4 space 3/4
space

TEACHER SUPPORT

Resources to make planning easy and give you more time to do what you do best—teach.

Teacher's Editions (K–8) are easy to use and filled with ideas and resources.

Teaching Resources (K–8) provide time-saving copying masters, such as dot and graph paper and vocabulary cards.

Intervention and Extension Copying Masters (3–8) provide extra support for pre-requisite skills and activities to expand chapter content.

School-Home Connection (K–8) includes information and activities to help parents reinforce instruction.

Problem of the Day Transparencies (1–8) provide a convenient way to present the *Problem of the Day*.

Answer and Solution Keys (6–8) give answers and solutions to all of the problems.

ASSESSMENT CHOICES

A wide range of assessment tools to measure performance before, during, and after instruction.

Teacher's Guides for Assessment (K–8) include an overview of assessment options, portfolio suggestions, and checklists.

Assessing Prior Knowledge Copying Masters (1–8) assess the prerequisite skills students need to begin a chapter.

Test Copying Masters (1–8) provide end-of-chapter tests, periodic "checkpoint" tests, and cumulative tests.

Performance Assessment Teacher's Guides (1–8) make assessment manageable with performance tasks, scoring rubrics, and student benchmark papers.

Math Advantage TestBanks (3–8), for use with the W.K. Bradford TestBuilder, let you print or customize *Math Advantage* tests—or create and print your own.

Computer Management System (1–8) allows you to score tests and print reports.

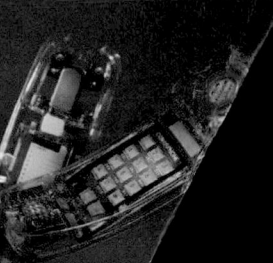

MANIPULATIVES OPTIONS

A great way to build your manipulative collection and meet the needs of your students.

Teacher Modeling Kits (K, 1, 2, 3, 4 & 5, 6–8) include overhead manipulatives, transparencies, and three-dimensional manipulatives to help you demonstrate concepts.

Overhead Manipulative Kits (K, 1 & 2, 3–5, 6–8) contain colorful, sturdy manipulatives to help you model concepts.

Core Kits (K, 1, 2, 3, 4 & 5, 6 & 7, 8) provide key manipulatives that help students work together to develop math concepts.

Build-A-Kit® Manipulatives (K–8) help you customize your classroom collection by adding just the manipulatives you need.

MATH ADVANTAGE
1 2 3
Number House
A KINDERGARTEN PROGRAM

- Teacher's Edition
- Pupil's Edition or Pupil's Edition Chapter Books
- Pupil's Edition Manipulatives Case
- Teacher's Guide for Assessment
- School-Home Connection
- Teaching Resources

- Practice Workbook • On My Own with Teacher's Edition
- Gameboards with Pieces
- Learning Center Cards
- Literature Big Books
- Movement Audiocassettes
- Transparency Package

- K.C. Koala Puppet
- Koala Counters
- Stanley's Sticker Stories CD-ROM
- Mighty Math Carnival Countdown CD-ROM
- Mighty Math Zoo Zillions CD-ROM

WITH 1...2...3 options to match your teaching style

24 Pupil's Editions
or
24 Sets of Pupil's Edition Chapter Books
or
Big Book of Pupil's Edition

 Visit The Harcourt Brace Internet Site at http://www.hbschool.com

HARCOURT BRACE

For more information, call: 1-800-225-5425.

PREPARATION FOR ALGEBRA

MATH
ADVANTAGE

MIDDLE SCHOOL BOOK I

HARCOURT
BRACE

Orlando • Atlanta • Austin • Boston • San Francisco • Chicago • Dallas • New York • Toronto • London

http://www.hbschool.com

All photographs by The Quarasan Group, Inc. All tab page photographs by FPG.

Table of Contents

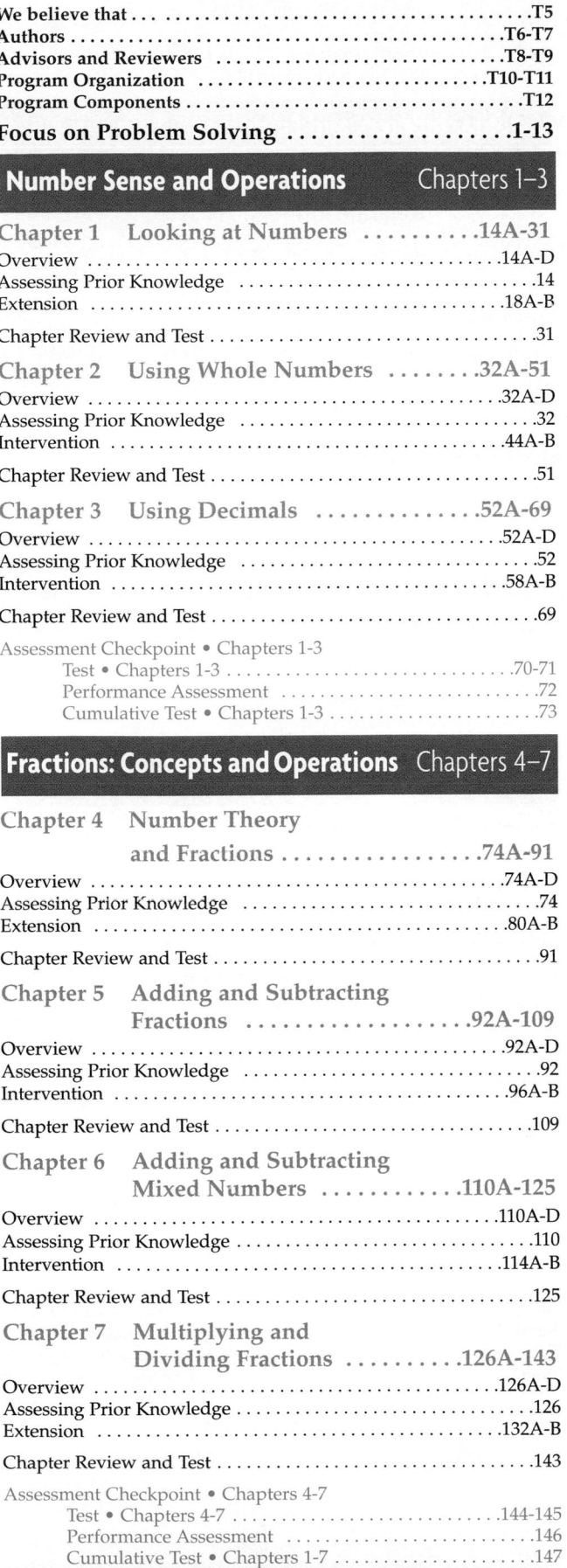

T**a**ble of C**o**ntents

We Believe That...

- **every child can be successful at learning mathematics.**

- instruction that includes a balance among conceptual understanding, connections to prior knowledge, skill proficiency, and problem-solving experiences best develops children's mathematical literacy.

- instruction that includes hands-on activities to develop concepts helps ✓ children engage in meaningful practice.

- problem solving is the focus of mathematics instruction and is best developed through teaching of strategies and practice in choosing and applying those strategies to solve relevant, engaging, and meaningful problems.

- children learn best when they understand what they are learning and why they are learning it.

- reasoning about mathematics, rather than memorizing rules and procedures, helps children make sense of mathematics.

- the communication skills of reading, representing, speaking, listening, and writing help children develop a strong conceptual understanding of mathematics.

- practice is a necessary component for building skill proficiency, and ongoing review of previously taught skills ensures proficiency in mathematical skills and procedures.

- children broaden their view of mathematics when they investigate mathematical situations drawn from real-life experiences, from other cultures, and from other subject areas.

- children learn best in an active environment in which they work together, think together, and communicate about the mathematics they are learning.

- children of all ages learn best when instruction links concrete experiences to pictorial representations and to abstract symbols.

- tools of mathematics—such as manipulatives, calculators and computers, and paper and pencil algorithms—linked to instruction help children discover the power of mathematics and provide multiple ways to present concepts and provide practice.

- a variety of assessments that are integrated into instruction evaluate how children think and what children can do rather than what they cannot do.

- instruction that involves all modalities and all types of intelligences helps all children learn mathematics.

- children whose primary language is not English will benefit from regular mathematics instruction that includes specific English-language acquisition strategies.

- a strong partnership between home and school helps students succeed in mathematics.

Authors

Grace M. Burton

Evan M. Maletsky

SENIOR AUTHORS

▶ **Grace M. Burton**
Chair, Department of Curricular Studies
Professor, School of Education
University of North Carolina at Wilmington
Wilmington, North Carolina

▶ **Evan M. Maletsky**
Professor of Mathematics
Montclair State University
Upper Montclair, New Jersey

George W. Bright

Sonia M. Helton

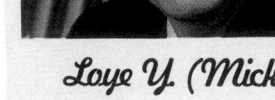

Loye Y. (Mickey)

AUTHORS

▶ **George W. Bright**
Professor of Mathematics Education
The University of North Carolina
   at Greensboro
Greensboro, North Carolina

▶ **Sonia M. Helton**
Professor of Childhood Education
Coordinator, College of Education
University of South Florida
St. Petersburg, Florida

▶ **Loye Y. (Mickey) Hollis**
Professor of Mathematics Education
Director of Teacher Education and
   Undergraduate Programs
University of Houston
Houston, Texas

Authors

AUTHORS

• • • • • • • • • • •

▶ **Howard C. Johnson**
Dean of the Graduate School
Associate Vice Chancellor
  for Academic Affairs
Professor, Mathematics and
  Mathematics Education
Syracuse University
Syracuse, New York

▶ **Joyce C. McLeod**
Visiting Professor
Rollins College
Winter Park, Florida

▶ **Evelyn M. Neufeld**
Professor, College of Education
San Jose State University
San Jose, California

▶ **Vicki Newman**
Classroom Teacher
McGaugh Elementary School
Los Alamitos Unified School
  District
Seal Beach, California

▶ **Terence H. Perciante**
Professor of Mathematics
Wheaton College
Wheaton, Illinois

▶ **Karen A. Schultz**
Associate Dean and
Director of Graduate Studies and
  Research
Research Professor, Mathematics
  Education
College of Education
Georgia State University
Atlanta, Georgia

▶ **Muriel Burger Thatcher**
Independent Mathematics
  Consultant
Mathematical Encounters
Pine Knoll Shores, North Carolina

Howard C. Johnson

Joyce C. McLeod

Evelyn M. Neufeld

Vicki Newman

Terence H. Perciante

Karen A. Schultz

Muriel Burger Thatcher

Advisors

Anne R. Biggins
Speech-Language Pathologist
Fairfax County Public Schools
Fairfax, Virginia

Carolyn Gambrel
Learning Disabilities Teacher
Fairfax County Public Schools
Fairfax, Virginia

Asa G. Hilliard III
Fuller E. Callaway
    Professor of Urban Education
Georgia State University
Atlanta, Georgia

Marsha W. Lilly
Secondary Mathematics
    Coordinator
Alief Independent School District
Alief, Texas

Clementine Sherman
Director, Division of USI
    Mathematics and Science
Dade County Public Schools
Miami, Florida

Judith Mayne Wallis
Elementary Language Arts/Social
    Studies/Gifted Coordinator
Alief Independent School District
Houston, Texas

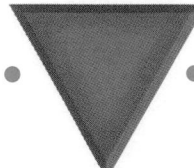

Reviewers

Tara Barca
Teacher
P.S. 127
Brooklyn, New York

Debbie Beck
Teacher
Gulf Gate Elementary
Sarasota, Florida

Carol Homberg Bley
Teacher
Thurgood Marshall Elementary
Newark, Delaware

Karla J. Bond
Teacher and Elementary
    Mathematics Coordinator
Ft. Zumwalt Public Schools/
    Forest Park Elementary
O'Fallon, Missouri

Patricia R. Claybaugh
Teacher
Scott Elementary
Evansville, Indiana

Barbara DeGroff
Teacher
Mountain Sky Junior High
Phoenix, Arizona

Nita L. Downey
ESL Teacher
Barrick Elementary
Houston, Texas

Richard Dutton
Director of Elementary Education
Denver Public Schools
Denver, Colorado

Catherine Fay
Teacher
South Davis Elementary
Orchard Park, New York

Kenneth Gay, Jr.
Teacher
Lopez Elementary
Seffner, Florida

Nancy S. Koven
Teacher
P.S. 114 Queens
Belle Harbor, New York

Debra Scrivens Latozas
K-12 Mathematics Coordinator
Bailey Lake/Clarkston
    Community Schools
Clarkston, Michigan

Chris Lechleitner
Teacher
John S. Irwin Elementary
Ft. Wayne, Indiana

Gwendolyn E. Long
Manager, Mathematics
    Instruction Practices
Chicago Public Schools
Chicago, Illinois

Robert McCain
Assistant Principal
Portland Middle School
Portland, Connecticut

Marie J. McKinney
Magnet/Technology Coordinator
IS 227
Brooklyn, New York

Leanne McNutt
Teacher
Crockett Elementary
Bryan, Texas

David H. Moore
Teacher
Tulip Grove Elementary
Hermitage, Tennessee

Judi Morlan
Teacher
Tarpon Springs Middle School
Tarpon Springs, Florida

Jessica Mowad
Principal
Glen Cove Elementary
El Paso, Texas

Nancy Egan Powell
Teacher
J.F. Kennedy Elementary
Kansas City, Kansas

Dawnellen Pybas
Teacher
Harvest Hills Elementary
Oklahoma City, Oklahoma

Barbara C. Rachbuch
Teacher and Team Leader
Heritage Middle School
Livingston, New Jersey

Joel Reed
Math/Science Curriculum
    Coordinator
Woodland Hills School District
Pittsburgh, Pennsylvania

Reviewers

Janet L. Reese
Teacher in Residence
Auburn University at Montgomery
Montgomery, Alabama

Jenny Reid
Teacher
Brookside Elementary #54
Indianapolis, Indiana

Frankie L. Robinson
Teacher
B.J. Ward Middle School
Bolingbrook, Illinois

Lillian Robison
Teacher
Heard Elementary
Dothan, Alabama

Susan Rosenbaum
Teacher
Longfellow Middle School
Norman, Oklahoma

Patricia Sapecky
Teacher
Mohawk Trails Elementary
Carmel, Indiana

Michael Steinberg
Teacher
P.S. 255-Barbara Reing School
Brooklyn, New York

Shirley Q. Tisdale
Teacher
Alexander Burger
Bronx, New York

Peggy Valentine
Teacher
Ft. Worth ISD
Ft. Worth, Texas

Terri Watson
Teacher
Thomas Haley Elementary
Irving, Texas

Kenneth West
Assistant Principal
Broward County Schools
Ft. Lauderdale, Florida

Field Test Teachers

Mary Lou Beasley
Southside Fundamental Middle
  St. Petersburg, Florida

Valerie Blevins
Bayside Middle
Virginia Beach, Virginia

Naida Brueland
Irving Middle
San Antonio, Texas

Shannon Coker
Virginia Beach Middle
Virginia Beach, Virginia

Christopher Coombs
Conniston Middle
West Palm Beach, Florida

Marcy DeBock
Clear Creek School
Buffalo, Wyoming

Rose DeForrest
Gulf Middle
Cape Coral, Florida

Gail Fein
Coronado Middle
Kansas City, Kansas

Oscar Herzer
Virginia Beach Middle
Virginia Beach, Virginia

Ed Killian
East Taylor Elementary
Johnstown, Pennsylvania

Alicia Kruska
Meadow Park Middle
Beavertown, Oregon

Donnette Parker
Decatur Middle
Decatur, Texas

Kit Parker
South Junior High
Boise, Idaho

Tracy Post
Plaza Park Middle
Evansville, Indiana

Joanne Prunty
Haviland Middle
Hyde Park, New York

Anna Reeve
DeLeon Middle
McAllen, Texas

Ann Sargent
Greenbriar Junior High
Parma, Ohio

Becky Scarborough
Landmark Middle
Jacksonville, Florida

Letha Silas
Middleton Middle
Tampa, Florida

Debra Sparks
Walker Elementary
Northport, Alabama

Greg Starke
Cole Middle School
Denver, Colorado

Rod Swerin
Meadow Park Middle
Beavertown, Oregon

Randy Vincent
Southside Fundamental Middle
St. Petersburg, Florida

Dolly Vogel
York Junior High
Conroe, Texas

Leslie Williams
Whittier Middle
Norman, Oklahoma

Program Organization

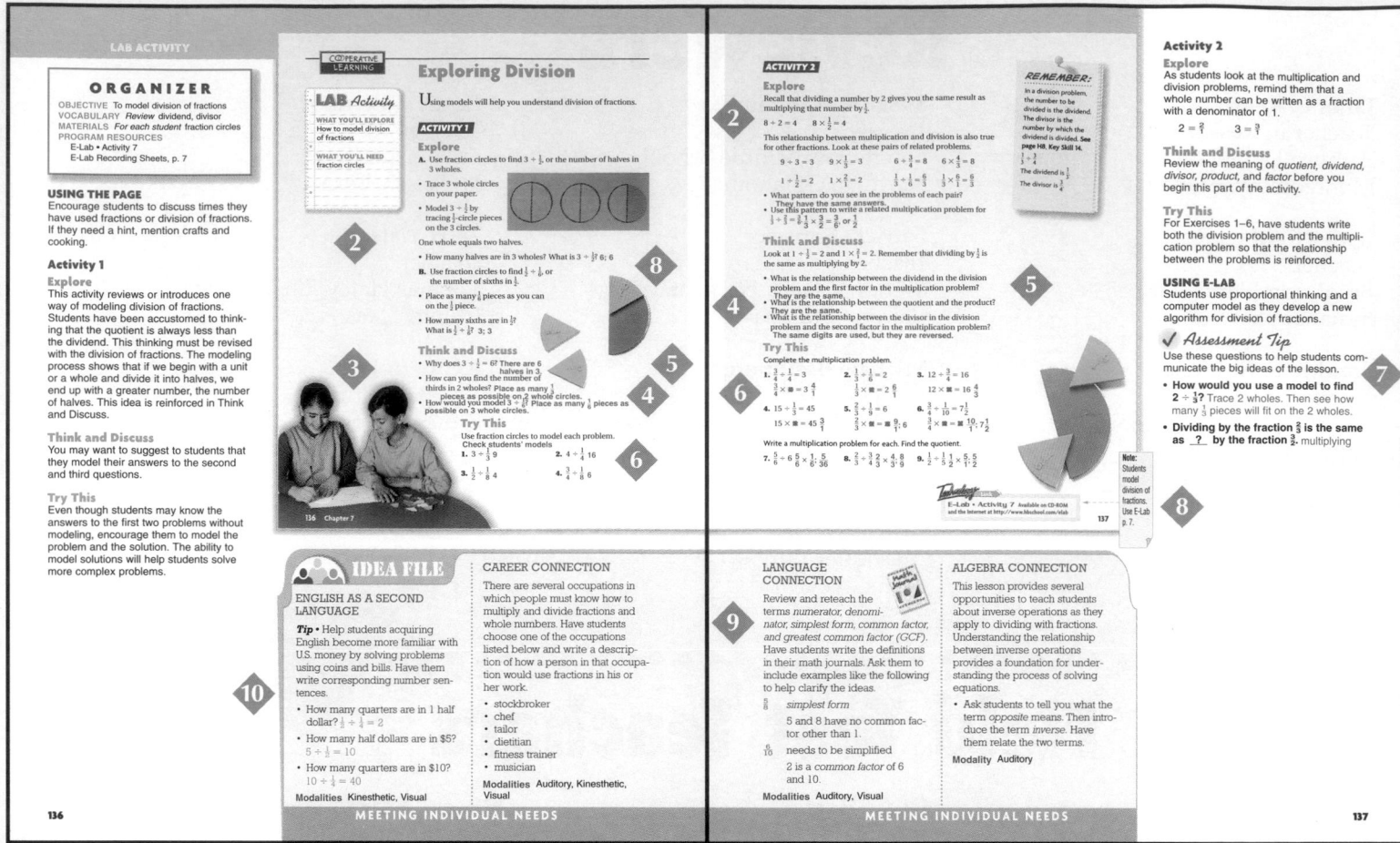

In **Middle School I**, students are introduced to many new concepts that go beyond basic computation and build a foundation for Algebra I. For the appropriate amount of instruction and practice, concepts are developed in either two-page or four-page lessons. Lab Activities provide hands-on experiences that introduce concepts or extend understanding.

Every lesson was written to focus on many of the following goals.

1. Problem solving is the focus of instruction.

2. Hands-on activities help students develop concepts.

3. Students learn best when instruction links concrete representations, pictorial representations, and abstract symbols.

4. Reasoning, rather than rules, helps students make sense of mathematics.

5. Communication skills help students develop a strong conceptual understanding.

6. Practice builds skill proficiency.

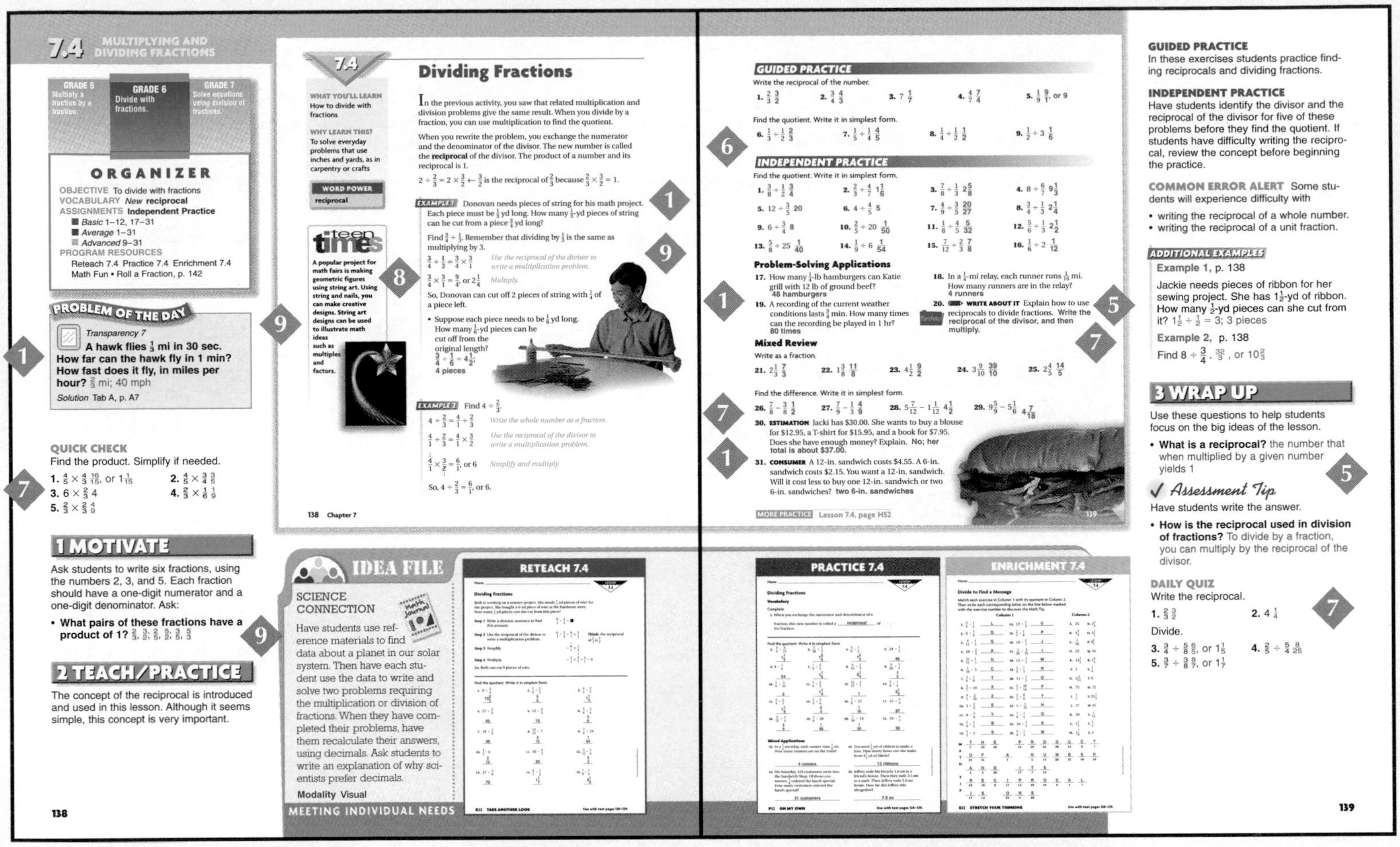

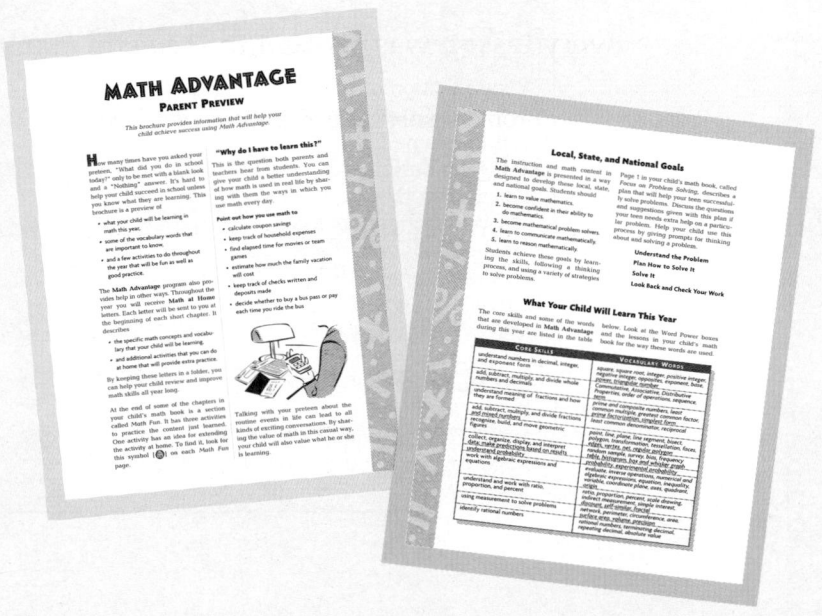

A variety of assessments integrated into instruction helps you evaluate how students think and what students *can* do.

Manipulatives, technology, mental computation, and paper-and-pencil algorithms provide multiple ways to present concepts and practice.

Mathematical situations drawn from real life, from other subject areas, and from other cultures help students see the need for mathematics.

Students benefit from instruction that includes specific content for English-language acquisition.

School-Home Connections

A strong partnership between school and home helps students succeed in mathematics.

At the beginning of the year, send the *Math Advantage Parent Preview* to each student's home. This describes the math content and some of the vocabulary words that will be covered during the school year.

A *Math at Home* letter, pictured on each chapter opener page, describes the content to be covered in the chapter and suggests activities to do at home.

Components Chart

Student Support	K	1	2	3	4	5	6	7	8
Pupil's Editions (K–8)	■	■	■	■	■	■	■	■	■
Big Book (K)	■								
Practice Workbook • On My Own™ (K–8)	■	■	■	■	■	■	■	■	■
Teacher's Edition also available									
Reteaching Workbook • Take Another Look (1–8)		■	■	■	■	■	■	■	■
Teacher's Edition also available									
Enrichment Workbook • Stretch Your Thinking (1–8)		■	■	■	■	■	■	■	■
Teacher's Edition also available									
Learning Center Cards (K–5)	■	■	■	■	■	■			
Gameboards with Pieces (K–2)	■	■	■						
Classroom Read-Aloud Library (1 & 2)		■	■						
Mighty Math™ Software (K–8)	■	■	■	■	■	■	■	■	■
Carnival Countdown™ (K–3)	■	■	■	■					
Zoo Zillions™ (K–3)	■	■	■	■					
Number Heroes™ (3–6)				■	■	■	■		
Calculating Crew™ (3–6)				■	■	■	■		
Cosmic Geometry™ (6–8)							■	■	■
Astro Algebra™ (6–8)							■	■	■
Stanley's Sticker Stories™ (K–2)	■	■	■						
E-Lab™ and E-Lab Recording Sheets (3–8)				■	■	■	■	■	■
Graph Links® Plus (1–6)		■	■	■	■	■	■		
Data ToolKit™ (5–8)						■	■	■	■
Harcourt Brace Internet Site (K–8)	■	■	■	■	■	■	■	■	■
Calculator Ten-Packs (K–8)	■	■	■	■	■	■	■	■	■

Teacher Support	K	1	2	3	4	5	6	7	8
Teacher's Editions (K–8)	■	■	■	■	■	■	■	■	■
Teaching Resources (K–8)	■	■	■	■	■	■	■	■	■
Intervention and Extension Copying Masters (3–8)				■	■	■	■	■	■
School-Home Connection (K–8)	■	■	■	■	■	■	■	■	■
Problem of the Day Transparencies (1–8)		■	■	■	■	■	■	■	■
Answer and Solution Keys (6–8)							■	■	■

Assessment Choices	K	1	2	3	4	5	6	7	8
Teacher's Guides for Assessment (K–8)	■	■	■	■	■	■	■	■	■
Assessing Prior Knowledge Copying Masters (1–8)		■	■	■	■	■	■	■	■
Test Copying Masters (1–8)		■	■	■	■	■	■	■	■
Performance Assessment Teacher's Guides (1–8)		■	■	■	■	■	■	■	■
Math Advantage TestBanks (3–8)				■	■	■	■	■	■
Computer Management System (1–8)		■	■	■	■	■	■	■	■

Manipulatives Options	K	1	2	3	4	5	6	7	8
Teacher Modeling Kits (K, 1, 2, 3, 4 & 5, 6–8)	■	■	■	■	■	■	■	■	■
Overhead Manipulatives Kits (K, 1 & 2, 3–5, 6–8)	■	■	■	■	■	■	■	■	■
Core Kits (K, 1, 2, 3, 4 & 5, 6 & 7, 8)	■	■	■	■	■	■	■	■	■
Build-A-Kit® Manipulatives (K–8)	■	■	■	■	■	■	■	■	■

1, 2, 3 Number House: A Kindergarten Program	K	1	2	3	4	5	6	7	8
Teacher's Edition	■								
Pupil's Edition or Chapter Books or Big Book	■								
Teacher's Guide for Assessment	■								
School-Home Connection	■								
Teaching Resources	■								
Practice Workbook • On My Own	■								
with Teacher's Edition	■								
Gameboards with Pieces	■								
Learning Center Cards	■								
Literature Big Books	■								
Movement Audiocassettes	■								
Transparency Package	■								
K.C. Koala Puppet	■								
Koala Counters	■								
Stanley's Sticker Stories CD-ROM	■								
Mighty Math Carnival Countdown CD-ROM	■								
Mighty Math Zoo Zillions CD-ROM	■								

PREPARATION FOR ALGEBRA

MATH

ADVANTAGE

Harcourt Brace & Company

Orlando • Atlanta • Austin • Boston • San Francisco • Chicago • Dallas • New York • Toronto • London

http://www.hbschool.com

▼▼ Senior Authors ▼▼

Grace M. Burton
Chair, Department of Curricular Studies
Professor, School of Education
University of North Carolina at Wilmington
Wilmington, North Carolina

Evan M. Maletsky
Professor of Mathematics
Montclair State University
Upper Montclair, New Jersey

▼▼ Authors ▼▼

George W. Bright
Professor of Mathematics Education
The University of North Carolina at Greensboro
Greensboro, North Carolina

Sonia M. Helton
Professor of Childhood Education
Coordinator, College of Education
University of South Florida
St. Petersburg, Florida

Loye Y. (Mickey) Hollis
Professor of Mathematics Education
Director of Teacher Education and Undergraduate Programs
University of Houston
Houston, Texas

Howard C. Johnson
Dean of the Graduate School
Associate Vice Chancellor for Academic Affairs
Professor, Mathematics and Mathematics Education
Syracuse University
Syracuse, New York

Joyce C. McLeod
Visiting Professor
Rolins College
Winter Park, Florida

Evelyn M. Neufeld
Professor, College of Education
San Jose State University
San Jose, California

Vicki Newman
Classroom Teacher
McGaugh Elementary School
Los Alamitos Unified School District
Seal Beach, California

Terence H. Perciante
Professor of Mathematics
Wheaton College
Wheaton, Illinois

Karen A. Schultz
Associate Dean and Director of Graduate Studies and Research
Research Professor, Mathematics Education
College of Education
Georgia State University
Atlanta, Georgia

Muriel Burger Thatcher
Independent Mathematics Consultant
Mathematical Encounters
Pine Knoll Shores, North Carolina

▼▼▼▼▼▼▼▼▼▼▼▼

Advisors

Anne R. Biggins
Speech-Language Pathologist
Fairfax County Public Schools
Fairfax, Virginia

Carolyn Gambrel
Learning Disabilities Teacher
Fairfax County Public Schools
Fairfax, Virginia

Asa G. Hilliard, III
Fuller E. Callaway Professor of Urban Education
Georgia State University
Atlanta, Georgia

Marsha W. Lilly
Secondary Mathematics Coordinator
Alief Independent School District
Alief, Texas

Clementine Sherman
Director, Division of USI Mathematics and Science
Dade County Public Schools
Miami, Florida

Judith Mayne Wallis
Elementary Language Arts/Social Studies/Gifted Coordinator
Alief Independent School District
Houston, Texas

iii

CONTENTS

NUMBER SENSE AND OPERATIONS — CHAPTERS 1–3

Key Skills

Key Skills

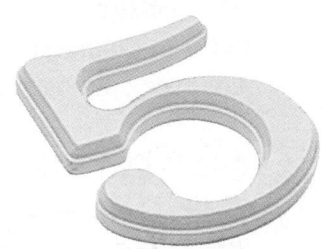

FRACTIONS: CONCEPTS AND OPERATIONS — CHAPTERS 4–7

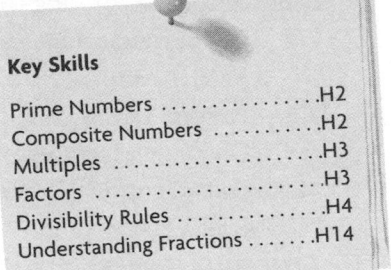

Key Skills

Mixed Numbers and Fractions . .H16
Renaming Mixed Numbers H18
Adding Mixed Numbers H18
Subtracting Mixed Numbers H19

Key Skills

Dividing Whole Numbers H8
Simplest Form of Fractions H15
Mixed Numbers and Fractions . .H16

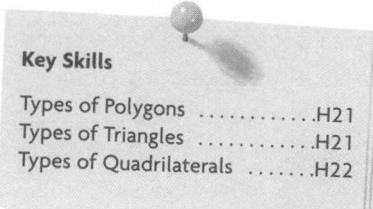

GEOMETRY   CHAPTERS 8–10

Key Skills

Types of Polygons H21
Types of Triangles H21
Types of Quadrilaterals H22

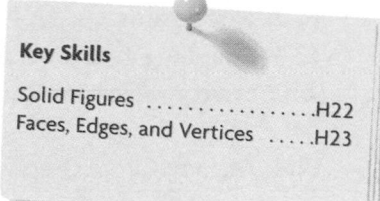

Key Skills

Range H27

Key Skills

Range H27

Key Skills

Mean H26
Median and Mode H26

Key Skills

Area H25

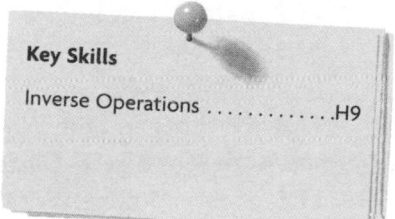

ALGEBRA: EQUATIONS AND RELATIONSHIPS   CHAPTERS 15–16

Key Skills

Inverse Operations H9

PROPORTIONAL REASONING   CHAPTERS 17–20

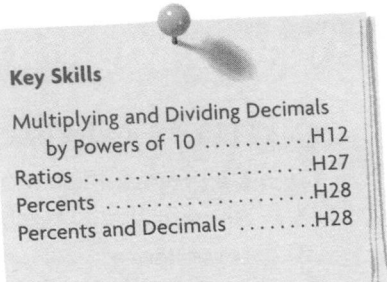

Key Skills

Inverse Operations H9
Writing a Fraction as a Decimal .. H19

Key Skills

Multiplying and Dividing Decimals
     by Powers of 10 .... H12
Ratios H27
Percents H28
Percents and Decimals H28

Key Skills

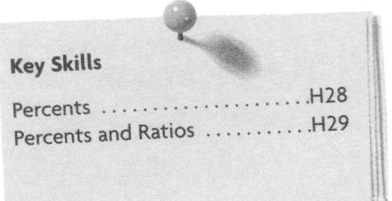

Percents H28
Percents and Ratios H29

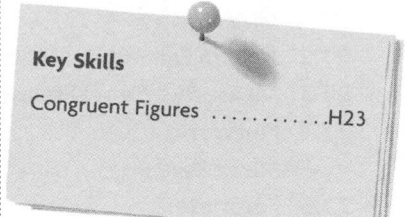

Key Skills
Congruent Figures H23

MEASUREMENT: ONE, TWO, AND THREE DIMENSIONS

Key Skills

Key Skills

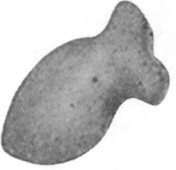

Key Skills

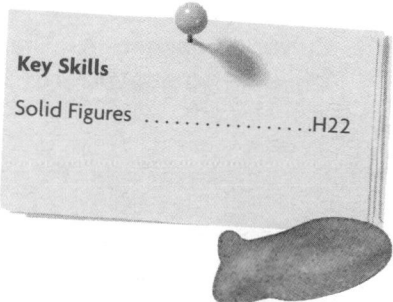

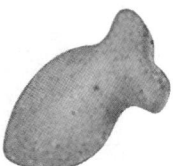
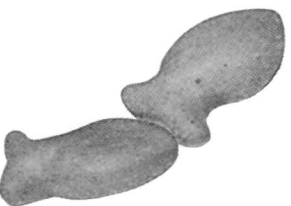

ALGEBRA: INTEGERS AND EQUATIONS   CHAPTERS 24–26

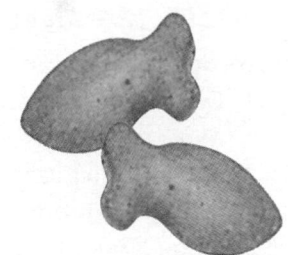

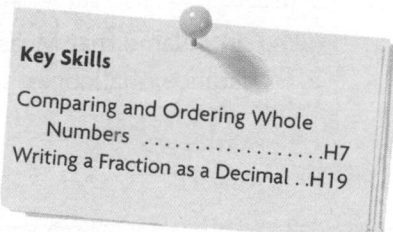

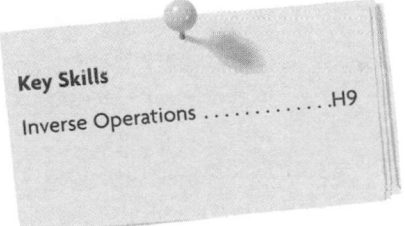

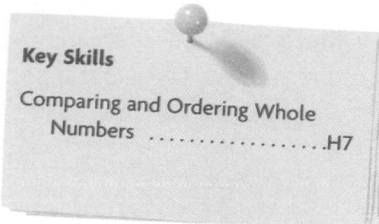

PATTERNS   CHAPTERS 27–28

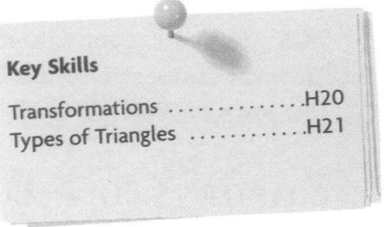

STUDENT HANDBOOK

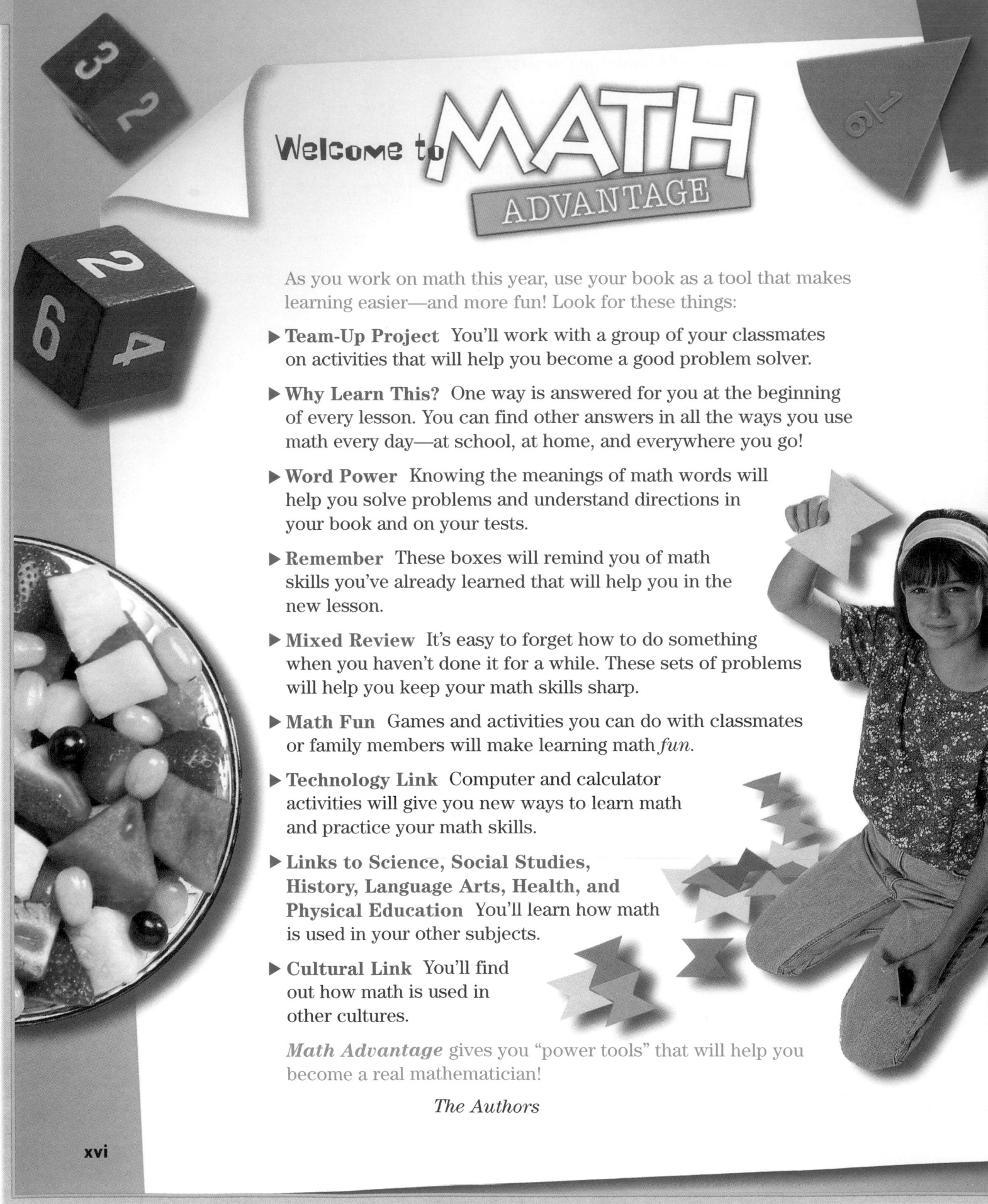

Welcome to MATH ADVANTAGE

As you work on math this year, use your book as a tool that makes learning easier—and more fun! Look for these things:

▶ **Team-Up Project** You'll work with a group of your classmates on activities that will help you become a good problem solver.

▶ **Why Learn This?** One way is answered for you at the beginning of every lesson. You can find other answers in all the ways you use math every day—at school, at home, and everywhere you go!

▶ **Word Power** Knowing the meanings of math words will help you solve problems and understand directions in your book and on your tests.

▶ **Remember** These boxes will remind you of math skills you've already learned that will help you in the new lesson.

▶ **Mixed Review** It's easy to forget how to do something when you haven't done it for a while. These sets of problems will help you keep your math skills sharp.

▶ **Math Fun** Games and activities you can do with classmates or family members will make learning math *fun*.

▶ **Technology Link** Computer and calculator activities will give you new ways to learn math and practice your math skills.

▶ **Links to Science, Social Studies, History, Language Arts, Health, and Physical Education** You'll learn how math is used in your other subjects.

▶ **Cultural Link** You'll find out how math is used in other cultures.

Math Advantage gives you "power tools" that will help you become a real mathematician!

The Authors

xvi

FOCUS ON PROBLEM SOLVING

Good problem solvers need to be good thinkers. They also need to know these strategies.

- Draw a Diagram
- Act It Out
- Make a Model
- Use a Formula
- Work Backward
- Find a Pattern
- Guess and Check
- Solve a Simpler Problem
- Make a Table or Graph
- Write an Equation
- Account for All Possibilities

This plan can help you think through a problem.

✓ **Understand** the problem.

Ask yourself...	Then try this.
What is the problem about?	Retell the problem in your own words.
What is the question?	Say the question as a fill-in-the-blank sentence.
What information is given?	List the information given in the problem.

✓ **Plan** how to solve it.

Ask yourself...	Then try this.
What strategies might I use?	List some strategies you can use.
About what will the answer be?	Predict what your answer will be. Make an estimate if it will help.

✓ **Solve** the problem.

Ask yourself...	Then try this.
How can I solve the problem?	Follow your plan and show your solution.
How can I write my answer?	Write your answer in a complete sentence.

✓ **Look Back** and check your answer.

Ask yourself...	Then try this.
How can I tell if my answer is reasonable?	Compare your answer to your estimate. Check your answer by redoing your work. Match your answer to the question.
How else might I have solved the problem?	Try using another strategy to solve the problem.

On the following pages, you can practice being a good problem solver. Each page reviews a different strategy that you can use throughout the year. These pages will help you recognize the kinds of problems that can be solved with each strategy. Think through each problem you work on and ask yourself questions as you **Understand, Plan, Solve, and Look Back.** Then be proud of your success!

1

USING THE PAGE

For more practice on this problem-solving strategy, see pp. 376–377

Draw a Diagram

Problem Solving
• Understand
• Plan
• Solve
• Look Back

Sarah and her brother Chris visit their aunt and cousin each Saturday. From their apartment they walk 4 blocks north and 5 blocks west to their aunt's house. They then continue 2 blocks south and 3 blocks east to their cousin's house. How many ways can Sarah and Chris return home by walking four blocks?

☑ **UNDERSTAND**   You must mark the starting point and map out the walk. You know the number of blocks and the direction walked to arrive at each stopping place.

☑ **PLAN**   Draw a diagram using graph paper. Draw a line to show the number of blocks walked. Use arrows to show the direction taken to get from one place to the next.

☑ **SOLVE**   Draw a diagram like the one at the right to show the location of their apartment, the aunt's house, and the cousin's house.

From their cousin's, sketch the 4 remaining blocks to home. Enlarge and identify the possible paths to home.

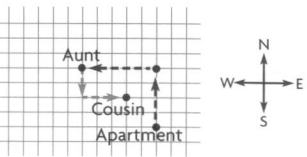

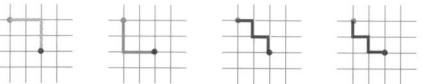

There are 4 possible ways.

☑ **LOOK BACK**   They are going to go south and east. The possible ways are EESS, SSEE, ESES, and SESE.

Try These

1. Freeway poles are placed at regular intervals. The distance from the first pole to the fifth pole is 300 ft. What is the distance from the twentieth pole to the twenty-ninth pole? **675 ft**

2. The last Friday of a month is the 25th day of that month. What day of the week is the first day of the month? **Tuesday**

3. There are 12 small square tables in the media center. Each table can seat only 1 person on each side. If the tables are pushed together to form one big rectangle, what arrangement will allow the most people to be seated? How many people can be seated? **1 × 12; 26 people**

2   Focus on Problem Solving

Act It Out

Problem Solving
- Understand
- Plan
- Solve
- Look Back

Eric found a toaster oven that toasts 2 slices of bread at once, but it toasts only one side at a time. Eric likes each side of his bread to be toasted for exactly one minute. How can he toast 3 slices of bread in 3 minutes?

☑ **UNDERSTAND**   You need to find a way to toast 3 slices of bread in three minutes. You know that each side takes 1 minute.

☑ **PLAN**   Place 3 pieces of paper in front of you to represent the bread. Label each side. You can use A1, A2, B1, B2, C1, and C2. Set up three steps, one for each minute.

☑ **SOLVE**   Step 1: Place slices A and B "in the toaster," with sides A1 and B1 facing up.

Step 2: Flip slice A1. Then replace B1 with C1.

Step 3: Replace A2 with B2. Then flip C1.

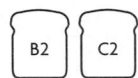

☑ **LOOK BACK**   If each step takes 1 minute, then all 3 slices have been toasted for 1 minute on each side.

Try These

1. A special matinee of a play costs $4.00 for adults and $1.00 for children under 10. If twice as many children attended as adults and the show received $72.00, how many adults attended the matinee?
12 adults

2. Zachary wants to buy a comic book for $0.85. He has ten coins in his pocket that add up to exactly the correct amount. What coins does Zachary have in his pocket?
Possible answer: 2 quarters, 3 dimes, 5 pennies

USING THE PAGE
For more practice on this problem-solving strategy, see pp. 334–335

USING THE PAGE

For more practice on this problem-solving strategy, see pp. 180–181

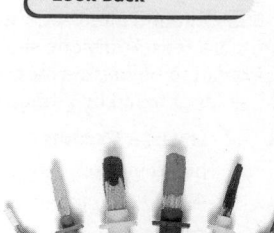

Problem Solving
• Understand
• Plan
• Solve
• Look Back

Make a Model

The figures below show three views of a cube, revealing six faces. This cube has a different color on each face. Determine which colors are opposite each other.

B: blue   R: red
Y: yellow   G: green
W: white   P: purple

☑ **UNDERSTAND** You must find the colors that are opposite each other. You know that certain colors cannot be opposites since they are touching.

☑ **PLAN** Use three cubes and label each face with color dots or mark as shown in the diagram above. Focus on one particular color and determine which colors could not be opposite it.

☑ **SOLVE** Given the information, place the known information on a grid. Use X's for "cannot be solutions." Use ✓'s for opposites. Opposites can occur only once in a column.

So, red is opposite purple. Green is opposite blue. Yellow is opposite white.

Make a cube with these opposites.

	Red	Blue	Green	Yellow	White	Purple
Red	X	X	X	X	X	
Blue	X	X	✓	X		
Green	X		X	X	X	X
Yellow	X	X	X	X		X
White	X		X	✓	X	
Purple	✓		X	X		X

☑ **LOOK BACK** Place your cube in each of the positions shown in the diagram at the top of the page. Make sure your cube matches in each position.

Try These

1. Erica and Amanda always disagree over who sits where in their family's 7-seat minivan. They like to sit in a variety of places. Since their mother and father always sit in the front 2 seats, how many different arrangements are possible for the remaining 5 seats? **20 arrangements**

2. Sam is making a plant stand using 4 cubes. He stacks the cubes, one on top of the other, and paints the outside of the stand red (not the bottom). How many faces of the original cubes are painted? **4 faces of bottom 3 cubes, 5 faces of top cube; total: 17 faces**

USING THE PAGE
For more practice on this problem-solving strategy, see pp. 36–37

Guess and Check

Handmade friendship bracelets use 20 beads. Handmade rings use 8 beads. Marne used a total of 176 beads to make 13 items. How many friendship bracelets did she make?

☑ **UNDERSTAND**   You must find the number of bracelets. You know the number of beads used in a bracelet and in a ring.

☑ **PLAN**   Guess a number of bracelets and rings. Check to see if the total number of beads is 176 and the number of items is 13.

☑ **SOLVE**   Record your first guess and total the number of beads.

Bracelets	Rings	Total Items	Number of Beads	
5	8	5 + 8 = 13	5 × 20 + 8 × 8 = 100 + 64 = 164	too low
8	5	8 + 5 = 13	8 × 20 + 5 × 8 = 160 + 40 = 200	too high
7	6	7 + 6 = 13	7 × 20 + 6 × 8 = 140 + 48 = 188	too high
6	7	6 + 7 = 13	6 × 20 + 7 × 8 = 120 + 56 = 176	correct

So, the number of bracelets is 6.

☑ **LOOK BACK**   Check that there are 13 items and 176 beads used.

bracelets: 6   bracelet beads: 6 × 20 = 120
rings: 7   ring beads: 7 × 8 = 56
total: 6 + 7 = 13 ✓   total beads: 120 + 56 = 176 ✓

Try These

1. Victor bought comics. He handed the clerk $20.00 and received $11.45 in change. He knows he bought more than 35 comics and each comic had the same price. How many books could he have bought and what did each cost? **Possible answers: 57 at $0.15 or 171 at $0.05**
2. The sum of two numbers is 49. Their difference is less than 10. What are all the possible pairs? **24 and 25, 23 and 26, 22 and 27, 21 and 28, 20 and 29**
3. Amy bought used books for $4.95. She paid $0.50 each for some books and $0.35 each for others. She bought fewer than 8 books at each price. How many did she buy for $0.50? **5 books**

USING THE PAGE
For more practice on this problem-solving strategy, see pp. 140–141

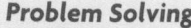

Problem Solving
- Understand
- Plan
- Solve
- Look Back

Work Backward

Jack, Jill, and Jodi collect baseball cards. Jack has 125 more cards in his collection than Jill. Jill has 75 cards fewer than Jodi. Jodi has 250 baseball cards. How many cards does Jack have?

☑ **UNDERSTAND** You need to find the number of cards Jack has in his collection. You know the number of cards Jodi has and how many more cards Jack has than Jill.

☑ **PLAN** Start from the end and work backward from what you know about Jodi's cards. Calculate Jill's and Jack's cards from this information.

☑ **SOLVE** Work from Jodi's collection. This is the only information you have without computing.

Jodi: 250 cards   250
Jill: 75 fewer than Jodi $250 - 75 = 175$
Jack: 125 more than Jill $175 + 125 = 300$

So, Jack has 300 baseball cards.

☑ **LOOK BACK** Check the original problem with the information.

Jack has 125 more cards than Jill: $175 + 125 = 300$

Jill has 75 cards fewer than Jodi: $250 - 75 = 175$

Try These

1. A driver travels east for 25 mi, turns north and goes another 15 mi. To return home over the same roads, in what direction and for how far will she have to travel? **south 15 mi, west 25 mi**

2. Mr. George ordered $50 of candy. He bought a box of deluxe caramels at $8 a box and 3 boxes of deluxe nuggets. What was the price per box for the deluxe nuggets? **$14**

3. A coin collection was distributed among four family members. Ryan received $\frac{1}{2}$ of the coins. Josephine received $\frac{1}{4}$ of the coins, and Lilly received $\frac{1}{5}$ of the coins. Beth got 100 coins. How many coins were in the collection? **2,000 coins**

Account for All Possibilities

Problem Solving
....
- **Understand**
- **Plan**
- **Solve**
- **Look Back**

A palindrome reads the same forward and backward. How many palindromes are there during a 24-hr day?

☑ **UNDERSTAND**   You must find the number of palindromes displayed in 24 hr You know what a palindrome is.

☑ **PLAN**   Make a list of the palindromes that appear within a 1-hr block of time. Be sure to account for A.M. and P.M.

☑ **SOLVE**   List all the possible palindromes within 1-hr blocks and look for a pattern:

1 o'clock	2 o'clock	3 o'clock	4 o'clock	5 o'clock	6 o'clock
1:01	2:02	3:03	4:04	5:05	6:06
1:11	2:12	3:13	4:14	5:15	6:16
1:21	2:22	3:23	4:24	5:25	6:26
1:31	2:32	3:33	4:34	5:35	6:36
1:41	2:42	3:43	4:44	5:45	6:46
1:51	2:52	3:53	4:54	5:55	6:56

7 o'clock	8 o'clock	9 o'clock	10 o'clock	11 o'clock	12 o'clock
7:07	8:08	9:09	10:01	11:11	12:21
7:17	8:18	9:19			
7:27	8:28	9:29			
7:37	8:38	9:39			
7:47	8:48	9:49			
7:57	8:58	9:59			

There are 6 palindromes in each single-digit hr. For each of 10, 11, and 12 o'clock, there is 1 palindrome. In 12 hr, the clock displays $(9 \times 6) + (1 \times 3)$, or 57 palindromes. For 24 hr, double that amount: $2 \times 57 = 114$.

So, the clock displays 114 palindromes during a 24-hr day.

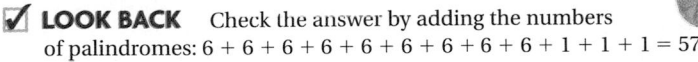

☑ **LOOK BACK**   Check the answer by adding the numbers of palindromes: $6 + 6 + 6 + 6 + 6 + 6 + 6 + 6 + 6 + 1 + 1 + 1 = 57$.

Try These

1. How many different 3-digit odd numbers can you make using the digits 3, 5, and 8? No digit should be used more than once in any number. **4 numbers**

2. How many ways can you make change for a quarter using only dimes, nickels, and pennies? **12 ways**

USING THE PAGE

For more practice on this problem-solving strategy, see pp. 504–505

Problem Solving
• Understand
• Plan
• Solve
• Look Back

Find a Pattern

John made a design using hexagons and triangles. The length of each side of the hexagons and of each side of the triangles is 1 in. What is the perimeter of the next figure in his design?

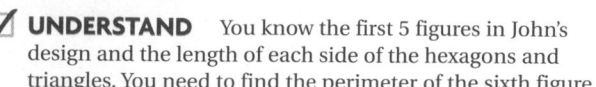 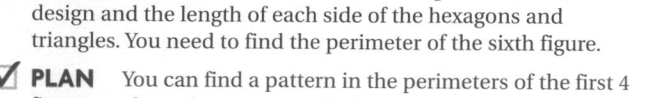 

☑ **UNDERSTAND**   You know the first 5 figures in John's design and the length of each side of the hexagons and triangles. You need to find the perimeter of the sixth figure.

☑ **PLAN**   You can find a pattern in the perimeters of the first 4 figures and use the pattern to find the perimeter of the next figure.

☑ **SOLVE**   Find the perimeter of the figures. Then find a pattern in the perimeters.

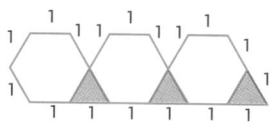

Figure ·	1	2	3	4	5
Perimeter	6	7	11	12	16

Think: $6 + 1 = 7$ The rule for the pattern is add 1, add 4,
$7 + 4 = 11$ add 1, add 4, and so on.
$11 + 1 = 12$
$12 + 4 = 16$ Sixth perimeter: $16 + 1 = 17$

So the perimeter of the sixth figure will be 17 ft.

☑ **LOOK BACK**   Sketch a picture of the sixth figure. Check the perimeter.

Perimeter = 17 in. ✓

Try These

1. The Computer Connection is planning to network all 10 of its members. The company will connect each member's computer with each of the other 9 members. How many cables will the hook-up require? **45 cables**

Find the pattern. Then find the next number or figure.

2. 1, 5, 9, 13, 17, . . . **21**   3. 1, 4, 16, 64, 256, . . **1,024**   4.

USING THE PAGE
For more practice on this problem-solving strategy, see pp. 306–307

Make a Table or Graph

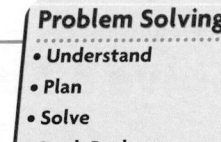

Problem Solving
• Understand
• Plan
• Solve
• Look Back

The Houston Public Library charges a fine of 10¢ a day for overdue books. The Fort Bend Library charges a fine of 50¢ for the first day and 5¢ for each additional day. On what day would an overdue book have the same fine at both libraries?

✓ **UNDERSTAND** You need to know on which day both libraries charge the same amount. You know what each library charges and the daily rate.

✓ **PLAN** Make a table showing the fines each library charges.

✓ **SOLVE** Identify each library's fine in the table with a money value.

Day	1	2	3	4	5	6	7	8	9
Houston Library	10¢	20¢	30¢	40¢	50¢	60¢	70¢	80¢	90¢
Fort Bend Library	50¢	55¢	60¢	65¢	70¢	75¢	80¢	85¢	90¢

So, on the 9th day both libraries would charge the same amount.

✓ **LOOK BACK** Is it reasonable that both libraries would both charge the same amount on the 9th day? Is there another way you could solve this problem?

Try These

1. Planes leave Houston Hobby Airport for Dallas Love Field every 45 min. The first plane leaves at 5:45 A.M. What is the departure time closest to 4:30 P.M.? **4:15 P.M.**

2. Sixth graders have light blue, white, and green tops for their uniforms and white, navy, light blue, and tan pants. How many different combinations of tops and pants do they have for their uniforms? **12 combinations**

3. The debate club has 10 members. Each member will debate each of the other members only once. How many debates will they have? **45 debates**

Technology Link

In *Data ToolKit,* you can make a table to help you solve the Try These exercises.

HARCOURT BRACE
Data ToolKit

Focus on Problem Solving 9

USING THE PAGE
For more practice on this problem-solving strategy, see pp. 198–199

Problem Solving
....
• Understand
• Plan
• Solve
• Look Back

Solve a Simpler Problem

At the end of each game, it is a tradition for each player to shake hands with every player on the opposing team. How many handshakes occur at the end of a game between two hockey teams of 20 players?

✓ **UNDERSTAND**   You know how many players are on each team. You need to find the total number of handshakes exchanged.

✓ **PLAN**   Examine the simplest case. Start with 1 player from each team. Then look at 2, 3, and 4 players from each team. Collect and record the data in a table. Look for a pattern.

✓ **SOLVE**   The simplest case involves one player per team shaking hands with the opponent.

Extend the pattern in the table to include 20 players on each team.

5 → 25 handshakes
6 → 36 handshakes
7 → 49 handshakes

20 → 400 handshakes

So, the total number of handshakes exchanged is 400.

	Players on Each Team	Number of Handshakes
	1	1
	2	4
	3	9
	4	16

✓ **LOOK BACK**   The table shows a pattern that relates to square numbers. So, the number of handshakes between two hockey teams each with 20 players is 20^2, or 400.

Try These

1. There were 10 players in the chess tournament. Each player plays every other player in one game. How many games need to be scheduled? **45 games**

2. The sum of the first 100 positive whole numbers is 5,050. What is the sum of the first 100 positive even whole numbers? **10,100**

Use a Formula

Problem Solving
....
• Understand
• Plan
• Solve
• Look Back

The first figure below is called a tetromino. It has a perimeter of 10 units. The second shape connects two tetrominoes and has a perimeter of 18 units. You can use the formula $P = 8n + 2$, where n is the number of connected tetrominoes, to find the perimeter of any connected tetromino shape.

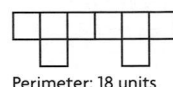

Perimeter: 10 units

Perimeter: 18 units

NOTE: Tetrominoes are always connected the same way.

If the perimeter of a connected tetromino shape is 58 units, how many connected tetromino shapes are there in the figure?

☑ **UNDERSTAND**   You are asked to find how many connected tetrominoes make a figure that has a perimeter of 58 units. You know the formula for finding the perimeter.

☑ **PLAN**   Since you know the value of P, you can replace P with that value in the formula and solve for n.

☑ **SOLVE**   Use the formula $P = 8n + 2$. Replace P with 58.

$$P = 8n + 2$$
$$58 = 8n + 2$$

Think: What number multiplied by 8, then increased by 2, gives 58?

$$8 \times 7 + 2 = 56 + 2 = 58$$

So, there are 7 connected tetrominoes in a figure with a perimeter of 58 units.

☑ **LOOK BACK**   You can also use the formula to find the perimeters of different connected tetromino shapes.

Number of Connected Tetrominos	1	2	3	4	5	6	7
Perimeter	8	18	26	34	42	50	58

Try These

1. Suppose the input number, n, comes from the set of whole numbers and the rule is $n^2 - 1$. Which of the following are output numbers?

80, 99, 224, 339, 481, 528
80, 99, 224, 528

2. A side of square B is 5 times the length of a side of square A. How many times greater is the area of square B than the area of square A? **25 times**

USING THE PAGE
For more practice on this problem-solving strategy, see pp. 406–407

USING THE PAGE
For more practice on this problem-solving strategy, see pp. 328–329

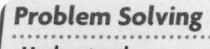
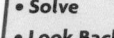

Problem Solving
• Understand
• Plan
• Solve
• Look Back

Write an Equation/Proportion

A recipe for chocolate chip cookies uses 2 eggs to make 15 cookies. If you want to make 60 cookies, how many eggs will you need?

☑ **UNDERSTAND**   You know the number of eggs needed for 15 cookies. You need to know how many eggs to use to make 60 cookies.

☑ **PLAN**   The same two things are compared in both cases. The number of eggs should be in proportion to the number of cookies.

☑ **SOLVE**   Use a proportion (equation) to compare the ratio of eggs to cookies.

$$\frac{\text{eggs}}{\text{cookies}} = \frac{\text{eggs}}{\text{cookies}}$$
$$\frac{2}{15} = \frac{n}{60}$$
$$15 \times n = 60 \times 2$$
$$15n = 120$$
$$\frac{15n}{15} = \frac{120}{15}$$
$$n = 8$$

Use the numbers you know. Let n represent the number of eggs needed for 60 cookies.

Find the cross products.
Divide each side by 15.

So, the 8 eggs are needed to make 60 cookies.

☑ **LOOK BACK**   You could also solve the problem by finding how many batches of 15 cookies make 60 cookies, and then multiplying the number of batches by 2 eggs.

$60 \div 15 = 4 \rightarrow 4$ batches $4 \times 2 = 8 \rightarrow 8$ eggs ✓

Try These

1. The perimeter of the triangle is 65 m. Find the unknown length. Write an equation to solve the problem.
$a + 13 + 26 = 65$; $a = 26$ m; 26 m
2. To park their car at Fun City, Savitri's parents paid $5.00 for the first hour and $1.50 for each additional hour. If they paid $14.00 for parking, how many hours were they at Fun City?
7 hr
3. Of the 180 bicycles sold last month at Cycle Villa, 30 were Mountain Climbers. Mr. Villa is planning to make a circle graph to show this information. How many degrees should the Mountain Climbers section of the graph be? **60 degrees**

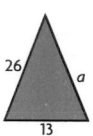

26   a

13

MIXED APPLICATIONS

Use the strategy of your choice to solve.

CHOOSE

1. A math book is lying open on a desk. What pages is it open to if the product of the page numbers of the facing pages is 7,832? **88 and 89**

2. A rectangle has an area of 2,000 ft². One side is 50 ft long. What is the perimeter of the rectangle? **180 ft**

3. Jacques arranged coins on a grid and left the coins in a pattern of 4 rows with 4 coins in each row. Each row had exactly one penny, one nickel, one dime, and one quarter. No row, either horizontal or vertical, had more than one coin of each kind. How had Jacques arranged the coins? **Possible answer: see below.**

4. Choose any three counting numbers less than 10. Write all the 2-digit numbers possible with these numbers. Do not repeat a digit. Find the sum of the 2-digit numbers. Divide by the sum of the three original numbers. Why is the quotient always 22? **All sums are multiples of 11 and 2.**

5. David made the following transactions on his bank account in one month:

   Deposit $120.00
   Check $35.28
   Check $63.97
   Deposit $85.00

   The balance at the end of the month was $326.79. What was the balance at the beginning of the month? **$221.04**

6. A target has three circles, one inside of the other. The bull's eye is 10 points, the middle circle is 7 points, and the outer circle is 5 points. If three darts are thrown and only two hit the target, what are the possible scores? **10, 12, 14, 15, 17, or 20 points**

7. The bells ring at 10:00 A.M., 10:40 A.M., 11:20 A.M., and 12:00 P.M. At which of these times will the bells ring?

   6:00 P.M.   8:00 P.M.   12:00 A.M.
   all times listed

8. How many cents are in *d* dimes? How many nickels are in *q* quarters? **10*d*; 5*q***

9. How many ways can you add 8 odd numbers to get a sum of 20? Each number may be used more than once. **11 ways**

10. A baked loaf of bread is about two times higher than the unbaked dough. A baked loaf is 5.5 in. high. How high was the unbaked dough? **2.75 in.**

11. Aaron is designing a studio and wants to change the square window. He will make it twice as long and half as tall. What happens to its area? **remains the same**

12. Write a problem that can be solved with more than one strategy. Exchange with a classmate and solve. **Problems will vary.**

13. David has 4 trophies. In how many different ways can he arrange them on his trophy shelf? **24 different ways**

Focus on Problem Solving   13

Additional Answer 3.			
P	N	D	Q
Q	D	N	P
N	P	Q	D
D	Q	P	N

PLANNING GUIDE

Looking at Numbers

BIG IDEA: Numbers can be represented in various equivalent forms using integers, fractions, decimals, and exponents.

Introducing the Chapter p. 14	Team-Up Project • Fascinating Facts p. 15		
OBJECTIVE	**VOCABULARY**	**MATERIALS**	**RESOURCES**
1.1 Whole Numbers and Decimals pp. 16–17 **Objective:** To use place value to express whole numbers and decimals **1 DAY**	*decimal point*	FOR EACH PAIR: place-value models, decimal models	■ Reteach, ■ Practice, ■ Enrichment 1.1; More Practice, p. H42 Math Fun: Reading, Writing, and Numbers, p. 30
EXTENSION • Other Number Systems, Teacher's Edition pp. 18A–18B			
1.2 Comparing and Ordering pp. 18-19 **Objective:** To compare and order whole numbers and decimals **1 DAY**			■ Reteach, ■ Practice, ■ Enrichment 1.2; More Practice, p. H42
1.3 Decimals and Fractions pp. 20-23 **Objective:** To write a decimal as a fraction and a fraction as a decimal **2 DAYS**			■ Reteach, ■ Practice, ■ Enrichment 1.3; More Practice, p. H42 Math Fun • Jeremy's Three, p. 30 *Calculator Crew • Nautical Number Line* Calculator Activities, p. H35
1.4 Exponents pp. 24-25 **Objective:** To represent numbers by using exponents **1 DAY**	*factors* **exponent** **base**		■ Reteach, ■ Practice, ■ Enrichment 1.4; More Practice, p. H43 Math Fun • The Numbers Are Falling, p. 30
Lab Activity: Squares and Square Roots pp. 26-27 **Objective:** To use a model to find squares and square roots	**square** **square root**	FOR EACH STUDENT: square tiles	🖥 E-Lab • Activity 1 E-Lab Recording Sheets, p. 1
1.5 Integers pp. 28-29 **Objective:** To order integers and to identify opposite integers **1 DAY**	**integers** **positive integers** **negative integers** **opposites**		■ Reteach, ■ Practice, ■ Enrichment 1.5; More Practice, p. H43
CHAPTER ASSESSMENT Chapter 1 Review p. 31			

NCTM CURRICULUM
STANDARDS FOR GRADES 5–8

- ✓ **Problem Solving** *Lessons 1.1, 1.2, 1.3, 1.4, 1.5*
- ✓ **Communication** *Lessons 1.1, 1.2, 1.3, 1.4, 1.5*
- ✓ **Reasoning** *Lessons 1.1, 1.2, 1.3, 1.4, 1.5*
- ✓ **Connections** *Lessons 1.1, 1.2, 1.3, 1.4, 1.5*
- ✓ **Number and Number Relationships** *Lessons 1.1, 1.2, 1.3, 1.4, 1.5*
- ✓ **Number Systems and Number Theory** *Lessons, 1.1, 1.2, 1.3, 1.4, 1.5*
- ✓ **Computation and Estimation** *Lessons 1.3, 1.4*
- ✓ **Patterns and Functions** *Lessons 1.2, 1.4*
- ✓ **Algebra** *Lessons 1.2, 1.4, 1.5*
- ✓ **Statistics** *Lesson 1.2*
- ☐ **Probability**
- ✓ **Geometry** *Lab Activity*
- ✓ **Measurement** *Lessons 1.2, 1.3, 1.5*

CHAPTER LEARNING GOALS

1A.1 To use place value to express and compare whole numbers and decimals.

1A.2 To write a decimal as a fraction and a fraction as a decimal.

1A.3 To use exponents to represent numbers.

1A.4 To order integers and identify opposite integers.

These goals for concepts, skills, and problem solving are assessed in many ways throughout the chapter. See the chart below for a complete listing of both traditional and informal assessment options.

Pretest Options
Pretest for Chapter Content, TCM pp. A5–A6, B155–B156
Assessing Prior Knowledge, TE p. 14; APK p. 3

Daily Assessment
Mixed Review, PE pp. 23, 25
Problem of the Day, TE pp. 16, 18, 20, 24, 28
Assessment Tip and Daily Quiz, TE pp. 17, 19, 23, 25, 27, 29

Formal Assessment
Chapter 1 Review, PE p. 31
Chapter 1 Test, TCM pp. A5–A6, B155–B156
Study Guide & Review, PE pp. 70–71
Test • Chapters 1–3 Test, TCM pp. A11–A12, B161–B162
Cumulative Review, PE p. 73
Cumulative Test, TCM pp. A13–A16, B163–B166

Performance Assessment
Team-Up Project Checklist, TE p. 15
What Did I Learn?, Chapters 1–3, TE p. 72
Interview/Task Test, PA p. 76

Portfolio
Suggested work samples:
Independent Practice, PE pp. 17, 19, 22, 25, 29
Math Fun/Cultural Connection, PE p. 30

Student Self-Assessment
How Did I Do?, TE p. 31, TGA p. 15
Portfolio Evaluation Form, TGA p. 24
Math Journal, TE pp. 14B, 27

Key
Assessing Prior Knowledge Copying Masters: APK
Test Copying Masters: TCM
Performance Assessment: PA
Teacher's Guide for Assessment: TGA
Teacher's Edition: TE
Pupil's Edition: PE

MATH JOURNAL
You may wish to use the following suggestions to have students write about the mathematics they are learning in this chapter:

PE page 22 • Talk About It

PE pages 17, 19, 23, 29 • Write About It

TE page 27 • Writing in Mathematics

TGA page 15 • How Did I Do?

In every lesson • record written answers to each Wrap-up Assessment Tip from the Teacher's Edition for test preparation.

VOCABULARY DEVELOPMENT
The terms listed at the right will be introduced or reinforced in this chapter. The boldfaced words in the list are new vocabulary terms in the chapter.

To help students develop an understanding of the words **exponent** and **base,** have them write a multiplication sentence in their Math Journal and label the numbers in the sentence with the words. Then have them tell in their own words how to use multiplication to find the value of a number with an exponent.

decimal point, pp. 16, 21
factors, pp. 24, 27, 31
exponent, pp. 24, 26, 31
base, pp. 24, 26
square, pp. 26, 27
square root, pp. 26, 27
integers, pp. 28, 29, 31
positive integers, p. 28
negative integers, p. 28
opposites, pp. 28, 31

MEETING INDIVIDUAL NEEDS

RETEACH
BELOW-LEVEL STUDENTS

Practice Activities, TE Tab B
Building A Number
(Use with Lesson 1.1.)

Interdisciplinary Connection, TE p. 26

Literature Link, TE p. 14C
Not for a Billion, Gazillion Dollars, Activity 1
(Use with Lesson 1.1.)

Technology
Calculating Crew • *Nautical Number Line*
(Use with Lesson 1.3.)

E-Lab • Activity 1
(Use with Lab Activity.)

PRACTICE
ON-LEVEL STUDENTS

Bulletin Board TE p. 14C
The Number Line Challenge
(Use with Lessons 1.1–1.5.)

Interdisciplinary Connection, TE p. 26

Literature Link, TE p. 14C
Not for a Billion, Gazillion Dollars, Activity 2
(Use with Lesson 1.1.)

Technology
Calculating Crew • *Nautical Number Line*
(Use with Lesson 1.3.)

E-Lab • Activity 1
(Use with Lab Activity.)

CHAPTER IDEA FILE

The activities on these pages provide additional practice and reinforcement of key chapter concepts. The chart above offers suggestions for their use.

LITERATURE LINK

Not for a Billion, Gazillion Dollars

by Paula Danzinger

Matthew learns important lessons about money management and friendship.

Activity 1—Matthew says he doesn't want to worry about money ". . . for a billion, gazillion dollars." Since *gazillion* is not a real word, have students write and name numbers that Matthew might have used to represent numbers more than a thousand.

Activity 2—Matthew's parents decide to keep a fraction of his allowance each week to help him pay back his debts. Tell students to assume Matthew gets $5 a week and his parents keep $\frac{1}{4}$ of it. Have them figure out how much allowance Matthew actually receives weekly and how long it will take him to save the $100 for the new computer program.

Activity 3—Matthew saved $100 in coins. Have students think about possible coin combinations and decide if it is possible for him to have saved 1,000, 15,000, or 800 coins to total $100.

BULLETIN BOARD

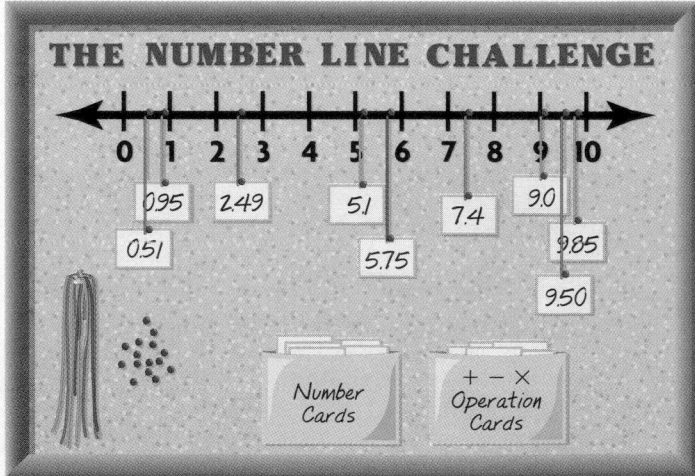

Place a variety of number cards in the pocket of the display. Students select numbers to plot on the number line and to use in different operations as they complete Chapter 1.

Activity 1—Each student selects 2 number cards and plots them on the number line, using tacks and string.

Activity 2—Each student chooses an operation card and completes that operation with 2 numbers from the number line.

Bulletin Board TE p. 14C
The Number Line Challenge
(Use with Lessons 1.1–1.5)

Interdisciplinary Connection, TE p. 26

 **Literature Link,** TE p. 14C
Not for a Billion, Gazillion Dollars, Activity 3
(Use with Lesson 1.1)

 Technology
Calculating Crew • Nautical Number Line
(Use with Lesson 1.3.)

E-Lab • Activity 1
(Use with Lab Activity.)

INTERDISCIPLINARY SUGGESTIONS

Connecting Looking at Numbers

Purpose To connect looking at numbers to other subjects students are studying

You may wish to connect *Looking at Numbers* to other subjects by using related interdisciplinary units.

Suggested Units

- The Language of Math: Writing Personal Dictionaries

- Packaging: How Numbers Are Used on the Things We Buy

- Number Games from Around the World

- Big and Small Numbers in the Universe

Modalities Auditory, Kinesthetic, Visual

Multiple Intelligences Mathematical/Logical, Visual/Spatial, Bodily/Kinesthetic

Technology

The purpose of using technology in the chapter is to provide reinforcement of skills and support in concept development.

E-Lab • Activity 1—Students discover the graphical pattern produced by repeatedly squaring or taking the square root of a number.

Calculating Crew • Nautical Number Line—The activities on this CD-ROM provide students with opportunities to practice identifying equivalent fractions for decimals.

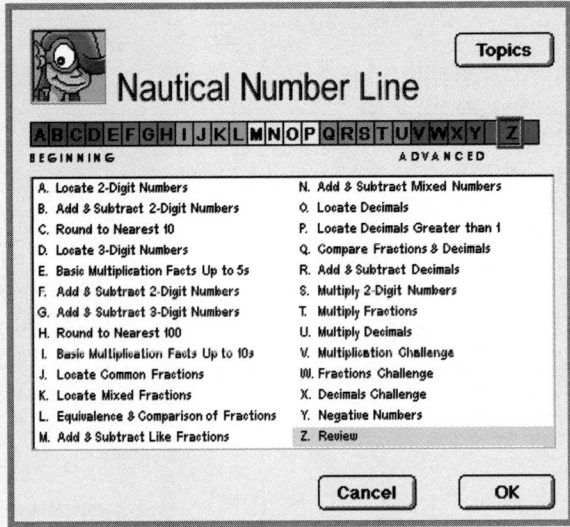

	Topics
Nautical Number Line	

A B C D E F G H I J K L M N O P Q R S T U V W X Y **Z**

BEGINNING   ADVANCED

A. Locate 2-Digit Numbers
B. Add & Subtract 2-Digit Numbers
C. Round to Nearest 10
D. Locate 3-Digit Numbers
E. Basic Multiplication Facts Up to 5s
F. Add & Subtract 2-Digit Numbers
G. Add & Subtract 3-Digit Numbers
H. Round to Nearest 100
I. Basic Multiplication Facts Up to 10s
J. Locate Common Fractions
K. Locate Mixed Fractions
L. Equivalence & Comparison of Fractions
M. Add & Subtract Like Fractions

N. Add & Subtract Mixed Numbers
O. Locate Decimals
P. Locate Decimals Greater than 1
Q. Compare Fractions & Decimals
R. Add & Subtract Decimals
S. Multiply 2-Digit Numbers
T. Multiply Fractions
U. Multiply Decimals
V. Multiplication Challenge
W. Fractions Challenge
X. Decimals Challenge
Y. Negative Numbers
Z. Review

Cancel   OK

Each Calculating Crew activity can be assigned by degree of difficulty. See Teacher's Edition page 23 for Grow Slide information.

 **Explorer Plus™ or fx–65 Calculator**—See suggestions for integrating the calculator in the PE Lesson p. 21. Additional ideas are provided in *Calculator Activities,* p. H35.

Visit our web site for additional ideas and activities.
http://www.hbschool.com

CHAPTER 1

Introducing the Chapter

Encourage students to talk about how they use fractions every day. Review the math content students will learn in the chapter.

WHY LEARN THIS?

The question "Why are you learning this?" will be answered in every lesson. Sometimes there is a direct application to everyday experiences. Other lessons help students understand why these skills are needed for future success in mathematics.

PRETEST OPTIONS

Pretest for Chapter Content See the *Test Copying Masters*, pages A5–A6, B155–B156 for a free-response or multiple-choice test that can be used as a pretest.

Assessing Prior Knowledge Use the copying master shown below to assess previously taught skills that are critical for success in this chapter.

Assessing Prior Knowledge Copying Masters, p. 3

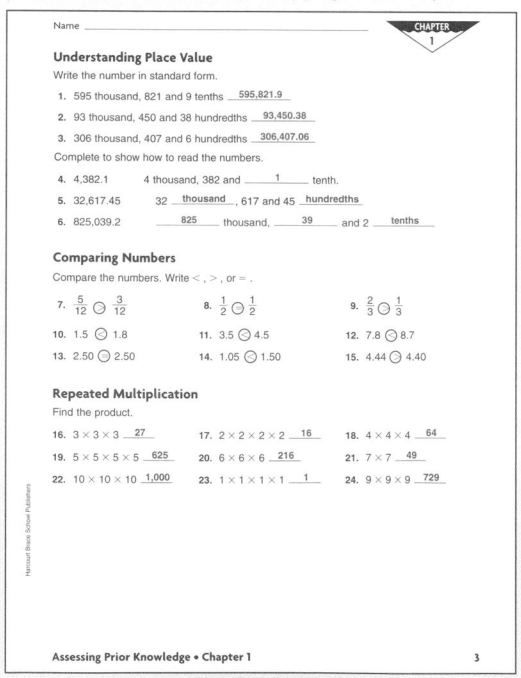

Troubleshooting

If students need help	Then use
• UNDERSTANDING PLACE VALUE	• Key Skills, PE p. H6
• COMPARING NUMBERS	• Practice Activities, TE Tab B, Activity 3A
• WITH MULTIPLICATION	• Key Skills, PE p. H8

CHAPTER 1
LOOKING AT NUMBERS

People have been digging up dinosaurs for nearly 200 years, but about 50 percent of all dinosaurs we know have been found in the last 20 years.

The smallest dinosaur, only 2 feet long, was called compsognathus. Its 3-inch head, about 12.5 percent of its body length, housed lots of small, sharp teeth. It first appeared about 145 million years ago in France and Germany.

It disappeared about 5 million years ago.

What's the largest dinosaur? In 1986, New Mexico scientists announced they had found the bones of Seismosaurus, "earth-shaker." Its tail bones suggested a creature 150 feet long, weighing 50 tons, or as much as ten large elephants.

We know by measuring their footprints that some dinosaur hunters could run as fast as 25 miles per hour. That's 0.416 miles per minute!

LOOK AHEAD

In this chapter you will solve problems that involve

• using place value of whole numbers and decimals

• comparing and ordering decimals

• changing decimals to fractions and fractions to decimals

• identifying and ordering integers

14   Chapter 1

 SCHOOL-HOME CONNECTION

You may wish to send to each student's family the *Math at Home* page. It includes an article that stresses the importance of communication about mathematics in the home, as well as helpful hints about homework and taking tests. The activity and extension provide extra practice that develops concepts and skills involving fractions and decimals.

Looking at Numbers: The Total Picture

Your child is beginning a chapter that teaches more about place value for whole numbers and decimals. Problems in this chapter deal with comparing and ordering numbers, using place value of whole numbers and decimals, seeing relationships between fractions and decimals, and using exponents and integers. Ask your child to share with you activities from this unit.

Here are an activity and an Extension that explore the size of 1,000,000. Use a calculator to help you answer the questions.

The Newest Millionaire

Imagine that you have just won a contest that has a grand prize of $1,000,000 in cash. The payout is to be made in one-dollar bills. You will need something to carry the money in when you pick it up. If you use a suitcase, what size will you need? How heavy will the suitcase be?

Data Bank
Each bill measures: 6 in. × 2.5 in.
A stack of twenty bills weigh about 0.7 oz.
A stack of twelve bills is about $\frac{1}{16}$ in. thick.
1 mile = 5,280 ft.

Extension • Round and Round We Go!

If the dollar bills were placed end to end, how far would the $1,000,000 reach? How far would it reach? Suppose your $1,000,000 was in hundred-dollar bills. How far would the hundred-dollar bills stretch? How much would they weigh?

For Your Information

Classrooms that are based on the standards differ from classes parents recall. Parents may recall exercises that, if done successfully, pleased the teachers, but you may not recollect why those exercises were learned. We may remember tests for which we memorized procedures to earn high scores, but we forget the mathematics that was learned to pass the test.

Parents can help their children see connections between their world and the skills and ideas being learned.

Condensed from "Mathematics Making a Living, Making a Life," National Council of Teachers of Mathematics

 It has been shown that getting a good night's sleep and eating breakfast help children learn better in school. They also help children do better on tests.

VOCABULARY
exponent—a number that shows how many times the base number is used as a factor
(base) x^2 (exponent)
square—the product of a number and itself
square root—two equal factors of a number

Math at Home

Team-Up Project

Fascinating Facts

What kind of facts interest you? Create a display of fascinating facts.

YOU WILL NEED: resource books such as, an almanac or encyclopedia

Plan • Work with a small group.

Act • Select four or five facts. Be sure your information is correct.
• Make a display of your facts.

Share • Name the type of numbers used for each fact in your display. Explain why it is appropriate.
• List all the different types of numbers used by the class. Identify the facts that could be expressed with another type of number.

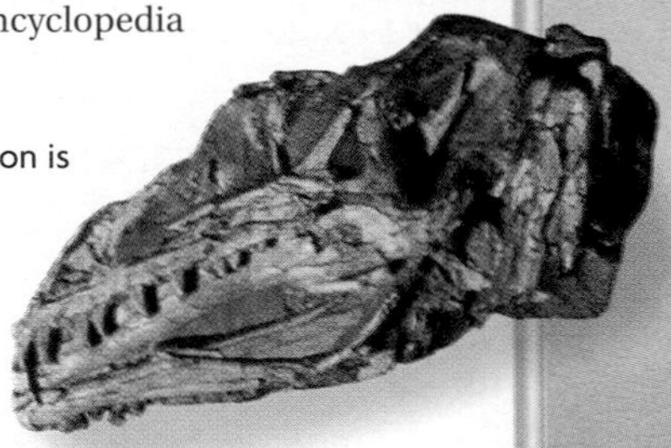

DID YOU

☑ Record at least four interesting facts?

☑ Check that the facts are correct?

☑ Make a display of your facts?

☑ Participate in the class discussion about the types of numbers that were used?

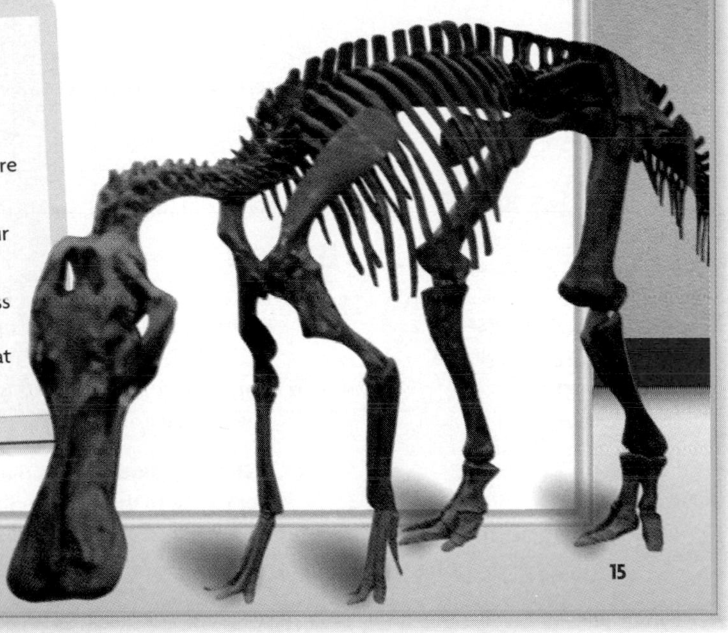

15

Project Checklist

Evaluate whether each group

☑ works cooperatively.

☑ recognizes various types of numbers.

☑ checks the facts.

☑ helps create the display.

Using the Project

Accessing Prior Knowledge Reading numbers

Purpose Students create a display of facts that use various number forms.

GROUPING small groups
MATERIALS 12 in. by 18 in. sheets of paper, drawing supplies, reference materials

Managing the Activity Students can find interesting facts in many places including magazines, reference books, and online. Encourage them to note their sources and to check the facts using more than one resource.

WHEN MINUTES COUNT

A shortened version of the activity can be completed in 20–30 minutes. Have each group locate a fascinating fact. Then lead a class discussion in which students list the different types of numbers used to express the facts.

A BIT EACH DAY

Day 1: Students work in groups to brainstorm about interesting facts or topics for their display.

Day 2: Students research and record a number of fascinating facts and check their information.

Day 3: Students select facts to use in the display.

Day 4: Students create their display.

Day 5: Groups present their displays. The class compiles a list of all of the types of numbers used.

The biggest and smallest bird egg sizes in the world are those of the ostrich **(7 inches, or 177 mm)** and Helena's hummingbird **($\frac{1}{2}$ inch or 12.7 mm)**.

The world record for rainfall in one year is **1,042 inches** in Cherrapunji, a town in India, in the foothills of the Himalaya.

HOME HAPPENINGS

Have students ask family members to share fascinating facts. Encourage them to check the facts.

GRADE 5	GRADE 6	GRADE 7
Read and write whole numbers and decimals in standard and expanded form.	Read and write whole numbers and decimals in standard and expanded form.	Explore decimal and binary numbers.

ORGANIZER

OBJECTIVE To use place value to express whole numbers and decimals

ASSIGNMENTS Independent Practice
- ■ *Basic* 1–17, odd; 18–20
- ■ *Average* 1–20
- ■ *Advanced* 5–17, odd; 18–20

PROGRAM RESOURCES
Reteach 1.1 Practice 1.1 Enrichment 1.1
Math Fun • Reading, Writing, and Numbers, p. 30
Extension • Other Number Systems, TE pp. 18A–18B

PROBLEM OF THE DAY

Transparency 1

Michael's number has six twos and five zeros. Each 2, except the one in the ten-thousandths place, is 100 times greater than the 2 to its right. Write his number in words. two million, twenty thousand, two hundred two and two hundred two ten-thousandths

Solution Tab A, p. A1

QUICK CHECK

Write in standard form.

1. 4,000 + 300 + 20 + 9 4,329
2. 70,000 + 500 + 80 + 7 70,587
3. 800,000 + 60 + 1 800,061
4. 30,000 + 9 30,009

1 MOTIVATE

MATERIALS: *For each pair* place-value models, decimal models

Have one student in each pair use place-value models to show the whole-number part of 21.509. Have the other student use decimal models to show the decimal part. Continue this activity, with student pairs changing roles for each number. Check students' models.

1. 709.08   2. 5,089.479   3. 80.3752

1.1

WHAT YOU'LL LEARN
How to use place value to express whole numbers and decimals

WHY LEARN THIS?
To understand and compare whole numbers and decimals used in population counts, prices, and sizes of bacteria

Whole Numbers and Decimals

Think about the numbers you see every day in school, in the newspaper, on television, and in books and magazines. How does place value help you understand these numbers?

Tiger Woods

The numbers you see every day are part of the decimal system. The decimal system uses ten digits: 0, 1, 2, 3, 4, 5, 6, 7, 8, and 9. The digits and the position of each digit in a number determine the number's value.

- Read each number on the place-value chart. What is the value of the digit 3 in each number? **3 thousands; 3 thousandths; 3 tens**

	PLACE VALUE									
Millions	Hundred Thousands	Ten Thousands	Thousands	Hundreds	Tens	Ones	Tenths	Hundredths	Thousandths	Ten-Thousandths
1	6	2	3	0	5	1				
							0	0	5	3	1
					3	2	4			

1,623,051 →
0.0531 →
32.4 →

When you read numbers, you use place value.

REMEMBER:
When reading a number with a decimal point, read the decimal point as "and." **See page H10.**
Read 8.2 as "eight and two tenths."

EXAMPLES

A. Write: 841,230 ← *standard form*
Read: eight hundred forty-one thousand, two hundred thirty

B. Write: 12.75 ← *standard form*
Read: twelve and seventy-five hundredths

- 💡 **CRITICAL THINKING** In the chart, what would be the next three places to the left of millions? **ten millions, hundred millions, billions**

IDEA FILE

ENGLISH AS A SECOND LANGUAGE

Tip • Many students acquiring English have trouble distinguishing the endings of words. In place value, it is important to clearly distinguish the *-th* that discriminates decimal places from whole-number places. Encourage students to exaggerate the ending sounds as they practice reading these numbers.

Modality Auditory

RETEACH 1.1

Name _____

Whole Numbers and Decimals

The numbers you use every day are part of the *decimal system*. To find a number's value, look at the digits and position of each digit. A place-value chart can help you.

	Place Value									
Millions	Hundred Thousands	Ten Thousands	Thousands	Hundreds	Tens	Ones	Tenths	Hundredths	Thousandths	Ten-Thousandths

decimal point

Sandy is asked to give the value of the digit 4 in the number 18.34.
- Sandy sees that the position of the digit 4 is two places to the right of the decimal point. 18.34
- Sandy uses the place-value chart to name the *position* of the digit. Counting two places to the right of the decimal point, he finds the digit 4 is in the hundredths place.
- Using both the digit and its position, Sandy discovers the digit's value is 4 hundredths, or 0.04.

Answer the questions about the value of each underlined digit. Use the place-value chart if necessary.

1. 4,082,113.7<u>2</u>3
   a. What is the position of the digit? **thousandths**
   b. What is the value of the digit? **3 thousandths; 0.003**

2. <u>5</u>8,166.7
   a. What is the position of the digit? **ten thousands**
   b. What is the value of the digit? **5 ten thousands; 50,000**

3. 30.1<u>9</u>4
   a. What is the position of the digit? **tenths**
   b. What is the value of the digit? **1 tenth; 0.1**

4. <u>9</u>,115,077.422
   a. What is the position of the digit? **millions**
   b. What is the value of the digit? **9 millions; 9,000,000**

Use with text pages 16–17.   TAKE ANOTHER LOOK R1

MEETING INDIVIDUAL NEEDS

GUIDED PRACTICE

Read the number. What is the value of the digit 6?

1. 34,065
   6 tens
2. 15,425.006
   6 thousandths
3. 2,654,000.25
   6 hundred thousands
4. 550.76
   6 hundredths

Write the number in words. **See Additional Answers, p. 31A.**

5. 10,040
6. 0.605
7. 6.5
8. 7,868
9. 863,042.46

Write the number in standard form.

10. seven million, twenty-four thousand, two hundred one **7,024,201**
11. eighty-four and seven hundred twenty-two thousandths **84.722**
12. six hundredths **0.06**
13. five thousand, forty-two **5,042**

INDEPENDENT PRACTICE

Write the value of the digit 4.

1. 34.1
   4 ones
2. 400
   4 hundreds
3. 11.394
   4 thousandths
4. 940,006
   4 ten thousands

Write in words the value of the underlined digit.

5. 120
   two tens
6. 5.0457 **seven ten-thousandths**
7. 800,927 **eight hundred thousands**
8. 345.69 **six tenths**

Write the number in words. **See Additional Answers, p. 31A.**

9. 46
10. 0.03
11. 1,500.1
12. 2.456
13. 1,037,804

Write the number in standard form.

14. one hundred twenty-five **125**
15. fifteen hundredths **0.15**
16. thirty-five and two thousand, three hundred sixty-nine ten-thousandths **35.2369**
17. eight hundred fifty thousand, two hundred forty-seven and fifty-six hundredths **850,247.56**

Problem-Solving Applications

18. The world's smallest cut diamond is 0.0009 in. in diameter and weighs 0.0012 carat. Write both numbers in words. **nine ten-thousandths; twelve ten-thousandths**
19. A movie studio announced that the box office sales for its new release reached nine million, four hundred fifty-six thousand, three hundred two in the first week. Write the number in standard form. **9,456,302**
20. ✏ **WRITE ABOUT IT** How do you know that 8.1 and 8.001 are not the same? **The ones have different place values.**

MORE PRACTICE Lesson 1.1, page H42

17

As students work through the examples on page 16, ask questions such as these: **What is the value of the digit 1 in Example A?** 1,000 **How many hundreds are in Example B?** zero hundreds

GUIDED PRACTICE
Complete these exercises orally with the entire class. Make sure that students understand the concept of place value before you assign the Independent Practice.

INDEPENDENT PRACTICE
To ensure student success, you may wish to have students complete Exercises 1, 5, 9, and 14 orally before you assign the remaining exercises.

ADDITIONAL EXAMPLE
Example, p. 16

Write: 7,204,805 ← standard form

Read: seven million, two hundred four thousand, eight hundred five

3 WRAP UP

Use these questions to help students focus on the big ideas of the lesson.

- **How does the value of the digit 6 in 672.48 compare to the value of the digit 6 in 9.061?** The value of the digit 6 in 672.48 is six hundred. The value of the digit 6 in 9.061 is six hundredths.

✓ *Assessment Tip*
Have students write the answer.

- **Describe how the value of each digit in a number compares to the value of the digit to its right.** It is 10 times greater.

DAILY QUIZ
Write the value of the digit 5.

1. 4.895 five thousandths
2. 50.3 five tens
3. 72.654 five hundredths
4. 7,057,213 five ten-thousands

Write the number in words.

5. 7.028 seven and twenty-eight thousandths
6. 904,000.4 nine hundred four thousand and four tenths

PRACTICE 1.1

Name _____

LESSON 1.1

Whole Numbers and Decimals

Write the value of the digit 6.

1. 26.8
2. 2.600
3. 78.056
4. 560.003

   6 ones **6 hundreds** **6 thousandths** **6 ten thousands**

Write in words the value of the underlined digit.

5. 571
6. 92,114.6
7. 458.035
8. 862.077

   7 tens; **2 thousands;** **5 thousandths;** **6 ten thou-**
   70 **2,000** **0.005** **sands; 60,000**

Write the number in words.

9. 12.7 __twelve and seven tenths__
10. 254.91 __two hundred fifty-four and ninety-one hundredths__
11. 4.814 __four and eight hundred fourteen thousandths__

Write the number in standard form.

12. three thousand, forty-nine
    3,049
13. sixty thousand, three hundred five
    60,305
14. twenty-eight hundredths
    0.28
15. seventy-five and six tenths
    75.6

Mixed Applications

16. Over a holiday weekend, a total of six thousand, four hundred eighty-seven cars passed through the midtown tunnel. Write the number in standard form.
    6,487
17. The zigot factory made nine million, six hundred twenty-two thousand, seventeen zigots during the month of June. Write the number in standard form.
    9,622,017
18. A rectangle has a length of 6 in. and a width of 4 in. Find the area of the rectangle in square inches.
    24 in.²
19. A scientist discovered that a plant grew 0.0027 cm over a 24-hr period. Use words to write the plant's growth.
    twenty-seven ten-thousandths

Use with text pages 16–17. **ON MY OWN** P1

ENRICHMENT 1.1

Name _____

LESSON 1.1

Riddle Fun

Match each number written in words in Column 1 with its standard form in Column 2. Then write each corresponding letter on the line below marked with the exercise number. You will not use every letter to solve the riddle.

Column 1		Column 2
1. two thousand and six tenths __L__		A. 46.008
2. three million, seventeen __O__		B. 12,000,000.050
3. forty-six and eight thousandths __A__		C. 0.2006
4. five hundred fourteen thousand and nine tenths __M__		E. 0.0046
5. twelve million and fifty thousandths __B__		H. 600,021
6. six hundred twenty-one ten-thousandths __U__		I. 2,000.6
7. two thousand six ten-thousandths __C__		M. 514,000.9
8. one million and one hundredth __T__		N. 12,050,000
9. forty-six ten-thousandths __E__		O. 3,000,017
10. five hundred nine ten-thousandths __Y__		T. 1,000,000.01
11. twelve million, fifty thousand __N__		U. 0.0621
12. six hundred thousand, twenty-one __H__		Y. 0.0509

What did one number say to another number?

Y	O	U		C	A	N		C	O	U	N	T
10	2	6		7	3	11		7	2	6	11	8

O	N		M	E
2	11		4	9

Use with text pages 16–17. **STRETCH YOUR THINKING** E1

17

EXTENSION

Other Number Systems

Use these strategies to extend understanding of other number systems.

Multiple Intelligences Logical/Mathematical

Modalities Auditory, Kinesthetic, Visual

Lesson Connection Students will relate various number systems to the decimal system, a place-value system.

Program Resources *Intervention and Extension Copying Masters,* p. 13

Challenge

TEACHER NOTES

ACTIVITY 1

Purpose To explore the ancient Egyptian number system as an example of a system that does not use place value

Discuss the following questions.

- Is the ancient Egyptian system a place-value number system? No. In this system one merely adds the value of each symbol. A heel bone means ten wherever it appears in the number.

- Is the ancient Egyptian system a base-ten system? Yes. Each symbol represents a power of 10.

WHAT STUDENTS DO

1. The ancient Egyptians used the following symbols.

| | ∩ | ℓ | ℓ̃ | ʃ |
1   10   100   1,000   10,000

The number 4,123 could be written as follows.

ℓℓℓℓ ∕∩∩∩||| or |||∩∩∩∕ ℓℓℓℓ

Write the number 32,132 in the Egyptian number system. (The order of the symbols does not matter.)

ʃʃʃ ℓ̃ℓ̃ ∕∩∩∩||

2. The names of the five symbols are scroll, flower, staff, heel bone, and crooked finger. Match these names with the numbers 1, 10, 100, 1,000, and 10,000. 1 : staff, 10 : heel bone, 100 : scroll, 1,000 : flower, 10,000 : crooked finger

3. Write the following numbers in the ancient Egyptian system.
   a. the number of students in your class Answers will vary.
   b. the current year Answers will vary.
   c. 1,000

   ℓ̃

   d. 999

   ℓℓℓℓℓℓℓℓℓ ∩∩∩∩∩∩∩∩∩ |||||||||

TEACHER NOTES

ACTIVITY 2

Purpose To compare the Roman numeral system with ancient Egyptian system and with the decimal system

Discuss the following questions.

- What is the main advantage of having symbols for 5, 50, and 500 in addition to those for powers of 10? Representing numbers is more efficient. For example, 50 = L, not XXXXX.

- What are the only two numbers formed by subtracting 1? IV, IX 10? XL, XC 100? CD, CM

- What is the advantage of being able to subtract 1, 10, or 100 in the Roman system? You can use fewer symbols. For example, 40 = XL, not XXXX.

Have students meet in small groups or with partners to discuss the given question. Have one student report the group's responses to the larger group.

WHAT STUDENTS DO

1. The Roman numeral system uses these symbols to name numbers:

I	V	X	L	C	D	M
1	5	10	50	100	500	1,000

To form other numbers, begin with the greatest symbol needed and add others (in decreasing value) from left to right.

Write 1,236 as M, CC, XXX, V, and I together: MCCXXXVI.

- Read these numbers: VII 7   XVIII 18   XXXIII 33   MCI 1,101

2. These six numbers can be written by using four identical symbols but are often shortened by using subtraction as shown.

4	9	40	90	400	900
IIII	VIIII	XXXX	LXXXX	CCCC	DCCCC
IV	IX	XL	XC	CD	CM
1 from 5	1 from 10	10 from 50	10 from 100	100 from 500	100 from 1,000

494 is CDXCIV and 3,949, is MMMCMXLIX.

- Read these numbers: XXIV 24   LIX 59   XLVI 46   XCV 95

- Write the Roman numerals for these numbers:

    22 XXII   137 CXXXVII   713 DCCXIII   2,817 MMDCCCXVII

3. How are the Roman and Egyptian number systems alike and different? Answers will vary.

Share Results

- Which of these two number systems did you find easier to use? Why? Answers will vary.

Intervention and Extension Copying Masters, p. 13

ORGANIZER

GRADE 5	GRADE 6	GRADE 7
Compare and order whole numbers and decimals.	Compare and order whole numbers and decimals.	Locate real numbers on a number line.

OBJECTIVE To compare and order whole numbers and decimals

ASSIGNMENTS Independent Practice
- *Basic* 1–18, even; 19–21
- *Average* 1–21
- *Advanced* 13–21

PROGRAM RESOURCES
Reteach 1.2 Practice 1.2 Enrichment 1.2
Math Fun • Jeremy's Two, p. 30

PROBLEM OF THE DAY

Transparency 1

Two numbers have the same seven digits in the same order. The leftmost place is in the millions place. The rightmost place is in the ten-thousandths place. The difference between the numbers is 1,975,427.4375. What are the two numbers? 1,975,625 and 197.5625

Solution Tab A, p. A1

QUICK CHECK

Write the number in standard form.

1. 200 + 7 = 207
2. 500 + 10 + 3 = 513
3. 1,000 + 50 + 4 = 1,054
4. 2,000 ┃ 900 + 60 + 8 ─ 2,968

1 MOTIVATE

To access students' knowledge ask:

- **How much time do you spend in math class each day? Do you think you spend more or less time than students in other countries?**

Copy the data shown below on the board, and have students write three questions about the information that will help them compare and order the data.

Average Time Spent in Math Class Per Day	
South Korea – 36 min	Switzerland – 50 min
France – 45 min	United States – 46 min

Comparing and Ordering

WHAT YOU'LL LEARN
How to compare and order whole numbers and decimals

WHY LEARN THIS?
To compare measurements or amounts of money

In large quantities plastic straws cost less than 1¢ each. Paper straws cost more, about 1.25¢ each. Some animal parks use paper straws that are biodegradable. They are relatively harmless to the park's animals if the straws are swallowed accidentally.

REMEMBER:
You can add a zero to the right of a decimal without changing its value. For example, 7.2 and 7.20 are equivalent.
See page H12, Key Skill 21.

Do you ever draw straws to determine who will be chosen from a group? Sam is drawing straws with his friends. The person who draws the shortest straw will have to help clean up after a party.

- The first straw is 5.5 cm, and the second straw is 5.7 cm. Which straw is longer?

You can use a number line to decide whether one number is greater than or less than another number.

Since 5.5 is to the left of 5.7, 5.5 is less than 5.7.
You can write 5.5 < 5.7, or you can write 5.7 > 5.5.
↑ means "less than"   ↑ means "greater than"

Another way to compare numbers is to use place value.

EXAMPLE 1 Compare 7.28 and 7.2. Use < or >.

7.28	7.2	*Compare the digits. Start at the left.*
7.28	7.2	*same number of ones*
7.28	7.2	*same number of tenths*
7.28	7.20	*Add a zero to 7.2 so that both decimals have the same number of places.*

Since 8 hundredths is greater than 0 hundredths, 7.28 > 7.2, or 7.2 < 7.28.

EXAMPLE 2 Use < to list the numbers in order from least to greatest. 12,010; 12,031; 12.1; 13,001.5

12.1 < 12,010   12,010 < 12,031   *Compare the numbers.*
    12,031 < 13,001.5

12.1 < 12,010 < 12,031 < 13,001.5   *List the numbers in order.*

- Use > to list the numbers in order from greatest to least.
  0.821, 3.821, 3.028 **3.821 > 3.028 > 0.821**

IDEA FILE

HISTORY CONNECTION

Have students make a time line from A.D. 1000 to 2000 with intervals of 100 years. Have them locate the year in which each of these items was invented and put the information on the time line. Students may wish to illustrate their work.

telescope—1608
pendulum clock—1641
computer—1943
telephone—1876
navigational compass—1086

Modalities Kinesthetic, Visual

RETEACH 1.2

Name _____

LESSON 1.2

Comparing and Ordering

Mindy is asked to list 18.3, 17.8, and 24.1 in order from greatest to least. She uses place value to compare the numbers.

Step 1 Mindy compares the first two numbers.
- She starts at the left. Both numbers have the digit 1 in the tens place.   18.3 ← > 17.8
- So, Mindy looks at the digits in the ones place. The first number has the digit 8 in the ones place, while the second number has the digit 7.   18.3 ← > 17.8
- Since 8 > 7, 18.3 > 17.8.

Step 2 Mindy compares the third number to the largest number so far, the first number.   24.1 ← → 18.3
- The third number has the digit 2 in the tens place, while the first number has the digit 1.
- Since 2 > 1, 24.1 > 18.3.

Using what she discovered, Mindy makes the list: 24.1 > 18.3 > 17.8.

List each set of numbers in order from greatest to least. Use >.
1. 32.8, 33.2, 38.1   2. 14.7, 14.07, 17.4   3. 22.43, 22.15, 22.6
   38.1 > 33.2 > 32.8   **17.4 > 14.7 > 14.07**   **22.6 > 22.43 > 22.15**
4. 9.03, 0.93, 3.09   5. 0.12, 0.01, 0.210   6. 8.56, 56.83, 38.6
   9.03 > 3.09 > 0.93   **0.210 > 0.12 > 0.01**   **56.83 > 38.6 > 8.56**

List each set of numbers in order from least to greatest. Use <.
7. 42.05, 45.02, 40.52   8. 19.7, 19.007, 19.07   9. 0.59, 0.95, 0.6
   40.52 < 42.05 < 45.02   **19.007 < 19.07 < 19.7**   **0.59 < 0.6 < 0.95**
10. 3.21, 2.31, 12.3   11. 6.14, 4.61, 4.16   12. 74.3, 37.4, 43.7
    2.31 < 3.21 < 12.3   **4.16 < 4.61 < 6.14**   **37.4 < 43.7 < 74.3**
13. 7.88, 8.88, 7.89   14. 8.241, 8.24, 8.2   15. 0.47, 0.93, 0.08
    7.88 < 7.89 < 8.88   **8.2 < 8.24 < 8.241**   **0.08 < 0.47 < 0.93**

R2 TAKE ANOTHER LOOK   Use with text pages 18–19.

GUIDED PRACTICE

Compare the numbers. Write <, >, or =.

1. 213 ● 21.3 >   **2.** 1.5 ● 1.7 <   **3.** 0.01 ● 0.12 <   **4.** 12.36 ● 12.3 >

5. 1,721.0 ● 1,721 =   **6.** 3.20 ● 32 <   **7.** 0.012 ● 0.12 <   **8.** 0.190 ● 0.19 =

Write the numbers in order from least to greatest. Use <.

9. 1.361, 1.351, 1.363
1.351 < 1.361 < 1.363

10. 8,621; 8,612; 8,613
8,612 < 8,613 < 8,621

11. 125.3, 124.32, 125.33
124.32 < 125.3 < 125.33

INDEPENDENT PRACTICE

Compare the numbers. Write <, >, or =.

1. 12.1 ● 12.2 <   **2.** 213 ● 223 <   **3.** 132.34 ● 122.34 >   **4.** 34.6 ● 34.60 =

5. 7.433 ● 7.432 >   **6.** 7.099 ● 7.999 <   **7.** 0.110 ● 0.1100 =   **8.** 45.678 ● 45.673 >

9. 644.56 ● 634.55 >   **10.** 99.088 ● 99.888 <   **11.** 133.23 ● 13.323 >   **12.** 457.8 ● 457.3 >

Write the numbers in order from least to greatest. Use <.

13. 4,556; 4,566; 4,555
4,555 < 4,556 < 4,566

14. 90, 90.22, 90.12
90 < 90.12 < 90.22

15. 5.004, 5.040, 5.4
5.004 < 5.040 < 5.4

Write the numbers in order from greatest to least. Use >.

16. 0.900, 0.090, 0.009
0.900 > 0.090 > 0.009

17. 126.3, 162.32, 126.33
162.32 > 126.33 > 126.3

18. 125.3, 124.32, 125.33
125.33 > 125.3 > 124.32

Problem-Solving Applications

For Problem 19, use the table.

Monthly Normal Rainfall for Seattle, Washington (in inches)												
Month	Jan	Feb	Mar	Apr	May	Jun	Jul	Aug	Sep	Oct	Nov	Dec
Rainfall	6.0	4.2	3.6	2.4	1.6	1.4	0.7	1.3	2.0	3.4	5.6	6.3

0.7 < 1.3 < 1.4 < 1.6 < 2.0 < 2.4 < 3.4 < 3.6 < 4.2 < 5.6 < 6.0 < 6.3

19. Joey's dad is planning the family's vacation to Seattle, Washington, for next year. He knows that some months have more rain than others. He would like to make sure the chance of rain is lower while they are there.
  a. Write in order from least to greatest the rainfall per month. Use <.
  b. Which month would be the best for a visit? Why? **July; because it has the least rainfall**

20. The batting averages of three players are 0.268, 0.280, and 0.265. Write the averages in order from greatest to least. Use >.
0.280 > 0.268 > 0.265

21. ▭ **WRITE ABOUT IT** Explain how you would compare 5,361 and 5,316. **Compare place values: 6 tens > 1 ten, so 5,361 > 5,316.**

MORE PRACTICE Lesson 1.2, page H42

19

In this lesson students use place value and number lines to compare and order whole numbers and decimals.

GUIDED PRACTICE
You may wish to complete these exercises as a whole-class activity.

INDEPENDENT PRACTICE
To ensure student success on Exercises 1–18, write <, >, and = on the board with their meanings.

ADDITIONAL EXAMPLES

Example 1, p. 18

Compare 6.93 and 6.9. Use < or >.

Since 3 hundredths is greater than 0 hundredths, 6.93 > 6.9 or 6.9 < 6.93.

Example 2, p. 18

Use > to list the numbers in order from greatest to least.

147; 1,471; 147.8; 147.83

1,471 > 147.83 > 147.8 > 147

3 WRAP UP

Use these questions to help students express the big ideas of the lesson.

• **Explain how you would compare 91.9 and 91.4 on a number line.** Locate the numbers on the number line. Since 91.4 is to the left of 91.9, 91.4 < 91.9.

✓ Assessment Tip
Have students write the answer.

• **Explain how you would determine the order of 4.5, 4, and 4.3 from greatest to least. Use >.** Annex a zero to the 4. Compare 4.5 to 4.0: 4.5 > 4.0. Compare 4.3 to 4.0: 4.3 > 4.0. List the numbers in order: 4.5 > 4.3 > 4.

DAILY QUIZ
Compare the numbers. Write >, <, or =.

1. 7,423 > 7,413   **2.** 8.047 < 8.0473
3. 5.474 < 6.474   **4.** 67.92 = 67.920

Use < to write the numbers in order from least to greatest.

5. 0.505, 0.5, 0.55 0.5 < 0.505 < 0.55
6. 27.6, 27.65, 27.62, 27 27 < 27.6 < 27.62 < 27.65

PRACTICE 1.2

Name _____

LESSON 1.2

Comparing and Ordering

Compare the numbers. Write <, >, or =.

1. 15.4 **>** 14.5   2. 5.67 **<** 5.76   3. 43.90 **=** 43.9

4. 7.91 **<** 9.17   5. 765.28 **>** 762.58   6. 0.234 **<** 2.304

Write the numbers in order from least to greatest. Use <.
7. 3,224; 2,432; 3,422   8. 88.5; 85.8; 58.8   9. 6.21; 6.02; 6.12

2,432 < 3,224 < 3,422   **58.8 < 85.8 < 88.5**   **6.02 < 6.12 < 6.21**

Write the numbers in order from greatest to least. Use >.
10. 0.005; 0.500; 0.050   11. 317.8; 318.7; 371.8   12. 16.04; 14.6; 16.4

0.500 > 0.050 > 0.005   **371.8 > 318.7 > 317.8**   **16.4 > 16.04 > 14.6**

Mixed Applications

For Problems 13–16, use the table. It shows the average points scored by the top eight NBA scorers for one basketball season.

TOP NBA SCORERS			
Player	Average	Player	Average
Charles Barkley	25.6	Shaquille O'Neal	23.4
Joe Dumars	23.5	Hakeem Olajuwon	26.1
Patrick Ewing	24.2	Karl Malone	27.0
Michael Jordan	32.6	Dominique Wilkins	29.9

13. Compare the averages of Shaquille O'Neal, Patrick Ewing, Joe Dumars, and Charles Barkley. Write the numbers in order from least to greatest. Use <.

23.4 < 23.5 < 24.2 < 25.6

14. Compare the averages of Karl Malone, Michael Jordan, Hakeem Olajuwon, and Dominique Wilkins. Write the numbers in order from greatest to least. Use >.

32.6 > 29.9 > 27.0 > 26.1

15. What is the value of the digit 2 in the greatest average listed in the table?

2 ones; 2

16. Use words to write Shaquille O'Neal's average for the season.

twenty-three and four tenths

P2 ON MY OWN   Use with text pages 18–19.

ENRICHMENT 1.2

Name _____

LESSON 1.2

Comparing Scoring Leaders

SCORING LEADERS	
Player	Average
Rick Barry	24.8
Elgin Baylor	27.4
Wilt Chamberlain	30.1
George Gervin	25.1
Bob Pettit	26.4
Oscar Robertson	25.7
Jerry West	27.0

The table above shows the scoring averages of some retired NBA scoring leaders. Use the table to answer the questions.

1. Which of the players listed in the table has the greatest average?

Wilt Chamberlain

2. Which of the players listed in the table has the lowest average?

Rick Barry

3. Use > to list the averages of Oscar Robertson, George Gervin, and Bob Pettit from greatest to least.

26.4 > 25.7 > 25.1

4. Use < to list all the averages shown in the table from least to greatest.

24.8 < 25.1 < 25.7 < 26.4 < 27.0 < 27.4 < 30.1

5. Dominique Wilkins's scoring average was 26.5. Where would he be placed in the list you made in Exercise 4?

after Bob Pettit (26.4) but before Jerry West (27.0)

6. Michael Jordan's scoring average was 32.3. Where would he be placed in the list you made in Exercise 4?

after Wilt Chamberlain (30.1)

7. Which player in the table had an average greater than 25.5 but less than 26.0?

Oscar Robertson

8. Which players in the table had an average greater than 26 but less than 28?

Elgin Baylor, Bob Pettit, Jerry West

E2 STRETCH YOUR THINKING   Use with text pages 18–19.

19

ORGANIZER

OBJECTIVE To write a decimal as a fraction and a fraction as a decimal

ASSIGNMENTS Independent Practice
- *Basic* 1–34, odd; 35–40; 41–47
- *Average* 1–40; 41–47
- *Advanced* 1–34, odd; 41–47

PROGRAM RESOURCES
Reteach 1.3 Practice 1.3 Enrichment 1.3
Math Fun • The Numbers Are Falling!, p. 30
Calculating Crew • *Nautical Number Line*

PROBLEM OF THE DAY

Transparency 1

From 4 o'clock to 5 o'clock Jacob, Lisa, and Chelsea took turns at the same computer game. Jacob played $\frac{1}{3}$ hour and Lisa played 0.25 hour. For how many minutes did Chelsea play the game? 25 min

Solution Tab A, p. A1

QUICK CHECK
Divide to change these fractions to decimals.

1. $\frac{1}{5}$ 0.2
2. $\frac{3}{5}$ 0.6
3. $\frac{2}{4}$ 0.5
4. $\frac{2}{5}$ 0.4
5. $\frac{3}{6}$ 0.5
6. $\frac{4}{8}$ 0.5

1 MOTIVATE

To help students access their knowledge about how they use fractions and decimals, ask them the following questions:

Are these numbers usually represented by fractions or by decimals?

- **scores for Olympic skaters** decimals
- **ingredients in a recipe** fractions
- **gallons of gasoline** decimals
- **distance of a race** decimals or fractions
- **value of a share of stock** fractions

1.3

WHAT YOU'LL LEARN
How to write a decimal as a fraction and a fraction as a decimal

WHY LEARN THIS?
To understand that decimals and fractions can be used to name the same amounts, such as the distance run

Sports Link

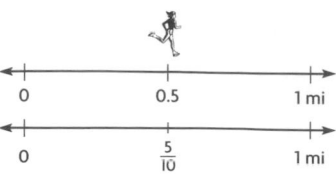

The distances of most races in the United States are measured in metric units. In the past these distances were measured in customary units. For example, instead of 100 m, racers ran 110 yd (100.6 m), and instead of 800 m, racers ran 880 yd (0.5 mi). How many meters does 880 yd equal? **804.8 m**

Decimals and Fractions

Jen ran five tenths of a mile. What is the distance Jen ran, expressed as a decimal? What is the distance Jen ran expressed as a fraction?

```
0        0.5      1 mi

0        5/10     1 mi
```

The number lines show that 0.5 and $\frac{5}{10}$ are equivalent. You can use place value to change a decimal to an equivalent fraction.

EXAMPLE 1 Use place value to change the decimal to a fraction.

A. 0.3 — *Identify the place value of the last digit. The 3 is in the tenths place.*

$\frac{3}{10}$ — *Use the value for the denominator.*

So, $0.3 = \frac{3}{10}$.

B. 0.42 — *Identify the place value of the last digit. The 2 is in the hundredths place.*

$\frac{42}{100}$ — *Use the value for the denominator.*

So, $0.42 = \frac{42}{100}$.

- Write 0.103 as a fraction. $\frac{103}{1,000}$

- When changing a decimal to a fraction, what would the denominator be if the last digit in a decimal has a place value of tenths? hundredths? thousandths? **10; 100; 1,000**

A calculator with fraction functions can be used to change a decimal to a fraction. For example, the calculator sequence below shows how to change 0.75 to a fraction.

0.75 [F◁▷D] [75/100] ← $0.75 = \frac{75}{100}$

RETEACH 1.3

Name _____

LESSON 1.3

Decimals and Fractions

Ron needs to change $\frac{3}{5}$ to a decimal. He remembers that he can use division to change a fraction to a decimal.

Step 1 Set up a division problem. Divide the numerator by the denominator.   5)3

Step 2 Place a decimal point after the numerator. Write a zero.   5)3.0

Step 3 Divide as you would with whole numbers.
Ron finds that $\frac{3}{5}$ = 0.6.

```
   0.6
5)3.0
   3.0
     0
```

Now Ron needs to change 0.364 to a fraction. He uses place value.

Step 1 Look at the place value of the last digit.

0.364
↑
thousandths

Step 2 Use the place value of the last digit as the denominator and the digits of the decimal as the numerator. $\frac{364}{1,000}$

Ron finds that $0.364 = \frac{364}{1,000}$.

Answer the questions to change the fraction to a decimal.

1. $\frac{1}{4}$
   a. What division problem will you use?   4)1
   b. What is the quotient?   0.25

2. $\frac{3}{8}$
   a. What division problem will you use?   8)3
   b. What is the quotient?   0.375

3. $\frac{17}{50}$
   a. What division problem will you use?   50)17
   b. What is the quotient?   0.34

4. $\frac{11}{20}$
   a. What division problem will you use?   20)11
   b. What is the quotient?   0.55

Use place value to change the decimal to a fraction.

5. 0.61 $\frac{61}{100}$
6. 0.92 $\frac{92}{100}$
7. 0.48 $\frac{48}{100}$
8. 0.137 $\frac{137}{1,000}$
9. 0.6897 $\frac{6,897}{10,000}$

Use with text pages 20–23.   **TAKE ANOTHER LOOK** R3

PRACTICE 1.3

Name _____

LESSON 1.3

Decimals and Fractions

Use place value to change the decimal to a fraction.

1. 0.5 $\frac{5}{10}$
2. 0.14 $\frac{14}{100}$
3. 0.06 $\frac{6}{100}$
4. 0.83 $\frac{83}{100}$
5. 0.29 $\frac{29}{100}$
6. 0.62 $\frac{62}{100}$
7. 0.317 $\frac{317}{1,000}$
8. 0.8053 $\frac{8,053}{10,000}$
9. 0.955 $\frac{955}{1,000}$
10. 0.125 $\frac{125}{1,000}$

Write as a decimal.

11. $\frac{7}{10}$ 0.7
12. $\frac{54}{100}$ 0.54
13. $\frac{1}{8}$ 0.125
14. $\frac{4}{5}$ 0.8
15. $\frac{3}{16}$ 0.1875
16. $\frac{91}{1,000}$ 0.091
17. $\frac{3}{4}$ 0.75
18. $\frac{1}{5}$ 0.2
19. $\frac{19}{10,000}$ 0.0019
20. $\frac{5}{16}$ 0.3125
21. $\frac{32}{1,000}$ 0.032
22. $\frac{5}{8}$ 0.625
23. $\frac{156}{10,000}$ 0.0156
24. $\frac{9}{16}$ 0.5625
25. $\frac{89}{10,000}$ 0.0089

Write < or >.

26. $\frac{3}{10}$ < $\frac{1}{2}$
27. $\frac{11}{20}$ < $\frac{7}{8}$
28. $\frac{91}{100}$ > $\frac{11}{16}$
29. 0.24 < $\frac{3}{4}$
30. 0.18 > $\frac{7}{50}$
31. 0.04 < $\frac{4}{10}$
32. 0.19 < $\frac{1}{5}$
33. 0.459 > $\frac{7}{20}$
34. 0.1 < $\frac{15}{16}$

Mixed Applications

35. Rhea needs to know the decimal equivalent of $\frac{3}{8}$ to find the cost of three sticks of gum. Change the fraction to a decimal.   0.375

36. Tia needs to know the decimal equivalent of $\frac{5}{8}$ to find the cost of five slices of pizza. Change the fraction to a decimal.   0.625

37. The batting averages of three players are 0.172, 0.223, and 0.194. Write the averages in order from greatest to least. Use >.   0.223 > 0.194 > 0.172

38. Todd and Jim drove home for the holidays. Todd drove $\frac{3}{8}$ of the distance. Jim drove $\frac{10}{16}$ of the distance. Who drove more miles? Explain.   Jim; $\frac{3}{8} < \frac{10}{16}$

Use with text pages 20–23.   **ON MY OWN** P3

You can change a fraction to an equivalent decimal by dividing the numerator by the denominator.

EXAMPLE 2 Use division to change the fraction to a decimal.

A. $\frac{1}{5}$ *Divide the numerator by the denominator.*

$$\begin{array}{r} 0.2 \\ 5\overline{)1.0} \\ \underline{-10} \\ 0 \end{array}$$

Place the decimal point.
Then divide as with whole numbers.

So, $\frac{1}{5} = 0.2$.

B. $\frac{1}{4}$ *Divide the numerator by the denominator.*

$$\begin{array}{r} 0.25 \\ 4\overline{)1.00} \\ \underline{-8}\downarrow \\ 20 \\ \underline{-20} \\ 0 \end{array}$$

Place the decimal point.
Then divide as with whole numbers.

So, $\frac{1}{4} = 0.25$.

You can also use a calculator to change a fraction to a decimal.

EXAMPLE 3 Change $\frac{1}{8}$ to a decimal.

Method 1

 1 [ ] 8 [F⇌D] [0.125]   *a calculator that has fraction functions*

Method 2

1 [÷] 8 [=] [0.125]   *a calculator that does not have fraction functions*

So, $\frac{1}{8} = 0.125$.

- Use a calculator to change $\frac{3}{8}$ to a decimal. **0.375**

GUIDED PRACTICE

Use place value to change the decimal to a fraction.

1. 0.4 $\frac{4}{10}$ 2. 0.03 $\frac{3}{100}$ 3. 0.32 $\frac{32}{100}$ 4. 0.425 $\frac{425}{1,000}$ 5. 0.7255 $\frac{7,255}{10,000}$

Use division to change the fraction to a decimal.

6. $\frac{3}{10}$ 0.3   7. $\frac{1}{2}$ 0.5   8. $\frac{23}{100}$ 0.23   9. $\frac{3}{4}$ 0.75   10. $\frac{3}{25}$ 0.12

 **Calculator Activities, page H35**

21

In this lesson students explore the relationship between fractions and decimals by using multiplication, division, and fraction functions on a calculator. Then they extend these skills to compare fractions to decimals.

COMMON ERROR ALERT Some students may have difficulty dividing a lesser number by a greater number. Remind them to place a decimal point after the dividend and to add zeros until the remainder is zero.

ADDITIONAL EXAMPLES

Example 1, p. 20

Use place value to change 0.7 to a fraction.

$0.7 = \frac{7}{10}$

In the number 0.32, identify the place value of the last digit. The 2 is in the hundredths place.

Example 2, p. 21

Use division to change $\frac{1}{2}$ to a decimal.

$\frac{1}{2} = 0.5$

Example 3, p. 21

Change $\frac{2}{5}$ to a decimal.

Method 1

 [2] [/] [5] [F⇌D] [0.4]

Method 2

[2] [÷] [5] [=] [0.4]

$\frac{2}{5} = 0.4$

ENRICHMENT 1.3

Name _____

Decimal Patterns

Find the next three terms in the pattern. Then write the rule for the pattern.

1. $\frac{1}{5}$, $\frac{2}{5}$, $\frac{3}{5}$, $\frac{4}{5}$, $\frac{5}{5}$, $\frac{6}{5}$, $\frac{7}{5}$, $\frac{8}{5}$, $\frac{9}{5}$

   0.20   0.40   0.60   0.80   1.00   1.20   **1.40**   **1.60**   **1.80**
   Add $\frac{1}{5}$ (0.20) to previous number.
   Rule: _____

2. $\frac{11}{2}$, $\frac{10}{2}$, $\frac{9}{2}$, $\frac{8}{2}$, $\frac{7}{2}$, $\frac{6}{2}$, $\frac{5}{2}$, $\frac{4}{2}$, $\frac{3}{2}$

   5.5   5.0   4.5   4.0   3.5   3.0   **2.5**   **2.0**   **1.5**
   Subtract $\frac{1}{2}$ (0.5) from previous number.
   Rule: _____

3. $\frac{3}{10}$, $\frac{6}{10}$, $\frac{9}{10}$, $\frac{12}{10}$, $\frac{15}{10}$, $\frac{18}{10}$, $\frac{21}{10}$, $\frac{24}{10}$, $\frac{27}{10}$

   0.3   0.6   0.9   1.2   1.5   1.8   **2.1**   **2.4**   **2.7**
   Add $\frac{3}{10}$ (0.3) to previous number.
   Rule: _____

4. $\frac{25}{4}$, $\frac{24}{4}$, $\frac{23}{4}$, $\frac{22}{4}$, $\frac{21}{4}$, $\frac{20}{4}$, $\frac{19}{4}$, $\frac{18}{4}$, $\frac{17}{4}$

   6.25   6.00   5.75   5.50   5.25   5.00   **4.75**   **4.50**   **4.25**
   Subtract $\frac{1}{4}$ (0.25) from previous number.
   Rule: _____

5. $\frac{44}{10}$, $\frac{40}{10}$, $\frac{36}{10}$, $\frac{32}{10}$, $\frac{28}{10}$, $\frac{24}{10}$, $\frac{20}{10}$, $\frac{16}{10}$, $\frac{12}{10}$, $\frac{8}{10}$

   4.4   4.0   3.6   3.2   2.8   2.4   2.0   **1.6**   **1.2**   **0.8**
   Subtract $\frac{4}{10}$ (0.4) from previous number.
   Rule: _____

6. $\frac{99}{100}$, $\frac{88}{100}$, $\frac{77}{100}$, $\frac{66}{100}$, $\frac{55}{100}$, $\frac{44}{100}$, $\frac{33}{100}$, $\frac{22}{100}$

   0.99   0.88   0.77   0.66   0.55   **0.44**   **0.33**   **0.22**
   Subtract $\frac{11}{100}$ (0.11) from previous number.
   Rule: _____

Use with text pages 20–23.   **STRETCH YOUR THINKING** **E3**

21

GUIDED PRACTICE

You may wish to have students complete these exercises in small groups. Encourage students with stronger skills to assist those with weaker skills.

INDEPENDENT PRACTICE

To assess students' understanding of the relationship between fractions and decimals, ask them which exercises they could change to a decimal by using place value. Exercises 13, 14, 19, and 20

💡 CRITICAL THINKING

How could you solve Exercises 15, 21, and 24 without dividing? by writing equivalent fractions with a denominator that is a multiple of 10

How could you solve Exercise 25 without changing the fractions to decimals? by changing $\frac{1}{2}$ to an equivalent fraction with a denominator of 4 and then by comparing $\frac{2}{4}$ to $\frac{3}{4}$

ADDITIONAL EXAMPLE

Example 4, p. 22

Lisa used $\frac{5}{8}$ yd of orange suede and $\frac{3}{5}$ yd of purple suede to make a vest. Did she use more orange or more purple suede to make the vest?

Compare $\frac{5}{8}$ and $\frac{3}{5}$.

$\frac{5}{8} > \frac{3}{5}$

So, she used more orange suede.

Comparing Fractions and Decimals

You can compare fractions that have the same denominators by looking at the numerators.

For example, to compare $\frac{1}{5}$ and $\frac{2}{5}$, you compare the numerators. $\frac{1}{5} < \frac{2}{5}$ since $1 < 2$.

When fractions do not have the same denominator, it is easier to compare them if you change them to decimals.

EXAMPLE 4 Students at Sumter Middle School wear their school colors, blue and yellow, to show school spirit. One day $\frac{2}{5}$ of the students wore blue and $\frac{1}{4}$ of the students wore yellow. Which color was worn by more students?

Compare $\frac{2}{5}$ and $\frac{1}{4}$.

$\frac{2}{5}$ ● $\frac{1}{4}$ *To compare the fractions, write each as a decimal.*

↓   ↓

0.4 ● 0.25   *Compare the decimals.*

0.4 > 0.25

Since $0.4 > 0.25$, $\frac{2}{5} > \frac{1}{4}$.

So, more students wore blue.

• Use decimals to compare $\frac{1}{8}$ and $\frac{2}{10}$. $0.125 < 0.2$, so $\frac{1}{8} < \frac{2}{10}$.

Talk About It

• Suppose you were comparing $\frac{1}{4}$ and $\frac{3}{4}$. Would you need to change to decimals? Explain. **Possible answer: No; $\frac{1}{4}$ and $\frac{3}{4}$ have the same denominator.**

• 💡 **CRITICAL THINKING** Explain how you would compare $\frac{5}{8}$ and 0.75.
Possible answer: by changing $\frac{5}{8}$ to a decimal and then comparing

INDEPENDENT PRACTICE

Use place value to change the decimal to a fraction.

1. 0.2 $\frac{2}{10}$ **2.** 0.8 $\frac{8}{10}$ **3.** 0.17 $\frac{17}{100}$ **4.** 0.09 $\frac{9}{100}$ **5.** 0.99 $\frac{99}{100}$ **6.** 0.45 $\frac{45}{100}$

7. 0.25 $\frac{25}{100}$ **8.** 0.568 $\frac{568}{1,000}$ **9.** 0.325 $\frac{325}{1,000}$ **10.** 0.875 $\frac{875}{1,000}$ **11.** 0.3525 $\frac{3,525}{10,000}$ **12.** 0.0823 $\frac{823}{10,000}$

Write as a decimal.

13. $\frac{2}{10}$ 0.2   **14.** $\frac{27}{100}$ 0.27   **15.** $\frac{3}{5}$ 0.6   **16.** $\frac{7}{8}$ 0.875   **17.** $\frac{15}{16}$ 0.9375   **18.** $\frac{4}{8}$ 0.5

22   Chapter 1

👥 IDEA FILE

CALCULATOR CONNECTION

An important aid for students to use as they study fractions and decimals is a calculator. Have them use the Explorer Plus™ or the fx-65 to

• change $\frac{1}{9}$, $\frac{2}{9}$, and $\frac{3}{9}$ to decimals and describe the pattern they discovered. The repeating decimal is the same as the numerator.

• repeat the process with $\frac{1}{11}$, $\frac{2}{11}$, and $\frac{3}{11}$. The repeating decimal is 9 times the numerator.

Modalities Kinesthetic, Visual

REVIEW

The scores for six finalists in a gymnastics event are shown below. Have students list the finalists in order, starting with the gymnast with the highest score.

Peggy	9.25	6th
Nan	9.725	2nd
Joan	9.705	3rd
Lin	9.75	1st
Kathy	9.3	4th
Judy	9.275	5th

Modalities Auditory, Visual

Write as a decimal.

19. $\frac{87}{10,000}$ **0.0087** **20.** $\frac{7}{100}$ **0.07** **21.** $\frac{1}{4}$ **0.25** **22.** $\frac{3}{8}$ **0.375** **23.** $\frac{7}{16}$ **0.4375** **24.** $\frac{2}{5}$ **0.4**

Write < or >.

25. $\frac{1}{2}$ ● $\frac{3}{4}$ **<** **26.** $\frac{1}{5}$ ● $\frac{3}{10}$ **<** **27.** $\frac{17}{20}$ ● $\frac{5}{8}$ **>** **28.** $\frac{1}{8}$ ● $\frac{17}{100}$ **<** **29.** $\frac{99}{100}$ ● $\frac{13}{16}$ **>**

30. $\frac{7}{16}$ ● $\frac{11}{20}$ **<** **31.** 0.7 ● $\frac{77}{100}$ **<** **32.** 0.18 ● $\frac{3}{50}$ **>** **33.** 0.335 ● $\frac{5}{16}$ **>** **34.** 0.09 ● $\frac{9}{10}$ **<**

Problem-Solving Applications

35. Hector needs to know the decimal equivalent of $\frac{1}{8}$ to find the cost of one slice of pizza. Change the fraction to a decimal. **0.125**

36. The decimal that is equivalent to $\frac{1}{5}$ is 0.2. What is the decimal that is equivalent to $\frac{2}{5}$? **0.4**

37. Al needs $\frac{3}{4}$ lb of plant food. The plant food shows decimal weights. What decimal weight should Al buy? **0.75 lb**

38. Jan ran $\frac{1}{8}$ mi. Karen ran 0.15 mi. Who ran the farthest? **Karen**

39. CRITICAL THINKING Use a calculator to change $\frac{1}{3}$ to a decimal. Then use paper and pencil to change $\frac{1}{3}$ to a decimal. What do you notice about the decimal? How is this decimal different from the decimal for $\frac{1}{2}$? **The 3 keeps repeating; the remainder is not 0.**

40. ◖▦▷ **WRITE ABOUT IT** Explain two ways to *Portfolio* change the fraction $\frac{4}{10}$ to a decimal. **Use place value; divide the numerator, 4, by the denominator, 10.**

Mixed Review

Multiply.

41. $2 \times 2 \times 2 \times 2$ **16** **42.** 11×11 **121** **43.** $6 \times 6 \times 6$ **216**

Find the area.

44. 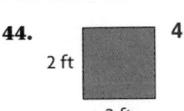 **4 ft²**
2 ft
2 ft

45. 1 in. ▭ **5 in.²**
5 in.

46. PLACE VALUE Use mental math to evaluate each expression.

a. $7,458,234 + 1,000$ **7,459,234**

b. $7,458,234 + 10$ **7,458,244**

c. $7,458,234 + 10,000$ **7,468,234**

47. LOGICAL REASONING Complete the following pattern.

$\frac{1}{4}$ → 0.25 $\frac{2}{4}$ → 0.50 $\frac{3}{4}$ → 0.75 $\frac{4}{4}$ → 1.00 $\frac{5}{4}$ → ? **1.25** $\frac{6}{4}$ → ? **1.50**

MORE PRACTICE Lesson 1.3, page H42

Technology **Link**

◉ In *Mighty Math Calculating Crew* the Nautical Number Line challenges you to find oceanic treasure by identifying equivalent fractions for decimals.

Use Grow Slide Level Q.

23

MATH CONNECTION

Measurement Bev used fractions to plan the size of each serving of food for her picnic, but the serving sizes of the food she bought are shown in decimals. Have students change the serving size of the items on her list to decimals.

hamburger—$\frac{1}{4}$ lb 0.25 lb

veggie salad—$\frac{2}{5}$ lb 0.4 lb

fruit salad—$\frac{3}{8}$ lb 0.375 lb

rolls—$\frac{2}{5}$ of a package
      0.4 of a package

Modalities Auditory, Visual

CAREER CONNECTION

Many careers involve the use of both decimals and fractions. Separate the class into small groups and have each group research one of the following occupations to find out how decimals and fractions are used.

• stockbroker
• engineer
• architect
• chef
• pharmacist

Modalities Auditory, Visual

MEETING INDIVIDUAL NEEDS

DEVELOPING ALGEBRAIC THINKING
Learning to work with variables is an important step in developing algebraic thinking. Ask students these questions:

• **If $x = y$, would you need to change $\frac{5}{x}$ and $\frac{3}{y}$ to decimals to compare them? Why or why not?** No, because if $x = y$, the denominators are the same so you compare the numerators.

3 WRAP UP

Use these questions to help students focus on the big ideas of the lesson.

• **Explain how to solve Exercise 33.** Possible answer: Change $\frac{5}{16}$ to the decimal 0.3125 by dividing 5 by 16. Then compare 0.335 to 0.3125.

• **Describe one situation in which you might have to change a fraction to a decimal or a decimal to a fraction.** Check students' answers.

✔ *Assessment Tip*
Have students write the answer.

• **Explain how to change a decimal to a fraction.** Possible answer: Identify the place value of the last digit and use that place value for the denominator. Make the digit a whole number and use that number for the numerator.

• **Explain how to change a fraction to a decimal.** Possible answer: Divide the denominator into the numerator.

DAILY QUIZ
Use place value to change the decimal to a fraction.

1. 0.7 $\frac{7}{10}$

2. 0.09 $\frac{9}{100}$

3. 0.411 $\frac{411}{1000}$

Change each fraction to a decimal. Write *a, b,* or *c.*

4. $\frac{9}{20}$ *a*   **a.** 0.45   **b.** 2.22   **c.** 0.9

5. $\frac{3}{5}$ *b*   **a.** 0.06   **b.** 0.6   **c.** 1.6

6. $\frac{7}{8}$ *a*   **a.** 0.875   **b.** 1.14   **c.** 0.87

GRADE 5 Model square and cubic numbers to find area and volume.	GRADE 6 Represent numbers by using exponents.	GRADE 7 Use exponents.

ORGANIZER

OBJECTIVE To represent numbers by using exponents

VOCABULARY *Review* factors

New **exponent, base**

ASSIGNMENTS **Independent Practice**
- *Basic* 1–23, even; 24–26; 27–33
- *Average* 1–26; 27–33
- *Advanced* 5–8, 16–23; 24–26; 27–33

PROGRAM RESOURCES
Reteach 1.4 Practice 1.4 Enrichment 1.4

PROBLEM OF THE DAY

Transparency 1

Replace the letters *a, b, c,* and *d* with the numbers 2, 3, 4, and 5 to make a true sentence.

$a^b + a^b = c^d$ $2^5 + 2^5 = 4^3$

Solution Tab A, p. A1

QUICK CHECK

Use mental math to find the product.

1. $10 \times 10 \times 10$ 1,000   **2.** $3 \times 3 \times 3$ 27
3. $1 \times 1 \times 1 \times 1 \times 1$ 1   **4.** 5×5 25
5. $2 \times 2 \times 2 \times 2$ 16   **6.** 4×4 16

1 MOTIVATE

To help students access their knowledge of exponents, pose the following problem:

- **A recording artist must sell 1,000,000 recordings in the United States to be awarded a gold record. How many times do you need to use 10 as a factor to reach 1,000,000?** 6 times

2 TEACH/PRACTICE

In this lesson students apply their understanding of multiplication to express numbers as exponents.

Exponents

WHAT YOU'LL LEARN
How to represent numbers by using exponents

WHY LEARN THIS?
To be able to write large numbers, such as 10,000, in a shortened form

WORD POWER
exponent
base

REMEMBER:

When you multiply two or more numbers to get a product, the numbers multiplied are called factors. **See page H3.**

$8 \times 3 \times 4 = 96$

The numbers 8, 3, and 4 are factors.

Large numbers can be hard to understand, but you can think of them in certain ways to make sense of them. For example, to understand how much $10,000 is, first you can ask, How many $10 bills would it take to make $10,000? Then think about powers of 10.

$10 \times 10 = 100$

$10 \times 10 \times 10 = 1,000$

$10 \times 10 \times 10 \times 10 = 10,000$

You can think of $10,000 as $10 \times 10 \times 10 \times 10$, or $10 \times 1,000$. So, there are one thousand $10 bills in $10,000.

Powers of numbers can be written with exponents. An **exponent** shows how many times a number called the **base** is used as a factor.

$$10^4 = 10 \times 10 \times 10 \times 10 = 10,000$$

Read 10^4 as "ten to the fourth power."

You use multiplication to find the value of a number with an exponent.

EXAMPLE 1 Find the value of 2^4.

The exponent, 4, tells you to use the base, 2, as a factor four times.

$2^4 = 2 \times 2 \times 2 \times 2 = 16$

So, the value of 2^4 is 16.

- Find the value of 8^2. $8^2 = 8 \times 8 = 64$

EXAMPLE 2 Express 81 by using an exponent and the base 3.

$81 = 9 \times 9 = 3 \times 3 \times 3 \times 3$ *Find the equal factors.*

$= 3^4$ *Write the base and the exponent.*

- Express 81 by using an exponent and the base 9. 9^2

📱 Calculator Activities, page H30

IDEA FILE

ENGLISH AS A SECOND LANGUAGE

Tip • Help students acquiring English relate the idea of base to the number on the bottom of an exponential expression, just as the base of a geometric solid is the foundation that the shape stands on. Have the whole class join in finding expressions that use the word *base* to mean "foundation" (the base of a pyramid, for example).

MEETING INDIVIDUAL NEEDS

RETEACH 1.4

Name _____

LESSON 1.4

Exponents

Powers of numbers can be written in exponent form. An *exponent* shows how many times a number called the *base* is used as a factor.

$10^5 = 10 \times 10 \times 10 \times 10 \times 10 = 100,000$
base   factors
10 used as a factor 5 times

$2^6 = 2 \times 2 \times 2 \times 2 \times 2 \times 2 = 64$
base   factors
2 used as a factor 6 times

Joan is asked to express 64 using an exponent and the base 4.

Step 1 The base is 4. So, Joan must find equal factors.
$8 \times 8 = 64$
$4 \times 2 \times 2 \times 4 = 64$
$4 \times 4 \times 4 = 64$

Step 2 Joan writes the base. Then she counts how many times it is used as a factor.
$4^3 \leftarrow$ used as a factor 3 times
base
So, $64 = 4^3$.

Write the factors in exponent form.

1. $4 \times 4 \times 4$ 4^3 2. $3 \times 3 \times 3 \times 3 \times 3$ 3^5
3. $10 \times 10 \times 10 \times 10 \times 10 \times 10$ 10^6 4. $30 \times 30 \times 30 \times 30$ 30^4

Express each number with an exponent and the base given.

5. 36, base 6 6^2 6. 32, base 2 2^5 7. 81, base 3 3^4
8. 625, base 5 5^4 9. 343, base 7 7^3 10. 128, base 2 2^7
11. 512, base 8 8^3 12. 1,000,000, base 10 10^6 13. 121, base 11 11^2

R4   TAKE ANOTHER LOOK   Use with text pages 24–25.

GUIDED PRACTICE

Tell how many zeros will be in the standard form of the number.

1. 10^3 3   **2.** 10^{12} 12   **3.** 10^{10} 10   **4.** 10^1 1   **5.** 10^2 2   **6.** 10^8 8

Write the equal factors. Then find the value.

7. 2^3 **8.** 5^2 **9.** 3^4 $3 \times 3 \times$ **10.** 9^3 **11.** 9^4 $9 \times 9 \times$ **12.** 1^4 $1 \times 1 \times$
$2 \times 2 \times 2$; 8 5×5; 25 3×3; 81 $9 \times 9 \times 9$; 729 9×9; 6,561 1×1; 1

INDEPENDENT PRACTICE

Write in exponent form.

1. $12 \times 12 \times 12$ 12^3 **2.** $1 \times 1 \times 1 \times 1 \times 1$ 1^5 **3.** $4 \times 4 \times 4 \times 4$ 4^4 **4.** $10 \times 10 \times 10$ 10^3

5. 7×7 7^2 **6.** 14×14 14^2 **7.** $2 \times 2 \times 2 \times 2 \times 2$ 2^5 **8.** $20 \times 20 \times 20$ 20^4

Find the value.

9. 4^5 1,024   **10.** 7^3 343   **11.** 1^{12} 1   **12.** 5^3 125   **13.** 2^3 8   **14.** 2^5 32

15. 14^1 14   **16.** 13^2 169   **17.** 10^8 100,000,000   **18.** 20^2 400   **19.** 2^{10} 1,024   **20.** 10^4 10,000

Express with an exponent and the given base.

21. 64, base 8 8^2 **22.** 216, base 6 6^3 **23.** 1,000; base 10 10^3

Problem-Solving Applications

24. Sandy has a job that pays 2¢ on the first day of work. Then, for each day after the first, she receives double the preceding day's wage. Using exponent form, write the number of cents she will receive on the sixth day. 2^6

25. Tomás had 10 stamps. He traded each stamp for 10 coins. Then he traded each coin for 10 baseball cards. How many baseball cards does he have? Use an exponent to write the answer. 10^3

26. ✏️ **WRITE ABOUT IT** Which is greater, 3^2 or 2^3? 3^2

Mixed Review

Describe the opposite of the action.

27. Take 2 steps forward. Take 2 steps backward.   **28.** Turn right. Turn left.   **29.** Jump up. Jump down.

Write the number in standard form.

30. ten thousand, one hundred ninety 10,190   **31.** two hundred thousand, twenty-four 200,024

32. **NUMBER SENSE** Dennis ran 3 mi farther than Horace. Together they ran 17 mi. How far did Dennis run? **10 mi**

33. **NUMBER SENSE** Naomi has 5 friends and 3 pizzas. If she gives them equal amounts, how much does each person get? $\frac{3}{5}$ of a pizza

MORE PRACTICE Lesson 1.4, page H43

25

GUIDED PRACTICE
Complete these exercises orally to ensure that all students understand the concepts.

INDEPENDENT PRACTICE
To ensure that students understand the basic concept of exponents, check their answers to Exercises 1–8.

💡**CRITICAL THINKING** Numbers with the exponent 3 are called cubes or cubed numbers; for example, 4^3 and 7^3 are cubed numbers. Why do you think the name *cubed* is used? The volume of a cube is equal to the length of one side of the cube raised to a power of 3.

ADDITIONAL EXAMPLES

Example 1, p. 24

Find the value of 3^3.

$3^3 = 3 \times 3 \times 3 = 27$

Example 2, p. 24

Express 16 by using an exponent and the base 2.

$16 = 2 \times 2 \times 2 \times 2 = 2^4$

3 WRAP UP

Use these questions to help students focus on the big ideas of the lesson.

- **How do you solve Exercise 20?** Possible answer: Multiply 10 by 10 by 10 by 10.

✓ *Assessment Tip*

Have students write the answer.

- **Explain how to find the value of a number with an exponent.** Possible explanation: Multiply the base number by itself as many times as the number of its exponent.

DAILY QUIZ
Find the value.

1. 10^5 100,000
2. 11^2 121
3. 3^4 81

Express with an exponent and the given base.

4. 1,331; base 11 11^3
5. 625; base 5 5^4

PRACTICE 1.4

Name _____   LESSON 1.4

Exponents

Vocabulary

Complete using *exponent* or *base*.

1. A(n) ___exponent___ shows how many times a number called the ___base___ is used as a factor.

Write in exponent form.

2. $5 \times 5 \times 5$ 5^4 3. $10 \times 10 \times 10 \times 10 \times 10$ 10^5 4. 18×18 18^2

Find the value.

5. 8^2 64   6. 10^6 1,000,000   7. 4^3 64   8. 1^{18} 1   9. 2^6 64
10. 6^4 1,296   11. 11^2 121   12. 10^3 1,000   13. 15^1 15   14. 30^2 900

Express with an exponent and the given base.

15. 125, base 5 5^3 16. 10,000, base 10 10^4 17. 256, base 4 4^4 18. 729, base 9 9^3

Mixed Applications

19. Rod has a job that pays $5.00 on the first day of work. Then, for each day after the first, he receives double the preceding day's wage. Using exponent form, write the number of dollars he will receive on the eighth day. 5^8

20. Max has 6 cartons. In each carton, he places 6 bags. In each of the bags, he places 6 cookies. How many cookies are contained in the cartons? Use an exponent to write the answer. 6^3

21. While exercising, Maria jogged $\frac{3}{8}$ mi and walked $\frac{5}{8}$ mi. Did she jog or walk the greater distance? Explain. walk; $\frac{3}{8} < \frac{5}{8}$

22. Three pieces of wire measure 28.3 cm, 23.8 cm, and 28.5 cm. Write the lengths in order from least to greatest. Use <. 23.8 cm < 28.3 cm < 28.5 cm

P4 ON MY OWN   Use with text pages 24–25.

ENRICHMENT 1.4

Name _____   LESSON 1.4

Puzzling Exponents

Complete the puzzle with the values.

Across	Down
1. 11^3	1. 25^3
2. 5^3	3. 6^3
3. 17^2	8. 30^2
4. 7^4	10. 6^4
5. 14^3	12. 2^{10}
6. 9^4	13. 7^3
7. 5^5	14. 20^2
8. 10^4	15. 12^3
9. 16^2	16. 24^2
10. 2^7	17. 3^7
11. 15^3	

E4 STRETCH YOUR THINKING   Use with text pages 24–25.

25

ORGANIZER

OBJECTIVE To use a model to find squares and square roots
VOCABULARY *New* **square, square root**
MATERIALS *For each student* square tiles
PROGRAM RESOURCES
E-Lab • Activity 1
E-Lab Recording Sheets, p. 1

USING THE PAGE
Discuss with students the meaning of arrays. Ask:

- **What objects in your environment show arrays?** Possible examples: panes of glass in windows, sheets of adhesive stamps, coin slots in a coin collector's folder

- **When did you use arrays in math class last year?** Possible answer: when learning multiplication and division

- **Have you ever modeled a square array?** Some students will have modeled square arrays when they learned to find the area of a square.

Activity 1
Explore
In this activity students model square numbers by using tiles to make square arrays. This hands-on experience gives students the opportunity to explore the relationship between square numbers and exponents before they proceed to finding squares on a calculator.

Think and Discuss
Discuss with students how they found the answer to each question. Some students will count individual tiles, some will count by fours and fives, and others will multiply 4×4 and 5×5.

Try This
Have students share with the class the models they make for 6^2, 7^2, and 8^2. Observe students as they use the calculator to find square numbers. Help them locate the x^2 key.

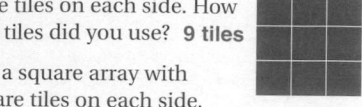

LAB *Activity*

WHAT YOU'LL EXPLORE
How to use a model to find squares and square roots

WHAT YOU'LL NEED
square tiles

WORD POWER
square
square root

26   Chapter 1

Squares and Square Roots

A **square** is the product of a number and itself. A square can be expressed with the exponent 2. Read 3^2 as "3 squared."

You can use square arrays to model square numbers.

ACTIVITY 1
Explore

- Make a square array with 3 square tiles on each side. How many tiles did you use? **9 tiles**

- Make a square array with 4 square tiles on each side.

- Make another square array with 5 square tiles on each side.

Think and Discuss

- How many tiles are in the square array with 4 square tiles on each side? Complete: $4^2 = \bullet$. **16 tiles; 16**

- How many tiles are in the square array with 5 square tiles on each side? Complete: $5^2 = \bullet$. **25 tiles; 25**

- 💡**CRITICAL THINKING** Suppose you have a square with n square tiles on each side. How many tiles would be in the square array? **n^2, or $n \times n$, tiles**

Try This

- Use square tiles to find 6^2, 7^2, and 8^2. **36; 49; 64**

You can use a calculator to find squares.

11 [x^2] [121]   *Enter the value of the base, 11. Then press* [x^2].

So, $11^2 = 121$.

- Use a calculator to find 15^2, 25^2, and 51^2. **225; 625; 2,601**

🖩 Calculator Activities, page H38

👥 **IDEA FILE**

ALTERNATIVE TEACHING STRATEGY

Have students find the value of these square numbers using a geoboard.

1. 2^2 **4**
2. 1^2 **1**
3. 4^2 **16**

Modalities Kinesthetic, Visual

INTERDISCIPLINARY CONNECTION

Have students find out more about how numbers are used to describe the solar system. Have them research answers to one or both of these questions:

- What is the greatest distance between two bodies in the solar system? Which two bodies did you choose?

- How do astronomers use decimals and exponents to describe very large distances?

Modality Visual

When you find the two equal factors of a number, you are finding the **square root** of the number. The symbol for a square root is $\sqrt{\ }$. Finding a square root is the opposite of finding a square.

Since $5^2 = 25$, $\sqrt{25} = 5$. Read $\sqrt{25}$ as "the square root of 25."

ACTIVITY 2

Explore

You can think about the sides of a square array to help you find the square root of a number.

- Make a square array with 16 square tiles. How many tiles are on each side of your square array? **4**

- Make a square array with 4 square tiles.

- Make another square array with 9 square tiles.

Think and Discuss

- How many tiles are on each side of your square array with 4 tiles? Complete: $\sqrt{4} = \bullet$. **2; 2**

- How many tiles are on each side of your square array with 9 tiles? Complete: $\sqrt{9} = \bullet$. **3; 3**

- How are squares and square roots different?
  Possible response: They are opposites.

- 💡**CRITICAL THINKING** Can you form a square with 10 tiles? 12 tiles? Explain. **No; no; you need the same number of tiles on each side to form a square.**

Try This

You can use a calculator to find square roots.

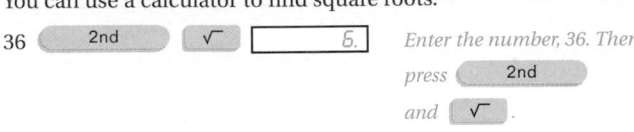

Enter the number, 36. Then press [2nd] *and* [√].

So, $\sqrt{36} = 6$.

- Use a calculator to find $\sqrt{121}$, $\sqrt{625}$, and $\sqrt{1,024}$. **11; 25; 32**

Technology Link
E-Lab • Activity 1 Available on CD-ROM and the Internet at http://www.hbschool.com/elab

Note: Students find patterns in square roots. Use E-Lab p. 1.

27

Activity 2

Explore

In this activity, students use square arrays to explore the concept of square roots. This hands-on experience leads students to find square roots on a calculator.

Think and Discuss

Ask:

- **How did you find the number of tiles on each side of the 4-tile array? on each side of the 9-tile array?** Some students will have counted individual tiles; others will have used mental math to find $2 \times 2 = 4$ and $3 \times 3 = 9$.

Try This

As students use the calculator to find square roots, make sure they are following the correct key sequence. Some students may need help locating the square and square-root keys.

USING E-LAB

Students discover the graphic pattern produced by repeatedly squaring or taking the square root of a number.

✓ Assessment Tip

Use these questions to help students express the big ideas of the lesson.

- **Use square tiles to find 9^2.** 81; check students' models.

- **Use a calculator to find the square root of 144.** 12

REVIEW

Have students find the numbers that have the same value. $4^2, 2^4 = 16$; $1^5, 1^7 = 1$; $8^2, 4^3, 2^6 = 64$

4^4	2^4	2^3	1^5
2^7	8^2	4^3	3^6
3^2	7^2	2^6	4^2
2^5	4^5	1^7	2^2

Modality Visual

GEOMETRY CONNECTION

Have students describe and illustrate the relationship between square root and the perimeter of a square. Possible answer: The square root of a number is equal to the length of one side of a square, and four times the length of one side of a square is equal to the perimeter.

Modalities Auditory, Visual

GRADE 5	GRADE 6	GRADE 7
Read integers on a thermometer.	Order integers and identify opposite integers.	Identify integers as a set of numbers.

ORGANIZER

OBJECTIVE To order integers and to identify opposite integers

VOCABULARY *New* **integers, positive integers, negative integers, opposites**

ASSIGNMENTS Independent Practice
- *Basic* 1–35, odd; 36–40
- *Average* 1–40
- *Advanced* 1–5, 16–40

PROGRAM RESOURCES
Reteach 1.5 Practice 1.5 Enrichment 1.5
Math Fun • The Numbers Are Falling!, p. 30

PROBLEM OF THE DAY

Transparency 1

One side of Jessica's square array is 2 tiles longer than a side of Dave's square array. Together, they used a total of 100 tiles. How many tiles are on each side of Dave's array? 6 tiles on a side

Solution Tab A, p. A1

QUICK CHECK
Name the opposite.

1. forward backward
2. left right
3. negative positive
4. greater less
5. increase decrease
6. smaller larger

1 MOTIVATE

On the board, draw a number line from 100 to 1,000 with intervals of 100. Have students label these numbers on the number line: 850, 500, 900, 450, 650, 700, 250, and 300. Then ask them:

- **Which is greater: 250 or 500?** 500
- **How do you know?** 500 is to the right of 250.
- **Which is less: 700 or 850?** 700
- **How do you know?** 700 is to the left of 850.

Repeat these questions for the other numbers on the number line.

1.5 ALGEBRA CONNECTION

Integers

WHAT YOU'LL LEARN
How to order integers and how to identify opposite integers

WHY LEARN THIS?
To compare temperatures and altitudes and find opposites

WORD POWER
integers
positive integers
negative integers
opposites

Science Link
Scientists use the Celsius scale instead of the Fahrenheit scale to measure temperatures. Water freezes at 32° on the Fahrenheit scale (32°F), which is 0° on the Celsius scale (0°C). Which temperature is colder, 2°F or ⁻2°C? 2°F

In a given year, the lowest temperature recorded for Atlantic City, New Jersey, was ⁻11°F. On a thermometer, is ⁻11°F above or below zero?

Just as a thermometer shows temperatures above and below zero, a number line can show numbers to the right and to the left of zero.

Integers can be shown on a number line. Integers greater than 0 are **positive integers**. Integers less than 0 are **negative integers**. The integer 0 is neither positive nor negative.

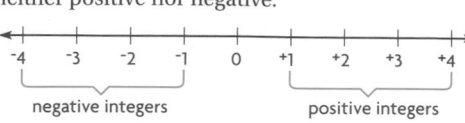

negative integers   positive integers

You can use a number line to compare integers. On a number line, each integer is greater than any integer to its left and less than any integer to its right.

EXAMPLES Compare the integers. Use < and > .

A. ⁺3 and ⁻4

⁻4 is to the left of ⁺3 on the number line.
So, ⁻4 < ⁺3.

⁺3 is to the right of ⁻4 on the number line.
So, ⁺3 > ⁻4.

B. ⁻4 and ⁻6

⁻6 is to the left of ⁻4 on the number line.
So, ⁻6 < ⁻4.

⁻4 is to the right of ⁻6 on the number line.
So, ⁻4 > ⁻6.

- Compare and order ⁻3, ⁺7, and 0 from least to greatest.
⁻3 < 0 < ⁺7

For every positive integer, there is an opposite negative integer. Integers that are **opposites** are the same distance from 0 on the number line, only in opposite directions.

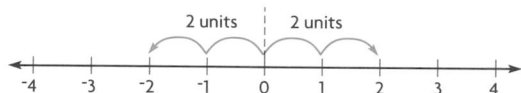

2 units   2 units

The numbers 2 and ⁻2 are both 2 units from 0. They are opposites.

IDEA FILE

SCIENCE CONNECTION

Have students label a number line from ⁻100°C to ⁺100°C with intervals of 10°C. Have them research the highest and lowest recorded temperatures for each of the seven continents and record the data on the number line.

Use TR p. R15.

Modalities Auditory, Kinesthetic, Visual

RETEACH 1.5

Name _____

Integers

A number line can help you compare integers. Remember, an integer is greater than any integers to its left and less than any integers to its right.

Larry is asked to compare ⁺2 and ⁻5.

- He finds ⁺2 and ⁻5 on the number line.
- Since ⁻5 is to the left of ⁺2, Larry knows that ⁻5 < ⁺2.
- Larry could have also written this relationship as ⁺2 > ⁻5.

Now Larry wants to compare and order ⁻4, ⁺3, and ⁻1 from least to greatest.

- He finds ⁻4, ⁺3, and ⁻1 on the number line.
- Since ⁻4 is the farthest left, it has the least value.
- Since ⁺3 is the farthest right, it has the greatest value.
- Larry knows that ⁻4 < ⁻1 < ⁺3.

Compare the integers. Refer to the above number line if it helps you. Use < or >.

1. ⁻1 < ⁺4   2. ⁺2 > ⁻4   3. ⁻6 < ⁺6   4. ⁺5 > ⁻2
5. ⁻3 < ⁺2   6. 0 > ⁻3   7. ⁺4 > ⁻1   8. ⁺3 > ⁻6

Order the integers from least to greatest. Use <.

9. ⁻2, ⁻3, ⁺5   10. ⁺2, 0, ⁻1   11. ⁺3, ⁻2, ⁻3
   ⁻3 < ⁻2 < ⁺5   ⁻1 < 0 < ⁺2   ⁻3 < ⁻2 < ⁺3
12. ⁺1, 0, ⁻1   13. ⁻6, ⁺5, ⁻4   14. ⁻2, ⁺1, ⁻6
   ⁻1 < 0 < ⁺1   ⁻6 < ⁻4 < ⁺5   ⁻6 < ⁻2 < ⁺1
15. ⁻5, ⁻6, ⁻4   16. ⁺2, 0, ⁺3   17. ⁺1, ⁻5, ⁺5
   ⁻6 < ⁻5 < ⁻4   0 < ⁺2 < ⁺3   ⁻5 < ⁺1 < ⁺5

Use with text pages 28–29.   TAKE ANOTHER LOOK R5

GUIDED PRACTICE

Compare the integers. Tell which is greater.

1. $^+2$ and $^-3$ $^+2$ **2.** $^-3$ and $^-6$ $^-3$ **3.** $^+3$ and $^+2$ $^+3$ **4.** $^-4$ and $^+3$ $^+3$ **5.** $^-4$ and $^-2$ $^-2$

Name the opposite of the given integer.

6. $^+5$ $^-5$ **7.** $^-10$ $^+10$ **8.** $^-3$ $^+3$ **9.** $^+15$ $^-15$ **10.** $^-8$ $^+8$

INDEPENDENT PRACTICE

Compare the integers. Use $<$ and $>$.

1. $^-3 \bullet ^+3$ $<$ **2.** $^-4 \bullet 0$ $<$ **3.** $^-2 \bullet ^-5$ $>$ **4.** $^-5 \bullet ^-3$ $<$ **5.** $^+15 \bullet ^+10$ $>$

6. $^-4 \bullet ^-2$ $<$ **7.** $^-10 \bullet ^-1$ $<$ **8.** $^+6 \bullet 0$ $>$ **9.** $^-6 \bullet ^-10$ $>$ **10.** $^+12 \bullet ^-11$ $>$

11. $^+6 \bullet ^+10$ $<$ **12.** $^-17 \bullet ^-3$ $<$ **13.** $^+22 \bullet ^+19$ $>$ **14.** $^-2 \bullet ^+20$ $<$ **15.** $^+13 \bullet ^-26$ $>$

Order the integers from least to greatest. Use $<$.

16. $0, ^-4, ^+7, ^-3$ $^-4 < ^-3 < 0 < ^+7$
17. $^-6, ^-3, ^+6, ^-4$ $^-6 < ^-4 < ^-3 < ^+6$
18. $^+5, ^-5, ^+2, ^-3$ $^-5 < ^-3 < ^+2 < ^+5$
19. $^+2, 0, ^-3, ^-10$ $^-10 < ^-3 < 0 < ^+2$
20. $^-3, ^+3, ^+7, ^-2$ $^-3 < ^-2 < ^+3 < ^+7$
21. $^-8, ^-2, ^+5, ^-6$ $^-8 < ^-6 < ^-2 < ^+5$
22. $^+8, ^+4, ^+2, ^-2$ $^-2 < ^+2 < ^+4 < ^+8$
23. $^+20, ^-11, ^+2, ^-9$ $^-11 < ^-9 < ^+2 < ^+20$

Name the opposite of the given integer.

24. $^-2$ $^+2$ **25.** $^+9$ $^-9$ **26.** $^+3$ $^-3$ **27.** $^-11$ $^+11$ **28.** $^+6$ $^-6$ **29.** $^+14$ $^-14$

30. $^-7$ $^+7$ **31.** $^-12$ $^+12$ **32.** $^-19$ $^+19$ **33.** $^-24$ $^+24$ **34.** $^-54$ $^+54$ **35.** $^-98$ $^+98$

Problem-Solving Applications

36. Juneau, Alaska, reports a temperature of $^-22°F$. Seattle, Washington, reports a temperature of $22°F$. Which city has the higher temperature? **Seattle, Washington**

37. The Smiths' house is 5 ft below sea level. The Jacksons' house is 3 ft above sea level. Whose house is closer to sea level? **the Jacksons' house**

38. The countdown for the space shuttle's launch is at minus 10 sec. What integer would represent 10 sec after the launch? **$^+10$**

39. On one space shuttle mission, a communications check is scheduled at $^-15$ sec. A fuel check is scheduled at $^-8$ sec. Which check will happen first? **communications check**

40. ✏️ **WRITE ABOUT IT** Write *sometimes*, *always*, or *never* for this statement: A negative integer is less than a positive integer. Explain your answer. **Always; on a number line, negative integers are always to the left of positive integers.**

MORE PRACTICE Lesson 1.5, page H43

29

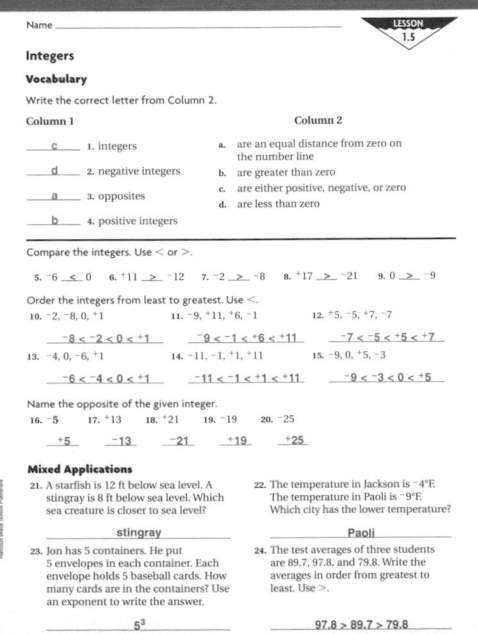

PRACTICE 1.5

Name _____ LESSON 1.5

Integers

Vocabulary

Write the correct letter from Column 2.

Column 1		Column 2
c	1. integers	a. are an equal distance from zero on the number line.
d	2. negative integers	b. are greater than zero
a	3. opposites	c. are either positive, negative, or zero
b	4. positive integers	d. are less than zero

Compare the integers. Use $<$ or $>$.

5. $^-6 \underline{<} 0$ 6. $^+11 \underline{>} ^-12$ 7. $^-2 \underline{>} ^-8$ 8. $^+17 \underline{>} ^-21$ 9. $0 \underline{>} ^-9$

Order the integers from least to greatest. Use $<$.

10. $^-2, ^-8, 0, ^+1$ $^-8 < ^-2 < 0 < ^+1$
11. $^-9, ^+11, ^+6, ^-1$ $^-9 < ^-1 < ^+6 < ^+11$
12. $^+5, ^-5, ^+7, ^-7$ $^-7 < ^-5 < ^+5 < ^+7$
13. $^-4, 0, ^-6, ^+1$ $^-6 < ^-4 < 0 < ^+1$
14. $^-11, ^-1, ^+1, ^+1$ $^-11 < ^-1 < ^+1 < ^+11$
15. $^-9, 0, ^+5, ^-3$ $^-9 < ^-3 < 0 < ^+5$

Name the opposite of the given integer.

16. $^-5$ $^+5$ 17. $^+13$ $^-13$ 18. $^+21$ $^-21$ 19. $^-19$ $^+19$ 20. $^-25$ $^+25$

Mixed Applications

21. A starfish is 12 ft below sea level. A stingray is 8 ft below sea level. Which sea creature is closer to sea level? **stingray**

22. The temperature in Jackson is $^-4°F$. The temperature in Paoli is $^-9°F$. Which city has the lower temperature? **Paoli**

23. Jon has 5 containers. He put 5 envelopes in each container. Each envelope holds 5 baseball cards. How many baseball cards are in the containers? Use an exponent to write the answer. **5^3**

24. The test averages of three students are 89.7, 97.8, and 79.8. Write the averages in order from greatest to least. Use $>$. **97.8 > 89.7 > 79.8**

Use with text pages 28–29. **ON MY OWN** P5

ENRICHMENT 1.5

Name _____ LESSON 1.5

Name That Floor

The floor selection keys found in a hotel's elevator are shown. Each of the following describes a ride in the elevator.

Use integers to show how each rider's position changed. Then name the floor on which each person left the elevator. The first one is done for you.

1. Sue entered the elevator on the seventh floor. She traveled up three floors and then down five floors before leaving the elevator.

$^+3; ^-5;$ **fifth floor**

2. Josh entered the elevator at the lobby. He went up six floors and then down two before leaving the elevator.

$^+6; ^-2;$ **third floor**

3. Rob got on the elevator on the twelfth floor. He rode down to the mezzanine and then up four floors before leaving the elevator.

$^-12, ^+4;$ **fourth floor**

4. Tim entered the elevator from the garage. He rode up eight floors, down two floors, and then up another three floors before exiting.

$^+8, ^-2, ^+3;$ **seventh floor**

5. Jon entered the elevator on the ninth floor. He rode up five floors, down six floors, and then down another two floors before leaving the elevator.

$^+5, ^-6, ^-2;$ **sixth floor**

6. Tina got on the elevator at the lobby. She rode up to the fourteenth floor, down six floors, up four floors, and then she exited the elevator.

$^+15, ^-6, ^+4;$ **twelfth floor**

7. Evan entered the elevator on the fifth floor. He rode down to the mezzanine, down to the garage, and up four floors before leaving.

$^-5, ^-2, ^+4;$ **second floor**

8. Carole got on the elevator on the first floor. She rode up six floors, down to the mezzanine, and then up two floors before exiting.

$^+6, ^-7, ^+2;$ **second floor**

Use with text pages 28–29. **STRETCH YOUR THINKING** E5

---

2 TEACH/PRACTICE

Before discussing the examples on page 28, you may wish to ask students this question.

• **In what situations are positive and negative integers used?** Possible answers: to describe heights and depths above and below sea level and above-ground and underground

GUIDED PRACTICE

You may wish to use a number line on the board to complete these exercises as a whole-class activity.

INDEPENDENT PRACTICE

In order to assess students' understanding of positive and negative integers, you may wish to have students complete Exercises 1–5 and then trade papers with a class-mate to check their answers.

ADDITIONAL EXAMPLE

Example, p. 28

Use a number line to compare the integers. Then write the answer using $>$ and $<$. $^+7$ and $^-6$

$^-6$ is to the left of $^+7$ on the number line.

So, $^-6 < ^+7$. $^+7$ is to the right of $^-6$ on the number line.

So, $^+7 > ^-6$.

3 WRAP UP

Use these questions to help students focus on the big ideas of the lesson.

• **What negative integers are greater than $^-5$?** $^-4, ^-3, ^-2, ^-1$

✔️ *Assessment Tip*

Have students write the answer.

• **Write two statements that describe the relationship between $^+6$ and $^-6$.** Possible answers: $^+6 > ^-6$; $^-6 < ^+6$; $^+6$ is the opposite of $^-6$; $^-6$ is the opposite of $^+6$.

DAILY QUIZ

Order the integers from least to greatest.

1. $^+8, ^-12, ^+12$ $^-12, ^+8, ^+12$
2. $^-18, ^-19, ^+1$ $^-19, ^-18, ^+1$
3. $^+8, ^-12, ^+12$ $^-12, ^+8, ^+12$

Name the opposite of the given integer.

4. $^-60$ $^+60$ **5.** $^+65$ $^-65$ **6.** $^+77$ $^-77$

29

USING THE PAGE

To provide additional practice activities related to Chapter 1: *Looking at Numbers*

READING, WRITING, AND NUMBERS

MATERIALS: *For each group* a set of number cards and word cards

Have students work in pairs or small groups to complete this activity.

In this activity, students use number and word cards to use place value to express whole numbers and decimals.

Multiple Intelligences Visual/Spatial, Logical/Mathematical, Verbal/Linguistic, Bodily/Kinesthetic, Interpersonal/Social, Intrapersonal/Introspective
Modalities Auditory, Kinesthetic, Visual

JEREMY'S TWO

Have students work in pairs or small groups to complete this activity.

In this activity, students analyze answers to determine whether they are correct.

Multiple Intelligences Logical/Mathematical, Verbal/Linguistic, Interpersonal/Social, Intrapersonal/Introspective
Modality Auditory

THE NUMBERS ARE FALLING!

MATERIALS: *For each student* blank number lines, TR p. R15

Have students work individually to complete this activity.

In this activity, students compare and order decimals, fractions, exponents, square roots, and integers on a number line.

Multiple Intelligences Visual/Spatial, Logical/Mathematical, Intrapersonal/Introspective
Modality Visual

MATH FUN!

READING, WRITING, AND NUMBERS

PURPOSE To practice using place value to express whole numbers and decimals (pages 16–17)
YOU WILL NEED a set of number cards and word cards

Work with a partner or small group. Make sure each player has a set of number cards. Take turns drawing a card from the word card pile. Place the card you draw face up, and read it aloud to the other players. Have each player form that number from his or her own set of number cards.

The player who forms the number first wins 10 points. The player who comes in second wins 5 points. A player who gets a number wrong loses 5 points. The first player to reach 100 points wins.

ten and four thousandths

one thousand two hundred fifty and four tenths

HOME NOTE Make a new set of word cards. Then play this game with your family.

 Portfolio

JEREMY'S TWO

PURPOSE To practice writing and comparing decimals and fractions (pages 20–23)

Jeremy just finished his math homework. He made two errors. Work with a partner or in a small group. Try to be the first group to find the two errors. The first team finished gets to make up four new problems for the other teams.

PROBLEMS	ANSWERS
1. Write 0.4 as a fraction.	$\frac{1}{5}$ **wrong**
2. Compare. $\frac{7}{14}$ ● 0.458	$>$
3. Compare. $\frac{7}{8}$ ● $\frac{77}{100}$	$<$ **wrong**
4. Write $\frac{13}{16}$ as a decimal.	0.8125

THE NUMBERS ARE FALLING!

PURPOSE To practice comparing and ordering decimals, fractions, exponents, square roots, and integers (pages 20–29)
YOU WILL NEED blank number line

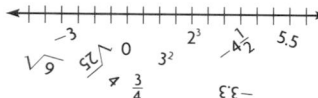

Help the numbers find their places on the number line.

Start by labeling the zero point. Find the value of any exponent or square root. Then find the location of each of the numbers. **Check students' number lines.**

CHAPTER 1 REVIEW

EXAMPLES

• Use place value to express numbers.
(pages 16–17)

Write: 1,956,450.0345 ← *standard form*

Read: one million, nine hundred fifty-six thousand, four hundred fifty and three hundred forty-five ten-thousandths

• Compare and order whole numbers and decimals. (pages 18–19)

9.3 ● 9.36	*Compare the digits.*
9.3 ● 9.36	*same number of ones*
9.3 ● 9.36	*same number of tenths*
9.30 ● 9.36	*Add a zero to 9.3.*

Since 0 hundredths is less than 6 hundredths, 9.3 < 9.36.

• Write decimals as fractions, and fractions as decimals. (pages 20–23)

0.09   *Identify the value of the last digit.*
        9 is in the hundredths place.

$\frac{9}{100}$ *Use that value to write the denominator.*

So, $0.09 = \frac{9}{100}$.

• Write numbers as exponents. (pages 24–25)

3^4 *Use 3 as a factor 4 times.*

$3 \times 3 \times 3 \times 3 = 81$

So, the value of 3^4 is 81.

• Order integers, and find opposite integers.
(pages 28–29)

$^-11$ ● $^+11$ *$^-11$ is to the left of $^+11$ on a number line.*

So, $^-11 < {}^+11$.

EXERCISES

Write the number in standard form.
1. two hundred fifty-five **255**
2. one thousand, six hundred four ten-thousandths **0.1604**
3. eight hundred twenty-one thousand, two hundred two and four tenths **821,202.4**

Compare the numbers. Write <, >, or =.
4. 113.3 ● 11.33 **>**   5. 2,281 ● 2,291 **<**
6. 82.46 ● 82.461 **<**   7. 820 ● 82 **>**
8. 4.5 ● 4.50 **=**   9. 0.0821 ● 0.821 **<**
10. Use < to list the numbers in order from least to greatest: 92.01, 920.1, 92.10.
92.01 < 92.10 < 920.1

Change the decimal to a fraction.
11. 0.1 $\frac{1}{10}$ 12. 0.003 $\frac{3}{1,000}$
Write as a decimal.
13. $\frac{3}{100}$ **0.03**   14. $\frac{3}{5}$ **0.6**
15. Jean has $\frac{2}{5}$ of a pizza and Joe has $\frac{1}{4}$. Who has more? **Jean**

16. **VOCABULARY** A(n) _?_ shows how many times a number called the base is used as a factor. **exponent**
Find the value.
17. 10^4 **10,000**   18. 9^2 **81**
19. 15^3 **3,375**   20. 1^{14} **1**

21. **VOCABULARY** Integers that are _?_ are an equal distance from zero on the number line. **opposites**
Write < or >.
22. $^-7$ ● $^+1$ **<**   23. 0 ● $^-2$ **>**
24. $^+4$ ● $^+3$ **>**   25. $^-5$ ● $^-10$ **>**

Chapter 1 Review **31**

USING THE PAGE

The Chapter 1 Review reviews using place value, comparing and ordering whole numbers and decimals, writing decimals as fractions, writing exponents, and ordering integers. Chapter objectives are provided, along with examples and practice exercises for each.

The Chapter 1 Review can be used as a review or a test. You may wish to place the Review in students' portfolios.

Assessment Checkpoint

For Performance Assessment in this chapter, see page 72, What Did I Learn?, items 1 and 4.

✔ Assessment Tip

Student Self-Assessment The How Did I Do? Survey helps students assess both what they have learned and how they learned it. Self-assessment helps students learn more about their own capabilities and develop confidence in themselves.

This survey is available as a copying master in the *Teacher's Guide for Assessment*, p. 15.

CHAPTER 1 TEST, page 1

Name _____

Choose the letter of the correct answer.

For questions 1–2, use the place-value chart below.

PLACE VALUE									
Millions	Hundred Thousands	Ten Thousands	Thousands	Hundreds	Tens	Ones	Tenths	Hundredths	Thousandths
1	4	2	6	3	2	1	.0	7	9

1. What is the value of the digit 4?
   A. 4 tens
   B. 4 thousands
   C. 4 ten thousands
   Ⓓ 4 hundred thousands

2. What is the value of the digit 7 in the place-value chart?
   A. 7 tenths
   Ⓑ 7 hundredths
   C. 7 thousandths
   D. 7 ten-thousandths

3. What is the value of the digit 8 in the number 407,836.21?
   Ⓐ 8 hundreds
   B. 8 thousands
   C. 8 ten thousands
   D. 8 hundred thousands

4. What is the value of the digit 5 in the number 1.275?
   A. 5 tens   B. 5 hundreds
   C. 5 hundredths Ⓓ 5 thousandths

5. What is the standard form of four million, sixty thousand, one hundred two?
   Ⓐ 4,060,102   B. 460,102
   C. 46,000,102   D. 4,006,102,000

6. Which shows six hundredths written in standard form?
   A. 0.006
   Ⓑ 0.06
   C. 0.60
   D. 600.0

7. Which is greater than 10.25?
   Ⓐ 10.4
   B. 1.40
   C. 0.14
   D. 10.04

8. Which shows the numbers written in order from least to greatest?
   A. 2.2, 2.02, 2.12
   B. 2.2, 2.12, 2.02
   C. 2.12, 2.02, 2.2
   Ⓓ 2.02, 2.12, 2.2

9. The batting averages of three players are 0.240, 0.229, and 0.252. Which shows these averages written in order from greatest to least?
   A. 0.229, 0.240, 0.252
   B. 0.252, 0.229, 0.240
   Ⓒ 0.252, 0.240, 0.229
   D. 0.240, 0.252, 0.229

10. A rabbit can run 56.3 km per hour, a coyote 69.2 km per hour, a giraffe 51.3 km per hour, and a greyhound 63.3 km per hour. Which of the four animals runs the fastest?
    A. rabbit
    Ⓑ coyote
    C. giraffe
    D. greyhound

Form A • Multiple-Choice   A5   Go on. ▶

CHAPTER 1 TEST, page 2

Name _____

11. How can you change the fraction $\frac{1}{6}$ to a decimal?
    A. Multiply 10 by $\frac{1}{6}$.
    B. Divide 10 by $\frac{1}{6}$.
    C. Divide 6 by 1.
    Ⓓ Divide 1 by 6.

12. What is 0.07 written as a fraction?
    A. $\frac{7}{10}$ B. $\frac{1}{7}$
    Ⓒ $\frac{7}{100}$ D. $\frac{1}{70}$

13. What is $\frac{2}{5}$ written as a decimal?
    A. 0.10   B. 0.20
    C. 0.25   Ⓓ 0.40

14. What decimal makes the following inequality true?
    $\frac{15}{25} <$ _?_
    A. 0.20   B. 0.45
    C. 0.50   Ⓓ 0.8

15. The lake was 1 mi away. Terry rode his bike 0.25 of the way. What is the distance Terry rode, expressed as a fraction?
    A. $\frac{1}{8}$ mi   B. $\frac{1}{5}$ mi
    Ⓒ $\frac{1}{4}$ mi   D. $\frac{1}{3}$ mi

16. Katie walked $\frac{3}{5}$ mi. What is the distance she walked, expressed as a decimal?
    A. 0.006 mi
    B. 0.06 mi
    Ⓒ 0.6 mi
    D. 6.0 mi

17. How many zeros are in the standard form of 10^6?
    A. 4   B. 5
    Ⓒ 6   D. 7

18. Which expression represents 5^2?
    A. 5 + 2
    B. 5 × 2
    Ⓒ 5 × 5
    D. 2 × 2 × 2 × 2 × 2

19. What is 4 × 4 × 4 × 4 × 4 written in exponent form?
    A. 4^3 B. 3^4
    Ⓒ 4^5 D. 5^4

20. What is the value of 6^3?
    A. 18   B. 36
    C. 42   Ⓓ 216

21. Which is less than 0?
    Ⓐ $^-2$ B. 0
    C. 4   D. 6

22. Which of these shows the integers written in order from least to greatest?
    A. 0, $^-4$, 2, 4
    B. 0, 2, $^-4$, 4
    Ⓒ $^-4$, 0, 2, 4
    D. $^-2$, 2, 0, $^-4$

23. Which of these shows the integers written in order from least to greatest?
    A. 1, $^-3$, $^-5$, 0
    B. 1, $^-5$, $^-3$, 0
    C. $^-3$, $^-5$, 1, 0
    Ⓓ $^-5$, $^-3$, 0, 1

24. What is the opposite of $^-4$?
    A. $^-\frac{1}{4}$ B. $^+\frac{1}{4}$
    Ⓒ $^+4$ D. $^+4^2$

Form A • Multiple-Choice   A6   ▶ Stop!

See *Test Book Copying Masters*, pp. A5–A6 for the Chapter Test in **multiple-choice** format and pp. B155–B156 for **free-response** format.

ADDITIONAL ANSWERS
Lesson 1.1, page 17

Guided Practice

5. ten thousand, forty

6. six hundred five thousandths

7. six and five tenths

8. seven thousand, eight hundred sixty-eight

9. eight hundred sixty-three thousand, forty-two and forty-six hundredths

Independent Practice

9. forty-six

10. three hundredths

11. one thousand, five hundred and one-tenth

12. two and four hundred fifty-six thousandths

13. one million, thirty-seven thousand, eight hundred four

2

PLANNING GUIDE

Using Whole Numbers

BIG IDEA: Mental-math and estimation experiences enhance number sense and often yield the most appropriate solutions to problems.

Introducing the Chapter p. 32

Team-Up Project • How My Calculator Works p. 33

OBJECTIVE	VOCABULARY	MATERIALS	RESOURCES
2.1 Mental-Math for Addition and Subtraction pp. 34-35 **Objective:** To use properties and mental math to find sums and differences **1 DAY**	Commutative Property Associative Property compensation		■ Reteach, ■ Practice, ■ Enrichment 2.1; More Practice, p. H43 Cultural Connection • Piñata Party, p. 50
2.2 Using Guess and Check to Add and Subtract Numbers pp. 36-37 **Objective:** To use the strategy guess and check to solve addition and subtraction problems with whole numbers **1 DAY**			■ Reteach, ■ Practice, ■ Enrichment 2.2; More Practice, p. H44 Problem Solving Think Along, TR p. R1
2.3 Mental Math for Multiplication pp. 38-39 **Objective:** To use properties to solve multiplication problems **1 DAY**	Identify Property of One Property of Zero Distributive Property		■ Reteach, ■ Practice, ■ Enrichment 2.3; More Practice, p. H44
2.4 Multiplication and Division pp. 40-43 **Objective:** To multiply and divide whole numbers **2 DAYS**			■ Reteach, ■ Practice, ■ Enrichment 2,4; More Practice, p. H44 Calculator Activities, p. H31 *Calculating Crew • Intergalactic Trader*
INTERVENTION • Multiplication and Division, Teacher's Edition pp. 44A–44B			
Lab Activity: Order of Operations pp. 44-45 **Objective:** To find a solution by using order of operations	order of operations	FOR EACH STUDENT: calculator	**E-Lab • Activity 2** E-Lab Recording Sheets, p. 2
2.5 Using Estimation pp. 46-49 **Objective:** To estimate sums, differences, products, and quotients **1 DAY**	clustering compatible numbers		■ Reteach, ■ Practice, ■ Enrichment 2.5; More Practice, p. H45

CHAPTER ASSESSMENT Chapter 2 Review p. 51

NCTM CURRICULUM
STANDARDS FOR GRADES 5–8

☑ **Problem Solving**
Lessons 2.1, 2.2, 2.3, 2.4, 2.5

☑ **Communication**
Lessons 2.1, 2.2, 2.3, 2.4, 2.5

☑ **Reasoning**
Lessons 2.1, 2.2, 2.3, 2.4, 2.5

☑ **Connections**
Lessons 2.1, 2.2, 2.3, 2.4, 2.5

☑ **Number and Number Relationships**
Lessons 2.1, 2.2, 2.3, 2.5

☑ **Number Systems and Number Theory**
Lessons 2.1, 2.2, 2.3, 2.4, 2.5

☑ **Computation and Estimation**
Lessons 2.1, 2.2, 2.3, 2.4, 2.5

☑ **Patterns and Functions**
Lessons 2.2, 2.4

☑ **Algebra**
Lessons 2.1, 2.2, 2.3, Lab Activity

☑ **Statistics**
Lessons 2.1, 2.2, 2.5

☐ **Probability**

☐ **Geometry**

☑ **Measurement**
Lessons 2.2, 2.4, 2.5

CHAPTER LEARNING GOALS

2A To solve problems that involve number theory and fractions.

2A.1 To use properties and mental math to find sums and differences.

2A.2 To solve problems by using the guess and check strategy.

2A.3 To multiply and divide whole numbers.

2A.4 To use estimation to find sums, differences, products, and quotients.

These goals for concepts, skills, and problem solving are assessed in many ways throughout the chapter. See the chart below for a complete listing of both traditional and informal assessment options.

Pretest Options
Pretest for Chapter Content, TCM pp. A7–A8, B157–B158
Assessing Prior Knowledge, TE p. 32; APK p. 4

Daily Assessment
Mixed Review, PE pp. 43, 49
Problem of the Day, TE pp. 34, 36, 38, 40, 46
Assessment Tip and Daily Quiz, TE pp. 35, 37, 39, 43, 45, 49

Formal Assessment
Chapter 2 Review, PE p. 51
Chapter 2 Test, TCM pp. A7–A8, B157–B158
Study Guide & Review, PE pp. 70–71
Test • Chapters 1–3 Test, TCM pp. A11–A12, B161–B162
Cumulative Review, PE p. 73
Cumulative Test, TCM pp. A13–A16, B163–B166

Performance Assessment
Team-Up Project Checklist, TE p. 33
What Did I Learn?, Chapters 1-3, TE p. 72
Interview/Task Test, PA p. 77

Portfolio
Suggested work samples:
Independent Practice, PE pp. 35, 39, 42–43, 48–49
Cultural Connection, PE p. 50

Student Self-Assessment
How Did I Do?, TGA p. 15
Portfolio Evaluation Form, TGA p. 24
Math Journal, TE pp. 32B

Key

Assessing Prior Knowledge Copying Masters: APK	Teacher's Guide for Assessment: TGA
Test Copying Masters: TCM	Teacher's Edition: TE
Performance Assessment: PA	Pupil's Edition: PE

MATH COMMUNICATION

MATH JOURNAL
You may wish to use the following suggestions to have students write about the mathematics they are learning in this chapter:

PE pages 38, 47 • Talk About It

PE pages 35, 37, 39, 43, 49 • Write About It

TE page 49 • Writing in Mathematics

TGA page 15 • How Did I Do?

In every lesson • record written answers to each Wrap-up Assessment Tip from the Teacher's Edition for test preparation.

VOCABULARY DEVELOPMENT
The terms listed at the right will be introduced or reinforced in this chapter. The boldfaced words in the list are new vocabulary terms in the chapter.

To help students develop an understanding of the order of operations, have them write the sentence "Please Excuse My Dear Aunt Sally" in their Math Journals. Then have them write the words that stand for the initial letters *P, E, M, D, A,* and *S.*

Commutative Property, pp. 34, 38
Associative Property, pp. 34, 38
compensation, p. 34
Identify Property of One, p. 38
Property of Zero, p. 38
order of operations, p. 44
clustering, p. 46
compatible numbers, p. 48

MEETING INDIVIDUAL NEEDS

RETEACH	PRACTICE
BELOW-LEVEL STUDENTS	**ON-LEVEL STUDENTS**

Practice Activities, TE Tab B *'Round We Go!* (Use with Lesson 2.5.)	**Practice Game,** TE p. 32C *'Round We Go!* (Use with Lesson 2.5.)
Interdisciplinary Connection, TE p. 45	**Interdisciplinary Connection,** TE p. 45
Literature Link, TE p. 32C *Where The Red Fern Grows,* Activity 1 (Use with Lesson 2.4.)	**Literature Link,** TE p. 32C *Where The Red Fern Grows,* Activity 2 (Use with Lesson 2.4.)
Technology *Calculating Crew* • *Nick Knack, Intergalactic Trader* (Use with Lesson 2.4.)	**Technology** *Calculating Crew* • *Nick Knack, Intergalactic Trader* (Use with Lesson 2.4.)
E-Lab • Activity 2 (Use with Lab Activity.)	**E-Lab** • Activity 2 (Use with Lab Activity.)

CHAPTER IDEA FILE

The activities on these pages provide additional practice and reinforcement of key chapter concepts. The chart above offers suggestions for their use.

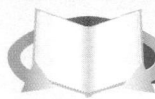

LITERATURE LINK

Where the Red Fern Grows

by Woodrow Wilson Rawls

Billy Coleman sees an advertisement for two coon-hound pups. He saves his allowance for two years to buy the dogs and through them learns about growing up and caring for others.

Activity 1—Have students find out the cost for a pet they would like to have. Then have students determine how many weeks it would take for them to save for a pet if they are given a certain allowance per week.

Activity 2—Have students find out how much it would cost in equipment and supplies to keep a pet for a year. Then have students determine whether or not they would be able to afford to buy, keep and maintain the pet based on the allowance used in Activity 1.

Activity 3—Have students find out which pet students would prefer to have in the classroom. Students should then calculate the cost per student for keeping the pet in the classroom for one school year.

PRACTICE GAME

'Round We Go!

Purpose To round whole numbers

Materials bean bags

About the Game Students in teams of eight to ten students form circles. The first player in each team tosses the bag to the player on the right and names a number between one and one hundred. The catcher rounds the number to the nearest ten and then tosses the bag to the right naming another number. Play continues in this way until the bag completes the circle and all players sit down. The first team seated wins.

Modalities Kinesthetic, Visual, Auditory

Multiple Intelligences
Logical/Mathematical,
Bodily/Kinesthetic,
Interpersonal/Social

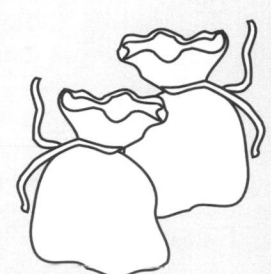

Practice Game, TE p. 32C
'Round We Go!
(Use with Lesson 2.5.)

Interdisciplinary Connection, TE p. 45

 **Literature Link,** TE p. 32C
Where The Red Fern Grows, Activity 3
(Use with Lesson 2.4.)

 Technology
Calculating Crew • Nick Knack,
Intergalactic Trader
(Use with Lesson 2.4.)

E-Lab • Activity 2
(Use with Lab Activity.)

INTERDISCIPLINARY SUGGESTIONS

Connecting Whole Numbers

Purpose To connect whole numbers to other subjects students are studying

You may wish to connect **Using Whole Numbers** to other subjects by using related interdisciplinary units.

Suggested Units

- Science Connection: Creatures Great and Small
- History Connection: Rulers of Monarchies
- Science Connection: Protecting Our Natural Resources
- Multicultural Connection: Recipes From Other Lands
- Math Connection: Logical Reasoning: Number Puzzles
- Social Studies Connection: Communicating With Our Elected Officials
- Consumer Connection: The Cost of Energy

Modalities Auditory, Kinesthetic, Visual

Multiple Intelligences Mathematical/ Logical, Bodily/Kinesthetic, Visual/Spatial

The purpose of using technology in the chapter is to provide reinforcement of skills and support in concept development.

 E-Lab • Activity 2—Students use reasoning skills and the concept of order of operations to examine, evaluate, and insert the correct operation in an expression in order to solve a problem.

Calculating Crew • *Nick Knack, Intergalactic Trader*—the activities on this CD-ROM provide students with opportunities to practice multiplying and dividing whole numbers.

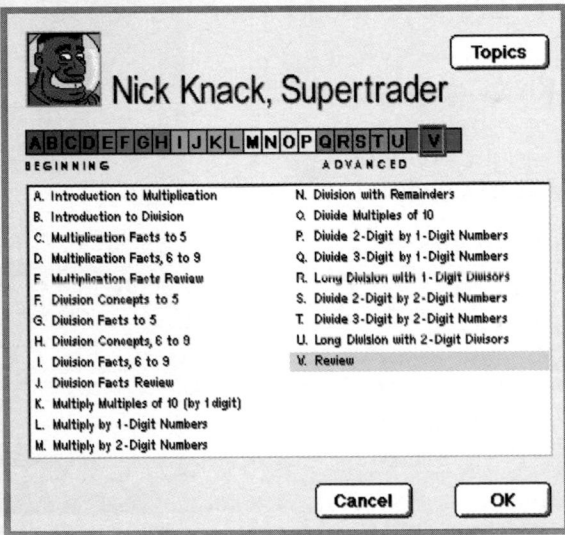

Each Calculating Crew activity can be assigned by degree of difficulty. See Teacher's Edition page 43 for Grow Slide information.

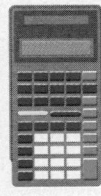 **Explorer Plus™ or fx–65 Calculator**—See suggestions for integrating the calculator in the PE Lesson p. 42. Additional ideas are provided in *Calculator Activities,* p. H31.

Visit our web site for additional ideas and activities.
http://www.hbschool.com

Introducing the Chapter

Encourage students to talk about times when they use mental math to add or subtract whole numbers. Review the math content students will learn in the chapter.

WHY LEARN THIS?

The question "Why are you learning this?" will be answered in every lesson. Sometimes there is a direct application to everyday experiences. Other lessons help students understand why these skills are needed for future success in mathematics.

PRETEST OPTIONS

Pretest for Chapter Content See the *Test Copying Masters*, pages A7–A8, B157–B158 for a free-response or multiple-choice test that can be used as a pretest.

Assessing Prior Knowledge Use the copying master shown below to assess previously taught skills that are critical for success in this chapter.

Assessing Prior Knowledge Copying Masters, p. 4

Name _____

CHAPTER 2

Multiplying by One-Digit Numbers
Find the product.

1. 24×2 = 48	2. 33×3 = 99	3. 19×4 = 76	4. 28×6 = 168	5. 39×7 = 273
6. 401×2 = 802	7. 214×3 = 642	8. 132×5 = 660	9. 274×8 = 2,192	10. 437×8 = 3,496

Dividing by One-Digit Numbers
Find the quotient.

11. $3\overline{)72}$ 24 12. $4\overline{)84}$ 21 13. $5\overline{)95}$ 19 14. $6\overline{)96}$ 16

15. $4\overline{)96}$ 24 16. $3\overline{)87}$ 29 17. $8\overline{)256}$ 32 18. $6\overline{)354}$ 59

Rounding
Round to the nearest 10.

19. 32 _30_ 20. 72 _70_ 21. 86 _90_ 22. 75 _80_ 23. 84 _80_

Round to the nearest 100.

24. 642 _600_ 25. 361 _400_ 26. 597 _600_ 27. 109 _100_ 28. 1,386 _1,400_

Round to the nearest 1,000.

29. 4,831 _5,000_ 30. 9,089 _9,000_ 31. 6,136 _6,000_ 32. 3,712 _4,000_ 33. 4,610 _5,000_

4 Assessing Prior Knowledge • Chapter 2

Troubleshooting

If students need help	Then use
• MULTIPLYING WHOLE NUMBERS	• Intervention • Multiplication and Division, TE pp. 44A–44B
• DIVIDING BY ONE-DIGIT NUMBERS	• Practice Activities, TE Tab B, Activity 6A
• ROUNDING	• Key Skills, PE p. H9

CHAPTER 2

USING WHOLE NUMBERS

LOOK AHEAD

In this chapter you will solve problems that involve

- using mental-math strategies for addition, subtraction, and multiplication

- using the strategy *guess and check*
- multiplying and dividing whole numbers
- using the order of operations

32 Chapter 2

SCHOOL-HOME CONNECTION

You may wish to send to each student's family the *Math at Home* page. It includes an article that stresses the importance of communication about mathematics in the home, as well as helpful hints about homework and taking tests. The activity and extension provide extra practice that develops concepts and skills involving whole numbers.

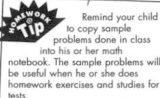

MATH AT HOME

Volume 6 No. 2
http://www.hbschool.com

The Whole Story: Uses of Whole Numbers

Your child is beginning a chapter that teaches more uses of whole numbers. Students will use mental math for addition, subtraction, and multiplication, and division of whole numbers. Share with your child some of the ways you use whole numbers in your daily life.

Here are an activity and an Extension for you to do with your child that use whole numbers to figure team standings in sports.

Pass the Puck!

In hockey, the standings of the teams are determined by a point system based on the number of wins and ties. A team's points are found by doubling the number of wins and adding the number of ties. Find the points for the following teams.

	WINS	TIES	POINTS
PHILADELPHIA	23	3	
FLORIDA	19	9	
DEVILS	20	3	
RANGERS	18	5	

Suppose a team has 50 points. What are two ways it might have earned that total?

Extension • Score a Goal!

Is it possible to have the same number of points but not the same number of wins? Give an example.

Is it possible to have fewer wins than another team but have more points? Give an example.

For Your Information

In a newspaper they are likely to find numbers such as **65K** and **$1.2 M**. Good number sense will tell them that 65K means 65,000 and $1.2 M means $1.2 million or 1,200,000.

In today's world, people need a strong sense of numbers to be able to understand the many uses of mathematics.

> Remind your child to copy sample problems done in class into his or her math notebook. The sample problems will be useful when he or she does homework exercises and studies for tests.

VOCABULARY

Commutative Property—the fact that numbers can be added in any order without changing their sum
$64 + 78 + 56 = 56 + 64 + 78$

Associative Property—the fact that addends can be grouped differently without changing their sum
$(64 + 78) + 56 = (56 + 64) + 78$

ANSWERS: Pass the Puck: Philadelphia – 49; Florida – 47; Devils – 43; Rangers – 41; Possible 50 points: a poster doubletons; 33 wins, 4 ties; 11 wins, 7 ties = 42 points; Net 25 wins, 0 ties = 40 points; 18 wins, 4 ties = 40 points; 17 wins, 4 ties = 38 points

2

Math at Home

Team-Up Project

How My Calculator Works

Suppose you met someone who did not know how to work your calculator. Make a guide for him or her.

YOU WILL NEED: calculator

Plan
- Work with a partner to write a guide for operating your calculator.

Act
- Select a calculator. Decide which operations to include in the guide.
- Figure out how to perform each operation and record your actions.
- Make a guide. Work out a sample problem to show each operation.

Share
- Try out your guide by exchanging calculators and guides with another group.

DID YOU
- ✓ Write a guide showing different operations?
- ✓ Use another group's guide?
- ✓ Discuss differences in different calculators?

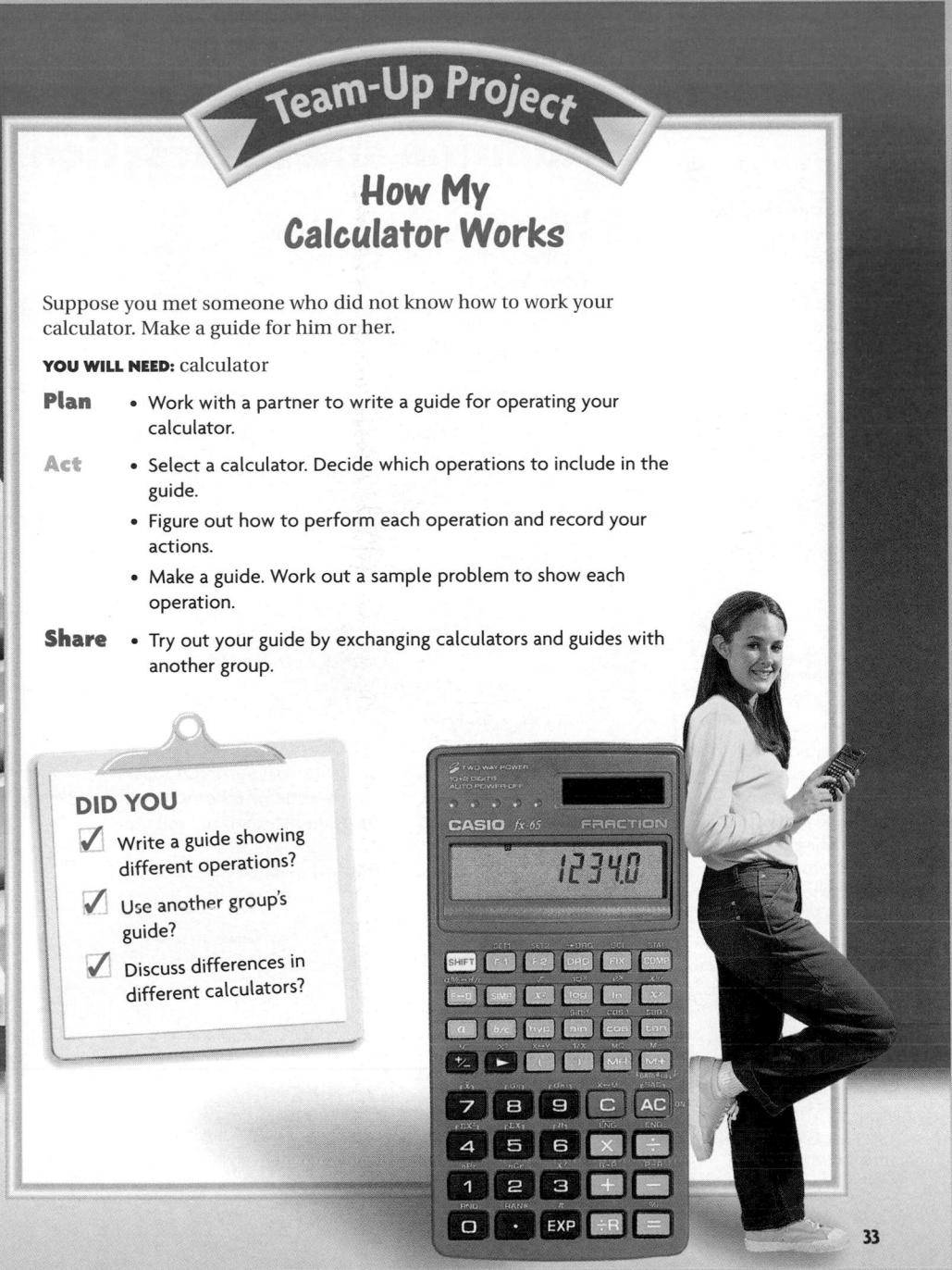

Project Checklist

Portfolio

Evaluate whether partners
- ✓ work cooperatively.
- ✓ record key strokes in order.
- ✓ follow the written guide.
- ✓ participate in the class discussion.

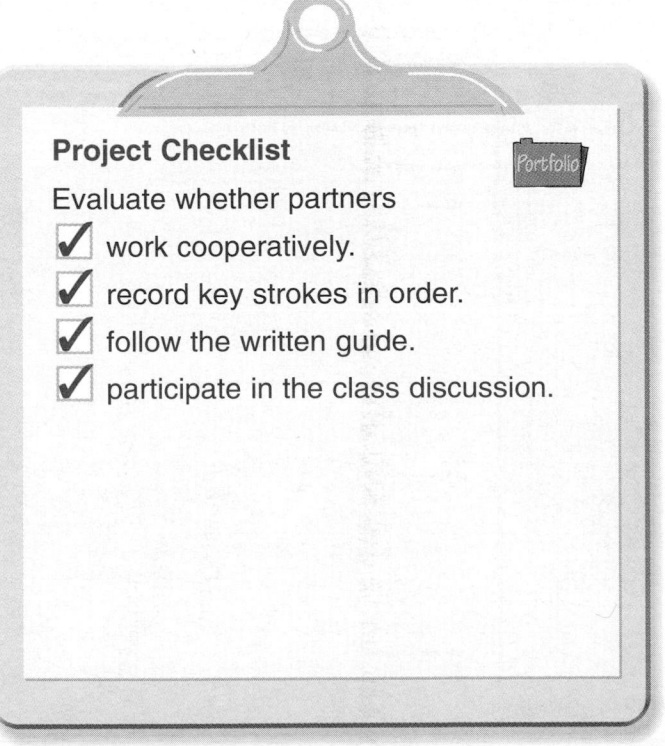

Using the Project

Accessing Prior Knowledge understanding of operations

Purpose to use whole numbers by writing guidebooks for operating calculators

GROUPING partners
MATERIALS calculators

Managing the Activity Help groups select which operations to include in their guides. Some students will include only basic operations; others may include other operations, memory keys, or grouping symbols.

 WHEN MINUTES COUNT

A shortened version of this activity can be completed in 15–20 minutes. Ask a student to describe the steps she uses to solve 235 1 895 on her calculator. Have other class members try the steps on their calculators. If someone has a calculator that requires different steps, have him describe them. Continue for other operations.

 A BIT EACH DAY

Day 1: Select partners, select a calculator, and decide which operations to include in the guide.

Day 2: Figure out how to perform each of the operations to be included in the guide. Make a list of the examples that will be worked out.

Day 3: Figure out the format for the guide. Write the guide.

Day 4: Exchange calculators and guides with another group. Be the learner. Follow the guide.

Day 5: Compare guides. Discuss differences in the way operations are performed on different calculators.

 ON THE ROAD

Have students ask people in various professions if they use specialized calculators. They may be surprised to learn that there are scientific calculators, business calculators, construction calculators, and many other specialized calculators.

GRADE 5	GRADE 6	GRADE 7
Add and subtract whole numbers.	Use properties and mental math to find sums and differences.	Estimate sums and differences.

ORGANIZER

OBJECTIVE To use properties and mental math to find sums and differences

VOCABULARY *New* **Commutative Property, Associative Property, compensation**

ASSIGNMENTS Independent Practice
- *Basic* 1–32, even; 33–36
- *Average* 1–36
- *Advanced* 13–36

PROGRAM RESOURCES
Reteach 2.1 Practice 2.1 Enrichment 2.1
Cultural Connection • Piñata Party, p. 50

PROBLEM OF THE DAY

Transparency 2

Subtract vertically and horizontally.

93	38	?	55
68	29	?	39
?	?	?	
25	9	16	

Solution Tab A, p. A2

QUICK CHECK

Use mental math to add or subtract.

1. 17 + 3 20
2. 42 − 2 40
3. 45 + 15 60
4. 25 + 25 50
5. 61 − 1 60
6. 28 − 3 25

1 MOTIVATE

Copy these exercises on the board: 15 + 16; 25 + 24; 49 + 50. Ask:

- **What strategy could you use to mentally find the sums?** Look for doubles: 15 + 15 = 30, 30 + 1 = 31; 25 + 25 = 50, 50 − 1 = 49; 50 + 50 = 100, 100 − 1 = 99.

Discuss with students other strategies they have learned that help them add or subtract by using mental math.

WHAT YOU'LL LEARN
How to use properties and mental math to find sums and differences

WHY LEARN THIS?
To find a sum or difference without paper and pencil

WORD POWER
Commutative
Property
Associative
Property
compensation

Language Link

One definition of *compensation* is "something given to balance something else." If you used compensation in the following problem, what would be given and where would it come from?

29 + 46

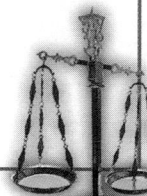

A 1 would be given to 29 and taken from 46.

Mental Math for Addition and Subtraction

How can you mentally find the total number of tickets sold by Marcus, Sandy, and Li?

One way to mentally find a sum is to use addition properties.

Number of Tickets Sold	
Marcus	64
Sandy	78
Li	56

Commutative Property

Numbers can be added in any order.

64 + 78 + 56 = 56 + 64 + 78

Associative Property

Addends can be grouped differently. The sum is always the same.

(64 + 78) + 56 = (56 + 64) + 78

EXAMPLE 1 Find the sum. 5 + 9 + 21 + 25

5 + 9 + 21 + 25 = 5 + 25 + 9 + 21   ← *Commutative Property*
    = (5 + 25) + (9 + 21) ← *Associative Property*
    = 30 + 30   ← *Use mental math.*
    = 60

So, the sum is 60.

A strategy you can use for addition or subtraction is **compensation**. You can change one number to a multiple of ten and then adjust the other addend to keep the balance.

EXAMPLE 2 Use compensation to solve.

A. 18 + 24

18 + 24 = (18 + 2) + (24 − 2)
    = 20 + 22
    = 42

Think: *18 + ■ = 20*
Add 2 and subtract 2.
Use mental math.

B. 43 − 31

43 − 31 = (43 − 1) − (31 − 1)
    = 42 − 30
    = 12

Think: *31 − ■ = 30*
Subtract 1 from each number.
Use mental math.

IDEA FILE

HISTORY CONNECTION

Have students find the number of years each of the longest-reigning female rulers in the world held the throne.

Ruler	Country	Reign
Victoria	England	1837–1901 64 yr
Wilhelmina	Netherlands	1890–1948 58 yr
Wu Chao	China	655–705 50 yr
Salote Tubou	Tonga	1918–1965 47 yr

Modality Visual

RETEACH 2.1

Name _____

LESSON 2.1

Mental Math for Addition and Subtraction

One way to find a sum mentally is to use addition properties.

Commutative Property
Numbers can be added in any order without changing the sum.
45 + 29 + 55 = 29 + 45 + 55
        Order has been changed.

Associative Property
Addends can be grouped differently. The sum is always the same.
(45 + 29) + 55 = 29 + (45 + 55)
        Grouping has been changed.

Zack wants to find the total of 8 + 14 + 22 + 6 using mental math. This is how he used addition properties to find the sum.

8 + 14 + 22 + 6 = 8 + 22 + 14 + 6   ← Commutative Property
            He switched 14 and 22.

    = (8 + 22) + (14 + 6)   ← Associative Property
            He grouped numbers that are easy to add mentally.

    = 30 + 20   ← He used mental math.
    = 50

Complete to find each sum.

1. 19 + 45 + 21 + 5

19 + 45 + 21 + 5 = 19 + __21__ + __45__ + 5   → ____ Commutative Property

    = __(19 + 21)__ + __(45 + 5)__   → ____ Associative Property

    = __40__ + __50__   → ____ Use mental math.

    = __90__

2. 34 + 17 + 23 + 6

34 + 17 + 23 + 6 = 34 + __6__ + __17__ + 23   → ____ Commutative Property

    = __(34 + 6)__ + __(17 + 23)__   → ____ Associative Property

    = __40__ + __40__   → ____ Use mental math.

    = __80__

Add. Use mental math.

3. 16 + 9 + 24   4. 33 + 26 + 17 + 44   5. 21 + 14 + 29 + 36
    __49__   __120__   __100__

R6   TAKE ANOTHER LOOK   Use with text pages 34–35.

GUIDED PRACTICE

Find the missing addend. Name the property used. Then find the value of the addition expression.

1. $15 + 4 + 5 = 15 + \blacksquare + 4$ **2.** $6 + 17 + 13 = 6 + (\blacksquare + 13)$ **3.** $24 + 11 + 16 = 24 + 16 + \blacksquare$
5; Comm.; 24   17; Assoc.; 36   11; Comm.; 51
4. $12 + 11 + 8 + 9 = 12 + \blacksquare + 11 + 9$ **5.** $24 + 6 + 7 + 23 = (24 + 6) + (7 + \blacksquare)$
8; Comm.; 40   23; Assoc.; 60

Use compensation to solve. Show your work.

6. $19 + 11 \; (19 + 1) +$ **7.** $38 + 12 \; (38 + 2) +$ **8.** $17 + 28 \; (17 + 3) +$ **9.** $16 + 35 \; (16 + 4) +$
$(11 - 1) = 20 + 10 = 30$ $(12 - 2) = 40 + 10 = 50$ $(28 - 3) = 20 + 25 = 45$ $(35 - 4) = 20 + 31 = 51$
10. $29 - 12 \; (29 - 2) -$ **11.** $32 - 11 \; (32 - 1) -$ **12.** $45 - 24 \; (45 - 4) -$ **13.** $56 - 35 \; (56 - 5) -$
$(12 - 2) = 27 - 10 = 17$ $(11 - 1) = 31 - 10 = 21$ $(24 - 4) = 41 - 20 = 21$ $(35 - 5) = 51 - 30 = 21$

INDEPENDENT PRACTICE

Use mental math to add.

1. $8 + 16 + 12$ **36**   **2.** $19 + 10 + 11$ **40**   **3.** $15 + 21 + 9$ **45**   **4.** $31 + 26 + 19$ **76**

5. $25 + 3 + 15 + 17$ **60**   **6.** $22 + 15 + 8 + 15$ **60**   **7.** $9 + 20 + 31 + 10$ **70**   **8.** $32 + 26 + 18 + 4$ **80**

9. $12 + 21 + 38 + 9$ **80**   **10.** $23 + 21 + 17 + 19$ **80**   **11.** $16 + 33 + 24 + 17$ **90**   **12.** $37 + 26 + 13 + 14$ **90**

Use compensation to add.

13. $17 + 24$ **41**   **14.** $19 + 23$ **42**   **15.** $14 + 27$ **41**   **16.** $18 + 26$ **44**   **17.** $16 + 28$ **44**

18. $28 + 24$ **52**   **19.** $48 + 32$ **80**   **20.** $29 + 32$ **61**   **21.** $26 + 37$ **63**   **22.** $37 + 33$ **70**

Use compensation to subtract.

23. $44 - 21$ **23**   **24.** $48 - 37$ **11**   **25.** $35 - 14$ **21**   **26.** $28 - 12$ **16**   **27.** $37 - 23$ **14**

28. $54 - 22$ **32**   **29.** $58 - 35$ **23**   **30.** $64 - 43$ **21**   **31.** $66 - 15$ **51**   **32.** $67 - 32$ **35**

Problem-Solving Applications

Use mental math to solve. For Problems 33–35, use the table below.

33. How many CDs were bought in all?
70 CDs
34. Whose purchases can be combined to total 30 CDs?
Brenda's and Ricardo's
35. If Nick gave 12 CDs to his cousin, how many would he have left? **11 CDs**

CDs Bought			
Brenda	Selena	Ricardo	Nick
12	17	18	23

36. ✏️➔ **WRITE ABOUT IT** Explain how to use compensation to add. **Change one addend to a multiple of ten, and adjust the other addend.**

MORE PRACTICE Lesson 2.1, page H43

35

As students work through Example 1, ask:

- **What is the purpose of the parentheses?** to group numbers

GUIDED PRACTICE

Have students complete Exercises 1–5 orally. Then have volunteers work Exercises 6–13 on the board. Discuss as a class.

INDEPENDENT PRACTICE

To make sure that students understand how to use compensation, check students' answers to Exercises 13–32.

ADDITIONAL EXAMPLES

Example 1, p. 34

Find the sum. $45 + 16 + 5 + 34$ 100

Example 2, p. 34

Use compensation to solve.

A. $29 + 14$
$29 + 14 = (29 + 1) + (14 - 1) = 30 + 13 = 43$

B. $73 - 44$
$73 - 44 = (73 - 3) - (44 - 3) = 70 - 41 = 29$

3 WRAP UP

Use these questions to help students focus on the big ideas of the lesson.

- **Use compensation and the table for Problems 33–35 to find how many more CDs Nick bought than Selena.**
6 CDs

✓ Assessment Tip

Have students write the answer.

- **What properties did you use to solve Independent Practice Exercise 1? Explain.** Possible answer: Commutative Property to reorder the addends and Associative Property to group the addends

DAILY QUIZ

Use mental math to add.

1. $43 + 12 + 7$ 62   **2.** $5 + 29 + 35$ 69

Use compensation to add or subtract.

3. $56 + 49$ 105   **4.** $72 + 17 = 89$
5. $76 - 44$ 32   **6.** $81 - 51 = 30$

35

2.2 USING WHOLE NUMBERS

GRADE 5	GRADE 6	GRADE 7
Use the strategy *guess and check* to solve problems, and choose appropriate strategies.	Use the strategy *guess and check* to solve problems, and choose appropriate strategies.	Use the strategy *guess and check* to solve problems, and choose appropriate strategies.

ORGANIZER

OBJECTIVE To use the strategy *guess and check* to solve addition and subtraction problems with whole numbers

ASSIGNMENTS Practice/Mixed Applications
- *Basic* 1–13
- *Average* 1–13
- *Advanced* 1–13

PROGRAM RESOURCES
Reteach 2.2 Practice 2.2 Enrichment 2.2
Problem-Solving Think Along, TR p. R1

PROBLEM OF THE DAY

Transparency 2

At noon on Monday, Latisha sets her watch to the correct time. If her watch loses one minute each hour and she does not correct it, what time will her watch show at noon on Wednesday? 11:12 At noon on what day will her watch show 10:00? Saturday

Solution Tab A, p. A2

QUICK CHECK

Complete.

1. 47 + _?_ = 76 29
2. _?_ − 72 = 22 94
3. 55 − 23 = _?_ 32
4. _?_ + 33 = 109 76
5. _?_ + 45 = 123 78
6. 87 − _?_ = 68 19

1 MOTIVATE

To stimulate discussion, present this problem to the class:

I am thinking of two consecutive counting numbers that have a sum of 43. Guess what they are.

Lead students to ask:

- **How many digits will each of the two numbers have? Why?**
- **Could one of the numbers have only 1 digit? Why?**

36

2.2

WHAT YOU'LL LEARN
How to use the strategy *guess and check* to solve addition and subtraction problems with whole numbers

WHY LEARN THIS?
To develop a systematic plan to check and revise good guesses

Problem Solving
- Understand
- Plan
- Solve
- Look Back

PROBLEM-SOLVING STRATEGY
Using Guess and Check to Add and Subtract

Sometimes it takes detective work to find the solution to a problem.

During an annual reading program, the sixth and seventh graders read 1,020 books. The sixth graders read 110 more books than the seventh graders. How many books did the sixth graders read? How many books did the seventh graders read?

☑ **UNDERSTAND** What are you asked to find? **number of books read by 6th graders and by 7th graders** What facts are given? **total number of books read; how many more read by 6th graders**

☑ **PLAN** What strategy will you use?

You can use the strategy *guess and check*. Make a guess that satisfies the first clue. Then check your guess to see if it satisfies the second clue.

☑ **SOLVE** How will you solve the problem?

Make a table to keep track of your guesses and checks for the first and second clues. Try to guess in an organized way so that your guesses get closer to the answer.

GUESS		CHECK	
Sixth Graders	Seventh Graders	Clue 1: The sum is 1,020.	Clue 2: The difference is 110.
500	520	1,020	20	← *too low*
600	420	1,020	180	← *too high*
565	455	1,020	110	← *satisfies both clues*

☑ **LOOK BACK** The difference for the first guess was too low. What changes were made for the second guess? **increased number for 6th graders; decreased number for 7th graders What if...** the sixth graders had read 120 more books than the seventh graders? How many books would the sixth graders have read? How many books would the seventh graders have read? **570 books; 450 books**

36 Chapter 2

IDEA FILE

CULTURAL CONNECTION

The oldest-known magic square is the *lo-shu*, which was discovered about 4,000 years ago in China. Some of the numbers from the *lo-shu* are shown below. Have students complete the square and find the sum. **15**

6	1	8
7	5	3
2	9	4

Modality Visual

MEETING INDIVIDUAL NEEDS

RETEACH 2.2

Name _____

Problem-Solving Strategy

Using Guess and Check to Add and Subtract

Looking for clues in a problem can help you find its answer. Then you can use the clues to help you guess and check the answer until you find the exact answer.

Valley Middle School is holding a canned food drive. Sixth-grade students have collected 150 more cans than seventh-grade students. Together, the students in both grades have collected a total of 530 cans. How many cans did the sixth graders collect? How many cans did the seventh graders collect?

Step 1: Think about what you know.
- You are asked to find the number of cans collected by each grade.
- You know the total number of cans collected and how many more cans the sixth graders collected than the seventh graders.

Step 2: Plan a strategy to solve.
- Use the *guess and check* strategy.
- Use these clues: total cans collected is 530; the difference between amounts collected by sixth and seventh graders is 150.

Step 3: Solve.
- Use a table to record your guesses and checks. Try to guess in an organized way to help you get closer to the exact answer.

GUESS		CHECK		
Sixth Graders	Seventh Graders	Clue 1: The sum is 530.	Clue 2: The difference is 150.	
330	200	330 + 200 = 530 ✓	330 − 200 = 130 ⊗	← Difference is too low.
350	180	350 + 180 = 530 ✓	350 − 180 = 170 ⊗	← Difference is too high.
340	190	340 + 190 = 530 ✓	340 − 190 = 150 ✓	← Both clues are satisfied.

Use the strategy *guess and check* with a table to help you solve.

1. In the problem above, what if the sixth graders had collected 120 more cans than the seventh graders? How many cans would each grade have collected?

   sixth grade: 325 cans; seventh grade: 205 cans

2. Tony collected 85 cans of either soup or fruit. He collected 15 more cans of soup than of fruit. How many cans of soup did he collect? How many cans of fruit did he collect?

   50 soup; 35 fruit

Use with text pages 36–37. **TAKE ANOTHER LOOK R7**

PRACTICE

Guess and check to solve. Make a table to record your guesses.

1. Douglas bought a total of 50 cans of orange and grape sodas. He bought 12 more cans of grape than of orange. How many of each kind did he buy? **31 grape, 19 orange**

2. The perimeter of a rectangular garden is 40 ft. The length is 6 ft more than the width. What is the length and width of the garden? *l* = **13 ft**, *w* = **7 ft**

3. Kim sold a total of 36 red and blue tickets. She sold 6 more red tickets than blue tickets. How many red and blue tickets did she sell? **21 red tickets, 15 blue tickets**

4. The Brown Bears soccer team played a total of 20 games. They won 4 more than they lost, and they tied in 2 games. How many games did they win? **11 games**

MIXED APPLICATIONS

CHOOSE

Choose a strategy and solve. **Choices of strategies will vary.**

Problem-Solving Strategies

- Guess and Check
- Solve a Simpler Problem
- Write an Equation
- Find a Pattern
- Act It Out
- Make a Table

5. Rosalia talks on the phone with Teresa, Ronald, and Jack. If each of the four friends has one conversation with each other friend, how many calls are made? **6 calls**

6. Ned spent a total of $23.45. He bought a ticket to a basketball game for $6.50, food for $6.95, and some T-shirts for $5.00 each. How many T-shirts did he buy? **2 T-shirts**

7. A bus travels 55 mi per hour. Abe starts out by bus at 8:00 A.M. How many miles does he travel if he arrives at his destination at 11:00 A.M.? **165 mi**

8. Tim has earned $2,100. To buy a used car, he needs twice that amount, plus $500. How much does the car cost? **$4,700**

9. Peter and his two brothers are collecting stamps. Peter has twice as many as his older brother, who has 21 stamps. Peter has three times as many as his younger brother. How many stamps do they have in all? **77 stamps**

10. Mira started an exercise routine and jogged 4 mi each day the first week. She increased her daily routine by 2 mi each week, but every third week she added only 1 mi. How many miles did she jog each day in week 5? **11 mi**

11. Martha saved $0.50 from her allowance one week. Then she saved $0.50 more each week than she had the week before. How many weeks did it take to save $10.50? **6 weeks**

12. Ira has 68 baseball cards. This is twice as many as Saul has, plus 2. How many cards does Saul have? **33 cards**

13. ✏️ **WRITE ABOUT IT** Write a problem that can be solved by guessing and checking. Have a partner use a table to solve. **Answers will vary.**

MORE PRACTICE Lesson 2.2, page H44

• **What digit might be in the tens place of each number?** 1, 3 or 2, 2

Continue to elicit questions from students to emphasize the importance of making careful guesses, not random guesses. Then have students find the answer. 21, 22

2 TEACH/PRACTICE

In this lesson, students learn to use the strategy *guess and check* to solve problems by looking for clues and tracking guesses and checks in a table.

PRACTICE
Have partners work through Problem 1 and complete the Problem-Solving Think Along, TR p. R1.

MIXED APPLICATIONS
To make sure students understand how to use problem-solving strategies, have students exchange papers and check a classmate's work. Emphasize that more than one strategy can be used to solve some problems.

3 WRAP UP

Use these questions to help students focus on the big ideas of the lesson.

• **Use *guess and check* to solve: A *Save the Planet* poster costs $4 less than a *Save the Planet* banner. Together, the poster and banner cost $20. What is the price of the poster?** $8 **the banner?** $12

✓ *Assessment Tip*
Have students write the answer.

• **Explain how you solved Problem 3.** Possible explanation: I made a table to keep track of my guesses and checks.

DAILY QUIZ
Guess and check to solve.

1. Juan paid $21 for two model airplanes. One airplane cost $7 more than the other. What was the cost of each airplane? $7, $14

2. The Lorenzos traveled 300 more miles during the first week of their camping trip than they did the second week. They traveled a total of 2,280 miles. How many miles did they travel each week? first week, 1,290 mi; second week, 990 mi

GRADE 5	GRADE 6	GRADE 7
Use multiplication properties.	**Use properties to solve multiplication problems.**	Multiply whole numbers and decimals.

ORGANIZER

OBJECTIVE To use properties to solve multiplication problems

VOCABULARY *Review* Identity Property of One, Property of Zero

ASSIGNMENTS **Independent Practice**
- *Basic* 1–24, even; 25–28
- *Average* 1–28
- *Advanced* 9–28

PROGRAM RESOURCES
Intervention • Multiplication and Division TE pp. 44A–44B
Reteach 2.3 Practice 2.3 Enrichment 2.3
Cultural Connection • Piñata Party, p. 50

PROBLEM OF THE DAY

Transparency 2

Replace the dots with the digits 0–9 to make a correct number sentence. Use each digit only once.

● × ● = 18   2 × 9 = 18
● × ● = 24   3 × 8 = 24
● × ● = 0   5 × 0 = 0
● × ● = 28   4 × 7 = 28
● × ● = 6   6 × 1 = 6

Solution Tab A, p. A2

QUICK CHECK
Use mental math to multiply.

1. 3 × 40 120   **2.** 5 × 50 250
3. 2 × 25 50   **4.** 4 × 60 240
5. 7 × 30 210   **6.** 6 × 70 420

1 MOTIVATE

Discuss with the class the advantages of using mental math to multiply. Lead students to understand that using mental math can be faster than using a calculator or paper and pencil and that they can use mental math when they don't have a calculator or paper and pencil.

Tell students about Mrs. Shakuntala Devi of India, who used mental math to multiply 7,686,369,774,870 by 2,465,099,745,779. She found the correct answer in only 28 seconds!

2.3

WHAT YOU'LL LEARN
How to use properties to solve multiplication problems

WHY LEARN THIS?
To mentally solve multiplication problems such as determining the number of party favors

WORD POWER
Distributive Property

REMEMBER:
Identity Property of One
The product of any factor and 1 is the factor.

17 × 1 = 17

Property of Zero
The product of any factor and zero is zero.

99 × 0 = 0

See page H6.

ALGEBRA CONNECTION
Mental Math for Multiplication

Marsha is making 32 party favors, with 12 treats in each favor. How can she use mental math to determine the number of treats she will need?

You can use the **Distributive Property** to make an easier problem.

$12 \times 32 = 12 \times (30 + 2)$
$= (12 \times 30) + (12 \times 2)$
$= 360 + 24$
$= 384$

So, Marsha will need 384 treats.

The Commutative and Associative Properties let you change the order of the factors or the way the factors are grouped so that you can mentally find the product.

EXAMPLE A party store has 8 party favor bags on each of 3 shelves. Each party favor bag contains 5 items. How many items are there altogether?

$8 \times 3 \times 5 = 8 \times 5 \times 3 \leftarrow$ *Commutative*
$= (8 \times 5) \times 3 \leftarrow$ *Associative*
$= 40 \times 3$
$= 120$

So, there are 120 items.

Talk About It
- Would you need to use the Commutative, Associative, or Distributive Property to solve $4 \times 25 \times 1$? Explain. **No; 4 × 25 = 100, and by the Identity Property of One, 100 × 1 = 100.**
- Would you need to use the Commutative, Associative, or Distributive Property to solve $9 \times (8 - 8)$? Explain. **No; 8 − 8 = 0. So, 9 × 0 = 0 by the Property of Zero.**

PROPERTIES

Commutative
Factors can be multiplied in any order without changing the product.

$6 \times 7 = 7 \times 6$

Associative
Factors can be grouped in any way without changing the product.

$(2 \times 9) \times 5 = 2 \times (9 \times 5)$

Distributive
A factor can be thought of as the sum of addends. Multiplying the sum by a number is the same as multiplying each addend by the number and then adding the products.

$3 \times 42 =$
$3 \times (40 + 2) =$
$(3 \times 40) + (3 \times 2)$

38 Chapter 2

IDEA FILE

CONSUMER CONNECTION

Present this problem to students:

- Suppose you were to take 2 showers, brush your teeth 3 times, and wash dishes 2 times today. How much water would you use? 106 gal

Activity	Water Used Each Time
Washing dishes	20 gal
Taking shower	30 gal
Brushing teeth	2 gal

Modality Auditory

MEETING INDIVIDUAL NEEDS

RETEACH 2.3

Name _____ LESSON 2.3

Mental Math Strategies for Multiplication

Toni is making cards for a craft show. She has 36 packages of 15 cards each. To find out how many cards she can create, Toni uses the Distributive Property and mental math.

$36 \times 15 = 36 \times (10 + 5)$ ← Use mental math to split 15 into numbers that are easier to multiply by 36.
$= (36 \times 10) + (36 \times 5)$ ← Use the Distributive Property.
$= 360 + 180$ ← Use mental math.
$= 540$

So, Toni can create 540 cards.

Complete to show how to use the Distributive Property to find each product.

1. $8 \times 14 = 8 \times (\underline{10} + 4)$
$= (8 \times \underline{10}) + (8 \times \underline{4})$
$= \underline{80} + \underline{32}$
$= \underline{112}$

2. $9 \times 34 = 9 \times (30 + \underline{4})$
$= (9 \times \underline{30}) + (9 \times \underline{4})$
$= \underline{270} + \underline{36}$
$= \underline{306}$

3. $3 \times 56 = 3 \times (\underline{50} + \underline{6})$
$= (3 \times \underline{50}) + (3 \times \underline{6})$
$= \underline{150} + \underline{18}$
$= \underline{168}$

4. $4 \times 63 = 4 \times (\underline{60} + \underline{3})$
$= (4 \times \underline{60}) + (4 \times \underline{3})$
$= \underline{240} + \underline{12}$
$= \underline{252}$

5. $13 \times 24 = 13 \times (\underline{20} + \underline{4})$
$= (13 \times \underline{20}) + (13 \times \underline{4})$
$= \underline{260} + \underline{52}$
$= \underline{312}$

6. $18 \times 55 = 18 \times (\underline{50} + \underline{5})$
$= (18 \times \underline{50}) + (18 \times \underline{5})$
$= \underline{900} + \underline{90}$
$= \underline{990}$

7. $19 \times 46 = 19 \times (\underline{40} + \underline{6})$
$= (\underline{19} \times \underline{40}) + (\underline{19} \times \underline{6})$
$= \underline{760} + \underline{114}$
$= \underline{874}$

8. $29 \times 54 = 29 \times (\underline{50} + \underline{4})$
$= (\underline{29} \times \underline{50}) + (\underline{29} \times \underline{4})$
$= \underline{1,450} + \underline{116}$
$= \underline{1,566}$

R8 TAKE ANOTHER LOOK   Use with text pages 38–39.

GUIDED PRACTICE

Use the Distributive Property to make an easier problem.

1. 4×33
$(4 \times 30) + (4 \times 3)$

2. 6×25
$(6 \times 20) + (6 \times 5)$

3. 7×24
$(7 \times 20) + (7 \times 4)$

4. 3×48
$(3 \times 40) + (3 \times 8)$

Use the Commutative and Associative Properties to rearrange the factors. **Possible answers given.**

5. $(2 \times 9) \times 5$
$9 \times (2 \times 5)$

6. $(8 \times 7) \times 5$
$7 \times (8 \times 5)$

7. $6 \times (5 \times 7)$
$(6 \times 5) \times 7$

8. $(12 \times 2) \times 4$
$12 \times (2 \times 4)$

9. $2 \times (11 \times 3)$
$(2 \times 3) \times 11$

INDEPENDENT PRACTICE

Find the missing factor.

1. $48 \times 5 = (40 \times 5) + (\blacksquare \times 5)$ **8**

2. $25 \times \blacksquare = (25 \times 30) + (25 \times 6)$ **36**

3. $55 \times 42 = (\blacksquare \times 40) + (55 \times 2)$ **55**

4. $6 \times 99 = (6 \times \blacksquare) + (6 \times 9)$ **90**

Use the Distributive Property and mental math to find the product.

5. 17×9 **153**
6. 28×8 **224**
7. 32×6 **192**
8. 16×21 **336**

Use mental math to find the product.

9. $9 \times 5 \times 2$ **90**
10. $4 \times 8 \times 2$ **64**
11. 3×21 **63**
12. $2 \times 6 \times 5$ **60**

13. $3 \times 10 \times 7$ **210**
14. 6×15 **90**
15. 12×6 **72**
16. $12 \times 4 \times 5$ **240**

17. 13×4 **52**
18. $25 \times 3 \times 2$ **150**
19. 8×28 **224**
20. 32×9 **288**

21. 48×2 **96**
22. 44×5 **220**
23. 110×4 **440**
24. 120×3 **360**

Problem-Solving Applications

Use mental math to solve.

25. A toy store has 6 boxes on each of 4 shelves. Each box has 10 items in it. How many items altogether are on the 4 shelves? **240 items**

26. For a fund-raiser, 25 students are making muffins. Each student bakes 16 muffins. How many muffins do they make altogether? **400 muffins**

27. LOGICAL REASONING Raoul sells three kinds of cookies. His average sales are 24, 26, and 28 cookies for the three kinds each day. For each kind, he wants to know the total he is likely to sell in the next 14 days. Write number sentences that use the properties of multiplication and addition to simplify his calculations. **Possible answer:** $24 \times 14 = (24 \times 10) + (24 \times 4)$, $26 \times 14 = (26 \times 10) + (26 \times 4)$, $28 \times 14 = (28 \times 10) + (28 \times 4)$

28. WRITE ABOUT IT What properties of multiplication can you use to help you with mental math? **Commutative, Associative, Distributive Properties, Property of Zero, and Identity Property of One**

MORE PRACTICE **Lesson 2.3, page H44**

39

---

2 TEACH/PRACTICE

Discuss the questions in Talk About It. Have students give other examples of the Property of Zero and the Identity of One.

GUIDED PRACTICE

Have volunteers complete Exercises 1–9 on the board. Discuss as a class.

INDEPENDENT PRACTICE

To ensure student success, answer Exercises 1, 5, 9, and 25 as a whole-class activity before you assign the remaining exercises.

CRITICAL THINKING Mia has three brothers under the age of 10. The product of their ages is 210. How old are they? **5, 6, and 7 years old**

ADDITIONAL EXAMPLE

Example, p. 38

Karen is planning to watch each of her 4 favorite videos 3 times over the weekend. Each video lasts 50 min. How much time will she spend watching videos? 600 min, or 10 hr

3 WRAP UP

Use these questions to help students focus on the big ideas of the lesson.

- **How do the Commutative and Associative Properties help you find the product $5 \times 3 \times 8$?** Possible answer: I can reorder the factors: $5 \times 3 \times 8 = 5 \times 8 \times 3$. Then I can group the factors so I can use mental math: $(5 \times 8) \times 3 = 40 \times 3 = 120$.

✔ Assessment Tip

Have students write the answer.

- **What property did you use to solve Exercise 23? Why?** Distributive Property; possible explanation: because it is easier to multiply 100 by 4, to multiply 10 by 4, and to add the products than it is to multiply 110 by 4.

DAILY QUIZ

Use mental math to find the product.

1. $7 \times 3 \times 3$ 63
2. $5 \times 7 \times 8$ 280
3. $5 \times 5 \times 5$ 125
4. $9 \times 2 \times 7$ 126
5. 220×3 660
6. 9×38 342

PRACTICE 2.3

Name _____
LESSON 2.3

Mental Math Strategies for Multiplication

Vocabulary

Match each property in Column A with its definition in Column B.

Column A

__c__ 1. Associative Property
__e__ 2. Commutative Property
__b__ 3. Distributive Property
__d__ 4. Identity Property of One
__a__ 5. Zero Property

Column B

a. The product of any factor and zero is zero.
b. Multiplying a factor's addends by a number, then adding the products, is the same as multiplying the factor by the number.
c. Factors can be grouped in any way without changing the product.
d. The product of any factor and 1 is the factor.
e. Factors can be multiplied in any order without changing the product.

Find the missing factor.

6. $56 \times 7 = (50 \times 7) + (\underline{\ 6\ } \times 7)$
7. $15 \times \underline{\ 48\ } = (15 \times 40) + (15 \times 8)$
8. $8 \times 93 = (8 \times \underline{\ 90\ }) + (8 \times 3)$
9. $63 \times 21 = (\underline{\ 63\ } \times 20) + (63 \times 1)$

Use the Distributive Property and mental math to find the product.

10. 18×6 **108**
11. 42×5 **210**
12. 17×4 **68**
13. 31×15 **465**
14. 86×3 **258**
15. 18×22 **396**
16. 47×3 **141**
17. 15×51 **765**

Use mental math to find the product.

18. $8 \times 3 \times 4$ **96**
19. $2 \times 7 \times 5$ **70**
20. $6 \times 9 \times 5$ **270**
21. $10 \times 4 \times 7$ **280**
22. 17×9 **153**
23. 61×6 **366**
24. 19×11 **209**
25. $4 \times 16 \times 5$ **320**

Mixed Applications

26. The Sports Shack has 8 boxes of baseballs on each of 5 shelves. Each box holds one dozen balls. How many baseballs are there in all?

480 baseballs

27. Bill has earned $87. To buy a new stereo, he needs twice that amount plus $50. How much does the stereo cost?

$224

P8 ON MY OWN
Use with text pages 38–39.

ENRICHMENT 2.3

Name _____
LESSON 2.3

Solve It

Use mental math to solve each problem in the Decoder Box. Find the product in the Tip Box. Each time the product appears, write the letter of that problem above it. When you have solved all the problems, you will discover the math tip.

Decoder Box

A $2 \times 9 \times 5 =$	90		N $13 \times 3 \times 2 =$	78
B $7 \times 14 =$	98		O $4 \times 11 \times 3 =$	132
C $120 \times 4 =$	480		P $9 \times 22 =$	198
D $6 \times 7 \times 10 =$	420		Q $5 \times 34 \times 4 =$	680
E $55 \times 3 =$	165		R $15 \times 4 \times 2 =$	120
F $6 \times 8 \times 5 =$	240		S $16 \times 7 =$	112
H $2 \times 13 \times 5 =$	130		T $25 \times 6 =$	150
I $7 \times 21 =$	147		U $88 \times 5 =$	440
K $29 \times 8 =$	232		V $25 \times 19 \times 4 =$	1,900
L $9 \times 2 \times 6 =$	108		Y $7 \times 8 \times 5 =$	280
M $8 \times 5 \times 4 =$	160		Z $9 \times 53 =$	477

Tip Box

F	A	C	T	O	R	S		C	A	N	B	E
240	90	480	150	132	120	112		480	90	78	98	165

M	U	L	T	I	P	L	I	E	D	I	N
160	440	108	150	147	198	108	147	165	420	147	78

A	N	Y	O	R	D	E	R
90	78	280	132	120	420	165	120

Now use the Decoder Box to help you find the answer to a riddle.
What is only useful when it's used up?

A	N		U	M	B	R	E	L	L	A
90	78		440	160	98	120	165	108	108	90

E8 STRETCH YOUR THINKING
Use with text pages 38–39.

39

ORGANIZER

OBJECTIVE To multiply and divide whole numbers

ASSIGNMENTS Independent Practice
- *Basic* 1–35, odd; 36–41; 42–49
- *Average* 1–41; 42–49
- *Advanced* 1–5, 11–15, 26–30; 36–41; 42–49

PROGRAM RESOURCES
Reteach 2.4 Practice 2.4 Enrichment 2.4
Calculating Crew • *Intergalactic Trader*
Cultural Connection • Piñata Party, p. 50

PROBLEM OF THE DAY

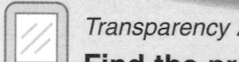

Transparency 2

Find the product. Compare the product with the first factor.

1. $13 \times 11 = $ _?_ 143
2. $72 \times 11 = $ _?_ 792
3. $326 \times 11 = $ _?_ 3,586
4. $6,045 \times 11 = $ _?_ 66,495

Solution Tab A, p. A2

QUICK CHECK

Complete the pattern.

1. 3×63 189
   30×63 1,890
   300×63 18,900

2. 6×41 246
   60×41 2,460
   600×41 24,600

3. 2×342 684
   20×342 6,840
   200×342 68,400

4. 4×52 208
   40×52 2,080
   400×52 20,800

1 MOTIVATE

Challenge students to write a math puzzle. For example:

An empty city bus leaves the bus station. At Stop 1, some people get on the bus. At Stop 2, twice as many people get on as did at Stop 1, and two people get off the bus. At Stop 3, six people get on, and three times the number that got off at Stop 2 get off. At Stop 4, three times the number of people get on the bus as get off. Now, there are thirty people on the bus. How many people got on at Stop 1? 10 people

---

2.4

WHAT YOU'LL LEARN
How to multiply and divide whole numbers

WHY LEARN THIS?
To solve problems such as finding the total amount earned by ticket sales

CULTURAL LINK

The doubling method below is known as the Russian Peasant Method.

$26 \times 35 = 910$

Halve	Double
26	35
13	70
6	~~140~~
3	280
1	+ 560
	910

Find the even numbers in the left column and cross out the numbers directly across from them in the right column. Add the remaining doubles.

Use this method to find 18×22.
396

40   Chapter 2

Multiplication and Division

Sometimes you can use shortcuts to multiply numbers. For example, to find 32×100, you know that you can just put two zeros after 32 to get 3,200. This activity will show you a different way to multiply two whole numbers when both of the factors are numbers between 10 and 20.

ACTIVITY

Find 16×19.

- Find the sum of the first factor and the ones digit of the second factor. $16 + 9 = 25$
- Add a zero to that sum.   250
- Find the product of the ones digits of the first and second factors. $6 \times 9 = 54$
- Find the sum. $250 + 54 = 304$

So, $16 \times 19 = 304$.

- Find 14×15.
  $14 + 5 = 19$; 190; $4 \times 5 = 20$; so, $14 \times 15 = 190 + 20$, or 210.

For most multiplication problems, you find the product by multiplying the two factors.

EXAMPLE 1   Find 132×24.

$$
\begin{array}{r}
132 \\
\times\ 24 \\
\hline
528 \\
+\ 2640 \\
\hline
3,168
\end{array}
$$

Multiply by the ones. 4×132
Multiply by the tens. 20×132
Use the zero as a placeholder. Add.

- Find 128×18.
  2,304

You can omit the zero placeholders when you multiply. Just be careful to line up your products correctly.

	Correct:	Wrong:
	132	132
	$\times\ 24$	$\times\ 24$
	528	528
	+264	+264

Sometimes you have to use multiplication to solve a problem.

EXAMPLE 2 A group is holding a concert to raise money to help clean up a local lake. Each ticket costs $125. The group has already sold 12,383 tickets. How much has it earned so far?

$$
\begin{array}{r}
12{,}383 \\
\times\quad 125 \\
\hline
61\ 915 \\
247\ 66 \\
+1\ 238\ 3 \\
\hline
1{,}547{,}875 \\
\end{array}
$$

Multiply by the ones. $5 \times 12{,}383$
Multiply by the tens. $20 \times 12{,}383$
Multiply by the hundreds. $100 \times 12{,}383$
Add.

So, the group has earned $1,547,875.

• How much will the group earn if it sells 15,225 tickets?
$1,903,125

GUIDED PRACTICE

Find the product.

1. $\begin{array}{r}13 \\ \times 14 \\ \hline 182\end{array}$	**2.** $\begin{array}{r}380 \\ \times 52 \\ \hline 19{,}760\end{array}$	**3.** $\begin{array}{r}2{,}382 \\ \times 12 \\ \hline 28{,}584\end{array}$	**4.** $\begin{array}{r}962 \\ \times 40 \\ \hline 38{,}480\end{array}$
5. $\begin{array}{r}849 \\ \times 413 \\ \hline 350{,}637\end{array}$	**6.** $\begin{array}{r}7{,}658 \\ \times 111 \\ \hline 850{,}038\end{array}$	**7.** $\begin{array}{r}1{,}453 \\ \times 721 \\ \hline 1{,}047{,}613\end{array}$	**8.** $\begin{array}{r}16{,}225 \\ \times 258 \\ \hline 4{,}186{,}050\end{array}$
9. $\begin{array}{r}123 \\ \times 12 \\ \hline 1{,}476\end{array}$	**10.** $\begin{array}{r}918 \\ \times 27 \\ \hline 24{,}786\end{array}$	**11.** $\begin{array}{r}342 \\ \times 521 \\ \hline 178{,}182\end{array}$	**12.** $\begin{array}{r}189 \\ \times 108 \\ \hline 20{,}412\end{array}$

Solving Division Problems

Liz is redesigning the school cafeteria to seat exactly 420 students. Each table in her design will seat 12 students. How many tables will she need?

You can use division to find the number of tables.

$$
\begin{array}{r}
35 \\
12\overline{)420} \\
-36\downarrow \\
\hline
60 \\
-60 \\
\hline
0
\end{array}
$$

Since 12 is greater than 4, the first digit of the quotient will be in the tens place. Divide the 42 tens.
Think: $12 \times 3 = 36$
Bring down the 0 ones. Divide the 60 ones.
Think: $12 \times 5 = 60$

So, Liz will need 35 tables in her design.

• What would you do if there were 424 students? How many tables would Liz need? **36 tables**

 Calculator Activities, page H31

In this lesson, students apply the properties they have learned by using an algorithm to multiply and divide whole numbers.

COMMON ERROR ALERT If students have difficulty lining up products correctly, they may want to continue to use zero as a placeholder or do their work on graph paper.

GUIDED PRACTICE
Have volunteers go to the board to find the products for Exercises 1–12. Discuss as a class.

ADDITIONAL EXAMPLES

Example 1, p. 40

Find 534×21. 11,214

Example 2, p. 41

The seating capacity of a stadium is 14,532. The stadium had sellout crowds for 135 events. How many people attended the events in all?

1,961,820 people

ENRICHMENT 2.4

Name _____

LESSON 2.4

Number Crossword

Solve each problem. Complete the puzzle with the answers.

Across		Down	
1. $1{,}685 \times 124 =$	208,940	1. $6{,}150 \div 3 =$	2,050
2. $65 \times 104 =$	6,760	4. $3{,}454 \div 22 =$	157
3. $2{,}549 \times 317 =$	808,033	5. $855 \div 19 =$	45
4. $596 \times 240 =$	143,040	9. $25{,}344 \div 36 =$	704
5. $99 \times 5 =$	495	10. $16{,}740 \div 54 =$	310
6. $6{,}058 \times 847 =$	5,131,126	11. $7{,}968 \div 16 =$	498
7. $351 \times 208 =$	73,008	12. $46{,}295 \div 47 =$	985
8. $872 \times 234 =$	204,048	13. $17{,}568 \div 36 =$	488
		14. $14{,}761 \div 29 =$	509
		15. $32{,}625 \div 87 =$	375
		16. $9{,}672 \div 93 =$	104
		17. $7{,}896 \div 12 =$	658

Use with text pages 40–43. STRETCH YOUR THINKING E9

Review with students how to interpret and use remainders. Explain that they can drop a remainder or increase a quotient by 1, depending on the situation. Then ask students these questions:

- **In Additional Example 3, why was the quotient increased by 1?** because 1 person was left over and it would not be safe to have 13 people on a minibus when the seating capacity is 12

- **Do you think that only 1 person will ride on the extra bus? Explain.** No; probably about half of the 13 people will ride on one bus, and the other half will ride on another bus.

Have students make up a problem in which they would drop the remainder. You may want to provide this example:

- **The instructions in a bow-making kit say to use 5 yd of ribbon to make each bow. How many bows could you make with 48 yd of ribbon?** 9 bows

INDEPENDENT PRACTICE

To ensure student success in multiplying and dividing whole numbers, have volunteers work exercises 1, 7, 11, and 21 on the board.

ADDITIONAL EXAMPLE

Example 3, p. 42

Exactly 2,449 people signed up for the *Kids in Science Exposition* at the Science Museum. Each minibus can transport 12 people to the museum. How many minibuses are needed for the exposition?

204 r1; 205 buses are needed for the exposition.

EXAMPLE 3 A math class is making a banner that is 290 ft long for a math fair. If the paper comes in 12-ft rolls, how many rolls of paper will the class need?

$$
\begin{array}{r}
24 \text{ r2} \\
12\overline{)290} \\
-24\downarrow \\
\hline
50 \\
-48 \\
\hline
2
\end{array}
$$

Since 12 is greater than 2, the first digit will be in the tens place. Divide the 29 tens.
Think: $12 \times 2 = 24$
Bring down the 0 ones. Divide the 50 ones.
Think: $12 \times 4 = 48$

So, the math class will need 25 rolls of paper.

Talk About It

- **CRITICAL THINKING** In Example 3, why will the math class need 25 rolls of paper instead of 24 rolls of paper? **With 24 rolls, the banner would be 2 ft too short.**

- When do you write a zero in the quotient? **when the difference is less than the divisor**

Another Method You can use some calculators to show the whole number remainder.

145 [÷R] 8 [=] [= 18 R1]

- How does a calculator show the remainder when you divide 145 by 8 using the [÷] key? **as 18.125**

In Example 3 the quotient had a remainder. You can express a remainder with an *r*, or you can express it as a fraction. The quotient in Example 3 can be expressed as 24 r2, or it can be expressed as $24\frac{2}{12}$.

- Find $13\overline{)254}$. Write the remainder as a fraction. $19\frac{7}{13}$
- Find $22\overline{)137}$. Write the remainder as a fraction. $6\frac{5}{22}$

INDEPENDENT PRACTICE

Find the product.

1. 27 $\times$ 11 **297**	**2.** 140 $\times$ 25 **3,500**	**3.** 364 $\times$ 30 **10,920**	**4.** 3,473 $\times$ 46 **159,758**	**5.** 298 $\times$ 89 **26,522**
6. 327 $\times$ 123 **40,221**	**7.** 5,233 $\times$ 238 **1,245,454**	**8.** 462 $\times$ 140 **64,680**	**9.** 5,470 $\times$ 240 **1,312,800**	**10.** 4,904 $\times$ 196 **961,184**

IDEA FILE

EXTENSION

Have students replace the boxes with the numbers 1–6. They can use each number only once.

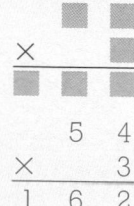

$$
\begin{array}{r}
5 \ 4 \\
\times \quad 3 \\
\hline
1 \ 6 \ 2
\end{array}
$$

Modality Visual

SCIENCE CONNECTION

The average amount of garbage generated by each person in the United States in one year includes 190 lb of plastic, 85 lb of glass, 72 lb of aluminum cans, and 24 lb of plastic containers.

- As a class project, students can calculate how much garbage is generated by all the students in the class. Have students make a pictograph showing the results.

- Have students find the total amount of each type of garbage generated by the number of people in their homes.

Modality Visual

Find the quotient.

11. $4\overline{)56}$ = 14 12. $8\overline{)432}$ = 54 13. $12\overline{)144}$ = 12 14. $4\overline{)412}$ = 103 15. $24\overline{)626}$ = 26 r2

16. $16\overline{)1,664}$ = 104 17. $14\overline{)7,044}$ = 503 r2 18. $19\overline{)1,068}$ = 56 r4 19. $20\overline{)6,020}$ = 301 20. $34\overline{)15,912}$ = 468

21. $18\overline{)2,250}$ = 125 22. $22\overline{)1,829}$ = 83 r3 23. $26\overline{)2,314}$ = 89 24. $68\overline{)24,820}$ = 365 25. $52\overline{)728}$ = 14

Find the quotient. Write the remainder as a fraction.

26. $5\overline{)49}$ $9\frac{4}{5}$ 27. $12\overline{)245}$ $20\frac{5}{12}$ 28. $34\overline{)855}$ $25\frac{5}{34}$ 29. $14\overline{)3,385}$ $241\frac{11}{14}$ 30. $36\overline{)7,225}$ $200\frac{25}{36}$

31. $15\overline{)386}$ $25\frac{11}{15}$ 32. $32\overline{)428}$ $13\frac{3}{8}$ 33. $18\overline{)5,720}$ $317\frac{7}{9}$ 34. $20\overline{)7,349}$ $367\frac{9}{20}$ 35. $45\overline{)2,270}$ $50\frac{4}{9}$

Problem-Solving Applications

36. Tasha's family lives in an apartment. Tasha's parents pay $725 each month for rent. How much rent do they pay in one year? **$8,700**

38. Lincoln Middle School had a car wash to raise money for the school. They charged $3 for every car. If they washed 23 cars, how much money did they earn? **$69**

40. Robbie's father earns $42,600 a year as an engineer. How much does he earn each month? **$3,550**

37. Bob's parents bought an entertainment center for $1,176. If they pay for it with 14 equal monthly payments, how much will each payment be? **$84**

39. Kristie used this calculator sequence:
298 ÷R 6 = = 49 R4
Rewrite the answer, with the remainder expressed as a fraction. **49$\frac{4}{6}$**

41. **WRITE ABOUT IT** Write a word problem that can be solved with multiplication or division. Exchange with a classmate and solve. **Check students' problems.**

Portfolio

Mixed Review

Round to the nearest thousand.

42. 3,789 **4,000** 43. 12,554 **13,000** 44. 8,432 **8,000**

Order the integers from least to greatest. Use < .

45. ⁻3, 4, ⁻2, 5
⁻3 < ⁻2 < 4 < 5

46. 0, ⁻1, 2, ⁻4
⁻4 < ⁻1 < 0 < 2

47. ⁻3, ⁻6, 7, 6
⁻6 < ⁻3 < 6 < 7

48. **COMPARING** Teresa is buying a gift for her mother. She has found two items that she likes, but she can buy only one. The blouse costs $15.99, and the perfume costs $15.89. She chooses the less expensive item. Which one does she buy? **the perfume**

49. **FIND A PATTERN** The first number in a pattern is 12. Each number after 12 is the sum of the previous number and 3. What is the fourth number in the pattern? **21**

Technology **Link**

In *Mighty Math Calculating Crew*, work with Captain Nick Knack in the *Intergalactic Trader* to multiply and divide whole numbers.

Use Grow Slide Level V.

MORE PRACTICE Lesson 2.4, page H44

43

---

3 WRAP UP

Use these questions to help students focus on the big ideas of the lesson.

- **An exhibitor at the Mayfield Stamp Collectors Exposition had 27 pages of rare antique stamps on display. Each page held 24 stamps. How many stamps were on display?** 648 stamps

✔ *Assessment Tip*

Have students write the answer.

- **Explain how you solved Exercise 25.** Possible answer: I divided 72 tens by 52; then I divided 208 ones by 52.

DAILY QUIZ

Find the product.

1. 34
   × 8
   272

2. 420
   × 13
   5,460

3. 4,216
   ×   51
   215,016

Find the quotient.

4. $5\overline{)180}$ = 36
5. $22\overline{)1,659}$ = 75 r9
6. $37\overline{)8,475}$ = 229 r2

CAREER CONNECTION

Have students find the number of miles walked in one month by each of these persons.

Person/Profession	Number of Miles Walked in One Year	
Police officer	1,632	136 mi
TV reporter	1,008	84 mi
Retail clerk	80	6.7 mi
Doctor	840	70 mi

Modality Visual

CORRELATIONS TO OTHER RESOURCES

CornerStone Mathematics ©1996 SkillsBank Corporation Level B, Using Whole Numbers, Lessons 5 and 7

INTERVENTION

MEETING INDIVIDUAL NEEDS

Multiplication and Division

Use these strategies to review the skills required for multiplying and dividing whole numbers.

Multiple Intelligences Logical/Mathematical

Modalities Auditory, Kinesthetic, Visual

Lesson Connection Students will multiply and divide in all the lessons in this chapter.

Program Resources *Intervention and Extension Copying Masters*, p. 1

Troubleshooting

WHAT STUDENTS NEED TO KNOW	HOW TO HELP	FOLLOW UP
Place value	Have students review the meaning of ones, tens, hundreds, thousands, ten thousands, hundred thousands, and millions. **Examples** The 7 in 37,425 stands for 7,000. The 2 in 37,425 stands for 20. The 3 in 3,452,106 stands for 3,000,000.	Other problems to solve: Give the value of the 3 in 19,382. 300 What digit in 65,626 stands for 60,000? the first digit See Key Skill 10, Student Handbook, page H6.
How to multiply by 1 and 0	Have students review the Identity Property of One and the Property of Zero. Identity Property of One: $a \times 1 = a$ $5 \times 1 = 5$ Property of Zero: $a \times 0 = 0$ $5 \times 0 = 0$	Other problems to solve: Find the product. 26×1 26 18×0 0 225×1 225 0×0 0 See Key Skill 9, Student Handbook, page H6.
That you can multiply two numbers in any order	Have students model the Commutative Property of Multiplication with cubes. **Example** Use grids to show 2×7 and 7×2. 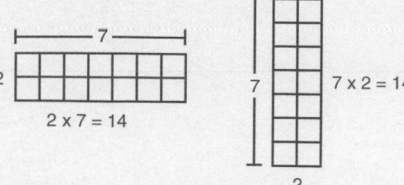	Other models to make: Use cubes to show that $6 \times 4 = 4 \times 6$. See Key Skill 9, Student Handbook, page H6.

WHAT STUDENTS NEED TO KNOW	HOW TO HELP	FOLLOW UP
That multiplication and division are closely related	Have students state the related multiplication and division facts. **Example** fact: $3 \times 7 = 21$ related facts: $7 \times 3 = 21$ $21 \div 7 = 3$ $21 \div 3 = 7$	Other problems to solve: Give the related facts for $18 \div 2 = 9$. $18 \div 9 = 2$, $2 \times 9 = 18$, $9 \times 2 = 18$ See Key Skill 15, Student Handbook, page H9.
Definition of terms	Have students state the factors and the product for a multiplication fact. Have students state the dividend, the divisor, and the quotient for a related division fact. **Examples** $2 \times 5 = 10$: The factors are 2 and 5, and the product is 10. $10 \div 5 = 2$: The dividend is 10, the divisor is 5, and the quotient is 2.	Other problems to solve: Name the product and factors in $2 \times 8 = 16$. 16; 2 and 8 Name the dividend, divisor, and quotient in $16 \div 8 = 2$. 16, 8, 2 See Key Skills 4 and 14, Student Handbook, pages H3 and H8.

dividend		divisor		quotient
12	÷	4	=	3
12	=	4	×	3
product		factor		factor

Intervention and Extension Copying Masters, p. 1

INTERVENTION STRATEGY

Name _____

CHAPTER 2

Multiplication and Division

Write the value of the 4 in each number.

1. 6,491   2. 34,285   3. 147,020   4. 9,341

   __400__   __4,000__   __40,000__   __40__

5. 4,382,099   6. 9,000,704   7. 5,489,156   8. 804,527

   __4,000,000__   __4__   __400,000__   __4,000__

Complete.

9. $73 \times$ __1__ $= 73$ 10. $6,420 \times 1 = 6,420$ 11. $298 \times$ __0__ $= 0$

12. $0 \times$ __0__ $= 0$ 13. $58 \times$ __71__ $= 71 \times 58$ 14. $203 \times 59 =$ __59__ $\times 203$

15. $19 \times 200 = 200 \times$ __19__   16. __184__ $\times 50,000 = 50,000 \times 184$ 17. $831 \times 1 = 1 \times$ __831__

Give the related multiplication and division facts for each fact.

18. $8 \times 2 = 16$ 19. $15 \div 3 = 5$

   __$2 \times 8 = 16$;__   __$15 \div 5 = 3$;__

   __$16 \div 8 = 2$; $16 \div 2 = 8$__   __$3 \times 5 = 15$; $5 \times 3 = 15$__

20. $56 \div 8 = 7$ 21. $8 \times 9 = 72$

   __$56 \div 7 = 8$;__   __$9 \times 8 = 72$,__

   __$7 \times 8 = 56$; $8 \times 7 = 56$__   __$72 \div 8 = 9$; $72 \div 9 = 8$__

Complete.

22. The answer in a multiplication problem is called the __product__.

23. Each of the numbers being multiplied is called a __factor__.

24. In the equation $21 \div 7 = 3$, the 7 is the __divisor__.

Intervention and Extension   1

ORGANIZER

OBJECTIVE To find a solution by using order of operations

VOCABULARY *New* **order of operations, algebraic operating system**

MATERIALS *For each student* calculator

PROGRAM RESOURCES
E-Lab • Activity 2
E-Lab Recording Sheets, p. 2

USING THE PAGE

In these activities, students will explore the order of operations. Discuss with students home activities in which doing things in order is very important. Possible responses: cooking, constructing models, taping television programs

Point out to students that it is also important to perform the operations in an expression in the correct order.

Activity 1

Explore

Many students will find the value of $3 + 2 \times 6 - 2$ by performing the operations in the order they are presented. After they have reviewed the correct order of operations, have them go back and find the value again.

Think and Discuss

After students evaluate $26 + 2 - 1 \times (24 \div 2)$, ask:

• **Was this expression easier to evaluate than the expressions you evaluated above? Why?** Most students will say yes, because the phrase that Drake used was helpful.

Try This

As students describe the order of operations they would use to find the value of the expression, have one student show each operation on the board:

$4^2 = 16$
$16 \times 3 = 48$
$120 - 14 = 106$
$106 + 48 = 154$

WORD POWER
order of operations
algebraic operating system

WHAT YOU'LL EXPLORE
How to find a solution using order of operations

WHAT YOU'LL NEED
calculator

Order of Operations

Marco's older brother, Drake, was working on his homework. At the top of the page, he wrote, "Please Excuse My Dear Aunt Sally." Marco didn't understand. He said, "We don't have an aunt named Sally."

Drake said, "You'll see why I wrote this on my paper."

ACTIVITY 1

Explore

Work with a partner.

• Use paper and pencil to find the value of $3 + 2 \times 6 - 2$. **13**

• How does your answer compare with your partner's answer? **Answers may vary.**

When you find the value of an expression with more than one operation, you need to use the **order of operations**.

1. Perform operations in parentheses.
2. Clear exponents.
3. Multiply and divide from left to right.
4. Add and subtract from left to right.

• Use the order of operations to evaluate $13 - 4 \div 2 + 6$. **17**

Think and Discuss

• Why do you think Marco's brother had "Please Excuse My Dear Aunt Sally" at the top of his homework page? **to help him remember the order of operations**
• What do the underlined letters in "Please Excuse My Dear Aunt Sally" represent? **parentheses, exponents, multiplication, division, addition, and subtraction**
• What order would you use to find the value of $26 + 2 - 1 \times (24 \div 2)$? Explain why the order is important. **parentheses, multiplication, addition, subtraction; so that the value of an expression is always the same**

Try This

• Tell the order in which you would perform each of the operations in the expression $120 - 14 + 4^2 \times 3$. Then find the value of the expression. **exponent, multiplication, subtraction, addition; 154**

 Calculator Activities, page H33

IDEA FILE

SOCIAL STUDIES CONNECTION

In a recent year, an average of 75,850 pieces of correspondence were sent to President Clinton each day. They included 2 times as many phone calls as letters, 100 times as many letters as telegrams, and 600 on-line messages. How many phone calls, letters, and telegrams did the President receive? phone calls, 50,000; letters, 25,000; telegrams, 250

Modality Visual

EXTENSION

Have students write an expression that equals 100 by using the number 5 five times. The expression may include any number of $+$, $-$, $\times$, and $\div$ signs as well as parentheses. $(5 \times 5 \times 5) - (5 \times 5)$ or $(5 + 5 + 5 + 5) \times 5$

Modality Visual

You can use a calculator to find the value of expressions with more than one operation. Some calculators use an **algebraic operating system (AOS)**. These calculators automatically follow the order of operations.

ACTIVITY 2

Explore

- Use your calculator to find the value of $8 \div 2 + 6 \times 3 - 4$. **Answers may vary.**
- Following the order of operations, use paper and pencil to find the value of $8 \div 2 + 6 \times 3 - 4$. **18**

- Exchange papers with your partner, and check each other's work.

Think and Discuss

- How does the value you calculated for $8 \div 2 + 6 \times 3 - 4$ compare with the value you got by using paper and pencil? Does your calculator use AOS? **Answers may vary.**

To find the value of an expression with a calculator that does not use AOS, follow the order of operations or use the memory keys.

Follow the order of operations to find the value of the expression $2 + 6 \times 3^2 - 4$.

3 × 3 × 6 + 2 − 4 = 52.

Use the memory keys to find the value of the expression $9^2 + 6 \div 2 \times 4$.

9 × 9 M+ 6 ÷ 2 × 4 M+ MRC 93.

- When you enter values into a calculator that does not have AOS, how do you know which values to enter first?
  Use the order of operations to identify order.
- Would you use memory keys or the order of operations to calculate the value of $4 \times 12 - 3 + 8 \times 2$? **Answers may vary.**

Try This

- A calculator shows the display [9.] as the value of $12 + 15 \div 3$. Does the calculator use AOS? Explain.
  No; 17 is the correct value, not 9.

Technology Link
E-Lab • Activity 2 Available on CD-ROM
and the Internet at http://www.hbschool.com/elab

45

Note:
Students evaluate expressions. Use E-Lab p. 2.

Activity 2
Explore
As students check each other's work, point out that whether they are using a calculator or pencil and paper, they need to follow the order of operations for the value of the expression to be correct.

Think and Discuss
When discussing the last question, have students explain their preferences. Some students may say they prefer to use memory keys because multiplication is used more than once in the expression. Others may say they prefer using the order of operations because there are fewer keys to remember.

Try This
Ask students:

- **How could you use memory keys to evaluate the expression?**

1 2 M+ 1 5 ÷ 3 M+ MR

- **How could you use the order of operations to evaluate the expression?**
  $15 \div 3 + 12$

USING E-LAB
Students use reasoning skills and the concept of order of operations to examine, evaluate, and insert the correct operation in an expression in order to solve a problem.

✔ Assessment Tip
Use these questions to help students communicate the big ideas of the lesson.

- **Find the value of the expression $7 + 8 - 2^3 \div 4$. Tell the order in which you performed the operations.** 13; exponents, division, addition, subtraction

- **A calculator shows the display 2 as the value of $16 \div (12 - 4)$. Does the calculator use AOS? Explain.** Yes; 2 is the correct answer when you follow the order of operations.

INTERDISCIPLINARY CONNECTION

One suggestion for an interdisciplinary unit for this chapter is Creatures Great and Small. In this unit you can have students

- write word problems about the five longest snakes in the world. Each problem should include two different operations.

Average Lengths of Snakes	
Reticulated python	35 ft
Anaconda	28 ft
Indian python	25 ft
Diamond python	21 ft
King cobra	19 ft

Modality Visual

CULTURAL CONNECTION

In Germany in the late 1800's, a man trained his horse Muhamed to add, subtract, multiply, and divide. To give an answer of 35, Muhamed would tap his left hoof 3 times and his right hoof 5 times. If Muhamed had known the order of operations, how would he have shown the value of the expression $9 + 12 \div 4$? He would have tapped his left hoof 1 time and his right hoof 2 times.

Modality Auditory

ORGANIZER

OBJECTIVE To estimate sums, differences, products, and quotients

VOCABULARY *New* **clustering, compatible numbers**

ASSIGNMENTS Independent Practice
- ■ *Basic* 1–35, even; 36–41; 42–51
- ■ *Average* 1–41; 42–51
- ■ *Advanced* 1–35, even; 36–41; 42–51

PROGRAM RESOURCES
Reteach 2.5 Practice 2.5 Enrichment 2.5

PROBLEM OF THE DAY

Transparency 2

Although not all the digits are shown, how do you know the quotients cannot be correct?

1. ■ 7)3, ■ ■ 5   2. 58)■, ■ ■ ■

1. 8 × 7 = 56, but there is a 5 in the units place in the dividend.

2. 306 × 58 = 17,748, but there are only four places in the dividend.

Solution Tab A, p. A2

QUICK CHECK

Round to the nearest ten and hundred.

1. 476 480; 500
2. 822 820; 800
3. 706 710; 700
4. 584 580; 600
5. 4,367 4,370; 4,400
6. 8,241 8,240; 8,200
7. 8,019 8,020; 8,000
8. 4,939 4,940; 4,900

1 MOTIVATE

To access students' prior knowledge of estimates, you may want to ask students the following questions:

- **Would you answer these questions with an exact number or an estimate?**
  exact − 1, 4, 6; estimate − 2, 3, 5, 7

2.5

WHAT YOU'LL LEARN
How to estimate sums, differences, products, and quotients

WHY LEARN THIS?
To solve problems that don't need an exact answer, such as estimating the distance a flying disc is thrown

WORD POWER
clustering
compatible
numbers

REMEMBER:
To round, look at the digit to the right of the place to which you are rounding.
- If that digit is 5 or greater, the digit being rounded increases by 1.
- If the digit to the right is less than 5, the digit being rounded remains the same. **See page H9.**

Using Estimation

Did you know that the longest throw of a flying disc was 656 ft 2 in., made by Scott Stokely? Mark, Kevin, and Deanna are throwing their disc, trying to equal the distance thrown by Scott Stokely. They wrote the distances of their throws in a table.

Name	Distance
Mark	145 ft
Kevin	148 ft
Deanna	151 ft

You can use rounding to find a sum when you don't need an exact answer.

$$
\begin{array}{rcl}
145 & \to & 150 \\
148 & \to & 150 \\
+151 & \to & +150 \\
\hline
& & 450
\end{array}
$$

Round each number to the nearest 10.

The estimate, 450, is not close to 656 ft, so the total distance is not about the same as the record.

You can use **clustering** to estimate a sum when all the addends are about the same.

EXAMPLE 1 Use clustering to estimate the sum.

$$
\begin{array}{r}
1,802 \\
2,182 \\
+1,999 \\
\end{array}
$$

The three addends are close to 2,000.

$3 \times 2,000 = 6,000$ *Multiply.*

So, the sum is about 6,000.

You can use rounding to estimate a difference.

EXAMPLE 2 Use rounding to estimate the difference.

$$
\begin{array}{r}
928 \\
-616 \\
\end{array}
$$

Round to the nearest hundred.	*Round to the nearest ten.*
$\begin{array}{r} 900 \\ -600 \\ \hline 300 \end{array}$	$\begin{array}{r} 930 \\ -620 \\ \hline 310 \end{array}$

So, the difference is 300 when you round to the nearest hundred, and it is 310 when you round to the nearest ten.

IDEA FILE

SCIENCE CONNECTION

Have students write word problems that can be solved using estimation. Have them use in their problems these data about the five most popular breeds of cats in the United States.

Breed	Number Registered
Persian	44,735
Maine coon	4,332
Siamese	3,025
Abyssinian	2,469
Exotic shorthair	1,610

Modality Visual

RETEACH 2.5

Name _____

LESSON 2.5

Using Estimation

You can use compatible numbers to estimate a quotient. **Compatible numbers** are helpful to use because they divide without a remainder, are close to the actual numbers, and are easy to compute mentally.

Oakdale Middle School is collecting recycled cans. The school has set a goal of collecting 2,788 cans. There are 38 homerooms in the school. About how many cans should each homeroom collect for the school to reach its goal?

Because you are asked "about how many," an estimate is appropriate for the answer. Use compatible numbers to estimate.

Step 1: Look at the actual numbers that make up the problem. Think about numbers that are close to the real numbers that will divide without a remainder.

$2,788 \div 38$ → $2,800 \div 40$

Step 2: Divide the compatible numbers.   → $2,800 \div 40 = 70$

So, each homeroom should collect about 70 cans.

Complete to show how compatible numbers are used to estimate the quotient.

1. $2,615 \div 47 \to 2,500 \div \underline{50} = \underline{50}$
2. $3,104 \div 62 \to 3,000 \div \underline{60} = \underline{50}$
3. $3,591 \div 88 \to \underline{3,600} \div 90 = \underline{40}$
4. $4,733 \div 74 \to 4,800 \div \underline{80} = \underline{60}$
5. $7,105 \div 77 \to \underline{7,200} \div 80 = \underline{90}$
6. $5,511 \div 62 \to 5,400 \div \underline{60} = \underline{90}$
7. $15,843 \div 381 \to 16,000 \div \underline{400} = \underline{40}$
8. $20,972 \div 287 \to 21,000 \div \underline{300} = \underline{70}$
9. $29,100 \div 307 \to \underline{30,000} \div 300 = \underline{100}$
10. $95,347 \div 795 \to 96,000 \div \underline{800} = \underline{120}$

Use compatible numbers to estimate the quotient. Possible estimates are given.

11. $434 \div 68$ __6__
12. $394 \div 5$ __80__
13. $448 \div 15$ __30__
14. $986 \div 102$ __10__
15. $627 \div 89$ __7__
16. $554 \div 63$ __9__
17. $293 \div 31$ __10__
18. $705 \div 97$ __7__
19. $1,246 \div 43$ __30__
20. $2,779 \div 28$ __100__
21. $3,896 \div 38$ __100__
22. $7,164 \div 78$ __90__

GUIDED PRACTICE

Use clustering to estimate the sum. **Possible estimates are given.**

	1.	2,940	2.	1,040	3.	4,480	4.	5,449
		3,100		1,300		4,100		4,869
		+2,834		+1,422		+3,967		+4,834
		9,000		**3,000**		**12,000**		**15,000**

Round to the nearest ten. Then find the sum or difference.

	5.	723	6.	420	7.	998	8.	375
		+819		+388		−114		−132
		1,540		**810**		**890**		**250**

Round to the nearest hundred. Then find the sum or difference.

	9.	184	10.	244	11.	667	12.	855
		+518		+238		−133		−268
		700		**400**		**600**		**600**

Estimating Products and Quotients

You can use rounding to estimate a product.

EXAMPLE 3   The school is holding a talent show in the cafeteria. The students set up 21 rows of 32 seats. About how many programs should the school print for the talent show?

Think: Are you asked to find an exact answer or an estimate?

$$32 \rightarrow 30$$
$$\times 21 \rightarrow \times 20$$

Estimate by rounding each factor to the nearest ten.

$$\begin{array}{r} 30 \\ \times 20 \\ \hline 600 \end{array}$$

Multiply.

So, the school will need to print about 600 programs.

• Estimate the product. 550 × 82 **Possible answers: 44,000; 48,000**

Talk About It

• When you rounded each factor in Example 3, did you round up to the next ten or down? **down**

• 💡**CRITICAL THINKING**   Would the exact answer be a number greater than the estimated answer or less than the estimated answer? How do you know? **Greater than; each of the two factors was rounded down.**

• Do you think it would be better for the estimated answer to be less than or greater than the exact answer? Explain. **Possible answer: greater than; to be sure that there are enough programs**

> **Science Link**
>
> In the vastness of space, distances between objects are enormous. Even within our own solar system, distances between planets are huge. Earth's orbit is about 151,000,000 km from the sun, and Mars's orbit is about 228,000,000 km from the sun. Rounding these distances to the nearest ten million, estimate the distance between Earth's orbit and Mars's orbit.

about 80,000,000 km

47

---

1. What is the world record for the long jump?
2. How many seashells are on the beach?
3. How many people in the United States play the guitar?
4. At what temperature does water freeze?
5. How many traffic lights are in Japan?
6. What size shoe do you wear?
7. How many snowflakes fell in Buffalo, New York, in 1996?

2 TEACH/PRACTICE

In this lesson, students broaden their knowledge of estimation to include not only rounding to estimate but clustering and using compatible numbers as well.

TEACHING TIP

Discuss with students the accuracy of the estimate 450 ft at the top of page 46.

• **If you rounded the addends to the nearest hundred, what would your estimate be?** 400 ft

• **Find the exact answer.** 444 ft

• **Which estimate is closer?** 450 ft

Lead students to conclude that rounding to a number with a lesser place value results in a more accurate estimate.

GUIDED PRACTICE

Complete Exercises 1–12 orally with the class. Make sure students understand how to make good estimates before you assign the Independent Practice.

ADDITIONAL EXAMPLES

Example 1, p. 46

Use clustering to estimate the sum.

5,889 The sum is about 18,000.
6,019
+5,999

Example 2, p. 46

Use rounding to estimate the difference.

785
−262

The difference is 500 when you round to the nearest hundred, and it is 530 when you round to the nearest ten.

Example 3, p. 47

Each of the 47 students who enter the Blair County Math Competition needs to solve 32 different problems. About how many problems will the sponsors of the competition need to write for the competitors? 1,500 problems

---

PRACTICE 2.5

Name _____

LESSON 2.5

Using Estimation

Vocabulary

1. Numbers that divide without a remainder, are close to the actual numbers, and are easy to compute mentally are called __compatible numbers__.

2. When all addends are about the same, you can use __clustering__ to estimate their sum.

Estimate the sum or difference. Possible estimates are given.

3.	2,489	4.	398	5.	4,723	6.	7,132	7.	5,401
	1,601		415		+2,198		6,594		+9,188
	+2,109		+368		**6,800**		+7,301		**14,500**
	6,000		**1,200**				**21,000**		

8.	478	9.	263	10.	5,877	11.	8,528	12.	8,903
	− 26		−211		−5,318		−6,491		−4,575
	450		**50**		**600**		**2,000**		**4,300**

Estimate the product. Possible estimates are given.

13.	53	14.	76	15.	72	16.	47	17.	660
	× 8		× 9		×28		×53		× 42
	400		**720**		**2,100**		**2,500**		**28,000**

18.	371	19.	68	20.	480	21.	375	22.	824
	× 78		×37		×192		×591		×693
	32,000		**2,800**		**100,000**		**240,000**		**560,000**

Estimate the quotient. Possible estimates are given.

23. 331 ÷ 5	24. 643 ÷ 9	25. 1,827 ÷ 59	26. 5,543 ÷ 77
70	**70**	**30**	**70**

27. 9,165 ÷ 28	28. 6,281 ÷ 875	29. 7,118 ÷ 614	30. 8,215 ÷ 897
300	**7**	**12**	**9**

Mixed Applications

31. There are 187 seats in a local movie theater. The theater has been sold out for the past 18 shows. How many people attended the shows?

__about 4,000 people__

32. Rhonda earned $371.09 this week. Her brother Rod earned $397.01. Who earned more money? How much more?

__Rod; $25.92 more__

P10 **ON MY OWN**   Use with text pages 46–49.

ENRICHMENT 2.5

Name _____

LESSON 2.5

Estimating Populations

POPULATION OF THE MIDDLE COLONIES: 1670 – 1750					
Colony	1670	1690	1710	1730	1750
Delaware	700	1,482	3,645	9,170	28,704
New Jersey	1,000	8,000	19,872	35,510	71,393
New York	5,754	13,909	21,625	48,594	76,696
Pennsylvania	—	11,450	24,450	51,707	119,666

The table shows how the population of the middle colonies changed from 1670 to 1750. Use the table to answer the questions. Estimate to the nearest thousand.

1. About how many people lived in either Delaware or New York in 1690?

__about 15,000 people__

2. About how many people lived in either Pennsylvania or New Jersey in 1730?

__about 88,000 people__

3. About how many more people lived in New York than in Delaware in 1710?

__about 18,000 more people__

4. About how many more people lived in Pennsylvania in 1750 than in 1690?

__about 109,000 more people__

5. About how many people lived in the middle colonies in 1670?

__about 8,000 people__

6. About how many people lived in the middle colonies in 1750?

__about 297,000 people__

7. About how many more people lived in the middle colonies in 1750 than in 1670?

__about 289,000 more people__

E10 **STRETCH YOUR THINKING**   Use with text pages 46–49.

47

CRITICAL THINKING
Discuss with students that sometimes it is better to overestimate than underestimate an answer. It depends upon the situation.

- **Would it be better to overestimate or underestimate the answer to Additional Example 3? Why? overestimate; so** there are enough problems for all the competitors to solve

- **What if the sponsors plan to use the money they make selling tickets for prize money for the competition? Should they overestimate or underestimate the amount of money they expect to make? Why? underestimate;** because they don't want to run short of prize money

ADDITIONAL EXAMPLE

Example 4, p. 48

Jeff made a total of $1,325 mowing lawns for 63 days this past summer. About how much did he earn each day?

Using rounding shows that Jeff earned about $22 each day. Using compatible numbers shows that Jeff earned about $20 each day.

When you estimate a quotient, you can use rounding or compatible numbers. **Compatible numbers** divide without a remainder, are close to the actual numbers, and are easy to compute mentally.

EXAMPLE 4 Loughman Middle School is starting a recycling program. The school has set a goal of collecting 1,545 lb of material to be recycled. The school has 36 homerooms. About how many pounds of recyclable material should the students in each homeroom collect?

Using rounding:

$1,545 \div 36$

$1,500 \div 40$ *Round the dividend and the divisor.*

$1,500 \div 40 = 37 \text{ r}20$ *Divide.*

So, the students in each homeroom should collect about 38 lb.

Using compatible numbers:

$1,545 \div 36$

$1,600 \div 40$ *4 is compatible with 16.*

$1,600 \div 40 = 40$ *Divide.*

So, the students in each homeroom should collect about 40 lb.

- Which estimate is easier to compute with mentally? **the one made with compatible numbers**

- Is $1,488 \div 36$ easier to estimate using rounding or compatible numbers? **Compatible numbers; $1,500 \div 30$ is easier to estimate than $1,500 \div 40$.**

INDEPENDENT PRACTICE

Estimate the sum or difference. **Possible estimates are given.**

1.	2.	3.	4.	5.
1,700	293	3,643	5,765	6,902
2,008	348	+4,211	5,948	+7,219
+2,324	+343	**8,000**	+6,324	**14,000**
6,000	**900**		**18,000**	

6.	7.	8.	9.	10.
389	152	3,556	9,123	4,687
− 43	−138	−3,339	−6,512	−1,022
350	**10**	**300**	**2,000**	**4,000**

48   Chapter 2

IDEA FILE

CULTURAL CONNECTION

Have students work in pairs to write problems using the populations per square mile of these cities. Then have students exchange papers and have their partners estimate the answers.

City	Population
Cairo	97,106
Hong Kong	237,501
New York City	11,480
Bangkok	58,379
Bombay	127,461

Modality Visual

CONSUMER CONNECTION

Have students round to the nearest ten to estimate the total annual energy cost for the use of these appliances. $400

Appliance	Annual Energy Cost
Washer/Dryer	$77
Color TV	$22
Range/Oven	$42
Refrigerator/Freezer	$239
Microwave oven	$7
Dishwasher	$12

Modality Visual

MEETING INDIVIDUAL NEEDS

Estimate the product. **Possible estimates are given.**

11.	36	12.	43	13.	59	14.	48	15.	940
	× 9		× 7		× 33		× 29		× 66
	360		**280**		**1,800**		**1,500**		**63,000**
16.	364	17.	53	18.	590	19.	482	20.	846
	× 12		× 41		× 335		× 299		× 562
	4,000		**2,000**		**180,000**		**150,000**		**480,000**

Estimate the quotient. **Possible estimates are given.**

21. 268 ÷ 5 **54**   **22.** 321 ÷ 4 **80**   **23.** 1,544 ÷ 28 **50**   **24.** 4,156 ÷ 64 **70**   **25.** 8,429 ÷ 39 **200**

26. 1,844 ÷ 22 **90**   **27.** 3,575 ÷ 56 **60**   **28.** 4,239 ÷ 670 **6**   **29.** 6,435 ÷ 529 **12**   **30.** 9,433 ÷ 309 **30**

31. 9,200 ÷ 8 **920**   **32.** 802 ÷ 23 **40**   **33.** 5,867 ÷ 18 **300**   **34.** 1,784 ÷ 178 **9**   **35.** 3,165 ÷ 211 **15**

Problem-Solving Applications Possible estimates are given.

36. Rudy had $64 to spend. He spent $18 on a haircut, $22 on shoes, and $7 for a movie ticket. About how much does Rudy have left? **about $10**

37. A garden shop sells annuals, shrubs, and trees. They have 78 annuals, 32 shrubs, and 18 trees. About how many plants do they have? **about 130 plants**

38. Maria collected souvenirs from each place she visited on her trip. She collected 156 colored leaves and 124 rocks and gems. About how many more colored leaves did she collect than rocks and gems? **about 40 more colored leaves**

39. The theater of the Natural Science and History Museum was filled to capacity for its 276 shows this fall. The theater holds 35 people. About how many people attended the shows? **about 12,000 people**

40. The Forest Service estimates that about 3,645 people visited the Cradle of Forestry in the past 31 days. About how many people visited the attraction each day? **about 120 people**

41. ▬▶ **WRITE ABOUT IT** Is it easier to use rounding or compatible numbers to estimate 756 ÷ 44? Explain. **It is easier to use compatible numbers, because they divide evenly and are easy to compute mentally.**

Mixed Review

Tell which numbers are equivalent.

42. 4.2, 4.25, 4.20
4.2, 4.20

43. 1.45, 1.450, 1.405
1.45, 1.450

44. 72.04, 72.40, 72.4
72.40, 72.4

Write as a decimal.

45. $\frac{1}{2}$ **0.5**   **46.** $\frac{7}{8}$ **0.875**   **47.** $\frac{3}{4}$ **0.75**   **48.** $\frac{1}{4}$ **0.25**   **49.** $\frac{5}{8}$ **0.625**

50. **RECYCLING** In a lifetime the average American will throw away 600 times his or her adult weight. How much trash is this for an adult who weighs 185 lb? **111,000 lb of trash**

51. **MENTAL MATH** Michelle spent $40 on a new pair of tennis shoes. She also spent $5 on socks and $15 on shorts. How much did she spend altogether? **$60**

MORE PRACTICE Lesson 2.5, page H45

49

CAREER CONNECTION

Journalism Have students write newspaper headlines using exact numbers and estimates. Two examples are given.

TEMPERATURES REACH A RECORD-BREAKING HIGH OF 102°F exact

MORE THAN 2,000 TURN OUT FOR MEMORIAL DAY PARADE estimate

Modality Visual

REVIEW

Have students find the value of this expression.

1 × 1 + 1 ÷ 1 − 1 + 1 ÷ 1 − 1 × (1 − 1) + 1 **3**

Modality Visual

💡**CRITICAL THINKING** Sue paid a total of about $100 for a pair of jeans, a T-shirt, and a vest. The vest cost about half as much as the jeans and about $22 more than the T-shirt. Estimate the price of each item. Possible answer: jeans, $60; T-shirt, $10; vest, $30

INDEPENDENT PRACTICE
To make sure students understand how to estimate, check answers to Exercises 1–4 before assigning the remaining exercises.

3 WRAP UP

Use these questions to help students focus on the big ideas of the lesson.

• **Three hundred forty-nine students completed the 6-mile *Walk for Hunger*. Round to the nearest ten to estimate the total mileage walked by the students.** 2,100 mi

✔ *Assessment Tip*

Have students write the answer.

• **Would the exact answer to Exercise 19 be greater than or less than the estimated answer? Why?** less than; because both factors were rounded up

DAILY QUIZ
Estimate the sum or difference. Possible estimates are given.

1.	756	**2.**	3,214	**3.**	4,893
	+229		−1,570		6,275
	1,000		1,600		+1,025
					12,000

Estimate the quotient. Possible estimates are given.

4. 415 ÷ 7 **60**   **5.** 4,765 ÷ 5 **900**

USING THE PAGE

Before the Lesson

Write the following problems on the overhead projector and have students discuss the different ways the problems can be solved using mental-math strategies and traditional methods.

- 12 + 16 + 9 + 8 45

- There are 163 apples and oranges. There are 29 more oranges than apples. How many of each fruit are there? 67 apples, 96 oranges

- 15 × 16 240

Cooperative Groups

Have students in each group read the paragraph about piñatas and work through the example. Then have each group solve Problems 1–4. When they finish, have groups compare their work and check the answers.

CULTURAL LINK

Mexico is a country in which many different cultures flourish. The Spanish culture, which was brought to Latin America by the conquistadors, has been blended with the native cultures already in place, including the Aztec, Maya, and Mixtec cultures.

CULTURAL CONNECTION

Piñata Party

A tradition for Mexican children at parties and during holidays is to try to break open a piñata. A piñata is a container made of paper or clay. The container is usually decorated like an animal and is filled with small candies. It is then hung from a tree or ceiling. Children take turns swinging a stick at the piñata to try and break it so the candy inside will fall to the ground. It is not as easy as it sounds, since the children are blindfolded while swinging at the piñata. Sometimes the piñata is hung by a pulley so it can be moved up and down while the children are trying to break it.

Charlie is having a birthday party at his home and decides to have a piñata. Inside the piñata he puts 17 cherry, 22 grape, 13 orange, and 28 lemon candies. How many candies are in the piñata?

$$17 + 22 + 13 + 28 = 17 + 13 + 22 + 28$$
$$= (17 + 13) + (22 + 28)$$
$$= 30 + 50$$
$$= 80$$

So, there are 80 candies in the piñata.

 Work Together

1. Charlie invites 17 guests to his party. Each of his guests gets an equal number of candies from the piñata. How many candies did each guest get? Will there be any candies left over? **4, yes**

2. Each of the 17 guests spends about $12 on a birthday gift for Charlie. How much money did Charlie's guests spend for gifts? **$204**

3. During the party, a game is played where teams of three people toss a ball as far as they can. On one team, Tina tosses the ball 63 ft, Jared tosses it 57 ft, and José tosses it 71 ft. Estimate the total distance the team tossed the ball. **about 190 ft**

4.  **WRITE ABOUT IT** Write a problem involving whole numbers about Charlie's party. Exchange with a partner and solve. **Check students' problems.**

CULTURAL LINK

The main food of most Mexicans is corn. Corn is used for several different types of food. The most popular is the tortilla. Tortillas are used to make tacos, enchiladas, burritos, or tostadas. They are then topped or filled with meats, cheeses, vegetables, and sauces.

CHAPTER 2 REVIEW

EXAMPLES

• **Use mental math to add and subtract.**
(pages 34–35)

$32 + 19 + 48 + 11 =$
$32 + 48 + 19 + 11 = \leftarrow$ *Commutative*
$(32 + 48) + (19 + 11) = \leftarrow$ *Associative*
$80 + 30 = 110 \leftarrow$ *mental math*

• **Use the strategy *guess and check* to solve addition and subtraction problems.**
(pages 36–37)

PROBLEM-SOLVING TIP: For help in solving problems by guessing and checking, see pages 5 and 36.

• **Use mental math to multiply.** (pages 38–39)
$22 \times 12 = 22 \times (10 + 2) \leftarrow$ *Distributive*
$= (22 \times 10) + (22 \times 2)$
$= 220 + 44$
$= 264$

• **Multiply and divide whole numbers.**
(pages 40–43)

$\begin{array}{r} 35 \\ 12\overline{)420} \\ -36\downarrow \\ \hline 60 \\ -60 \\ \hline 0 \end{array}$
The first digit is in the tens place.

Bring down the 0 ones.

• **Estimate sums, differences, products, and quotients.** (pages 46–49)

$\begin{array}{r} 521 \\ -328 \end{array}$ *Round to the nearest hundred.* $\rightarrow \begin{array}{r} 500 \\ -300 \\ \hline 200 \end{array}$

EXERCISES

1. **VOCABULARY** When you change one number to a multiple of ten and adjust the other addend, you are using __?__.
compensation

Find the sum or difference by using mental math.

2. $28 + 45 + 32$ **105** 3. $29 + 84$ **113** 4. $93 - 26$ **67**

5. Jane has a bag with 48 marbles. She has 12 more red marbles than yellow ones. How many does Jane have of each color?

6. The sum of Dustin's and Nick's ages is 30. Nick is 6 years older. What are their ages?
5. **30 red marbles, 18 yellow marbles**
6. **Dustin, 12 years old; Nick, 18 years old**

Use mental math to multiply.

7. 9×510 **4,590** 8. $22 \times 2 \times 0$ **0**
9. 24×5 **120** 10. $12 \times 6 \times 5$ **360**
11. $15 \times 6 \times 2$ **180** 12. $5 \times 9 \times 3$ **135**

Find the product.

13. $\begin{array}{r} 64 \\ \times 23 \\ \hline 1,472 \end{array}$ 14. $\begin{array}{r} 2,813 \\ \times 85 \\ \hline 239,105 \end{array}$ 15. $\begin{array}{r} 332 \\ \times 132 \\ \hline 43,824 \end{array}$

Find the quotient.

16. $8\overline{)584}$ **73** 17. $92\overline{)5,996}$ **65 r16**
18. $12\overline{)527}$ **43 r11** 19. $58\overline{)3,962}$ **68 r18**

20. Write the quotients for Exercises 18 and 19 using fractions for the remainders.
$43\frac{11}{12}; 68\frac{9}{29}$

Estimate. **Answers may vary.**

21. $\begin{array}{r} 8,271 \\ +7,834 \\ \hline 16,000 \end{array}$ 22. $\begin{array}{r} 1,802 \\ -959 \\ \hline 800 \end{array}$ 23. $\begin{array}{r} 832 \\ -487 \\ \hline 300 \end{array}$

24. $\begin{array}{r} 536 \\ \times 54 \\ \hline \end{array}$ **25,000** 25. $6\overline{)431}$ **70** 26. $8\overline{)789}$ **100**

USING THE PAGE

The Chapter 2 Review reviews using mental math to add, subtract, and multiply, using the strategy *guess and check* to add and subtract, multiplying and dividing whole numbers, and estimating sums, differences, products, and quotients. Chapter objectives are provided, along with examples and practice exercises for each.

Portfolio The Chapter 2 Review can be used as a review or a test. You may wish to place the Review in students' portfolios.

Assessment Checkpoint

For Performance Assessment in this chapter, see page 72, What Did I Learn?, items 2 and 5.

✔ Assessment Tip

Student Self-Assessment The How Did I Do? Survey helps students assess both what they have learned and how they learned it. Self-assessment helps students learn more about their own capabilities and develop confidence in themselves.

A self-assessment survey is available as a copying master in the *Teacher's Guide for Assessment*, p. 15.

CHAPTER 2 TEST, page 1

Name _____

Choose the letter of the correct answer.

For questions 1–4, use mental math.

1. $7 + 18 + 13$
A. 28 B. 31
C. 35 **D. 38**

2. $29 + 41 + 31 + 9$
A. 99 B. 100
C. 110 D. 113

3. $53 - 29$
A. 21 B. 22
C. 23 **D. 24**

4. $62 - 17$
A. 27 B. 38
C. 40 **D. 45**

5. A rectangular box has a perimeter of 50 ft. The box is 5 ft longer than it is wide. What is the width of the box?
A. 10 ft B. 15 ft
C. 20 ft D. 40 ft

6. Barry sold a total of 42 green or red banners. He sold 10 more green banners than red banners. How many green banners did he sell?
A. 16 banners
B. 22 banners
C. 26 banners
D. 32 banners

7. The Rogers hockey team played a total of 20 games. They tied in 3 games and lost 1 more game than they won. How many games did they win?
A. 7 games
B. 8 games
C. 9 games

D. 10 games

8. Ms. Lee baked a total of 60 pies. She baked a dozen more apple pies than lemon pies. How many apple pies did she bake?
A. 18 pies B. 24 pies
C. 36 pies D. 48 pies

9. Which property of multiplication is being used below?
$7 \times 25 = (7 \times 20) + (7 \times 5)$
A. Distributive Property
B. Property of One
C. Commutative Property
D. Associative Property

For questions 10–11, use mental math to find the product.

10. 4×44
A. 156 B. 160 **C. 176** D. 180

11. $4 \times 10 \times 6$
A. 240 B. 260 C. 360 D. 400

12. Find the missing factor:
$27 \times 43 = (27 \times \blacksquare) + (27 \times 3)$
A. 34 **B. 40** C. 43 D. 74

13. Rob is solving the multiplication problem below. What mistake, if any, has Rob made?
$\begin{array}{r} 126 \\ \times 23 \\ \hline 378 \\ 252 \end{array}$
A. His work has no mistake.
B. The product 3×126 is 370.
C. The products are lined up incorrectly.
D. He did not multiply by the hundreds.

Form A • Multiple-Choice   A7   Go on. ▶

CHAPTER 2 TEST, page 2

Name _____

14. In this division problem, where would you put the first digit of the quotient?
$17\overline{)1,840}$
A. in the ones place
B. in the tens place
C. in the hundreds place
D. in the thousands place

15. $\begin{array}{r} 120 \\ \times 35 \\ \hline \end{array}$
A. 960
B. 3,200
C. 4,100
D. not here

16. $\begin{array}{r} 437 \\ \times 226 \\ \hline \end{array}$
A. 97,862
B. 98,762
C. 106,662
D. 107,662

17. $6\overline{)324}$
A. 50 r4 B. 52 r2
C. 53 **D. 54**

18. $20\overline{)4,060}$
A. 203 B. 213
C. 230 D. 233

19. Clea's parents bought a computer for $1,530.00. They are paying for it in 18 monthly payments. How much is each payment?
A. $77.50
B. $85.00
C. $95.00
D. $127.50

20. The Stone family spends $125 on groceries each week. How much does the family spend on groceries in one year? Hint: 1 yr = 52 wk
A. $650 B. $2,500
C. $5,250 **D. $6,500**

21. Choose the best estimate of the sum.
$\begin{array}{r} 5,199 \\ 4,206 \\ +3,749 \\ \hline \end{array}$
A. 11,000 B. 12,000
C. 13,000 D. 15,000

22. Choose the best estimate of the quotient.
$1,765 \div 28$
A. 0.6 **B. 60**
C. 600 D. 6,000

23. The music store sold 589 CDs of waterfall sounds and 209 audiotapes of the sounds. About how many more CDs were sold than audiotapes?
A. about 200 more CDs
B. about 300 more CDs
C. about 400 more CDs
D. about 500 more CDs

24. The movie theater at the waterfall park had 213 shows this summer. The theater, which holds 480 people, was full for every show. About how many people attended the shows?
A. about 10,000 people
B. about 10,200 people
C. about 80,000 people
D. about 100,000 people

Form A • Multiple-Choice   A8   ▶ Stop!

See *Test Copying Masters*, pp. A7–A8 for the Chapter Test in **multiple-choice** format and pp. B157–B158 for **free-response** format.

PLANNING GUIDE

Using Decimals

BIG IDEA: Efficient computation involves understanding the connection among basic facts, and place value in algorithms.

Introducing the Chapter p. 52		Team-Up Project • Selecting Computer Programs p. 53	
OBJECTIVE	**VOCABULARY**	**MATERIALS**	**RESOURCES**
3.1 Adding Decimals pp. 54-55 **Objective:** To add decimals and estimate sums **1 DAY**		FOR EACH PAIR: decimal squares	■ Reteach, ■ Practice, ■ Enrichment 3.1; More Practice, p. H45 Math Fun • Decimal Puzzles, p. 68
3.2 Subtracting Decimals pp. 56-57 **Objective:** To subtract decimals **1 DAY**			■ Reteach, ■ Practice, ■ Enrichment 3.2; More Practice, p. H45 Math Fun • Three in a Row, P. 68
 INTERVENTION • Placing the Decimal Point, Teacher's Edition pp. 58A–58B			
3.3 Multiplying Decimals pp. 58-61 **Objective:** To multiply decimals **2 DAYS**			■ Reteach, ■ Practice, ■ Enrichment 3.3; More Practice, p. H46 Math Fun • $1,000 Winner, p. 68
Lab Activity: Exploring Division of Decimals pp. 62-63 **Objective:** To use a model to divide decimals		FOR EACH STUDENT: decimal squares, colored pencils, scissors; 10 quarters, 10 dimes, 10 nickels	🖥 **E-Lab • Activity 3** E-Lab Recording Sheets, p. 3
3.4 Dividing Decimals pp. 64-67 **Objective:** To divide a decimal by a whole number and to divide a decimal by a decimal **2 DAYS**		FOR EACH STUDENT: calculator	■ Reteach, ■ Practice, ■ Enrichment 3.4; More Practice, p. H46 💿 ***Number Heroes •*** *Quizzo*
CHAPTER ASSESSMENT Chapter 3 Review p. 69			
ASSESSMENT CHECKPOINT Review • Chapter 1–3 pp. 70–71 Performance Assessment • Chapter 1–3 p. 72 Cumulative Review • Chapter 1–3 p. 73			

NCTM CURRICULUM
STANDARDS FOR GRADES 5–8

✓ **Problem Solving**
Lessons 3.1, 3.2, 3.3, 3.4

✓ **Communication**
Lessons 3.1, 3.2, 3.3, 3.4

✓ **Reasoning**
Lessons 3.1, 3.2, 3.3, 3.4

✓ **Connections**
Lessons 3.1, 3.2, 3.3, 3.4

✓ **Number and Number Relationships**
Lessons 3.1, 3.2, 3.3, 3.4

✓ **Number Systems and Number Theory**
Lessons 3.1, 3.2, 3.3, 3.4

✓ **Computation and Estimation**
Lessons 3.1, 3.2, 3.3, 3.4

✓ **Patterns and Functions**
Lessons 3.3, 3.4

☐ **Algebra**

☐ **Statistics**

☐ **Probability**

✓ **Geometry**
Lesson 3.3, Lab Activity

✓ **Measurement**
Lessons 3.2, 3.3, 3.4

CHAPTER LEARNING GOALS

3A To solve problems that involve number theory and fractions.

3A.1 To add and subtract decimals.

3A.2 To multiply decimals.

3A.3 To divide decimals.

These goals for concepts, skills, and problem solving are assessed in many ways throughout the chapter. See the chart below for a complete listing of both traditional and informal assessment options.

Pretest Options
Pretest for Chapter Content, TCM pp. A9–A10, B159–B160
Assessing Prior Knowledge, TE p. 52; APK p. 5

Daily Assessment
Mixed Review, PE pp. 57, 61
Problem of the Day, TE pp. 54, 56, 58, 64
Assessment Tip and Daily Quiz, TE pp. 55, 57, 61, 63, 67

Formal Assessment
Chapter 3 Review, PE p. 69
Chapter 3 Test, TCM pp. A9–A10, B159–B160
Study Guide & Review, PE pp. 70–71
Test • Chapters 1-3 Test, TCM pp. A11–A12, B161–B162
Cumulative Review, PE p. 73
Cumulative Test, TCM pp. A13–A16, B163–B166

Performance Assessment
Team-Up Project Checklist, TE p. 53
What Did I Learn?, Chapters 1–3, TE p. 72
Interview/Task Test, PA p. 78

Portfolio
Suggested work samples:
Independent Practice, PE pp. 55, 57, 60–61, 67
Math Fun, PE p. 68

Student Self-Assessment
How Did I Do?, TGA p. 15
Portfolio Evaluation Form, TGA p. 24
Math Journal, TE pp. 52B, 67

Key
Assessing Prior Knowledge Copying Masters: APK	Teacher's Guide for Assessment: TGA
Test Copying Masters: TCM	Teacher's Edition: TE
Performance Assessment: PA	Pupil's Edition: PE

MATH JOURNAL
You may wish to use the following suggestions to have students write about the mathematics they are learning in this chapter:

PE pages 59, 65 • Talk About It

PE pages 55, 57, 61, 67 • Write About It

TE page 67 • Writing in Mathematics

TGA page 15 • How Did I Do?

In every lesson • record written answers to each Wrap-up Assessment Tip from the Teacher's Edition for test preparation.

VOCABULARY DEVELOPMENT
There are no new words introduced in this chapter but mathematics vocabulary is reinforced visually and verbally. Encourage students to review the mathematics vocabulary in their journals.

MEETING INDIVIDUAL NEEDS

RETEACH	PRACTICE
BELOW-LEVEL STUDENTS	**ON-LEVEL STUDENTS**

BELOW-LEVEL STUDENTS

Practice Activities, TE Tab B
Name the Mystery Number
(Use with Lessons 3.1–3.4)

Interdisciplinary Connection, TE p. 61

 Literature Link, TE p. 52C
King of the Wind, Activity 1
(Use with Lesson 3.4.)

 Technology
Number Heroes • *Quizzo*
(Use with Lessons 3.2, 3.4.)

E-Lab • Activity 3
(Use with Lab Activity.)

ON-LEVEL STUDENTS

Practice Game, TE p. 52C
Decimal Challenge
(Use with Lessons 3.1–3.4.)

Interdisciplinary Connection, TE p. 61

 Literature Link, TE p. 52C
King of the Wind, Activity 2
(Use with Lesson 3.4.)

 Technology
Number Heroes • *Quizzo*
(Use with Lessons 3.2, 3.4.)

E-Lab • Activity 3
(Use with Lab Activity.)

CHAPTER IDEA FILE

The activities on these pages provide additional practice and reinforcement of key chapter concepts. The chart above offers suggestions for their use.

LITERATURE LINK

King of the Wind

by Marguerite Henry

This book tells the story of a thoroughbred horse, Sham and his stable boy, Agba. Together, the two travel all over the world winning horse races and learning about many of life's cruelties.

Activity 1—In 1973, Secretariat ran the Kentucky Derby in 1 minute, 59 seconds, the track record for this 1.25 mile long race. Have students calculate the rate of speed for Secretariat.

Activity 2—Have students research and compare the top speeds of five different animals. Then have them calculate which animal would win the Kentucky Derby and how much sooner that animal would finish than all the others.

Activity 3—Have students estimate the speeds of different modes of transportation such as walking, biking, riding in a car, and roller blading. Using these different figures, have students compare how long it would take them to get from their own homes to the house of a friend who lives at least a mile away.

PRACTICE GAME

Decimal Challenge

Purpose To add, subtract, and multiply decimals

Materials game cards (TR p. R80); spinner (TR p. R67), prepared as illustrated below; calculator

About the Game Players in small groups stack the cards face down. A facilitator spins for an operation and turns up 2 cards. Other players race to perform the operation while the facilitator checks the answers. The first player to answer correctly earns 1 point. Another player becomes the facilitator and repeats the process. Play continues until all players have served as the facilitator. The player with the most points wins.

Modalities Visual, Tactile

Multiple Intelligences
Interpersonal/Social,
Logical/Mathematical,
Bodily/Kinesthetic

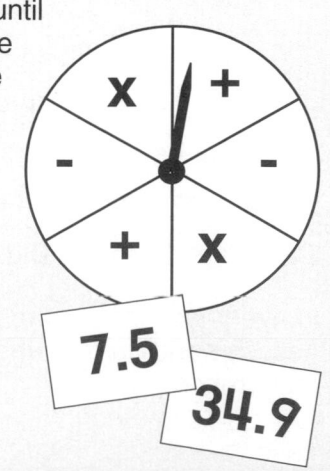

Practice Game, TE p. 52C
Decimal Challenge
(Use with Lessons 3.1–3.4.)

Interdisciplinary Connection, TE p. 61

Literature Link, TE p. 52C
King of the Wind, Activity 3
(Use with Lessons 3.3, 3.4.)

Technology
Number Heroes • *Quizzo*
(Use with Lessons 3.2, 3.4.)

E-Lab • Activity 3
(Use with Lab Activity.)

The purpose of using technology in the chapter is to provide reinforcement of skills and support in concept development.

E-Lab • Activity 3—Students explore multiplication of decimals by using a computer model.

Number Heroes • *Quizzo*—The activities on this CD-ROM provide students with opportunities to practice decimal division.

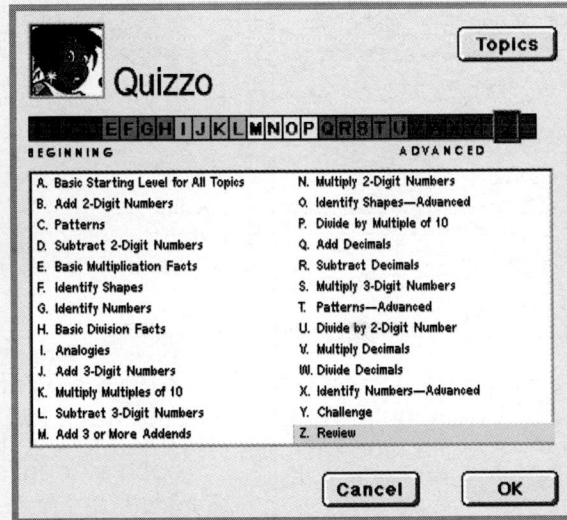

Each Number Heroes activity can be assigned by degree of difficulty. See Teacher's Edition pages 57 and 67 for Grow Slide information.

INTERDISCIPLINARY SUGGESTIONS

Connecting Decimals

Purpose To connect decimals to other subjects students are studying

You may wish to connect **Using Decimals** to other subjects by using related interdisciplinary units.

Suggested Units

- Cultural Connection: Measurement in Other Countries
- Consumer Connection: Home Entertainment
- Social Studies Connection: Mass Transit Systems
- History Connection: The Faces on Our Currency
- Science Connection: Where Does Our Garbage Go?

Modalities Auditory, Kinesthetic, Visual

Multiple Intelligences Mathematical/ Logical, Bodily/Kinesthetic, Visual/Spatial

Visit our web site for additional ideas and activities.
http://www.hbschool.com

Introducing the Chapter

Encourage students to talk about why the decimal point is important. Review the math content students will learn in the chapter.

WHY LEARN THIS?

The question "Why are you learning this?" will be answered in every lesson. Sometimes there is a direct application to everyday experiences. Other lessons help students understand why these skills are needed for future success in mathematics.

PRETEST OPTIONS

Pretest for Chapter Content See the *Test Copying Masters*, pages A9–A10, B159–B160 for a free-response or multiple-choice test that can be used as a pretest.

Assessing Prior Knowledge Use the copying master shown below to assess previously taught skills that are critical for success in this chapter.

Assessing Prior Knowledge Copying Masters, p. 5

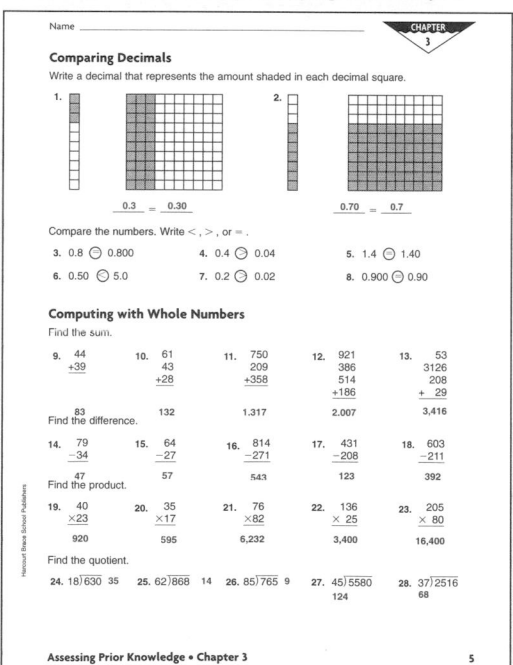

Troubleshooting

If students need help	Then use
• COMPARING DECIMALS	• Key Skills, PE p. H11
• COMPUTING WITH WHOLE NUMBERS	• Practice Activities, TE Tab A, Activity 8A

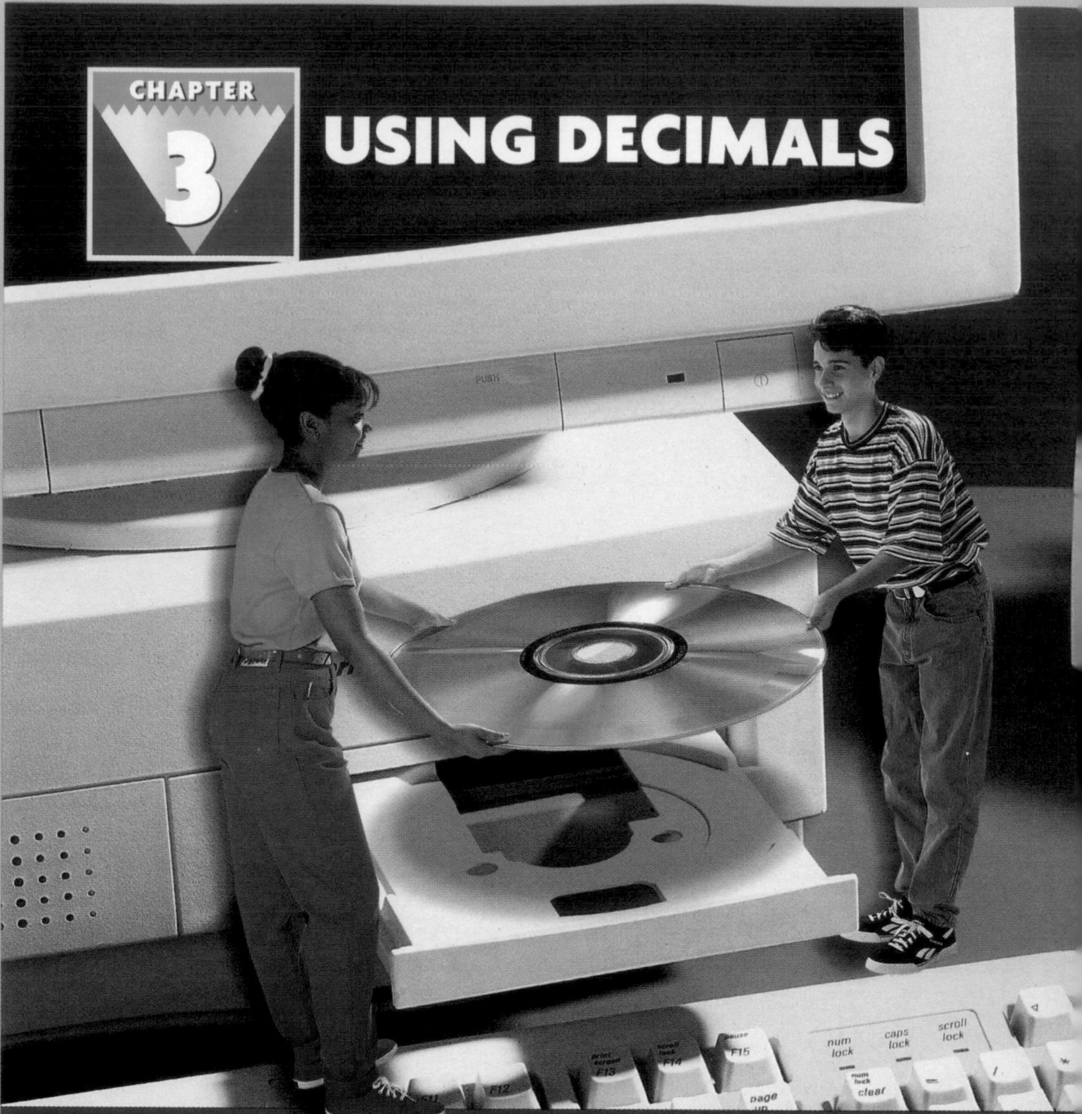

LOOK AHEAD

In this chapter you will solve problems that involve

• adding, subtracting, multiplying, and dividing decimals

52 Chapter 3

SCHOOL-HOME CONNECTION

You may wish to send to each student's family the *Math at Home* page. It includes an article that stresses the importance of communication about mathematics in the home, as well as helpful hints about homework and taking tests. The activity and extension provide extra practice that develops concepts and skills involving decimals.

Math at Home

Team-Up Project

Selecting Computer Programs

Suppose your class has been given $500 to buy computer programs. Make a list of the programs you would select. Find their total cost. Remember, you can not spend over $500.

YOU WILL NEED: magazines, catalogs

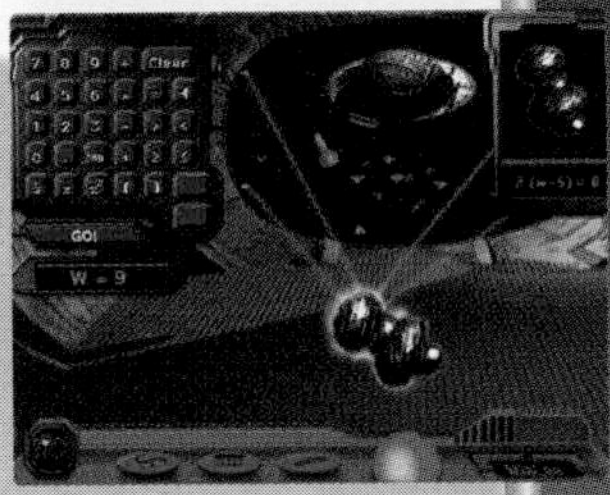

Plan
- Work with a partner to select computer programs. Find their total cost.

Act
- Look in magazines or catalogs to find computer programs and their cost.
- Figure out which computer programs you would buy with $500. Make a list of the programs and their cost.
- Find the total cost of the programs.
- Will you have any money left? If so, how much?

Share
- Compare your list and total cost with that of other groups. Which group spent the most money without going over $500?

DID YOU
- ☑ Select computer programs?
- ☑ List the programs and their cost?
- ☑ Find the total cost?
- ☑ Compare lists with other groups?

Project Checklist

Portfolio

Evaluate whether partners
- ☑ work cooperatively.
- ☑ participate in selecting programs.
- ☑ help calculate total cost and ensure that the $500 limit is not exceeded.

Team-Up Project

Using the Project

Accessing Prior Knowledge adding and subtracting money

Purpose to use decimals

GROUPING partners

MATERIALS computer magazines, catalogs, newspaper ads

Managing the Activity You may want to limit the programs from which students may select to educational programs. Encourage students to spend as much of the $500 as they can without going over.

WHEN MINUTES COUNT

A shortened version of this activity can be completed in 15-20 minutes. Write a list of programs and their prices on the board. Have groups of students decide which they would buy with $250.

A BIT EACH DAY

Day 1: Students select partners and begin looking for programs.

Day 2: Students continue selecting programs and considering possible purchases.

Day 3: Students narrow their selections to a group with a total cost of $500 or less.

Day 4: Partners compare their choices and total costs.

Day 5: The class works together to make a final selection for the whole class.

ON THE ROAD

Students may want to visit a computer store or other store that sells computer programs to shop for best prices or find out more about the programs.

ORGANIZER

OBJECTIVE To add decimals and estimate sums

ASSIGNMENTS Independent Practice
- ■ *Basic* 1–26, odd; 27–29
- ■ *Average* 1–29
- ■ *Advanced* 1–4, 23–26; 27–29

PROGRAM RESOURCES
Reteach 3.1 Practice 3.1 Enrichment 3.1
Math Fun • Decimal Puzzles, p. 68

PROBLEM OF THE DAY

Transparency 3

Replace each ♥ with a digit from 0–9 to make a true number sentence.

♥.♥ ♥ ♥ + ♥ ♥.♥ ♥ + ♥.♥ = 22.815

Possible answer: 0.725 + 13.69 + 8.4 = 22.815

Solution Tab A, p. A3

QUICK CHECK
Find the sum.

1. 47 + 62 109
2. 98 + 56 154
3. 78 + 4 + 29 111
4. 8 + 46 + 54 108
5. 34 + 52 + 29 115
6. 342 + 87 + 224 653

1 MOTIVATE

MATERIALS: *For each pair* decimal squares

Organize students into pairs. Review with them how to use decimal squares to add decimals. Display and discuss this example.

1.5 + 0.7 = 2.2

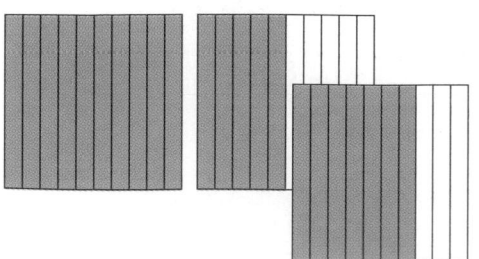

3.1 Adding Decimals

WHAT YOU'LL LEARN
How to add decimals and estimate sums

WHY LEARN THIS?
To find the total cost of a purchase or determine whether you have enough money for a purchase

Shana has $9.50 and is eating out for dinner. She is looking at her bill to see whether she has enough money to order a dessert for $1.25. How can she determine whether she has enough money without finding the exact total?

One way is to estimate the sum by using rounding.

$1.05 + $4.89 + $1.19 + $1.25
↓   ↓   ↓   ↓
1 + 5 + 1 + 1 = 8 ← The total with dessert will be about $8.

So, Shana has enough money to buy a dessert.

Have you ever noticed how decimals are arranged on a cash register receipt? When you add decimals, you align the decimal points.

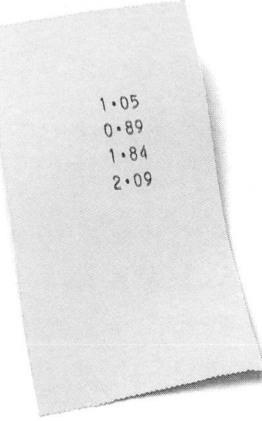

1·05
0·89
1·84
2·09

EXAMPLE 1 Margo is buying school supplies. The items and the prices are listed below. Find the total cost.

| poster board | $1.05 | construction paper | $0.89 |
| markers | $1.84 | glue | $2.09 |

$1.05   *Align the decimal points.*
0.89
1.84
+ 2.09
─────
$5.87   *Place the decimal point. Then add.*

So, the total cost of the supplies is $5.87.

REMEMBER:
You can add zeros to the right of a decimal without changing its value.
2.75 = 2.750
17 = 17.0
See page H12, Key Skill 21.

EXAMPLE 2 Add. 3.24 + 1.5 + 0.425

3.24		3.240	*Align the decimal points.*
1.5	or	1.500	*You can use zeros as placeholders.*
+ 0.425		+ 0.425	
		5.165	*Add.*

So, the sum is 5.165.

• Find the sum. 1.2 + 0.812 + 33 **35.012**

IDEA FILE

HISTORY CONNECTION

When Carl placed four coins on the table, he saw the faces of these famous political leaders: Thomas Jefferson, George Washington, Abraham Lincoln, and Franklin Roosevelt. What is the value of his coins? $0.41

• Have students find out what images were on the coins before those of the Presidents. Have them make a display that shows their findings.

Modalities Auditory, Kinesthetic, Visual

RETEACH 3.1

Name _____

LESSON 3.1

Adding Decimals

Ian is buying school supplies. Find the total cost.

| Binder | $3.49 | Paper | $2.79 |
| Pen | $0.88 | Highlighter | $0.98 |

Step 1: List the items in a column. Remember to align the decimal points.
$3.49
0.88
2.79
+ 0.98

Step 2: Write the decimal point for the answer. Place it directly under the other points.

Step 3: Add as you would whole numbers. Remember to regroup if needed.
$3.49
0.88
2.79
+ 0.98
─────
$8.14

The total cost of Ian's school supplies is $8.14.

Solve.

1. Rita is buying ingredients to make a birthday cake for her brother.

| Sugar | $1.79 | Flour | $0.89 |
| Eggs | $1.09 | Vanilla | $2.59 |

What is the total cost of the items? __$6.36__

2. Tom and Maria are going to have a car wash to raise money for the Red Cross. They bought these items.

| Bucket | $3.29 | Window cleaner | $1.09 |
| Washing solution | $4.09 | Paper towels | $0.59 |

How much did they spend? __$9.06__

Find the sum.

3. 2.89 + 1.65 + 3.86 __8.4__
4. 4.62 + 7.89 + 9.17 __21.68__
5. 2.891 + 3.006 + 2.861 __8.758__
6. 18.21 + 6.85 + 2.77 __27.83__
7. 1.0791 + 4.0689 __5.148__
8. 2.478 + 6.811 + 7.222 __16.511__

Use with text pages 54–55.

TAKE ANOTHER LOOK R11

GUIDED PRACTICE

Round to the nearest whole number.

1. 0.2 **0** **2.** 1.55 **2** **3.** 1.75 **2** **4.** 5.42 **5** **5.** 10.65 **11** **6.** 20.51 **21**

Estimate the sum. **Possible estimates are given.**

7. 2.4 + 12.9 **15** **8.** 0.18 + 4.37 + 6.82 **11** **9.** 7.561 + 3.019 + 1.775 **13**

Find the sum.

10. 0.33 + 0.07
0.4
11. 1.05 + 1.45
2.5
12. 2.8 + 3.55 + 1.4
7.75
13. 22.5 + 11.62 + 12
46.12

INDEPENDENT PRACTICE

Estimate the sum. **Possible estimates are given.**

1. 0.3 + 0.6 **1** **2.** 1.4 + 2.7 **4** **3.** 7.1 + 4.5 **12** **4.** 0.77 + 3.24 **4**

5. 4.09 + 5.89 **10** **6.** 6.95 + 2.44 **9** **7.** 0.392 + 1.021 **1** **8.** 9.356 + 6.811 **16**

Find the sum.

9. 0.28 + 7.52 **7.8** **10.** 3.95 + 4.16 **8.11** **11.** 6.03 + 0.29 **6.32** **12.** 5.41 + 3.20 **8.61**

13. 0.431 + 1.549 + 2.017
3.997
14. 9.365 + 40.271 + 1.537
51.173
15. 3.1428 + 90.6218
93.7646
16. 12.5 + 8.64
21.14
17. 0.94 + 1.237
2.177
18. $21.18 + $2
$23.18
19. $420.52 + $35.48
$456.00
20. 2.4 + 46 + 8.15
56.55
21. 0.21 + 8.99 + 53.614
62.814
22. $6 + $118.59 + $0.35
$124.94

23.
211.32
946.8
+ 63.754
1,221.874

24.
2.49
10.9521
+45.1
58.5421

25.
90.2
100.38
1.341
+1,082.6219
1,274.5429

26.
1.0981
96.235
62.1
+11.74
171.1731

Problem-Solving Applications

27. The computer club has raised $120.95 for new software. A math program costs $39.50, a game costs $24.75, and a reading program costs $66.29. About how much more money does the club need to pay for the software? **about $10**

28. Mr. Ross made a list of computer hardware he found on sale. A mouse pad was $4.75, a 10-pack of diskettes was $5.89, and a color printer was $249.79. How much money will he need to buy the hardware? **$260.43**

29. ▣ **WRITE ABOUT IT** Explain how to add two or more decimals. Show an example in your explanation. **Check students' explanations and examples.**

Portfolio

55

PRACTICE 3.1

Name _____
LESSON 3.1

Adding Decimals
Estimate the sum. **Possible estimates are given.**
1. 1.6 + 3.3 __5__   2. 0.2 + 0.9 __1__   3. 8.1 + 7.5 __16__   4. 5.16 + 0.58 __6__
5. 9.14 + 8.68 __18__   6. 7.58 + 3.25 __11__   7. 0.291 + 1.833 __2__   8. 4.693 + 9.711 __15__

Find the sum.
9. 0.34 + 8.19 __8.53__   10. 6.92 + 3.55 __10.47__   11. 0.418 + 3.291 __3.709__   12. 8.93 + 2.60 __11.53__
13. 4.89 + 2.45 __7.34__   14. 0.68 + 7.12 __7.8__   15. 1.681 + 2.899 __4.58__   16. 7.86 + 4.20 __12.06__
17. 1.246 + 2.081 __3.327__   18. 59.328 + 1.294 __60.622__   19. 5.0804 + 23.7381 __28.8185__
20. 446.09   21. 8.71   22. 23.75   23. 3.056
    811.36   13.99   873.33   28.1174
    +73.52   +67.2   +2,586.02   1,691.396
    __1,330.97__ __89.9__   __3,483.10__ + 44.21
                                           __1,766.7794__

Mixed Applications
24. The Garden Club has raised $213.42 for a new garden in the park. Plants for the garden cost $178.95, fencing costs $87.99, and fertilizer costs $44.15. About how much more money does the club need to cover the costs? __about $100__
25. Ryan read 8 pages of a book today. He plans to increase the number of pages he reads each day by 4. How many pages will he read four days from now? __24 pages__
26. Beth needs soccer equipment. Shoes cost $39.95, shorts cost $19.95, and socks cost $2.79. How much money will she need to buy the equipment? __$62.69__
27. This year, 432 students signed up to play intramural sports. Each team needs 18 players. How many teams are there? __24 teams__

Use with text pages 54–55.   **ON MY OWN** P11

ENRICHMENT 3.1

Name _____
LESSON 3.1

Shopping at the School Store

SALE FLYER			
Bookcover	$0.25	Note Cards	$0.98
Eraser	$0.15	Pen	$0.79
Paper	$1.89	Set of Markers	$3.95
Gym Shorts	$4.59	School Sweatshirt	$14.79
Highlighter	$1.29	School T-shirt	$5.98

The Heartsville School Store is having a back-to-school sale. Use the sale flyer above to solve.

1. Alana bought five different items. The total cost was $13.09. What were the items?
__bookcover, note cards, highlighter, school T-shirt, and gym shorts__

2. Zack bought two items. The total cost was more than $25, but less than $30. What did he buy? How much did he spend?
__2 school sweatshirts; $29.58__

3. Rico bought 2 each of three different items. They cost $2.38. What were the items?
__pens, erasers, and bookcovers__

4. Adam spent $11.22 for four different items. What did he buy?
__pen, paper, set of markers, and gym shorts__

5. Lonnie spent more than $5.00 but less than $5.50 on six items. What did Lonnie buy?
__eraser, pen, bookcover, note cards, paper, and highlighter__

6. Donna bought 3 each of two different items. She spent a total of $3.69. What did Donna buy?
__bookcovers and note cards__

7. Name two combinations of five items. Each combination should have a total cost greater than $3, but less than $4. You may include more than one of some items.
__Possible answers: 1 bookcover, 2 erasers, 1 paper, 1 highlighter; 3 pens, 1 note cards, 1 bookcover__

8. Name two combinations of four different items. Each combination should have a total cost greater than $16, but less than $20.
__Possible answers: school sweatshirt, gym shorts, bookcover, and eraser; gym shorts, school T-shirt, set of markers, and paper__

Use with text pages 54–55.   **STRETCH YOUR THINKING** E11

---

2 TEACH/PRACTICE

Throughout this lesson, emphasize the importance of aligning the decimal points and using zeros as placeholders.

GUIDED PRACTICE
After students complete the exercises, go over them orally before assigning the Independent Practice.

INDEPENDENT PRACTICE
To assess student understanding of estimating sums and adding decimals, check students' answers to Exercises 1–9.

💡 **CRITICAL THINKING** How many zeros can be placed after the last digit to the right of a decimal point without changing the value of the decimal? **an unlimited number**

ADDITIONAL EXAMPLES

Example 1, p. 54

At the Westward Ho Gift Shop, Jenny bought a hat for $8.75, spurs for $7.99, and postcards for $0.60. How much did she spend? **$17.34**

Example 2, p. 54

Add. 0.367 + 2.4 + 1.59 **4.357**

3 WRAP UP

Use these questions to help students focus on the big ideas of the lesson.

- **At the flea market, Tara bought an old comic book for $0.75, a used gumball machine for $8.50, and a camera for $15. How much did she spend in all?** $24.25

✓ *Assessment Tip*

Have students write the answer.

- **How does using zeros as placeholders help you add decimals with different place values?** Possible answer: by helping to keep the place values aligned

DAILY QUIZ
Estimate the sum.

1. 4.56 + 0.77 **6**
2. 3.428 + 21.76 + 0.22 **25**

Find the sum.

3. 4.9 + 18.8 **23.7** **4.** 0.7 + 2.045 **2.745**
5. 0.45 + 3.2 + 45.683 **49.333**

55

GRADE 5	GRADE 6	GRADE 7
Use models to subtract decimals.	Subtract decimals.	Estimate differences of whole numbers and decimals.

ORGANIZER

OBJECTIVE To subtract decimals
ASSIGNMENTS Independent Practice
- ■ *Basic* 1–18, even; 19–27; 28–37
- ■ *Average* 1–27; 28–37
- ■ *Advanced* 9–27; 28–37

PROGRAM RESOURCES
Reteach 3.2 Practice 3.2 Enrichment 3.2
Math Fun • Three in a Row, p. 68

PROBLEM OF THE DAY

Transparency 3

Carrie and her four friends noticed that the change each got from a $10.00 bill had the same digits as the amount spent. Carrie spent $4.55 and got $5.45 in change. Each of her friends spent different amounts. How much did each spend? What was the change? $9.05, $0.95; $1.85, $8.15; $2.75, $7.25; $3.65, $6.35

Solution Tab A, p. A3

QUICK CHECK

Find the sum.

1. 19.4 + 0.8   20.2
2. 7.07 + 11.6   18.67
3. 0.63 + 21.06   21.69
4. 5.07 + 4.5   9.57
5. 55.5 + 5.5   61
6. 3.09 + 3.09   6.18

1 MOTIVATE

Organize students into pairs. Review with students how to use a number line to subtract decimals. Discuss this example.

1.3 − 0.7 = 0.6

Repeat this activity with 0.9 − 0.2 and 1.3 − 0.8. 0.7, 0.5

3.2

WHAT YOU'LL LEARN
How to subtract decimals

WHY LEARN THIS?
To know how much change to expect when you make a purchase

Subtracting Decimals

How does knowing how to subtract decimals help you determine if you received the correct amount of change from a purchase?

Andrew bought new basketball shoes for $53.49. He gave the cashier $55.00. How much change should Andrew get from the cashier?

$55.00
− 53.49   *Align the decimal points.*

$55.00
− 53.49
$ 1.51   *Place the decimal point.*
          Subtract.

So, Andrew should get $1.51 in change from the cashier.

- Suppose Andrew gave the cashier $60.00. How much change should he get? **$6.51**

Sometimes adding zeros to the right of a decimal is helpful when subtracting.

EXAMPLE Find the difference. 101.2 − 8.73

101.20   *Align the decimal points.*
− 8.73   *Add a zero to 101.2.*

101.20   *Place the decimal point.*
− 8.73
 92.47   *Subtract.*

- How can you check your subtraction? **by adding**

Consumer Link

When you pay for an item, you can use a strategy to avoid getting a pocketful of change or to avoid getting pennies. For example, if the total is $7.04, you can give the cashier $10.04 and get three $1 bills in change. Mark bought a magazine for $2.49. He paid the cashier $3.09. How much change should he get back? **$0.60**

GUIDED PRACTICE

Copy the problem. Place the decimal point in the difference.

1. 9.7 − 3.01 = 669
   6.69
2. 37.5 − 0.19 = 3731
   37.31

Find the difference.

3. $4.38 − $2.26
   $2.12
4. $54.99 − $12.81
   $42.18
5. $80.93 − $60.51
   $20.42
6. 3.2 − 2.6
   0.6
7. $6.18 − $5.55
   $0.63
8. 10.712 − 1.53
   9.182

IDEA FILE

SOCIAL STUDIES CONNECTION

Have students write subtraction word problems using the data below.

Public Transit System	Passengers per Year
Bombay	1.407 billion
Moscow	2.426 billion
New York City	1.006 billion
Paris	1.156 billion
Tokyo	1.694 billion

Modality Visual

RETEACH 3.2

Name _____

LESSON 3.2

Subtracting Decimals

If you know how to subtract decimals, you can make sure you receive the correct change when making a purchase.

Joe bought a new baseball glove for $69.95. He gave the clerk four $20 bills. How much change should he get from the cashier?

Step 1: Write the amount of money Joe gave the cashier.
four $20 bills = $80.00

Step 2: Write the price of the glove below it. Be sure to align the decimal points.
$80.00
− 69.95

Step 3: Carry the decimal point down to the answer. Place it directly under the other decimal points. Subtract.
$80.00
− 69.95
$10.05

Joe should receive $10.05.

Complete to solve.

1. Renee bought a pair of shoes for $22.95. She gave the clerk a $20 bill and a $10 bill.
   a. How much money did Renee give the clerk?   **$30.00**
   b. How much do the shoes cost?   **$22.95**
   c. What subtraction problem will you use to find her change?   **$30.00 − 22.95**
   d. How much change should Renee receive?   **$7.05**

Solve.

2. Paula is starting a new exercise program. She wants to walk 5.5 km every day. She walks 2.8 km to and from school. How much farther must she walk each day to reach her goal?   **2.7 km farther**

3. Chen is making muffins. He uses 0.35 L of milk for the batch. If he began with 1.50 L of milk, how much is left?   **1.15 L**

4. Jerry is making hamburgers. He started with 2.75 lb of hamburger and has used 1.6 lb. How much is left?   **1.15 lb**

5. Pat has 4.50 m of plywood. She will use 2.95 m to make bookshelves. How much will be left?   **1.55 m**

R12   TAKE ANOTHER LOOK   Use with text pages 56–57.

MEETING INDIVIDUAL NEEDS

INDEPENDENT PRACTICE

Find the difference.

1. 9.3 − 6.1 **3.2**   **2.** 11.2 − 1.7 **9.5**   **3.** 5.95 − 3.26 **2.69**   **4.** $12.41 − $8.04 **$4.37**

5. 4.981 − 3.235 **1.746**   **6.** 7.306 − 5.872 **1.434**   **7.** 6.0781 − 2.3195 **3.7586**   **8.** 18.6902 − 9.2571 **9.4331**

9. 12.304 − 6.92 **5.384**   **10.** 16.9278 − 5.051 **11.8768**   **11.** 17.45 − 4.207 **13.243**   **12.** 38.124 − 2.3819 **35.7421**

13.
$$\begin{array}{r} 3 \\ -\ 2.894 \\ \hline 0.106 \end{array}$$

14.
$$\begin{array}{r} 48.28 \\ -\ 16.1790 \\ \hline 32.1010 \end{array}$$

15.
$$\begin{array}{r} 982.3024 \\ -\ 890.5196 \\ \hline 91.7828 \end{array}$$

Find the missing number.

16. 2.45 + ■ = 5.262 **2.812**   **17.** 61.21 + ■ = 88.163 **26.953**   **18.** 3.22 + ■ = 7.24 **4.02**

19. 48.900 − ■ = 10.208 **38.692**   **20.** 7.560 − ■ = 4.211 **3.349**   **21.** 100.25 − ■ = 0.82 **99.43**

Problem-Solving Applications

22. Jake's batting average is 0.325. Last year it was 0.235. What is the difference between his batting average last year and his batting average this year? **0.090**

23. Jan bought 3.3 lb of potatoes. Paul bought a 5-lb bag of potatoes. How many more pounds of potatoes did Paul buy than Jan? **1.7 lb**

24. Valerie is balancing her checkbook. She has a balance of $482.79. She writes a check for $22.18. What is the new balance in her checkbook? **$460.61**

25. Mariah bought a video for $32.09. How much change will she get back if she gives the cashier $40.09? **$8**

26. Arnold drove 108.2 mi to his aunt's house. On the way back, he took a different route and drove 92.6 mi. How much shorter was his route back? **15.6 mi**

27. ✏ **WRITE ABOUT IT** Write a word problem that can be solved by subtracting decimals. Exchange with a classmate and solve. **Problems will vary.**

Portfolio

Mixed Review

Multiply.

28. 32 × 15 **480**   **29.** 149 × 53 **7,897**   **30.** 201 × 90 **18,090**   **31.** 698 × 712 **496,976**

Use the Distributive Property to find the product.

32. 25 × 8 (20 × 8) + (5 × 8) = **200**   **33.** 35 × 6 (30 × 6) + (5 × 6) = **210**   **34.** 55 × 9 (50 × 9) + (5 × 9) = **495**   **35.** 89 × 4 (80 × 4) + (9 × 4) = **356**

36. **CONSUMER MATH** Tanya has $20.00 for grocery shopping. Milk costs $1.45, bread costs $1.05, cheese costs $4.50, and juice costs $5.00. How much more money does she need to get 2 of each product? **$4.00**

37. **PATTERNS** Sean delivers 18 newspapers after school. He plans to increase his deliveries by 6 newspapers a week. How many newspapers will he be delivering at the end of 4 weeks? **42 newspapers**

PRACTICE 3.2

Name _____

LESSON 3.2

Subtracting Decimals

Find the difference.

1. 8.7 − 4.2   2. 13.2 − 5.9   3. 5.41 − 1.36   4. $15.93 − $7.08
    4.5   **7.3**   **4.05**   **$8.85**

5. 5.962 − 1.748   6. 4.036 − 2.751   7. 8.1163 − 3.0948
    4.214   **1.285**   **5.0215**

8. 17.1053 − 12.7559   9. 15.082 − 4.19   10. 18.1429 − 6.204
    4.3494   **10.892**   **11.9389**

11. 14.16 − 6.385   12. 45.324 − 7.0871   13. 16.076 − 3.28
    7.775   **38.2369**   **12.796**

14.
$$\begin{array}{r} 5 \\ -\ 3.218 \\ \hline 1.782 \end{array}$$
15.
$$\begin{array}{r} 54.08 \\ -\ 29.7561 \\ \hline 24.3239 \end{array}$$
16.
$$\begin{array}{r} 14 \\ -\ 5.39 \\ \hline 8.61 \end{array}$$
17.
$$\begin{array}{r} 631.5039 \\ -\ 420.1417 \\ \hline 211.3622 \end{array}$$

18.
$$\begin{array}{r} 21 \\ -\ 9.45 \\ \hline 11.55 \end{array}$$
19.
$$\begin{array}{r} 89.01 \\ -\ 67.56 \\ \hline 21.45 \end{array}$$
20.
$$\begin{array}{r} 25.8 \\ -\ 17.226 \\ \hline 8.574 \end{array}$$
21.
$$\begin{array}{r} 71.043 \\ -\ 58.6492 \\ \hline 12.3938 \end{array}$$

Find the missing number.

22. 1.83 + **0.98** = 2.81   23. 10.13 + **4.96** = 15.09

24. 12.379 + **11.021** = 23.4   25. 16.007 − **20.576** = 36.583

Mixed Applications

26. Mel uses 2.8 lb from a 10-lb bag of potatoes to make hash browns. How many pounds of potatoes are left?

    7.2 lb

27. The distance from Joan's house to school is 6.3 mi. The distance from Rick's house to school is 1.9 mi. How much closer to the school is Rick's house?

    4.4 mi closer

28. Zack has $6.00 for grocery shopping. Juice costs $2.29, bread costs $1.29, and eggs cost $1.09. How much more money does he need to get 2 of each item?

    $3.34 more

29. Carla does 22 sit-ups a day. She plans to increase her exercise by 4 sit-ups a day. How many sit-ups will she be doing at the end of 3 days?

    34 sit-ups

P12 ON MY OWN   Use with text pages 56–57.

ENRICHMENT 3.2

Name _____

LESSON 3.2

Pattern Practice

Identify the subtraction rule that was used to create each pattern. Then name the next three decimals.

1. 94.8, 94.3, 92.8, 90.3, **86.8** , **82.3** , **76.8** . . .
   Rule: _____ Subtract 0.5, then 1.5, then 2.5, and so on.

2. 78.26, 77.15, 74.93, 71.6, **67.16** , **61.61** , **54.95** . . .
   Rule: _____ Subtract 1.11, then 2.22, then 3.33, and so on.

3. 52.3, 50.2, 48, 45.7, **43.3** , **40.8** , **38.2** . . .
   Rule: _____ Subtract 2.1, then 2.2, then 2.3, then 2.4, and so on.

4. 27.5, 26, 23.5, 22, **19.5** , **18** , **15.5** . . .
   Rule: _____ Subtract 1.5, then 2.5, then 1.5, and so on.

5. 64.23, 58.53, 53.73, 48.03, **43.23** , **37.53** , **32.73** . . .
   Rule: _____ Subtract 5.7, then 4.8, then 5.7, and so on.

6. 103.26, 93.27, 84.39, 76.62, **69.96** , **64.41** , **59.97** . . .
   Rule: _____ Subtract 9.99, then 8.88, then 7.77, and so on.

7. 815.634, 803.289, 748.968, 736.623, **682.302** , **669.957** , **615.636** . . .
   Rule: _____ Subtract 12.345, then 54.321, then 12.345, and so on.

8. 128.003, 121.757, 113.558, 107.312, **99.113** , **92.867** , **84.668** . . .
   Rule: _____ Subtract 6.246, then 8.199, then 6.246, and so on.

9. 38.15, 35.25, 29.85, 26.95, **21.55** , **18.65** , **13.25** . . .
   Rule: _____ Subtract 2.9, then 5.4, then 2.9, and so on.

10. Make up two pattern problems of your own. Exchange papers with a classmate and solve. **Check students' work.**

_____

E12 STRETCH YOUR THINKING   Use with text pages 56–57.

As students broaden their decimal skills to include subtraction, their consumer skills will continue to grow.

GUIDED PRACTICE

Have volunteers complete Exercises 1–8 on the board. Discuss as a class.

INDEPENDENT PRACTICE

To assess students' understanding of subtracting decimals, check the answers to Exercises 1–26.

COMMON ERROR ALERT

Some students may neglect to use zeros as placeholders when subtracting decimals. Emphasize the importance of using zeros as placeholders so all the decimals have the same place value.

TEACHING TIPS

Have students use *Calculating Crew* and visit Wanda Wavelet at *Nautical Number Line* to practice adding and subtracting decimals on a number line. See Grow Slide Level R.

To practice adding and subtracting decimals using algorithms, have students visit *Number Heroes* and join Star Brilliant at *Quizzo.* See Grow Slide Levels Q and R.

ADDITIONAL EXAMPLE

Example, p. 56

Find the difference. 76.8 − 67.34 9.46

3 WRAP UP

Use these questions to help students focus on the big ideas of the lesson.

- **The gas tank in Max's car holds 16 gal. When he had the tank filled, he got 15.3 gal. How close was the tank to being empty?** 0.7 gal

✔ *Assessment Tip*

Have students write the answer.

- **Gary says that 24.3 − 17.56 = 6.86. Explain what he did wrong.** Possible explanation: He forgot to place a zero after the 3.

DAILY QUIZ

Find the difference.

1. 8.5 − 7.8 0.7   **2.** 5.76 − 3.2 2.56

3. 6.3 − 4.22 2.08   **4.** 51.3 − 5.13 46.17

5. 145.89 − 65.036 80.854

INTERVENTION

MEETING INDIVIDUAL NEEDS

Placing the Decimal Point

Use these strategies to review the skills required for placing the decimal point when multiplying and dividing decimals.

Multiple Intelligences Logical/Mathematical

Modalities Auditory, Kinesthetic, Visual

Lesson Connection Students will multiply and divide decimals in Lessons 3.3 and 3.4.

Program Resources *Intervention and Extension Copying Masters*, p. 2

Troubleshooting

WHAT STUDENTS NEED TO KNOW	HOW TO HELP	FOLLOW UP
How to multiply whole numbers mentally by powers of 10	Have students multiply mentally by powers of 10. **Example** 20 × 1,000 = 20,000 — 1 zero, 3 zeros, 4 zeros. Add the numbers of zeros in the factors. 1 + 3 = 4 Be sure there are 4 zeros in the product.	Other problems to solve: 3 × 10,000 30,000 12 × 1,000 12,000 450 × 100 45,000 See Key Skill 22, Student Handbook, page H12.
How to divide whole numbers mentally by powers of 10	Have students divide mentally by powers of 10. **Example** 38,000 ÷ 100 = 380 — 3 zeros, 2 zeros, 1 zero. Subtract the numbers of zeros. 3 − 2 = 1. Be sure there is 1 zero in the quotient.	Other problems to solve: 500 ÷ 10 50 34,000 ÷ 1,000 34 170,000 ÷ 100 1,700 See Key Skill 22, Student Handbook, page H12.
How to multiply and divide mentally	Have students combine the technique of counting zeros with basic multiplication and division facts. **Examples** 200 × 5,000 = 1,000,000 — 2 zeros, 3 zeros, 2 × 5 2 + 3 = 5 zeros. 32,000 ÷ 80 = 400 — 3 zeros, 1 zero, 32 ÷ 8 3 − 1 = 2 zeros.	Other problems to solve: 300 × 400 120,000 500 × 400 200,000 5,600 ÷ 80 70 48,000,000 ÷ 6,000 8,000

WHAT STUDENTS NEED TO KNOW	HOW TO HELP	FOLLOW UP
That estimating the product can help you place the decimal point when you multiply decimals	Have students estimate the product of decimals and then use the estimate to place the decimal point. **Example** Use an estimate to place the decimal point in the product. $3.1 \times 47.89 = 148459$ **Think:** $3 \times 50 = 150$, so the product is close to 150. Place the decimal point after the 8. The answer is 148.459.	Other problems to solve: Use an estimate to place the decimal point. $5.9 \times 13.7 = 8083$ 80.83 $73{,}000 \times 1.07 = 78110$ 78,110
That estimating the quotient can help you place the decimal point when you divide decimals.	Have students estimate a quotient, and then use the estimate to place the decimal point. **Example** Use an estimate to place the decimal point in the quotient. $750.3 \div 20.5 = 36600$ **Think:** $800 \div 20 = 40$, so the answer is about 40. Place the decimal point after the 36. The answer is 36.600, or 36.6.	Other problems to solve: Use an estimate to place the decimal point. $392.7 \div 10.2 = 3850$ 38.5 $579.156 \div 2.89 = 200400$ 200.4

Intervention and Extension Copying Masters, p. 2

INTERVENTION STRATEGY

Name _____   CHAPTER 3

Placing the Decimal Point

Multiply.

1. $56 \times 1{,}000$	2. 11×100	3. $205 \times 10{,}000$	4. 9×100
56,000	1,100	2,050,000	900

5. 600×300	6. $40 \times 2{,}000$	7. $5 \times 4{,}000$	8. $3 \times 20{,}000$
180,000	80,000	20,000	60,000

Divide.

9. $36{,}000 \div 10$	10. $5{,}000 \div 100$	11. $403{,}000 \div 1{,}000$	12. $300{,}000 \div 10{,}000$
3,600	50	403	30

13. $42{,}000 \div 700$	14. $360 \div 60$	15. $50{,}000 \div 50$	16. $120{,}000 \div 300$
60	6	1,000	400

Estimate whether the answer is *more than 10 or less than 10*

17. 3.4×5.17	18. $21.8 \div 3.06$	19. $274.1 \div 4.9$	20. 19.7×1.14
more than 10	less than 10	more than 10	more than 10

Estimate each answer. Then place the decimal point.
Possible estimates are given.

21. $5.6 \times 20.5 = 1148$	22. $30.4 \times 27.3 = 82992$
120; 114.8	900; 829.92

23. $85.69 \div 1.9 = 45100$	24. $220.4 \div 30.4 = 72500$
40; 45.1	7; 7.25

25. $5.84 \times 720 = 42048$	26. $197.824 \div 0.88 = 22480$
4,200; 4,204.8	200; 224.8

2   Intervention and Extension

58B

3.3

Multiplying Decimals

| GRADE 5 Find patterns in decimal factors and products. | GRADE 6 Multiply decimals. | GRADE 7 Multiply whole numbers and decimals. |

ORGANIZER

OBJECTIVE To multiply decimals
ASSIGNMENTS Independent Practice
- *Basic* 1–39, even; 40–47; 48–55
- *Average* 1–47; 48–55
- *Advanced* 12–47; 48–55

PROGRAM RESOURCES
Reteach 3.3 Practice 3.3 Enrichment 3.3
Math Fun • $1,000 Winner!, p. 68

PROBLEM OF THE DAY

Transparency 3

The sum of two decimal numbers is 9.3. Their difference is 4.3, and their product is 17.00. What are the numbers? 2.5 and 6.8

Solution Tab A, p. A3

QUICK CHECK
Find the product.

1. 4 × 23 92
2. 3 × 15 45
3. 5 × 759 3,795
4. 6 × 2,747 16,482
5. 32 × 48 1,536
6. 467 × 26 12,142

1 MOTIVATE

Discuss with students the relationship between multiplication and addition. Have them write a multiplication expression for this addition expression:

4.7 + 4.7 + 4.7 + 4.7 + 4.7. 5 × 4.7

- **What does the first factor represent?** the number of times the addend is repeated
- **What does the second factor represent?** the repeated addend
- **Which expression do you think would be easier to solve?** 5 × 4.7 **Why?** Possible answer: because it takes less time to do the computation

Have students write multiplication expressions for these addition expressions:

1. 4.9 + 4.9 + 4.9 + 4.9 + 4.9 5 × 4.9
2. 7.89 + 7.89 + 7.89 + 7.89 4 × 7.89
3. 36.08 + 36.08 + 36.08 3 × 36.08

58

WHAT YOU'LL LEARN
How to multiply decimals

WHY LEARN THIS?
To find the total amount of money you will earn for working a number of hours

CULTURAL LINK

You can send a letter anywhere in the world. All you need is a stamp. Every country has some type of postal service. The country with the greatest number of post offices is India. The postage rate to send a letter from the U.S. to any country across an ocean is $0.60 for the first 0.5 ounce and $0.40 for each additional ounce. How much would it cost to send a letter with photos that weighs 3 ounces to India? $1.60

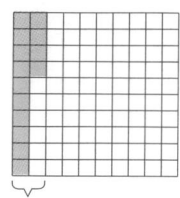
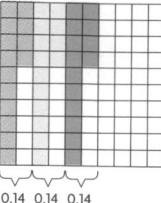

58 Chapter 3

You can use a model to find the product of a decimal and a whole number.

ACTIVITY

WHAT YOU'LL NEED: colored pencils, decimal squares

- To find 3 × 0.14, shade 0.14, or 14 squares, three times. Use a different color each time.
- Count the number of shaded squares. What is 3 × 0.14? **0.42**

0.14   0.14 0.14 0.14

- Use a decimal square to find 2 × 0.16. **Check students' models; 0.32.**

Sometimes, when the factors are larger, as in 8 × 1.52, it is easier to use paper and pencil to find the product of a whole number and a decimal.

EXAMPLE 1 Mark collects stamps. He bought 9 different stamps at $1.12 each. How much did he spend?

Estimate the products.

Multiply as with whole numbers.

Since the estimate is 9, place the decimal point after the 10.

$1.12 × 9 → $1 × 9 = 9

```
  1 1
  $1.12          $1.12
×    9          ×    9
───────        ───────
 $10 08         $10.08
```

So, Mark paid $10.08 for 9 stamps.

- Suppose Mark had paid $0.82 for each stamp. How much would he have paid for 9 stamps? **$7.38**

- **CRITICAL THINKING** How does an estimate help you determine whether the product of two factors is reasonable? **Possible answer: An estimate is easier to calculate and is close to the product.**

RETEACH 3.3

Name _____

LESSON 3.3

Multiplying Decimals

Marty worked 26.5 hours this week. He earns $6.40 per hour. How much money did Marty earn this week?
To solve, you need to find the product 26.5 × 6.40.

Step 1: Multiply as you would with whole numbers.
```
  $6.40
× 26.5
──────
169600
```

Step 2: Count the number of decimal places in the factors.
```
 $6.40  →  2 decimal places
× 26.5  →  1 decimal place
──────
169600
```

Step 3: Starting at the right side of the answer, count over that number of places. This is where the decimal point is placed.
```
   $6.40
 × 26.5
────────
$169.600
```

Marty earned $169.60 this week.

Complete to solve.
1. Sean worked 34 hours this week. He earns $9.25 an hour. How much money did Sean earn this week?
   a. What multiplication problem will you use? _34 × $9.25_
   b. How many decimal places are in the factors? _2_
   c. How much did Sean earn? _$314.50_
2. Denise earns $4.65 an hour. Last week she worked 17 hours. How much did she earn last week?
   a. What multiplication problem will you use? _17 × $4.65_
   b. How many decimal places are in the factors? _2_
   c. How much did Denise earn? _$79.05_

Solve.
3. Tyler has 3 pieces of wire. Each piece is 18.64 cm long. If the pieces are placed end-to-end, what distance will they cover? _55.92 cm_
4. The chef bought 12 packages of rice. Each package weighed 0.75 kg. How much did they weigh in all? _9 kg_

Use with text pages 58–61.   **TAKE ANOTHER LOOK R13**

PRACTICE 3.3

Name _____

LESSON 3.3

Multiplying Decimals

Estimate the product. **Possible estimates are given.**
1. 6.3 × 0.75 _6_
2. 9.7 × 48.8 _490_
3. 5.96 × 62.15 _360_
4. 37.6 × 8.3 _320_
5. 32.08 × 7.3 _210_
6. 428.9 × 5.6 _2,400_
7. 897.35 × 5.3 _4,500_
8. 186.45 × 9.6 _2,000_

Place the decimal point in the product.
9. 6.17 × 8.2 = 50594 _50.594_
10. 24.01 × 8.51 = 2043251 _204.3251_
11. 8.94 × 5.27 = 471138 _47.1138_
12. 8.04 × 1.7 = 13668 _13.668_
13. 19.6 × 5.8 = 11368 _113.68_
14. 30.7 × 8.33 = 255731 _255.731_

Find the product.
15. 5 × 0.9 _4.5_
16. 9 × 1.2 _10.8_
17. 4 × 3.47 _13.88_
18. $18.93 × 7 _$132.51_
19. $5.55 × 9 _$49.95_
20. 5 × 2.89 _14.45_
21. $31.82 × 4 _$127.28_
22. 4.61 × 8 _36.88_
23. $2.49 × 6 _$14.94_
24. 35.98 × 6.3 _226.674_
25. 73.02 × 9.1 _664.482_
26. 8.5 × 16.03 _136.255_

Mixed Applications
27. The Flower Club is selling plants for $7.95 each. If the club sells a total of 285 plants, how much money will they make? _$2,265.75_
28. Ronnie is selling handmade baby sweaters. She charges $29.50 for one sweater. If she sells 32 sweaters, how much money will she make? _$944.00_
29. This year 592 students signed up to play basketball. Each team needs 8 players. How many teams are there? _74 teams_
30. There are 104 students in the school band. They are traveling in cars to a performance. Each car will transport 4 band members. How many cars are needed? _26 cars_

Use with text pages 58–61.   **ON MY OWN P13**

Estimate the product. **Possible estimates are given.**

1. 3.62×7 **28** **2.** 2.15×8 **16** **3.** 4.04×5 **20** **4.** 6.82×9 **63**

Copy the problem. Use an estimate to place the decimal point in the product.

5. $9 \times 5.4 = 486$
   48.6
6. $7 \times 4.1 = 287$
   28.7
7. $22 \times 0.55 = 121$
   12.1
8. $7.32 \times 3 = 2196$
   21.96
9. $0.82 \times 5 = 41$
   4.1
10. $32.5 \times 6 = 195$
   195., or 195

Find the product.

11. 0.33×3 **0.99**
12. 0.42×2 **0.84**
13. 1.25×4 **5**

14. 3.23×8 **25.84**
15. 1.3×5 **6.5**
16. 28.84×7
   201.88

Multiplying a Decimal and a Decimal

You can use decimal squares or paper and pencil to find the product of two decimals.

EXAMPLE 2 Find 0.2×0.6.

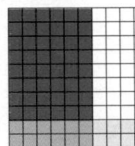

Shade 6 columns blue for 0.6.

Shade 2 rows yellow for 0.2.

The area in which the shading overlaps shows the product, or 0.2 of 0.6.

$$\begin{array}{r} 0.2 \\ \times\, 0.6 \\ \hline 0.12 \end{array}$$

So, $0.2 \times 0.6 = 0.12$.

• Find 0.5×0.3. **0.15**

Talk About It

• How do you determine the number of rows and columns to shade to find the product of 2 decimals using a decimal square? **The first factor is rows, the second factor is columns.**
• Look at the model. Is the product greater or less than either factor? **The product is less than either factor.**
• How is the product 0.2×0.6 different from the product 2×6? **The whole-number product is greater than either factor; the decimal product is less than either factor.**
• Look at the number of decimal places in the factors and in the product in Example 2. Write a rule for placing the decimal point in the product when multiplying decimals. **Add the number of decimal places in the factors, and then count that total number of places from the right in the product.**

59

Students begin this lesson by modeling decimal multiplication. They then make the transition to multiplying without models.

COMMON ERROR ALERT Some students may neglect to estimate a product before finding the exact product. Emphasize the importance of estimation as a way to avoid making mistakes when placing a decimal point in a product.

GUIDED PRACTICE
Complete Exercises 1–4 orally with the class. Then have volunteers work Exercises 11–13 on the board. Discuss as a class.

TEACHING TIP
To help students better understand the concept discussed in the first two questions of Talk About It, have them try this example: **Find 0.5 × 0.8. What fraction in simplest form is equal to 0.5?** $\frac{1}{2}$ **What is half of 0.8?** 0.4 **So, 0.5 × 0.8 = 0.4.** Repeat this activity with 0.5×0.4 and with 0.5×0.6. 0.2, 0.3

ADDITIONAL EXAMPLES

Example 1, p. 58

Once each year Jeff replaces the batteries in the smoke detectors in his home. Each battery costs $2.89. How much will Jeff spend on batteries for 6 smoke detectors? $17.34

Example 2, p. 59

Find 0.4×0.7.

Shade 4 columns blue for 0.4.

Shade 7 rows yellow for 0.7.

The area in which the shading overlaps shows the product, or 0.4 of 0.7.

$0.4 \times 0.7 = 0.28$

Remind students to place a zero before the decimal point if the product is less than 1.

Point out to students that when you multiply with money, sometimes the product has a decimal with a value of thousandths. If a zero is in the thousandths place, such as in Example 3, they can drop the zero so the product makes sense. If the decimal in the thousandths place is greater than 0, they can round the decimal to the nearest hundredth.

INDEPENDENT PRACTICE

To ensure student success, you may want to complete Exercises 1, 9, 12, 33, and 40 as a class before assigning the rest of the exercises.

💡CRITICAL THINKING **Tammy wrote to 13 people while she was away at camp. She spent a total of $3.44 for 20-cent postcard stamps and 32-cent letter stamps. How many postcards did she write? How many letters did she write?**
6 postcards; 7 letters

ADDITIONAL EXAMPLES

Example 3, p. 60

Roland needs 2.5 lb of pecans to make pies for the Boy Scout bake sale. Pecans sell for $5.40 per lb. How much will Roland pay for pecans? $13.50

Example 4, p. 60

Find the product. 0.036×0.45

$$
\begin{array}{r}
0.036 \leftarrow 3 \text{ decimal places} \\
\times\ 0.45 \leftarrow 2 \text{ decimal places} \\
\hline
180 \qquad \text{Multiply as with whole numbers.} \\
144 \qquad \\
\hline
0.01620 \qquad \text{Place the decimal point.}
\end{array}
$$

Since the answer must have five decimal places, you need one zero to the left of 1.

$0.036 \times 0.45 = 0.01620$ or 0.0162

You can place the decimal point in a product by estimating. Another method of placing the decimal point is to add the numbers of decimal places in the factors and then count that total number of places from the right in the product.

$$
\begin{array}{r}
0.8 \leftarrow \textit{1 decimal place} \\
\times 0.11 \leftarrow \textit{2 decimal places} \\
\hline
0.088 \leftarrow \textit{1 + 2, or 3 decimal places}
\end{array}
$$

EXAMPLE 3 George works 37.5 hr per week at the bank. He earns $7.70 an hour. How much does he earn in a week?

$$
\begin{array}{r}
7.70 \leftarrow 2 \text{ decimal places} \\
\times 37.5 \leftarrow 1 \text{ decimal place} \\
\hline
3850 \quad \textit{Multiply as with whole numbers.} \\
5390 \quad \\
2310 \quad \textit{Place the decimal point.} \\
\hline
288.750 \leftarrow 3 \text{ decimal places}
\end{array}
$$

So, George earns $288.75 in a week.

When you multiply decimals, sometimes you have to use zeros as placeholders.

EXAMPLE 4 Find the product 0.042×0.073.

$$
\begin{array}{r}
0.042 \leftarrow 3 \text{ decimal places} \\
\times 0.073 \leftarrow 3 \text{ decimal places} \\
\hline
126 \quad \textit{Multiply as with whole numbers.} \\
294 \quad \\
\hline
0.003066 \quad \textit{Place the decimal point.} \\
\textit{Since the answer must have 6 decimal places,} \\
\textit{you need 2 zeros to the left of 3.}
\end{array}
$$

So, $0.042 \times 0.073 = 0.003066$.

• Find the product. 9.73×0.84 **8.1732**

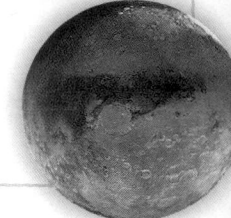

INDEPENDENT PRACTICE

Estimate the product. Possible estimates are given.

1. 0.95×4.2 **4**   **2.** 28.32×9.81 **280**   **3.** 51.42×8.16 **400**   **4.** 99.65×2.42 **200**

5. 4.82×20.19 **100**   **6.** 541.28×6.95 **3,500**   **7.** 603.95×9.1 **5,400**   **8.** 992.06×8.8 **9,000**

Copy the problem. Place the decimal point in the product.

9. $4.01 \times 5.6 = 22456$ **22.456**   **10.** $6.37 \times 2.91 = 185367$ **18.5367**   **11.** $20.4 \times 9.52 = 194208$ **194.208**

👥 IDEA FILE

CONSUMER CONNECTION

Have students estimate the total amount of money they spend on video rentals in one year. Tell them to multiply the average number of videos they rent per month by 12. Then have them multiply that number by the average video rental fee they pay. Have students compare estimates with those of their classmates.

Modality Visual

CAREER CONNECTION

Copy this ad on the board. Ask students: Which job pays the most per week? How much does it pay? lawn mowing; $72.75

PART-TIME SUMMER JOBS!	
Car Washing $4.95/hr 14 hours a week	**Lawn Mowing** $4.85/hr 15 hours a week
Bagging Groceries $5.05/hr 13 hours a week	

Modality Visual

Find the product.

12. 2×0.6 **1.2**

13. 5×2.4 **12**

14. 8×1.6 **12.8**

15. $\$3.20 \times 4$ **\$12.80**

16. 2×0.12 **0.24**

17. 8×0.11 **0.88**

18. $\$1.21 \times 5$ **\$6.05**

19. 4×2.62 **10.48**

20. 6×3.35 **20.1**

21. $\$8.46 \times 2$ **\$16.92**

22. 9.35×7 **65.45**

23. 6.04×8 **48.32**

24. 0.2×0.4 **0.08**

25. 6.3×0.9 **5.67**

26. 0.21×2.1 **0.441**

27. 3.21×4.5 **14.445**

28. 6.15×2.4 **14.76**

29. 4.08×1.35 **5.5080**

30. 6.21×0.95 **5.8995**

31. 35.42×2.33 **82.5286**

32. 24.63×1.09 **26.8467**

33. 29.147×5.61 **163.51467**

34. 0.189×2.09 **0.39501**

35. 2.354×1.92 **4.51968**

36. 118.001×0.37 **43.66037**

37. 148.9×0.006 **0.8934**

38. $1,200.5 \times 8.2$ **9,844.10**

39. $8,116.9 \times 1.402$ **11,379.8938**

Problem-Solving Applications

40. Band members are selling cookies for $2.15 a dozen. If they sell 55 dozen, how much money will they make? **$118.25**

41. Jennifer bought 4.8 lb of tomatoes at $1.25 per pound. How much did the tomatoes cost? **$6**

42. Keith has a storage room that is 5.2 m wide. He has 3 pieces of furniture that are each 1.6 m wide. Is there enough room for all of them? Explain. **yes; $1.6 \times 3 = 4.8$**

43. Joanne brought 4.5 bottles of soda to the town picnic. Each bottle will fill 10 cups. How many cups can she fill with soda? **45 cups**

44. The middle school is having a car wash for a fund-raiser. They charge $5.25 for each car. If they wash 45 cars, how much money will they earn? **$236.25**

45. Michael is making an apple pie. The recipe calls for 2.25 lb of apples. Apples are $1.60 per pound. How much will he pay for the apples? **$3.60**

46. Robbie wants 2 CDs that cost $8.99 each. He earns $5.00 a week mowing lawns. About how many weeks will he have to work to earn enough to buy the CDs? **about 4 weeks**

47. ▬ **WRITE ABOUT IT** Explain how you know how many decimal places to put in the product. **by adding the numbers of decimal places in the factors**

Mixed Review

48. $240 \div 6$ **40**

49. $320 \div 80$ **4**

50. $210 \div 70$ **3**

51. $600 \div 15$ **40**

Use mental math or compensation to solve.

52. $15 + 26 + 5$ **46**

53. $44 + 35 + 6$ **85**

54. $24 - 16$ **8**

55. **DECISION MAKING** Chris saves $5 a week. He has saved $150 toward a new bike that costs $200. Should he continue saving, or should he put $100 down and pay $20 a month for 6 months? Explain. **He should continue saving; the cost will be less.**

56. **SPORTS** This year 240 students play after-school sports. Each team needs 15 players. How many teams are there? **16 teams**

MORE PRACTICE Lesson 3.3, page H46

61

SCIENCE CONNECTION

Have students research what is in landfills in their state. Then have them compare their findings with what is in a typical landfill in the United States. You may wish to point out that the decimals sum to 1, which represents the whole (or entire) landfill.

Contents of Typical U.S. Landfills	
Metal	0.08
Plastic	0.24
Paper	0.3
Other Trash	0.19
Food and Yard Waste	0.11
Rubber and Leather	0.08

Modality Visual

INTERDISCIPLINARY CONNECTION

One suggestion for an interdisciplinary unit in this chapter is Measurement in Other Countries. Tell students that in the United States, land is frequently measured in acres. In Japan, land is measured in *tans* and in Italy it is measured in *quadratos*. A *tan* is equal to 0.2449 acres, and a *quadrato* is equal to 1.25 acres. If you owned 7 *tans* of land in Japan and 6 *quadratos* in Italy, how many acres of land would you own? 9.2143 acres

Modalities Auditory, Visual

DEVELOPING ALGEBRAIC THINKING

The use of variables is an important part of algebraic thinking. To enhance students' comfort working with variables, have them write an addition expression and a multiplication expression for Problem 43. Let $y = 4.5$. $y + y + y + y + y + y + y + y + y + y$; $10y$

TEACHING TIPS

Have students use **Number Heroes** and visit Star Brilliant at *Quizzo* to practice multiplying decimals using algorithms. See Grow Slide Level V.

To practice multiplying decimals on a number line, have students use **Calculating Crew** and visit the *Nautical Number Line*. See Grow Slide Level U.

3 WRAP UP

Use these questions to help students focus on the big ideas of the lesson.

- **The sales consultant for Beauty Bonus Hair Care Products gave nine 2.25 oz samples of shampoo to her new customers. How many oz of shampoo did she give to her new customers?** 20.25 oz

✓ *Assessment Tip*

Have students write the answer.

- **How does estimation help you multiply decimals?** Possible answer: It shows you where to place the decimal point.

DAILY QUIZ

Find the product.

1. 3×0.6 1.8

2. 5×3.3 16.5

3. 7.3×3.4 24.82

4. 2.18×5.6 12.208

5. 0.053×0.086 0.004558

ORGANIZER

OBJECTIVE To use a model to divide decimals

MATERIALS *For each student* decimal squares, colored pencils, scissors

PROGRAM RESOURCES
E-Lab • Activity 3
E-Lab Recording Sheets, p. 3

USING THE PAGE

Discuss situations in which students may need to divide decimals, such as sharing money with other people or cutting ribbon into equal lengths.

Activity 1

Explore

In this activity students use decimal squares to model dividing decimals by whole numbers. This hands-on experience will help students visualize the concept of equal groups.

Think and Discuss

After discussing the questions, have students summarize what they discovered about the relationship between the quotient of a whole-number dividend and the quotient of a tenths dividend. The quotient of a whole-number dividend is 10 times greater.

Try This

Observe students as they model each exercise. Ask questions that encourage students to verbalize their solutions.

- **In Exercise 2, how many decimal squares did you put in each group?** none **Why?** because 3 cannot be divided equally into 5 groups

- **How many tenths did you put in each group?** 7 tenths

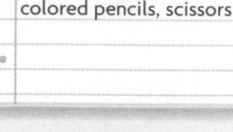

COOPERATIVE LEARNING

Exploring Division of Decimals

LAB *Activity*

WHAT YOU'LL EXPLORE
How to use a model to divide decimals

WHAT YOU'LL NEED
decimal squares, colored pencils, scissors

REMEMBER:
1 tenth (0.1) = 1 column

See page H11.

You can shade and cut apart decimal squares to divide a decimal by a whole number.

ACTIVITY 1

Explore

Work with a partner to find 3.6 ÷ 3.

- Shade 3.6 decimal squares.

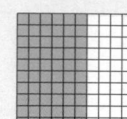

- Divide the shaded squares into 3 equal groups. Cut out each of the 6 tenths in the partially shaded decimal square.

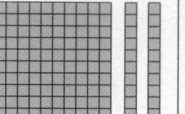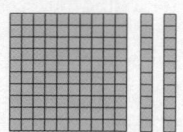

- What decimal names each group? What is the quotient?
1.2; 1.2
- Use decimal squares to find 1.2 ÷ 4. Use scissors to cut out each shaded tenth, and divide the tenths into 4 equal groups.
Check students' models; 0.3.

Think and Discuss

- Find 36 ÷ 3. How is the quotient the same as for 3.6 ÷ 3? How is it different? **12; it has the same digits; it is 10 times as great.**
- Find 12 ÷ 4. How is the quotient the same as for 1.2 ÷ 4? How is it different? **3; it has the same digits; it is 10 times as great.**

Try This

- Use decimal squares to find the quotients.

1. 4.4 ÷ 4 **1.1**   **2.** 3.5 ÷ 5 **0.7**   **3.** 1.5 ÷ 3 **0.5**

62   Chapter 3

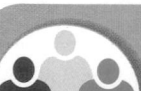 IDEA FILE

ALTERNATIVE TEACHING STRATEGY

Show students how to draw diagrams to divide decimals. Have students draw diagrams to find the quotient of 1.6 ÷ 8. 0.2; Check students' diagrams.

Modality Visual

 CALCULATOR CONNECTION

Have students use calculators to find the pattern in these exercises. Then have them write three exercises that use the pattern and have a classmate find the quotient. The quotient is 10 times greater each time.

1. 7.8 ÷ 0.78 10
   7.8 ÷ 0.078 100
   7.8 ÷ 0.0078 1,000

2. 18.24 ÷ 1.824 10
   18.24 ÷ 0.1824 100
   18.24 ÷ 0.01824 1,000

Modality Visual

You can shade and cut apart decimal squares to divide a decimal by a decimal.

ACTIVITY 2

Explore

Work with a partner to find $3.6 \div 1.2$.

- Shade 3.6 decimal squares.

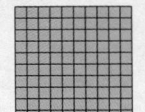  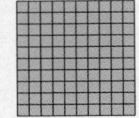 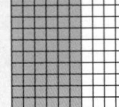

- Use scissors to cut apart the 6 tenths.

- Divide the shaded squares and shaded tenths into equal groups of 1.2. How many groups of 1.2 are in 3.6? What is the quotient? **3; 3**

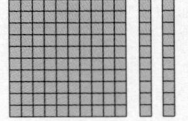 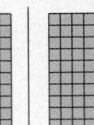 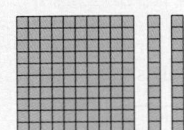 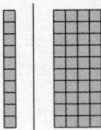 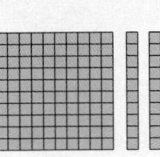 

- Use decimal squares to find $3.2 \div 1.6$. **Check students' models; 2.**
- How many groups of 1.6 are in 3.2? **2**

Think and Discuss

- Find $36 \div 12$. How is the quotient the same as for $3.6 \div 1.2$? How is the problem different from $3.6 \div 1.2$? **3; it has the same digits; the divisor and dividend are each 10 times as great.**
- You know that $3.6 \div 12 = 0.3$ and $3.6 \div 1.2 = 3$. What do you think $3.6 \div 0.12$ equals? **30**

Try This

Use decimal squares to find the quotients.

1. $7.8 \div 1.3$ **6**   **2.** $5.6 \div 0.8$ **7**   **3.** $1.56 \div 0.52$ **3**

4. $5.5 \div 1.1$ **5**   **5.** $3.6 \div 0.9$ **4**   **6.** $11.2 \div 1.4$ **8**

E–Lab • Activity 3 Available on CD-ROM and the Internet at http://www.hbschool.com/elab

63

Note:
Students explore multiplication. Use E-Lab p. 3.

Activity 2

Explore

Unlike Activity 1, in which the number of decimal models that should go in each group was being determined, in this activity the number of groups is being determined.

Think and Discuss

Have students explain how they know $3.6 \div 0.12 = 30$. Possible explanation: by using a pattern: $3.6 \div 12 = 0.3$, $3.6 \div 1.2 = 3$, so, $3.6 \div 0.12 = 30$

Try This

- **Which exercises could you solve by using mental math? Explain.** Most students will say Exercise 2: $56 \div 8 = 7$, so $5.6 \div 0.8 = 7$. Some students may also say Exercise 3: $156 \div 52 = 3$, so $1.56 \div 0.52 = 3$.

Have students model all six exercises even if they were able to find the answer by using mental math. The conceptual understanding they develop in the process of modeling will help them later on when they are presented with more complex problems.

USING E-LAB

Students explore multiplication of decimals by using a computer model.

✓ Assessment Tip

Use these questions to help students communicate the big ideas of the lesson. Check students' models.

- **Draw a model to show the division equation.**

1. $0.6 \div 2 = 0.3$ **2.** $0.9 \div 0.1 = 9$

CONSUMER CONNECTION

MATERIALS: *For each student*
10 quarters, 10 dimes, 10 nickels

Display this model for dividing decimals:

$$\$1.00 \div \$0.25 = 4$$

$0.25 $0.25 $0.25 $0.25

Have students use money to find the quotient for $\$0.30 \div \$0.05$ and for $\$1.60 \div \$0.80$. **6; 2; Check students' models.**

Modalities Auditory, Kinesthetic, Visual

EXTENSION

Have students shade decimal squares to show the square roots of these numbers. One example is shown.

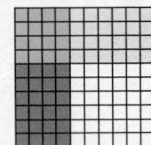

0.16 0.4

1. 0.09 0.3   **2.** 0.01 0.1
3. 0.04 0.2   **4.** 0.25 0.5
5. 0.81 0.9   **6.** 0.36 0.6

Modalities Kinesthetic, Visual

GRADE 5
Find patterns in decimal division.

GRADE 6
Divide a decimal by a whole number and divide a decimal by a decimal.

GRADE 7
Divide whole numbers and decimals.

ORGANIZER

OBJECTIVE To divide a decimal by a whole number and to divide a decimal by a decimal

MATERIALS *For each student* calculator

ASSIGNMENTS Independent Practice
- *Basic* 1–34, even; 35–42
- *Average* 1–42
- *Advanced* 11–42

PROGRAM RESOURCES
Intervention • Placing the Decimal Point, TE pp. 58A–58B
Reteach 3.4 Practice 3.4 Enrichment 3.4
Number Heroes • *Quizzo*
Math Fun • $1,000 Winner!, p. 68

PROBLEM OF THE DAY

Transparency 3

Lashonda and Mark each have the same number of coins. Lashonda has $8.25 in quarters. Mark has all dimes. How much more money does Lashonda have than Mark? $4.95 more

Solution Tab A, p. A3

QUICK CHECK

Find the quotient.

1. 168 ÷ 2 **84**
2. 364 ÷ 4 **91**
3. 209 ÷ 11 **19**
4. 1,632 ÷ 51 **32**

1 MOTIVATE

To stimulate discussion, you may wish to tell students they can use estimation to help them divide decimals. They can check the reasonableness of a quotient by estimating. Discuss this example with the class:

63.6 ÷ 8.2 = 2.05

64 ÷ 8 = 8 ← **Round or use compatible numbers to estimate the quotient.**

• **Is the quotient 2.05 reasonable?** No, the quotient should be about 8.

Dividing Decimals

WHAT YOU'LL LEARN
How to divide a decimal by a whole number and how to divide a decimal by a decimal

WHY LEARN THIS?
To find the cost of one item when you know the cost of several of the same item

Sports Link

Millions of people around the world play tennis for recreation. Tennis is played on a rectangular court with a net across the middle. The total length of a tennis court is 23.7 m. What is the length on each side of the net?

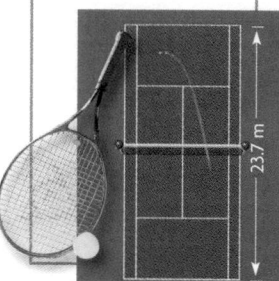

23.7 m

11.85 m

Jacob is going to take pictures on the class field trip. He bought a 3-pack of film for $9.93. What was the cost of each roll of film?

You can use division to find the cost. Dividing a decimal by a whole number is like dividing with whole numbers.

$$3)\overline{9.93}$$ *Place a decimal point above the decimal point in the dividend.*

```
   3.31
3)9.93
  -9↓
   09
   -9↓
    03
    -3
     0
```
Divide as with whole numbers.

Each roll of film cost $3.31.

You have to put a zero in the quotient when the divisor is greater than the dividend.

EXAMPLE 1 Felicia purchased a container of 3 tennis balls for $2.49. How much did she pay for each tennis ball?

$$3)\overline{2.49}$$ *Place a decimal point above the decimal point in the dividend.*

```
   0.83
3)2.49
  -0↓
   24
  -24↓
    09
    -9
     0
```
Since the divisor, 3, is greater than the 2 in the dividend, put a zero in the quotient in the ones place. Divide as with whole numbers.

So, Felicia paid $0.83 for each tennis ball.

• Find the quotient. 6.48 ÷ 8 **0.81**

RETEACH 3.4

Name _____

LESSON 3.4

Dividing Decimals

Mary has saved $0.35 each day. She now has a total of $9.10. How long has Mary been saving?

To solve, you need to divide 9.10 by 0.35.

Step 1: Make the divisor a whole number by multiplying the divisor and the dividend by a multiple of 10.

$0.35)\overline{9.10}$ →(×100 ×100)→ $35)\overline{910}$

Step 2: Place the decimal point in the quotient directly above the decimal point in the dividend. Divide as you would with whole numbers.

```
      26.
35)910
   70
   210
  -210
     0
```

Since the remainder is 0, the answer is a whole number. You do not need to show the decimal point.

Mary has been saving for 26 days.

Complete to solve.

1. Dusty bought 7 movie tickets for his friends. The tickets cost a total of $34.65. How much does each friend owe him for the cost of one ticket?

   a. What division problem will you use to solve? **$34.65 ÷ 7**

   b. Do you need to multiply to make the divisor a whole number? **no**

   c. How do you decide where to place the decimal point in the quotient? **Place it directly above the decimal point in the dividend.**

   d. How much does each friend owe? **$4.95**

Solve.

2. Steve is saving for a tape player that costs $80.85. Each week he puts aside $7.35. How long will it take Steve to save enough money to buy the tape player? **11 weeks**

3. Marty is cutting strips of fabric to decorate the classroom. Each strip is to be 1.5 in. wide. He has 23.25 in. of material. How many strips can he cut? **15.5 strips**

PRACTICE 3.4

Name _____

LESSON 3.4

Dividing Decimals

Rewrite the problem so that the divisor is a whole number.

1. 8.5 ÷ 2.3 **85 ÷ 23**
2. 6.4 ÷ 1.3 **64 ÷ 13**
3. 9.1 ÷ 0.15 **910 ÷ 15**
4. 33.17 ÷ 6.8 **331.7 ÷ 68**

Complete.

5. 56.8 ÷ 0.8 = 568 ÷ **8**
6. 7.21 ÷ 0.03 = **721** ÷ 3
7. 4.12 ÷ 2.3 = 41.2 ÷ **23**

Find the quotient.

8. 36.9 ÷ 3 **12.3**
9. 22.4 ÷ 7 **3.2**
10. 37.5 ÷ 5 **7.5**
11. 89.6 ÷ 8 **11.2**
12. 14)78.4 **5.6**
13. 40)6.8 **0.17**
14. 13)150.8 **11.6**
15. 70)23.8 **0.34**
16. 5.32 ÷ 0.7 **7.6**
17. 1.88 ÷ 0.4 **4.7**
18. 2.12 ÷ 0.2 **10.6**
19. 5.4 ÷ 0.08 **67.5**
20. 7.54)24.882 **3.3**
21. 12.6)806.4 **64**
22. 0.91)6.734 **7.4**
23. 10.9)81.75 **7.5**
24. 2.9)0.3335 **0.115**
25. 0.18)64.296 **357.2**
26. 12.3)84.87 **6.9**
27. 8.7)53.244 **6.12**

Mixed Applications

28. Ryan buys 6 lb of mixed nuts for $16.74. How much does 1 lb of nuts cost? **$2.79**

29. Jon buys golf balls for $12.79. He gives the clerk a $20 bill. How much change does Jon get? **$7.21**

30. James is saving $15.75 every week to buy a plane ticket that costs $141.75. In how many weeks will he have enough money saved? **9 weeks**

31. Michele jogged $1\frac{1}{5}$ mi this week. She plans to increase the distance by $\frac{2}{5}$ mi every week. How far will she jog 5 weeks from now? **4 mi**

GUIDED PRACTICE

Copy the problem. Place the decimal point in the quotient.

1. $4.92 \div 2 = 246$ **2.46**	**2.** $9.6 \div 3 = 32$ **3.2**	**3.** $2.96 \div 4 = 74$ **0.74**
4. $41.5 \div 5 = 83$ **8.3**	**5.** $121.2 \div 12 = 101$ **10.1**	**6.** $57.75 \div 15 = 385$ **3.85**
7. $32.5 \div 5 = 65$ **6.5**	**8.** $825.9 \div 3 = 2753$ **275.3**	**9.** $224.6 \div 4 = 5615$ **56.15**

Divide.

1.97 **10.** $7.88 \div 4$	**4.61** **11.** $9.22 \div 2$	**0.893** **12.** $8.93 \div 10$	**20.54** **13.** $102.7 \div 5$
2.22 **14.** $4.44 \div 2$	**0.83** **15.** $2.49 \div 3$	**6.1** **16.** $30.5 \div 5$	**3.06** **17.** $33.66 \div 11$
0.322 **18.** $5\overline{)1.61}$	**4.935** **19.** $2\overline{)9.87}$	**0.395** **20.** $20\overline{)7.9}$	**7.4** **21.** $16\overline{)118.4}$

Dividing Decimals by Decimals

ACTIVITY

WHAT YOU'LL NEED: calculator

• Use a calculator to find the following quotients.

• Look for a pattern. Try to predict the last quotient in each set.

Set A	Set B
$0.52 \div 0.02 = $ ■ **26**	$0.768 \div 0.016 = $ ■ **48**
$5.2 \div 0.2 = $ ■ **26**	$7.68 \div 0.16 = $ ■ **48**
$52 \div 2 = $ ■ **26**	$76.8 \div 1.6 = $ ■ **48**
$520 \div 20 = $ ■ **26**	$768 \div 16 = $ ■ **48**

When the decimal point is moved the same number of places in the dividend and the divisor, the quotient remains unchanged.

Talk About It

• Describe the pattern that helped you predict the last quotient in each set.

• For each set above, write another division problem that has the same quotient. **Possible problems: 5,200 ÷ 200; 7,680 ÷ 160**

• Look at $5.2 \div 0.2$. Multiply both 5.2 and 0.2 by 10. How do the quotients of $5.2 \div 0.2$ and $52 \div 2$ compare? How does multiplying the divisor and the dividend by 10 affect the quotient? **52 ÷ 2; the quotients are the same; the quotient stays the same.**

• 🔦 **CRITICAL THINKING** Look at $7.68 \div 0.16$. Multiply both 7.68 and 0.16 by 100. How do the quotients of $7.68 \div 0.16$ and $768 \div 16$ compare? How does multiplying the divisor and the dividend by 100 affect the quotient? **768 ÷ 16; the quotients are the same; the quotient stays the same.**

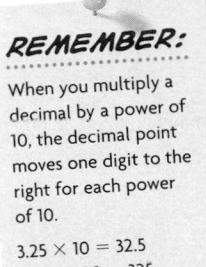

REMEMBER:

When you multiply a decimal by a power of 10, the decimal point moves one digit to the right for each power of 10.

$3.25 \times 10 = 32.5$
$3.25 \times 100 = 325$
$3.25 \times 1,000 = 3,250$

See page H12.

65

Have students estimate the quotient for these problems and decide whether it is reasonable.

1. $18.06 \div 8.6 = 10$ $18 \div 9 = 2$, so the quotient is not reasonable.

2. $46.2 \div 8.4 = 5.5$ $45 \div 9 = 5$, so the quotient is reasonable.

2 TEACH/PRACTICE

In this lesson students apply their whole-number division skills to divide decimals. Discuss the examples and the Talk About It questions with the class to focus on decimal concepts such as zeros in the dividend and quotient and using powers of ten to divide.

GUIDED PRACTICE

Complete Exercises 1–9 orally as a class and then have volunteers solve Exercises 10–21 on the board. Discuss as a class.

COMMON ERROR ALERT Some students may write remainders for Exercises 12–13. Remind them to add zeros to the dividend until the remainder is zero.

TEACHING TIP

If students need additional practice multiplying by powers of 10, have them use the Review in the Idea File on p. 66.

ADDITIONAL EXAMPLE

Example 1, p. 64

The sale price of golf balls is 4 for $3.40.

What is the price of one golf ball? $0.85

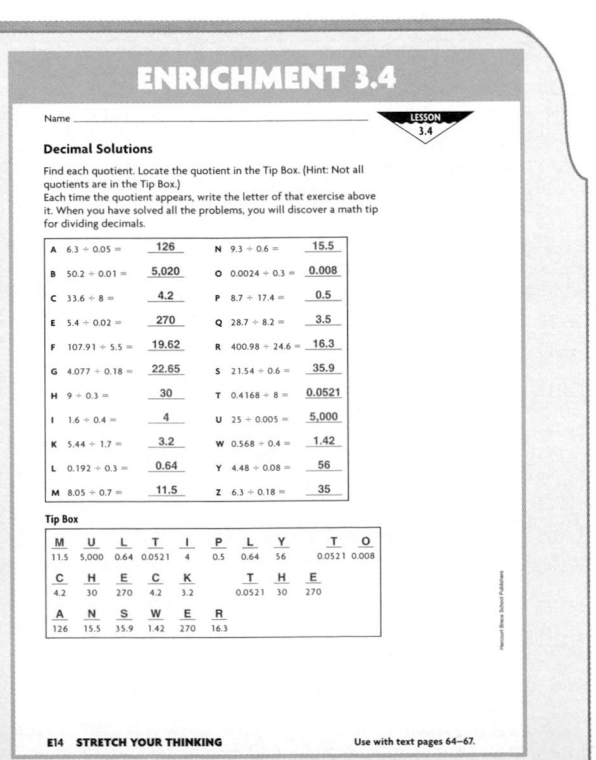

ENRICHMENT 3.4

Name _____

LESSON 3.4

Decimal Solutions

Find each quotient. Locate the quotient in the Tip Box. (Hint: Not all quotients are in the Tip Box.)
Each time the quotient appears, write the letter of that exercise above it. When you have solved all the problems, you will discover a math tip for dividing decimals.

A $6.3 \div 0.05 = $	**126**	**N** $9.3 \div 0.6 = $	**15.5**	
B $50.2 \div 0.01 = $	**5,020**	**O** $0.0024 \div 0.3 = $	**0.008**	
C $33.6 \div 8 = $	**4.2**	**P** $8.7 \div 17.4 = $	**0.5**	
E $5.4 \div 0.02 = $	**270**	**Q** $28.7 \div 8.2 = $	**3.5**	
F $107.91 \div 5.5 = $	**19.62**	**R** $400.98 \div 24.6 = $	**16.3**	
G $4.077 \div 0.18 = $	**22.65**	**S** $21.54 \div 0.6 = $	**35.9**	
H $9 \div 0.3 = $	**30**	**T** $0.4168 \div 8 = $	**0.0521**	
I $1.6 \div 0.4 = $	**4**	**U** $25 \div 0.005 = $	**5,000**	
K $5.44 \div 1.7 = $	**3.2**	**W** $0.568 \div 0.4 = $	**1.42**	
L $0.192 \div 0.3 = $	**0.64**	**Y** $4.48 \div 0.08 = $	**56**	
M $8.05 \div 0.7 = $	**11.5**	**Z** $6.3 \div 0.18 = $	**35**	

Tip Box

M	U	L	T	I	P	L	Y		T	O
11.5	5,000	0.64	0.0521	4	0.5	0.64	56		0.0521	0.008

C	H	E	C	K		T	H	E
4.2	30	270	4.2	3.2		0.0521	30	270

A	N	S	W	E	R
126	15.5	35.9	1.42	270	16.3

E14 STRETCH YOUR THINKING Use with text pages 64–67.

65

CRITICAL THINKING
It costs Stella $0.25 to make one dozen seashell magnets. If she sells two magnets for $0.25, how many magnets will she need to sell to make a $5.00 profit? **4 dozen**

ADDITIONAL EXAMPLES

Example 2, p. 66

Divide. 39.8 ÷ 0.4 **99.5**

Example 3, p. 66

Mr. Baker drives the 4.75-mile shuttle bus run between the parking lot and the airport. He drove a total of 99.75 miles the day before the Super Bowl. How many trips did he make? **21 trips**

A picture can last a lifetime. So, before you take that snapshot of your friends or family, read these tips to improve your picture:

• Move in close to your subject to fill the viewfinder.

• Try to keep the backgrounds clean and simple.

• When you take a picture of a scenic view, include family or friends.

To divide a decimal by a decimal, first multiply the divisor and the dividend by a power of 10 to change the divisor to a whole number.

$$0.5)\overline{12.55} \rightarrow 5)\overline{125.5}$$
$$0.5 \times 10 = 5$$
$$12.55 \times 10 = 125.5$$

• What power of 10 would you multiply the divisor and the dividend by in the problem 425.7 ÷ 0.12 to change the divisor to a whole number? **by 100**

EXAMPLE 2 Divide. 22.8 ÷ 0.8

$$0.8)\overline{22.8}$$

Make the divisor a whole number by multiplying the divisor and dividend by 10.
$$0.8 \times 10 = 8 \qquad 22.8 \times 10 = 228$$

Place the decimal point in the quotient.
Divide as with whole numbers.

```
    28.5
8.)228.0
  −16↓
    68
   −64↓
     40
    −40
      0
```

Place a zero in the tenths place in the dividend, and continue to divide.

So, 22.8 ÷ 0.8 = 28.5

• Think about 55.8 ÷ 0.18. To change the divisor to a whole number, you multiply by 100. What does the dividend, 55.8, become? Explain. **5,580; when you multiply 55.8 by 100, you add a zero so that you can move the decimal point 2 places.**

You can use division of decimals to solve problems that involve money.

EXAMPLE 3 Each member of the class gave Mark $0.75 to buy a camera for the class. Mark received a total of $21.75. How many students are in the class?

$$0.75)\overline{21.75}$$

Make the divisor a whole number by multiplying the divisor and dividend by 100.
$$0.75 \times 100 = 75 \qquad 21.75 \times 100 = 2,175$$

```
     29.
75.)2175.
  −150↓
    675
   −675
      0
```

Place the decimal point in the quotient. Divide as with whole numbers.

Since the remainder is 0, the answer is a whole number, and you do not need to show the decimal point.

So, there are 29 students in the class.

IDEA FILE

CALCULATOR CONNECTION

The average amount of time shoppers spend in different sections of the grocery store is shown below. Have students use their calculators to change the time to minutes, expressed to the nearest tenth.

Produce	181 sec	3.0 min
Bread	42.3 sec	0.7 min
Seafood	40.5 sec	0.7 min
Meat	154 sec	2.6 min
Baby food	130.9 sec	2.2 min

Modalities Kinesthetic, Visual

REVIEW

Have students name the power of ten they would multiply by to make the decimal a whole number.

1. 04.21 10^2
2. 6.325 10^3
3. 349.2 10^1
4. 0.9385 10^4
5. 124.01 10^2
6. 0.304 10^3
7. 3.3 10^1
8. 44.4444 10^4

Modality Auditory

CORRELATION TO OTHER RESOURCES

CornerStone Mathematics
©1996 SkillsBank Corporation
Level B, Using Decimals, Lesson 4

MEETING INDIVIDUAL NEEDS

INDEPENDENT PRACTICE

Rewrite the problem so that the divisor is a whole number.

1. 9.6 ÷ 1.6
96 ÷ 16

2. 5.5 ÷ 1.1
55 ÷ 11

3. 6.3 ÷ 0.18
630 ÷ 18

4. 48.24 ÷ 2.4
482.4 ÷ 24

Complete.

5. 48.4 ÷ 0.4 = 484 ÷ ■ **4**

6. 8.19 ÷ 0.09 = ■ ÷ 9 **819**

7. 3.57 ÷ 2.1 = 35.7 ÷ ■ **21**

8. 3.846 ÷ 64.1 = ■ ÷ 641
38.46

9. 239 ÷ 0.075 = ■ ÷ 75
239,000

10. 84.36 ÷ 0.2812 = ■ ÷ 2812
843,600

Find the quotient.

11. 21.6 ÷ 3 **7.2**

12. 12.8 ÷ 4 **3.2**

13. 80.1 ÷ 9 **8.9**

14. 90.3 ÷ 6 **15.05**

15. 11)109.01 **9.91**

16. 12)286.8 **23.9**

17. 90)10.8 **0.12**

18. 60)12.6 **0.21**

19. 2.75 ÷ 0.5 **5.5**

20. 1.26 ÷ 0.2 **6.3**

21. 13.2 ÷ 0.06 **220**

22. 42.5 ÷ 0.05 **850**

23. 3.2)2.24 **0.7**

24. 2.8)4.48 **1.6**

25. 4.2)3.78 **0.9**

26. 8.2)229.6 **28**

27. 0.38)13.3 **35**

28. 0.55)2.42 **4.4**

29. 6.41)135.892 **21.2**

30. 2.48)1.3392 **0.54**

31. 49.3)201.144 **4.08**

32. 38.2)469.86 **12.3**

33. 29.1)186.24 **6.4**

34. 18.2)378.56 **20.8**

Problem-Solving Applications

35. Cynthia bought 4 lb of apples and paid $2.40. How much did 1 lb of apples cost? **$0.60**

36. Sal served equal amounts of punch to 9 guests. She used a total of 94.5 oz of punch. How many ounces did each guest receive? **10.5 oz**

37. Emilio bought 5 movie tickets for his friends. The tickets cost a total of $28.75. How much does each friend owe him for the cost of one ticket? **$5.75**

38. Jonelle is saving $4.95 every week to buy a video that costs $29.70, including tax. How many weeks will she have to save?
6 weeks

39. A new parking garage at the mall is 16.8 m high. Each floor is 4.2 m high. How many floors are there? **4 floors**

40. Mel knows that 5 lb of mixed nuts sell for $12.50. He buys only 1 lb. How much does he pay? **$2.50**

41. **CRITICAL THINKING** Michael divided 4.25 by 0.25 and got a quotient of 0.17. Explain what he did wrong. What is the correct quotient? **He didn't multiply the dividend by 100; 17.**

42. **WRITE ABOUT IT** When you divide a decimal by a decimal, what do you do to the divisor and the dividend? **Multiply by a power of 10 that will make the divisor a whole number.**

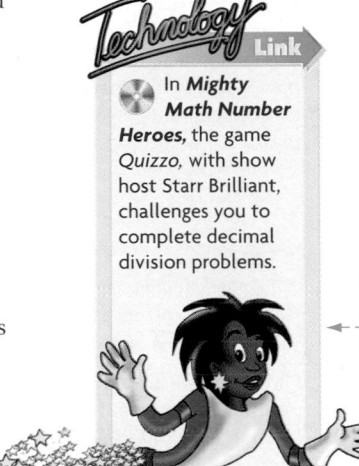

Technology Link

In *Mighty Math Number Heroes,* the game *Quizzo,* with show host Starr Brilliant, challenges you to complete decimal division problems.

Use Grow Slide Level Z.

MORE PRACTICE Lesson 3.4, page H46

67

USING THE PAGE

To provide additional practice activities related to Chapter 3: *Using Decimals*

DECIMAL PUZZLES

Have students work individually to complete this activity.

In this activity, students practice adding decimals by completing a diagram so that the sum of every three numbers in a line is the same.

Multiple Intelligences Visual/Spatial, Logical/Mathematical, Intrapersonal/Introspective
Modality Visual

THREE IN A ROW

MATERIALS: *For each group* 1-in. graph paper, colored markers or squares of paper marked with each players name

Have students work in pairs to complete this activity.

In this activity, students practice subtracting decimals by writing subtraction problems that can be answered on a gameboard.

Multiple Intelligences Visual/Spatial, Logical/Mathematical, Verbal/Linguistic, Interpersonal/Social, Intrapersonal/Introspective
Modalities Auditory, Visual

$1,000 WINNER!

MATERIALS: *For each group* 6-section spinner numbered 1–6

Have students work in pairs or small groups to complete this activity.

In this activity, students practice multiplying and dividing decimals by performing the operation corresponding to each spin on a game spinner.

Multiple Intelligences Visual/Spatial, Logical/Mathematical, Bodily/Kinesthetic, Interpersonal/Social, Intrapersonal/Introspective
Modalities Kinesthetic, Visual

DECIMAL PUZZLES

PURPOSE To practice adding decimals (pages 54–55)

Copy the diagram. Then place each number listed below in one of the circles so that the sum of every three numbers in a line is the same.

2.1	2.2	2.3	2.4	2.5
2.6	2.7	2.8	2.9	

What is the sum? **7.5**

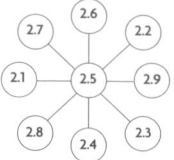

THREE IN A ROW

PURPOSE To practice subtracting decimals (pages 56–57)
YOU WILL NEED paper, squares of paper marked with each player's name

Copy the cost grid. In turn, each player makes up a subtraction problem that has an answer shown on the grid using the supply list. Each player has $3.50 to spend. The player covers the answer with a marker. If it is incorrect, the player loses a turn. The player to get three in a row wins the game.

$2.52	$0.32	$3.14
$2.61	$3.13	$2.82
$3.18	$2.25	$0.57

School Supplies			
Pencil	$0.32	Highlighter	$1.25
Pen	$0.98	Notebook	$2.93
Eraser	$0.36	Folder	$0.37
Paper	$0.89	Scissors	$3.18
Ruler	$0.68	Protractor	$0.93

$1,000 WINNER!

PURPOSE To practice multiplying and dividing decimals (pages 58–67)
YOU WILL NEED 6-section spinner numbered 1–6

Play with a partner. Take turns spinning the pointer. Each player begins with $100. Use the chart to tell what to do with the number you spin. (For example, if you spin a 1, multiply your money by 5.) If you lose all your money, you're out. The first person to get $1,000, or the only one left with any money, is the winner.

If You Spin	What To Do
1	× 5
2	× 0.55
3	× 50.5
4	÷ 5
5	÷ 0.55
6	÷ 50.5

HOME NOTE

Play this game with your family.

CHAPTER 3 REVIEW

EXAMPLES

• Estimate and add decimals. (pages 54–55)

```
  1.996
  2.400    Use zeros as placeholders.
+ 3.260
  7.656    Align the decimals.
           Add.
```

• Subtract decimals. (pages 56–57)

$92.9 - 18.76$

```
  92.90    Add a zero to 92.9.
- 18.76
  74.14    Place the decimal point.
           Subtract.
```

• Multiply decimals. (pages 58–61)

```
  0.0281   ← 4 decimal places
×    9.2   ← 1 decimal place
    562
   2529
  0.25852  ← 5 decimal places
```

• Divide decimals. (pages 64–67)

```
0.81)1.863   Make the divisor a whole number.
             0.81 × 100 = 81
             1.863 × 100 = 186.3

       2.3   Divide as with whole numbers.
81.)186.3
   -162 ↓
    243
   -243
      0
```

EXERCISES

Estimate the sum.

1. 0.87 + 2.4 **3**   **2.** 3.2 + 0.7+ 1.2 **5**
3. 3.27 + 19.9 **23**   **4.** 0.5 + 7.9 + 4.1 **13**

Add.

5. 3.9 + 7.2 **11.1**   **6.** 480.7 + 0.824 **481.524**
7. 28.7 + 0.45 + 3.2 **32.35**   **8.** $341.59 + $10.52 **$352.11**
9. Ross found a glove for $28.99, a helmet for $21.50, and shoes for $35.00. How much money will he need to buy these items? **$85.49**

Find the difference.

10. 3.2 − 1.5 **1.7**   **11.** 91.06 − 81.95 **9.11**
12. 9.638 − 9.284 **0.354**   **13.** 231.72 − 27.926 **203.794**
14. 45.2 − 3.78 **41.42**   **15.** 28 − 17.224 **10.776**
16. Paul bought a basketball for $22.98. He gave the cashier $25.00. How much change should he get from the cashier? **$2.02**

Find the product.

17. 0.15 × 5 **0.75**   **18.** 48.23 × 7 **337.61**
19. 1.96 × 9 **17.64**   **20.** 9.07 × 75.3 **682.971**
21. 0.14 × 0.26 **0.0364**   **22.** 5.7 × 0.98 **5.586**
23. Stephanie works 15 hr per week at a theme park. She earns $5.60 an hour. How much does she earn in one week? **$84.00**

Find the quotient.

24. 7)16.1 **2.3**   **25.** 9)43.65 **4.85**   **26.** 15)10.8 **0.72**
27. 8.45 ÷ 0.5 **16.9**   **28.** 7.8 ÷ 0.12 **65**
29. Ms. Pegg bought a 25-pack of diskettes for $11.25 for her computer class. How much did she pay for each diskette? **$0.45**
30. A new building is 32.8 ft high. Each floor is 8.2 ft high. How many floors are there? **4 floors**

Chapter 3 Review **69**

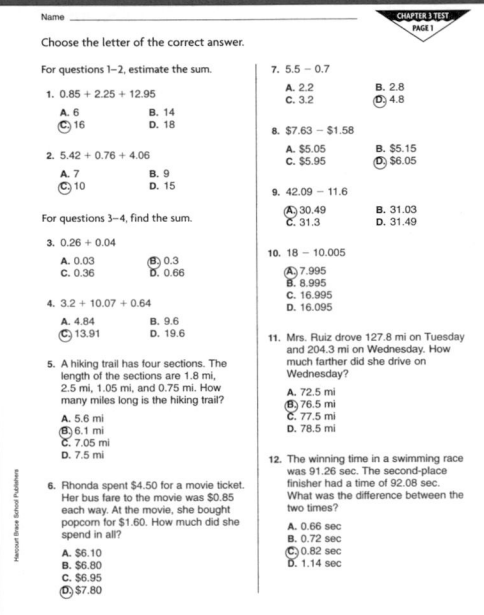

USING THE PAGE

The Chapter 3 Review reviews estimating, adding, subtracting, multiplying, and dividing decimals. Chapter objectives are provided, along with examples and practice exercises for each.

 The Chapter 3 Review can be used as a review or a test. You may wish to place the Review in students' portfolios.

Assessment Checkpoint

For Performance Assessment in this chapter, see page 72, What Did I Learn?, items 3 and 6.

✔ Assessment Tip

Student Self-Assessment The How Did I Do? Survey helps students assess both what they have learned and how they learned it. Self-assessment helps students learn more about their own capabilities and develop confidence in themselves.

A self-assessment survey is available as a copying master in the *Teacher's Guide for Assessment*, p. 15.

CHAPTER 3 TEST, page 1

Name _____

Choose the letter of the correct answer.

For questions 1–2, estimate the sum.

1. 0.85 + 2.25 + 12.95
A. 6   B. 14
C. 16   D. 18

2. 5.42 + 0.76 + 4.06
A. 7   B. 9
C. 10   D. 15

For questions 3–4, find the sum.

3. 0.26 + 0.04
A. 0.03   B. 0.3
C. 0.36   D. 0.66

4. 3.2 + 10.07 + 0.64
A. 4.84   B. 9.6
C. 13.91   D. 19.6

5. A hiking trail has four sections. The length of the sections are 1.8 mi, 2.5 mi, 1.05 mi, and 0.75 mi. How many miles long is the hiking trail?
A. 5.6 mi
B. 6.1 mi
C. 7.05 mi
D. 7.5 mi

6. Rhonda spent $4.50 for a movie ticket. Her bus fare to the movie was $0.85 each way. At the movie, she bought popcorn for $1.60. How much did she spend in all?
A. $6.10
B. $6.80
C. $6.95
D. $7.80

7. 5.5 − 0.7
A. 2.2   B. 2.8
C. 3.2   D. 4.8

8. $7.63 − $1.58
A. $5.05   B. $5.15
C. $5.95   D. $6.05

9. 42.09 − 11.6
A. 30.49   B. 31.03
C. 31.3   D. 31.49

10. 18 − 10.005
A. 7.995
B. 8.995
C. 16.995
D. 16.095

11. Mrs. Ruiz drove 127.8 mi on Tuesday and 204.3 mi on Wednesday. How much farther did she drive on Wednesday?
A. 72.5 mi
B. 76.5 mi
C. 77.5 mi
D. 78.5 mi

12. The winning time in a swimming race was 91.26 sec. The second-place finisher had a time of 92.08 sec. What was the difference between the two times?
A. 0.66 sec
B. 0.72 sec
C. 0.82 sec
D. 1.14 sec

Form A • Multiple-Choice   A9   Go on. ▶

CHAPTER 3 TEST, page 2

Name _____

13. Use the decimal square to find 3 × 0.12.

A. 0.036   B. 0.36
C. 3.6   D. 36.0

14. How many decimal places are there in the product 0.055 × 0.07?
A. 2 decimal places
B. 3 decimal places
C. 4 decimal places
D. 5 decimal places

15. 5 × 0.7
A. 0.035   B. 0.35
C. 35   D. not here

16. $2.75 × 8
A. $16.60   B. $22.00
C. $22.50   D. not here

17. 0.3 × 0.4
A. 0.12   B. 1.12
C. 1.2   D. not here

18. 0.27 × 4.2
A. 0.1134   B. 1.134
C. 1.34   D. not here

19. Brittany charges $3.50 per hour for baby-sitting. If she baby-sits for 4.5 hr, how much will she earn?
A. $14.00   B. $14.75
C. $15.25   D. $15.75

20. Craig is planning a cookout for 27 students. He needs 0.25 lb of hamburger for each student. How many pounds of hamburger does he need in all?
A. 6.75 lb   B. 12.5 lb
C. 16.5 lb   D. 67.5 lb

21. Which of these numbers should you multiply by to make the divisor in the division problem a whole number?
0.65)2.145
A. 1   B. 10
C. 100   D. not here

22. 15.3 ÷ 9
A. 0.17   B. 1.7
C. 1.77   D. 17

23. 0.18)3.96
A. 0.022   B. 0.22
C. 2.2   D. 22

24. 13.44 ÷ 0.24
A. 0.56   B. 5.06
C. 5.6   D. 56

Form A • Multiple-Choice   A10   ▶ Stop!

See *Test Copying Masters*, pp. A9–A10 for the Chapter Test in **multiple-choice** format and pp. B159–B160 for **free-response** format.

69

STUDY GUIDE & REVIEW

✔ *Assessment Checkpoint*

USING THE PAGES

This Study Guide reviews the following content:

Chapter 1 Looking at Numbers, pp. 14–31

Chapter 2 Using Whole Numbers, pp. 32–51

Chapter 3 Using Decimals, pp. 52–69

CHAPTERS
1-3

STUDY GUIDE & REVIEW

VOCABULARY CHECK

1. A(n) __?__ shows how many times a number called the base is used as a factor. (page 24) **exponent**

2. Integers that are __?__ are an equal distance from zero on the number line. (page 28) **opposites**

EXAMPLES

- **Use place value to express numbers.** (pages 16–17)

  Write: 1,900,050.0325 *standard form*

  Read: one million, nine hundred thousand, fifty and three hundred twenty-five ten-thousandths

EXERCISES

Write the number in standard form.

3. five hundred twenty thousand, ten and eight thousandths **520,010.008**

4. two million, four hundred thousand, twenty-six and six tenths **2,400,026.6**

- **Write decimals as fractions, and fractions as decimals.** (pages 20–23)

  0.08 *Identify the value of the last digit. 8 is in the hundredths place.*

  $\frac{8}{100}$ *Use that value for the denominator.*

  So, $0.08 = \frac{8}{100}$.

Use place value to change the decimal to a fraction.

5. 0.2 $\frac{2}{10}$

6. 0.004 $\frac{4}{1,000}$

7. 0.65 $\frac{65}{100}$

8. 0.015 $\frac{15}{1,000}$

Write as a decimal.

9. $\frac{5}{100}$ **0.05**

10. $\frac{2}{5}$ **0.4**

- **Write numbers using exponents.** (pages 24–25)

  4^3 *Use 4 as a factor 3 times.*

  $4 \times 4 \times 4 = 64$ *So,* $4^3 = 64$

Find the value.

11. 8^2 **64**

12. 10^3 **1,000**

13. 1^{11} **1**

14. 20^3 **8,000**

- **Multiply and divide whole numbers.** (pages 40–43)

  $$\begin{array}{r} 107 \ \text{r}11 \\ 16)\overline{1723} \\ -16 \\ \hline 12 \\ -0 \\ \hline 123 \\ -112 \\ \hline 11 \end{array}$$

  Divide.
  Multiply.
  Subtract.
  Repeat as necessary.

  ← *Remainder*

Find the product.

15. 7,654 × 93 **711,822**

16. 342 × 125 **42,750**

Find the quotient.

17. 7)716 **102 r2**

18. 84)5,056 **60 r16**

See *Test Copying Masters*, pp. A11–A12 for copying masters of the Test • Chapters 1–3 in **multiple-choice** format and pp. B161–B162 for **free-response** format.

TEST • CHAPTERS 1–3, p. 1

Name _____

Choose the letter of the correct answer.

1. What is the value of the digit 3 in the number 37,529.04?
   - A. 3 tens
   - B. 3 hundreds
   - C. 3 thousands
   - (D.) 3 ten thousands
   1-A.1

2. Which is less than 7.46?
   - (A.) 7.39
   - B. 7.50
   - C. 17.2
   - D. 70.5
   1-A.1

3. What is $\frac{5}{8}$ written as a decimal?
   - A. 0.375
   - B. 0.58
   - (C.) 0.625
   - D. 0.75
   1-A.2

4. Mark skated $\frac{4}{5}$ mi. What is the distance he skated, expressed as a decimal?
   - A. 0.66 mi
   - B. 0.75 mi
   - (C.) 0.80 mi
   - D. 5.6 mi
   1-A.2

5. How many zeros are there in the standard form of 10^4?
   - A. 3
   - (B.) 4
   - C. 5
   - D. 6
   1-A.3

6. What is the value of 3^4?
   - A. 12
   - B. 36
   - (C.) 81
   - D. 243
   1-A.3

7. Which of these shows the integers written in order from least to greatest?
   - A. 9, 0, ⁻3, ⁻6
   - B. 0, ⁻3, ⁻6, 9
   - C. ⁻6, 9, ⁻3, 0
   - (D.) ⁻6, ⁻3, 0, 9
   1-A.4

8. What is the opposite of ⁺16?
   - (A.) ⁻16
   - B. $\frac{-1}{16}$
   - C. $\frac{-1}{16}$
   - D. ⁺4²
   1-A.4

9. Julie and Wanda scored a total of 38 points in a basketball game. Wanda scored 8 more points than Julie. How many points did Julie score?
   - (A.) 15 points
   - B. 16 points
   - C. 23 points
   - D. 30 points
   2-A.2

10. Kane made 54 treats for the bake sale. He made 12 brownies and made 6 more chocolate chip cookies than he did sugar cookies. How many sugar cookies did he make?
    - A. 12 cookies
    - (B.) 18 cookies
    - C. 24 cookies
    - D. 36 cookies
    2-A.2

11. Find the missing factor.
    $16 \times \underline{\ ?\ } = (16 \times 60) + (16 \times 4)$
    - A. 4
    - B. 8
    - C. 16
    - (D.) 64
    2-A.3

12. 350 × 26
    - A. 2,800
    - B. 6,300
    - (C.) 9,100
    - D. 72,100
    2-A.3

Form A • Multiple-Choice   A11   Go on. ▶

TEST • CHAPTERS 1–3, p. 2

Name _____

13. 35)5,090
    - A. 115 r45
    - B. 135 r15
    - (C.) 145 r15
    - D. 154 r45
    2-A.3

14. The distance between two cities is 2,470 mi round trip. A pilot made the trip 9 times in one month. How many miles did the pilot travel?
    - (A.) 22,230 mi
    - B. 23,200 mi
    - C. 23,320 mi
    - D. 24,330 mi
    2-A.3

15. Choose the best estimate of the difference.
    8,462
    − 2,248
    - A. 6,000
    - B. 6,100
    - C. 6,200
    - (D.) 6,300
    2-A.4

16. Choose the best estimate of the product.
    781
    × 412
    - A. 280,000
    - (B.) 320,000
    - C. 350,000
    - D. 400,000
    2-A.4

17. Find the sum.
    0.48 + 3.32 + 2.55
    - A. 4.75
    - B. 5.25
    - (C.) 6.35
    - D. 8.67
    3-A.1

18. Each member of a relay team ran 100 yd. The times for each member were 12.23 sec, 12.35 sec, 11.96 sec, and 11.89 sec. What was the total time for the relay team?
    - A. 47.33 sec
    - (B.) 48.43 sec
    - C. 49.34 sec
    - D. 50.44 sec
    3-A.1

19. 6.5093 − 2.2508
    - A. 4.1513
    - (B.) 4.2585
    - C. 4.3350
    - D. 4.3422
    3-A.1

20. Alani threw a shot put 12.83 m on his first try and 13.09 m on his second try. How much farther was his second throw?
    - (A.) 0.26 m
    - B. 0.92 m
    - C. 1.26 m
    - D. 1.83 m
    3-A.1

21. 8 × 4.73
    - A. 32.64
    - B. 35.46
    - (C.) 37.84
    - D. 38.48
    3-A.2

22. Heather is performing a chemistry experiment using 8 test tubes. She needs 2.50 mL of acid in each test tube. How much acid does she need?
    - A. 10.5 mL
    - B. 16.0 mL
    - C. 16.5 mL
    - (D.) 20.0 mL
    3-A.2

23. 2.6)2.08
    - A. 0.52
    - (B.) 0.80
    - C. 0.82
    - D. 1.10
    3-A.3

24. 53.6 ÷ 8
    - (A.) 6.7
    - B. 7.6
    - C. 7.8
    - D. 8.6
    3-A.3

Form A • Multiple-Choice   A12   ▶ Stop!

• **Estimate sums, differences, products, and quotients.** (pages 46–49)

$$518 \quad \textit{Round to the} \quad \rightarrow \quad 500$$
$$-329 \quad \textit{nearest hundred.} \quad \quad -300$$
$$\overline{} \quad \quad \quad \quad \quad \quad \quad \quad \overline{200}$$

Estimate. Answers may vary.

19. 7,261
   +5,842
   ___
   13,000

20. 2,401
   − 779
   ___
   1,600

21. 615 × 49 **30,000**

22. 895 ÷ 9 **100**

• **Add, subtract, multiply, and divide decimals.** (pages 54–67)

$$\begin{array}{r} 2.1 \\ 0.8\overline{)1.68} \\ \underline{1\,6} \\ 8 \\ \underline{8} \\ 0 \end{array}$$

Make the divisor a whole number to place decimal point.
$0.8 \times 10 = 8 \quad 1.68 \times 10 = 16.8$
Divide as with whole numbers.

Add.

23. 38.7 + 0.869
   39.569

24. 17.45 + 8.318
   25.768

Find the difference.

25. 85.06 − 75.98
   9.08

26. 224.62 − 27.948
   196.672

Find the product.

27. 63.5 × 9
   571.5

28. 8.95 × 0.8
   7.16

Find the quotient.

29. 30.45 ÷ 15 **2.03**

30. 0.5)75.5 **151**

PROBLEM-SOLVING APPLICATIONS

Solve. Explain your method. **Explanations will vary.**

31. Roger started a worm farm with 10 worms. Each month there are 10 times as many worms as the month before. Using the exponent form, write the number of worms Roger will have in 10 mo. (pages 24–25) **10^{10}**

32. Edward has 20 marbles. He has 10 more blue marbles than red ones. How many does Edward have of each color? (pages 36–37) **15 blue marbles, 5 red marbles**

33. The PTA has $140.50 to spend on picnic supplies. They bought hot dogs for $24.75, rolls for $11.45, drinks for $48.96, and cookies for $25.29. About how much does the PTA have left? (pages 54–55) **about $30**

34. In a 24-hour auto race, a car drove 3,057.6 mi. How many miles per hour is this? (pages 64–67) **127.4 mi per hr**

✓ *Assessment Checkpoint*

PORTFOLIO SUGGESTIONS

Portfolio

The portfolio illustrates the growth, talents, achievements, and reflections of the mathematics learner. You may wish to have students spend a short time selecting work samples for their portfolios and complete the Portfolio Summary Guide in the *Teacher's Guide for Assessment*, page 23. Students may wish to address the following questions:

• **What new understanding of math have I developed in the past several weeks?**

• **What growth in understanding or skills can I see in my work?**

• **What can I do to improve my understanding of math ideas?**

• **What would I like to learn more about?**

For information about how to organize, share, and evaluate portfolios, see the *Teacher's Guide for Assessment* pages 20–22.

✔ Assessment Checkpoint

USING PERFORMANCE ASSESSMENT

With the *Performance Assessment Teacher's Guide,*

- have students work individually or in pairs as an alternative to formal assessment.
- use the evaluation checklist to evaluate items 1-3.
- use the rubic and model papers below to evaluate items 4-6.
- use the coping masters to evaluate interview/task tests on pp. 76-78.

SCORING RUBRIC FOR PROBLEM-SOLVING TASKS, ITEMS 4–6

Levels of Performance

Level 3 Accomplishes the purposes of the task. Student gives clear explanations, shows understanding of mathematical ideas and processes, and computes accurately.

Level 2 Purposes of the task not fully achieved. Student demonstrates satisfactory but limited understanding of the mathematical ideas and processes.

Level 1 Purposes of the task not accomplished. Student shows little evidence of understanding the mathematical ideas and processes and makes computational and/or procedural errors.

CHAPTERS
1–3 **WHAT DID I LEARN?**

✏ **Write About It**

For items 1-3, see Evaluation Checklist in the *Performance Assessment Guide.* Explanations will vary.

1. Show how to change $\frac{3}{4}$ into a decimal. Explain your method.
(pages 20–23) Check students' explanations; 0.75

2. Choose a strategy and then explain how to solve the problem.
(pages 36–37) Check students' explanations; *guess and check is a possible strategy;* 30 pencils, 18 pens
Andy had 12 more pencils than pens. If Andy had a total of 48 pens and pencils, how many pens and how many pencils did Andy have?

3. Find the quotient of 34.6 ÷ 0.4. Explain your method and show your work. (pages 64–67)
Check students' explanations; 86.5

✔ Performance Assessment

Choose a strategy and solve. Explain your method.

Problem-Solving Strategies
- Find a Pattern
- Write an Equation
- Make a Table
- Make a Model
- Act It Out
- Make a Graph

Choice of strategies will vary. Explanations will vary.

4. Tabby had 10 pennies. Each week she received 10 times as many pennies. If this pattern continues, how many pennies will Tabby have after five weeks? Write the total using exponent form. (pages 24–25)
100,000; 10^5

5. The Broncos soccer team played a total of 24 games. They won 5 more games than they lost, and they tied in 3 games. How many games did they win? How many did they lose? (pages 36–37) **win: 13, lose: 8**

6. Shawna wants to buy film. She can buy 3 rolls of 24 exposures for $9.89, or she can buy single rolls of 36 exposures for $3.98 each. Which is the better deal? Explain why you think so. (pages 64–67)
Check students' explanations; 36 exposure roll; 24 exposures are about $0.14 per exposure while 36 exposures are about $0.11 per exposure

24 Exposures	3 Rolls for $9.89
36 Exposures	$3.98 each

STUDENT WORK SAMPLES for Item 4

Level 3

Tabby = 10
1ˢᵗ 10^1 = 10
2ⁿᵈ 10^2 = 100
3ᴿᵈ 10^3 = 1000
4ᵗʰ 10^4 = 10,000
5ᵗʰ 10^5 = 100,000

Tabby would have 100,000 pennies.

Level 2

1ˢᵗ
Tabby = 10
Pennies = 10
Pennies = 10
Pennies = 10
Pennies = 10

10
100
1000
10000
100,000

Level 1

(student table of powers of ten)

10×4 =48 200
2nd 200 5th

1ˢᵗ tabby = 1 = 10^0 10^0 10^0 10^0 10^0
2nd 2 = 10^1 10^2 10^1 10^1 10^1
3ʳᵈ 3 = 10^2 10^2 10^2 10^2 10^2
4ᵗʰ 4 = 10^3 10^4 10^4 10^4 10^4
5ᵗʰ 5 = 10^4 10^4 10^4 10^4 10^4
6ᵗʰ 6 = 10^5 10^5 10^5 10^5 10^5
7ᵗʰ 7 = 10^6 10^6 10^6 10^6 10^6
8ᵗʰ 8 = 10^7 10^7 10^7 10^7 10^7
9ᵗʰ 9 = 10^8 10^8 10^8 10^8 10^8
10ᵗʰ 10 = 10^9 10^9 10^9 10^9 10^9

10×4 =40 (×)5 = 200

The student showed the pattern of exponents and multiples of ten in a list. The answer is correct and the total is shown in exponent form.

The student showed an understanding of the pattern of multiples of ten but did not use exponents.

The student showed organization in her listing of the powers of ten. However, her table is not explained and her answer is not correct.

CUMULATIVE REVIEW

Solve the problem. Then write the letter of the correct answer.

1. Which is *five hundred five thousand, five hundred fifty and fifty-five ten thousandths* in standard form? (pages 16–17)

A. 55,055.55
B. 505,550.055
C. 505,550.0055
D. 505,000,550.55 **1A.1**

2. Compare. 9.45 ● 9.452 (pages 18–19)

A. <
B. >
C. =
D. ≥ **1A.1**

3. Change 0.005 to a fraction. (pages 20–23)

A. $\frac{5}{10}$
B. $\frac{5}{100}$
C. $\frac{5}{1,000}$
D. $\frac{5}{10,000}$ **1A.2**

4. Find the value of 10^5. (pages 24–25)

A. 50
B. 105
C. 10,000
D. 100,000 **1A.3**

5. Compare. ⁻5 ● 0 (pages 28–29)

A. <
B. >
C. =
D. ≥ **1A.4**

6. Use mental math to add. 13 + 26 + 47 + 54 (pages 34–35)

A. 100
B. 130
C. 140
D. 150 **2A.1**

7. 6,478 × 82 (pages 40–43)

A. 531,196
B. 494,486
C. 64,780
D. 79 **2A.3**

8. 24⟌4916 (pages 40–43)

A. 24 r20
B. 204
C. 204 r20
D. 117,984 **2A.3**

9. 42.6 + 0.968 (pages 54–55)

A. 0.1394
B. 4.3568
C. 43.568
D. 4,356.8 **3A.1**

10. 36.1 − 18.298 (pages 56–57)

A. 1.7802
B. 17.802
C. 17.998
D. 1,780.2 **3A.1**

11. 8.085 × 3.5 (pages 58–61)

A. 2,829.75
B. 282.975
C. 28.2975
D. 2.31 **3A.2**

12. 36.12 ÷ 4.2 (pages 64–67)

A. 0.086
B. 8.6
C. 860
D. 151.704 **3A.3**

CUMULATIVE REVIEW & TEST

✔ *Assessment Checkpoint*

USING THE PAGE

This review is in multiple-choice format to provide the students with practice for standardized testing. More in-depth Cumulative Tests are provided in free-response and multiple-choice formats. See *Test Copying Masters*, pages A13–A16 for the multiple-choice test and pages B163–B166 for the free-response test.

CUMULATIVE TEST
FREE-RESPONSE FORMAT
CHAPTERS 1–3, PAGES 1-4

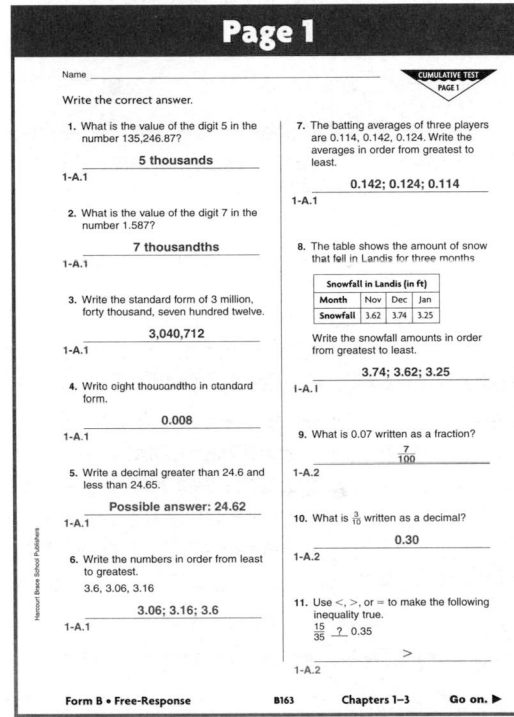

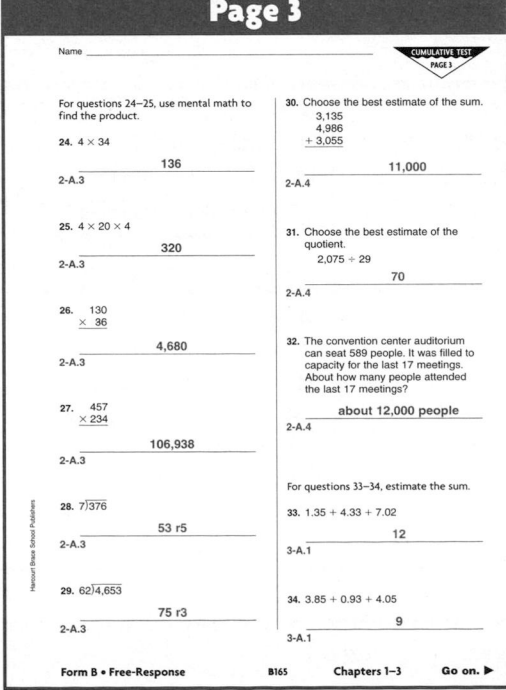

Number Theory and Fractions

BIG IDEA: Factors, multiples, and primes relate directly to computational algorithms and real-world problem solving.

Introducing the Chapter p. 74		Team-Up Project • Foods From Different Cultures p. 75	
OBJECTIVE	**VOCABULARY**	**MATERIALS**	**RESOURCES**
4.1 Multiples and Factors pp. 76-77 **Objective:** To find multiples and factors and to tell whether a number is prime or composite **1 DAY**	*multiple* **prime number** **composite number**		■ Reteach, ■ Practice, ■ Enrichment 4.1; More Practice, p. H46 Cultural Connection • Holiday Trip to Japan, p. 90
4.2 Prime Factorization pp. 78-79 **Objective:** To write a composite number as the product of prime numbers **1 DAY**	**prime factorization**		■ Reteach, ■ Practice, ■ Enrichment 4.2; More Practice, p. H47
EXTENSION • Primes and Composites, Teacher's Edition pp. 80A–80B			
4.3 Least Common Multiple and Greatest Common Factor pp. 80-83 **Objective:** To find the LCM and GCF of numbers **2 DAYS**	**least common multiple (LCM)** **greatest common factor (GCF)**		■ Reteach, ■ Practice, ■ Enrichment 4.3; More Practice, p. H47
Lab Activity: Equivalent Fractions pp. 84-85 **Objective:** To use fraction bars to find equivalent fractions		FOR EACH STUDENT: fraction bars	🖥 **E-Lab • Activity 4** E-Lab Recording Sheets, p. 4
4.4 Fractions in Simplest Form pp. 86-87 **Objective:** To write fractions in simplest form **1 DAY**	**simplest form**		■ Reteach, ■ Practice, ■ Enrichment 4.4; More Practice, p. H47 💿 **Number Heroes •** *Fraction Fireworks*
4.5 Mixed Numbers and Fractions pp. 88-89 **Objective:** To write fractions as mixed numbers and mixed numbers as fractions **1 DAY**	**mixed number**		■ Reteach, ■ Practice, ■ Enrichment 4.5; More Practice, p. H48 🖩 Calculator Activities, p. H31
CHAPTER ASSESSMENT Chapter 4 Review p. 91			

NCTM CURRICULUM
STANDARDS FOR GRADES 5–8

- ✔ **Problem Solving** *Lessons 4.1, 4.2, 4.3, 4.4, 4.5*
- ✔ **Communication** *Lessons 4.1, 4.2, 4.3, 4.4, 4.5*
- ✔ **Reasoning** *Lessons 4.1, 4.2, 4.3, 4.4, 4.5*
- ✔ **Connections** *Lessons 4.1, 4.2, 4.3, 4.4, 4.5*
- ✔ **Number and Number Relationships** *Lessons 4.1, 4.2, 4.3, 4.4, 4.5*
- ✔ **Number Systems and Number Theory** *Lessons 4.1, 4.2, 4.3, 4.4, 4.5*
- ✔ **Computation and Estimation** *Lessons 4.1, 4.2, 4.3, Lab Activity, 4.4, 4.5*
- ✔ **Patterns and Functions** *Lessons 4.1, 4.2, 4.3, 4.4*
- ✔ **Algebra** *Lessons 4.2, 4.4*
- ✔ **Statistics** *Lesson 4.1*
- ☐ **Probability**
- ✔ **Geometry** *Lesson 4.5*
- ✔ **Measurement** *Lessons 4.3, 4.5*

CHAPTER LEARNING GOALS

4A To solve problems that involve number theory and fractions

4A.1 To identify factors and multiples of a number and tell whether a number is prime or composite

4A.2 To write a composite number as the product of prime numbers

4A.3 To find the least common multiple and greatest common factor

4A.4 To write fractions in simplest form

4A.5 To write fractions as mixed numbers and mixed numbers as fractions

These goals for concepts, skills, and problem solving are assessed in many ways throughout the chapter. See the chart below for a complete listing of both traditional and informal assessment options.

Pretest Options
Pretest for Chapter Content, TCM pp. A17–A18, B167–B168
Assessing Prior Knowledge, TE p. 74; APK p. 6

Daily Assessment
Mixed Review, PE pp. 83, 87
Problem of the Day, TE pp. 76, 78, 80, 86, 88
Assessment Tip and Daily Quiz, TE pp. 77, 79, 83, 85, 87, 89

Formal Assessment
Chapter 4 Review, PE p. 91
Chapter 4 Test, TCM pp. A17–A18, B167–B168
Study Guide & Review, PE pp. 144–145
Test • Chapters 4–7 Test, TCM pp. A25–A26, B175–B176
Cumulative Review, PE p. 147
Cumulative Test, TCM pp. A27–A30, B177–B180

Performance Assessment
Team-Up Project Checklist, TE p. 75
What Did I Learn?, Chapters 4–7, TE p. 146
Interview/Task Test, PA p. 79

Portfolio
Suggested work samples:
Independent Practice, PE pp. 77, 79, 83, 87, 89
Cultural Connection, PE p. 90

Student Self-Assessment
How Did I Do?, TGA p 15
Portfolio Evaluation Form, TGA p. 24
Math Journal, TE pp. 74B, 84

Key

Assessing Prior Knowledge Copying Masters: APK	Teacher's Guide for Assessment: TGA
Test Copying Masters: TCM	Teacher's Edition: TE
Performance Assessment: PA	Pupil's Edition: PE

MATH JOURNAL
You may wish to use the following suggestions to have students write about the mathematics they are learning in this chapter:

PE page 80 • Talk About It

PE pages 77, 79, 83, 87, 89 • Write About It

TE pages 84 • Writing in Mathematics

TGA page 15 • How Did I Do?

In every lesson • record written answers to each Wrap-up Assessment Tip from the Teacher's Edition for test preparation.

VOCABULARY DEVELOPMENT
The terms listed at the right will be introduced or reinforced in this chapter. The boldfaced terms in the list are new vocabulary terms in the chapter.

So that students understand the terms rather than just memorize definitions, have them construct their own definitions and pictorial representations and record those in their Math Journals.

multiple, pp. 76, 80, 81
prime number, pp. 76, 78
composite number, pp. 76, 78
prime factorization, p. 78
least common multiple (LCM), p. 80
greatest common factor (GCF), pp. 81, 82, 86
simplest form, p. 86
mixed number, p. 88

MEETING INDIVIDUAL NEEDS

RETEACH	PRACTICE
BELOW-LEVEL STUDENTS	**ON-LEVEL STUDENTS**
Practice Activities, TE Tab B *Switching Digits* (Use with Lessons 4.4–4.5)	**Bulletin Board** TE p. 74C *A Forest of Factors* (Use with Lessons 4.1–4.5)
Interdisciplinary Connection, TE p. 76	**Interdisciplinary Connection,** TE p. 76
Literature Link, TE p. 74C *The School Play,* Activity 1 (Use with Lessons 4.1–4.5)	**Literature Link,** TE p. 74C *The School Play,* Activity 2 (Use with Lessons 4.1–4.5)
Technology *Number Heroes* • *Fraction Fireworks* (Use with Lesson 4.4.)	**Technology** *Number Heroes* • *Fraction Fireworks* (Use with Lesson 4.4.)
E-Lab • Activity 4 (Use with Lab Activity.)	**E-Lab** • Activity 4 (Use with Lab Activity.)

CHAPTER IDEA FILE

The activities on these pages provide additional practice and reinforcement of key chapter concepts. The chart above offers suggestions for their use.

 **LITERATURE LINK**

The School Play

by Gary Soto

A class of sixth graders puts on a play for the school and realizes that making mistakes is not the end of the world.

Activity 1—Mrs. Bunnin's sixth graders are rehearsing a play. They have 3 weeks to rehearse. Have students make up the schedule for the class that adds up to 1 hour of practice per school day.

Activity 2—In the story, each student who got at least 12 words out of 15 right on the spelling test was given a speaking part in the class play. Have students count their classmates and make up a number of students who could have done well on the test. They then think about how to represent the fraction of the class who would have speaking parts.

Activity 3—Ask each student to think of a favorite movie and assign the key roles to classmates. Have students write the fraction of the girls who would have roles and the fraction of the boys who would have roles. Students compare their fractions to see who was able to cast the most students.

BULLETIN BOARD

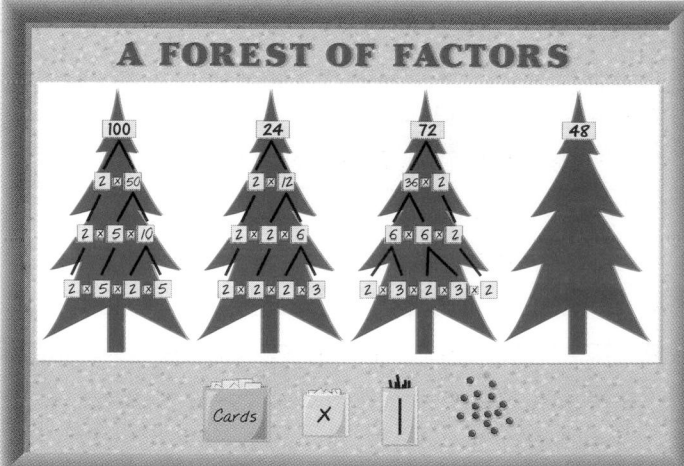

Vary daily the numbers at the top of each tree to review factorization as you complete Chapters 4–7.

Activity 1—Students create a factor tree for a number posted on a pine tree.

Activity 2—Students complete 3 factor trees and, if possible, write out the prime factorization in exponent form.

Bulletin Board TE p. 74C
A Forest of Factors
(Use with Lessons 4.1–4.5)

Interdisciplinary Connection, TE p. 76

 **Literature Link,** TE p. 74C
The School Play, Activity 3
(Use with Lessons 4.1–4.5)

Technology
Number Heroes • *Fraction Fireworks*
(Use with Lesson 4.4.)

E-Lab • Activity 4
(Use with Lab Activity.)

INTERDISCIPLINARY SUGGESTIONS

Connecting Number Theory and Fractions

Purpose To connect number theory and fractions to other subjects students are studying.

You may wish to connect ***Number Theory and Fractions*** to other subjects by using related interdisciplinary units.

Suggested Units

- Math in History: The Sieve of Eratosthenes
- Party Planning: Making Numbers Come Out Even
- Language Arts: Make a Book of Number Riddles
- Artistic Factors: Design Spiraling Pinwheels

Modalities Auditory, Kinesthetic, Visual

Multiple Intelligences Mathematical/ Logical, Visual/Spatial, Bodily/Kinesthetic

---

Technology

The purpose of using technology in the chapter is to provide reinforcement of skills and support in concept development.

E-Lab • Activity 4—Students use a tool to find factors.

Number Heroes • *Fraction Fireworks*—The activities on this CD-ROM provide students with opportunities to practice finding equivalent fractions.

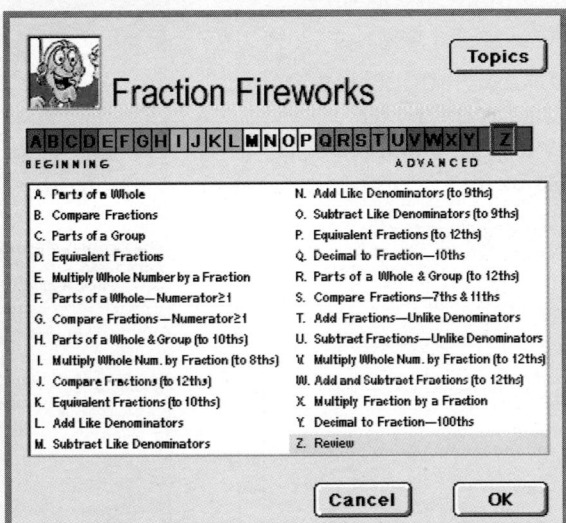

| Topics |
Fraction Fireworks

A B C D E F G H I J K L M N O P Q R S T U V W X Y Z
BEGINNING   ADVANCED

A. Parts of a Whole
B. Compare Fractions
C. Parts of a Group
D. Equivalent Fractions
E. Multiply Whole Number by a Fraction
F. Parts of a Whole—Numerator≥1
G. Compare Fractions—Numerator≥1
H. Parts of a Whole & Group (to 10ths)
I. Multiply Whole Num. by Fraction (to 8ths)
J. Compare Fractions (to 12ths)
K. Equivalent Fractions (to 10ths)
L. Add Like Denominators
M. Subtract Like Denominators

N. Add Like Denominators (to 9ths)
O. Subtract Like Denominators (to 9ths)
P. Equivalent Fractions (to 12ths)
Q. Decimal to Fraction—10ths
R. Parts of a Whole & Group (to 12ths)
S. Compare Fractions—7ths & 11ths
T. Add Fractions—Unlike Denominators
U. Subtract Fractions—Unlike Denominators
V. Multiply Whole Num. by Fraction (to 12ths)
W. Add and Subtract Fractions (to 12ths)
X. Multiply Fraction by a Fraction
Y. Decimal to Fraction—100ths
Z. Review

| Cancel | | OK |

Each Number Heroes activity can be assigned by degree of difficulty. See Teacher's Edition page 87 for Grow Slide information.

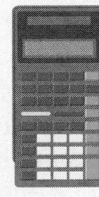 **Explorer Plus™ or fx–65 Calculator**—See suggestions for integrating the calculator in the PE Lesson p. 88. Additional ideas are provided in *Calculator Activities,* p. H31.

Visit our web site for additional ideas and activities.
http://www.hbschool.com

Introducing the Chapter

Encourage students to talk about equivalent fractions. Review the math content students will learn in the chapter.

WHY LEARN THIS?

The question "Why are you learning this?" will be answered in every lesson. Sometimes there is a direct application to everyday experiences. Other lessons help students understand why these skills are needed for future success in mathematics.

PRETEST OPTIONS

Pretest for Chapter Content See the *Test Copying Masters*, pages A17–A18, B167–B168 for a free-response or multiple-choice test that can be used as a pretest.

Assessing Prior Knowledge Use the copying master shown below to assess previously taught skills that are critical for success in this chapter.

Assessing Prior Knowledge Copying Masters, p. 6

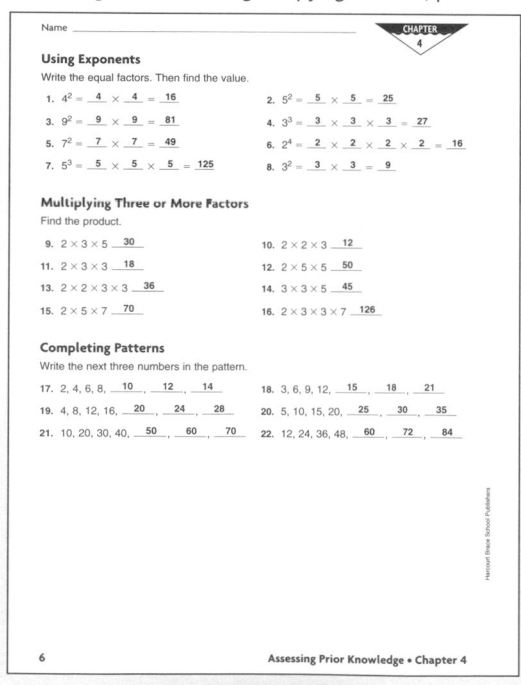

Troubleshooting

If students need help	Then use
• USING EXPONENTS	• More Practice, Lesson 1.4 PE p. H43
• MULTIPLYING WHOLE NUMBERS	• Practice Activities, TE Tab A, Activity 8A
• COMPLETING PATTERNS	• Focus on Problem Solving, PE p. 8

CHAPTER 4
NUMBER THEORY AND FRACTIONS

LOOK AHEAD

In this chapter you will solve problems that involve

• multiples and factors
• prime factorization

• the least common multiple and the greatest common factor
• equivalent fractions, fractions in simplest form, and mixed numbers

74 Chapter 4

SCHOOL-HOME CONNECTION

You may wish to send to each student's family the *Math at Home* page. It includes an article that stresses the importance of communication about mathematics in the home, as well as helpful hints about homework and taking tests. The activity and extension provide extra practice that develops concepts and skills involving least common multiples and greatest common factors.

Bits and Pieces: Number Theory and Fractions

In this chapter your child will learn key number-theory concepts—such as prime numbers, factors, and multiples—that will make the study of fractions easier and more meaningful. Ask your child what he or she has learned in class.

Here is an activity that involves finding a group of specially defined numbers. The Extension looks at a relationship that's useful for writing fractions in simplest form.

Is It Perfect?

A number is called perfect if all its factors, other than the number itself, add up to the number. For example, 6 is a perfect number.

The factors of 6 (numbers that can be multiplied to equal 6) are 1, 2, 3, 6.

Add all the factors except 6: $1 + 2 + 3 = 6$.

Since the factors less than the number itself add up to equal 6, 6 is perfect.

Determine which numbers less than or equal to 30 are perfect numbers.

Extension • Guess the Rule

The following pairs of numbers are said to be relatively prime:

4 and 9; 12 and 25; 16 and 49; 10 and 27; 15 and 169.

The following pairs of numbers are not relatively prime:

10 and 15; 4 and 8; 16 and 24; 21 and 42; 49 and 21.

What do you think *relatively prime* means? How is this concept related to fractions?

For Your Information

Parents should take an interest in their child's homework. However, parents need to be careful not to cross the line and do the work for the child. If the child brings into class a completed assignment that indicates he or she had no difficulty, the teacher cannot accurately gauge the child's progress. Furthermore, doing the homework for your child will not help him or her gain self confidence.

Encourage your child to use pictures, diagrams, or manipulatives to help clarify problems on an assignment. For example, in looking for fractions equivalent to $\frac{1}{2}$, the child could draw pictures of regions or use counters.

Remind your child to be sure that all answers on a test are clearly indicated and properly labeled. The teacher should not have to hunt around the test paper or work paper to find the answers.

Team-Up Project

Foods From Different Cultures

Suppose you want to prepare lunch for your class. Your lunch is to represent the food of another culture. Make a menu and a shopping list of what you need. Include the amount needed of each item.

YOU WILL NEED: magazines, recipe books

Plan
- Work with a small group to make a menu. Make a shopping list of supplies needed to provide lunch for your class.

Act
- Decide what foods to serve. Make a menu.
- Look in books or magazines to find recipes.
- List the amount of each item you need to buy.

Share
- Show your menu. Describe your lunch and tell what culture your foods are from.
- Explain how you made your supply list.

DID YOU
- ✓ Make a menu?
- ✓ Make a list of the amounts of each item needed?
- ✓ Share your work with the class?

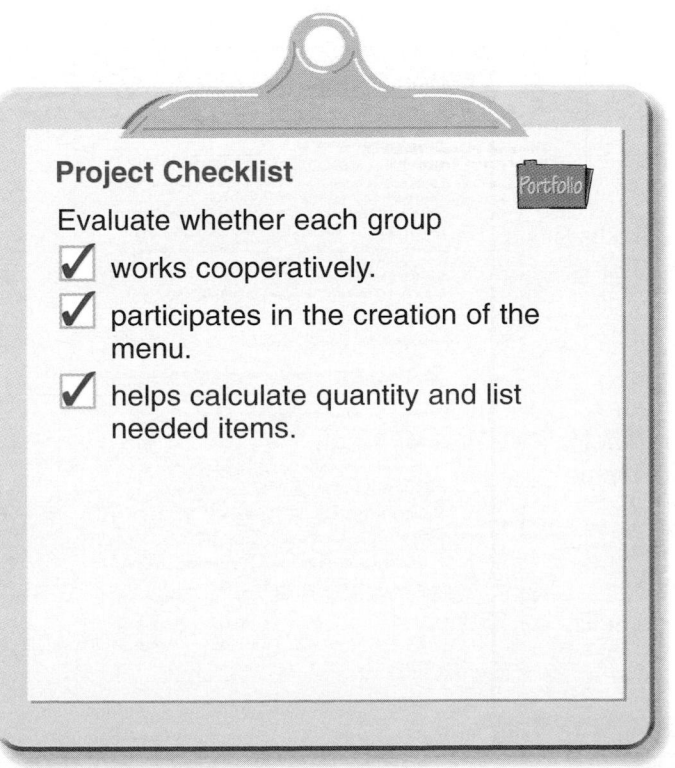

Project Checklist

Evaluate whether each group
- ✓ works cooperatively.
- ✓ participates in the creation of the menu.
- ✓ helps calculate quantity and list needed items.

Using the Project

Accessing Prior Knowledge multiplying or adding whole numbers and fractions

Purpose to add fractions

GROUPING small groups
MATERIALS recipes, cookbooks

Managing the Activity You may want to provide information about traditional foods from countries your class has studied or will study. Ask students to describe any foods from these countries with which they are familiar.

WHEN MINUTES COUNT

A shortened version of this activity can be completed in 15–20 minutes. Have students make a menu and list supplies needed to provide lunch for the entire class.

A BIT EACH DAY

Day 1: Students form groups and select a type of food to serve at their lunch.

Day 2: Students make a menu of the foods they want to serve and find recipes.

Day 3: Students count the number of students in the class. They make a list of the amount of each ingredient they need to make food for the entire class.

Day 4: Groups describe their lunches and present their menus to the class.

Day 5: Groups exchange menus and shopping lists. They take turns explaining how they determined shopping list quantities.

HOME HAPPENINGS

Encourage students to ask about the history of foods that are traditionally eaten in their family.

GRADE 5	GRADE 6	GRADE 7
Model factors of numbers to identify prime and composite numbers.	Find multiples and factors and identify numbers as prime or composite.	Use multiples and factors to add and subtract fractions.

ORGANIZER

OBJECTIVE To find multiples and factors and to tell whether a number is prime or composite

VOCABULARY *Review* multiple
New **prime number, composite number**

ASSIGNMENTS Independent Practice
- *Basic* 1–40, even; 41–44
- *Average* 1–44
- *Advanced* 10–40, even; 41–44

PROGRAM RESOURCES
Reteach 4.1 Practice 4.1 Enrichment 4.1
Cultural Connection • Holiday Trip to Japan, p. 90

PROBLEM OF THE DAY

Transparency 4

The sum of the ages of Mr. and Mrs. Olsen and their two children is 108. Their ages are sets of twin prime numbers. What are their ages?
NOTE: Twin primes are two prime numbers whose difference is 2. 11 + 13 + 41 + 43 = 108

Solution Tab A, p. A4

QUICK CHECK
Write the missing numbers.

1. 1 × ___ = 9 9
   ___ × 3 = 9 3
2. ___ × 15 = 15 1
   5 × ___ = 15 3
3. 1 × 18 = ___ 18
   ___ × 9 = 18 2
4. ___ × 16 = 16 1
   2 × ___ = 16 8

1 MOTIVATE

To access students' understanding of multiples and factors, ask:

- **How many different rectangles can you make on grid paper using 16 squares? Describe your rectangles.** 3 rectangles—1 by 16; 2 by 8; 4 by 4

- **How many rectangles can you make using 11 squares?** 1 rectangle—1 by 11

- **How many rectangles can you make using 9 squares?** 2 rectangles—1 by 9, 3 by 3

4.1

WHAT YOU'LL LEARN
How to find multiples and factors and how to tell whether a number is prime or composite

WHY LEARN THIS?
To use patterns of multiples in making designs and developing schedules, such as for exercise sessions

WORD POWER
prime number
composite number

REMEMBER:
Multiples of a number are the products when a number is multiplied by 1, 2, 3, 4, and so on.

When you multiply two or more numbers to get a product, the numbers multiplied are called factors. **See page H3.**

Multiples and Factors

Albert has football practice every third day during November. His first day of practice is November 3. On what other dates in November will he have football practice?

S	M	T	W	Th	F	S
					1	2
3	4	5	6	7	8	9
10	11	12	13	14	15	16
17	18	19	20	21	22	23
24	25	26	27	28	29	30

The dates of Albert's practices are multiples of 3.

To find multiples of any number, multiply the number by the counting numbers 1, 2, 3, 4, and so on. The first six multiples of 3 are shown below.

0 3 6 9 12 15 18
3×1 3×2 3×3 3×4 3×5 3×6

- What are the first five multiples of 9? **9, 18, 27, 36, 45**

Remember that when you multiply, you are using factors.

$3 \times 6 = 18 \leftarrow$ 3 and 6 are factors of 18.

Some numbers, such as 5, have only two factors: 1 and the number itself. Numbers with only two factors are called **prime numbers**.

$5 = 5 \times 1 \leftarrow$ factors of 5: 1 and 5
$7 = 7 \times 1 \leftarrow$ factors of 7: 1 and 7

Numbers that have more than two factors are called **composite numbers**.

$6 = 1 \times 6$
$6 = 2 \times 3$ $\leftarrow$ factors of 6: 1, 2, 3, and 6

The numbers 0 and 1 are neither prime nor composite.

EXAMPLES Tell whether each number is prime or composite.

A. 12
factors: 1, 2, 3, 4, 6, 12

12 is composite.

B. 23
factors: 1, 23

23 is prime.

C. 63
factors: 1, 3, 7, 9, 21, 63

63 is composite.

- Starting with 2, what are the first ten prime numbers?
2, 3, 5, 7, 11, 13, 17, 19, 23, 29

IDEA FILE

INTERDISCIPLINARY CONNECTION

One suggestion for an interdisciplinary unit for this chapter is Math in History: The Sieve of Eratosthenes. You can have students

- research when and where Eratosthenes lived and the mathematical discoveries he made.

- explain how his system of sifting for primes works.

Modalities Auditory, Kinesthetic, Visual

RETEACH 4.1

Name _____

Multiples and Factors

You multiply numbers called *factors* to get a product.
- Some numbers have only two factors, the number itself and 1. These special numbers are called **prime numbers**.
- Numbers with more than two factors are called **composite numbers**.

List the factors of 24. Then tell whether 24 is prime or composite.
Step 1 Any number multiplied by 1 equals that number. So, one pair of factors is 1 and 24.
Step 2 24 is even, so it is divisible by 2. Another pair of factors is 2 and 12.
Step 3 Is 24 divisible by 3? Yes. So, another pair of factors is 3 and 8.
Step 4 Check to see if the number is divisible by 4, 5, and 6. Another pair of factors is 4 and 6.

You have found the following factors of 24:
1, 2, 3, 4, 6, 8, 12, 24.
Since 24 has more than one pair of factors, it is a composite number.

Complete to find the factors of the number. Then tell whether the number is prime or composite.
1. 16
Any number multiplied by __1__ equals that number. So, one pair of factors is 1 and __16__. Since the number is even, it is divisible by __2__. So, another pair of factors is __2__ and __8__. A third pair of factors is __4__ and __4__. Since 16 has more than two factors, it is a __composite__ number.

Write P for *prime* or C for *composite*.
2. 12 __C__ 3. 21 __C__ 4. 13 __P__ 5. 18 __C__ 6. 29 __P__
7. 41 __P__ 8. 63 __C__ 9. 15 __C__ 10. 19 __P__ 11. 72 __C__

Use with text pages 76–77.   **TAKE ANOTHER LOOK** R15

GUIDED PRACTICE

Write the first three multiples.

1. 4 4, 8, 12
2. 7 7, 14, 21
3. 9 9, 18, 27
4. 13 13, 26, 39

5. 27 27, 54, 81
6. 50 50, 100, 150
7. 19 19, 38, 57
8. 14 14, 28, 42

Write the factors. Tell whether the number is *prime* or *composite*.

9. 12 1, 2, 3, 4, 6, 12; composite
10. 11 1, 11; prime
11. 15 1, 3, 5, 15; composite
12. 23 1, 23; prime

INDEPENDENT PRACTICE

Name the first four multiples.

1. 25 25, 50, 75, 100
2. 10 10, 20, 30, 40
3. 21 21, 42, 63, 84
4. 15 15, 30, 45, 60
5. 11 11, 22, 33, 44
6. 16 16, 32, 48, 64

Find the missing multiple or multiples.

7. 8, 16, 24, __?__, __?__ 32, 40
8. __?__, 24, 36, 48 12
9. __?__, 14, 21, 28, __?__ 7, 35

Write the factors.

10. 9 1, 3, 9
11. 16 1, 2, 4, 8, 16
12. 12 1, 2, 3, 4, 6, 12
13. 37 1, 37
14. 34 1, 2, 17, 34

15. 18 1, 2, 3, 6, 9, 18
16. 23 1, 23
17. 42 1, 2, 3, 6, 7, 14, 21, 42
18. 121 1, 11, 121
19. 41 1, 41

20. 27 1, 3, 9, 27
21. 54 1, 2, 3, 6, 9, 18, 27, 54
22. 31 1, 31
23. 77 1, 7, 11, 77
24. 84 1, 2, 3, 4, 6, 7, 12, 14, 21, 28, 42, 84

Write P for *prime* or C for *composite*.

25. 31 P
26. 14 C
27. 9 C
28. 20 C
29. 33 C
30. 11 P

31. 16 C
32. 29 P
33. 30 C
34. 51 C
35. 48 C
36. 100 C

37. How do you know a number is prime? **It has only two factors, 1 and itself.**

38. What is a composite number? **a number that has more than two factors**

39. List the prime numbers between 20 and 45. **23, 29, 31, 37, 41, 43**

40. List the composite numbers from 80 through 90. **80, 81, 82, 84, 85, 86, 87, 88, 90**

Problem-Solving Applications

41. Jasmine's mother filled her car's gasoline tank every eighth day in June, beginning on June 8. How many times did she fill it in June? On what dates? **3 times; 8, 16, 24**

42. Juan has a violin lesson every fourth day during September. His first lesson is on September 4. What are the dates of his other lessons in September? **8, 12, 16, 20, 24, 28**

43. Can a composite number have prime numbers as factors? Explain. **Yes. Composite numbers can have 2, 3, 5, 7, and so on as factors.**

44. ◀▶ **WRITE ABOUT IT** Give an example to show how a factor of a number and a multiple of that number are related. **3 is a factor of 9, 18 is a multiple of 9, and 3 is a factor of 18**

MORE PRACTICE Lesson 4.1, page H46

77

2 TEACH/PRACTICE

In this lesson students use multiples of a number to determine a boy's football schedule for one month. Students broaden their understanding of number theory as they explore the properties of prime and composite numbers.

GUIDED PRACTICE

Encourage students to use mental math to solve Exercises 1–8. Some students may want to draw rectangles using squares on grid paper to help them find the factors for Exercises 9–12.

INDEPENDENT PRACTICE

Students may choose to use a calculator to help them solve some exercises. If they have difficulty with Exercises 7–9, suggest that they look for a pattern in each set of numbers.

ADDITIONAL EXAMPLE

Example, p. 76

Tell whether each number is prime or composite.

1. 71
factors: 1, 71;
71 is prime.

2. 14
factors: 1, 2, 7, 14;
14 is composite.

3 WRAP UP

Use these questions to help students focus on the big ideas of the lesson.

• **Can a factor of a number be a multiple of the same number? Explain.** Yes, because every number has 1 and itself as factors, and 1 times any number is a multiple of that number.

✔ *Assessment Tip*

Have students write the answer.

• **Can a prime number have composite numbers as factors? Explain.** No. A prime number has only two factors: 1 and the number itself.

DAILY QUIZ

Tell whether the number is *prime* or *composite*.

1. 36 composite
2. 127 prime
3. 43 prime
4. 38 composite
5. 55 composite

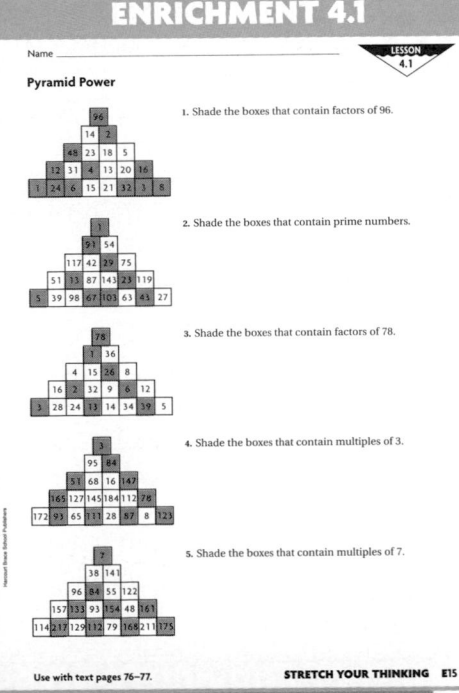

PRACTICE 4.1

Name _____

LESSON 4.1

Multiples and Factors

Vocabulary

Complete.

1. Numbers that have more than two factors are called _____ **composite numbers**

2. Numbers with only two factors are called _____ **prime numbers**

Name the first four multiples.

3. 12   4. 20   5. 18   6. 13
12, 24, 36, 48   20, 40, 60, 80   18, 36, 54, 72   13, 26, 39, 52

Find the missing multiples.

7. 6, 12, **18**, **24**   8. 14, 28, **42**, **56**, 70   9. **13**, 26, 39, **52**, 65

Write the factors.

10. 28   11. 51   12. 37   13. 72 **1, 2, 3, 4, 6, 8, 9**
1, 2, 4, 7, 14, 28   1, 3, 17, 51   1, 37   12, 18, 24, 36, 72

Write P for prime or C for composite.

14. 54   **C**   15. 53   **P**   16. 42   **C**

Mixed Applications

17. Fran has 36 calendars. Each person in her office can have 4 calendars. How many people work in Fran's office? Is that number a factor of 36? **9; yes**

18. Shawna is making a poster that measures 14 in. by 20 in. What is the area of the poster? **280 in.²**

19. Pete played tennis every sixth day in July, beginning on July 6. How many times did he play in July? On what dates? **5 times; July 6, July 12, July 18, July 24, July 30**

20. Grace wants to make equal-size packages of pens from 32 pens. In how many ways can she do that? How many pens would be in each package? **5 ways: packages of 1, 2, 4, 8, or 16**

Use with text pages 76–77.   ON MY OWN P15

ENRICHMENT 4.1

Name _____

LESSON 4.1

Pyramid Power

1. Shade the boxes that contain factors of 96.

2. Shade the boxes that contain prime numbers.

3. Shade the boxes that contain factors of 78.

4. Shade the boxes that contain multiples of 3.

5. Shade the boxes that contain multiples of 7.

Use with text pages 76–77.   STRETCH YOUR THINKING E15

77

GRADE 5	GRADE 6	GRADE 7
Model factors of numbers to identify prime and composite numbers.	Use two methods to find the prime factorization of composite numbers.	Use prime factorization to find LCM and GCF.

ORGANIZER

OBJECTIVE To write a composite number as the product of prime numbers

VOCABULARY New **prime factorization**

ASSIGNMENTS Independent Practice
- *Basic* 1–33, odd; 34–39
- *Average* 1–39
- *Advanced* 11–33, odd; 34–39

PROGRAM RESOURCES
Reteach 4.2 Practice 4.2 Enrichment 4.2
Extension • Primes and Composites,
TE pp. 80A–80B

PROBLEM OF THE DAY

Transparency 4

Which product is greater?

$95 \times 21 = \underline{\ ?\ }$ $57 \times 35 = \underline{\ ?\ }$

The products are the same

Solution Tab A, p. A4

QUICK CHECK

Write the number in standard form.

1. 4^3 $4 \times 4 \times 4$
2. 3^2 3×3
3. 6^4 $6 \times 6 \times 6 \times 6$
4. 2^5 $2 \times 2 \times 2 \times 2 \times 2$
5. 7^7 $7 \times 7 \times 7 \times 7 \times 7 \times 7 \times 7$

1 MOTIVATE

To access students' knowledge of factoring, ask:

- **How many different ways can you stack 12 videos if every stack has the same number of videos? Use counters to model the problem.** 6 ways: 1 stack of 12, 2 stacks of 6, 3 stacks of 4, 4 stacks of 3, 6 stacks of 2, 12 stacks of 1

- **What are the factors of 12?** 1, 2, 3, 4, 6, 12

- **Which factors are prime numbers?** 2, 3

- **Which factors are composite numbers?** 4, 6, 12

4.2

Prime Factorization

WHAT YOU'LL LEARN
How to write a composite number as the product of prime numbers

WHY LEARN THIS?
To compare the prime factors of two numbers and find the factors they have in common

WORD POWER
prime factorization

Computer Link

One of the largest known prime numbers to date has 420,921 digits. Suppose you just wanted to type out the largest prime number on your computer. If you could type 80 digits per line (without commas) and 60 lines per page, about how many pages would it take to display this number?

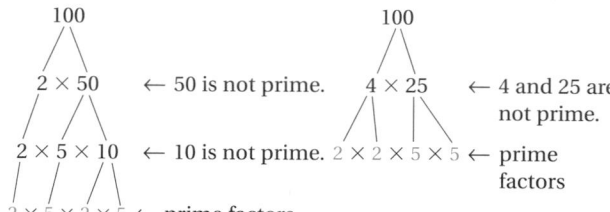

about 88 pages

You have learned that prime numbers have only two factors. The prime numbers less than 50 are listed below.

2, 3, 5, 7, 11, 13, 17, 19, 23, 29, 31, 37, 41, 43, 47

You have also learned that composite numbers have more than two factors. A composite number can be written as the product of prime factors. This is called the **prime factorization** of the number.

You can divide to find the prime factors of a number.

EXAMPLE 1 Find the prime factorization of 36.

$2\lfloor 36$
$2\lfloor 18$
$3\lfloor 9$
$3\lfloor 3$
$\quad 1$

Repeatedly divide by the smallest possible prime factor until the quotient is 1.

$2 \times 2 \times 3 \times 3$ *List the prime numbers you divided by. These are the prime factors.*

So, the prime factorization of 36 is $2 \times 2 \times 3 \times 3$, or $2^2 \times 3^2$.

- What is the prime factorization of 24? $2 \times 2 \times 2 \times 3$, or $2^3 \times 3$

Another Method You can use a factor tree to find the prime factors of a composite number. Different trees are possible but the prime factors are always the same.

EXAMPLE 2 Find the prime factorization of 100.
Choose any two factors of 100. Continue until only prime factors are left.

100
2×50 ← 50 is not prime.
$2 \times 5 \times 10$ ← 10 is not prime.
$2 \times 5 \times 2 \times 5$ ← prime factors

100
4×25 ← 4 and 25 are not prime.
$2 \times 2 \times 5 \times 5$ ← prime factors

So, the prime factorization of 100 is $2 \times 2 \times 5 \times 5$, or $2^2 \times 5^2$.

- What is the prime factorization of 66? $2 \times 3 \times 11$

78 Chapter 4

IDEA FILE

EXTENSION

Two prime numbers with a difference of 2 are called twin primes. For example, 17 and 19 are prime numbers, and $19 - 17 = 2$, so 17 and 19 are twin primes.

Ask:

- **Are the prime factors of 35 and 21 twin primes?** The prime factors of 35 are 5 and 7, which are twin primes. The prime factors of 21 are 3 and 7, which are not twin primes.

Modality Auditory

RETEACH 4.2

Name _____ LESSON 4.2

Prime Factorization

When you write a composite number as the product of prime factors, you have found the **prime factorization** of the number. A factor tree can help you find the prime factors of a composite number.
What is the prime factorization of 36?

Step 1 Choose any two factors of 36. Draw two lines from 36. Write one factor at the end of each line.

Step 2 Look at the factors. Are they prime? composite? Since they are composite, you must continue to find more factors.

Step 3 Look at the new bottom row of factors. Are they prime? composite? Since they are prime, you have found the prime factorization of 36.

You can write the prime factorization of 36 as $2 \times 2 \times 3 \times 3$, or $2^2 \times 3^2$.

Complete each factor tree to find the prime factors of the number.

1. 70
Prime factorization:
$2 \times \underline{5} \times \underline{7}$

2. 63
Prime factorization:
$3 \times \underline{3} \times \underline{7}$ or $3^2 \times \underline{7}$

3. 54
Prime factorization:
$\underline{2} \times 3 \times \underline{3} \times 3$ or $2 \times \underline{3^3}$

4. 200
Prime factorization:
$\underline{2} \times \underline{2} \times \underline{2} \times \underline{5} \times \underline{5}$
or $\underline{2^3} \times \underline{5^2}$

R16 TAKE ANOTHER LOOK Use with text pages 78–79.

GUIDED PRACTICE

Tell which of the prime numbers 2, 3, 5, and 7 are factors of the number.

1. 8 **2**
2. 24 **2, 3**
3. 30 **2, 3, 5**
4. 42 **2, 3, 7**
5. 100 **2, 5**

Draw a factor tree to show the prime factorization of the number. **Check students' factor trees.**

6. 12 **2 × 2 × 3**
7. 65 **5 × 13**
8. 16 **2 × 2 × 2 × 2**
9. 66 **2 × 3 × 11**
10. 42 **2 × 3 × 7**

Write the prime factorization in exponent form.

11. $2 \times 2 \times 2$ **2^3**
12. $2 \times 2 \times 7$ **$2^2 \times 7$**
13. $2 \times 3 \times 3 \times 3$ **2×3^3**
14. $2 \times 7 \times 7$ **2×7^2**
15. $3 \times 3 \times 3 \times 3$ **3^4**

INDEPENDENT PRACTICE

Use division to find the prime factors. Write the prime factorization.

1. 24 **2 × 2 × 2 × 3**
2. 36 **2 × 2 × 3 × 3**
3. 15 **3 × 5**
4. 50 **2 × 5 × 5**
5. 40
6. 93 **3 × 31**
7. 26 **2 × 13**
8. 38 **2 × 19**
9. 75 **3 × 5 × 5**
10. 88

See below for answers for exercises 5, 10, 15, 20, 25.

Use a factor tree to find the prime factors. Write the prime factorization in exponent form. **Check students' factor trees.**

11. 9 **3 × 3; 3^2**
12. 18 **2 × 3 × 3; 2×3^2**
13. 32 **2^5**
14. 49 **7 × 7; 7^2**
15. 27
16. 20 **2 × 2 × 5; $2^2 \times 5$**
17. 52 **2 × 2 × 13; $2^2 \times 13$**
18. 45 **3 × 3 × 5; $3^2 \times 5$**
19. 76 **2 × 2 × 19; $2^2 \times 19$**
20. 99
21. 56 **2 × 2 × 2 × 7; $2^3 \times 7$**
22. 48 **2 × 2 × 2 × 3; $2^4 \times 3$**
23. 72 **2 × 2 × 2 × 3 × 3; $2^3 \times 3^2$**
24. 64 **2 × 2 × 2 × 2 × 2; 2^6**
25. 84

Solve for *n* to complete the prime factorization.

26. $2 \times n \times 5 = 20$ **2**
27. $44 = 2 \times 2 \times n$ **11**
28. $n \times 3 \times 7 = 42$ **2**
29. $75 = 3 \times 5 \times n$ **5**
30. $n \times 13 = 26$ **2**
31. $3 \times n \times 7 = 105$ **5**
32. $2^n = 32$ **5**
33. $78 = 2 \times 3 \times n$ **13**

Problem-Solving Applications

34. The prime factors of a number are the three smallest prime numbers. No factor is repeated. What are the factors? What is the number? **2, 3, 5; 30**

35. **CRITICAL THINKING** The prime factorization of 25 is 5^2. Without dividing or using a factor tree, tell the prime factorization of 75. **3×5^2**

36. A number, *n*, is a prime factor of both 15 and 50. What is *n*? **n = 5**

37. A number, *a*, is a prime factor of both 12 and 60. What is *a*? **a = 3, or 2**

38. There are two numbers between 100 and 250 that have 3, 5, and 7 as prime factors. What are the numbers? **105 and 210**

39. ✏ **WRITE ABOUT IT** Do the prime factors of a number differ depending on which factors you choose first? Explain. **No. The order of factors does not affect the value of their product.**

5. **2 × 2 × 2 × 5**
10. **2 × 2 × 2 × 11**
15. **3 × 3 × 3; 3^3**
20. **3 × 3 × 11; $3^2 \times 11$**
25. **2 × 2 × 3 × 7; $2^2 \times 3 \times 7$**

MORE PRACTICE Lesson 4.2, page H47

PRACTICE 4.2

Name _____

LESSON 4.2

Prime Factorization

Vocabulary

1. Write *true* or *false*. Prime factorization is writing a composite number as the product of composite factors. ___**false**___

Use division to find the prime factors. Write the prime factorization.

2. 28
3. 50
4. 76
5. 108

2 × 2 × 7   **2 × 5 × 5**   **2 × 2 × 19**   **2 × 2 × 3 × 3 × 3**

Use a factor tree to find the prime factors. Write the prime factorization in exponent form. **Check students' factor trees.**

6. 55
7. 120
8. 92

5 × 11   **2 × 2 × 2 × 3 × 5; $2^3 \times 3 \times 5$**   **2 × 2 × 23; $2^2 \times 23$**

Solve for *n* to complete the prime factorization.

9. $n \times 17 = 51$ ___3___
10. $3^n \times 2 = 18$ ___2___
11. $2 \times 2 \times 2 \times n = 40$ ___5___

Mixed Applications

12. Beth has 6 coins in her pocket. The total amount is $1.00. She has no pennies. What are the coins?

___3 quarters, 2 dimes, 1 nickel or 5 dimes and 1 half-dollar or 1 half-dollar, 1 quarter, 1 dime, 3 nickels___

13. There are 24 students in Mrs. Garcia's class. She wants to divide the class evenly into groups of at least 4 students. Write the ways in which she can divide the class.

___2 groups of 12, 4 groups of 6, 6 groups of 4, or 3 groups of 8___

14. The two prime factors of a number are greater than 12 but less than 25. The number is greater than 300 but less than 400. What is the number?

___323; 17 and 19, or 391; 17 and 23___

15. The three prime factors of a number are all less than 10. The number is greater than 100. If no factor is repeated, what is the number? What are the factors?

___105; 3, 5, and 7___

P16   **ON MY OWN**   Use with text pages 78–79.

ENRICHMENT 4.2

Name _____

LESSON 4.2

Figure the Prime Factors

Shade one box in each row to show the prime factors of the given number.

1. 84
2. 225
3. 56
4. 144
5. 234
6. 96
7. 208
8. 340
9. 456

E16   **STRETCH YOUR THINKING**   Use with text pages 78–79.

This lesson offers students two methods of finding the prime factorization of a number: dividing by prime factors and using a factor tree.

GUIDED PRACTICE

In these exercises, students demonstrate their understanding of prime factorization. Check Exercises 6–10 to verify students' understanding of factor trees.

INDEPENDENT PRACTICE

Encourage students to check their answers by using mental math or a calculator.

ADDITIONAL EXAMPLES

Example 1, p. 78

Divide to find the prime factorization of 56.

$56 \div 2 = 28 \to 28 \div 2 = 14 \to 14 \div 2 = \to 7 \div 7 = 1$

The prime factorization of 56 is $2 \times 2 \times 2 \times 7$, or $2^3 \times 7$.

Example 2, p. 78

Use a factor tree to find the prime factorization of 42.

42 / 7 × 6 / 7 × 2 × 3

The prime factorization of 42 is $7 \times 3 \times 2$.

3 WRAP UP

Use these questions to help students focus on the big ideas of the lesson.

- **To find the prime factorization of a number, do you prefer the method shown in Example 1 or the method shown in Example 2? Why?** Possible answer: Example 2, because drawing a diagram helps me understand the problem.

✔ *Assessment Tip*

Have students write the answer.

- **How do you know when you have finished the prime factorization of a number?** All the factors are prime numbers.

DAILY QUIZ

Write the prime factorization in exponent form.

1. 36 **$2^2 \times 3^2$**
2. 54 **2×3^3**
3. 120 **$2^3 \times 3 \times 5$**
4. 63 **$3^2 \times 7$**

Solve for *y*.

5. $90 = 2 \times 3^y \times 5$ **y = 2**
6. $25 = 5^y$ **y = 2**

EXTENSION

Primes and Composites

Use these strategies to extend understanding of prime and composite numbers.

Multiple Intelligences Logical/Mathematical, Visual/Spatial

Modalities Auditory, Kinesthetic, Visual

Lesson Connection Students will extend their understanding of factors, primes, and composites.

Materials Graph paper

Program Resources *Intervention and Extension Copying Masters*, p. 14

Challenge

TEACHER NOTES

ACTIVITY 1

Purpose To explore a method of finding all prime numbers less than a given number

Vocabulary Review

- **prime number**—a whole number greater than 1 whose only factors are itself and 1

- **composite number**—a whole number greater than 1 with more than two whole-number factors

Discuss the following questions.

- Why is 1 not circled? It is not prime.

- In the sieve to 100, why were the multiples of 11 already crossed out by the time you circled 11? They are all multiples of smaller numbers, too.

WHAT STUDENTS DO

1. The Sieve of Eratosthenes is a method of finding prime numbers. The following sieve finds all primes up to 30.

Sieve of Eratosthenes

Here are the steps for making such a sieve.

a. Cross out 1.
b. Circle 2, a prime. Then cross out all multiples of 2.
c. Circle 3, and then cross out all multiples of 3. (Some will already be crossed out.)
d. Circle 5, the next number after 3 that is not already crossed out. Cross out multiples of 5.
e. Continue until every number is either circled or crossed out. The circled numbers are prime.

Now write the numbers 1–100 on graph paper. Make a sieve to find all the prime numbers under 100. There are 25 prime numbers under 100: 2, 3, 5, 7, 11, 13, 17, 19, 23, 29, 31, 37, 41, 43, 47, 53, 59, 61, 67, 71, 73, 79, 83, 89, and 97.

2. As you look at your sieve, what visual patterns do you see? Did you write 10 rows of 10, 20 rows of 5, or some other grid? How do the patterns on your sieve compare with those of your classmates?

Answers will vary.

TEACHER NOTES

ACTIVITY 2

Purpose To expand understanding of prime numbers by exploring twin primes

Discuss the following questions.

- Are 2 and 3 twin primes? No, they differ by 1.

- Look at the numbers that are between the twin primes in each pair, starting with the pair 5 and 7. What do these numbers have in common? They are all multiples of 6.

Have students meet in small groups or with partners to discuss the given questions. Have one student report the group's responses to the larger group.

WHAT STUDENTS DO

1. Prime numbers that differ by 2 are called **twin primes**. The first twin primes are 3 and 5.

   List all the twin primes under 100. Use the sieve you constructed in Activity 1. 3 and 5, 5 and 7, 11 and 13, 17 and 19, 29 and 31, 41 and 43, 59 and 61, 71 and 73.

2. Copy and complete the table. Then construct a bar graph showing the frequency of twin primes.

Twin Primes Under 100				
Interval	1–25	26–50	51–75	76–100
Frequency	4	2	2	0

Check students' graphs.

Share Results

- Why are there no even primes other than 2? All other even numbers are multiples of 2 and so are composite numbers.

- Why do you think the frequency of twin primes decreases as numbers get larger? The frequency of primes decreases.

Intervention and Extension Copying Masters, p. 14

ORGANIZER

GRADE 5	GRADE 6	GRADE 7
Find the LCM and GCF of numbers.	Find the LCM and GCF of numbers.	Use the LCM and GCF of numbers to compute fractions.

OBJECTIVE To find the LCM and GCF of numbers

VOCABULARY *New* **least common multiple (LCM), greatest common factor (GCF)**

ASSIGNMENTS Independent Practice
- ■ *Basic* 1–29, even; 30–33; 34–44
- ■ *Average* 1–33; 34–44
- ■ *Advanced* 13–33; 34–44

PROGRAM RESOURCES
Reteach 4.3 Practice 4.3 Enrichment 4.3

PROBLEM OF THE DAY

Transparency 4

Which number in each group does not belong with the rest? Why?

1. 19 15 20 18 21 **15; because the other numbers are consecutive numbers**
2. 426 792 158 345 620 **345; because the other numbers are even**
3. 36 56 63 84 49 **36; because the other numbers are multiples of 7**
4. 24 39 51 16 67 **67; because the other numbers are composites**

Solution Tab A, p. A4

QUICK CHECK
Use mental math to complete the pattern.

1. 5, 10, 15, __, __, __, __ **20, 25, 30, 35**
2. 7, 14, 21, __, __, __, __ **28, 35, 42, 49**
3. 15, 30, 45, __, __, __, __ **60, 75, 90, 105**
4. 9, 18, 27, __, __, __, __ **36, 45, 54, 63**
5. 11, 22, 33, __, __, __, __ **44, 55, 66, 77**

1 MOTIVATE

To illustrate the concepts involved in LCM and GCF, draw a number line on the board, labeled 1–20. Have students count by twos and write a 2 above each number they land on. Then have them count by threes. Ask:

4.3

WHAT YOU'LL LEARN
How to find the LCM and GCF of numbers

WHY LEARN THIS?
To solve everyday problems such as finding how many people can share things equally

WORD POWER
least common multiple (LCM)
greatest common factor (GCF)

Consumer Link
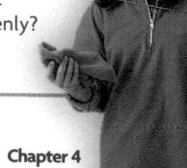
Hot dogs and buns aren't the only consumer products that don't always work out evenly. Suppose your older cousin is changing the oil in her car. Her car needs five quarts of oil every time it is changed, but the cheapest way to buy oil is by the gallon. How many gallons of oil should she buy, and how many oil changes will it take for everything to work out evenly?

80 Chapter 4

5 gallons; 4 oil changes

LCM and GCF

You have learned that you can find multiples of a number by multiplying the number by 1, 2, 3, 4, and so on.

This number line shows multiples of 4 and 6.

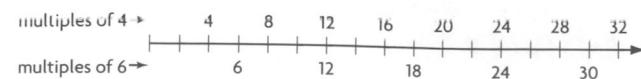
multiples of 4 → 4 8 12 16 20 24 28 32
multiples of 6 → 6 12 18 24 30

The multiples, such as 12 and 24, that are shown on each side of the number line are called common multiples. The smallest of the common multiples is called the **least common multiple**, or **LCM**.

Talk About It

- On the number line above, which multiples of 4 are also multiples of 6? **12 and 24**
- What is the least common multiple, or LCM, of 4 and 6? **12**
- What is another common multiple of 4 and 6? **Possible Answer: 36**
- 💡 **CRITICAL THINKING** Is there a greatest common multiple? Explain. **No. You can always continue multiplying.**

You can use the LCM to solve problems.

EXAMPLE 1 Frank is buying hot dogs for a class picnic. Hot dogs are sold in packages of 10. Hot-dog buns are sold in packages of 8. What is the smallest number of hot dogs and buns Frank can buy to have an equal number of each?

10: 10, 20, 30, 40, 50, 60, 70, 80, 90   *List the multiples.*
8: 8, 16, 24, 32, 40, 48, 56, 64, 72, 80   *Find the common multiples.*

The LCM of 10 and 8 is 40.   *Find the LCM.*

So, Frank needs 40 hot dogs and 40 buns.

- How many packages does Frank have to buy to get 40 hot dogs? to get 40 buns? **4 packages; 5 packages**
- What is the LCM of 3 and 7? **21**

RETEACH 4.3

Name _____

LCM and GCF

The greatest common factor, or GCF, of two numbers is the largest common factor of both numbers. You can use prime factors to find the GCF of two numbers.

What is the GCF of 28 and 36?

Step 1 Use factor trees to find the prime factors of the numbers.
28 = 2 × 2 × 7
36 = 2 × 2 × 3 × 3

Step 2 Find the prime factors that are in both trees.   2 and 2

Step 3 Multiply the common factors.   2 × 2 = 4

Using this method, you discover that the GCF of 28 and 36 is 4.

To find the least common multiple, or (LCM), of two numbers, you can list the multiples of each number. The smallest number in both lists is the LCM.

To find the LCM for 8 and 12:
8 → 8, 16, 24, 32, . . .
12 → 12, 24, 36, 48, . . .
The LCM for 8 and 12 is 24.

Complete to find the GCF of 12 and 72. Factor trees may vary.

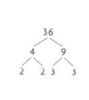

1. Use factor trees to find the prime factors of the numbers.
2. Find the common prime factors. **2, 2, and 3**
3. Multiply the common factors. **2 × 2 × 3 = 12**
4. The GCF of 12 and 72 is **12**

Find the LCM for each pair of numbers.

5. 9 and 12 **36**
6. 20 and 15 **60**

Use with text pages 80–83.   **TAKE ANOTHER LOOK   R17**

PRACTICE 4.3

Name _____

LCM and GCF

Vocabulary

Complete.

1. The smallest of the common multiples is called the **least common multiple, or LCM**
2. The largest of the common factors is called the **greatest common factor, or GCF**

Find the LCM for each set of numbers.

3. 12, 18	4. 7, 14	5. 16, 20	6. 4, 5, 6	7. 2, 6, 7
36	**14**	**80**	**60**	**42**

Find the GCF for each set of numbers.

8. 15, 45	9. 6, 14	10. 24, 40	11. 8, 12, 52	12. 16, 24, 32
15	**2**	**8**	**4**	**8**

Find the common prime factors. Then find the GCF.

13. 15, 50	14. 16, 24	15. 49, 70	16. 45, 108	17. 18, 36
5; 5	**2 × 2 × 2; 8**	**7; 7**	**3 × 3; 9**	**2 × 3 × 3; 18**

Mixed Applications

18. Walt has 51 football stickers and 68 baseball stickers. He will put them on cards that all have the same number of stickers. What is the greatest number of cards Walt can make? **17 cards**

19. Hot dogs come in packages of 8, and rolls come in packages of 6. What is the smallest number of each that you can buy so there are no extras? **3 packages of hot dogs, 4 packages of rolls**

20. Maya has 65 dimes and 104 pennies. She will put them in packages that are all the same. What is the greatest number of packages she can make? **13 packages**

21. Rob jogged on the 4th, 8th, and 12th of the month. If he continues jogging in this pattern, what will be the next two days that Rob jogs? **16th and 20th**

Use with text pages 80–83.   **ON MY OWN   P17**

EXAMPLE 2 Country Flavor granola snacks are sold in 6-oz, 9-oz, and 18-oz packages. What is the least number of ounces you can buy to have equal amounts of the different sizes?

6: 6, 12, 18, 24, 30, 36, 42, 48, 54, 60   *List multiples of 6, 9, and*
9: 9, 18, 27, 36, 45, 54, 63, 72, 81, 90   *18. Find the common*
18: 18, 36, 54, 72, 90, 108, 126   *multiples.*

The LCM of 6, 9, and 18 is 18.   *Find the LCM.*

So, you would need to buy 18 oz of each size.

• Since you need 18 oz, how many of each size would you need to buy? **three 6-oz packages, two 9-oz packages, and one 18-oz package**

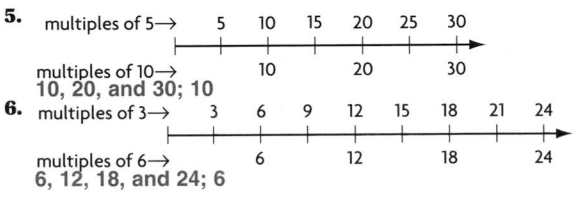

GUIDED PRACTICE

List the first three multiples of each number in the pair.

3: 3, 6, 9;	7: 7, 14, 21;	11: 11, 22, 33;	15: 15, 30, 45;
1. 3, 6	**2.** 7, 14	**3.** 11, 8	**4.** 15, 12
6: 6, 12, 18	14: 14, 28, 42	8: 8, 16, 24	12: 12, 24, 36

Name the common multiples on the number line. Name the LCM.

5. multiples of 5→   5   10   15   20   25   30

multiples of 10→   10   20   30
10, 20, and 30; 10

6. multiples of 3→   3   6   9   12   15   18   21   24

multiples of 6→   6   12   18   24
6, 12, 18, and 24; 6

Find the LCM of each pair of numbers.

7. 3, 7 **21**   **8.** 2, 3 **6**   **9.** 6, 9 **18**   **10.** 8, 20 **40**

Greatest Common Factor

Factors shared by two or more numbers are called common factors. The largest of the common factors is called the **greatest common factor**, or **GCF**.

To find the GCF of two or more numbers, list all the factors of each number, find the common factors, and then find the greatest common factor.

12: 1, 2, 3, 4, 6, 12   The common factors are 1, 2, 3, 6.
18: 1, 2, 3, 6, 9, 18   The GCF of 12 and 18 is 6.

• Why are you able to find a greatest common factor but not a greatest common multiple? **Factors are limited to a certain few, but multiples are limitless.**

81

ENRICHMENT 4.3

Name _____

LESSON 4.3

Multiple Relationships

Fill in each blank with a number from the box below. Then, identify the relationship described. The first two are done for you. You will not use every number in the box.

108	62	51	8	51	19	120	36
180	95	9	15	16	310	192	25
75	30	94	60	27	64	12	35

1. 17 is to __51__ and 85 as 19 is to 57 and 95. What is the relationship?
   17 is the GCF of 51 and 85; 19 is the GCF of 57 and 95

2. __42__ is to 6 and 14 as 48 is to __12__ and 16. What is the relationship?
   42 is the LCM of 6 and 14; 48 is the LCM of 12 and 16

3. 45 is to __9__ and 15 as 120 is to 24 and 30. What is the relationship?
   45 is the LCM of 9 and 15; 120 is the LCM of 24 and 30

4. 36 is to 108 and 180 as __64__ is to 192 and 320. What is the relationship?
   36 is the GCF of 108 and 180; 64 is the GCF of 192 and 320

5. 47 is to 94 and 235 as 62 is to 124 and __310__. What is the relationship?
   47 is the GCF of 94 and 235; 62 is the GCF of 124 and 310

6. 105 is to 15 and 35 as __108__ is to 12 and 27. What is the relationship?
   105 is the LCM of 15 and 35; 108 is the LCM of 12 and 27

7. 12 is to 24 and 36 as __25__ is to 50 and 75. What is the relationship?
   12 is the GCF of 24 and 36; 25 is the GCF of 50 and 75

Use with text pages 80–83.   **STRETCH YOUR THINKING**   E17

• **Did you land on any of the same numbers both times?** yes **Which ones?** 6, 12, 18

Repeat the activity with the numbers 4 and 5.

2 TEACH/PRACTICE

This lesson emphasizes real-life situations in which students need to determine the LCM or the GCF. Have them come up with as many examples as possible. Display the examples in the classroom.

GUIDED PRACTICE

These exercises provide practice in finding multiples and LCMs. Suggest that students use mental math to help them find the answers.

ADDITIONAL EXAMPLES

Example 1, p. 80

Evelyn is making pumpkin pies for the school bake sale. Piecrusts are sold in packages of three. Pie filling is sold in 4-can packages. What is the least number of piecrusts and cans of pie filling Evelyn can buy to have the same number of each?

12 piecrusts and 12 cans of pie filling

How many packages of each should she buy?

4 packages of piecrusts and 3 packages of pie filling

Example 2, p. 81

Thrift Market sells apple juice in 8-oz containers, pineapple juice in 12-oz containers, and grape juice in 16-oz containers. Jeff wants to have the same amount of each to make some punch. What is the least number of ounces he should buy of each flavor of fruit juice?

48 oz

How many containers of each flavor should he buy? 6 containers of apple juice, 4 containers of pineapple juice, and 3 containers of grape juice

81

CRITICAL THINKING When Tina arranged the seating for her party at tables for 2, 3, or 4 people, there was always 1 person left without a place to sit. When she arranged the seating at tables for 5, everyone had a place to sit. What is the least possible number of people there could have been at Tina's party? 25 people

COMMON ERROR ALERT Some students may confuse *LCM* with *GCF*. Write *least common multiple* and *greatest common factor* on the board. Discuss the meaning of *least, greatest, multiple,* and *factor.* Then have them complete this problem.

• **What is the GCF and LCM of 12 and 18?**

Factors	Number	Multiples
1, 2, 3, 4, 6, 12	12	12, 24, 36, 48, 60, 72
1, 2, 3, 6, 9, 18	18	18, 36, 54, 72
Common factors: 1, 2, 3, 6		Common multiples: 36, 72, . . .
GCF 6		LCM 36

INDEPENDENT PRACTICE
In these exercises, students demonstrate their understanding of LCM and GCF. Have students explain how they solved Exercises 7–12 and 24–29. Their responses should indicate that the procedure is the same no matter how many numbers are in a set.

ADDITIONAL EXAMPLES

Example 3, p. 82

Leon is making bouquets of red and white carnations for a party. He wants the number of each color to be the same in every bouquet. He has 16 red carnations and 24 white carnations. What is the greatest number of bouquets he can make without any flowers left over?

He can make 8 bouquets, each with 3 white and 2 red carnations.

Example 4, p. 82

Use prime factors to find the GCF of 30 and 12.

$$30 = 2 \times 5 \times 3$$
$$12 = 2 \times 2 \times 3$$

The prime factors are 2, 3, and 5.

The common prime factors are 2 and 3.

$$2 \times 3 = 6$$

The GCF of 30 and 12 is 6.

The GCF can be used to solve problems.

EXAMPLE 3 Emma is packaging items to give her friends. She has 45 pencils and 36 stickers. All packages have to contain the same number of each item. What is the greatest number of packages she can make without any items left over?

You can find the greatest number of packages by finding the GCF of 45 and 36.

45: 1, 3, 5, 9, 15, 45   *List the factors.*
36: 1, 2, 3, 4, 6, 9, 12, 18, 36   *Find the common factors.*

The GCF of 45 and 36 is 9.   *Find the GCF.*

So, Emma can make 9 packages without any items left over.

• **CRITICAL THINKING** How many pencils and how many stickers will be in each package? **5 pencils and 4 stickers**

Another Method To find the GCF of two numbers, you can use their prime factors. List the prime factors, find the common prime factors, and then find their product.

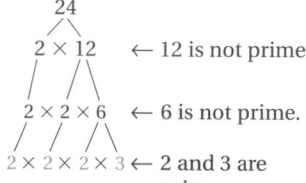

18
2 × 9 ← 9 is not prime.
2 × 3 × 3 ← 2 and 3 are prime.

24
2 × 12 ← 12 is not prime.
2 × 2 × 6 ← 6 is not prime.
2 × 2 × 2 × 3 ← 2 and 3 are prime.

The prime factors of 18 are $2 \times 3 \times 3$.
The prime factors of 24 are $2 \times 2 \times 2 \times 3$.
The common prime factors are 2 and 3.
Find the product of the common factors: $2 \times 3 = 6$.
The GCF of 18 and 24 is 6.

EXAMPLE 4 Use prime factors to find the GCF of 54 and 72.

54:   $2 \times 3 \times 3 \times 3$ *Find the prime factors.*
72:   $2 \times 2 \times 2 \times 3 \times 3$

2, 3, and 3   *Find the common prime factors.*

$2 \times 3 \times 3 = 18$ *Multiply the common factors.*

So, the GCF of 54 and 72 is 18.

• What is the GCF of 36 and 60? **12**

IDEA FILE

HISTORY CONNECTION

One of the most famous math books, *Elements,* was written in about 300 B.C. by the Greek mathematician Euclid. In this book he showed different methods for finding the greatest common factor.

• Have students find out more about Euclid and his book and share with the class what they learn.

Modalities Auditory, Visual

ENGLISH AS A SECOND LANGUAGE

Tip • Display this example in the classroom. Each day have a different student explain how the vocabulary words are related to the numbers.

factors → 5 × 3 = 15 ← multiple

factors → 2 × 3 = 6 ← multiple

common factor → 3

greatest common factor → 3

Find the LCM for each set of numbers.

1. 3, 4 **12**   **2.** 8, 12 **24**   **3.** 1, 6 **6**   **4.** 5, 6 **30**   **5.** ▢, ▢ **20**   **6.** 7, 9 **63**

7. 4, 14 **28**   **8.** 9, 15 **45**   **9.** 6, 8 **24**   **10.** 1, 3, 7 **21**   **11.** 3, 5, 12 **60**   **12.** 2, 8, 10 **40**

Find the common prime factors. Then find the GCF.

13. 8, 24   **14.** 32, 128   **15.** 12, 20   **16.** 9, 15   **17.** 24, 30
    2 × 2 × 2; 8   2 × 2 × 2 × 2 × 2; 32   2 × 2; 4   3; 3   2 × 3; 6

Find the GCF for each set of numbers.

18. 6, 9 **3**   **19.** 4, 20 **4**   **20.** 9, 24 **3**   **21.** 12, 20 **4**   **22.** 16, 18 **2**   **23.** 5, 8 **1**

24. 9, 21 **3**   **25.** 16, 40 **8**   **26.** 3, 5, 10 **1**   **27.** 15, 20, 35 **5**   **28.** 12, 15, 24 **3**   **29.** 60, 72 **12**

Problem-Solving Applications

For Problems 30–31, use the following information.

Peter will distribute toothpaste samples with pamphlets about dental care. The toothpaste samples come in packages of 15. The pamphlets come in packages of 20.

30. What is the smallest number of toothpaste samples and pamphlets that he needs without having any left over? **60 of each**

31. How many packages of each does he need? **4 packages of toothpaste samples, 3 packages of pamphlets**

32. Rochelle has 36 markers and 48 erasers. She will put them in packages that are all the same. What is the greatest number of packages she can make? **12 packages**

33. ✏️ **WRITE ABOUT IT** What does *common* in *least common multiple* and *greatest common factor* mean? Give an example of each using 18 and 24. **It means that two or more numbers have an identical multiple or factor. For 18 and 24, the least common multiple is 72 and the greatest common factor is 6.**

Mixed Review

Find the value of *n*.

34. $15 \div 3 = n$ **5**   **35.** $16 \div 4 = n$ **4**   **36.** $24 \div n = 6$ **4**   **37.** $6 \times n = 30$ **5**   **38.** $n \times 8 = 56$ **7**

Solve.

39. $0.16 + 3.85$ **4.01**   **40.** $12.5 - 4.68$ **7.82**   **41.** 14.85×9.2 **136.62**   **42.** $70 \div 3.5$ **20**

43. **FRACTIONS** Sharon shared one huge cookie with six friends. What fractional part of the cookie did each person get? $\frac{1}{7}$ **of the cookie**

44. **PATTERNS** Ned delivered newspapers on the 3rd, 11th, and 19th of the month. If he continues this pattern, when will he make his next delivery? **on the 27th**

3 WRAP UP

Use these questions to help students focus on the big ideas of the lesson.

- **Terry wants to plant a pansy-and-geranium garden. Pansies are sold in pots of 6. Geraniums are sold in pots of 4. What is the least number of pansies and geraniums she should buy to have an equal number of each?** 12 **How many pots of each should she buy?** 2 pots of pansies, 3 pots of geraniums

- **David is passing out treats at his sister's birthday party. He has 14 cookies and 35 sticks of gum. Does he have enough to divide equally among each of 7 guests? Explain.** Yes. The GCF of 14 and 35 is 7. Each guest will get 2 cookies and 5 sticks of gum.

✔ *Assessment Tip*

Have students write the answer.

- **Explain the difference between the LCM and the GCF.** The LCM is the least number that is a multiple of two or more numbers. The GCF is the greatest number that is a factor of two or more numbers.

DAILY QUIZ

Find the LCM of each pair of numbers.

1. 12, 16 48   **2.** 12, 20 60
3. 4, 12 12   **4.** 10, 12 60

Find the GCF of each pair of numbers.

5. 18, 30 6   **6.** 8, 28 4
7. 4, 26 2   **8.** 20, 50 10

HOME CONNECTION

Students can play this game with someone at home.

1. Roll two number cubes labeled 1–6.
2. Find and record the LCM of the numbers rolled.
3. After five turns, the winner is the player with the greater number of least common multiples that are 10 or less.

Modalities Kinesthetic, Visual

CONSUMER CONNECTION

Present this problem to students.

Suppose you are making egg sandwiches for the sixth-grade brunch. Eggs come in cartons of 12, and English muffins come in packages of 10.

- **What is the least number of eggs and English muffins you can buy to have an equal number of each?** 60

- **How many cartons and packages should you buy?** 5 cartons, 6 packages

Modality Auditory

ORGANIZER

OBJECTIVE To use fraction bars to find equivalent fractions

MATERIALS *For each student* fraction bars

PROGRAM RESOURCES
E-Lab • Activity 4
E-Lab Recording Sheets, p. 4

USING THE PAGE

Discuss with students how they use equivalent fractions in their everyday lives. Guide the discussion toward such topics as adapting recipes for a large crowd, judging distances in sporting events, and participating in activities at scheduled times.

Activity 1

Explore

In this activity students model equivalent fractions with fraction bars. This hands-on experience will allow students to make a smooth transition to the algorithm for finding equivalent fractions presented in Activity 2. Observe how students position the fraction bars they are comparing. It is important that the fraction bars be aligned on the left and right.

Think and Discuss

To assess students' understanding of the activity, you may wish to ask them the following questions.

- **What fraction bars were not equivalent to $\frac{1}{2}$?** $\frac{1}{3}, \frac{1}{5}, \frac{1}{7}, \frac{1}{9}$

- **How did you know?** Accept reasonable explanations. Have one student demonstrate how many fourths are equivalent to $\frac{6}{8}$.

Try This

Exercises 1 and 2 are presented in the same language as Activity 1. However, in Exercises 3 and 4 students are presented with a different representation of equivalent fractions. Ask a volunteer to read Exercise 3 aloud and then to put it into the format of Exercises 1 and 2. Explain to students that to solve these exercises, they will follow the same procedure they followed in Activity 1. Encourage students to work together to solve Exercise 5.

Equivalent Fractions

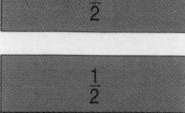

LAB Activity

WHAT YOU'LL EXPLORE
How to use fraction bars to find equivalent fractions

WHAT YOU'LL NEED
fraction bars

Fractions can be written in different ways to name the same amount or part. One way to find equivalent fractions is to use fraction bars.

ACTIVITY 1

Explore

Work with a partner to find how many eighths are equivalent to $\frac{1}{2}$.

- Use the $\frac{1}{2}$ fraction bar.

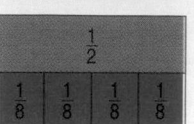

- Place $\frac{1}{8}$ bars along the $\frac{1}{2}$ bar until the lengths are equal.

Since 4 of the $\frac{1}{8}$ bars are equivalent to the $\frac{1}{2}$ bar, $\frac{1}{2} = \frac{4}{8}$.

- Use the fraction bars to find other fractions that are equivalent to $\frac{1}{2}$.

Think and Discuss $\frac{1}{4}, 2; \frac{1}{6}, 3; \frac{1}{10}, 5; \frac{1}{12}, 6$

- Which other fraction bars did you use to find fractions equivalent to $\frac{1}{2}$? How many of each bar did you use?

- Explain how you would use fraction bars to find how many fourths are equivalent to $\frac{6}{8}$. **Use 6 of the $\frac{1}{8}$ bars. Lay $\frac{1}{4}$ bars along the $\frac{1}{8}$ bars until the lengths are equal. Count the $\frac{1}{4}$ bars.**

Try This

Use fraction bars to solve.

1. How many twelfths are equivalent to $\frac{3}{4}$? **9**

2. How many tenths are equivalent to $\frac{4}{5}$? **8**

3. $\frac{5}{6} = \frac{\blacksquare}{12}$ **10**

4. $\frac{\blacksquare}{12} = \frac{2}{3}$ **8**

5. Find as many fractions as you can that are equivalent to $\frac{1}{4}$.
Possible answers: $\frac{2}{8}, \frac{3}{12}, \frac{4}{16}$

IDEA FILE

ENGLISH AS A SECOND LANGUAGE

Tip • Discuss these word fractions to help students understand the concept of number fractions.

A room is part of a house. $\frac{room}{house}$

A flower is part of a garden. $\frac{flower}{garden}$

A tree is part of a forest. $\frac{tree}{forest}$

A letter is part of a word. $\frac{letter}{word}$

Modality Auditory

WRITING IN MATHEMATICS

Have students solve these riddles. Encourage them to document the problem-solving strategies they used.

1. The difference between my numerator and denominator is 12. I am equivalent to $\frac{1}{2}$. What fraction am I? $\frac{12}{24}$

2. The product of my numerator and denominator is 35. I am equivalent to $\frac{20}{28}$. what fraction am I? $\frac{5}{7}$

Modality Visual

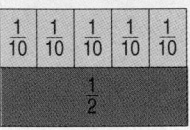

You can also use multiplication and division to find equivalent fractions.

Explore

Yes. The six $\frac{1}{12}$ bars form the same length as the two $\frac{1}{4}$ bars.

• The fractions $\frac{2}{4}$ and $\frac{6}{12}$ are modeled below. Are they equivalent? Explain.

| $\frac{1}{4}$ | $\frac{1}{4}$ |

| $\frac{1}{12}$ | $\frac{1}{12}$ | $\frac{1}{12}$ | $\frac{1}{12}$ | $\frac{1}{12}$ | $\frac{1}{12}$ |

What number can you use for ■ to show that $\frac{2 \times ■}{4 \times ■} = \frac{6}{12}$? **$\frac{3}{3}$**

• The fractions $\frac{5}{10}$ and $\frac{1}{2}$ are modeled below. Are they equivalent? Explain. **Yes. The $\frac{1}{2}$ bar is the same length as the five $\frac{1}{10}$ bars.**

| $\frac{1}{10}$ | $\frac{1}{10}$ | $\frac{1}{10}$ | $\frac{1}{10}$ | $\frac{1}{10}$ |

| $\frac{1}{2}$ |

What number can you use for ■ to show that $\frac{5 \div ■}{10 \div ■} = \frac{1}{2}$? **$\frac{5}{5}$**

Think and Discuss

• Multiplying the numerator and denominator by 3 is like multiplying by $\frac{3}{3}$. What does $\frac{3}{3}$ equal? **1 whole**

• Suppose you multiply the numerator and the denominator of $\frac{2}{3}$ by 5. What equivalent fraction does that make? **$\frac{10}{15}$**

• Suppose you divide both the numerator and denominator of $\frac{6}{12}$ by 2. What equivalent fraction does that make? **$\frac{3}{6}$**

• What whole numbers can you use for ■ to get a fraction equivalent to $\frac{1}{3}$? $\frac{1}{3} \times \frac{■}{■}$ **Possible answers: 2, 3, 4**

• What whole numbers can you use for ■ to get a fraction equivalent to $\frac{12}{16}$? $\frac{12}{16} \div \frac{■}{■}$ **2, 4**

Try This

Copy and complete.

1. $\frac{2 \times ■}{3 \times ■} = \frac{6}{9}$ **3**
2. $\frac{4 \times ■}{5 \times ■} = \frac{16}{20}$ **4**
3. $\frac{30 \div ■}{35 \div ■} = \frac{6}{7}$ **5**
4. $\frac{15 \div ■}{24 \div ■} = \frac{5}{8}$ **3**

Complete each number sentence.

5. $\frac{9}{12} = \frac{■}{4}$ **3**
6. $\frac{3}{24} = \frac{■}{8}$ **1**
7. $\frac{17}{34} = \frac{■}{2}$ **1**

Technology **Link**
E-Lab • **Activity 4** Available on CD-ROM and the Internet at http://www.hbschool.com/elab

85

Note: Students find factors. Use E-Lab p. 4.

Activity 2

Explore

Point out to students the relationship between the fraction-bar models and the algorithms for finding equivalent fractions. You may wish to ask:

• **Why did we multiply in the first example and divide in the second example?** We multiplied by 3 because three $\frac{1}{12}$ bars are equivalent to one $\frac{1}{4}$ bar. We divided by 5 because five $\frac{1}{10}$ bars are equivalent to one $\frac{1}{2}$ bar.

Think and Discuss

Have students work in small groups to solve these problems. Encourage discussion.

Try This

Although some students may be able to solve these problems using mental math, suggest that they write out all the steps they used to complete the problems. This will allow you to assess their understanding of the process. Emphasize to students that they need to understand the process so that they can solve problems with more complex numbers.

USING E-LAB

Students use a tool to find factors.

✓ *Assessment Tip*

Use these questions to help students communicate the big ideas of the lesson.

• **How many fourths are equivalent to $\frac{9}{12}$?** 3

• **How could you show that $\frac{3}{5} = \frac{6}{10}$ using multiplication or division?** Multiply the numerator 3 and the denominator 5 by 2.

EXTENSION

Have students complete and explain these patterns.

1. $\frac{1}{3} = \frac{2}{6} = \frac{3}{9} = \frac{4}{12} = $ ____ = ____ = ____ **$\frac{5}{15}, \frac{6}{18}, \frac{7}{21}$**

2. $\frac{1}{2} = \frac{4}{8} = \frac{2}{4} = \frac{8}{16} = $ ____ = ____ = ____ **$\frac{4}{8}, \frac{16}{32}, \frac{8}{16}$**

3. $\frac{32}{64} = \frac{8}{16} = \frac{16}{32} = $ ____ = ____ = ____ **$\frac{4}{8}, \frac{8}{16}, \frac{2}{4}$**

Modality Visual

REVIEW

Have students write a fraction that is commonly used to represent each amount.

1. 30 minutes $\frac{1}{2}$ hour

2. 8 ounces $\frac{1}{2}$ pound

3. 15 minutes $\frac{1}{4}$ hour

4. 6 eggs $\frac{1}{2}$ dozen

5. 45 minutes $\frac{3}{4}$ of an hour

Modality Visual

GRADE 5
Use fraction bars to picture fractions in simplest form.

GRADE 6
Write fractions in simplest form.

GRADE 7
Write fraction sums, differences, products, and quotients in simplest form.

ORGANIZER

OBJECTIVE To write fractions in simplest form
VOCABULARY *New* **simplest form**
ASSIGNMENTS Independent Practice
- *Basic* 1–25, odd; 26–29; 30–37
- *Average* 1–29; 30–37
- *Advanced* 1–6, 13–25, odd; 26–29; 30–37

PROGRAM RESOURCES
Reteach 4.4 Practice 4.4 Enrichment 4.4
Number Heroes* • *Fraction Fireworks

PROBLEM OF THE DAY

Transparency 4

Write these everyday items in "simplest form."

1. 50 pennies $\frac{1}{2}$ dollar
2. 24 eggs 2 dozen eggs
3. 8 oz of butter $\frac{1}{2}$ lb of butter
4. 2 qts of ice cream $\frac{1}{2}$ gal of ice cream
5. 36 in. of material 1 yd of material

Solution Tab A, p. A4

QUICK CHECK

Use mental math to find the missing numbers.

1. $72 \div 8 = a$; $a \div 3 = b$ a = 9, b = 3
2. $150 \div 15 = a$; $a \div 5 = b$ a = 10, b = 2
3. $36 \div 9 = a$; $a \div 2 = b$; $b \div 2 = c$ a = 4, b = 2, c = 1
4. $105 \div 5 = a$; $a \div 3 = b$; $b \div 7 = c$ a = 21, b = 7, c = 1

1 MOTIVATE

To give students the opportunity to show what they know about simplest form, have them describe the results of this survey.

One hundred dentists were surveyed:

- $\frac{4}{8}$ of the dentists use A-1 Toothpaste.
- $\frac{2}{4}$ of the dentists use Smiley Toothpaste.

The same number of dentists used each brand of toothpaste because $\frac{4}{8}$ and $\frac{2}{4}$ are equivalent fractions.

4.4

WHAT YOU'LL LEARN
How to write fractions in simplest form

WHY LEARN THIS?
To use fractions in the form that is easiest to understand and most commonly used

WORD POWER
simplest form

You probably know that automobile odometers measure distance, but did you know that they divide miles into tenths, while many highway signs divide miles into half miles and quarter miles? When you learn to drive, you will need to remember that $\frac{5}{10}$ mi = $\frac{1}{2}$ mi, $\frac{2}{10}$ mi or $\frac{3}{10}$ mi is about $\frac{1}{4}$ mi, and $\frac{7}{10}$ mi or $\frac{8}{10}$ mi is about $\frac{3}{4}$ mi.

Fractions in Simplest Form

In the Lab Activity, you learned how to write equivalent fractions.

When the numerator and denominator of an equivalent fraction have no common factor other than 1, the equivalent fraction is in **simplest form**.

EXAMPLE 1 Write $\frac{6}{18}$ in simplest form.

6: 1, 2, 3, 6   *Find the common factors of 6 and 18.*
18: 1, 2, 3, 6, 9, 18

$\frac{6}{18} = \frac{6 \div 3}{18 \div 3} = \frac{2}{6}$ *Divide the numerator and denominator by a common factor until the fraction is in simplest form.*

$\frac{2}{6} = \frac{2 \div 2}{6 \div 2} = \frac{1}{3}$

So, $\frac{1}{3}$ is the simplest form of $\frac{6}{18}$.

- Explain how you know that $\frac{1}{3}$ is in simplest form. **The only common factor of 1 and 3 is 1.**

You can use the GCF to write fractions in simplest form.

EXAMPLE 2 Write $\frac{16}{24}$ in simplest form.

16: 1, 2, 4, 8, 16   *Find the GCF of 16 and 24.*
24: 1, 2, 3, 4, 6, 8, 12, 24

$\frac{16}{24} = \frac{16 \div 8}{24 \div 8} = \frac{2}{3}$ *Divide the numerator and denominator by the GCF.*

So, $\frac{2}{3}$ is the simplest form of $\frac{16}{24}$.

- What is the simplest form of $\frac{5}{10}$? $\frac{1}{2}$

GUIDED PRACTICE

Write the common factors for the numerator and denominator.

1. $\frac{4}{8}$ 1, 2, 4   2. $\frac{9}{24}$ 1, 3   3. $\frac{8}{18}$ 1, 2   4. $\frac{12}{54}$ 1, 2, 3, 6

Write the fraction in simplest form.

5. $\frac{4}{32}$ $\frac{1}{8}$ 6. $\frac{14}{21}$ $\frac{2}{3}$ 7. $\frac{9}{54}$ $\frac{1}{6}$ 8. $\frac{48}{54}$ $\frac{8}{9}$

9. $\frac{8}{22}$ $\frac{4}{11}$ 10. $\frac{6}{18}$ $\frac{1}{3}$ 11. $\frac{9}{30}$ $\frac{3}{10}$ 12. $\frac{32}{48}$ $\frac{2}{3}$

IDEA FILE

CORRELATIONS TO OTHER RESOURCES

💾 **To help students practice finding the LCM and the GCF** 💿 **and reducing fractions to simplest form, use *Fraction Attraction*, which provides practice opportunities at a virtual amusement park.**

Available for Mac and Windows in disk and CD-ROM formats.

💾 **CornerStone Mathematics** 💿 **©1996 SkillsBank Corporation Level B, Using Fractions and Percents, Lesson 1**

RETEACH 4.4

Name _____

LESSON 4.4

Fractions in Simplest Form

When the numerator and denominator of a fraction have no common factor other than 1, the fraction is in **simplest form**. You can use a GCF to write a fraction in simplest form.
What is the simplest form of $\frac{32}{56}$?

Step 1 Find the GCF of 32 and 56 by listing the factors of each. The GCF is 8.
32: 1, 2, 4, 8, 16, 32
56: 1, 2, 4, 7, 8, 14, 28, 56

Step 2 Divide the numerator and denominator by the GCF
$\frac{32}{56} = \frac{32 \div 8}{56 \div 8} = \frac{4}{7}$

So, $\frac{4}{7}$ is the simplest form of $\frac{32}{56}$.

Complete to find the simplest form of $\frac{88}{104}$.

1. Find the GCF of 88 and 104.   88: 1, 2, 4, 8, 11, 22, 44, 88
   The GCF is __8__.   104: 1, 2, 4, 8, 13, 26, 52, 104

2. Divide the numerator and denominator by the GCF. $\frac{88 \div 8}{104 \div 8} = \frac{11}{13}$
3. So, $\frac{11}{13}$ is the simplest form of $\frac{88}{104}$.

Complete to find the simplest form of $\frac{78}{120}$.

4. Find the GCF of 78 and 120.   78: 1, 2, 3, 6, 13, 26, 39, 78
   The GCF is __6__.   120: 1, 2, 3, 4, 5, 6, 8, 10, 12, 15, 20, 24, 30, 40, 60, 120

5. Divide the numerator and denominator by the GCF. $\frac{78 \div 6}{120 \div 6} = \frac{13}{20}$
6. So, $\frac{13}{20}$ is the simplest form of $\frac{78}{120}$.

Find the GCF of the pair of numbers. Then write the fraction in simplest form.

7. 4, 10; $\frac{4}{10}$ 8. 8, 12; $\frac{8}{12}$ 9. 18, 36; $\frac{18}{36}$ 10. 21, 60; $\frac{21}{60}$
   2; $\frac{2}{5}$ 4; $\frac{2}{3}$ 18; $\frac{1}{2}$ 3; $\frac{7}{20}$

R18 TAKE ANOTHER LOOK   Use with text pages 86–87.

INDEPENDENT PRACTICE

Write the common factors and the GCF of the numerator and denominator.

1. $\frac{1}{17}$ 1; 1 **2.** $\frac{9}{24}$ 1, 3; 3 **3.** $\frac{6}{27}$ 1, 3; 3 **4.** $\frac{9}{63}$ 1, 3, 9; 9 **5.** $\frac{10}{35}$ 1, 5; 5 **6.** $\frac{16}{40}$
1, 2, 4,
8; 8

Write the fraction in simplest form.

7. $\frac{4}{24}$ $\frac{1}{6}$ **8.** $\frac{9}{12}$ $\frac{3}{4}$ **9.** $\frac{6}{48}$ $\frac{1}{8}$ **10.** $\frac{12}{16}$ $\frac{3}{4}$ **11.** $\frac{10}{18}$ $\frac{5}{9}$ $\frac{3}{4}$ **12.** $\frac{15}{20}$

13. $\frac{18}{90}$ $\frac{1}{5}$ **14.** $\frac{28}{42}$ $\frac{2}{3}$ **15.** $\frac{21}{33}$ $\frac{7}{11}$ **16.** $\frac{24}{30}$ $\frac{4}{5}$ **17.** $\frac{42}{60}$ $\frac{7}{10}$ **18.** $\frac{28}{98}$
$\frac{2}{7}$

19. Choose the fraction that is the simplest form of $\frac{21}{24}$.

  a. $\frac{3}{4}$ **b.** $\frac{7}{8}$ **c.** $\frac{2}{3}$ b

20. Choose the fraction that is the simplest form of $\frac{48}{120}$.

  a. $\frac{1}{8}$ **b.** $\frac{4}{10}$ **c.** $\frac{2}{5}$ c

Write the missing number.

21. $\frac{2}{12} = \frac{1}{\blacksquare}$ 6 **22.** $\frac{\blacksquare}{36} = \frac{2}{9}$ 8 **23.** $\frac{21}{24} = \frac{7}{\blacksquare}$ 8 **24.** $\frac{40}{\blacksquare} = \frac{5}{8}$ 64 **25.** $\frac{9}{\blacksquare} = \frac{3}{4}$ 12

Problem-Solving Applications

26. Clint has 6 apple muffins, 2 corn muffins, and 4 bran muffins. What fraction of the muffins are bran? Write the fraction in simplest form. $\frac{1}{3}$

27. The numerator of a fraction is 12. The GCF of the numerator and denominator is 4. What is the denominator? any multiple of 4

28. Some calculators have a [SIMP] key that can be used to simplify fractions. What fraction would this key sequence give? 10 [/] 15 [SIMP] [=] $\frac{2}{3}$

29. ▭▶ **WRITE ABOUT IT** When do you know that a fraction is in simplest form? when 1 is the GCF of the numerator and the denominator

Mixed Review

Solve.

30. $(6 \div 3) + 2$ 4 **31.** $(4 \div 2) + 1$ 3 **32.** $(12 \div 4) + 2$ 5

Solve.

33. 3.5×0.01 **0.035** **34.** 4.1×0.2 **0.82** **35.** 0.5×1.2 **0.6**

36. **TRAVEL** Mrs. Garcia rents a car for 5 days at $20.95 per day and $0.15 per mile. She travels 315 mi. What is her total cost? **$152.00**

37. **MEASUREMENT** Mr. Bell needs a fence around a rectangular garden that is 8 ft by 10 ft. The fence will cost $11.50 per yard. What will be the total cost? **$138.00**

MORE PRACTICE Lesson 4.4, page H47

Technology Link

In *Mighty Math Number Heroes*, you can practice finding equivalent fractions by playing the game *Fraction Fireworks*.

Use Grow Slide Level Z.

87

---

2 TEACH/PRACTICE

In this lesson, students apply their understanding of factors and GCF to write fractions in simplest form.

GUIDED PRACTICE

Before starting these exercises, review the definition of common factors as they relate to fractions.

INDEPENDENT PRACTICE

After students finish Exercises 1–25, ask:

- **In Exercises 1–6, which fraction has only one common factor?** $\frac{1}{17}$

💡**CRITICAL THINKING** To write $\frac{18}{27}$ in simplest form, Karen wrote $\frac{6}{9}$ and Bryan wrote $\frac{2}{5}$. Analyze their answers. Karen's answer is not in simplest form. The denominator in Bryan's answer should be 3.

ADDITIONAL EXAMPLES

Example 1, p. 86

Use common factors to simplify $\frac{36}{48}$. $\frac{3}{4}$

Example 2, p. 86

Use the GCF to write $\frac{20}{25}$ in simplest form.

The GCF of 20 and 25 is 5. $\frac{20}{25} = \frac{20 \div 5}{25 \div 5} = \frac{4}{5}$. So, $\frac{4}{5}$ is the simplest form of $\frac{20}{25}$.

3 WRAP UP

Use these questions to help students focus on the big ideas of the lesson.

- **Denny has 15 action videos, 11 drama videos, and 9 musical videos. What part of his video collection is action videos? Write the answer in simplest form.** $\frac{3}{7}$

✓ *Assessment Tip*

Have students write the answer.

- **To write a fraction in simplest form, is the method shown in Example 1 or Example 2 more efficient? Explain.** Example 2, because you only have to divide once

DAILY QUIZ

Write the fraction in simplest form.

1. $\frac{21}{24}$ $\frac{7}{8}$ **2.** $\frac{12}{16}$ $\frac{3}{4}$ **3.** $\frac{24}{60}$ $\frac{2}{5}$

4. $\frac{12}{18}$ $\frac{2}{3}$ **5.** $\frac{55}{60}$ $\frac{11}{12}$ **6.** $\frac{40}{50}$ $\frac{4}{5}$

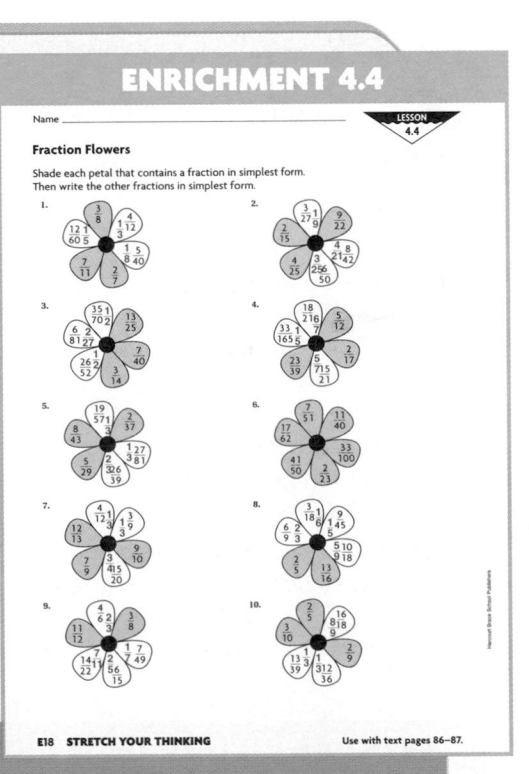

PRACTICE 4.4

Name _____
LESSON 4.4

Fractions in Simplest Form

Vocabulary

Complete.

1. When the numerator and denominator of a fraction have no common factor other than 1, the fraction is in ____simplest form____.

Write the common factors and the GCF of the numerator and denominator.

2. $\frac{8}{32}$ **3.** $\frac{10}{50}$ **4.** $\frac{2}{13}$ **5.** $\frac{14}{49}$ **6.** $\frac{1}{19}$
1, 2, 4, 8; 8 1, 2, 5, 10; 10 1; 1 1, 7; 7 1; 1

7. $\frac{12}{18}$ **8.** $\frac{25}{40}$ **9.** $\frac{15}{40}$ **10.** $\frac{9}{54}$ **11.** $\frac{6}{33}$
1, 2, 3, 6; 6 1, 5, 25; 25 1, 5; 5 1, 3, 9; 9 1, 3; 3

Write the fraction in simplest form.

12. $\frac{9}{36}$ **13.** $\frac{15}{50}$ **14.** $\frac{11}{121}$ **15.** $\frac{15}{36}$ **16.** $\frac{14}{28}$
$\frac{1}{4}$ $\frac{3}{10}$ $\frac{1}{11}$ $\frac{5}{12}$ $\frac{1}{2}$

17. $\frac{30}{66}$ **18.** $\frac{63}{72}$ **19.** $\frac{27}{81}$ **20.** $\frac{25}{65}$ **21.** $\frac{12}{42}$
$\frac{5}{11}$ $\frac{7}{8}$ $\frac{1}{3}$ $\frac{5}{13}$ $\frac{2}{7}$

Write the missing number.

22. $\frac{36}{72} = \frac{1}{\boxed{2}}$ **23.** $\frac{50}{75} = \frac{2}{3}$ **24.** $\frac{17}{85} = \frac{1}{5}$ **25.** $\frac{63}{84} = \frac{3}{\boxed{4}}$ **26.** $\frac{2}{3} = \frac{64}{96}$

Mixed Applications

27. Ryan has 6 Yankees baseball cards, 5 Dodgers cards, and 7 Braves cards. What fraction of the cards are Yankees cards? Write the fraction in simplest form. $\frac{1}{3}$

28. Wanda has 5 containers of orange juice, 11 of apple juice, and 4 of grape juice. What fraction of the containers are *not* orange juice? Write the fraction in simplest form. $\frac{3}{4}$

29. Ron rents a car for 6 days at $22.95 per day and $0.15 per mile. He travels 420 miles. What is his total cost? **$200.70**

30. Betty rents a VCR for 1 week at $5.95 per day. She also rents 3 tapes, each for $4.95 per week. What is her total cost? **$56.50**

P18 ON MY OWN   Use with text pages 86–87.

ENRICHMENT 4.4

Name _____
LESSON 4.4

Fraction Flowers

Shade each petal that contains a fraction in simplest form. Then write the other fractions in simplest form.

E18 STRETCH YOUR THINKING   Use with text pages 86–87.

87

ORGANIZER

OBJECTIVE To write fractions as mixed numbers and mixed numbers as fractions

VOCABULARY *New* **mixed number**

ASSIGNMENTS Independent Practice

- *Basic* 1–43, even; 44–47
- *Average* 1–47
- *Advanced* 1–12, 25–43, even; 44–47

PROGRAM RESOURCES

Reteach 4.5 Practice 4.5 Enrichment 4.5

PROBLEM OF THE DAY

Transparency 4

Brad had 100 coins, all of which were different from Chelsea's 50 coins. Brad and Chelsea took their coins to the bank and exchanged them for bills. Each got the same three bills, with no coins left over. What bills did they receive? What coins did each start with?

Possible answer: Each received one $5 bill and two $1 bills. Brad had 10 quarters, 40 dimes, and 50 pennies; Chelsea had 10 half dollars and 40 nickels.

Solution Tab A, p. A4

QUICK CHECK

Is the fraction closer to 0, $\frac{1}{2}$, or 1?

1. $\frac{5}{6}$ 1   **2.** $\frac{1}{8}$ 0   **3.** $\frac{4}{9}$ $\frac{1}{2}$

4. $\frac{1}{16}$ 0   **5.** $\frac{3}{5}$ $\frac{1}{2}$ **6.** $\frac{9}{10}$ 1

1 MOTIVATE

Have students do this activity in small groups.

1. Fold a sheet of paper in half; then unfold it.
   What fraction does the whole sheet of paper represent? $\frac{2}{2}$
2. Refold the paper as in Step 1, and then fold the paper in half again and unfold it.
   Now what fraction does the paper represent? $\frac{4}{4}$
3. **What whole number do all of these fractions represent?** 1

WHAT YOU'LL LEARN
How to write fractions as mixed numbers and mixed numbers as fractions

WHY LEARN THIS?
So you can express amounts, such as slices of pizza, as either a fraction or a mixed number

WORD POWER
mixed number

Real-Life Link

Although U.S. money is based on a decimal system, some terms for coins name fractional parts of a dollar, such as quarter and half dollar. Suppose you want to buy something that costs $3.50. You have 9 quarters and 3 half dollars. Do you have enough money? Explain.

Yes, because 9 quarters and 3 half dollars are $3.75.

Mixed Numbers and Fractions

A pizza store sells pizza by the slice. The pizzas are cut into fourths. You want to buy 11 slices to share with your friends. How many pizzas is this?

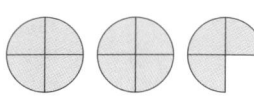

11 pieces, or $\frac{11}{4}$

Look at the model above. Since 11 pieces equal 2 whole pizzas and $\frac{3}{4}$ of another pizza, $\frac{11}{4} = 2\frac{3}{4}$.

A fraction like $\frac{11}{4}$ is greater than 1 because the numerator is greater than the denominator. A fraction greater than 1 can be written as a mixed number. A **mixed number** is a whole number and a fraction.

EXAMPLE 1 Write $\frac{17}{3}$ as a mixed number.

$$\frac{17}{3} \rightarrow 3)\overline{17} \begin{array}{c} 5\frac{2}{3} \\ \underline{-15} \\ 2 \end{array}$$

Divide the numerator by the denominator.
Write the remainder as a fraction in simplest form. Use the remainder as the numerator and the divisor as the denominator.

So, $\frac{17}{3} = 5\frac{2}{3}$.

Another Method You can use a calculator to change a fraction to a mixed number.

17 [b/c] 3 [SHIFT] [ab/c↔d/c] [5⅔]

You can also write a mixed number as a fraction.

EXAMPLE 2 Write $3\frac{2}{5}$ as a fraction.

$$3\frac{2}{5} \rightarrow \frac{(3 \times 5) + 2}{5} = \frac{17}{5}$$

Multiply the whole number by the denominator and add the numerator. Use the same denominator.

So, $3\frac{2}{5} = \frac{17}{5}$.

- Write $2\frac{4}{7}$ as a fraction. $\frac{18}{7}$

Calculator Activities, page H31

IDEA FILE

CAREER CONNECTION

Stockbroker Have students bring the financial section of the newspaper to school. Discuss how fractions are used in the stock market. Each student can choose a company from the stock exchange and check the price of the stock every day for a week. At the end of the week, students can report on the performance of the stocks they chose.

Modalities Auditory, Kinesthetic, Visual

RETEACH 4.5

Name _____

LESSON 4.5

Mixed Numbers and Fractions

A mixed number is made up of two parts: a whole number and a fraction.
- You can rewrite a mixed number as an equivalent fraction. This type of fraction will have a numerator that is greater than the denominator.
- You can also rewrite a fraction with a numerator greater than the denominator as a mixed number.

Write $4\frac{2}{3}$ as an equivalent fraction.
Step 1 Multiply the whole number by the denominator of the fraction.
$4 \times 3 = 12$
Step 2 Add the numerator. This sum is the numerator of the equivalent fraction.
$12 + 2 = 14$
Step 3 Use the same denominator to write the equivalent fraction.
$\frac{14}{3}$

Write $\frac{19}{5}$ as a mixed number.
Step 1 Divide the numerator by the denominator.
$19 \div 5 = 3 r4$
Step 2 Use the remainder as the numerator in the fraction part of the mixed number. The denominator is the divisor.
$3\frac{4}{5}$ ← remainder ← divisor

Complete.
1. Write $7\frac{3}{5}$ as a fraction.

Step 1 Multiply the whole number by the denominator of the fraction.
$7 \times \boxed{5} = \boxed{35}$

Step 2 Add the ___numerator___. This sum is the ___numerator___ of the equivalent fraction.
$\boxed{35} + \boxed{3} = \boxed{38}$

Step 3 Use the same ___denominator___ to write the equivalent fraction.
$\frac{\boxed{38}}{\boxed{5}}$

Write the mixed number as a fraction.
2. $5\frac{1}{3}$ $\frac{16}{3}$ 3. $4\frac{1}{5}$ $\frac{21}{5}$ 4. $5\frac{2}{7}$ $\frac{37}{7}$ 5. $6\frac{1}{3}$ $\frac{19}{3}$ 6. $7\frac{4}{5}$ $\frac{39}{5}$

Write the fraction as a mixed number.
7. $\frac{17}{3}$ $5\frac{2}{3}$ 8. $\frac{19}{6}$ $3\frac{1}{6}$ 9. $\frac{37}{8}$ $4\frac{5}{8}$ 10. $\frac{52}{9}$ $5\frac{7}{9}$ 11. $\frac{69}{11}$ $6\frac{3}{11}$

Use with text pages 88–89.   TAKE ANOTHER LOOK R19

GUIDED PRACTICE

Write the fraction as a mixed number.

1. $\frac{3}{2}$ $1\frac{1}{2}$ 2. $\frac{5}{3}$ $1\frac{2}{3}$ 3. $\frac{7}{2}$ $3\frac{1}{2}$ 4. $\frac{9}{5}$ $1\frac{4}{5}$ 5. $\frac{13}{4}$ $3\frac{1}{4}$

Write the mixed number as a fraction.

6. $1\frac{1}{4}$ $\frac{5}{4}$ 7. $1\frac{3}{5}$ $\frac{8}{5}$ 8. $2\frac{2}{3}$ $\frac{8}{3}$ 9. $3\frac{1}{3}$ $\frac{10}{3}$ 10. $2\frac{4}{7}$ $\frac{18}{7}$

INDEPENDENT PRACTICE

Write the fraction as a mixed number or a whole number.

1. $\frac{7}{4}$ $1\frac{3}{4}$ 2. $\frac{9}{2}$ $4\frac{1}{2}$ 3. $\frac{11}{2}$ $5\frac{1}{2}$ 4. $\frac{23}{4}$ $5\frac{3}{4}$ 5. $\frac{27}{3}$ 9 6. $\frac{19}{5}$ $3\frac{4}{5}$

7. $\frac{31}{6}$ $5\frac{1}{6}$ 8. $\frac{18}{11}$ $1\frac{7}{11}$ 9. $\frac{29}{7}$ $4\frac{1}{7}$ 10. $\frac{32}{4}$ 8 11. $\frac{75}{16}$ $4\frac{11}{16}$ 12. $\frac{50}{9}$ $5\frac{5}{9}$

13. $\frac{31}{9}$ $3\frac{4}{9}$ 14. $\frac{41}{6}$ $6\frac{5}{6}$ 15. $\frac{6}{5}$ $1\frac{1}{5}$ 16. $\frac{19}{3}$ $6\frac{1}{3}$ 17. $\frac{22}{4}$ $5\frac{1}{2}$ 18. $\frac{25}{7}$ $3\frac{4}{7}$

Write the mixed number as a fraction.

19. $3\frac{2}{3}$ $\frac{11}{3}$ 20. $6\frac{1}{2}$ $\frac{13}{2}$ 21. $5\frac{1}{3}$ $\frac{16}{3}$ 22. $1\frac{9}{10}$ $\frac{19}{10}$ 23. $4\frac{1}{9}$ $\frac{37}{9}$ 24. $10\frac{2}{3}$ $\frac{32}{3}$

25. $9\frac{1}{4}$ $\frac{37}{4}$ 26. $2\frac{3}{8}$ $\frac{19}{8}$ 27. $4\frac{9}{11}$ $\frac{53}{11}$ 28. $8\frac{4}{9}$ $\frac{76}{9}$ 29. $6\frac{3}{5}$ $\frac{33}{5}$ 30. $9\frac{7}{8}$ $\frac{79}{8}$

31. $7\frac{1}{4}$ $\frac{29}{4}$ 32. $1\frac{3}{5}$ $\frac{8}{5}$ 33. $7\frac{1}{2}$ $\frac{15}{2}$ 34. $2\frac{1}{6}$ $\frac{13}{6}$ 35. $2\frac{1}{12}$ $\frac{25}{12}$ 36. $4\frac{3}{8}$ $\frac{35}{8}$

Tell which are equivalent numbers in each set.

37. $9\frac{1}{2}, \frac{17}{2}, \frac{19}{2}, 9\frac{1}{3}$ $9\frac{1}{2}$ and $\frac{19}{2}$ 38. $\frac{13}{5}, 5\frac{1}{5}, 4\frac{1}{5}, \frac{26}{5}$ $5\frac{1}{5}$ and $\frac{26}{5}$ 39. $\frac{28}{3}, \frac{13}{3}, 8\frac{1}{3}, 9\frac{1}{3}$ $\frac{28}{3}$ and $9\frac{1}{3}$

Write the missing number.

40. $\frac{40}{9} = 4\frac{\blacksquare}{9}$ 4 41. $\frac{\blacksquare}{3} = 3\frac{2}{3}$ 11 42. $\frac{\blacksquare}{6} = 9\frac{1}{6}$ 55 43. $8\frac{3}{5} = \frac{43}{\blacksquare}$ 5

Problem-Solving Applications

44. Renee has $1\frac{3}{4}$ yd of fabric. Does she have enough for a pillow cover that requires $\frac{3}{2}$ yd of fabric? **Yes. The** Explain. **pillow cover requires only $1\frac{1}{2}$ yd of fabric.**

45. Don drinks $\frac{5}{3}$ glasses of orange juice. His sister drinks $1\frac{2}{3}$ glasses of orange juice. Do they drink the same amount? Explain. **yes; $\frac{5}{3} = 1\frac{2}{3}$**

46. Rick changed $3\frac{1}{4}$ to a fraction. He used this method: $3\frac{1}{4} = 3 \times \frac{4}{4} + \frac{1}{4} = \frac{12}{4} + \frac{1}{4} = \frac{13}{4}$ Explain his method. **Possible answer: change whole number to a fraction; find the sum of fractions.**

47. ✏ **WRITE ABOUT IT** Can any fraction be changed to a mixed number? Explain. **No. The numerator must be greater than the denominator.**

MORE PRACTICE Lesson 4.5, page H48

89

Students begin the lesson by relating fractions and mixed numbers to a real-life model. Then they proceed to explore the procedures for writing mixed numbers as fractions and fractions as mixed numbers.

GUIDED PRACTICE

Have students explain how they solved Exercises 5 and 10. Their explanations should demonstrate an understanding of the algorithms they used.

INDEPENDENT PRACTICE

Students may choose to use calculators for some of the exercises.

ADDITIONAL EXAMPLES

Example 1, p. 88

Write $\frac{7}{2}$ as a mixed number. $3\frac{1}{2}$

Example 2, p. 88

Write $1\frac{3}{5}$ as a fraction. $\frac{8}{5}$

3 WRAP UP

Use these questions to help students focus on the big ideas of the lesson.

• **What operations do you use to write a fraction as a mixed number?** division and subtraction

• **What operations do you use to write a mixed number as a fraction?** multiplication and addition

✓ *Assessment Tip*

Have students write the answers.

Compare the numerator and denominator in

• **a fraction less than 1.** The numerator is less than the denominator.

• **a fraction greater than 1.** The numerator is greater than the denominator.

DAILY QUIZ

Write the fraction as a mixed number or a whole number.

1. $\frac{12}{5}$ $2\frac{2}{5}$ 2. $\frac{10}{7}$ $1\frac{3}{7}$ 3. $\frac{49}{9}$ $5\frac{4}{9}$

Write the mixed number as a fraction.

4. $2\frac{3}{5}$ $\frac{13}{5}$ 5. $1\frac{7}{8}$ $\frac{15}{8}$ 6. $9\frac{1}{3}$ $\frac{28}{3}$

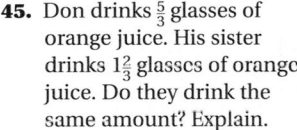

PRACTICE 4.5

Name _____

LESSON 4.5

Mixed Numbers and Fractions

Vocabulary

Complete.

1. A whole number and a fraction is called a(n) ___ **mixed number** ___.

Write the fraction as a mixed number or a whole number.

2. $\frac{20}{5}$ 4 3. $\frac{19}{4}$ $4\frac{3}{4}$ 4. $\frac{22}{7}$ $3\frac{1}{7}$ 5. $\frac{39}{10}$ $3\frac{9}{10}$ 6. $\frac{19}{10}$ $1\frac{9}{10}$

7. $\frac{75}{15}$ 5 8. $\frac{44}{13}$ $3\frac{5}{13}$ 9. $\frac{50}{7}$ $7\frac{1}{7}$ 10. $\frac{63}{21}$ 3 11. $\frac{41}{8}$ $5\frac{1}{8}$

Write the mixed number as a fraction.

12. $6\frac{2}{7}$ $\frac{44}{7}$ 13. $4\frac{6}{11}$ $\frac{50}{11}$ 14. $9\frac{2}{3}$ $\frac{29}{3}$ 15. $11\frac{1}{5}$ $\frac{56}{5}$ 16. $2\frac{2}{3}$ $\frac{8}{3}$

Tell which are equivalent numbers in each set.

17. $8\frac{1}{5}, \frac{81}{5}, 41\frac{1}{5}, 41\frac{1}{5}$ $8\frac{1}{5}$ and $\frac{41}{5}$ 18. $\frac{22}{3}, 8\frac{1}{3}, \frac{19}{3}, 6\frac{1}{3}$ $\frac{19}{3}$ and $6\frac{1}{3}$

Write the missing number.

19. $\frac{54}{11} = 4\frac{10}{11}$ 20. $\frac{97}{12} = 8\frac{1}{12}$ 21. $\frac{5}{\blacksquare} = \frac{68}{9}$ 22. $9\frac{1}{\blacksquare} = \frac{19}{\blacksquare}$

Mixed Applications

23. Jon jogs $\frac{7}{4}$ of a mile. Kathy jogs $1\frac{3}{4}$ of a mile. Do they jog the same distance? Explain. **Yes; $\frac{7}{4} = 1\frac{3}{4}$.**

24. Vicky has 32 pens. Each person in her office can take 4 pens. How many people work in her office? **8 people**

25. Terri is making a quilt. She needs $4\frac{1}{3}$ yd of fabric. If she buys $1\frac{1}{3}$ yd, will she have enough to make the quilt? Explain. **Yes. The quilt requires only $1\frac{1}{3}$ yd of fabric.**

26. Chris has a dozen pieces of fruit. She has 4 oranges, 5 apples, and some bananas. What fraction of the fruit is bananas? Write the fraction in simplest form. **$\frac{1}{4}$**

Use with text pages 88–89. ON MY OWN P19

ENRICHMENT 4.5

Name _____

LESSON 4.5

Fraction Squares

Shade the squares that show fractions equivalent to the mixed number in the center.

Use with text pages 88–89. STRETCH YOUR THINKING E19

OBJECTIVE To practice solving problems by using multiples and factors

USING THE PAGE

Before the Lesson

Write the following statement on the over-head projector. Twenty-two out of every 100 people in Japan live in rural areas. Have the students make a factor tree to find the prime factors of 22. Since the first two numbers are 2 and 11, which are prime numbers, the prime factorization is 2 × 11. Have students define the following terms in a chart and add examples:

prime numbers
composite numbers
factors
common multiples
LCM

Cooperative Groups

Have the cooperative groups read the infor-mation about Japan and answer the ques-tions under the chart. Then have them work together to solve Problems 1–2. When they finish, ask all the groups to post their work and solutions on a large piece of paper. They can compare and discuss the results. When all groups have posted their work, have students label their work with terms from the definition chart.

CULTURAL CONNECTION

COOPERATIVE LEARNING

Holiday Trip to Japan

This fall, Hideki and his family are taking a trip to Japan to visit his grandparents, who grow rice on a farm on the island of Kyushu. They will arrive at the end of the rice harvest in time to celebrate Labor Thanksgiving Day. Labor Thanksgiving Day is a national holiday in Japan to honor work of all kinds.

Since it will be Hideki's first visit to Japan, he went to the library to read about Japan. He found that $\frac{1}{6}$ of the people in Japan are farmers.

On a hundreds chart, circle the positive multiples of 6 to show the number of farmers for every 100 people in Japan.

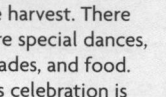

> **CULTURAL LINK**
>
> Long ago, Japanese farmers had a celebration after the rice harvest. There were special dances, parades, and food. This celebration is now known as Labor Thanksgiving Day.

```
 1  2  3  4  5 (6) 7  8  9 10 11(12)13 14 15 16 17(18)19 20
21 22 23(24)25 26 27 28 29(30)31 32 33 34 35(36)37 38 39 40
41(42)43 44 45 46 47(48)49 50 51 52 53(54)55 56 57 58 59(60)
61 62 63 64 65(66)67 68 69 70 71(72)73 74 75 76 77(78)79 80
81 82 83(84)85 86 87 88 89(90)91 92 93 94 95(96)97 98 99 100
```

So, about 16 of every hundred people in Japan are farmers.

• What pattern do you notice with the circled numbers?
  The numbers end in 6, 2, 8, 4, 0 and then the pattern repeats.
• Are the circled numbers prime or composite? Explain.
  They are composite, since they all have more than two factors.
• Suppose $\frac{1}{7}$ of the people in Japan were farmers. Would there be more or fewer farmers for every 100 people in Japan? **fewer**

[Portfolio] Work Together

1. Hideki has two uncles in Japan who are also farmers. One uncle has rice fields of 3 acres each. Another uncle has rice fields of 4 acres each. His grandfather has rice fields of 2 acres each. What is the least number of rice fields each of them would have to plant so they would have the same number of acres planted? **1st uncle: 4; 2nd uncle: 3; grandfather: 6**

2. Hideki takes $400 out of his savings account for the trip. He plans to spend $80 on gifts to take to his relatives in Japan. What fraction of the total does he plan to spend? Write the fraction in simplest form.
$\frac{1}{5}$

90 Cultural Connection

CHAPTER 4 REVIEW

EXAMPLES

• **Determine whether a number is prime or composite.** (pages 76–77)

$3 = 1 \times 3$ $10 = 1 \times 10$
                  2×5

factors of 3: 1 and 3   factors of 10: 1, 2, 5, 10
3 is prime.   10 is composite.

• **Write a composite number as the product of prime numbers.** (pages 78–79)

Find the prime factors of 44.

```
      44
     /  \
    2 × 22   ← 22 is not prime.
   / \  / \
  2 × 2 × 11  ← 2² × 11
```

• **Find the LCM and GCF of numbers.** (pages 80–83)

Find the LCM and GCF for the pair of numbers 9 and 12.

LCM	GCF
9: 9, 18, 27, 36, 45	9: 1, 3, 9
12: 12, 24, 36, 48	12: 1, 2, 3, 4, 6, 12

The LCM is 36.   The GCF is 3.

• **Write fractions in simplest form.** (pages 86–87)

$\frac{6}{8} = \frac{6 \div 2}{8 \div 2} = \frac{3}{4}$ *Divide by the GCF.*

• **Write mixed numbers as fractions and fractions as mixed numbers.** (pages 88–89)

$1\frac{2}{3} \rightarrow \frac{3 \times 1 + 2}{3} = \frac{5}{3}$

$1\frac{2}{3} = \frac{5}{3}$

$\frac{4}{3} \rightarrow 3\overline{)4} \quad 1\frac{1}{3}$
$\qquad\qquad \frac{-3}{1}$

EXERCISES

1. **VOCABULARY** Numbers with only two factors are called _?_. **prime numbers**

Tell whether the number is prime or composite.

2. 7 **prime**   3. 9 **composite**
4. 12 **composite**   5. 26 **composite**
6. 11 **prime**   7. 37 **prime**
8. 49 **composite**   9. 71 **prime**

10. **VOCABULARY** When a composite number is written as the product of prime factors, this is called _?_. **prime factorization**

Find the prime factors of the number.

11. 9 3^2 12. 8 2^3
13. 14 2×7 14. 18 2×3^2
15. 12 $2^2 \times 3$ 16. 33 3×11
17. 50 2×5^2 18. 49 7^2

19. **VOCABULARY** The largest of the common factors is called the _?_. **greatest common factor**

Find the LCM and GCF for each pair of numbers.

20. 3, 9 **9, 3**   21. 2, 6 **6, 2**
22. 6, 4 **12, 2**   23. 10, 15 **30, 5**
24. 8, 12 **24, 4**   25. 9, 27 **27, 9**
26. 15, 25 **75, 5**   27. 12, 15 **60, 3**

Write the fraction in simplest form.

28. $\frac{2}{8}$ $\frac{1}{4}$ 29. $\frac{3}{9}$ $\frac{1}{3}$
30. $\frac{6}{21}$ $\frac{2}{7}$ 31. $\frac{9}{24}$ $\frac{3}{8}$
32. $\frac{4}{10}$ $\frac{2}{5}$ 33. $\frac{15}{20}$ $\frac{3}{4}$

Write each fraction as a mixed number and each mixed number as a fraction.

34. $\frac{8}{5}$ $1\frac{3}{5}$ 35. $\frac{11}{7}$ $1\frac{4}{7}$ 36. $1\frac{1}{3}$ $\frac{4}{3}$
37. $2\frac{3}{7}$ $\frac{17}{7}$ 38. $\frac{9}{2}$ $4\frac{1}{2}$ 39. $2\frac{1}{6}$ $\frac{13}{6}$
40. $3\frac{2}{5}$ $\frac{17}{5}$ 41. $\frac{7}{3}$ $2\frac{1}{3}$ 42. $\frac{10}{7}$ $1\frac{3}{7}$

Chapter 4 Review   91

CHAPTER 4 REVIEW & TEST

USING THE PAGE
The Chapter 4 Review reviews prime and composite numbers, finding the LCM and GCF of numbers, writing fractions in simplest form, and writing mixed numbers as fractions and fractions as mixed numbers. Chapter objectives are provided, along with examples and practice exercises for each.

Portfolio The Chapter 4 Review can be used as a review or a test. You may wish to place the Review in students' portfolios.

Assessment Checkpoint

For Performance Assessment in this chapter, see page 146, What Did I Learn?, items 1 and 5.

✔ Assessment Tip

Student Self-Assessment The How Did I Do? Survey helps students assess both what they have learned and how they learned it. Self-assessment helps students learn more about their own capabilities and develop confidence in themselves.

A self-assessment survey is available as a copying master in the *Teacher's Guide for Assessment*, p. 15.

CHAPTER 4 TEST, page 1

Name _____   CHAPTER 4 TEST
                                         PAGE 1

Choose the letter of the correct answer.

1. Which of these numbers is a prime number?
   A. 27   B. 29
   C. 33   D. 39

2. What is the next multiple of 12?
   12, 24, 36, _?_
   A. 40   B. 42
   C. 44   D. not here

3. What are the factors of 49?
   A. 1, 7
   B. 1, 3, 7, 13, 49
   C. 1, 7, 49
   D. 3, 7, 13

4. Which of these numbers is a composite number?
   A. 2   B. 17
   C. 51   D. 47

5. Herbert has a game every seventh day during November. His first game is on November 7. How many games does he have in November?
   A. 3 games
   B. 4 games
   C. 5 games
   D. 7 games

6. In February Dorothy runs 2 mi every third day. She begins on February 3. How many miles does she run in February?
   A. 9 mi   B. 12 mi
   C. 18 mi   D. 24 mi

7. What is the missing factor?

   A. 2   B. 3
   C. 4   D. 5

8. What is the missing factor?

   A. 2   B. 4
   C. 6   D. 8

9. What is the prime factorization of 39?
   A. 3 × 13   B. 3 × 3 × 7
   C. $3^2 \times 2$ D. 3 × 6 × 3

10. Solve for n to complete the prime factorization.
    $2^n = 16$
    A. 2   B. 4
    C. 6   D. 8

11. What is the least common multiple of 3 and 4?
    3: 3 6 9 12 15 18 21 24
    4: 4 8 12 16 20 24 28 31
    A. 7   B. 8
    C. 12   D. 24

Form A • Multiple-Choice   A17   Go on. ▶

CHAPTER 4 TEST, page 2

Name _____   CHAPTER 4 TEST
                                         PAGE 2

12. Use the factor trees to find the greatest common factor of 16 and 12.

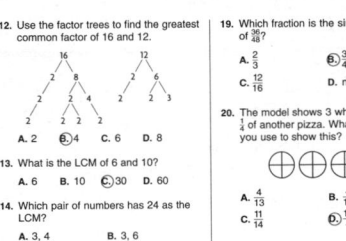

    A. 2   B. 4
    C. 6   D. 8

13. What is the LCM of 6 and 10?
    A. 6   B. 10   C. 30   D. 60

14. Which pair of numbers has 24 as the LCM?
    A. 3, 4   B. 3, 6
    C. 8, 12   D. 24, 48

15. What is the GCF of 24 and 40?
    A. 2   B. 4   C. 8   D. 12

16. Harry has 40 erasers and 60 pencils to put in packages. All the packages will contain the same number of erasers and the same number of pencils. What is the greatest number of packages he can make?
    A. 10   B. 12   C. 15   D. 20

17. Which fraction is the simplest form of $\frac{8}{12}$?
    A. $\frac{2}{4}$ B. $\frac{2}{3}$
    C. $\frac{4}{6}$ D. not here

18. Which fraction is the simplest form of $\frac{10}{12}$?
    A. $\frac{4}{5}$ B. $\frac{5}{7}$
    C. $\frac{6}{8}$ D. not here
    E. $\frac{7}{8}$

19. Which fraction is the simplest form of $\frac{9}{12}$?
    A. $\frac{2}{3}$ B. $\frac{3}{4}$
    C. $\frac{12}{16}$ D. not here

20. The model shows 3 whole pizzas and $\frac{1}{4}$ of another pizza. What fraction could you use to show this?

    A. $\frac{4}{13}$ B. $\frac{4}{11}$
    C. $\frac{11}{14}$ D. $\frac{13}{4}$

21. What is $2\frac{1}{2}$ written as a fraction?
    A. $\frac{2}{5}$ B. $\frac{3}{2}$
    C. $\frac{5}{2}$ D. $\frac{12}{4}$

22. What is $\frac{25}{7}$ written as a mixed number?
    A. $5\frac{2}{7}$ B. 3
    C. $3\frac{1}{7}$ D. $3\frac{3}{7}$

23. What is the missing number?
    $5\frac{3}{12} = \frac{\blacksquare}{12}$
    A. 36   B. 48
    C. 63   D. not here

24. What is the missing number?
    $\frac{\blacksquare}{4} = 4\frac{3}{4}$
    A. 11   B. 15
    C. 16   D. 19

Form A • Multiple-Choice   A18   ▶ Stop!

See *Test Copying Masters*, pp. A17–A18 for the Chapter Test in **multiple-choice** format and pp. B167–B168 for **free-response** format.

91

5

PLANNING GUIDE

Adding and Subtracting Fractions

BIG IDEA: Addition and subtraction of fractions can be used to solve many real-world problems.

Introducing the Chapter p. 92		Team-Up Project • Drawing People p. 93	
OBJECTIVE	**VOCABULARY**	**MATERIALS**	**RESOURCES**
5.1 Adding and Subtracting Like Fractions pp. 94-95 **Objective:** To add and subtract fractions with like denominators **1 DAY**			■ Reteach, ■ Practice, ■ Enrichment 5.1; More Practice, p. H48 Math Fun • The Pizza Runner, p. 108
INTERVENTION • Equivalent Fractions and Fractions in Simplest Form, Teacher's Edition pp. 96A–96B			
Lab Activity: Addition and Subtraction of Unlike Fractions pp. 96-97 **Objective:** To use fraction bars to add and subtract fractions with unlike denominators		FOR EACH STUDENT: fraction bars	💻 **E-Lab • Activity 5** E-Lab Recording Sheets, p. 5 Math Fun • Fraction Riddles, p. 108
5.2 Adding and Subtracting Unlike Fractions pp. 98-99 **Objective:** To use diagrams to add and subtract unlike fractions **1 DAY**			■ Reteach, ■ Practice, ■ Enrichment 5.2; More Practice, p. H48 Math Fun • Roll It Out, p.108
5.3 Adding Unlike Fractions pp. 100-101 **Objective:** To add unlike fractions **1 DAY**	least common denominator (LCD)		■ Reteach, ■ Practice, ■ Enrichment 5.3; More Practice, p. H49 🖩 Calculator Activities, p. H34
5.4 Subtracting Unlike Fractions pp. 102-103 **Objective:** To subtract unlike fractions **1 DAY**			■ Reteach, ■ Practice, ■ Enrichment 5.4; More Practice, p. H49 💿 *Number Heroes • Fraction Fireworks*
5.5 Estimating Sums and Differences pp. 104-107 **Objective:** To estimate sums and differences of fractions **2 DAYS**			■ Reteach, ■ Practice, ■ Enrichment 5.5; More Practice, p. H49

CHAPTER ASSESSMENT Chapter 5 Review p. 109

NCTM CURRICULUM
STANDARDS FOR GRADES 5–8

- ✓ **Problem Solving** *Lessons 5.1, 5.2, 5.3, 5.4, 5.5*
- ✓ **Communication** *Lessons 5.1, 5.2, 5.3, 5.4, 5.5*
- ✓ **Reasoning** *Lessons 5.1, 5.2, 5.3, 5.4, 5.5*
- ✓ **Connections** *Lessons 5.1, 5.2, 5.3, 5.4, 5.5*
- ✓ **Number and Number Relationships** *Lessons 5.1, 5.2, 5.5*
- ✓ **Number Systems and Number Theory** *Lessons 5.1, 5.2, 5.3, 5.4*
- ✓ **Computation and Estimation** *Lessons 5.1, 5.2, 5.3, 5.4, 5.5*
- ☐ **Patterns and Functions**
- ☐ **Algebra**
- ☐ **Statistics**
- ☐ **Probability**
- ☐ **Geometry**
- ✓ **Measurement** *Lessons 5.1, 5.2, 5.3, 5.4, 5.5*

ASSESSMENT OPTIONS

CHAPTER LEARNING GOALS

5A To add and subtract fractions

5A.1 To add and subtract like fractions

5A.2 To add and subtract unlike fractions

5A.3 To estimate sums and differences of fractions

These goals for concepts, skills, and problem solving are assessed in many ways throughout the chapter. See the chart below for a complete listing of both traditional and informal assessment options.

Pretest Options
Pretest for Chapter Content, TCM pp. A19–A20, B169–B170
Assessing Prior Knowledge, TE p. 92; APK p. 7

Daily Assessment
Mixed Review, PE pp. 103, 107
Problem of the Day, TE pp. 94, 98, 100, 102, 104
Assessment Tip and Daily Quiz, TE pp. 95, 97, 99, 101, 103, 107

Formal Assessment
Chapter 5 Review, PE p. 109
Chapter 5 Test, TCM pp. A19–A20, B169–B170
Study Guide & Review, PE pp. 144–145
Test • Chapters 4–7, TCM pp. A25–A26, B175–B176
Cumulative Review, PE p. 147
Cumulative Test, TCM pp. A27–A30, B177–B180

Performance Assessment
Team-Up Project Checklist, TE p. 93
What Did I Learn?, Chapters 4–7, TE p. 146
Interview/Task Test, PA p. 80

Portfolio
Suggested work samples:
Independent Practice, PE pp. 95, 99, 101, 106
Math Fun, PE p. 108

Student Self-Assessment
How Did I Do?, TGA p. 15
Portfolio Evaluation Form, TGA p. 24
Math Journal, TE pp. 92B, 96

Key

Assessing Prior Knowledge Copying Masters: APK	Teacher's Guide for Assessment: TGA
Test Copying Masters: TCM	Teacher's Edition: TE
Performance Assessment: PA	Pupil's Edition: PE

MATH COMMUNICATION

MATH JOURNAL
You may wish to use the following suggestions to have students write about the mathematics they are learning in this chapter:

PE page 106 • Talk About It

PE pages 95, 99, 101, 103, 107 • Write About It

TE pages 96, 107 • Writing in Mathematics

TGA page 15 • How Did I Do?

In every lesson • record written answers to each Wrap-up Assessment Tip from the Teacher's Edition for test preparation.

VOCABULARY DEVELOPMENT
The terms listed at the right will be introduced or reinforced in this chapter. The boldfaced terms in the list are new vocabulary terms in the chapter.

So that students understand the terms rather than just memorize definitions, have them construct their own definitions and pictorial representations and record those in their Math Journals.

least common denominator (LCD), pp. 100, 102

MEETING INDIVIDUAL NEEDS

Practice Activities, TE Tab B
Getting Close to the Answer
(Use with Lessons 5.1–5.4.)

Interdisciplinary Connection, TE p. 97

Literature Link, TE p. 92C
At Last I Kill a Buffalo, Activity 1
(Use with Lessons 5.1–5.4.)

Technology
Number Heroes • *Fraction Fireworks*
(Use with Lesson 5.4.)

E-Lab • Activity 5
(Use with Lab Activity.)

Practice Game, TE p. 92C
Fast Fractions
(Use with Lessons 5.1, 5.2, 5.4.)

Interdisciplinary Connection, TE p. 97

Literature Link, TE p. 92C
At Last I Kill a Buffalo, Activity 2
(Use with Lessons 5.1–5.4.)

Technology
Number Heroes • *Fraction Fireworks*
(Use with Lesson 5.4.)

E-Lab • Activity 5
(Use with Lab Activity.)

CHAPTER IDEA FILE

The activities on these pages provide additional practice and reinforcement of key chapter concepts. The chart above offers suggestions for their use.

LITERATURE LINK

At Last I Kill a Buffalo

by Luther Standing Bear

A young Sioux boy goes on his first buffalo hunt and learns what it takes to be a man.

Activity 1—In the story, the boy rides in quietly and surprises the herd of buffalo. He can cover 10 mi. a day. Have students calculate what fraction of the day it will take for the boy to get to the buffalo if the buffalo are 5 mi. away.

Activity 2—The buffalo herd has 50 buffalo. One boy kills 2, another kills 4, and a third kills 5. Have students express the number each boy has killed as a fraction of the herd. Add these to find the total fraction killed.

Activity 3—Have students research an endangered species and find out what fraction of that species dies each year. Students should compare their numbers with those of classmates and figure out which species have the highest and the lowest number of deaths.

PRACTICE GAME

Fast Fractions

Purpose To add like and unlike fractions

Materials 2 number cubes, one numbered 1–6, the other 7–12

About the Game In pairs, Player A rolls both number cubes and records the value of the 1–6 cube as the numerator of a fraction and the value of the 7–12 cube as the denominator. Then Player B rolls both number cubes to make a second fraction. Players compete to add the 2 fractions. The first player to get a correct answer scores 1 point. After 10 rounds the player with more points wins.

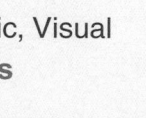

Modalities Kinesthetic, Visual

Multiple Intelligences
Visual/Spatial,
Logical/Mathematical

$$\frac{2}{7} + \frac{4}{10}$$

Practice Game, TE p. 92C
Fast Fractions
(Use with Lessons 5.1, 5.2, 5.4.)

Interdisciplinary Connection, TE p. 97

Literature Link, TE p. 92C
At Last I Kill a Buffalo, Activity 3
(Use with Lessons 5.1–5.4)

Technology
Number Heroes • *Fraction Fireworks*
(Use with Lesson 5.4.)

E-Lab • Activity 5
(Use with Lab Activity.)

INTERDISCIPLINARY SUGGESTIONS

Connecting Adding and Subtracting Fractions

Purpose To connect computation with fractions to other subjects students are studying

You may wish to connect **Adding and Subtracting Fractions** to other subjects by using related interdisciplinary units.

Suggested Units

• Consumer Connection: Fractions: Who Needs Them?

• History Connection: Egyptian Fractions

• Science Connection: Measuring in Small Quantities

• Multicultural Connection: Recipes from Other Lands

• Art Connection: Geometric Designs and Fractional Parts

Modalities Auditory, Kinesthetic, Visual

Multiple Intelligences Mathematical/Logical, Bodily/Kinesthetic, Visual/Spatial

The purpose of using technology in the chapter is to provide reinforcement of skills and support in concept development.

E-Lab • Activity 5—Students use the concepts of this lesson, reasoning skills, and a computer tool for finding multiples to add fractions with unlike denominators.

Number Heroes • *Fraction Fireworks*—The activities on this CD-ROM provide students with opportunities to practice finding the correct sum or difference of fractions.

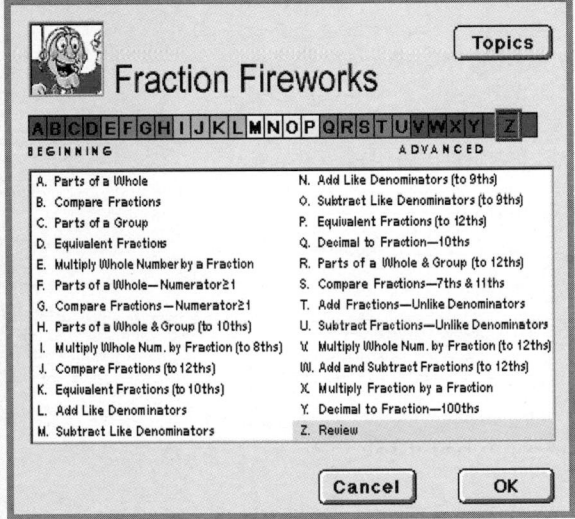

```
                                          Topics
Fraction Fireworks
A B C D E F G H I J K L M N O P Q R S T U V W X Y  Z
BEGINNING                                ADVANCED

A. Parts of a Whole            N. Add Like Denominators (to 9ths)
B. Compare Fractions           O. Subtract Like Denominators (to 9ths)
C. Parts of a Group            P. Equivalent Fractions (to 12ths)
D. Equivalent Fractions        Q. Decimal to Fraction—10ths
E. Multiply Whole Number by a Fraction   R. Parts of a Whole & Group (to 12ths)
F. Parts of a Whole—Numerator≥1          S. Compare Fractions—7ths & 11ths
G. Compare Fractions—Numerator≥1         T. Add Fractions—Unlike Denominators
H. Parts of a Whole & Group (to 10ths)   U. Subtract Fractions—Unlike Denominators
I. Multiply Whole Num. by Fraction (to 8ths)  V. Multiply Whole Num. by Fraction (to 12ths)
J. Compare Fractions (to 12ths)          W. Add and Subtract Fractions (to 12ths)
K. Equivalent Fractions (to 10ths)       X. Multiply Fraction by a Fraction
L. Add Like Denominators                 Y. Decimal to Fraction—100ths
M. Subtract Like Denominators            Z. Review

                          Cancel      OK
```

Each Number Heroes activity can be assigned by degree of difficulty. See Teacher's Edition page 103 for Grow Slide information.

Explorer Plus™ or fx–65 Calculator—Ideas for integrating the calculator in the PE Lesson p. 100 are provided in the *Calculator Handbook,* p. H34.

Visit our web site for additional ideas and activities.
http://www.hbschool.com

CHAPTER 5

Introducing the Chapter

Encourage students to talk about when they might add or subtract fractions. Review the math content students will learn in the chapter.

WHY LEARN THIS?

The question "Why are you learning this?" will be answered in every lesson. Sometimes there is a direct application to everyday experiences. Other lessons help students understand why these skills are needed for future success in mathematics.

PRETEST OPTIONS

Pretest for Chapter Content See the *Test Copying Masters*, pages A19–A20, B169–B170 for a free-response or multiple-choice test that can be used as a pretest.

Assessing Prior Knowledge Use the copying master shown below to assess previously taught skills that are critical for success in this chapter.

Assessing Prior Knowledge Copying Masters, p. 7

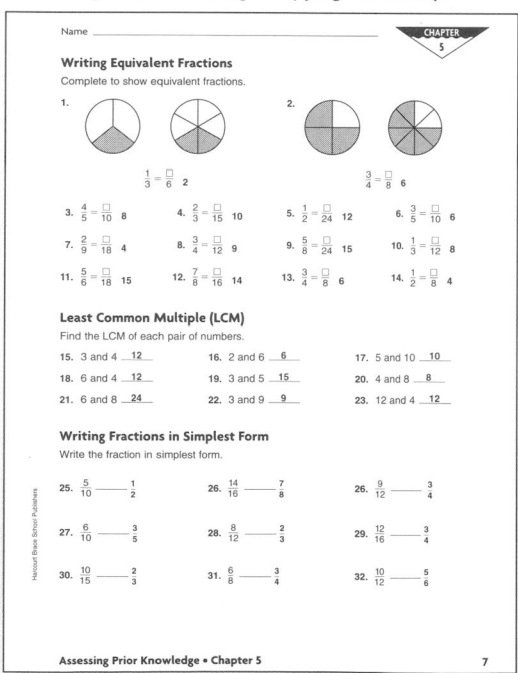

Troubleshooting

If students need help	Then use
• WRITING EQUIVALENT FRACTIONS	• Practice Activities, TE Tab B, Activity 16A
• FINDING LCM	• Key Skills, PE p. H5
• WRITING FRACTIONS IN SIMPLEST FORM	• Intervention • Equivalent Fractions, TE pp. 96A–96B

CHAPTER 5
ADDING AND SUBTRACTING FRACTIONS

LOOK AHEAD

In this chapter you will solve problems that involve

• adding and subtracting like and unlike fractions

• estimating sums and differences of fractions

92   Chapter 5

SCHOOL-HOME CONNECTION

You may wish to send to each student's family the *Math at Home* page. It includes an article that stresses the importance of communication about mathematics in the home, as well as helpful hints about homework and taking tests. The activity and extension provide extra practice that develops concepts and skills involving adding and subtracting fractions.

Fractions: Making Sum Difference

Your child is beginning a chapter that takes students farther in their work with fractions. In this chapter students will solve problems that involve adding and subtracting fractions and estimating sums and differences of fractions. Share with your child situations in which you need to add and subtract fractions for example, in building, or cooking.

Here is an activity for you to do with your child that involves addition and subtraction of fractions in a real-life situation. The Extension asks you and your child to solve a well-known mathematical puzzle. Good luck!

Something from the Oven

A baker planning to make several batches of cookies would have to have enough of the ingredients on hand. This activity simulates that situation. Find three different cookie recipes. Add the amounts of sugar needed for the three recipes. Find the amount of flour needed for all three. Do the recipes call for more sugar or flour? How much more?

Extension • On the Road Again!

Solve this: A farmer died, leaving 17 cows to his 3 sons. His will said that $\frac{1}{2}$ of the cows must go to the eldest son, $\frac{1}{3}$ of the cows to the middle son, and $\frac{1}{9}$ of the cows to the youngest son. But you cannot take $\frac{1}{2}$, $\frac{1}{3}$, or $\frac{1}{9}$ of 17 cows without killing some of the cows. A neighbor said he would lend the sons one of his cows. This solved the problem, since the eldest son got $\frac{1}{2}$ of 18, or 9, the middle son got $\frac{1}{3}$ of 18, or 6, and the youngest got $\frac{1}{9}$ of 18, or 2. That totaled 17 cows, so the neighbor took his cow and left. Can you figure out why the neighbor's plan worked?

For Your Information

Although there have been changes in the way mathematics is taught, much of what children study today is familiar to their parents. Basic facts, adding, subtracting, multiplying, dividing, whole numbers, fractions, decimals, percentages, and proportions remain a key part of a child's mathematics experience. However, math today also deals with statistics, probability, algebraic thinking, and geometric relationships and properties to prepare students for the demands of the present and the future.

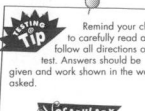

Remind your child to carefully read and follow all directions on a test. Answers should be given and work shown in the way asked.

VOCABULARY

least common denominator (LCD)—the LCM (smallest of the common multiples of two or more numbers) of two or more denominators

Math at Home

Team-Up Project

Drawing People

Suppose you want to draw a picture of a baby, a young child, a teenager and an adult. What fraction of the height of each drawing is the head? What fraction is the body? What fraction are the legs?

YOU WILL NEED: tape measure or yardstick

Plan
- Work with a small group to measure the length from head to shoulder, shoulder to waist , and waist to floor of people of various ages. Also measure their height.

Act
- Use a tape measure or yardstick. Make a table to record your data.
- Measure several people of different ages.
- Write fractions comparing the lengths measured to the height of each person.
- Use the fractions to draw a baby, young child, teenager, and adult.

Share
- Discuss your findings with other groups. See if the fractions are the same for all age groups.
- Compare drawings and discuss methods you used to draw the figures.

DID YOU
- ☑ Measure people of different ages?
- ☑ Use fractions to draw people of different ages?
- ☑ Share your data and findings?

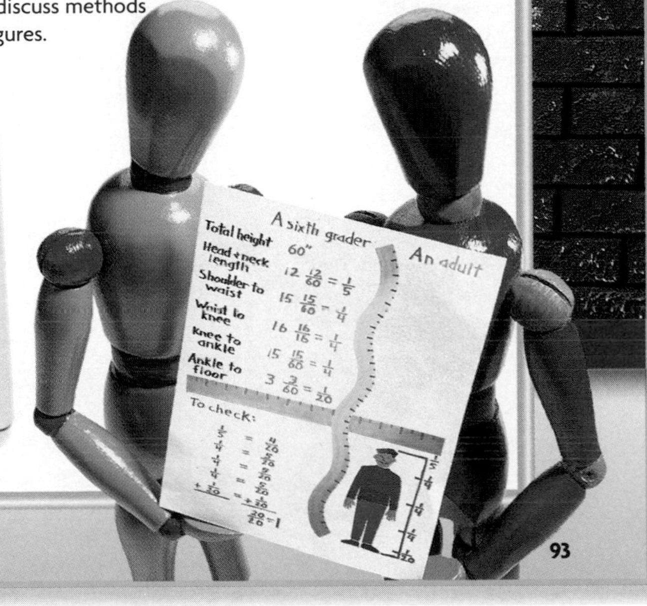

93

Project Checklist

Portfolio

Evaluate whether each group
- ☑ works cooperatively.
- ☑ measures accurately.
- ☑ writes and compares fractions correctly.

Using the Project

Accessing Prior Knowledge write and compare fractions

Purpose to add and subtract fractions

GROUPING small groups
MATERIALS yardsticks, meter sticks, or tape measures

Managing the Activity Help students find fractions that are easy to visualize. Encourage them to find fractions with denominators of eight or less that are close to their measurements.

WHEN MINUTES COUNT

A shortened version of this activity can be completed in 15–20 minutes. Have students measure each other, write fractions, and use their fractions to make drawings of themselves.

A BIT EACH DAY

Day 1: Students form groups, create tables on which to record their data, and speculate about how the fraction of head length to total height might be different for young people and adults.

Day 2: Students make and record measurements.

Day 3: Students write fractions for the measurements they made and make generalizations about the fraction of total height that is head, body, and legs for various ages.

Day 4: Students use their fractions to make drawings.

Day 5: Students compare their fractions and drawings with those of other groups. They discuss how the fractions change as humans grow and how to use fractions to make drawings.

HOME HAPPENINGS

Have students measure family members as part of the data collection process.

5.1

GRADE 5	GRADE 6	GRADE 7
Add and subtract fractions with like denominators.	Add and subtract fractions with like denominators.	Add and subtract fractions.

ORGANIZER

OBJECTIVE To add and subtract fractions with like denominators

ASSIGNMENTS Independent Practice
- *Basic* 1–30, odd; 31–36
- *Average* 1–36
- *Advanced* 11–36

PROGRAM RESOURCES
Reteach 5.1 Practice 5.1 Enrichment 5.1
Math Fun • The Pizza Runner, p. 108

PROBLEM OF THE DAY

Transparency 5
Write the next two terms of the pattern.

1. $\frac{1}{6} = \frac{1}{6}$
$\frac{2}{6} = \frac{1}{3}$
$\frac{3}{6} = \frac{1}{2}$ $\frac{4}{6} = \frac{2}{3}, \frac{5}{6} = \frac{5}{6}$

2. $\frac{1}{8} = \frac{1}{8}$
$\frac{2}{8} = \frac{1}{4}$
$\frac{3}{8} = \frac{3}{8}$ $\frac{4}{8} = \frac{1}{2}, \frac{5}{8} = \frac{5}{8}$

3. $\frac{1}{10} = \frac{1}{10}$
$\frac{2}{10} = \frac{1}{5}$
$\frac{3}{10} = \frac{3}{10}$ $\frac{4}{10} = \frac{2}{5}, \frac{5}{10} = \frac{1}{2}$

Solution Tab A, p. A5

QUICK CHECK

Write as a mixed number in simplest form.

1. $\frac{13}{3}$ $4\frac{1}{3}$
2. $\frac{11}{2}$ $5\frac{1}{2}$
3. $\frac{30}{7}$ $4\frac{2}{7}$
4. $\frac{10}{4}$ $2\frac{1}{2}$
5. $\frac{34}{6}$ $5\frac{2}{3}$
6. $\frac{75}{9}$ $8\frac{1}{3}$

1 MOTIVATE

To help students access their prior knowledge of fractions, ask:

- **How are fractions used in music?**
Musical notes represent fractions that indicate how long a tone is held.

- **Can you name these notes?**

whole note, half note, quarter note, eighth note

- **How many half notes are equal to a whole note? Explain.** 2 half notes; 2 halves equal 1 whole ($\frac{1}{2} + \frac{1}{2} = 1$).

5.1

WHAT YOU'LL LEARN
How to add and subtract fractions with like denominators

WHY LEARN THIS?
To find the total amount of time spent on an activity, such as reading a favorite book

REMEMBER:
When the numerator and denominator of an equivalent fraction have no common factor other than 1, the equivalent fraction is in simplest form. **See page H15.**
$\frac{3}{6} = \frac{1}{2}$
The factor of 1 is 1. The factors of 2 are 1 and 2. The common factor is 1.

Adding and Subtracting Like Fractions

$\mathbf{K}$im reads his library book for $\frac{2}{6}$ hr on Tuesday and $\frac{3}{6}$ hr on Saturday. How long does Kim spend reading his library book?

You can use models to add and subtract like fractions. The model below shows $\frac{2}{6} + \frac{3}{6}$.

$\frac{2}{6}$ $\frac{3}{6}$ $\frac{2}{6} + \frac{3}{6} = \frac{5}{6}$

Kim spends $\frac{5}{6}$ hr reading his library book.

You can add like fractions without models.

EXAMPLE 1 Add. $\frac{7}{9} + \frac{5}{9}$

$\frac{7}{9} + \frac{5}{9} = \frac{12}{9}$ *Add the numerators. Write the sum over the denominator.*

$= \frac{12 \div 3}{9 \div 3} = \frac{4}{3}$, or $1\frac{1}{3}$ *Write the fraction in simplest form. Write the answer as a fraction or a mixed number.*

- What is $\frac{9}{15} + \frac{12}{15}$ in simplest form? $\frac{7}{5}$, or $1\frac{2}{5}$

The method for subtracting like fractions is similar to the method for adding.

EXAMPLE 2 Subtract. $\frac{5}{8} - \frac{1}{8}$

$\frac{5}{8} - \frac{1}{8} = \frac{4}{8}$ *Subtract the numerators. Write the difference over the denominator.*

$= \frac{4 \div 4}{8 \div 4} = \frac{1}{2}$ *Write the answer in simplest form.*

So, $\frac{5}{8} - \frac{1}{8} = \frac{1}{2}$.

- What is $\frac{5}{6} - \frac{1}{6}$? $\frac{4}{6}$, or $\frac{2}{3}$

IDEA FILE

EXTENSION

Have students double this recipe by adding fractional amounts.

Snack Mix
$\frac{1}{4}$ c raisins $\frac{1}{2}$ c
$\frac{1}{6}$ c sunflower seeds $\frac{1}{3}$ c
$\frac{1}{3}$ c mixed nuts $\frac{2}{3}$ c
$\frac{3}{8}$ c dried apples $\frac{3}{4}$ c

Will the original recipe fit in a 2-c bowl? yes

What about the doubled recipe? no

Modalities Auditory, Visual

RETEACH 5.1

Name _____

Adding and Subtracting Like Fractions

Fractions that have the same denominator are *like fractions*. When adding or subtracting like fractions, you work with the numerators.

Add. $\frac{5}{7} + \frac{2}{7}$	Subtract. $\frac{5}{6} - \frac{1}{6}$
Step 1 Add the numerators.	**Step 1** Subtract the numerators.
$5 + 2 = 7$	$5 - 1 = 4$
Step 2 Write the sum of the numerators over the denominator.	**Step 2** Write the difference of the numerators over the denominator.
$\frac{5}{7} + \frac{2}{7} = \frac{7}{7}$	$\frac{5}{6} - \frac{1}{6} = \frac{4}{6}$
Step 3 Write the fraction in simplest form.	**Step 3** Write the fraction in simplest form.
$\frac{7}{7} = 1$	$\frac{4}{6} = \frac{4 \div 2}{6 \div 2} = \frac{2}{3}$
$\frac{5}{7} + \frac{2}{7} = 1$	$\frac{5}{6} - \frac{1}{6} = \frac{2}{3}$

Complete to find each sum or difference.

1. Add. $\frac{3}{4} + \frac{3}{4}$
Add the ___numerators___ $3 + $ __3__ $ = $ **6**
Write the sum over the ___denominator___ $\frac{6}{4}$
Write the fraction in ___simplest form___ $1\frac{1}{2}$
$\frac{3}{4} + \frac{3}{4} = 1\frac{1}{2}$

2. Subtract. $\frac{7}{10} - \frac{1}{10}$
Subtract the ___numerators___ $7 - $ __1__ $ = $ **6**
Write the difference over the ___denominator___ $\frac{6}{10}$
Write the fraction in ___simplest form___ $\frac{3}{5}$
$\frac{7}{10} - \frac{1}{10} = \frac{3}{5}$

Find each sum or difference. Write the answer in simplest form.

3. $\frac{11}{12} - \frac{5}{12}$ $\frac{1}{2}$
4. $\frac{5}{6} + \frac{1}{6}$ 1
5. $\frac{3}{8} + \frac{7}{8}$ $1\frac{1}{4}$
6. $\frac{8}{9} - \frac{2}{9}$ $\frac{2}{3}$

R20 TAKE ANOTHER LOOK   Use with text pages 94–95.

GUIDED PRACTICE

Find the sum or difference. Use a model to help you find the answer. **Check students' models.**

1. $\frac{1}{5} + \frac{2}{5}$ **$\frac{3}{5}$**   2. $\frac{2}{4} - \frac{1}{4}$ **$\frac{1}{4}$**   3. $\frac{1}{3} + \frac{1}{3}$ **$\frac{2}{3}$**   4. $\frac{2}{6} + \frac{3}{6}$ **$\frac{5}{6}$**   5. $\frac{3}{7} - \frac{1}{7}$ **$\frac{2}{7}$**

Find the sum or difference. Write the answer in simplest form.

6. $\frac{4}{5} - \frac{3}{5}$ **$\frac{1}{5}$**   7. $\frac{3}{8} + \frac{3}{8}$ **$\frac{3}{4}$**   8. $\frac{2}{3} + \frac{1}{3}$ **1**   9. $\frac{4}{6} - \frac{3}{6}$ **$\frac{1}{6}$**   10. $\frac{11}{12} - \frac{5}{12}$ **$\frac{1}{2}$**

INDEPENDENT PRACTICE

Find the sum or difference. Write the answer in simplest form.

1. $\frac{2}{5} + \frac{1}{5}$ **$\frac{3}{5}$**   2. $\frac{4}{9} + \frac{2}{9}$ **$\frac{2}{3}$**   3. $\frac{1}{4} + \frac{2}{4}$ **$\frac{3}{4}$**   4. $\frac{3}{4} + \frac{1}{4}$ **1**   5. $\frac{1}{6} + \frac{4}{6}$ **$\frac{5}{6}$**

6. $\frac{4}{7} - \frac{1}{7}$ **$\frac{3}{7}$**   7. $\frac{9}{10} - \frac{7}{10}$ **$\frac{1}{5}$**   8. $\frac{6}{10} - \frac{4}{10}$ **$\frac{1}{5}$**   9. $\frac{6}{7} - \frac{4}{7}$ **$\frac{2}{7}$**   10. $\frac{3}{8} - \frac{2}{8}$ **$\frac{1}{8}$**

11. $\frac{9}{12} + \frac{2}{12}$ **$\frac{11}{12}$**   12. $\frac{9}{10} - \frac{8}{10}$ **$\frac{1}{10}$**   13. $\frac{10}{20} + \frac{1}{20}$ **$\frac{11}{20}$**   14. $\frac{9}{12} - \frac{5}{12}$ **$\frac{1}{3}$**   15. $\frac{8}{9} + \frac{3}{9}$ **$1\frac{2}{9}$**

16. $\frac{9}{17} + \frac{5}{17}$ **$\frac{14}{17}$**   17. $\frac{9}{14} - \frac{5}{14}$ **$\frac{2}{7}$**   18. $\frac{14}{20} + \frac{11}{20}$ **$1\frac{1}{4}$**   19. $\frac{9}{16} - \frac{5}{16}$ **$\frac{1}{4}$**   20. $\frac{4}{15} + \frac{5}{15}$ **$\frac{3}{5}$**

21. $\frac{5}{8} + \frac{1}{8}$ **$\frac{3}{4}$**   22. $\frac{5}{6} - \frac{3}{6}$ **$\frac{1}{3}$**   23. $\frac{8}{15} - \frac{3}{15}$ **$\frac{1}{3}$**   24. $\frac{11}{15} - \frac{8}{15}$ **$\frac{1}{5}$**   25. $\frac{7}{10} + \frac{9}{10}$ **$1\frac{3}{5}$**

26. $\frac{7}{12} - \frac{5}{12}$ **$\frac{1}{6}$**   27. $\frac{10}{10} - \frac{7}{10}$ **$\frac{3}{10}$**   28. $\frac{6}{8} + \frac{6}{8}$ **$1\frac{1}{2}$**   29. $\frac{16}{25} + \frac{9}{25}$ **1**   30. $\frac{11}{12} - \frac{5}{12}$ **$\frac{1}{2}$**

Problem-Solving Applications

33. **$\frac{4}{5}$ of the lumber**

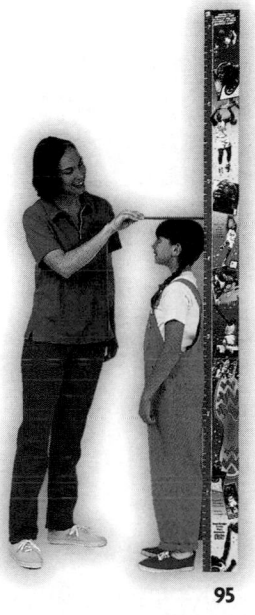

31. Ms. Johnson measures the height of her children on their birthdays. Lauren grew $\frac{2}{12}$ ft since her last birthday. John grew $\frac{5}{12}$ ft since his last birthday. How much more did John grow than Lauren? **$\frac{3}{12}$, or $\frac{1}{4}$ ft**

$\frac{3}{4}$ of the pizza; $\frac{1}{4}$ of the pizza

32. Brad ate $\frac{3}{8}$ of a small pizza, and Jimmy ate $\frac{3}{8}$ of the same pizza. How much of the pizza did they eat? How much was left over?

33. Tori keeps pieces of lumber in his garage. Of the lumber, $\frac{1}{5}$ is pine and $\frac{3}{5}$ is oak. How much of the lumber is pine and oak?

34. Jerry's CD collection is $\frac{5}{7}$ oldies, $\frac{1}{7}$ rock, and $\frac{1}{7}$ jazz. How much of his collection is not oldies music? **$\frac{2}{7}$ of his collection**

35. Erica has filled $\frac{3}{4}$ of an album with postcards. One fourth of the postcards are foreign postcards. What part of the album contains other postcards? **$\frac{2}{4}$**

36. **WRITE ABOUT IT** How do you add or subtract like fractions? **Write the sum or difference of the numerators over the denominator. Then simplify the fraction.**

MORE PRACTICE Lesson 5.1, page H48

95

---

2 TEACH/PRACTICE

A real-life setting is the springboard for this lesson. Then students relate fraction circles to the algorithms for adding and subtracting fractions with like denominators.

GUIDED PRACTICE
To make sure students understand how to add fractions with like denominators, have them work with a partner to complete Exercises 1–10.

INDEPENDENT PRACTICE
To assess students' understanding of the concepts, check answers to Exercises 1–5 before assigning the remaining exercises.

DEVELOPING ALGEBRAIC THINKING
Write an equation for Problem 31 using variables. Let y = the total number of feet.
$y = \frac{5}{12} - \frac{2}{12}$ $y = \frac{3}{12}$, or $\frac{1}{4}$

ADDITIONAL EXAMPLES

Example 1, p. 94
Add $\frac{4}{5}$ and $\frac{3}{5}$. $\frac{4}{5} + \frac{3}{5} = \frac{7}{5}$, or $1\frac{2}{5}$

Example 2, p. 94
Subtract $\frac{1}{6}$ from $\frac{5}{6}$. $\frac{5}{6} - \frac{1}{6} = \frac{4}{6}$, or $\frac{2}{3}$

3 WRAP UP

Use these questions to help students focus on the big ideas of the lesson.

• **What operations would you use to write the answer to $\frac{5}{8} + \frac{7}{8}$ as a mixed number?** addition, division

• **What operations would you use to write the answer to $\frac{7}{9} - \frac{4}{9}$ in simplest form?** subtraction, division

✔ Assessment Tip
Have students write the answer.

• **What is the difference between adding and subtracting like fractions?** The only difference is the operation.

DAILY QUIZ
Find the correct answer in simplest form. Choose *a, b,* or *c*.

1. $\frac{8}{9} + \frac{4}{9}$ a. $1\frac{1}{3}$ b. $1\frac{3}{9}$ c. $\frac{9}{12}$ a

2. $\frac{7}{8} - \frac{3}{8}$ a. $\frac{10}{8}$ b. $\frac{1}{2}$ c. $\frac{4}{8}$ b

3. $\frac{3}{5} - \frac{1}{5}$ a. $\frac{1}{5}$ b. $\frac{4}{5}$ c. $\frac{2}{5}$ c

4. $\frac{5}{8} + \frac{1}{8}$ a. $\frac{6}{8}$ b. $\frac{3}{4}$ c. $\frac{1}{2}$ b

5. $\frac{2}{5} + \frac{4}{5}$ a. $\frac{6}{10}$ b. $\frac{5}{6}$ c. $1\frac{1}{5}$ c

---

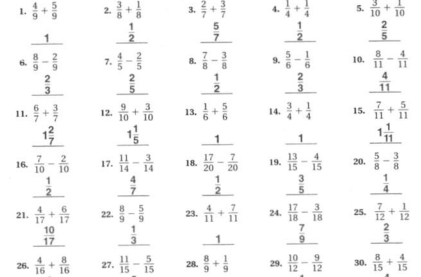

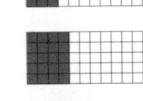
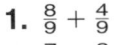
95

INTERVENTION

MEETING INDIVIDUAL NEEDS

Equivalent Fractions and Fractions in Simplest Form

Use these strategies to review the skills required for finding equivalent fractions and to simplify fractions.

Multiple Intelligences Logical/Mathematical, Visual/Spatial

Modalities Auditory, Kinesthetic, Visual

Lesson Connection Students will find equivalent fractions and write fractions in simplest form throughout this chapter.

Program Resources *Intervention and Extension Copying Masters*, p. 3

Troubleshooting

WHAT STUDENTS NEED TO KNOW	HOW TO HELP	FOLLOW UP
That a fraction has a numerator and a denominator	Have students identify the numerator and denominator. **Example** $\dfrac{5}{8}$ ← numerator ← denominator	Of the fractions $\frac{2}{3}$ and $\frac{5}{2}$, which has the larger numerator? the larger denominator? $\frac{5}{2}$; $\frac{2}{3}$
How to draw and shade a model of a fraction	Have students model a fraction by drawing vertical lines to divide a rectangle into equal parts. **Example** Show $\frac{2}{5}$ by shading. 	Other models to make: $\frac{1}{4}$, $\frac{3}{4}$, $\frac{2}{8}$, $\frac{4}{5}$ Check students' models. See Key Skill 25, Student Handbook, page H14.
How to show equivalent fractions by drawing and shading models	Have students model equivalent fractions by drawing horizontal lines in the model of one fraction. **Example** Show that $\frac{2}{5} = \frac{4}{10}$. Divide the model for $\frac{2}{5}$ in half by drawing a horizontal line. 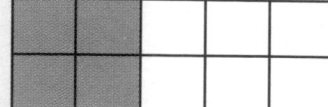	Other models to make: $\frac{2}{3}$ and $\frac{4}{6}$ $\frac{3}{5}$ and $\frac{6}{10}$ Check students' models.

WHAT STUDENTS NEED TO KNOW	HOW TO HELP	FOLLOW UP
How to find equivalent fractions by multiplying the numerator and denominator by the same number	Have students write several fractions equivalent to a given fraction. **Example** Write three fractions equivalent to $\frac{2}{3}$. Possible answers: $\frac{4}{6}$, $\frac{6}{9}$, $\frac{8}{12}$	Other problems to solve: Write three fractions equivalent to $\frac{3}{4}$. $\frac{6}{8}$, $\frac{9}{12}$, $\frac{12}{16}$ See Key Skill 26, Student Handbook, page H14.
That a fraction is in simplest form if the numerator and denominator have no common factors other than 1	Have students list all the factors of the numerator and denominator to determine whether the fraction is in simplest form. **Example** $\frac{6}{15}$ ← factors are 1, 2, **3**, 6 ← factors are 1, **3**, 5, 15 3 is a common factor, so $\frac{6}{15}$ is not in simplest form.	Other problems to solve: Tell whether each fraction is in simplest form. $\frac{4}{9}$ yes $\frac{6}{10}$ no $\frac{8}{12}$ no $\frac{3}{10}$ yes See Key Skill 27, Student Handbook, page H15.
How to divide the numerator and denominator of a fraction by a common factor until the fraction is in simplest form	Have students write fractions in simplest form. **Examples** $\frac{14}{35} = \frac{14 \div 7}{35 \div 7} = \frac{2}{5}$ $\frac{12}{18} = \frac{12 \div 2}{18 \div 2} = \frac{6}{9} \rightarrow \frac{6 \div 3}{9 \div 3} = \frac{2}{3}$	Other fractions to simplify: $\frac{10}{12}$ $\frac{1}{2}$ $\frac{8}{12}$ $\frac{2}{3}$ $\frac{12}{15}$ $\frac{4}{5}$ $\frac{5}{20}$ $\frac{1}{4}$ See Key Skill 27, Student Handbook, page H15.

Intervention and Extension Copying Masters, p. 3

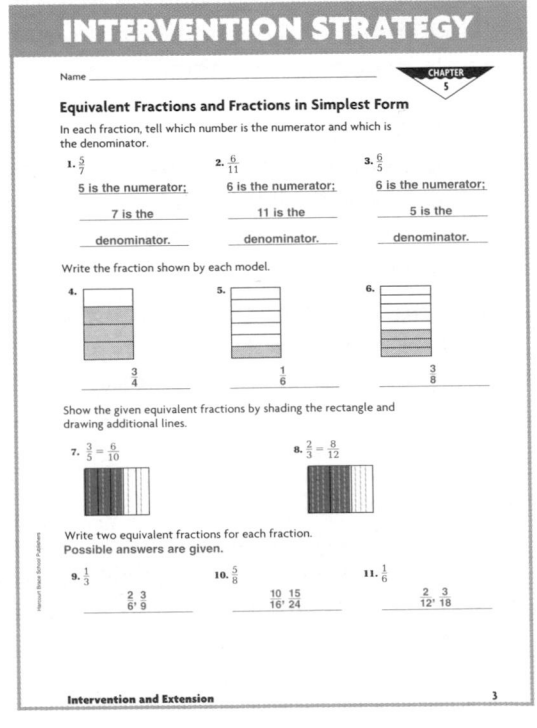

ORGANIZER

OBJECTIVE To use fraction bars to add and subtract fractions with unlike denominators
VOCABULARY *Review* least common multiple (LCM)
MATERIALS *For each student* fraction bars
PROGRAM RESOURCES
E-Lab • Activity 5
E-Lab Recording Sheets, p. 5

USING THE PAGE
Read and discuss the first three paragraphs. Explain to students that you can rename two things with one different name. For example, you can rename *cats* and *dogs* with the name *pets.*

- **What name could you use to rename *stamps, writing paper,* and *envelopes?* mailing supplies**

Point out that sometimes you can rename one of the two things with the name of the other. For example, of *literature* and *novel,* you can rename *novel* with the name *literature.*

- **Of *sports* and *golf,* which name can you use to rename the other?** *sports*

Renaming fractions is similar to renaming things. You rename fractions based on what they have in common.

Activity 1
Explore
In this activity, students use fraction bars to add unlike fractions. Observe students as they experiment with different fraction bars, making sure their second row of fraction bars fits exactly across their first row.

Think and Discuss
Ask students how fraction bars could be used to model adding $\frac{1}{2}$ and $\frac{1}{4}$. Most students will suggest using fourths bars or eighths bars.

- **Why do both fourths bars and eighths bars work?** because 4 and 8 are both common multiples of the denominators of $\frac{1}{2}$ and $\frac{1}{4}$

Try This
Discuss Exercise 3 with the class. Some students will say the answer is $\frac{9}{10}$; others will say $\frac{18}{20}$. Have students explain how they found their answers. Make sure students understand that $\frac{9}{10}$ was found by using the LCM and that $\frac{18}{20}$ was found by using a common denominator.

LAB Activity

WHAT YOU'LL EXPLORE
How to use fraction bars to add and subtract fractions with unlike denominators

WHAT YOU'LL NEED
fraction bars

Addition and Subtraction of Unlike Fractions

You have 5 pencils and your best friend has 3 pens. How many pencils are there altogether? How many more pencils do you have than your friend?

To add or subtract, you must have things with the same name, such as 5 pencils and 3 pencils. Your sum or difference has the same name, 8 pencils or 2 pencils.

The denominators of fractions do not always have the same name. Fractions with different denominators are called *unlike fractions.* You can use fraction bars to rename the denominators before adding.

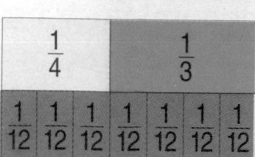

ACTIVITY 1

REMEMBER:
The LCM is the smallest of the common multiples of two or more numbers. **See page H5.**

2: 2, 4, 6, 8, 10, 12, . . .

3: 3, 6, 9, 12, 15, . . .

The LCM of 2 and 3 is 6. The LCM can be used to write common denominators of two or more fractions.

Explore
Work with a partner to find $\frac{1}{4} + \frac{1}{3}$.

- Use fraction bars to show both fractions.

- Which fraction bars fit exactly across $\frac{1}{4}$ and $\frac{1}{3}$? Think about the LCM of 4 and 3. **twelfths**

- What is $\frac{1}{4} + \frac{1}{3}$? **$\frac{7}{12}$**

$\frac{1}{4}$	$\frac{1}{3}$

$\frac{1}{4}$	$\frac{1}{3}$

$\frac{1}{12}$	$\frac{1}{12}$	$\frac{1}{12}$	$\frac{1}{12}$	$\frac{1}{12}$	$\frac{1}{12}$	$\frac{1}{12}$

Think and Discuss
- Look at the model for $\frac{1}{4} + \frac{1}{3}$. What do you know about $\frac{1}{4}$ and $\frac{3}{12}$? about $\frac{1}{3}$ and $\frac{4}{12}$? **They are equivalent; they are equivalent.**

- How are the denominators of $\frac{1}{4}$, $\frac{1}{3}$, and $\frac{1}{12}$ related? (HINT: Think about common multiples.) **12 is the LCM of 3 and 4.**

Try This
Use fraction bars to find the sum.

1. $\frac{1}{3} + \frac{1}{4}$ **$\frac{7}{12}$**   2. $\frac{1}{2} + \frac{1}{3}$ **$\frac{5}{6}$**   3. $\frac{1}{2} + \frac{2}{5}$ **$\frac{9}{10}$**

IDEA FILE

WRITING IN MATHEMATICS

Have students write a word problem that can be solved with these fraction bars.

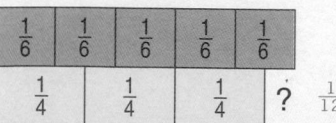

$\frac{1}{6}$	$\frac{1}{6}$	$\frac{1}{6}$	$\frac{1}{6}$	$\frac{1}{6}$

$\frac{1}{4}$	$\frac{1}{4}$	$\frac{1}{4}$	?	$\frac{1}{12}$

Possible answer: Ann bought 5 slices of a pizza. Each slice was $\frac{1}{6}$ of the pizza. After she gave Peter, Mary, and Steve each $\frac{1}{4}$ of the whole pizza, how much did she have left?

Modality Visual

REVIEW

In this activity students model five different fractions equivalent to $\frac{1}{2}$.

1. Fold a sheet of paper in half, unfold it, then shade 1 part. What part is shaded? $\frac{1}{2}$

2. Fold the paper in half twice and unfold it. Now what part is shaded? $\frac{2}{4}$

3. Repeat step 2 for three, four, and five folds. How are the fractions related? They are equivalent.

4. What is special about the first fraction? It is in simplest form.

Modalities Auditory, Kinesthetic, Visual

Fraction bars can also be used to subtract unlike fractions.

ACTIVITY 2

Explore

Work with a partner to find $\frac{1}{2} - \frac{1}{3}$.

- Use fractions bars to show $\frac{1}{2}$ and $\frac{1}{3}$.

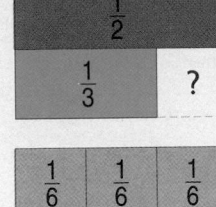

- Which fraction bars fit exactly across both $\frac{1}{2}$ and $\frac{1}{3}$? Think about the LCM. **sixths**

- Compare $\frac{3}{6}$ and $\frac{2}{6}$. How much more is $\frac{3}{6}$ than $\frac{2}{6}$? **$\frac{1}{6}$ more**

- What is $\frac{3}{6} - \frac{2}{6}$? What is $\frac{1}{2} - \frac{1}{3}$? **$\frac{1}{6}$; $\frac{1}{6}$**

Think and Discuss

- How are the denominators of $\frac{1}{2}$, $\frac{2}{3}$, and $\frac{1}{6}$ related? (HINT: Think about common multiples.) **The LCM of 2 and 3 is 6.**
- Look at the model of $\frac{3}{4} - \frac{1}{3}$. Which fraction bars do you think will fit exactly across $\frac{3}{4}$ and $\frac{1}{3}$? Explain. **Twelfths; 12 is the LCM of 4 and 3.**

- Which fraction bars do you think would fit exactly across $\frac{1}{2} - \frac{1}{4}$? **fourths or eighths**

Try This

Use fraction bars to subtract.

1. $\frac{3}{4} - \frac{1}{3}$ **$\frac{5}{12}$**   2. $\frac{2}{5} - \frac{1}{10}$ **$\frac{3}{10}$**   3. $\frac{1}{3} - \frac{1}{4}$ **$\frac{1}{12}$**

4. $\frac{1}{2} - \frac{2}{5}$ **$\frac{1}{10}$**   5. $\frac{1}{2} - \frac{5}{12}$ **$\frac{1}{12}$**   6. $\frac{1}{4} - \frac{1}{6}$ **$\frac{1}{12}$**

Technology Link

E-Lab • Activity 5 Available on CD-ROM and the Internet at http://www.hbschool.com/elab

97

Note: Students add fractions. Use E-Lab p. 5.

Activity 2

Explore

After students find the fraction bars that fit exactly across $\frac{1}{2}$ and $\frac{1}{3}$, ask:

- **How many sixths are in $\frac{1}{2}$?** 3 sixths
- **How many sixths are in $\frac{1}{3}$?** 2 sixths
- **How would you classify the fractions $\frac{1}{2} = \frac{3}{6}$ and $\frac{1}{3} = \frac{2}{6}$?** They are two sets of equivalent fractions.

Think and Discuss

After students test their predictions for $\frac{3}{4}$ and $\frac{1}{3}$, discuss the results. Then ask:

- **What is the difference between $\frac{3}{4}$ and $\frac{1}{3}$?** $\frac{5}{12}$

Try This

Before students begin, ask:

- **How do Exercises 2 and 5 differ from Exercises 1, 3, 4, and 6?** In Exercises 2 and 5, the LCM is the denominator of one of the fractions.

USING E-LAB

Students use the concepts of this lesson, reasoning skills, and a computer tool for finding multiples to add fractions with unlike denominators.

✓ *Assessment Tip*

Use these questions to help students communicate the big ideas of the lesson.

- **Which of these exercises do you need to rename before you add or subtract?** Exercises 1 and 3

  1. $\frac{3}{4} + \frac{2}{3}$ 2. $\frac{4}{9} - \frac{2}{9}$
  3. $\frac{5}{8} - \frac{1}{4}$ 4. $\frac{4}{5} + \frac{2}{5}$

- **What fraction bars would you use to model $\frac{2}{3} + \frac{3}{4}$?** thirds, fourths, twelfths
- **Use fraction bars to solve $\frac{4}{5} - \frac{1}{3}$.** $\frac{7}{15}$

INTERDISCIPLINARY CONNECTION

One suggestion for an interdisciplinary unit for this chapter is Fractions: Who Needs Them? In this unit you can have students

- become consumer sleuths. Have them look for ways fractions are useful to shoppers.
- make a list of fractions and then write a brief paragraph explaining why we need fractions.

Modalities Auditory, Visual

EXTENSION

When the denominator of one fraction is a multiple of the denominator of another fraction, the denominators are *compatible*. Have students use fraction bars to solve problems that have compatible denominators.

1. $\frac{2}{3} - \frac{1}{6}$ **$\frac{1}{2}$**
2. $\frac{4}{5} + \frac{1}{10}$ **$\frac{9}{10}$**
3. $\frac{3}{7} - \frac{1}{14}$ **$\frac{5}{14}$**
4. $\frac{3}{8} + \frac{1}{4}$ **$\frac{5}{8}$**

Modalities Auditory, Kinesthetic, Visual

5.2

GRADE 5
Use fraction bars to add unlike fractions.

GRADE 6
Use diagrams to add and subtract unlike fractions.

GRADE 7
Add and subtract fractions.

ORGANIZER

OBJECTIVE To use diagrams to add and subtract unlike fractions

ASSIGNMENTS Independent Practice
- *Basic* 1–15, even; 16–20
- *Average* 1–20
- *Advanced* 4–20

PROGRAM RESOURCES
Reteach 5.2 Practice 5.2 Enrichment 5.2
Intervention • Equivalent Fractions and Fractions in Simplest Form, TE
p. 96A–96B

PROBLEM OF THE DAY

Transparency 5

How many minutes are in $\frac{1}{2}$ hr?
30 min **in $\frac{1}{3}$ hr?** 20 min **in $\frac{1}{4}$ hr?** 15 min
in $\frac{1}{5}$ hr? 12 min **in $\frac{1}{6}$ hr?** 10 min

Solution Tab A, p. A5

QUICK CHECK
Write the fraction in simplest form.

1. $\frac{8}{18}$ $\frac{4}{9}$
2. $\frac{9}{36}$ $\frac{1}{4}$
3. $\frac{16}{28}$ $\frac{4}{7}$
4. $\frac{15}{20}$ $\frac{3}{4}$
5. $\frac{11}{22}$ $\frac{1}{2}$
6. $\frac{10}{40}$ $\frac{1}{4}$

1 MOTIVATE

Present this problem to students:

- **Jane is designing a closet organizer. She will use $\frac{1}{8}$ of her closet for shoes, $\frac{1}{2}$ for pants and tops, $\frac{1}{8}$ for sweaters, and $\frac{1}{4}$ for dresses. Draw a diagram of Jane's closet.** Check students' diagrams.

2 TEACH/PRACTICE

In this lesson students will draw and use diagrams to help them add and subtract unlike fractions.

COMMON ERROR ALERT If students find it difficult to draw accurate diagrams that represent fractions, using graph paper can be very helpful.

5.2

WHAT YOU'LL LEARN
How to use diagrams to add and subtract unlike fractions

WHY LEARN THIS?
To understand how to solve problems involving addition and subtraction of fractions by drawing a diagram

Geography Link

According to the diagram below, about $\frac{1}{2}$ of all the earth's water is in the Pacific Ocean. About $\frac{1}{4}$ of all the water on earth is in the Atlantic Ocean, and about $\frac{1}{5}$ of the earth's water is in the Indian Ocean. How much of the earth's water is not part of these three oceans? **about $\frac{1}{20}$**

- Others
- Indian Ocean
- Atlantic Ocean
- Pacific Ocean
- All earth's water

Adding and Subtracting Unlike Fractions

In the Lab Activity, you learned how to use fraction bars to add and subtract fractions. You can also use diagrams.

Ulises spent $\frac{1}{5}$ hr putting the clean dishes away. He spent $\frac{1}{2}$ hr cleaning the kitchen after dinner. What fractional part of an hour did Ulises spend on these chores?

Use the diagram at the right to find $\frac{1}{5} + \frac{1}{2}$.

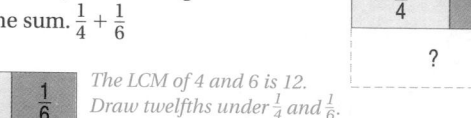

The diagram shows that $\frac{1}{5} + \frac{1}{2} = \frac{7}{10}$.

To draw a diagram of an addition or subtraction problem, think about the LCM of the denominators and about equivalent fractions.

EXAMPLE 1 Complete the diagram to find the sum. $\frac{1}{4} + \frac{1}{6}$

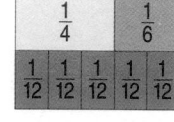

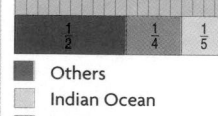

The LCM of 4 and 6 is 12. Draw twelfths under $\frac{1}{4}$ and $\frac{1}{6}$.
Think: $\frac{1}{4} = \frac{3}{12}$ and $\frac{1}{6} = \frac{2}{12}$.

So, $\frac{1}{4} + \frac{1}{6} = \frac{5}{12}$.

EXAMPLE 2 Draw a diagram to find the difference. $\frac{1}{2} - \frac{1}{5}$

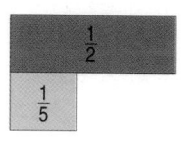

The LCM of 2 and 5 is 10.
Think: $\frac{1}{2} = \frac{5}{10}$ and $\frac{1}{5} = \frac{2}{10}$.
Draw $\frac{1}{2}$ as $\frac{5}{10}$ and $\frac{1}{5}$ as $\frac{2}{10}$.

Think: *How much greater is $\frac{5}{10}$ than $\frac{2}{10}$?*
$\frac{5}{10} - \frac{2}{10} = \frac{3}{10}$
So, $\frac{1}{2} - \frac{1}{5} = \frac{3}{10}$.

IDEA FILE

CORRELATION TO OTHER RESOURCES

CornerStone Mathematics ©1996 SkillsBank Corporation **Level B, Using Fractions and Percents, Lessons 2 and 3**

***Fraction Attraction* Published by Sunburst** Students build their understanding of the concepts of fractions.

RETEACH 5.2

Name _____

LESSON 5.2

Adding and Subtracting Unlike Fractions

Fractions with different denominators are *unlike fractions*. You can use a diagram to help you add unlike fractions. When drawing the diagram, think about the least common multiple, or LCM of the denominators.

Copy and complete the diagram of $\frac{1}{6} + \frac{1}{3}$.

Step 1: The LCM of 6 and 3 is 6. Draw sixths under $\frac{1}{6}$ and $\frac{1}{3}$. Remember, $\frac{1}{3} = \frac{2}{6}$.

Step 2: Count the number of sixths you drew. There are 3 sixths.

Step 3: Write the fraction in simplest form, if necessary. $\frac{3}{6} = \frac{1}{2}$

So, $\frac{1}{6} + \frac{1}{3} = \frac{1}{2}$.

Complete each diagram to help you find the sum.

1. a. $\frac{1}{5}$ equals how many tenths? $\frac{2}{10}$
   b. $\frac{1}{5} + \frac{3}{10} = \frac{1}{2}$

2. a. $\frac{1}{4}$ equals how many eighths? $\frac{2}{8}$
   b. $\frac{5}{8} + \frac{1}{4} = \frac{7}{8}$

3. a. $\frac{1}{4}$ equals how many twelfths? $\frac{3}{12}$
   b. $\frac{7}{12} + \frac{1}{4} = \frac{10}{12}$, or $\frac{5}{6}$

4. a. $\frac{1}{5}$ equals how many tenths? $\frac{2}{10}$
   b. $\frac{1}{5} + \frac{7}{10} = \frac{9}{10}$

5. a. $\frac{1}{3}$ equals how many ninths? $\frac{3}{9}$
   b. $\frac{5}{9} + \frac{1}{3} = \frac{8}{9}$

Use with text pages 98–99.

TAKE ANOTHER LOOK R21

GUIDED PRACTICE

Copy and complete the diagram to find the sum or difference.

1. $\frac{1}{4} + \frac{1}{2}$ $\frac{3}{4}$

| $\frac{1}{4}$ | $\frac{1}{2}$ |

2. $\frac{1}{3} - \frac{1}{6}$ $\frac{1}{6}$

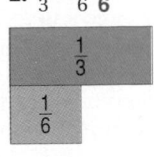

| $\frac{1}{3}$ |
| $\frac{1}{6}$ |

3. $\frac{1}{2} + \frac{3}{10}$ $\frac{8}{10}$, or $\frac{4}{5}$

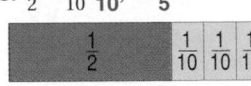

| $\frac{1}{2}$ | $\frac{1}{10}$ | $\frac{1}{10}$ | $\frac{1}{10}$ |

INDEPENDENT PRACTICE

Write the addition or subtraction problem and then solve.

1.
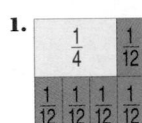
$\frac{1}{4} + \frac{1}{12} = \frac{4}{12}$, or $\frac{1}{3}$

2.
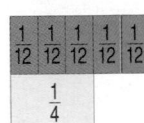
$\frac{5}{12} - \frac{1}{4} = \frac{2}{12}$, or $\frac{1}{6}$

3.
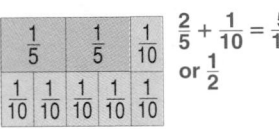
$\frac{2}{5} + \frac{1}{10} = \frac{5}{10}$, or $\frac{1}{2}$

Draw a diagram to find each sum or difference. **Check students' diagrams.**

4. $\frac{7}{12} - \frac{1}{4}$ $\frac{4}{12}$, or $\frac{1}{3}$

5. $\frac{1}{2} + \frac{1}{6}$ $\frac{4}{6}$, or $\frac{2}{3}$

6. $\frac{3}{5} - \frac{1}{2}$ $\frac{1}{10}$

7. $\frac{3}{4} + \frac{1}{6}$ $\frac{11}{12}$

8. $\frac{2}{3} - \frac{1}{2}$ $\frac{1}{6}$

9. $\frac{5}{12} + \frac{1}{3}$ $\frac{9}{12}$, or $\frac{3}{4}$

10. $\frac{7}{12} - \frac{1}{3}$ $\frac{3}{12}$, or $\frac{1}{4}$

11. $\frac{9}{10} - \frac{3}{5}$ $\frac{3}{10}$

12. $\frac{3}{8} + \frac{1}{4}$ $\frac{5}{8}$

13. $\frac{1}{3} + \frac{1}{6}$ $\frac{3}{6}$, or $\frac{1}{2}$

14. $\frac{5}{8} - \frac{1}{4}$ $\frac{3}{8}$

15. $\frac{5}{6} - \frac{1}{4}$ $\frac{7}{12}$

Problem-Solving Applications

16. Jackie's mom made pudding cups for the party. Of the cups she made, $\frac{5}{8}$ are chocolate and $\frac{1}{4}$ are butterscotch. How many more cups are chocolate than butterscotch? **$\frac{3}{8}$ more of the cups are chocolate.**

17. Jane needs $\frac{1}{4}$ yd of blue fabric and $\frac{2}{3}$ yd of purple fabric for her sewing project. How much fabric does she need for her sewing project? $\frac{11}{12}$ yd

18. Kerri used $\frac{1}{2}$ gal of water to water the house plants. He put $\frac{1}{3}$ gal of water in the dog's water bowl. How much water did Kerri use for the house plants and the dog's bowl? $\frac{5}{6}$ gal

19. Erin walks $\frac{1}{2}$ mi to school. Gene walks $\frac{3}{10}$ mi to school. How much farther does Erin walk than Gene? $\frac{2}{10}$ mi, or $\frac{1}{5}$ mi farther

20. ✏️ **WRITE ABOUT IT** Write a word problem that can be solved by the picture in Exercise 2. **Possible word problem: Jane has $\frac{5}{12}$ yd of fabric. She uses $\frac{1}{4}$ yd for a vest. How much fabric does she have left?**

MORE PRACTICE Lesson 5.2, page H48

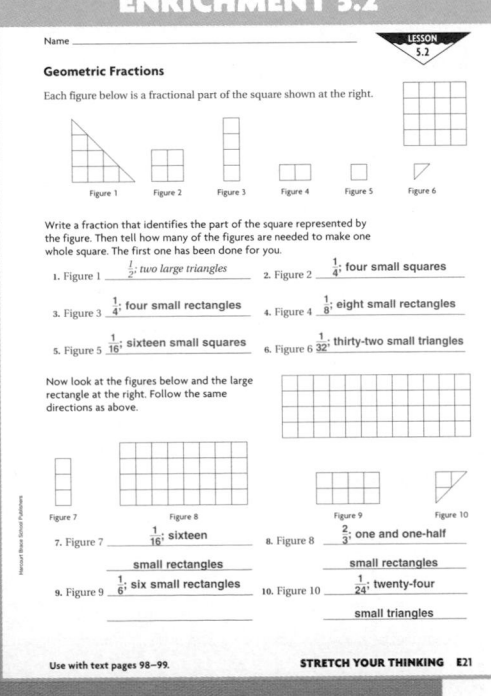

99

GUIDED PRACTICE

Have students work with a partner to complete Exercises 1–3, to make sure students understand how to add and subtract unlike fractions.

INDEPENDENT PRACTICE

To assess students' understanding of the concepts, check answers to Exercises 1–3 before assigning the remaining exercises.

ADDITIONAL EXAMPLES

Example 1, p. 98

Complete the diagram to find the sum of $\frac{2}{5}$ and $\frac{1}{2}$.

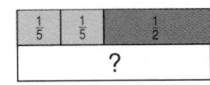 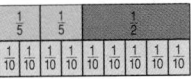

$\frac{2}{5} + \frac{1}{2} = \frac{9}{10}$

Example 2, p. 98

Draw a diagram to solve $\frac{3}{4} - \frac{2}{3}$.

$\frac{3}{4} - \frac{2}{3} = \frac{1}{12}$

3 WRAP UP

Use these questions to help students focus on the big ideas of the lesson.

- Ryan used $\frac{1}{2}$ gal of paint on his entertainment center and $\frac{1}{4}$ gal on bookshelves. Draw a diagram to find the answers to these questions. Check students' diagrams.

- How much paint did he use? $\frac{3}{4}$ gal

- If $\frac{7}{8}$ gal of paint was in the can when he started, how much paint did he have left? $\frac{1}{8}$ gal

✔ Assessment Tip

Have students write the answer.

- **How does drawing a diagram help you add and subtract fractions?** Possible answer: By drawing fraction models, you are better able to see equivalent fractions.

DAILY QUIZ

Draw a diagram to find the sum or difference. Check students' diagrams.

1. $\frac{3}{8} + \frac{1}{2}$ $\frac{7}{8}$

2. $\frac{9}{10} - \frac{2}{5}$ $\frac{1}{2}$

3. $\frac{5}{6} - \frac{1}{10}$ $\frac{11}{15}$

4. $\frac{1}{3} + \frac{1}{2}$ $\frac{5}{6}$

5. $\frac{3}{4} - \frac{1}{3}$ $\frac{5}{12}$

6. $\frac{2}{5} - \frac{3}{10}$ $\frac{1}{10}$

PRACTICE 5.2

Name _____ LESSON 5.2

Adding and Subtracting Unlike Fractions

Write the addition or subtraction problem and then solve.

1.
| $\frac{1}{5}$ | $\frac{1}{5}$ | $\frac{1}{10}$ |
| $\frac{1}{10}$ | | |

$\frac{3}{5} + \frac{1}{10} = \frac{7}{10}$

2.
| $\frac{1}{8}$ | $\frac{1}{8}$ | $\frac{1}{8}$ | $\frac{1}{8}$ | $\frac{1}{8}$ |
| $\frac{1}{4}$ | | | | |

$\frac{5}{8} - \frac{1}{4} = \frac{3}{8}$

3.
| $\frac{1}{3}$ | $\frac{1}{9}$ |
| $\frac{1}{9}$ | $\frac{1}{9}$ | $\frac{1}{9}$ |

$\frac{2}{3} + \frac{1}{9} = \frac{7}{9}$

Draw a diagram to find each sum or difference. Check students' diagrams.

4. $\frac{1}{2} - \frac{1}{6}$ $\frac{1}{3}$

5. $\frac{3}{10} + \frac{2}{5}$ $\frac{7}{10}$

6. $\frac{5}{6} - \frac{1}{4}$ $\frac{7}{12}$

7. $\frac{2}{9} + \frac{1}{3}$ $\frac{5}{9}$

8. $\frac{7}{8} - \frac{1}{4}$ $\frac{5}{8}$

9. $\frac{1}{3} + \frac{1}{2}$ $\frac{5}{6}$

10. $\frac{7}{10} - \frac{3}{5}$ $\frac{1}{10}$

11. $\frac{1}{3} + \frac{1}{2}$ $\frac{5}{6}$

12. $\frac{8}{9} - \frac{2}{3}$ $\frac{2}{9}$

Mixed Applications

13. Spencer's backyard measures 12 yd by 8 yd. What is the area of his backyard? **96 sq yd**

14. Chris eats $\frac{1}{8}$ of a pizza. Ryan eats $\frac{3}{4}$ of the same pizza. How much of the pizza do they eat? How much is left? **$\frac{7}{8}$ eaten; $\frac{1}{8}$ left**

15. Joan walked $\frac{1}{2}$ mi to her friend's house. She then walked $\frac{1}{4}$ mi to school. What is the total distance that Joan walked? **$\frac{9}{20}$ mi**

16. In Mr. Sims's math class, $\frac{2}{5}$ of the students are wearing blue shirts and $\frac{3}{10}$ of the students are wearing white shirts. What part of the class is wearing *neither* blue nor white shirts? **$\frac{3}{10}$ of the class**

Use with text pages 98–99. ON MY OWN **P21**

ENRICHMENT 5.2

Name _____ LESSON 5.2

Geometric Fractions

Each figure below is a fractional part of the square shown at the right.

Figure 1 · Figure 2 · Figure 3 · Figure 4 · Figure 5 · Figure 6

Write a fraction that identifies the part of the square represented by the figure. Then tell how many of the figures are needed to make one whole square. The first one has been done for you.

1. Figure 1 $\frac{1}{2}$: *two large triangles*

2. Figure 2 $\frac{1}{4}$: four small squares

3. Figure 3 $\frac{1}{4}$: four small rectangles

4. Figure 4 $\frac{1}{8}$: eight small rectangles

5. Figure 5 $\frac{1}{16}$: sixteen small squares

6. Figure 6 $\frac{1}{32}$: thirty-two small triangles

Now look at the figures below and the large rectangle at the right. Follow the same directions as above.

Figure 7 · Figure 8 · Figure 9 · Figure 10

7. Figure 7 $\frac{1}{16}$: sixteen small rectangles

8. Figure 8 $\frac{2}{3}$: one and one-half small rectangles

9. Figure 9 $\frac{1}{6}$: six small rectangles

10. Figure 10 $\frac{1}{24}$: twenty-four small triangles

Use with text pages 98–99. STRETCH YOUR THINKING **E21**

GRADE 5	GRADE 6	GRADE 7
Use LCD to add unlike fractions.	Add unlike fractions.	Add and subtract fractions.

ORGANIZER

OBJECTIVE To add unlike fractions
VOCABULARY *New* **least common denominator (LCD)**
ASSIGNMENTS Independent Practice
- *Basic* 1–27, odd; 28–32
- *Average* 1–32
- *Advanced* 9–32

PROGRAM RESOURCES
Reteach 5.3 Practice 5.3 Enrichment 5.3
Math Fun • Fraction Riddles, p. 108

PROBLEM OF THE DAY

Transparency 5
Add horizontally and vertically.

$\frac{1}{4}$	$\frac{1}{12}$	?	$\frac{1}{3}$
$\frac{1}{3}$	$\frac{1}{6}$	?	$\frac{1}{2}$
? $\frac{7}{12}$	? $\frac{1}{4}$	?	$\frac{5}{6}$

Solution Tab A, p. A5

QUICK CHECK
Find the LCM of each pair of numbers.

1. 7, 14 14   **2.** 3, 6 6   **3.** 8, 12 24
4. 6, 4 12   **5.** 12, 15 60   **6.** 20, 25 100

1 MOTIVATE

Have students compare the denominators in these exercises to answer the questions.

1. $\frac{5}{8} + \frac{1}{8}$ **2.** $\frac{2}{3} + \frac{5}{9}$ **3.** $\frac{3}{4} + \frac{4}{5}$

- **Would Exercise 1 or Exercise 2 be easier to solve? Why?** Exercise 1; because the denominators are the same

- **Would Exercise 2 or Exercise 3 be easier to solve? Why?** Exercise 2; because the LCM of the denominators is one of the denominators

2 TEACH/PRACTICE

In this lesson, students use equivalent fractions to add unlike fractions.

5.3

WHAT YOU'LL LEARN
How to add unlike fractions

WHY LEARN THIS?
To solve problems such as finding the total distance traveled by bike

WORD POWER
least common denominator (LCD)

> **Sports Link**
>
> In football there are four 15-min quarters, in hockey there are three 20-min periods, and in basketball there are two 24-min halves. Last Sunday you watched $\frac{1}{4}$ of a football game, $\frac{1}{3}$ of a hockey game, and $\frac{1}{2}$ of a basketball game. Is the sum of the fractions of the three events you watched more or less than one whole event?
>
>

The sum of the fractions, $\frac{13}{12}$, is more than one whole event.

Adding Unlike Fractions

When you use models or pictures to add unlike fractions, you rename the fractions so that they have a common denominator. To add fractions without models or pictures, you can write equivalent fractions by using the **least common denominator**, or **LCD**. The LCD is the LCM of the denominators.

EXAMPLE 1 You and a friend have decided to meet at the neighborhood recreation center at 4:15 P.M. Both of you will ride your bikes directly from home. You will ride $\frac{1}{2}$ mi and your friend will ride $\frac{3}{5}$ mi. What is the total distance the two of you will ride? Find $\frac{1}{2} + \frac{3}{5}$.

$$\begin{aligned} \frac{1}{2} &= \frac{1 \times 5}{2 \times 5} = \frac{5}{10} \\ + \frac{3}{5} &= \frac{3 \times 2}{5 \times 2} = \frac{6}{10} \end{aligned}$$

The LCM of 2 and 5 is 10, so the LCD of $\frac{1}{2}$ and $\frac{3}{5}$ is tenths. Multiply to write equivalent fractions using the LCD.

$$\begin{aligned} \frac{1}{2} &= \frac{5}{10} \\ + \frac{3}{5} &= \frac{6}{10} \\ \hline &\frac{11}{10}, \text{ or } 1\frac{1}{10} \end{aligned}$$

Add the numerators. Write the sum over the denominator.

Write the answer as a fraction or as a mixed number.

So, you and your friend will ride a total of $1\frac{1}{10}$ mi.

- What if you ride $\frac{1}{2}$ mi and your friend rides $\frac{3}{4}$ mi? What is the total number of miles the two of you ride? $\frac{5}{4}$ **mi, or** $1\frac{1}{4}$ **mi**

- **CRITICAL THINKING** When is the LCD of two fractions equal to the product of the denominators? **When the denominators have 1 as their GCF.**

EXAMPLE 2

A.
$$\begin{aligned} \frac{2}{3} &= \frac{2 \times 4}{3 \times 4} = \frac{8}{12} \\ + \frac{1}{4} &= \frac{1 \times 3}{4 \times 3} = \frac{3}{12} \\ \hline &\frac{11}{12} \end{aligned}$$

B.
$$\begin{aligned} \frac{1}{2} &= \frac{1 \times 3}{2 \times 3} = \frac{3}{6} \\ + \frac{5}{6} &= \frac{5 \times 1}{6 \times 1} = \frac{5}{6} \\ \hline &\frac{8}{6} = \frac{4}{3}, \text{ or } 1\frac{1}{3} \end{aligned}$$

C.
$$\begin{aligned} \frac{3}{5} &= \frac{3 \times 3}{5 \times 3} = \frac{9}{15} \\ + \frac{1}{3} &= \frac{1 \times 5}{3 \times 5} = \frac{5}{15} \\ \hline &\frac{14}{15} \end{aligned}$$

D.
$$\begin{aligned} \frac{3}{4} &= \frac{3 \times 3}{4 \times 3} = \frac{9}{12} \\ + \frac{4}{6} &= \frac{4 \times 2}{6 \times 2} = \frac{8}{12} \\ \hline &\frac{17}{12}, \text{ or } 1\frac{5}{12} \end{aligned}$$

 Calculator Activities, page H34

IDEA FILE

EXTENSION

Have students complete this magic square and find the sum.

$\frac{2}{15}$	$\frac{7}{15}$	$\frac{2}{5}$
$\frac{3}{5}$	$\frac{1}{3}$	$\frac{1}{15}$
$\frac{4}{15}$	$\frac{1}{5}$	$\frac{8}{15}$

The sum is $\frac{15}{15}$, or 1.
Modality Visual

RETEACH 5.3

Name _____

LESSON 5.3

Adding Unlike Fractions

Remember, fractions with different denominators are *unlike fractions*.
- To add unlike fractions, you first write equivalent fractions, using the least common denominator of all the fractions.
- The least common denominator, or LCD, is the LCM of the denominators.

Mabel has $\frac{1}{4}$ lb of pink gumdrops and $\frac{2}{3}$ lb of licorice gumdrops. How many total pounds does she have?
Step 1: Find the LCM of the denominators. The LCM of 4 and 3 is 12. So, the LCD of $\frac{1}{4}$ and $\frac{2}{3}$ is 12.
Step 2: Multiply to write equivalent fractions, using the LCD.
$$\frac{1}{4} = \frac{1 \times 3}{4 \times 3} = \frac{3}{12}$$
$$\frac{2}{3} = \frac{2 \times 4}{3 \times 4} = \frac{8}{12}$$
Step 3: Add the numerators. Write the sum over the denominator.
$$\frac{3}{12} + \frac{8}{12} = \frac{11}{12}$$ Remember, keep the denominator the same.
Step 4: Write the answer as a fraction or as a mixed number in simplest form. $\frac{11}{12}$ is already in simplest form.
Mabel has $\frac{11}{12}$ lb of gumdrops.

Complete to find each sum. Write the answer as a fraction or a mixed number in simplest form.

1. Find $\frac{2}{5} + \frac{3}{10}$.
The LCM of 5 and 10 is 10.
So, $\frac{2}{5} + \frac{3}{10} = \frac{4}{10} + \frac{3}{10} = \frac{7}{10}$.

2. Find $\frac{7}{8} + \frac{2}{3}$.
The LCM of 8 and 3 is 24.
So, $\frac{7}{8} + \frac{2}{3} = \frac{21}{24} + \frac{16}{24} = \frac{37}{24} = 1\frac{13}{24}$.

3. Find $\frac{1}{4} + \frac{5}{6}$.
The LCM of 4 and 6 is 12.
So, $\frac{1}{4} + \frac{5}{6} = \frac{3}{12} + \frac{10}{12} = \frac{13}{12} = 1\frac{1}{12}$.

4. Find $\frac{1}{6} + \frac{3}{8}$.
The LCM of 6 and 8 is 24.
So, $\frac{1}{6} + \frac{3}{8} = \frac{4}{24} + \frac{13}{24}$.

R22   TAKE ANOTHER LOOK   Use with text pages 100–101.

GUIDED PRACTICE

Name the LCM. Then write equivalent fractions with the LCD.

1. $\frac{1}{2} + \frac{3}{8}$ 8; $\frac{4}{8} + \frac{3}{8}$
2. $\frac{7}{10} + \frac{1}{5}$ 10; $\frac{7}{10} + \frac{2}{10}$
3. $\frac{1}{3} + \frac{1}{8}$ 24; $\frac{8}{24} + \frac{3}{24}$
4. $\frac{5}{6} + \frac{1}{4}$ 12; $\frac{10}{12} + \frac{3}{12}$

Find the sum. Write your answer in simplest form.

5. $\frac{1}{3} + \frac{1}{4}$ $\frac{7}{12}$
6. $\frac{4}{9} + \frac{1}{3}$ $\frac{7}{9}$
7. $\frac{2}{5} + \frac{2}{4}$ $\frac{9}{10}$
8. $\frac{2}{3} + \frac{3}{4}$ $\frac{17}{12}$, or $1\frac{5}{12}$

INDEPENDENT PRACTICE

Write equivalent fractions with the LCD.

1. $\frac{5}{6} + \frac{2}{3}$ $\frac{5}{6} + \frac{4}{6}$
2. $\frac{1}{4} + \frac{1}{2}$ $\frac{1}{4} + \frac{2}{4}$
3. $\frac{1}{5} + \frac{7}{15}$ $\frac{3}{15} + \frac{7}{15}$
4. $\frac{1}{4} + \frac{1}{3}$ $\frac{3}{12} + \frac{4}{12}$
5. $\frac{1}{4} + \frac{5}{8}$ $\frac{2}{8} + \frac{5}{8}$
6. $\frac{1}{2} + \frac{2}{3}$ $\frac{3}{6} + \frac{4}{6}$
7. $\frac{3}{4} + \frac{4}{5}$ $\frac{15}{20} + \frac{16}{20}$
8. $\frac{2}{3} + \frac{7}{8}$ $\frac{16}{24} + \frac{21}{24}$
9. $\frac{3}{8} + \frac{1}{2}$ $\frac{3}{8} + \frac{4}{8}$
10. $\frac{4}{5} + \frac{1}{7}$ $\frac{28}{35} + \frac{5}{35}$
11. $\frac{3}{4} + \frac{1}{6}$ $\frac{9}{12} + \frac{2}{12}$
12. $\frac{5}{6} + \frac{1}{3}$ $\frac{5}{6} + \frac{2}{6}$

Find the sum. Write your answer in simplest form.

13. $\frac{1}{6} + \frac{2}{3}$ $\frac{5}{6}$
14. $\frac{1}{4} + \frac{2}{3}$ $\frac{11}{12}$
15. $\frac{1}{2} + \frac{3}{10}$ $\frac{4}{5}$
16. $\frac{1}{3} + \frac{1}{2}$ $\frac{5}{6}$
17. $\frac{3}{4} + \frac{5}{8}$ $\frac{11}{8}$, or $1\frac{3}{8}$
18. $\frac{3}{8} + \frac{1}{3}$ $\frac{17}{24}$
19. $\frac{1}{6} + \frac{5}{12}$ $\frac{7}{12}$
20. $\frac{7}{12} + \frac{2}{3}$
21. $\frac{5}{6} + \frac{1}{4}$
22. $\frac{3}{8} + \frac{1}{12}$ $\frac{11}{24}$
23. $\frac{3}{5} + \frac{3}{10}$ $\frac{9}{10}$
24. $\frac{3}{5} + \frac{3}{4}$
25. $\frac{3}{4} + \frac{3}{20}$ $\frac{9}{10}$
26. $\frac{3}{8} + \frac{3}{20}$ $\frac{21}{40}$
27. $\frac{3}{5} + \frac{1}{3}$ $\frac{14}{15}$

20. $\frac{15}{12}$, or $1\frac{1}{4}$ 21. $\frac{13}{12}$, or $1\frac{1}{12}$ 24. $\frac{27}{20}$, or $1\frac{7}{20}$

Problem-Solving Applications

28. Mary is helping her mother bake cookies. The recipe calls for $\frac{1}{2}$ c of brown sugar and $\frac{3}{4}$ c of white sugar. What is the total amount of sugar Mary's mother needs? $\frac{5}{4}$ c, or $1\frac{1}{4}$ c of sugar

29. Craig feeds his pets each night. He feeds his big dog $\frac{1}{2}$ can of dog food and his small dog $\frac{1}{3}$ can of dog food. How much dog food does he need each night? $\frac{5}{6}$ can

30. Sara is landscaping her yard. She will plant trees in $\frac{1}{10}$ of the yard and flowers in $\frac{1}{5}$ of the yard. How much of Sara's yard will she plant with trees and flowers? $\frac{3}{10}$ of the yard

31. Last week James practiced the tuba for $\frac{5}{6}$ hr on Monday, $\frac{1}{3}$ hr on Wednesday, and $\frac{1}{4}$ hr on Friday. How many hours did he practice last week? $\frac{17}{12}$ hr, or $1\frac{5}{12}$ hr

32. **WRITE ABOUT IT** Explain how to add fractions with unlike denominators. **Rename the fractions by using the LCD, and then add the numerators, writing the sum over the denominator. Then write the answer in simplest form**

MORE PRACTICE Lesson 5.3, page H49

101

GUIDED PRACTICE
To make sure students understand how to find the LCD, complete Exercises 1–8 as a class and discuss.

INDEPENDENT PRACTICE
To assess students' understanding of the concepts, check answers to Exercises 1–4 before assigning the remaining exercises.

CRITICAL THINKING How can you make this problem easier to solve: $\frac{3}{5} + \frac{4}{7} + \frac{2}{5} + \frac{2}{7}$? Add the fractions with like denominators first: $\frac{3}{5} + \frac{2}{5} = \frac{5}{5}$, or 1; $\frac{4}{7} + \frac{2}{7} = \frac{6}{7}$; $1 + \frac{6}{7} = 1\frac{6}{7}$.

DEVELOPING ALGEBRAIC THINKING

- If *a* and *b* are the denominators of two fractions, will $a \times b$ always be a common denominator of the fractions? yes

- Will $a \times b$ always be the LCD of the fractions? no

ADDITIONAL EXAMPLES

Example 1, p. 100

Find the sum of $\frac{3}{4}$ and $\frac{3}{5}$. $\frac{27}{20}$, or $1\frac{7}{20}$

Example 2, p. 100

Find the sums: $\frac{3}{8} + \frac{1}{4}$ $\frac{5}{8}$, $\frac{1}{3} + \frac{1}{2}$ $\frac{5}{6}$, $\frac{3}{4} + \frac{3}{5}$ $1\frac{7}{20}$, $\frac{1}{6} + \frac{5}{12}$ $\frac{7}{12}$

3 WRAP UP

Use these questions to help students focus on the big ideas of the lesson.

- Roger ate $\frac{1}{3}$ c of Fruity Cereal and $\frac{1}{4}$ c of Great Grain Cereal for breakfast. How much cereal did he eat? $\frac{7}{12}$ c

✓ Assessment Tip

Have students write the answer.

- When you add fractions, is it more efficient to use a common denominator or the LCD to write equivalent fractions? Why? Students' explanations should demonstrate their understanding that using the LCD will result in lesser, more manageable numbers and reduce the need to simplify answers.

DAILY QUIZ

Find the sum. Write the answer in simplest form.

1. $\frac{1}{3} + \frac{5}{8}$ $\frac{23}{24}$
2. $\frac{5}{8} + \frac{5}{12}$ $1\frac{1}{24}$
3. $\frac{1}{6} + \frac{1}{2}$ $\frac{2}{3}$
4. $\frac{1}{5} + \frac{7}{10}$ $\frac{9}{10}$
5. $\frac{2}{3} + \frac{3}{4}$ $1\frac{5}{12}$
6. $\frac{1}{3} + \frac{5}{6}$ $1\frac{1}{6}$

PRACTICE 5.3

Name _____ LESSON 5.3

Adding Unlike Fractions

Vocabulary

1. To add unlike fractions, you write equivalent fractions by using the _____

least common denominator

Write equivalent fractions with the LCD.

2. $\frac{3}{8} + \frac{1}{2}$ $\frac{3}{8} + \frac{4}{8}$
3. $\frac{3}{4} + \frac{1}{6}$ $\frac{9}{12} + \frac{2}{12}$
4. $\frac{1}{8} + \frac{1}{4}$ $\frac{7}{8} + \frac{2}{8}$
5. $\frac{2}{3} + \frac{4}{5}$ $\frac{10}{15} + \frac{12}{15}$
6. $\frac{7}{10} + \frac{2}{5}$ $\frac{7}{10} + \frac{8}{10}$
7. $\frac{7}{12} + \frac{2}{3}$ $\frac{7}{12} + \frac{8}{12}$
8. $\frac{5}{9} + \frac{1}{3}$ $\frac{8}{9} + \frac{3}{9}$
9. $\frac{11}{15} + \frac{1}{3}$ $\frac{11}{15} + \frac{9}{15}$

Find the sum. Write the answer in simplest form.

10. $\frac{3}{9} + \frac{2}{9}$ $\frac{5}{9}$
11. $\frac{1}{7} + \frac{1}{2}$ $\frac{9}{14}$
12. $\frac{5}{3} + \frac{5}{12}$ $\frac{13}{12}$, or $1\frac{1}{12}$
13. $\frac{3}{10} + \frac{1}{2}$ $\frac{4}{5}$
14. $\frac{1}{3} + \frac{7}{9}$ $\frac{10}{9}$, or $1\frac{1}{9}$
15. $\frac{4}{5} + \frac{1}{6}$ $\frac{29}{30}$
16. $\frac{3}{8} + \frac{1}{3}$ $\frac{17}{24}$
17. $\frac{4}{9} + \frac{1}{2}$ $\frac{17}{18}$
18. $\frac{1}{10} + \frac{1}{4}$ $\frac{7}{20}$
19. $\frac{1}{3} + \frac{1}{2}$ $\frac{1}{2}$

Mixed Applications

20. Ron is mixing salad dressing. He combines $\frac{1}{2}$ c oil with $\frac{1}{4}$ c vinegar. What is the total amount of oil and vinegar he uses? $\frac{3}{4}$ c

21. Marie is making a birdhouse. She needs $\frac{5}{6}$ ft of plywood for the roof and $\frac{5}{6}$ ft of plywood for the sides. How much plywood does she need for the birdhouse? $1\frac{2}{3}$ ft

22. The Garden Club sold plants to raise money for a new park. Each plant sold for $6. The club sold 417 plants. How much money did the club raise? $2,502

23. Marla is paid $4 per hour for baby-sitting. Last month, Marla baby-sat for 22 hr. How much money did she earn? $88.00

P22 ON MY OWN   Use with text pages 100–101.

ENRICHMENT 5.3

Name _____ LESSON 5.3

Sum It Up

The shaded portion of each figure below models a fraction. Use the figures to find each sum. Write your answer in simplest form.

A   B   C

D   E   F

1. A + C = $\frac{5}{8}$
2. B + F = $\frac{21}{32}$
3. D + E = $\frac{15}{16}$
4. C + F = $\frac{19}{32}$
5. A + B = $\frac{11}{16}$
6. C + E = $\frac{3}{4}$
7. B + D = $\frac{3}{4}$
8. A + F = $\frac{23}{32}$
9. D + A = $\frac{13}{16}$
10. C + B = $\frac{9}{16}$
11. E + A = $\frac{7}{8}$
12. F + D = $\frac{25}{32}$
13. C + D = $\frac{11}{16}$
14. B + E + C = $1\frac{1}{16}$
15. F + A + D = $1\frac{5}{32}$
16. A + B + C = $\frac{15}{16}$
17. D + E + F = $1\frac{9}{32}$
18. A + F + B = $1\frac{1}{32}$

E22 STRETCH YOUR THINKING   Use with text pages 100–101.

101

GRADE 5	GRADE 6	GRADE 7
Use the LCD to subtract unlike fractions.	Subtract unlike fractions.	Add and subtract fractions.

ORGANIZER

OBJECTIVE To subtract unlike fractions

ASSIGNMENTS Independent Practice
- ■ *Basic* 1–20, even; 21–24; 25–34
- ■ *Average* 1–24; 25–34
- ■ *Advanced* 11–24; 25–34

PROGRAM RESOURCES
Reteach 5.4 Practice 5.4 Enrichment 5.4
Math Fun • Roll It Out, p. 108
Number Heroes • *Fraction Fireworks*

PROBLEM OF THE DAY

Transparency 5

Subtract horizontally and vertically.

$\frac{8}{9}$	$\frac{1}{3}$	? $\frac{5}{9}$
$\frac{1}{2}$	$\frac{1}{6}$	? $\frac{1}{3}$
? $\frac{7}{18}$	? $\frac{1}{6}$	? $\frac{2}{9}$

Solution Tab A, p. A5

QUICK CHECK

Write the missing numbers to show an equivalent fraction.

1. $\frac{4}{5} = \frac{4 \times ?}{5 \times ?} = \frac{20}{25}$ 5, 5
2. $\frac{5}{8} = \frac{? \times 5}{? \times 5} = \frac{?}{40}$ 5, 8, 25
3. $\frac{4}{7} = \frac{4 \times 4}{7 \times 4} = \frac{?}{?}$ 16, 28
4. $\frac{3}{4} = \frac{? \times 3}{4 \times ?} = \frac{9}{?}$ 3, 3, 12
5. $\frac{2}{9} = \frac{2 \times ?}{? \times 2} = \frac{4}{18}$ 2, 9, 18

1 MOTIVATE

Read and discuss Teen Times. Have students draw and label a diagram to show what part of their money they save and what part they spend.

- **How does the amount of money you save compare with the amount the average teenager saves? How does the amount of money you spend compare with the amount the average teenager spends?** Answers will vary but should accurately reflect the diagrams and include language such as *more than, less than, about the same.*

5.4

WHAT YOU'LL LEARN
How to subtract unlike fractions

WHY LEARN THIS?
To solve problems such as finding the difference in portions of earnings spent

teen times

The average teenager spends about $\frac{2}{3}$ of the money earned through part-time jobs—baby-sitting and lawn mowing—on personal items, such as clothes, entertainment, and snacks. About $\frac{1}{4}$ of the money is saved for college or for buying a car.

Subtracting Unlike Fractions

You have learned how to add unlike fractions. Do you think you can use a similar method to subtract unlike fractions?

Monica works part time for her uncle. She plans to spend $\frac{2}{3}$ of her earnings on school clothes and $\frac{1}{5}$ of her earnings on school supplies. What is the difference between what she spends on clothes and what she spends on school supplies?

Find $\frac{2}{3} - \frac{1}{5}$.

$$\frac{2}{3} = \frac{2 \times 5}{3 \times 5} = \frac{10}{15}$$
$$-\frac{1}{5} = \frac{1 \times 3}{5 \times 3} = \frac{3}{15}$$

The LCD of $\frac{2}{3}$ and $\frac{1}{5}$ is fifteenths. Multiply to find the equivalent fractions using the LCD.

$$\frac{2}{3} = \frac{10}{15}$$
$$-\frac{1}{5} = \frac{3}{15}$$
$$\overline{\frac{7}{15}}$$

Subtract the numerators. Write the difference over the denominator.

So, Monica spends $\frac{7}{15}$ more of her earnings on clothes.

- Suppose Monica spends $\frac{1}{3}$ of her earnings on clothes. What is the difference between what she spends on clothes and what she spends on school supplies? $\frac{2}{15}$ **more on clothes**

EXAMPLES

A.
$$\frac{5}{6} = \frac{5 \times 3}{6 \times 3} = \frac{15}{18}$$
$$-\frac{7}{9} = \frac{7 \times 2}{9 \times 2} = \frac{14}{18}$$
$$\overline{\frac{1}{18}}$$

B.
$$\frac{5}{12} = \frac{5}{12}$$
$$-\frac{1}{4} = \frac{1 \times 3}{4 \times 3} = \frac{3}{12}$$
$$\overline{\frac{2}{12} = \frac{1}{6}}$$

GUIDED PRACTICE

Use the LCD to rewrite the problem with equivalent fractions.

1. $\frac{4}{5} - \frac{1}{3}$ $\frac{12}{15} - \frac{5}{15}$
2. $\frac{2}{3} - \frac{1}{2}$ $\frac{4}{6} - \frac{3}{6}$
3. $\frac{5}{7} - \frac{1}{2}$ $\frac{10}{14} - \frac{7}{14}$
4. $\frac{2}{3} - \frac{2}{5}$ $\frac{10}{15} - \frac{6}{15}$

Subtract. Write the answer in simplest form.

5. $\frac{7}{9} - \frac{1}{6}$ $\frac{11}{18}$
6. $\frac{3}{4} - \frac{1}{6}$ $\frac{7}{12}$
7. $\frac{3}{4} - \frac{3}{8}$ $\frac{3}{8}$
8. $\frac{2}{5} - \frac{1}{3}$ $\frac{1}{15}$

IDEA FILE

ALTERNATIVE TEACHING STRATEGY

Some students will have difficulty subtracting fractions. These students may have more success if they use a method with which they are already familiar. You may wish to have students use number lines to solve subtraction problems.

Modalities Kinesthetic, Visual

RETEACH 5.4

Name _____

LESSON 5.4

Subtracting Unlike Fractions

Remember, unlike fractions have *different* denominators. When subtracting unlike fractions, change them to equivalent fractions with the same denominator. Then follow the same steps you use to subtract like fractions.

Alonso used $\frac{3}{4}$ c sugar and $\frac{1}{8}$ c honey in a cookie recipe. How much more sugar than honey did he use?

Find $\frac{3}{4} - \frac{1}{8}$

Step 1: Find the LCD of $\frac{3}{4}$ and $\frac{1}{8}$. THINK: What is the least multiple of 4 and 8? It is 8.

Step 2: Multiply to find equivalent fractions, using the LCD. $\frac{3}{4} = \frac{3 \times 2}{4 \times 2} = \frac{6}{8}$ $\frac{1}{8} = \frac{1 \times 1}{8 \times 1} = \frac{1}{8}$

Step 3: Follow the same steps you use to subtract like fractions: Subtract the numerators and write the difference over the same denominator. $\frac{6}{8} - \frac{1}{8} = \frac{5}{8}$ ← Remember, keep the denominator the same.

Step 4: Write the answer in simplest form. $\frac{5}{8}$ is already in simplest form.

Alonso used $\frac{5}{8}$ c more sugar.

1. Complete to find $\frac{9}{10} - \frac{5}{6}$.

Step 1: The LCD of $\frac{9}{10}$ and $\frac{5}{6}$ is ___30___.

Step 2: $\frac{9}{10} = \frac{9 \times 3}{10 \times 3} = \frac{27}{30}$ $\frac{5}{6} = \frac{5 \times 5}{6 \times 5} = \frac{25}{30}$

Step 3: $\frac{27}{30} - \frac{25}{30} = \frac{2}{30}$

Step 4: In simplest form. $\frac{2}{30} = \frac{1}{15}$

Find each difference. Write the answer in simplest form.

2. $\frac{5}{7} - \frac{1}{2}$ $\frac{3}{14}$
3. $\frac{11}{12} - \frac{7}{8}$ $\frac{1}{24}$
4. $\frac{4}{5} - \frac{1}{2}$ $\frac{3}{10}$
5. $\frac{3}{4} - \frac{2}{3}$ $\frac{1}{12}$

Use with text pages 102–103.   TAKE ANOTHER LOOK **R23**

Subtract. Write the answer in simplest form.

1. $\frac{1}{3} - \frac{1}{4} \quad \frac{1}{12}$
2. $\frac{1}{2} - \frac{2}{5} \quad \frac{1}{10}$
3. $\frac{5}{6} - \frac{1}{4} \quad \frac{7}{12}$
4. $\frac{5}{9} - \frac{1}{3} \quad \frac{2}{9}$
5. $\frac{7}{9} - \frac{1}{2} \quad \frac{5}{18}$

6. $\frac{5}{7} - \frac{1}{3} \quad \frac{3}{14}$
7. $\frac{3}{8} - \frac{1}{4} \quad \frac{1}{8}$
8. $\frac{4}{5} - \frac{1}{3} \quad \frac{7}{15}$
9. $\frac{1}{4} - \frac{1}{6} \quad \frac{1}{12}$
10. $\frac{7}{8} - \frac{3}{4} \quad \frac{1}{8}$

11. $\frac{8}{9} - \frac{2}{3} \quad \frac{2}{9}$
12. $\frac{7}{10} - \frac{1}{5} \quad \frac{1}{2}$
13. $\frac{5}{6} - \frac{1}{8} \quad \frac{17}{24}$
14. $\frac{1}{3} - \frac{1}{7} \quad \frac{4}{21}$
15. $\frac{3}{5} - \frac{1}{4} \quad \frac{7}{20}$

16. $\frac{5}{6} - \frac{5}{9} \quad \frac{5}{18}$
17. $\frac{4}{5} - \frac{3}{10} \quad \frac{1}{2}$
18. $\frac{3}{4} - \frac{1}{6} \quad \frac{7}{12}$
19. $\frac{3}{10} - \frac{1}{4} \quad \frac{1}{20}$
20. $\frac{7}{8} - \frac{2}{3} \quad \frac{5}{24}$

Problem-Solving Applications

21. Barnie and his family walked each day of their vacation. On the first day, they walked $\frac{5}{12}$ mi. On the second day, they walked $\frac{1}{4}$ mi. How much farther did they walk on the first day than on the second? $\frac{2}{12}$ mi, or $\frac{1}{6}$ mi farther

22. Barnie and his family stopped to pick strawberries. Together they picked $\frac{3}{4}$ qt. Barnie's brother, Carl, ate $\frac{1}{12}$ qt before they left the strawberry field. How much was left? $\frac{8}{12}$ qt, or $\frac{2}{3}$ qt

23. When Barnie and his family left for their trip, they had a full tank of gas. When they reached their first stop, they had only $\frac{1}{8}$ tank. How much gas had they used? $\frac{7}{8}$ tank

24. ✏ **WRITE ABOUT IT** How do you subtract fractions with unlike denominators? **Write equivalent fractions with the LCD, and then subtract the numerators. Use the same denominator. Then write the answer in simplest form.**

Mixed Review

Compare the value of the numerator with the value of the denominator. Write *much less than, about the same as,* or *about $\frac{1}{2}$.*

25. $\frac{1}{5}$ **much less than**
26. $\frac{7}{8}$ **about the same as**

27. $\frac{4}{9}$ **about $\frac{1}{2}$**
28. $\frac{2}{15}$ **much less than**

Write the fraction that is in simplest form.

29. $\frac{2}{4}, \frac{1}{2}, \frac{3}{6}, \frac{6}{12} \quad \frac{1}{2}$
30. $\frac{4}{6}, \frac{20}{30}, \frac{6}{9}, \frac{2}{3} \quad \frac{2}{3}$

31. $\frac{6}{8}, \frac{9}{12}, \frac{3}{4}, \frac{18}{24} \quad \frac{3}{4}$
32. $\frac{3}{18}, \frac{7}{21}, \frac{2}{7}, \frac{3}{6} \quad \frac{2}{7}$

33. **CONSUMER** Beth is buying a dozen doughnuts for the scout troop. She plans to buy $\frac{1}{3}$ glazed, $\frac{1}{3}$ cake, and $\frac{1}{3}$ jelly. How many of each kind is she getting? **4 of each kind**

34. **SPORTS** Eric skated 500 m in 37.14 sec. Andre skated the same distance in 37.139 sec. Who won the race? **Andre**

MORE PRACTICE Lesson 5.4, page H49

Technology Link

In *Mighty Math Number Heroes,* you can create fireworks in the game *Fraction Fireworks* by finding the correct sum or difference of fractions.

Use Grow Slide Level W.

103

---

2 TEACH/PRACTICE

In this lesson students use equivalent fractions to subtract unlike fractions.

GUIDED PRACTICE

To make sure students understand how to subtract unlike fractions, complete Exercises 1–8 as a class.

INDEPENDENT PRACTICE

These exercises check students' abilities to find the LCD, to subtract unlike fractions, and to write their answers in simplest form.

ADDITIONAL EXAMPLE

Example, p. 102

$$\frac{3}{5} = \frac{3 \times 3}{5 \times 3} = \frac{9}{15}$$
$$- \frac{1}{3} = \frac{1 \times 5}{3 \times 5} = \frac{5}{15}$$
$$\frac{4}{15}$$

3 WRAP UP

Use these questions to help students focus on the big ideas of the lesson.

- **Marilyn's house is $\frac{5}{8}$ mi from the elementary school and $\frac{3}{4}$ mi from the middle school. Which school is she closer to?** the elementary school

- **How much closer? $\frac{1}{8}$ mi**

✓ *Assessment Tip*

Have students write the answer.

- **When Pam solved Problem 22, she wrote the equation $\frac{1}{12} - \frac{3}{4} = ?$. Explain why her equation is wrong.** Pam's equation should be $\frac{3}{4} - \frac{1}{12}$ because she is subtracting a lesser number, $\frac{1}{12}$, from a greater number, $\frac{3}{4}$.

DAILY QUIZ

Subtract. Write the answer in simplest form.

1. $\frac{5}{6} - \frac{1}{3} \quad \frac{1}{2}$
2. $\frac{9}{10} - \frac{2}{5} \quad \frac{1}{2}$
3. $\frac{1}{3} - \frac{1}{6} \quad \frac{1}{6}$
4. $\frac{9}{10} - \frac{7}{8} \quad \frac{1}{40}$
5. $\frac{5}{6} - \frac{5}{8} \quad \frac{5}{24}$
6. $\frac{1}{5} - \frac{1}{10} \quad \frac{1}{10}$

PRACTICE 5.4

Name _____

LESSON 5.4

Subtracting Unlike Fractions

Subtract. Write the answer in simplest form.

1. $\frac{1}{2} - \frac{1}{5} \quad \frac{3}{10}$
2. $\frac{6}{7} - \frac{1}{4} \quad \frac{17}{28}$
3. $\frac{9}{10} - \frac{3}{5} \quad \frac{3}{10}$
4. $\frac{7}{8} - \frac{1}{2} \quad \frac{3}{8}$
5. $\frac{3}{4} - \frac{5}{8} \quad \frac{1}{8}$

6. $\frac{4}{5} - \frac{1}{3} \quad \frac{7}{15}$
7. $\frac{5}{8} - \frac{1}{10} \quad \frac{21}{40}$
8. $\frac{1}{2} - \frac{1}{6} \quad \frac{1}{3}$
9. $\frac{7}{10} - \frac{1}{4} \quad \frac{9}{20}$
10. $\frac{5}{6} - \frac{1}{3} \quad \frac{1}{2}$

11. $\frac{11}{12} - \frac{1}{4} \quad \frac{2}{3}$
12. $\frac{9}{10} - \frac{1}{6} \quad \frac{11}{15}$
13. $\frac{3}{4} - \frac{1}{12} \quad \frac{2}{3}$
14. $\frac{7}{8} - \frac{1}{3} \quad \frac{11}{21}$
15. $\frac{4}{5} - \frac{1}{6} \quad \frac{19}{30}$

16. $\frac{3}{4} - \frac{1}{2} \quad \frac{1}{4}$
17. $\frac{3}{8} - \frac{3}{4} \quad \frac{7}{24}$
18. $\frac{3}{5} - \frac{1}{4} \quad \frac{8}{15}$
19. $\frac{13}{14} - \frac{2}{7} \quad \frac{9}{14}$
20. $\frac{1}{3} - \frac{1}{5} \quad \frac{2}{15}$

21. $\frac{7}{10} - \frac{2}{5} \quad \frac{3}{10}$
22. $\frac{4}{7} - \frac{1}{3} \quad \frac{5}{21}$
23. $\frac{7}{12} - \frac{1}{4} \quad \frac{1}{3}$
24. $\frac{7}{15} - \frac{2}{5} \quad \frac{1}{15}$
25. $\frac{5}{6} - \frac{1}{3} \quad \frac{1}{2}$

26. $\frac{7}{9} - \frac{1}{3} \quad \frac{4}{9}$
27. $\frac{2}{3} - \frac{2}{7} \quad \frac{8}{21}$
28. $\frac{5}{8} - \frac{1}{3} \quad \frac{7}{24}$
29. $\frac{3}{4} - \frac{1}{9} \quad \frac{5}{9}$
30. $\frac{5}{6} - \frac{1}{2} \quad \frac{1}{3}$

Mixed Applications

31. Ron picks $\frac{3}{4}$ qt of strawberries. He gives $\frac{1}{3}$ qt to a neighbor. What quantity of strawberries does he have left? $\frac{5}{12}$ qt

32. Josh hikes $\frac{11}{12}$ mi before noon and $\frac{5}{6}$ mi after noon. How much farther does Josh hike before noon than after? $\frac{1}{12}$ mi

33. The auditorium is being set up for a concert. Each row has 32 chairs. How many rows will be needed to seat 832 people? **26 rows**

34. Tickets to the concert cost $5. A total of 825 tickets were sold. How much money did the orchestra raise? **$4,125**

Use with text pages 102–103.   ON MY OWN   P23

ENRICHMENT 5.4

Name _____

LESSON 5.4

Think Like an Egyptian

Ancient Egyptians used only *unit fractions* to represent parts of a whole. (Unit fractions have 1 as the numerator.) This is how the ancient Egyptians wrote fractions by using symbols.

$= \frac{1}{2}$ $= \frac{1}{6}$ $= \frac{1}{8}$

$= \frac{1}{11}$ $= \frac{1}{12}$

Look at the fractions above. Then use Egyptian symbols to represent these fractions.

1. $\frac{1}{4}$ 2. $\frac{1}{10}$ 3. $\frac{1}{5}$ 4. $\frac{1}{7}$

Not every fraction has a 1 as a numerator. How did the Egyptians represent an amount like $\frac{3}{4}$? They showed it as a sum of two unit fractions.

$\frac{1}{2} + \frac{1}{4} = \frac{3}{4}$

Tell what fractions are being subtracted. Then find the difference. Write your answer using symbols as an ancient Egyptian would have.

5. $\frac{1}{2} - \frac{1}{3} = \frac{1}{6}$
6. $\frac{1}{4} - \frac{1}{8} = \frac{1}{8}$
7. $\frac{1}{3} - \frac{1}{4} = \frac{1}{12}$

8. $\frac{1}{4} - \frac{1}{5} = \frac{1}{20}$
9. $\frac{1}{2} - \frac{1}{5} = \frac{3}{10}$
10. $\frac{1}{4} - \frac{1}{10} = \frac{3}{20}$

Use with text pages 102–103.   STRETCH YOUR THINKING   E23

103

GRADE 5
Estimate sums and differences of fractions.

GRADE 6
Estimate sums and differences of fractions.

GRADE 7
Estimate sums and differences of fractions.

ORGANIZER

OBJECTIVE To estimate sums and differences of fractions

ASSIGNMENTS Independent Practice
- *Basic* 1–31, odd; 32–38; 39–51
- *Average* 1–38; 39–51
- *Advanced* 17–38; 39–51

PROGRAM RESOURCES
Reteach 5.5 Practice 5.5 Enrichment 5.5

PROBLEM OF THE DAY

Transparency 5

Mike's and Kay's numbers are both less than 1. The digit in Mike's numerator is the same as the digit in Kay's denominator. Kay's number is $\frac{1}{10}$ greater than Mike's. What are Mike's and Kay's numbers? Mike's number is $\frac{2}{5}$ and Kay's is $\frac{1}{2}$.

Solution Tab A, p. A5

QUICK CHECK

Which number is about half the given number? Write *a, b,* or *c.*

1. 8	**a.** 5	**b.** 2	**c.** 7 a
2. 10	**a.** 8	**b.** 2	**c.** 4 c
3. 9	**a.** 3	**b.** 5	**c.** 7 b
4. 5	**a.** 1	**b.** 2	**c.** 4 b
5. 7	**a.** 3	**b.** 6	**c.** 1 a

1 MOTIVATE

To access students' prior knowledge of estimates, ask:

- **When is it helpful to estimate sums and differences? Give an example of each time.** Possible answers: (1) to check the reasonableness of answers, (2) when an exact answer isn't necessary, and (3) to predict about what an exact answer will be

Explain to students that they estimate sums and differences of fractions for the same reasons.

104

5.5

WHAT YOU'LL LEARN
How to estimate sums and differences of fractions

WHY LEARN THIS?
To estimate answers to problems when exact answers are not needed

104 Chapter 5

Estimating Sums and Differences

Sometimes when you solve problems that involve adding and subtracting fractions, you do not need to know the exact answer. One way to find the estimated answer is to use a number line.

Look at the number line below. Is $\frac{1}{8}$ closer to 0, $\frac{1}{2}$, or 1?

$$0 \quad \frac{1}{8} \quad \frac{2}{8} \quad \frac{3}{8} \quad \frac{1}{2} \quad \frac{5}{8} \quad \frac{6}{8} \quad \frac{7}{8} \quad 1$$

Since $\frac{1}{8}$ is closer to 0, you would round $\frac{1}{8}$ to 0.

To estimate with fractions less than 1, round them to 0, $\frac{1}{2}$, or 1.

Round $\frac{1}{9}$ to 0. The numerator is much less than the denominator.

Round $\frac{3}{8}$ to $\frac{1}{2}$. The numerator is about half the denominator.

Round $\frac{4}{5}$ to 1. The numerator is about the same as the denominator.

EXAMPLE 1 Gloria likes to hike and jog. She jogged $\frac{3}{4}$ mi on Monday and $\frac{4}{10}$ mi on Tuesday. What is a good estimate of the distance Gloria jogged on the two days?

Estimate. $\frac{3}{4} + \frac{4}{10}$

$$\begin{array}{rl} \frac{3}{4} & \to 1 \quad \textit{Round each fraction.} \\ + \frac{4}{10} & \to \frac{1}{2} \\ \hline & 1\frac{1}{2} \quad \textit{Add.} \end{array}$$

So, Gloria jogged about $1\frac{1}{2}$ mi.

- Suppose Gloria jogs the same distance every Monday and Tuesday. What is a good estimate of the number of miles she jogs in 2 weeks? **3 mi**

- **CRITICAL THINKING** In a fraction, if the numerator is more than half the denominator, is the fraction greater than $\frac{1}{2}$? **yes**

When you estimate the sum of more than two fractions, you follow the same rules of rounding.

EXAMPLE 2 One week Gloria couldn't hike or jog because it rained. On Monday it rained $\frac{3}{5}$ in., on Tuesday it rained $\frac{4}{10}$ in., and on Thursday it rained $\frac{4}{5}$ in. What is a good estimate of the amount of rain for the three days?

Estimate. $\frac{3}{5} + \frac{4}{10} + \frac{4}{5}$

$$\begin{array}{ll}\frac{3}{5} \rightarrow \frac{1}{2} & \textit{Round each fraction.} \\ \frac{4}{10} \rightarrow \frac{1}{2} & \\ +\frac{4}{5} \rightarrow 1 & \\ \hline \quad 1\frac{2}{2}, \text{ or } 2 & \textit{Add.} \\ & \textit{Write the answer as a whole number.}\end{array}$$

So, there was about 2 in. of rain for the three days.

- Suppose it rained $\frac{6}{10}$ in. on Friday also. What is a good estimate of the amount of rain for the four days? **about $2\frac{1}{2}$ in.**

- Gloria and her niece hiked $\frac{3}{4}$ mi on Wednesday, $\frac{7}{10}$ mi on Friday, $\frac{1}{5}$ mi on Saturday, and $\frac{4}{10}$ mi on Sunday. What is a good estimate for the total number of miles Gloria and her niece hiked? **about $2\frac{1}{2}$ mi**

Geography Link

The Appalachian Trail runs 2,159 mi along the ridges of the Appalachian Mountains, from Springer Mountain, Georgia, to Mount Katahdin, Maine. The average time it takes to hike the full length of the trail, is about 4–6 months. Suppose you started at the Georgia end early in May but hiked only about $\frac{1}{4}$ of the trail by the end of June. Then you hiked about $\frac{2}{5}$ of the trail in July and August. How much of the trail do you estimate you would still need to hike?

about $\frac{1}{2}$ of the trail

GUIDED PRACTICE

Use the number line to round each fraction to 0, $\frac{1}{2}$, or 1.

$$0 \quad \frac{1}{12} \quad \frac{1}{6} \quad \frac{1}{4} \quad \frac{1}{3} \quad \frac{5}{12} \quad \frac{1}{2} \quad \frac{7}{12} \quad \frac{2}{3} \quad \frac{3}{4} \quad \frac{5}{6} \quad \frac{11}{12} \quad 1$$

1. $\frac{11}{12}$ **1**   2. $\frac{7}{12}$ **$\frac{1}{2}$**   3. $\frac{1}{6}$ **0**   4. $\frac{1}{4}$ **0 or $\frac{1}{2}$**

Round each fraction. Write *about 0, about $\frac{1}{2}$, or about 1.*

5. $\frac{1}{9}$ 6. $\frac{10}{11}$ 7. $\frac{2}{6}$ 8. $\frac{7}{9}$ 9. $\frac{4}{7}$
   about 0   **about 1**   **about $\frac{1}{2}$**   **about 1**   **about $\frac{1}{2}$**

Estimate the sum. **Possible estimates given.**

10. $\frac{3}{8} + \frac{1}{5}$ **$\frac{1}{2}$**   11. $\frac{2}{9} + \frac{9}{11}$ **1**   12. $\frac{2}{3} + \frac{5}{9}$ 13. $\frac{7}{8} + \frac{9}{10}$ **2**   12. **$1\frac{1}{2}$, or 2**

14. $\frac{2}{3} + \frac{9}{10}$ **2**   15. $\frac{7}{8} + \frac{1}{2}$ **$1\frac{1}{2}$**   16. $\frac{1}{16} + \frac{5}{8}$ **$\frac{1}{2}$**   17. $\frac{1}{2} + \frac{3}{4}$ **$1\frac{1}{2}$, or 2**

18. $\frac{5}{6} + \frac{2}{5} + \frac{1}{8}$ **$1\frac{1}{2}$**   19. $\frac{7}{9} + \frac{1}{5} + \frac{7}{8}$ **2**   20. $\frac{3}{10} + \frac{5}{8} + \frac{4}{9}$ **$1\frac{1}{2}$**

105

2 TEACH/PRACTICE

Students begin the lesson with two different strategies to help them round fractions: (1) using a number line and (2) comparing the numerator and denominator. Then they use their rounding skills to estimate sums and differences of fractions.

GUIDED PRACTICE

Have volunteers complete Exercises 1–4 at the board, using a number line. Then complete the remaining exercises orally as a class and discuss.

ADDITIONAL EXAMPLES

Example 1, p. 104

Roger bought $\frac{3}{4}$ lb of broccoli and $\frac{1}{8}$ lb of cauliflower. About how many pounds of vegetables did he buy?

Estimate: $\frac{3}{4} + \frac{1}{8}$

Round each fraction.
$$\begin{array}{r}\frac{3}{4} \rightarrow 1 \\ +\frac{1}{8} \rightarrow 0 \\ \hline\end{array}$$

Add.   1

Roger bought about 1 lb of vegetables.

Example 2, p. 105

Erica used $\frac{5}{8}$ yd of lace, $\frac{3}{8}$ yd of ribbon, and $\frac{7}{8}$ yd of braid to trim the pillows she made for the craft show. About how many yards of trim did she use?

Estimate: $\frac{5}{8} + \frac{3}{8} + \frac{7}{8}$

Round each fraction.
$$\begin{array}{r}\frac{5}{8} \rightarrow \frac{1}{2} \\ \frac{3}{8} \rightarrow \frac{1}{2} \\ +\frac{7}{8} \rightarrow 1 \\ \hline\end{array}$$

Add. $1\frac{2}{2}$, or 2

Erica used about 2 yd of trim.

ENRICHMENT 5.5

Name _____   **LESSON 5.5**

You Pose the Problem

A mistake was made at the printer. This page should have contained eight word problems that require estimation of a sum or a difference of fractions. However, instead of the problems, the answers were printed.

Now you must create a word problem for each answer. Be sure that your problem requires estimating a sum or a difference. Exchange papers with a classmate. Check each other's work. **Check students' work.**

1. _____   2. _____

Answer: $\frac{1}{2}$ c flour   Answer: 1 yd fabric

3. _____   4. _____

Answer: 2 mi   Answer: 1 c sugar

5. _____   6. _____

Answer: $\frac{1}{2}$ in.   Answer: 1 mi

7. _____   8. _____

Answer: $\frac{1}{2}$ the flowers   Answer: 2 ft wire

E24 STRETCH YOUR THINKING   Use with text pages 104–107.

TEACHING TIP

Rounding fractions to 0, $\frac{1}{2}$, or 1 works for estimating a sum or difference most of the time, but not all the time. Try estimating the sum $\frac{1}{8} + \frac{9}{40} + \frac{3}{10}$. Since all the addends round to 0, the estimated sum would be about 0. The exact answer, however, is $\frac{13}{20}$, or about $\frac{1}{2}$, making the estimate of 0 not reasonable. A similar situation occurs when all the fractions in a problem round to 1. In all these situations, it's a good idea to bend the rules for rounding fractions so that the estimates make sense.

INDEPENDENT PRACTICE

To assess students' understanding of the concepts, complete Exercises 1–6 orally as a class before assigning the remaining exercises.

💡 **CRITICAL THINKING** **Bill is helping his father fence a field, the sides of which are $\frac{3}{5}$ mi, $\frac{4}{7}$ mi, $\frac{9}{10}$ mi and $\frac{8}{10}$ mi. About how many miles of fence do they need?** about 3 mi

Draw a picture of the field to help you solve the problem.

COMMON ERROR ALERT If students need additional practice using a number line to round fractions, they can make up their own problems and use number lines. Before students begin Exercises 5–9 in Guided Practice, remind them to round to 0 if the numerator is much less than the denominator, to $\frac{1}{2}$ if the numerator is about half the denominator, and to 1 if the numerator is about the same as the denominator. To check students' understanding, discuss Exercise 18 as a class.

ADDITIONAL EXAMPLES

Example 3, p. 106

Estimate. $\frac{4}{5} - \frac{1}{3}$

$\frac{4}{5} \to 1$

$- \frac{1}{3} \to \frac{1}{2}$

$\quad\quad \frac{1}{2}$

$\frac{4}{5} - \frac{1}{3}$ is about $\frac{1}{2}$.

Example 4, p. 106

Joan is covering two shelves with contact paper. One shelf is $\frac{3}{4}$ yd long and the other is $\frac{5}{8}$ yd long. How much more paper will she need for the longer shelf? about $\frac{1}{2}$ yd

106

You can also estimate the differences of fractions by rounding to the nearest 0, $\frac{1}{2}$, or 1.

EXAMPLE 3 Estimate. $\frac{2}{3} - \frac{5}{9}$

$\frac{2}{3} \to 1$　　*Round each fraction.*

$- \frac{5}{9} \to \frac{1}{2}$

$\quad\quad \frac{1}{2}$　　*Subtract.*

So, $\frac{2}{3} - \frac{5}{9}$ is about $\frac{1}{2}$.

• Estimate. $\frac{7}{8} - \frac{5}{6}$ **about 0**

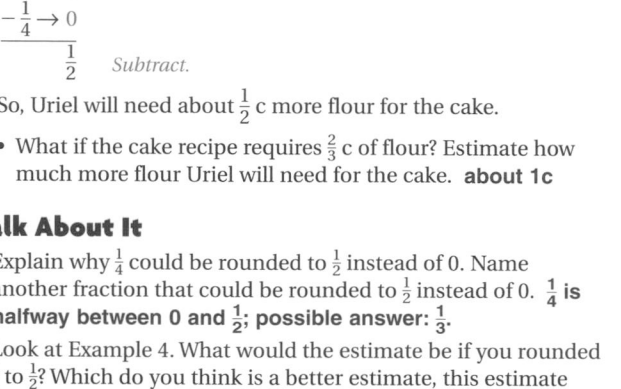

EXAMPLE 4 Uriel is making a pie and a cake for a party. The cake recipe requires $\frac{5}{8}$ c of flour, and the pie recipe requires $\frac{1}{4}$ c of flour. What is a good estimate of how much more flour Uriel will need for the cake than the pie?

Estimate. $\frac{5}{8} - \frac{1}{4}$

$\frac{5}{8} \to \frac{1}{2}$　　*Round each fraction.*

$- \frac{1}{4} \to 0$

$\quad\quad \frac{1}{2}$　　*Subtract.*

So, Uriel will need about $\frac{1}{2}$ c more flour for the cake.

• What if the cake recipe requires $\frac{2}{3}$ c of flour? Estimate how much more flour Uriel will need for the cake. **about 1c**

Talk About It

• Explain why $\frac{1}{4}$ could be rounded to $\frac{1}{2}$ instead of 0. Name another fraction that could be rounded to $\frac{1}{2}$ instead of 0. **$\frac{1}{4}$ is halfway between 0 and $\frac{1}{2}$; possible answer: $\frac{1}{3}$.**

• Look at Example 4. What would the estimate be if you rounded $\frac{1}{4}$ to $\frac{1}{2}$? Which do you think is a better estimate, this estimate or $\frac{1}{2}$? Explain. **0; possible answer: $\frac{1}{2}$, since 0 does not give you an idea of how much flour is needed.**

INDEPENDENT PRACTICE

Round each fraction. Write *about 0, about $\frac{1}{2}$,* or *about 1.*

1. $\frac{2}{3}$ about 1　**2.** $\frac{2}{9}$ about 0　**3.** $\frac{5}{8}$ about $\frac{1}{2}$　**4.** $\frac{13}{14}$ about 1　**5.** $\frac{3}{12}$ about 0　**6.** $\frac{5}{12}$ about $\frac{1}{2}$

👥 **IDEA FILE**

LANGUAGE ARTS CONNECTION

Discuss with students that many words in the English language have Latin origins. The word *fraction* comes from the Latin word *frangere*, which means "to break." Encourage students to research the origins of other math terms and to report to the class what they discover.

Modality Auditory

EXTENSION

Have students identify which numbers are estimates and which are exact numbers.

1. Bob watched TV for about $\frac{1}{2}$ hr. estimate

2. The milk carton is almost half empty. estimate

3. He grew $1\frac{5}{8}$ in. last year. exact

4. More than $\frac{3}{4}$ of the class attended the party. estimate

5. One-fifth of people's fingers are thumbs. exact

6. The recipe calls for $\frac{1}{3}$ c of sugar. exact

Modality Auditory

Estimate the sum or difference. **Possible estimates are given.**

7. $\frac{2}{3} + \frac{1}{9}$ 1

8. $\frac{10}{12} - \frac{7}{10}$ 0

9. $\frac{7}{9} - \frac{5}{9}$ $\frac{1}{2}$

10. $\frac{5}{16} + \frac{4}{9}$ 1

11. $\frac{7}{8} - \frac{1}{4}$ 1

12. $\frac{5}{9} - \frac{1}{2}$ 0

13. $\frac{1}{3} + \frac{6}{7}$ $1\frac{1}{2}$

14. $\frac{12}{15} - \frac{3}{7}$ $\frac{1}{2}$

15. $\frac{7}{8} + \frac{3}{4}$ 2

16. $\frac{11}{12} + \frac{4}{7}$ $1\frac{1}{2}$

17. $\frac{11}{12} - \frac{1}{9}$ 1

18. $\frac{13}{15} + \frac{8}{9}$ 2

19. $\frac{3}{5} + \frac{6}{7}$ $1\frac{1}{2}$

20. $\frac{11}{14} - \frac{1}{3}$ $\frac{1}{2}$

21. $\frac{8}{10} - \frac{10}{18}$ $\frac{1}{2}$

22. $\frac{7}{8} + \frac{1}{3}$ $1\frac{1}{2}$

23. $\frac{15}{16} - \frac{4}{5}$ 0

24. $\frac{4}{5} - \frac{2}{7}$ 1

25. $\frac{6}{10} + \frac{1}{5}$ $\frac{1}{2}$

26. $\frac{9}{12} + \frac{10}{18}$ $1\frac{1}{2}$

27. $\frac{1}{9} + \frac{5}{8}$ $\frac{1}{2}$

28. $\frac{8}{9} - \frac{1}{6}$ 1

29. $\frac{5}{6} + \frac{7}{8}$ 2

30. $\frac{3}{13} + \frac{2}{15}$ 0

31. $\frac{5}{6} - \frac{1}{8}$ $\frac{1}{2}$

Problem-Solving Applications

32. Paula planted zinnia and marigold seeds in her flower garden. Of the seeds that sprouted, $\frac{1}{8}$ were zinnias and $\frac{3}{7}$ were marigolds. About what portion of the sprouts were zinnias or marigolds? **about $\frac{1}{2}$ of the sprouts**

33. In May, $\frac{13}{15}$ of Ed's flowers bloomed. During June, $\frac{3}{8}$ of the flowers bloomed. Estimate what fraction more of the flowers bloomed in May than in June. **About $\frac{1}{2}$ more of the flowers bloomed in May.**

34. Lori is baking muffins. For each batch she needs $\frac{1}{2}$ c of milk, $\frac{3}{4}$ c of water, and $\frac{1}{3}$ c of apple juice. She wants to make three batches. About how many cups of liquid will she use? **about 6 c**

35. Larry has three fabric samples measuring $\frac{5}{6}$ yd, $\frac{1}{4}$ yd, and $\frac{2}{3}$ yd. He estimates the total length to be about $1\frac{1}{2}$ yd. Is his estimate reasonable? Explain. **No; he has about $2\frac{1}{2}$ or 3 yd of fabric**

36. Kathi bicycled $\frac{7}{8}$ mi to the post office. From there she bicycled $\frac{3}{5}$ mi to the park. She then bicycled $\frac{6}{10}$ mi home. About how many miles did she bicycle? **about 2 mi**

37. Rogelio read $\frac{1}{3}$ hr on Monday, $\frac{3}{4}$ hr on Tuesday, and $\frac{1}{8}$ hr on Friday. He says he read about $1\frac{1}{2}$ hr. Is he right? Explain. **Yes; $\frac{1}{2} + 1 + 0 \approx 1\frac{1}{2}$.**

38. ✏️ **WRITE ABOUT IT** Explain how to round fractions less than 1. **Compare the numerator with the denominator to decide whether the fraction is closer to 0, $\frac{1}{2}$, or 1.**

Mixed Review

Write as a fraction.

39. $1\frac{3}{4}$ $\frac{7}{4}$

40. $2\frac{1}{4}$ $\frac{9}{4}$

41. $2\frac{5}{8}$ $\frac{21}{8}$

42. $3\frac{2}{3}$ $\frac{11}{3}$

43. $3\frac{7}{6}$ $\frac{25}{6}$

44. $5\frac{1}{5}$ $\frac{26}{5}$

Write the GCF of each pair of numbers.

45. 21, 9 **3**

46. 8, 64 **8**

47. 25, 10 **5**

48. 18, 24 **6**

49. 36, 16 **4**

50. **MENTAL MATH** The band is setting up chairs in the school auditorium for the concert. They can put 25 chairs in each row. How many rows will they need to seat 350 people? **14 rows**

51. **MONEY** The band held a concert to raise money for a trip. Each ticket cost $5. They sold 345 tickets. How much money did they raise? **$1,725**

MORE PRACTICE Lesson 5.5, page H49

Use these questions to help students focus on the big ideas of the lesson.

- **As of Wednesday, only $\frac{1}{3}$ of the class had purchased tickets for the dance. By Friday, $\frac{3}{4}$ of the class had their tickets. What part of the class bought their tickets between Wednesday and Friday?** about half the class

✔ *Assessment Tip*

Have students write the answer.

- **Explain how to estimate the sum of fractions.** Round each fraction to 0, $\frac{1}{2}$, or 1 and then add.

DAILY QUIZ

Estimate the sum or difference.

1. $\frac{9}{16} + \frac{5}{8}$ about 1

2. $\frac{9}{10} - \frac{3}{8}$ about $\frac{1}{2}$

3. $\frac{7}{8} + \frac{5}{12}$ about $1\frac{1}{2}$

4. $\frac{5}{7} + \frac{6}{11}$ about $1\frac{1}{2}$

5. $\frac{8}{10} - \frac{2}{5}$ about $\frac{1}{2}$

6. $\frac{3}{5} - \frac{1}{6}$ about $\frac{1}{2}$

WRITING IN MATHEMATICS

Have students rewrite each statement to include an appropriate estimate.

1. Only 5 eggs are in the carton. About $\frac{1}{2}$ dozen eggs are in the carton.

2. The show lasts for 55 min. The show lasts about 1 hr.

3. She is 62 in. tall. She is about 5 ft tall.

4. She walked the dog every day for 6 days. She walked the dog every day for about 1 wk.

Modalities Auditory, Visual

ALTERNATIVE TEACHING STRATEGY

Have students make five fraction circles (TR pp. R19–R20). They should shade the first two fraction circles to show fractions that round to 0, the next fraction circle to show a fraction that rounds to $\frac{1}{2}$, and the last two fraction circles to show fractions that round to 1. Then have students write each fraction. Check students' work.

Modalities Kinesthetic, Visual

USING THE PAGE
To provide additional practice activities related to Chapter 5: *Adding and Subtracting Fractions*

THE PIZZA RUNNER

MATERIALS: *For each group* number cubes

Have students work in pairs or small groups to complete this activity.

In this activity, students practice adding fractions by determining what portion of a pizza a boy eats in a day.

Multiple Intelligences Logical/Mathematical, Bodily/Kinesthetic, Interpersonal/Social, Intrapersonal/Introspective
Modalities Auditory, Kinesthetic

FRACTION RIDDLES

Have students work individually to complete this activity.

In this activity, students practice adding and subtracting fractions by solving and writing fraction riddles.

Multiple Intelligences Logical/Mathematical, Intrapersonal/Introspective
Modality Auditory

ROLL IT OUT

MATERIALS: *For each group* 2 number cubes

Have students work in pairs or small groups to complete this activity.

In this activity, students create fractions less than one by rolling number cubes and then adding or subtracting the fractions to try to get a given fraction.

Multiple Intelligences Logical/Mathematical, Verbal/Linguistic, Bodily/Kinesthetic, Interpersonal/Social, Intrapersonal/Introspective
Modalities Auditory, Kinesthetic, Visual

MATH FUN!

The Pizza Runner

PURPOSE To practice adding like fractions (pages 94–95)
YOU WILL NEED number cube

Cooper loves pizza and eats it for breakfast, lunch and dinner. His parents say that if he eats a whole pizza before dinner, he will have to run a mile to work it off.

Work with a partner or small group. Take turns rolling the number cube. Each player gets two rolls per turn. The first roll shows how many pieces Cooper ate for breakfast.

The second roll shows how many pieces he ate for lunch. Show each roll as a fraction of the pizza. Find the sum of the fractions and compare it to 1. A whole pizza has 8 slices.

Does Cooper have to run? If he does, you earn a point. The first player to earn 5 points wins the game.

FRACTION RIDDLES

PURPOSE To practice finding equivalent fractions (pages 96–99)

See if you can identify the numbers from the information that is given.

1. I am a fraction in simplest form. If you multiply my numerator and denominator by the first prime number, the result is $\frac{10}{12}$. Who am I? $\frac{5}{6}$

2. I am a fraction greater than 1. If you divide my numerator and denominator by the first composite number, the result is $\frac{7}{3}$. Who am I? $\frac{28}{12}$

$\frac{1}{2}$ $\frac{3}{4}$ $\frac{1}{4}$

Roll It Out

Portfolio

PURPOSE To practice adding and subtracting fractions (pages 100–103)
YOU WILL NEED two number cubes

$\frac{3}{4}$ $\frac{5}{6}$ $\frac{2}{3}$ $\frac{7}{10}$ $\frac{1}{2}$ $\frac{1}{6}$ $\frac{0}{0}$ $\frac{1}{1}$ $\frac{1}{4}$ $\frac{1}{3}$

Work with a partner or small group. Each player should copy the fractions onto a sheet of paper.

Take turns rolling the number cubes. Use the numbers to make a fraction less than 1. Roll again and make another fraction. Add or subtract. Can you make one of the

fractions shown? If so, cross if off and roll again. If you can't, it is the next player's turn. The first player to cross out all the fractions wins.

HOME NOTE Make a new set of fractions and play this game with your family.

CHAPTER 5 REVIEW

EXAMPLES

• Add and subtract like fractions. (pages 94–95)

$\frac{5}{6} - \frac{1}{6} = \frac{4}{6}$ *Subtract the numerators.*
Write the difference over the denominator.

$= \frac{2}{3}$ *Write in simplest form.*

• Add unlike fractions. (pages 98–101)

$\frac{1}{3} = \frac{4}{12}$ *Write equivalent fractions using the LCD.*
$+\frac{1}{4} = \frac{3}{12}$
$\overline{\quad\quad \frac{7}{12}}$ *Add the numerators.*
Write the sum over the denominator.

10. $\frac{17}{12}$, or $1\frac{5}{12}$ 11. $\frac{3}{2}$, or $1\frac{1}{2}$ 12. $\frac{7}{6}$, or $1\frac{1}{6}$

• Subtract unlike fractions. (pages 98–99, 102–103)

$\frac{5}{6} = \frac{20}{24}$ *Write equivalent fractions using the LCD.*
$-\frac{3}{8} = \frac{9}{24}$
$\overline{\quad\quad \frac{11}{24}}$ *Subtract the numerators.*
Write the difference over the denominator.

• Estimate the sum or difference. (pages 104–107)

$\frac{6}{7} \rightarrow 1$ *Round each fraction.*
$-\frac{4}{9} \rightarrow \frac{1}{2}$
$\overline{\quad\quad \frac{1}{2}}$ *Subtract.*

25. 1 or $\frac{1}{2}$

27. about $1\frac{1}{2}$ dollars

EXERCISES

Add or subtract. Write in simplest form.

1. $\frac{7}{8} + \frac{3}{8}$ **$\frac{5}{4}$, or $1\frac{1}{4}$**   2. $\frac{8}{9} - \frac{2}{9}$ **$\frac{2}{3}$**

3. $\frac{13}{14} - \frac{3}{14}$ **$\frac{5}{7}$**   4. $\frac{6}{7} + \frac{5}{7}$ **$\frac{11}{7}$, or $1\frac{4}{7}$**

5. A blueberry-pecan muffin recipe requires $\frac{3}{4}$ c of whole-wheat flour. If Doug has $\frac{1}{4}$ c, how much more whole-wheat flour does he need? **$\frac{1}{2}$ c**

6. **VOCABULARY** The ? is the LCM of the denominators. **least common denominator**

Add. Write in simplest form.

7. $\frac{1}{2} + \frac{1}{3}$ **$\frac{5}{6}$**   8. $\frac{3}{4} + \frac{1}{6}$ **$\frac{11}{12}$**   9. $\frac{2}{5} + \frac{2}{4}$ **$\frac{9}{10}$**

10. $\frac{2}{3} + \frac{3}{4}$ 11. $\frac{5}{6} + \frac{2}{3}$ 12. $\frac{1}{3} + \frac{5}{6}$

13. Sara has $\frac{5}{6}$ yd of ribbon and Mark has $\frac{3}{4}$ yd of ribbon. How much ribbon do Sara and Mark have altogether? **$\frac{19}{12}$ yd, or $1\frac{7}{12}$ yd**

Subtract. Write in simplest form.

14. $\frac{3}{4} - \frac{1}{3}$ **$\frac{5}{12}$**   15. $\frac{7}{8} - \frac{1}{4}$ **$\frac{5}{8}$**   16. $\frac{5}{6} - \frac{2}{9}$ **$\frac{11}{18}$**

17. $\frac{8}{9} - \frac{1}{6}$ **$\frac{13}{18}$**   18. $\frac{7}{8} - \frac{5}{6}$ **$\frac{1}{24}$**   19. $\frac{3}{4} - \frac{5}{12}$ **$\frac{1}{3}$**

20. Eric practiced $\frac{3}{4}$ hr on Monday and $\frac{2}{5}$ hr on Saturday. How much longer did Eric practice on Monday? **$\frac{7}{20}$ hr**

Estimate the sum or difference.

21. $\frac{7}{12} + \frac{1}{4}$ **1**   22. $\frac{3}{4} - \frac{1}{3}$ **1**   23. $\frac{2}{9} - \frac{1}{7}$ **0**

24. $\frac{4}{5} + \frac{3}{8}$ **$1\frac{1}{2}$**   25. $\frac{4}{5} - \frac{1}{4}$ 26. $\frac{2}{5} + \frac{2}{11}$ **$\frac{1}{2}$**

27. Nick has $\frac{4}{10}$ of a dollar and Narong has $\frac{4}{5}$ of a dollar. About how much money do the two boys have altogether?

USING THE PAGE

The Chapter 5 Review reviews adding and subtracting like and unlike fractions and estimating sums or differences. Chapter objectives are provided, along with examples and practice exercises for each.

 The Chapter 5 Review can be used as a review or a test. You may wish to place the Review in students' portfolios.

Assessment Checkpoint

For Performance Assessment in this chapter, see page 146, What Did I Learn?, items 2 and 6.

✔ Assessment Tip

Student Self-Assessment The How Did I Do? Survey helps students assess both what they have learned and how they learned it. Self-assessment helps students learn more about their own capabilities and develop confidence in themselves.

A self-assessment survey is available as a copying master in the *Teacher's Guide for Assessment,* p. 15.

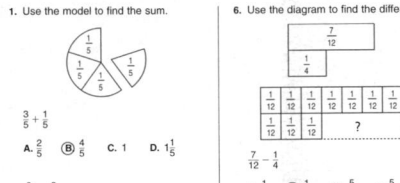

CHAPTER 5 TEST, page 1

Name _____

Choose the letter of the correct answer.

1. Use the model to find the sum.

$\frac{3}{5} + \frac{1}{5}$

A. $\frac{2}{5}$ **B.** $\frac{4}{5}$ C. 1   D. $1\frac{1}{5}$

2. $\frac{9}{10} - \frac{3}{10}$

A. $\frac{3}{5}$ B. $\frac{5}{6}$ C. $\frac{8}{7}$ D. $1\frac{1}{5}$

3. $\frac{5}{9} + \frac{5}{9}$

A. $\frac{1}{3}$ B. $\frac{4}{7}$ C. $\frac{1}{2}$ **D.** $\frac{2}{3}$

4. A recipe calls for $\frac{1}{4}$ c of brown sugar and $\frac{1}{4}$ c of white sugar. How much sugar is needed in the recipe?

A. $\frac{1}{4}$ c   B. $\frac{3}{8}$ c   **C.** $\frac{1}{2}$ c   D. $\frac{3}{4}$ c

5. Use the diagram to find the sum.

$\frac{1}{3} + \frac{1}{6}$

A. $\frac{1}{6}$ **B.** $\frac{1}{2}$ C. $\frac{2}{3}$ D. $\frac{3}{4}$

6. Use the diagram to find the difference.

$\frac{7}{12} - \frac{1}{4}$

A. $\frac{1}{4}$ **B.** $\frac{1}{3}$ C. $\frac{5}{12}$ D. $\frac{5}{6}$

For questions 7–8, find the sum or difference.

7. $\frac{3}{10} + \frac{1}{2}$

A. $\frac{1}{3}$ B. $\frac{1}{2}$ C. $\frac{3}{5}$ **D.** $\frac{4}{5}$

8. $\frac{3}{4} - \frac{1}{8}$

A. $\frac{1}{4}$ B. $\frac{1}{2}$ **C.** $\frac{5}{8}$ D. $\frac{2}{3}$

9. What is the LCM of $\frac{2}{3}$ and $\frac{3}{8}$?

A. 11   B. 12   C. 18   **D.** 24

10. What LCM would you use to write this problem with equivalent fractions?

$\frac{5}{9} - \frac{1}{6}$

A. 12   B. 15   **C.** 18   D. 30

11. $\frac{2}{3} + \frac{1}{3}$

A. $\frac{3}{8}$ B. $\frac{11}{15}$ C. $\frac{3}{4}$ D. $\frac{11}{12}$

Form A • Multiple-Choice   A19   Go on. ▶

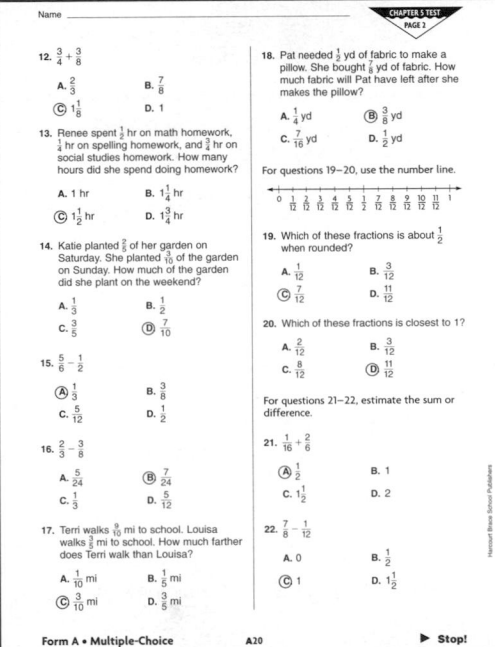

CHAPTER 5 TEST, page 2

Name _____

12. $\frac{3}{4} + \frac{3}{8}$

A. $\frac{2}{3}$ B. $\frac{7}{8}$ **C.** $1\frac{1}{8}$ D. 1

13. Renee spent $\frac{1}{3}$ hr on math homework, $\frac{1}{2}$ hr on spelling homework, and $\frac{3}{4}$ hr on social studies homework. How many hours did she spend doing homework?

A. 1 hr   B. $1\frac{1}{4}$ hr   **C.** $1\frac{1}{2}$ hr   D. $1\frac{3}{4}$ hr

14. Katie planted $\frac{2}{8}$ of her garden on Saturday. She planted $\frac{1}{5}$ of the garden on Sunday. How much of the garden did she plant on the weekend?

A. $\frac{1}{3}$ B. $\frac{1}{2}$ C. $\frac{3}{5}$ **D.** $\frac{7}{10}$

15. $\frac{5}{6} - \frac{1}{2}$

A. $\frac{1}{3}$ B. $\frac{3}{8}$ C. $\frac{5}{12}$ D. $\frac{1}{2}$

16. $\frac{2}{3} - \frac{3}{8}$

A. $\frac{5}{24}$ **B.** $\frac{7}{24}$ C. $\frac{5}{3}$ D.

17. Terri walks $\frac{7}{10}$ mi to school. Louisa walks $\frac{2}{5}$ mi to school. How much farther does Terri walk than Louisa?

A. $\frac{1}{10}$ mi   B. $\frac{3}{10}$ mi   **C.** $\frac{3}{10}$ mi   D. $\frac{7}{8}$ mi

18. Pat needed $\frac{1}{3}$ yd of fabric to make a pillow. She bought $\frac{5}{8}$ yd of fabric. How much fabric will Pat have left after she makes the pillow?

A. $\frac{1}{4}$ yd   **B.** $\frac{3}{8}$ yd   C. $\frac{7}{16}$ yd   D. $\frac{1}{2}$ yd

For questions 19–20, use the number line.

19. Which of these fractions is about $\frac{1}{2}$ when rounded?

A. $\frac{1}{12}$ B. $\frac{3}{12}$ **C.** $\frac{7}{12}$ D. $\frac{11}{12}$

20. Which of these fractions is closest to 1?

A. $\frac{2}{12}$ B. $\frac{3}{12}$ C. $\frac{8}{12}$ **D.** $\frac{11}{12}$

For questions 21–22, estimate the sum or difference.

21. $\frac{1}{16} + \frac{2}{6}$

A. $\frac{1}{2}$ B. 1   C. $1\frac{1}{2}$ D. 2

22. $\frac{7}{8} - \frac{1}{2}$

A. 0   B. $\frac{1}{2}$ **C.** 1   D. $1\frac{1}{2}$

Form A • Multiple-Choice   A20   ▶ Stop!

See *Test Copying Masters,* pp. A19–A20 for the Chapter Test in **multiple-choice** format and pp. B169–B170 for **free-response** format.

PLANNING GUIDE

Adding and Subtracting Mixed Numbers

BIG IDEA: Addition and subtraction of mixed numbers can be used to solve many real-world problems.

Introducing the Chapter p. 110

Team-Up Project • Precise Measurements p. 111

OBJECTIVE	VOCABULARY	MATERIALS	RESOURCES
6.1 Adding Mixed Numbers pp. 112-113 **Objective:** To use visual models to add mixed numbers **1 DAY**	*mixed number*		■ Reteach, ■ Practice, ■ Enrichment 6.1; More Practice, p. H50 Cultural Connection • Soccer and Cricket, p. 124 ⊙ *Calculating Crew • Nautical Number Line*
INTERVENTION • Comparing and Ordering Fractions, Teacher's Edition pp. 114A–114B			
Lab Activity: Subtracting Mixed Numbers pp. 114-115 **Objective:** To use fraction bars to subtract mixed numbers		FOR EACH PAIR: fraction bars	🖳 **E-Lab • Activity 6** E-Lab Recording Sheets, p. 6
6.2 Subtracting Mixed Numbers pp. 116-117 **Objective:** To subtract mixed numbers **2 DAYS**			■ Reteach, ■ Practice, ■ Enrichment 6.2; More Practice, p. H50 Cultural Connection • Soccer and Cricket, p. 124
6.3 Adding and Subtracting Mixed Numbers pp. 118-121 **Objective:** To add and subtract mixed numbers **2 DAYS**			■ Reteach, ■ Practice, ■ Enrichment 6.3; More Practice, p. H50 Cultural Connection • Soccer and Cricket, p. 124 🖩 Calculator Activities, p. H34
6.4 Estimating Sums and Differences pp. 122-123 **Objective:** To estimate sums and differences of mixed numbers **1 DAY**			■ Reteach, ■ Practice, ■ Enrichment 6.4; More Practice, p. H51
CHAPTER ASSESSMENT Chapter 6 Review p. 125			

NCTM CURRICULUM
STANDARDS FOR GRADES 5–8

- ✔ **Problem Solving** *Lessons 6.1, 6.2, 6.3, 6.4*
- ✔ **Communication** *Lessons 6.1, 6.2, 6.3, 6.4*
- ✔ **Reasoning** *Lessons 6.1, 6.2, 6.3, 6.4*
- ✔ **Connections** *Lessons 6.1, 6.2, 6.3, 6.4*
- ✔ **Number and Number Relationships** *Lessons 6.1, 6.2, 6.3, 6.4*
- ✔ **Number Systems and Number Theory** *Lessons 6.1, Lab Activity, 6.2*
- ✔ **Computation and Estimation** *Lessons 6.1, Lab Activity, 6.2*
- ☐ **Patterns and Functions**
- ☐ **Algebra**
- ☐ **Statistics**
- ☐ **Probability**
- ☐ **Geometry**
- ✔ **Measurement** *Lessons 6.1, 6.2*

CHAPTER LEARNING GOALS

6A To solve problems that involve adding and subtracting mixed numbers

6A.1 To add and subtract mixed numbers

6A.2 To estimate sums and differences of mixed numbers

These goals for concepts, skills, and problem solving are assessed in many ways throughout the chapter. See the chart below for a complete listing of both traditional and informal assessment options.

Pretest Options
Pretest for Chapter Content, TCM pp. A21–A22, B171–B172
Assessing Prior Knowledge, TE p. 110; APK p. 8

Daily Assessment
Mixed Review, PE pp. 121, 123
Problem of the Day, TE pp. 112, 116, 118, 122
Assessment Tip and Daily Quiz, TE pp. 113, 117, 121, 123

Formal Assessment
Chapter 6 Review, PE p. 125
Chapter 6 Test, TCM pp. A21–A22, B171–B172
Study Guide & Review, PE pp. 144–145
Test • Chapters 4-7, TCM pp. A25–A26, B175–B176
Cumulative Review, PE p. 147
Cumulative Test, TCM pp. A27–A30, B177–B180

Performance Assessment
Team-Up Project Checklist, TE p. 111
What Did I Learn?, Chapters 4–7, TE p. 146
Interview/Task Test, PA p. 81

Portfolio
Suggested work samples:
Independent Practice, PE pp. 113, 117, 120, 123
Cultural Connection, PE p. 124

Student Self-Assessment
How Did I Do?, TGA p. 15
Portfolio Evaluation Form, TGA p. 24
Math Journal, TE pp. 110B, 112, 114

Key
Assessing Prior Knowledge Copying Masters: APK
Test Copying Masters: TCM
Performance Assessment: PA
Teacher's Guide for Assessment: TGA
Teacher's Edition: TE
Pupil's Edition: PE

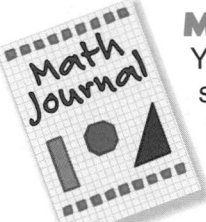

MATH JOURNAL
You may wish to use the following suggestions to have students write about the mathematics they are learning in this chapter:

PE page 119 • Talk About It

PE pages 113, 117, 121, 123 • Write About It

TE pages 112, 114 • Writing in Mathematics

TGA page 15 • How Did I Do?

In every lesson • record written answers to each Wrap-up Assessment Tip from the Teacher's Edition for test preparation.

VOCABULARY DEVELOPMENT
mixed number, pp. 112, 114, 116, 118, 119, 120, 121, 122

There are no new words introduced in this chapter but mathematics vocabulary is reinforced visually and verbally. Encourage students to review the mathematics vocabulary in their journals.

MEETING INDIVIDUAL NEEDS

RETEACH
BELOW-LEVEL STUDENTS

Practice Activities, TE Tab B
Spinning For a Divisor
(Use with Lessons 6.2–6.3.)

Interdisciplinary Connection, TE p. 121

Literature Link, TE p. 110C
Too Soon a Woman, Activity 1
(Use with Lessons 6.1–6.4.)

Technology
Calculating Crew • *Nautical Number Line*
(Use with Lesson 6.1.)

E-Lab • Activity 6
(Use with Lab Activity.)

PRACTICE
ON-LEVEL STUDENTS

Practice Game, TE p. 110C
The 100-Mile Race
(Use with Lessons 6.1–6.3.)

Interdisciplinary Connection, TE p. 121

Literature Link, TE p. 110C
Too Soon a Woman, Activity 2
(Use with Lessons 6.1–6.4.)

Technology
Calculating Crew • *Nautical Number Line*
(Use with Lesson 6.1.)

E-Lab • Activity 6
(Use with Lab Activity.)

CHAPTER IDEA FILE

The activities on these pages provide additional practice and reinforcement of key chapter concepts. The chart above offers suggestions for their use.

LITERATURE LINK

Too Soon a Woman

by Dorothy M. Johnson

This story tells of a family's wagon trip through the plains in search of a better life.

Activity 1—Pa leaves to go in search of food. He says he will be back in 4 days, but he does not return for $6\frac{1}{2}$ days. Have students write and solve an equation to tell how much longer Pa was gone than he expected.

Activity 2—Mary finds a 10-lb mushroom in the woods and decides to eat part of it. Have students calculate the fraction of the mushroom that is left if Mary eats $\frac{1}{10}$ of it. How might the mushroom be divided among the other 4 people?

Activity 3—Students should determine what fraction of a pound of butter it would take to make the cookies. Have students form groups and add up the butter that would be needed to make everyone's cookies. Finally, they should determine what fraction of a pound would be used.

PRACTICE GAME

The 100-Mile Race

Purpose To add and subtract mixed numbers

Materials number cube (TR, p. 71), game board (TR, p. 85)

About the Game In small groups, players take turns rolling the number cube and moving around the game board. For each space they land on, players add to or subtract their score from the number rolled, depending on whether there is a plus sign or minus sign in the space. If a player's score goes below 0, he or she starts the next turn with 0 points. The first player to score 100 points wins.

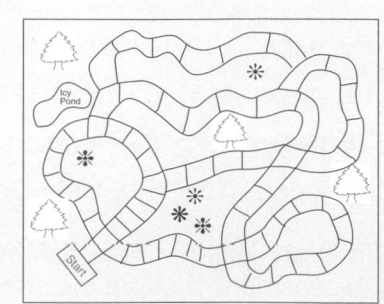

Modalities
Kinesthetic, Visual

Multiple Intelligences
Logical/Mathematical,
Bodily/Kinesthetic,
Interpersonal/Social

Practice Game, TE p. 110C
The 100-Mile Race
(Use with Lessons 6.1–6.3.)

Interdisciplinary Connection, TE p. 121

 **Literature Link,** TE p. 110C
Too Soon a Woman, Activity 3
(Use with Lessons 6.1–6.4.)

 Technology
Calculating Crew • Nautical Number Line
(Use with Lesson 6.1.)

E-Lab • Activity 6
(Use with Lab Activity.)

INTERDISCIPLINARY SUGGESTIONS

Connecting Mixed Numbers

Purpose To connect using mixed numbers to other subjects students are studying

You may wish to connect *Adding and Subtracting Mixed Numbers* to other subjects by using related interdisciplinary units.

Materials assortment of measuring devices and tools

Suggested Units

- Metric and Customary: Measuring with Decimals and Fractions
- Cooking with Math
- Fractions in the Stock Market
- Crafts and Fractions

Modalities Auditory, Kinesthetic, Visual

Multiple Intelligences Visual/Spatial, Logical/Mathematical, Bodily/Kinesthetic

Technology

The purpose of using technology in the chapter is to provide reinforcement of skills and support in concept development.

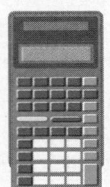 **E-Lab • Activity 6**—Students use visual thinking, reasoning, and a fraction model to subtract mixed numbers.

Calculating Crew • Nautical Number Line—The activities on this CD-ROM provide students with opportunities to practice adding mixed numbers.

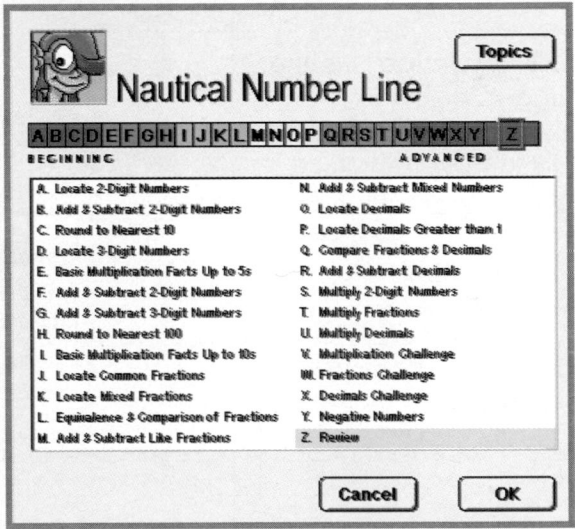

Each Calculating Crew activity can be assigned by degree of difficulty. See Teacher's Edition page 113 for Grow Slide information.

Explorer Plus™ or fx–65 Calculator—Ideas for integrating the calculator in the PE Lesson p. 118 are provided in the *Calculator Handbook,* p. H34.

Visit our web site for additional ideas and activities.
http://www.hbschool.com

Introducing the Chapter

Encourage students to talk about adding and subtracting mixed numbers. Review the math content students will learn in the chapter.

WHY LEARN THIS?

The question "Why are you learning this?" will be answered in every lesson. Sometimes there is a direct application to everyday experiences. Other lessons help students understand why these skills are needed for future success in mathematics.

PRETEST OPTIONS

Pretest for Chapter Content See the *Test Copying Masters*, pages A21–A22, B171–B172 for a free-response or multiple-choice test that can be used as a pretest.

Assessing Prior Knowledge Use the copying master shown below to assess previously taught skills that are critical for success in this chapter.

Assessing Prior Knowledge Copying Masters, p. 8

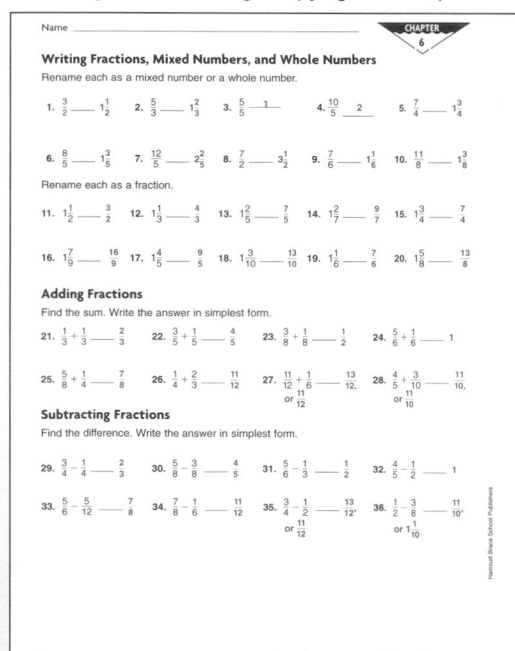

Troubleshooting

If students need help	Then use
• WRITING MIXED NUMBERS AS FRACTIONS	• Practice Activities, TE Tab A, Activity 18A
• ADDING FRACTIONS	• Key Skills, PE p. H17
• SUBTRACTING FRACTIONS	• Practice Activities, TE Tab B, Activity 20A

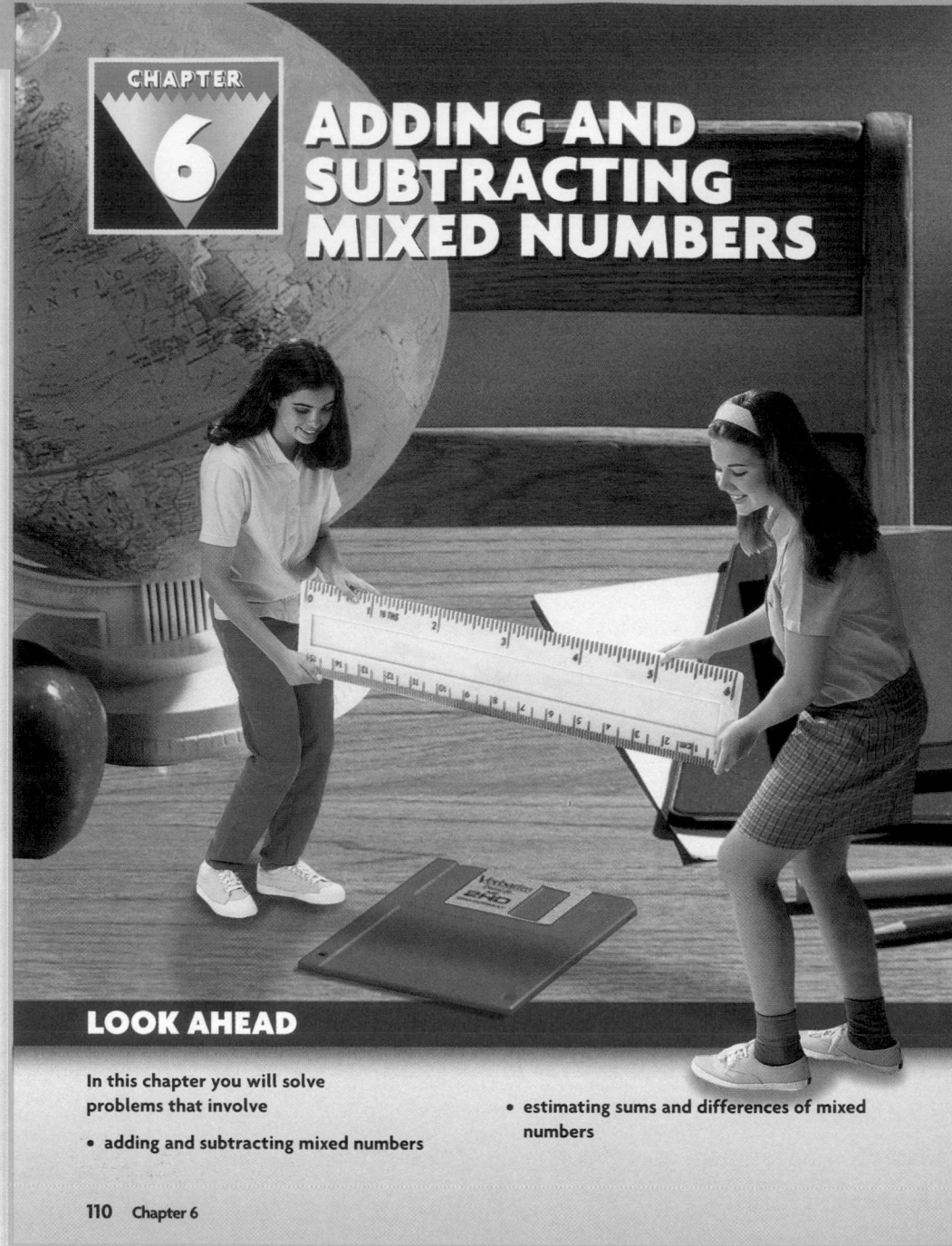

CHAPTER 6 ADDING AND SUBTRACTING MIXED NUMBERS

LOOK AHEAD

In this chapter you will solve problems that involve

• adding and subtracting mixed numbers

• estimating sums and differences of mixed numbers

110 Chapter 6

SCHOOL-HOME CONNECTION

You may wish to send to each student's family the *Math at Home* page. It includes an article that stresses the importance of communication about mathematics in the home, as well as helpful hints about homework and taking tests. The activity and extension provide extra practice that develops concepts and skills involving mixed numbers.

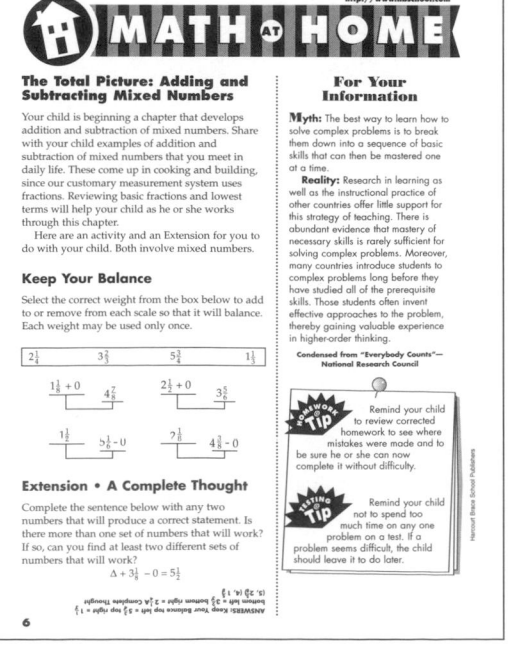

Math at Home

Team-Up Project

Precise Measurements

Have you ever needed a precise measurement? Find 6 objects in the classroom and compare them to the length and width of a pencil box that is $8\frac{1}{2}$ in. by $1\frac{3}{4}$ in. Make a chart to show the comparisons.

YOU WILL NEED: ruler

Plan
- Work with a small group. Find and measure six objects. Compare the length and width of each object to the length and width of the pencil box.

Act
- Find six objects in the classroom to measure.
- Measure the length and width of each object to the nearest $\frac{1}{8}$ in. Record the data.
- Find the difference between the length and width of each object and the length and width of the pencil box.
- Make a table to show the comparisons.

Share
- Compare your chart with other groups. Tell how you found the differences.

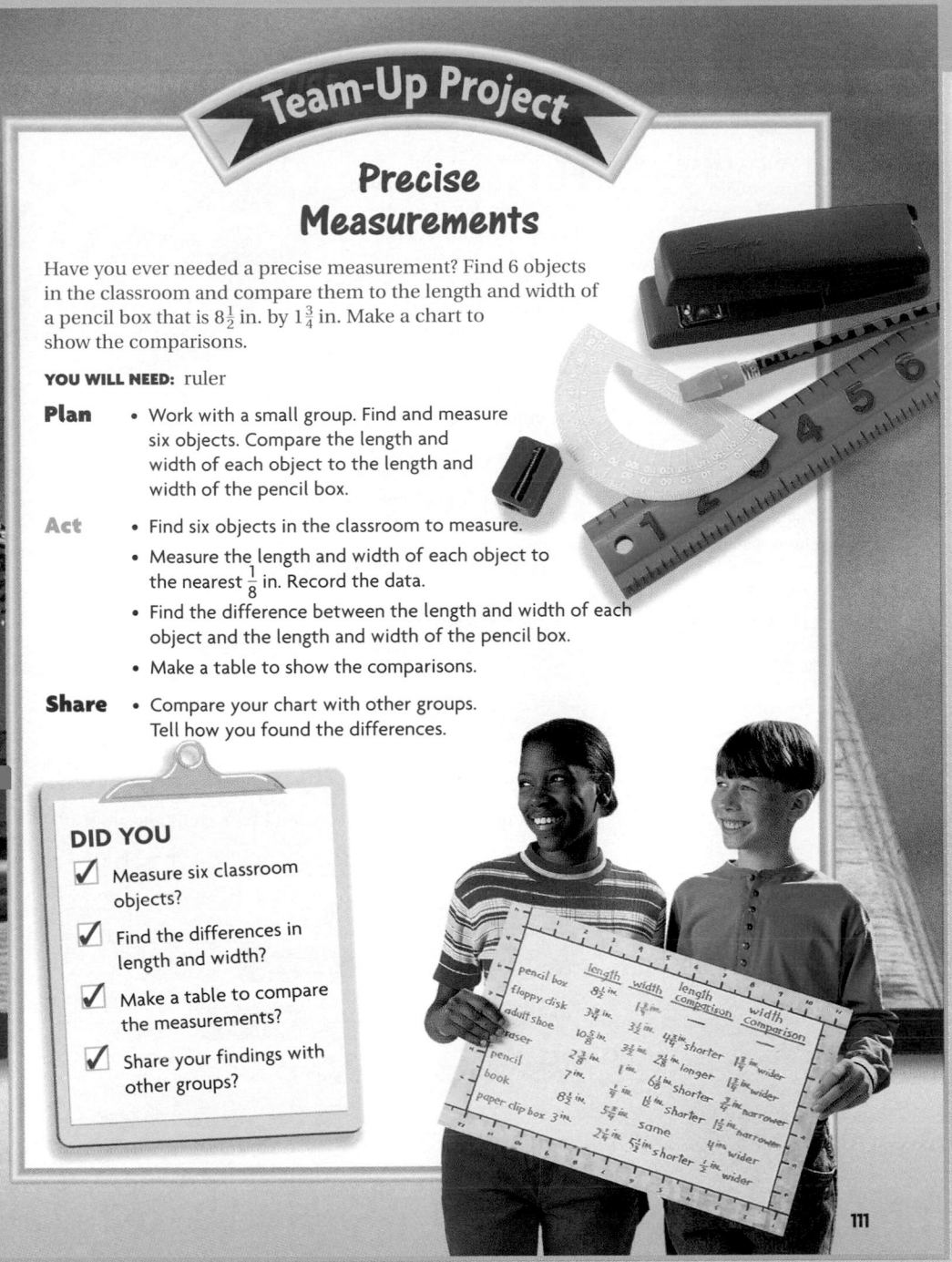

DID YOU
- ✓ Measure six classroom objects?
- ✓ Find the differences in length and width?
- ✓ Make a table to compare the measurements?
- ✓ Share your findings with other groups?

111

Project Checklist

Portfolio

Evaluate whether each group
- ✓ works cooperatively to find and measure lengths and widths of objects.
- ✓ adds and subtracts mixed numbers to compare objects.
- ✓ completes a chart showing comparisons in sizes.

Using the Project

Accessing Prior Knowledge fractions of units

Purpose to add and subtract mixed numbers

GROUPING small groups

MATERIALS rulers, yardsticks, various classroom objects to measure

Managing the Activity If students have not added and subtracted mixed numbers before, have them line up a paper template measuring $8\frac{1}{2}$ in. $\times$ $1\frac{3}{4}$ in. with a ruler and mark the ruler. They can count on or count back in appropriate units as they compare 6 objects with the ruler marks.

WHEN MINUTES COUNT

A shortened version of this activity can be completed in 15–20 minutes. Divide the class into three groups and give each group one object to measure and compare in length and width with the pencil box.

A BIT EACH DAY

Day 1: Students form groups and find 6 objects in the classroom to measure.

Day 2: Students measure the objects and record them in a chart.

Day 3: Students add or subtract mixed numbers to find the differences between the objects and the pencil box.

Day 4: Students continue to measure for comparisons and complete their charts.

Day 5: Group members compare their charts with other groups and find how many objects were measured.

ON THE ROAD

Hardware stores, science stores, and sports stores are good places to find various tools for measuring length. Students may want to visit these or other places that have measuring tools.

111

GRADE 5
Use visual models to add mixed numbers.
GRADE 6
Use visual models to add mixed numbers.
GRADE 7
Add and subtract mixed numbers.

ORGANIZER

OBJECTIVE To use visual models to add mixed numbers

VOCABULARY *Review* mixed number

ASSIGNMENTS Independent Practice
- ■ *Basic* 1–6, odd; 7–10
- ■ *Average* 1–10
- ■ *Advanced* 3–10

PROGRAM RESOURCES

Intervention • Comparing and Ordering Fractions, TE pp. 114A–114B

Reteach 6.1 Practice 6.1 Enrichment 6.1

Calculating Crew • Nautical Number Line

Cultural Connection • Soccer and Cricket, p. 124

PROBLEM OF THE DAY

Transparency 6

Melissa rides the bus $1\frac{2}{3}$ mi north and $3\frac{1}{4}$ mi east to get to school. Brandon rides his bike $2\frac{3}{4}$ mi south and $2\frac{1}{6}$ mi west to get to the same school. Who travels a longer distance to school? They travel the same distance, $4\frac{11}{12}$ mi.

Solution Tab A, p. A6

QUICK CHECK
Find the LCD.

1. $\frac{3}{4}, \frac{7}{12}$ 12
2. $\frac{2}{3}, \frac{1}{9}$ 9
3. $\frac{3}{5}, \frac{1}{4}$ 20
4. $\frac{1}{6}, \frac{1}{5}$ 30
5. $\frac{5}{12}, \frac{1}{8}$ 24
6. $\frac{2}{9}, \frac{5}{6}$ 18

1 MOTIVATE

To interest students in the real-life applications of mixed numbers, read and discuss the Career Link.

- **How might costume designers use mixed numbers?** Possible answer: to measure fabric

- **How might stage designers use mixed numbers?** Possible answer: to measure wood for constructing sets

Discuss with students how mixed numbers may have been used in making some of their favorite movies.

Adding Mixed Numbers

WHAT YOU'LL LEARN
How to use visual models to add mixed numbers

WHY LEARN THIS?
To solve problems about measurement involving addition of mixed numbers

REMEMBER:
A mixed number is a whole number and a fraction combined.
See page H16.
$4\frac{2}{5}$ is a mixed number.

Have you ever assembled a model airplane? A series of diagrams showing the steps makes the assembly easier. Diagrams can also help you understand the steps for adding mixed numbers.

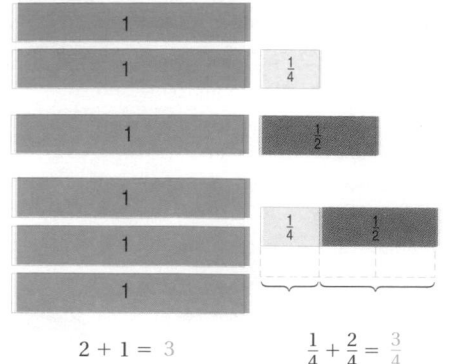

Mark uses $2\frac{1}{4}$ ft of lumber to build a chair and $1\frac{1}{2}$ ft of lumber to build a stool. How much lumber did he use?

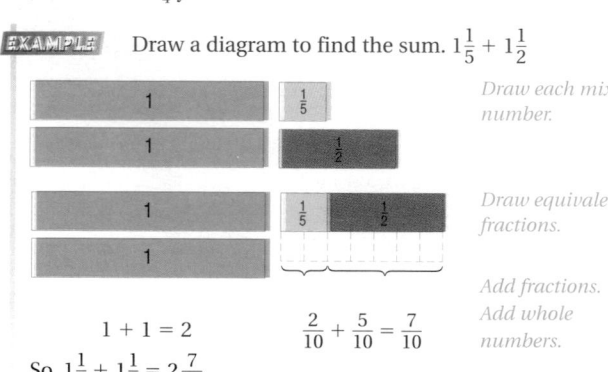

Complete the diagram to find $2\frac{1}{4} + 1\frac{1}{2}$.

Combine whole numbers. Combine fractions. Draw equivalent fractions with the LCD, fourths.

$2 + 1 = 3$ $\frac{1}{4} + \frac{2}{4} = \frac{3}{4}$

Add fractions. Add whole numbers.

So, Mark used $3\frac{3}{4}$ yd of lumber to build the chair and stool.

EXAMPLE Draw a diagram to find the sum. $1\frac{1}{5} + 1\frac{1}{2}$

Draw each mixed number.

Draw equivalent fractions.

Add fractions. Add whole numbers.

$1 + 1 = 2$ $\frac{2}{10} + \frac{5}{10} = \frac{7}{10}$

So, $1\frac{1}{5} + 1\frac{1}{2} = 2\frac{7}{10}$.

IDEA FILE

ENGLISH AS A SECOND LANGUAGE

Tip • At the end of each lesson in this chapter, have students make a list of math words they have used and then have them vote to choose 3–5 that they think are the most important. Have them nominate the hardest words or concepts and explain why they are difficult. You may wish to have them enter the words in their math journals.

Modality Auditory

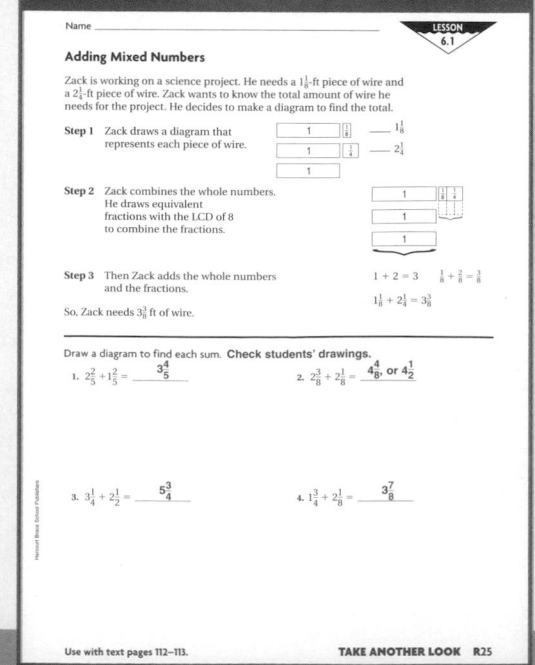

RETEACH 6.1

Name _____

Adding Mixed Numbers

Zack is working on a science project. He needs a $1\frac{1}{8}$-ft piece of wire and a $2\frac{1}{4}$-ft piece of wire. Zack wants to know the total amount of wire he needs for the project. He decides to make a diagram to find the total.

Step 1 Zack draws a diagram that represents each piece of wire.

Step 2 Zack combines the whole numbers. He draws equivalent fractions with the LCD of 8 to combine the fractions.

Step 3 Then Zack adds the whole numbers and the fractions.

$1 + 2 = 3$ $\frac{1}{8} + \frac{2}{8} = \frac{3}{8}$

$1\frac{1}{8} + 2\frac{1}{4} = 3\frac{3}{8}$

So, Zack needs $3\frac{3}{8}$ ft of wire.

Draw a diagram to find each sum. **Check students' drawings.**

1. $2\frac{2}{5} + 1\frac{2}{5}$ = $3\frac{4}{5}$
2. $2\frac{3}{8} + 2\frac{1}{8}$ = $4\frac{4}{8}$, or $4\frac{1}{2}$
3. $3\frac{1}{4} + 2\frac{1}{2}$ = $5\frac{3}{4}$
4. $1\frac{3}{8} + 2\frac{1}{2}$ = $3\frac{7}{8}$

Use with text pages 112–113.   TAKE ANOTHER LOOK R25

GUIDED PRACTICE

Write the addition problem shown by the diagram. Then find the sum.

1.
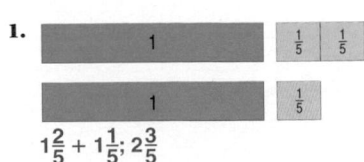

$1\frac{2}{5} + 1\frac{1}{5}; 2\frac{3}{5}$

2. $2\frac{1}{2} + 1\frac{1}{4}; 3\frac{3}{4}$
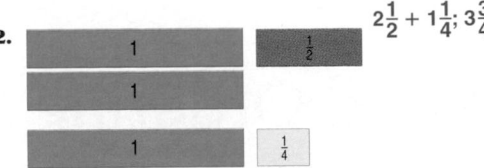

Draw a diagram to find each sum. Write the answer in simplest form. **Check students' diagrams.**

3. $2\frac{3}{5} + 1\frac{1}{5}$ $3\frac{4}{5}$

4. $1\frac{1}{3} + 1\frac{1}{3}$ $2\frac{2}{3}$

5. $1\frac{1}{12} + 2\frac{1}{4}$ $3\frac{4}{12}$, or $3\frac{1}{3}$

6. $1\frac{1}{2} + 1\frac{1}{6}$ $2\frac{4}{6}$, or $2\frac{2}{3}$

INDEPENDENT PRACTICE

Write the addition problem shown by the diagram. Then find the sum. Write the answer in simplest form.

1.

$1\frac{1}{5} + 1\frac{3}{5} = 2\frac{4}{5}$

2.
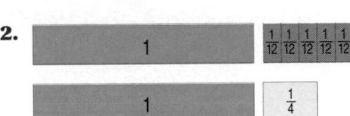

$1\frac{5}{12} + 1\frac{1}{4} = 2\frac{8}{12}$, or $2\frac{2}{3}$

Draw a diagram to find each sum. Write the answer in simplest form. **Check students' diagrams.**

3. $1\frac{1}{8} + 1\frac{5}{8}$ $2\frac{6}{8}$, or $2\frac{3}{4}$

4. $2\frac{1}{8} + 2\frac{1}{4}$ $4\frac{3}{8}$

5. $2\frac{1}{8} + 1\frac{3}{8}$ $3\frac{4}{8}$, or $3\frac{1}{2}$

6. $1\frac{1}{3} + 1\frac{1}{6}$ $2\frac{3}{6}$, or $2\frac{1}{2}$

Problem-Solving Applications

For Problems 7–9, use a diagram to solve.

7. Hank helped Madison paint scenery for the school play. He painted for $1\frac{1}{6}$ hr on Monday and $2\frac{1}{3}$ hr on Saturday. How many hours did he spend painting? **$3\frac{1}{2}$ hr**

8. The students rehearsed for the school play for $2\frac{1}{6}$ hr on Tuesday and $3\frac{1}{4}$ hr on Saturday. How many hours did the students rehearse? **$5\frac{5}{12}$ hr**

9. Joey used $2\frac{1}{8}$ yd of fabric for his costume, and Tami used $4\frac{1}{2}$ yd of fabric for her costume. How many yards of fabric did Joey and Tami use? **$6\frac{5}{8}$ yd**

10. **WRITE ABOUT IT** How is adding mixed numbers different from adding fractions? **Possible answer: With mixed numbers you have to add whole numbers.**

Portfolio

MORE PRACTICE Lesson 6.1, page H50

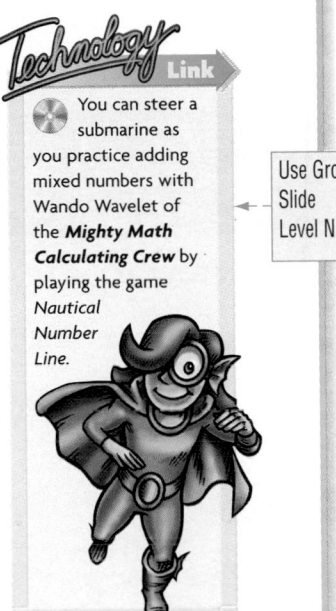

Technology **Link**

You can steer a submarine as you practice adding mixed numbers with Wando Wavelet of the *Mighty Math Calculating Crew* by playing the game *Nautical Number Line.*

Use Grow Slide Level N.

113

The models students draw for adding mixed numbers may look like the fraction bars shown in the book, or they may be fraction circles or squares.

GUIDED PRACTICE
Complete these exercises orally with the whole class. Make sure students understand how to use visual models to add mixed numbers.

INDEPENDENT PRACTICE
To make sure students understand how to use visual models to add mixed numbers, check students' answers for Exercises 1–9.

DEVELOPING ALGEBRAIC THINKING
Using variables is an important concept in algebra. To show students that they already use variables, ask them to write an equation for Problem 8. Tell them to let x = the total number of hours the students rehearsed. $x = 2\frac{1}{6} + 3\frac{1}{4}$

ADDITIONAL EXAMPLE
Example, p.112

Draw a diagram to find the sum. $1\frac{1}{4} + 1\frac{1}{6}$

Check students' diagrams; $2\frac{5}{12}$

3 WRAP UP

Use this question to help students focus on the big ideas of the lesson.

• **Was it easier to draw a diagram for Exercise 5 or Exercise 6? Explain.**
Exercise 5; it didn't require drawing equivalent fractions.

✓ *Assessment Tip*
Have students write the answer.

• **A set designer worked $1\frac{1}{4}$ hr in the morning and $2\frac{1}{2}$ hr in the afternoon. How many hours did she work in all? Draw a diagram to solve.** $3\frac{3}{4}$ hr; check students' diagrams.

DAILY QUIZ
Draw a diagram to find the sum. Write the answer in simplest form.

1. $1\frac{1}{6} + 2\frac{5}{6}$ 4

2. $1\frac{1}{8} + 1\frac{3}{8}$ $2\frac{1}{2}$

3. $2\frac{1}{3} + 1\frac{1}{9}$ $3\frac{5}{9}$

4. $1\frac{1}{5} + 1\frac{1}{10}$ $2\frac{3}{10}$

PRACTICE 6.1

Name _____

LESSON 6.1

Adding Mixed Numbers

Write the addition problem shown by the diagram. Then find the sum. Write the answer in simplest form.

1. $1\frac{1}{5} + 1\frac{2}{5} = 2\frac{3}{5}$

2. $2\frac{1}{4} + 1\frac{1}{4} = 3\frac{2}{4}$

3. $2\frac{5}{12} + 1\frac{1}{4} = 3\frac{9}{12}$, or $3\frac{3}{4}$

4. $1\frac{1}{5} + 2\frac{2}{5} = 3\frac{3}{5}$

Draw a diagram to find each sum. Write the answer in simplest form. Check students' diagrams.

5. $1\frac{3}{5} + 1\frac{1}{5}$ $2\frac{4}{5}$

6. $2\frac{2}{8} + 1\frac{1}{4}$ $3\frac{5}{8}$

7. $2\frac{1}{8} + 1\frac{3}{8}$ $3\frac{4}{8}$, or $3\frac{1}{2}$

8. $3\frac{1}{4} + 1\frac{1}{4}$ $4\frac{4}{8}$

9. $2\frac{3}{8} + 1\frac{1}{2}$ $3\frac{7}{8}$

10. $2\frac{5}{8} + 1\frac{1}{6}$ $3\frac{5}{6}$

Mixed Applications

For Problems 11–14, use a diagram to solve.

11. Ryan uses $2\frac{1}{4}$ m of wire for his science project. Sean uses $1\frac{1}{6}$ m of wire for his project. How much wire do Ryan and Sean use? $3\frac{1}{2}$ m

12. The school band practices $2\frac{1}{4}$ hr on Friday and $3\frac{1}{4}$ hr on Saturday. How many hours does the band practice? $5\frac{3}{4}$ hr

13. The Garden Club held a bake sale. The club sold 42 cakes for $9.50 each, 188 cookies for $0.20 each, and 124 cupcakes for $0.45 each. How much did the club raise? $492.40

14. The wagot factory makes 7,400 wagots every hour. At this rate, how many wagots are produced during an 8-hour shift? 59,200 wagots

Use with text pages 112–113. **ON MY OWN P25**

ENRICHMENT 6.1

Name _____

LESSON 6.1

Addition Patterns

Write the next three terms. Then identify the rule.

1. 10; $14\frac{1}{5}$; $18\frac{2}{5}$; $22\frac{3}{5}$; $26\frac{4}{5}$; 31; $35\frac{1}{5}$
Rule: _____ Add $4\frac{1}{5}$ to previous term.

2. 1; $2\frac{1}{4}$; $3\frac{1}{2}$; $4\frac{3}{4}$; 6; $7\frac{1}{4}$; $8\frac{1}{2}$
Rule: _____ Add $1\frac{1}{4}$ to previous term.

3. 5; $7\frac{1}{5}$; $9\frac{2}{5}$; $11\frac{3}{5}$; $13\frac{4}{5}$; 16; $18\frac{1}{5}$
Rule: _____ Add $2\frac{1}{5}$ to previous term.

4. 1; $2\frac{3}{8}$; $3\frac{3}{4}$; $5\frac{1}{8}$; $6\frac{1}{2}$; $7\frac{7}{8}$; $9\frac{1}{4}$
Rule: _____ Add $1\frac{3}{8}$ to previous term.

5. 4; $7\frac{2}{9}$; $10\frac{4}{9}$; $13\frac{6}{9}$; $16\frac{8}{9}$; $20\frac{1}{9}$; $23\frac{3}{9}$
Rule: _____ Add $3\frac{2}{9}$ to previous term.

6. 2; $4\frac{3}{10}$; $6\frac{6}{10}$; $8\frac{9}{10}$; $11\frac{1}{5}$; $13\frac{1}{2}$; $15\frac{4}{5}$
Rule: _____ Add $2\frac{3}{10}$ to previous term.

7. 3; $4\frac{1}{3}$; $5\frac{2}{3}$; 7; $8\frac{1}{3}$; $9\frac{2}{3}$; 11
Rule: _____ Add $1\frac{1}{3}$ to previous term.

8. 3; $3\frac{1}{6}$; $5\frac{1}{3}$; $7\frac{1}{2}$; $9\frac{2}{3}$; $11\frac{5}{6}$; 14
Rule: _____ Add $2\frac{1}{6}$ to previous term.

9. 8; $9\frac{1}{8}$; $10\frac{1}{4}$; $11\frac{3}{8}$; $12\frac{1}{2}$; $13\frac{5}{8}$; $14\frac{3}{4}$
Rule: _____ Add $1\frac{1}{8}$ to previous term.

10. 2; $14\frac{3}{4}$; $17\frac{1}{4}$; $20\frac{1}{4}$; 23; $25\frac{3}{4}$; $28\frac{1}{4}$
Rule: _____ Add $2\frac{3}{4}$ to previous term.

Use with text pages 112–113. **STRETCH YOUR THINKING E25**

113

INTERVENTION

Comparing and Ordering Fractions

Use these strategies to review the skills required for comparing and ordering fractions.

Multiple Intelligences Logical/Mathematical, Visual/Spatial

Modalities Auditory, Kinesthetic, Visual

Lesson Connection Students will compare fractions in Lessons 6.2 and 6.3.

Program Resources *Intervention and Extension Copying Masters*, p. 4

 *Troubleshooting*

WHAT STUDENTS NEED TO KNOW	HOW TO HELP	FOLLOW UP
How to compare fractions with like denominators	Have students compare fractions with like denominators by comparing their numerators. **Examples** $\frac{5}{8} < \frac{7}{8}$ because $5 < 7$. $\frac{7}{12} > \frac{6}{12}$ because $7 > 6$. 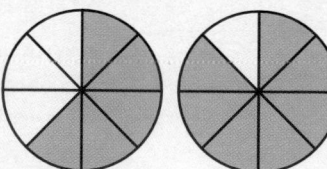 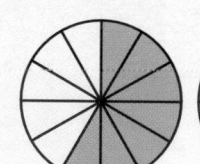	Other problems to solve: Use < or > to complete. $\frac{1}{4}$ $\frac{3}{4}$ (<), $\frac{6}{9}$ $\frac{7}{9}$ (<), $\frac{3}{10}$ $\frac{1}{10}$ (>)
How to write an equivalent fraction with a given denominator	Have students find the missing numerator. **Example** $\frac{5}{8} = \frac{?}{24}$ **Think:** To get 24, you multiply the denominator, 8, by 3. So, multiply the numerator, 5, by 3. $\frac{5}{8} = \frac{5 \times 3}{8 \times 3} = \frac{15}{24}$	Other problems to solve: Complete. $\frac{2}{2} = \frac{?}{12}$ (12), $\frac{1}{5} = \frac{?}{10}$ (2) $\frac{3}{4} = \frac{?}{16}$ (12)

WHAT STUDENTS NEED TO KNOW	HOW TO HELP	FOLLOW UP
How to compare fractions with unlike denominators	Have students compare two fractions with unlike denominators by rewriting them with a common denominator. Point out that any common multiple of the denominators will do. **Example** Compare $\frac{1}{3}$ and $\frac{2}{5}$. $\frac{1}{3} = \frac{5}{15}$ and $\frac{2}{5} = \frac{6}{15}$ $\frac{5}{15} < \frac{6}{15}$, so $\frac{1}{3} < \frac{2}{5}$	Other fractions to compare: $\frac{5}{7}$ and $\frac{2}{3}$ $(\frac{5}{7} > \frac{2}{3})$ $\frac{3}{5}$ and $\frac{1}{2}$ $(\frac{3}{5} > \frac{1}{2})$ See Key Skill 28, Student Handbook, page H15.
How to arrange three or more fractions in order from least to greatest	Have students list a group of fractions in order by comparing two fractions at a time. It may be helpful to first rewrite all the fractions with a common denominator. **Example** List $\frac{1}{2}$, $\frac{3}{8}$, and $\frac{1}{3}$ in order from least to greatest. A common multiple of 2, 8, and 3 is 24. $\frac{1}{2} = \frac{12}{24}$, $\frac{3}{8} = \frac{9}{24}$, and $\frac{1}{3} = \frac{8}{24}$. Since $\frac{8}{24} < \frac{9}{24} < \frac{12}{24}$, the fractions in order are: $\frac{1}{3}$, $\frac{3}{8}$, and $\frac{1}{2}$.	Other problems to solve: List $\frac{3}{4}$, $\frac{2}{3}$, and $\frac{5}{8}$ in order from least to greatest. $(\frac{5}{8}, \frac{2}{3}, \frac{3}{4})$ See Key Skill 29, Student Handbook, page H16.

Intervention and Extension Copying Masters, p. 4

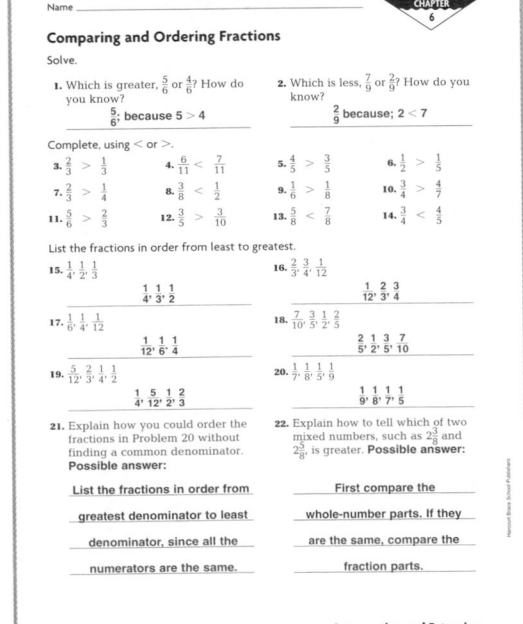

ORGANIZER

OBJECTIVE To use fraction bars to subtract mixed numbers

MATERIALS *For each pair* fraction bars

PROGRAM RESOURCES
E-Lab • Activity 6
E-Lab Recording Sheets, p. 6

USING THE PAGE

To help students understand that the ability to manipulate fractions is important in everyday life, ask them the following questions.

- **Why might you need to subtract mixed numbers when hiking, cooking, sewing, recycling, and traveling?**
  Possible answers: to find out how much farther you have to travel; to make a smaller amount of a recipe; to find yardage

- **In what other situations might you need to subtract mixed numbers?**

Activity 1

Explore

In this activity, students model three different mixed-number subtraction problems. In the first problem, they rename fractions in order to subtract. In the second problem, they rename a mixed number in order to subtract. And in the third problem, they rename fractions and a mixed number in order to subtract. These hands-on experiences will help students visualize the process of subtracting mixed numbers.

Think and Discuss

Discuss the first question. Have students use fraction bars to solve the problem. $2\frac{3}{8} - 1\frac{1}{8} = 1\frac{1}{4}$ After students respond to the second question, ask:

- **Which mixed number do you need to rename?** $3\frac{2}{5}$

Subtracting Mixed Numbers

You can use fraction bars to subtract mixed numbers.

ACTIVITY

Explore

A. Work with a partner to find $2\frac{2}{5} - 1\frac{3}{10}$.

- Use fraction bars to model $2\frac{2}{5}$.

- Since you are subtracting tenths, think about the LCD for $\frac{2}{5}$ and $\frac{3}{10}$. Change $\frac{2}{5}$ to $\frac{4}{10}$.

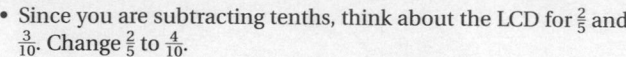

- Subtract $1\frac{3}{10}$ from $2\frac{4}{10}$.

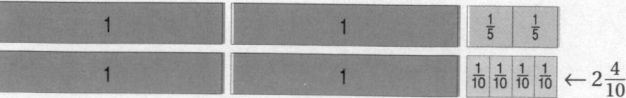

- What is $2\frac{2}{5} - 1\frac{3}{10}$? $1\frac{1}{10}$

Sometimes you need to regroup to subtract mixed numbers.

B. Work with a partner to find $2\frac{1}{4} - 1\frac{3}{4}$.

- Use fraction bars to model $2\frac{1}{4}$.

- Here is another way to model $2\frac{1}{4}$.

- From which model can you subtract $1\frac{3}{4}$? the second one
- Subtract $1\frac{3}{4}$ from $1\frac{5}{4}$. What is $2\frac{1}{4} - 1\frac{3}{4}$? $\frac{2}{4}$, or $\frac{1}{2}$

114 Chapter 6

IDEA FILE

WRITING IN MATHEMATICS

Have students write a word problem that could be solved with these fraction bars.

Possible answer: Carrie used $1\frac{1}{8}$ yd of a $2\frac{5}{16}$-yd length of material to make a backpack. How much material is left? $1\frac{3}{16}$ yd

Modality Visual

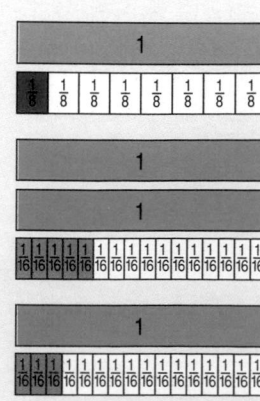

C. Work with a partner to find $2\frac{1}{6} - 1\frac{5}{12}$.

- Use fraction bars to model $2\frac{1}{6}$.

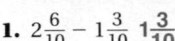

$\leftarrow 2\frac{1}{6}$

- Here is another way to model $2\frac{1}{6}$.

$\leftarrow 1\frac{7}{6}$

- From which model can you subtract $1\frac{5}{12}$? **neither**
- Since you are subtracting twelfths, think of the LCD for $\frac{1}{6}$ and $\frac{5}{12}$. Change the sixths to twelfths.

$\leftarrow 1\frac{14}{12}$

- Subtract $1\frac{5}{12}$ from $1\frac{14}{12}$. What is $2\frac{1}{6} - 1\frac{5}{12}$? $\frac{9}{12}$, or $\frac{3}{4}$

Think and Discuss

- Think about $2\frac{3}{8} - 1\frac{1}{8}$. Do you need to rename before you subtract? Explain. **No; $\frac{3}{8}$ is greater than $\frac{1}{8}$, and the denominators are the same.**
- Think about $3\frac{2}{5} - 1\frac{4}{5}$. Do you need to rename before you subtract? Explain. **Yes; $\frac{2}{5}$ is less than $\frac{4}{5}$.**
- Think about $5\frac{1}{2} - 3\frac{5}{6}$. Do you need to rename before you subtract? Explain. **Yes; $\frac{1}{2}$ is less than $\frac{5}{6}$, and the denominators are different.**

Try This

Use fraction bars to subtract. **Check students' work.**

1. $2\frac{6}{10} - 1\frac{3}{10}$ $1\frac{3}{10}$
2. $1\frac{2}{5} - \frac{1}{5}$ $1\frac{1}{5}$
3. $1\frac{3}{4} - \frac{1}{8}$ $1\frac{5}{8}$
4. $2\frac{2}{3} - 1\frac{1}{2}$ $1\frac{1}{6}$
5. $3\frac{3}{8} - \frac{3}{4}$ $2\frac{5}{8}$
6. $2\frac{2}{3} - 1\frac{3}{4}$ $\frac{11}{12}$
7. $3\frac{1}{2} - 1\frac{3}{5}$ $1\frac{9}{10}$
8. $3 - 2\frac{1}{2}$ $\frac{1}{2}$
9. $2 - 1\frac{3}{8}$ $\frac{5}{8}$

Technology Link
E–Lab • Activity 6 Available on CD-ROM
and the Internet at http://www.hbschool.com/elab

Note: Students use fraction bars. Use E-Lab p. 6.

115

GRADE 5	GRADE 6	GRADE 7
Use fraction bars to subtract mixed numbers.	Subtract mixed numbers.	Add and subtract mixed numbers.

ORGANIZER

OBJECTIVE To subtract mixed numbers

ASSIGNMENTS Independent Practice
- Basic 1–14, odd; 15–18
- Average 1–18
- Advanced 7–18

PROGRAM RESOURCES
Reteach 6.2 Practice 6.2 Enrichment 6.2
Cultural Connection • Soccer and Cricket, p. 124

PROBLEM OF THE DAY

 Transparency 6

Write the next 4 numbers. HINT: What is being subtracted in each sequence?

1. $10, 8\frac{3}{4}, 7\frac{1}{2}, 6\frac{1}{4}, \underline{?}, \underline{?}, \underline{?}, \underline{?}$
   $5, 3\frac{3}{4}, 2\frac{1}{2}, 1\frac{1}{4}$

2. $9, 7\frac{7}{8}, 6\frac{3}{4}, 5\frac{5}{8}, \underline{?}, \underline{?}, \underline{?}, \underline{?}$
   $4\frac{1}{2}, 3\frac{3}{8}, 2\frac{1}{4}, 1\frac{1}{8}$

3. $11\frac{7}{10}, 10\frac{2}{5}, 9\frac{1}{10}, 7\frac{4}{5}, \underline{?}, \underline{?}, \underline{?}, \underline{?}$
   $6\frac{1}{2}, 5\frac{1}{5}, 3\frac{9}{10}, 2\frac{3}{5}$

Solution Tab A, p. A6

QUICK CHECK

Complete to show equivalent fractions.

1. $1\frac{2}{3} = 1\frac{?}{6}$ 4
2. $2\frac{3}{4} = 2\frac{?}{8}$ 6
3. $2\frac{4}{3} = 2\frac{8}{?}$ 6
4. $3\frac{5}{2} = 3\frac{?}{4}$ 10
5. $1\frac{6}{5} = 1\frac{?}{10}$ 12
6. $2\frac{4}{3} = 2\frac{?}{12}$ 16

1 MOTIVATE

To help students access their prior knowledge of adding and subtracting mixed numbers, pose this problem to the class.

- **Joe delivers meals to homebound senior citizens every Tuesday and Thursday. He drives $2\frac{3}{8}$ mi on his Tuesday route and $1\frac{5}{8}$ mi on his Thursday route. How much longer is his Tuesday route?** $\frac{3}{4}$ mi

Copy these number sentences on the board.

$$1\frac{5}{8} - 2\frac{3}{8} \qquad 2\frac{3}{8} - 1\frac{5}{8}$$

WHAT YOU'LL LEARN
How to subtract mixed numbers

WHY LEARN THIS?
To understand how to use diagrams to solve problems that involve subtracting mixed numbers

Subtracting Mixed Numbers

You can use diagrams to subtract mixed numbers.

Liliana tutors younger students after school. She tutored for $2\frac{1}{4}$ hr on Monday and for $1\frac{3}{4}$ hr on Wednesday. How much longer did she tutor on Monday than on Wednesday? Complete the diagram to find $2\frac{1}{4} - 1\frac{3}{4}$.

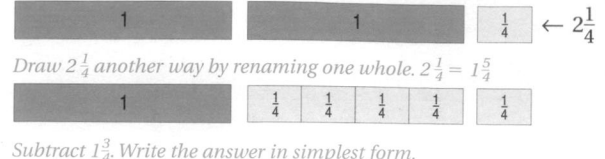

Draw $2\frac{1}{4}$ another way by renaming one whole. $2\frac{1}{4} = 1\frac{5}{4}$

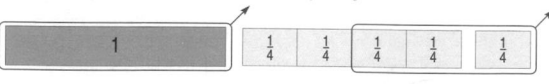

Subtract $1\frac{3}{4}$. Write the answer in simplest form.

$$2\frac{1}{4} - 1\frac{3}{4} = 1\frac{5}{4} - 1\frac{3}{4} = \frac{2}{4}, \text{ or } \frac{1}{2}$$

So, Liliana tutored $\frac{1}{2}$ hr longer on Monday.

EXAMPLE Draw a diagram to find the difference. $2\frac{1}{3} - 1\frac{3}{6}$

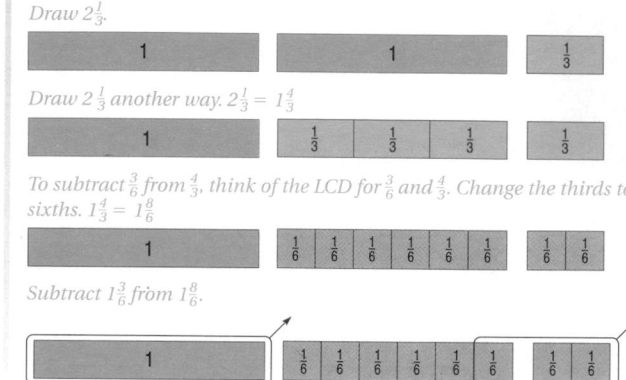

Draw $2\frac{1}{3}$.

Draw $2\frac{1}{3}$ another way. $2\frac{1}{3} = 1\frac{4}{3}$

To subtract $\frac{3}{6}$ from $\frac{4}{3}$, think of the LCD for $\frac{3}{6}$ and $\frac{4}{3}$. Change the thirds to sixths. $1\frac{4}{3} = 1\frac{8}{6}$

Subtract $1\frac{3}{6}$ from $1\frac{8}{6}$.

So, $2\frac{1}{3} - 1\frac{3}{6} = \frac{5}{6}$.

teen times

In many schools, "peer teachers" help other students who may have trouble with certain subjects. Find out if your school has a peer teacher program, and then volunteer to teach fellow students who may need your help.

IDEA FILE

EXTENSION

Have students work in pairs. One student writes a mixed number. The other student writes another mixed number that can be subtracted from the first mixed number. The first student then draws a diagram to solve the problem. Have students change roles and continue this activity until each student has solved three problems.

Modalities Kinesthetic, Visual

RETEACH 6.2

Name _____

LESSON 6.2

Subtracting Mixed Numbers

Ryan jogs $3\frac{3}{5}$ hr each week. So far this week, he has jogged $2\frac{4}{5}$ hr. How much longer must he jog to reach his goal? Drawing a diagram can help answer this question.

Step 1 Draw a diagram to show $3\frac{3}{5}$. ← $3\frac{3}{5}$

Step 2 You need to regroup to subtract $\frac{4}{5}$ from $\frac{3}{5}$. Draw a diagram to show $3\frac{3}{5}$ another way. ← $2\frac{8}{5}$

Step 3 Subtract $2\frac{4}{5}$ from $2\frac{8}{5}$.

So, $3\frac{3}{5} - 2\frac{4}{5} = \frac{4}{5}$.

Ryan needs to jog $\frac{4}{5}$ hr to reach his goal.

Draw a diagram to find each difference. Write the answer in simplest form. **Check students' diagrams.**

1. $4\frac{1}{6} - 2\frac{5}{6} = \underline{1\frac{1}{3}}$
2. $3\frac{2}{5} - 1\frac{4}{5} = \underline{1\frac{3}{5}}$
3. $5\frac{5}{12} - 3\frac{7}{12} = \underline{1\frac{5}{6}}$
4. $4\frac{1}{8} - 1\frac{3}{8} = \underline{2\frac{3}{4}}$

R26 TAKE ANOTHER LOOK Use with text pages 116–117.

GUIDED PRACTICE

For Exercises 1–6, match the mixed number with the diagram.

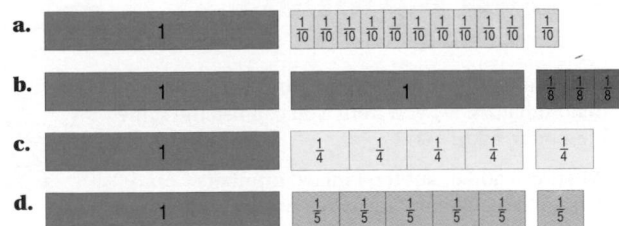

a. | 1 | 1/10 1/10 1/10 1/10 1/10 1/10 1/10 1/10 1/10 1/10 | 1/10

b. | 1 | 1 | 1/8 1/8 1/8

c. | 1 | 1/4 1/4 1/4 1/4 | 1/4

d. | 1 | 1/5 1/5 1/5 1/5 | 1/5

1. $2\frac{1}{5}$ d   **2.** $1\frac{5}{4}$ c   **3.** $1\frac{11}{10}$ a   **4.** $1\frac{6}{5}$ d   **5.** $2\frac{1}{4}$ c   **6.** $2\frac{3}{8}$ b

Draw a diagram to find each difference. Write the answer in simplest form. **Check students' diagrams.**

7. $2\frac{2}{3} - 1\frac{1}{3}$ $1\frac{1}{3}$ **8.** $3\frac{1}{6} - 2\frac{5}{6}$ $\frac{1}{3}$ **9.** $4\frac{1}{6} - 3\frac{1}{3}$ $\frac{5}{6}$ **10.** $5\frac{1}{4} - 1\frac{3}{8}$ $3\frac{7}{8}$

INDEPENDENT PRACTICE

Write each mixed number by renaming one whole.

1. $2\frac{1}{2}$ $1\frac{3}{2}$ **2.** $2\frac{3}{4}$ $1\frac{7}{4}$ **3.** $3\frac{2}{3}$ $2\frac{5}{3}$ **4.** $3\frac{5}{6}$ $2\frac{11}{6}$ **5.** $5\frac{5}{8}$ $4\frac{13}{8}$ **6.** $4\frac{5}{12}$ $3\frac{17}{12}$

Draw a diagram to find each difference. Write the answer in simplest form. **Check students' diagrams.**

7. $2\frac{3}{5} - 1\frac{1}{5}$ $1\frac{2}{5}$ **8.** $2\frac{1}{3} - 1\frac{2}{3}$ $\frac{2}{3}$ **9.** $2\frac{3}{4} - 1\frac{1}{4}$ $1\frac{1}{2}$ **10.** $2\frac{1}{5} - 1\frac{7}{10}$ $\frac{1}{2}$

11. $2\frac{1}{3} - 2\frac{1}{6}$ $\frac{1}{6}$ **12.** $3\frac{1}{12} - 1\frac{5}{6}$ $1\frac{1}{4}$ **13.** $3\frac{5}{6} - 1\frac{1}{6}$ $2\frac{2}{3}$ **14.** $3\frac{2}{3} - 2\frac{3}{4}$ $\frac{11}{12}$

Problem-Solving Applications

For Problems 15–17, draw a diagram to solve. Write the answer in simplest form. **Check students' diagrams.**

15. During morning break, the class drank $2\frac{3}{4}$ qt of fruit punch and $1\frac{1}{4}$ qt of apple juice. How much more of fruit punch than apple juice did the class drink? $1\frac{1}{2}$ qt

16. Karen had $3\frac{3}{8}$ yd of cotton fabric. She used $2\frac{5}{8}$ yd for a skirt. How much fabric was left? $\frac{3}{4}$ yd

17. Joel's father drives $4\frac{2}{3}$ mi to work. His mother drives $2\frac{3}{4}$ mi to work. How much farther does Joel's father drive to work than his mother? $1\frac{11}{12}$ mi

18. ✏ **WRITE ABOUT IT** When you draw a diagram to show subtraction of mixed numbers, how do you know when to regroup before you subtract? **when the fraction of the first mixed number is less than the fraction of the second mixed number**

MORE PRACTICE Lesson 6.2, page H50

117

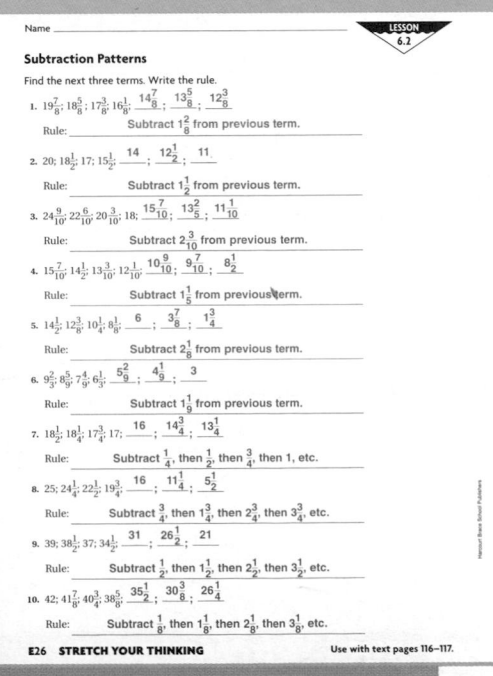

PRACTICE 6.2

Name _____

LESSON 6.2

Subtracting Mixed Numbers

Write each mixed number by renaming one whole.

1. $3\frac{1}{2}$ $\frac{3}{2}$ 2. $5\frac{3}{4}$ $\frac{7}{4}$ 3. $2\frac{7}{8}$ $1\frac{15}{8}$ 4. $4\frac{6}{8}$ $3\frac{11}{6}$ 5. $3\frac{5}{12}$ $2\frac{17}{12}$

6. $4\frac{3}{10}$ $3\frac{13}{10}$ 7. $2\frac{4}{9}$ $1\frac{13}{9}$ 8. $3\frac{5}{8}$ $2\frac{11}{8}$ 9. $5\frac{3}{7}$ $4\frac{10}{7}$ 10. $4\frac{3}{11}$ $3\frac{14}{11}$

Draw a diagram to find each difference. Write the answer in simplest form. Check students' diagrams.

11. $4\frac{3}{5} - 2\frac{2}{5}$ $2\frac{2}{5}$ 12. $3\frac{1}{3} - 1\frac{2}{3}$ $1\frac{2}{3}$ 13. $6\frac{2}{5} - 4\frac{3}{5}$ $1\frac{4}{5}$

14. $6\frac{3}{8} - 2\frac{1}{8}$ $4\frac{2}{8}$ 15. $3\frac{1}{2} - 2\frac{1}{4}$ $1\frac{1}{4}$ 16. $5\frac{3}{4} - 1\frac{1}{8}$ $4\frac{5}{8}$

17. $5\frac{1}{4} - 3\frac{2}{3}$ $1\frac{1}{2}$ 18. $4\frac{3}{5} - 2\frac{1}{10}$ $2\frac{1}{2}$ 19. $4\frac{1}{4} - 3\frac{1}{2}$ $1\frac{1}{4}$

20. $4\frac{1}{8} - 2\frac{3}{4}$ $1\frac{3}{8}$ 21. $5\frac{1}{9} - 3\frac{3}{9}$ $1\frac{4}{9}$ 22. $3\frac{1}{2} - 1\frac{4}{5}$ $1\frac{7}{10}$

Mixed Applications

For Problems 23–26, draw a diagram to solve. Write the answer in simplest form. Check students' diagrams.

23. Kate has $5\frac{3}{8}$ m of wire. She uses $2\frac{5}{8}$ m for a school project. How much wire is left?

$3\frac{1}{2}$ m

24. The school cafeteria has $5\frac{1}{4}$ qt of grape juice and $2\frac{3}{8}$ qt of apple juice. How much more grape juice than apple juice is there?

$2\frac{7}{8}$ qt

25. Michele read for $2\frac{1}{4}$ hr on Monday and $1\frac{1}{2}$ hr on Tuesday. How much time did she spend reading?

$3\frac{3}{4}$ hr

26. Carey works at a local theater during the summer. She is paid $4.75 per hour and works 6 hr each day. If she works 4 days each week, how much does she earn each week?

$114

P26 ON MY OWN

Use with text pages 116–117.

ENRICHMENT 6.2

Name _____

LESSON 6.2

Subtraction Patterns

Find the next three terms. Write the rule.

1. $19\frac{7}{8}$; $18\frac{5}{8}$; $17\frac{3}{8}$; $16\frac{1}{8}$; $14\frac{7}{8}$, $13\frac{5}{8}$, $12\frac{3}{8}$

   Rule: ___ Subtract $1\frac{2}{8}$ from previous term.

2. 20; $18\frac{1}{2}$; 17; $15\frac{1}{2}$; 14 , $12\frac{1}{2}$, 11

   Rule: ___ Subtract $1\frac{1}{2}$ from previous term.

3. $24\frac{9}{10}$; $22\frac{6}{10}$; $20\frac{3}{10}$; 18; $15\frac{7}{10}$, $13\frac{4}{10}$, $11\frac{1}{10}$

   Rule: ___ Subtract $2\frac{3}{10}$ from previous term.

4. $15\frac{7}{10}$; $14\frac{1}{2}$; $13\frac{3}{10}$; $12\frac{1}{10}$; $10\frac{9}{10}$, $9\frac{7}{10}$, $8\frac{1}{2}$

   Rule: ___ Subtract $1\frac{1}{5}$ from previous term.

5. $14\frac{1}{2}$; $12\frac{3}{8}$; $10\frac{1}{4}$; $8\frac{1}{8}$; 6 , $3\frac{7}{8}$, $1\frac{3}{4}$

   Rule: ___ Subtract $2\frac{1}{8}$ from previous term.

6. $9\frac{5}{9}$; $8\frac{3}{9}$; $7\frac{4}{9}$; $6\frac{1}{9}$; $5\frac{2}{9}$, $4\frac{4}{9}$, 3

   Rule: ___ Subtract $1\frac{1}{9}$ from previous term.

7. $18\frac{1}{2}$; $18\frac{1}{4}$; $17\frac{3}{4}$; 17; 16 , $14\frac{3}{4}$, $13\frac{1}{4}$

   Rule: ___ Subtract $\frac{1}{4}$, then $\frac{1}{2}$, then $\frac{3}{4}$, then 1, etc.

8. 25; $24\frac{1}{4}$; $22\frac{3}{4}$; $19\frac{1}{2}$; 16 , $11\frac{1}{2}$, $5\frac{1}{2}$

   Rule: ___ Subtract $\frac{3}{4}$, then $1\frac{1}{4}$, then $2\frac{3}{4}$, then $3\frac{3}{4}$, etc.

9. 39; $38\frac{1}{2}$; 37; $34\frac{1}{2}$; 31 , $26\frac{1}{2}$, 21

   Rule: ___ Subtract $\frac{1}{2}$, then $1\frac{1}{2}$, then $2\frac{1}{2}$, then $3\frac{1}{2}$, etc.

10. 42; $41\frac{7}{8}$; $40\frac{1}{2}$; $38\frac{3}{8}$; $35\frac{1}{2}$, $30\frac{3}{8}$, $26\frac{1}{4}$

    Rule: ___ Subtract $\frac{1}{8}$, then $1\frac{3}{8}$, then $2\frac{1}{8}$, then $3\frac{3}{8}$, etc.

E26 STRETCH YOUR THINKING

Use with text pages 116–117.

- **Which number sentence would you use to solve the problem?** $2\frac{3}{8} - 1\frac{5}{8}$

- **Why?** Possible answer: because $2\frac{3}{8}$ is greater than $1\frac{5}{8}$

- **Will you need to rename a mixed number to solve the problem?** yes

- **Why?** because $\frac{3}{8}$ is less than $\frac{5}{8}$

2 TEACH/PRACTICE

As students work through the example, focus attention on the second and third steps. Ask:

- **Why do you need to draw $2\frac{1}{3}$ another way?** because $\frac{1}{3}$ is less than $\frac{3}{6}$

- **Why do you need to change the thirds to sixths?** because 6 is the LCD of $\frac{4}{3}$ and $\frac{3}{6}$

GUIDED PRACTICE

Have volunteers work Exercises 7–10 on the board. Discuss the answers with the whole class.

INDEPENDENT PRACTICE

To make sure that students understand how to rename fractions before subtracting, check students' answers to Exercises 1–6.

ADDITIONAL EXAMPLE

Example, p. 116

Draw a diagram to find the difference. $2\frac{2}{5} - 1\frac{9}{10}$ Check students' diagrams; $\frac{1}{2}$.

3 WRAP UP

Use these questions to help students focus on the big ideas of the lesson.

- **Exit 19 is at the $3\frac{5}{8}$-mi mark on the highway. Exit 18 is at the $1\frac{3}{4}$-mi mark. What is the distance between the exits? Draw a diagram to solve.** $1\frac{7}{8}$ mi; check students' diagrams.

✔ *Assessment Tip*

Have students write the answer.

- **What did you need to do to solve Exercise 10 that you didn't need to do to solve Exercise 9?** rename one whole number and a fraction

DAILY QUIZ

Draw a diagram to find each difference.

1. $3\frac{2}{5} - 1\frac{1}{5}$ $2\frac{1}{5}$ **2.** $2\frac{1}{8} - 1\frac{3}{8}$ $\frac{3}{4}$

3. $2\frac{2}{3} - 1\frac{1}{9}$ $1\frac{5}{9}$ **4.** $2\frac{1}{4} - 1\frac{1}{3}$ $\frac{11}{12}$

117

GRADE 5	GRADE 6	GRADE 7
Rename to subtract mixed numbers.	Add and subtract mixed numbers.	Add and subtract mixed numbers.

ORGANIZER

OBJECTIVE To add and subtract mixed numbers

ASSIGNMENTS Independent Practice
- *Basic* 1–36, even; 37–41; 42–53
- *Average* 1–41; 42–53
- *Advanced* 1–5, 17–21, 27–31; 37–41; 42–53

PROGRAM RESOURCES
Reteach 6.3 Practice 6.3 Enrichment 6.3

PROBLEM OF THE DAY

Transparency 6

Add vertically and horizontally.

$1\frac{2}{3}$	$2\frac{1}{4}$	?	$3\frac{11}{12}$
$3\frac{1}{6}$	$4\frac{1}{2}$	?	$7\frac{2}{3}$
?	?	?	
$4\frac{5}{6}$	$6\frac{3}{4}$	$11\frac{7}{12}$	

Solution Tab A, p. A6

QUICK CHECK

Add or subtract. Write the answer in simplest form.

1. $\frac{7}{16} + \frac{1}{4}$ $\frac{11}{16}$ **2.** $\frac{9}{10} - \frac{5}{6}$ $\frac{1}{15}$

3. $\frac{3}{4} - \frac{2}{5}$ $\frac{7}{20}$ **4.** $\frac{1}{3} + \frac{3}{4}$ $1\frac{1}{12}$

5. $\frac{3}{7} + \frac{2}{9}$ $\frac{41}{63}$ **6.** $\frac{2}{3} - \frac{5}{8}$ $\frac{1}{24}$

1 MOTIVATE

Organize the class into pairs. Have one student in each pair draw a diagram to find $2\frac{3}{8} + 1\frac{1}{4}$. Then have the student explain the diagram to his or her partner. Have partners change roles and repeat this activity for $2\frac{2}{3} - 1\frac{1}{9}$.

2 TEACH/PRACTICE

In this lesson students move from using models for adding and subtracting mixed numbers to using an algorithm. Work through the examples with the class, and discuss each step in solving the problems.

6.3

WHAT YOU'LL LEARN
How to add and subtract mixed numbers

WHY LEARN THIS?
To solve problems such as finding a total amount of time worked

REMEMBER:

To write a fraction as a mixed number, divide the numerator by the denominator. The quotient is the whole-number part of the mixed number. Use the remainder as the new numerator and the divisor as the denominator.
See page H16.

$$\frac{13}{3} \rightarrow 3\overline{)13} \quad \begin{array}{r} 4\frac{1}{3} \\ \underline{-12} \\ 1 \end{array}$$

Adding and Subtracting Mixed Numbers

Think about how you add fractions with unlike denominators. Do you think you can use the same method to add $1\frac{2}{3}$ and $2\frac{1}{4}$?

When you add or subtract mixed numbers with unlike fractions, you can write equivalent fractions by using the LCD.

Andi works with her father every Friday and Saturday. Andi's father pays her for the number of hours she works. This week Andi worked $3\frac{2}{5}$ hr on Friday and $2\frac{1}{2}$ hr on Saturday. What is the total number of hours Andi worked?

Find $3\frac{2}{5} + 2\frac{1}{2}$.

$$\begin{array}{r} 3\frac{2}{5} = 3\frac{4}{10} \\ +2\frac{1}{2} = 2\frac{5}{10} \\ \hline 5\frac{9}{10} \end{array}$$

The LCD of $\frac{2}{5}$ and $\frac{1}{2}$ is tenths.
Write equivalent fractions, using the LCD.
Add fractions.
Add whole numbers.

So, Andi worked $5\frac{9}{10}$ hr.

• Is the time Andi worked closer to 5 hr or 6 hr? Explain. **6 hr; $5\frac{9}{10}$ hr is closer to 6 hr than 5 hr.**

Sometimes you need to rewrite a sum.

EXAMPLE 1 Jared, Andi's brother, likes to compete with his sister. He worked $4\frac{2}{3}$ hr on Friday and $3\frac{3}{4}$ hr on Saturday. What is the total number of hours he worked?

Find $4\frac{2}{3} + 3\frac{3}{4}$.

$$\begin{array}{r} 4\frac{2}{3} = 4\frac{8}{12} \\ +3\frac{3}{4} = 3\frac{9}{12} \\ \hline 7\frac{17}{12} = 7 + 1\frac{5}{12} = 8\frac{5}{12} \end{array}$$

Write equivalent fractions, using the LCD.
Add fractions.
Add whole numbers.
Rename the fraction as a mixed number. Rewrite the sum.

So, Jared worked $8\frac{5}{12}$ hr.

 Calculator Activities page H34

IDEA FILE

ALTERNATIVE TEACHING STRATEGY

Have students use the eighths ruler (see TR p. 23) to add and subtract these mixed numbers.

1. $2\frac{3}{8} - 1\frac{1}{8}$ $1\frac{1}{4}$

2. $1\frac{3}{4} + 3\frac{5}{8}$ $5\frac{3}{8}$

3. $5\frac{1}{2} - 1\frac{3}{8}$ $4\frac{1}{8}$

Modalities Kinesthetic, Visual

RETEACH 6.3

Name _____

LESSON 6.3

Adding and Subtracting Mixed Numbers

Mrs. Ruiz buys $4\frac{1}{2}$ lb of apples. She uses $1\frac{2}{3}$ lb to bake apple tarts. How many pounds of apples are left?

Step 1 The LCD of $\frac{1}{2}$ and $\frac{2}{3}$ is 6. Rename the fractions using the LCD.

$$4\frac{1}{2} = 4\frac{3}{6}$$
$$-1\frac{2}{3} = 1\frac{4}{6}$$

Step 2 Since you can't subtract $\frac{4}{6}$ from $\frac{3}{6}$, rename $4\frac{3}{6}$.
Think: $4\frac{3}{6} = 3 + \frac{6}{6} + \frac{3}{6} = 3\frac{9}{6}$

$$4\frac{1}{2} = 4\frac{3}{6} = 3\frac{9}{6}$$
$$-1\frac{2}{3} = 1\frac{4}{6} = 1\frac{4}{6}$$
$$\overline{2\frac{5}{6}}$$

Now, subtract the fractions. Then subtract the whole numbers.

So, $4\frac{1}{2} - 1\frac{2}{3} = 2\frac{5}{6}$

Mrs. Ruiz has $2\frac{5}{6}$ pounds of apples left.

Find the difference. Write the answer in simplest form.

1. $6\frac{2}{5}$ $-3\frac{3}{10}$ $\overline{2\frac{1}{10}}$	2. $4\frac{1}{6}$ $-2\frac{3}{4}$ $\overline{1\frac{5}{12}}$	3. $9\frac{3}{7}$ $-5\frac{1}{2}$ $\overline{3\frac{13}{14}}$
4. $11\frac{1}{3}$ $-7\frac{5}{9}$ $\overline{3\frac{7}{9}}$	5. $10\frac{3}{8}$ $-3\frac{1}{2}$ $\overline{6\frac{7}{8}}$	6. $6\frac{1}{10}$ $-4\frac{4}{5}$ $\overline{1\frac{3}{10}}$
7. $9\frac{1}{4}$ $-4\frac{3}{8}$ $\overline{4\frac{7}{8}}$	8. $12\frac{2}{5}$ $-6\frac{1}{2}$ $\overline{5\frac{9}{10}}$	9. $8\frac{1}{3}$ $-5\frac{5}{6}$ $\overline{2\frac{1}{6}}$

Use with text pages 118–121.

TAKE ANOTHER LOOK R27

Talk About It

- When should you rename the fractions before adding mixed numbers? Give an example. **when the denominators are not the same; possible example: $1\frac{1}{2} + 2\frac{1}{3}$**

- When can you rename a fraction as a mixed number? **when the numerator is larger than the denominator**

- 💡 **CRITICAL THINKING** How would you rename the mixed number $2\frac{3}{2}$? **Rename $\frac{3}{2}$ as $1\frac{1}{2}$, and then add it to the whole number 2 to get $3\frac{1}{2}$.**

GUIDED PRACTICE

Rename the fraction, using the LCD. Rewrite each problem.

1. $1\frac{1}{3} + 1\frac{1}{2}$ $1\frac{2}{6} + 1\frac{3}{6}$
2. $2\frac{2}{3} + 1\frac{1}{5}$ $2\frac{10}{15} + 1\frac{3}{15}$
3. $3\frac{3}{7} + 2\frac{2}{3}$ $3\frac{9}{21} + 2\frac{14}{21}$
4. $6\frac{1}{6} + 4\frac{1}{2}$ $6\frac{1}{6} + 4\frac{3}{6}$
5. $3\frac{3}{4} + 2\frac{1}{3}$ $3\frac{9}{12} + 2\frac{4}{12}$
6. $2\frac{4}{5} + 4\frac{1}{4}$ $2\frac{16}{20} + 4\frac{5}{20}$

Rename the fraction as a mixed number. Write the new mixed number.

7. $1\frac{5}{3}$ $1\frac{2}{3}; 2\frac{2}{3}$
8. $2\frac{4}{3}$ $1\frac{1}{3}; 3\frac{1}{3}$
9. $2\frac{7}{4}$ $1\frac{3}{4}; 3\frac{3}{4}$
10. $3\frac{13}{7}$ $1\frac{6}{7}; 4\frac{6}{7}$

11. $2\frac{7}{5}$ $1\frac{2}{5}; 3\frac{2}{5}$
12. $4\frac{8}{7}$ $1\frac{1}{7}; 5\frac{1}{7}$
13. $3\frac{12}{7}$ $1\frac{5}{7}; 4\frac{5}{7}$
14. $6\frac{15}{11}$ $1\frac{4}{11}; 7\frac{4}{11}$

15. $3\frac{18}{4}$ $4\frac{2}{4}; 7\frac{2}{4}$
16. $1\frac{12}{11}$ $1\frac{1}{11}; 2\frac{1}{11}$
17. $5\frac{18}{3}$ $6; 11$
18. $2\frac{12}{4}$ $3; 5$

Subtracting Mixed Numbers

You can subtract mixed numbers without using diagrams or models.

EXAMPLE 2 Tyler's mother bought $3\frac{3}{4}$ yd of fabric on Wednesday and $2\frac{3}{8}$ yd on Saturday. How much more fabric did she buy on Wednesday?

Find $3\frac{3}{4} - 2\frac{3}{8}$.

$$3\frac{3}{4} = 3\frac{6}{8}$$ *Write equivalent fractions, using the LCD.*
$$-2\frac{3}{8} = 2\frac{3}{8}$$
$$\overline{\phantom{-2\frac{3}{8}=}\,1\frac{3}{8}}$$ *Subtract the fractions.*
Subtract the whole numbers.

So, Tyler's mother bought $1\frac{3}{8}$ yd more on Wednesday.

- What if Tyler's mother bought $1\frac{2}{3}$ yd of fabric on Saturday? How many more yards did she buy on Wednesday? **$2\frac{1}{12}$ yd more**

119

💡 **CRITICAL THINKING** On Sunday, Bryan ran $1\frac{1}{4}$ mi and Sally ran $\frac{3}{4}$ mi. Starting on Monday, Sally will increase the distance she runs by $\frac{1}{4}$ mi each day and Bryan will increase the distance he runs by $\frac{1}{8}$ mi each day. On what day will they run the same distance? **Thursday** Explain how you solved the problem. Possible answer: by making a chart

Dana rented three science fiction videos. *Space Chase* is $\frac{3}{4}$ hour longer than *Doomsday*. *Doomsday* is $\frac{1}{4}$ hour longer than *It's Here*. The combined running time for the three videos is $6\frac{1}{2}$ hours. How long is each video? *Space Chase* is $2\frac{3}{4}$ hours long, *Doomsday* is 2 hours long, and *It's Here* is $1\frac{3}{4}$ hours long.

ADDITIONAL EXAMPLES

Example 1, p. 118

A group of hikers walked $4\frac{3}{4}$ mi before stopping for refreshments. They walked another $3\frac{5}{8}$ mi before setting up camp. How many miles did they hike in all? $8\frac{3}{8}$ mi

Example 2, p. 119

The river by Camp Somerset is $1\frac{3}{8}$ mi shorter than the $3\frac{1}{2}$-mi-long river by Camp Hope. How long is the river by Camp Somerset? $2\frac{1}{8}$ mi

---

PRACTICE 6.3

Name _____

LESSON 6.3

Adding and Subtracting Mixed Numbers

Rename the fractions using the LCD. Rewrite the problem.

1. $1\frac{1}{3} + 2\frac{5}{6}$ $1\frac{2}{6} + 2\frac{5}{6}$
2. $2\frac{3}{4} + 1\frac{1}{2}$ $2\frac{3}{4} + 1\frac{2}{4}$
3. $5\frac{5}{6} + 3\frac{1}{4}$ $5\frac{10}{12} + 3\frac{3}{12}$
4. $6\frac{1}{5} + 2\frac{3}{4}$ $6\frac{4}{20} + 2\frac{15}{20}$

Rename the fraction as a mixed number. Write the new mixed number.

5. $5\frac{10}{3}$ $1\frac{1}{3}; 6\frac{1}{3}$
6. $6\frac{7}{4}$ $1\frac{3}{4}; 7\frac{3}{4}$
7. $8\frac{9}{5}$ $1\frac{4}{5}; 9\frac{4}{5}$
8. $5\frac{13}{10}$ $1\frac{3}{10}; 6\frac{3}{10}$

Tell whether you must rename to subtract. Write *yes* or *no*.

9. $8\frac{3}{4} - 6\frac{1}{2}$ **Yes**
10. $4\frac{1}{5} - 2\frac{7}{10}$ **Yes**
11. $7\frac{1}{4} - 2\frac{3}{5}$ **Yes**
12. $5\frac{3}{8} - 3\frac{5}{3}$ **Yes**

Find the sum. Write the answer in simplest form.

13. $3\frac{1}{2} + 2\frac{3}{10}$ $5\frac{1}{2}$
14. $1\frac{3}{8} + 4\frac{1}{2}$ $5\frac{7}{8}$
15. $5\frac{5}{8} + 2\frac{3}{4}$ $9\frac{1}{2}$
16. $1\frac{7}{9} + 1\frac{2}{3}$ $3\frac{4}{9}$

Find the difference. Write the answer in simplest form.

17. $4\frac{2}{3} - 1\frac{1}{2}$ $3\frac{1}{6}$
18. $5\frac{1}{8} - 3\frac{1}{4}$ $2\frac{11}{20}$
19. $3\frac{1}{2} - 1\frac{4}{9}$ $1\frac{8}{9}$
20. $4\frac{3}{8} - 2\frac{1}{2}$ $2\frac{1}{8}$

Mixed Applications

21. Carol bought a share of stock for $22\frac{1}{8}$ dollars. A week later the stock increased $1\frac{3}{4}$ dollars. What is her share of stock worth now? $23\frac{7}{8}$ dollars

22. Bob practices the piano $6\frac{1}{4}$ hr each week. This week he has practiced $4\frac{3}{5}$ hr. How many more hours will Bob practice this week? $1\frac{17}{20}$ hr

23. At the Olympics, a souvenir vendor makes $185.50 in sales in the morning and $290.75 in the afternoon. How much money in all does the vendor make that day? $476.25

24. A newspaper company surveyed 1,000 people about their major health concern. The major concern of 656 people was keeping fit. How many people did not list keeping fit as a major concern? 344 people

Use with text pages 118–121. **ON MY OWN** P27

ENRICHMENT 6.3

Name _____

LESSON 6.3

Puzzling Fractions

Match each Exercise in Column 1 with its sum or difference in Column 2. Then, to discover the Math Tip in the box, write each corresponding letter above the line marked with the Exercise number.

Column 1

1. $1\frac{1}{3} + 2\frac{3}{4}$ H
2. $5\frac{1}{2} - 3\frac{1}{4}$ O
3. $2\frac{5}{6} + 4\frac{8}{9}$ A
4. $6\frac{2}{3} - 5\frac{1}{2}$ S
5. $4\frac{5}{6} + 1\frac{2}{3}$ D
6. $5\frac{3}{10} - 1\frac{4}{5}$ L
7. $2\frac{1}{4} + 5\frac{1}{3}$ U
8. $6\frac{3}{4} - 3\frac{1}{4}$ B
9. $1\frac{1}{5} + 2\frac{1}{2}$ N
10. $7\frac{11}{12} - 5\frac{1}{6}$ W
11. $3\frac{3}{8} - 1\frac{1}{4}$ E
12. $6\frac{2}{5} + 1\frac{7}{20}$ P
13. $8\frac{5}{12} - 1\frac{5}{6}$ I
14. $5\frac{1}{4} + 1\frac{1}{2}$ T
15. $9\frac{3}{5} - 3\frac{2}{15}$ M
16. $4\frac{7}{10} - 2\frac{13}{20}$ F
17. $4\frac{2}{9} + 3\frac{1}{3}$ R

Column 2

A. $7\frac{1}{9}$
B. $3\frac{7}{12}$
D. $6\frac{1}{2}$
E. $2\frac{1}{8}$
F. $2\frac{1}{20}$
H. $4\frac{1}{12}$
I. $6\frac{7}{12}$
L. $3\frac{1}{2}$
M. $6\frac{8}{15}$
N. $3\frac{12}{35}$
O. $2\frac{1}{4}$
P. $7\frac{3}{4}$
Q. $7\frac{7}{9}$
S. $1\frac{1}{6}$
T. $6\frac{7}{12}$
U. $7\frac{1}{12}$
W. $2\frac{3}{4}$

Math Tip

A	N	S	W	E	R	S		S	H	O	U	L	D
3	9	4	10	11	17	14		4	1	2	7	6	5

							B	E		I	N		
							8	11		13	9		

S	I	M	P	L	E	S	T		F	O	R	M	
4	13	15	12	6	11	4	14		16	2	17	15	

Use with text pages 118–121. **STRETCH YOUR THINKING** E27

119

INDEPENDENT PRACTICE

You may wish to check students' answers to Exercises 1–5 to make sure that they understand how to rename fractions before they proceed to the remaining exercises.

DEVELOPING ALGEBRAIC THINKING

To accustom students to working with variables, ask:

• **If *x* is a whole number, will you always need to rename *x* to subtract a mixed number from it? Explain.** Yes; before you can subtract a fraction from another number, both numbers must be fractions.

ADDITIONAL EXAMPLES

Example 3, p. 120

Find the difference.

$6 - 3\frac{2}{3}$ $2\frac{1}{3}$

Example 4, p. 120

Find the difference.

$4\frac{3}{12} - 2\frac{5}{6}$ $1\frac{5}{12}$

Sometimes a whole number needs to be renamed before you can subtract.

EXAMPLE 3 Find the difference. $5 - 1\frac{4}{5}$

$$5 = 4\frac{5}{5}$$ *Since you are subtracting fifths, rename 5 as $4\frac{5}{5}$.*

$$-1\frac{4}{5} = 1\frac{4}{5}$$ *Subtract the fractions.*
Subtract the whole numbers.
$$\overline{3\frac{1}{5}}$$

So, $5 - 1\frac{4}{5} = 3\frac{1}{5}$.

• What if you wanted to find $8 - 3\frac{4}{7}$? What would you rename 8 so you could subtract $3\frac{4}{7}$? **$7\frac{7}{7}$**

Sometimes a mixed number needs to be renamed before you can subtract.

EXAMPLE 4 Find the difference. $4\frac{1}{6} - 2\frac{7}{9}$

$$4\frac{1}{6} = 4\frac{3}{18}$$ *The LCD of $\frac{1}{6}$ and $\frac{7}{9}$ is 18.*
Write equivalent fractions, using the LCD.
$$-2\frac{7}{9} = 2\frac{14}{18}$$

$$4\frac{1}{6} = 4\frac{3}{18} = 3\frac{21}{18}$$ *Since you can't subtract $\frac{14}{18}$ from $\frac{3}{18}$, rename $4\frac{3}{18}$.*
$4\frac{3}{18} = 3 + \frac{18}{18} + \frac{3}{18} = 3\frac{21}{18}$
$$-2\frac{7}{9} = 2\frac{14}{18} = 2\frac{14}{18}$$ *Subtract the fractions.*
Subtract the whole numbers.
$$\overline{1\frac{7}{18}}$$

So, $4\frac{1}{6} - 2\frac{7}{9} = 1\frac{7}{18}$.

• 💡 **CRITICAL THINKING** What if you wanted to find $4\frac{3}{8} - 2\frac{1}{12}$? What equivalent fractions would you write using the LCD? **$4\frac{9}{24}$ and $2\frac{2}{24}$**

INDEPENDENT PRACTICE

Rename the fractions, using the LCD. Rewrite the problem.

1. $1\frac{1}{4} + 1\frac{3}{8}$ $1\frac{2}{8} + 1\frac{3}{8}$ **2.** $3\frac{1}{4} + 1\frac{2}{3}$ $3\frac{3}{12} + 1\frac{8}{12}$ **3.** $2\frac{1}{2} + 3\frac{2}{5}$ $2\frac{5}{10} + 3\frac{4}{10}$ **4.** $2\frac{5}{6} + 4\frac{4}{9}$ $2\frac{15}{18} + 4\frac{8}{18}$ **5.** $2\frac{1}{4} + 3\frac{2}{5}$ $2\frac{5}{20} + 3\frac{8}{20}$

Rename the fraction as a mixed number. Write the new mixed number.

6. $2\frac{3}{2}$ $1\frac{1}{2}$; $3\frac{1}{2}$ **7.** $1\frac{7}{4}$ $1\frac{3}{4}$; $2\frac{3}{4}$ **8.** $3\frac{7}{5}$ $1\frac{2}{5}$; $4\frac{2}{5}$ **9.** $4\frac{11}{6}$ $1\frac{5}{6}$; $5\frac{5}{6}$ **10.** $2\frac{11}{7}$ $1\frac{4}{7}$; $3\frac{4}{7}$ **11.** $5\frac{13}{10}$ $1\frac{3}{10}$; $6\frac{3}{10}$

Tell whether you must rename to subtract. Write *yes* or *no*.

12. $2\frac{1}{4} - 1\frac{3}{8}$ **yes** **13.** $3\frac{1}{4} - 1\frac{2}{3}$ **yes** **14.** $7\frac{1}{2} - 3\frac{2}{5}$ **no** **15.** $5\frac{5}{6} - 2\frac{7}{9}$ **no** **16.** $4\frac{1}{4} - 2\frac{2}{5}$ **yes**

👥 IDEA FILE

CULTURAL CONNECTION

Tell students that the January harvest festival in South India is called Pongal. A popular rice dish served in India goes by the same name. Have students find this recipe in a cookbook and share it with the class. Discuss how fractions and mixed numbers are used in the recipe.

Modalities Auditory, Visual

GEOGRAPHY CONNECTION

Have students write one addition problem and one subtraction problem using this data. Check students' problems.

Annual Rainfall	
Houston, TX	$44\frac{3}{4}$ in.
Phoenix, AZ	$7\frac{1}{10}$ in.
Atlanta, GA	$48\frac{3}{5}$ in.
Chicago, IL	$33\frac{1}{3}$ in.

Modality Visual

Find the sum. Write the answer in simplest form.

17. $1\frac{1}{4} + 1\frac{1}{8}$ **$2\frac{3}{8}$**   **18.** $2\frac{1}{4} + 4\frac{1}{3}$ **$6\frac{7}{12}$**   **19.** $4\frac{1}{2} + 3\frac{4}{5}$ **$8\frac{3}{10}$**   **20.** $6\frac{5}{6} + 5\frac{7}{9}$ **$12\frac{11}{18}$**   **21.** $7\frac{3}{4} + 3\frac{2}{5}$ **$11\frac{3}{20}$**

22. $2\frac{3}{4} + 4\frac{5}{12}$ **$7\frac{1}{6}$**   **23.** $6\frac{1}{9} + 7\frac{1}{3}$ **$13\frac{4}{9}$**   **24.** $3\frac{2}{7} + 8\frac{1}{3}$ **$11\frac{13}{21}$**   **25.** $5\frac{5}{6} + 4\frac{2}{9}$ **$10\frac{1}{18}$**   **26.** $4\frac{5}{7} + 3\frac{1}{2}$ **$8\frac{3}{14}$**

Find the difference. Write the answer in simplest form.

27. $6\frac{1}{2} - 3\frac{1}{5}$ **$3\frac{3}{10}$**   **28.** $7\frac{5}{6} - 2\frac{5}{9}$ **$5\frac{5}{18}$**   **29.** $4\frac{1}{3} - 2\frac{1}{4}$ **$2\frac{1}{12}$**   **30.** $3\frac{1}{4} - 1\frac{1}{6}$ **$2\frac{1}{12}$**   **31.** $7\frac{1}{4} - 4\frac{3}{5}$ **$2\frac{13}{20}$**

32. $12\frac{1}{4} - 10\frac{3}{4}$ **$1\frac{1}{2}$**   **33.** $5\frac{1}{2} - 3\frac{7}{10}$ **$1\frac{4}{5}$**   **34.** $12\frac{1}{9} - 7\frac{1}{3}$ **$4\frac{7}{9}$**   **35.** $11\frac{1}{4} - 9\frac{7}{8}$ **$1\frac{3}{8}$**   **36.** $15\frac{1}{6} - 7\frac{2}{3}$ **$7\frac{1}{2}$**

Problem-Solving Applications

37. Todd's mother bought a share of stock for $25\frac{3}{8}$ dollars. A week later the stock increased $2\frac{3}{4}$ dollars. What is her share of stock worth now? **$28\frac{1}{8}$ dollars**

38. Robin practices the guitar $5\frac{1}{4}$ hr a week. This week she has practiced $3\frac{5}{12}$ hr. How many more hours will Robin practice this week? **$1\frac{5}{6}$ hr**

39. Sally's father drives $19\frac{1}{3}$ mi to work in the morning. On the way home from work, he picks up Sally's little brother from day care. He drives $22\frac{4}{9}$ mi on the way home. How many miles does he drive to and from work? **$41\frac{7}{9}$ mi**

40. Pete lives $1\frac{3}{4}$ mi from school. Tony lives $1\frac{2}{3}$ mi from school. How much farther from school does Pete live than Tony? **$\frac{1}{12}$ mi**

41. **WRITE ABOUT IT** Explain how subtracting or adding mixed numbers with unlike denominators differs from subtracting or adding mixed numbers with like denominators. **You must write unlike fractions in mixed numbers as like fractions before you can add or subtract.**

Mixed Review

Round each fraction to 0, $\frac{1}{2}$, or 1.

42. $\frac{2}{9}$ **0**   **43.** $\frac{10}{13}$ **1**   **44.** $\frac{9}{20}$ **$\frac{1}{2}$**   **45.** $\frac{2}{15}$ **0**   **46.** $\frac{22}{27}$ **1**   **47.** $\frac{17}{36}$ **$\frac{1}{2}$**

Write the equivalent fraction that is in simplest form.

48. $\frac{1}{2}, \frac{5}{10}, \frac{4}{8}, \frac{6}{12}$ **$\frac{1}{2}$**   **49.** $\frac{8}{12}, \frac{2}{3}, \frac{12}{18}, \frac{18}{27}$ **$\frac{2}{3}$**   **50.** $\frac{8}{18}, \frac{16}{36}, \frac{4}{9}, \frac{20}{45}$ **$\frac{4}{9}$**   **51.** $\frac{30}{36}, \frac{20}{24}, \frac{25}{30}, \frac{5}{6}$ **$\frac{5}{6}$**

52. DECIMALS The scouts held a bake sale to raise money for the homeless shelter. They sold 244 cookies for $0.15 each, 12 cakes for $15.50 each, and 145 brownies for $0.30 each. How much money did the scouts raise? **$266.10**

53. In the United States, 1,500 aluminum cans are recycled every second. At this rate, how many cans are recycled in a minute? in an hour? in a day? **90,000 cans; 5,400,000 cans; 129,600,000 cans**

MORE PRACTICE Lesson 6.3, page H50   121

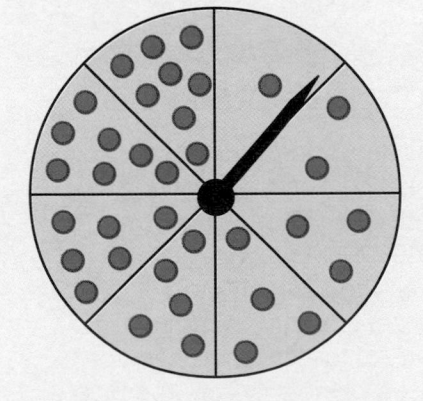

Use these questions to help students focus on the big ideas of the lesson.

• Wendy was $60\frac{1}{4}$ in. tall when she started sixth grade. She grew $2\frac{3}{8}$ in. during the school year. How tall was she by the end of the school year? $62\frac{5}{8}$ in.

✓ Assessment Tip

Have students write the answer.

• **Did you rename to subtract in Exercise 32? Explain.** Yes; $\frac{1}{4}$ is less than $\frac{3}{4}$.

DAILY QUIZ

Find the sum. Write the answer in simplest form.

1. $1\frac{1}{8} + 4\frac{3}{4}$ **$5\frac{7}{8}$**   **2.** $5\frac{2}{3} + 3\frac{5}{6}$ **$9\frac{1}{2}$**

Find the difference. Write the answer in simplest form.

3. $2\frac{2}{3} - 1\frac{5}{12}$ **$1\frac{1}{4}$**   **4.** $7 - 4\frac{5}{9}$ **$2\frac{4}{9}$**

5. $3\frac{1}{3} - 1\frac{3}{4}$ **$1\frac{7}{12}$**   **6.** $6\frac{1}{2} - 5\frac{4}{5}$ **$\frac{7}{10}$**

INTERDISCIPLINARY CONNECTION

One suggestion for an interdisciplinary unit for this chapter is Fraction Games. In this unit you can have students use what they have learned in this chapter to design a game that involves adding and subtracting fractions. Suggest a theme like Pizza Party, Competing on Time, or Driving Around. Students may design a game board and their own game pieces, or they may describe how fraction bars can be used to play the game.

Modalities Kinesthetic, Visual

GRADE 5	GRADE 6	GRADE 7
Solve problems using mixed numbers.	Estimate sums and differences of mixed numbers.	Estimate sums and differences.

ORGANIZER

OBJECTIVE To estimate sums and differences of mixed numbers

ASSIGNMENTS Independent Practice
- *Basic* 1–15, odd; 19; 20–29
- *Average* 1–19; 20–29
- *Advanced* 11–19; 20–29

PROGRAM RESOURCES
Reteach 6.4 Practice 6.4 Enrichment 6.4

PROBLEM OF THE DAY

Transparency 6

Complete the Magic Square. The magic sum is $1\frac{1}{4}$.

Solution Tab A, p. A6

$\frac{1}{2}$	$\frac{7}{12}$	$\frac{1}{6}$
$\frac{1}{12}$	$\frac{5}{12}$	$\frac{3}{4}$
$\frac{2}{3}$	$\frac{1}{4}$	$\frac{1}{3}$

QUICK CHECK

Round the fractions to 0, $\frac{1}{2}$, or 1.

1. $\frac{4}{5}$ 1
2. $\frac{3}{8}$ $\frac{1}{2}$
3. $\frac{2}{9}$ 0
4. $\frac{2}{5}$ $\frac{1}{2}$
5. $\frac{7}{12}$ $\frac{1}{2}$
6. $\frac{8}{9}$ 1

1 MOTIVATE

Copy this grocery list on the board, and then ask students these questions.

- **Which items weigh about 0 lb?**
  Swiss cheese

- **Which items weigh about $\frac{1}{2}$ lb?**
  tuna salad, potato salad

- **Which items weigh about 1 lb?**
  ham, potato salad

Grocery List
$\frac{1}{8}$ lb Swiss cheese
$\frac{7}{8}$ lb ham
$\frac{7}{16}$ lb tuna salad
$\frac{3}{4}$ lb potato salad

2 TEACH/PRACTICE

Students begin the lesson by using a ruler to round mixed numbers. Then they apply their rounding skills to estimating the sums and differences of mixed numbers.

6.4

WHAT YOU'LL LEARN
How to estimate sums and differences of mixed numbers

WHY LEARN THIS?
To estimate answers to problems such as total length

REMEMBER:
To estimate with a fraction less than 1, round it to 0, $\frac{1}{2}$, or 1.
See page 104.

Round $\frac{1}{9}$ to 0.

Round $\frac{3}{8}$ to $\frac{1}{2}$.

Round $\frac{4}{5}$ to 1.

Science Link

House cats, tigers, and lions are all closely related. There are differences among cats, especially in size. The average lion is $3\frac{1}{2}$ ft tall. The average house cat is $\frac{3}{4}$ ft tall. About how much taller is the lion?

about $2\frac{1}{2}$ ft taller

Estimating Sums and Differences

Look at the ruler below. Is $2\frac{5}{8}$ in. closer to 2 in., $2\frac{1}{2}$ in., or 3 in.? $2\frac{1}{2}$ in.

inches 1 2 3

To estimate sums and differences of mixed numbers, round each mixed number to the nearest whole number or to $\frac{1}{2}$.

EXAMPLE 1 Arnold has $9\frac{1}{4}$ yd of rope and Sherry has $6\frac{7}{8}$ yd of rope. About how much rope do Arnold and Sherry have together?

$9\frac{1}{4} + 6\frac{7}{8}$ *Round each mixed number.*
↓ ↓ **Think:** $\frac{1}{4}$ *rounds to 0;* $\frac{7}{8}$ *rounds to 1.*
$9 + 7 = 16$ *Add.*

So, Arnold and Sherry have about 16 yd of rope.

- What if Arnold has $10\frac{3}{4}$ yd? About how much rope do they have together? **about 18 yd**

EXAMPLE 2 Teri is going to build houses for Kevin's dog and cat. Teri needs $14\frac{4}{9}$ ft of lumber for the dog's house and $11\frac{1}{6}$ ft of lumber for the cat's house. About how much more feet of lumber does Teri need for the dog's house?

$14\frac{4}{9} - 11\frac{1}{6}$ *Round each mixed number.*
↓ ↓ **Think:** $\frac{4}{9}$ *rounds to* $\frac{1}{2}$, $\frac{1}{6}$ *rounds to 0.*
$14\frac{1}{2} - 11 = 3\frac{1}{2}$ *Subtract.*

So, Teri needs about $3\frac{1}{2}$ more feet of lumber for the dog's house.

- What if Teri builds a birdhouse? She needs $1\frac{2}{3}$ ft of lumber to build it. About how many more feet of lumber will Teri use to build the cat's house than the birdhouse? **about 9 more feet of lumber**

122 Chapter 6

IDEA FILE

SCIENCE CONNECTION

Distances in our solar system are measured in astronomical units (AU). One AU is equal to 92,900,000 mi, the distance from the Earth to the Sun. Have students round to the nearest half AU the distance each of these planets is from the Sun.

Jupiter, $5\frac{1}{5}$ AU 5 AU

Saturn, $9\frac{3}{5}$ AU $9\frac{1}{2}$ AU

Pluto, $39\frac{4}{5}$ AU 40 AU

Neptune, $30\frac{1}{10}$ AU 30 AU

Modalities Auditory, Visual

RETEACH 6.4

Name _____

LESSON 6.4

Estimating Sums and Differences

To estimate sums and differences of mixed numbers, round mixed numbers to the nearest whole number or $\frac{1}{2}$.

Rita is working on two sewing projects. She needs $4\frac{3}{5}$ yd of material for one project and $3\frac{7}{8}$ yd of material for another project. About how much material does Rita need?

Step 1 Round each mixed number. $4\frac{3}{5}$ $3\frac{7}{8}$
  Think: $\frac{3}{5}$ rounds to $\frac{1}{2}$. $\frac{7}{8}$ rounds to 1.
  $4\frac{3}{5} \longrightarrow 4\frac{1}{2}$ $3\frac{7}{8} \longrightarrow 4$

Step 2 Add the rounded addends. $4\frac{1}{2} + 4 = 8\frac{1}{2}$

So, Rita needs about $8\frac{1}{2}$ yd of material.

Estimate the sum or difference. Possible estimates are given.

1. $5\frac{1}{7} - 2\frac{5}{6} =$ __2__
2. $8\frac{3}{4} + 1\frac{1}{15} =$ __10__
3. $11\frac{2}{9} - 4\frac{7}{8} =$ __6__
4. $8\frac{11}{12} - 6\frac{1}{6} =$ __3__
5. $5\frac{3}{7} + 2\frac{2}{3} =$ __$8\frac{1}{2}$__
6. $9\frac{3}{8} + 4\frac{1}{5} =$ __$13\frac{1}{2}$__
7. $5\frac{4}{9} + 1\frac{7}{8} =$ __$7\frac{1}{2}$__
8. $7\frac{5}{12} - 3\frac{7}{8} =$ __$3\frac{1}{2}$__
9. $14\frac{9}{10} - 9\frac{1}{6} =$ __6__
10. $7\frac{1}{2} + 1\frac{3}{9} =$ __$8\frac{1}{2}$__

R28 TAKE ANOTHER LOOK Use with text pages 122–123.

GUIDED PRACTICE

Round the fractions to 0, $\frac{1}{2}$, or 1. Then rewrite the problem. **Possible answers are given.**

1. $2\frac{1}{4} + 1\frac{1}{8}$ 2 + 1   **2.** $4\frac{3}{8} - 1\frac{1}{3}$ $4\frac{1}{2} - 1$ **3.** $2\frac{1}{2} + 4\frac{4}{5}$ $2\frac{1}{2} + 5$ **4.** $5\frac{1}{6} + 2\frac{7}{9}$ 5 + 3   **5.** $9\frac{7}{12} - 5\frac{1}{5}$

$9\frac{1}{2} - 5$

Estimate the sum or difference. **Possible estimates are given.**

6. $3\frac{1}{4} + 1\frac{5}{8}$ $4\frac{1}{2}$ **7.** $6\frac{3}{16} - 4\frac{1}{6}$ 2   **8.** $4\frac{11}{12} - 3\frac{1}{5}$ 2   **9.** $8\frac{7}{8} + 5\frac{5}{9}$ $14\frac{1}{2}$ **10.** $10\frac{3}{4} - 2\frac{2}{7}$ 9

INDEPENDENT PRACTICE

Estimate the sum or difference. **Possible estimates are given.**

1. $4\frac{3}{4} + 1\frac{1}{6}$ 6   **2.** $12\frac{1}{12} + 6\frac{7}{8}$ 19   **3.** $7\frac{9}{16} - 4\frac{2}{9}$ $3\frac{1}{2}$ **4.** $10\frac{9}{11} + 2\frac{3}{16}$ 13   **5.** $8\frac{7}{12} - 6\frac{3}{14}$ $2\frac{1}{2}$

6. $6\frac{5}{12} - 4\frac{7}{8}$ $1\frac{1}{2}$ **7.** $8\frac{2}{7} + 4\frac{13}{15}$ 13   **8.** $6\frac{7}{16} - 3\frac{3}{17}$ $3\frac{1}{2}$ **9.** $8\frac{6}{11} + 8\frac{7}{16}$ 17   **10.** $9\frac{2}{11} - 6\frac{7}{15}$ $2\frac{1}{2}$

11. $3\frac{1}{6} + 1\frac{1}{3}$ 4   **12.** $2\frac{7}{12} + 12\frac{1}{10}$ $14\frac{1}{2}$ **13.** $8\frac{2}{9} - 2\frac{7}{18}$ $5\frac{1}{2}$ **14.** $10\frac{1}{2} + 2\frac{7}{8}$ $13\frac{1}{2}$ **15.** $12\frac{7}{15} - 10\frac{7}{18}$ 2

Problem-Solving Applications Possible estimates are given.

16. Jill is working on a project for a craft show. She uses $1\frac{7}{8}$ yd of blue ribbon and $10\frac{5}{12}$ yd of purple ribbon. For her next project, she wants to use same amount of ribbon but only one color of ribbon. What is a good estimate of the amount of ribbon Jill will need? **about $12\frac{1}{2}$ yd**

17. Mr. Smith's last water bill showed that he was using $22\frac{15}{16}$ gal of water a day. One faucet leaked $2\frac{3}{4}$ gal of water a day. Now that the faucet has been fixed, estimate how many gallons of water a day Mr. Smith will use. **about 20 gal**

18. Mr. Kelly had $60\frac{3}{16}$ ft of plastic piping. He installed $28\frac{7}{12}$ ft in the bathroom. About how much piping does he have left?
about $31\frac{1}{2}$ ft

19. ✏ **WRITE ABOUT IT** Explain how to estimate sums or differences of mixed numbers. **Round the mixed number to the nearest whole number. Then add or subtract.**

Mixed Review

Multiply.

20. 0.5×0.25 **0.125**   **21.** 0.75×0.2 **0.15**   **22.** 0.5×0.75 **0.375**   **23.** 0.125×0.2 **0.025**

Find the sum or difference. Write the answer in simplest form.

24. $\frac{5}{6} - \frac{1}{4}$ $\frac{7}{12}$ **25.** $\frac{3}{5} + \frac{2}{3}$ $1\frac{4}{15}$ **26.** $\frac{4}{5} - \frac{1}{2}$ $\frac{3}{10}$ **27.** $\frac{1}{6} + \frac{3}{8}$ $\frac{13}{24}$

28. James travels 60 mi a day to and from work. He works 5 days a week. His car gets 25 mi per gallon of gasoline. How many gallons does James use every week?
12 gal

29. Marci bought 3 CDs for $36.97. The first CD cost $12.99 and the second cost $10.99. How much did the third CD cost?
$12.99

💡**CRITICAL THINKING** One share of KidStuf stock is currently priced at $24\frac{1}{8}$. The price dropped $2\frac{7}{8}$ points yesterday and $1\frac{5}{8}$ points the day before. About what was the stock price three days ago? about $28\frac{1}{2}$

ADDITIONAL EXAMPLES

Example 1, p. 122

Betsy bought $3\frac{1}{8}$ lb of Muenster cheese and $4\frac{7}{8}$ lb of mozzarella cheese at the delicatessen. About how much cheese did she buy in all?

about 8 lb

Example 2, p. 122

The guests drank $4\frac{5}{12}$ qt of orange juice and $3\frac{5}{6}$ qt of tomato juice at the annual school brunch. About how much more orange juice did they drink?

about $\frac{1}{2}$ qt more

3 WRAP UP

Use these questions to help students focus on the big ideas of the lesson.

- **Roger sold $1\frac{5}{12}$ qt of the $4\frac{1}{6}$ qt of raspberries he picked. About how many quarts of raspberries does he have left?** about $2\frac{1}{2}$ qt

✔ *Assessment Tip*

Have students write the answer.

- **Explain how you solved Exercise 16.** Possible answer: by rounding $1\frac{7}{8}$ to 2 and $10\frac{5}{12}$ to $10\frac{1}{2}$, then adding 2 to $10\frac{1}{2}$, to get the sum of $12\frac{1}{2}$.

DAILY QUIZ

Estimate the sum or difference. Possible estimates are given.

1. $3\frac{2}{3} + 7\frac{5}{9}$ $11\frac{1}{2}$ **2.** $9\frac{1}{4} - 2\frac{8}{9}$ 6
3. $12\frac{4}{9} + 5\frac{1}{8}$ $17\frac{1}{2}$ **4.** $8\frac{5}{8} - 4\frac{3}{8}$ 4
5. $5\frac{1}{6} - 2\frac{5}{12}$ $2\frac{1}{2}$ **6.** $9\frac{4}{5} - 3\frac{1}{3}$ 7

CO⊙PERATIVE
LEARNING

OBJECTIVE To practice solving problems involving mixed numbers

USING THE PAGE

Before the Lesson

Display a dollar bill, a nickel, a dime, a quarter, and a half-dollar. Have the students write fractions representing each of the coins' relationship to the dollar bill.

$\frac{1}{20}$, $\frac{1}{10}$, $\frac{1}{4}$, $\frac{1}{2}$

Write "1 hour" on the overhead projector. Have students write the fractional part of an hour for each of the following number of minutes.

40, 15, 30, 45, and 50. $\frac{2}{3}$, $\frac{1}{4}$, $\frac{1}{2}$, $\frac{3}{4}$, $\frac{5}{6}$

Cooperative Groups

Have students work together in groups to read the first part of the lesson and answer the questions under the example. Then have them solve Problems 1–3.

When they finish, ask them to write a schedule with 4 after-school activities. Have them list the time it takes to complete each activity using mixed numbers. Ask groups to write a subtraction problem and an addition problem with the mixed numbers. Have groups exchange the problems and solve them.

Cricket is a game played with a hard ball and a bat with flat sides. There are two teams of 11 players each. The game is similar to baseball, but there are two pitching and batting positions in cricket instead of one. The field has two wickets (like goal posts) that are 22 yd apart. Fair play is important to the game. The expression, "It's not cricket," implies something is not fair.

CULTURAL CONNECTION

Soccer and Cricket

Soccer is much more popular in England than in the United States. Thousands of people in England regularly attend soccer matches, much as thousands of Americans regularly attend football games. The fans are very enthusiastic at a soccer match while supporting their favorite team.

Another popular sport in England is cricket. Cricket somewhat resembles the American sport of baseball.

Anthony lives in London, England, and enjoys playing cricket and soccer. Before a soccer match, Anthony wants to make two banners. One banner will be $3\frac{1}{2}$ m long and the other will be $4\frac{1}{5}$ m long. How long will the combined length of the banners be?

$$3\frac{1}{2} = 3\frac{5}{10}$$
$$+ 4\frac{1}{5} = 4\frac{2}{10}$$
$$\overline{\phantom{+4\frac{1}{5}=} 7\frac{7}{10}}$$

So, the combined length of the banners will be $7\frac{7}{10}$ m.

• How much longer is the larger banner than the smaller banner? $\frac{7}{10}$ m

Work Together

1. Anthony played two soccer matches in one day. The first match lasted $1\frac{1}{2}$ hr with a 30-min break. The second match lasted $1\frac{3}{4}$ hr with a 45-min break. How long did the matches and breaks take? Write your answer as a mixed number. $4\frac{1}{2}$ hr

2. Anthony plays cricket in the spring, summer, and early fall. He spends 10 hr a week practicing cricket. He spends $3\frac{2}{5}$ hr practicing the drums every week. How much longer does he play cricket than practice the drums? $6\frac{3}{5}$ hr longer

3. Debbie and Anthony take the London Underground to Piccadilly Circus. The ride on the Underground takes $\frac{1}{2}$ hr. They spend 2 hr shopping, 1 hr at Buckingham Palace, and 45 min sightseeing. The trip back takes 40 min. How much time did the trip take? $4\frac{11}{12}$ hr

Piccadilly Circus is one of London's most popular tourist spots. In this context, the word *circus* means "circle" or "circular." Piccadilly Circus is a busy traffic circle in London with many special shops where visitors can buy souvenirs. Nearby is Buckingham Palace, the Queen's London residence.

REVIEW

EXAMPLES

EXERCISES

• **Use diagrams to add and subtract mixed numbers.** (pages 112–117)

Check students' diagrams

Draw a model to find $1\frac{3}{5} - 1\frac{1}{5}$.
Draw $1\frac{3}{5}$.

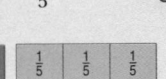

Subtract $1\frac{1}{5}$.

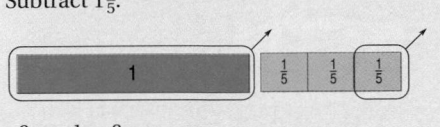

$1\frac{3}{5} - 1\frac{1}{5} = \frac{2}{5}$

Draw a diagram to find each sum or difference. Write the answer in simplest form.

1. $2\frac{3}{4} + 3\frac{1}{8}$ $5\frac{7}{8}$
2. $1\frac{3}{8} - 1\frac{1}{8}$ $\frac{1}{4}$
3. $2\frac{3}{8} + 1\frac{1}{2}$ $3\frac{7}{8}$
4. $2\frac{5}{6} - 1\frac{3}{4}$ $1\frac{1}{12}$
5. $1\frac{1}{6} + 3\frac{2}{3}$ $4\frac{5}{6}$
6. $2\frac{2}{10} - 1\frac{3}{5}$ $\frac{6}{10}$, or $\frac{3}{5}$
7. $2\frac{1}{2} + 2\frac{1}{5}$ $4\frac{7}{10}$
8. $1\frac{3}{6} - 1\frac{1}{3}$ $\frac{1}{6}$

9. Pat lives $1\frac{2}{5}$ mi from school. Kat lives $2\frac{3}{10}$ mi from school. How much closer to school does Pat live than Kat? $\frac{9}{10}$ **mi**

• **Add and subtract mixed numbers.** (pages 118–121)

$2\frac{1}{6} = 2\frac{1}{6}$
$-1\frac{1}{2} = 1\frac{3}{6}$

Write equivalent fractions by using the LCD.

$2\frac{1}{6} = 2\frac{1}{6} = 1\frac{7}{6}$
$-1\frac{1}{2} = 1\frac{3}{6} = 1\frac{3}{6}$
$\frac{4}{6}$, or $\frac{2}{3}$

Rename the mixed number on top.
$2\frac{1}{6} = 1 + \frac{6}{6} + \frac{1}{6}$
$= 1\frac{7}{6}$
Subtract the fractions. Subtract the whole numbers.

Find the sum or difference.

10. $2\frac{2}{7} + 3\frac{3}{7}$ $5\frac{5}{7}$
11. $7\frac{3}{4} - 5\frac{1}{3}$ $2\frac{5}{12}$
12. $8\frac{1}{3} - 3\frac{1}{8}$ $5\frac{5}{24}$
13. $11\frac{5}{6} - 6\frac{2}{3}$ $5\frac{1}{6}$
14. $1\frac{2}{4} + 5\frac{1}{8}$ $6\frac{5}{8}$
15. $6\frac{3}{5} + 2\frac{1}{3}$ $8\frac{14}{15}$
16. $3\frac{1}{4} - 2\frac{1}{2}$ $\frac{3}{4}$
17. $4\frac{3}{8} + 2\frac{3}{4}$ $7\frac{1}{8}$

18. Allie grew $2\frac{2}{3}$ in. during the fifth grade and $3\frac{1}{4}$ in. during the sixth grade. How much did Allie grow during the two school years? $5\frac{11}{12}$ **in.**

• **Estimate the sum or difference.** (pages 122–123)

$3\frac{2}{5} - 1\frac{3}{11}$ *Round each mixed number.*
↓   ↓
$3\frac{1}{2} - 1 = 2\frac{1}{2}$ *Subtract.*

25. about $1\frac{1}{2}$ ft taller

Estimate the sum or difference.

19. $9\frac{4}{5} - 2\frac{7}{9}$ 7
20. $6\frac{1}{4} + 3\frac{2}{9}$ 9
21. $4\frac{9}{20} + 1\frac{4}{5}$ $6\frac{1}{2}$
22. $3\frac{3}{7} + 1\frac{1}{14}$ $4\frac{1}{2}$
23. $4\frac{3}{4} - 1\frac{1}{8}$ 4
24. $3\frac{2}{4} - 1\frac{6}{7}$ $1\frac{1}{2}$

25. Barry is $5\frac{5}{8}$ ft tall. Kevin is $4\frac{1}{4}$ ft tall. About how much taller is Barry than Kevin?

USING THE PAGE

The Chapter 6 Review reviews adding and subtracting mixed numbers and estimating sums or differences. Chapter objectives are provided, along with examples and practice exercises for each.

 The Chapter 6 Review can be used as a review or a test. You may wish to place the Review in students' portfolios.

Assessment Checkpoint

For Performance Assessment in this chapter, see page 146, What Did I Learn?, items 3 and 7.

✔ Assessment Tip

Student Self-Assessment The How Did I Do? Survey helps students assess both what they have learned and how they learned it. Self-assessment helps students learn more about their own capabilities and develop confidence in themselves.

A self-assessment survey is available as a copying master in the *Teacher's Guide for Assessment,* p. 15.

CHAPTER 6 TEST, page 1

Name _____

CHAPTER 6 TEST
PAGE 1

Choose the letter of the correct answer.

For questions 1–2, choose the addition problem shown by the diagram.

1.
A. $3\frac{1}{7} + 1\frac{1}{7}$ B. $3\frac{2}{7} + 1\frac{1}{7}$
C. $3\frac{3}{7} + 1$ D. $3 + 1\frac{3}{7}$

2.
A. $\frac{3}{4} + \frac{7}{8}$ B. $1\frac{2}{8} + 2\frac{1}{4}$
C. $2\frac{1}{4} + 1\frac{3}{8}$ D. $2\frac{1}{8} + 1\frac{3}{4}$

For questions 3–4, draw a diagram to help you find the sum.

3. $3\frac{1}{9} + 2\frac{1}{3}$
A. $5\frac{2}{9}$ B. $5\frac{1}{3}$ C. $5\frac{1}{2}$ D. $5\frac{2}{3}$

4. $1\frac{1}{2} + 2\frac{1}{4}$
A. $3\frac{1}{2}$ B. $3\frac{3}{4}$ C. $3\frac{5}{8}$ D. $4\frac{1}{4}$

5. Which diagram matches the mixed number $2\frac{1}{3}$?
A.   B.
C.   D.

6. Use the diagram to find the difference.

$3\frac{1}{4} - 1\frac{1}{2}$
A. $1\frac{3}{4}$ B. $1\frac{7}{8}$
C. $2\frac{1}{4}$ D. $2\frac{3}{4}$

For questions 7–8, draw a diagram to help you find the difference. Answers are in simplest form.

7. $4\frac{4}{5} - 3\frac{2}{5}$
A. $\frac{2}{5}$ B. $1\frac{1}{5}$
C. $1\frac{2}{5}$ D. $1\frac{2}{5}$

8. $5\frac{5}{12} - 3\frac{1}{2}$
A. $1\frac{5}{6}$ B. $1\frac{11}{12}$
C. $2\frac{1}{12}$ D. $2\frac{2}{5}$

9. $3\frac{3}{8} + 1\frac{3}{8}$
A. $2\frac{1}{8}$ B. $4\frac{1}{2}$
C. $4\frac{3}{4}$ D. not here

10. $4\frac{3}{5} + 2\frac{5}{6}$
A. $6\frac{1}{2}$ B. $7\frac{1}{2}$
C. $7\frac{2}{3}$ D. not here

Form A • Multiple-Choice   A21   Go on. ▶

CHAPTER 6 TEST, page 2

Name _____

CHAPTER 6 TEST
PAGE 2

11. A costume requires $2\frac{1}{2}$ yd of fabric for the jacket and $1\frac{1}{8}$ yd for the pants. How much fabric is required in all?
A. $3\frac{5}{8}$ yd B. $3\frac{4}{5}$ yd
C. 4 yd D. $4\frac{3}{8}$ yd

12. Last week the basketball team practiced for $6\frac{1}{2}$ hr. It usually practices for $9\frac{1}{4}$ hr. How many hours less did the team practice last week?
A. $2\frac{3}{4}$ hr B. $2\frac{7}{8}$ hr
C. $3\frac{1}{4}$ hr D. $3\frac{3}{4}$ hr

13. In March it rained $5\frac{3}{10}$ in. In April it rained $3\frac{1}{5}$ in. How much more did it rain in March than in April?
A. $1\frac{1}{5}$ in. B. $1\frac{1}{2}$ in.
C. $2\frac{1}{10}$ in. D. $2\frac{1}{2}$ in.

14. For a class party, $2\frac{3}{4}$ gal of apple juice and $1\frac{1}{2}$ gal of ginger ale were combined to make punch. How many gallons of punch were made?
A. $1\frac{1}{4}$ gal B. $3\frac{7}{8}$ gal
C. 4 gal D. $4\frac{1}{4}$ gal

15. Which will give the best estimate of the sum?
$5\frac{7}{8} + 2\frac{3}{4}$
A. 5 + 2 B. 6 + 2
C. 5 + 3 D. 6 + 3

16. Which will give the best estimate of the difference?
$9\frac{1}{6} - 2\frac{9}{10}$
A. 9 – 3 B. 9 – 2
C. 10 – 2 D. 10 – 3

For questions 17–18, choose the best estimate for each sum or difference.

17. $3\frac{2}{9} + 2\frac{1}{7}$
A. about 1 B. about 5
C. about 7 D. about 8

18. $7\frac{5}{8} - 2\frac{1}{12}$
A. about $4\frac{1}{2}$ B. about $5\frac{1}{2}$
C. about $6\frac{1}{2}$ D. about 10

For questions 19–20, choose the best estimate for the word problem.

19. At a fruit stand, the Raleigh family bought $4\frac{1}{4}$ lb of cooking apples and $2\frac{3}{10}$ lb of apples to eat. About how many pounds of apples did the family buy altogether?
A. about 6 lb B. about 7 lb
C. about 9 lb D. about 10 lb

20. On a hike, the scout troop hiked $13\frac{5}{9}$ mi on the first day. They hiked $9\frac{1}{10}$ mi on the second day. About how much farther did they hike the first day than the second day?
A. about $3\frac{1}{2}$ mi B. about 5 mi
C. about 5 mi D. about 21 mi

Form A • Multiple-Choice   A22   ▶ Stop!

See *Test Copying Masters,* pp. A21–A22 for the Chapter Test in **multiple-choice** format and pp. B171–B172 for **free-response** format.

125

PLANNING GUIDE

Multiplying and Dividing Fractions

BIG IDEA: Multiplication and division of fractions and mixed numbers can be used to solve many real-world problems.

Introducing the Chapter p. 126	Team-Up Project • Taking Stock p. 127		
OBJECTIVE	**VOCABULARY**	**MATERIALS**	**RESOURCES**
7.1 Multiplying with Fractions pp. 128-131 **Objective:** To multiply fractions **2 DAYS**	*factor product simplest form GCF*	FOR EACH STUDENT: paper, ruler, 2 different-color pencils or markers	■ Reteach, ■ Practice, ■ Enrichment 7.1; More Practice, p. H51 Math Fun • Egyptian Symbols, p. 142
EXTENSION • Reciprocals, Teacher's Edition pp. 132A–132B			
7.2 Simplifying Factors pp. 132–133 **Objective:** To multiply fractions after simplifying each factor **1 DAY**			■ Reteach, ■ Practice, ■ Enrichment 7.2; More Practice, p. H51
7.3 Mixed Numbers pp. 134–135 **Objective:** To multiply fractions and mixed numbers **1 DAY**			■ Reteach, ■ Practice, ■ Enrichment 7.3; More Practice, p. H52 Math Fun • Cooking Up a Treat, p. 142 ⊙ ***Number Heroes •*** *Fraction Fireworks*
Lab Activity: Exploring Division pp. 136–137	*dividend divisor*	FOR EACH STUDENT: fraction circles	⊡ **E-Lab • Activity 7** E-Lab Recording Sheets, p. 7
7.4 Dividing Fractions pp. 138-139 **Objective:** To divide with fractions **1 DAY**	**reciprocal**		■ Reteach, ■ Practice, ■ Enrichment 7.4; More Practice, p. H52 Math Fun • Roll a Fraction, p. 142
7.5 Problem-Solving Strategy: Work Backward by Dividing Mixed Numbers pp. 140-141 **Objective:** To divide mixed numbers when using the strategy *work backward* **1 DAY**			■ Reteach, ■ Practice, ■ Enrichment 7.5; More Practice, p. H52 Problem-Solving Think Along, TR p. R1
CHAPTER ASSESSMENT Chapter 7 Review p. 143			
ASSESSMENT CHECKPOINT Review • Chapter 4–7 pp. 144–145 Performance Assessment • Chapter 4–7 p. 146 Cumulative Review • Chapter 4–7 p. 147			

NCTM CURRICULUM
STANDARDS FOR GRADES 5–8

✓ **Problem Solving**
Lessons 7.1, 7.2, 7.3, 7.4, 7.5

✓ **Communication**
Lessons 7.1, 7.2, 7.3, 7.4, 7.5

✓ **Reasoning**
Lessons 7.1, 7.2, 7.5

✓ **Connections**
Lessons 7.1, 7.2, 7.4

✓ **Number and Number Relationships**
Lesson 7.2

✓ **Number Systems and Number Theory**
Lesson 7.2

✓ **Computation and Estimation**
Lessons 7.1, 7.2, 7.3, 7.4, 7.5

✓ **Patterns and Functions**
Lab Activity

✓ **Algebra**
Lab Activity

☐ **Statistics**

☐ **Probability**

☐ **Geometry**

☐ **Measurement**

CHAPTER LEARNING GOALS

7A.1 To simplify factors and multiply fractions and mixed numbers

7A.2 To divide fractions

7A.3 To solve problems by working backward

These goals for concepts, skills, and problem solving are assessed in many ways throughout the chapter. See the chart below for a complete listing of both traditional and informal assessment options.

Pretest Options
Pretest for Chapter Content, TCM pp. A23–A24, B173–B174
Assessing Prior Knowledge, TE p. 126; APK p. 9

Daily Assessment
Mixed Review, PE pp. 131, 139
Problem of the Day, TE pp. 128, 132, 134, 138, 140
Assessment Tip and Daily Quiz, TE pp. 131, 133, 135, 137, 139, 141

Formal Assessment
Chapter 7 Review, PE p. 143
Chapter 7 Test, TCM pp. A23–A24, B173–B174
Study Guide & Review, PE pp. 144–145
Test • Chapters 4–7, TCM pp. A25–A26, B175–B176
Cumulative Review, PE p. 147
Cumulative Test, TCM pp. A27–A30, B177–B180

Performance Assessment
Team-Up Project Checklist, TE p. 127
What Did I Learn?, Chapters 4–7, TE p. 146
Interview/Task Test, PA p. 82

Portfolio
Suggested work samples:
Independent Practice, PE pp. 131, 133, 135, 139
Math Fun, PE p. 142

Student Self-Assessment
How Did I Do?, TGA p. 15
Portfolio Evaluation Form, TGA p. 24
Math Journal, TE pp. 126B, 130

Key
Assessing Prior Knowledge Copying Masters: APK
Test Copying Masters: TCM
Performance Assessment: PA
Teacher's Guide for Assessment: TGA
Teacher's Edition: TE
Pupil's Edition: PE

MATH COMMUNICATION

MATH JOURNAL
You may wish to use the following suggestions to have students write about the mathematics they are learning in this chapter:

PE page 132 • Talk About It

PE pages 131, 133, 135, 139, 141 • Write About It

TE page 130 • Writing in Mathematics

TGA page 15 • How Did I Do?

In every lesson • record written answers to each Wrap-up Assessment Tip from the Teacher's Edition for test preparation.

VOCABULARY DEVELOPMENT
The terms listed at the right will be introduced or reinforced in this chapter. The boldfaced terms in the list are new vocabulary terms in the chapter.

factor, pp. 129, 137
product, p. 129
simplest form, p. 129
GCF, p. 129
dividend, p. 137
divisor, p. 137
reciprocal, p. 138

So that students understand the terms rather than just memorize definitions, have them construct their own definitions and pictorial representations and record those in their Math Journals.

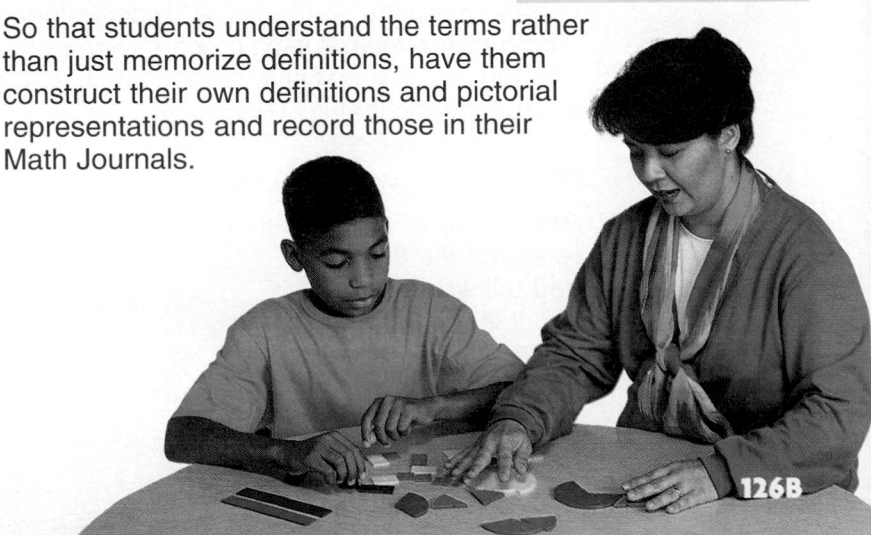

MEETING INDIVIDUAL NEEDS

RETEACH	**PRACTICE**
BELOW-LEVEL STUDENTS	**ON-LEVEL STUDENTS**
Practice Activities, TE Tab B *Finding Averages* (Use with Lessons 7.1–7.3.)	**Practice Game,** TE p. 126C *Two Many Fractions* (Use with Lessons 7.1–7.5.)
Interdisciplinary Connection, TE p. 140	**Interdisciplinary Connection,** TE p. 140
Literature Link, TE p. 126C *Island of the Blue Dolphins,* Activity 1 (Use with Lesson 7.1.)	**Literature Link,** TE p. 126C *Island of the Blue Dolphins,* Activity 2 (Use with Lesson 7.1.)
Technology *Number Heroes* • *Fraction Fireworks* (Use with Lesson 7.3.)	**Technology** *Number Heroes* • *Fraction Fireworks* (Use with Lesson 7.3.)
E-Lab • Activity 7 (Use with Lab Activity.)	**E-Lab** • Activity 7 (Use with Lab Activity.)

CHAPTER IDEA FILE

The activities on these pages provide additional practice and reinforcement of key chapter concepts. The chart above offers suggestions for their use.

 **LITERATURE LINK**

Island of the Blue Dolphins
by Scott O'Dell

This is the story of a young Indian girl named Karana, who is left behind on her island when her tribe leaves for a better place. Alone, she learns to take care of herself and begins to appreciate her heritage.

Activity 1—The island Karana lives on is 1 league × 2 leagues. Have students look up the definition of *league.* Then have them find the area, in square miles, that the island would have if it were rectangular.

Activity 2—The fathom is a unit of depth typically used to measure water. Have students look up the length of a *fathom.* Dolphins can swim at depths of 300 meters of less. Have students figure out how many fathoms that is. Then have them research and figure out how deep, in fathoms, an anglerfish or other deep-sea fish can swim.

Activity 3—Students should research the term *knot* and tell why sailors use it instead of *nautical miles per hour.*

PRACTICE GAME

Two Many Fractions

Purpose To practice dividing fractions

Materials cards numbered 0–9 for each group of four students

Each student draws this diagram:

 ×

About the Game The cards are mixed and placed facedown in a pile. The top four cards are turned over. Each student writes the four numbers in the boxes of his or her choice and then multiplies the fractions. A point is awarded to the player(s) with the product closest to 2. A score of 3 points wins the game.

Students can also play the game by making a division problem. Have students change the diagram and find the quotient closest to 2.

Modalities Kinesthetic, Visual

Multiple Intelligences Visual/Spatial, Logical/Mathematical

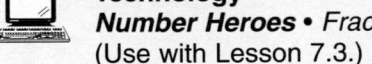

Practice Game, TE p. 126C
Two Many Fractions
(Use with Lessons 7.1–7.5.)

Interdisciplinary Connection, TE p. 140

Literature Link, TE p. 126C
Island of the Blue Dolphins, Activity 3
(Use with Lesson 7.1.)

Technology
Number Heroes • *Fraction Fireworks*
(Use with Lesson 7.3.)

E-Lab • Activity 7
(Use with Lab Activity.)

INTERDISCIPLINARY SUGGESTIONS

Connecting Fractions

Purpose To connect using fractions to other subjects students are studying

You may wish to connect **Multiplying and Dividing Fractions** to other subjects by using related interdisciplinary units.

Suggested Units

• Metric and Customary: Measuring with Decimals and Fractions

• Cooking with Math

• Fractions in the Stock Market

• Crafts and Fractions

Modalities Auditory, Kinesthetic, Visual

Multiple Intelligences Visual/Spatial, Logical/Mathematical, Bodily/Kinesthetic

The purpose of using technology in the chapter is to provide reinforcement of skills and support in concept development.

E-Lab • Activity 7—Students use proportional thinking and a computer model as they develop a new algorithm for division of fractions.

Number Heroes • *Fraction Fireworks*— The activities on this CD-ROM provide students with opportunities to practice showing products of fractions.

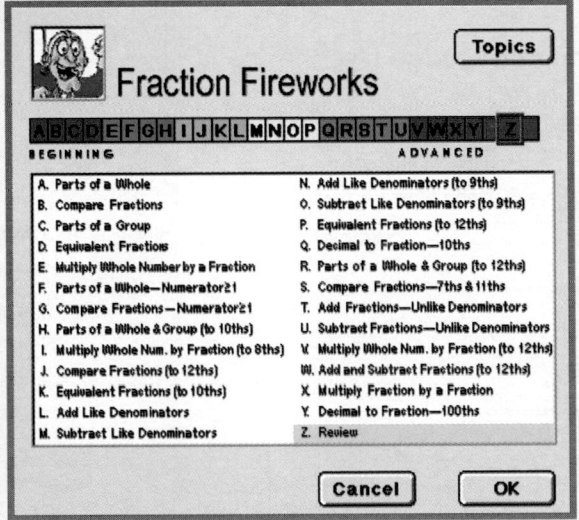

Each Number Heroes activity can be assigned by degree of difficulty. See Teacher's Edition page 135 for Grow Slide information.

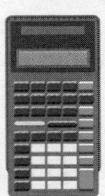
Explorer Plus™ or fx–65 Calculator—See suggestions for integrating the calculator in the PE lesson p. 135. Additional ideas are provided in the *Calculator Handbook,* p. H34.

Visit our web site for additional ideas and activities.
http://www.hbschool.com

CHAPTER 7

Introducing the Chapter

Encourage students to talk about multiplying and dividing fractions. Review the math content students will learn in the chapter.

WHY LEARN THIS?

The question "Why are you learning this?" will be answered in every lesson. Sometimes there is a direct application to everyday experiences. Other lessons help students understand why these skills are needed for future success in mathematics.

PRETEST OPTIONS

Pretest for Chapter Content See the *Test Copying Masters*, pages A23–A24, B173–B174 for a free-response or multiple-choice test that can be used as a pretest.

Assessing Prior Knowledge Use the copying master shown below to assess previously taught skills that are critical for success in this chapter.

Assessing Prior Knowledge Copying Masters, p. 9

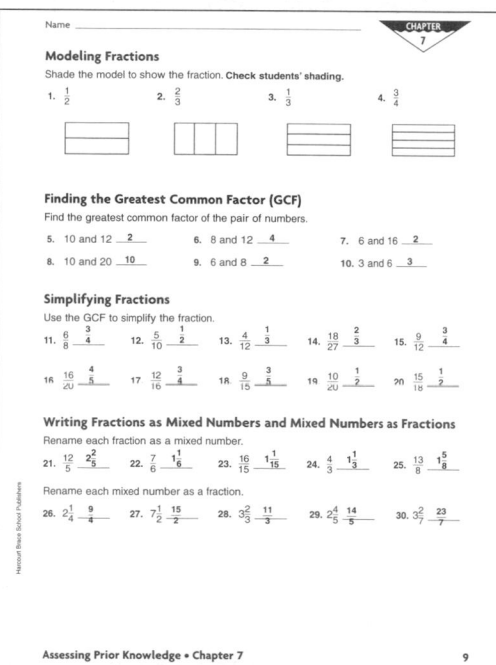

Troubleshooting

If students need help	Then use
• MODELING FRACTIONS	• Key Skills, PE p. H14
• FINDING GCF	• Practice Activities, TE Tab A, Activity 13B
• SIMPLIFYING FRACTIONS	• Key Skills, PE p. H15
• WRITING FRACTIONS/ MIXED NUMBERS	• Key Skills, PE p. H16

126

CHAPTER 7

MULTIPLYING AND DIVIDING FRACTIONS

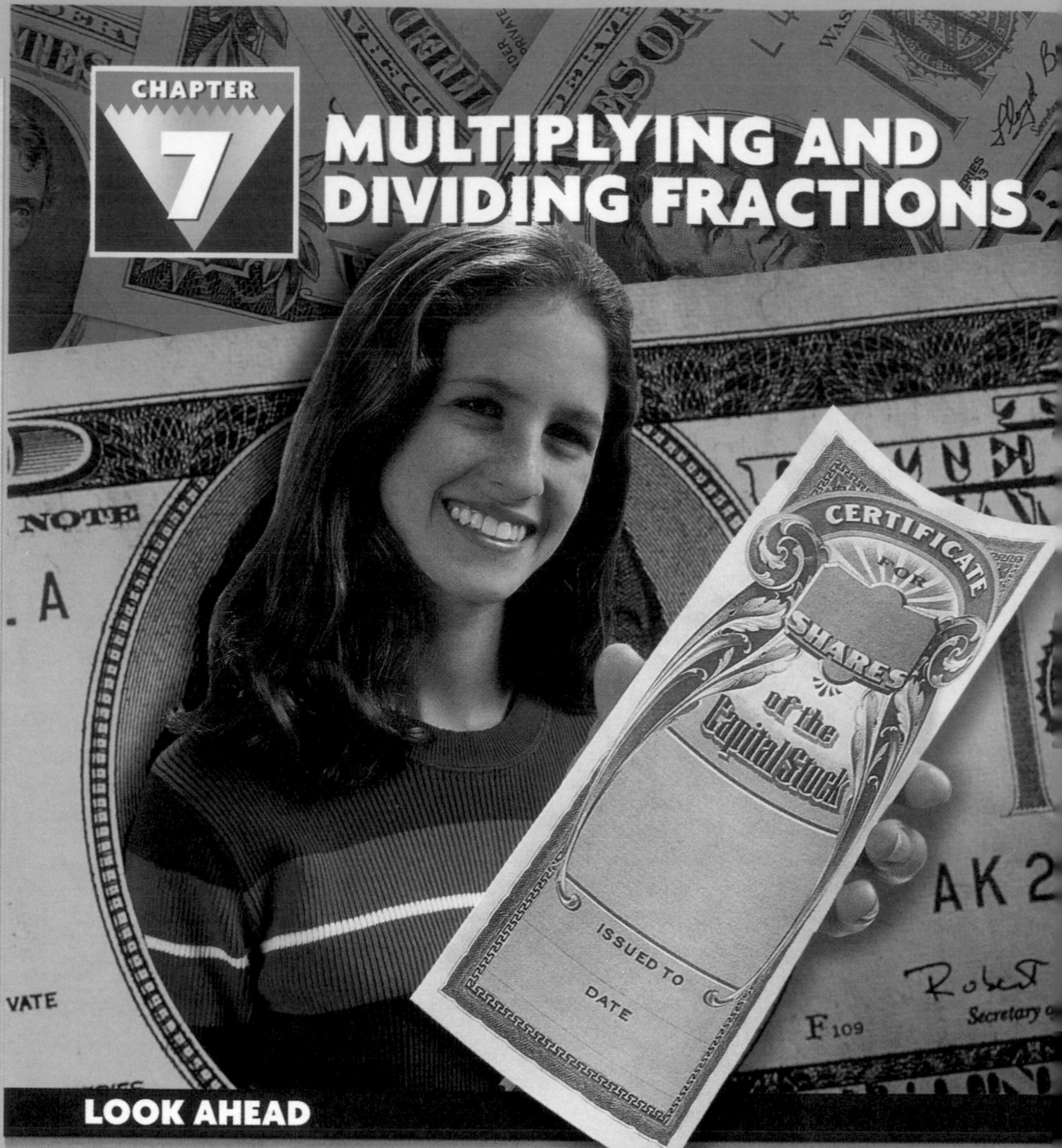

LOOK AHEAD

In this chapter you will solve problems that involve

- multiplying fractions and mixed numbers
- dividing fractions and mixed numbers

126 Chapter 7

SCHOOL-HOME CONNECTION

You may wish to send to each student's family the *Math at Home* page. It includes an article that stresses the importance of communication about mathematics in the home, as well as helpful hints about homework and taking tests. The activity and extension provide extra practice that develops concepts and skills involving fractions.

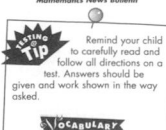

Math at Home

Taking Stock

Suppose that you have $1,000 to invest in the stock market. You want to pick stocks that you think will increase in value.

YOU WILL NEED: daily newspaper

Plan
- Work with a small group to invest $1,000 in stocks you think will increase in value over a 2-week period of time.

Act
- Choose 4–5 stocks from different companies.
- Decide how many shares of stock to buy with $1,000.
- Make a table that shows the name of each company you chose, the original price of each share of stock, the number of shares you purchased, and the days of the week (Monday–Friday) for 2 weeks.
- Look at the newspaper each day and record the daily price of each of your stocks under the corresponding day.
- At the end of 2 weeks, compare the final value of your stocks to your original $1,000 investment. Are your stocks worth more or less than when you started the project?

Share
- Present your results to the class.

DID YOU
- ✔ Select 4–5 companies for your investment?
- ✔ Decide how many shares to buy with $1,000?
- ✔ Make a table to record the daily price changes?

Project Checklist

Evaluate whether each group
- ✔ cooperatively selects companies to "invest" in.
- ✔ decides how many shares of stock to "purchase" for each company.
- ✔ makes a table and records daily information about each stock.
- ✔ compares final value of stocks with original investment.
- ✔ makes a presentation of the results to the class.

Using the Project

Accessing Prior Knowledge Multiplying a whole number and a fraction

Purpose The purpose of the project is for students to buy shares of stock, keep track of how the prices of the shares change, and determine how much money they made or lost.

GROUPING small groups
MATERIALS newspapers

Managing the Activity To help students understand how to read the prices of stocks, put this list on the chalkboard.

$\frac{1}{8} = \$0.125$ $\frac{1}{4} = \$0.25$

$\frac{3}{8} = \$0.375$ $\frac{1}{2} = \$0.50$

$\frac{5}{8} = \$0.625$ $\frac{3}{4} = \$0.75$

$\frac{7}{8} = \$0.875$

WHEN MINUTES COUNT

A shortened version of the activity can be completed in 20–30 minutes. Have each group look at the stock listings in the newspaper and pick 3 companies. For each company, have groups determine the cost of 100 shares at the closing price and the dollar value of the listed change in price.

A BIT EACH DAY

Day 1: Students work in groups to decide which stocks to buy and how many shares to buy.

Day 2: Students make a table and record the information needed for each stock.

Days 3–8: Students record the daily price of each stock.

Day 9: Students compare the final value of their stocks to the original investment and determine how much they made or lost.

Day 10: Groups present their results.

HOME HAPPENINGS

Suggest that students discuss the project with their families and share what they've learned about how to determine the price of stock.

GRADE 5
Explore multiplication of fractions.

GRADE 6
Multiply fractions.

GRADE 7
Solve equations using multiplication of fractions.

ORGANIZER

OBJECTIVE To multiply fractions
VOCABULARY *Review* factor, product, simplest form, GCF
MATERIALS *For each student* paper, ruler, 2 different-colored pencils or markers
ASSIGNMENTS Independent Practice
- *Basic* 1–29, odd; 30–33; 34–43
- *Average* 1–33; 34–43
- *Advanced* 1–15; 30–33; 34–43

PROGRAM RESOURCES
Extension • Reciprocals, TE pp. 132A–132B
Reteach 7.1 Practice 7.1 Enrichment 7.1
Math Fun • *Egyptian Symbols*, p. 142

PROBLEM OF THE DAY

Transparency 7

In a jump-rope marathon, Cara will earn and donate to charity $5 for each half hour or fraction of a half hour that she jumps rope. How much money will Cara earn if she jumps rope for 175 min? $30

Solution Tab A, p. A7

QUICK CHECK
Solve.

1. $\frac{1}{2}$ of 12 = __?__ 6
2. $\frac{1}{2}$ of 3 = __?__ $1\frac{1}{2}$
3. $\frac{1}{3}$ of 6 = __?__ 2
4. $\frac{1}{3}$ of 12 = __?__ 4
5. $\frac{1}{10}$ of 30 = __?__ 3
6. $\frac{1}{4}$ of 8 = __?__ 2

1 MOTIVATE

To help students visualize finding a part of a part, help them draw a model of the following situation:

Marcus had a dozen cookies. He took 6 of the cookies to school in his lunch. He shared $\frac{1}{3}$ of the 6 cookies with a friend. What part of the dozen cookies did he share with his friend? $\frac{1}{6}$

7.1

WHAT YOU'LL LEARN
How to multiply fractions

WHY LEARN THIS?
To solve everyday problems such as sharing part of a pizza or candy bar

Multiplying with Fractions

Joshua had $\frac{3}{4}$ of a candy bar. He gave his sister $\frac{1}{2}$ of what he had. To find what part of the whole candy bar Joshua gave his sister, find $\frac{1}{2}$ of $\frac{3}{4}$, or $\frac{1}{2} \times \frac{3}{4}$.

ACTIVITY

WHAT YOU'LL NEED: paper, ruler, 2 different-colored pencils or markers

Make a model to help find the product $\frac{1}{2} \times \frac{3}{4}$. Use a piece of paper to represent a whole.

- Fold the paper into 4 equal parts as shown below. Each part represents $\frac{1}{4}$. Color 3 parts to represent $\frac{3}{4}$.

 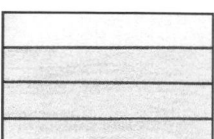

- Fold the paper in half so that you divide each of the fourths into 2 equal parts. Using the other color, shade 1 of the halves.

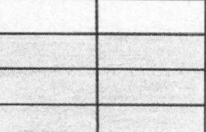 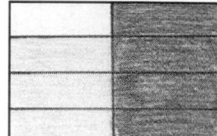

Of the 8 parts, 3 are shaded twice. These 3 parts represent $\frac{1}{2} \times \frac{3}{4}$.

The model shows that $\frac{1}{2} \times \frac{3}{4} = \frac{3}{8}$.

- Is $\frac{3}{8}$ larger or smaller than the $\frac{3}{4}$ you started with? $\frac{3}{8}$ is smaller than $\frac{3}{4}$
- Make a model to find $\frac{1}{4} \times \frac{5}{8}$. **Check students' models.**
- What is the product of $\frac{1}{4} \times \frac{5}{8}$? $\frac{5}{32}$

128 Chapter 7

RETEACH 7.1

Name _____

Multiplying with Fractions

Bo knows that $\frac{2}{3}$ of the students in his class play soccer. Of those students, $\frac{1}{6}$ are in the school band. Bo wants to know what fraction of his class play soccer and are in the school band.

Step 1 Write a multiplication sentence. $\frac{2}{3} \times \frac{1}{6} = \blacksquare$

Step 2 Multiply the numerators. Multiply the denominators. $\frac{2 \times 1}{3 \times 6} = \frac{2}{18}$

Step 3 Divide the numerator and denominator by the GCF, 2. $= \frac{2 \div 2}{18 \div 2}$

Step 4 Write the product in simplest form. $= \frac{1}{9}$

So, $\frac{1}{9}$ of Bo's class play soccer and are in the school band.

Complete to find each product.

1. $\frac{3}{4} \times \frac{2}{9} = \frac{3 \times \boxed{2}}{\boxed{4} \times 9}$
$= \frac{6}{36}$
$= \frac{6 \div \boxed{6}}{36 \div \boxed{6}}$
$= \frac{1}{6}$

2. $\frac{1}{8} \times \frac{2}{3} = \frac{1 \times 2}{8 \times \boxed{3}}$
$= \frac{2}{24}$
$= \frac{2 \div \boxed{2}}{24 \div \boxed{2}}$
$= \frac{1}{12}$

3. $\frac{5}{7} \times \frac{3}{10} = \frac{5 \times \boxed{3}}{7 \times 10}$
$= \frac{15}{70}$
$= \frac{15 \div 5}{70 \div 5} = \frac{3}{14}$

4. $\frac{4}{9} \times \frac{3}{8} = \frac{\boxed{4} \times 3}{9 \times \boxed{8}}$
$= \frac{12}{72}$
$= \frac{12 \div 12}{72 \div 12} = \frac{1}{6}$

Use with text pages 128–131. **TAKE ANOTHER LOOK** R29

PRACTICE 7.1

Name _____

Multiplying with Fractions

Make a model to find the product.

1. $\frac{1}{2} \times 6$ 3
2. $\frac{2}{5} \times \frac{1}{2}$ $\frac{1}{5}$
3. $\frac{1}{8} \times \frac{1}{2}$ $\frac{1}{16}$
4. $10 \times \frac{1}{2}$ 5
5. $\frac{1}{2} \times \frac{1}{3}$ $\frac{1}{6}$

Find the product. Write it in simplest form.

6. $\frac{1}{4} \times \frac{1}{6}$ $\frac{1}{24}$
7. $\frac{1}{5} \times \frac{1}{2}$ $\frac{1}{10}$
8. $\frac{3}{8} \times \frac{1}{4}$ $\frac{3}{32}$
9. $\frac{3}{4} \times \frac{1}{5}$ $\frac{3}{20}$
10. $\frac{4}{5} \times \frac{1}{2}$ $\frac{2}{5}$
11. $\frac{1}{4} \times \frac{8}{9}$ $\frac{2}{9}$
12. $\frac{3}{7} \times \frac{2}{3}$ $\frac{2}{7}$
13. $\frac{5}{9} \times \frac{9}{10}$ $\frac{1}{2}$
14. $\frac{5}{6} \times \frac{2}{5}$ $\frac{1}{3}$
15. $\frac{5}{8} \times \frac{2}{5}$ $\frac{4}{7}$

Complete the multiplication sentence.

16. $\frac{5}{6} \times \frac{\boxed{1}}{3} = \frac{5}{18}$
17. $\frac{4}{5} \times \frac{2}{\boxed{7}} = \frac{8}{35}$
18. $\frac{2}{3} \times \frac{\boxed{4}}{7} = \frac{8}{21}$
19. $\frac{9}{\boxed{10}} \times \frac{10}{11} = \frac{9}{11}$
20. $\frac{3}{4} \times \frac{\boxed{1}}{2} = \frac{3}{8}$
21. $\frac{5}{13} \times \frac{2}{\boxed{5}} = \frac{2}{13}$

Mixed Applications

22. Sally knows that $\frac{2}{5}$ of the students in her class are in the school band. Of those students, $\frac{3}{4}$ own their instruments. What fraction of Sally's class own their own instruments? $\frac{3}{10}$

23. Ryan takes $\frac{1}{5}$ hr to walk to the baseball field. He spends $\frac{1}{2}$ of that time walking down his street. What part of an hour does Ryan spend walking down his street? How many minutes is this? $\frac{1}{10}$ hour; 6 minutes

24. Bill chose a number, added 2, multiplied the sum by 4 and divided by 8. The final number was 4. What number had Bill chosen? 6

25. A tour bus travels 19.5 mi each day taking people on sightseeing trips. How many miles does the tour bus travel during a 6-day week? 117 mi

Use with text pages 128–131. **ON MY OWN** P29

As you model multiplication of fractions, look for a relationship between the factors and the product.

EXAMPLE 1 Find $\frac{2}{3}$ of $\frac{1}{2}$, or $\frac{2}{3} \times \frac{1}{2}$.

Fold a piece of paper into 2 equal parts. Color 1 part to show $\frac{1}{2}$.

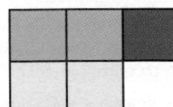

Fold the paper into thirds. Color $\frac{2}{3}$ of the paper.

Of the 6 parts, 2 are shaded twice. These parts represent $\frac{2}{3} \times \frac{1}{2}$.

So, $\frac{2}{3} \times \frac{1}{2} = \frac{2}{6}$.

- **CRITICAL THINKING** Compare the numerator and denominator of the product with the numerators and denominators of the factors. What relationship do you see? $\frac{2}{3} \times \frac{1}{2} = \frac{2 \times 1}{3 \times 2} = \frac{2}{6}$

In the Activity and Example 1, you can see this relationship:

$$\frac{\text{numerator} \times \text{numerator}}{\text{denominator} \times \text{denominator}} = \frac{\text{numerator}}{\text{denominator}}$$

↑ factor ↑ factor = ↑ product

You can use this relationship to multiply fractions without a model.

EXAMPLE 2 Use the relationship above to find $\frac{1}{3} \times \frac{3}{4}$. Write the product in simplest form.

$\frac{1}{3} \times \frac{3}{4} = \frac{1 \times 3}{3 \times 4}$ *Multiply the numerators.*
Multiply the denominators.

$= \frac{3}{12}$

$= \frac{3 \div 3}{12 \div 3}$ *Divide the numerator and the denominator by the GCF, 3.*

$= \frac{1}{4}$ *Write the product in simplest form.*

So, $\frac{1}{3} \times \frac{3}{4} = \frac{1}{4}$.

- Example 2 shows that $\frac{1}{3} \times \frac{3}{4} = \frac{1}{4}$. Explain why the product is less than the factor $\frac{3}{4}$. **because $\frac{1}{3}$ is less than 1**

Language Link

Fractions are an everyday part of our language. For example, if someone tells you it is quarter to two, you know that the time is 1:45, or 15 min before 2:00. Make a list of other everyday expressions that include fractions.

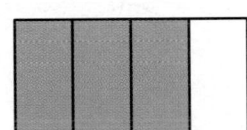

Expressions will vary.

REMEMBER:

To write a fraction in simplest form, divide the numerator and the denominator by the greatest common factor (GCF). **See page H15.**

$\frac{4}{10} = \frac{4 \div 2}{10 \div 2}$ GCF:2
$= \frac{2}{5}$

129

ENRICHMENT 7.1

Name _____

LESSON 7.1

Multiply to Find a Message

Match each exercise in Column 1 with its product in Column 2. Then write each corresponding letter on the line below marked with the exercise number to discover the Math Tip.

Column 1

1. $15 \times \frac{2}{5}$ **H**
2. $\frac{5}{7} \times 14$ **S**
3. $\frac{2}{3} \times \frac{1}{6}$ **J**
4. $\frac{3}{8} \times \frac{4}{21}$ **W**
5. $12 \times \frac{2}{3}$ **A**
6. $\frac{1}{8} \times \frac{2}{7}$ **O**
7. $13 \times \frac{1}{26}$ **D**
8. $\frac{3}{10} \times \frac{5}{9}$ **R**
9. $\frac{1}{4} \times \frac{20}{23}$ **B**
10. $18 \times \frac{2}{3}$ **L**
11. $\frac{7}{20} \times \frac{5}{14}$ **T**
12. $\frac{1}{2} \times \frac{1}{5}$ **Y**
13. $\frac{3}{12} \times \frac{2}{9}$ **F**

14. $24 \times \frac{1}{6}$ **U**
15. $\frac{6}{7} \times 21$ **P**
16. $\frac{3}{10} \times 5$ **I**
17. $\frac{9}{11} \times \frac{1}{3}$ **C**
18. $\frac{13}{25} \times \frac{5}{6}$ **Q**
19. $\frac{4}{5} \times 35$ **E**
20. $\frac{9}{10} \times \frac{2}{3}$ **V**
21. $\frac{1}{10} \times 9$ **K**
22. $\frac{3}{7} \times 35$ **M**
23. $\frac{5}{9} \times \frac{18}{25}$ **N**
24. $\frac{1}{5} \times 5$ **Z**
25. $\frac{7}{10} \times 3$ **G**
26. $\frac{1}{2} \times \frac{6}{7}$ **X**

Column 2

A. 8   N. $\frac{2}{7}$
B. $\frac{5}{23}$ O. $\frac{1}{28}$
C. $\frac{3}{11}$ P. 18
D. $\frac{1}{2}$ Q. $\frac{13}{30}$
E. 28   R. $\frac{1}{6}$
F. $\frac{1}{18}$ S. 10
G. $2\frac{1}{10}$ T. $\frac{1}{8}$
H. 6   U. 4
I. $1\frac{1}{2}$ V. $\frac{3}{5}$
J. $\frac{1}{9}$ W. $\frac{1}{14}$
K. $\frac{9}{10}$ X. $\frac{3}{7}$
L. 12   Y. $\frac{1}{10}$
M. 15   Z. $1\frac{2}{3}$

M A __ F __ R __ A __ C __ T __ I __ O __ N __
A   5   13   8   17   11   16   6   23

A T __ G __ R __ E __ A __ T __ E __ R __   T __ H __ A __ N __
H   25   8   19   5   11   19   8   11   1   5   23

T O __ N __ E __   C __ A __ N __   B __ E __
I   6   23   19   17   5   23   9   19

W __ R __ I __ T __ T __ E __ N __   A __ S __   A __
    4   8   16   11   11   19   23   5   2   5

P M __ I __ X __ E __ D __   N __ U __ M __ B __ E __ R __
    22   16   26   19   7   23   14   22   9   19   8

Use with text pages 128–131.   **STRETCH YOUR THINKING** E29

---

2 TEACH/PRACTICE

Use the overhead with grids of different colors to model the Activity and Example 1. Help students picture the overlap of the two colors as the product of the two fractions.

TEACHING TIP
Students usually understand that $\frac{1}{2}$ of a quantity, an amount of money, or a region, is less than the original amount. However, the multiplication of whole numbers that students have done often creates a mind-set that the product can never be less than one of the factors. This mind-set is difficult to change when students begin to work with fractions. Some students will question the fact that a product can be smaller than one of its factors.

CRITICAL THINKING **Aaron did the problem $\frac{3}{4} \times \frac{2}{3}$, and his answer was $\frac{5}{12}$. What mistake did he make?** He added the numerators instead of multiplying them.

ADDITIONAL EXAMPLES

Example 1, p. 129

Find $\frac{1}{3}$ of $\frac{3}{4}$, or $\frac{1}{3} \times \frac{3}{4}$.

Fold a piece of paper into 4 equal parts. Color 3 parts to show $\frac{3}{4}$.

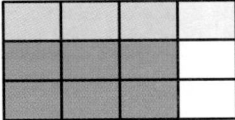

Fold the paper into thirds. Color $\frac{1}{3}$ of the paper.

Of the 12 parts, 3 are shaded twice. This part represents $\frac{1}{3} \times \frac{3}{4}$. So, $\frac{1}{3} \times \frac{3}{4} = \frac{3}{12}$, or $\frac{1}{4}$.

Example 2, p. 129

Use the relationship shown in the middle of page 129 to find $\frac{2}{5} \times \frac{1}{2}$. Write the product in simplest form.

Multiply the numerators. Multiply the denominators.

$\frac{2}{5} \times \frac{1}{2} = \frac{2 \times 1}{5 \times 2} = \frac{2}{10}$

Divide the numerator and the denominator by the GCF, 2. Write the product in simplest form.

$\frac{2 \div 2}{10 \div 2} = \frac{1}{5}$

So, $\frac{2}{5} \times \frac{1}{2} = \frac{1}{5}$.

129

GUIDED PRACTICE

For Exercises 5–8, you may want to use a calculator to find the product after students make an estimate. Allow variation in the estimates, but ask how students got them. Some students may need a model to determine the answer.

For Exercises 9–12, remind students that simplest form relates to equivalent fractions. If necessary, review writing equivalent fractions.

For Exercises 11 and 12, students will need to know how to write fractions greater than one. Explain that the simplest form of the answer may be left as a fraction or changed to a mixed number.

ADDITIONAL EXAMPLES

Example 3, p. 130

Mrs. Hector's class wants to order pizzas. Each pizza has 8 slices. Each student plans to eat 3 slices of pizza. There are 24 students in the class. Estimate the number of pizzas they need to order. Answers will vary, but $\frac{3}{8}$ is about $\frac{1}{2}$, so 12 pizzas would be a reasonable estimate.

Example 4, p. 130

Mrs. Hector insists that the class determine exactly how many pizzas to order. Find $24 \times \frac{3}{8}$. 9 pizzas

---

Multiplying Fractions and Whole Numbers

You can use the same method to multiply a whole number by a fraction.

EXAMPLE 3 Christina is on the track team at school. She practices every afternoon after school. She runs 9 times around the $\frac{3}{4}$-mi track. What is a reasonable estimate of the distance she runs?

$9 \times \frac{3}{4} \leftarrow \frac{3}{4}$ is about 1.   *Round the fraction to the nearest whole number.*

$9 \times \frac{3}{4} \approx 9 \times 1$

$9 \times 1 = 9$ *Multiply.*

So, she runs about 9 mi.

• What is a reasonable estimate of $\frac{2}{3} \times 11$? **11**

EXAMPLE 4 Christina's coach wants to know the exact distance she runs. Find $9 \times \frac{3}{4}$.

$9 \times \frac{3}{4} = \frac{9}{1} \times \frac{3}{4}$ *Write the whole number as a fraction.*

$= \frac{9 \times 3}{1 \times 4}$ *Multiply the numerators.*
    Multiply the denominators.

$= \frac{27}{4}$, or $6\frac{3}{4}$ *Write the answer as a fraction or as a mixed number.*

So, Christina runs exactly $6\frac{3}{4}$ mi.

GUIDED PRACTICE

Make a model to find the product. **Check students' models.**

1. $\frac{1}{2} \times \frac{3}{4}$ **$\frac{3}{8}$**   **2.** $\frac{3}{4} \times \frac{1}{2}$ **$\frac{3}{8}$**   **3.** $\frac{1}{3} \times \frac{5}{8}$ **$\frac{5}{24}$**   **4.** $\frac{1}{2} \times \frac{1}{4}$ **$\frac{1}{8}$**

Estimate the product.

5. $\frac{7}{9} \times 3$ **3**   **6.** $5 \times \frac{7}{8}$ **5**   **7.** $4 \times \frac{5}{7}$ **4**   **8.** $\frac{1}{2} \times \frac{3}{4}$ **$\frac{1}{2}$**

Find the product. Write the answer in simplest form.   **11.** $\frac{12}{7}$, or $1\frac{5}{7}$

9. $\frac{3}{4} \times \frac{2}{5}$ **$\frac{3}{10}$**   **10.** $\frac{2}{5} \times \frac{7}{8}$ **$\frac{7}{20}$**   **11.** $2 \times \frac{6}{7}$ **12.** $\frac{2}{3} \times 16$ **$\frac{32}{3}$, or $10\frac{2}{3}$**

13. $\frac{2}{3} \times 4$ **$\frac{8}{3}$, or $2\frac{2}{3}$**   **14.** $9 \times \frac{2}{3}$ **6**   **15.** $\frac{5}{6} \times \frac{2}{3}$ **$\frac{5}{9}$**   **16.** $\frac{3}{5} \times \frac{5}{6}$ **$\frac{1}{2}$**

Sports Link

To earn a letter in some sports, you have to play for your team a certain fraction of the time. Suppose a basketball player has to play in $\frac{3}{4}$ of all the game quarters to earn a letter. If the team plays 20 games, how many quarters must the player participate in to earn a letter?

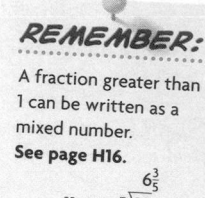

60 quarters

REMEMBER:

A fraction greater than 1 can be written as a mixed number.
See page H16.

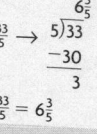

$\frac{33}{5} \rightarrow 5\overline{)33}$
$\underline{-30}$
3

$\frac{33}{5} = 6\frac{3}{5}$

IDEA FILE

WRITING IN MATHEMATICS

Have students write a journal entry that describes why models are useful for learning to multiply fractions.

Encourage them to use illustrations to reinforce their arguments.

Modality Visual

ALTERNATIVE TEACHING STRATEGY

MATERIALS: fraction strips

Some students may better understand multiplying fractions by fractions if they use fraction strips as models.

For $\frac{1}{3} \times \frac{3}{4}$, use three $\frac{1}{4}$ strips, put them in 3 equal groups, and pick up one group. product: $\frac{1}{4}$

Have students model other problems.

Modalities Kinesthetic, Visual

Make a model to help you find the product.

1. $\frac{1}{2} \times \frac{3}{4}$ **$\frac{3}{8}$**   2. $\frac{1}{2} \times 4$ **2**   3. $\frac{3}{5} \times \frac{1}{2}$ **$\frac{3}{10}$**   4. $\frac{1}{2} \times \frac{1}{2}$ **$\frac{1}{4}$**   5. $7 \times \frac{1}{2}$ **$\frac{7}{2}$, or $3\frac{1}{2}$**

Find the product. Write it in simplest form. **Answers for Exercises 21 and 22 see below.**

6. $\frac{1}{3} \times \frac{2}{3}$ **$\frac{2}{9}$**   7. $\frac{1}{2} \times \frac{1}{2}$ **$\frac{1}{4}$**   8. $\frac{3}{4} \times \frac{1}{4}$ **$\frac{3}{16}$**   9. $\frac{1}{5} \times \frac{2}{3}$ **$\frac{2}{15}$**   10. $\frac{1}{4} \times \frac{2}{7}$ **$\frac{1}{14}$**

11. $\frac{4}{5} \times \frac{7}{8}$ **$\frac{7}{10}$**   12. $\frac{2}{9} \times \frac{3}{4}$ **$\frac{1}{6}$**   13. $\frac{2}{5} \times \frac{5}{7}$ **$\frac{2}{7}$**   14. $\frac{1}{8} \times \frac{4}{5}$ **$\frac{1}{10}$**   15. $\frac{5}{9} \times \frac{3}{10}$ **$\frac{1}{6}$**

16. $\frac{4}{9} \times \frac{3}{5}$ **$\frac{4}{15}$**   17. $\frac{6}{7} \times \frac{7}{8}$ **$\frac{3}{4}$**   18. $\frac{2}{3} \times 21$ **14**   19. $24 \times \frac{1}{12}$ **2**   20. $\frac{1}{8} \times 16$ **$\frac{2}{1}$, or 2**

21. $\frac{5}{6} \times 5$ 22. $13 \times \frac{3}{5}$ 23. $5 \times \frac{1}{9}$ **$\frac{5}{9}$**   24. $14 \times \frac{7}{12}$ **$\frac{49}{6}$, or $8\frac{1}{6}$**   25. $\frac{2}{5} \times 14$ **$\frac{28}{5}$, or $5\frac{3}{5}$**

Complete the multiplication sentence.

26. $\frac{1}{5} \times \frac{\blacksquare}{3} = \frac{4}{15}$ **4**   27. $\frac{\blacksquare}{4} \times \frac{3}{8} = \frac{9}{32}$ **3**   28. $\frac{2}{\blacksquare} \times \frac{3}{5} = \frac{2}{5}$ **3**   29. $\frac{\blacksquare}{4} \times \frac{2}{3} \times \frac{\blacksquare}{2} = \frac{1}{2}$
2, 3, or 3, 2

Problem-Solving Applications

30. Sandra takes $\frac{1}{2}$ hr to walk to school. She spends $\frac{1}{2}$ of that time walking down her street. What part of an hour does Sandra spend walking down her street? How many minutes is this? **$\frac{1}{4}$ hr; 15 min**

31. In a stock market game, Roberto bought 150 shares of stock at $\frac{5}{8}$ per share. How much money did 150 shares cost? **$93\frac{3}{4}$, or $93.75**

32. There are 144 registered voters in Booker County. In the last election, $\frac{3}{4}$ of them voted. How many voters voted in the last election? **108 voters**

33. ▭ **WRITE ABOUT IT** Explain how you can multiply two fractions. **Use a model; multiply the numerators and the denominators, and write the answer in simplest form.**

Mixed Review

Find the greatest common factor (GCF) of each pair of numbers.

34. 2 and 4 **2**   35. 4 and 8 **4**   36. 6 and 9 **3**   37. 15 and 10 **5**

Find the sum. Write it in simplest form.

38. $\frac{1}{12} + \frac{7}{12}$ **$\frac{2}{3}$**   39. $\frac{3}{4} + \frac{1}{6}$ **$\frac{11}{12}$**   40. $2\frac{2}{7} + 3\frac{3}{7}$ **$5\frac{5}{7}$**   41. $1\frac{1}{2} + 2\frac{2}{5}$ **$3\frac{9}{10}$**

42. **NUMBER SENSE** Cindy chose a number, added 5, multiplied the sum by 3, subtracted 10, and doubled the result. Her final number was 28. What number had she chosen? **3**

43. **ELAPSED TIME** School begins at 8:30 A.M. Albert must be there 15 min early. The trip to school takes 35 min. When should Albert leave for school? **7:40 A.M.**

21. **$\frac{25}{6}$, or $4\frac{1}{6}$**   22. **$\frac{39}{5}$, or $7\frac{4}{5}$**

CULTURAL CONNECTION

The ancient Egyptians thought fractions were troublesome. They wrote every fraction except $\frac{2}{3}$ as the sum of fractions with the numerator 1, or *unit fractions*. For example, $\frac{3}{4}$ was written as $\frac{1}{4} + \frac{1}{2}$. Greek mathematicians used the same method, and even seventeenth-century Russian documents show fractions as sums of unit fractions.

SCIENCE CONNECTION

Fractions are used very little in science. One time when they are used, however, is when it is necessary to convert temperatures from Celsius to Fahrenheit. The formulas for these conversions are

$°F = (\frac{9}{5} \times °C) + 32$

$°C = \frac{5}{9} \times (°F - 32)$

• Have students convert three temperatures from Fahrenheit to Celsius.

Modalities Auditory, Visual

INDEPENDENT PRACTICE
Exercises require students to use a model to find a product, to multiply a fraction by a fraction or a whole number, and to express the product in simplest form.

3 WRAP UP

Use these questions to help students focus on the big ideas of the lesson.

• **What would you say to a friend who missed class and asked you for help in multiplying $\frac{3}{5} \times \frac{2}{3}$?** Multiply the numerators and the denominators.

• **What does it mean to write your answer in simplest form?** There are no common factors, other than 1, in the numerator and denominator.

✓ Assessment Tip

Have students write the answer.

• **How does the product of two whole numbers differ from the product of two fractions?** The product of two whole numbers is always a whole number. The product of two fractions may be a fraction or a whole number.

DAILY QUIZ

1. Make a model to find the product. $\frac{1}{3} \times \frac{3}{5} = ?$ Check students' models; $\frac{1}{5}$.

2. Write $\frac{12}{15}$ in simplest form. $\frac{4}{5}$

Multiply. Write the answer in simplest form.

3. $\frac{4}{5} \times \frac{3}{7}$ $\frac{12}{35}$

4. $\frac{4}{5} \times \frac{3}{4}$ $\frac{3}{5}$

5. Write a problem that can be answered by finding $8 \times \frac{1}{3}$. Check students' answers.

EXTENSION

MEETING INDIVIDUAL NEEDS

Reciprocals

Use these strategies to extend understandings about reciprocals.

Multiple Intelligences Logical/Mathematical

Modalities Auditory, Kinesthetic, Visual

Lesson Connection Students will find reciprocals of different kinds of numbers, and use reciprocals to divide fractions.

Materials Calculators

Program Resources *Intervention and Extension Copying Masters*, p. 15

Challenge

TEACHER NOTES

ACTIVITY 1

Purpose To extend understanding of reciprocals

Vocabulary Review

- **reciprocals**—two numbers whose product is 1

Discuss the following questions.

- How do you write a whole number as a fraction? Write the whole number over 1.

- How do you write a mixed number as a fraction? Write it as a fraction with the numerator greater than the denominator.

- How do you write a decimal as a fraction? Write it as a fraction with denominator 10, 100, 1,000, and so on. So, $1.75 = 1\frac{75}{100} = \frac{175}{100} = \frac{7}{4}$.

- You can write 0 as $\frac{0}{1}$. Does this mean the reciprocal of 0 is $\frac{1}{0}$? No. $\frac{1}{0}$ means $1 \div 0$, and division by zero is not allowed. So zero has no reciprocal.

WHAT STUDENTS DO

1. For a fraction, you find the reciprocal by interchanging the numerator and denominator. So, the reciprocal of $\frac{2}{3}$ is $\frac{3}{2}$. But what about the reciprocals of whole numbers? mixed numbers? decimals?

   Find the reciprocal of each of the following numbers.

   $4 = \frac{4}{1}$; reciprocal is $\frac{1}{4}$

   $2\frac{1}{3} = \frac{7}{3}$; reciprocal is $\frac{3}{7}$

   $0.5 = \frac{5}{10} = \frac{1}{2}$; reciprocal is $\frac{2}{1}$ or 2

   $1.75 = 1\frac{75}{100} = \frac{7}{4}$; reciprocal is $\frac{4}{7}$

   $\frac{5}{8}$ reciprocal is $\frac{8}{5}$

2. Multiply each of the numbers in Problem 1 by its reciprocal. What is the product? A number times its reciprocal is always 1.

3. Rewrite each division problem as a multiplication problem. Then write the answer in simplest form.

   a. $7 \div 2\frac{3}{4}$

   $\frac{7}{1} \div \frac{11}{4}$

   $\frac{7}{1} \times \frac{4}{11} = \frac{28}{11} = 2\frac{6}{11}$

   b. $4\frac{1}{2} \div 1.5$

   $\frac{9}{2} \div \frac{3}{2}$

   $\frac{9}{2} \times \frac{2}{3} = 3$

4. Does every number have a reciprocal? Explain. Every number except zero has a reciprocal. Zero has no reciprocal. Zero times its reciprocal would have to equal 1, but zero times any number is always zero.

TEACHER NOTES

ACTIVITY 2

Purpose To use a calculator to understand reciprocals

Discuss the following questions.

- What is true about the reciprocals of large numbers, such as 1,000 or 20,000? The reciprocals are near 0.

- What is true about the reciprocals of small numbers, such as $\frac{1}{500}$ and $\frac{1}{10,000}$? The reciprocals are quite large.

- What numbers have reciprocals that are near 1? numbers near 1

Have students meet in small groups or with partners to discuss the three questions. Have one student report the group's responses to the larger group.

WHAT STUDENTS DO

1. There are two ways to find a reciprocal on your calculator.

   a. Use the reciprocal key, $\boxed{1/x}$. To find the reciprocal of 2.5, enter 2.5 $\boxed{1/x}$. You get 0.4.

   b. If your calculator has no $\boxed{1/x}$ key, divide 1 by the number to find the reciprocal. Enter 1 $\boxed{\div}$ 2.5 $\boxed{=}$. You get 0.4.

   Find the reciprocal of each number by using your calculator.

   | 3 0.3333333 | 1.1 0.9090909 | 1.001 0.999009 |
   | 1.5 0.6666666 | 1.01 0.990099 | |

   - Describe reciprocals of numbers close to 1. The reciprocals are close to 1.

   - What is the reciprocal of 1? 1

2. Use your calculator to find the reciprocals of small numbers.

   - What is the reciprocal of 0.00005? 20,000

   - What can you say about the reciprocals of numbers close to 0? They are greater numbers.

3. What does your calculator display if you try to find the reciprocal of 0? Most give an error message, "E," because 0 has no reciprocal.

Share Results

- Explain what reciprocals have to do with division. Dividing by a number is the same as multiplying by its reciprocal.

- What is the only number with no reciprocal? 0

- Discuss what you learned about reciprocals of numbers close to 1 and close to 0. near 1; large

Intervention and Extension Copying Masters, p. 15

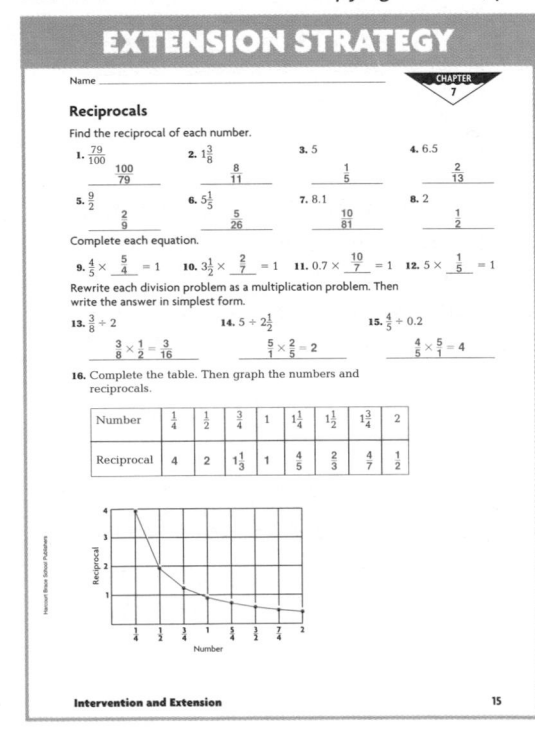

GRADE 5	GRADE 6	GRADE 7
Multiply a fraction by a fraction.	Simplify fraction factors before multiplying.	Solve equations involving multiplication of fractions.

ORGANIZER

OBJECTIVE To multiply fractions after simplifying each factor

ASSIGNMENTS Independent Practice
- ■ *Basic* 1–32, even; 33–37
- ■ *Average* 1–37
- ■ *Advanced* 9–37

PROGRAM RESOURCES
Reteach 7.2 Practice 7.2 Enrichment 7.2

PROBLEM OF THE DAY

Transparency 7

Use the digits 1–9 to make as many pairs of equivalent fractions as you can. You may use each number only once in each pair. Possible answers: $\frac{1}{2}, \frac{4}{8}, \frac{1}{4}, \frac{2}{8}, \frac{1}{3}, \frac{2}{6}, \frac{2}{3}, \frac{6}{9}$

Solution Tab A, p. A7

QUICK CHECK

Write in simplest form.

1. $\frac{9}{12}$ $\frac{3}{4}$
2. $\frac{6}{8}$ $\frac{3}{4}$
3. $\frac{10}{12}$ $\frac{5}{6}$
4. $\frac{7}{14}$ $\frac{1}{2}$
5. $\frac{12}{20}$ $\frac{3}{5}$
6. $\frac{8}{6}$ $\frac{4}{3}$, or $1\frac{1}{3}$

1 MOTIVATE

Ask students to name any fraction. Make up a second fraction, making sure that it has a common factor in either the numerator or the denominator. For example, suppose a student says $\frac{12}{23}$. Choose a fraction with 23 in the numerator or 12 in the denominator, such as $\frac{5}{12}$. Then immediately write the product, $\frac{5}{23}$. After a few examples, ask if anyone can explain how you found the product so quickly.

2 TEACH/PRACTICE

Help students understand that they may simplify factors before they multiply but that if they forget or do not recognize a common factor, the answer will be the same if they simplify the product.

7.2

WHAT YOU'LL LEARN
To simplify factors before multiplying fractions and find the simplest form faster

WHAT YOU'LL LEARN
To simplify factors before multiplying fractions and find the simplest form faster

WHY LEARN THIS?
To make the multiplying of fractions easier

Science Link

Fractions are used very little in science. One time when they are used, however, is when it is necessary to convert temperatures from Celsius to Fahrenheit. The formulas for these conversions are:

°F = $\frac{9}{5}$ × °C + 32°

°C = $\frac{5}{9}$ × (°F − 32°)

Suppose the temperature is 100°C. What is this in °F? **212°F**

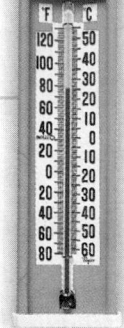

Simplifying Factors

You can often simplify fractions before you multiply.

Cheryl has $\frac{2}{3}$ of a case of soda left from a picnic. If she gives $\frac{1}{4}$ of it away, what part of the case will she give away?

Think: What is $\frac{1}{4}$ of $\frac{2}{3}$?

Find $\frac{1}{4} \times \frac{2}{3}$.

$\frac{1}{4} \times \frac{2}{3} \leftarrow$ The GCF of 2 and 4 is 2.   *Look for a numerator and denominator with common factors. Find the greatest common factor (GCF).*

$\frac{1}{\underset{2}{4}} \times \frac{\overset{1}{2}}{3} \quad \begin{array}{l} \leftarrow 2 \div 2 = 1 \\ \leftarrow 4 \div 2 = 2 \end{array}$ *Divide the numerator and denominator by the GCF, 2.*

$\frac{1}{\underset{2}{4}} \times \frac{\overset{1}{2}}{3} = \frac{1 \times 1}{2 \times 3} = \frac{1}{6}$ *Multiply.*

So, Cheryl gave away $\frac{1}{6}$ of a case of soda.

Talk About It

- Find the product of $\frac{1}{4}$ and $\frac{2}{3}$ without using the GCF. $\frac{2}{12}$

- Would you have to simplify your answer if you didn't use the GCF? Explain. **Yes; the factors are not in simplest form.**

- 💡**CRITICAL THINKING** When you use the GCF, do you need to simplify the fraction after you multiply? Explain. **No; the product is in simplest form.**

EXAMPLE Find $\frac{2}{3} \times \frac{3}{10}$.

$\frac{2}{3} \times \frac{3}{10} = \frac{2}{\underset{1}{3}} \times \frac{\overset{1}{3}}{10} = \frac{1}{5}$ *The GCF of 3 and 3 is 3. The GCF of 2 and 10 is 2.*

So, $\frac{2}{3} \times \frac{3}{10} = \frac{1}{5}$.

- Look at the problem $\frac{5}{8} \times \frac{4}{5}$. To completely simplify the factors, would you use one GCF or two? Explain your answer. **Two; the GCF of 5 and 5 is 5, and the GCF of 4 and 8 is 4.**

IDEA FILE

ALTERNATIVE TEACHING STRATEGY

Review simplifying fractions by dividing the numerator and denominator by the same number. Extend this concept to multiplication of fractions by writing the product $\frac{2}{3} \times \frac{1}{2}$ as $\frac{(2 \times 1)}{(3 \times 2)}$ and checking for common factors before multiplying. Students should understand that a number can be divided by 1 without changing it, and that dividing a number or fraction by $\frac{2}{2}$, $\frac{3}{3}$, or $\frac{n}{n}$ is the same as dividing by 1.

RETEACH 7.2

Name _____

LESSON 7.2

Simplifying Factors

Tim has $\frac{5}{8}$ of a case of baseballs left over from the season. If he gives $\frac{3}{4}$ of it away, what part of it will he give away?

Step 1 What is $\frac{5}{8}$ of $\frac{3}{4}$? $\frac{5}{8} \times \frac{3}{4} = \blacksquare$
Find $\frac{5}{8} \times \frac{3}{4}$.

Step 2 Look for a numerator and denominator with common factors. Find the greatest factor (GCF). $\frac{5}{8} \times \frac{3}{4}$ The GCF of 3 and 6 is 3.

Step 3 Divide the numerator and denominator by the GCF, 3. $\frac{5}{8} \times \frac{\overset{1}{3}}{4}$ $3 \div 3 = 1$ $6 \div 3 = 2$

Step 4 Multiply. $\frac{5}{8} \times \frac{1}{4} = \frac{5 \times 1}{2 \times 4} = \frac{5}{8}$

So, Tim gave away $\frac{5}{8}$ of the case of baseballs.

Use GCFs to simplify the factors so that the answer is in simplest form.

1. $\frac{5}{6} \times \frac{2}{3}$ 2. $\frac{3}{4} \times \frac{5}{12}$ 3. $\frac{5}{9} \times \frac{3}{10}$ 4. $\frac{3}{4} \times \frac{6}{7}$
   $\frac{5}{9}$ _____ $\frac{5}{16}$ _____ $\frac{1}{6}$ _____ $\frac{9}{14}$

5. $\frac{4}{5} \times \frac{3}{9}$ 6. $\frac{5}{8} \times \frac{12}{25}$ 7. $\frac{2}{3} \times \frac{9}{10}$ 8. $\frac{3}{4} \times \frac{8}{15}$
   $\frac{1}{5}$ _____ $\frac{2}{3}$ _____ $\frac{3}{10}$ _____ $\frac{1}{10}$

9. $\frac{3}{5} \times \frac{5}{9}$ 10. $\frac{2}{3} \times \frac{7}{16}$ 11. $\frac{4}{5} \times \frac{15}{6}$ 12. $\frac{5}{8} \times \frac{4}{9}$
   $\frac{1}{3}$ _____ $\frac{7}{24}$ _____ $\frac{3}{4}$ _____ $\frac{5}{2}$

13. $\frac{1}{8} \times \frac{2}{16}$ 14. $\frac{7}{12} \times \frac{3}{14}$ 15. $\frac{9}{11} \times \frac{5}{18}$ 16. $\frac{7}{12} \times \frac{4}{7}$
   $\frac{1}{64}$ _____ $\frac{1}{12}$ _____ $\frac{5}{22}$ _____ $\frac{1}{3}$

R30 TAKE ANOTHER LOOK   Use with text pages 132–133.

GUIDED PRACTICE

Tell each GCF you would use to simplify the fractions.

1. $\frac{5}{8} \times \frac{8}{16}$ **8**

2. $\frac{8}{9} \times \frac{3}{16}$ **8, 3**

3. $\frac{2}{9} \times \frac{3}{4}$ **2, 3**

4. $\frac{4}{3} \times \frac{1}{2}$ **2**

5. $\frac{1}{8} \times \frac{4}{5}$ **4**

6. $\frac{2}{3} \times \frac{5}{6}$ **2**

7. $\frac{2}{15} \times \frac{5}{6}$ **2, 5**

8. $6 \times \frac{2}{3}$ **3**

INDEPENDENT PRACTICE

Tell each GCF you would use to simplify the fractions.

1. $\frac{1}{2}, \frac{2}{3}$ **2**

2. $\frac{1}{5}, \frac{5}{16}$ **5**

3. $\frac{14}{27}, \frac{6}{7}$ **7, 3**

4. $\frac{3}{4}, \frac{8}{9}$ **3, 4**

5. $\frac{1}{4}, \frac{4}{5}$ **4**

6. $\frac{4}{20}, \frac{5}{8}$ **4, 5**

7. $\frac{2}{9}, \frac{9}{16}$ **2, 9**

8. $\frac{3}{5}, \frac{4}{6}$ **3**

Use GCFs to simplify the factors. Write the new problem.

9. $\frac{3}{4} \times \frac{4}{7}$ **$\frac{3}{1} \times \frac{1}{7}$**

10. $\frac{1}{2} \times \frac{6}{7}$ **$1 \times \frac{3}{7}$**

11. $\frac{3}{5} \times \frac{5}{6}$ **$1 \times \frac{1}{2}$**

12. $\frac{9}{12} \times \frac{6}{18}$ **$\frac{1}{2} \times \frac{1}{2}$**

13. $\frac{5}{8} \times \frac{12}{25}$ **$\frac{1}{2} \times \frac{3}{5}$**

14. $\frac{4}{7} \times \frac{5}{16}$ **$\frac{1}{7} \times \frac{5}{4}$**

15. $\frac{3}{5} \times \frac{5}{6}$ **$\frac{1}{5} \times \frac{1}{2}$**

16. $\frac{4}{5} \times \frac{5}{16}$ **$1 \times \frac{1}{4}$**

Use GCFs to simplify the factors so that the answer is in simplest form.

17. $\frac{2}{3} \times \frac{1}{6}$ **$\frac{1}{9}$**

18. $\frac{5}{6} \times \frac{3}{8}$ **$\frac{5}{16}$**

19. $\frac{7}{10} \times \frac{5}{7}$ **$\frac{1}{2}$**

20. $\frac{4}{5} \times 5$ **4**

21. $\frac{8}{9} \times \frac{3}{4}$ **$\frac{2}{3}$**

22. $\frac{6}{7} \times \frac{5}{8}$ **$\frac{15}{28}$**

23. $\frac{4}{7} \times \frac{21}{28}$ **$\frac{3}{7}$**

24. $\frac{9}{10} \times \frac{5}{18}$ **$\frac{1}{4}$**

25. $\frac{3}{10} \times \frac{5}{9}$ **$\frac{1}{6}$**

26. $\frac{3}{13} \times \frac{13}{3}$ **1**

27. $\frac{5}{6} \times \frac{7}{10}$ **$\frac{7}{12}$**

28. $\frac{4}{5} \times \frac{5}{8}$ **$\frac{1}{2}$**

29. $42 \times \frac{1}{6}$ **7**

30. $\frac{8}{9} \times 27$ **24**

31. $\frac{4}{9} \times 45$ **20**

32. $81 \times \frac{1}{3}$ **27**

Problem-Solving Applications

33. On Saturday, 39 teenagers went into the skateboard shop. Of those teens, $\frac{1}{3}$ bought skateboards. How many teens bought skateboards? **13 teens**

34. At Longwood School, $\frac{3}{5}$ of the students in sixth grade are boys. Of those boys, $\frac{1}{3}$ are on the honor roll. What fraction of the sixth-grade boys are on the honor roll? **$\frac{1}{5}$ of the sixth-grade boys**

35. The Hightown Apartments are being remodeled. So far, $\frac{3}{7}$ of them have been repainted, and another $\frac{2}{7}$ have been recarpeted. How many of the 343 apartments have been repainted or recarpeted? **245 apartments**

36. Mark took a survey and found that $\frac{4}{5}$ of the skateboard riders wore helmets. Of those, $\frac{1}{2}$ wore knee pads. What fraction of the skateboard riders used both helmets and knee pads? **$\frac{2}{5}$ of the riders**

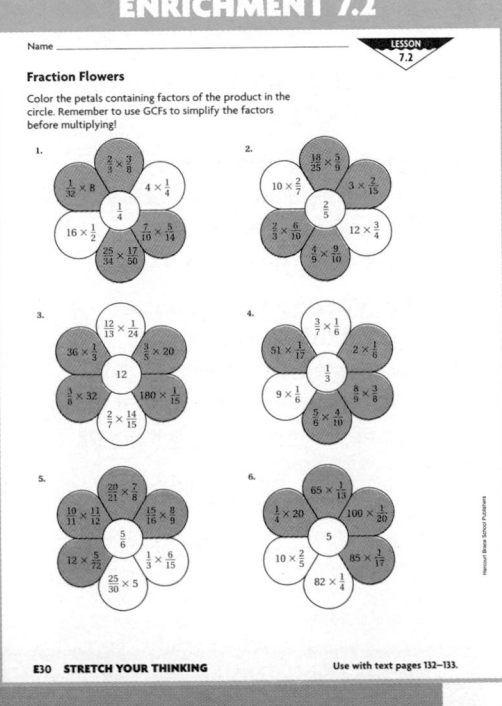

37. ✏ **WRITE ABOUT IT** What are the advantages of simplifying before multiplying fractions? **You do not have to simplify the product; multiplying the factors is easier.**

MORE PRACTICE Lesson 7.2, page H51

133

GUIDED PRACTICE

In these exercises students practice the skill of finding the GCFs of two fractions.

INDEPENDENT PRACTICE

As students work Exercises 17–32, observe them to determine whether they have developed the algorithm for finding the product. An incorrect answer may be caused by an error in simplifying or by a computational error.

COMMON ERROR ALERT Some students may want to divide a common factor into the two numbers in the numerators or in the denominators. Remind students that they must use both a numerator and a denominator each time they simplify.

ADDITIONAL EXAMPLE

Example, p. 132

Find $\frac{5}{6} \times \frac{2}{5}$.

The GCF of 6 and 2 is 2. The GCF of 5 and 5 is 1.

$$\frac{5}{6} \times \frac{2}{5} = \frac{\overset{1}{5}}{\underset{3}{6}} \times \frac{\overset{1}{2}}{\underset{1}{5}} = \frac{1}{3}$$

So, $\frac{5}{6} \times \frac{2}{5} = \frac{1}{3}$.

3 WRAP UP

Use these questions to help students focus on the big ideas of the lesson.

- **Why is the product the same whether you multiply and then simplify or simplify and then multiply?** Answers will vary but should include a statement about common factors.

✓ *Assessment Tip*

Have students write the answer.

- **Why does $\frac{3}{4} \times \frac{8}{5} = \frac{3}{1} \times \frac{2}{5}$?** You can change 4 to 1 and 8 to 2 because they have a common factor.

DAILY QUIZ

Simplify factors. Then multiply.

1. $\frac{3}{5} \times \frac{1}{3}$ **$\frac{1}{5}$**

2. $\frac{7}{12} \times \frac{3}{14}$ **$\frac{1}{8}$**

3. $\frac{6}{7} \times \frac{7}{8}$ **$\frac{3}{4}$**

4. $20 \times \frac{3}{10}$ **6**

5. $\frac{1}{2} \times \frac{8}{10}$ **$\frac{2}{5}$**

6. $\frac{2}{3} \times \frac{9}{12}$ **$\frac{1}{2}$**

PRACTICE 7.2

LESSON 7.2

Name _____

Simplifying Factors

Tell each GCF you would use to rewrite the fractions.

1. $\frac{3}{4}, \frac{8}{5}$ **2.** $\frac{8}{9}, \frac{11}{16}$ **3.** $\frac{7}{10}, \frac{5}{21}$ **4.** $\frac{9}{11}, \frac{2}{3}$

   ___4___   ___8___   ___5, 7___   ___3___

5. $\frac{6}{25}, \frac{5}{18}$ **6.** $\frac{12}{13}, \frac{1}{30}$ **7.** $\frac{4}{15}, \frac{10}{17}$ **8.** $\frac{7}{9}, \frac{27}{28}$

   ___6, 5___   ___6___   ___5___   ___7, 9___

Use GCFs to simplify the factors. Write the new problem.

9. $\frac{3}{4} \times \frac{8}{9}$ **10.** $\frac{1}{6} \times \frac{6}{15}$ **11.** $\frac{5}{8} \times \frac{8}{9}$ **12.** $\frac{3}{8} \times \frac{4}{12}$

  $\frac{1}{1} \times \frac{2}{3}$ $\frac{1}{1} \times \frac{1}{5}$ $\frac{1}{1} \times \frac{5}{9}$ $\frac{1}{2} \times \frac{1}{4}$

13. $\frac{2}{5} \times \frac{4}{9}$ **14.** $\frac{3}{8} \times \frac{4}{12}$ **15.** $\frac{3}{5} \times \frac{10}{9}$ **16.** $\frac{5}{6} \times \frac{20}{36}$

  $\frac{2}{5} \times \frac{1}{5}$ $\frac{1}{2} \times \frac{1}{4}$ $\frac{1}{1} \times \frac{2}{5}$ $\frac{1}{6} \times \frac{4}{9}$

Use GCFs to simplify the factors so that the answer is in simplest form.

17. $\frac{5}{6} \times \frac{3}{10}$ **18.** $\frac{7}{10} \times \frac{2}{5}$ **19.** $\frac{1}{2} \times \frac{12}{13}$ **20.** $18 \times \frac{2}{6}$

   $\frac{1}{4}$ $\frac{7}{25}$ $\frac{6}{13}$ ___6___

21. $\frac{4}{5} \times 30$ **22.** $\frac{9}{10} \times \frac{2}{3}$ **23.** $\frac{7}{8} \times 24$ **24.** $\frac{9}{11} \times \frac{22}{27}$

   ___24___ $\frac{3}{5}$ ___21___ $\frac{2}{3}$

Mixed Applications

25. The boy's locker room is being remodeled. On Monday, $\frac{5}{6}$ of the lockers were replaced. On Tuesday, $\frac{4}{5}$ of the lockers were replaced. How many of the 504 lockers have been replaced?

   ___336 lockers___

26. At the Smith School, $\frac{3}{4}$ of the students in the sixth grade play a team sport. Of those, $\frac{2}{3}$ are in the school band. What fraction of the sixth graders play a team sport and are in the band? $\frac{1}{2}$

27. Jon chose a number, added 6, multiplied by 4, and subtracted 2. The result was 70. What's the number?

   ___12___

28. On Monday the band rehearsed for $2\frac{1}{4}$ hr. On Wednesday they rehearsed for $3\frac{1}{2}$ hr. How many hours did the band rehearse? $5\frac{3}{4}$ hr

P30 **ON MY OWN**   Use with text pages 132–133.

ENRICHMENT 7.2

LESSON 7.2

Name _____

Fraction Flowers

Color the petals containing factors of the product in the circle. Remember to use GCFs to simplify the factors before multiplying!

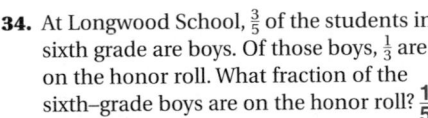

E30 **STRETCH YOUR THINKING**   Use with text pages 132–133.

133

ORGANIZER

OBJECTIVE To multiply fractions and mixed numbers

ASSIGNMENTS Independent Practice
- *Basic* 1–20, odd; 21–26
- *Average* 1–26
- *Advanced* 1–10; 21–26

PROGRAM RESOURCES
Reteach 7.3 Practice 7.3 Enrichment 7.3
Math Fun • Cooking Up a Treat, p. 142
Number Heroes • *Fraction Fireworks*

PROBLEM OF THE DAY

Transparency 7

It took André 1 min to fill his aquarium $\frac{1}{3}$ full. How long will it take him to fill the aquarium $\frac{3}{4}$ full?
$2\frac{1}{4}$ min

Solution Tab A, p. A7

QUICK CHECK

Write the mixed number as a fraction.

1. $1\frac{1}{4}$ $\frac{5}{4}$
2. $2\frac{2}{5}$ $\frac{12}{5}$
3. $2\frac{1}{2}$ $\frac{5}{2}$
4. $1\frac{5}{6}$ $\frac{11}{6}$
5. $2\frac{2}{1}$
6. $3\frac{1}{3}$ $\frac{10}{3}$

1 MOTIVATE

Show this model of $1\frac{1}{2}$. Ask:

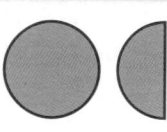

- **How would you show $\frac{1}{2}$ of $1\frac{1}{2}$?** Take $\frac{1}{2}$ of the whole and $\frac{1}{2}$ of $\frac{1}{2}$.

2 TEACH/PRACTICE

Provide opportunities for students to make the connection between multiplying fractions and multiplying with mixed numbers. The only new part of this process is that mixed numbers must be expressed as fractions before you proceed as usual.

7.3

WHAT YOU'LL LEARN
How to multiply fractions and mixed numbers

WHY LEARN THIS?
To solve everyday problems that involve measurements given as mixed numbers, such as $3\frac{1}{3}$ yd and $1\frac{1}{2}$ mi

REMEMBER:

To change a mixed number to a fraction, multiply the whole number by the denominator and add the numerator to the product. Write the sum as the new numerator. Use the same denominator.
See page H16.

$2\frac{3}{4} = \frac{(2 \times 4) + 3}{4}$

$= \frac{11}{4}$

Mixed Numbers

You can multiply a fraction and a mixed number.

Tony took $1\frac{1}{3}$ dozen cookies to his family reunion. His relatives ate $\frac{1}{2}$ of them. What part of a dozen did they eat?

Think: What is $\frac{1}{2}$ of $1\frac{1}{3}$? Find $\frac{1}{2} \times 1\frac{1}{3}$.

$\frac{1}{2} \times 1\frac{1}{3} = \frac{1}{2} \times \frac{4}{3}$ *Write the mixed number as a fraction.*

$= \frac{1}{2} \times \frac{\overset{2}{\cancel{4}}}{3}$ *Use the GCF to simplify.*

$= \frac{1}{1} \times \frac{2}{3} = \frac{2}{3}$ *Multiply.*

So, the relatives ate $\frac{2}{3}$ dozen cookies.

- Jackie took $2\frac{1}{2}$ gallons of ice water to the reunion. The relatives drank $\frac{3}{4}$ of it. How much did they drink? $1\frac{7}{8}$ **gallons**

You can use the same method to multiply two mixed numbers.

EXAMPLE After lunch at the reunion, Cami rode her bike $1\frac{3}{5}$ mi. Tim rode $1\frac{1}{2}$ times as far. How far did Tim ride?

Think: Tim rode $1\frac{1}{2} \times 1\frac{3}{5}$ mi.

$1\frac{1}{2} \times 1\frac{3}{5} = \frac{3}{2} \times \frac{8}{5}$ *Write the mixed numbers as fractions.*

$= \frac{3}{2} \times \frac{\overset{4}{\cancel{8}}}{5}$ *Use the GCF to simplify.*

$= \frac{3}{1} \times \frac{4}{5}$ *Multiply.*

$= \frac{12}{5}$, or $2\frac{2}{5}$

So, Tim rode $2\frac{2}{5}$ mi.

You can use a calculator to multiply mixed numbers, such as $3\frac{3}{4} \times 1\frac{1}{2}$.

3 [UNIT] 3 [/] 4 [×] 1 [UNIT]

1 [/] 2 [=]   45/8

- What happens to the product when you press the key [Ab/c] ? **The product is changed to a mixed number.**

IDEA FILE

CORRELATION TO OTHER RESOURCES

- **CornerStone Mathematics** ©1996 SkillsBank Corporation Level B, Using Fractions and Percents, Lessons 4 and 5

- *Fraction Attraction* **Published by Sunburst** Students complete activities that provide conceptual development and skill in working with fractions.

RETEACH 7.3

Name _____

LESSON 7.3

Mixed Numbers

Barbara bought $2\frac{1}{4}$ dozen donuts. Her family ate $\frac{2}{3}$ of them. How many dozen did her family eat?

Step 1 What is $\frac{2}{3}$ of $2\frac{1}{4}$? $\frac{2}{3} \times 2\frac{1}{4} = \blacksquare$
Find $\frac{2}{3} \times 2\frac{1}{4}$.

Step 2 Write the mixed number as a fraction. $\frac{2}{3} \times 2\frac{1}{4} = \frac{2}{3} \times \frac{9}{4}$

Step 3 Use the GCF to simplify. The GCF of 2 and 4 is 2. The GCF of 3 and 9 is 3. $= \frac{1}{1} \times \frac{3}{2}$

Step 4 Multiply. $= \frac{1}{1} \times \frac{3}{2} = \frac{3}{2} = 1\frac{1}{2}$

So, Barbara's family ate $1\frac{1}{2}$ dozen.

Find the product. Write it in simplest form.

1. $\frac{2}{3} \times 1\frac{3}{5}$ 2. $4\frac{1}{2} \times 6\frac{2}{3}$ 3. $2\frac{1}{5} \times 3\frac{1}{8}$
   $\frac{2}{3}$ 30 $6\frac{7}{8}$

4. $\frac{8}{9} \times 3\frac{3}{4}$ 5. $1\frac{3}{7} \times 1\frac{5}{9}$ 6. $2\frac{1}{3} \times 1\frac{5}{7}$
   $3\frac{1}{3}$ $2\frac{2}{9}$ 4

7. $\frac{5}{8} \times 4\frac{4}{5}$ 8. $2\frac{3}{4} \times 1\frac{3}{5}$ 9. $1\frac{1}{3} \times 5\frac{3}{4}$
   3      $4\frac{2}{5}$ $7\frac{2}{3}$

10. $2\frac{4}{9} \times 1\frac{1}{2}$ 11. $1\frac{2}{5} \times \frac{5}{7}$ 12. $3\frac{5}{6} \times 2\frac{1}{4}$
   $3\frac{2}{3}$ 7 $8\frac{5}{8}$

13. $1\frac{1}{2} \times \frac{9}{10}$ 14. $\frac{4}{7} \times 2\frac{3}{8}$ 15. $\frac{5}{7} \times 4\frac{1}{5}$
   $1\frac{1}{35}$ $1\frac{5}{14}$ 3

Use with text pages 134–135.   TAKE ANOTHER LOOK   R31

GUIDED PRACTICE

Rewrite the problem by changing each mixed number to a fraction.

1. $2\frac{1}{4} \times \frac{1}{7}$ $\frac{9}{4} \times \frac{1}{7}$

2. $\frac{2}{9} \times 2\frac{3}{8}$ $\frac{2}{9} \times \frac{19}{8}$

3. $1\frac{2}{3} \times 3\frac{2}{5}$ $\frac{5}{3} \times \frac{17}{5}$

4. $4\frac{1}{3} \times 1\frac{6}{11}$ $\frac{13}{3} \times \frac{17}{11}$

Find the product.

5. $\frac{3}{4} \times 1\frac{1}{2}$ $\frac{9}{8}$, or $1\frac{1}{8}$

6. $\frac{1}{2} \times 2\frac{1}{3}$ $\frac{7}{6}$, or $1\frac{1}{6}$

7. $1\frac{1}{2} \times 1\frac{1}{2}$ $\frac{9}{4}$, or $2\frac{1}{4}$

8. $1\frac{2}{5} \times 2\frac{1}{4}$ $\frac{63}{20}$, or $3\frac{3}{20}$

INDEPENDENT PRACTICE

Rewrite the problem by changing each mixed number to a fraction.

1. $4\frac{3}{4} \times \frac{1}{8}$ $\frac{19}{4} \times \frac{1}{8}$

2. $\frac{2}{7} \times 6\frac{3}{5}$ $\frac{2}{7} \times \frac{33}{5}$

3. $1\frac{1}{3} \times 3\frac{3}{4}$ $\frac{4}{3} \times \frac{15}{4}$

4. $2\frac{1}{3} \times 1\frac{6}{7}$ $\frac{7}{3} \times \frac{13}{7}$

Find the product. Write it in simplest form.

5. $3\frac{1}{2} \times \frac{4}{5}$ $\frac{14}{5}$, or $2\frac{4}{5}$

6. $1\frac{1}{3} \times 3\frac{3}{4}$ 5

7. $3\frac{2}{3} \times 1\frac{5}{8}$ $\frac{143}{24}$, or $5\frac{23}{24}$

8. $\frac{1}{5} \times 3\frac{1}{3}$ $\frac{2}{3}$

9. $4\frac{2}{3} \times 1\frac{3}{4}$ $\frac{49}{6}$, or $8\frac{1}{6}$

10. $1\frac{3}{8} \times 4\frac{2}{3}$ $\frac{77}{12}$, or $6\frac{5}{12}$

11. $5\frac{1}{2} \times \frac{1}{6}$ $\frac{11}{12}$

12. $\frac{1}{2} \times 3\frac{1}{7}$ $\frac{11}{7}$, or $1\frac{4}{7}$

13. $4\frac{1}{6} \times 3\frac{3}{5}$ 15

14. $1\frac{3}{4} \times \frac{1}{3}$ $\frac{7}{12}$

15. $10\frac{1}{5} \times 8\frac{1}{3}$ 85

16. $\frac{1}{5} \times 5\frac{5}{6}$ $\frac{7}{6}$, or $1\frac{1}{6}$

17. $3\frac{1}{3} \times 2\frac{1}{7}$ $\frac{50}{7}$, or $7\frac{1}{7}$

18. $3\frac{1}{4} \times \frac{2}{5}$ $\frac{13}{10}$, or $1\frac{3}{10}$

19. $\frac{7}{8} \times 7\frac{3}{7}$ $\frac{13}{2}$, or $6\frac{1}{2}$

20. $9\frac{3}{8} \times 4\frac{4}{5}$ 45

Problem-Solving Applications

21. Richard and Hector walked to the picnic. Richard walked $1\frac{3}{5}$ mi. Hector walked $2\frac{1}{2}$ times as far as Richard. How far did Hector walk? **4 mi**

22. Mr. Jackson has $1\frac{2}{3}$ c of fruit. The fruit is $\frac{1}{4}$ grapes. How many cups of grapes does he have? $\frac{5}{12}$ **c**

23. Lakeshia is training for a track meet. She walks $5\frac{3}{4}$ mi every day. How many miles does she walk in one week? $40\frac{1}{4}$ **mi**

24. CALCULATOR Use a fraction calculator to show the key sequence you would use to find $\frac{1}{2} \times 2\frac{1}{4}$. **Sequences will vary depending on calculator.**

25. CALCULATOR What product would be the result of this key sequence?

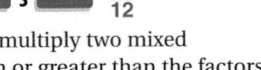 $\frac{14}{12}$

26. ✏️ **WRITE ABOUT IT** When you multiply two mixed numbers, is the product less than or greater than the factors? Give an example. **Possible example:** $3\frac{2}{3} \times 1\frac{2}{5} = 5\frac{2}{15}$; **greater**

MORE PRACTICE Lesson 7.3, page H52

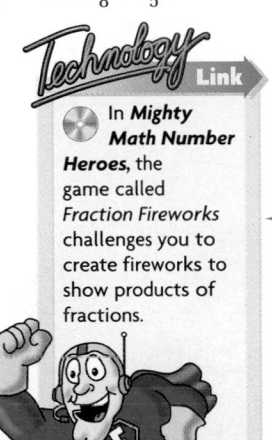

In *Mighty Math Number Heroes*, the game called *Fraction Fireworks* challenges you to create fireworks to show products of fractions.

Use Grow Slide Level X.

135

GUIDED PRACTICE

To check students' conceptual understanding of mixed numbers and fractions, work through Exercises 1 and 5.

💡**CRITICAL THINKING** **How are mixed numbers and fractions greater than 1 alike?** They both represent quantities greater than 1. **How are they different?** Mixed numbers show the whole numbers, fractions do not.

INDEPENDENT PRACTICE

You may want to point out to students that a product can be a fraction, a whole number, or a mixed number. Use Exercises 5, 6, and 8 as examples.

ADDITIONAL EXAMPLE

Example, p. 134

The recipe for Thirst Ender drink requires $1\frac{1}{2}$ cups of sugar for each gallon of water. How much sugar would you add to make 6 gallons of Thirst Ender?

$6 \times 1\frac{1}{2} = 6 \times \frac{3}{2} = 9$ cups

3 WRAP UP

Use these questions to help students focus on the big idea of the lesson.

- **If someone phoned you to ask how to multiply either two mixed numbers or a mixed number and a fraction, how would you explain it?** Check students' answers.

✓ *Assessment Tip*

Have students write the answer.

- **Without multiplying, tell whether the product $\frac{2}{3} \times \frac{3}{4}$ is a fraction, a whole number, or a mixed number. Explain how you know.** The answer is a fraction. Both factors are fractions less than 1.

DAILY QUIZ

Write each number as a fraction.

1. $5\frac{2}{5}$ $\frac{27}{5}$

2. $7\frac{1}{7}$ $\frac{50}{7}$

Find the product. Simplify if needed.

3. $\frac{2}{3} \times 2\frac{2}{3}$ $\frac{16}{9}$, or $1\frac{7}{9}$

4. $2\frac{1}{2} \times 3\frac{3}{4}$ $7\frac{5}{8}$, or $9\frac{3}{8}$

5. $1\frac{7}{9} \times 3\frac{3}{4}$ $\frac{20}{3}$, or $6\frac{2}{3}$

135

PRACTICE 7.3

Name _____

LESSON 7.3

Mixed Numbers

Rewrite the problem by changing each mixed number to a fraction.

1. $3\frac{1}{2} \times \frac{6}{14}$ $\frac{7}{2} \times \frac{6}{14}$

2. $\frac{3}{8} \times 2\frac{2}{11}$ $\frac{3}{8} \times \frac{24}{11}$

3. $6\frac{2}{5} \times 1\frac{2}{5}$ $\frac{32}{5} \times \frac{10}{5}$

4. $2\frac{1}{2} \times \frac{4}{10}$ $\frac{5}{2} \times \frac{4}{10}$

5. $2\frac{1}{3} \times 2\frac{4}{5}$ $\frac{15}{7} \times \frac{14}{5}$

6. $1\frac{2}{3} \times \frac{2}{5}$ $\frac{10}{8} \times \frac{2}{5}$

7. $\frac{2}{9} \times 2\frac{1}{4}$ $\frac{9}{9} \times \frac{9}{4}$

8. $1\frac{3}{10} \times \frac{15}{26}$ $\frac{13}{10} \times \frac{15}{26}$

Find the product. Write it in simplest form.

9. $2\frac{1}{2} \times 1\frac{1}{3}$ $3\frac{1}{3}$

10. $3\frac{1}{2} \times 2\frac{1}{3}$ 8

11. $8\frac{3}{4} \times \frac{2}{5}$ $3\frac{1}{2}$

12. $3\frac{3}{5} \times 1\frac{1}{5}$ 4

13. $3\frac{1}{3} \times 2\frac{2}{5}$ 8

14. $1\frac{3}{4} \times \frac{3}{14}$ $\frac{3}{8}$

15. $4\frac{2}{5} \times \frac{10}{11}$ 4

16. $\frac{6}{7} \times 2\frac{1}{10}$ $1\frac{4}{5}$

17. $3\frac{1}{2} \times 1\frac{1}{4}$ $4\frac{3}{8}$

18. $3\frac{1}{2} \times 1\frac{1}{5}$ $4\frac{1}{5}$

19. $4\frac{3}{8} \times \frac{1}{2}$ $2\frac{3}{16}$

20. $6\frac{1}{2} \times \frac{5}{8}$ $4\frac{1}{16}$

21. $2\frac{1}{4} \times 3\frac{1}{9}$ $7\frac{1}{5}$

22. $9\frac{3}{5} \times 1\frac{1}{4}$ 12

23. $\frac{3}{5} \times 1\frac{2}{3}$ 1

24. $12\frac{1}{2} \times 1\frac{1}{2}$ $18\frac{3}{4}$

25. $1\frac{1}{5} \times \frac{1}{4}$ $\frac{3}{10}$

26. $3\frac{1}{4} \times 1\frac{5}{8}$ $6\frac{7}{8}$

27. $2\frac{2}{5} \times \frac{9}{10}$ $\frac{9}{10}$

28. $5\frac{1}{4} \times 1\frac{1}{2}$ $7\frac{5}{8}$

Mixed Applications

29. Marsha had $2\frac{1}{2}$ c of mixed nuts. The nuts are $\frac{1}{4}$ peanuts. How many cups of peanuts does she have? $\frac{2}{4}$ c

30. Suzanne is practicing for the track meet. She runs $2\frac{1}{4}$ times around the track. A lap of the track is $\frac{1}{4}$ mi. How many miles does she run? $\frac{9}{16}$ mi

31. In March, it rained a total of 5.3 in. The normal rainfall for March is 3.9 in. How many inches above normal was the rainfall?   1.4 in.

32. Ty bought a 7.5 lb roast. He wants it to be medium well. The directions say it should roast for 25 min per lb. How long should he roast the meat?   187.5 min or $3\frac{1}{8}$ hr

Use with text pages 134–135.   **ON MY OWN** P31

ENRICHMENT 7.3

Name _____

LESSON 7.3

Fraction Analogies

Read each analogy. Explain how each pair of numbers are related. The first one is done for you.

1. $2\frac{1}{3}$ is to $3\frac{1}{2}$ as $4\frac{3}{8}$ is to $6\frac{9}{16}$.

Think: What number times $2\frac{1}{3}$ equals $3\frac{1}{2}$?

$3\frac{1}{2}$ is the product of $2\frac{1}{3}$ and $1\frac{1}{2}$, just as $6\frac{9}{16}$ is the product of $4\frac{3}{8}$ and $1\frac{1}{2}$.

2. $1\frac{2}{5}$ is to $2\frac{9}{20}$ as $3\frac{1}{2}$ is to $6\frac{1}{8}$.

$2\frac{9}{20}$ is the product of $1\frac{2}{5}$ and $1\frac{3}{4}$, as $6\frac{1}{8}$ is the product of

$3\frac{1}{2}$ and $1\frac{3}{4}$.

3. $4\frac{3}{8}$ is to $1\frac{3}{32}$ as $5\frac{3}{4}$ is to $1\frac{7}{16}$.

$1\frac{3}{32}$ is the product of $4\frac{3}{8}$ and $\frac{1}{4}$, as $1\frac{7}{16}$ is the product of

$5\frac{3}{4}$ and $\frac{1}{4}$.

4. $3\frac{1}{2}$ is to $9\frac{1}{4}$ as $4\frac{7}{8}$ is to $12\frac{3}{16}$.

$9\frac{1}{4}$ is the product of $3\frac{1}{2}$ and $2\frac{1}{4}$, as $12\frac{3}{16}$ is the product of

$4\frac{7}{8}$ and $2\frac{1}{4}$.

5. $5\frac{1}{4}$ is to $1\frac{1}{15}$ as $6\frac{3}{4}$ is to $1\frac{7}{20}$.

$1\frac{1}{15}$ is the product of $5\frac{1}{4}$ and $\frac{1}{5}$, as $1\frac{7}{20}$ is the product of

$6\frac{3}{4}$ and $\frac{1}{5}$.

6. $9\frac{5}{6}$ is to $11\frac{1}{16}$ as $3\frac{1}{4}$ is to $3\frac{21}{32}$.

$11\frac{1}{16}$ is the product of $9\frac{5}{6}$ and $1\frac{1}{8}$, as $3\frac{21}{32}$ is the product of

$3\frac{1}{4}$ and $1\frac{1}{8}$.

7. $2\frac{5}{12}$ is to $3\frac{5}{8}$ as $4\frac{8}{9}$ is to $6\frac{4}{9}$.

$3\frac{5}{8}$ is the product of $2\frac{5}{12}$ and $1\frac{1}{3}$, as $6\frac{4}{9}$ is the product of

$4\frac{8}{9}$ and $1\frac{1}{3}$.

Use with text pages 134–135.   **STRETCH YOUR THINKING** E31

ORGANIZER

OBJECTIVE To model division of fractions
VOCABULARY *Review* dividend, divisor
MATERIALS *For each student* fraction circles
PROGRAM RESOURCES
  E-Lab • Activity 7
  E-Lab Recording Sheets, p. 7

USING THE PAGE

Encourage students to discuss times they have used fractions or division of fractions. If they need a hint, mention crafts and cooking.

Activity 1

Explore

This activity reviews or introduces one way of modeling division of fractions. Students have been accustomed to thinking that the quotient is always less than the dividend. This thinking must be revised with the division of fractions. The modeling process shows that if we begin with a unit or a whole and divide it into halves, we end up with a greater number, the number of halves. This idea is reinforced in Think and Discuss.

Think and Discuss

You may want to suggest to students that they model their answers to the second and third questions.

Try This

Even though students may know the answers to the first two problems without modeling, encourage them to model the problem and the solution. The ability to model solutions will help students solve more complex problems.

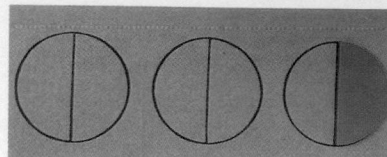

COOPERATIVE LEARNING

Exploring Division

Using models will help you understand division of fractions.

LAB *Activity*

WHAT YOU'LL EXPLORE
How to model division of fractions

WHAT YOU'LL NEED
fraction circles

ACTIVITY 1

Explore

A. Use fraction circles to find $3 \div \frac{1}{2}$, or the number of halves in 3 wholes.

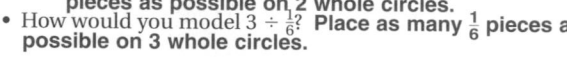

* Trace 3 whole circles on your paper.
* Model $3 \div \frac{1}{2}$ by tracing $\frac{1}{2}$-circle pieces on the 3 circles.

One whole equals two halves.

* How many halves are in 3 wholes? What is $3 \div \frac{1}{2}$? **6; 6**

B. Use fraction circles to find $\frac{1}{2} \div \frac{1}{6}$, or the number of sixths in $\frac{1}{2}$.

* Place as many $\frac{1}{6}$ pieces as you can on the $\frac{1}{2}$ piece.
* How many sixths are in $\frac{1}{2}$? What is $\frac{1}{2} \div \frac{1}{6}$? **3; 3**

Think and Discuss

* Why does $3 \div \frac{1}{2} = 6$? **There are 6 halves in 3.**
* How can you find the number of thirds in 2 wholes? **Place as many $\frac{1}{3}$ pieces as possible on 2 whole circles.**
* How would you model $3 \div \frac{1}{6}$? **Place as many $\frac{1}{6}$ pieces as possible on 3 whole circles.**

Try This

Use fraction circles to model each problem. **Check students' models**

1. $3 \div \frac{1}{3}$ **9**
2. $4 \div \frac{1}{4}$ **16**
3. $\frac{1}{2} \div \frac{1}{8}$ **4**
4. $\frac{3}{4} \div \frac{1}{8}$ **6**

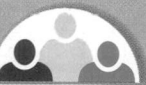

IDEA FILE

ENGLISH AS A SECOND LANGUAGE

Tip • Help students acquiring English become more familiar with U.S. money by solving problems using coins and bills. Have them write corresponding number sentences.

* How many quarters are in 1 half dollar? $\frac{1}{2} \div \frac{1}{4} = 2$
* How many half dollars are in $5? $5 \div \frac{1}{2} = 10$
* How many quarters are in $10? $10 \div \frac{1}{4} = 40$

Modalities Kinesthetic, Visual

CAREER CONNECTION

There are several occupations in which people must know how to multiply and divide fractions and whole numbers. Have students choose one of the occupations listed below and write a description of how a person in that occupation would use fractions in his or her work.

* stockbroker
* chef
* tailor
* dietitian
* fitness trainer
* musician

Modalities Auditory, Kinesthetic, Visual

ACTIVITY 2

Explore

Recall that dividing a number by 2 gives you the same result as multiplying that number by $\frac{1}{2}$.

$$8 \div 2 = 4 \qquad 8 \times \frac{1}{2} = 4$$

This relationship between multiplication and division is also true for other fractions. Look at these pairs of related problems.

$$9 \div 3 = 3 \qquad 9 \times \frac{1}{3} = 3 \qquad 6 \div \frac{3}{4} = 8 \qquad 6 \times \frac{4}{3} = 8$$

$$1 \div \frac{1}{2} = 2 \qquad 1 \times \frac{2}{1} = 2 \qquad \frac{1}{3} \div \frac{1}{6} = \frac{6}{3} \qquad \frac{1}{3} \times \frac{6}{1} = \frac{6}{3}$$

- What pattern do you see in the problems of each pair?
  They have the same answers.
- Use this pattern to write a related multiplication problem for
  $\frac{1}{3} \div \frac{2}{3} = \frac{3}{6} \cdot \frac{1}{3} \times \frac{3}{2} = \frac{3}{6}, \text{ or } \frac{1}{2}$

> **REMEMBER:**
> In a division problem, the number to be divided is the dividend. The divisor is the number by which the dividend is divided. **See page H8, Key Skill 14.**
> $\frac{1}{3} \div \frac{3}{4}$
> The dividend is $\frac{1}{3}$.
> The divisor is $\frac{3}{4}$.

Think and Discuss

Look at $1 \div \frac{1}{2} = 2$ and $1 \times \frac{2}{1} = 2$. Remember that dividing by $\frac{1}{2}$ is the same as multiplying by 2.

- What is the relationship between the dividend in the division problem and the first factor in the multiplication problem?
  They are the same.
- What is the relationship between the quotient and the product?
  They are the same.
- What is the relationship between the divisor in the division problem and the second factor in the multiplication problem?
  The same digits are used, but they are reversed.

Try This

Complete the multiplication problem.

1. $\frac{3}{4} \div \frac{1}{4} = 3$
   $\frac{3}{4} \times \blacksquare = 3 \, \frac{4}{1}$

2. $\frac{1}{3} \div \frac{1}{6} = 2$
   $\frac{1}{3} \times \blacksquare = 2 \, \frac{6}{1}$

3. $12 \div \frac{3}{4} = 16$
   $12 \times \blacksquare = 16 \, \frac{4}{3}$

4. $15 \div \frac{1}{3} = 45$
   $15 \times \blacksquare = 45 \, \frac{3}{1}$

5. $\frac{2}{3} \div \frac{1}{9} = 6$
   $\frac{2}{3} \times \blacksquare = \blacksquare \, \frac{9}{1}; 6$

6. $\frac{3}{4} \div \frac{1}{10} = 7\frac{1}{2}$
   $\frac{3}{4} \times \blacksquare = \blacksquare \, \frac{10}{1}; 7\frac{1}{2}$

Write a multiplication problem for each. Find the quotient.

7. $\frac{5}{6} \div 6 \, \frac{5}{6} \times \frac{1}{6}; \frac{5}{36}$

8. $\frac{2}{3} \div \frac{3}{4} \, \frac{2}{3} \times \frac{4}{3}; \frac{8}{9}$

9. $\frac{1}{2} \div \frac{1}{5} \, \frac{1}{2} \times \frac{5}{1}; \frac{5}{2}$

Technology Link

E-Lab • Activity 7 Available on CD-ROM
and the Internet at http://www.hbschool.com/elab

Note: Students model division of fractions. Use E-Lab p. 7.

137

Activity 2

Explore

As students look at the multiplication and division problems, remind them that a whole number can be written as a fraction with a denominator of 1.

$$2 = \frac{2}{1} \qquad 3 = \frac{3}{1}$$

Think and Discuss

Review the meaning of *quotient, dividend, divisor, product,* and *factor* before you begin this part of the activity.

Try This

For Exercises 1–6, have students write both the division problem and the multiplication problem so that the relationship between the problems is reinforced.

USING E-LAB

Students use proportional thinking and a computer model as they develop a new algorithm for division of fractions.

✓ *Assessment Tip*

Use these questions to help students communicate the big ideas of the lesson.

- **How would you use a model to find $2 \div \frac{1}{3}$?** Trace 2 wholes. Then see how many $\frac{1}{3}$ pieces will fit on the 2 wholes.

- **Dividing by the fraction $\frac{2}{3}$ is the same as __?__ by the fraction $\frac{3}{2}$.** multiplying

---

LANGUAGE CONNECTION

Math Journal

Review and reteach the terms *numerator, denominator, simplest form, common factor,* and *greatest common factor (GCF)*. Have students write the definitions in their math journals. Ask them to include examples like the following to help clarify the ideas.

$\frac{5}{8}$ *simplest form*

5 and 8 have no common factor other than 1.

$\frac{6}{10}$ needs to be simplified

2 is a *common factor* of 6 and 10.

Modalities Auditory, Visual

ALGEBRA CONNECTION

This lesson provides several opportunities to teach students about inverse operations as they apply to dividing with fractions. Understanding the relationship between inverse operations provides a foundation for understanding the process of solving equations.

- Ask students to tell you what the term *opposite* means. Then introduce the term *inverse*. Have them relate the two terms.

Modality Auditory

ORGANIZER

GRADE 5	GRADE 6	GRADE 7
Multiply a fraction by a fraction.	Divide with fractions.	Solve equations using division of fractions.

OBJECTIVE To divide with fractions
VOCABULARY *New* **reciprocal**
ASSIGNMENTS Independent Practice
- *Basic* 1–12, 17–31
- *Average* 1–31
- *Advanced* 9–31

PROGRAM RESOURCES
Reteach 7.4 Practice 7.4 Enrichment 7.4
Math Fun • Roll a Fraction, p. 142

PROBLEM OF THE DAY

Transparency 7
A hawk flies $\frac{1}{3}$ mi in 30 sec. How far can the hawk fly in 1 min? How fast does it fly, in miles per hour? $\frac{2}{3}$ mi; 40 mph

Solution Tab A, p. A7

QUICK CHECK
Find the product. Simplify if needed.

1. $\frac{4}{5} \times \frac{4}{3}$ $\frac{16}{15}$, or $1\frac{1}{15}$ 2. $\frac{4}{5} \times \frac{3}{4}$ $\frac{3}{5}$
3. $6 \times \frac{2}{3}$ 4   4. $\frac{2}{3} \times \frac{1}{6}$ $\frac{1}{9}$
5. $\frac{2}{3} \times \frac{2}{3}$ $\frac{4}{9}$

1 MOTIVATE

Ask students to write six fractions, using the numbers 2, 3, and 5. Each fraction should have a one-digit numerator and a one-digit denominator. Ask:

- **What pairs of these fractions have a product of 1?** $\frac{2}{3}, \frac{3}{2}. \frac{2}{5}, \frac{5}{2}. \frac{3}{5}, \frac{5}{3}$

2 TEACH/PRACTICE

The concept of the reciprocal is introduced and used in this lesson. Although it seems simple, this concept is very important.

7.4

Dividing Fractions

WHAT YOU'LL LEARN
How to divide with fractions

WHY LEARN THIS?
To solve everyday problems that use inches and yards, as in carpentry or crafts

WORD POWER
reciprocal

teen times

A popular project for math fairs is making geometric figures using string art. Using string and nails, you can make creative designs. String art designs can be used to illustrate math ideas such as multiples and factors.

In the previous activity, you saw that related multiplication and division problems give the same result. When you divide by a fraction, you can use multiplication to find the quotient.

When you rewrite the problem, you exchange the numerator and the denominator of the divisor. The new number is called the **reciprocal** of the divisor. The product of a number and its reciprocal is 1.

$$2 \div \frac{2}{3} = 2 \times \frac{3}{2} \leftarrow \frac{3}{2} \text{ is the reciprocal of } \frac{2}{3} \text{ because } \frac{2}{3} \times \frac{3}{2} = 1.$$

EXAMPLE 1 Donovan needs pieces of string for his math project. Each piece must be $\frac{1}{3}$ yd long. How many $\frac{1}{3}$-yd pieces of string can he cut from a piece $\frac{3}{4}$ yd long?

Find $\frac{3}{4} \div \frac{1}{3}$. Remember that dividing by $\frac{1}{3}$ is the same as multiplying by 3.

$\frac{3}{4} \div \frac{1}{3} = \frac{3}{4} \times \frac{3}{1}$ *Use the reciprocal of the divisor to write a multiplication problem.*

$\frac{3}{4} \times \frac{3}{1} = \frac{9}{4}$, or $2\frac{1}{4}$ *Multiply.*

So, Donovan can cut off 2 pieces of string with $\frac{1}{4}$ of a piece left.

- Suppose each piece needs to be $\frac{1}{6}$ yd long. How many $\frac{1}{6}$-yd pieces can he cut off from the original length? $\frac{3}{4} \div \frac{1}{6} = 4\frac{1}{2}$; 4 pieces

EXAMPLE 2 Find $4 \div \frac{2}{3}$.

$4 \div \frac{2}{3} = \frac{4}{1} \div \frac{2}{3}$ *Write the whole number as a fraction.*

$\frac{4}{1} \div \frac{2}{3} = \frac{4}{1} \times \frac{3}{2}$ *Use the reciprocal of the divisor to write a multiplication problem.*

$\frac{\overset{2}{4}}{1} \times \frac{3}{\underset{1}{2}} = \frac{6}{1}$, or 6 *Simplify and multiply.*

So, $4 \div \frac{2}{3} = \frac{6}{1}$, or 6.

IDEA FILE

SCIENCE CONNECTION

Have students use reference materials to find data about a planet in our solar system. Then have each student use the data to write and solve two problems requiring the multiplication or division of fractions. When they have completed their problems, have them recalculate their answers, using decimals. Ask students to write an explanation of why scientists prefer decimals.

Modality Visual

RETEACH 7.4

Name _____

LESSON 7.4

Dividing Fractions

Beth is working on a science project. She needs $\frac{2}{3}$-yd pieces of wire for the project. She bought a 6-yd piece of wire at the hardware store. How many $\frac{2}{3}$-yd pieces can she cut from this piece?

Step 1 Write a division sentence to find this amount. $\frac{6}{1} \div \frac{2}{3} = \blacksquare$

Step 2 Use the reciprocal of the divisor to write a multiplication problem. $\frac{6}{1} \div \frac{2}{3} = \frac{6}{1} \times \frac{3}{2}$ **Think:** the reciprocal of $\frac{2}{3}$ is $\frac{3}{2}$.

Step 3 Simplify. $= \frac{6}{1} \times \frac{3}{2}$

Step 4 Multiply. $= \frac{3}{1} \times \frac{3}{1} = \frac{9}{1} = 9$

So, Beth can cut 9 pieces of wire.

Find the quotient. Write it in simplest form.

1. $8 \div \frac{3}{4}$ $10\frac{2}{3}$
2. $\frac{5}{9} \div \frac{2}{3}$ $\frac{5}{6}$
3. $\frac{4}{5} \div \frac{2}{3}$ $1\frac{1}{5}$
4. $27 \div \frac{3}{5}$ 45
5. $12 \div \frac{4}{5}$ 15
6. $\frac{5}{8} \div \frac{3}{4}$ $\frac{5}{6}$
7. $18 \div \frac{3}{8}$ 48
8. $\frac{14}{15} \div 7$ $\frac{2}{15}$
9. $\frac{3}{8} \div 24$ $\frac{1}{64}$
10. $\frac{4}{5} \div 6$ $\frac{2}{15}$
11. $16 \div \frac{4}{5}$ 20
12. $\frac{5}{12} \div \frac{5}{8}$ $\frac{2}{3}$
13. $27 \div \frac{3}{8}$ 72
14. $\frac{4}{7} \div \frac{1}{2}$ $1\frac{1}{7}$
15. $\frac{7}{8} \div \frac{1}{4}$ $3\frac{1}{2}$

R32 TAKE ANOTHER LOOK   Use with text pages 138–139.

GUIDED PRACTICE

Write the reciprocal of the number.

1. $\frac{2}{3}$ $\frac{3}{2}$ 2. $\frac{3}{4}$ $\frac{4}{3}$ 3. 7 $\frac{1}{7}$ 4. $\frac{4}{7}$ $\frac{7}{4}$ 5. $\frac{1}{9}$ $\frac{9}{1}$, or 9

Find the quotient. Write it in simplest form.

6. $\frac{1}{3} \div \frac{1}{2}$ $\frac{2}{3}$ 7. $\frac{1}{5} \div \frac{1}{4}$ $\frac{4}{5}$ 8. $\frac{1}{4} \div \frac{1}{2}$ $\frac{1}{2}$ 9. $\frac{1}{2} \div 3$ $\frac{1}{6}$

INDEPENDENT PRACTICE

Find the quotient. Write it in simplest form.

1. $\frac{3}{8} \div \frac{1}{2}$ $\frac{3}{4}$ 2. $\frac{2}{3} \div \frac{4}{7}$ $1\frac{1}{6}$ 3. $\frac{7}{8} \div \frac{1}{3}$ $2\frac{5}{8}$ 4. $8 \div \frac{6}{7}$ $9\frac{1}{3}$

5. $12 \div \frac{3}{5}$ 20 6. $4 \div \frac{4}{5}$ 5 7. $\frac{4}{9} \div \frac{3}{5}$ $\frac{20}{27}$ 8. $\frac{3}{4} \div \frac{1}{3}$ $2\frac{1}{4}$

9. $6 \div \frac{3}{4}$ 8 10. $\frac{2}{5} \div 20$ $\frac{1}{50}$ 11. $\frac{1}{8} \div \frac{4}{5}$ $\frac{5}{32}$ 12. $\frac{5}{6} \div \frac{1}{3}$ $2\frac{1}{2}$

13. $\frac{5}{8} \div 25$ $\frac{1}{40}$ 14. $\frac{1}{9} \div 6$ $\frac{1}{54}$ 15. $\frac{7}{12} \div \frac{2}{3}$ $\frac{7}{8}$ 16. $\frac{1}{6} \div 2$ $\frac{1}{12}$

Problem-Solving Applications

17. How many $\frac{1}{4}$-lb hamburgers can Katie grill with 12 lb of ground beef? **48 hamburgers**

18. In a $\frac{1}{4}$-mi relay, each runner runs $\frac{1}{16}$ mi. How many runners are in the relay? **4 runners**

19. A recording of the current weather conditions lasts $\frac{3}{4}$ min. How many times can the recording be played in 1 hr? **80 times**

20. 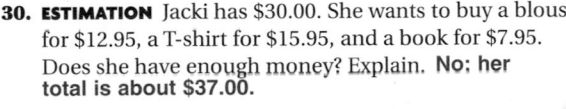 **WRITE ABOUT IT** Explain how to use reciprocals to divide fractions. **Write the reciprocal of the divisor, and then multiply.**

Mixed Review

Write as a fraction.

21. $2\frac{1}{3}$ $\frac{7}{3}$ 22. $1\frac{3}{8}$ $\frac{11}{8}$ 23. $4\frac{1}{2}$ $\frac{9}{2}$ 24. $3\frac{9}{10}$ $\frac{39}{10}$ 25. $2\frac{4}{5}$ $\frac{14}{5}$

Find the difference. Write it in simplest form.

26. $\frac{7}{8} - \frac{3}{8}$ $\frac{1}{2}$ 27. $\frac{7}{9} - \frac{1}{3}$ $\frac{4}{9}$ 28. $5\frac{7}{12} - 1\frac{1}{12}$ $4\frac{1}{2}$ 29. $9\frac{5}{9} - 5\frac{1}{6}$ $4\frac{7}{18}$

30. **ESTIMATION** Jacki has $30.00. She wants to buy a blouse for $12.95, a T-shirt for $15.95, and a book for $7.95. Does she have enough money? Explain. **No; her total is about $37.00.**

31. **CONSUMER** A 12-in. sandwich costs $4.55. A 6-in. sandwich costs $2.15. You want a 12-in. sandwich. Will it cost less to buy one 12-in. sandwich or two 6-in. sandwiches? **two 6-in. sandwiches**

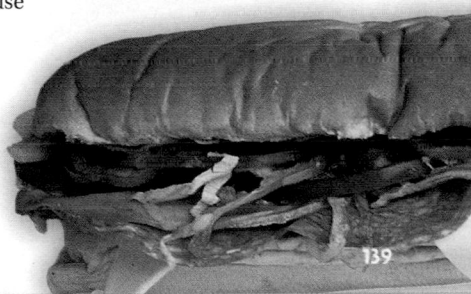

139

MORE PRACTICE Lesson 7.4, page H52

GUIDED PRACTICE
In these exercises students practice finding reciprocals and dividing fractions.

INDEPENDENT PRACTICE
Have students identify the divisor and the reciprocal of the divisor for five of these problems before they find the quotient. If students have difficulty writing the reciprocal, review the concept before beginning the practice.

COMMON ERROR ALERT Some students will experience difficulty with

• writing the reciprocal of a whole number.
• writing the reciprocal of a unit fraction.

ADDITIONAL EXAMPLES

Example 1, p. 138

Jackie needs pieces of ribbon for her sewing project. She has $1\frac{1}{2}$-yd of ribbon. How many $\frac{1}{2}$-yd pieces can she cut from it? $1\frac{1}{2} \div \frac{1}{2} = 3$; 3 pieces

Example 2, p. 138

Find $8 \div \frac{3}{4}$. $\frac{32}{3}$, or $10\frac{2}{3}$

3 WRAP UP

Use these questions to help students focus on the big ideas of the lesson.

• **What is a reciprocal?** the number that when multiplied by a given number yields 1

✓ *Assessment Tip*

Have students write the answer.

• **How is the reciprocal used in division of fractions?** To divide by a fraction, you can multiply by the reciprocal of the divisor.

DAILY QUIZ

Write the reciprocal.

1. $\frac{2}{3}$ $\frac{3}{2}$ 2. 4 $\frac{1}{4}$

Divide.

3. $\frac{3}{4} \div \frac{5}{8}$ $\frac{6}{5}$, or $1\frac{1}{5}$ 4. $\frac{2}{5} \div \frac{5}{4}$ $\frac{8}{25}$

5. $\frac{3}{7} \div \frac{3}{8}$ $\frac{8}{7}$, or $1\frac{1}{7}$

PRACTICE 7.4

Name _____

LESSON 7.4

Dividing Fractions

Vocabulary

Complete.

1. When you exchange the numerator and denominator of a fraction, this new number is called a __reciprocal__ of the fraction.

Find the quotient. Write it in simplest form.

2. $\frac{4}{5} \div \frac{8}{15}$ $1\frac{1}{2}$ 3. $\frac{7}{10} \div \frac{1}{2}$ $1\frac{2}{5}$ 4. $\frac{5}{6} \div \frac{1}{2}$ $1\frac{2}{3}$ 5. $24 \div \frac{1}{2}$ 48

6. $9 \div \frac{1}{6}$ 54 7. $\frac{7}{9} \div \frac{2}{3}$ $1\frac{1}{6}$ 8. $\frac{9}{10} \div \frac{2}{5}$ $2\frac{1}{4}$ 9. $\frac{9}{20} \div \frac{3}{4}$ $\frac{3}{5}$

10. $\frac{5}{8} \div \frac{5}{16}$ 2 11. $\frac{5}{6} \div \frac{2}{3}$ $1\frac{1}{4}$ 12. $\frac{12}{21} \div \frac{4}{7}$ 1 13. $\frac{5}{8} \div \frac{1}{4}$ $2\frac{1}{2}$

14. $\frac{3}{4} \div \frac{2}{3}$ $1\frac{1}{8}$ 15. $\frac{5}{9} \div \frac{5}{6}$ $\frac{2}{3}$ 16. $\frac{7}{8} \div 12$ $\frac{1}{16}$ 17. $15 \div \frac{5}{9}$ 27

18. $\frac{5}{12} \div \frac{3}{4}$ $\frac{5}{9}$ 19. $\frac{3}{8} \div 18$ $\frac{1}{48}$ 20. $\frac{7}{10} \div 14$ $\frac{1}{20}$ 21. $24 \div \frac{4}{5}$ 30

Mixed Applications

22. In a $\frac{1}{2}$-mi relay, each runner runs $\frac{1}{8}$ mi. How many runners are on the team? **4 runners**

23. Zoe used $\frac{1}{3}$ yd of ribbon to make a bow. How many bows can she make from $4\frac{1}{3}$ yd of fabric? **13 ribbons**

24. On Saturday, 124 customers went into the Sandwich Shop. Of these customers, $\frac{1}{4}$ ordered the lunch special. How many customers ordered the lunch special? **31 customers**

25. Jeffrey rode his bicycle 1.6 mi to a friend's house. Then they rode 2.5 mi to a park. Then Jeffrey rode 3.8 mi home. How far did Jeffrey ride altogether? **7.9 mi**

P32 ON MY OWN

Use with text pages 138–139.

ENRICHMENT 7.4

Name _____

LESSON 7.4

Divide to Find a Message

Match each exercise in Column 1 with its quotient in Column 2. Then write each corresponding letter on the line below marked with the exercise number to discover the Math Tip.

Column 1		Column 2	
1. $\frac{3}{4} \div \frac{1}{2}$ L	14. $12 \div \frac{1}{6}$ U	A. 25	N. $2\frac{3}{4}$
2. $6 \div \frac{1}{9}$ Q	15. $\frac{5}{7} \div \frac{1}{8}$ P	B. $1\frac{3}{4}$	O. $3\frac{1}{2}$
3. $\frac{8}{11} \div \frac{1}{3}$ G	16. $18 \div \frac{2}{3}$ J	C. $\frac{1}{50}$	P. $2\frac{2}{7}$
4. $10 \div \frac{2}{5}$ A	17. $\frac{3}{20} \div \frac{3}{10}$ I	D. 22	Q. 54
5. $\frac{11}{12} \div \frac{1}{2}$ N	18. $15 \div \frac{1}{3}$ W	E. $13\frac{1}{2}$	R. $1\frac{4}{5}$
6. $\frac{1}{10} \div 5$ C	19. $\frac{3}{5} \div \frac{1}{3}$ R	F. 1	S. $\frac{1}{4}$
7. $\frac{3}{4} \div \frac{1}{7}$ T	20. $11 \div \frac{1}{2}$ D	G. $2\frac{2}{11}$	T. 6
8. $\frac{5}{7} \div 10$ X	21. $\frac{5}{7} \div \frac{10}{14}$ F	H. 75	U. 72
9. $\frac{4}{5} \div \frac{2}{3}$ Z	22. $\frac{2}{3} \div \frac{8}{9}$ Y	I. $\frac{1}{2}$	V. $21\frac{1}{3}$
10. $3 \div \frac{2}{7}$ E	23. $5 \div \frac{1}{15}$ H	J. 27	W. 45
11. $8 \div \frac{3}{5}$ V	24. $7\frac{1}{2} \div \frac{1}{4}$ O	K. 50	X. $\frac{1}{14}$
12. $\frac{7}{8} \div \frac{1}{2}$ B	25. $10 \div \frac{1}{5}$ K	L. $1\frac{1}{2}$	Y. $\frac{3}{4}$
13. $\frac{1}{2} \div 2$ E	26. $\frac{5}{8} \div \frac{1}{3}$ M	M. $1\frac{7}{8}$	Z. 4

M T   H   E   P   R   O   D   U   C   T
   7   23   10   15   24   20   14   6   7

A O   F   A   N   U   M   B   E   R
T 24 21   4   5   14   26   12   10   19

H A   N   D   I   T   S
   4   5   10   17   7   13

T R   E   C   I   P   R   O   C   A   L
I 19 10 17 15 19 24 6   4   1

P I   S   O   N   E
   17 13   24   5   10

E32 STRETCH YOUR THINKING

Use with text pages 138–139.

139

GRADE 5	GRADE 6	GRADE 7
Solve a problem using the strategy *work backward*, and choose appropriate strategies.	Solve a problem using the strategy *work backward*, and choose appropriate strategies.	Solve a problem using the strategy *work backward*, and choose appropriate strategies.

ORGANIZER

OBJECTIVE To divide mixed numbers
ASSIGNMENTS Practice/Mixed Applications
- ■ *Basic* 1–12
- ■ *Average* 1–12
- ■ *Advanced* 1–12

PROGRAM RESOURCES
Reteach 7.5 Practice 7.5 Enrichment 7.5
Problem-Solving Think Along, TR p. R1

PROBLEM OF THE DAY

Transparency 7

Ming Li ran 90 ft from first base to second base. Each stride was $3\frac{4}{5}$ ft long. About how many strides did Ming Li take? 24 strides ($90 \div 3\frac{4}{5} = 23\frac{13}{19}$ but she could not take a fraction of a stride.)

Solution Tab A, p. A7

QUICK CHECK
Find *n*.

1. $3 \times 8 = n$ 24
2. $4 \times n = 12$ 3
3. $16 \div n = 2$ 8
4. $n \times 7 = 35$ 5
5. $12 \times n = 6$ $\frac{1}{2}$

1 MOTIVATE

Have students work with partners. Tell them that they are to think of two actions that are opposites. Have pairs demonstrate their opposite actions to the class. For example: One student opens the door; the other closes the door. After they have finished, make a list of opposites.

2 TEACH/PRACTICE

Students should discover that division with mixed numbers uses the same algorithm as division with fractions except that the mixed number must first be written as a fraction. Some students will need to be reminded to use the reciprocal.

7.5

WHAT YOU'LL LEARN
How to work backward by dividing mixed numbers

WHY LEARN THIS?
To solve problems such as finding the length or width of a box when you know the area

Problem Solving
- • **Understand**
- • **Plan**
- • **Solve**
- • **Look Back**

PROBLEM-SOLVING STRATEGY
Work Backward by Dividing Mixed Numbers

You can divide fractions that are greater than 1.

Melissa is making a rectangular box. She has $4\frac{1}{2}$ ft² of colored paper to cover the top. The width of the box must be $1\frac{1}{2}$ ft. What is the greatest length the box can have?

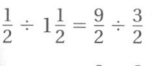

✓ **UNDERSTAND**   What are you asked to find? **the greatest length of the box**
What facts are given? **the area of the paper and width of the box**

✓ **PLAN**   What strategy will you use?

You can *work backward* to find the length.

The area of a rectangle is found by multiplying the length and the width. Since you know the area and the width, you can divide to find the length.

✓ **SOLVE**   How will you solve the problem?

Area equals length times width: $4\frac{1}{2} = \blacksquare \times 1\frac{1}{2}$

To find the length, work backward by dividing: $4\frac{1}{2} \div 1\frac{1}{2} = \blacksquare$

$4\frac{1}{2} \div 1\frac{1}{2} = \frac{9}{2} \div \frac{3}{2}$ *Write the mixed numbers as fractions.*

$= \frac{9}{2} \times \frac{2}{3}$ *Use the reciprocal of the divisor to write a multiplication problem.*

$= \frac{\overset{3}{9}}{\underset{1}{2}} \times \frac{\overset{1}{2}}{\underset{1}{3}}$ *Simplify the factors.*

$= \frac{3}{1}$, or 3   *Multiply.*

So, she can make the box 3 ft long.   **Multiply 3 ft by $1\frac{1}{2}$ ft to see if the product is $4\frac{1}{2}$ ft².**

✓ **LOOK BACK**   How can you check your answer?

What if . . . the width of the box is $\frac{3}{4}$ ft? How long can the box be? **6 ft**

140   Chapter 7

IDEA FILE

INTERDISCIPLINARY CONNECTION

One suggestion for an interdisciplinary unit for this chapter is Metric and Customary: Measuring with Decimals and Fractions. In this unit you can have students

- • compare units of measure from the two systems. Use units for length, volume, mass, and weight.
- • measure several objects using both systems.

Modalities Auditory, Kinesthetic, Visual

RETEACH 7.5

Name _____

LESSON 7.5

Work Backward by Dividing Mixed Numbers

The area of a rectangular rug is 57 sq ft. The width of the rug is $6\frac{1}{3}$ ft. What is the length of the rug?

Think: You can work backward to find the length. You find the area of a rectangle by multiplying the length times the width. Since you know the area and the width, you can divide to find the length.

Step 1 Write a multiplication sentence to represent the problem. $57 = \blacksquare \times 6\frac{1}{3}$

Step 2 Work backward by dividing to find the length. $57 \div 6\frac{1}{3} = \blacksquare$

Write the mixed number as a fraction. $\frac{57}{1} \div \frac{19}{3}$

Use the reciprocal of the divisor to write a multiplication problem. $\frac{57}{1} \times \frac{3}{19}$

Simplify the factors. $\frac{\overset{3}{57}}{1} \times \frac{3}{\underset{1}{19}}$

Multiply. $\frac{9}{1}$, or 9

So, the length of the rug is 9 ft.

You can check your answer by multiplying. $9 \times 6\frac{1}{3}$
$\frac{3}{1} \times \frac{19}{3}$
$3 \times 19 = 57$

Work backward to solve.

1. The area of a rectangular room is $187\frac{1}{2}$ sq ft. The width of the room is $12\frac{1}{2}$ ft. What is the length of the room?   **15 ft**

2. Jim has a $2\frac{1}{2}$ lb supply of dog treats. He gives his dog $\frac{1}{4}$ lb of treats each day. How many days will his supply last?   **10 days**

Use with text pages 140–141.   TAKE ANOTHER LOOK   R33

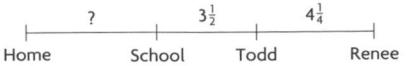

Work backward to solve.

1. Ms. Jones leaves home to pick up Amy at school. Then she travels $3\frac{1}{2}$ mi to pick up Todd and $4\frac{1}{4}$ mi to pick up Renee. Ms. Jones drives a total of $12\frac{1}{2}$ mi. What is the distance from home to school? **$4\frac{3}{4}$ mi**

```
|-------|--------|--------|--------|
    ?        3½       4¼
  Home    School    Todd    Renee
```

2. Terry is trying to decrease the amount of time he spends talking on the phone. This week he talked for $2\frac{1}{2}$ hr, which is $\frac{2}{3}$ of the time he talked last week. How long did he talk on the phone last week? **$3\frac{3}{4}$ hr**

3. Gerald has $8\frac{3}{4}$ yd of fabric. This is 7 times the amount he needs to make 1 costume for the school play. How much fabric does he need for each costume? **$1\frac{1}{4}$ yd**

4. Marsha wants to plant a rectangular garden. She wants it to have an area of $18\frac{3}{4}$ ft². The length is 6 ft. How wide will the garden have to be? **$3\frac{1}{8}$ ft**

MIXED APPLICATIONS

Choose a strategy and solve. **Choices of strategies will vary.**

> **Problem-Solving Strategies**
> - Work Backward
> - Guess and Check
> - Draw a Diagram
> - Write an Equation
> - Use a Formula
> - Make a Table

5. A library shelf is 8 ft wide. A set of sports books fills $\frac{1}{4}$ of the shelf space. Each book in the set is $1\frac{1}{2}$ in. thick. How many books are in the set? **16 books**

6. Lisa left home and walked $1\frac{1}{4}$ mi north. Then she walked $\frac{3}{4}$ mi west, $\frac{1}{3}$ mi north, $1\frac{1}{4}$ mi east, and $1\frac{2}{3}$ mi south. How far is she from her home? **$\frac{1}{2}$ mi**

7. Tim's family is driving $10\frac{1}{2}$ hr to Rompa to visit relatives. They will stop every $3\frac{1}{2}$ hr to rest. How many times will they stop to rest during the trip? **2 times**

8. Caren has a $1\frac{1}{2}$-lb supply of cat treats. Each day, she gives her cat $\frac{1}{8}$ lb of the treats. How many days will her supply last? **12 days**

9. Quincy went to see two one-act plays. With an intermission of $\frac{1}{4}$ hr, the evening lasted $2\frac{1}{2}$ hr. The first play was $1\frac{1}{4}$ hr long. How long did the second play last? **1 hr**

10. Joel chose a number and multiplied by $1\frac{1}{2}$. The product was $10\frac{1}{2}$. What was the number? **7**

11. Tommy has a rectangular toolbox with a base 21 in. long and 9 in. wide. He wants to cover the bottom of the toolbox with felt material. How many square inches of felt does he need to cover the bottom of the toolbox? **189 in.²**

12. **✏ WRITE ABOUT IT** Describe the steps needed to use the work backward strategy. **Identify the operation used. Use the inverse operation to solve.**

MORE PRACTICE Lesson 7.5, page H52

141

141

USING THE PAGE

To provide additional practice activities related to Chapter 7: *Multiplying and Dividing Fractions*

EGYPTIAN SYMBOLS

Have students work individually to complete this activity.

In this activity, students use Egyptian symbols to practice writing and multiplying fractions.

Multiple Intelligences Visual/Spatial, Logical/Mathematical, Interpersonal/Social, Intrapersonal/Introspective
Modalities Auditory, Visual

COOKING UP A TREAT

MATERIALS: *For each student* calculator

Have students work individually to complete this activity.

In this activity, students multiply mixed numbers to adjust the measurements of a recipe in order to make a given number of bread loaves.

Multiple Intelligences Logical/Mathematical, Intrapersonal/Introspective
Modality Auditory

ROLL A FRACTION

MATERIALS: *For each group* three number cubes, calculator

Have students work in pairs or small groups to complete this activity.

In this activity, students roll number cubes and use the numbers to create fractions less than one. Students then write division problems with the fractions, trying to get the largest quotient.

Multiple Intelligences Logical/Mathematical, Verbal/Linguistic, Bodily/Kinesthetic, Interpersonal/Social, Intrapersonal/Introspective
Modalities Auditory, Kinesthetic, Visual

ADDITIONAL ANSWERS

Egyptian Symbols

1. $\frac{1}{15}$, $\frac{3}{4}$, $\frac{1}{20}$;
2. $\frac{2}{3}$, $\frac{1}{2}$, $\frac{1}{3}$;

EGYPTIAN SYMBOLS

PURPOSE To practice multiplying fractions (pages 128–131)

Ancient Egyptians used stick symbols for ones, heel bones for tens, and scrolls for hundreds. In Egyptian fractions the symbol ◇ represented 1 as the numerator. Some fractions had their own symbols.

1	∩	ℓ
1	10	100

▢ for $\frac{1}{2}$, ✕ for $\frac{1}{4}$, ⚘ for $\frac{2}{3}$, and ⚘ for $\frac{3}{4}$

EXAMPLE: $\frac{1}{6}$ $\frac{1}{10}$ $\frac{1}{12}$

Write the fractions for the symbols. Find the product. Write the product in symbols.

1.

2. ⚘ ▢

COOKING UP A TREAT

PURPOSE To practice multiplying mixed numbers (pages 134–135)
YOU WILL NEED calculator

A common use of fractions in the home is in cooking. Suppose you need to make bread for a family meal that will celebrate a cultural holiday. Adjust the recipe for injera, the national bread of Ethiopia, to make the given number of loaves.

1. 30 loaves
See below.

2. 120 loaves

INJERA (makes about 15 small, round loaves)	
5 tbsp all-purpose flour	$1\frac{1}{2}$ c water
3 c pancake mix	$3\frac{1}{2}$ c club soda
$\frac{1}{4}$ tsp baking soda	

 HOME NOTE Choose a family recipe, and adjust the ingredients to make a larger amount.

ROLL A FRACTION

Portfolio

PURPOSE To practice dividing fractions (pages 138–139)
YOU WILL NEED three number cubes, calculator

Work with a partner or small group. Take turns rolling three number cubes. Use the numbers rolled to make fractions less than 1. Each student writes a division problem with two of the fractions. The person with the greatest quotient earns one point. Use a

calculator to check your work. The first player to get 10 points wins the game.

EXAMPLE: Player 1 makes the problem $\frac{4}{5} \div \frac{3}{5} = 1\frac{1}{3}$. Player 2 makes $\frac{3}{4} \div \frac{4}{5} = \frac{15}{16}$. Player 1 gets a point, because $1\frac{1}{3}$ is greater than $\frac{15}{16}$.

142 Math Fun!

Cooking Up A Treat

1. 10 tbsp all-purpose flour
   6 c pancake mix
   $\frac{1}{2}$ tsp baking soda
   3 c water
   7 c club soda

2. 40 tbsp all-purpose flour
   24 c pancake mix
   2 tsp baking soda
   12 c water
   28 c club soda

CHAPTER 7 REVIEW

EXAMPLES

EXERCISES

- **Multiply fractions.** (pages 128–131)

$\frac{3}{7} \times \frac{2}{3} = \frac{6}{21}$ *Multiply the numerators.*
Multiply the denominators.

$= \frac{2}{7}$ *Write in simplest form.*

Find the product. Write it in simplest form.

1. $\frac{1}{6} \times \frac{3}{5}$ $\frac{1}{10}$ **2.** $\frac{2}{3} \times \frac{4}{7}$ $\frac{8}{21}$

3. $16 \times \frac{5}{12}$ $6\frac{2}{3}$ **4.** $\frac{3}{8} \times 10$ $3\frac{3}{4}$

- **Simplify fraction factors.** (pages 132–133)

$\frac{5}{7} \times \frac{3}{10} = \frac{5}{7} \times \frac{3}{10}$ *Divide by the GCF.*

$= \frac{3}{14}$ *Multiply.*

Use the GCF to simplify the factors. Then find the product.

5. $\frac{2}{3} \times \frac{1}{6}$ $\frac{1}{9}$ **6.** $\frac{8}{9} \times \frac{3}{4}$ $\frac{2}{3}$ **7.** $\frac{6}{7} \times \frac{5}{8}$ $\frac{15}{28}$

8. $\frac{3}{4} \times 12$ 9 **9.** $6 \times \frac{2}{3}$ 4 **10.** $16 \times \frac{5}{6}$

10. $\frac{40}{3}$, or $13\frac{1}{3}$

- **Multiply fractions and mixed numbers.**
(pages 134–135)

$3\frac{1}{2} \times \frac{4}{5} = \frac{7}{2} \times \frac{4}{5}$ *Write the mixed number as a fraction.*

$= \frac{7}{2} \times \frac{4}{5}$ *Simplify the factors.*

$= \frac{14}{5}$, or $2\frac{4}{5}$ *Multiply.*

Find the product. Write it in simplest form.

11. $1\frac{2}{3} \times \frac{3}{4}$ $\frac{5}{4}$, or $1\frac{1}{4}$ **12.** $\frac{4}{15} \times 4\frac{3}{8}$ $\frac{7}{6}$, or $1\frac{1}{6}$

13. $1\frac{1}{3} \times 3\frac{3}{4}$ $\frac{5}{1}$, or 5 **14.** $1\frac{7}{8} \times 4\frac{2}{3}$ $\frac{35}{4}$, or $8\frac{3}{4}$

15. $1\frac{1}{2} \times \frac{3}{4}$ $\frac{9}{8}$, or $1\frac{1}{8}$ **16.** $4\frac{1}{2} \times 2\frac{1}{3}$ $\frac{21}{2}$, or $10\frac{1}{2}$

- **Divide fractions.** (pages 138–139)

$\frac{3}{4} \div \frac{1}{8} = \frac{3}{4} \times \frac{8}{1}$ *Write a multiplication problem using the reciprocal of the divisor. Simplify and multiply.*

$\frac{3}{4} \times \frac{8}{1}$

$= \frac{6}{1}$, or 6

17. VOCABULARY When the numerator and the denominator of a fraction are exchanged, the new fraction is called the _?_ of the original fraction. **reciprocal**

Find the quotient. Write it in simplest form.

18. $\frac{6}{7} \div \frac{3}{5}$ **19.** $\frac{3}{4} \div \frac{1}{3}$ **20.** $\frac{1}{3} \div \frac{4}{5}$ $\frac{5}{12}$

21. $9 \div \frac{3}{8}$ 24 **22.** $8 \div \frac{6}{7}$ **23.** $\frac{4}{5} \div 4$ $\frac{1}{5}$

- **Use the strategy** *work backward* **to divide mixed numbers.** (pages 140–141)

PROBLEM-SOLVING TIP: For help in solving problems by working backward, see pages 6 and 140.

18. $\frac{10}{7}$, or $1\frac{3}{7}$ **19.** $\frac{9}{4}$, or $2\frac{1}{4}$

22. $\frac{28}{3}$, or $9\frac{1}{3}$ **24.** $1\frac{3}{5}$ times faster

24. Terry came in third place running $1\frac{3}{4}$ mph in a three-legged race at school. After practicing, he now runs $2\frac{4}{5}$ mph. How many times faster does he run now?

25. Jami chose a number and multiplied by $2\frac{1}{4}$. The product was $7\frac{7}{8}$. What number did she choose? $3\frac{1}{2}$

Chapter 7 Review   **143**

CHAPTER 7 REVIEW & TEST

USING THE PAGE

The Chapter 7 Review reviews multiplying and dividing fractions. Chapter objectives are provided, along with examples and practice exercises for each.

The Chapter 7 Review can be used as a review or a test. You may wish to place the Review in students' portfolios.

Assessment Checkpoint

For Performance Assessment in this chapter, see page 146, What Did I Learn?, items 4 and 8.

✓ Assessment Tip

Student Self-Assessment The How Did I Do? Survey helps students assess both what they have learned and how they learned it. Self-assessment helps students learn more about their own capabilities and develop confidence in themselves.

A self-assessment survey is available as a copying master in the *Teacher's Guide for Assessment*, p. 15.

CHAPTER 7 TEST, page 1

Name _____

CHAPTER 7 TEST
PAGE 1

Choose the letter of the correct answer.

1. Use the model to help you find the product.

   $\frac{1}{3} \times \frac{1}{2}$

Ⓐ $\frac{1}{6}$ **B.** $\frac{1}{3}$ **C.** $\frac{5}{12}$ **D.** $\frac{1}{2}$

2. Which multiplication sentence shows a reasonable estimate for $1\frac{4}{5} \times 5$?

A. $\frac{4}{5} \times 5 = 4$
B. $1 \times 5 = 5$
Ⓒ $2 \times 5 = 10$
D. $4 \times 5 = 20$

3. $\frac{1}{2} \times \frac{1}{2}$

A. $\frac{1}{8}$ Ⓑ $\frac{1}{4}$ **C.** $\frac{11}{16}$ **D.** not here

4. $\frac{2}{5} \times \frac{3}{4}$

Ⓐ $\frac{3}{10}$ **B.** $\frac{1}{3}$ **C.** $\frac{1}{2}$ **D.** not here

5. $3 \times \frac{6}{8}$

A. $\frac{2}{5}$ Ⓑ $2\frac{1}{4}$ **C.** $2\frac{2}{5}$ **D.** not here

6. $\frac{2}{3} \times \frac{7}{10}$

A. $\frac{3}{10}$ **B.** $\frac{1}{3}$ **C.** $\frac{13}{30}$ Ⓓ not here

7. In which of these multiplication problems could you use two GCFs to simplify fractions?

A. $\frac{3}{4} \times \frac{1}{12}$ **B.** $\frac{1}{6} \times \frac{1}{12}$
C. $\frac{2}{3} \times \frac{3}{5}$ Ⓓ $\frac{3}{5} \times \frac{5}{9}$

For questions 8–9, choose the GCF you should use to simplify the fractions.

8. $\frac{3}{7} \times \frac{5}{12}$

Ⓐ 3 **B.** 5 **C.** 7 **D.** 12

9. $\frac{5}{6} \times \frac{4}{7}$

Ⓐ 2 **B.** 4 **C.** 5 **D.** 6

10. Solve. Use GCFs to simplify the factors. Answers are in simplest form.

   $\frac{3}{8} \times \frac{4}{9}$

A. $\frac{1}{12}$ **B.** $\frac{1}{9}$ Ⓒ $\frac{1}{6}$ **D.** $\frac{1}{3}$

11. What is the first step in finding this product?

   $2\frac{1}{2} \times 1\frac{1}{3}$

A. Find the GCF.
B. Estimate the product using whole numbers.
C. Round the mixed numbers to whole numbers.
Ⓓ Write the mixed numbers as fractions.

12. $4\frac{1}{2} \times \frac{2}{3}$

A. $1\frac{1}{3}$ **B.** $1\frac{1}{2}$
C. $2\frac{2}{3}$ Ⓓ 3

Form A • Multiple-Choice   A23   Go on. ▶

CHAPTER 7 TEST, page 2

Name _____

CHAPTER 7 TEST
PAGE 2

13. $1\frac{4}{5} \times 1\frac{1}{3}$

A. $1\frac{9}{10}$ Ⓑ $2\frac{2}{5}$
C. $2\frac{8}{15}$ **D.** $2\frac{4}{5}$

14. $4\frac{1}{6} \times 2\frac{2}{5}$

A. $6\frac{17}{30}$ **B.** $8\frac{1}{15}$
Ⓒ 10 **D.** $12\frac{1}{2}$

15. A hiking trail around a lake is $6\frac{3}{4}$ mi long. Frederico has hiked $\frac{1}{3}$ of the way. How far has he hiked?

Ⓐ $2\frac{1}{4}$ mi   **B.** $2\frac{3}{4}$ mi
C. $3\frac{1}{2}$ mi   **D.** $4\frac{1}{2}$ mi

16. Some friends made 4 batches of chocolate chip cookies. They used $1\frac{2}{3}$ c of chocolate chips in each batch. How many cups of chips did they use to make all the cookies?

A. $4\frac{2}{3}$ c   **B.** $5\frac{1}{2}$ c
Ⓒ $6\frac{2}{3}$ c   **D.** $7\frac{1}{3}$ c

17. What is the reciprocal of $\frac{3}{4}$?

A. $\frac{1}{4}$ **B.** 1
C. $\frac{3}{8}$ Ⓓ $\frac{4}{3}$

18. How can you rewrite this division problem as a multiplication problem?

   $\frac{4}{5} \div \frac{2}{3}$

A. $\frac{4}{5} \times \frac{2}{3}$ Ⓑ $\frac{4}{5} \times \frac{3}{2}$
C. $\frac{5}{4} \times \frac{2}{3}$ **D.** $\frac{5}{4} \times \frac{3}{2}$

19. $\frac{7}{8} \div \frac{1}{2}$

A. $\frac{7}{16}$ **B.** $1\frac{1}{3}$
Ⓒ $1\frac{3}{4}$ **D.** $2\frac{2}{7}$

20. $8 \div \frac{4}{5}$

A. $\frac{1}{10}$ **B.** $\frac{1}{6}$
C. 6 Ⓓ 10

21. The area of a rectangular rug is $19\frac{1}{4}$ ft². The rug is $5\frac{1}{2}$ ft long. What is the width?

Ⓐ $3\frac{1}{2}$ ft   **B.** $3\frac{3}{4}$ ft
C. $4\frac{1}{4}$ ft   **D.** $4\frac{1}{2}$ ft

22. Gerri bought a piece of salmon that weighs $2\frac{1}{4}$ lb. She wants each serving to be $\frac{1}{4}$ lb. How many servings can she make?

A. 5 servings   **B.** 9 servings
C. 10 servings   Ⓓ 11 servings

23. Claude has a piece of wood $5\frac{1}{2}$ ft long. It is 3 times the length he needs to make a small shelf. How long will the shelf be?

A. $1\frac{5}{6}$ ft   Ⓑ $1\frac{7}{8}$ ft
C. $2\frac{1}{4}$ ft   **D.** $2\frac{3}{8}$ ft

24. A DJ plays songs for $12\frac{1}{2}$ min without interruption. Each song lasts $2\frac{1}{2}$ min. How many songs does the DJ play?

A. 4 songs   Ⓑ 5 songs
C. $5\frac{1}{2}$ songs   **D.** 6 songs

Form A • Multiple-Choice   A24   ▶ Stop!

See *Test Copying Masters*, pp. A23–A24 for the Chapter Test in **multiple-choice** format and pp. B173–B174 for **free-response** format.

143

CHAPTERS 4-7

✓ *Assessment Checkpoint*

USING THE PAGES

This Study Guide reviews the following content:

Chapter 4 Number Theory and Fractions, pp. 74–91

Chapter 5 Adding and Subtracting Fractions, pp. 92–109

Chapter 6 Adding and Subtracting Mixed Numbers, pp. 110–125

Chapter 7 Multiplying and Dividing Fractions, pp. 126–143

CHAPTERS 4-7 STUDY GUIDE & REVIEW

VOCABULARY CHECK

1. Numbers such as 5 that have only two factors 1 and 5 are _?_ numbers. (page 76) **prime**

2. The _?_ of a composite number is when it is written as the product of prime factors. Example: $36 = 2 \times 2 \times 3 \times 3$. (page 78) **prime factorization**

3. The _?_ is the LCM of the denominators. (page 100) **least common denominator (LCD)**

4. When you rewrite a fraction problem as a multiplication problem to solve it, you use the _?_ of the divisor. (page 138) **reciprocal**

EXAMPLES

- **Find the LCM and GCF of numbers.** (pages 80–83)

Find the LCM and GCF for 8 and 12.

LCM = 24   GCF = 4
8: 8, 16, **24**   8: 1, 2, **4**
12: 12, **24**   12: 1, 2, 3, **4**, 6, 12

- **Write fractions in simplest form.** (pages 86–87)

Write $\frac{12}{24}$ in simplest form.

$\frac{12}{24} = \frac{12 \div 12}{24 \div 12} = \frac{1}{2}$ *Divide by the GCF.*

- **Write mixed numbers as fractions and fractions as mixed numbers.** (pages 88–89)

Write $2\frac{3}{5}$ as a fraction.

$2\frac{3}{5} \rightarrow \frac{(5 \times 2) + 3}{5} = \frac{13}{5}$

Write $\frac{17}{4}$ as a mixed number.

$\begin{array}{r} 4 \\ 4\overline{)17} \\ -16 \\ \hline 1 \end{array}$ *Divide numerator by denominator*
Write the remainder as a fraction.
$\frac{17}{4} = 4\frac{1}{4}$

EXERCISES

Find the LCM for each pair of numbers.

5. 5, 10 **10**   **6.** 9, 15 **45**

Find the GCF for each pair of numbers.

7. 5, 10 **5**   **8.** 9, 15 **3**

Write the fraction in simplest form.

9. $\frac{9}{21}$ **$\frac{3}{7}$**   **10.** $\frac{20}{40}$ **$\frac{1}{2}$**   **11.** $\frac{50}{100}$ **$\frac{1}{2}$**

12. $\frac{6}{8}$ **$\frac{3}{4}$**   **13.** $\frac{7}{21}$ **$\frac{1}{3}$**   **14.** $\frac{15}{40}$ **$\frac{3}{8}$**

Write the mixed number as a fraction.

15. $1\frac{2}{3}$ **$\frac{5}{3}$**   **16.** $3\frac{1}{7}$ **$\frac{22}{7}$**   **17.** $5\frac{2}{5}$ **$\frac{27}{5}$**

Write the fraction as a mixed number.

18. $\frac{16}{7}$ **$2\frac{2}{7}$**   **19.** $\frac{9}{5}$ **$1\frac{4}{5}$**   **20.** $\frac{14}{3}$ **$4\frac{2}{3}$**

21. $\frac{17}{12}$ **$1\frac{5}{12}$**   **22.** $\frac{9}{4}$ **$2\frac{1}{4}$**   **23.** $\frac{7}{2}$ **$3\frac{1}{2}$**

See *Test Copying Masters*, pp. A25–A26 for copying masters of the Test • Chapters 4–7 in **multiple-choice** format and pp. B175–B176 for **free-response** format.

TEST • CHAPTERS 4-7, p. 1

Name _____

Choose the letter of the correct answer.

1. What are the factors of 25?
   (A) 1, 5, 25   B. 1, 2, 5, 25
   C. 1, 25   D. 1, 2, 5, 10, 25
   4-A.1

2. What is the prime factorization of 58?
   (A) 2 × 29   B. 2 × 4 × 8
   C. 2^2 × 29   D. 2 × 5 × 8
   4-A.2

3. What is the LCM of 8 and 18?
   A. 8   B. 26
   C. 36   (D) 72
   4-A.3

4. Which fraction is the simplest form of $\frac{10}{16}$?
   A. $\frac{4}{8}$ (B) $\frac{5}{8}$
   C. $\frac{5}{9}$ D. $\frac{6}{10}$
   4-A.4

5. What is $4\frac{3}{5}$ written as a fraction?
   A. $\frac{3}{5}$ B. $\frac{4}{5}$
   C. $\frac{12}{5}$ (D) $\frac{23}{5}$
   4-A.5

6. What is the missing number?
   $\frac{\Box}{9} = 3\frac{7}{9}$
   A. 7   B. 19   C. 27   (D) 34
   4-A.5

7. $\frac{4}{11} + \frac{9}{11}$
   A. $\frac{5}{11}$ B. $\frac{12}{11}$
   (C) $\frac{13}{11}$ D. $13\frac{1}{11}$
   5-A.1

8. Mika's box of pencils was full. She used $\frac{2}{7}$ of the pencils and gave away $\frac{3}{7}$ to her friends. What fraction of the pencils did she have remaining?
   A. $\frac{1}{7}$ (B) $\frac{2}{7}$
   C. $\frac{3}{7}$ D. $\frac{5}{7}$
   5-A.1

9. $\frac{4}{5} - \frac{2}{3}$
   A. $\frac{1}{15}$ (B) $\frac{2}{15}$
   C. $\frac{4}{15}$ D. $\frac{2}{3}$
   5-A.2

10. $\frac{7}{10} - \frac{3}{5}$
    (A) $\frac{1}{10}$ B. $\frac{4}{10}$
    C. $\frac{2}{5}$ D. $\frac{4}{5}$
    5-A.2

11. Which of these fractions is about $\frac{1}{2}$ when rounded?
    A. $\frac{2}{9}$ (B) $\frac{4}{9}$
    C. $\frac{7}{9}$ D. $\frac{8}{9}$
    5-A.3

12. Which of these fractions is closest to 1?
    (A) $\frac{7}{8}$ B. $\frac{2}{11}$
    C. $\frac{5}{32}$ D. $1\frac{7}{8}$
    5-A.3

Form A • Multiple-Choice   A25   Go on. ▶

TEST • CHAPTERS 4-7, p. 2

Name _____

13. Choose the addition problem shown by the diagram.
    A. $1\frac{1}{3} + 1\frac{2}{3}$ B. $2\frac{1}{3} + 1\frac{1}{3}$
    (C) $2\frac{1}{3} + 1\frac{2}{3}$ D. $2\frac{2}{3} + 1\frac{1}{3}$
    6-A.1

14. Use the diagram to find the difference.
    $2\frac{3}{5} - 1\frac{4}{5}$
    A. $\frac{1}{5}$ (B) $\frac{4}{5}$
    C. $1\frac{1}{5}$ D. $1\frac{4}{5}$
    6-A.1

15. Which will give the best estimate of the sum?
    $3\frac{1}{7} + 4\frac{5}{6}$
    A. 4 + 4   (B) 3 + 5
    C. 4 + 5   D. 6 + 7
    6-A.1

16. Cara and Jean went to the market. Jean drove $4\frac{1}{5}$ mi and Cara drove $8\frac{1}{5}$ mi. About how many more miles did Cara drive than Jean?
    (A) about 4 mi   B. about 5 mi
    C. about 6 mi   D. about 7 mi
    6-A.2

17. It rained $8\frac{4}{5}$ in. during June. The average rainfall for June is $6\frac{1}{10}$ in. About how much more did it rain above the average?
    A. about 1 in.   B. about 2 in.
    C. about $2\frac{1}{7}$ in.   (D) about 3 in.
    6-A.2

18. $\frac{4}{7} \times \frac{1}{2}$
    (A) $\frac{2}{7}$ B. $\frac{4}{9}$ C. $\frac{5}{9}$ D. $\frac{5}{14}$
    7-A.1

19. Van needed to hang 3 bird feeders, using wire. He used $\frac{8}{9}$ yd for each bird feeder. How many yards of wire did he use in all?
    A. $\frac{24}{27}$ yd   B. $1\frac{2}{3}$ yd
    C. $2\frac{6}{9}$ yd   (D) $2\frac{2}{3}$ yd
    7-A.1

20. $\frac{3}{8} \div \frac{1}{3}$
    A. $\frac{1}{8}$ (B) $1\frac{1}{8}$ C. $1\frac{5}{8}$ D. $2\frac{1}{8}$
    7-A.2

21. $12 \div \frac{3}{5}$
    A. $5\frac{1}{3}$ B. $10\frac{3}{5}$ C. $15\frac{4}{5}$ (D) 20
    7-A.2

22. A relay team finished in $6\frac{2}{3}$ min. Each of the 4 runners on the team had the same time. How long did each person run?
    A. $\frac{2}{3}$ min   B. 1 min
    (C) $1\frac{2}{3}$ min   D. $1\frac{3}{4}$ min
    7-A.2

Form A • Multiple-Choice   A26   ▶ Stop!

✓ *Assessment Checkpoint*

PORTFOLIO SUGGESTIONS

The portfolio illustrates the growth, talents, achievements, and reflections of the mathematics learner. You may wish to have students spend a short time selecting work samples for their portfolios and complete the Portfolio Summary Guide in the *Teacher's Guide for Assessment*, page 23. Students may wish to address the following questions:

- **What new understanding of math have I developed in the past several weeks?**
- **What growth in understanding or skills can I see in my work?**
- **What can I do to improve my understanding of math ideas?**
- **What would I like to learn more about?**

For information about how to organize, share, and evaluate portfolios, see the *Teacher's Guide for Assessment*, pages 20-22.

- **Add and subtract unlike fractions.**

(pages 98–103)

$\frac{3}{4} = \frac{21}{28}$ *Write equivalent fractions using the LCD.*

$-\frac{3}{7} = \frac{12}{28}$ *Subtract the numerators.*

$\frac{9}{28}$ *Write the difference over the denominator.*

Add. Write the answer in simplest form.

24. $\frac{2}{3} + \frac{1}{5}$ $\frac{13}{15}$ **25.** $\frac{1}{3} + \frac{2}{6}$ $\frac{2}{3}$

26. $\frac{3}{4} + \frac{3}{8}$ $\frac{9}{8}$, or $1\frac{1}{8}$ **27.** $\frac{5}{9} + \frac{5}{6}$ $\frac{25}{18}$, or $1\frac{7}{18}$

Subtract. Write the answer in simplest form.

28. $\frac{3}{8} - \frac{1}{12}$ $\frac{7}{24}$ **29.** $\frac{3}{5} - \frac{1}{2}$ $\frac{1}{10}$

- **Add and subtract mixed numbers.**

(pages 118–121)

$3\frac{1}{6} = 3\frac{1}{6} = 2\frac{7}{6}$ *Write equivalent fractions. Rename as needed.*

$-1\frac{1}{2} = 1\frac{3}{6} = 1\frac{3}{6}$ *Subtract fractions. Subtract whole numbers.*

$1\frac{4}{6} = 1\frac{2}{3}$ *Write in simplest form.*

Add. Write the answer in simplest form.

30. $2\frac{3}{5} + 3\frac{1}{3}$ $5\frac{14}{15}$ **31.** $5\frac{5}{6} + 4\frac{2}{3}$ $10\frac{1}{2}$

Subtract. Write in simplest form.

32. $8\frac{2}{3} - 3\frac{3}{8}$ $5\frac{5}{12}$ **33.** $7\frac{1}{4} - 6\frac{3}{8}$ $\frac{7}{8}$

- **Multiply fractions and mixed numbers.**

(pages 134–135)

$\frac{1}{2} \times 1\frac{1}{3} = \frac{1}{2} \times \frac{4}{3} =$ *Write mixed numbers as fractions.*

  Use the GCF to simplify.

$\frac{1}{\overset{}{2}} \times \frac{\overset{2}{4}}{3} = \frac{2}{3}$ *Multiply numerators. Multiply denominators.*

Multiply. Write the answer in simplest form.

34. $\frac{3}{4} \times \frac{2}{3}$ $\frac{1}{2}$ **35.** $\frac{2}{9} \times 1\frac{1}{2}$ $\frac{1}{3}$

36. $2\frac{1}{3} \times 3$ 7 **37.** $2\frac{3}{5} \times 2\frac{7}{5}$ $8\frac{21}{25}$

- **Divide fractions.** (pages 138–139)

$6 \div \frac{2}{3} = \frac{6}{1} \div \frac{2}{3}$ *Write the whole number as a fraction.*

$= \frac{6}{1} \times \frac{3}{2}$ *Use the reciprocal of the divisor to write a multiplication problem.*

$\frac{\overset{3}{6}}{1} \times \frac{3}{\overset{}{2}} = \frac{9}{1} = 9$ *Simplify. Multiply.*

Divide. Write the answer in simplest form.

38. $\frac{1}{2} \div \frac{1}{3}$ $\frac{3}{2}$, or $1\frac{1}{2}$ **39.** $\frac{3}{4} \div \frac{9}{16}$ $\frac{4}{3}$, or $1\frac{1}{3}$

40. $5 \div \frac{10}{11}$ $\frac{11}{2}$, or $5\frac{1}{2}$ **41.** $\frac{6}{7} \div 3$ $\frac{2}{7}$

PROBLEM-SOLVING APPLICATIONS

Solve. Explain your method.

42. A shelf is 6 ft wide. Tom's cassettes fill up $\frac{1}{6}$ of the shelf. Each cassette is $\frac{3}{4}$ in. thick. How many cassettes does Tom have on the shelf? (pages 134–135) **16 cassettes**

43. This week Alicia exercised for 3 hr. This is $1\frac{1}{2}$ times the amount she exercised last week. How long did she exercise last week? (pages 138–139) **2 hr**

Review • Chapters 4–7   **145**

PERFORMANCE ASSESSMENT

✔ *Assessment Checkpoint*

With the *Performance Assessment Teacher's Guide,*

- have students work individually or in pairs as an alternative to formal assessment.
- use the evaluation checklist to evaluate items 1–4.
- use the rubric and model papers below to evaluate items 5–8.
- use the copying masters to evaluate interview/task tests on pp. 79–82.
- find additional performance assessment for chapters 1–7 on pp. 9–24.

SCORING RUBRIC FOR PROBLEM SOLVING TASKS, ITEMS 5–8

Levels of Performance

Level 3   Accomplishes the purposes of the task. Student gives clear explanations, shows understanding of mathematical ideas and processes, and computes accurately.

Level 2   Purposes of the task not fully achieved. Student demonstrates satisfactory but limited understanding of the mathematical ideas and processes.

Level 1   Purposes of the task not accomplished. Student shows little evidence of understanding the mathematical ideas and processes and makes computational and/or procedural errors.

STUDENT WORK SAMPLES for Item 7

CHAPTERS 4–7 WHAT DID I LEARN?

 Write About It   For items 1–4, see Evaluation Checklist in the *Performance Assessment Guide*. Explanations will vary.

1. Show the steps you would use to find the prime factorization of 48. Explain your method. (pages 78–79) **Check students' explanations; $2^4 \times 3$**

2. Show how to find the difference. Explain your method. (pages 100–101)

   $\frac{7}{8} - \frac{2}{3} = n$ **Check students' explanations; $n = \frac{5}{24}$**

3. Show the steps to find the difference. $5\frac{3}{8} - 2\frac{5}{6}$ (pages 116–117) **Check students' explanations; $2\frac{13}{24}$**

4. Explain the steps and then solve the problem.

   Angela has $2\frac{1}{4}$ yd of ribbon. She plans to use it on 6 packages. She uses the same amount of ribbon on each package. How much ribbon is used for each package? (pages 138–139) **Check students' explanations; $\frac{3}{8}$ yd**

✔ Performance Assessment

Choose a strategy and solve. Explain your method.

> **Problem-Solving Strategies**
> - Draw a Diagram
> - Write an Equation
> - Make a Table
> - Make a Model
> - Find a Pattern
> - Work Backward

Choices of strategies will vary. Explanations will vary.

5. The Olympic Summer Games are held every 4 years. They were held in 1996. How many times will they be held between the years 2000 and 2050? List the years in which they will be held. (pages 76–77)

6. Michael estimated that it would take him $2\frac{1}{2}$ hr to do his homework. As it turned out, he finished his homework in $1\frac{3}{4}$ hr. How much less time did it take him than expected? (pages 118–121) **$\frac{3}{4}$ hr**

7. Mrs. Esparza's car holds 10 gal of gas. On Friday her tank was $\frac{4}{5}$ full. On Saturday she used 5 gal. How much gas was left in the tank? (pages 128–129) **3 gal**

8. Kristin is baking bread. She uses $2\frac{1}{3}$ c of flour for $\frac{1}{4}$ recipe. How much flour does the whole recipe call for? (pages 140–141) **$9\frac{1}{3}$ c**

5. **13 times; 2000, 2004, 2008, 2012, 2016, 2020, 2024, 2028, 2032, 2036, 2040, 2044, 2048**

Level 3	Level 2	Level 1

 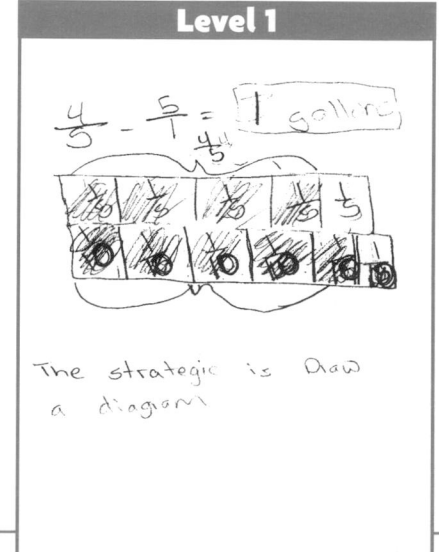

The student modeled the problem accurately. She correctly interpreted the diagram stating that there was $\frac{3}{10}$ of a tank of gas left.

The student modeled the problem correctly but interpreted her diagram incorrectly. The student's answer, $\frac{3}{5}$, was incorrect.

The student made a diagram but was unable to show how 10 gallons related to $\frac{4}{5}$ of a tank. His computation is not explained and his answer is incorrect.

CUMULATIVE REVIEW

Solve the problem. Then write the letter of the correct answer.

1. Compare. 126.5 ● 12.65 (pages 18–19)

- **A.** <
- **B.** > (circled)
- **C.** =
- **D.** ≥ **1A.1**

2. Write $\frac{1}{1,000}$ as a decimal. (pages 20–23)

- **A.** 0.001 (circled)
- **B.** 0.01
- **C.** 0.1
- **D.** 1,000 **1A.2**

3. Find the value of 12^3. (pages 24–25)

- **A.** 36
- **B.** 123
- **C.** 144
- **D.** 1,728 (circled) **1A.3**

4. 224 × 136 (pages 40–43)

- **A.** 2,240
- **B.** 29,244
- **C.** 30,464 (circled)
- **D.** 31,464 **2A.3**

Choose the best estimate. (pages 46–49)

5.   6,324
      + 2,731

- **A.** 3,000
- **B.** 8,000
- **C.** 9,000 (circled)
- **D.** 10,000 **2A.4**

6. 316.54 − 63.856 (pages 56–57)

- **A.** 2.52684
- **B.** 252.684 (circled)
- **C.** 252.696
- **D.** 25,268.4 **3A.1**

7. $\frac{5}{9} + \frac{2}{3}$ (pages 94–97)

- **A.** $\frac{7}{12}$
- **B.** $\frac{1}{9}$
- **C.** $\frac{8}{9}$
- **D.** $1\frac{2}{9}$ (circled) **5A.1**

8. $\frac{2}{5} - \frac{1}{3}$ (pages 98–99, 102–103)

- **A.** $\frac{1}{15}$ (circled)
- **B.** $\frac{2}{15}$
- **C.** $\frac{1}{3}$
- **D.** $\frac{1}{2}$ **5A.1**

9. $4\frac{3}{4} + 2\frac{5}{6}$ (pages 118–121)

- **A.** $6\frac{4}{5}$
- **B.** $6\frac{7}{12}$
- **C.** $7\frac{7}{12}$ (circled)
- **D.** $7\frac{3}{4}$ **6A.1**

10. $4\frac{3}{5} - 3\frac{2}{3}$ (pages 118–121)

- **A.** $\frac{3}{5}$
- **B.** $\frac{14}{15}$ (circled)
- **C.** $1\frac{1}{2}$
- **D.** $1\frac{14}{15}$ **6A.1**

11. $\frac{3}{4} \times \frac{3}{4}$ (pages 128–131)

- **A.** $\frac{9}{16}$ (circled)
- **B.** $\frac{3}{5}$
- **C.** 1
- **D.** $2\frac{1}{9}$ **7A.1**

12. $\frac{4}{7} \div 2$ (pages 138–139)

- **A.** $\frac{4}{7}$
- **B.** $\frac{2}{7}$ (circled)
- **C.** $1\frac{1}{7}$
- **D.** $3\frac{1}{2}$ **7A.2**

Cumulative Review   147

✔ *Assessment Checkpoint*

USING THE PAGE

This review is in multiple-choice format to provide the students with practice for standardized testing. More in-depth Cumulative Tests are provided in free-response and multiple-choice formats. See *Test Copying Masters*, pages A27–A30 for the multiple-choice test and pages B177–B180 for the free-response test.

CUMULATIVE TEST
FREE-RESPONSE FORMAT
CHAPTERS 1–7, PAGES 1–4

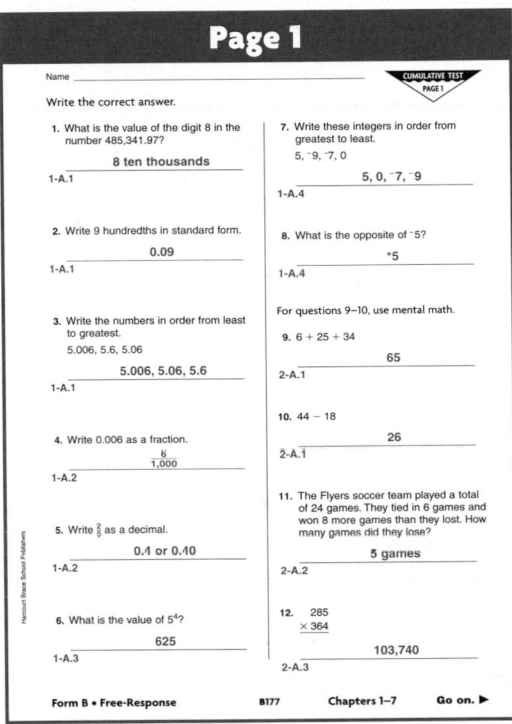

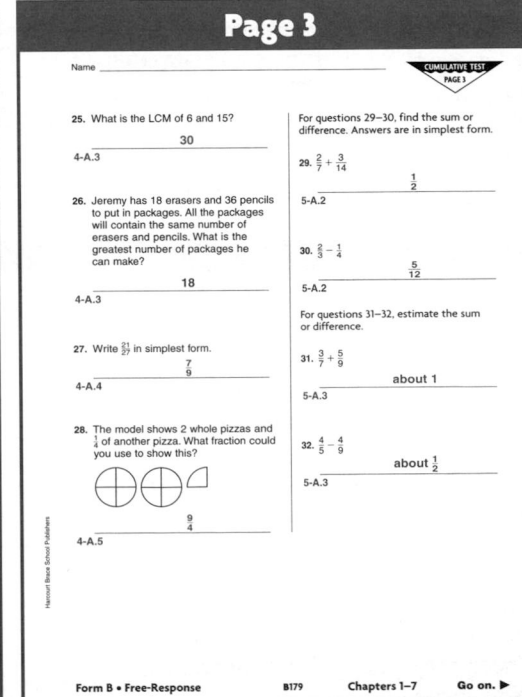

PLANNING GUIDE

Geometric Figures

BIG IDEA: Relationships between objects can be described.

Introducing the Chapter p. 148		Team-Up Project • Catalog Geometry p. 149	
OBJECTIVE	**VOCABULARY**	**MATERIALS**	**RESOURCES**
8.1 Points, Lines, and Planes pp. 150-151 **Objective:** To use terms to describe geometric figures **1 DAY**	point line plane line segment ray		■ Reteach, ■ Practice, ■ Enrichment 8.1; More Practice, p. H53 Cultural Connection • Geometry in Spanish Buildings, p. 164
8.2 Classifying Lines pp. 152-153 **Objective:** To name the different types of lines **1 DAY**	parallel intersecting perpendicular	FOR EACH STUDENT: one blank sheet of paper, cut in half	■ Reteach, ■ Practice, ■ Enrichment 8.2; More Practice, p. H53 Cultural Connection • Geometry in Spanish Buildings, p. 164
8.3 Angles pp. 154-155 **Objective:** To name angles and measure them **1 DAY**	vertex *acute angle right angle obtuse angle straight angle*	FOR EACH STUDENT: protractor	■ Reteach, ■ Practice, ■ Enrichment 8.3; More Practice, p. H53 *Cosmic Geometry • Amazing Angles*
8.4 Constructing Congruent Segments and Angles pp. 156-159 **Objective:** To construct congruent line segments and angles **2 DAYS**	congruent	FOR EACH STUDENT: compass, straightedge	■ Reteach, ■ Practice, ■ Enrichment 8.4; More Practice, p. H54
Lab Activity: Bisecting a Line Segment pp. 160-161 **Objective:** To bisect a line segment	bisect	FOR EACH STUDENT: compass, straightedge	E-Lab • Activity 8 E-Lab Recording Sheets, p. 8
INTERVENTION • Classifying Triangles and Quadrilaterals, Teacher's Edition pp. 164A–164B			
8.5 Polygons pp. 162-163 **Objective:** To name and classify polygons **1 DAY**	polygon regular polygon	FOR EACH PAIR: geoboard, dot paper	■ Reteach, ■ Practice, ■ Enrichment 8.5; More Practice, p. H54 Cultural Connection • Geometry in Spanish Buildings, p. 164

CHAPTER ASSESSMENT   Chapter 8 Review   p. 165

NCTM CURRICULUM

STANDARDS FOR GRADES 5–8

- ✓ **Problem Solving** *Lessons 8.1, 8.2, 8.3, 8.4, 8.5*
- ✓ **Communication** *Lessons 8.1, 8.2, 8.3, 8.4, 8.5*
- ✓ **Reasoning** *Lessons 8.1, 8.2, 8.3, 8.4, 8.5*
- ✓ **Connections** *Lessons 8.1, 8.2, 8.3, 8.4, 8.5*
- ☐ **Number and Number Relationships**
- ☐ **Number Systems and Number Theory**
- ☐ **Computation and Estimation**
- ☐ **Patterns and Functions**
- ☐ **Algebra**
- ☐ **Statistics**
- ☐ **Probability**
- ✓ **Geometry** *Lessons 8.1, 8.2, 8.3, 8.4, Lab Activity, 8.5*
- ✓ **Measurement** *Lessons 8.3, 8.4, 8.5*

CHAPTER LEARNING GOALS

8A.1 To identify and describe points, line, and planes

8A.2 To classify lines and angles and measure angles

8A.3 To construct congruent line segments and angles

8A.4 To identify polygons by the number of sides and angles

These goals for concepts, skills, and problem solving are assessed in many ways throughout the chapter. See the chart below for a complete listing of both traditional and informal assessment options.

Pretest Options
Pretest for Chapter Content, TCM pp. A31–A32, B181–B182
Assessing Prior Knowledge, TE p. 148; APK p. 10

Daily Assessment
Mixed Review, PE pp. 153, 159
Problem of the Day, TE pp. 150, 152, 154, 156, 162
Assessment Tip and Daily Quiz, TE pp. 151, 153, 155, 159, 161, 163

Formal Assessment
Chapter 8 Review, PE p. 165
Chapter 8 Test, TCM pp. A31–A32, B181–B182
Study Guide & Review, PE pp. 202–203
Test • Chapters 8–10, TCM pp. A37–A38, B187–B188
Cumulative Review, PE p. 205
Cumulative Test, TCM pp. A39–A42, B189–B192

Performance Assessment
Team-Up Project Checklist, TE p. 149
What Did I Learn?, Chapters 8–10, TE p. 204
Interview/Task Test, PA p. 83

Portfolio
Suggested work samples:
Independent Practice, PE pp. 151, 153, 155, 158, 163
Cultural Connection, PE p. 164

Student Self-Assessment
How Did I Do?, TGA p. 15
Portfolio Evaluation Form, TGA p. 24
Math Journal, TE pp. 148B, 161

Key
Assessing Prior Knowledge Copying Masters: APK
Test Copying Masters: TCM
Performance Assessment: PA
Teacher's Guide for Assessment: TGA
Teacher's Edition: TE
Pupil's Edition: PE

MATH COMMUNICATION

MATH JOURNAL
You may wish to use the following suggestions to have students write about the mathematics they are learning in this chapter:

PE pages 152, 157 • Talk About It

PE pages 151, 153, 155, 159, 163 • Write About It

TE page 161 • Writing in Mathematics

TGA page 15 • How Did I Do?

In every lesson • record written answers to each Wrap-up Assessment Tip from the Teacher's Edition for test preparation.

VOCABULARY DEVELOPMENT
The terms listed at the right will be introduced or reinforced in this chapter. The boldfaced terms in the list are new vocabulary terms in the chapter.

So that students understand the terms rather than just memorize definitions, have them construct their own definitions and pictorial representations and record those in their Math Journals.

point, p. 150
line, p. 150
plane, p. 150
line segment, pp. 150, 156, 157, 160, 162
ray, pp. 150, 154
parallel, p. 152
intersecting, p. 152
perpendicular, p. 152
vertex, p. 154
acute angle, p. 154
right angle, p. 154
obtuse angle, p. 154
straight angle, p. 154
congruent, pp. 156, 157, 158
bisect, p. 160
polygon, p. 162
regular polygon, p. 162

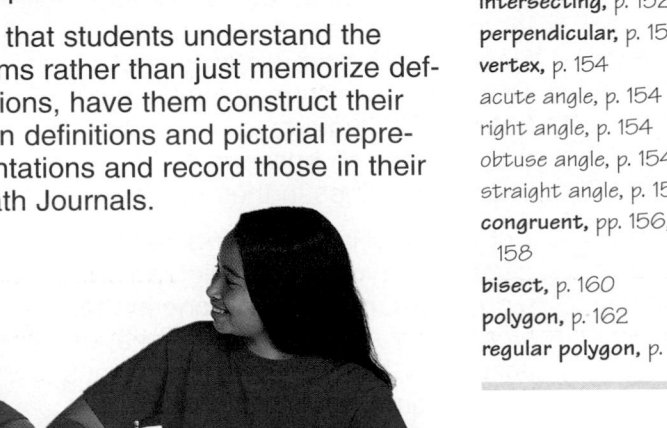

MEETING INDIVIDUAL NEEDS

MEETING INDIVIDUAL NEEDS

MEETING INDIVIDUAL NEEDS

RETEACH
BELOW-LEVEL STUDENTS

Practice Activities, TE Tab B
Greatest Answer—Least Answer
(Use with Lessons 8.1–8.5.)

Interdisciplinary Connection, TE p. 150

 Literature Link, TE p. 148C
A New Way of Life, Activity 1
(Use with Lesson 8.5.)

 Technology
Cosmic Geometry • Amazing Angles
(Use with Lesson 8.3.)

E-Lab • Activity 8
(Use with Lab Activity.)

PRACTICE
ON-LEVEL STUDENTS

Bulletin Board, TE p. 148C
Geometric Figures
(Use with Lesson 8.5.)

Interdisciplinary Connection, TE p. 150

 Literature Link, TE p. 148C
A New Way of Life, Activity 2
(Use with Lesson 8.5.)

 Technology
Cosmic Geometry • Amazing Angles
(Use with Lesson 8.3.)

E-Lab • Activity 8
(Use with Lab Activity.)

CHAPTER IDEA FILE

The activities on these pages provide additional practice and reinforcement of key chapter concepts. The chart above offers suggestions for their use.

LITERATURE LINK

A New Way of Life
by Paul Rutledge

This essay explains adjustments that Vietnamese refugees settling in the United States have to make. The author briefly tells about the differences in family life, education, language, and customs, and points out that being new to a country is not easy.

Activity 1—Students imagine that a refugee has come to their school. To help the new classmate, students label geometric figures in the classroom.

Activity 2—Tell students to imagine inviting a new student home. Have students draw a map and write directions from school to their house. Tell students to make sure they distinguish between parallel and intersecting lines in their directions.

Activity 3—Have students use different polygons to design a floor plan for a family like the one in the story. They should draw the shapes accurately and defend their reasons for the configurations.

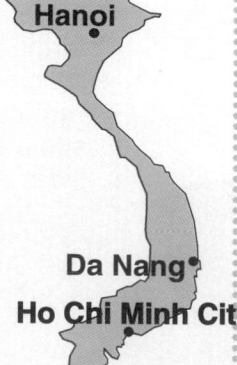

Hanoi

Da Nang

Ho Chi Minh City

BULLETIN BOARD

GEOMETRIC FIGURES		
Description	**Example**	**Observed**
Line Segment		
Ray		
Plane		
Parallel lines		
Intersecting lines		
Perpendicular lines		
Acute angle		
Obtuse angle		
Straight angle		
Right angle		

Preview a variety of geometric figures by adding new figures to the bulletin board as you introduce Chapters 8–10.

Activity 1—Students find examples of each figure in the classroom and tally them on a chart.

Activity 2—Students cut pictures from magazines of geometric figures, label, and add to bulletin board.

Activity 3—Students trace the figures and label the line segments and angles.

Bulletin Board, TE p. 148C
Geometric Figures
(Use with Lesson 8.5.)

Interdisciplinary Connection, TE p. 150

Literature Link, TE p. 148C
A New Way of Life, Activity 3
(Use with Lesson 8.5.)

Technology
Cosmic Geometry • *Amazing Angles*
(Use with Lesson 8.3.)

E-Lab • Activity 8
(Use with Lab Activity.)

INTERDISCIPLINARY SUGGESTIONS

Connecting Geometric Figures

Purpose To connect geometric figures to other subjects students are studying

You may wish to connect **Geometric Figures** to other subjects by using related interdisciplinary units.

Suggested Units

- Science Connection: Weather Phenomena

- Language Arts Connection: Communicating Without Words

- History Connection: Long-Standing Towers

- Multicultural Connection: Flags Around the World

- Art Connection: Origami—The Japanese Art of Paper Folding

Modalities Auditory, Kinesthetic, Visual

Multiple Intelligences Mathematical/Logical, Bodily/Kinesthetic, Visual/Spatial

The purpose of using technology in the chapter is to provide reinforcement of skills and support in concept development.

E-Lab • Activity 8—Students use reasoning and visual thinking skills to expand the concept of bisection to include symmetry and to apply the symmetry concept to two-dimensional plane figures as they look for symmetry lines in the figures.

***Cosmic Geometry* •** *Amazing Angles*—The activities on this CD-ROM provide students with opportunities to practice estimating angle measurements and classifying angles.

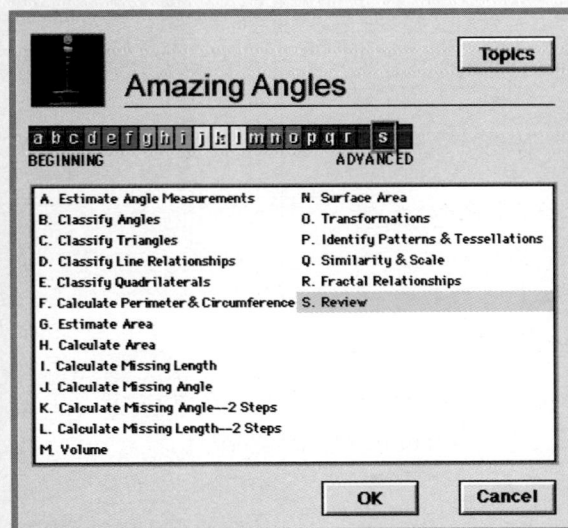

Each Cosmic Geometry activity can be assigned by degree of difficulty. See Teacher's Edition page 155 for Grow Slide information.

Visit our web site for additional ideas and activities.
http://www.hbschool.com

Introducing the Chapter

Encourage students to talk about geometric figures. Review the math content students will learn in the chapter.

WHY LEARN THIS?

The question "Why are you learning this?" will be answered in every lesson. Sometimes there is a direct application to everyday experiences. Other lessons help students understand why these skills are needed for future success in mathematics.

PRETEST OPTIONS

Pretest for Chapter Content See the *Test Copying Masters*, pages A31–A32, B181–B182 for a free-response or multiple-choice test that can be used as a pretest.

Assessing Prior Knowledge Use the copying master shown below to assess previously taught skills that are critical for success in this chapter.

Assessing Prior Knowledge Copying Masters, p. 10

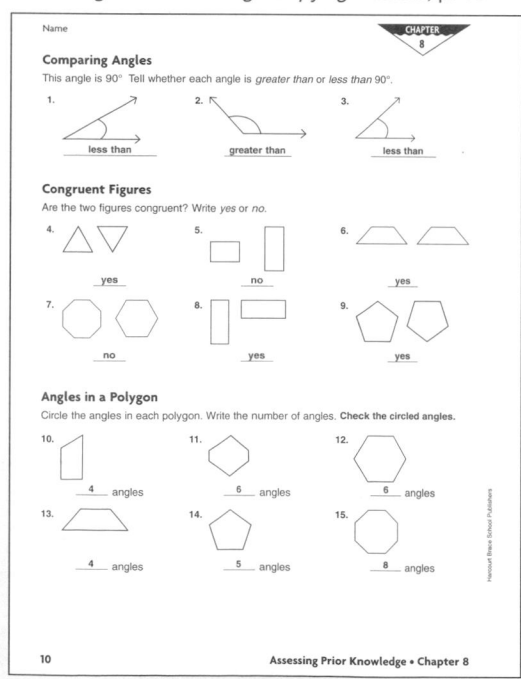

Troubleshooting

If students need help	Then use
• COMPARING ANGLES TO 90°	• protractors and have students draw 90° angles
• IDENTIFYING CONGRUENT FIGURES	• Key Skills, PE p. H23
• IDENTIFYING NUMBER OF ANGLES IN POLYGONS	• Key Skills, PE p. H21

148

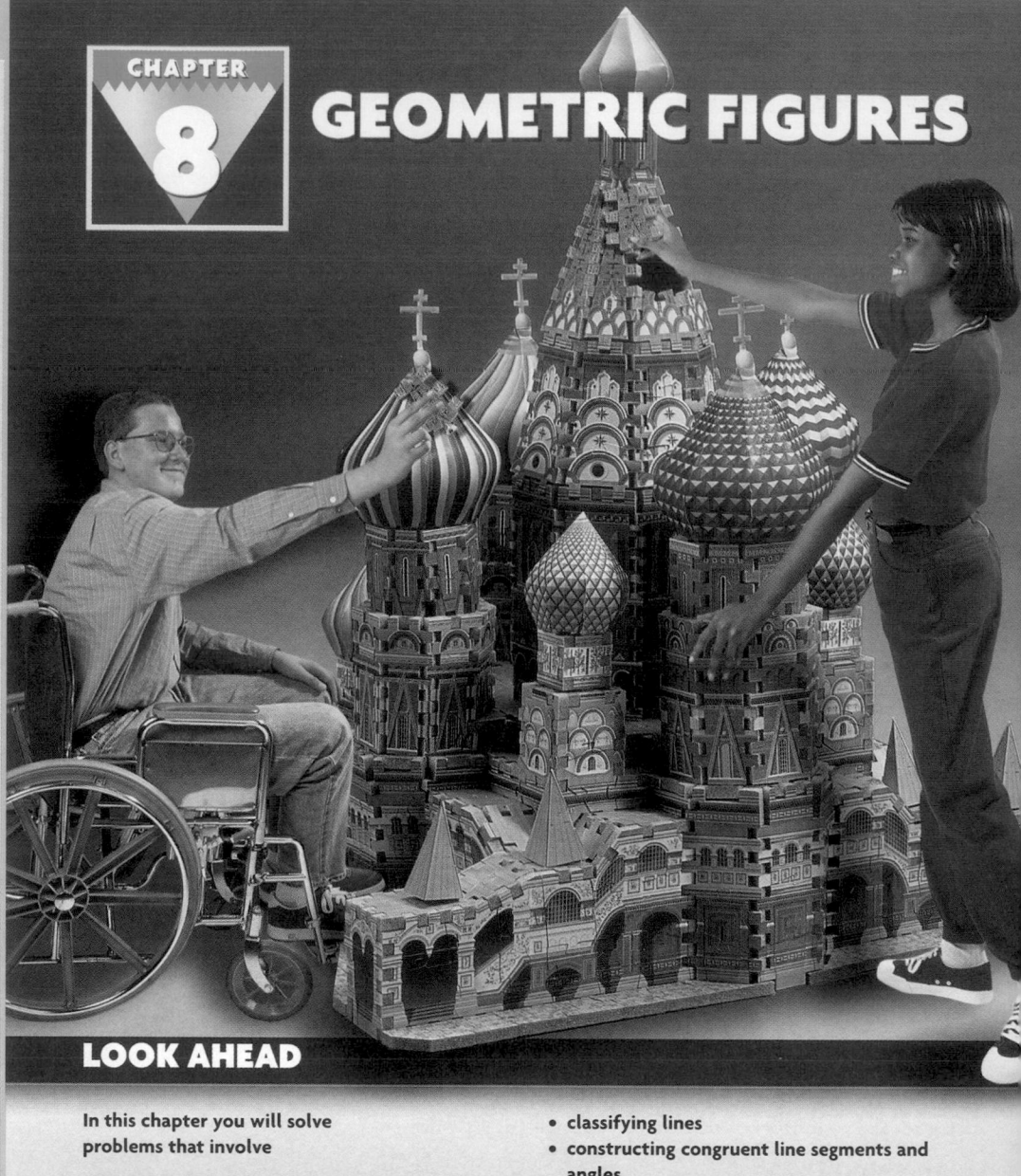

CHAPTER 8 GEOMETRIC FIGURES

LOOK AHEAD

In this chapter you will solve problems that involve

• constructing more complex geometric figures from basic geometric shapes

• classifying lines
• constructing congruent line segments and angles
• recognizing a polygon by the number of sides and type of angles

148   Chapter 8

 SCHOOL-HOME CONNECTION

You may wish to send to each student's family the *Math at Home* page. It includes an article that stresses the importance of communication about mathematics in the home, as well as helpful hints about homework and taking tests. The activity and extension provide extra practice that develops concepts and skills involving geometric figures.

Math at Home

Catalog Geometry

You can find examples of geometric figures in many everyday items. Find some items in a catalog. Outline examples of geometric figures in the items.

YOU WILL NEED: catalog

Plan
- Work with a small group to make a display showing geometric figures in catalog items. Use the list of geometric elements and see how many of them you can find.

Act
- Use catalogs that can be cut. Find some items that show geometric figures.
- Cut out the items and paste them on a sheet of paper.
- Outline and label the geometric figures that are in the items.

Share
- Join with other groups to make a display.

GEOMETRIC FIGURES

- **POINTS**
  midpoints, endpoints
- **LINES**
  parallel, intersecting, diagonal, perpendicular, segments, rays
- **ANGLES**
  acute, obtuse, right
- **POLYGONS**
  triangle, quadrilateral, pentagon, hexagon, octagon
- **TRIANGLES**
  right, isosceles, scalene, equilateral
- **QUADRILATERALS**
  parallelograms, trapezoids, squares, rectangles, rhombuses
- **CIRCLES**
  radius, diameter, circumference

DID YOU
- ✓ Find examples of geometric figures in catalog items?
- ✓ Identify geometric figures in each item?
- ✓ Make a display of your work?

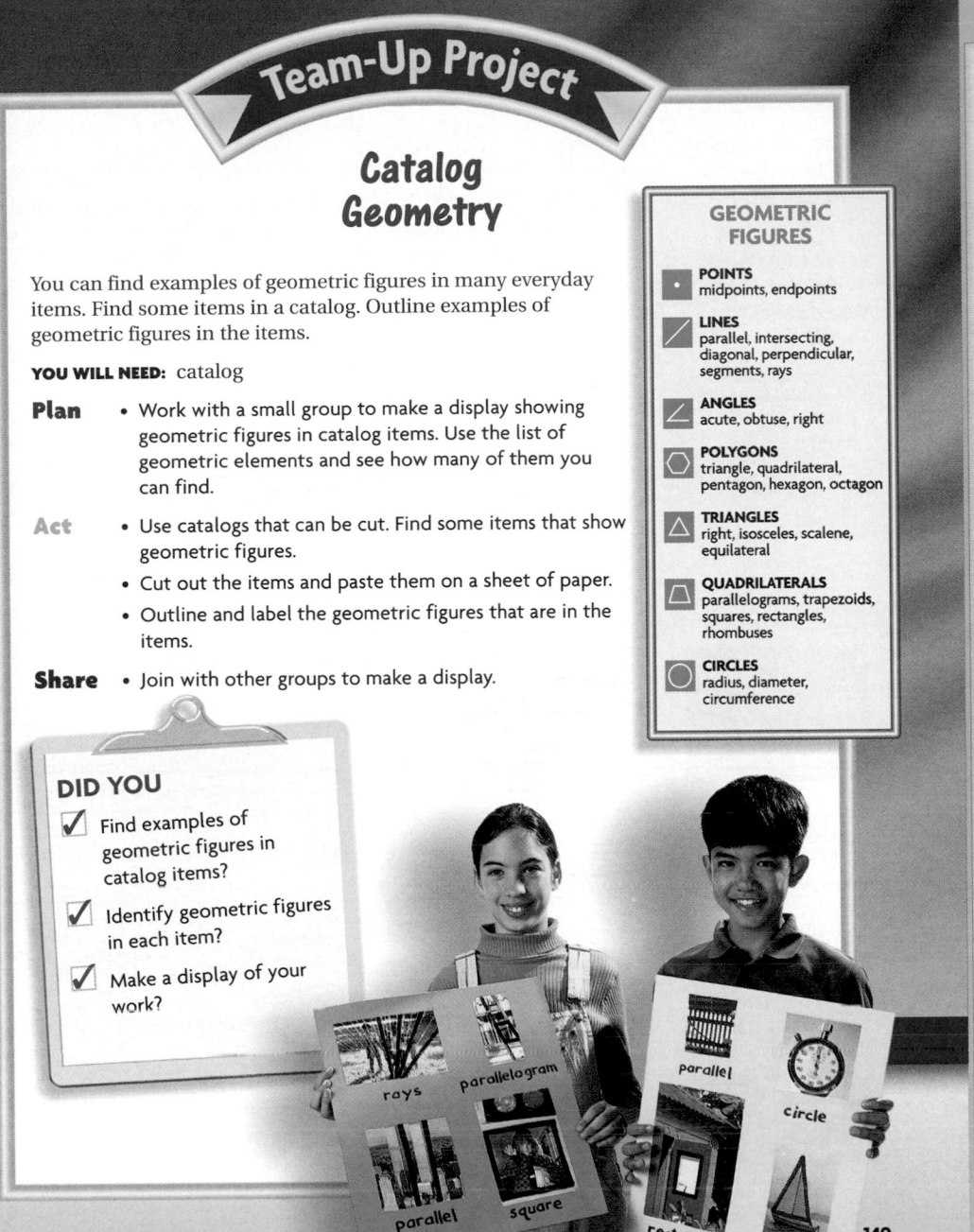

rays   parallelogram   parallel   circle

parallel   square   rectangle   triangle

149

Project Checklist

Evaluate whether each group
- ✓ works cooperatively.
- ✓ helps locate examples of geometric design.
- ✓ is able to identify geometric elements.

Using the Project

Accessing Prior Knowledge identifying geometric figures, spatial visualization

Purpose to identify lines, angles, and geometric figures in everyday items

GROUPING small groups

MATERIALS catalogs or newspapers, paper, glue

Managing the Activity Encourage students to identify angles as right, obtuse, or acute and estimate the measures.

WHEN MINUTES COUNT

A shortened version of this activity can be completed in 15–20 minutes. Ask students to identify lines, angles, and geometric figures used in the design of items shown on the chapter opener.

A BIT EACH DAY

Day 1: Students form groups and organize bringing in catalogs.

Day 2: Students select items from the catalogs and cut them out.

Day 3: Students paste the selected items on sheets of paper.

Day 4: Students identify and outline geometric elements.

Day 5: The class creates a display.

HOME HAPPENINGS

Ask students to bring in catalogs that can be cut up for the activity. Students can challenge a family member to see who can name the most geometric elements in a piece of home furniture.

ORGANIZER

OBJECTIVE To use terms to describe geometric figures

VOCABULARY *New* **point, line, line segment, ray, plane**

ASSIGNMENTS Independent Practice
- ■ *Basic* 1–14
- ■ *Average* 1–14
- ■ *Advanced* 5–14

PROGRAM RESOURCES
Reteach 8.1 Practice 8.1 Enrichment 8.1
Cultural Connection • Geometry in Spanish
  Buildings, p. 164

PROBLEM OF THE DAY

Transparency 8

Kate and four other students each made a poster of a different geometric figure. Tim did not draw the ray. Marty's figure has no symbol. Joe's figure needs three points. Lyda's and Tim's figures go forever. What figure did each student's poster show? Joe's, a plane; Marty's, a point; Lyda's, a ray; Tim's, a line; Kate's, a line segment

Solution Tab A, p. A8

QUICK CHECK

NOTE: Draw the figures on the board.

Name the geometric figure.

1. •————•
line segment

2. •
point

3. •————▸
ray

4. ◂——•——•——▸
   X   Y
line

5.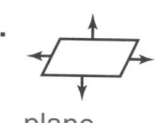
plane

1 MOTIVATE

Copy these symbols on the board. Ask:

• **What do these symbols represent?**

right turn

first aid

information

Points, Lines, and Planes

WHAT YOU'LL LEARN
How to use terms to describe geometric figures

WHY LEARN THIS?
To describe objects around you more clearly

WORD POWER
point
line
plane
line segment
ray

Science Link

Because the Earth is tilted on its axis, the sun's rays at any one time are much more direct at some places on Earth than at others. Look at the diagram above. If the Earth's axis were not tilted, where would the direct rays of the sun strike the Earth all year?

at the equator

The building blocks of geometry are points, lines, and planes.

• Fold a sheet of paper, open it up, and mark a location on the crease. Use geometric terms to describe the paper, the crease, and the marked location. **plane, line, point**

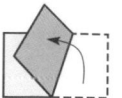  

Geometric figures are named and pictured in special ways.

A **point** is an exact location.	• P point P Use the name of the point.
A **line** is a straight path that goes on forever in opposite directions. It has no endpoints.	line AB, $\overleftrightarrow{AB}$ line BA, $\overleftrightarrow{BA}$ Use two points on the line to name the line.
A **plane** is a flat surface that goes on forever in all directions.	•N •M •L plane LMN Use three points that are not on a line to name the plane.

Parts of a line can be named by using named points on the line

A **line segment** is part of a line. It has two endpoints.	line segment XY, $\overline{XY}$ X——Y line segment YX, $\overline{YX}$
A **ray** is part of a line. It begins at its endpoint and goes on forever in only one direction.	ray JK, $\overrightarrow{JK}$ J——K—▸

A flagpole and a pencil are examples of line segments. A light beam from a laser can represent a ray.

IDEA FILE

INTERDISCIPLINARY CONNECTION

One suggestion for an interdisciplinary unit for this chapter is Weather Phenomena. Have students

• name the geometric figures suggested by each of these weather-condition symbols.

 lightning 1 line, 1 ray

 snow drifting 2 rays

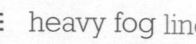 heavy fog lines

Modalities Auditory, Visual

RETEACH 8.1

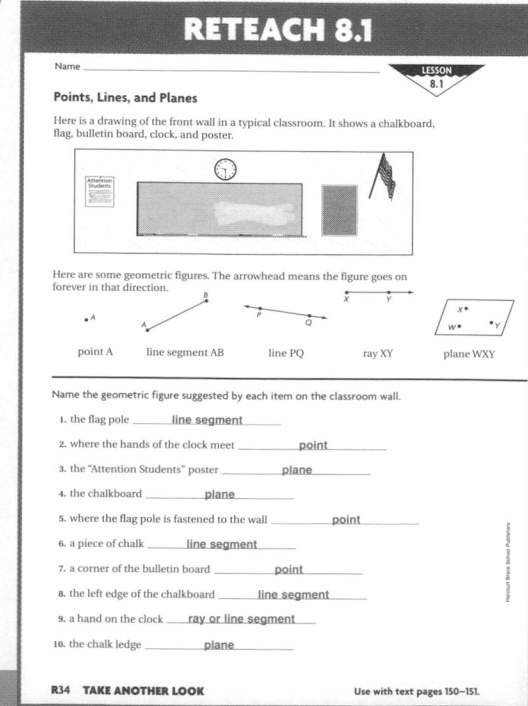

Points, Lines, and Planes

Here is a drawing of the front wall in a typical classroom. It shows a chalkboard, flag, bulletin board, clock, and poster.

Here are some geometric figures. The arrowhead means the figure goes on forever in that direction.

point A   line segment AB   line PQ   ray XY   plane WXY

Name the geometric figure suggested by each item on the classroom wall.

1. the flag pole ___ line segment
2. where the hands of the clock meet ___ point
3. the "Attention Students" poster ___ plane
4. the chalkboard ___ plane
5. where the flag pole is fastened to the wall ___ point
6. a piece of chalk ___ line segment
7. a corner of the bulletin board ___ point
8. the left edge of the chalkboard ___ line segment
9. a hand on the clock ___ ray or line segment
10. the chalk ledge ___ plane

R34 TAKE ANOTHER LOOK   Use with text pages 150–151.

GUIDED PRACTICE

Name the geometric figure.

1.
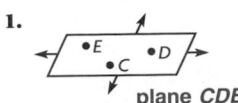
plane *CDE*

2. •_F_   point *F*

3. ←•——•→
A   B
line *AB* or line *BA*;
$\overleftrightarrow{AB}$, $\overleftrightarrow{BA}$

4. •——•→
A   B
ray *AB*; $\overrightarrow{AB}$

5. ←•——•→
S   T
line *ST* or *TS*; $\overleftrightarrow{ST}$ or $\overleftrightarrow{TS}$

6. •——•
P   Q
line segment *PQ* or *QP*;
$\overline{PQ}$ or $\overline{QP}$

INDEPENDENT PRACTICE

For Exercises 1–4, use the figure below. **All possible answers are given.**

P   Q   R

1. Name three different segments. $\overline{PQ}$, $\overline{QR}$, $\overline{PR}$

2. Name six different rays. $\overrightarrow{PQ}$, $\overrightarrow{QR}$, $\overrightarrow{PR}$, $\overrightarrow{RQ}$, $\overrightarrow{QP}$, $\overrightarrow{RP}$

3. Two names for the line are $\overleftrightarrow{PR}$ and $\overleftrightarrow{RP}$. Give four more names for the line. $\overleftrightarrow{PQ}$, $\overleftrightarrow{QR}$, $\overleftrightarrow{QP}$, $\overleftrightarrow{RQ}$

4. Give another name for ray *RQ*. $\overrightarrow{RP}$

Problem-Solving Applications

Name the geometric figure that is suggested.

5. railroad tracks through a town **lines**

6. path between bus stops on a road **line segment**

7. a star in the sky **point**

8. the short or long hand on a clock **ray**

For Problems 9–12, name the geometric figure suggested by each part of the map.

9. the three towns on the map **points**

10. the two interstate highways **lines**

11. the route from Pompton Lakes to Parsippany **line segment**

12. Is a line a good model to use to describe Route 202? Explain. **No; it is not straight.**

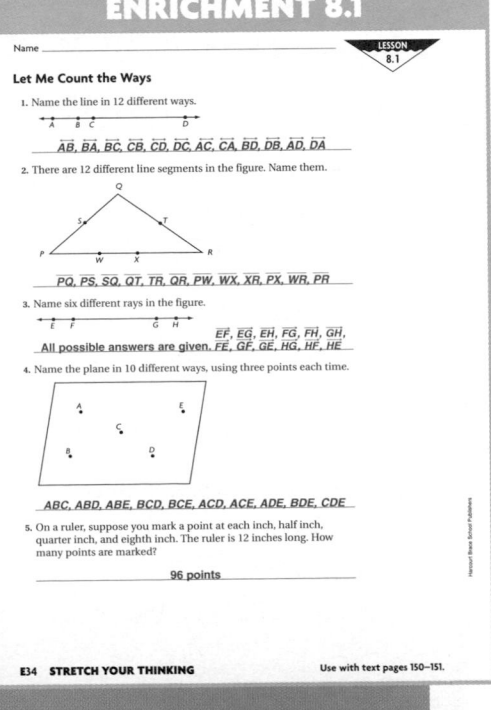
MAP
Pompton Lakes
I-287   202
Parsippany   Mountain View
I-80

13. Draw a map of your neighborhood. Label your house with point *A*. Include other landmarks, such as parks and buildings. Use geometric terms to describe the landmarks and routes from one landmark to another. **Maps and descriptions will vary.**

14. ✏ **WRITE ABOUT IT** Can a line segment be part of a ray? Explain. **Yes. Any two points of the ray can be the endpoints of a line segment.**

MORE PRACTICE Lesson 8.1, page H53

151

• **Why are these symbols used instead of words?**

Explain to students that symbols are used in geometry for the same reasons.

2 TEACH/PRACTICE

Read and discuss the chart on page 150. Have students suggest objects in their environment that represent geometric figures. Check students' answers.

COMMON ERROR ALERT Some students may name rays with the endpoint last. Point out that the endpoint of a ray should always be named first.

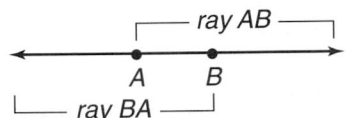
ray AB
←————•——•————→
A   B
ray BA

GUIDED PRACTICE

To be sure students understand the terms used to describe geometric figures, discuss Exercises 1–6 with the entire class.

INDEPENDENT PRACTICE

To assure student success, have students complete Exercises 1, 5, and 9 orally before you assign the rest of the exercises. Check students' maps.

3 WRAP UP

Use these questions to help students focus on the big ideas of the lesson.

• **What geometric figure is suggested by a tabletop?** plane **a laser beam?** ray **a straight bike trail?** line **a pencil?** line segment **a city location on a map?** point

✓ *Assessment Tip*

Have students write the answer.

• **How does a line differ from a line segment?** A line goes on forever in opposite directions; a line segment has two endpoints and is part of a line.

DAILY QUIZ

NOTE: Draw the figure on the board. Use the figure to name at least one point, one line, one line segment, and one ray.

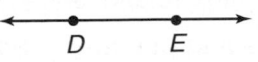
←————•——•————→
D   E

All possible answers: point *D*; point *E*; line segment *DE*; line segment *ED*; ray *DE*; ray *ED*; line *DE*; line *ED*

151

GRADE 5 Use terms to describe line relationships.	GRADE 6 Name the different types of lines.	GRADE 7 Find congruent line segments and angles.

ORGANIZER

OBJECTIVE To name the different types of lines

VOCABULARY *New* **parallel, intersecting, perpendicular**

ASSIGNMENTS Independent Practice
- ■ *Basic* 1–15
- ■ *Average* 1–15
- ■ *Advanced* 1–15

PROGRAM RESOURCES
Reteach 8.2 Practice 8.2 Enrichment 8.2
Cultural Connection • Geometry in Spanish
Buildings, p. 164

PROBLEM OF THE DAY

Transparency 8

Thad describes his drawing with these symbols: $\overleftrightarrow{AB} \parallel \overleftrightarrow{CD}$, $\overleftrightarrow{AC} \perp \overleftrightarrow{AB}$, and $\overleftrightarrow{AB} \perp \overleftrightarrow{BD}$. What symbols can he use to describe $\overleftrightarrow{AC}$ and $\overleftrightarrow{BD}$? $\overleftrightarrow{AC}$ and $\overleftrightarrow{BD}$ are parallel to each other: $\overleftrightarrow{AC} \parallel \overleftrightarrow{BD}$.

Solution Tab A, p. A8

QUICK CHECK

Tell whether the statements are true or false. If false, make it true.

1. A line segment is part of a line. true
2. Line *AB* is the same as line *BA*. true
3. Ray AB is the same as ray BA. False; ray *AB* has *A* as an endpoint, and ray *BA* has *B* as an endpoint.
4. A line goes on forever in one direction. False; a line goes on forever in both directions.
5. A line has two endpoints. False; a line has no endpoints.

1 MOTIVATE

Use this activity to assess students' prior knowledge of line relationships.

1. Fold one sheet of paper in half length-wise, then repeat the process. Open the paper. **Describe the relationship between the fold lines.** They are parallel. **How do you know?** They never cross.

WHAT YOU'LL LEARN
How to name the different types of lines

WHY LEARN THIS?
To help you understand directions to a place

WORD POWER
parallel
intersecting
perpendicular

Social Studies Link

Stone Mountain, located in Georgia, is 825 ft high. Visitors can hike to the top or ride there on an aerial tramway. There are two cars on the tramway at all times. Each car is suspended from a thick cable. Do you think the cables are parallel lines or intersecting lines? Explain.

The cables are parallel to keep the cars from colliding with each other.

Classifying Lines

A line is a straight path that goes on forever in opposite directions. The chart below shows some line relationships.

GEOMETRIC FIGURES

Parallel lines are lines in a plane that are always the same distance apart. They never intersect and have no common points.	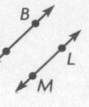	Line *AB* is parallel to line *ML*. $\overleftrightarrow{AB} \parallel \overleftrightarrow{ML}$
Intersecting lines are lines that cross at exactly one point.	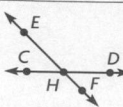	Line *EF* intersects line *CD* at point *H*.
Perpendicular lines are lines that intersect to form 90° angles, or right angles.	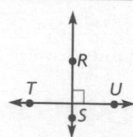	Line *RS* is perpendicular to line *TU*. $\overleftrightarrow{RS} \perp \overleftrightarrow{TU}$

Line segments can also be parallel, intersecting, or perpendicular.

Examples of lines and line segments can be found in this picture of a covered bridge.

Talk About It

- In the picture of the covered bridge, what type of line segments are represented by $\overline{AB}$ and $\overline{CD}$? **parallel**

- In the same picture, which figures represent intersecting lines? $\overleftrightarrow{IJ}$ and $\overleftrightarrow{KL}$; $\overline{GH}$ and $\overline{EF}$

IDEA FILE

CULTURAL CONNECTION

Have students find pictures of the national flag for each of these countries and classify the stripes on each flag as parallel, intersecting, or perpendicular.

Tanzania	parallel
Tonga	perpendicular
Switzerland	perpendicular
Thailand	parallel
Jamaica	intersecting
South Africa	intersecting

Modality Visual

RETEACH 8.2

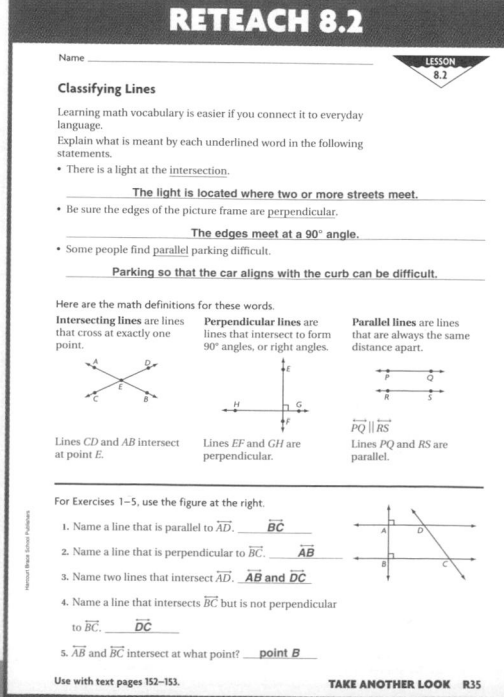

GUIDED PRACTICE

Name the type of lines.

1.

intersecting

2.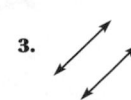

perpendicular

3.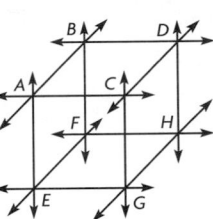

parallel

Make and label a drawing for Exercises 4–5. **Check students' drawings.**

4. Line *KL* is perpendicular to line *MN*.   5. Line *CD* is parallel to line *EF*.

INDEPENDENT PRACTICE

The figure at the right shows 12 lines drawn on the edges of a cube.

1. Name all the lines that are parallel to $\overleftrightarrow{AC}$. **$\overleftrightarrow{BD}$, $\overleftrightarrow{EG}$, $\overleftrightarrow{FH}$**

2. Name all the lines that intersect $\overleftrightarrow{CD}$. **$\overleftrightarrow{AC}$, $\overleftrightarrow{BD}$, $\overleftrightarrow{CG}$, $\overleftrightarrow{DH}$**

3. Name all the lines that are perpendicular to and intersect $\overleftrightarrow{DB}$. **$\overleftrightarrow{AB}$, $\overleftrightarrow{CD}$, $\overleftrightarrow{BF}$, $\overleftrightarrow{DH}$**

4. Name all the lines that are parallel to $\overleftrightarrow{EF}$. **$\overleftrightarrow{AB}$, $\overleftrightarrow{GH}$, $\overleftrightarrow{CD}$**

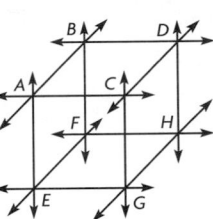

Problem-Solving Applications

5. Tina is describing to a friend where she lives. Her friend knows 3rd Street intersects Oak Street. Tina tells her that she lives on 4th Street which is parallel to 3rd Street and intersects Oak Street. Draw a map showing 3rd Street, 4th Street, and Oak Street. **Check students' maps.**

6. ⬤▶ **WRITE ABOUT IT** How are perpendicular and intersecting lines the same? different? **Both lines cross at one point; perpendicular lines form right angles, but intersecting lines might not.**

Mixed Review

Tell whether each angle is greater than or less than 90°.

7.   less than

8. greater than

9.  greater than

10. 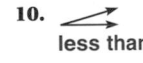 less than

Multiply.

11. $2\frac{1}{2} \times 3\frac{1}{3}$ **$\frac{25}{3}$, or $8\frac{1}{3}$**

12. $4\frac{3}{8} \times 1\frac{1}{5}$ **$\frac{21}{4}$, or $5\frac{1}{4}$**

13. $6\frac{2}{3} \times 2\frac{1}{4}$ **15**

14. Damon Clark is paid $350 a week. This week he spent $60 on groceries, saved $110, and paid a bill. He has $100 left. How much was his bill? **$80**

15. Maureen has 3 stamps to trade. She trades each stamp for 3 coins. How many coins does she get? Express the answer with an exponent. **3^2 coins**

MORE PRACTICE Lesson 8.2, page H53

153

---

2. Fold another sheet of paper in half lengthwise. Now fold it in half again in the other direction. Open the paper. **Describe the relationship between the fold lines.** Possible answers: They are perpendicular; they are intersecting. **How do you know?** They form a right angle.

2 TEACH/PRACTICE

Read and discuss page 152. Have students use relationships between lines to describe locations in their community. Possible answer: Main Street is perpendicular to Jackson Avenue.

GUIDED PRACTICE

Complete these exercises as a whole-class activity. Make sure students understand how to name different types of lines before you assign the Independent Practice.

INDEPENDENT PRACTICE

To ensure student success, you may wish to have students complete Exercises 1, 5, 9, and 11 orally before you assign the remaining exercises.

3 WRAP UP

Use these questions to help students focus on the big ideas of the lesson.

- **What is the relationship between two lines that intersect to form 90° angles?** They are perpendicular.

✓ *Assessment Tip*

Have students write the answer.

- **What are parallel lines?** lines that never intersect and have no common point

DAILY QUIZ

NOTE: Draw the figure on the board.

Use the figure to tell whether the statements are true or false.

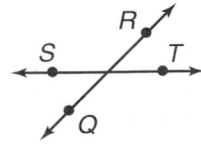

1. Line *QR* is parallel to line *ST*. false
2. Line *QR* intersects line *ST*. true
3. Line *QR* is perpendicular to line *ST*. false

PRACTICE 8.2

Name _____

LESSON 8.2

Classifying Lines

Vocabulary

Complete.

1. _____Intersecting_____ lines are lines that cross at exactly one point.

2. _____Perpendicular_____ lines are lines that intersect to form 90° angles, or right angles.

3. _____Parallel_____ lines are lines in a plane that are always the same distance apart. They never intersect and have no common points.

The figure at the right shows 12 lines drawn on the edges of a rectangular prism.   **All possible answers are given.**

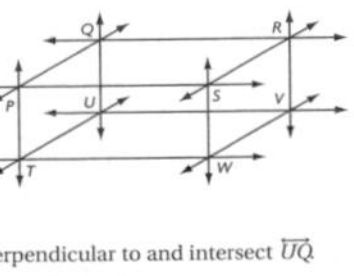

4. Name all the lines that are parallel to $\overleftrightarrow{TW}$.
   _____$\overleftrightarrow{UV}$, $\overleftrightarrow{PS}$, $\overleftrightarrow{QR}$_____

5. Name all the lines that intersect $\overleftrightarrow{PQ}$.
   _____$\overleftrightarrow{PS}$, $\overleftrightarrow{QR}$, $\overleftrightarrow{PT}$, $\overleftrightarrow{QU}$_____

6. Name all the lines that are perpendicular to and intersect $\overleftrightarrow{UQ}$.
   _____$\overleftrightarrow{PQ}$, $\overleftrightarrow{QR}$, $\overleftrightarrow{TU}$, $\overleftrightarrow{UV}$_____

Write *true* or *false*. Change any false statement into a true statement.

7. Intersecting lines can be parallel.
   False; parallel lines never intersect.

8. Parallel lines are never perpendicular.
   true

Mixed Applications

9. A ladder leans against a wall. Do the ladder and wall suggest parallel, intersecting, or perpendicular lines?
   _____intersecting lines_____

10. At Cheese City, sandwiches are made with $2\frac{1}{2}$ oz of cheese. At House of Cheese, they use $1\frac{3}{4}$ oz. How much more cheese do they put in a sandwich at Cheese City?
   $\frac{3}{4}$ oz

Use with text pages 152–153.   ON MY OWN   P35

ENRICHMENT 8.2

Name _____

LESSON 8.2

Space Project

In a plane, two lines are either parallel or intersecting. But *in space*, there are three possibilities: parallel, intersecting, and *skew* (SKYOO).

Look at the lines that form the edges of the cube. $\overline{AB}$ and $\overline{DE}$ are skew lines. They don't intersect, and they are not parallel. Also, $\overline{BC}$ and $\overline{DE}$ are skew lines.

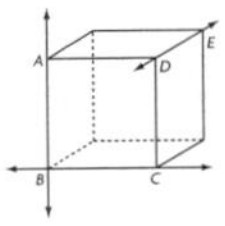

Look at your classroom.
- There is a line where the front wall meets the ceiling. Call it line 1.
- There is a line where the front wall meets the floor. Call it line 2.
- There is a line where the right side wall meets the ceiling. Call it line 3.

Are the lines parallel, intersecting, or skew?

1. lines 1 and 2   2. lines 2 and 3   3. lines 1 and 3
   _____parallel_____   _____skew_____   _____intersecting_____

For Exercises 4–10, use the figure at the right.
Tell whether the lines are parallel, intersecting, or skew.

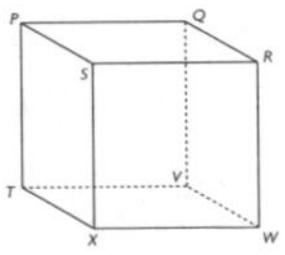

4. $\overline{SR}$ and $\overline{TX}$ _____skew_____

5. $\overline{VW}$ and $\overline{QR}$ _____parallel_____

6. $\overline{SP}$ and $\overline{QP}$ _____intersecting_____

7. $\overline{PT}$ and $\overline{XW}$ _____skew_____

8. $\overline{TV}$ and $\overline{RW}$ _____skew_____

9. $\overline{TV}$ and $\overline{SR}$ _____parallel_____

10. $\overline{PS}$ and $\overline{QV}$ _____skew_____

Use with text pages 152–153.   STRETCH YOUR THINKING   E35

153

GRADE 5	GRADE 6	GRADE 7
Name rays and angles.	Name and measure angles.	Find congruent line segments and angles.

ORGANIZER

OBJECTIVE To name angles and to measure them

VOCABULARY *New* vertex

ASSIGNMENTS Independent Practice
- ■ *Basic* 1–17
- ■ *Average* 1–17
- ■ *Advanced* 1–17

PROGRAM RESOURCES
Reteach 8.3 Practice 8.3 Enrichment 8.3
Cosmic Geometry • Amazing Angles

PROBLEM OF THE DAY

Transparency 8

The measure of Marlon's angle is twice that of Amee's and half that of Nate's. The sum of the measures of Amee's and Trisha's angles is equal to the sum of the measures of Marlon's and Nate's angles. The sum of the measures of all the angles is equal to the measure of a straight angle. What is the measure of each student's angle? Marlon's, 30°; Nate's, 60°; Amee's, 15°; Trisha's, 75°

Solution Tab A, p. A8

QUICK CHECK

NOTE: Draw the figures on the board.

Tell whether the geometric figure is a ray.

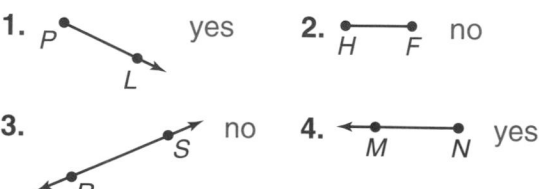

1. P L — yes   2. H F — no
3. S R — no   4. M N — yes

1 MOTIVATE

MATERIALS: *For each student* protractor

Use this activity to review the elements of a protractor.

- **How many scales are on a protractor?** 2 scales

WHAT YOU'LL LEARN
How to name angles and how to measure them

WHY LEARN THIS?
To use angles to build models and make art projects

WORD POWER
vertex

Science Link

The Earth's axis is tilted 23.5° from a line perpendicular to the plane of its orbit. The diagram on page 150 shows how the sun's direct rays fall on the Tropic of Cancer in June. The diagram below shows how the sun's direct rays fall on the Tropic of Capricorn in December. How many degrees from the equator are the Tropics of Cancer and Capricorn? **23.5°**

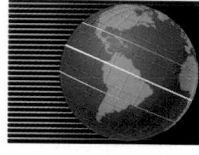

Angles

You know that a ray begins at a point and goes on forever in only one direction. An angle is formed by two rays with a common endpoint, the **vertex** of the angle. An angle can be named by three letters, a point from each side and the vertex as the middle letter. It can also be named with a single letter, its vertex.

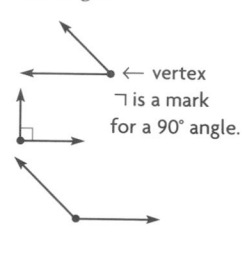

∠*DEF*, ∠*FED*, or ∠*E*

Angles are measured in degrees. The number of degrees determines the type of the angle.

The measurement of an acute angle is less than 90°.

The measurement of a right angle is 90°.

← vertex
⌐ is a mark for a 90° angle.

The measurement of an obtuse angle is more than 90° and less than 180°.

The measurement of a straight angle is 180°.

You can find the number of degrees in an angle by using a protractor. A protractor has a top and bottom scale. Either scale can be used to measure angles.

EXAMPLE Find the measure of ∠*ZYX*. Since the angle opens from the left, use the top scale of the protractor, so that 0° is on $\overrightarrow{YZ}$.

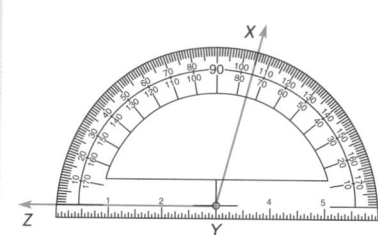

Place the center mark of the protractor on the vertex of the angle.

Place the base of the protractor along ray YZ.

Read the top scale where $\overrightarrow{YX}$ crosses to find the number of degrees in the angle.

- There are 105° in the angle. What kind of angle is it? **obtuse**

IDEA FILE

HISTORY CONNECTION

More than 3,000 years ago, the Babylonians developed the concept of degrees when they divided a circle into 360°. Why do you think they chose 360°? because they observed that the sun took about 360 days to travel in a circle How does the protractor we use today relate to the discovery by the Babylonians? It is half of a circle, measuring 180°.

Modality Visual

RETEACH 8.3

Name _____

LESSON 8.3

Angles

The table describes four types of angles. If you get confused between *acute* and *obtuse*, think of the *A* in *Acute*.

Acute — the capital A forms an acute angle.

Angle	Measure	Example
acute	less than 90°	
right	90°	
obtuse	greater than 90° but less than 180°	
straight	180°	

Measure the angle. Write the measure and *acute, right, obtuse,* or *straight*.

1. 110°, obtuse   2. 90°, right   3. 130°, obtuse
4. 55°, acute   5. 50°, acute   6. 180°, straight
7. 90°, right   8. 125°, obtuse   9. 25°, acute

R36   TAKE ANOTHER LOOK   Use with text pages 154–155.

GUIDED PRACTICE

Write *right, acute, obtuse,* or *straight* for each angle measure.

1. 90°
right

2. 36°
acute

3. 180°
straight

4. 100°
obtuse

5. 89°
acute

Name each angle. Tell whether it is *acute, right,* or *obtuse.*

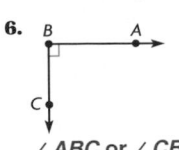

6.
∠ABC or ∠CBA or ∠B, right

7.
∠DEF or ∠FED or ∠E, acute

8.
∠GHI or ∠IHG or ∠H, obtuse

INDEPENDENT PRACTICE

Measure the angle. Write the measure and *acute, right, obtuse,* or *straight.*

1. 77°, acute

2. 180°, straight

3. 116°, obtuse

4. 90°, right

5. 162°, obtuse

6. 18°, acute

Problem-Solving Applications

At 3:00, the hands of a clock form a right angle. For Problems 7–14, name the type of angle formed at each of the given times.

7. 2:00
acute

8. 3:00
right

9. 6:00
straight

10. 5:30
acute

11. 10:25
obtuse

12. 11:45
acute

13. 9:00
right

14. 11:08
acute

15. CRITICAL THINKING One acute angle and one obtuse angle share the vertex, *C.* The two angles form a straight line. What is the sum of the measures of the two angles? **180°**

16. CRITICAL THINKING Does the measure of an angle depend on the lengths of the rays that are drawn for its sides? Explain.

17. ✏️ **WRITE ABOUT IT** Describe the difference between an acute angle and an obtuse angle. **acute angle: < 90°; obtuse angle: > 90°, < 180°**

16. No; the measure of an angle is the measurement from one ray to another ray with a common vertex.

Technology Link
In *Mighty Math Cosmic Geometry,* you can go to *Amazing Angles* to practice estimating angle measurements and classifying angles

Use Grow Slide Levels A and B.

PRACTICE 8.3

Name _____
LESSON 8.3

Angles

Vocabulary

Complete.

1. The measure of a ___**right**___ angle is 90°.

2. The measure of a ___**straight**___ angle is 180°.

3. The measure of an ___**obtuse**___ angle is more than 90° and less than 180°.

4. The measure of an ___**acute**___ angle is less than 90°.

5. The ___**vertex**___ of an angle is formed by two rays with a common endpoint.

Measure the angle. Write the measure and *acute, right, obtuse,* or *straight.*

6. 40°, acute

7. 125°, obtuse

8. 90°, right

9. 110°, obtuse

10. 180°, straight

11. 33°, acute

Mixed Applications

12. The hands of a clock indicate that the time is about 4:15. Do the hands form an angle that is obtuse, right, or acute?
___**acute**___

13. Mario can wash a car in about ⅓ hr. How many cars can he wash in 1⅔ hr?
___**5 cars**___

P36 **ON MY OWN**
Use with text pages 154–155.

ENRICHMENT 8.3

Name _____
LESSON 8.3

Angles in 3-D

An angle is formed by two rays or lines that lie in a plane. You can also talk about the angle formed when two planes intersect. This angle can be acute, right, or obtuse.

Think of the two halves of a gameboard as two intersecting planes. As you unfold the gameboard, the angle goes from acute, to right, to obtuse.

Answer these questions about real-life situations that involve angles formed by intersecting planes.

1. When you play chess, the gameboard is unfolded to what type of angle?
___**straight**___

2. What does it mean when a door is "ajar"? What kind of angle is formed between the door and the wall?
___It is open just a little; acute.___

3. An adjustable outdoor lounge chair has several settings. What types of angles can be formed by the seat and back of the chair?
___The settings will result in obtuse angles and possibly a right or straight angle.___

4. A roof line forms an angle made by the two sides of the roof. Why might such an angle be acute? obtuse? Discuss.
___The angle can vary simply for the sake of appearance, or the variations may serve a function. A steep roof, forming an acute angle, might be appropriate in areas where heavy snow falls.___

E36 **STRETCH YOUR THINKING**
Use with text pages 154–155.

---

2 TEACH/PRACTICE

Read and discuss with students page 154. To stimulate discussion, you may wish to ask students to identify different types of angles in the classroom.

GUIDED PRACTICE

Complete these exercises as a whole-class activity. Make sure students understand how to name and measure angles before you assign the Independent Practice.

INDEPENDENT PRACTICE

To ensure that students understand the concepts in this lesson, you may wish to check their answers to Exercises 1–6.

ADDITIONAL EXAMPLE

Example, p. 154

Find the measure of the angle. Since the angle opens from the right, use the top scale of the protractor so that 0° is on ray *BC.*

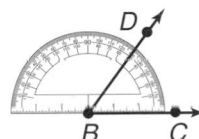

There are 50° in the angle. What kind of angle is it? acute angle

3 WRAP UP

Use these questions to help students focus on the big ideas of the lesson.

• **How do you know the angle in Independent Practice Exercise 6 is acute?** because it is less than 90°

✓ *Assessment Tip*

Have students write the answer.

• **What is an angle?** Possible answer: It is two rays with a common endpoint called the vertex.

DAILY QUIZ

For each time, name the angle formed by the hands of a clock.

1. 10:00 acute

2. 4:15 acute

3. 9:00 right

4. 6:00 straight

5. 3:55 obtuse

6. 8:45 acute

155

GRADE 5	GRADE 6	GRADE 7
Measure angles.	Construct congruent line segments and angles.	Construct congruent line segments and angles.

ORGANIZER

OBJECTIVE To construct congruent line segments and angles

VOCABULARY *New* **congruent**

MATERIALS *For each student* compass, straightedge

ASSIGNMENTS Independent Practice
- *Basic* 1–9, odd; 10–12; 13–22
- *Average* 1–12; 13–22
- *Advanced* 7–12; 13–22

PROGRAM RESOURCES
Reteach 8.4 Practice 8.4 Enrichment 8.4

PROBLEM OF THE DAY

Transparency 8

How many squares are there in all? 15 squares

Solution Tab A, p. A8

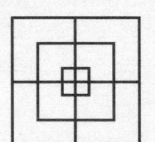

QUICK CHECK

Write *right, acute, obtuse,* or *straight* for each angle measure.

1. 50° acute
2. 88° acute
3. 145° obtuse
4. 95° obtuse
5. 180° straight
6. 90° right

1 MOTIVATE

Use this activity to show students the importance of accurate measurement of geometric figures. Copy these line segments on the board. NOTE: The segments are the same length.

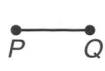

- **Which line segment appears to be longer?** Most students will say line segment *FG*.

Have volunteers measure the length of each line segment. They will discover that the line segments are actually the same length. Explain to students that something that appears to be different than it actually is, is called an illusion.

156

WHAT YOU'LL LEARN
How to construct congruent line segments and angles

WHY LEARN THIS?
To make drawings for construction projects or art projects in school or in future jobs

WORD POWER
congruent

teen times

The compass that you use to draw circles and arcs is a tool used in occupations such as architecture, carpentry, and engineering.

Constructing Congruent Segments and Angles

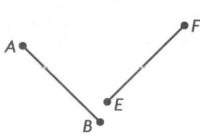

All line segments of the same length are said to be **congruent**.

We usually compare by measuring with a ruler, but we can also use a compass to see if two line segments are congruent.

ACTIVITY

WHAT YOU'LL NEED: compass

- Trace $\overline{EF}$ and $\overline{AB}$ on your paper.
- Place the compass point on point *E*. Open the compass to the length of $\overline{EF}$.
- Use the compass to show that $\overline{AB}$ has the same length.
- What is the relationship between the length of $\overline{AB}$ and the length of $\overline{EF}$? **The lengths are equal.**

You can use a compass and a straightedge to construct congruent line segments.

EXAMPLE 1

Trace $\overline{CD}$. Construct a line segment congruent to $\overline{CD}$.

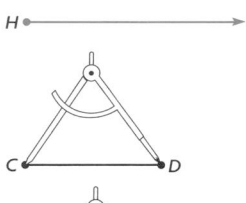

Draw a ray that is longer than $\overline{CD}$. Label the endpoint H.

Place the compass point on point C. Open the compass to the length of $\overline{CD}$.

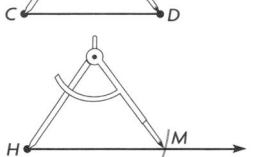

Use the same opening, but with the compass point on H.

Draw an arc that intersects the ray. Label the intersection point M.

$\overline{CD} \cong \overline{HM} \leftarrow$ segment CD is congruent to segment HM.

156 Chapter 8

RETEACH 8.4

Name _____

LESSON 8.4

Constructing Congruent Segments and Angles

Here are two constructions you can do with a compass and a straightedge.

1. Construct a line segment congruent to $\overline{AB}$.

Step 1 Draw a ray longer than $\overline{AB}$. Call the endpoint C.

Step 2 Measure $\overline{AB}$ using your compass. Put the point on A, and open the compass so the pencil is on B.

Step 3 Use the same opening as in Step 2. Put the compass point on C. Draw an arc intersecting the ray. Label the intersection point D. $\overline{AB} \cong \overline{CD}$.

2. Construct an angle congruent to ∠ABC.

Step 1 With the compass point on B, draw an arc through ∠ABC.

Step 2 Draw $\overline{DE}$. Use the same opening as in Step 1. With the compass point on D, draw an arc just like the one you drew in Step 1.

Step 3 Use the compass to measure the arc in ∠ABC.

Step 4 Using the same opening, locate point F on the arc.

Step 5 Draw $\overline{DF}$. ∠ABC ≅ ∠FDE.

1. Use a compass and a straightedge to construct a line segment congruent to $\overline{PQ}$. Check students' drawings.

2. Use a compass and a straightedge to construct an angle congruent to ∠RST. Check students' drawings.

Use with text pages 156–159.

TAKE ANOTHER LOOK R37

PRACTICE 8.4

Name _____

LESSON 8.4

Constructing Congruent Segments and Angles

Vocabulary

Complete.

1. All line segments of the same length are said to be ___congruent___.

2. Two angles that have the same measure in degrees are ___congruent___ angles.

Use a protractor to measure the angles in each pair. Tell whether they are congruent. Write each measure and *yes* or *no*.

3. 105°, yes
4. 22°, yes
5. 120°, 60°, no

6. In the space at the right, use a compass and a straightedge to construct an angle congruent to the one below. Check students' drawings.

Mixed Applications

7. Look at the four sides of an ordinary window pane. What do you notice about them?
Opposite sides are congruent.

8. In a science experiment on seeds, Claudia measured the sprout from her seed to be 0.25 in. Paul measured the sprout from his seed as $\frac{1}{8}$ in. Whose seed had sprouted more? Explain.
Claudia's; $0.25 > \frac{1}{8}$

Use with text pages 156–159.

ON MY OWN P37

Talk About It

- In Example 1, why must the ray be longer than $\overline{CD}$? __
  so you can find the part of it that is the same as $\overline{CD}$
- What are some examples of congruent line segments in your classroom? **Possible answers: new pencils, chair legs, lines on notebook paper**
- What are some careers in which people must know how to construct congruent figures? **Possible answers: architecture, engineering, graphic design**

GUIDED PRACTICE

Use a compass to determine if the line segments in each pair are congruent. Write *yes* or *no*.

1.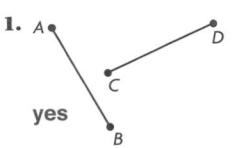

yes

2.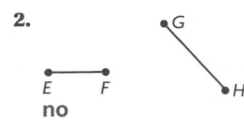

no

Use a compass and a straightedge to construct a line segment congruent to each figure. **Check students' constructions.**

3. **4.** **5.**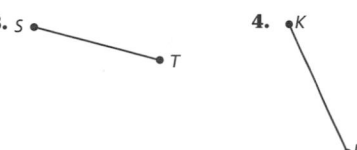

6. Use a ruler to draw a line segment *XY* that is 6 cm long. Then construct a line segment congruent to *XY*. **Check students' constructions.**

Construct Congruent Angles

Remember that angles are measured in degrees. All angles that have the same measure in degrees are congruent angles.

Look at the picture of the handmade quilt, at the right. When the quilt was made, pieces of fabric were cut by using congruent line segments and congruent angles. The pieces were then sewn together to make the pattern that you see.

- In what other handmade items might congruent angles be used?
  Possible answers: clothing, furniture

157

---

2 TEACH/PRACTICE

As students construct congruent line segments and angles in this lesson, they will see how congruent figures are used in the designs of quilts and rugs.

TEACHING TIP
Students may not be familiar with the term *arc*. Explain that an arc is part of a circle. If they were to draw a continuous arc, they would make a circle.

COMMON ERROR ALERT Students may inadvertently change the opening of the compass when they are constructing a congruent line segment or angle. If this occurs, have them measure the original line segment or angle again and adjust the size of the newly constructed line segment or angle as needed.

GUIDED PRACTICE
Have students complete Exercises 1–5. Then you may want to have students exchange their work with a classmate for checking. Make sure students are proficient at these skills before you assign the Independent Practice.

CRITICAL THINKING **If angle *PQR* is congruent to angle *BCD*, and angle *BCD* is congruent to angle *MNP*, is angle *PQR* congruent to angle *MNP*?** yes

If the rays in angle *PQR* are longer than the rays in angle *FGH*, could angles *PQR* and *FGH* be congruent? Explain. Yes; angles are congruent if they have the same angle measure, regardless of the lengths of the rays. Also, rays go on forever.

ENRICHMENT 8.4

Name _____

LESSON 8.4

Measure Up

Use a ruler and a protractor to measure the sides and angles of each figure. List all pairs of congruent sides or angles for each figure.

1. $\overline{AD} \cong \overline{BC}, \angle A \cong \angle B, \angle D \cong \angle C$

2. $\overline{EJ} \cong \overline{GH}, \overline{EF} \cong \overline{GF}, \angle J \cong \angle H,$ $\angle E \cong \angle G$

3. $\overline{KN} \cong \overline{LM}, \overline{KL} \cong \overline{NM}, \angle N \cong \angle L,$ $\angle K \cong \angle M$

4. $\overline{SR} \cong \overline{RQ}, \angle S \cong \angle R$

5. None

6. $\overline{XY} \cong \overline{XZ}, \angle Y \cong \angle Z$

Use with text pages 156–159.   **STRETCH YOUR THINKING** E37

157

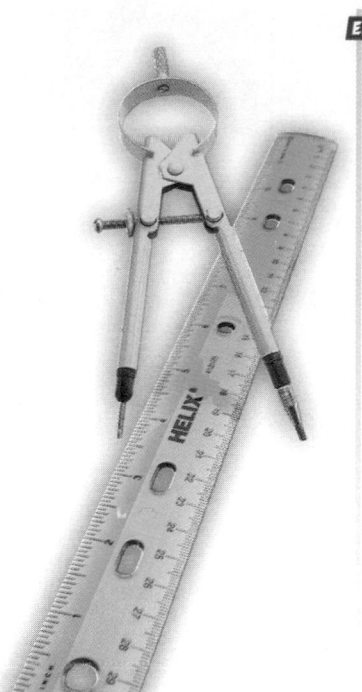

You used a compass and a straightedge to construct congruent line segments. You can use the same tools to construct congruent angles.

EXAMPLE 2 Trace ∠*ABC*. Construct an angle congruent to ∠*ABC*.

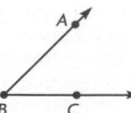

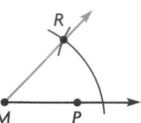

Draw $\overrightarrow{MP}$.

Draw an arc through ∠ABC.

Use the same compass opening to draw an arc through $\overrightarrow{MP}$.

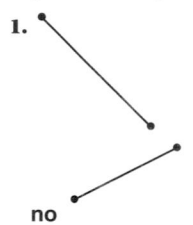
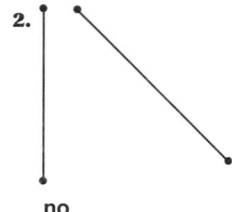

Use the compass to measure the arc in ∠ABC.

Use the same compass opening to locate point R.

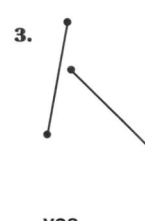

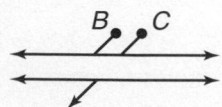

Draw $\overrightarrow{MR}$.

So, ∠*ABC* ≅ ∠*RMP*.

• Draw an obtuse angle. Use a compass and straightedge to copy the angle. **Check students' drawings.**

INDEPENDENT PRACTICE

Use a compass to determine if the line segments in each pair are congruent. Write *yes* or *no*.

1.

no

2.

no

3.

yes

IDEA FILE

EXTENSION

Have students guess whether either ray extends below the horizontal lines. They can check their guess by using a straightedge. The ray with endpoint *C* extends below the horizontal lines.

Modality Visual

LANGUAGE ARTS CONNECTION

A semaphore code is a set of signals used by sailors and railroad workers. Letters and numbers are represented by the positions of flags. Have students research the signals that stand for the letters in their first names. Have them construct congruent angles for each letter. Check students' constructions.

M   A   T   H

Modality Visual

Use a protractor to measure the angles in each pair. Tell whether they are congruent. Write each measure and *yes* or *no*.

4.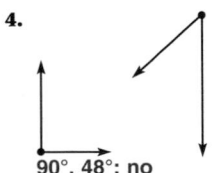

90°, 48°; no

5.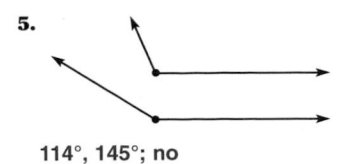

114°, 145°; no

6.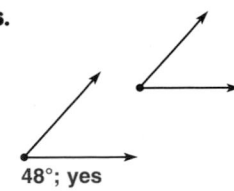

48°; yes

Use a compass and a straightedge to construct a congruent line segment or angle. **Check students' drawings.**

7.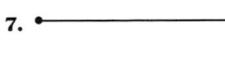

8.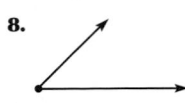

9.

Problem-Solving Applications

10. Construct a line segment that has a length equal to the length of $\overline{AB}$ plus the length of $\overline{CD}$. **Students' line segments should be $2\frac{1}{2}$ in.**

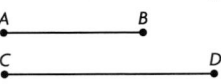

11. Construct an angle that has a measure equal to the measure of $\angle ABC$ plus the measure of $\angle DEF$. **Students' angles should have a measure of 130°.**

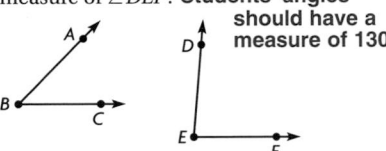

12. ▬ **WRITE ABOUT IT** Explain how to construct congruent angles. **Check students' explanations.**

Portfolio

Mixed Review

Tell how many sides each figure has. Name the figure.

13. △ 3; triangle

14. □ 4; square

15. ⬡ 6; hexagon

16. ⬠ 5; pentagon

Divide. Write the answer in simplest form.

17. $3\frac{1}{2} \div \frac{3}{4}$ $4\frac{2}{3}$

18. $2\frac{1}{4} \div 1\frac{1}{2}$ $1\frac{1}{2}$

19. $1\frac{1}{4} \div 1\frac{1}{8}$ $1\frac{1}{9}$

20. $3\frac{1}{5} \div 2\frac{2}{3}$ $1\frac{1}{5}$

21. Rochelle's car gets 15 mi per gallon of gasoline. She lives 180 mi from Brooke's house. How many gallons of gasoline does she use going to and from Brooke's house? **24 gal**

22. Juanita has $\frac{1}{2}$ yd of fabric. She uses only $\frac{1}{3}$ yd. How much is left? **$\frac{1}{6}$ yd**

INDEPENDENT PRACTICE

To ensure student success, you may wish to complete Exercises 1, 4, and 10 as a whole-class activity, before you assign the remaining exercises.

3 WRAP UP

Use these questions to help students focus on the big ideas of the lesson.

- **What are congruent line segments?** line segments with the same length
- **What are congruent angles?** angles with the same measure in degrees

✓ *Assessment Tip*

Have students write the answer.

- **Can a 90° angle facing one direction be congruent to a 90° angle facing in the opposite direction? Explain.** Yes. Possible explanation: angles are congruent if they have the same angle measure; it doesn't matter what direction they face.

DAILY QUIZ

NOTE: Draw these figures on the board.

Use a compass and a straightedge to construct a congruent line segment or angle. Check students' drawings.

1.

2.

3.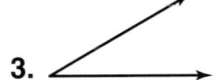

4.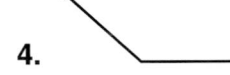

HISTORY CONNECTION

The Leaning Tower of Pisa was constructed in Tuscany, Italy, about 800 years ago. An uneven settling of the ground on which it was constructed caused it to lean. It now leans about 17 feet. Ask students what type of angle they think the tower forms with the ground. acute Have students research the tower to find how much further it is expected to lean and what is being done to correct the situation.

Modality Auditory

SCIENCE CONNECTION

Have students find photographs of spiderwebs and trace the webs. Have them examine the webs for angles. Ask them to measure and label each angle. You may wish to display their "geowebs."

Modality Visual

MEETING INDIVIDUAL NEEDS

159

ORGANIZER

OBJECTIVE To bisect a line segment
VOCABULARY *New* **bisect**
MATERIALS *For each student* compass,
straightedge
PROGRAM RESOURCES
E-Lab • Activity 8
E-Lab Recording Sheets, p. 8

USING THE PAGE

In this activity students are introduced to bisecting line segments—a skill they will use as they broaden their geometric experiences.

Explore

Observe students as they draw an arc through line segment *RS*. Make sure they open the compass a little more, not less, than half the distance from *R* to *S*. Point out that they need to draw a longer arc to bisect a line segment than they do to construct a line segment. If any students accidentally change the opening of the compass before they draw the second arc, have them place the compass point back on point *R* and adjust the opening to match the arc they drew. If the arcs don't intersect, have students use their compass to make longer arcs.

Think and Discuss

Discuss the first four questions. Then have students measure line segments *LW* and *LR*, and ask:

• **Does line segment *SD* bisect line segment *WR*?** yes

• **How do you know?** because line segments *LW* and *LR* are equal

• **Are line segments *LW* and *LR* perpendicular?** no

Point out to students that the angles formed by a line segment that bisects another line segment are not always right angles.

COOPERATIVE LEARNING

Bisecting a Line Segment

LAB *Activity*

WHAT YOU'LL EXPLORE
How to bisect a line segment

WHAT YOU'LL NEED
compass
straightedge

WORD POWER
bisect

You can use a compass and straightedge to bisect a line segment. When you **bisect** a line segment, you divide it into two equal parts.

ACTIVITY

Explore

Work with a partner to bisect a segment.

Draw a segment $\overline{RS}$.

R •————————• S

• Place the compass point on point *R*. Open the compass to a little more than half the distance from *R* to *S*. Draw an arc through $\overline{RS}$ as shown.

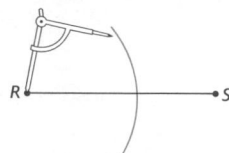

• Keep the same compass opening. Place the compass point on point *S*. Draw an arc as shown. Label the points *T* and *U* where the arcs intersect.

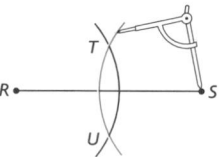

• Use a straightedge to draw a line through *T* and *U*. Label the point *P* where $\overleftrightarrow{TU}$ intersects $\overline{RS}$.

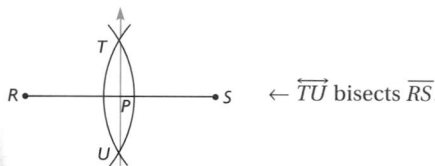

$\leftarrow \overleftrightarrow{TU}$ bisects $\overline{RS}$.

160 Chapter 8

IDEA FILE

REVIEW

Present this problem to students: When Bev adjusts her recliner, it forms an angle with the floor. She adjusts the recliner to a 90° angle to watch TV, a 20° angle to take a nap, and a 65° angle to read a book. Sometimes she raises the footrest to a 30° angle. Describe the angles Bev makes with her recliner. 90°—right angle; the others are acute angles.

Modality Auditory

ART CONNECTION

MATERIALS: graph paper, compass, straightedge

Have students make and color this quilt-square design. Check students' constructions.

1. Draw a 20-unit × 20-unit square.
2. Bisect the four line-segments that make up the square.
3. Connect the midpoints of each line segment to form another square.
4. Repeat steps 2 and 3 twice to make two more smaller squares within squares.

Modalities Auditory, Kinesthetic, Visual

Think and Discuss

- $\overleftrightarrow{TU}$ bisects $\overline{RS}$. What does that mean? **$\overleftrightarrow{TU}$ divides $\overline{RS}$ into two equal parts.**

- Look at $\overleftrightarrow{TU}$ and $\overline{RS}$. Would you describe them as parallel or intersecting? **intersecting**

- $\angle TPS$, $\angle SPU$, $\angle UPR$ and $\angle RPT$ are right angles. How many degrees are in each angle? **90°**

- 💡**CRITICAL THINKING** $\overleftrightarrow{TU}$ and $\overline{RS}$ intersect at point P to form four right angles. Besides intersection, what other relationship do you see between $\overleftrightarrow{TU}$ and $\overline{RS}$? **They are perpendicular.**

- Look at the figure at the right. How could you determine whether $\overline{SD}$ bisects $\overline{WR}$? **by using your compass to measure $\overline{LW}$ and $\overline{LR}$**

Try This

Tell whether $\overline{JK}$ bisects the other line segment. Write *yes* or *no*.

1.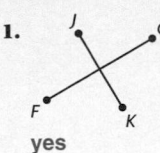
yes

2.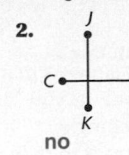
no

3.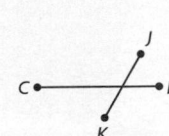
no

4.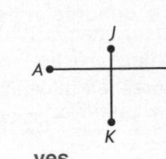
yes

5. Draw a line segment. Label it CD. Then construct a segment AB that bisects $\overline{CD}$.

6. Draw a line segment. Label it JK. Then construct a segment LM that bisects $\overline{JK}$.
For Exercises 5–6, check students' constructions.

Technology **Link**
E-Lab • Activity 8 Available on CD-ROM and the Internet at http://www.hbschool.com/elab

Note: Students bisect figures. Use E-Lab p. 8.

161

	GRADE 5	GRADE 6	GRADE 7
	Classify quadrilaterals.	Name and classify types of polygons.	Classify and compare triangles.

ORGANIZER

OBJECTIVE To name and classify polygons

VOCABULARY *New* **polygon, regular polygon**

ASSIGNMENTS Independent Practice
- ■ *Basic* 1–13
- ■ *Average* 1–13
- ■ *Advanced* 5–13

PROGRAM RESOURCES
Intervention • Classifying Triangles and Quadrilaterals, TE pp.164A–164B
Reteach 8.5 Practice 8.5 Enrichment 8.5

PROBLEM OF THE DAY

Transparency 8

Jason used 4 clock faces to draw regular polygons with 3 to 6 sides. One vertex of each polygon was on the dot for 12:00. The other vertices of his triangle were at 20 minutes past and 20 minutes to the hour. Where were the vertices of his other polygons? square: 12:00, 15 past, 30 past, 45 past; pentagon: 12:00, 12 past, 24 past, 36 past, 48 past; hexagon: 12:00, 10 past, 20 past, 30 past, 40 past, 50 past

Solution Tab A, p. A8

QUICK CHECK

Sketch the following figures on the board.

Do the line segments or angles appear to be congruent? If they don't appear to be congruent, explain why.

1.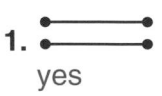
yes

2.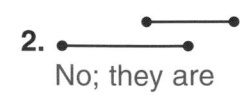
No; they are different lengths.

3.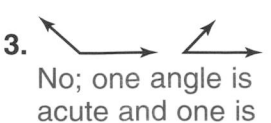
No; one angle is acute and one is obtuse.

4.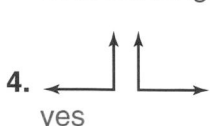
yes

1 MOTIVATE

Review with students the relationship

8.5

Polygons

WHAT YOU'LL LEARN
How to name polygons by the number of sides and angles

WHY LEARN THIS?
To recognize different polygons that you see in patterns and designs

WORD POWER
polygon
regular polygon

REMEMBER:
Triangles are classified by their angles and by their sides.
See page H21.

Classified by Sides
Equilateral
Isosceles
Scalene

Classified by Angles
Acute
Right
Obtuse

Activity item #3:
Possible answers: square, rectangle; square, rectangle, parallelogram; square, rectangle, parallelogram

Line segments are used to form other geometric figures. A **polygon** is a closed plane figure formed by three or more line segments. Polygons are named by the number of their sides and angles.

COMMON POLYGONS	
Name	**Sides and Angles**
Triangle	3
Quadrilateral	4
Pentagon	5
Hexagon	6
Octagon	8

ACTIVITY

WHAT YOU'LL NEED: geoboard and dot paper

- Work with a partner. Use the geoboard to show as many different triangles as you can. Include isosceles, scalene, right, acute, and obtuse triangles. Record your triangles on dot paper.
Check students' triangles.
- Use the geoboard to show as many different quadrilaterals as you can. Include a square, rectangle, parallelogram, and trapezoid. Record your quadrilaterals on dot paper.
Check students' quadrilaterals.
- Look at your quadrilaterals. Which have right angles? Which have both pairs of opposite sides parallel? In which are both pairs of opposite sides congruent?

- **CRITICAL THINKING** Show other types of polygons such as pentagons, hexagons, and octagons. Record them on dot paper.
Check students' polygons.
Some of the polygons you made may be regular polygons. A **regular polygon** has all sides congruent and all angles congruent.

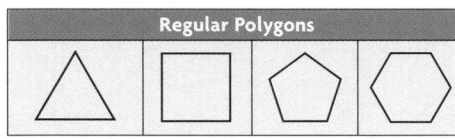
Regular Polygons

- Look at the regular polygons above. What types of angles do you see in each polygon? **triangle: acute; quadrilateral: right; pentagon: obtuse; hexagon: obtuse**

162 Chapter 8

IDEA FILE

CULTURAL CONNECTION

The art of paper folding called origami originated in Japan. Origami designs incorporate many geometric concepts, such as congruent angles and regular polygons. Have students find pictures of origami designs and identify the geometric concepts in them.

Modality Visual

RETEACH 8.5

Name _____

LESSON 8.5

Polygons

A polygon has three or more sides. The sides are line segments that meet at their endpoints only. The first figure is a polygon. The other two figures are not polygons.

This polygon has 5 sides.   Neither figure is a polygon.

You name a polygon by the number of sides.

Number of Sides	Name	Number of Angles
3	Triangle	3
4	Quadrilateral	4
5	Pentagon	5
6	Hexagon	6
8	Octagon	8

In a polygon, the number of sides is equal to the number of angles.

A regular polygon has all sides congruent and all angles congruent.

Not regular.
The angles are congruent but the sides are not.

Not regular.
The sides are congruent but the angles are not.

A regular polygon.
All the sides and all the angles are congruent.

Name each polygon. Then tell whether the polygon is regular. Write *yes* or *no*.

1. pentagon; yes   2. hexagon; no   3. triangle; yes   4. quadrilateral; no

R38 TAKE ANOTHER LOOK   Use with text pages 162–163.

GUIDED PRACTICE

Cut out a triangular piece of paper.

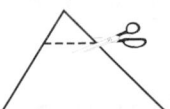

1. How many sides does the triangle have? How many vertices? **3; 3**

2. Cut off one vertex. Tell how many sides and vertices the polygon has. Name the polygon. **4 of each; quadrilateral**

3. Cut another vertex. Tell how many sides and vertices the polygon has. Name the polygon. **5 of each; pentagon**

4. Predict the number of sides and vertices if another vertex is cut. **6 of each**

INDEPENDENT PRACTICE

Name the quadrilateral. Tell whether both pairs of opposite sides are parallel. Write *yes* or *no*.

1.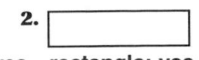
parallelogram; yes

2.
rectangle; yes

3.
trapezoid; no

4.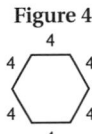
square; yes

For Exercises 5–8, use Figures 1–4. All measurements are in inches.

Figure 1	Figure 2	Figure 3	Figure 4
 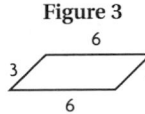

5. Which figure is a hexagon? **Figure 4**

6. Which figures have only obtuse angles? **Figures 2 and 4**

7. Is Figure 1 an equilateral, scalene, or isosceles triangle? **scalene**

8. Which figures appear to be regular polygons? **Figures 2 and 4**

Problem-Solving Applications

9. Tell how many triangles and how many quadrilaterals are in the figure at the right. **12 triangles; 11 quadrilaterals**

10. Draw a square. Then divide it into two congruent triangles. What type of triangles are they? **isosceles, or right**

11. Janice's mother bought a table that seats six people, one at each side. What shape is the table? **hexagon**

12. All parallelograms have opposite sides parallel. Are squares and rectangles parallelograms? Explain. **Yes; opposite sides are parallel.**

13. ◀▶ **WRITE ABOUT IT** Explain the difference between a regular polygon and a polygon that is not regular. **Regular: all sides and angles are congruent; not regular: no sides or angles need to be congruent.**

Portfolio

MORE PRACTICE **Lesson 8.5, page H54**

163

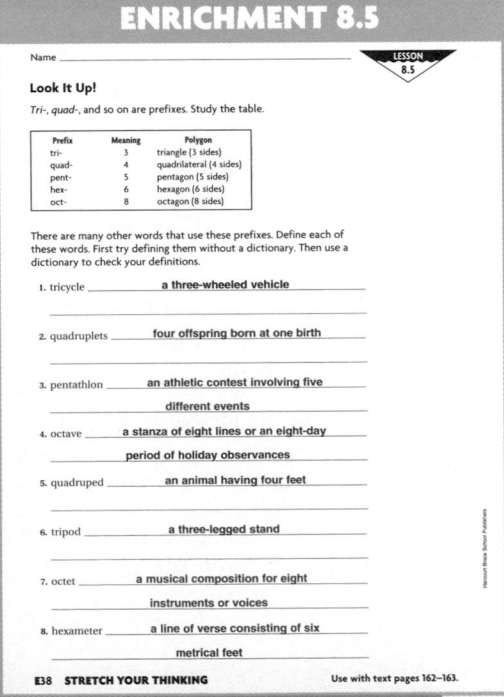

---

between the number of sides and angles that make up polygons and the prefixes in polygon names. Ask:

- **What geometric figure begins with the prefix *tri-*?** triangle
- **How many sides and angles does a triangle have?** 3 sides and 3 angles
- **What other words use the prefix *tri-* to represent 3?** Possible answer: tricycle—a 3-wheel vehicle

Repeat the questions above for the other polygons named in the chart.

2 TEACH/PRACTICE

In this lesson students identify the properties of polygons they model on geoboards.

GUIDED PRACTICE

Complete these exercises as a whole-class activity. Make sure students can name polygons by the number of sides and angles before you assign the Independent Practice.

INDEPENDENT PRACTICE

You may wish to check students' answers to Exercises 1–4 before assigning the remainder of the Independent Practice.

3 WRAP UP

Use these questions to help students focus on the big ideas of the lesson.

- **What polygons have more than 4 sides and 4 angles?** pentagon, hexagon, octagon
- **What polygon has fewer than 4 sides and 4 angles?** triangle

✓ *Assessment Tip*

Have students write the answer.

- **How do you know that the quadrilateral in Independent Practice Exercise 3 is a trapezoid?** because it has only one pair of parallel sides

DAILY QUIZ

NOTE: Sketch these figures on the board.

Name the polygon.

1.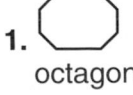
octagon

2.
quadrilateral

3.
pentagon

4.
hexagon

163

INTERVENTION

Classifying Triangles and Quadrilaterals

Use these strategies to review the skills required to classify triangles and quadrilaterals.

Multiple Intelligences Logical/Mathematical, Visual/Spatial, Bodily/Kinesthetic

Modalities Auditory, Kinesthetic, Visual

Lesson Connection Students will classify polygons in Lesson 8.5.

Materials Graph paper, scissors

Program Resources *Intervention and Extension Copying Masters,* p. 5

Troubleshooting

WHAT STUDENTS NEED TO KNOW	HOW TO HELP	FOLLOW UP
Triangles can be classified by their angles.	Have students draw on graph paper several examples each of a right triangle, an acute triangle, and an obtuse triangle. Check students' triangles. **Examples** This is a right triangle because it has a right angle. This is an acute triangle because each angle is acute. This is an obtuse triangle because it has one obtuse angle.	Other problems to solve: Classify the triangle as right, acute, or obtuse, given the three angles. 40°, 70°, 70° acute 30°, 100°, 50° obtuse 40°, 80°, 60° acute 25°, 90°, 65° right See Key Skill 40, Student Handbook, page H21.
Triangles can be classified by their sides.	Have students draw and cut out an example of a scalene, an isosceles, and an equilateral triangle. Check students' triangles. **Examples** This is an equilateral triangle because all three sides have the same length. This is an isosceles triangle because two sides have the same length. This is a scalene triangle because no sides have the same length.	Other problems to solve: Classify the triangle as equilateral, isosceles, or scalene, given the lengths of the sides. 3 in., 11 in., 9 in. scalene 18 m, 18 m, 18 m equilateral 7 cm, 5 cm, 7 cm isosceles See Key Skill 40, Student Handbook, page H21.

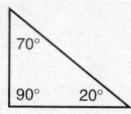
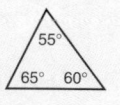
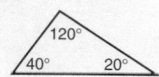
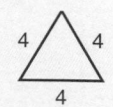
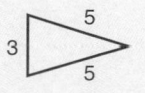

WHAT STUDENTS NEED TO KNOW	HOW TO HELP	FOLLOW UP
Triangles can be classified by a combination of their angles and their sides.	Have students identify a triangle by a combination of its angles and sides. **Examples** 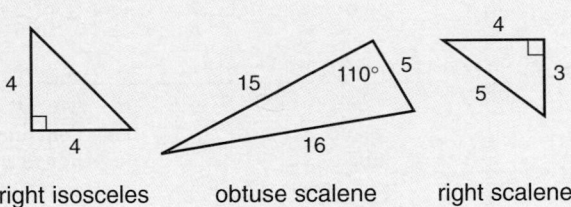 right isosceles obtuse scalene right scalene	Other activities: Draw an obtuse isosceles triangle. Check students' drawings. Draw several equilateral triangles. What kinds of angles does an equilateral triangle always have? Check students' drawings; acute angles.
Some quadrilaterals have special names.	Have students discuss and develop descriptions of each of the following special quadrilaterals. Checks students' descriptions.  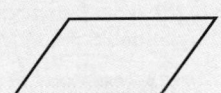 **trapezoid:** 1 pair of parallel sides **parallelogram:** 2 pairs of parallel sides **rhombus:** all sides equal 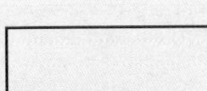  **rectangle:** opposite sides equal; 4 right angles **square:** all sides equal; 4 right angles	Questions to ask: Which quadrilateral has only 1 pair of parallel sides? trapezoid Which quadrilaterals have 2 pairs of parallel sides? parallelogram, rhombus, rectangle, square Which quadrilaterals have 4 right angles? rectangle, square See Key Skill 41, Student Handbook, page H22.

Intervention and Extension Copying Masters, p. 5

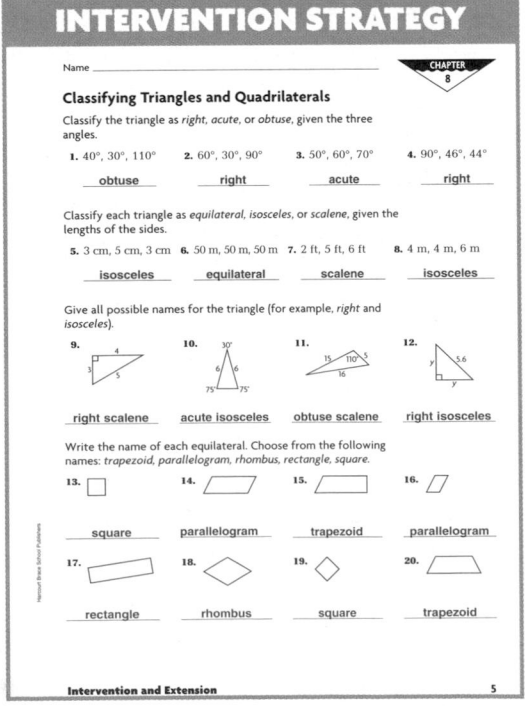

164B

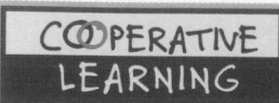

COOPERATIVE LEARNING

OBJECTIVE To recognize and draw geometric figures

USING THE PAGE

Before the Lesson
Display pictures of interesting buildings in the community. Ask students to identify the different geometric figures they see in the pictures. Have them use the terms listed on the student page. Ask student volunteers to make a chart listing the terms followed by examples. The chart can be used as a guide when they solve the problems.

Cooperative Groups
Have the students work in cooperative groups to solve the problems in Problems 1–4. Ask each group to find examples of geometric figures in the photo and to complete Problems 1–4.

CULTURAL LINK

There are many festivals and celebrations in Spain. One of the most colorful is the Tomato Festival in Valencia. Tomatoes are tossed about to test the juicy quality of the new tomato harvest.

CULTURAL CONNECTION

COOPERATIVE LEARNING

Geometry in Spanish Buildings

Marisol did a geometry project about her native land of Spain. She made a bulletin board display to show geometric figures in buildings. She used pictures of buildings in Spain. Then students circled examples of geometric figures they found in the pictures. One of the pictures Marisol brought in was of the cathedral in Segovia. Identify geometric shapes in the figure. Use the words from the chart below.

point	line segment
plane	congruent line segments
angle	perpendicular lines
right angle	intersecting lines
congruent angles	isosceles triangle

Portfolio **Work Together** Use the picture of the windmill in La Mancha.

1. What congruent geometric figures form the blades of the windmill? **squares**

2. Draw examples of perpendicular, congruent, parallel, and intersecting lines found in the windmill blades.
   Check students' drawings.

3. The blades of the windmill are constructed with a long support that bisects several smaller supports. Draw one of the smaller supports and construct its perpendicular bisector.
   Check students' constructions.

4. What geometric figure best describes the ground the windmill rests on? **plane**

CULTURAL LINK

Visitors to Spain can tour buildings that are hundreds of years old. There are more than 1,000 castles and fortresses scattered across the country.

CHAPTER 8 REVIEW

EXAMPLES

• **Identify points, lines, and planes.**
(pages 150–151)

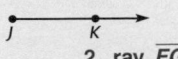
$\overrightarrow{JK}$, or ray JK

2. ray, $\overrightarrow{FG}$ 3. line segment, $\overline{CD}$

• **Classify lines.** (pages 152–153)

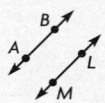
$\overleftrightarrow{AB} \parallel \overleftrightarrow{ML}$ Lines AB and
ML are parallel.

4. perpendicular lines
5. $\overleftrightarrow{JK}$ or $\overleftrightarrow{HI}$
6. $\overleftrightarrow{HJ}, \overleftrightarrow{HI}, \overleftrightarrow{JK}, \overleftrightarrow{KI}$

• **Identify and measure angles.** (pages 154–155)

The measure of an acute angle is less than 90°.

The measure of a right angle is 90°.

• **Construct congruent line segments and congruent angles.** (pages 156–159)

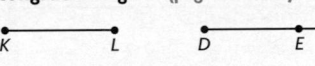

• **Identify polygons.** (pages 162–163)

triangle   quadrilateral

This hexagon has 6 congruent sides and 6 congruent angles.

EXERCISES

1. **VOCABULARY** A(n) _?_ is part of a line and has two endpoints . **line segment**

Identify each figure. Give the symbol.
2. 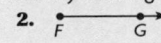   3.

4. **VOCABULARY** Right angles are formed when _?_ intersect.

For Exercises 5–6, use the diagram at the right.
Possible answers are given.
5. $\overrightarrow{HJ} \perp$ _?_.
6. $\overleftrightarrow{KH}$ intersects _?_.

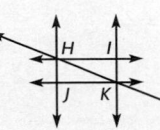

7. **VOCABULARY** In an angle, the common endpoint of the rays is the _?_. **vertex**

Trace the angle and measure it. Write *acute*, *right*, or *obtuse*.
8. 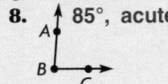 85°, acute   9. 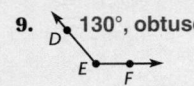 130°, obtuse

10. **VOCABULARY** Line segments are _?_ if they have the same length. **congruent**

Trace the figure. Then construct one that is congruent. **Check students' constructions.**
11. 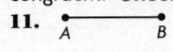   12.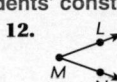

13. **VOCABULARY** A _?_ has all sides congruent and all angles congruent. **regular polygon**
Name the polygon.
14. 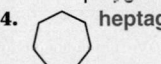 heptagon 15. 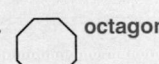 octagon

Chapter 8 Review **165**

USING THE PAGE
The Chapter 8 Review reviews identifying points, lines, and planes; classifying lines; identifying and measuring angles; constructing congruent line segments and congruent angles; identifying polygons. Chapter objectives are provided, along with examples and practice exercises for each.

Portfolio The Chapter 8 Review can be used as a review or a test. You may wish to place the Review in students' portfolios.

Assessment Checkpoint
For Performance Assessment in this chapter, see page 204, What Did I Learn?, items 1 and 4.

✔ **Assessment Tip**
Student Self-Assessment The How Did I Do? Survey helps students assess both what they have learned and how they learned it. Self-assessment helps students learn more about their own capabilities and develop confidence in themselves.

A self-assessment survey is available as a copying master in the *Teacher's Guide for Assessment*, p. 15.

CHAPTER 8 TEST, page 1

Choose the letter of the correct answer.

1. A flat surface that goes on forever in all directions is called a _?_.
A. point   B. line
C. ray   (D) plane

2. What is the symbol for this figure?
A. $\overleftrightarrow{SR}$ B. $\overline{RS}$
(C) $\overrightarrow{RS}$ D. $\overrightarrow{RS}$

3. What geometric figure is suggested by the point at the end of a pencil?
(A) point   B. line
C. ray   D. plane

4. What does the symbol $\overline{XY}$ refer to?
A. plane XY   (B) line segment XY
C. line XY   D. ray XY

5. Which pair of lines is described by the statement $\overleftrightarrow{AB} \perp \overleftrightarrow{CD}$?

6. Which statement about pairs of lines is *not* true?
A. Intersecting lines are sometimes perpendicular.
B. Parallel lines are never intersecting lines.
C. Perpendicular lines form right angles.
(D) Parallel lines are sometimes intersecting lines.

For questions 7–8, use the figure.

7. Which of these lines intersects $\overleftrightarrow{EF}$ and is perpendicular to it?
(A) $\overleftrightarrow{AB}$ B. $\overleftrightarrow{GH}$
C. $\overleftrightarrow{CD}$ D. not here

8. What is the relationship between $\overleftrightarrow{CD}$ and $\overleftrightarrow{EF}$?
A. They are parallel.
(B) They are intersecting but not perpendicular.
C. They are perpendicular.
D. not here

9. An angle that measures 45° is classified as a(n) _?_.
A. obtuse angle   B. right angle
(C) acute angle   D. straight angle

10. At which of these times do the hands of a clock form a right angle?
A. 1:00   (B) 3:00
C. 5:00   D. 6:00

Form A • Multiple-Choice   A31   Go on. ▶

CHAPTER 8 TEST, page 2

11. Which could be the measure of the angle shown below?
A. 60°   B. 90°
(C) 125°   D. 180°

12. What is the measure of a straight angle?
A. less than 90°
B. 90°
C. more than 90° but less than 180°
(D) 180°

13. Two segments that are congruent have the same _?_.
(A) length   B. endpoints
C. angles   D. not here

14. Which angle appears to be congruent to ∠EFG?

15. What construction do the figures below represent?
(A) congruent line segments
B. bisected line segments
C. congruent angles
D. perpendicular line segments

16. What construction do the figures below represent?
A. congruent line segments
B. bisected line segments
(C) congruent angles
D. parallel line segments

17. If a figure is a parallelogram, then it must be a _?_.
A. rectangle   (B) quadrilateral
C. trapezoid   D. square

18. What is the name for this polygon?
A. pentagon   B. trapezoid
C. hexagon   (D) octagon

19. How many angles does a pentagon have?
(A) 5   B. 6   C. 7   D. not here

20. A triangle that is a regular polygon is a(n) _?_.
(A) equilateral triangle
B. right triangle
C. scalene triangle
D. not here

Form A • Multiple-Choice   A32   ▶ Stop!

See *Test Copying Masters*, pp. A31–A32 for the Chapter Test in **multiple-choice** format and pp. B181–B182 for **free-response** format.

Symmetry and Transformations

BIG IDEA: Geometric figures can be transformed through slides, turns, and flips.

Introducing the Chapter p. 166		Team-Up Project • Letter Chains p. 167	
OBJECTIVE	**VOCABULARY**	**MATERIALS**	**RESOURCES**
9.1 Symmetry and Congruence pp. 168-171 **Objective:** To identify line symmetry and rotational symmetry **2 DAYS**	line symmetry rotational symmetry	FOR EACH STUDENT: paper, scissors, dark crayon	■ Reteach, ■ Practice, ■ Enrichment 9,1; More Practice, p. H54 Math Fun • It's All in the Alphabet, p. 182
Lab Activity: Symmetry and Tangrams pp. 172-173 **Objective:** To use tangrams to build symmetric figures		FOR EACH STUDENT: tangram pieces or tangram pattern	E-Lab • Activity 9 E-Lab Recording Sheets, p. 9 Math Fun • Puzzle-Grams, p. 182
9.2 Transformations pp. 174-177 **Objective:** To use translations, rotations, and reflections to transform geometric shapes **2 DAYS**	transformation translation rotation reflection		■ Reteach, ■ Practice, ■ Enrichment 9.2; More Practice, p. H55 Math Fun • Puzzle-Grams, p. 182
EXTENSION • Transformations on a Coordinate Plane, Teacher's Edition pp. 178A–178B			
9.3 Tessellations pp. 178-179 **Objective:** To use polygons to make a tessellation **1 DAY**	tessellation		■ Reteach, ■ Practice, ■ Enrichment 9.3; More Practice, p. H55 **Cosmic Geometry •** *Tessellation Creation Station* Math Fun • Alphabet Tessellations, p. 182
9.4 Problem-Solving Strategy: Making a Model pp. 180-181 **Objective:** To use the strategy *make a model* to solve problems that involve tessellations **1 DAY**			■ Reteach, ■ Practice, ■ Enrichment 9.4; More Practice, p. H55 Problem-Solving Think Along TR p. R1
CHAPTER ASSESSMENT Chapter 9 Review p. 183			

NCTM CURRICULUM
STANDARDS FOR GRADES 5–8

✓	**Problem Solving** *Lessons 9.1, 9.2, 9.3, 9.4*
✓	**Communication** *Lessons 9.1, 9.2, 9.3, 9.4*
✓	**Reasoning** *Lessons 9.1, 9.2, 9.3, 9.4*
✓	**Connections** *Lessons 9.1, 9.2, 9.3, 9.4*
☐	**Number and Number Relationships**
☐	**Number Systems and Number Theory**
✓	**Computation and Estimation** *Lessons 9.1, 9.3*
✓	**Patterns and Functions** *Lessons 9.1, Lab Activity, 9.2, 9.3, 9.4.*
☐	**Algebra**
☐	**Statistics**
☐	**Probability**
✓	**Geometry** *Lessons 9.1, Lab Activity, 9.2, 9.3, 9.4.*
✓	**Measurement** *Lessons 9.1, 9.2, 9.3*

ASSESSMENT OPTIONS

CHAPTER LEARNING GOALS

9A.1 To identify line symmetry and rotational symmetry

9A.2 To identify and use transformations of geometric shapes

9A.3 To use polygons to make tessellations

9A.4 To use a model to solve a problem

These goals for concepts, skills, and problem solving are assessed in many ways throughout the chapter. See the chart below for a complete listing of both traditional and informal assessment options.

Pretest Options
Pretest for Chapter Content, TCM pp. A33–A34, B183–B184
Assessing Prior Knowledge, TE p. 166; APK p. 11

Daily Assessment
Mixed Review, PE p. 171
Problem of the Day, TE pp. 168, 174, 178, 180
Assessment Tip and Daily Quiz, TE pp. 171, 173, 177, 179, 181

Formal Assessment
Chapter 9 Review, PE p. 183
Chapter 9 Test, TCM pp. A33–A34, B183–B184
Study Guide & Review, PE pp. 202–203
Test • Chapters 8–10, TCM pp. A37–A38, B187–B188
Cumulative Review, PE p. 205
Cumulative Test, TCM pp. A39–A42, B189–B192

Performance Assessment
Team-Up Project Checklist, TE p. 167
What Did I Learn?, Chapters 8–10, TE p. 204
Interview/Task Test, PA p. 84

Portfolio
Suggested work samples:
Independent Practice, PE pp. 170, 176
Math Fun, PE p. 182

Student Self-Assessment
How Did I Do?, TGA p. 15
Portfolio Evaluation Form, TGA p. 24
Math Journal, TE pp. 166B, 177

Key
Assessing Prior Knowledge Copying Masters: APK	Teacher's Guide for Assessment: TGA
Test Copying Masters: TCM	Teacher's Edition: TE
Performance Assessment: PA	Pupil's Edition: PE

MATH COMMUNICATION

MATH JOURNAL
You may wish to use the following suggestions to have students write about the mathematics they are learning in this chapter:

PE pages 175, 178 • Talk About It

PE pages 171, 177, 181 • Write About It

TE page 177 • Writing in Mathematics

TGA page 15 • How Did I Do?

In every lesson • record written answers to each Wrap-up Assessment Tip from the Teacher's Edition for test preparation.

VOCABULARY DEVELOPMENT
The terms listed at the right will be introduced or reinforced in this chapter. The boldfaced terms in the list are new vocabulary terms in the chapter.

So that students understand the terms rather than just memorize definitions, have them construct their own definitions and pictorial representations and record those in their Math Journals.

line symmetry, pp. 168, 173
rotational symmetry, pp. 169, 173
transformation, pp. 174, 176
translation, pp. 174, 176
rotation, pp. 174, 176
reflection, pp. 174, 176
tessellation, p. 178

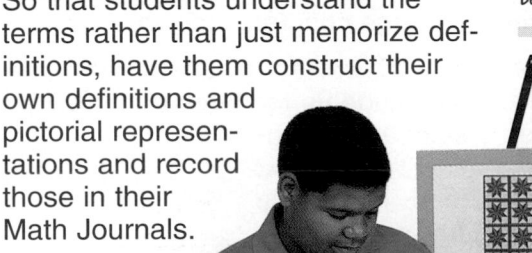

MEETING INDIVIDUAL NEEDS

RETEACH

BELOW-LEVEL STUDENTS

Practice Activities, TE Tab B
Multiplying Tenths
(Use with Lessons 9.1–9.4.)

Interdisciplinary Connection, TE p. 178

 Literature Link, TE p. 166C
Where the Sidewalk Ends, Activity 1
(Use with Lesson 9.3.)

 Technology
Cosmic Geometry • *Tessellation Creation Station*
(Use with Lesson 9.3.)

E-Lab • Activity 9
(Use with Lab Activity.)

PRACTICE

ON-LEVEL STUDENTS

Practice Game, TE p. 166C
Tessellation Terms
(Use with Lessons 9.3–9.4.)

Interdisciplinary Connection, TE p. 178

 Literature Link, TE p. 166C
Where the Sidewalk Ends, Activity 2
(Use with Lesson 9.3.)

 Technology
Cosmic Geometry • *Tessellation Creation Station*
(Use with Lesson 9.3.)

E-Lab • Activity 9
(Use with Lab Activity.)

CHAPTER IDEA FILE

The activities on these pages provide additional practice and reinforcement of key chapter concepts. The chart above offers suggestions for their use.

 ## LITERATURE LINK

Where the Sidewalk Ends

by Shel Silverstein

In this collection of poems, Silverstein uses humor and rhyme to address the trials and tribulations of childhood.

Activity 1—Have students sketch a hopscotch pattern like one that is often drawn on a sidewalk in chalk. Then have them tell whether this pattern tessellates a plane.

Activity 2—Have students draw a pattern using shapes of their choice, such as hexagons, squares, triangles, and/or pentagons that could be used to tessellate a large concrete sidewalk.

Activity 3—Sidewalks are sometimes made out of bricks. Have students research and draw brick patterns such as herringbone, basket weave, and running bond. Then have students invent their own pattern using brick shapes.

PRACTICE GAME

Tessellation Terms

Purpose To identify symmetry, transformations, and tessellations

Materials game cards (TR p. R86)

About the Game In a group of 12, a player shuffles cards and deals one to each player. Players with pictures on their cards try to find players with matching word cards as quickly as possible. The pairs of players who come in first, second, and third place receive prizes. Then 6 new players replace the winners. Play continues as necessary to include all students.

Modalities Auditory, Kinesthetic, Visual

Multiple Intelligences Verbal/Linguistic, Visual/Spatial, Bodily/ Kinesthetic

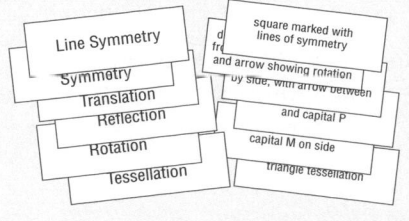

Interdisciplinary Connection, TE p. 178

 **Literature Link,** TE p. 166C
Where the Sidewalk Ends, Activity 3
(Use with Lesson 9.3.)

 Technology
Cosmic Geometry • *Tessellation Creation Station*
(Use with Lesson 9.3.)

E-Lab • Activity 9
(Use with Lab Activity.)

INTERDISCIPLINARY SUGGESTIONS

Connecting Symmetry and Transformations

Purpose To connect symmetry and congruence to other subjects students are studying

You may wish to connect **Symmetry and Transformations** to other subjects by using related interdisciplinary units.

Suggested Units

• Art Connection: Geometric Patterns in Art

• History Connection: Tangrams

• Multicultural Connection: Tribal Masks

• Writing Connection: Words in Motion

• Career Connection: Gift Wrap Designer

• Science Connection: Patterns in Nature

• Language Arts Connection: Symmetrical Words

Modalities Auditory, Kinesthetic, Visual

Multiple Intelligences Mathematical/Logical, Bodily/Kinesthetic, Visual/Spatial

The purpose of using technology in the chapter is to provide reinforcement of skills and support in concept development.

E-Lab • **Activity 9**—Students use visual thinking and reasoning skills as they translate, rotate, and reflect a plane figure to a position generated by the computer until the figures match.

Cosmic Geometry • *Tessellation Creation Station*—The activities on this CD-ROM provide students with opportunities to practice making tessellations.

Each Cosmic Geometry activity can be assigned by degree of difficulty. See Teacher's Edition page 179 for Grow Slide information.

Visit our web site for additional ideas and activities.
http://www.hbschool.com

Introducing the Chapter

Encourage students to talk about the symmetry they see around them. Review the math content students will learn in the chapter.

WHY LEARN THIS?

The question "Why are you learning this?" will be answered in every lesson. Sometimes there is a direct application to everyday experiences. Other lessons help students understand why these skills are needed for future success in mathematics.

PRETEST OPTIONS

Pretest for Chapter Content See the *Test Copying Masters*, pages A33–A34, B183–B184 for a free-response or multiple-choice test that can be used as a pretest.

Assessing Prior Knowledge Use the copying master shown below to assess previously taught skills that are critical for success in this chapter.

Assessing Prior Knowledge Copying Masters, p. 11

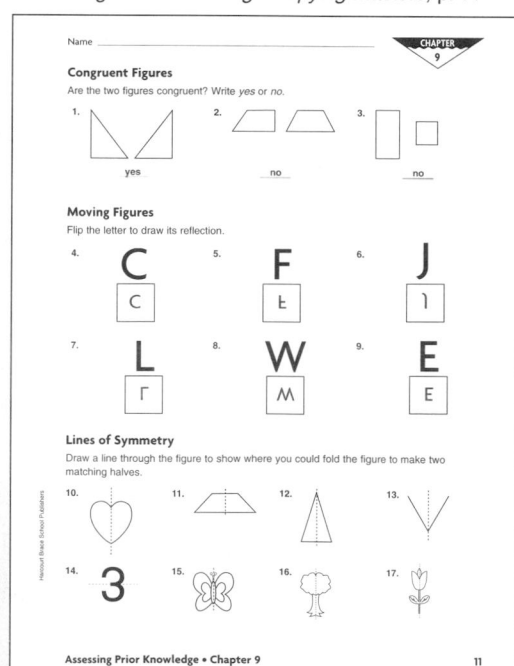

Troubleshooting

If students need help	Then use
• IDENTIFYING CONGRUENT FIGURES	• Key Skills, PE p. H23
• REFLECTING FIGURES	• Key Skills, PE p. H20
• IDENTIFYING LINES OF SYMMETRY	• Glossary, PE pp. H87–H96

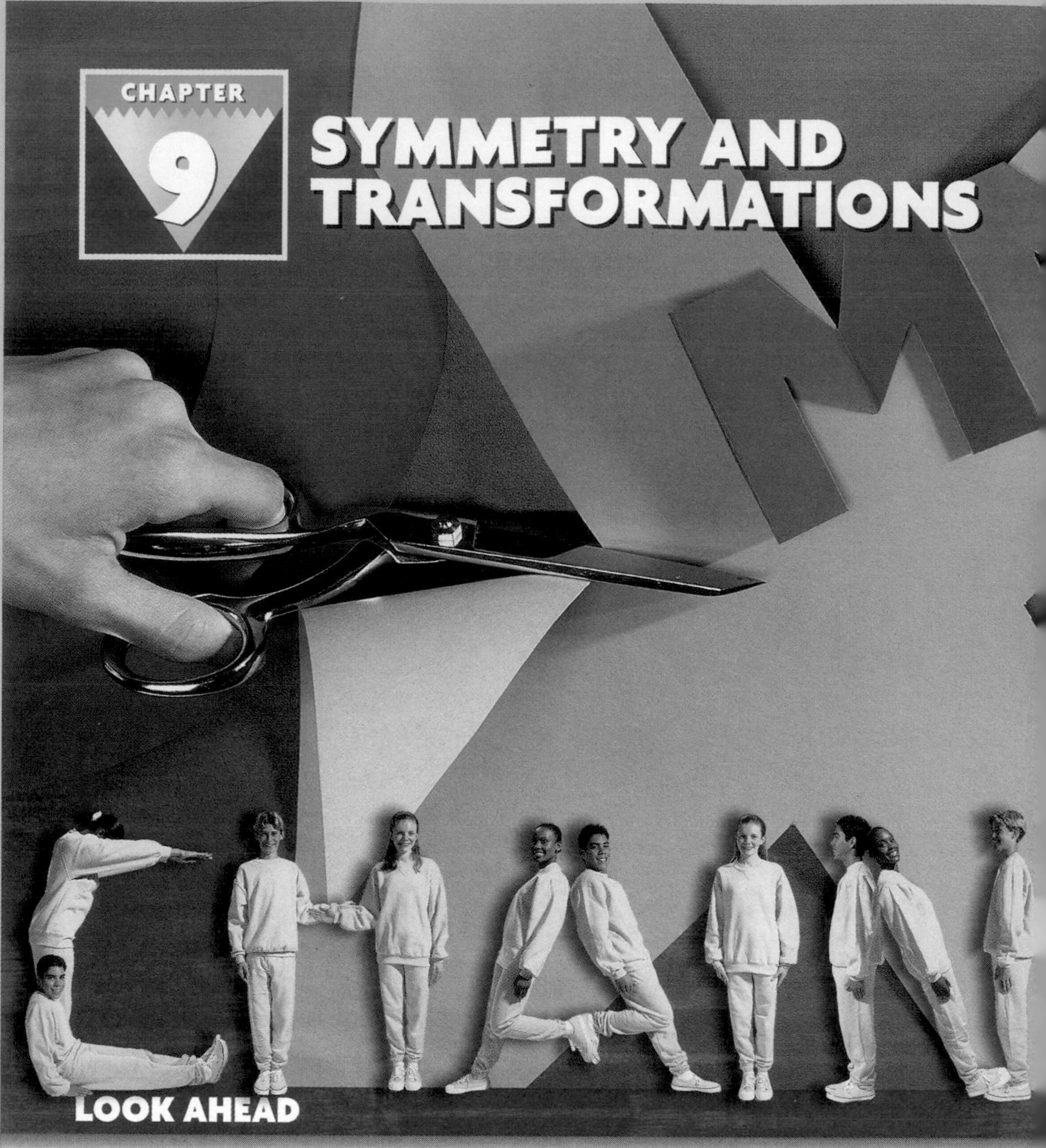

CHAPTER 9 SYMMETRY AND TRANSFORMATIONS

LOOK AHEAD

In this chapter you will solve problems that involve

- symmetry and congruence of figures
- transformations of plane figures
- tessellations

SCHOOL-HOME CONNECTION

You may wish to send to each student's family the *Math at Home* page. It includes an article that stresses the importance of communication about mathematics in the home, as well as helpful hints about homework and taking tests. The activity and extension provide extra practice that develops concepts and skills involving congruent figures and symmetry.

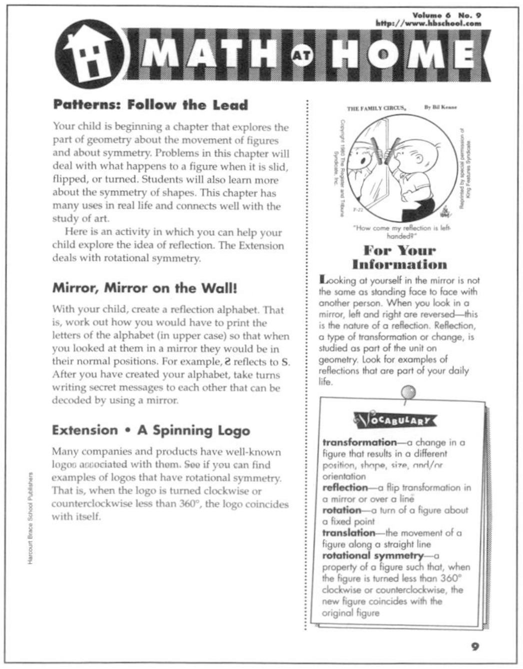

Patterns: Follow the Lead

Your child is beginning a chapter that explores the part of geometry about the movement of figures and about symmetry. Problems in this chapter will deal with what happens to a figure when it is slid, flipped, or turned. Students will also learn more about the symmetry of shapes. This chapter has many uses in real life and connects well with the study of art.

Here is an activity in which you can help your child explore the idea of reflection. The Extension deals with rotational symmetry.

Mirror, Mirror on the Wall!

With your child, create a reflection alphabet. That is, work out how you would have to print the letters of the alphabet (in upper case) so that when you looked at them in a mirror they would be in their normal positions. For example, Ƨ reflects to S. After you have created your alphabet, take turns writing secret messages to each other that can be decoded by using a mirror.

Extension • A Spinning Logo

Many companies and products have well-known logos associated with them. See if you can find examples of logos that have rotational symmetry. That is, when the logo is turned clockwise or counterclockwise less than 360°, the logo coincides with itself.

For Your Information

Looking at yourself in the mirror is not the same as standing face to face with another person. When you look in a mirror, left and right are reversed—this is the nature of a reflection. Reflection, a type of transformation or change, is studied as part of the unit on geometry. Look for examples of reflections that are part of your daily life.

VOCABULARY

transformation—a change in a figure that results in a different position, shape, size, and/or orientation.
reflection—a flip transformation in a mirror or over a line
rotation—a turn of a figure about a fixed point
translation—the movement of a figure along a straight line
rotational symmetry—a property of a figure such that, when the figure is turned less than 360° clockwise or counterclockwise, the new figure coincides with the original figure

Math at Home

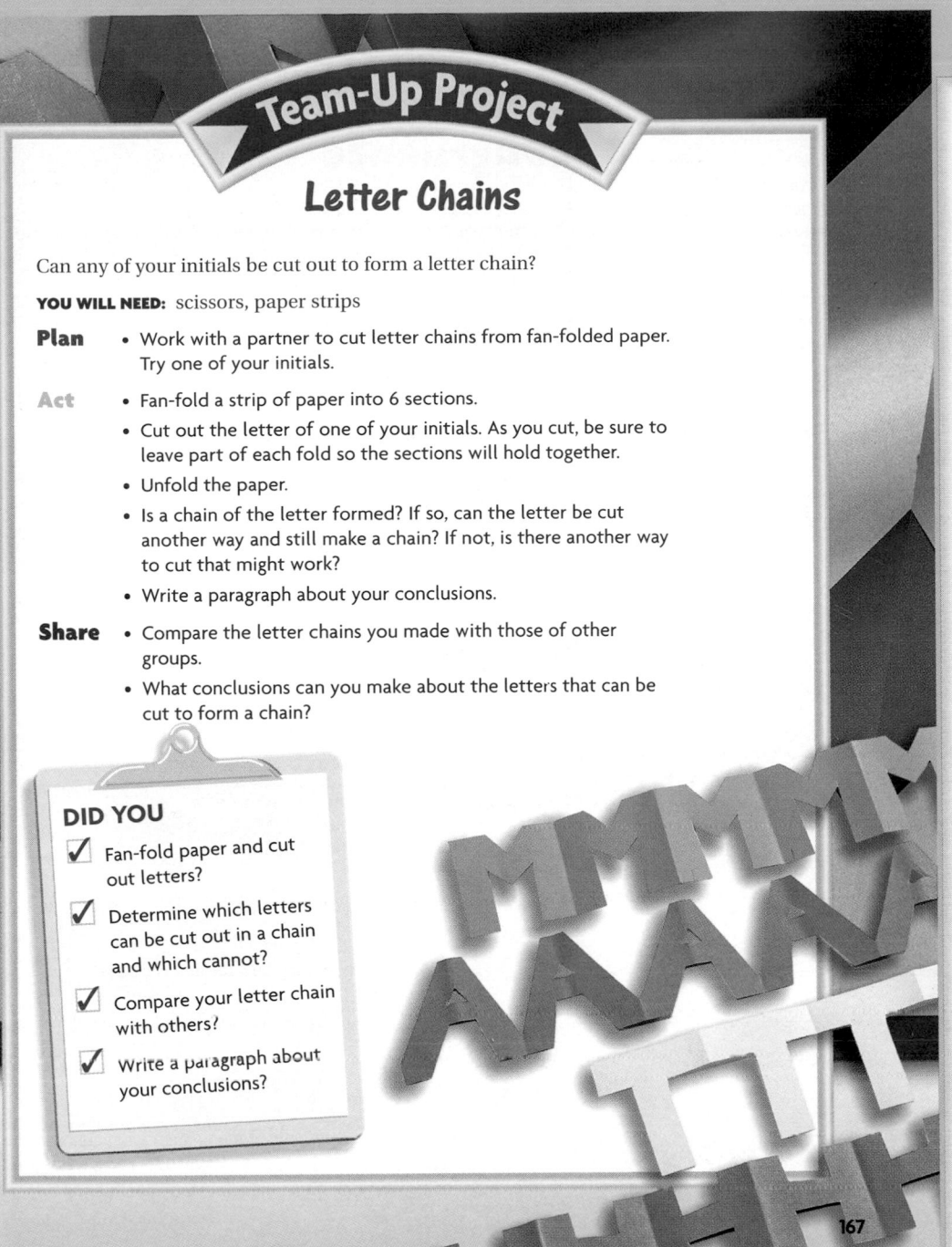

Team-Up Project

Letter Chains

Can any of your initials be cut out to form a letter chain?

YOU WILL NEED: scissors, paper strips

Plan
- Work with a partner to cut letter chains from fan-folded paper. Try one of your initials.

Act
- Fan-fold a strip of paper into 6 sections.
- Cut out the letter of one of your initials. As you cut, be sure to leave part of each fold so the sections will hold together.
- Unfold the paper.
- Is a chain of the letter formed? If so, can the letter be cut another way and still make a chain? If not, is there another way to cut that might work?
- Write a paragraph about your conclusions.

Share
- Compare the letter chains you made with those of other groups.
- What conclusions can you make about the letters that can be cut to form a chain?

DID YOU
- ✔ Fan-fold paper and cut out letters?
- ✔ Determine which letters can be cut out in a chain and which cannot?
- ✔ Compare your letter chain with others?
- ✔ Write a paragraph about your conclusions?

167

Project Checklist

Portfolio

Evaluate whether partners
- ✔ work cooperatively.
- ✔ cut letters from fan-folded paper.
- ✔ recognize the transformations illustrated in the letter chains.
- ✔ understand that figures with a horizontal line of symmetry can form chains.
- ✔ write a paragraph about their conclusions.

Using the Project

Accessing Prior Knowledge using spatial visualization, symmetry, and transformations

Purpose to study symmetry and transformations in chains cut from fan-folded paper

GROUPING partners
MATERIALS scissors, paper

Managing the Activity You may want to assign groups extra letters so that the class as a whole considers all letters. Notebook paper or magazine pages cut in half the long way fold well for this project.

WHEN MINUTES COUNT

A shortened version of this activity can be completed in 15–20 minutes. Ask students to explain how each of the chains on the student page was cut from the fan-folded paper. Have them describe how each letter is transformed in unfolding and explain why some letters can be cut in chains and others cannot.

A BIT EACH DAY

Day 1: Students select partners and make sketches to show how they will cut out letters so they remain attached at the folds.

Day 2: Students fan-fold paper and cut out letters.

Day 3: Students figure out which letters can make letter chains and which cannot.

Day 4: Partners share their letter chains.

Day 5: Students make generalizations about the characteristics of figures that can be cut into chains and write a paragraph about their conclusions.

HOME HAPPENINGS

Tell students to show a younger child how to make a chain of paper dolls.

GRADE 5	GRADE 6	GRADE 7
Identify and draw lines of symmetry.	Identify line symmetry and rotational symmetry.	Identify line symmetry and rotational symmetry.

ORGANIZER

OBJECTIVE To identify line symmetry and rotational symmetry

VOCABULARY *New* **line symmetry, rotational symmetry**

ASSIGNMENTS Independent Practice
- *Basic* 1–20, odd; 21–24; 25–33
- *Average* 1–24; 25–33
- *Advanced* 9–24; 25–33

PROGRAM RESOURCES
Reteach 9.1 Practice 9.1 Enrichment 9.1
Math Fun • It's All in the Alphabet, p. 182

PROBLEM OF THE DAY

Transparency 9

Not only letters but also whole words can have lines of symmetry. Write at least five words and draw their lines of symmetry.

DECIDED TOT MAAM EXCEED

Some other words are *HAH, HUH, TAT, TUT, MUM, DICE, CODE, DECODE, DID, DEED, BID, BODE, BOX, EXCEEDED, BEECH.*

Solution Tab A, p. A9

QUICK CHECK
Find the product.

1. $\frac{1}{4} \times 360$ 90
2. $\frac{1}{9} \times 360$ 40
3. $\frac{1}{2} \times 360$ 180
4. $\frac{1}{3} \times 360$ 120
5. $\frac{1}{8} \times 360$ 45
6. $\frac{1}{6} \times 360$ 60

1 MOTIVATE

Use this activity to review congruent angles. Draw these line segments on the board.

- **Is $\overline{KL}$ congruent to $\overline{PQ}$?** no
- **How do you know?** because $\overline{PQ}$ is longer than $\overline{KL}$

WHAT YOU'LL LEARN
How to identify line symmetry and rotational symmetry

WHY LEARN THIS?
To recognize and make symmetric and congruent designs as used in art and clothing

WORD POWER
line symmetry
rotational symmetry

REMEMBER:
Regular polygons have all sides congruent and all angles congruent.
See page H21.

Symmetry and Congruence

Symmetry is an interesting part of geometry that can be found all around you. You can find many examples of symmetry in nature and in manufactured objects.

ACTIVITY

WHAT YOU'LL NEED: paper, scissors, dark crayon

- Fold the paper in half. Use the crayon to write your name in cursive along the fold line.

- Fold the paper along the fold line so your name is inside. Use the handle of the scissors to make a rubbing of your name.

- Unfold the paper. Your name appears on the other half of the paper.

The design you made has line symmetry. A figure has **line symmetry** if it can be folded or reflected so that the two parts of the figure match, or are congruent.

- Where is the line of symmetry on the design you made above? **horizontally across the middle**
- How many lines of symmetry does the design have? **one**

Some figures have several lines of symmetry.

EXAMPLE 1 Find all the lines of symmetry in the regular pentagon.

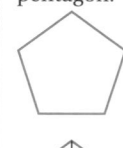

Trace the figure and cut it out.

Fold it in half in different ways.

If the halves match, then the fold is a line of symmetry.

Count the different lines of symmetry.

The figure has five lines of symmetry.

- Do all pentagons have five lines of symmetry? Explain. **No; only regular pentagons have five lines of symmetry.**

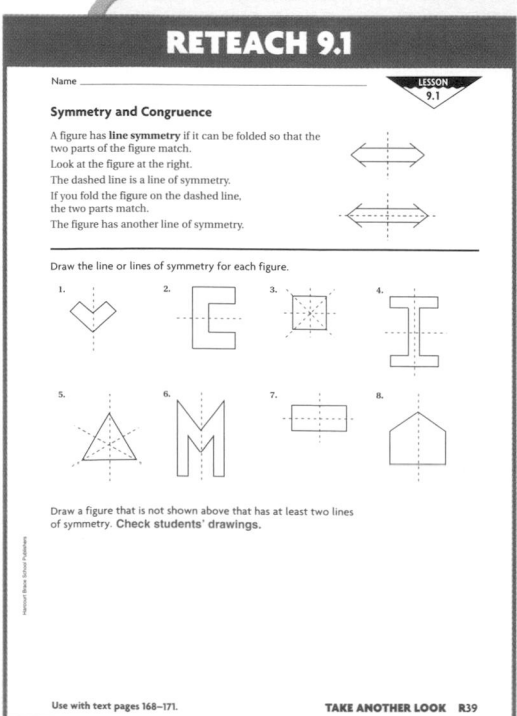

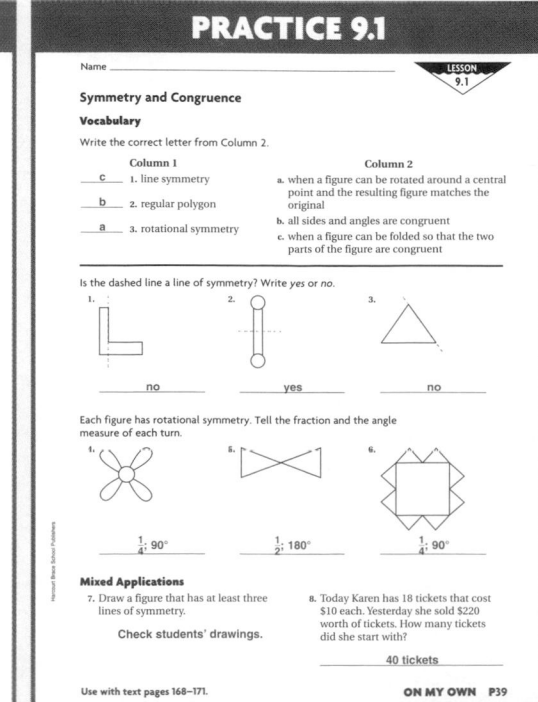

If a figure has line symmetry, its two halves are congruent, or the same size and same shape, when the figure is folded.

EXAMPLE 2 Look at the figure. Are the halves on either side of the line congruent?

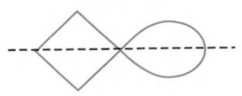

Trace the figure.
Fold along the dashed line.
The two halves are the same size and shape.

So, the two halves are congruent.

GUIDED PRACTICE

Trace the figure. Draw the lines of symmetry.
Check students' drawings.

1.    2. 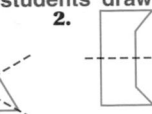   3. 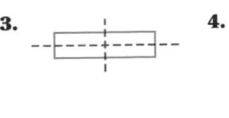   4.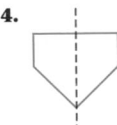

Is the dashed line a line of symmetry? Write *yes* or *no*.

5. 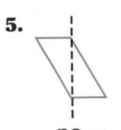   6.    7. 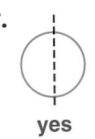   8.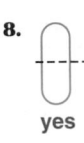

no   no   yes   yes

Rotational Symmetry

Another type of symmetry involves turning the figure instead of folding. A figure has **rotational symmetry** when it can be rotated less than 360° around a central point, or point of rotation and still match the original figure.

The figure at the right has rotational symmetry. If you turn it $\frac{1}{4}$ turn, or 90°, around the point of rotation, it will match the original figure.

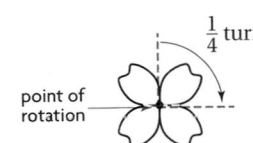

$\frac{1}{4}$ turn, or 90°

point of rotation

- 💡 **CRITICAL THINKING** How many times would you have to turn the figure 90° to get it back to its original position? **4 times**

169

- Is $\overline{ST}$ congruent to $\overline{KL}$? yes
- **How do you know?** because they are the same length

Draw these angles on the board.

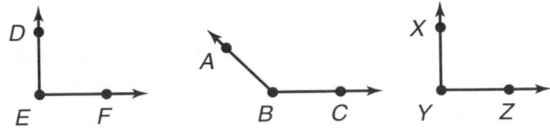

- Is ∠**ABC** congruent to ∠**DEF**? no
- **How do you know?** because ∠ABC is greater than ∠DEF
- Is ∠**XYZ** congruent to ∠**DEF**? yes
- **How do you know?** because the angles are the same size

Explain to students that in this lesson they will learn about congruency in other figures.

2 TEACH/PRACTICE

Students begin this lesson by making a design with one line of symmetry. Have them work in pairs or small groups as they explore figures with more than one line of symmetry in Examples 1 and 2.

COMMON ERROR ALERT
Some students may incorrectly assume that all diagonal lines are lines of symmetry. Have them experiment with a variety of figures to disprove this idea.

GUIDED PRACTICE
Have students work with a partner to trace the figures and practice drawing lines of symmetry in these exercises.

ADDITIONAL EXAMPLES

Example 1, p. 168

Find all the lines of symmetry in this figure.

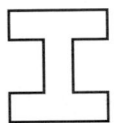 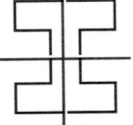

The figure has 2 lines of symmetry.

Example 2, p. 169

Look at the figure. Are the halves on both sides of the line congruent? yes

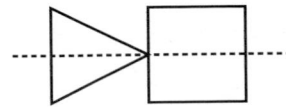

TEACHING TIP

Have students explore rotational symmetry with the figures on TR p. R29 before beginning the Independent Practice exercises. Students should follow these steps:

1. Cut out the figures.
2. Mark a point of rotation on each figure.
3. Put one figure on a sheet of paper.
4. Place your pencil point at the point of rotation and turn the figure until it looks like the original figure.
5. Mark dashed lines, points, and an arrow to show where the rotation begins and ends. (See Example 3 in your book.)

COMMON ERROR ALERT

If some students have difficulty determining the fraction and the angle measure of each turn, have them follow these steps:

1. Write the number of turns needed for a 360° rotation as the denominator.
2. Write 1 as the numerator.
3. Multiply the fraction by 360°.

INDEPENDENT PRACTICE

To assess students' understanding of the concepts, check students' drawings for Exercises 1–4 before assigning the remaining exercises.

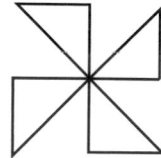
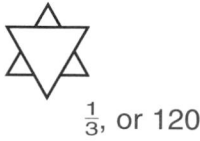
EXAMPLE 3 Does the figure below have rotational symmetry?

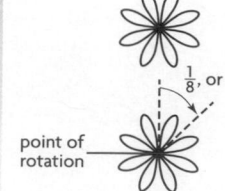

point of rotation

$\frac{1}{8}$, or 45°

Trace the figure.

Place your pencil point at the point of rotation.

Rotate the figure until it looks like the original figure.

The figure has rotational symmetry.

EXAMPLE 4 The figure below has rotational symmetry. What are the fraction and the angle measure of each turn?

Trace the figure.

Place your pencil point in the center of the figure.

$\frac{1}{2}$, or 180°

Rotate the figure until it looks like the original figure.

Identify the fraction and the angle measure of the turn.

So, the figure matches itself when turned $\frac{1}{2}$ turn, or 180°.

- What if you rotated the figure $\frac{1}{4}$ turn? What would it look like? Would it match the original figure?

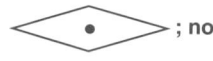 ; no

Real-Life Link

When you were younger, did you ever use a Spirograph® and a colored pencil to draw "flowers"? A Spirograph® works on the principle of rotational symmetry. It guides you to repeat a figure after a rotation of a certain number of degrees. If you repeat a figure 6 times around a full turn, how many degrees is each rotation? 60°

INDEPENDENT PRACTICE

Trace the figure. Draw all lines of symmetry. **Check students' drawings.**

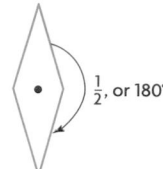

1.   2.   3.   4.

IDEA FILE

ART CONNECTION

Most people have heard that no two snowflakes are alike. Wilson Bently was so intrigued by the variety of patterns in snowflakes that he took photographs of more than 5,000 snowflake patterns.

- Have students make their own snowflake patterns and share with the class the rotational symmetry of their patterns.

Modalities Kinesthetic, Visual

Is the dashed line a line of symmetry? Write *yes* or *no*.

5. 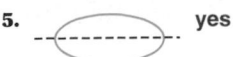 yes   6.  yes   7.  no   8.  yes

9.  no   10.  no   11.  yes   12.  yes

Tell whether each figure has rotational symmetry. Write *yes* or *no*.

13.  yes   14. yes   15. no   16. yes

Each figure has rotational symmetry. Tell the fraction and angle measure of each turn.

17.  $\frac{1}{2}$, 180°   18. $\frac{1}{4}$, 90°   19.  $\frac{1}{2}$, 180°   20.  $\frac{1}{8}$, 45°

Problem-Solving Applications

21. Draw a figure that has at least two lines of symmetry. **Check students' drawings.**

22. Does a circle have line symmetry? rotational symmetry? Explain. **Yes; yes; diameters divide it equally; any turn around the center matches the original.**

23. **CRITICAL THINKING** Brenda has a square cake. She cuts it into pieces along all the lines of symmetry of the square. How many pieces does she have? **8 pieces**

24. **WRITE ABOUT IT** Why must a rotation be less than 360° to show rotational symmetry? **Any figure will look like the original if turned 360°.** Portfolio

Mixed Review

Trace the letter. Then flip it over from top to bottom and draw the new figure.

25. A ∀   26. T ⊥   27. ∩ U   28. M W

Identify the type of lines.

29. ⟷ parallel   30. ✕ intersecting   31. ⊕ perpendicular and intersecting

32. Two hamsters use the exercise wheel 9 times in one day. Hamster A uses it twice as often as Hamster B. How many times does each use it? **Hamster A: 6; Hamster B: 3**

33. Rolanda needs enough punch for 20 people to have 2 c each. She knows that 8 fl oz = 1 c, and 4 c = 1 qt. How many quarts of punch should she buy? **10 qt**

MORE PRACTICE Lesson 9.1, page H54

171

CRITICAL THINKING Can a figure have line symmetry but not rotational symmetry? Give an example. Yes. Possible answer:

Can a figure have rotational symmetry but not line symmetry? no

3 WRAP UP

Use these questions to help students focus on the big ideas of the lesson.

• **How do you know whether a figure has line symmetry?** Possible answer: if the figure can be folded so that the two parts are congruent

✓ Assessment Tip

Have students write the answer.

• **How do you know whether a figure has rotational symmetry?** Possible answer: if the a figure can be rotated less than 360° around a central point and match the original figure

DAILY QUIZ

Tell whether the dashed line is a line of symmetry. Write *yes* or *no*.

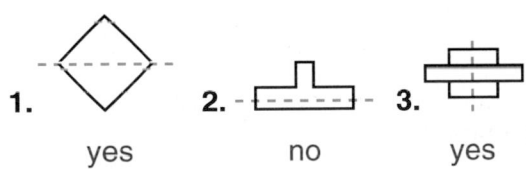

1. yes   2. no   3. yes

Tell whether each figure has rotational symmetry. Write *yes* or *no*.

4. no   5. yes   6. no

171

ORGANIZER

OBJECTIVE To use tangrams to build symmetric figures

MATERIALS *For each student* tangram pieces

PROGRAM RESOURCES
E-Lab • Activity 9
E-Lab Recording Sheets, p. 9
Math Fun • Puzzle-Grams, p. 182

USING THE PAGE

Introduce students to tangrams with a class discussion.

• **Have you ever put a jigsaw puzzle together?**

• **How many pictures can you make from one jigsaw puzzle?**

You can make thousands of pictures from just one tangram puzzle. Ever since tangrams were introduced in China years ago, they have been a popular leisure-time activity for people around the world.

Explore

Activity 1

As students begin to assemble the puzzle, point out that the pieces of the original tangram can be placed in any position to create the new shape.

Activity 2

Observe students as they assemble the puzzle and draw lines of symmetry. If they are not sure of the lines of symmetry, have them fold the shape so that the two parts are congruent.

Activity 3

Encourage students to be creative as they use all the tangram pieces to build a shape. Discuss the different possibilities of shapes: polygons, animals, people, buildings, airplanes, and others.

C○○PERATIVE LEARNING

Symmetry and Tangrams

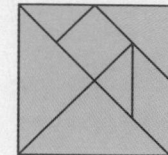

LAB *Activity*

WHAT YOU'LL EXPLORE
How to use tangrams to build symmetric figures

WHAT YOU'LL NEED
tangram pieces

Many people enjoy solving puzzles. The tangram is a puzzle made of seven polygons. The seven pieces are cut from a square. The square is not the only shape that can be made with the tangram pieces.

Work with a partner to do the following activities.

Explore

ACTIVITY 1

• Use the tangram pieces to build the shape at the right.

• Trace the outline of the shape you made.

• Draw the line or lines of symmetry. How many lines of symmetry are there? **one**

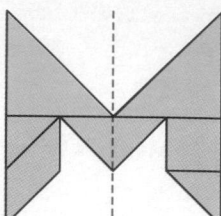

ACTIVITY 2

• Use five tangram pieces to build the shape at the right.

• Trace the outline of the shape you made.

• Draw the line or lines of symmetry. How many lines of symmetry are there? **two**

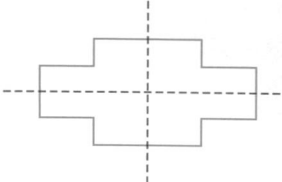

Possible arrangement:

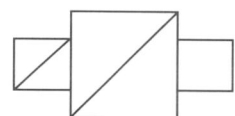

172　Chapter 9

IDEA FILE

HISTORY CONNECTION

No one knows who invented tangrams. However, we do know that Chinese tangrams were made of ivory, bone, or wood, inlaid with decorative carving. In Europe, tangrams were made on sets of cards decorated with pictures of the things the tangrams were meant to represent.

EXTENSION

In the nineteenth century, people wanted to solve tangram puzzles that had curved shapes, so they created circular tangrams. Circular tangrams are made up of two pieces from one circle and five pieces from another circle.

• Have students make shapes from the circular tangram shown below and identify which shapes have rotational or line symmetry.

Modalities Kinesthetic, Visual

MEETING INDIVIDUAL NEEDS

ACTIVITY 3

- Use all seven of the tangram pieces to build another shape with line symmetry.
- Trace the outline of your shape, and exchange with your partner. Build your partner's shape with tangram pieces.
- Trace the new shape you built, and draw the line or lines of symmetry.

Think and Discuss

- Which of the tangram pieces are congruent? **two large triangles; two small triangles**
- Can the two small triangles be arranged to form a shape congruent to another piece? Which piece? **yes; either of the two quadrilaterals, and the medium triangle**
- Which of the tangram pieces has more than one line of symmetry? How many lines of symmetry are there? **the square; four**
- Which of the tangram pieces have rotational symmetry? **both quadrilaterals**
- Do any of the shapes you built have rotational symmetry? If so, which ones? **Answers will vary.**

Try This

Use more than one tangram piece to make each shape. Trace the outline of each shape you make, and draw the line of symmetry. If a shape has no lines of symmetry, write *none*. **Make sure students' shapes have more than one piece.**

1.
2.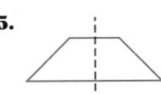
3.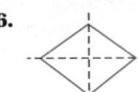

4.
   none
5.
6.

Use two or more tangram pieces to make each shape. Tell whether the shape has *line symmetry, rotational symmetry,* or *both.*

7.

   line symmetry
8.

   both
9.

   line symmetry

E-Lab • Activity 9 Available on CD-ROM and the Internet at http://www.hbschool.com/elab

Note: Students transform figures. Use E-Lab p. 9.

173

Think and Discuss
Discuss each question. Have students use tangram pieces to model their answers.

Try This
When students finish each set of exercises, have them explain and demonstrate their solutions.

USING E-LAB
Students use visual thinking and reasoning skills as they translate, rotate, and reflect a plane figure to a position generated by the computer until the figures match.

✔ *Assessment Tip*

Students write the answer.

Use these questions to help students communicate the big ideas of the lesson.

- **How do you know that the shape you made in Exercise 1 has line symmetry?** Possible answer: because it has two parts that are congruent

- **What are the fraction and the angle measure of the turn needed to make the shape in Exercise 8 look like the original shape?** $\frac{1}{2}$, 180°

LANGUAGE ARTS CONNECTION

The origin of the word *tangram* is unknown, but some believe it is a combination of the word *Tang*, the name of a great Chinese dynasty, and the Greek suffix *-gram*, which means "writing or drawing." Others believe it is a variation of the obsolete English word *tangram*, which means "puzzle." The Chinese call the tangram "the board of wisdom."

GRADE 5	GRADE 6	GRADE 7
Draw translations, rotations, and reflections on a coordinate grid.	Use translations, rotations, and reflections to transform geometric shapes.	Identify and draw translations, rotations, and reflections.

ORGANIZER

OBJECTIVE To use translations, rotations, and reflections to transform geometric shapes

VOCABULARY *New* **transformation, translation, rotation, reflection**

ASSIGNMENTS Independent Practice
- *Basic* 1–16, even; 17–20; 21–30
- *Average* 1–20; 21–30
- *Advanced* 5–20; 21–30

PROGRAM RESOURCES
Reteach 9.2 Practice 9.2 Enrichment 9.2
Extension • Transformations on a
Coordinate Plane, TE pp. 178A–178B

PROBLEM OF THE DAY

Transparency 9

Rashan and two classmates each moved one of these letters. Louis did not flip or slide his letter. Meg did not use reflection to move her letter. Which letter did each move and what transformation did each use? Louis turned the letter *P* (rotation). Meg slid the letter *G* (translation). So Rashan flipped the letter *F* (reflection).

Solution Tab A, p. A9

QUICK CHECK

Tell whether each figure has rotational symmetry. Write *yes* or *no*.

1. yes

2. no

3. yes

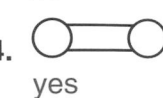
4. yes

1 MOTIVATE

Use this activity to assess students' understanding of transformations in a real-life setting. Sketch these pictures on the board.

9.2

WHAT YOU'LL LEARN
How to use translations, rotations, and reflections to transform geometric shapes

WHY LEARN THIS?
To solve jigsaw puzzles or cut out patterns from fabric for clothing

WORD POWER
transformation
translation
rotation
reflection

teen times

Have you ever heard the old expression "The mirror doesn't lie"? It means that the way you see yourself in a mirror is the way others see you. The next time you look at your reflection in a mirror, raise your right hand and wave. The reflected "you" will be waving its left hand.

Transformations

When you turn a figure around a point of rotation, it doesn't change in size or shape. A movement that doesn't change the size or shape of a figure is a rigid **transformation**. Three types of these transformations are described below.

A **translation** is the movement of a figure along a straight line. Only the location of the figure changes.

Turning a figure around a point is called a **rotation**. Both the position and the location of the figure change.

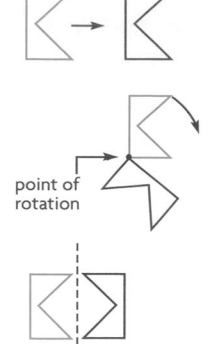

point of rotation

Flipping a figure over a line is called a **reflection**. Both the position and the location of the figure change.

line of reflection

EXAMPLE 1 Tell whether the transformation is a *translation*, a *rotation*, or a *reflection*.

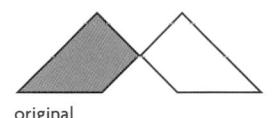

original quadrilateral

Trace the original quadrilateral on the left.

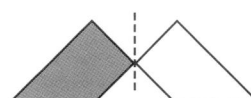

Move your tracing by translating, rotating, or reflecting it to determine how it was moved.

This transformation of the figure is a reflection.

• What position would the original quadrilateral have after being transformed with a translation?

Possible answer:

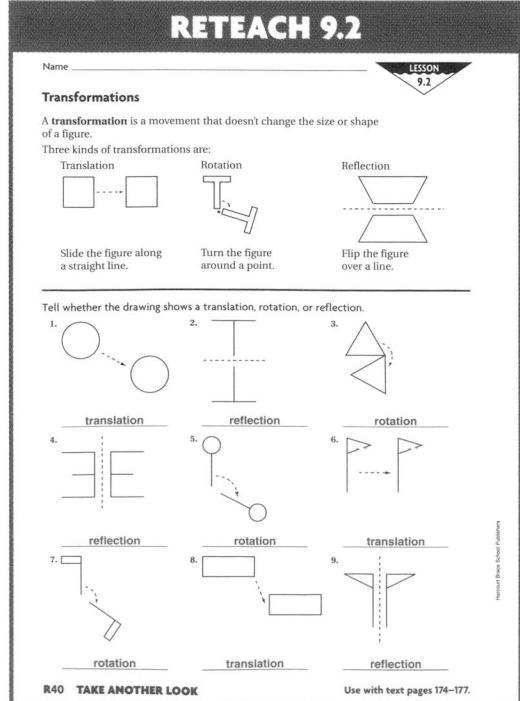
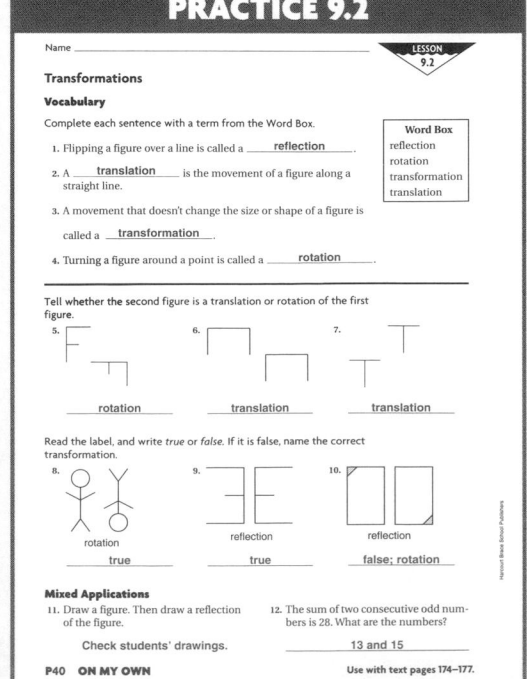

A point of rotation can be on or outside a figure.

EXAMPLE 2 Draw a 90° clockwise rotation of the figure at the right about the point shown.

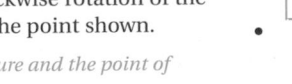

Trace the figure and the point of rotation.

Place your pencil on the point of rotation.

Rotate the figure clockwise 90°.

Trace the figure in its new location.

• How would the rotation look if you turned the figure 180° clockwise? **Check students' work.**

Talk About It

• Look at Examples 1 and 2. Is the new figure similar or congruent to the original figure? Explain. **Congruent; the size and shape of the figure did not change.**

• 💡 **CRITICAL THINKING** If a figure is rotated 360°, how does its new position compare to its old position? **They are the same.**

GUIDED PRACTICE

Tell which type of transformation the second figure is of the first figure. Write *translation, rotation,* or *reflection.*

1.  **2.**   **3.**  **4.**

rotation   translation   translation   rotation or reflection

Trace the figure, line, and point. Draw a reflection about the line and then rotate that figure 90° clockwise.   **Check students' drawings.**

5.  **6.**  **7.**  **8.**

Read the label, and write *true* or *false.*

9. **10.** **11.** **12.**

translation   rotation   reflection   translation
false   true   false   true

175

   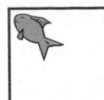

Frame 1 Frame 2 Frame 3 Frame 4

Tell students that the pictures represent individual frames in a movie. When they are combined with other frames and run through a projector, the image of a fish swimming through water will appear on the screen.

• **What type of transformation describes how the picture of the fish changes from one frame to the next?** from Frame 1 to Frame 2: slide or translation; from Frame 2 to Frame 3: turn or rotation; from Frame 3 to Frame 4: flip or reflection

2 TEACH/PRACTICE

Discuss the examples of transformations at the top of page 174. Have students give real-life examples of each type of transformation. Possible answers—translation: a baseball player sliding into first base; rotation—a Ferris wheel going around; reflection—the image you see in a car's rearview mirror

GUIDED PRACTICE

Complete Exercises 1–4 as a class to make sure students understand the different types of transformations.

ADDITIONAL EXAMPLES

Example 1, p. 174

Tell whether the transformation is a *translation,* a *rotation,* or a *reflection.*

original figure

The transformation of the figure is a reflection.

Example 2, p. 175

Draw a 45° clockwise rotation of the figure shown.

INDEPENDENT PRACTICE

To assess students' understanding of the concepts, check answers to Exercises 1–4 before assigning the remaining exercises.

ADDITIONAL EXAMPLE

Example 3, p. 176

What moves were made to transform the figure into each new position?

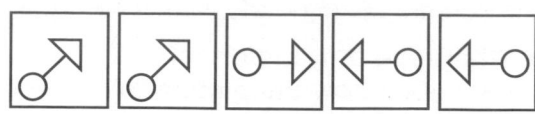

start   first   second   third   fourth

The transformations made were a translation, a rotation, a reflection, and a translation.

You can use more than one transformation to change a figure's position and location. Patterns in material, wallpaper, and draperies are examples of figures being transformed more than one way.

EXAMPLE 3 What moves were made to transform the block letter *E* into each new position?

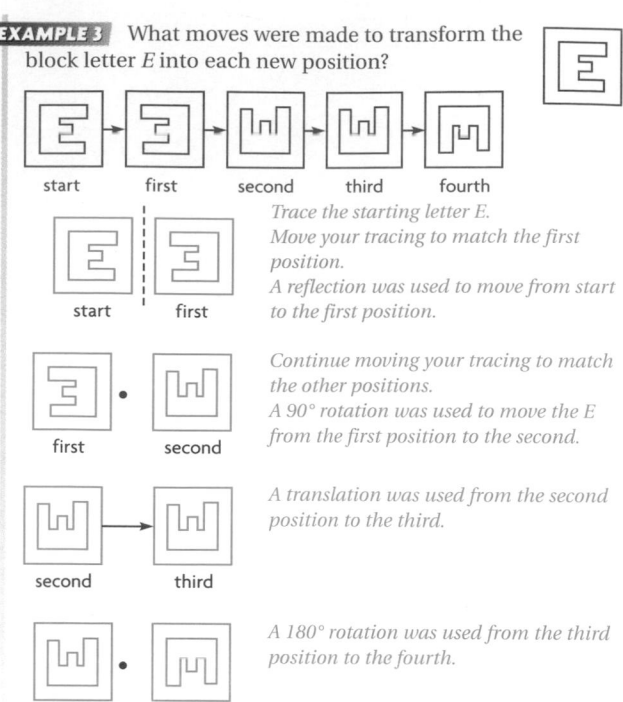

start   first   second   third   fourth

Trace the starting letter E. Move your tracing to match the first position.
A reflection was used to move from start to the first position.

start   first

Continue moving your tracing to match the other positions.
A 90° rotation was used to move the E from the first position to the second.

first   second

A translation was used from the second position to the third.

second   third

A 180° rotation was used from the third position to the fourth.

third   fourth

So, the transformations made were a reflection, a rotation, a translation, and a rotation.

Real-Life Link

You know that the mirror image of an object—the reflection of an object in a mirror—is reversed left to right from the actual object. The mirror image of the word *Ambulance* is written on the front of an ambulance so the word can be read in the rear-view mirror of a car. Choose any word, and write its mirror image.

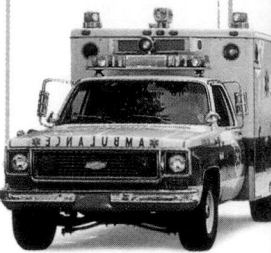

Answer should be a mirror image of a word chosen by the student.

INDEPENDENT PRACTICE

Tell whether the second figure is a *translation* or a *rotation* of the first figure.

1. B B
translation

2. △ ▷
rotation

3. ☐ ◇
rotation

4. 6 translation
6

IDEA FILE

CAREER CONNECTION

Gift-Wrap Designer Gift-wrap designers use transformations to create a wide range of gift-wrap designs. Have students use transformations to design their own gift wrap. Brainstorm design ideas with the class. Some students may want their design to reflect a special occasion or a favorite sport or hobby. Others may want to create a design for a special friend or family member.

Modalities Kinesthetic, Visual

Trace the figure, the line, and point. Draw a reflection about the line and then rotate that figure 90° clockwise. **Check students' drawings.**

5.
6.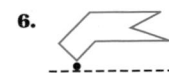
7.
8.

Read the label, and write *true* or *false*. If it is false, name the correct transformation.

9.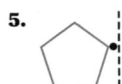
reflection
false; rotation

10.
reflection **true**

11. rotation **true**

12.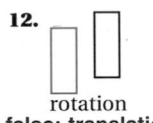
rotation
false; translation

Tell what moves were made to transform each figure into its next position.

13.
rotation, translation, reflection or rotation

14. 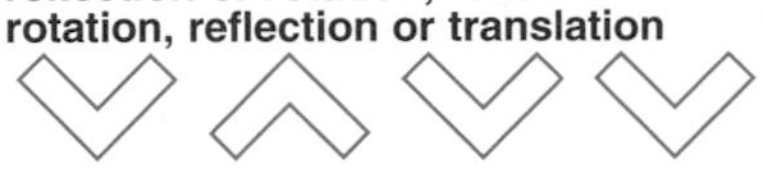 **reflection or rotation, reflection or rotation, reflection or translation**

15.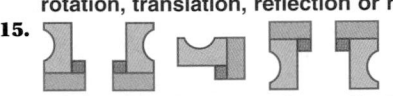
reflection, rotation, rotation, reflection

16. 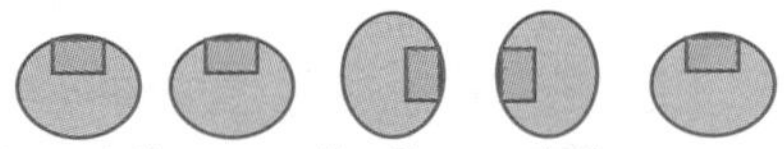 **translation or reflection, rotation, reflection or rotation, rotation**

Problem-Solving Applications

17. **LOGICAL REASONING** Simone was looking in the mirror. She blinked her left eye. Which eye did her reflection blink? **right**

18. Draw a figure. Then draw a translation, a rotation, and a reflection of the figure. **Check students' drawings.**

19. Reflect the word *bob* vertically. What word appears? **pop**

20. ⬛ **WRITE ABOUT IT** Can a reflection also be a rotation? Explain. **Yes; some figures can rotate halfway and form a reflection.**

Mixed Review

Use a protractor to measure the angle. **Approximate angle measures are given.**

21.  90°
22.  120°
23.  45°
24.  120°

Tell whether the angle is *acute, right,* or *obtuse.*

25. 100° **obtuse**
26. 45° **acute**
27. 90° **right**
28. 160° **obtuse**

29. **LOGICAL REASONING** A cat was tangled in 50 in. of yarn. He freed himself from 10 in. with each try. How many tries did he take to get free? **5 tries**

30. Today Katrina has 10 tickets that cost $20 each. Yesterday she sold $400 worth of tickets. How many tickets did she start with? **30 tickets**

MORE PRACTICE Lesson 9.2, page H55

177

WRITING CONNECTION

Math Journal

Challenge students to find as many words as they can that would look the same before and after a reflection. One example is shown below.

M M
O O
W W

Modality Visual

ART CONNECTION

Have students draw transformations of a figure and use the shapes to make a mobile. One example is shown below.

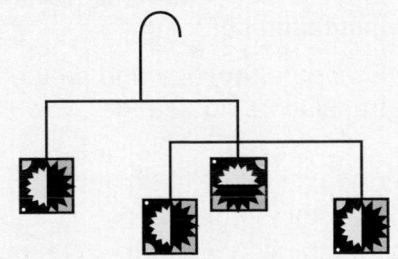

Modalities Kinesthetic, Visual

MEETING INDIVIDUAL NEEDS

💡 **CRITICAL THINKING** How could you move the letters *n* and *d* to look like other letters? Name the transformations and the new letters. *n*: rotation, *u; d*: rotation, *p*; reflection, *b*, or *q*

3 WRAP UP

Use these questions to help students focus on the big ideas of the lesson.

• **Fred designed this stencil. What moves did he make to transform each hockey stick into each new position?** rotation, reflection, rotation

✓ *Assessment Tip*

Have students write the answer.

• **How do you know the figure in Independent Practice Exercise 9 is not a translation?** Possible answer: because in a translation, only the location of the figure changes, not the position

DAILY QUIZ

Read the label and write *true* or *false*. If it is false, name the correct transformation.

1.
rotation
true

2.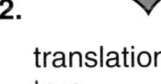
translation
true

3.
reflection
false; rotation or translation

4.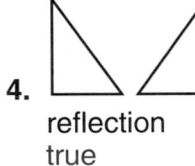
reflection
true

177

EXTENSION

Transformations on a Coordinate Plane

Use these strategies to extend understanding of transformations on a coordinate plane.

Multiple Intelligences Logical/Mathematical, Visual/Spatial

Modalities Auditory, Kinesthetic, Visual

Lesson Connection Students will extend their understanding of transformations.

Materials 0.5 cm coordinate grid, Quadrant 1 TR pp. R61–R62

Program Resources *Intervention and Extension Copying Masters*, p. 16

Challenge

TEACHER NOTES

ACTIVITY 1

Purpose To explore translations on a coordinate plane

Vocabulary Review

- **coordinates**—the numbers that make up an ordered pair

- **x-coordinate**—the first number of an ordered pair

- **y-coordinate**—the second number of an ordered pair

Discuss the following questions.

- How would you describe a translation from point *S* to point *T*? 2 units to the right and 2 units down

- Compare the perimeter and area of rectangle *WXYZ* before and after the translation. no change

WHAT STUDENTS DO

1. The coordinates of point *S* are (1,3). This means you find point *S* like this:

   a. Start at the origin.

   b. Move 1 unit right.

   c. Move 3 units up.

   - What are the coordinates of point *T*? (3,1)

   - Which point has an *x*-coordinate of 3? point *T*

2. On a sheet of graph paper, graph triangle *PQR*:*P* = (1,2), *Q* = (1,4), and *R* = (5,2). Now translate the triangle 4 units to the right. What are the new coordinates of points *P, Q,* and *R*? *P* = (5,2), *Q* = (5,4), *R* = (9,2)

   - For each point, by how much did the *x*-coordinate change? the *y*-coordinate? increased by 4; not at all

   - Compare the size and shape of triangle *PQR* before and after the translation. no change

3. Graph rectangle *WXYZ* on a sheet of graph paper: *W* = (2,2), *X* = (2,4), *Y* = (5,4), and *Z* = (5,2). Translate the rectangle 3 units up.

   - Describe how the translation changed the coordinates of *W, X, Y,* and *Z*. The *x*-coordinates did not change. The *y*-coordinates increased by 3.

   - Compare the size and shape of rectangle *WXYZ* before and after the translation. no change

TEACHER NOTES

ACTIVITY 2

Purpose To explore reflections on a coordinate plane

Discuss the following questions.

- Where is the line of symmetry for the figure consisting of triangle *ABC* and its reflection? the vertical line through (6,0)

- Consider the figure consisting of rectangle *EFGH* and its reflection. How many lines of symmetry does this figure have? 2

- Why does reflection change the orientation of the triangle, but not the orientation of the rectangle? Possible answer: The rectangle is being reflected in a line that is parallel to one of its lines of symmetry. If the rectangle were at an angle to the line of reflection, its orientation would have changed.

Have students meet in small groups or with partners to discuss the given questions. Have one student report the group's responses to the larger group.

WHAT STUDENTS DO

1. Graph triangle *ABC*: *A* = (0,0), *B* = (5,0), and *C* = (5,3). Then draw a vertical line through (6,0). Reflect triangle *ABC* about that line. Draw the new triangle.

   - What are the new coordinates of points *A*, *B*, and *C*? *A* = (12,0), *B* = (7,0), and *C* = (7,3)

   - What do you notice? The size of the triangle does not change; the orientation does.

2. Graph rectangle *EFGH*. *E* = (1,0), *F* = (5,0), *G* = (5,3), and *H* = (1,3). Draw a horizontal line through (0,4). Reflect the rectangle about that line.

   - What are the new coordinates of points *E*, *F*, *G*, and *H*? *E* = (1,8), *F* = (5,8), *G* = (5,5), and *H* = (1,5)

   - Does the rectangle's size or shape change when it is reflected? no

Share Results

- How do the coordinates change when a figure is translated to the right? The *x*-coordinates increase.

- Discuss the effects of translation on the physical properties of a figure: size, shape, location, and so on. Only the location changes. Size and shape do not change.

- How are reflection and translation in the coordinate plane similar? How are they different? Reflection and translation can both change the location of a figure, but not its size and shape. Reflection can also change the orientation of a figure.

Intervention and Extension Copying Masters, p. 16

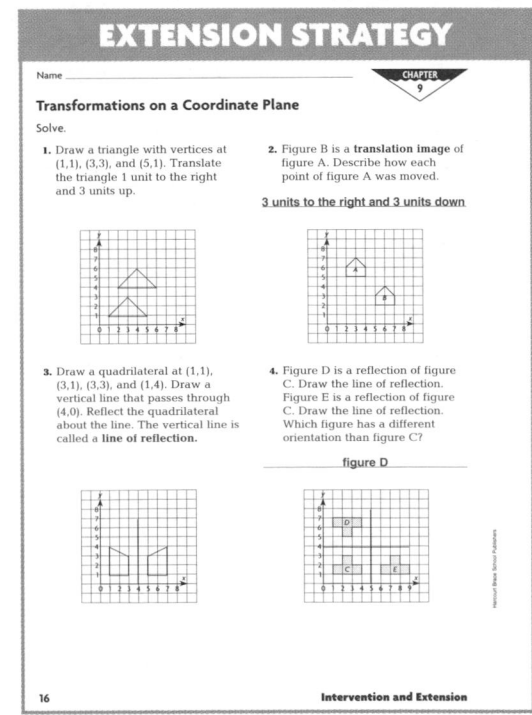

ORGANIZER

OBJECTIVE To use polygons to make tessellations

VOCABULARY *New* **tessellation**

ASSIGNMENTS Independent Practice
- ■ *Basic* 1–17
- ■ *Average* 1–17
- ■ *Advanced* 1–17

PROGRAM RESOURCES
Reteach 9.3 Practice 9.3 Enrichment 9.3
Math Fun • Alphabet Tessellations, p. 182
Cosmic Geometry • Tessellation Creation Station

PROBLEM OF THE DAY

  Transparency 9

Which term does not belong with the others?

1. intersecting, perpendicular, protractor, parallel protractor
2. rotation, translation, reflection, tessellation tessellation
3. acute, vertex, obtuse, right vertex
4. straight, scalene, equilateral, isosceles straight

Solution Tab A, p. A9

QUICK CHECK

Name the transformation. Write *translation, rotation,* or *reflection.*

1. H ⊥ rotation
2. ZƧ reflection
3. V∧ rotation
4. (figure) translation

1 MOTIVATE

MATERIALS: *For each student* protractor

Review with students how to measure angles. Copy on the board the angles shown.

9.3

WHAT YOU'LL LEARN
How to use polygons to make a tessellation

WHY LEARN THIS?
To understand which polygons will form tessellations so you can use them in art and construction projects

WORD POWER
tessellation

Art Link

The term *tessellation* is from the Latin word *tessella,* "a small square tile used in Roman mosaics." M. C. Escher, a modern Dutch artist, used tessellations in his work. Some of his art uses two or even three different repeating patterns. In the example below, how many different patterns form a tessellation? **three**

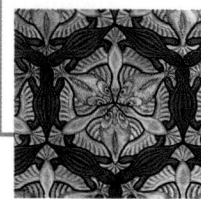

Tessellations

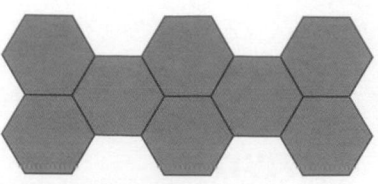

The design at the right can be seen in nature and in manufactured objects. Where have you seen this design before?
Possible answers: honeycomb, tile designs
A repeating arrangement of shapes that completely covers a plane, with no gaps and no overlaps, is called a **tessellation**. A hexagon was used in the tessellation above.

ACTIVITY

WHAT YOU'LL NEED: pattern blocks, colored pencils or markers, protractor

Work with a partner to make a tessellation.

- Choose a pattern block shape to use for your tessellation.
- Design your tessellation. Remember that the shapes must fit together without overlapping or leaving gaps.
- Record your tessellation. Color it to make a pleasing design.
- Choose a vertex inside your design. Measure each of the angles around the vertex. What is the sum of the angle measures? **360°**

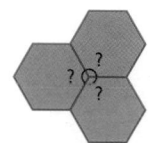

Talk About It

- Will each of the six pattern block shapes form a tessellation? Explain. **Yes; any of these blocks of the same shape can be arranged without gaps or overlaps.**
- Look at the designs you and your classmates made. What is the sum of the angles where vertices meet? **360°**

GUIDED PRACTICE

Trace each polygon. Then cut out several of each shape. Tell whether the polygon forms a tessellation. Write *yes* or *no.*

1. △ yes 2. □ yes 3. ⯃ no 4. ⬠ no

IDEA FILE

INTERDISCIPLINARY CONNECTION

One suggestion for an interdisciplinary unit for this chapter is Geometric Patterns in Art. In this unit you can have students

- find photographs of mosaics and identify the tessellation patterns used to create the designs.
- research the works of M.C. Escher, the artist who made tessellation a popular form of art.

Modality Visual

RETEACH 9.3

Name _____

LESSON 9.3

Tessellations

An arrangement of shapes that completely covers a plane, with no gaps and no overlaps, is called a **tessellation**.
The design at the right shows a tessellation.
There are no gaps or overlaps.
The shape used to make the tessellation is a square.
There are four angles at the vertex in the middle of the design.
The measure of each angle is 90°. The sum of the angle measures is 360°.

Trace each shape. Then cut out several of each shape. Tell whether the shape forms a tessellation. Write *yes* or *no.*

1. yes 2. no 3. yes
4. no 5. yes 6. yes

Use with text pages 178–179.   **TAKE ANOTHER LOOK** R41

INDEPENDENT PRACTICE

Trace the polygon several times, and cut out your tracings. Tell whether the polygon forms a tessellation. Write *yes* or *no*.

1.  yes
2.  yes
3. no
4.  yes

Find the measures of the angles that surround the circled vertex. Then find the sum of the measures.

5. 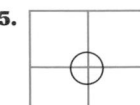 each angle: 90°; sum: 360°

6. each angle: 60°; sum: 360°

7. each angle: 90°; sum: 360°

Problem-Solving Applications

For Problems 8–10, use the figure at the right.

8. What polygons are at the circled vertex? **triangles and a hexagon**
9. What is the sum of the measures of the angles at the circled vertex? **360°**

10. Trace the design. Use the design to make a tessellation by connecting several more of the same pattern. **Check students' work.**

For Problems 11–13, use the figure at the right.

11. Which polygons are at the circled vertex? **two squares, a triangle, and a hexagon**
12. What is the sum of the measures of the angles at the circled vertex? **360°**

13. Trace the design. Then continue the pattern. Does it form a tessellation? **no**

14. Draw a design of your own that forms a tessellation. **Check students' drawings.**
15. Use two equilateral triangles to form a diamond. Then draw a tessellation using diamonds and half-diamonds. **Check students' drawings.**

16. **CRITICAL THINKING** Can any parallelogram be used to make a tessellation? Explain. **Yes; the sum of the measures of the four angles is 360°.**
17. **WRITE ABOUT IT** Explain how you know when shapes form a tessellation. **They cover a plane with no gaps or overlaps, and the angles at each vertex have a sum of 360°.**

vertex

vertex

Technology **Link**

In *Mighty Math Cosmic Geometry,* you can practice making tessellations in the *Tessellation Station.*

MORE PRACTICE Lesson 9.3, page H55

179

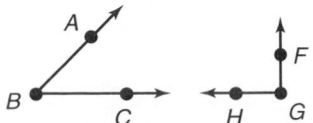

• **Which scale will you use to measure** ∠*ABC*? **Why?** the bottom scale; because ∠*ABC* opens on the right

2 TEACH/PRACTICE

Encourage students to use what they know about transformations to make tessellations.

GUIDED PRACTICE

Have students work with a partner to determine which of the polygons in Exercises 1–4 forms a tessellation.

INDEPENDENT PRACTICE

To assess students' understanding of the concepts, check students' work for Exercises 1–4 before assigning the remaining exercises.

3 WRAP UP

Use these questions to help students focus on the big ideas of the lesson.

• **What is a tessellation?** an arrangement of shapes that covers a plane, with no gaps or overlaps

✔ *Assessment Tip*

Have students write the answer.

• **How do you know that the polygon in Independent Practice Exercise 3 does not form a tessellation?** There are gaps.

DAILY QUIZ

Draw the polygon, trace it several times, and cut out your tracings. Tell whether the polygon forms a tessellation. Write *yes* or *no*.

1. yes
2. yes

Find the measures of the circled vertex. Then find the sum of the measures.

3. each angle: 90° sum: 360°
4. each angle: 45° sum: 360°

179

ORGANIZER

OBJECTIVE To use the strategy *make a model* to solve problems that involve tessellations

ASSIGNMENTS Practice/Mixed Applications
- *Basic* 1–10
- *Average* 1–10
- *Advanced* 1–10

PROGRAM RESOURCES
Reteach 9.4 Practice 9.4 Enrichment 9.4
Problem-Solving Think Along, TR p. R1

PROBLEM OF THE DAY

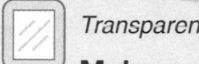 *Transparency 9*

Make each statement true by writing *All* or *Some*.

1. __?__ equilateral triangles are isosceles triangles. **All**
2. __?__ isosceles triangles have a right angle. **Some**
3. __?__ scalene triangles have an obtuse angle. **All**
4. __?__ triangles with all acute angles are equilateral triangles. **Some**
5. __?__ triangles have at least two acute angles. **All**

Solution Tab A, p. A9

QUICK CHECK

NOTE: Sketch these drawings on the board or put them on an overhead transparency. Tell what moves were made to transform each figure into each new position. Write *translation, rotation,* or *reflection*.

original

1.
reflection or rotation, reflection or rotation, rotation

original

2.
translation, rotation, reflection

180

Making a Model

WHAT YOU'LL LEARN
How to use the strategy *make a model* to solve problems that involve tessellations

WHY LEARN THIS?
To solve problems that involve woodworking, decorating, or craft projects

Problem Solving
- Understand
- Plan
- Solve
- Look Back

Nikki is designing a tile mosaic for the top of her table. The shape Nikki uses must tessellate a plane. She wants to use the shape at the right. Can she use this shape for her design?

You can make a model to solve problems like this.

☑ **UNDERSTAND** What are you asked to find?
whether the shape she wants to use will tessellate a plane
What information is given?
the shape she wants to use

☑ **PLAN** What strategy will you use?

You can use the strategy *make a model*. Trace the shape Nikki wants to use. Cut out several copies of the shape. Use the paper shapes to see whether the shape will tessellate a plane.

☑ **SOLVE** How will you solve the problem?

Begin by moving the shapes around to see if they fit together. The pieces must not overlap or leave any gaps when you place them together.

You may also check to see if the shapes will cluster around a point. If they do, check to see if the cluster can be drawn again and again. If it can, the shape will tessellate a plane.

☑ **LOOK BACK** If the figures cluster around a point, the sum of the measures of the angles is 360°. What is the sum of the measures of the angles of the figure Nikki is using? **360°**

What if . . . Nikki wants to use right triangles to make her mosaic? Will a right triangle tessellate a plane? Make a model to test your ideas. **Yes; check students' models.**

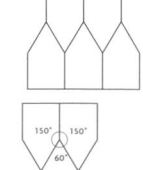

180 Chapter 9

IDEA FILE

CAREER CONNECTION

Book Cover Designer Present this problem to students: Marion is designing a cover for a computer-graphics book. She wants her design to be computer graphics of patterns that tessellate.

- Have students draw a design she could use.

Modalities Kinesthetic, Visual

RETEACH 9.4

Name _____

LESSON 9.4

Problem-Solving Strategy

Making a Model

Larry is designing a tile mosaic for a floor. The shape he uses must tessellate a plane. He wants to use the shape at the right. Can he use this shape for his design?

To solve this problem, you can make a model.

Trace and cut out several of the shapes Larry wants to use. Use the cutouts to see if the shape will tessellate a plane.

Move the shapes around to see if they fit together without gaps or overlaps.

Check that the sum of the angle measures around a point is 360°.
The shape Larry wants to use will tessellate the plane.

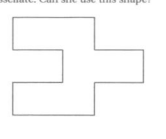

150° 150°
60°

Make a model to solve.

1. Erika is making a design from the shape below. She wants the design to tessellate. Can she use this shape?

2. For a tabletop design, Shawn uses octagons and squares. Will his design tessellate a plane?

yes

yes

R42 TAKE ANOTHER LOOK   Use with text pages 180–181.

PRACTICE

Make a model to solve.

1. Denise is making a design from the shape below. She wants the design to tessellate a plane. Can she use this figure? **yes**

2. Will this shape tessellate a plane? Explain. **Yes; copies of the shape fit together without gaps or overlaps.**

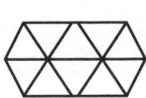

3. For a tile design, Bill used squares and equilateral triangles. Will his design tessellate a plane? **yes**

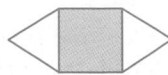

4. Draw a figure that does not tessellate a plane. **Check students' drawings.**

MIXED APPLICATIONS

CHOOSE

Choose a strategy and solve. **Choices of strategies will vary.**

> **Problem-Solving Strategies**
> - Work Backward
> - Guess and Check
> - Write an Equation
> - Use a Formula
> - Make a Model
> - Act It Out

5. Colin sat to the left of Gregory. Gregory sat across from June. Heather did not sit next to June, but next to Howie, who sat across from Colin. Did Heather sit to the right or left of Howie? **Heather is left of Howie.**

6. Maria spent $40.00 at the mall. She bought a present for $12.45, a concert ticket for $15.95, and two hats that were the same price. How much did each hat cost? **$5.80**

7. Mr. Rodriguez bought a washing machine for $450. He made a down payment of $130 and paid the rest in equal payments of $80 a month. How many months did it take to pay for the washing machine? **4 months**

8. Pedro is stocking toys on the shelves in the storeroom. The shelves are 0.5 m apart, and the bottom shelf is 1.5 m from the floor. There are 6 shelves. How far from the floor is the top shelf? **4 m**

9. Marissa wants to use stones shaped like the one below to make a walk around her garden. She wants to make sure the shape will tessellate. Can she use this shape? **yes**

10. ✏ **WRITE ABOUT IT** Explain how you can decide whether a figure will tessellate a plane. **Make a model to see if there are any gaps or if pieces overlap.**

MORE PRACTICE Lesson 9.4, page H55

181

---

1 MOTIVATE

Review with students patterns that show tessellations.

- **Which of these patterns tessellate? How do you know?**

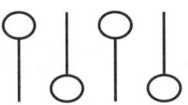

the first pattern; because the arrangement has no gaps or overlaps

2 TEACH/PRACTICE

Discuss the problem and four-step procedure on page 180.

PRACTICE

Have students work with a partner to complete Problem 1. Instruct them to use the Problem-Solving Think Along, TR p. R1 as they work through the problem.

MIXED APPLICATIONS

Emphasize that more than one strategy can be used to solve some problems.

3 WRAP UP

Use these questions to help students focus on the big ideas of the lesson.

- **How do you know that the shape in Problem 1 tessellates?** It fits together without gaps or overlaps.

✔ *Assessment Tip*

Have students write the answer.

- **How do you know the figure you drew for Problem 4 does not tessellate?** When the shapes are moved around, they overlap.

DAILY QUIZ

Make a model to solve.

1. Which bricks could you use to build a wall in which the bricks tessellate? a

a.   b.

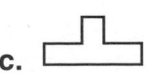

c.

2. Which floor-tile pattern tessellates? b

181

USING THE PAGE

To provide additional practice activities related to Chapter 9: *Symmetry and Transformations*

IT'S ALL IN THE ALPHABET

MATERIALS: *For each group* sets of 2-in. letters, construction paper

Have students work in pairs or small groups to complete this activity.

In this activity, students determine whether letters of the alphabet have line symmetry or rotational symmetry.

Multiple Intelligences Visual/Spatial, Logical/Mathematical, Verbal/Linguistic, Bodily/Kinesthetic, Interpersonal/Social, Intrapersonal/Introspective
Modalities Auditory, Kinesthetic, Visual

PUZZLE-GRAMS

MATERIALS: *For each pair* tangram patterns, tagboard or construction paper, scissors, straightedge

Have students work in pairs to complete this activity.

In this activity, students use translations, rotations, and reflections to make shapes.

Multiple Intelligences Visual/Spatial, Logical/Mathematical, Bodily/Kinesthetic, Interpersonal/Social, Intrapersonal/Introspective
Modalities Kinesthetic, Visual

ALPHABET TESSELLATIONS

MATERIALS: *For each student* tagboard, construction paper, scissors, straightedge, colored pencils

Have students work individually to complete this activity.

In this activity, students make tessellating designs from letters of the alphabet.

Multiple Intelligences Visual/Spatial, Logical/Mathematical, Interpersonal/Social, Intrapersonal/Introspective
Modality Visual

MATH FUN!

Portfolio | IT'S ALL IN THE ALPHABET

PURPOSE To practice identifying symmetry (pages 168–171)
YOU WILL NEED set of 2-in. letters, construction paper

A B C D E F G H I J K L M N O P Q R S T U V W X Y Z

Work with a partner or team. Divide your construction paper into two sections. Label the sections:

1. line symmetry
2. rotational symmetry

Under each section, draw the letters that have the given symmetry.

Test your answers. Cut each letter out. Fold it in half different ways. Do they have the same size and shape? Try rotating the letter until it looks like itself again.

PUZZLE-GRAMS

PURPOSE To practice using translations, rotations, and reflections (pages 172–177)
YOU WILL NEED tangram pattern, tagboard or construction paper

 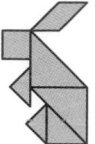 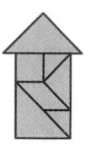

Use the tangram pattern to make your own set of tangram pieces. Decide how many polygons will be in your puzzle-gram. Use translations, rotations, and reflections of your pieces to make a new shape such as a house, a person, or an animal. Trace the shape. Give the outline and the puzzle pieces you used to a classmate to solve.

 Take your new puzzle-gram home. Challenge your family to create new puzzles.

ALPHABET TESSELLATIONS

PURPOSE To practice making tessellations out of letters (pages 178–179)
YOU WILL NEED tagboard, construction paper, scissors, straightedge, colored pencils

Look at the alphabet at the top of the page. Can you find a letter that will tessellate? Try out your idea. Redraw your letter on the tagboard. Cut out the letter and trace a tessellating design. Then look at your tessellation pattern. See how the shapes look at different angles. Color them.

182 Math Fun!

CHAPTER 9 REVIEW

EXAMPLES

- **Identify line symmetry and rotational symmetry.** (pages 168–171)

line symmetry   rotational symmetry

- **Transform geometric shapes.** (pages 174–177)

  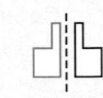

translation   rotation   reflection

- **Use polygons to make a tessellation.** (pages 178–179)

An octagon will not tessellate.

A right triangle will tessellate.

- **Use the strategy *make a model* to solve problems.** (pages 180–181)

PROBLEM-SOLVING TIP: For help in solving problems by making a model, see pages 4 and 180.

Yes; check students' models.

EXERCISES

1. **VOCABULARY** A figure has __?__ if it can be folded or reflected so that the two parts of the figure match or are congruent. **line symmetry**

Trace the figure. Draw the lines of symmetry.

2. 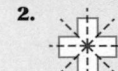   3.    4.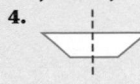

The figure has rotational symmetry. Identify the fraction and the angle measure of the turn.

5.  $\frac{1}{4}$, or 90°   6. $\frac{1}{2}$, or 180°

7. **VOCABULARY** A __?__ results when a figure is flipped over a line. **reflection**

Tell whether the second figure is a *reflection*, *translation*, or *rotation* of the first.

8. 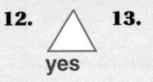   9. 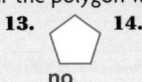   10.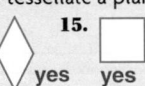
   rotation   reflection   rotation

11. **VOCABULARY** A __?__ is an arrangement of shapes that completely covers a plane with no gaps and no overlaps. **tessellation**

Tell whether the polygon will tessellate a plane.

12. yes   13. no   14. yes   15. yes

16. Tim wants to make a quilt by using quadrilaterals. Can he use any kind of quadrilateral? Make a model to test your answer. **Yes; check students' models.**

17. Will the shape tessellate a plane? Make a model to test your answer.

Chapter 9 Review **183**

USING THE PAGE

The Chapter 9 Review reviews identifying line symmetry and rotational symmetry; transforming geometric shapes; using polygons to make a tessellation; using the strategy *make a model* to solve problems. Chapter objectives are provided, along with examples and practice exercises for each.

The Chapter 9 Review can be used as a review or a test. You may wish to place the Review in students' portfolios.

Assessment Checkpoint

For Performance Assessment in this chapter, see page 204, What Did I Learn?, items 2 and 5.

✔ Assessment Tip

Student Self-Assessment The How Did I Do? Survey helps students assess both what they have learned and how they learned it. Self-assessment helps students learn more about their own capabilities and develop confidence in themselves.

A self-assessment survey is available as a copying master in the *Teacher's Guide for Assessment*, p. 15.

CHAPTER 9 TEST, page 1

CHAPTER 9 TEST, page 2

See *Test Copying Masters*, pp. A33–A34 for the Chapter Test in **multiple-choice** format and pp. B183–B184 for **free-response** format.

183

PLANNING GUIDE

BIG IDEA: Geometric figures in three-dimensions can be described, identified, classified, and constructed.

Introducing the Chapter p. 184		Team-Up Project • Box Patterns p. 185	
OBJECTIVE	**VOCABULARY**	**MATERIALS**	**RESOURCES**
10.1 Solid Figures pp. 186-189 **Objective:** To identify solid figures **1 DAY**	polyhedron base lateral face vertex	FOR EACH GROUP: solid figures	■ Reteach, ■ Practice, ■ Enrichment 10.1; More Practice, p. H56 Cultural Connection • Turkish Foods, p. 200
10.2 Faces, Edges, and Vertices pp. 190-191 **Objective:** To name and count the faces, edges, and vertices of prisms and pyramids **1 DAY**	vertex edge face		■ Reteach, ■ Practice, ■ Enrichment 10.2; More Practice, p. H56 ◉ *Calculating Crew •* *Dr. Gee's 3-D Lab*
10.3 Building Solids pp. 192-193 **Objective:** To build models of prisms from nets **1 DAY**	net	FOR EACH STUDENT: 4-in. x 6-in. index card, inch ruler, scissors, tape	■ Reteach, ■ Practice, ■ Enrichment 10.3; More Practice, p. H56
10.4 Two-Dimensional Views of Solids pp. 194-195 **Objective:** To draw and identify different views of a solid **1 DAY**		FOR EACH STUDENT: cylinder	■ Reteach, ■ Practice, ■ Enrichment 10.4; More Practice, p. H57 Cultural Connection • Turkish Foods, p. 200
Lab Activity: Different Views of Solids pp. 196-197 **Objective:** To learn how different views of a solid figure differ		FOR EACH STUDENT: centimeter cubes, centimeter graph paper	🖥 **E-Lab • Activity 10** E-Lab Recording Sheets, p. 10 Cultural Connection • Turkish Foods, p. 200
EXTENSION • Perspective Drawings, Teacher's Edition pp. 198A–198B			
10.5 Problem-Solving Strategy: Solve a Simpler Problem pp. 198-199 **Objective:** To solve a difficult problem by looking at a related simpler problem **1 DAY**			■ Reteach, ■ Practice, ■ Enrichment 10.5; More Practice, p. H57 Problem-Solving Think Along, TR p. R1

CHAPTER ASSESSMENT   Chapter 10 Review   p. 201

ASSESSMENT CHECKPOINT   Review • Chapter 8–10   pp. 202–203
Performance Assessment • Chapter 8–10   p. 204
Cumulative Review • Chapter 8–10   p. 205

NCTM CURRICULUM
STANDARDS FOR GRADES 5–8

- ☑ **Problem Solving**
  Lessons 10.1, 10.2, 10.3, 10.4, 10.5
- ☑ **Communication**
  Lessons 10.1, 10.2, 10.3, 10.4, 10.5
- ☑ **Reasoning**
  Lessons 10.1, 10.2, 10.3, 10.4, 10.5
- ☑ **Connections**
  Lessons 10.1, 10.2, 10.3, 10.4, 10.5
- ☑ **Number and Number Relationships**
  Lesson 10.5
- ☑ **Number Systems and Number Theory**
  Lesson 10.5
- ☑ **Computation and Estimation**
  Lessons 10.2, 10.3, 10.5
- ☑ **Patterns and Functions**
  Lessons 10.3, 10.4
- ☑ **Algebra**
  Lesson 10.5
- ☐ **Statistics**
- ☐ **Probability**
- ☑ **Geometry**
  Lessons 10.1, 10.2, 10.3, 10.4, 10.5
- ☑ **Measurement**
  Lesson 10.3

ASSESSMENT OPTIONS

CHAPTER LEARNING GOALS

10A.1 To identify solid figures and their parts

10A.2 To identify nets for solid figures and different points of view

10A.3 To solve problems by using the strategy *solving a simpler problem*

These goals for concepts, skills, and problem solving are assessed in many ways throughout the chapter. See the chart below for a complete listing of both traditional and informal assessment options.

Pretest Options
Pretest for Chapter Content, TCM pp. A35–A36, B185–B186
Assessing Prior Knowledge, TE p. 184; APK p. 12

Daily Assessment
Mixed Review, PE pp. 189, 193
Problem of the Day, TE pp. 186, 190, 192, 194, 198
Assessment Tip and Daily Quiz, TE pp. 189, 191, 193, 195, 197, 199

Formal Assessment
Chapter 10 Review, PE p. 201
Chapter 10 Test, TCM pp. A35–A36, B185–B186
Study Guide & Review, PE pp. 202–203
Test • Chapters 8–10, TCM pp. A37–A38, B187–B188
Cumulative Review, PE p. 205
Cumulative Test, TCM pp. A39–A42, B189–B192

Performance Assessment
Team-Up Project Checklist, TE p. 185
What Did I Learn?, Chapters 8–10, TE p. 204
Interview/Task Test, PA p. 85

Portfolio
Suggested work samples:
Independent Practice, PE pp. 188, 191, 193, 195
Cultural Connection, PE p. 200

Student Self-Assessment
How Did I Do?, TGA p. 15
Portfolio Evaluation Form, TGA p. 24
Math Journal, TE pp. 184B, 188

Key
Assessing Prior Knowledge Copying Masters: APK
Test Copying Masters: TCM
Performance Assessment: PA
Teacher's Guide for Assessment: TGA
Teacher's Edition: TE
Pupil's Edition: PE

MATH COMMUNICATION

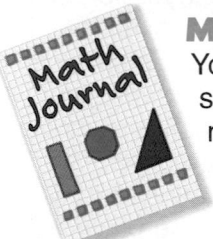

MATH JOURNAL
You may wish to use the following suggestions to have students write about the mathematics they are learning in this chapter:

PE pages 186, 190 • Talk About It

PE pages 189, 191, 193, 195, 199 • Write About It

TE page 188 • Writing in Mathematics

TGA page 15 • How Did I Do?

In every lesson • record written answers to each Wrap-up Assessment Tip from the Teacher's Edition for test preparation.

VOCABULARY DEVELOPMENT
The terms listed at the right will be introduced or reinforced in this chapter. The boldfaced terms in the list are new vocabulary terms in the chapter.

So that students understand the terms rather than just memorize definitions, have them construct their own definitions and pictorial representations and record those in their Math Journals.

polyhedron, pp. 186, 187
base, pp. 186, 187, 198
lateral face, pp. 186, 187
vertex, pp. 186, 187, 198
edge, pp. 190, 198
face, pp. 190, 192
net, p. 192

MEETING INDIVIDUAL NEEDS

RETEACH
BELOW-LEVEL STUDENTS

Practice Activities, TE Tab B
Problems for Products
(Use with Lessons 10.1–10.5.)

Interdisciplinary Connection, TE p. 189

Literature Link, TE p. 184C
The Phantom Tollbooth, Activity 1
(Use with Lesson 10.2.)

Technology
Calculating Crew • *Dr. Gee's 3D Lab*
(Use with Lesson 10.2.)

E-Lab • Activity 10
(Use with Lab Activity.)

PRACTICE
ON-LEVEL STUDENTS

Practice Game, TE p. 184C
Match Up
(Use with Lessons 10.1–10.4.)

Interdisciplinary Connection, TE p. 189

Literature Link, TE p. 184C
The Phantom Tollbooth, Activity 2
(Use with Lesson 10.2.)

Technology
Calculating Crew • *Dr. Gee's 3D Lab*
(Use with Lesson 10.2)

E-Lab • Activity 10
(Use with Lab Activity.)

CHAPTER IDEA FILE

The activities on these pages provide additional practice and reinforcement of key chapter concepts. The chart above offers suggestions for their use.

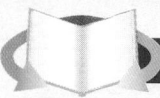

LITERATURE LINK

The Phantom Tollbooth
by Norton Juster

Milo travels through a toll booth where he meets characters who teach him valuable lessons. One of the places he visits is Digitopolis, where some inhabitants have many faces on their heads.

Activity 1—Have students tell the name of the character with four triangular faces. Then have them name four moods that might be used with these faces.

Activity 2—Milo describes Dodecahedron as somewhat like a cube that's had all its corners cut off. Have students think about how many faces Dodecahedron would have if he were really shaped this way, and what he would be named.

Activity 3—Some drawings of Dodecahedron show him as a twelve-sided regular polyhedron in which all of the edges and angles are identical. Have students determine the shapes of Dodecahedron's faces and whether or not there are other regular polyhedrons with faces the same shape.

PRACTICE GAME

Match Up

Purpose To practice identifying solid figures

Materials game cards (TR p. R87)

About the Game In small groups, players shuffle the game cards and lay them out in 4 rows of 5. A player turns up 2 cards. If the 2 cards match a picture of a solid to its name, the player keeps the cards and takes another turn. If the cards do not match, the player turns the cards face down, and the next player takes a turn. Play continues until players have matched all the cards. The player with the most pairs wins.

Modalities
Kinesthetic, Visual

Multiple Intelligences
Verbal/Linguistic,
Visual/Spatial,
Bodily/Kinesthetic

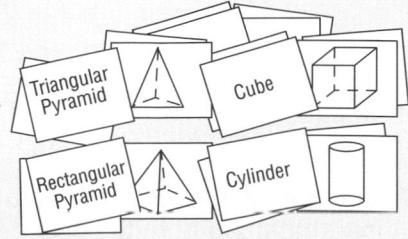

Interdisciplinary Connection, TE p. 189

Literature Link, TE p. 184C
The Phantom Tollbooth, Activity 3
(Use with Lesson 10.2.)

Technology
Calculating Crew • *Dr. Gee's 3D Lab*
(Use with Lesson 10.2.)

E-Lab • Activity 10
(Use with Lab Activity.)

The purpose of using technology in the chapter is to provide reinforcement of skills and support in concept development.

E-Lab • Activity 10—Students explore the relationship between solids and cross-sectional views by slicing solids.

Calculating Crew • *Dr. Gee's 3D Lab*—The activities on this CD-ROM provide students with opportunities to practice identifying solids based on their faces, vertices, and edges.

INTERDISCIPLINARY SUGGESTIONS

Connecting Solid Figures

Purpose To connect solid figures to other subjects students are studying

You may wish to connect **Solid Figures** to other subjects by using related interdisciplinary units.

Suggested Units

- The Seven Wonders of the Ancient World
- Record-Breaking Kids
- Writing Connection: Graphic Organizers
- Algebra Connection: Who Discovered Formulas?
- Art Connection: Silhouettes
- Career Connection: Masonry

Modalities Auditory, Kinesthetic, Visual

Multiple Intelligences Mathematical/ Logical, Bodily/Kinesthetic, Visual/Spatial

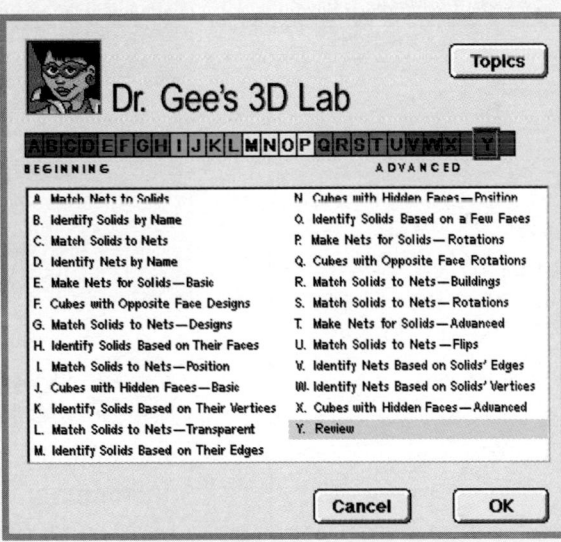

Each Calculating Crew activity can be assigned by degree of difficulty. See Teacher's Edition page 191 for Grow Slide information.

Visit our web site for additional ideas and activities.
http://www.hbschool.com

CHAPTER 10

Introducing the Chapter

Encourage students to talk about solid figures. Review the math content students will learn in the chapter.

WHY LEARN THIS?
The question "Why are you learning this?" will be answered in every lesson. Sometimes there is a direct application to everyday experiences. Other lessons help students understand why these skills are needed for future success in mathematics.

PRETEST OPTIONS

Pretest for Chapter Content See the *Test Copying Masters*, pages A35–A36, B185–B186 for a free-response or multiple-choice test that can be used as a pretest.

Assessing Prior Knowledge Use the copying master shown below to assess previously taught skills that are critical for success in this chapter.

Assessing Prior Knowledge Copying Masters, p. 12

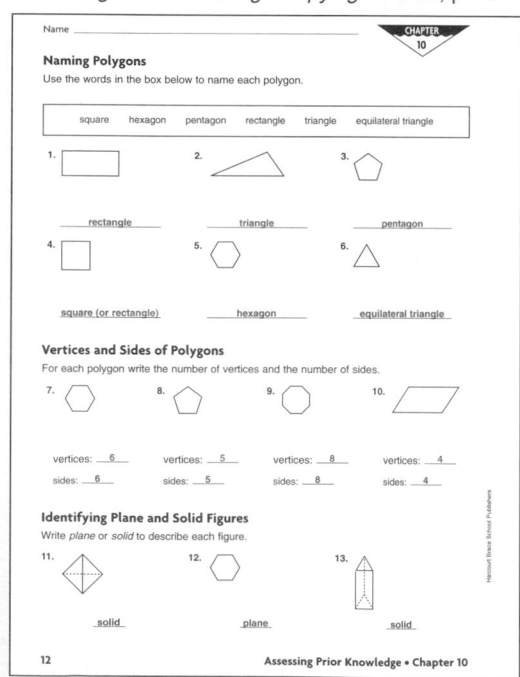

Troubleshooting

If students need help	Then use
• IDENTIFYING POLYGONS	• Key Skills, PE p. H21
• FINDING THE NUMBER OF SIDES AND VERTICES	• Key Skills, PE p. H23
• IDENTIFYING PLANE AND SOLID FIGURES	• Key Skills, PE p. H22

184

CHAPTER 10 SOLID FIGURES

LOOK AHEAD

In this chapter you will solve problems that involve

• classifying solid figures

• naming and counting edges, faces, and vertices in solid figures
• identifying views of solid figures
• drawing two-dimensional views of solid figures

184   Chapter 10

SCHOOL-HOME CONNECTION

You may wish to send to each student's family the *Math at Home* page. It includes an article that stresses the importance of communication about mathematics in the home, as well as helpful hints about homework and taking tests. The activity and extension provide extra practice that develops concepts and skills involving solid figures.

MATH AT HOME

Solid Figures: The Added Dimension

Your child is beginning a chapter that deals with classification of solid figures. Students will name and count edges, faces, and vertices of figures, and will identify different views. With your child, look for examples of three-dimensional shapes around your house. Discuss how the shapes are alike and how they are different, as well as characteristics of the shapes.

Here is an activity for you and your child that focuses on constructing cubes of different sizes. In the Extension you will build different rectangular prisms.

Cube It!

For this activity you will need a collection of small cubes, such as sugar cubes. One cube will stand for the smallest cube you can build using the sugar cubes. Your job is to figure out how many sugar cubes are needed to build the next larger cube. Once you have constructed the second cube, try to build the next largest size cube possible using the sugar cubes. What characteristics must the cubes you build have?

Extension • Box It!

Take 36 sugar cubes. Using all 36 cubes, build a rectangular prism (the shape of a box). Write down its measurements. Then use the same cubes to build a different rectangular prism, again using all 36. Write down its measurements also. See if you can do this at least two more times.

For Your Information

Spatial visualization, the ability to "see" a shape and relationships among shapes even if the figures are not present, is an important part of geometry. Children can improve their spatial visualization ability by building and changing shapes both flat and solid. Parents can help by providing their children with building materials such as Legos™, Construx™, K'nex™, and Tinkertoys™. These materials allow children to explore relationships and discover characteristics and properties of shapes and figures.

Remind your child that when asked for a written explanation, he or she should answer in complete sentences and be sure all parts of the question are answered.

VOCABULARY

polygon—a closed figure formed by three or more line segments that are polygons
polyhedron—a three-dimensional figure in which all the surfaces are flat and polygons
lateral face—a face that is not a base of a polyhedron
prism—a three-dimensional figure with two parallel and congruent faces that are polygons
pyramid—a three-dimensional figure with triangular faces and one base that is a polygon

ANSWER: Cube It! The number of sugar cubes needed always equals the length cubed (23 or 6 cubes total, 33 or 9 cubes total).

10

Math at Home

Team-Up Project

Box Patterns

Find out how a flat pattern is used to make a box.

YOU WILL NEED: cardboard box (any size), scissors

Plan
- Work with a partner. Locate an empty box that can be torn apart.

Act
- Measure the length, width, and height of the box. Record the dimensions.
- Figure out where the box is taped or glued together and cut or pull it apart.
- Record the measurements for each section of the flattened box.
- Make a poster to show the pattern of the box and the dimensions for each section. On the reverse side of the poster make a three-dimensional drawing of the assembled box and give its measurements.

Share
- Present your poster to the class.

DID YOU

- ☑ Measure the length, width, and height of the box?
- ☑ Find the measurements for each section of the flattened box?
- ☑ Make a poster to show the pattern of the box and the three-dimensional shape of the box?

185

Project Checklist

Evaluate whether partners

- ☑ work cooperatively.
- ☑ measure accurately.
- ☑ understand how three-dimensional boxes are formed from flat patterns.

Using the Project

Accessing Prior Knowledge making linear measurements, working with two-dimensional shapes

Purpose to see how a flat pattern is used to make a box

GROUPING partners
MATERIALS boxes, rulers

Managing the Activity You may want to provide a variety of boxes for students to disassemble.

WHEN MINUTES COUNT

A shortened version of this activity can be completed in 15–20 minutes. Give students a copy of the net from TR p. R42. Have students determine the dimensions of the rectangular prism.

A BIT EACH DAY

Day 1: Students select partners, locate a box, and measure its length, width, and height.

Day 2: Students observe how the box is constructed and cut or tear it apart and unfold it to discover the flat pattern.

Day 3: Students measure the dimensions of each section and write them on the flattened box.

Day 4: Partners make a poster using their flattened box.

Day 5: The class views the posters and tries to guess the shapes and dimensions of the boxes when they are assembled.

HOME HAPPENINGS

Encourage students to work with a family member to see how many different patterns they can find for constructing a box.

ORGANIZER

OBJECTIVE To identify solid figures
VOCABULARY *New* **polyhedron, base, lateral face, vertex**
ASSIGNMENTS Independent Practice
- *Basic* 1–17, even; 18–21; 22–31
- *Average* 1–21; 22–31
- *Advanced* 10–21; 22–31
PROGRAM RESOURCES
Reteach 10.1 Practice 10.1 Enrichment 10.1
Cultural Connection • Turkish Foods, p. 200

PROBLEM OF THE DAY

 Transparency 10

Complete the comparisons.

1. Sphere is to no flat faces as __?__ is to all flat faces. polyhedron

2. __?__ is to all triangular faces as rectangular prism is to all rectangular faces. Triangular pyramid

3. Square is to __?__ as circle is to cone. square pyramid

4. Circle is to __?__ as triangle is to triangular prism. cylinder

5. Circle is to square as sphere is to __?__. cube

Solution Tab A, p. A10

QUICK CHECK

NOTE: Draw the figures on the board.

Name the figure.

1.
square

2.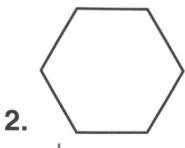
hexagon

3.
circle

4.
pentagon

186

WHAT YOU'LL LEARN
How to identify solid figures

WHY LEARN THIS?
To describe the shapes of real-life objects such as food containers

WORD POWER
polyhedron
lateral face
base
vertex

Solid figures come in many sizes and shapes. What shapes are used most often in packaging food? **Possible answer: boxes and cans** At a grocery store, many items are sold in boxes. Most boxes have flat, rectangular faces.

Talk About It

- Name items likely to be packaged in the four boxes shown. **Possible answer: cereal, cake mix, detergent, pasta**
- How many rectangular faces does each box have? **six rectangular faces**

- Which faces on each box are congruent? **opposite faces**

Another familiar shape in a grocery store is a can.

- Name items likely to be packaged in the four cans shown. **Possible answer: tuna, soup, juice, vegetables**
- Which two parts of each can are congruent? **the circular faces**

- How would you describe the parts that are not circles? **They are curved surfaces.**

Some foods are packaged in other shapes of containers.

- Name a food for each of these shapes. **Possible answer: bread, soda, milk**
- Describe the shape of any package you have seen that is different from any shown here. **Possible answer: perfume bottle**

In geometry many of these shapes are given special names.

A **polyhedron** is a solid figure with flat faces that are polygons.

A prism is a polyhedron with two congruent, parallel bases. Its **lateral faces** are rectangles. A prism is named for the shape of its **bases.**

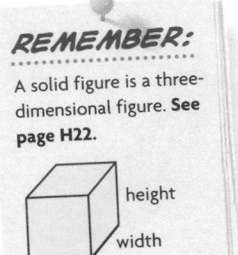

REMEMBER:

A solid figure is a three-dimensional figure. **See page H22.**

height
width
length

186 Chapter 10

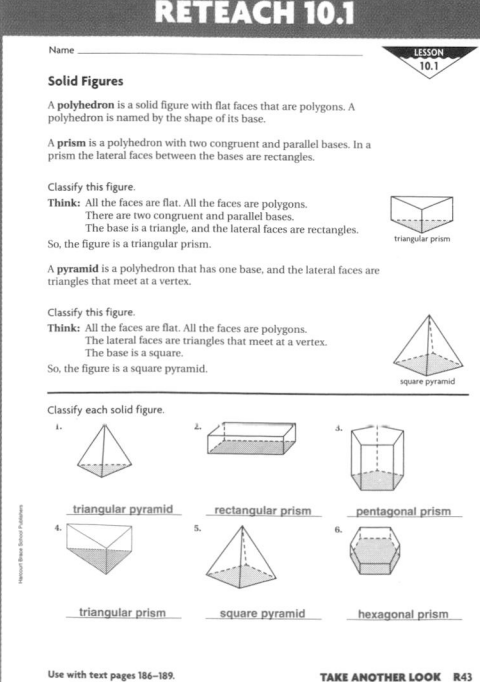

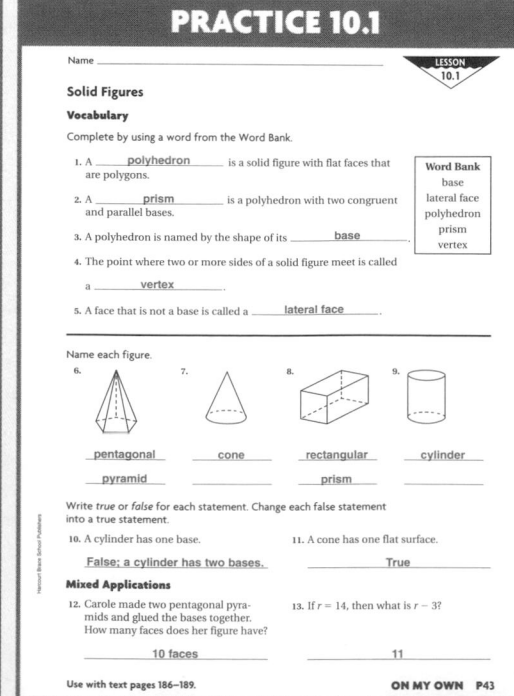

Boxes are usually rectangular prisms that have rectangular bases.
Other polygons can be used as bases as well.

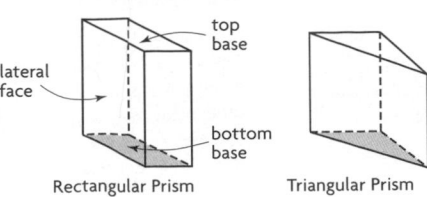

Rectangular Prism   Triangular Prism   Hexagonal Prism

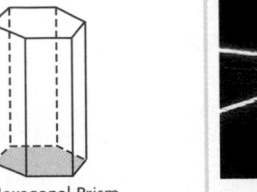

In science the word *prism* refers to a triangular prism that can separate rays of white light into the colors of the visible spectrum. Violet has the shortest wavelength (about 0.0000001 m). Just outside the visible spectrum is ultraviolet, the rays that cause sunburn. Ultraviolet has a wavelength of about 0.00000001 m. Is this wavelength longer or shorter than that of violet? **shorter**

EXAMPLE 1 Classify this solid figure.

All the faces are flat and are polygons, so the figure is a polyhedron.

The rectangular lateral faces indicate that the figure is a prism.

Two of the faces are congruent pentagons, so they must be the bases.

Since the lateral faces are rectangles, the figure is a pentagonal prism.

• What is the name of a prism that has squares as its bases? **square prism**

A cylinder has two flat, circular bases and a curved lateral surface. Most cans are examples of cylinders.

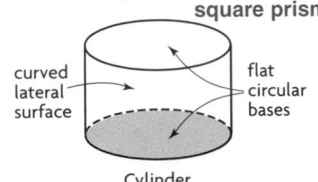

curved lateral surface — flat circular bases

Cylinder

• **CRITICAL THINKING** How are a cylinder and a prism similar? How are they different? **Both have two congruent, parallel bases; the cylinder does not have polygons as faces.**

If you connect a single point on the top base of a cylinder to all the points around the bottom base, you form a cone.

A cone has one flat circular base, a curved lateral surface, and a **vertex.**

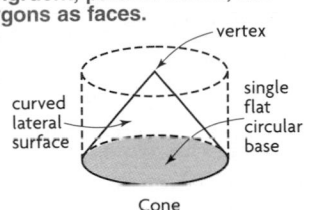

vertex
curved lateral surface — single flat circular base

Cone

• **CRITICAL THINKING** How are a cone and a cylinder similar? How are they different? **Both have a flat circular base and a curved lateral surface; a cone has one base and a vertex, and a cylinder has two bases.**

187

ENRICHMENT 10.1

Name _____

LESSON 10.1

Cross-Figure Puzzle

Use the names of the figures below to complete the puzzle.

Across

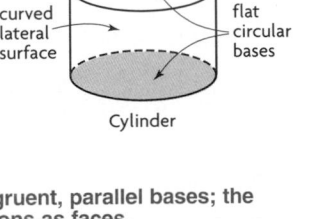

1. P E N T A G O N
   S Q U A R E
2. T R I A N G L E
3. R E C T A N G U L A R P R I S M
4. P Y R A M I D
   C O N E
5.

Down

6.   1.   7.   8.

9. How are the figures shown for 6 Across and 7 Down similar? How are they different?

Both have a circular base and a curved lateral surface;

a cone has 1 base and a vertex.

Use with text pages 186–189.   **STRETCH YOUR THINKING**   E43

MATERIALS: *For each group* solid figures

Organize the class into small groups, and have each group sort and classify solid figures. Have students begin by separating the solid figures into figures with flat surfaces and figures with curved surfaces.

• **What are the figures with flat surfaces called?** polyhedrons
• **What word describes the shape of the faces of the polyhedrons?** *polygonal*
• **What are the figures with curved surfaces called?** cones, cylinders
• **Are their bases polygons?** no
• **What are polyhedrons with two or more rectangular faces and two congruent and parallel bases called?** prisms
• **What are polyhedrons with three or more triangular faces with a common vertex called?** pyramids

Next, have students separate the polyhedrons into figures with two or more rectangular faces and figures with three or more triangular faces.

Begin by reading and discussing page 186, including the definition of *polyhedron*.

• **Which group of containers on page 186 would you classify as polyhedrons?** the boxes in the first group
• **Why wouldn't you classify the other groups of containers as polyhedrons?** because they have curved faces that aren't polygons

COMMON ERROR ALERT If students have difficulty visualizing the hidden faces of the pictures of solid figures in the book, have solid figures available for them to refer to.

ADDITIONAL EXAMPLE

Example, p. 187

NOTE: Draw the figure on the board.

Classify this solid figure. hexagonal prism

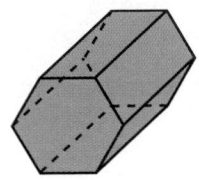

187

TEACHING TIPS

Read and discuss the questions about cylinders and cones on page 186.

- **How many bases does a cylinder have?** 2 bases **How many bases does a cone have?** 1 base
- **What shape are the bases of a cone and cylinder?** circular
- **Describe the lateral surfaces of cones and cylinders.** They are curved.
- **What is unique about a cone?** Its curved surface has a vertex.
- **Are any cones or cylinders polyhedrons?** no

Proceed to page 188 to explore pyramids. Read and discuss Example 2.

- **How many bases does a pyramid have?** 1 base
- **How many lateral faces does a pyramid have?** the same number as the number of sides of its base
- **What shape are the lateral faces of a pyramid?** triangular
- **Are all pyramids polyhedrons?** yes

Have students use the program **Cosmic Geometry** and visit the *Geometry Academy* to learn more about polygons and polyhedrons.

GUIDED PRACTICE

Complete Exercises 1–9 orally as a class. Make sure students understand how to identify solid figures before you assign the Independent Practice.

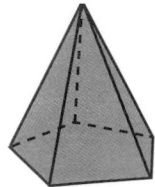
188

A pyramid is related to a prism as a cone is related to a cylinder.

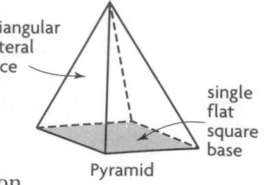

EXAMPLE 2 Classify this solid figure.

All the faces are flat and are polygons, so the figure is a polyhedron.

The triangular lateral faces with a common vertex indicate that it is a pyramid.

The one square face is the base.

The solid figure is a square pyramid.

- What is the name of a pyramid that has a pentagon as its base? **pentagonal pyramid**

REMEMBER:
A pyramid is a solid figure whose one base is a polygon and whose other faces are triangles that have a common vertex. See page H23

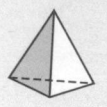

GUIDED PRACTICE

Is the figure a polyhedron? Explain your thinking. **Explanations should state that polyhedrons have flat faces that are polygons.**

1. 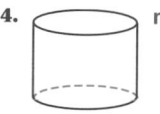 yes 2.  no 3. 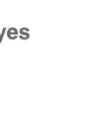 yes

4. 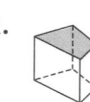 no 5.  yes 6. 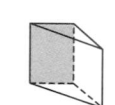 yes

7. Which figures above are prisms? Name them. 1. rectangular prism; 6. hexagonal prism
8. Which figures above are pyramids? Name them. 3. triangular pyramid; 5. square pyramid
9. Name the other figures. 2. cone; 4. cylinder

INDEPENDENT PRACTICE

Write *base* or *lateral face* to identify the shaded face.

1. base 2. base 3. lateral face 4. lateral face

IDEA FILE

WRITING IN MATHEMATICS

Have students make a web that can be used to classify solid figures. You can start them off with the beginning of the web below.

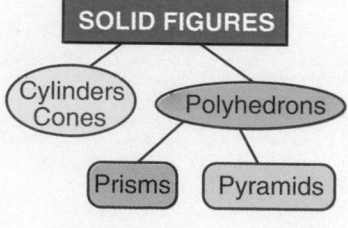

Modality Visual

CAREER CONNECTION

Masonry On May 17, 1996, in a charity event for *Habitat for Humanity*, Travis McGee laid 1,494 bricks in just 60 minutes. That's almost 1 brick every 2 seconds!

- Have students interview masons to find out how they use geometry in their work.

Modalities Auditory, Kinesthetic

MEETING INDIVIDUAL NEEDS

Name each figure.

5.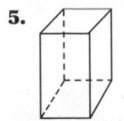
rectangular prism

6.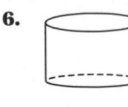
cylinder

7.
triangular prism

8.
pentagonal pyramid

9. 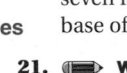 **cone**

Write *true* or *false* for each statement. Change each false statement into a true statement.

10. A cone has two vertices. **False; a cone has one vertex.**

11. A cylinder has two bases. **true**

12. A cone has no flat surfaces. **False; the base of a cone is flat.**

13. The bases of a cylinder are congruent. **true**

14. The lateral surface of a cylinder is curved. **true**

15. A cone can be cut from a solid cylinder. **true**

16. If a cylinder is placed on its lateral surface, it can roll in a straight line. **true**

17. A cylinder is a polyhedron. **False; a cylinder is not a polyhedron.**

Problem-Solving Applications

18. Jerome made a clay paperweight in the shape of a pentagonal prism. How many faces does the paperweight have? **seven faces**

19. Jessica carved a wooden pyramid with seven faces. How many sides does the base of the pyramid have? **six sides**

20. Beth made two square pyramids and glued the congruent bases together to make an ornament. How many faces does her ornament have? **eight faces**

21. **WRITE ABOUT IT** Name three household items that are shaped like rectangular prisms. **Answers will vary.**
Portfolio

Mixed Review

Tell how many sides and how many vertices each polygon has. Then name the polygon.

22.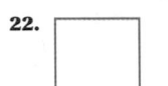
4, 4; square

23.
6, 6; hexagon

24.
3, 3; triangle

25.
4, 4; rectangle

26.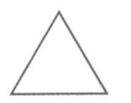
5, 5; pentagon

Tell whether the two figures are similar. Write *yes* or *no*.

27.  yes

28. 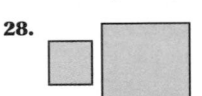 yes

29.  no

30. MENTAL MATH Mark has completed $\frac{2}{3}$ of an 18-week course. How many weeks has he completed? **12 weeks**

31. ALGEBRA If $x = 5$, then what is $x + 6$? **11**

MORE PRACTICE Lesson 10.1, page H56

189

INDEPENDENT PRACTICE

Some students may have difficulty visualizing the shaded faces of the figures pictured in these exercises. Have these students use chalk to mark the shaded faces on corresponding solid figures.

3 WRAP UP

Use these questions to help students focus on the big ideas of the lesson.

- **How are a pyramid and a cone alike?** Each has one base and a vertex.

- **How are they different?** A cone has a circular base and a curved surface; a pyramid has a polygonal base and triangular sides.

✓ *Assessment Tip*

Have students write the answer.

- **How are a prism and a pyramid alike?** Possible answers: Both are solid figures. Both are polyhedrons.

- **How are they different?** Possible answers: A pyramid has triangular lateral faces with a common vertex. A prism has rectangular lateral faces with no common vertex. A pyramid has only one base. A prism has two congruent and parallel bases.

DAILY QUIZ

NOTE: Draw the figures on the board.

Name each figure.

1.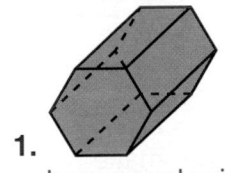
hexagonal prism

2.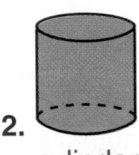
cylinder

3.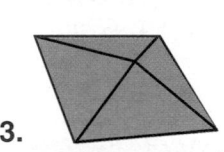
square pyramid

4.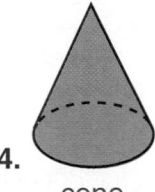
cone

INTERDISCIPLINARY CONNECTION

One suggestion for an interdisciplinary unit for this chapter is Seven Wonders of the Ancient World.

- Have small groups of students research the seven wonders. Have each group choose one and find out more about it.

- Have the groups build models of the wonders they have chosen and describe the solid figures used to construct the models.

Modalities Auditory, Kinesthetic, Visual

CULTURAL CONNECTION

Molds of solid figures are frequently used to construct sand castles. In 1988, students and faculty of Ellon Academy near Aberdeen, Scotland, built a sand castle 5.2 miles long.

- Have students make a list of the solid figures they would use to build a sand castle. Then have them design their sand castles.

Modalities Auditory, Kinesthetic, Visual

GRADE 5	GRADE 6	GRADE 7
Identify faces, edges, and vertices of prisms and pyramids.	Name and count the faces, edges, and vertices of prisms and pyramids.	Find patterns in the number of faces, edges, and vertices of prisms and pyramids.

ORGANIZER

OBJECTIVE To name and count the faces, edges, and vertices of prisms and pyramids

VOCABULARY *Review* vertex, edge, face

ASSIGNMENTS **Independent Practice**
- *Basic* 1–8, odd; 9–12
- *Average* 1–12
- *Advanced* 6–12

PROGRAM RESOURCES
Reteach 10.2 Practice 10.2 Enrichment 10.2
Calculating Crew • Dr. Gee's 3D Lab

PROBLEM OF THE DAY

Transparency 10

Fred, Lee, and Ann have drawn different figures. Fred's figure has as many vertices as Lee's has faces. Lee's figure has twice as many vertices and twice as many edges as Ann's. Ann's figure has as many faces as it has vertices. Fred's figure has one more face than Ann's. What figure has each drawn? Ann, triangular pyramid; Fred, triangular prism; Lee, rectangular prism

Solution Tab A, p. A10

QUICK CHECK

Write the number of sides that make up the figure.

1. triangle 3 sides   **2.** pentagon 5 sides
3. square 4 sides   **4.** hexagon 6 sides
5. rectangle 4 sides   **6.** octagon 8 sides

1 MOTIVATE

Use this activity to review the terms *face, edge,* and *vertex*. Draw these figures on the board.

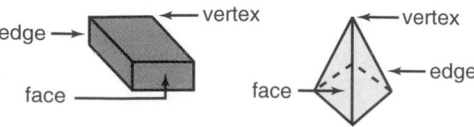

Have someone label the figures with each term as you discuss the definitions of the terms with the class.

10.2

Faces, Edges, and Vertices

WHAT YOU'LL LEARN
How to name and count the faces, edges, and vertices of prisms and pyramids

WHY LEARN THIS?
To understand the qualities of prisms and pyramids that make them good shapes for packaging

REMEMBER:

Vertices are named as points.

Edges are named as line segments.

Faces are named as polygons.

See page H23.

teen times

One of the things that makes some gemstones beautiful is the shape of their crystals, which are prisms. Amethyst, for example, forms six-sided crystals, and olivine crystals are four-sided.

Jamie wants to paint this box. No two faces that touch will have the same color. What is the least number of colors she needs?

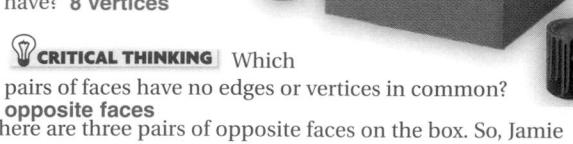

Talk About It
- How many faces does the box have?
  6 faces
- The faces meet to form edges. How many edges does the box have?
  12 edges
- The edges meet to form vertices. How many vertices does the box have? **8 vertices**

- **CRITICAL THINKING** Which pairs of faces have no edges or vertices in common?
  opposite faces

There are three pairs of opposite faces on the box. So, Jamie needs three colors.

EXAMPLE 1 Name the vertices, edges, and faces of this triangular pyramid.

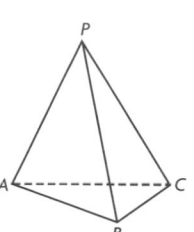

Vertices: points *A, B, C,* and *P*; 4 vertices
Edges: line segments *AB, BC, AC, PA, PB,* and *PC*; 6 edges
Faces: triangles *APB, BPC, APC,* and *ABC*; 4 faces

EXAMPLE 2 Count the vertices, edges, and faces of this triangular prism.

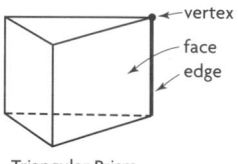
Triangular Prism

Vertices: 3 on each base, 6 vertices in all
Edges: 3 on each base, 3 lateral edges, 9 edges in all
Faces: 2 bases, 3 lateral faces, 5 faces in all

IDEA FILE

ALGEBRA CONNECTION

Swiss mathematician Leonard Fuller developed this formula about the number of edges (E), faces (F), and vertices (V) of polyhedrons:

$$E = F + V - 2$$

- Have students use the formula to find the number of edges of an octagonal pyramid and of a hexagonal prism. 16 edges, 18 edges

Modalities Auditory, Visual

RETEACH 10.2

Name _____

LESSON 10.2

Faces, Edges, and Vertices

Look at the triangular pyramid at the right. The faces meet to form edges. The edges meet to form a vertex. This triangular pyramid has 4 faces. The names of the faces are triangles *ABC, ABD, DBC,* and *CAD.* There are 6 edges. The names of the edges are line segments *AB, BC, CA, AD, BD,* and *CD.* Point *A* is the one vertex.

Name the edges, vertices, and faces for each figure.

1.

Edges: line segments *LM, MN, NO, LO, LP, MQ, NR, OS, PQ, QR, RS, SP*

Vertices: points *L, M, N, O, P, Q, R, S*

Faces: rectangles *LMNO, PQRS, PLOS, PQML, QRNM, SRNO*

2.

Edges: line segments *JK, KL, LM, MN, NJ, JO, KP, LQ, MR, NS, SO, OP, PQ, QR, RS*

Vertices: points *J, K, L, M, N, O, P, Q, R, S*

Faces: pentagons *JKLMN, OPQRS;* rectangles *JKPO, KLQP, LMRQ, MNSR, NJOS*

GUIDED PRACTICE

For Exercises 1–7, use the figure at the right.

1. Name the vertices.
   A, B, C, D, E, F
2. line segments *AB, BC, CA, DE, EF, FD, CF, AD, BE*
   Name the edges.
3. Name the faces.
   ABC, DEF, CFEB, ADEB, ADFC
4. Count the vertices.
   6 vertices
5. Count the edges.
   9 edges
6. Count the faces.
   5 faces
7. Which of the words below describe the figure? **prism, polyhedron**
   prism   pyramid   polyhedron

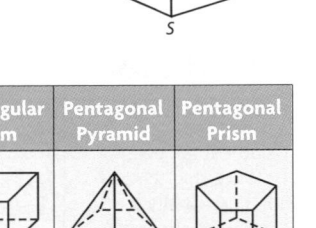

INDEPENDENT PRACTICE

For Exercises 1–5, use the figure at the right.

1. Name the vertices.
   N, M, R, P, S
2. Name the edges. **line segments SP, PN, NR, RS, MN, MR, MS, MP**
3. Name the faces.
   MRS, MSP, MPN, MNR, RSPN
4. Name the figure.
   rectangular pyramid
5. Write *polyhedron* or *not a polyhedron* to describe the figure.
   polyhedron

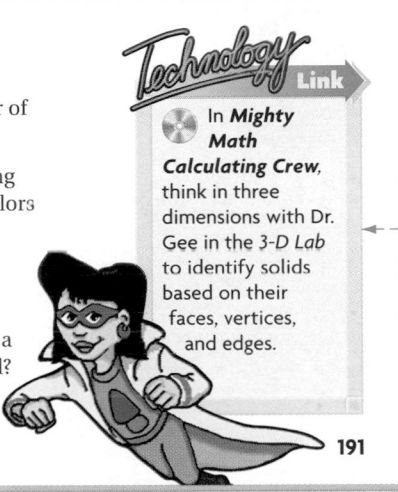

Copy and complete the table.

		Triangular Pyramid	Triangular Prism	Rectangular Pyramid	Rectangular Prism	Pentagonal Pyramid	Pentagonal Prism
6.	Number of faces	? 4	? 5	? 5	? 6	? 6	? 7
7.	Number of vertices	? 4	? 6	? 5	? 8	? 6	? 10
8.	Number of edges	? 6	? 9	? 8	? 12	? 10	? 15

Problem-Solving Applications

9. Maria wants to paint a triangular pyramid, with no two touching faces the same color. What is the least number of colors she needs? **4 colors**

10. Jon paints a pentagonal pyramid so that no two touching faces are the same color. What is the least number of colors he needs? **4 colors**

11. Mark made a prism and a pyramid that have 7 faces each. What figures did he make? **pentagonal prism, hexagonal pyramid**

12. **WRITE ABOUT IT** What is the least number of faces a prism can have? What is the least number for a pyramid? **prism: 5 faces; pyramid: 4 faces**

MORE PRACTICE Lesson 10.2, page H56

Technology Link

In *Mighty Math Calculating Crew*, think in three dimensions with Dr. Gee in the *3-D Lab* to identify solids based on their faces, vertices, and edges.

Use Grow Slide Levels V and W.

191

---

2 TEACH/PRACTICE

Read and discuss page 190, focusing on the similarities and differences between prisms and pyramids.

GUIDED PRACTICE

Complete the exercises orally with the class. Make sure students understand how to name and count the faces, edges, and vertices.

INDEPENDENT PRACTICE

To assure that students understand the concepts, check their answers to Exercises 1–4 before assigning the remaining exercises.

ADDITIONAL EXAMPLES

Example 1, p. 190

Draw a pentagonal pyramid and name the vertices, edges, and faces. Check students' drawings.

Example 2, p. 190

Draw a pentagonal prism and count the vertices, edges, and faces. Check students' drawings.

3 WRAP UP

Use these questions to help students focus on the big ideas of the lesson.

- **Jane's ice sculpture has 6 rectangular faces, 8 vertices, and 12 edges. Name the figure.** rectangular prism

✓ *Assessment Tip*

Have students write the answer.

- **Why does a triangular prism have more vertices and edges than a triangular pyramid?** because a triangular prism has two bases and a triangular pyramid has one base

DAILY QUIZ

NOTE: Sketch the figure on the board.

Name the vertices, edges, and faces of this figure. Name the figure.

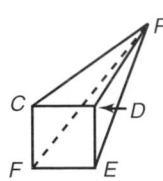

vertices: points *C, D, E, F, P*; edges: line segments *CD, DE, EF, FC, CP, DP, EP, FP*; faces: triangles *CPD, DPE, EPF, FPC,* and square *CDEF*; figure: square pyramid

191

PRACTICE 10.2

LESSON 10.2

Name _____

Faces, Edges, and Vertices

For Exercises 1–8, use the figure at the right.

1. Name the vertices.
   _A, B, C, D, E, F_

2. Name the edges.
   _AB, AC, BC, BE, CD, AF,_
   _FE, ED, DF_

3. Name the faces.
   ABC, DEF, AFEB, BEDC,
   CDFA

4. Name the figure.
   triangular prism

5. Is the figure a polyhedron? Explain why.
   Yes; it is a solid figure with
   flat faces that are polygons.

Complete the table.

		Pentagonal Pyramid	Rectangular Prism	Pentagonal Prism	Triangular Pyramid	Triangular Prism	Rectangular Pyramid
6.	Number of edges	10	12	15	6	9	8
7.	Number of faces	6	6	7	4	5	5
8.	Number of vertices	6	8	10	4	6	5

Mixed Applications

9. Scott wants to paint a rectangular prism so no two faces have the same color. What is the least number of colors he needs?
   6 colors

10. A survey showed that 2/5 of the 160 sixth-grade students play soccer. How many students is that?
   64 students

ENRICHMENT 10.2

LESSON 10.2

Name _____

Ornament Creations

A group of students combined solid figures to create unusual ornaments. Read about each student's creation. Then answer the questions that follow.

1. For her ornament, Ricki glued together 2 rectangular prisms. Her creation has 6 faces and 8 vertices.
   - How many edges does her creation have?
     12 edges
   - Sketch Ricki's ornament.

   Prisms could also be glued end-to-end.

2. Seth glued together 2 rectangular pyramids. His creation has 10 faces and 16 edges.
   - How many vertices does Seth's ornament have?
     8 vertices
   - Sketch Seth's ornament.

   Pyramids could also be glued base-to-base.

3. Tonya combined a triangular prism and a triangular pyramid. Her creation has 7 faces and 12 edges.
   - How many vertices does Tonya's creation have?
     7 vertices
   - Sketch Tonya's ornament.

4. Rusty combined a pentagonal prism and a pentagonal pyramid. His creation has 11 faces and 11 vertices.
   - How many edges does his ornament have?
     20 edges
   - Sketch Rusty's ornament.

GRADE 5	GRADE 6	GRADE 7
Identify nets for solid figures.	Build models of prisms from nets.	Make nets for solid figures.

ORGANIZER

OBJECTIVE To build models of prisms from nets

VOCABULARY *New* **net**

MATERIALS *For each student* 4-in. × 6-in. index card, inch ruler, scissors, tape

ASSIGNMENTS Independent Practice
- Basic 1–10; 11–18
- Average 1–10; 11–18
- Advanced 1–10; 11–18

PROGRAM RESOURCES
Reteach 10.3 Practice 10.3 Enrichment 10.3

PROBLEM OF THE DAY

Transparency 10

How many arrangements of five squares can you make in which at least one side of each square is touching another? How many of these arrangements can you fold into a five-sided box? 12 arrangements; 8 arrangements

Solution Tab A, p. A10

QUICK CHECK

Write the number of faces of the figure.

1. rectangular prism 6 faces
2. square prism 6 faces
3. pentagonal prism 7 faces
4. triangular prism 5 faces

1 MOTIVATE

MATERIALS: *For each group* scissors, tape, small six-sided box

Use this activity to assess students' prior knowledge of nets. Organize the class into cooperative groups and have students make as few cuts as possible along the edges of a box so it will lie flat.

- **What is another name for a flat box?** a net

Have students share their nets with the class and discuss how different nets can form the same shape.

10.3

WORD POWER
net

Sports Link

Look at the soccer ball in the photograph below. Notice that the ball has essentially, a spherical shape. Yet its separate parts or faces appear to be regular polygons. What are these polygons?

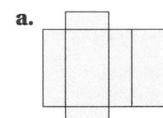

regular pentagons and regular hexagons

Building Solids

You can build solid figures by joining faces. The faces can be cut from paper, taped together, and then folded to form the solid.

ACTIVITY

WHAT YOU'LL NEED: 4-in. × 6-in. index card, inch ruler, scissors, tape

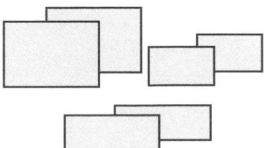

Look at the rectangular prism at the right.

- How many faces does the prism have? **6**

- What are the dimensions of the faces?
Two are 2 in. × 3 in., two are 1 in. × 3 in., and two are 1 in. × 2 in.
Follow these steps to make a pattern for the prism.

Step 1: Draw the faces on the index card.

Step 2: Cut out the six rectangles.

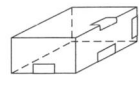

Step 3: Tape the pieces together to form the prism.

Step 4: Remove the tape from some of the edges so that the pattern lies flat.

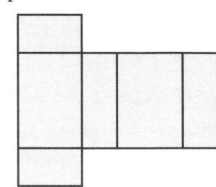

An arrangement of the rectangles that folds to form the prism is called a **net** for the prism.

Which arrangement is a net for the prism? **a**

a.   b.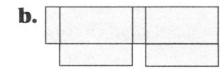

IDEA FILE

ART CONNECTION

MATERIALS: *For each student* TR pp. 42–46, tagboard, ruler, scissors, markers, tape

Have students follow these instructions to construct a building.

- Copy onto tagboard the net you want for your building. Don't copy the tabs.
- Decide which faces will become each side of the building and decorate them.
- Cut and tape the net to form your building.

Modalities Kinesthetic, Visual

RETEACH 10.3

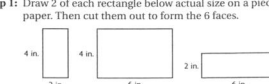

Name _____

Building Solids

You can build solid figures by joining faces. The faces can be cut from paper, taped together, and then folded to form the solid figure.
Look at the rectangular prism at the right.
There are 6 faces. The dimensions of the faces that form the front and back are 4 in. × 6 in.
The dimensions of the faces that form the top and bottom are 2 in. × 6 in.
The dimensions of the faces that form the sides are 2 in. × 4 in.

Follow these steps to build the figure.
Step 1: Draw 2 of each rectangle below actual size on a piece of paper. Then cut them out to form the 6 faces.

Step 2: Tape the pieces together to form the prism.
Step 3: Cut the tape between some of the edges so that the pattern lies flat. This flat arrangement of rectangles is a net for the prism.

Use the rectangular prism at the right for Problems 1–5.
1. How many faces does the prism have?
6 faces
2. What are the dimensions of the faces that form the front and back?
3 in. × 5 in.
3. What are the dimensions of the faces that form the sides?
5 in. × 5 in.
4. What are the dimensions of the faces that form the top and bottom?
3 in. × 5 in.
5. Use paper and tape to build this solid figure. Check students' prisms.

Use with text pages 192–193.   **TAKE ANOTHER LOOK R45**

GUIDED PRACTICE

Cut six 2 in. × 2 in. squares from an index card. Arrange the squares as shown below. Is the arrangement a net for a cube? Write *yes* or *no*.

1. 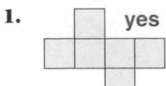 **yes**
2. 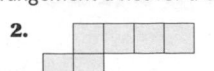 **no**
3. 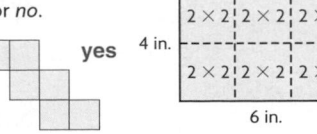 **yes**

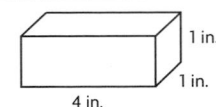
4 in.
6 in.

INDEPENDENT PRACTICE

1. Use the rectangular prism at the right. Draw the faces on an index card.   **Check students' drawings.**
2. Cut out the faces, and arrange them to form a net for the prism. Tape the pieces together to form a prism. **Check students' prisms**

1 in.
1 in.
4 in.

Will the arrangement of squares fold to form a cube? Write *yes* or *no*.

3.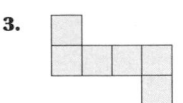
4.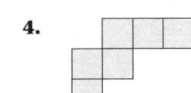
5.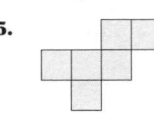
6.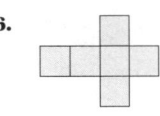

yes   **no**   **yes**   **yes**

Problem-Solving Applications

Melanie made a net for a prism that is 2 in. high and has two of these triangles as the bases.

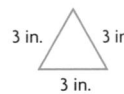
3 in.   3 in.
3 in.

7. How many faces does her prism have? **5**

8. Draw a net she could have made. **Check students' nets.**

9. What are the dimensions of the lateral faces? **3 in. × 2 in.**

10. ✏ **WRITE ABOUT IT** How would the net change if Melanie used two 3-in. squares for the bases? **The triangles would become squares, and one more 3-in. × 2-in. rectangle would be added.**

Mixed Review

Draw each polygon. **Check students' drawings.**

11. right triangle
12. square
13. rectangle
14. pentagon

Draw a rectangle. Then draw its image after each transformation. **Check students' drawings.**

15. reflected horizontally
16. rotated 90°

17. **NUMBER SENSE** If the LCM of two numbers is 4 and the sum of the numbers is 6, what are the numbers? **2 and 4**

18. **DATA** A survey showed that $\frac{1}{3}$ of the 120 people at a movie were between 12 and 16 years old. How many is this? **40 people**

MORE PRACTICE Lesson 10.3, page H56

193

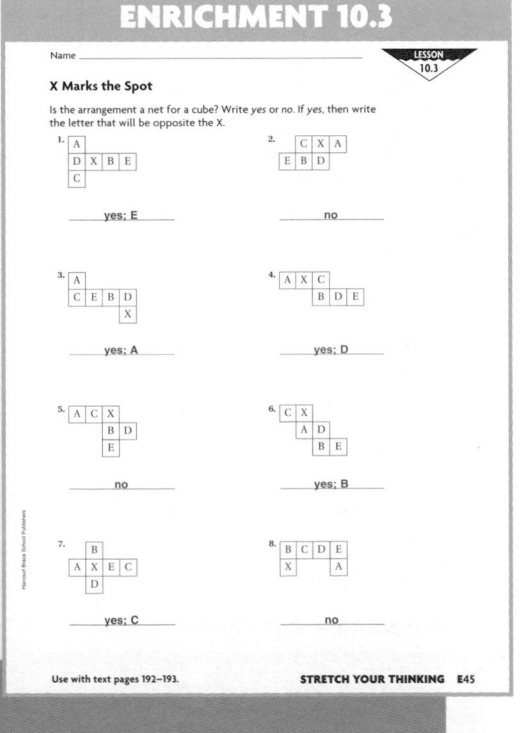

In this activity students explore nets by constructing rectangular prisms. Have students do this activity in small groups. Point out again that a rectangular prism has more than one net. When students finish, have them share their nets with other members of their group. Have each group report to the class how many different nets the group created that could be used to make the prism.

GUIDED PRACTICE

Complete these exercises as a whole-class activity to make sure students understand how to build models of prisms.

INDEPENDENT PRACTICE

To ensure that students understand the concepts, check their answers to Exercises 1–6 before assigning the remaining exercises.

3 WRAP UP

Use these questions to help students focus on the big ideas of the lesson.

- **How do nets for a cube differ from nets for a rectangular prism?** All the faces of a net for a cube are congruent. Only some of the faces of a net for a rectangular prism are congruent.

✓ *Assessment Tip*

Have students write the answer.

- **What is a net?** Possible answer: an arrangement of plane figures that fold to form a solid figure

DAILY QUIZ

NOTE: Draw the figures on the board.

1. Name the net that will form a cube. Write *a* or *b*. **b**

a.   b.

2. Name the net that will form a rectangular prism. Write *a* or *b*. **b**

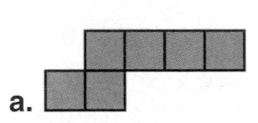 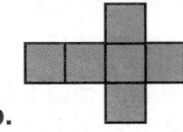
a.   b.

193

GRADE 5	GRADE 6	GRADE 7
Identify solid figures from different views.	Draw and identify different views of a solid.	Draw three-dimensional figures.

ORGANIZER

OBJECTIVE To draw and identify different views of a solid

ASSIGNMENTS Independent Practice
- *Basic* 1–11, odd; 12–13
- *Average* 1–13
- *Advanced* 4–13

PROGRAM RESOURCES
Reteach 10.4 Practice 10.4 Enrichment 10.4
Cultural Connections • Turkish Foods, p. 200
Extension • Perspective Drawings, TE
pp. 196A–196B

PROBLEM OF THE DAY

Transparency 10

What is the least number of toothpicks you can move to change 629 to 538?

629

2 toothpicks

Solution Tab A, p. A10

QUICK CHECK

Name the figure.

	Faces	Vertices	Edges	
1.	4	4	6	triangular pyramid
2.	7	10	15	pentagonal prism
3.	5	6	9	triangular prism
4.	6	6	10	pentagonal pyramid

1 MOTIVATE

Ask students to sketch this problem:

Jeff, Lisa, and Peggy are looking at a solid figure made from cubes. Jeff says there are 3 cubes in the figure. Lisa says 2, and Peggy says 6.

- **What view of the figure is each of them looking at?** Possible answer: Peggy: top or bottom; Jeff: front or back; Lisa: side

---

10.4

WHAT YOU'LL LEARN
How to draw and identify different views of a solid

WHY LEARN THIS?
To draw models of real-life objects

Two-Dimensional Views of Solids

Kris wants to show how objects appear from different views. She decides to show how her pet, Cleo, appears when you look at him from the top, side, and front.

top view   side view   front view

You can draw different views of a solid.

ACTIVITY

WHAT YOU'LL NEED: cylinder

- Look at the top of the cylinder.
  Draw the top view.
  The top view should be a circle.
- Look at the front of the cylinder.
  Draw the front view.
  The front view should be a rectangle.
- Look at a side of the cylinder.
  Draw the side view.
  The side view should be a rectangle.

EXAMPLE Identify the solid that has these views.

⊠	top view	*The top view shows that the base is square and that the sides come together at a point.*
△	front view	*The front and side views show that the solid has triangular sides.*
△	side view	

So, this solid is a square pyramid.

teen times

Have you ever seen an aerial view of a city? Aerial views can also be called top views. Today, many of these views come from pictures taken by satellites.

IDEA FILE

ART CONNECTION

Students can apply their understanding of different views of solids to a common form of art known as *silhouette*. Have students work in pairs to draw silhouettes of their heads and shoulders from back, front, and side views. Have them trace the shadows of their heads and shoulders on gray paper taped to the board. Then have them cut around the outlines of their silhouettes and mount them on white paper.

Modalities Kinesthetic, Visual

MEETING INDIVIDUAL NEEDS

RETEACH 10.4

Name _____

LESSON 10.4

Two-Dimensional Views of Solids

Solid figures can look different when you view them from the front, top, and bottom.
Look at the triangular prism at the right.
If you look at the prism only from the front, you would see a triangle.

If you look at the prism only from the top, you would see a rectangle.

If you look at the prism only from the bottom, you would see a rectangle.

Name the figure you would see from each view of the solid figure.

1. Front: _____triangle_____
   Top: _____triangle_____
   Bottom: _____triangle_____

2. Front: _____square_____
   Top: _____square_____
   Bottom: _____square_____

3. Front: _____triangle_____
   Top: _____pentagon_____
   Bottom: _____pentagon_____

R46 TAKE ANOTHER LOOK   Use with text pages 194–195.

GUIDED PRACTICE

Name each solid that has the given top view. Refer to the solids in the box.

   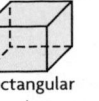  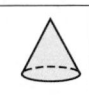

| triangular pyramid | triangular prism | rectangular pyramid | rectangular prism | pentagonal pyramid | hexagonal prism | cylinder | cone |

1. ⊠

rectangular pyramid

2. ◯

cylinder

3.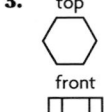

pentagonal pyramid

4. △

triangular pyramid

INDEPENDENT PRACTICE

Name the solid figure that has the given views.

1.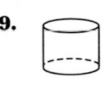
top
△ ◯
front bottom

cone

2. ⊠
top
△ △
front side

rectangular pyramid

3. top
⬡
front side
▭ ▭

hexagonal prism

For Exercises 4–7, use the solids at the top of the page.

4. Which solid has rectangles in all of its views? **rectangular prism**

5. Which solid has triangles in all of its views? **triangular pyramid**

6. Which figure has a pentagon in one of its views? **pentagonal pyramid**

7. Which solids have circles in some of their views? **cone and cylinder**

Draw the front, top, and bottom views of each solid. **See Additional Answers, p. 205A.**

8. **9.** **10.** **11.**

Problem-Solving Applications

12. Draw the top, front, and side views of an object in your classroom. **Check students' drawings.**

13. ✎ **WRITE ABOUT IT** Name objects at home that have a rectangle when you view the object from the top. **Possible answer: cereal box**

2 TEACH/PRACTICE

Shadows of solid figures will broaden students' understanding of different views of solids.

GUIDED PRACTICE

Complete Exercises 1–4 orally with the class to ensure that students understand how to find different views of a solid.

INDEPENDENT PRACTICE

To ensure that students understand the concepts, check their answers to Exercises 1–3 before assigning the remaining exercises.

ADDITIONAL EXAMPLE

Example, p. 194

These are three views of a solid figure. Name the figure.

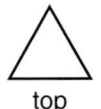 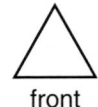 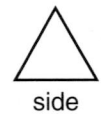
top front side

Since all three views are triangles, the solid is a triangular pyramid.

3 WRAP UP

Use these questions to help students focus on the big ideas of the lesson.

- **What figures always have the shape of a triangle for their side views?** triangular pyramids and cones

✓ *Assessment Tip*

Have students write the answer.

- **Is the top view of a prism always the same as the bottom view? Explain.** Yes; because the bases of prisms are congruent

DAILY QUIZ

NOTE: Draw the figures on the board.

Draw the top, bottom, and side views of each solid.

1.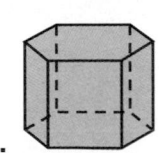

top and bottom: hexagon; side: rectangle

2.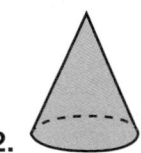

top and bottom: circle; side: triangle

ORGANIZER

OBJECTIVE To learn how different views of a solid figure differ
MATERIALS *For each student* centimeter cubes, centimeter graph paper
PROGRAM RESOURCES
E-Lab • Activity 10
E-Lab Recording Sheets, p. 10

USING THE PAGE

• **Did you ever hear the expression** *bird's-eye view*? **What does it mean?** Possible answer: what things on the ground look like from the sky

Explain to students that in this lesson they will look at solid figures from a bird's-eye view and from other views as well.

Activity 1

Explore

Observe students as they build the solid. Discuss the drawing of the top view.

• **Does the drawing of the top view show only the cube at the very top? Explain.** No; it shows all of the cubes you can see from the top.

Point out to students that the same principle applies to the front view and the side view.

Think and Discuss

Discuss the questions on page 192.

• **Which views show how wide the solid is?** the top view and the front view

Try This

When students finish Exercises 1–4, repeat the questions from Think and Discuss for each exercise.

COOPERATIVE LEARNING

Different Views of Solids

LAB *Activity*

WHAT YOU'LL EXPLORE
How the views of a solid figure differ

WHAT YOU'LL NEED
centimeter cubes
centimeter graph paper

How do you think this solid would look if you viewed it from the top?

ACTIVITY 1

Explore

• Use centimeter cubes to build the solid at the right.

• The top view of the solid is shown. Draw the front view and the side view on graph paper. **Check students' drawings.**

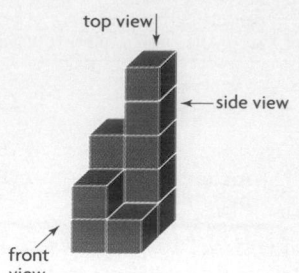
top view
side view
front view

top view

Think and Discuss

• How many cubes did it take to build the solid? **11 cubes**

• How many cubes do you see in the top view? the front view? the side view? **4 cubes; 8 cubes; 7 cubes**

• Which views show how high the solid is? **front view and side view**

Try This   See Additional Answers, p. 205A.

Each solid is made with 10 cubes. On graph paper, draw a top view, a front view, and a side view for each solid.

1.    **2.**

3.    **4.**

196   Chapter 10

IDEA FILE

REVIEW

Have students identify the solid figure that has the same shape as each of these items.

1. videotape **rectangular prism**
2. tornado **cone**
3. water pipe **cylinder**
4. building block **square prism or cube**
5. tepee **cone**

Modalities Auditory, Visual

CULTURAL CONNECTION

Ten students from Nakamura Elementary School in Yokohama, Japan, built a triangle of 5,525 empty cans in just 30 minutes. Have students build their own triangle by stacking 6 empty cans. Then have them draw on graph paper the front, top, and side views of the triangle. Let each square represent one can.

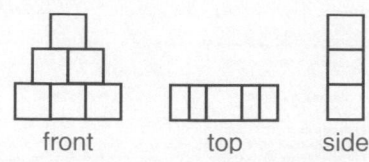
front   top   side

Modalities Kinesthetic, Visual

MEETING INDIVIDUAL NEEDS

Explore

- Build this 2-cm cube.
- Draw a top view, a front view, and a side view on graph paper.
- All three views are the same 2 × 2 squares. Build a different solid whose three views are the same. **Solids may vary.**

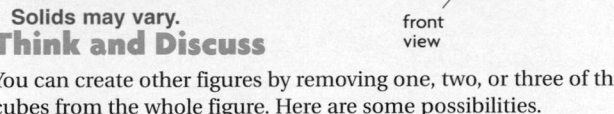
top view
side view
front view

Think and Discuss

You can create other figures by removing one, two, or three of the cubes from the whole figure. Here are some possibilities.

a. 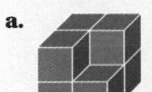   b.    c. 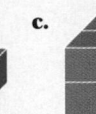   d.

Which of the solids above match the three views given?

1.

top   front   side   c

2.
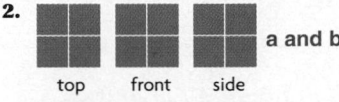
top   front   side   a and b

3.
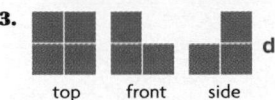
top   front   side   d

Try This   Models and drawings will vary.

- Build another 2-cm cube.
- Create a different solid by removing just three of the eight cubes.
- Draw the three views for your figure.

Technology Link
E–Lab • Activity 10 Available on CD-ROM
and the Internet at http://www.hbschool.com/elab

Note: Students slice solids. Use E-Lab p. 10.

197

Activity 2

Explore

Observe students as they build the 2-cm cube and draw the views. When they proceed to build other solids with three identical views, have them draw on graph paper the top, side, and front views of these solids and share their drawings with the class.

Think and Discuss

- **Which views show how high the solid is?** front view and side view
- **Which views show how wide the solid is?** top view and front view

Try This

When students finish the activity, ask:

- **Can you tell how many cubes are in a solid by looking at just one view? Explain.** No; you need to look at all the views.

USING E-LAB

Students explore the relationship between solids and cross-sectional views by slicing solids.

✓ *Assessment Tip*

Use these questions to help students communicate the big ideas of the lesson.

- **How many cubes did it take to build the solid labeled *c* on page 197?** 6 cubes
- **Which views show how high the 2-cm cube at the top of page 197 is?** front view and side view

---

EXTENSION

To extend the content of the lesson, ask students:

- How many faces directly face each other in the solids labeled *a, b, c,* and *d* on page 197?
  a: 18 faces
  b: 12 faces
  c: 14 faces
  d: 10 faces

Modality Visual

ALGEBRA CONNECTION

Have students write equations to show how many cubes were removed from the 8-cube figure at the top of page 197 to make the number of cubes in the solids labeled *a, b, c,* and *d.* Let *x* equal the number of cubes removed.
a: $8 - x = 7$, $x = 1$ cube;
b: $8 - x = 6$, $x = 2$ cubes;
c: $8 - x = 6$, $x = 2$ cubes;
d: $8 - x = 5$, $x = 3$ cubes

Modality Visual

EXTENSION

Perspective Drawings

Use these strategies to extend understanding of solid figures by making perspective drawings.

Multiple Intelligences Logical/Mathematical, Visual/Spatial

Modalities Auditory, Kinesthetic, Visual

Lesson Connection Students will make perspective drawings of solid figures.

Materials Straightedge

Program Resources *Intervention and Extension Copying Masters,* p. 17

Challenge

TEACHER NOTES

ACTIVITY 1

Purpose To learn how to draw in one-point perspective

Vocabulary Review

• **cube**—a rectangular prism with each face a square

Discuss the following questions.

• Is the rear face the same size as the front face? Discuss. The rear face is drawn smaller, although all faces of a cube are actually the same size.

• Why would you decide to include—or not to include—the dashed "hidden" lines? Possible answer: to suggest whether the material is opaque

• What is the result of placing the vanishing point above the box? below? The viewer looks down onto the box; the viewer looks up from below the box.

WHAT STUDENTS DO

1. Draw in one-point perspective a cube like the one shown. Follow these steps.

   a. Draw a square. Connect the four vertices to a *vanishing point.*

   b. Draw the top rear edge of the cube, parallel to the top front edge. Draw the right rear edge parallel to the right front edge.

   c. Draw the remaining visible edges of the cube.

   d. (Optional) Draw the "hidden" edges, using dashed lines. Erase the segments to the vanishing point.

      • Are the top left and top right edges of a cube parallel? Are they drawn parallel? Yes; no, if extended on paper, they would meet, at a vanishing point

2. Draw several cubes. Experiment with various vanishing points: left, right, above, and below the cube.

3. Draw a table in one-point perspective.

   • How do you find the vanishing point for the one-point perspective drawing? by extending the edges

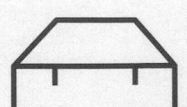

TEACHER NOTES

WHAT STUDENTS DO

ACTIVITY 2

Purpose To learn how to draw in two-point perspective

Discuss the following questions.

• If you were making a drawing of the school building as viewed from the front, would you use one-point perspective or two-point perspective? What if the view were from an angle so that the viewer faced the corner where the front and side meet? one-point; two-point

In two-point perspective, the table should look something like this.

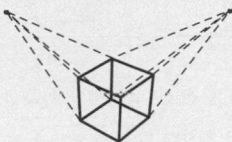

Have students meet in small groups or with partners to discuss the given questions. Have one student report each group's responses to the larger group.

1. Draw in two-point perspective a cube like the one shown. Follow these steps.

   a. Mark two vanishing points. They should lie along a horizontal line.

   b. Draw a vertical line segment for the front edge of the cube. Connect the ends of the line segment to the vanishing points.

   c. Draw the left and right vertical edges, and connect the endpoints to the vanishing points.

   d. Draw the remaining visible edges of the cube.

   e. (Optional) Draw the hidden lines, using dashed lines. Erase the lines to the vanishing point.

   • Which edge is closest to the viewer? the front edge, which was drawn first

2. On a separate sheet of paper, draw a table in two-point perspective.

Share Results

• Discuss when to use one-point perspective and when to use two-point perspective. Use one-point when the viewer faces a side, two-point when the viewer faces a corner.

• How can you find the vanishing point or points? by extending the edges

Intervention and Extension Copying Masters, p. 17

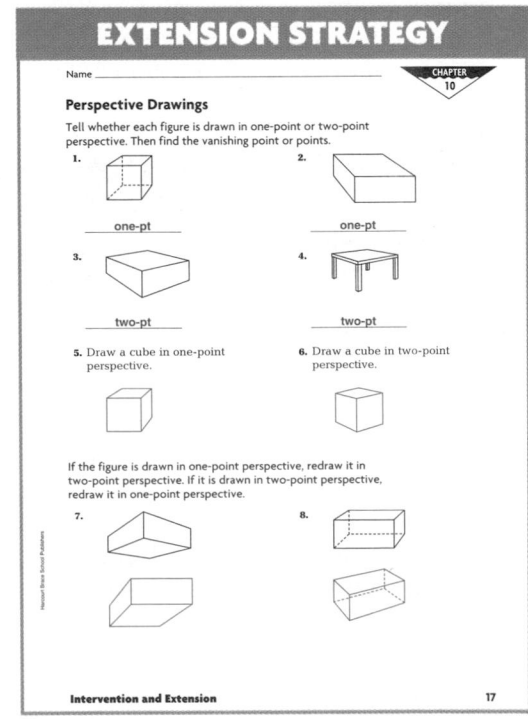

GRADE 5	GRADE 6	GRADE 7
Solve a problem by using the strategy *solve a simpler problem*, and choose appropriate strategies.	Solve a problem by using the strategy *solve a simpler problem*, and choose appropriate strategies.	Solve a problem by using the strategy *solve a simpler problem*, and choose appropriate strategies.

ORGANIZER

OBJECTIVE To solve a difficult problem by looking at a related simpler problem

ASSIGNMENTS Practice/Mixed Applications
- *Basic* 1–12
- *Average* 1–12
- *Advanced* 1–12

PROGRAM RESOURCES
Reteach 10.5 Practice 10.5 Enrichment 10.5
Problem-Solving Thinking Along, TR p. R1

PROBLEM OF THE DAY

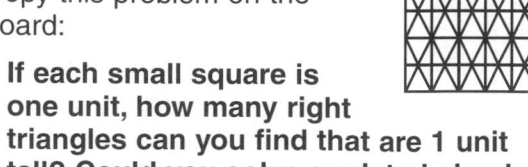

Transparency 10

Ms. Courtney told nine students to make three straight lines with four students in each line. How did they do it? They stood in the shape of a triangle.

Solution Tab A, p. A10

QUICK CHECK

Write the number of edges the base of the figure has.

1. pentagonal prism   5 edges
2. hexagonal prism   6 edges
3. triangular prism   3 edges
4. square prism   4 edges
5. rectangular prism   4 edges
6. octagonal prism   8 edges

1 MOTIVATE

Copy this problem on the board:

- **If each small square is one unit, how many right triangles can you find that are 1 unit tall? Could you solve a related simpler problem to find the answer? How?** yes; by finding the number of triangles in one square

- **How will solving a related simpler problem help you find the answer?** Possible answer: by making the more difficult problem easier to understand

- **What is the answer?** 40 right triangles

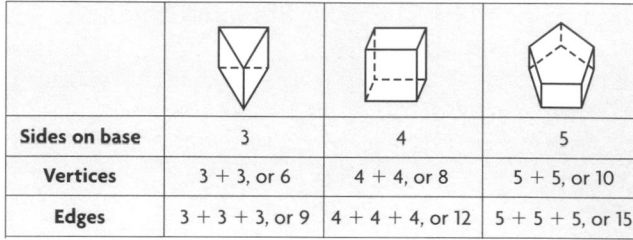

10.5 PROBLEM-SOLVING STRATEGY
Solve a Simpler Problem

WHAT YOU'LL LEARN
How to solve a difficult problem by looking at a related simpler problem

WHY LEARN THIS?
To find a simpler problem in tasks that seem too difficult to solve

> **Problem Solving**
> • Understand
> • Plan
> • Solve
> • Look Back

Brett is building models of prisms, using balls of clay for the vertices and straws for the edges. How many balls of clay and how many straws will he need to make a prism whose bases have 12 sides each?

If a problem seems difficult to solve, it sometimes helps to think of a similar, but simpler, problem.

☑ **UNDERSTAND** What are you asked to find? **how many balls of clay and how many straws will be needed** What facts are given? **The prism bases have 12 sides each.**

☑ **PLAN** What strategy will you use?

You can *solve a simpler problem* by thinking about the numbers of vertices and edges on prisms whose bases have 3, 4, and 5 sides. Then use these numbers to try to find a pattern.

☑ **SOLVE** How will you solve the problem?

Make a table to organize the data for prisms whose bases have 3, 4, and 5 sides. Show the numbers of sides, vertices, and edges.

Sides on base	3	4	5
Vertices	3 + 3, or 6	4 + 4, or 8	5 + 5, or 10
Edges	3 + 3 + 3, or 9	4 + 4 + 4, or 12	5 + 5 + 5, or 15

From the pattern in the table, you can see that for a prism whose bases have 12 sides each, the number of vertices is 12 + 12, or 24, and the number of edges is 12 + 12 + 12, or 36.

So, Brett will need 24 balls of clay for the vertices and 36 straws for the edges.

☑ **LOOK BACK** What other strategy could you use to solve this problem? *draw a diagram*

What if . . . the prism had bases with 20 sides each? How many vertices and how many edges would it have? **vertices: 20 + 20 = 40; edges: 20 + 20 + 20 = 60**

IDEA FILE

EXTENSION

Have students solve by solving a simpler problem: **Melissa drew a 30-row triangle, using the pattern below. How many exes were in the last row?** 59 exes **Explain the simpler problem you used to find the answer.** Add each row number to the previous row number.

Row 1: x
Row 2: x x x
Row 3: x x x x x
Row 4: x x x x x x x

Modality Visual

RETEACH 10.5

Name _____

LESSON 10.5

Problem-Solving Strategy

Solve a Simpler Problem

Rhea is building models of prisms, using balls of clay for vertices and straws for edges. How many balls of clay and how many straws will she need to make a prism whose base has 8 sides?

Sometimes it helps you solve a more difficult problem by solving a similar, but simpler, problem first.

You can *solve a simpler problem* by thinking about the numbers of vertices and edges on prisms whose bases have 3, 4, and 5 sides.

Number of sides on base	3	4	5
Number of vertices	3 + 3, or 6	4 + 4, or 8	5 + 5, or 10
Number of edges	3 + 3 + 3, or 9	4 + 4 + 4, or 12	5 + 5 + 5, or 15

From the pattern in the table, you can see that for a prism whose base has 8 sides, the number of vertices is 8 + 8, or 16, and the number of edges is 8 + 8 + 8, or 24.

So, Rhea will need 16 balls of clay and 24 straws.

Solve by first solving a simpler problem.

1. Suppose Rhea is building a model of a prism whose base has 16 sides. How many balls of clay and how many straws will she need? 32 balls of clay; 48 straws

2. Greg wants to make a model of a pyramid whose base has 5 sides. He will use balls of clay for the vertices and toothpicks for the edges. How many toothpicks will he need? 10 toothpicks

3. Terri is making edible prisms. She uses cheese cubes for vertices and carrot sticks for edges. How many cheese cubes and carrot sticks will she need to make a prism whose base has 14 sides? 28 cheese cubes and 42 carrot sticks

Use with text pages 198–199.   TAKE ANOTHER LOOK R47

MEETING INDIVIDUAL NEEDS

PRACTICE

Solve by first solving a simpler problem.

1. Aaron wants to make a model of a prism whose bases have 10 sides each. He will use balls of clay for the vertices and straws for the edges. How many balls of clay will he need? **20 balls of clay**

2. Look back at Problem 1. How many straws will Aaron need? How many faces will his prism have? **30 straws; 12 faces**

3. Amy wants to make a model of a pyramid whose base has 10 sides. She will use balls of clay for the vertices and toothpicks for the edges. How many balls of clay will she need? **11 balls of clay**

4. Look back at Problem 3. How many toothpicks will Amy need? How many faces will her pyramid have? **20 toothpicks; 11 faces**

MIXED APPLICATIONS

CHOOSE

Choose a strategy and solve. **Choices of strategies will vary.**

Problem-Solving Strategies
- Work Backward
- Guess and Check
- Draw a Diagram
- Write an Equation
- Use a Formula
- Make a Table

5. Mr. Dressel went to a garden shop. He spent $8.59 for a rosebush, $4.95 for fertilizer, and $9.95 for a sprinkler. He had $6.51 when he returned home. How much money did he take with him when he went to the garden shop? **$30.00**

6. T.J. used 5 gal of gasoline to make a 130-mi trip. How much gasoline will he need to make a 780-mi trip? **30 gal**

7. After the party $\frac{3}{4}$ of the cake was left. Sarah divided the cake equally among the 6 guests. How much of the cake did each guest get? $\frac{1}{8}$ **of the cake**

8. Mike wants to put a $3\frac{1}{2}$-ft table against a 13-ft wall. If he centers the table, how far will it be from each end of the wall? **$4\frac{3}{4}$ ft**

9. Heidi baked cookies after school. She gave half to her neighbor and divided the rest equally among the 4 people in her family. Each person in her family got 3 cookies. How many cookies did Heidi bake in all? **24 cookies**

10. Joni bought 3 model airplanes and a tube of glue at the hobby shop. The glue cost $2.59. The total bill was $20.59. What was the average price Joni paid for each of the models? **$6.00**

11. Jason is building a model of a prism. He used 24 toothpicks as edges to make his model. How many sides did the base of his prism have? **8 sides**

12. **✏ WRITE ABOUT IT** Look back at Problem 1. Explain how you used the strategy *solve a simpler problem*. **Explanations will vary.**

MORE PRACTICE Lesson 10.5, page H57

199

Name _____

LESSON 10.5

Problem-Solving Strategy

Solve a Simpler Problem

Solve by first solving a simpler problem.

1. Jon is building models of edible prisms. He uses gumdrops for vertices and licorice for edges. How many gumdrops and pieces of licorice will he need to make a prism whose bases have 8 sides?

16 gumdrops and
24 pieces of licorice

2. Carol wants to make a model of a prism whose base has 9 sides. She will use balls of clay for the vertices and straws for the edges. How many balls of clay and straws will she need? How many faces will her prism have?

18 balls of clay; 27 straws
11 faces

3. Chloe used 30 toothpicks as edges to make a model for a prism. How many sides did the base of her prism have? How many vertices?

10 sides; 20 vertices

4. Dan used 12 balls of clay as vertices to make a model for a prism. How many sides did the base of his prism have? How many edges?

6 sides; 18 edges

Mixed Applications

Solve.

CHOOSE A STRATEGY

- Work Backward • Draw a Diagram • Use a Formula • Guess and Check • Write an Equation
Choices of strategies will vary.

5. Alyssa used 8 gal of gasoline to make a 192-mi trip. How much gasoline will she need to make a 360-mi trip?

15 gal

6. Peter has $4.20 in dimes and quarters. He has 21 coins. How many coins of each kind does he have?

7 dimes; 14 quarters

7. Zack bought muffins at the bakery. He gave half of the muffins to a friend and divided the rest equally among the six members of his family. If each person in his family got 3 muffins, how many muffins did Zack buy?

36 muffins

8. Dennis bought 4 books and a bookmark at the bookstore. The bookmark cost $1.25. The total bill was $21.25. What was the average price Dennis paid for each book?

$5.00

Use with text pages 198–199.

ON MY OWN P47

Name _____

LESSON 10.5

Number Patterns

Name the next three terms. Then identify the rule used to form the pattern.

1. 28, 32, 40, 52, __68__ __88__ __112__
   Rule: Add 4, then 8, then 12, then 16, and so on.

2. 132, 127, 117, 102, __82__ __57__ __27__
   Rule: Subtract 5, then 10, then 15, then 20, and so on.

3. 43, 55, 77, 109, __151__ __203__ __265__
   Rule: Add 12, then 22, then 32, then 42, and so on.

4. 154, 147, 133, 112, __84__ __49__ __7__
   Rule: Subtract 7, then 14, then 21, then 28, and so on.

5. 63, 72, 90, 126, __198__ __342__ __630__
   Rule: Add 9, then 18, then 36, then 72, and so on.

6. 117, 109, 125, 117, 133, __125__ __141__ __133__
   Rule: Subtract 8, then add 16.

7. 203, 181, 161, 143, __127__ __113__ __101__
   Rule: Subtract 22, then 20, then 18, then 16, and so on.

8. 97, 112, 129, 148, __169__ __192__ __217__
   Rule: Add 15, then 17, then 19, then 21, and so on.

9. 101, 121, 140, 158, __175__ __191__ __206__
   Rule: Add 20, then 19, then 18, then 17, and so on.

10. 200, 175, 152, 131, __112__ __95__ __80__
    Rule: Subtract 25, then 23, then 21, then 19, and so on.

Use with text pages 198–199.

STRETCH YOUR THINKING E47

In this lesson students use simpler related problems to solve more-difficult geometry problems.

PRACTICE
Have partners work through Problem 1 together and complete the Problem-Solving Think Along, TR p. R1.

MIXED APPLICATIONS
To make sure students understand how to use problem-solving strategies, have students exchange papers and check a classmate's work. Emphasize that more than one strategy can be used to solve some problems.

DEVELOPING ALGEBRAIC THINKING
The ability to work with variables is extremely important in developing algebraic thinking. To promote comfort using variables, have students write equations to show how to find the number of vertices and edges for a prism given the number of sides of the base.

3 WRAP UP

Use these questions to help students focus on the big ideas of the lesson.

- **How many faces will Brett's prism have?** 14 faces

✓ *Assessment Tip*

Have students write the answer.

- **What related simpler problem did you use to solve Problem 3?** Possible answer: I found the number of balls of clay needed for a prism whose base has 3 sides.

DAILY QUIZ
Solve by first solving a simpler problem.

1. Eduardo's kite design has two 8-sided bases, streamers attached to each vertex, and wooden dowels for the edges. How many streamers does he need to make the kite? How many dowels does he need? 16 streamers, 24 dowels

2. Kim's plan for the frame of her model Egyptian pyramid has a base with 12 sides. She needs metal fasteners at each vertex to attach the wire strips she will use for edges. How many metal fasteners will she need? How many faces will her pyramid have? 13 metal fasteners, 13 faces

199

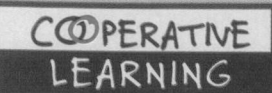

OBJECTIVE To identify and draw solid figures

USING THE PAGE

Before the Lesson

Have a chart on the wall or overhead projector with the solid figures reviewed in this chapter. Ask student volunteers to identify and describe the different figures. Stack cubes in two layers. Have students remove a few of the cubes and view the changes from the front, side, and top of the arrangement.

Cooperative Groups

Have the students work together in cooperative groups to answer the questions above the Work Together and to complete Problems 1–3.

► CULTURAL LINK

Turkey has its own pyramid-shaped buildings formed by nature. These strange rock formations in Anatolia were produced by volcanic eruptions millions of years ago. Houses and hotels have been carved out of this soft rock. Churches, houses, and schools in the ancient, underground cities of Cappadocia sheltered thousands of people. Even horses were stabled underground. Underground escape routes, some of them several miles long, led from the city.

200

CULTURAL CONNECTION

COOPERATIVE LEARNING

Turkish Foods

Nursel (NOOR·sel) likes to help at her aunt's deli and grocery store. Her aunt sells Turkish foods, pastries, and spices. Many people buy food from the store to give as gifts or to take to parties. Her aunt has special gift boxes and containers for packing some of the foods, pastries, and spices. Identify the shapes of the containers. Which shape would best hold spices, pastries, and yogurt?

Box *A* is a rectangular prism and could hold the baklava. Box *B* is a hexagonal prism and could hold the spices. Box *C* is a cylinder and could hold the yogurt.

• Which of the boxes are polyhedrons? Explain. *A* and *B*. **A polyhedron is a solid figure with flat faces that are polygons.**

Work Together

1. Nursel's favorite dessert is lokma, round doughnuts in syrup. Her aunt makes the lokma in different sizes so she can stack ten of them from large to small. What shapes of boxes could be used to package one of these stacks? **Possible answers: cube, pyramid, cone**

2. Tightly rolled cheese pastries called sigara boregi are packaged in a polyhedron box with two triangular bases and three other faces. Draw a picture of the box and identify which polyhedron it is. **Check students' drawings. triangular prism**

3. Doner kebab is the most popular food prepared at the deli. Slices of spiced grilled mutton are set in buttered pita bread. These Turkish sandwiches are placed in square boxes on the counter as shown below. Nursel handed her aunt the two top left boxes from the stack. Draw a side, top, and front view of the stack after she removed the two. **Check students' drawings.**

► CULTURAL LINK

Turkish foods are served in restaurants and homes all over the world. A typical meal might include eggplant or tomatoes stuffed with rice and meat. Circassian chicken with walnut sauce is another favorite. Kebabs with pepper, onion, and cubes of lamb or fish are also served in many homes. Yogurt is another international favorite.

CHAPTER 10 REVIEW

EXAMPLES

- **Classify solid figures.** (pages 186–189)

Name the figure. Then write *polyhedron* or *not a polyhedron.*

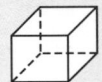 rectangular prism
polyhedron

- **Name and count faces, edges, and vertices of solid figures.** (pages 190–191)

4 faces: *ABD, BCD, DCA, ABC*
6 edges: line segments *AD,*
BD, CD, AB, BC, CA
4 vertices: *A, B, C, D*

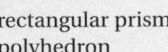

- **Build models of prisms.** (pages 192–193)

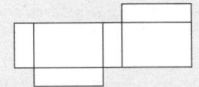
will make a prism

won't make a prism

- **Draw different views of solid figures.**
(pages 194–195)

top   front   side

- **Use the strategy** *solve a simpler problem.*
(pages 198–199)

PROBLEM-SOLVING TIP: For help in solving problems by solving a simpler problem, see pages 10 and 198.

EXERCISES

Name the figure. Then write *polyhedron* or *not a polyhedron.*

1.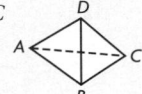
triangular prism
polyhedron

2. cylinder
not poly.

3.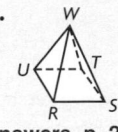
rectangular
pyramid
polyhedron

Name and count the faces, edges, and vertices.

4. 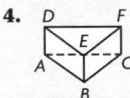 **5.** 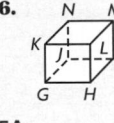 **6.**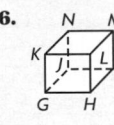

See Additional Answers, p. 205A.

Will the net fold to form a cube? Write *yes* or *no.*

7. yes

8. no

9. What are the dimensions of the faces of this prism? 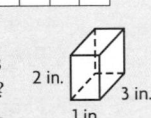 2 in.
3 in.
1 in.

9. 2 faces: 2 in. × 1 in.; 2 faces: 1 in. × 3 in.; 2 faces: 2 in. × 3 in.

Draw the top, front, and side views of the solid.
See Additional Answers, p. 205A.
10. 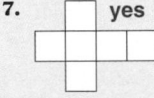 **11.** 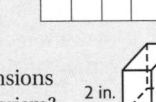 **12.**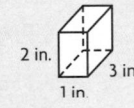

13. Name a solid figure that has circles in some of its views. **cylinder or cone**

14. Mari made a model of a prism whose bases have 9 sides each. How many edges and how many faces did her model have? **27 edges; 11 faces**

15. Matt made a model of a pyramid with 16 edges. How many sides did the base of his pyramid have? **8 sides**

Chapter 10 Review   **201**

USING THE PAGE

The Chapter 10 Review reviews solid figures. Chapter objectives are provided, along with examples and practice exercises for each.

The Chapter 10 Review can be used as a review or a test. You may wish to place the Review in students' portfolios.

Assessment Checkpoint

For Performance Assessment in this chapter, see page 204, What Did I Learn?, items 3 and 6.

✔ Assessment Tip

Student Self-Assessment The How Did I Do? Survey helps students assess both what they have learned and how they learned it. Self-assessment helps students learn more about their own capabilities and develop confidence in themselves.

A self-assessment survey is available as a copying master in the *Teacher's Guide for Assessment*, p. 15.

CHAPTER 10 TEST, page 1

Name _____

CHAPTER 10 TEST
PAGE 1

Choose the letter of the correct answer.

1. What question should you ask to find out if a solid figure is a polyhedron?
   A. How many bases are there?
   B. Are all the faces polygons?
   C. Which faces are congruent?
   D. Is there more than 1 vertex?

2. Which statement about the bases of pyramids and prisms is true?
   A. Both pyramids and prisms have 2 bases.
   B. Both pyramids and prisms have only 1 base.
   C. Pyramids have 1 base; prisms have 2 bases.
   D. Prisms have 1 base; pyramids have 2 bases.

For questions 3–5, use this group of figures.

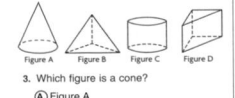
Figure A   Figure B   Figure C   Figure D

3. Which figure is a cone?
   A. Figure A
   B. Figure B
   C. Figure C
   D. Figure D

4. Which figure is a triangular prism?
   A. Figure A
   B. Figure B
   C. Figure C
   D. Figure D

5. How many of these figures are polyhedrons?
   A. 4 of them
   B. 3 of them
   C. 2 of them
   D. not here

For questions 6–8, use the figure below.

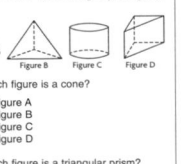

6. How many faces does this solid figure have?
   A. 4   B. 5   C. 6   D. 7

7. How many edges does this figure have?
   A. 5   B. 6   C. 8   D. not here

8. Which names a vertex of this figure?
   A. △ABC
   B. BC
   C. CA
   D. point D

9. A piece of cheese is cut in the shape of a triangular prism. How many edges does it have?
   A. 5   B. 6   C. 8   D. 9

10. Horace built a rectangular prism, using toothpicks for edges and miniature marshmallows for vertices. How many marshmallows did he use?
    A. 5   B. 6   C. 8   D. 12

11. What solid figure can you make from this net?
    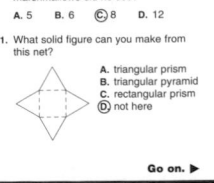
    A. triangular prism
    B. triangular pyramid
    C. rectangular prism
    D. not here

Form A • Multiple-Choice   A35   Go on. ▶

CHAPTER 10 TEST, page 2

Name _____

CHAPTER 10 TEST
PAGE 2

12. What solid figure can you make from this net?
    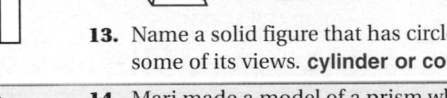
    A. rectangular prism
    B. rectangular pyramid
    C. triangular prism
    D. cube

13. How many faces of this prism have the dimensions 2 in. × 4 in.?
    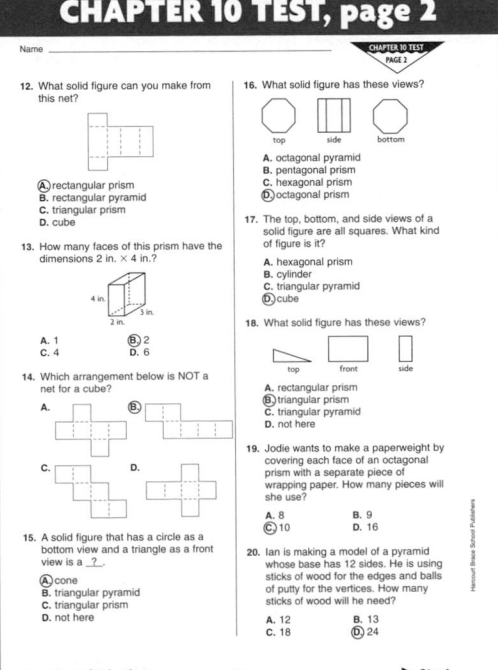 4 in.
    2 in.   3 in.
    A. 1   B. 2
    C. 4   D. 6

14. Which arrangement below is NOT a net for a cube?
    A.
    B.
    C.   D.

15. A solid figure that has a circle as a bottom view and a triangle as a front view is a __?__.
    A. cone
    B. triangular pyramid
    C. triangular prism
    D. not here

16. What solid figure has these views?
    top   side   bottom
    A. octagonal pyramid
    B. pentagonal prism
    C. hexagonal prism
    D. octagonal prism

17. The top, bottom, and side views of a solid figure are all squares. What kind of figure is it?
    A. hexagonal prism
    B. cylinder
    C. triangular pyramid
    D. cube

18. What solid figure has these views?
    top   front   side
    A. rectangular prism
    B. triangular prism
    C. triangular pyramid
    D. not here

19. Jodie wants to make a paperweight by covering each face of an octagonal prism with a separate piece of wrapping paper. How many pieces will she use?
    A. 8   B. 9
    C. 10   D. 16

20. Ian is making a model of a pyramid whose base has 12 sides. He is using sticks of wood for the edges and balls of putty for the vertices. How many sticks of wood will he need?
    A. 12   B. 13
    C. 18   D. 24

Form A • Multiple-Choice   A36   ▶ Stop!

See *Test Copying Masters*, pp. A35–A36 for the Chapter Test in **multiple-choice** format and pp. B185–B186 for **free-response** format.

201

STUDY GUIDE & REVIEW

CHAPTERS 8–10

✓ *Assessment Checkpoint*

USING THE PAGES
This Study Guide reviews the following content:

Chapter 8 Geometric Figures, pp. 148–165

Chapter 9 Symmetry and Transformations, pp. 166–183

Chapter 10 Solid Figures, pp. 184–201

CHAPTERS 8–10 STUDY GUIDE & REVIEW

VOCABULARY CHECK

1. When two lines intersect to form right angles, or 90° angles, they are _?_ to each other. (page 152) **perpendicular**

2. If a figure can be folded or reflected so that its two parts match, or are congruent, that figure has _?_. (page 168) **line symmetry**

3. A cone's curved lateral surface comes to a point at the cone's single _?_. (page 187) **vertex**

EXAMPLES

• **Classify lines.** (pages 152–153)

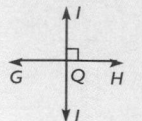

$\overleftrightarrow{GH}$ intersects $\overleftrightarrow{IJ}$ at point Q.
$\overleftrightarrow{GH} \perp \overleftrightarrow{IJ}$

• **Identify and measure angles.** (pages 154–155)

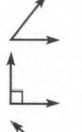

An acute angle has a measurement less than 90°.
A right angle has a measurement of 90°.
An obtuse angle has a measurement greater than 90° and less than 180°.

• **Identify line symmetry and rotational symmetry.** (pages 168–171)

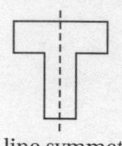 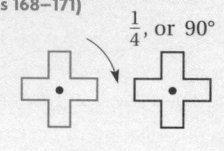

$\frac{1}{4}$, or 90°

line symmetry   rotational symmetry

EXERCISES

For Exercises 4–5, use the diagram below.

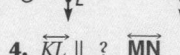

4. $\overleftrightarrow{KL} \parallel \_?\_ \overleftrightarrow{MN}$ 5. $\overleftrightarrow{KL} \_?\_ \overleftrightarrow{OP} \perp$

Trace the angle and measure it. Write *acute, right,* or *obtuse.*

6. 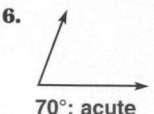   7.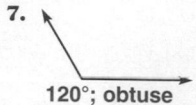

70°; acute   120°; obtuse

8. Trace the figure. Draw the line(s) of symmetry.

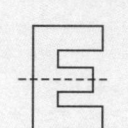 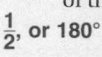

9. The figure has rotational symmetry. Identify the fraction and angle measure of the turn.

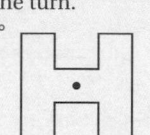

$\frac{1}{2}$, or 180°

202 Review • Chapters 8–10

See *Test Copying Masters,* pp. A37–A38 for copying masters of the Test • Chapters 8–10 in **multiple-choice** format and pp. B187–B188 for **free-response** format.

202

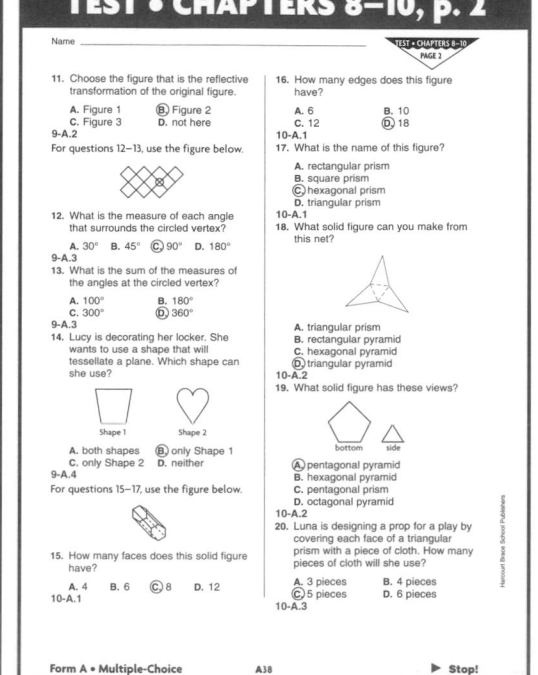

- **Use polygons to make a tessellation.**
  (pages 178–179)

  A hexagon will tessellate.

Tell whether the polygon will tessellate a plane. Write *yes* or *no*.

10.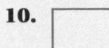

yes

11.

no

12.

no

- **Classify solid figures.** (pages 186–189)

  Name the figure. Then write *polyhedron* or *not a polyhedron*.

  rectangular pyramid; polyhedron

Name the figure. Then write *polyhedron or not a polyhedron.*

13.

cylinder;
not a polyhedron

14.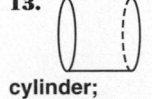

rectangular prism;
polyhedron

- **Name and count faces, edges, and vertices of solid figures.** (pages 190–191)

  5 faces: *DEF, ABC, EBCF, FCAD, DABE*
  9 edges: *DE, EF, FD, AB, BC, CA, BE, CF, AD*
  6 vertices: *A, B, C, D, E, F*

  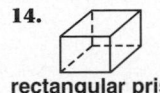

Name and count the faces, edges, and vertices. **See Additional Answers, p. 203A.**

15.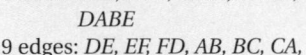

4 faces, 6 edges,
4 vertices

16.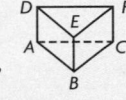

6 faces, 12 edges,
8 vertices

PROBLEM-SOLVING APPLICATIONS

Choose a strategy to solve.

17. Name the geometric figure that is suggested by the small or large hand of a clock. Can a line segment be part of this geometric figure? Explain. (pages 150–151)
ray; yes: any two points of the ray can be the endpoints of a line segment

19. Jervey wants to make a quilt entirely from octagonal pieces of cloth. Will he be able to do this? Why or why not? (pages 180–181)
No; octagons will not tessellate a plane.

18. Monica wanted to order a table that could seat eight people, but wasn't a quadrilateral. She also wanted each person to have equal space. What shape should she order? (pages 162–163) **regular octagon**

20. Susan made a model of a prism. Its base has 5 sides. What kind of prism is her model? How many vertices, edges, and faces does it have? (pages 190–191)
a pentagonal prism; 10 vertices; 15 edges; 7 faces

Portfolio

PORTFOLIO SUGGESTIONS
The portfolio illustrates the growth, talents, achievements, and reflections of the mathematics learner. You may wish to have students spend a short time selecting work samples for their portfolios and complete the Portfolio Summary Guide in the *Teacher's Guide for Assessment*, page 23. Students may wish to address the following questions:

- **What new understanding of math have I developed in the past several weeks?**

- **What growth in understanding or skills can I see in my work?**

- **What can I do to improve my understanding of math ideas?**

- **What would I like to learn more about?**

For information about how to organize, share, and evaluate portfolios, see the *Teacher's Guide for Assessment*, pages 20–22.

ADDITIONAL ANSWERS
Grade 6, Unit 3, Chapters 1–10

PE Assessment, Study Guide & Review

15. faces: *GHI, GJH, JHI, GJI*
    edges: *GJ, HI, JI, GH, JH, GI*
    vertices: *G, J, I, H*

16. faces: *PQLK, QRML, ORMN, POML, KNML, PORQ*
    edges: *PO, QR, OR, PQ, KN, ML, NM, KL, QL, RM, PK, ON*
    vertices: *P, O, R, Q, K, N, M, L*

CHAPTERS 8–10

✔ *Assessment Checkpoint*

With the *Performance Assessment Teacher's Guide,*

• have students work individually or in pairs as an alternative to formal assessment.

• use the evaluation checklist to evaluate items 1–3.

• use the rubric and model papers below to evaluate items 4–6.

• use the copying masters to evaluate interview/task tests on pp. 83–85.

SCORING RUBRIC FOR PROBLEM SOLVING TASKS, ITEMS 4–6

Levels of Performance

Level 3 Accomplishes the purposes of the task. Student gives clear explanations, shows understanding of mathematical ideas and processes, and computes accurately.

Level 2 Purposes of the task not fully achieved. Student demonstrates satisfactory but limited understanding of the mathematical ideas and processes.

Level 1 Purposes of the task not accomplished. Student shows little evidence of understanding the mathematical ideas and processes and makes computational and/or procedural errors.

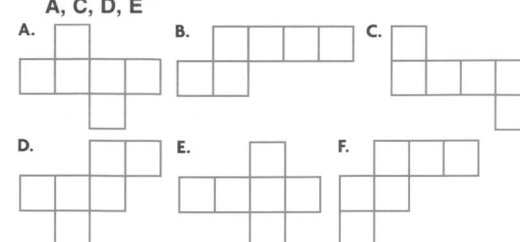

CHAPTERS 8–10 **WHAT DID I LEARN?**

For items 1–3, see Evaluation Checklist in the *Performance Assessment Guide.* Explanations will vary.

✏ Write About It

1. Explain how a regular pentagon is different from any other pentagon. Draw a regular pentagon. (pages 162–163) **Check students' explanations.**

2. Write the letters of the alphabet in all capital letters. Draw all the lines of symmetry on each letter. Explain. (pages 168–171) **Check students' drawings.**

3. Which of the nets below can be folded into a cube? Explain. (pages 192–193)
A, C, D, E

A.   B.   C.

D.   E.   F.

```
A  B  C  D  E  F  G  H
I  J  K  L  M  N  O  P
Q  R  S  T  U  V  W  X
Y  Z
```

✔ Performance Assessment

Choose a strategy and solve. Explain your method.

Problem-Solving Strategies
• Find a Pattern • Make a Model • Use a Table
• Draw a Diagram • Act It Out • Solve a Simpler Problem

Choice of strategies will vary. Explanations will vary.

4. Draw $\overline{AB}$ parallel to $\overline{CD}$. Then draw $\overline{EF}$ perpendicular to $\overline{AB}$. What is the relationship between $\overline{EF}$ and $\overline{AB}$?
(pages 156–159)
Check students' drawings; $\overline{EF} \perp \overline{AB}$.

5. Amy wants to design a tessellation screen saver for her computer. Choose a polygon that Amy could use. Then tessellate a plane using that polygon and draw at least 3 rows of the tessellation. (pages 178–179)
Check students' tessellations; use a square, rectangle, hexagon, or any triangle.

6. Tony is creating a model of a pyramid on his computer. If the base has 8 sides, how many vertices will his pyramid have? How many edges? How many faces?
(pages 190–191)
9 vertices, 16 edges, 9 faces.

STUDENT WORK SAMPLES for Item 5

Level 3	Level 2	Level 1

The student chose a shape that tessellates, drew the design correctly, and clearly explains her work.

The student showed an understanding of a tessellating design, but some triangles are smaller than others and do not match the tessellation pattern.

The student tried to tessellate with L-shaped figures but incorrectly drew shapes of different sizes and left gaps.

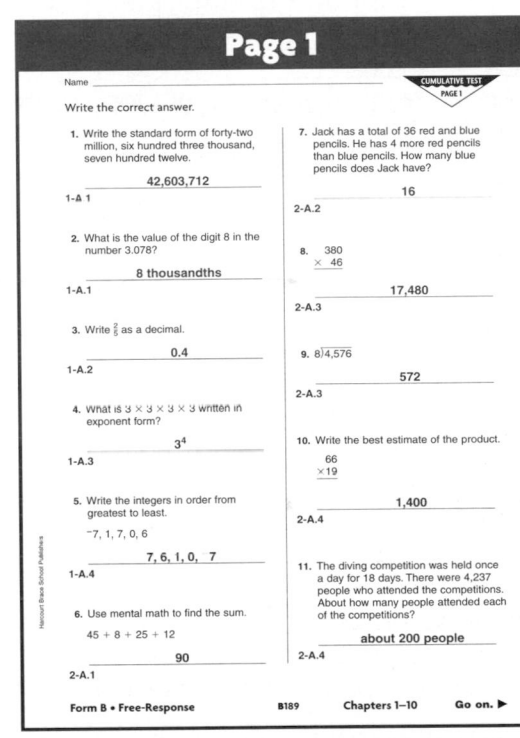
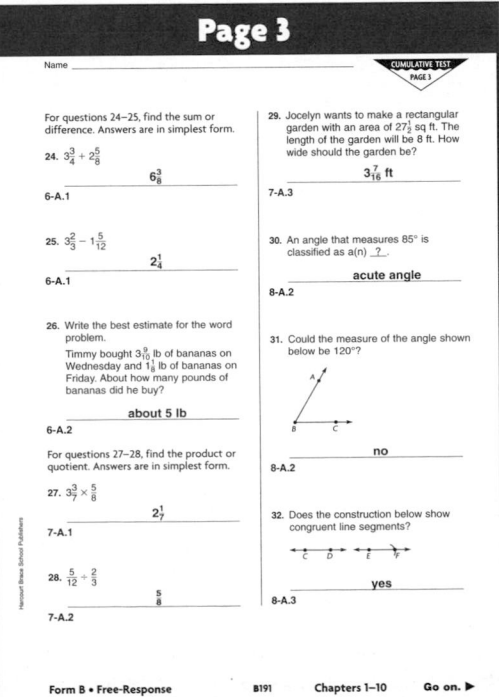
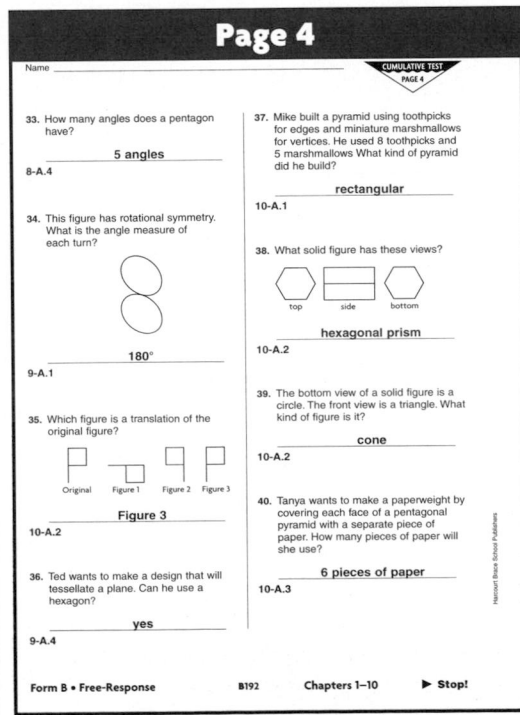

CUMULATIVE REVIEW

Solve the problem. Then write the letter of the correct answer.

1. Which of the following is the standard form for two hundred thirty and seventy-one ten-thousandths? (pages 16–17)

A. 230.0071 B. 230.071
C. 230.71 D. 2,300.0071 **1A.1**

2. 7.306×4.8 (pages 58–61)

A. 35.0688 B. 350.688
C. 3,506.88 D. 350,688 **3A.2**

For Exercises 3–6, choose the correct answer that is in simplest form.

3. $\frac{1}{9} + \frac{2}{3}$ (pages 98–99)

A. $\frac{1}{4}$ B. $\frac{7}{9}$
C. $\frac{14}{18}$ D. $1\frac{1}{9}$ **5A.2**

4. $3\frac{5}{7} - 1\frac{4}{5}$ (pages 118–121)

A. $\frac{60}{35}$ B. $1\frac{32}{35}$
C. $2\frac{1}{7}$ D. $2\frac{32}{35}$ **6A.1**

5. $\frac{4}{9} \times \frac{3}{7}$ (pages 128–131)

A. $\frac{12}{72}$ B. $\frac{12}{56}$
C. $\frac{4}{21}$ D. $1\frac{1}{27}$ **7A.1**

6. $\frac{6}{8} \div 2$ (pages 138–139)

A. $\frac{3}{8}$ B. $\frac{7}{16}$
C. $\frac{3}{4}$ D. $1\frac{1}{2}$ **7A.2**

7. Identify the figure. (pages 150–151)

A. angle XYZ B. ray XY
C. line segment XY D. line XY **8A.1**

8. What type of angle is shown? (pages 154–155)

A. acute B. straight
C. right D. obtuse **8A.2**

9. What is the name of this polygon? (pages 162–163)

A. heptagon B. hexagon
C. octagon D. pentagon **8A.4**

For Exercises 10–11, tell what kind of transformation the second figure is of the first. (pages 174–177)

10. R Я

A. translation B. reflection
C. rotation D. tessellation **9A.2**

11. ⊤ · ⊥

A. translation B. reflection
C. tessellation D. rotation **9A.2**

12. What is the name of this solid figure? (pages 186–189)

A. rectangular pyramid B. cone
C. triangular pyramid D. cube **10A.1**

✓ Assessment Checkpoint

USING THE PAGE
This review is in multiple-choice format to provide the students with practice for standardized testing. More in-depth Cumulative Tests are provided in free-response and multiple-choice formats. See *Test Copying Masters*, pages A39–A42 for the multiple-choice test and pages B189–B192 for the free-response test.

CUMULATIVE TEST
FREE-RESPONSE FORMAT
CHAPTERS 1-10, PAGES 1-4

ADDITIONAL ANSWERS
Lesson 10.4, page 195

8. top   front   bottom   9. top   front   bottom

10. top   front   bottom   11. top   front   bottom

Chapter 10 Lab, page 196

Try This

1. top   front   side   2. top   front   bottom

3. top   front   side   4. top   front   side

Chapter 10 Review, page 201

4. 5 faces: *ABED, BEFC, ADFC, ABC, DFE;*
   9 edges: *AB, DE, EF, BC, DF, AC, DA, EB, FC;*
   6 vertices: *A, B, C, D, E, F*

5. 5 faces: *UWR, WRS, WTS, WUT, UTSR;*
   8 edges: *WU, WR, WS, WT, RS, ST, TU, UR;*
   5 vertices: *R, S, T, U, W*

6. 6 faces: *KLHG, LHIM, NMIJ, KNJG, GHIJ, KLMN;*
   12 edges: *KL, LH, HG, GK, KN, NJ, JG, NM, MI, IJ,*
   ML, HI;
   8 vertices: *G, H, I, J, K, L, M, N*

10. top   front   side

11. top   front   side

12. top   front   side

Organizing Data

BIG IDEA: Statistics is a tool for solving problems. Analyzing statistical data helps you make predictions and decisions.

Introducing the Chapter p. 206		Team-Up Project • What Was the Question? p. 207	
OBJECTIVE	**VOCABULARY**	**MATERIALS**	**RESOURCES**
11.1 Defining the Problem pp. 208-209 **Objective:** To define decisions to be made from a survey **2 DAYS**			■ Reteach, ■ Practice, ■ Enrichment 11.1; More Practice, p. H57
11.2 Choosing a Sample pp. 210-211 **Objective:** To identify samples **1 DAY**	sample population random sample		■ Reteach, ■ Practice, ■ Enrichment 11.2; More Practice, p. H58
11.3 Bias in Surveys pp. 212-213 **Objective:** To determine whether a sample or question in a survey is biased **1 DAY**	biased		■ Reteach, ■ Practice, ■ Enrichment 11.3; More Practice, p. H58 Math Fun • What's Your Bias?, p. 220
Lab Activity: Questions and Surveys pp. 214-215 **Objective:** To write survey questions and to conduct a survey			E-Lab • Activity 11 E-Lab Recording Sheets, p. 11 Math Fun • Family Heritage Survey, p. 220
11.4 Collecting and Organizing Data pp. 216-219 **Objective:** To record and organize data collected in a survey **2 DAYS**	tally table frequency table cumulative frequency range		■ Reteach, ■ Practice, ■ Enrichment 11.4 More Practice, p. H58 Math Fun • What Kind of Pet Do You Have?, p. 220 Data ToolKit

INTERVENTION • Range and Intervals, Teacher's Edition pp. 220A–220B

CHAPTER ASSESSMENT Chapter 11 Review   p. 221

NCTM CURRICULUM
STANDARDS FOR GRADES 5–8

- ✔ **Problem Solving** *Lessons 11.1, 11.2, 11.3, 11.4*
- ✔ **Communication** *Lessons 11.1, 11.2, 11.3, 11.4*
- ✔ **Reasoning** *Lessons 11.1, 11.2, 11.3, 11.4*
- ✔ **Connections** *Lessons 11.1, 11.2, 11.3, 11.4*
- ✔ **Number and Number Relationships** *Lesson 11.4*
- ✔ **Number Systems and Number Theory** *Lessons 11.2, 11.4*
- ✔ **Computation and Estimation** *Lessons 11.2, 11.4*
- ☐ **Patterns and Functions**
- ☐ **Algebra**
- ✔ **Statistics** *Lessons 11.1, 11.2, 11.3, 11.4*
- ☐ **Probability**
- ☐ **Geometry**
- ☐ **Measurement**

ASSESSMENT OPTIONS

CHAPTER LEARNING GOALS

11A.1 To identify information needed to make decisions

11A.2 To identify sample sizes and types of samples when conducting a survey

11A.3 To determine whether a sample or question in a survey is biased

11A.4 To use and organize data from a survey

These goals for concepts, skills, and problem solving are assessed in many ways throughout the chapter. See the chart below for a complete listing of both traditional and informal assessment options.

Pretest Options
Pretest for Chapter Content, TCM pp. A43–A44, B193–B194
Assessing Prior Knowledge, TE p. 206; APK p. 13

Daily Assessment
Mixed Review, PE pp. 209, 219
Problem of the Day, TE pp. 208, 210, 212, 216
Assessment Tip and Daily Quiz, TE pp. 209, 211, 213, 215, 219

Formal Assessment
Chapter 11 Review, PE p. 221
Chapter 11 Test, TCM pp. A43–A44, B193–B194
Study Guide & Review, PE pp. 272–273
Test • Chapters 11–14, TCM pp. A53–A54, B203–B204
Cumulative Review, PE p. 275
Cumulative Test, TCM pp. A55–A58, B205–B208

Performance Assessment
Team-Up Project Checklist, TE p. 207
What Did I Learn?, Chapters 11–14, TE p. 272
Interview/Task Test, PA p. 86

Portfolio
Suggested work samples:
Independent Practice, PE pp. 209, 211, 213, 218
Math Fun, PE p. 220

Student Self-Assessment
How Did I Do?, TGA p. 15
Portfolio Evaluation Form, TGA p. 24
Math Journal, TE pp. 206B, 208

Key
Assessing Prior Knowledge Copying Masters: APK
Test Copying Masters: TCM
Performance Assessment: PA
Teacher's Guide for Assessment: TGA
Teacher's Edition: TE
Pupil's Edition: PE

MATH COMMUNICATION

MATH JOURNAL
You may wish to use the following suggestions to have students write about the mathematics they are learning in this chapter:

PE page 208 • Talk About It

PE pages 209, 211, 213, 219 • Write About It

TE page 208 • Writing in Mathematics

TGA page 15 • How Did I Do?

In every lesson • record written answers to each Wrap-up Assessment Tip from the Teacher's Edition for test preparation.

VOCABULARY DEVELOPMENT
The terms listed at the right will be introduced or reinforced in this chapter. The boldfaced terms in the list are new vocabulary terms in the chapter.

So that students understand the terms rather than just memorize definitions, have them construct their own definitions and pictorial representations and record those in their Math Journals.

sample, p. 210
population, p. 210
random sample, p. 210
biased, p. 212
tally table, p. 216
frequency table, pp. 216, 217
cumulative frequency, pp. 216, 217
range, pp. 216, 218

MEETING INDIVIDUAL NEEDS

RETEACH
BELOW-LEVEL STUDENTS

Practice Activities, TE Tab B
Racing to the Finish Line
(Use with Lessons 11.1–11.4.)

Interdisciplinary Connection, TE p. 218

 Literature Link, TE p. 206C
Little Farm in the Ozarks, Activity 1
(Use with Lesson 11.4.)

 Technology
Data ToolKit
(Use with Lesson 11.4.)

E-Lab • Activity 11
(Use with Lab Activity.)

PRACTICE
ON-LEVEL STUDENTS

Bulletin Board, TE p. 206C
Read It and Record It
(Use with Lessons 11.1–11.4.)

Interdisciplinary Connection, TE p. 218

 Literature Link, TE p. 206C
Little Farm in the Ozarks, Activity 2
(Use with Lesson 11.4.)

 Technology
Data ToolKit
(Use with Lesson 11.4.)

E-Lab • Activity 11
(Use with Lab Activity.)

CHAPTER IDEA FILE

The activities on these pages provide additional practice and reinforcement of key chapter concepts. The chart above offers suggestions for their use.

LITERATURE LINK

Little Farm in the Ozarks
by Roger Lea MacBride

This story tells of Rose Wilder's experiences in her new home in the Missouri Ozarks.

Activity 1—Rose's first experience is in a general store. She is amazed by all there is to purchase there. Have students choose four kinds of candy and tally how many classmates like each kind.

Activity 2—Rose wins a school spelling bee. Have students survey their classmates to find out how many spelling bees each has participated in. Then have students chart the percentage of those who have been in one spelling bee, two, three, and so on. Ask students how they think the distribution will change over the years.

Activity 3—Rose's parents kept a record of their money to make ends meet. Have students keep a record of their families' food purchases for a week. Then have students make a table of food categories and the money spent on each category. Next, have students compare their families' budgets with those of other classmates.

BULLETIN BOARD

READ IT AND RECORD IT

NUMBER OF BOOKS WE'VE READ			
Our Class	Week 1	Week 2	Week 3
Girls	ⅢⅡ ⅢⅡ Ⅰ		
Boys	ⅢⅡ ⅢⅡ Ⅲ		

Use this bulletin board to practice collecting and organizing data as you complete chapters 11–14. Each day for a three-week period, have students put a mark on the tally table for each book they read.

Activity 1 Students make a bar graph showing how many books boys read each week, how many girls read, and how many boys and girls read.

Activity 2 Students make a multiple line graph showing all the data from the completed tally table.

Bulletin Board, TE p. 206C
Read It and Record It
(Use with Lessons 11.1–11.4.)

Interdisciplinary Connection, TE p. 218

Literature Link, TE p. 206C
Little Farm in the Ozarks, Activity 3
(Use with Lesson 11.4.)

Technology
Data ToolKit
(Use with Lesson 11.4.)

E-Lab • Activity 11
(Use with Lab Activity.)

INTERDISCIPLINARY SUGGESTIONS

Connecting Data

Purpose To connect organizing data to other subjects students are studying

You may wish to connect *Organizing Data* to other subjects by using related interdisciplinary units.

Suggested Units

- Technology Connection: Kids and Computers
- History Connection: Ancient Artifacts
- Multicultural Connection: Where Does Chocolate Come From?
- Writing Connection: Planning a Survey
- Literature Connection: Best-Selling Paperbacks
- Art Connection: America's Favorite Colors
- Career Connection: Publishing
- Language Connection: Long Words
- Music Connection: Popular Songs

Modalities Auditory, Kinesthetic, Visual

Multiple Intelligences Mathematical/Logical, Bodily/Kinesthetic, Visual/Spatial

The purpose of using technology in the chapter is to provide reinforcement of skills and support in concept development.

E-Lab • Activity 11—Students analyze experiment results to discover that when doing an experiment or conducting a survey, the results often depend on the samples that you take.

Data ToolKit
This program provides a spreadsheet and graphing tool that will help students organize data they collect, choose appropriate graphs for their data, compare graphs they chose, and find measures of central tendency.

Visit our web site for additional ideas and activities.
http://www.hbschool.com

Introducing the Chapter

Encourage students to talk about how they organize information. Review the math content students will learn in the chapter.

WHY LEARN THIS?

The question "Why are you learning this?" will be answered in every lesson. Sometimes there is a direct application to everyday experiences. Other lessons help students understand why these skills are needed for future success in mathematics.

PRETEST OPTIONS

Pretest for Chapter Content See the *Test Copying Masters*, pages A43–A44, B193–B194 for a free-response or multiple-choice test that can be used as a pretest.

Assessing Prior Knowledge Use the copying master shown below to assess previously taught skills that are critical for success in this chapter.

Assessing Prior Knowledge Copying Masters, p. 13

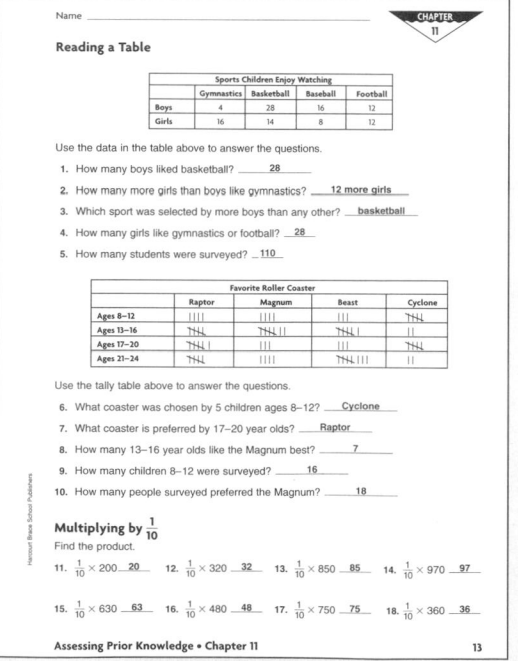

Troubleshooting

If students need help	Then use
• READING A TABLE	• Focus on Problem Solving, PE p. 9
• MULTIPLYING BY $\frac{1}{10}$	• Practice Activities, TE Tab B, Activity 19A

ORGANIZING DATA

LOOK AHEAD

In this chapter you will solve problems that involve

- defining a problem
- choosing a sample
- writing a survey and determining whether it is biased
- collecting and organizing data

206   Chapter 11

SCHOOL-HOME CONNECTION

You may wish to send to each student's family the *Math at Home* page. It includes an article that stresses the importance of communication about mathematics in the home, as well as helpful hints about homework and taking tests. The activity and extension provide extra practice that develops concepts and skills involving organizing data.

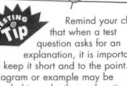

Math at Home

What Was The Question?

Surveys appear in newspapers, magazines, on the radio, and on television. Use the results of a survey to write possible survey questions.

YOU WILL NEED: newspapers, magazines

Plan • Work with a partner to write possible questions for survey results and three questions you can answer using data from survey results.

Act • Look and listen for survey results.

• Write questions that might have been asked to get those results. Write three questions you can answer using the data.

• Display your questions and the survey results.

Share • Present your survey results and questions to the class.

• Try to think of other possible questions for each survey.

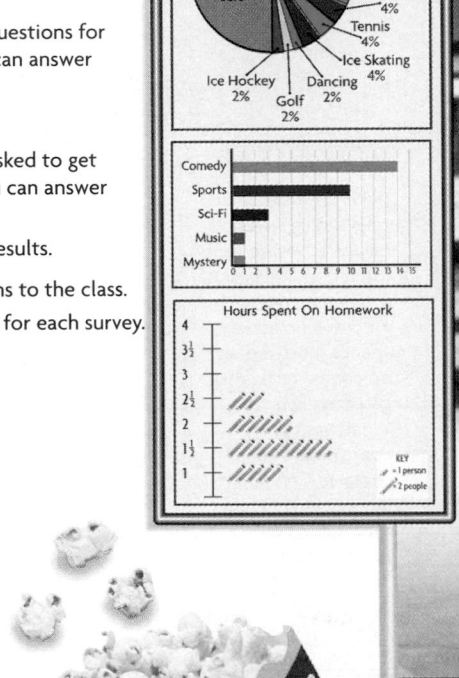

SURVEYS

Basketball 9%
Football 24%
Volleyball 7%
Baseball 6%
Soccer 36%
Lacrosse 4%
Tennis 4%
Ice Skating 4%
Ice Hockey 2%
Dancing 2%
Golf 2%

Comedy
Sports
Sci-Fi
Music
Mystery
0 1 2 3 4 5 6 7 8 9 10 11 12 13 14 15

Hours Spent On Homework
4
3½
3
2½
2
1½
1
KEY
= 1 person
= 2 people

DID YOU

☑ Find survey results?

☑ Write possible survey questions?

☑ Write possible questions you can answer from the data?

☑ Share your survey with the class?

207

Project Checklist

Evaluate whether partners

☑ work cooperatively.

☑ identify survey data.

☑ write appropriate questions.

Using the Project

Accessing Prior Knowledge reading graphs and tables, identifying appropriate data

Purpose to write questions for survey data

GROUPING partners

MATERIALS magazines, newspapers, radios, TVs, on-line services

Managing the Activity You may want to gather magazines such as *Zillions* or *US News* that often present survey data.

WHEN MINUTES COUNT

A shortened version of this activity can be completed in 15–20 minutes. Ask students to write survey questions for the survey data shown in the chapter opener.

A BIT EACH DAY

Day 1: Students find partners and brainstorm about where they might find survey data.

Day 2: Students locate survey data.

Day 3: Students write survey questions.

Day 4: Students write three questions they can answer using data from the surveys.

Day 5: Students share their questions and suggest other possible questions.

ON THE ROAD

Have students ask friends and family what survey questions they have answered.

ORGANIZER

OBJECTIVE To define decisions to be made from a survey

ASSIGNMENTS Independent Practice
- *Basic* 1–4; 5–14
- *Average* 1–4; 5–14
- *Advanced* 1–4; 5–14

PROGRAM RESOURCES
Reteach 11.1 Practice 11.1 Enrichment 11.1

PROBLEM OF THE DAY

Transparency 11

Frank used centimeter cubes to make a solid 3 cubes wide, 5 cubes long, and 4 cubes high. How many more cubes does he need if he wants to make the solid

a. twice as wide, half as high, but with the same length? He does not need any more cubes.

b. twice as wide, twice as long, and twice as high? He needs 420 more cubes.

Solution Tab A, p. A11

QUICK CHECK
Divide. Round the answer to the nearest whole number.

1. 530 ÷ 10 53
2. 890 ÷ 10 89
3. 70 ÷ 10 7
4. 425 ÷ 10 43
5. 234 ÷ 10 23
6. 78 ÷ 10 8

1 MOTIVATE

Discuss with students their previous experience with surveys.

- **What is a survey?**
- **Who conducts surveys?**
- **Why do they conduct surveys?**
- **Describe any survey you have participated in.**

11.1 Defining the Problem

WHAT YOU'LL LEARN
How to define a problem by deciding what information you need

WHY LEARN THIS?
To make decisions from information you get from a survey

A sixth-grade class is planning an overnight camping trip. The students in charge of the food need to buy breakfast cereal for the trip. They plan to survey everyone going on the trip, because they have to make some decisions. What decisions do they have to make? What information do they need?

The students wrote down the decisions they have to make.

a. the amount of cereal to take
b. the type of cereal to take

To make the decisions, the students need this information:

1. How many people are going on the trip?
2. How much cereal does the average person eat at breakfast?
3. What type of cereal do the people going on the trip like?

Talk About It

- What other decisions can you think of that the students should make? **Possible answer: what to take for people who don't like cereal, how much milk for the cereal**
- What method would you use to gather the information needed? **Possible answer: Have everyone going on the trip answer questions 2 and 3.**

Before you take a survey to gather information, you have to define the problem. You can do this by deciding what information you need to make decisions.

EXAMPLE Another group of students is in charge of transportation for the camping trip. What is one decision they have to make? What information do they need?

Here is a possible decision for them to make:

whether to use cars or buses for transportation

The students need to get this information:

1. How many people are going on the trip?
2. How many people can ride in each car or in each bus?
3. How much money will be needed to buy gasoline for the cars or to rent the buses?

Consumer Link

In 1993, American television showed 1.3 million advertisements for cereal, at a cost of $762 million. Only the automobile industry spends more money on commercials than cereal makers do. What was the approximate amount cereal makers spent per day in 1993 for their television advertisements?

about $2 million

IDEA FILE

WRITING IN MATHEMATICS

To help students organize their ideas for planning a survey, have them make a flowchart that helps them organize the process.

Example:

> Survey
> Writing Questions
> Collecting Information
> Analyzing Data
> Reaching Conclusions

Modalities Auditory, Visual

RETEACH 11.1

Name _____

LESSON 11.1

Defining the Problem

The PTO is going to sell ice cream at the school carnival. The PTO needs to make some decisions about the sale.
- the amount of ice cream to have available for the sale
- how much to charge for the ice cream
- how to keep the ice cream frozen at the carnival

These are some decisions that the PTO will *not* have to make.
- what rides will be at the carnival
- what type of booths will be at the carnival
- the date and time of the carnival

While the decisions in the second list do apply to the carnival, they do *not* apply to the selling of ice cream. Those decisions have to be made by the people running the carnival.

Solve.
1. The local baseball league will be selling hot dogs at the carnival. Circle the decisions that the baseball league has to make.
   - where the carnival will be held
   - ○ how much to charge for a hot dog
   - ○ how to cook the hot dogs
   - the date of next year's carnival
   - ○ the number of hot dogs to buy for the carnival

2. At the carnival, owners of a local restaurant plan to give out coupons for a free meal. Circle the decisions that the restaurant owners need to make about the coupons.
   - how much to pay their workers
   - ○ how many coupons to bring to the carnival
   - ○ who will hand out the coupons
   - the hours the restaurant is open
   - ○ dates the coupons can be used

3. The organizers of the carnival want to charge admission to the event. Circle the decisions they have to make about admission tickets.
   - ○ how many tickets will be available
   - where people will park their cars
   - ○ how admission tickets will be sold
   - who will be in charge of advertising the carnival
   - ○ whether adults and children must pay the same amount for a ticket

R48 TAKE ANOTHER LOOK

Use with text pages 208–209.

GUIDED PRACTICE

Solve.

1. Wendy and Jo are in charge of buying beverages for a sixth-grade party. Which of these decisions will they have to make?
  a. which beverages to buy
  b. the location of the party
  c. the amount of beverages to buy
  a and c

2. For the decisions you chose in Problem 1, tell what information is needed to make the decisions. **Possible answers: the number of people attending, the types of beverages people like, the amount each person will drink**

INDEPENDENT PRACTICE

Solve.

1. The Student Council is going to sell popcorn at the next basketball game. Which of these decisions will they have to make?
  a. the amount of popcorn to make
  b. the price they should charge for the popcorn
  c. the name of the team their school is playing
  a and b

2. Which of the following questions should the Student Council answer before they sell popcorn?
  a. How many games did their school win this season?
  b. What is the average number of people that attend each game?
  c. How much are people willing to pay for popcorn?
  b and c

Problem-Solving Applications

3. A group of students is in charge of music for a dance. What decisions do they need to make? What information do they need to make the decisions? **See Additional Answers, p. 221A.**

4. ▶ **WRITE ABOUT IT** Describe a situation in which you had to gather information in order to make a decision. Tell what information you needed. **Descriptions will vary.**

Mixed Review

Find the product.

5. $\frac{1}{10} \times 30$ **3**   **6.** $\frac{1}{10} \times 70$ **7**   **7.** $\frac{1}{10} \times 320$ **32**   **8.** $\frac{1}{10} \times 240$ **24**

Name the figure.

9. cone

10.  rectangular prism

11.  cylinder

12.  rectangular pyramid

13. **CHOOSE A STRATEGY** Walt has 160 stamps in his collection. He has 3 times as many national stamps as international stamps. How many of each does he have? **120 national stamps; 40 international stamps**

14. Suzanne bought 3 cans of lemonade, 4 cans of punch, 3 cans of soda, and 10 cans of juice. What fraction of the cans were punch? juice? **$\frac{1}{5}$ punch; $\frac{1}{2}$ juice**

MORE PRACTICE Lesson 11.1, page H57

209

---

2 TEACH/PRACTICE

The lesson begins with students identifying the information they need to make decisions for a survey.

GUIDED PRACTICE

To make sure students understand how to define a problem by making decisions, discuss the exercises as a class.

INDEPENDENT PRACTICE

To assess students' understanding of the concepts, check answers to Exercises 1–2.

ADDITIONAL EXAMPLE

Example, p. 208

Students at Jefferson Middle School want to start a school newspaper. What is one decision they need to make? What information do they need in order to make it? Check students' answers.

3 WRAP UP

Use these questions to help students focus on the big ideas of the lesson.

- **The local fire chief is planning a fire-prevention program for middle schools. What is one decision she needs to make to plan her program? What information does she need to make this decision?** Possible decision: whether she needs a different program at each grade level; possible information: What was covered in previous fire-prevention programs?

✔ *Assessment Tip*

Have students write the answer.

- **What do you need to do to make decisions for a survey?** Possible answer: Define what you want to find out.

DAILY QUIZ

Solve.

1. Members of the Collectors' Club are planning an exhibit of their collections. What is one decision they need to make? Possible answer: the date the exhibit will be held

2. What information do they need to make this decision? Possible answer: what events are already scheduled

209

GRADE 5 Graph survey results.	GRADE 6 Identify samples.	GRADE 7 Choose an appropriate sample for a survey.

ORGANIZER

OBJECTIVE To identify samples
VOCABULARY *New* **sample, population, random sample**
ASSIGNMENTS Independent Practice
■ *Basic* 1–10
■ *Average* 1–10
■ *Advanced* 1–10
PROGRAM RESOURCES
Reteach 11.2 Practice 11.2 Enrichment 11.2

PROBLEM OF THE DAY

Transparency 11

Dana's survey showed that 3 out of 8 students preferred vegetarian pizza and 1 out of 6 students preferred cheese pizza. How many more of the 72 students surveyed by Dana liked vegetarian pizza than liked cheese pizza? 15 more students liked vegetarian pizza.

Solution Tab A, p. A11

QUICK CHECK

Find $\frac{1}{10}$ of the given number. Round the answer to the nearest whole number.

1. 750 75   2. 922 92   3. 75 8
4. 104 10   5. 85 9   6. 2,456 246

1 MOTIVATE

Present this problem to students to introduce sampling:

• **Suppose you are in charge of the quality-control department for a manufacturer of game pieces. You want to check 1 out of every 10 game pieces for manufacturing defects. How many game pieces will you check in a production run of 536?** $\frac{1}{10}$ of 536

• **What is the answer?** 53.6

• **Would you check 53.6 game pieces?** No

• **What would you do?** Round 53.6 to 54, and check 54 game pieces.

11.2

Choosing a Sample

For the sixth-grade camping trip, the game committee is going to take a survey to find out which game most students want to play. How many students should they ask?

First, you need to look at the population. A **population** is a particular group of people, such as sixth graders. If the population is large, you might survey a **sample**, or a part of the population, to represent the population.

EXAMPLE 1 There are 340 people going on the camping trip. The game committee wants to survey a sample of this population. How many people should the committee survey?

The size of a sample depends on the size of the population. For a population of 340 students, a good sample is about 1 out of every 10, or 34 students.

• Suppose you are surveying 1 person out of every 10 for a sample. How many students would you survey out of 521? **52 students**

A sample can be chosen randomly. It is a **random sample** if every individual in the given population had an equal chance of being selected.

EXAMPLE 2 Ann is surveying the students at her school to find out their favorite sport. Will each of the following methods give a random sample of all the students? Explain.

A. At a basketball game, Ann can survey 1 out of every 10 students at the game who are from her school.

B. Ann can use a computer to select 92 students from a mixed-up list of the 920 students in the school. She can then survey those students.

Method A will not produce a random sample since some of the students at Ann's school probably don't go to basketball games. These students would not have an equal chance of being selected for the sample.

Method B will give a random sample since every student will have an equal chance of being selected.

Real-Life Link

The most popular board game, Monopoly®, was invented by Charles B. Darrow in 1934. Today Monopoly is played throughout the world. The game is made in 43 countries and in 26 languages. Take a survey of your friends to find out how many play Monopoly.

IDEA FILE

ART CONNECTION

Present this problem to students:

Did you know that blue is the color favored by most Americans? It is followed by red, green, white, pink, purple, and orange. Suppose you want to know the favorite color of the students in our school.

• **How would you choose a random sample for a survey?** Possible answer: Survey 1 out of every 10 students registered at school.

Modality Auditory

RETEACH 11.2

Name _____

Choosing a Sample

A **population** is a particular group of people. A **sample** is a part of a population. It is a **random sample** if every person in the population has an equal chance of being selected.

1. Carole is surveying members of her community to find out their favorite restaurants. Will each of the following methods produce a random sample of her community? Explain.

a. She stands outside a local mall and asks 1 out of every 10 people leaving the mall to name his or her favorite restaurant.

   No, because all members of her community might not shop at the mall.

b. She calls every tenth name listed in a telephone directory of her community.

   If all members of the community are included in the directory, then this will produce a random sample.

2. Evan is on a committee to determine if the 650 players in the Tri-Town Soccer League want to extend the season. Will each of the following methods produce a random sample of all the players? Explain.

a. Evan surveys the captain of each of the league's 40 teams.

   No; the sample is too small and does not include players who are not team captains.

b. Evan surveys the 65 oldest players in the league.

   No; the sample does not include younger members of the league.

c. Evan selects every tenth name on a list of the 650 league members. Then he surveys these 65 players.

   Yes; every individual in the population has an equal chance of being selected.

Use with text pages 210–211.   TAKE ANOTHER LOOK   R49

MEETING INDIVIDUAL NEEDS

GUIDED PRACTICE

Tell how many you would survey from each group if you survey 1 out of every 10 people. **Possible answer given.**

1. 420 students
   42 students
2. 110 teachers
   11 teachers
3. 280 boys
   28 boys
4. 630 girls
   63 girls
5. 390 voters
   39 voters
6. 860 doctors
   86 doctors
7. 510 drivers
   51 drivers
8. 330 athletes
   33 athletes

Tell whether each is a random sample. Write *yes* or *no*. Then explain.

9. Kerry wanted to know the most popular movies in the sixth grade. She surveyed 30 boys out of the 300 students in the sixth grade. **No; it excludes girls.**

10. Albert wanted to find out what snack foods students in his school prefer. He surveyed 1 out of every 10 students entering the school. **Yes; each student had an equal chance of being chosen.**

INDEPENDENT PRACTICE

Eric wants to find out if Park School cafeteria has a good selection of school lunches. There are 570 sixth, seventh, and eighth graders in the school.

1. How many people should he survey?
   possible answer: 57 people
2. Whom should Eric survey? **sixth, seventh, and eighth graders**
3. How can Eric get a random sample?
   Possible response: by randomly choosing students at lunch
4. Is a random sample of only sixth graders fair? Explain. **No; the sample should also include seventh and eighth graders.**

Problem-Solving Applications

Tell whether a random sample was chosen. Explain.

5. The news station wanted to survey a large number of voters about the election. As voters left the voting booth, 4 out of every 50 were surveyed. **Yes; all voters had equal chances of selection.**

6. Vanessa wanted to ask shoppers to name their favorite store. She surveyed 1 out of every 10 shoppers in Tony's Marketplace. **No; only shoppers from Tony's Marketplace were included.**

For Problems 7–9, use the table at the right.

7. The table shows the results of a survey Rob took at school. How many students did he survey? **100 students**

8. Rob surveyed 1 out of every 10 students as they left school. Was this a random sample of the students at school? **yes**

9. Suppose Rob surveyed 200 students. About how many would you expect to choose rock music? **about 56 students**

10. ✏ **WRITE ABOUT IT** Choose a topic you could find out about by conducting a survey. Then tell how you would choose a random sample. **Topics will vary; by surveying 1 out of every 10 people without excluding any part of the population.**

Favorite Music	
Type of Music	Number of Students
Pop	30
Rock	28
Country	27
Rap	15

Portfolio

MORE PRACTICE **Lesson 11.2, page H58**

211

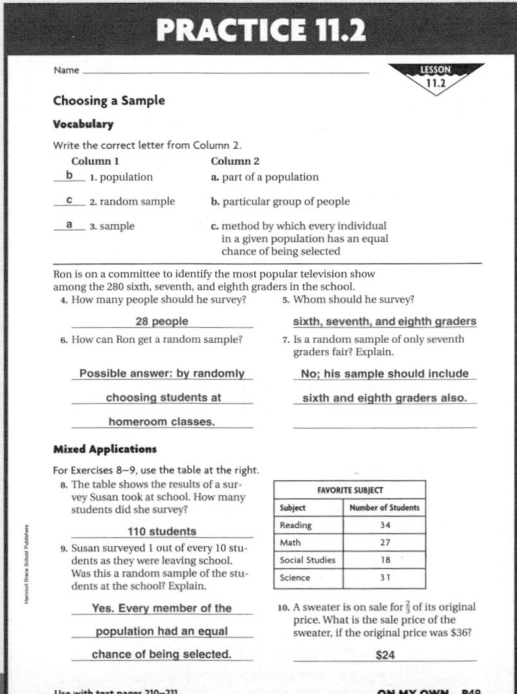
---

2 TEACH/PRACTICE

Reinforce the vocabulary words presented in the lesson as you go over the examples.

GUIDED PRACTICE

To reinforce how to identify samples, complete the exercises orally.

INDEPENDENT PRACTICE

To assess students' understanding of the concepts, check answers to Exercises 1–4.

ADDITIONAL EXAMPLES

Example 1, p. 210

Pizza will be served to 185 students. The cook is conducting a survey to find out what toppings students want on their pizza. How many students should be surveyed? A sample size should be about 1 out of every 10, so 19 of 185 students should be surveyed.

Example 2, p. 210

Dana is surveying students at the middle school. Which of the following methods produce a random sample? Explain.

a. A survey of 25 out of the 254 seventh graders.

b. A survey of 1 out of every 10 students registered at the middle school. Sample b because every student will have an equal chance of being selected.

3 WRAP UP

Use these questions to help students focus on the big ideas of the lesson.

- **A store surveyed 1 out of every 10 adult customers in a day. Will this method produce a random sample? Explain.** No. It excluded children.

✔ *Assessment Tip*

Have students write the answer.

- **What is a random sample?** Possible answer: a selection from a population in which every individual has an equal chance of being selected

DAILY QUIZ

Lyn is conducting a survey to find out how many of the 749 16- to 18-year-old licensed drivers in Miami wear seat belts.

1. How many drivers should be in the survey? 75
2. How can she get a random sample? Possible answer: by surveying 25 licensed drivers from each age group

211

GRADE 5
Graph survey results.

GRADE 6
Determine whether a sample or question in a survey is biased.

GRADE 7
Determine whether a sample is biased.

ORGANIZER

OBJECTIVE To determine whether a sample or question in a survey is biased

VOCABULARY *New* biased

ASSIGNMENTS Independent Practice
- *Basic* 1–14
- *Average* 1–14
- *Advanced* 1–14

PROGRAM RESOURCES
Reteach 11.3 Practice 11.3 Enrichment 11.3
Math Fun • What's Your Bias, p. 220

PROBLEM OF THE DAY

Transparency 11

In a survey, 90 students were in favor of a field trip to a nearby factory, 48 were against it, and 12 couldn't make up their minds. What fraction of the students were in favor of the trip? $\frac{90}{150}$, **or** $\frac{3}{5}$, **of the students were in favor of the trip.**

Solution Tab A, p. A11

QUICK CHECK
Is this sample large enough to represent the given population? Write *yes* or *no*.

1. 5 out of 520 librarians no
2. 90 out of 867 pilots yes
3. 10 out of 89 dancers yes
4. 8 out of 77 senior citizens yes

1 MOTIVATE

Read and discuss Teen Times to begin the lesson.

2 TEACH/PRACTICE

Encourage students to refer to the Random Sample Checklist in the Idea File as they do the exercises in this lesson.

GUIDED PRACTICE
To make sure students can determine whether a sample or question in a survey is biased, discuss the exercises as a class.

212

11.3

Bias in Surveys

WHAT YOU'LL LEARN
How to determine whether a sample or a question in a survey is biased

WHY LEARN THIS?
To understand what is a good sample and a good question for a poll or survey

WORD POWER
biased

A group of students wants to find out the favorite sport at North Middle School. Dwayne surveyed 15 students on the hockey team. Do you think the sample is a good one?

When you collect data from a survey, your sample should be large enough to represent the whole population, and every individual in that population should have an equal chance of being selected.

A sample is **biased** if individuals in the population are not represented in the sample. Dwayne's sample is biased since it only contains students on the hockey team.

EXAMPLE 1 Regina wants to survey sixth-grade students to find out the number of weekend hours students play video games. There are 450 students in the sixth grade. Which of the following sampling methods would be biased? Tell how you know.

A. Randomly survey 45 sixth-grade boys.

B. Randomly survey 45 sixth graders who do not play video games.

C. Randomly survey 45 sixth-grade students.

D. Randomly survey 4 sixth-grade students.

Choices A, B, and D are biased. Choice A excludes girls, choice B excludes students who play video games, and choice D calls for too few students to be representative of the population of 450.

Sometimes, the questions are biased. Questions that lead to a specific response or exclude a group are biased.

EXAMPLE 2 Is the following question biased?

Do you agree with the well-known and recognized Shoe Association that expensive shoes are more comfortable?

This question is biased since it leads you to agree with the Shoe Association.

- Rewrite the question so that it is not biased.
  Possible question: Which shoes do you think are more comfortable, Brand A or Brand B?

teen times

Some magazines, such as *Zillions®* include surveys. In a recent issue, these results were given for the number of hours kids play video games in a week. (The percent is based on the number of kids surveyed.)

Hours	Percent
0	28%
1–2	27%
3–4	17%
5–6	8%
7–8	6%
9+	14%

212   Chapter 11

IDEA FILE

HOME CONNECTION

HOME NOTE Students may use this checklist for reference in class or at home.

Random Sample Checklist

Is the sample large enough to represent the whole population?

Is any part of the population excluded from the sample?

Do the questions lead to a specific response?

Modalities Auditory, Visual

RETEACH 11.3

Name _____

LESSON 11.3

Bias in Surveys

A sample is **biased** if any individual in the population is not represented in the sample.

Linda wants to survey 125 sixth-grade students to find out the number of hours they spend reading each week. She makes a list of four sampling methods she could use.
- randomly survey 5 students
- randomly survey 15 sixth-grade boys
- randomly survey 13 students at her school
- randomly survey 15 sixth-grade students

She decided that the first method would not include a large-enough sample. The second method would not include sixth-grade girls. The third method was biased because it would not be especially for sixth graders. Linda decided to use the last method. It would contain a large-enough sample, and every member of the sixth grade would have an equal chance of being selected.

1. Marian wants to survey the 58 sixth-grade teachers at Oak Park School to find out the average number of hours sixth-grade teachers spend grading papers. She makes a list of four sampling methods she could use. Circle the sampling method that is not biased. Then explain how the other three methods are biased.
   a. randomly survey all teachers who have taught for more than 10 years
   b. randomly survey 10 male teachers at the school
   (c. randomly survey 8 sixth-grade teachers)
   d. randomly survey 8 math teachers at the school
   Method a excludes teachers who have taught less than 10 years.
   Method b excludes female teachers. Method d excludes

   teachers who teach subjects other than math.

2. Dan wants to survey the 230 members of a golf club to find out the average number of hours they play golf each week. He makes a list of four sampling methods he could use. Circle the sampling method that is not biased. Then explain how the other three methods are biased.
   a. randomly survey all club members who have won tournaments
   (b. randomly survey 30 club members)
   c. randomly survey all members who have ever shot a hole-in-one
   d. randomly survey 25 members under the age of 30
   Method a excludes members who have not won tournaments. Method c excludes members who never shot a hole-in-one.

   Method d excludes members over the age of 30.

R50 **TAKE ANOTHER LOOK**   Use with text pages 212–213.

GUIDED PRACTICE

Whitney is surveying his math class of 40 students to find out if more students pack a lunch or get the school lunch. Tell whether the sampling method is *biased* or *not biased*. Explain.

1. Randomly survey 10 girls. **biased; excludes boys**
2. Randomly survey students in the lunch line. **biased; excludes the others**
3. Randomly survey 10 students in the class. **not biased; equal chance**
4. Randomly survey 2 students in the class. **biased; sample not large enough**

Determine whether the question is biased.

5. Do you agree with the fruit industry spokesperson that green apples are good? **biased**
6. Did you like the movie *James and the Giant Peach*? **not biased**

INDEPENDENT PRACTICE

Tell whether the sampling method is *biased* or *not biased*. Explain.

The 500 Orange County teachers are doing a survey to find out how much time is spent correcting papers.

1. Randomly survey 50 male teachers. **biased; excludes female teachers**
2. Randomly survey English teachers. **biased; excludes the other teachers**
3. Randomly survey 10 teachers. **biased; sample not large enough**
4. Randomly survey 50 teachers. **not biased; all have equal chance**

The Good Soup Company is conducting a survey to find out what kind of soup consumers like best.

5. Randomly survey 2 out of 100 shoppers. **biased; sample not large enough**
6. Randomly survey 10 out of every 100 shoppers. **not biased; all have equal chance**
7. Randomly survey 20 out of 100 women shoppers. **biased; excludes men**
8. Randomly survey people who buy Good Soup. **biased; does not include people who buy other brands**

Determine whether the question is biased.

9. Do you agree with the President that breakfast is important? **biased**
10. Do you agree with the head chef that people like salad more than soup? **biased**
11. What is your favorite song? **not biased**
12. Is rock music your favorite type of music? **not biased**

Problem-Solving Applications

For Problems 13–14, use the following information.

The 600 students at Major Middle School are being surveyed to find out their choice of a mascot.

13. If 60 students are surveyed, do equal numbers of boys and girls have to be chosen? Explain. **No; boys and girls are randomly selected.**
14. **✏ WRITE ABOUT IT** Write a survey question that is biased. Then rewrite the question so it is not biased. **Questions will vary.**

Portfolio

MORE PRACTICE Lesson 11.3, page H58

213

ADDITIONAL EXAMPLES

Example 1, p. 212

The 121 7th-grade students are going to a museum. They are going to conduct a survey to find out what type of museum students want to visit. Which of the following sampling methods would you use? Tell why.

a. Randomly survey 5 7th-grade students.
b. Randomly survey 12 students who are at least 13 years old.
c. Randomly survey 12 7th-grade students.

Choice c is the only method that is not biased.

Example 2, p. 212

Is the following question biased?

Do you agree with a successful discount computer company that its computers have the best prices? This question is biased because it leads you to agree with the computer company.

3 WRAP UP

Use these questions to help students focus on the big ideas of the lesson.

- **How can Independent Practice Exercise 1 be rewritten so the sample is not biased?** Possible answer: Randomly sample 25 male and 25 female teachers.

✔ Assessment Tip

Have students write the answer.

- **What is a biased question?** Possible answer: a question that leads to a specific response

DAILY QUIZ

Tell whether the sampling method is biased or unbiased. Explain.

Sarah is doing a survey of 75 chess players to find out their favorite game strategy.

1. Randomly survey 8 members of the Chess Club. biased; excludes chess players who are not members of the Chess Club
2. Randomly survey 8 chess players. Not biased; all have an equal chance.

213

ORGANIZER

OBJECTIVE To write survey questions and to conduct a survey

PROGRAM RESOURCES
E-Lab • Activity 11
E-Lab Recording Sheets, p. 11
Math Fun • Family Heritage Survey, p. 220

USING THE PAGE

Students now have the background to develop and conduct a survey. They begin the lesson by exploring the elements of good survey questions. Then they proceed to conduct their own survey.

Activity 1

Explore

Read and discuss the guidelines for writing good survey questions. Encourage students to refer to these guidelines as they analyze the questions for the survey about cafeteria food. When students are ready to begin writing questions for the software company's survey, remind them to be sure their questions don't lead the respondents to a specific response.

Activity 2

Explore

As students plan their survey, review with them the criteria for a random sample:

1. The sample must be large enough to represent the total population.

2. No part of the population should be excluded from the sample.

You may wish to have students conduct their survey after school and bring the results to class the next day. This will give them the opportunity to survey everyone in their random sample.

Think and Discuss

When students compare their survey results with the results of other groups who chose the same topic, have them also compare the population surveyed.

• **Do the results from a parent population differ from those of a student population?** Check students' answers.

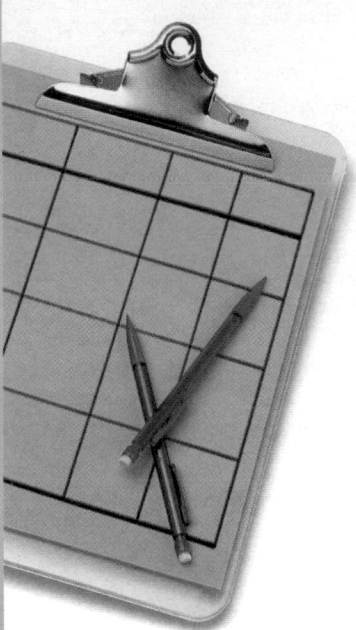

LAB *Activity*

WHAT YOU'LL EXPLORE
How to write survey questions and how to conduct a survey

Questions and Surveys

You have learned how to define a problem, identify random samples, and determine whether a sample or question is biased. Now you can use what you know to write survey questions and conduct a survey.

ACTIVITY 1

Explore

Many companies use surveys to identify consumers' preferences. A lot of time is spent writing the questions for these surveys so that the data gathered are accurate and consistent. When you write a survey question, you should keep the following in mind.

Questions should

1. use vocabulary that is easily understood.
2. be clear and concise.
3. have the same meaning for everyone.
4. result in one clear response per person.

• The questions below were written for a survey about cafeteria food. Work with a partner to review the questions and decide how they could be improved.

> **Do you eat broccoli and peas, do you prefer salad without dressing, and do you like to drink milk for lunch?**
> long; more than one response
> **Since hamburgers have more fat than turkey sandwiches, and mayonnaise has a lot of calories, wouldn't a turkey sandwich with mayonnaise be more healthful?**
> long; unclear

• A software company is thinking about opening a new store. The company wants to conduct a survey to find out if people would shop at this store, and how often. Write two survey questions the company can use. **Check students' questions.**

• Switch papers with a partner. Check your partner's questions. Then work together to rewrite the questions, if necessary. **Check students' questions.**

214 Chapter 11

IDEA FILE

CORRELATION TO OTHER RESOURCES

Kids, Cash, Creative Money Making by J. Lamancusa (McGraw-Hill, 1993).
This book provides a wealth of information for students interested in starting their own businesses. Have this book available to students as they conduct a survey to determine what type of business they would like to start.

CULTURAL CONNECTION

In the sixteenth century, cocoa was introduced in Spain. To take away the bitter taste of the cocoa, sugar and milk were added. Today, chocolate is the most popular candy in the world.

• Have students choose a random sample for a survey to find out the most popular candy among their teachers, parents, and adult neighbors.

Modality Auditory

MEETING INDIVIDUAL NEEDS

Explore

- Work in a group. Select one of the following topics for a survey:

    1. favorite sport
    2. favorite television show
    3. favorite type of snack food

- Decide what your target group, or population, is, such as middle-school students, teachers, or parents and neighbors. What is the size of the population? How many people do you need to survey to have a good sample?

- Prepare three questions for your survey. With your group, check and revise the survey questions so that they are not biased and are written according to the list on page 214.

- Make a recording sheet for your data. Here is a sample recording sheet for collecting data.

Person Number	Question 1	Question 2	Question 3
1			
2			
3			
4			

- Survey a random sample of the population you chose.

Think and Discuss Answers will vary.

- What other questions could you have written?

- Compare your survey results with those of another group that chose the same topic. Are your results about the same? How do that group's questions compare with your group's questions?

Try This

Write a paragraph describing how you carried out your survey. Do you think your sample was random? Explain.

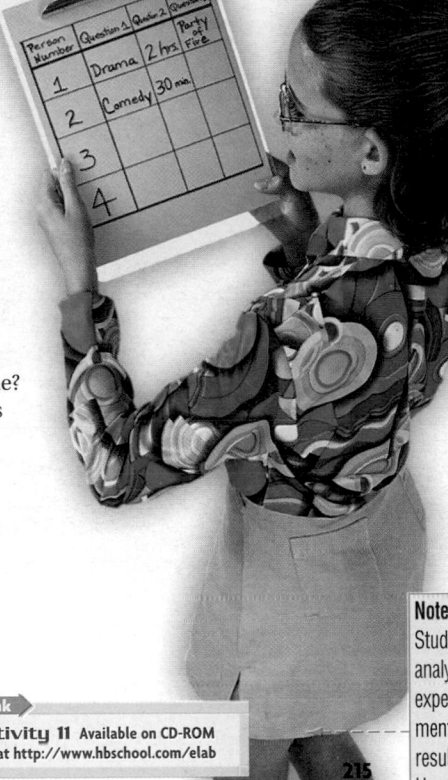

Technology Link
E–Lab • Activity 11 Available on CD-ROM and the Internet at http://www.hbschool.com/elab

Note: Students analyze experiment results. Use E-Lab p. 11.

Try This

You may want to display students' descriptions of how they administered their survey along with their survey results on a bulletin board. Alternatively, you may want to compile students' work into a Survey Portfolio to share with classroom visitors.

USING E-LAB

Students analyze experiment results to discover that when doing an experiment or conducting a survey, the results often depend on the sample that you take.

✔ *Assessment Tip*

Use these questions to help students communicate the big ideas of the lesson.

- **Mayville Department Store wants to conduct a survey to determine the quality of its customer service. What are two survey questions the company can use?** Check students' questions.

- **This question was written for the favorite television show survey: "Do you like to watch game shows or do you prefer evening shows, and would cartoons be better?" Rewrite this question so it complies with the guidelines on page 214.** Check students' questions.

CAREER CONNECTION

Publishing Editors who work for publishing companies analyze and edit manuscript according to guidelines similar to those used for survey questions. Interview an editor who works for a newspaper, magazine, or book publisher to learn more about his or her job.

Modality Auditory

LITERATURE CONNECTION

The all-time best-selling children's paperback is *Charlotte's Web* by E. B. White. Have students write two biased survey questions to find out what the most popular book is among their classmates. Then have them exchange papers and rewrite the questions so they are not biased.

Modality Visual

GRADE 5	GRADE 6	GRADE 7
Graph survey results.	Record and organize data collected in a survey.	Organize and display results of a survey.

ORGANIZER

OBJECTIVE To record and organize data collected in a survey

VOCABULARY *New* **tally table, frequency table, cumulative frequency, range**

ASSIGNMENTS Independent Practice
- ■ *Basic* 1–12; 13–18
- ■ *Average* 1–12; 13–18
- ■ *Advanced* 1–12; 13–18

PROGRAM RESOURCES
Intervention • Range and Intervals, TE p. 220A–220B
Reteach 11.4 Practice 11.4 Enrichment 11.4
Data Tool Kit
Math Fun • What Kind of Pet Do You Have?, p. 220

PROBLEM OF THE DAY

Transparency 11

Anne's line plot of ages of students has a range of 5. Each age has twice as many x's as the previous one. If the last age has 16 x's, how many students did Anne include in her data? 31 students

Solution Tab A, p. A11

QUICK CHECK

Is this a good question for a survey of people who take walks or run? Write *yes* or *no* and tell why.

1. Do you like to walk in the morning on weekends, jog on weekend afternoons, or walk and jog in the summer? No. The question is too long and involved.

2. When do you walk or run during the day? Yes. The question is unbiased.

3. Do you agree with running champions that it's more fun to run in the evening? No. The question is leading.

1 MOTIVATE

Introduce the lesson with a discussion of the Favorite Snack Foods survey at the top of the page. Ask students:

11.4

WHAT YOU'LL LEARN
How to record and organize data collected in a survey

WHY LEARN THIS?
To make it easy to understand the data you collect

WORD POWER
tally table
frequency table
cumulative frequency
range

Collecting and Organizing Data

Mike conducted a survey to find out his classmates' favorite type of snack food. He recorded the data in a tally table. A **tally table** is a table that has categories that allow you to record each piece of data as it is collected.

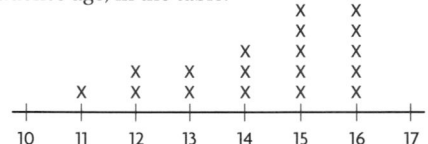

FAVORITE SNACK FOODS

Snack	Tally
Fruit	|||| ||||
Cereal	||||
Chips	|||| ||||
Cookies	|||| |||

• How can you determine the number of students who prefer each type of snack food? **by counting the number of tally marks**

Another method for recording data is by using a line plot.

EXAMPLE 1 A movie theater surveyed 50 students to find out if they go to the movies at least once a month. The age of each student who said *yes* was recorded. Use the data below to make a line plot.

Ages of Students Who Said *Yes*

12	15	11	16	15	16	14	12	14
16	16	15	13	16	15	13	15	14

Step 1: Draw a horizontal line.
Step 2: On your line, write the numerical values for the ages, using vertical tick marks.
Step 3: Plot the data by placing an x on your line plot for each value, or student's age, in the table.

```
                                    x       x
                                    x       x
                            x       x       x
                    x   x   x       x       x
            x   x   x   x   x       x       x
         ───┼───┼───┼───┼───┼───┼───┼───┼───
           10  11  12  13  14  15  16  17
```

RETEACH 11.4

Name _____

LESSON 11.4

Collecting and Organizing Data

A local fitness center wants to survey 50 adults to find out if they exercise daily. The age of each person who said yes was recorded. Use the data below to make a line plot.

Ages of Adults Who Said *Yes*

28	33	45	25	50	42	33
25	48	31	37	28	25	50
42	29	45	50	38	31	29
38	52	40	33	25	37	52

Step 1: Draw a horizontal line.
Step 2: On your line, write the numerical values for the ages, using vertical tick marks.
Step 3: Plot the data by drawing an x on the line plot for each value, or person's age, in the table.

```
    x
    x   x       x           x
    x   x   x   x   x   x   x   x   x   x
  ──┼───┼───┼───┼───┼───┼───┼───┼───┼───┼──
   25  28  29  31  33  37  38  40  42  45  48  50  52
```

For Exercises 1–2, use the data to make a line plot.
1. A local bookstore wants to survey 50 students to find out if they buy at least one book a month. The surveyor recorded the age of each student who said yes.

Age of Students Who Said *Yes*

13	12	16	15	11	14	15
11	16	14	14	13	15	11
14	15	11	12	16	11	13

```
                x   x
        x   x   x   x
        x   x   x   x   x
        x   x   x   x   x
      ──┼───┼───┼───┼───┼──
       11  12  13  14  15  16
```

2. The 20 students in science class take a test. Here are their results.

Science Test Scores

76	82	85	95	98
92	78	76	90	85
88	95	74	78	76
74	85	92	82	88

```
        x
  x x x x x           x   x x x
──┼───┼───┼───┼───┼───┼───┼───┼───┼──
 74  76  78  82  85  88  90  92  95  98
```

Use with text pages 216–219.

TAKE ANOTHER LOOK R51

PRACTICE 11.4

Name _____

LESSON 11.4

Collecting and Organizing Data

Vocabulary

Complete each sentence by writing a term from the Word Bank.

Word Bank
cumulative frequency
frequency table
range
tally table

1. A ____tally table____ is a table that has categories that allow you to record each piece of data as it is collected.

2. A running total of the number of people surveyed is called ____cumulative frequency____.

3. The difference between the greatest number and least number in a set of numbers is the ____range____.

4. A ____frequency table____ gives a total for each category or group in a set of data.

For Exercises 5–6, use the data in the box at the right.
5. Make a line plot.

Students' Heights (cm)

160	129	158	155	136	128
128	159	142	147	148	144
152	133	135	136	162	158
139	160	128	139	159	144
155	147	136	148	162	133

```
 x   x   x   x   x   x   x
──┼───┼───┼───...───┼───┼──
128 129 133 135 136 139 142 144 147 148 152 155 158 159 160 162
```

6. Find the range. ____34____

Mixed Applications

7. Make a line plot for reading test scores of 98, 75, 82, 100, 96, 100, 81, 78, 100, 92, 84, 86, 78, 100, 78. Then determine the range, and make a frequency table that uses intervals. Check students' tables.

range = 25

8. Rick surveyed his classmates about their color preferences. He collected this data: 6 prefer red, 8 prefer blue, 2 prefer white, 5 prefer green, and 4 prefer white. Organize this information in a tally table and in a frequency table. Check students' tables.

Use with text pages 216–219.

ON MY OWN P51

A frequency table helps you organize the data from a tally table or a line plot. A **frequency table** gives you the total for each category or group. You can add a cumulative frequency column to the table. **Cumulative frequency** keeps a running total of the number of people surveyed.

EXAMPLE 2 A new radio station surveyed people in a community to determine the type of music they preferred. Use the data below to make a frequency table. What type of music do you think this radio station will play?

TYPES OF MUSIC PREFERRED	
Music Type	**Tally**
Classical	ʮʮ ʮ
Country	ʮ ʮ ʮ ʮ ʮ ʮ
Rock and roll	ʮ |||
Pop	ʮ ʮ ʮ ʮ

List the music categories in one column. Put the total for each category in the frequency column and the running total in the cumulative frequency column.

Music Type	Frequency	Cumulative Frequency	
Classical	15	15	
Country	30	45	← 15 + 30 = 45
Rock and roll	9	54	← 45 + 9 = 54
Pop	20	74	← 54 + 20 = 74

So, they will probably play country music since 30 out of 74 people surveyed chose this music type.

• **CRITICAL THINKING** What part of the frequency table tells you the size of the sample surveyed? **the last entry in the cumulative frequency column**

GUIDED PRACTICE

Use the data in the table at the right.
See Additional Answers, p. 221A.
1. Make a tally table.
2. Make a line plot.
3. Make a cumulative frequency table.

Miles Jogged in One Week						
20	21	26	35	38	26	40
28	33	44	32	42	28	20
42	32	24	28	35	28	20

217

One of the best-known composers of classical music is Wolfgang Amadeus Mozart. A musical prodigy, he started composing at five years of age. His first major opera was performed 9 years later in 1770. How old was he when his first opera was performed?

14 years old

• **What snacks are in the Favorite Snack Foods survey shown in the book?** fruit, cereal, chips, cookies

• **Which snack is most popular?** fruit

• **Many Americans prefer candy, chips, and cookies for snacks? What important snack are they missing?** fruit

• **Why is it important?** Possible answer: It is a nutritious snack.

2 TEACH/PRACTICE

Students should be familiar with tally tables, so a brief review of how a tally table is used will suffice.

• **Why are tally marks grouped in fives?** so you can easily skip-count by fives

• **What is the advantage of recording data from a survey in a tally table?** Possible answer: You can record survey participants' responses individually.

GUIDED PRACTICE
Have students work in pairs to complete Exercises 1–3. Discuss as a class.

ADDITIONAL EXAMPLES

Example 1, p. 216

The Kids Can Help Club conducted a survey to determine how much time middle-school students spend doing volunteer work in an average week. Use the data to make a line plot.

Time Spent on Volunteer Work				
1 hr	2 hrs	3 hrs	2 hrs	4 hrs
3 hrs	1 hr	4 hrs	3 hrs	2 hrs
4 hrs	3 hrs	2 hrs	4 hrs	1 hr
2 hrs	4 hrs	3 hrs	4 hrs	3 hrs

Check students' line plots.

Example 2, p. 217

The Clothes Closet Company conducted a survey to find out what method of shopping people prefer. Use the data to make a frequency table. What method of selling do you think the company will focus on in the future?

PREFERRED SHOPPING METHODS	
Shopping Method	**Tally**
Mall	ʮ ʮ ʮ ʮ ʮ |||
Catalog	ʮ ʮ ||
Television	ʮ ʮ ʮ	
Internet	ʮ |||

Check students' tables; mall shopping

217

ENRICHMENT 11.4

Name _____

LESSON 11.4

Video Game Survey

Marla surveyed sixth-grade students about the number of hours per week they spend playing video games. She organized her data in the graph below. Use the graph to answer the questions.

NUMBER OF HOURS SPENT PLAYING VIDEO GAMES PER WEEK

1. How many students did Marla survey?
   50 students
2. What is the range of her data?
   6
3. What fraction of the students surveyed spend 4 hours a week playing video games?
   $\frac{1}{10}$ of the students
4. What fraction of the students surveyed spend less than 2 hours a week playing video games?
   $\frac{2}{5}$ of the students
5. What fraction of the students surveyed spend more than 3 hours a week playing video games?
   $\frac{7}{25}$ of the students
6. For every 1 student who responded "6 hours," there were 2 students who responded "5 hours." What other pair of responses has a similar relationship?
   (1:2 ratio) 3 hr : 0 hr; 6 hr : 1 hr
7. Suppose Marla continued her survey and polled 100 additional classmates. Based on her first survey, how many students would likely respond "0 hours"?
   42 students
8. Suppose Marla wanted to expand her survey and ask another question about video game play. What are two possible questions she might ask?
   Possible answers: What is your favorite game? What type of system do you have?

Use with text pages 216–219.   **STRETCH YOUR THINKING** E51

Example 3, p. 218

Use the jogging data from the chart at the bottom of page 217 to make a frequency table with intervals.

$44 - 20 = 24 \leftarrow$ Find range.

$3 \times 8 = 24 \leftarrow$ Use range to determine intervals.

Make 3 intervals that include 8 consecutive distances. Tally the number of people who jogged each distance. Record the tallies for each interval in the Tally column. Then record the numerical value for each interval in the Frequency column.

Distance (mi)	Tally	Frequency
20–27	⊣⊣⊣ ‖	7
28–35	⊣⊣⊣ ‖‖	9
36–44	⊣⊣⊣	5

Range and Intervals

A country music radio station took a survey to find out the ages of its listeners. The survey results are shown below.

Ages of Country Music Listeners									
14	16	35	38	43	28	36	43	41	27
21	12	27	33	18	24	19	30	29	35

The age of the youngest listener is 12. The age of the oldest listener is 43. Instead of listing each age, you can use the range to determine intervals for the data. The **range** is the difference between the greatest number and the least number in a set of data.

EXAMPLE 3   Use the data above to make a frequency table with intervals.

$43 - 12 = 31$ *Find the range.*

Since $4 \times 8 = 32$, which is close to 31, make 4 intervals that include 8 consecutive ages.   *Use the range to determine intervals.*

From the data, tally each age in the appropriate interval and record the value for each row in the frequency column.

COUNTRY MUSIC LISTENERS		
Age Group	**Tally**	**Frequency**
12–19	⼁⼁⼁	5
20–27	‖‖	4
28–35	⼁⼁⼁ ⼁	6
36–43	⼁⼁⼁	5

• **CRITICAL THINKING** What age groups could you have if you made 2 intervals? Explain. **12–27, 28–43; since $2 \times 16 = 32$, you would have 2 intervals that include 16 ages in each interval.**

INDEPENDENT PRACTICE

For Exercises 1–2, use the table at the right.

1. Copy and complete the table.

2. What is the size of the sample? **80**

Ability	Frequency	Cumulative Frequency
Beginner	20	? **20**
Intermediate	26	? **46**
Advanced	34	? **80**

IDEA FILE

LANGUAGE CONNECTION

Tell students: The longest word in the Spanish language is *superextra-ordinarisimo.* What do you think it means? *extraordinary* Make a tally table and frequency table to find the frequency of each vowel in the word. Check students' tables. a: 2; e: 2; i: 3; o: 2; u: 1

Modalities Auditory, Visual

INTERDISCIPLINARY CONNECTION

One suggestion for an interdisciplinary unit for this chapter is Kids and Computers. In this unit, you can have students

• conduct a survey to find out how many students in their school have computers at home.

• display the results of their survey in a frequency table using intervals.

Modalities Auditory, Kinesthetic, Visual

Norma asked the students in her math class about their favorite meal. For Exercises 3–4, use the tally table.

3. Organize the data into a frequency table. Include a cumulative frequency column. **See Additional Answers, p. 221A.**
4. What is the size of the sample? **28 people**

FAVORITE MEALS	
Meals	Tally
Pizza	⫴⫴ ⫴⫴
Tuna	⫴⫴ I
Spaghetti	⫴⫴ III
Meatloaf	IIII

For Exercises 5–8, use the data in the box at the right.

5. Make a line plot. **Check students' line plots.**
6. Find the range. **36**
7. How many heights would be in each interval if you made 4 intervals? **9 heights**
8. Make a cumulative frequency table using intervals of 9 for the heights. **See Additional Answers, p. 221A.**

Students' Heights (cm)					
160	130	142	153	164	160
161	162	132	155	140	130
150	145	140	138	166	155
154	155	160	160	155	158

Problem-Solving Applications

9. Sheri surveyed her classmates about their cereal preferences and collected the following data: 4 classmates like Hi-Fiber, 6 like Hi-Sugar, 12 like Munchy, and 8 like Crunchy. Organize this information in a tally table and a frequency table. **See Additional Answers, p. 221A.**
10. Make a line plot for history test scores of 55, 85, 94, 62, 88, 98, 77, 75, 63, 54, 60, 82, 85, 89, 91, 93, 75, 70, 66, 84, 88, 82, 71, 73. Then determine the range and make a frequency table using intervals. **range: 44; possible intervals: 50–59, 60–69, 70–79, 80–89, and 90–99**
11. Make a frequency table of each vowel using the following sentence: *When I go to school, I like math class the most.* **See Additional Answers, p. 221A.**
12. ◀■▶ **WRITE ABOUT IT** Explain why information on a frequency table can be more helpful than information on a tally table. **A frequency table shows totals. You don't have to count.**

Mixed Review

Tell whether you should use a line graph, circle graph, or bar graph.

13. high temperatures for one week **line graph**
14. soccer team's heights **bar graph**

Tell how many faces each figure has.

15. rectangular prism **6 faces**
16. triangular pyramid **4 faces**

17. **RATES** Lisa rode her bike a total of 36 km at a rate of 9 km per hour. How long did she ride? **4 hr**
18. **TIME** Manuel spent $4\frac{1}{2}$ hr mowing lawns. He finished at 4:45 P.M. When did he start? **12:15 P.M.**

Technology Link

In *Data ToolKit,* you can use data to make a frequency table and then show this data as a line plot.

HARCOURT BRACE
Data ToolKit

MORE PRACTICE Lesson 11.4, page H58

219

INDEPENDENT PRACTICE

To assure students' success, complete Exercises 1–4 as a whole class before assigning the remaining exercises.

💡**CRITICAL THINKING** In what situation might you be unable to use intervals in a frequency table? when the range for data is small

3 WRAP UP

Use these questions to help students focus on the big ideas of the lesson.

• **How did you determine the range and intervals for Independent Practice Problem 10?** Possible answer: To find the range, I subtracted the lowest test score from the highest test score. To determine the intervals, I found two factors that almost equal the range.

✓ *Assessment Tip*

Have students write the answer.

Make a tally table and a frequency table showing frequency of vowels in the following sentence:

• **All the King's horses and all the King's men couldn't put Humpty Dumpty together again.**

	Tally	Frequency	Cumulative Frequency
a	⫴⫴	5	5
e	⫴⫴ I	6	11
i	III	3	14
o	III	3	17
u	IIII	4	21

DAILY QUIZ

Refer to the frequency table you made for Independent Practice Exercise 9 to answer these questions.

1. What is the cumulative frequency for each cereal? Hi-Fiber: 4; Hi-Sugar: 10; Munchy: 22; Crunchy: 30
2. How many students were in the survey? 30

Refer to the frequency table you made for Independent Practice Exercise 11 to answer these questions.

3. What is the range of the data? 3
4. How many times did vowels appear in the sentence? 13 times

219

INTERVENTION

Range and Intervals

Use these strategies to review the skills required to find the range in a set of numbers and to then use the range to determine intervals.

Multiple Intelligences Logical/Mathematical, Visual/Spatial, Bodily/Kinesthetic

Modalities Auditory, Kinesthetic, Visual

Lesson Connection Students will use range and intervals in Lesson 11.4.

Program Resources *Intervention and Extension Copying Masters,* p. 6

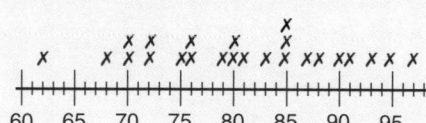

Troubleshooting

WHAT STUDENTS NEED TO KNOW	HOW TO HELP	FOLLOW UP
How to organize data by using a line plot	Have students construct a line plot to organize a set of test scores. **Example** The 24 test scores are: 93, 88, 70, 62, 72, 72, 87, 80, 97, 68, 76, 79, 85, 76, 85, 91, 80, 81, 75, 85, 83, 70, 95, 90. The line plot would look like this. 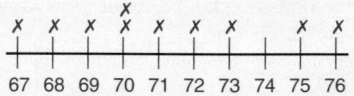	Another line plot to construct: The heights (in in.) of 10 baseball players are: 72, 75, 70, 71, 76, 68, 70, 73, 67, 69. Make a line plot.
The *range* is the difference between the greatest number and the least number in a set of numbers.	Have students find the range by subtracting the least number from the greatest number in the set of data. **Examples** In the set of test scores above, the range is 97−62, or 35. The tallest student in the class is 71 in. tall, and the shortest student is 58 in. tall. So the range is 71 − 58, or 13 in.	Other problems to solve: Find the range of the baseball players' heights above. 9 in. Find the range for these weights: 185 lb, 194 lb, 220 lb, 170 lb, and 190 lb. 50 lb See Key Skill 51, Student Handbook, page H27.

WHAT STUDENTS NEED TO KNOW	HOW TO HELP	FOLLOW UP

How to organize data into intervals

Have students group the data on test scores from the previous page into 4 intervals and make a frequency table.

Example

The range of test scores on the previous page is 35. To divide the data into 4 intervals, divide 35 by 4. $35 \div 4 = 8.7$

You could use 4 intervals of 9 or 10 values.

Range	Tally	Frequency
60–69	||	2
70–79	⫽⫽ |||	8
80–89	⫽⫽ ||||	9
90–99	⫽⫽	5

Another problem to solve:

Group the following data into 3 intervals, and make a frequency table.

24, 20, 79, 63, 27, 38, 55, 41, 26, 33, 62, 48

Range: $79 - 20 = 59$

$59 \div 3 \approx 20$

20–39	40–59	60–79
6	3	3

See Key Skill 51, Student Handbook, page H27.

You can change an interval to make it smaller or larger.

Have students vary the intervals and reconstruct the frequency table for the above test scores. Try it with intervals of 3, 5, 8, and 15, for example.

Example

The frequency table for intervals of 15 might look like this.

Range	Tally	Frequency
56–70	||||	4
71–85	⫽⫽ ⫽⫽ |||	13
86–100	⫽⫽ ||	7

Another problem to solve:

Regroup the data from above, using 4 intervals.

20–34	5
35–49	3
50–64	3
65–79	1

See Key Skill 51, Student Handbook, page H27.

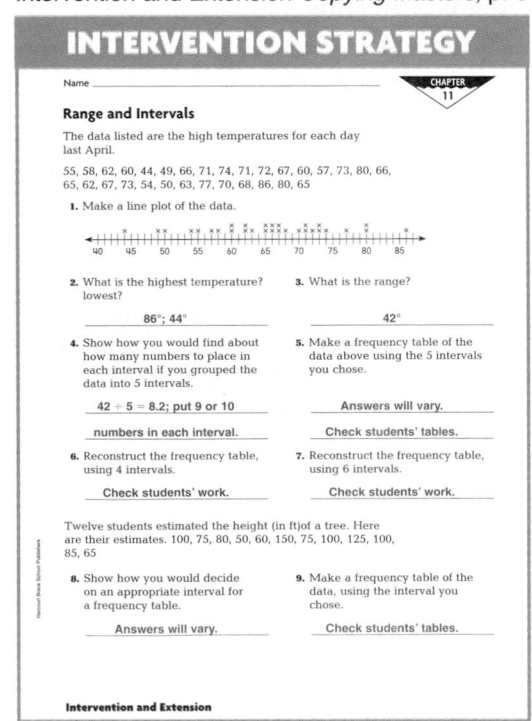

Intervention and Extension Copying Masters, p. 6

INTERVENTION STRATEGY

Name _____ CHAPTER 11

Range and Intervals

The data listed are the high temperatures for each day last April.

55, 58, 62, 60, 44, 49, 66, 71, 74, 71, 72, 67, 60, 57, 73, 80, 66, 65, 62, 67, 73, 54, 50, 63, 77, 70, 68, 86, 80, 65

1. Make a line plot of the data.

2. What is the highest temperature? lowest?

86°; 44°

3. What is the range?

42°

4. Show how you would find about how many numbers to place in each interval if you grouped the data into 5 intervals.

$42 \div 5 = 8.2$; put 9 or 10 numbers in each interval.

5. Make a frequency table of the data above using the 5 intervals you chose.

Answers will vary. Check students' tables.

6. Reconstruct the frequency table, using 4 intervals.

Check students' work.

7. Reconstruct the frequency table, using 6 intervals.

Check students' work.

Twelve students estimated the height (in ft) of a tree. Here are their estimates. 100, 75, 80, 50, 60, 150, 75, 100, 125, 100, 85, 65

8. Show how you would decide on an appropriate interval for a frequency table.

Answers will vary.

9. Make a frequency table of the data, using the interval you chose.

Check students' tables.

Intervention and Extension

USING THE PAGE

To provide additional practice activities related to Chapter 11: *Organizing Data*

WHAT'S YOUR BIAS?

Have students work in small groups to complete this activity.

In this activity, students write biased survey questions, exchange questions with another group, and identify the bias in the other group's questions.

Multiple Intelligences Verbal/Linguistic, Interpersonal/Social, Intrapersonal/Introspective
Modality Auditory

FAMILY HERITAGE SURVEY

Have students work individually to complete this activity.

In this activity, students write survey questions, survey a target group, and create a recording sheet of the data.

Multiple Intelligences Verbal/Linguistic, Interpersonal/Social, Intrapersonal/Introspective
Modality Auditory

WHAT KIND OF PET DO YOU HAVE?

MATERIALS: *For each group* poster paper, markers

Have students work in small groups to complete this activity.

In this activity, students conduct a survey about the kinds of pets people have and use the results to create a frequency and a cumulative frequency table.

Multiple Intelligences Visual/Spatial, Logical/Mathematical, Verbal/Linguistic, Interpersonal/Social, Intrapersonal/Introspective
Modalities Auditory, Kinesthetic, Visual

What's your bias?

PURPOSE To practice determining whether a sample or question in a survey is biased (pages 212–213)

Work in a small group. Brainstorm a list of 5–10 survey topics. Pick one survey topic. Then write some biased questions you could ask to gather information about the topic.

Trade survey questions with another group. Identify the source of bias in the questions. Explain why you think the questions are biased or not biased. **Answers will vary.**

 ## Family Heritage Survey

PURPOSE To practice writing survey questions and conducting a survey (pages 214–215)

Work in a small group. Create a survey question to find out about cultural heritage. Make sure your question is clear and has the same meaning for everyone.

Choose your target group. Will you survey teachers, families, neighbors, friends, or students in your own or in another classroom? Survey your target group. Make a recording sheet for your data. Share your results with the class.

 Ask as many family members such as aunts and uncles as you can about their cultural heritage. Identify countries and cities of origin on a world map.

What kind of pet do you have?

PURPOSE To practice recording and organizing data (pages 216–219)
YOU WILL NEED poster paper, markers

Work in a small group. Conduct a Pet Survey. Ask each person what kind of pet he or she has. Make a frequency table and a cumulative frequency table from the data you collect. Why can information in a frequency table be more helpful than information from a tally table?
Possible answer: It helps you organize your data from a tally table.

Share your results with the class. Compare tables. Are they exactly alike? Why or why not? **Answers will vary.**

Make up some fun pet questions of your own that you can ask in a survey. **Answers will vary.**

CHAPTER 11 REVIEW

EXAMPLES

Decide what information you need.
(pages 208–209)

Decision:
sporting events to have on field day

Information needed:
What kind of equipment and facilities does the school have?
How many students are participating?

Identify samples. (pages 210–211)

Each of 300 students had an equal chance of being surveyed about exercise and 30 of them were chosen. Was this a random sample? Explain.

Yes; each student had an equal chance of being part of the sample.

Determine whether a survey is biased.
(pages 212–213)

A toy company wants to survey 50 of its customers. Which sample is not biased?
a random sample of 50 people in one toy store
a random sample of 50 people who have bought the company's toys

Choice B is not biased.

Collect and organize data. (pages 216–219)

Cumulative frequency keeps a running total of the number of people surveyed.

Age Group	Frequency	Cumulative Frequency
25–30	6	6
31–36	4	10

EXERCISES

Solve.

1. Ted's class is in charge of planning a sixth-grade field trip. Which of these decisions will the students have to make? **a**
   a. the kind of transportation they will use
   b. where the seventh-grade class wants to go

2. What information might you need to decide how many chaperones to have on a field trip?

   2. Possible answer: how many students per chaperone

If you survey 1 out of every 10 people, how many would you survey from each group?

3. 350 students   4. 520 residents

   3. 35 students   4. 52 residents

5. To determine the favorite snack at school, Al surveyed the students in his class. Is this a random sample of the school? Explain. **No; his sample should include other classes.**

6. **VOCABULARY** A sample is _?_ if any type of individual in the population is not represented by the sample. **biased**

7. A restaurant randomly surveys 10 out of every 100 customers in a day about the quality of its service. Is this sample biased or not biased? **not biased**

Determine if the following question is biased.

8. Which do you prefer: hot, delicious apple pie or cold pudding? **biased**

For Exercises 9 and 10, use the following data.

Ages of Persons Surveyed					
23	18	14	16	21	13
13	26	25	24	14	22

Check students' displays.

9. Make a line plot.

10. Make a cumulative frequency table with intervals.

USING THE PAGE

The Chapter 11 Review reviews deciding what information is needed, identifying samples, determining whether a survey is biased, and collecting and organizing data. Chapter objectives are provided, along with examples and practice exercises for each.

 The Chapter 11 Review can be used as a review or a test. You may wish to place the Review in students' portfolios.

Assessment Checkpoint

For Performance Assessment in this chapter, see page 274, What Did I Learn?, items 1 and 5.

✔ Assessment Tip

Student Self-Assessment The How Did I Do? Survey helps students assess both what they have learned and how they learned it. Self-assessment helps students learn more about their own capabilities and develop confidence in themselves.

A self-assessment survey is available as a copying master in the *Teacher's Guide for Assessment*, p. 15.

CHAPTER 11 TEST, page 1

Name _____

CHAPTER 11 TEST
PAGE 1

Choose the letter of the correct answer.

1. A class is planning a book fair. Which decision can the students make?
   A. How much will each book cost?
   B. On what date will the fair take place?
   C. Do other schools have book fairs?
   D. How many books will be sold?

2. What information do you need to help decide whether to have a bake sale or a car wash to raise money for a class trip?
   A. Which activity has good past results?
   B. How many people run a bake sale?
   C. Which day is best for a bake sale?
   D. Where is the class trip?

3. The sixth-grade trip has 1 chaperone for 6 students. What is necessary to know to find how many chaperones are needed?
   A. number of miles they will travel
   B. number of students going
   C. number of buses
   D. number of places they will visit

4. A school survey is conducted about choosing new chairs for the library. Which is a useful student survey question?
   A. How much do different chairs cost?
   B. How often are the chairs used?
   C. Which chairs are most comfortable?
   D. What is the budget?

For questions 5–6, find the number of people you should survey if you survey 1 out of every 10 people for a sample.

5. 500 students
   A. 5 students   B. 25 students
   C. 50 students   D. 110 students

6. 330 voters
   A. 3 voters   B. 13 voters
   C. 30 voters   **D.** 33 voters

7. The librarian wants to survey students to see what magazines they prefer. How can she get a random sample?
   A. Choose students in the library.
   B. Ask teachers to choose students.
   C. Survey students in one classroom.
   D. Randomly survey students as they enter the school.

8. A ranger surveyed 1 out of every 6 hikers about the trails. Did he choose a random sample? Explain why or why not.
   A. Yes; all hikers had an equal chance.
   B. No; not enough hikers were selected.
   C. No; the sample did not have non-hikers.
   D. No; only adults should be surveyed.

For questions 9–10, use the table. As voters left the voting booth, 1 out of every 10 voters was asked how he or she voted on Question 1.

Question 1	Number of Votes
Yes	32
No	28

9. How many voters were surveyed?
   A. 28 voters   **B.** 32 voters
   C. 50 voters   D. 60 voters

10. Suppose 300 voters were surveyed. How many would you expect to have voted yes?
    A. 32 voters   B. 128 voters
    C. 160 voters   D. 200 voters

Form A • Multiple-Choice   A43   Go on. ▶

CHAPTER 11 TEST, page 2

Name _____

CHAPTER 11 TEST
PAGE 2

11. A bank randomly surveys 11 out of every 110 women customers. Is this sample biased? Explain why or why not.
    A. Yes; it excludes men.
    B. Yes; it is not large enough.
    C. Yes; only customers are surveyed.
    D. No; all have an equal chance.

12. Which of the following questions in a sports survey is biased?
    A. Who is your favorite player?
    B. Do you prefer an exciting sport like basketball or a sport like baseball?
    C. How often do you attend games?
    D. What sports do you play?

For questions 13–14, 500 students at Walden Middle School are being surveyed to find out their choice of a new name for the school.

13. How could a random sample be chosen?
    A. Choose the first 50 students whose names begin with A.
    B. Ask each teacher to choose 5 students.
    C. Choose the first 50 students to arrive at school.
    D. Use the computer to randomly choose 50 students.

14. Which question is biased?
    A. Do you have a favorite on the list of suggested names?
    B. Do you agree with the principal that Scott School is the best name?
    C. Do you prefer the name of a famous person or a geographical name?
    D. Can you suggest other names?

For questions 15–16, use the data showing the ages of newspaper readers who took part in a survey.

Ages of Newspaper Readers					
34	47	38	42	24	36
61	52	39	22	50	42
21	67	15	59	71	19

15. What is the range?
    A. 15   B. 50
    C. 56   D. 71

16. If you want to arrange the data into 4 age intervals, how many ages should be put in each interval?
    A. 9   B. 11
    C. 12   **D.** 14

For questions 17–18, use the table.
Meredith surveyed students in her P.E. class.

Favorite Breakfasts	Tally
Cereal	llll l
Bagel	llll ll
Pancakes	llll lll
Eggs	llll

17. What is the size of the sample?
    A. 24 students   B. 25 students
    C. 27 students   D. 30 students

18. Meredith plans to add a column to the table showing the numerical value for the tally marks. How should she label the new column?
    A. Number Tally
    B. Frequency
    C. Cumulative Frequency
    D. Range

Form A • Multiple-Choice   A44   ▶ Stop!

See *Test Copying Masters,* pp. A43–A44 for the Chapter Test in **multiple-choice** format and pp. B193–B194 for **free-response** format.

221

ADDITIONAL ANSWERS
Lesson 11.1, page 209

Independent Practice

3. Possible Answer: What type of music do they want to play? How long is the dance? How long does each song take to play? How many songs do they have time to play?

Lesson 11.4, page 217

Guided Practice

1.

Miles Jogged in One Week	
Number of Miles	Tally
20	///
21	/
24	/
26	//
28	////
32	//
33	/
35	//
38	/
40	/
42	//
44	/

2.

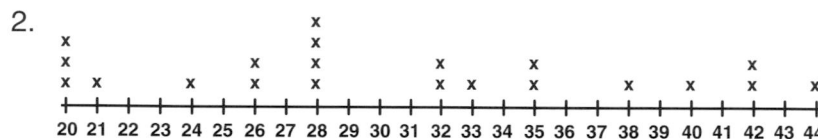

```
                        x
 x                      x
 x           x          x
 x  x        x    x          x    x          x
 x  x    x   x    x          x    x    x     x    x    x    x
 +--+--+--+--+--+--+--+--+--+--+--+--+--+--+--+--+--+--+--+--+--+--+--+--+
20 21 22 23 24 25 26 27 28 29 30 31 32 33 34 35 36 37 38 39 40 41 42 43 44
```

3.

Miles Jogged in One Week		
Number of Miles	Frequency	Cumulative Frequency
20	3	3
21	1	4
24	1	5
26	2	7
28	4	11
32	2	13
33	1	14
35	2	16
38	1	17
40	1	18
42	2	20
44	1	21

Lesson 11.4, page 219

Independent Practice

3.

Favorite Meals		
Meals	Frequency	Cumulative Frequency
Pizza	10	10
Tuna	6	16
Spaghetti	8	24
Meatloaf	4	28

8.

Students' Heights (cm)		
Heights (cm)	Frequency	Cumulative Frequency
130–139	4	4
140–149	4	8
150–159	8	16
160–169	8	24

9.

Favorite Cereals		
Cereal	Tally	Frequency
Hi-Fiber	////	4
Hi-Sugar	ЖН /	6
Munchy	ЖН ЖН //	12
Crunchy	ЖН ///	8

11.

Number of Vowels in Sentence	
Vowels	Frequency
a	2
e	3
i	3
o	5
u	0

PLANNING GUIDE

Displaying Data

BIG IDEA: Information can be collected and displayed as graphs.

Introducing the Chapter p. 222	Team-Up Project • Create the Data p. 223		
OBJECTIVE	**VOCABULARY**	**MATERIALS**	**RESOURCES**
12.1 Using Graphs to Display Data pp. 224-227 **Objective:** To display data in a bar graph, a line graph, and a stem-and-leaf plot **2 DAYS**	stem-and-leaf plot		■ Reteach, ■ Practice, ■ Enrichment 12.1; More Practice, p. H59 Cultural Connection • Zimbabwean Exports, p. 236
12.2 Histograms pp. 228-229 **Objective:** To make histograms **1 DAY**	histogram		■ Reteach, ■ Practice, ■ Enrichment 12.2; More Practice, p. H59
12.3 Graphing Two or More Sets of Data pp. 230-231 **Objective:** To graph two or more sets of data **1 DAY**			■ Reteach, ■ Practice, ■ Enrichment 12.3; More Practice, p. H59
EXTENSION • Using Spreadsheets for Data, Teacher's Edition pp. 232A–232B			
Lab Activity: Exploring Circle Graphs pp. 232-233 **Objective:** To make a circle graph from a bar graph		FOR EACH STUDENT: 8 $\frac{1}{2}$-in. x 11-in. or larger paper, markers, scissors, tape	💻 **E-Lab • Activity 12** E-Lab Recording Sheets, p. 12
12.4 Circle Graphs pp. 234-235 **Objective:** To make and read a circle graph **2 DAYS**		FOR EACH PAIR: protractor, centimeter ruler	■ Reteach, ■ Practice, ■ Enrichment 12.4; More Practice, p. H60 Cultural Connection • Zimbabwean Exports, p. 236 💾 **Data ToolKit**
CHAPTER ASSESSMENT Chapter 12 Review p. 237			

NCTM CURRICULUM
STANDARDS FOR GRADES 5–8

✓ **Problem Solving**
Lessons 12.1, 12.2, 12.3, 12.4

✓ **Communication**
Lessons 12.1, 12.2, 12.3, 12.4

✓ **Reasoning**
Lessons 12.1, 12.2, 12.3, 12.4

✓ **Connections**
Lessons 12.1, 12.2, 12.3, 12.4

✓ **Number and Number Relationships**
Lessons 12.1, 12.3, 12.4

✓ **Number Systems and Number Theory**
Lessons 12.2, 12.3, 12.4

✓ **Computation and Estimation**
Lessons 12.1, 12.3, 12.4

✓ **Patterns and Functions**
Lesson 12.3

☐ **Algebra**

✓ **Statistics**
Lessons 12.1, 12.2, 12.3, 12.4

☐ **Probability**

✓ **Geometry**
Lesson 12.4

✓ **Measurement**
Lessons 12.1, 12.3, 12.4

ASSESSMENT OPTIONS

CHAPTER LEARNING GOALS

12A.1 To display data in a bar graph, a line graph, and a stem-and-leaf plot

12A.2 To make histograms and circle graphs

12A.3 To graph two or more sets of data

These goals for concepts, skills, and problem solving are assessed in many ways throughout the chapter. See the chart below for a complete listing of both traditional and informal assessment options.

Pretest Options
Pretest for Chapter Content, TCM pp. A45–A47, B195–B197
Assessing Prior Knowledge, TE p. 222; APK p. 14

Daily Assessment
Mixed Review, PE pp. 227, 235
Problem of the Day, TE pp. 224, 228, 230, 234
Assessment Tip and Daily Quiz, TE pp. 227, 229, 231, 233, 235

Formal Assessment
Chapter 12 Review, PE p. 237
Chapter 12 Test, TCM pp. A45–A47, B195–B197
Study Guide & Review, PE pp. 272–273
Test • Chapters 11–14, TCM pp. A53–A54, B203–B204
Cumulative Review, PE p. 275
Cumulative Test, TCM pp. A55–A58, B205–B208

Performance Assessment
Team-Up Project Checklist, TE p. 223
What Did I Learn?, Chapters 11–14, TE p. 274
Interview/Task Test, PA p. 87

Portfolio
Suggested work samples:
Independent Practice, PE pp. 227, 229, 231, 235
Cultural Connection, PE p. 236

Student Self-Assessment
How Did I Do?, TGA p. 15
Portfolio Evaluation Form, TGA p. 24
Math Journal, TE pp. 222B, 226

Key

Assessing Prior Knowledge Copying Masters: APK	Teacher's Guide for Assessment: TGA
Test Copying Masters: TCM	Teacher's Edition: TE
Performance Assessment: PA	Pupil's Edition: PE

MATH COMMUNICATION

MATH JOURNAL
You may wish to use the following suggestions to have students write about the mathematics they are learning in this chapter:

PE page 226 • Talk About It

PE pages 227, 229, 231, 235 • Write About It

TE page 226 • Writing in Mathematics

TGA page 15 • How Did I Do?

In every lesson • record written answers to each Wrap-up Assessment Tip from the Teacher's Edition for test preparation.

VOCABULARY DEVELOPMENT
The terms listed at the right will be introduced or reinforced in this chapter. The boldfaced terms in the list are new vocabulary terms in the chapter.

stem-and-leaf plot, pp. 224, 226
histogram, p. 228

So that students understand the terms rather than just memorize definitions, have them construct their own definitions and pictorial representations and record those in their Math Journals.

MEETING INDIVIDUAL NEEDS

MEETING INDIVIDUAL NEEDS

RETEACH
BELOW-LEVEL STUDENTS

Practice Activities, TE Tab B
Multiples Game
(Use with Lessons 12.1–12.4.)

Interdisciplinary Connection, TE p. 234

 Literature Link, TE p. 222C
Exploring the Titanic, Activity 1
(Use with Lessons 12.1–12.4.)

 Technology
Data ToolKit
(Use with Lesson 12.4.)

E-Lab • Activity 12
(Use with Lab Activity.)

PRACTICE
ON-LEVEL STUDENTS

Practice Game, TE p. 222C
Oranges for Sale
(Use with Lesson 12.1.)

Interdisciplinary Connection, TE p. 234

 **Literature Link,** TE p. 222C
Exploring the Titanic, Activity 2
(Use with Lessons 12.1–12.4.)

 Technology
Data ToolKit
(Use with Lesson 12.4.)

E-Lab • Activity 12
(Use with Lab Activity.)

CHAPTER IDEA FILE

The activities on these pages provide additional practice and reinforcement of key chapter concepts. The chart above offers suggestions for their use.

 **LITERATURE LINK**

Exploring the *Titanic*
by Robert D. Ballard

This is a true account of the author's expedition to find the wreck of the *Titanic.*

Activity 1—A total of 1,503 people died when the *Titanic* sank. Have students create a chart comparing this figure with the number of students in their class, in their school, and in their town.

Activity 2—The miniature sub *Alvin,* used to explore the wreck, can descend at a maximum rate of 100 ft per minute. Have students draw a chart comparing this rate with that of an athlete running a 4 minute mi and of a car going 55 mi per hour.

Activity 3—The depth of the ocean at the *Titanic* wreck is 12,460 ft Students should research other shipwrecks or oceanic features. Then have students draw a chart or graph to compare their depths.

PRACTICE GAME

Oranges for Sale

Purpose To practice displaying data on a line graph and interpreting the results

Materials game board (TR p. R88), game directions and scoring guide (TR p. R89), colored pencils

About the Game Players in pairs graph earnings for orange sales over a period of 1 year. Players interpret the results, make comparisons, and use a scoring guide to determine who has the most points.

Modalities Visual, Kinesthetic

Multiple Intelligences Interpersonal/Social, Logical/Mathematical, Bodily/Kinesthetic, Visual/Spatial

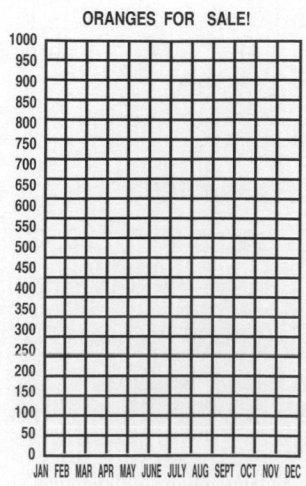

Practice Game, TE p. 222C
Oranges for Sale
(Use with Lesson 12.1.)

Interdisciplinary Connection, TE p. 234

 **Literature Link,** TE p. 222C
Exploring the Titanic, Activity 3
(Use with Lessons 12.1–12.4.)

 Technology
Data ToolKit
(Use with Lesson 12.4.)

E-Lab • Activity 12
(Use with Lab Activity.)

INTERDISCIPLINARY SUGGESTIONS

Connecting Data Displays

Purpose To connect displays of data to other subjects students are studying

You may wish to connect **Displaying Data** to other subjects by using related interdisciplinary units.

Suggested Units

• Science Connection: Our Natural Resources

• Multicultural Connection: African-American Elected Officials

• Consumer Connection: Advertising

• Technology Connection: Media at Home

• Geography Connection: Great Waterfalls

Modalities Auditory, Kinesthetic, Visual

Multiple Intelligences Mathematical/ Logical, Bodily/Kinesthetic, Visual/Spatial

The purpose of using technology in the chapter is to provide reinforcement of skills and support in concept development.

E-Lab • Activity 12—Students use a computer tool to make circle graphs which they then interpret in a variety of ways.

Data ToolKit

This program provides a spreadsheet and graphing tool that will help students organize data they collect, choose appropriate graphs for their data, compare graphs they chose, and find measures of central tendency.

Visit our web site for additional ideas and activities.
http://www.hbschool.com

CHAPTER 12

Introducing the Chapter

Encourage students to talk about using graphs to display data. Review the math content students will learn in the chapter.

WHY LEARN THIS?
The question "Why are you learning this?" will be answered in every lesson. Sometimes there is a direct application to everyday experiences. Other lessons help students understand why these skills are needed for future success in mathematics.

PRETEST OPTIONS

Pretest for Chapter Content See the *Test Copying Masters*, pages A45–A47, B195–B197 for a free-response or multiple-choice test that can be used as a pretest.

Assessing Prior Knowledge Use the copying master shown below to assess previously taught skills that are critical for success in this chapter.

Assessing Prior Knowledge Copying Masters, p. 14

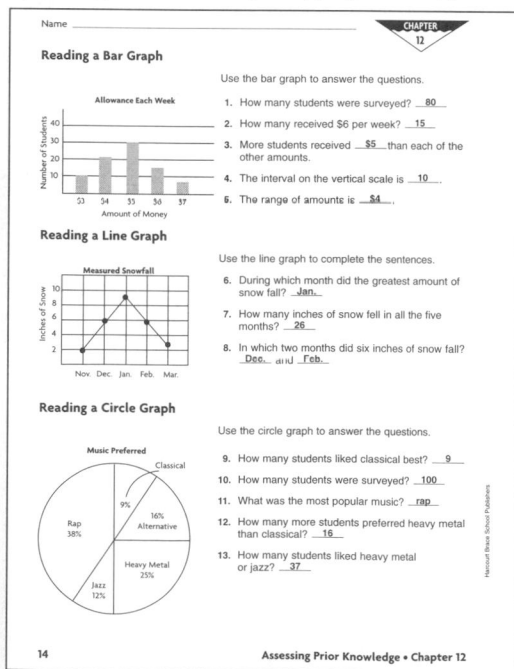

Troubleshooting

If students need help	Then use
• READING A BAR GRAPH	• *Data ToolKit* software
• READING A LINE GRAPH	• *Data ToolKit* software
• READING A CIRCLE GRAPH	• *Data ToolKit* software

CHAPTER 12 — DISPLAYING DATA

LOOK AHEAD

In this chapter you will solve problems that involve

• displaying data in bar graphs, stem-and-leaf plots, line graphs, histograms, and circle graphs

• displaying two or more sets of data in one graph

222 Chapter 12

 SCHOOL-HOME CONNECTION

You may wish to send to each student's family the *Math at Home* page. It includes an article that stresses the importance of communication about mathematics in the home, as well as helpful hints about homework and taking tests. The activity and extension provide extra practice that develops concepts and skills involving displaying data.

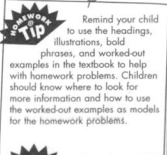

Displaying Data: A Picture Is Worth 1,000 Words

Your child is beginning a chapter on displaying data in different types of graphs and charts. Children are expected to be able to read and analyze the graphs (bar, line, histogram, circle, stem, and leaf) to answer questions and solve problems. Look for graphs in newspapers and magazines to discuss with your child.

Here is an activity to do with your child that involves displaying data. The Extension suggests conducting a survey to collect data that you can display.

Data by the Bag

Purchase two bags of M & Ms™. Sort one bag of candy by color. Construct a histogram (a bar graph of frequencies) to show how many yellow, red, orange, brown, green, and blue candies are in the bag. Find out what fraction of the total each color is. Make a circle graph to show this. Then sort the second bag. Make a double bar graph to show how the color groups in the two bags compare. Discuss the graphs.

Note: You can use instead of M & Ms™ anything that comes in several colors, shapes, or sizes.

Extension • Playing Favorites

Conduct a survey of at least 20 people to find out which color in the M & Ms™ package is their favorite. Graph the results.

For Your Information

Your daily newspaper uses many graphs—bar graphs, pictographs, line graphs, circle graphs, tables, and charts. Graphs show information to the reader more clearly and quickly than words can.

The ability to read and understand graphs is an important skill. Beginning in kindergarten, children are taught how to make, read, analyze, and choose graphs.

Parents can help their children feel sure of these skills by sharing and discussing graphs they come across in daily life.

Remind your child to use the headings, illustrations, bold phrases, and worked-out examples in the textbook to help with homework problems. Children should know where to look for more information and how to use the worked-out examples as models for the homework problems.

Remind your child that when asked to construct a graph on a test to be sure to label the graph and axes, and watch the scale they are using. Also, one needs to be sure the graph selected communicates the data properly.

12

Create The Data

Survey results sometimes do not show the data used to create them. Is there only one possible set of data?

YOU WILL NEED: newspapers, magazines

Plan
- Work with a partner to create a data table for a display in which totals are not shown.

Act
- Find survey results in which you are not given totals.
- Make your own data that could have been used for the survey results.
- Display your data set in a table with the survey results.

Share
- Trade displays with another group. Make your own data for their display.
- Compare your data with that of the other group. Check to be sure both sets of data are appropriate for the display.

1. In a recent survey the most popular gum flavors reported were:

   1. mint
   2. bubble-gum
   3. grape
   4. cinnamon
   5. fruit

   Mint gum was preferred 3:1 to fruit-flavored gum.

2. Here are the survey results for an ideal vacation as voted on by sixth-graders:

   Mountains 7%
   Beach 15%
   Amusement Park 78%

DID YOU

- ✓ Make your own data for your group's display?
- ✓ Make your own data for another group's display?
- ✓ Compare the two sets of data for each display and check that they are appropriate?

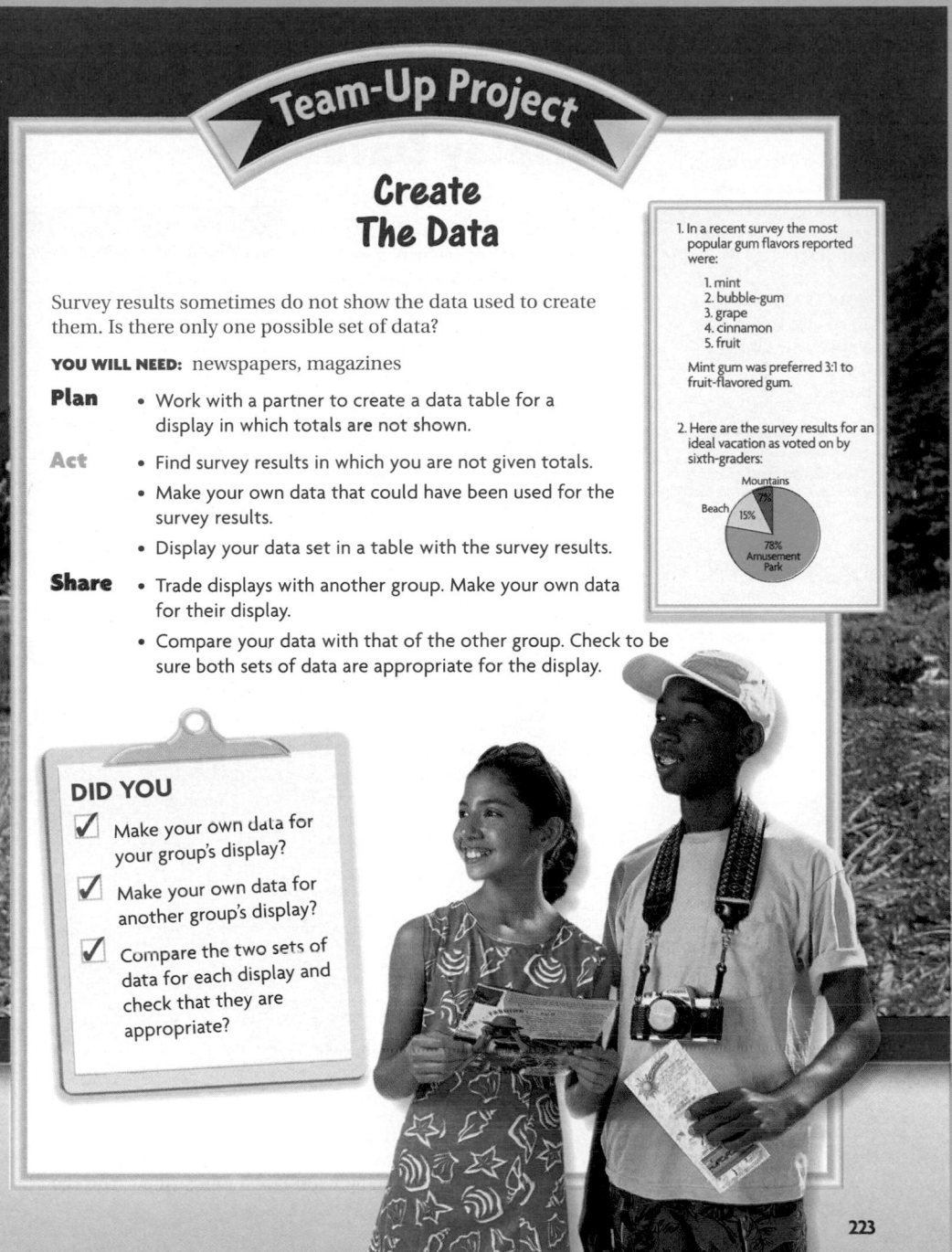

223

Project Checklist

Portfolio

Evaluate whether partners
- ✓ work cooperatively.
- ✓ create appropriate data.
- ✓ recognize that a display may fit more than one set of data.

Using the Project

Accessing Prior Knowledge making tables, using fractions and percents, working backwards

Purpose to create data for a display in which totals are not shown

GROUPING partners

MATERIALS magazines, newspapers, other resources where graphs and other data displays may be found

Managing the Activity Point out that circle graphs often give percents or fractions rather than specific data totals. Ratios and other comparisons describe data without giving totals.

WHEN MINUTES COUNT

A shortened version of this activity can be completed in 15–20 minutes. Have students make up at least two possible sets of data for each of the survey results shown in the chapter opener.

A BIT EACH DAY

Day 1: Students select partners and start looking for an appropriate data display.

Day 2: Students find a data display and set up a table on which totals can be recorded.

Day 3: Students make up data.

Day 4: Partners exchange displays and make up data.

Day 5: Partners compare data for each display.

ON THE ROAD

Ask students to watch for examples of comparisons (such as, 3 to 1) and percents (such as, 56% said yes) that are not accompanied by exact totals.

12.1

ORGANIZER

OBJECTIVE To display data in a bar graph, a line graph, and a stem-and-leaf plot

VOCABULARY *New* **stem-and-leaf plot**

ASSIGNMENTS **Independent Practice**

■ *Basic* 1–9; 10–16
■ *Average* 1–9; 10–16
■ *Advanced* 1–9; 10–16

PROGRAM RESOURCES

Reteach 12.1 Practice 12.1 Enrichment 12.1
Cultural Connection • Zimbabwean Exports, p. 236

PROBLEM OF THE DAY

Transparency 12

Maurie made a stem-and-leaf plot of his relatives' ages. He is the second-youngest of 4 teenagers, all 2 years apart in age. His mother is 3 times as old as he is and 24 years younger than her father. How does Maurie show his grandfather's age in the plot? His grandfather is 69 years old; check students' stem-and-leaf plots.

Solution Tab A, p. A12

QUICK CHECK

Complete the pattern.

1. 0; 150; 300; _?_ ; _?_ ; _?_ ; _?_
   450; 600; 750; 900

2. 0; 500; 1,000; _?_ ; _?_ ; _?_ ; _?_
   1,500; 2,000; 2,500; 3,000

3. 0; 125; 250; _?_ ; _?_ ; _?_ ; _?_
   375; 500; 625; 750

4. 1,000; 1,500; 2,000; _?_ ; _?_ ; _?_ ;
   _?_ 2,500; 3,000; 3,500; 4,000

1 MOTIVATE

Use this activity to review the parts of a graph. Draw a graph on the board without labels. Have a volunteer label the parts of the graph as you discuss them with the class.

224

12.1

WHAT YOU'LL LEARN
How to display data in a bar graph, a line graph, and a stem-and-leaf plot

WHY LEARN THIS?
To be able to organize and visually display data for a report or project

WORD POWER
stem-and-leaf plot

REMEMBER:
A frequency table shows a total in each row, one total for each category.
See page 217.

224 Chapter 12

Using Graphs to Display Data

Hailey's class is conducting a survey on the favorite type of vacation. Hailey organized the data in a frequency table. She wants to make a graph for a class presentation.

FAVORITE VACATION	
Vacation Spot	Frequency
Amusement Park	12
Beach	10
Mountains	4
Camping	10

ACTIVITY

• Work in a small group to make a graph for the data above. **Graphs may vary.**

• Tell which type of graph you made and why you chose it. **Answers may vary.**

A bar graph is a good graph to show data grouped by category.

EXAMPLE 1 Leslie is working on a report about transportation. She collected data about places that have the highest and lowest gas cost per person for a year. Use her data to make a bar graph.

YEARLY GAS COST	
Place	Cost
Wyoming	$734
Montana	$675
New York	$354
Washington, D.C.	$365

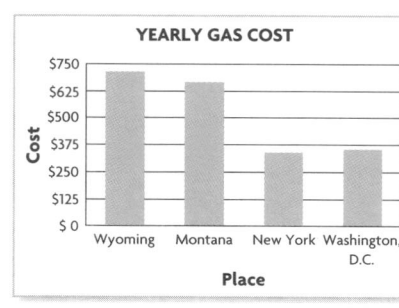

Use a scale from $0 to $750 with equal intervals.

Use bars of equal width, and leave equal space between them.

Title the graph and both axes.

• Why do you think the data sets for favorite vacation and yearly cost for gas are appropriate as bar graphs? **The data are categorical.**

RETEACH 12.1

LESSON 12.1

Name _____

Using Graphs to Display Data

Bar graphs and line graphs are useful for displaying data.
A bar graph shows categorical data.
• Use a scale with equal intervals on the vertical axis.
• Use bars of equal width with equal space between them on the horizontal axis.
• Name each bar.
• Title the graph and both axes.

A line graph shows change over time.
• Use a scale with equal intervals on the vertical axis.
• Mark a point for each piece of data, and connect the points.
• Title the graph and both axes.

NUMBER OF RAINY DAYS LAST WINTER

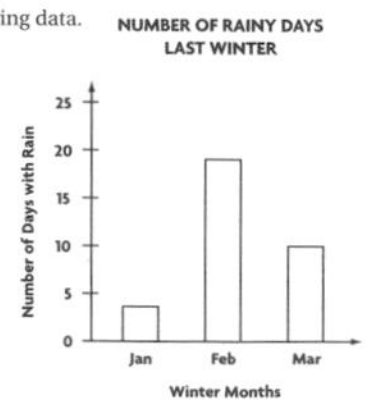

Complete.

1. The vertical axis on the bar graph shows equal intervals of ___5___ days.

2. The vertical axis on the line graph shows equal intervals of ___15___ degrees.

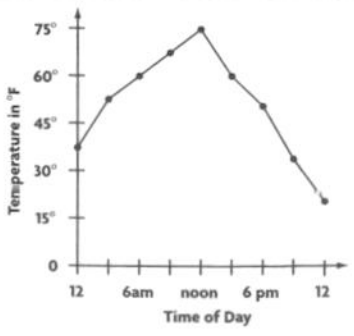

For Exercises 3–5, use the bar graph above.

3. Which month was the wettest? ___Feb___

4. How many rainy days were there last January? ___4 days___

5. How many more rainy days were there in February than in March? ___9 days___

For Exercises 6–7, use the line graph above.

6. What was the highest temperature? at what time? ___75°; at noon___

7. By how many degrees did the temperature fall between noon and 6 PM? ___25°___

R52 TAKE ANOTHER LOOK Use with text pages 224–227.

PRACTICE 12.1

LESSON 12.1

Name _____

Using Graphs to Display Data

Vocabulary

Complete.

1. Stem-and-leaf plots order the data from ___least___ to ___greatest___.

Make an appropriate graph for the set of data. Check students' graphs.

2.
Month	May	June	July	Aug	Sept
High Temperature (in °F)	56	71	86	93	79

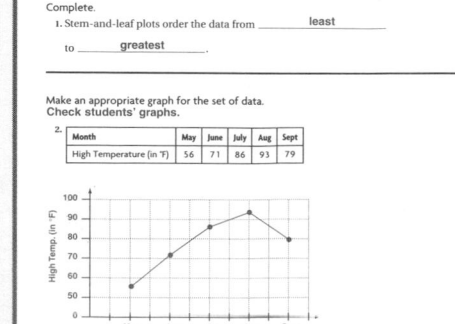

Mixed Applications

3. Make a stem-and-leaf plot for Janet's math test scores: 95, 83, 78, 90, 75, 85, 91, 98, 80.

Janet's Math Test Scores

```
7 | 5 8
8 | 0 3 5
9 | 0 1 5 8
```
Key: 8|3 = 83

4. Mrs. Lopez bought some $0.05-pencils and $0.03-erasers for her class. She bought 17 items in all and spent $.69. How many of each did she buy?

___9 pencils and 8 erasers___

P52 ON MY OWN Use with text pages 224–227.

A bar graph and a line graph are often used to display the same data. However, a line graph is best used when the data show change over time.

EXAMPLE 2 Mr. Henry Phillips works for a small business. He has to report to the business owners about the amount of profit the company has made over the last 6 months. Use Mr. Phillips' data to make a line graph.

PROFIT REPORT						
Month	Nov	Dec	Jan	Feb	Mar	Apr
Profit	$4,500	$5,750	$6,000	$7,500	$8,000	$8,400

Use a scale from $0 to $9,000 with equal intervals of $1,500.

Mark a point for each month.

Connect points.

Title the graph and both axes.

- **CRITICAL THINKING** Suppose the data included a seventh month with a profit of $12,000. How would you change the vertical scale on the line graph? **Possible answer: by showing intervals of $2,000 up to $12,000**

GUIDED PRACTICE

1. Use the data in the table below to determine the numerical scale for a bar graph.

AMOUNT OF SPORTS PARTICIPATION								
Student	A	B	C	D	E	F	G	H
Weeks	32	28	8	16	5	2	20	12

Possible answer: 4, 8, 12, 16, 20, 24, 28, 32

Tell whether a bar graph or a line graph is more appropriate.

2. a city's temperature readings for one week **line graph**

3. sales made by several salespersons on April 1 **bar graph**

225

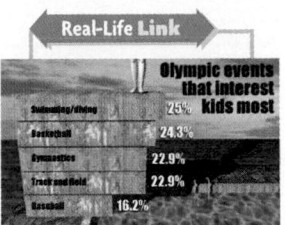

Real-Life Link

Olympic events that interest kids most

Swimming/diving 25%
Basketball 24.3%
Gymnastics 22.9%
Track and field 22.9%
Baseball 16.2%

The use of graphs in print media has become much more common since *USA Today* hit the streets in 1985. Many of the stories in this national newspaper use colorful graphs to help readers understand important data. Many local newspapers include graphs in their news stories, too. Look at a copy of today's paper, and record the types of graphs used and the number of times each is used.

Technology **Link**

In *Data ToolKit,* you can use data to make a table and then a bar graph or a line graph.

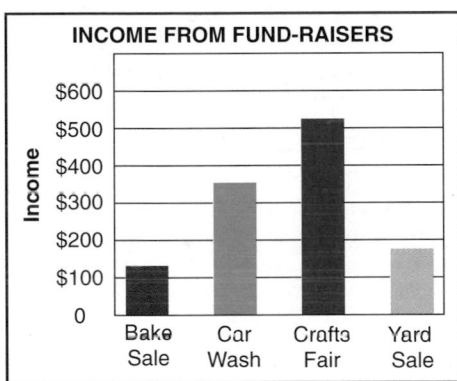

Data ToolKit

2 TEACH/PRACTICE

In this lesson students learn to choose and make the most appropriate graph to display real-world data. Read and discuss the example for each type of graph. Point out to students that the scale on a graph always begins with 0 and ends with a number as great as or greater than the greatest number in the data. The scale has equal intervals, which are usually multiples of 2, 5, or 10, depending on the data.

ADDITIONAL EXAMPLES

Example 1, p. 224

The student council president is working on a report about fund-raisers. She collected data about the income from fund-raisers held at her school. Use her data to make a bar graph.

INCOME FROM FUND-RAISERS	
Bake Sale	$109.00
Car Wash	$358.00
Crafts Fair	$520.00
Yard Sale	$179.00

Example 2, p. 225

The Music Club is working on a report about summer concerts. The club collected data about the student attendance of last year's summer concerts. Use the data to make a line graph.

CONCERT ATTENDANCE–1998	
Month	Number of Students
May	47
June	106
July	120
August	190

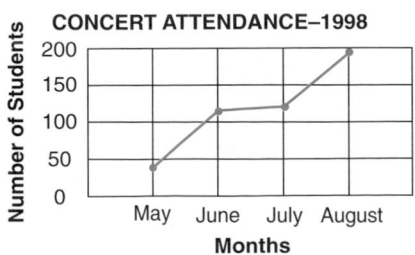

ENRICHMENT 12.1

Name _____

LESSON 12.1

Bar None

Use a yardstick or meterstick to measure the following distances, and complete the bar graph.

A. the width of your math book
B. the height of your desk
C. the length of this sheet of paper
D. the length of your pencil
E. the width of your desk
F. the height of the doorknob on the classroom door

Be sure to title the graph and both axes, and select appropriate intervals on the vertical axis. **Check students' graphs.**

Title: _____

A   B   C   D   E   F

E52   **STRETCH YOUR THINKING**   Use with text pages 224–227.

GUIDED PRACTICE

Discuss Exercises 1–3 with the class to make sure students understand how to determine numerical scales for a bar graph and a line graph.

INDEPENDENT PRACTICE

Have students evaluate the scales they plan to use for their graphs before they begin to make the graphs. Have students ask themselves questions such as "Will the scale easily fit on the page? Are the intervals equal?" Some students may need to try different intervals to find the scale that works best with each set of data.

ADDITIONAL EXAMPLE

Example 3, p. 226

Jeff conducted a survey of local stores to find the prices of their most popular videos. Use the data from the table to make a stem-and-leaf plot.

$14	$19	$31	$24	$24	$29
$29	$28	$19	$32	$18	$29

Stem	Leaves
1	4 8 9 9
2	4 4 8 9 9 9
3	1 2

Stem-and-Leaf Plots

You can use a **stem-and-leaf plot** to organize data when you want to see each item in the data. For a stem-and-leaf plot, choose the stems first and then write the leaves. In the example below, the tens digits appear vertically in order from least to greatest as stems. The ones digits appear horizontally in order from least to greatest as leaves.

EXAMPLE 3 In a recent tower-building competition with cards, the following numbers of levels were reached without the cards falling. Use the data to make a stem-and-leaf plot.

CARD-STACKING COMPETITION					
21	18	32	47	50	33
19	21	11	54	31	18
33	42	21	29	16	12

According to the *Guinness Book of World Records*, Bryan Berg of Boston, Massachusetts, holds the current record for the highest card tower. Berg's record is 102 levels, in a tower 19 ft $6\frac{1}{2}$ in. high.

11	12	16	18	18	19	*First, group data by tens digits.*
21	21	21	29			*Then, order data from least to*
31	32	33	33			*greatest.*
42	47					
50	54					

Card-Stacking Competition

Stem	Leaves	
1	1 2 6 8 8 9	*Use tens digits as stems.*
2	1 1 1 9	*Use ones digits as leaves.*
3	1 2 3 3	*Write leaves in increasing order.*
4	2 7	
5	0 4	The entry 4|2 means 42 levels.

• How is this type of display useful in determining each of the levels achieved in the competition? **Possible answer: Each item in the data is used in the display.**

Talk About It

• How many levels are shown by the second stem and its third leaf? **21 levels**

• How many levels are shown by the fifth stem and its first leaf? **50 levels**

• Suppose you added 55 to the data. Where would it appear on your stem-and-leaf plot? **fifth stem and third leaf**

• In your own words, describe how to make a stem-and-leaf plot. **Check students' responses.**

IDEA FILE

WRITING IN MATHEMATICS

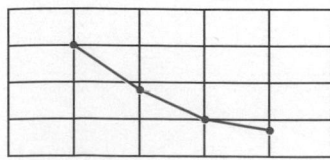

Have students describe two situations this graph could represent. Possible answer: the change in temperature from June to December

Modality Visual

SOCIAL STUDIES CONNECTION

Explain to students that some people live in countries that have populations smaller than those of some large cities. The populations of some of those countries are shown below. Have students make a stem-and-leaf plot using these data. Check students' graphs.

Country	Population (in thousands)
San Marino	24
Liechtenstein	31
Nauru	10
Palau	17

Modality Visual

Choose the appropriate graph for the data. Then make the graph.
See Additional Answers, p. 237A.

1.

NAME	Mary	Tom	Carol	Jim	Tony	Patty
Height (in inches)	35	48	58	60	66	40

bar graph

2.

MONTH	Jan	Feb	Mar	Apr	May	Jun
Average Temperature (in °F)	65	72	75	77	80	85

line graph

3.

MATH TEST SCORES											
60	65	66	70	71	80	81	92	59	77	75	71
80	82	84	93	95	98	80	82	75	75	77	84

stem-and-leaf plot

Problem-Solving Applications

A school record shows that six classes have the following numbers of students: 24 students, 28 students, 32 students, 33 students, 34 students, and 26 students.

4. Which graph, a bar or a line graph, would be better for the data? **bar graph**

5. Make a graph of the data. **See Additional Answers, p. 237A.**

Joe's bank balances for June through December were: $210, $350, $600, $400, $1,000, $750, and $900.

6. What kind of graph could Joe make to see the increases and decreases in his account over six months? **line graph**

7. Make a graph of Joe's data. **See Additional Answers, p. 237A.**

8. Make a stem-and-leaf plot for Mrs. Green's golf scores: 95, 92, 88, 88, 85, 90, 92, 90, 88, 85, 82, 82, 85, 84, 80, 82, 83, 77. **See Additional Answers, p. 237A.**

9. ▬ **WRITE ABOUT IT** Make up a problem in which you could use a stem-and-leaf plot to show the data. **Answers will vary.**

Mixed Review

10. Make a frequency table for the data at the right. **See Additional Answers, p. 237A.**

STUDENT AGES							
10	15	18	20	15	11	13	14
9	10	15	14	18	19	15	11

If you survey 1 out of 10, how many would you survey for each group?

11. 100 teachers **10**

12. 240 boys **24**

13. 620 students **62**

14. 530 girls **53**

15. **SPORTS** Gary was at the 10-yd line when he lost 4 yd. How far was it then to the 50-yd line? **44 yd**

16. **MEASUREMENT** Natalie uses $2\frac{1}{4}$ c of flour to bake a loaf of bread. How much flour will she need for 8 loaves? **18 c**

MORE PRACTICE Lesson 12.1, page H59

227

WRITING IN MATHEMATICS

Have pairs of students write an appropriate scale for each set of data. Ask them to explain in their journals why they chose their scale. Possible answers are given.

1. 58; 385; 555; 444; 101; 422
0; 100; 200; 300; 400; 500; 600

2. 78; 51; 29; 70; 38; 91
0; 25; 50; 75; 100

3. 1,088; 1,195; 789; 2,450; 2,001
0; 500; 1,000; 1,500; 2,000; 2,500

Modality Visual

CULTURAL CONNECTION

Read and discuss the data shown in the table about African-American Elected Officials. Have students choose and make an appropriate graph to show the data. Check students' graphs.

African-American Elected Officials			
Year	1970	1980	1990
Number	1,469	4,890	7,355

Modality Visual

💡**CRITICAL THINKING** Both graphs below show the same data. Why do you think they look different? Possible answer: because the intervals in the scale on the second graph are larger than the intervals in the scale on the first graph.

Make up a scale for each graph.
Possible answer: 0, 1, 2, 3; 0, 2, 4, 6

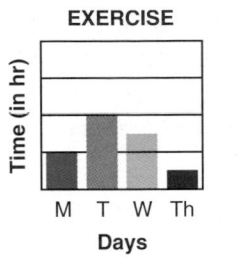

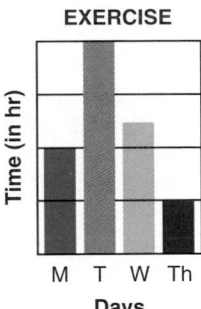

3 WRAP UP

Use these questions to help students focus on the big ideas of the lesson.

• **What type of graph would you make to show the ages of the members of your family? Why?** stem-and-leaf plot; because a stem-and-leaf plot shows each item in a set of data

✔ *Assessment Tip*

Have students write the answer.

• **For Independent Practice Problem 4, why is a bar graph more appropriate than a line graph for showing the data?** because the data show categories, not change over time

DAILY QUIZ

Choose an appropriate graph for the data.

1. an increase in sales from January to May line graph

2. the number of miles walked by each participant in a walkathon stem-and-leaf plot

3. the five most popular CDs of sixth-graders bar graph

12.2

Histograms

GRADE 5
Make line graphs.

GRADE 6
Make histograms.

GRADE 7
Show distribution of data with a histogram.

ORGANIZER

OBJECTIVE To make histograms
VOCABULARY *New* **histogram**
ASSIGNMENTS **Independent Practice**
■ *Basic* 1–10
■ *Average* 1–10
■ *Advanced* 1–10
PROGRAM RESOURCES
Reteach 12.2 Practice 12.2 Enrichment 12.2

PROBLEM OF THE DAY

Transparency 12

Draw the circles and replace the *?*'s with the numbers 1–10. The sum of the numbers in the outer circle is 4 times the sum of the 3 odd numbers in the inner circle. The sum of the inner-circle numbers is 11.

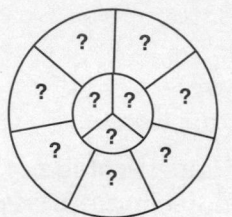
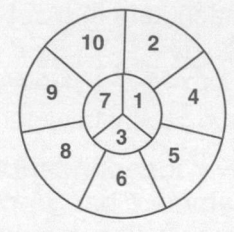

Solution Tab A, p. A12

WHAT YOU'LL LEARN
How to make histograms

WHY LEARN THIS?
To graph groups of data when you want to show the number of times data occur within intervals

WORD POWER
histogram

REMEMBER:
You can use the range to find intervals.

See page 218.

2	8	12	15
15	20	25	30
35	40	42	44

Range: 44 − 2 = 42
5 intervals: 42 ÷ 5 ≈ 9
Use 5 intervals of 9 or 10 values.

0–9, 10–19, 20–29, 30–39, 40–49

A **histogram** is a bar graph that shows the frequency, or the number of times, data occur within intervals. The bars in a histogram are connected, rather than separated.

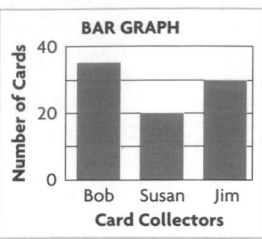
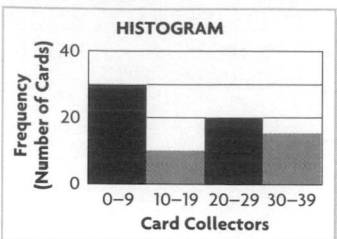

- In which graph do you find more information about the individual card collectors? **bar graph**

- In which graph do you find more information about card collectors in general? **histogram**

EXAMPLE The table below shows the number of runs each player scored in one season. Use the data to make a histogram.

RUNS SCORED BY TEXAS RANGERS

108	115	110	76	21	46	58
24	78	24	25	37	36	38

First, make a frequency table with intervals of 25. Start with 0.

Interval	0–24	25–49	50–74	75–99	100–124
Frequency	3	5	1	2	3

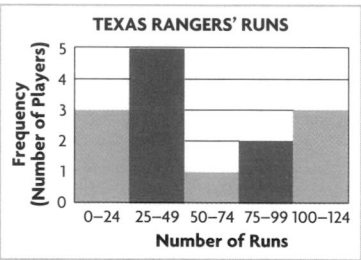

Title the graph and label the scales and axes.

Graph the number of players for each interval.

QUICK CHECK

Name an appropriate graph for the data.

1. a decrease in concert attendance
   line graph
2. the heights of all the sixth graders
   stem-and-leaf plot
3. the ten most successful movies
   bar graph

1 MOTIVATE

Copy the results of this survey on the board. Ask:

VIDEO PURCHASES

13	6	3	9	4	11	10	12	5	3
7	18	8	2	3	17	5	12	7	6

IDEA FILE

GEOGRAPHY CONNECTION

Have students research the names, heights, and locations of the ten highest waterfalls in the world. Then have them make a histogram and a bar graph showing the data. Discuss with students how the graphs are alike and how they are different. Check students' graphs.

Modality Visual

RETEACH 12.2

Name _____

Histograms

A histogram is a type of bar graph. Histograms are different from bar graphs in two ways.
- The bars are side-by-side, not spaced apart.
- Each bar refers to an interval, not a single item.

All of the students in Mr. Higgins' physical education class ran the 100-yard dash. This histogram shows the students' times to the nearest second.

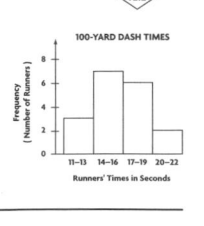

Complete.

1. The number of runners with times of 17–19 sec was ___6___

2. ___3___ students had times of 11–13 sec.

There were 18 students who recorded how many baskets they made in 30 sec. Their results are in the following table.

Student	1	2	3	4	5	6	7	8	9
Baskets made	5	8	17	13	5	19	7	7	

Student	10	11	12	13	14	15	16	17	18
Baskets made	8	9	6	14	7	16	12	12	10

3. Use the data above to complete the histogram.

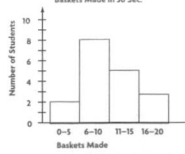

Use with text pages 228–229. **TAKE ANOTHER LOOK** R53

MEETING INDIVIDUAL NEEDS

GUIDED PRACTICE

Tell whether a bar graph or a histogram is more appropriate.

1. number of voters at different intervals of time **histogram**

2. heights of tallest buildings in the United States **bar graph**

3. favorite rock groups of sixth-grade students **bar graph**

For Exercises 4–5, use the frequency table at the right.

4. Make a histogram. **See Additional Answers, p. 237A.**

5. Tell how you could change the way the histogram looks. **Change the intervals.**

10-MILE RACE	
Minutes	Runners
0–49	10
50–99	40
100–149	20

INDEPENDENT PRACTICE

Tell whether a bar graph or a histogram is more appropriate.

1. 100 scores on a science test **histogram**

2. ages of 75 ice-skating competitors **histogram**

3. frequency of certain girls' names **bar graph**

For Exercises 4–5, use the table at the right. **See Additional Answers, p. 237A.**

4. Make a histogram.

5. How would the number of members change in each age group if you changed the histogram to show four age groups? **The number of members in the different age groups would increase.**

MEMBERS IN THE COMMUNITY BAND					
Age	20–29	30–39	40–49	50–59	60–69
Members	10	12	18	6	8

Problem-Solving Applications

For Problems 6–8, use the histogram at the right.

6. What type of table might have been used before making the histogram? **Possible answer: frequency table**

7. Make a frequency table that corresponds to the histogram. **See Additional Answers, p. 237A.**

8. How would the histogram change if the intervals were 7–9:59, 10–12:59, and 1–3:59? **Bars would be taller; would have more numbers on the vertical axis.**

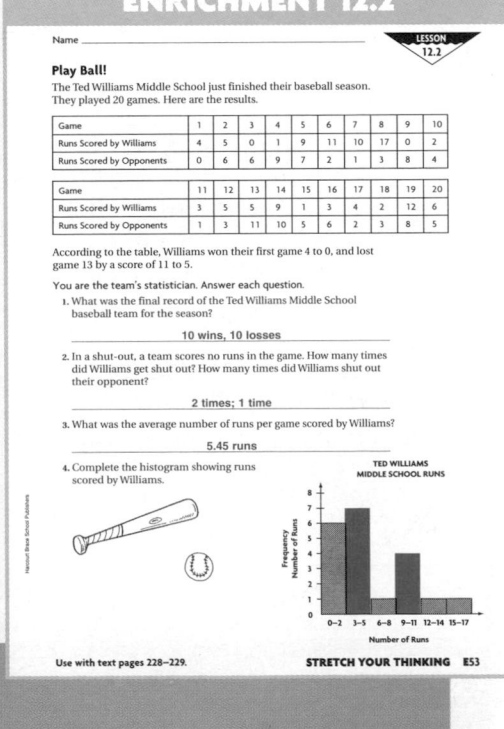

9. Write a problem based on data that can be displayed in a histogram. **Answers will vary.**

10. ✏️ **WRITE ABOUT IT** Explain the difference between a bar graph and a histogram. **A histogram shows data in intervals, and a bar graph shows categories.**

MORE PRACTICE Lesson 12.2, page H59

229

• **How could you display these data? What would the intervals be?**
Histogram: 0–4; 5–9; 10–14; 15–19

2 TEACH/PRACTICE

Focus students' attention on the difference between bar graphs and histograms.

GUIDED PRACTICE

Complete Exercises 1–5 as a whole class.

INDEPENDENT PRACTICE

The exercises ask students to distinguish the uses of bar graphs and histograms.

ADDITIONAL EXAMPLE

Example, p. 228

Use the data to make a histogram.

Books Read by Club Members									
13	53	59	36	75	50	44	64	12	57
19	39	8	61	66	17	48	42	79	48

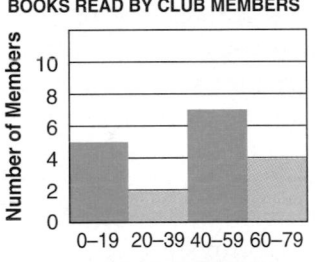

3 WRAP UP

Use these questions to help students focus on the big ideas of the lesson.

• **How would you label the *axes* of the histogram for Independent Practice Exercise 2?** *x*-axis = Ice-Skaters' Ages; *y*-axis = Number of Ice-Skaters

✓ *Assessment Tip*

Have students write the answer.

• **Why is a histogram more appropriate than a bar graph for displaying the data in Independent Practice Exercise 1?** The data can be shown in intervals but cannot be grouped by category.

DAILY QUIZ

Tell whether a bar graph or a histogram is more appropriate.

1. the number of teenagers in the four largest cities in Japan bar graph

2. the 100 distances in a long-jump competition histogram

3. the number of candy bars sold by students in a school fund-raiser histogram

229

PRACTICE 12.2

Name _____

LESSON 12.2

Histograms

Vocabulary

Write the letter for the correct graph.

1. bar graph c
2. histogram a
3. line graph b

Tell whether a bar graph or a histogram is more appropriate.

4. frequency of fish caught at different times of day **histogram**

5. average monthly telephone bill for a year **bar graph**

6. number of shoppers in a store during 3 different time intervals **histogram**

For Exercises 7–8, use the table below.

Campers at Day Camp				
Age	5–7	8–10	11–13	14–16
Number	6	11	18	9

7. Make a histogram. **Check students' graphs.**

8. How would the number of campers in each group change if you used 5 groups instead of 4? **number of campers per group would decrease**

CAMPERS AT DAY CAMP

Mixed Applications

For Problems 9 and 10, use the histogram at the right.

9. During what time period did the most flights arrive? **9:00–10:59**

10. Chen baby-sits for $2.50 per hour. If he baby-sat 6 hr last week and 3 hr this week, how much did he make? **$22.50**

FLIGHT ARRIVALS

Use with text pages 228–229.

ON MY OWN P53

ENRICHMENT 12.2

Name _____

LESSON 12.2

Play Ball!

The Ted Williams Middle School just finished their baseball season. They played 20 games. Here are the results.

Game	1	2	3	4	5	6	7	8	9	10
Runs Scored by Williams	4	5	0	1	9	11	10	17	0	2
Runs Scored by Opponents	0	6	6	9	7	2	1	3	8	4

Game	11	12	13	14	15	16	17	18	19	20
Runs Scored by Williams	3	5	5	9	1	3	4	2	12	6
Runs Scored by Opponents	1	3	11	10	5	6	2	3	8	5

According to the table, Williams won their first game 4 to 0, and lost game 13 by a score of 11 to 5.

You are the team's statistician. Answer each question.

1. What was the final record of the Ted Williams Middle School baseball team for the season? **10 wins, 10 losses**

2. In a shut-out, a team scores no runs in the game. How many times did Williams get shut out? How many times did Williams shut out their opponent? **2 times; 1 time**

3. What was the average number of runs per game scored by Williams? **5.45 runs**

4. Complete the histogram showing runs scored by Williams.

TED WILLIAMS MIDDLE SCHOOL RUNS

Use with text pages 228–229.

STRETCH YOUR THINKING E53

12.3

	GRADE 5	GRADE 6	GRADE 7
	Choose the appropriate graph for displaying a set of data.	Graph two or more sets of data.	Use appropriate graphs.

ORGANIZER

OBJECTIVE To graph two or more sets of data

ASSIGNMENTS Independent Practice
- ■ *Basic* 1–8
- ■ *Average* 1–8
- ■ *Advanced* 1–8

PROGRAM RESOURCES
Reteach 12.3 Practice 12.3 Enrichment 12.3
Extension • Using Spreadsheets for Data, TE
pp. 232A–232B

PROBLEM OF THE DAY

Transparency 12

On the first day of each month, Irene records on a multiple-line graph the times of the sunrise and the sunset. In January, sunrise is 7:54 A.M. and sunset is 5:17 P.M.; in February, the times are 7:40 A.M. and 5:50 P.M.; and in March, 7:05 A.M. and 6:23 P.M. Compare the 2 lines. The sunrise line will go down, and the sunset line will go up.

Solution Tab A, p. A12

QUICK CHECK

Tell whether a bar graph or a line graph is more appropriate.

1. the depths of the Great Lakes
   bar graph
2. the changes in your height from Grade 1 to Grade 6 line graph
3. a decline in the sales of analog clocks line graph

1 MOTIVATE

MATERIALS: two transparencies of TR p. R60

Use the data from the Cost of Jeans graph to make two graphs. On one transparency, draw a graph showing the data about girls' jeans; on the other, draw a graph showing the data about boys' jeans. Have students write five questions using each graph separately or both graphs together.

12.3

WHAT YOU'LL LEARN
How to graph two or more sets of data

WHY LEARN THIS?
To be able to show two or more sets of data on one graph

Geography Link

Temperature readings for cities on either side of the United States-Mexico border are often given in both Fahrenheit and Celsius. Suppose the temperature in both Laredo, Texas, and Nuevo Laredo, Mexico, on the United States-Mexico border, is given as 30° on a nice summer day. Is this Celsius or Fahrenheit?

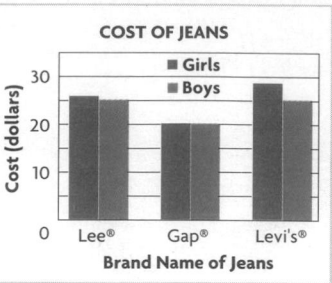

Celsius

Possible answer: for accuracy and so you don't connect the wrong points

Graphing Two or More Sets of Data

Two or more different sets of data can be graphed on one graph. The multiple-bar graph below shows two sets of data.

COST OF JEANS

(bar graph: Girls, Boys — Cost (dollars) vs Brand Name of Jeans: Lee®, Gap®, Levi's®)

- Why does the graph need a key? **to indicate which bars show cost of jeans for girls and which show cost for boys**

EXAMPLE Use the data below to make a multiple-line graph.

AVERAGE DAILY TEMPERATURES

	Mon	Tue	Wed	Thu	Fri
High	82°F	78°F	75°F	64°F	70°F
Low	70°F	67°F	65°F	54°F	62°F

AVERAGE DAILY TEMPERATURES

(line graph: Temperature (°F) vs Day — High, Low)

This means there is a break in the scale.

Determine an appropriate scale.

Mark a point for each high temperature, and connect the points.

Mark a point for each low temperature, and connect the points.

Title the graph and both axes. Include a key.

- 💡 **CRITICAL THINKING** When you make a multiple-line graph, why should you graph one set of data at a time?

IDEA FILE

HOME CONNECTION

Have students use the data below to make a multiple-bar graph. Check students' graphs.

PERCENT OF U.S. HOUSEHOLDS WITH ELECTRONIC MEDIA

	Phone	Radio	VCR
1980	92%	99%	3%
1990	92%	99%	71%

Modality Visual

RETEACH 12.3

Name _____

LESSON 12.3

Graphing Two or More Sets of Data

At Fox Run School, the sixth grade and seventh grade classes are competing to see which class can collect the most aluminum cans. The bar graphs show the results for each of the first three weeks.

GRADE 6 WEEKLY TOTALS   **GRADE 7 WEEKLY TOTALS**

To compare the two classes' results more easily, you can show all the data in one graph. The following graph is a double-bar graph.

WEEKLY TOTALS

Key: Grade 6, Grade 7

The key tells which bar represents the sixth grade and which, the seventh grade. A bar graph that shows two or more sets of data is called a multiple-bar graph.

For Exercises 1–3, use the multiple-bar graph above.

1. How do you know which bar refers to which grade? **Read the key.**

2. Which grade collected more cans in week 1? week 2? week 3? **Grade 7; Grade 7; Grade 6**

3. Which grade collected more cans all together? **Grade 6**

GUIDED PRACTICE

1. Make a multiple-bar graph for the following data, from a survey of 100.

NUMBERS OF PETS				
	Dogs	Cats	Birds	Fish
Adult Household	24	10	2	1
Children in Household	38	16	4	5

See Additional Answers, p. 237A.

2. Make a multiple-line graph for the following data, which compare scores from two seasons.

BASKETBALL SCORES					
Game	1	2	3	4	5
1996	60	66	45	50	55
1997	64	75	82	80	71

See Additional Answers, p. 237B.

INDEPENDENT PRACTICE

1. Make a multiple-bar graph showing the following price comparisons for some regular and nonfat dairy products.
See Additional Answers, p. 237B.

COSTS OF DAIRY PRODUCTS				
	Milk	Yogurt	Cottage Cheese	Ice Cream
Regular	$1.20	$0.89	$1.75	$1.99
Nonfat	$1.40	$0.99	$2.05	$2.20

2. Make a multiple-line graph for the following data, which compare some average stock prices.
See Additional Answers, p. 237B.

STOCK PRICES				
	Jan	Feb	Mar	Apr
1995	$44	$53	$64	$38
1996	$48	$55	$62	$38

Problem-Solving Applications

Lisette is recording band membership in her school for the past three years. For Problems 3–5, use the table at the right. **3. multiple-line graph; changes over time**

3. Which type of graph would you use for the data? Explain.

4. Make a graph for Lisette's data. **See Additional Answers, p. 237B.**

BAND MEMBERSHIP									
	Sep	Oct	Nov	Dec	Jan	Feb	Mar	Apr	May
1995	48	48	45	44	44	44	44	42	38
1996	50	51	53	53	52	52	51	50	51
1997	55	55	54	51	50	48	48	44	44

5. How many lines are on your graph? Why? **3 lines; 1 line for each year's data**

6. A survey shows how many boys and girls from each classroom are in the band. What type of graph would you use? Explain. **multiple-bar graph; shows comparisons of categories**

7. Suppose you find the predicted high and low temperatures for your city for one week. What type of graph would you make? **multiple-line graph**

8. ✏️ **WRITE ABOUT IT** Tell when you would use a multiple-bar or multiple-line graph. **to compare two or more sets of data**

---

2 TEACH/PRACTICE

Read and discuss p. 230 as a class. Have students identify what two sets of data are shown in each graph.

GUIDED PRACTICE
You may want to have students work with a partner to complete the exercises. Then discuss them as a class.

INDEPENDENT PRACTICE
Check answers to Exercises 1 and 2 before assigning Problems 3–8.

ADDITIONAL EXAMPLE
Example, p. 230

Use the data to make a graph.

BASKETBALL GAME ATTENDANCE				
	1993	1994	1995	1996
Fifth Graders	65	70	72	80
Sixth Graders	50	49	59	89

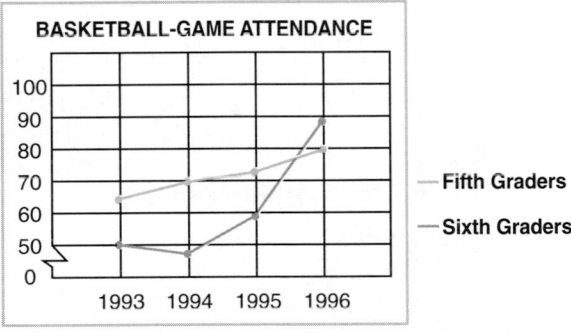

BASKETBALL-GAME ATTENDANCE

— Fifth Graders
— Sixth Graders

3 WRAP UP

Use these questions to help students focus on the big ideas of the lesson.

- **Why does a multiple-bar graph need a key?** Possible answer: to indicate which bar shows which set of data

✔️ *Assessment Tip*

Have students write the answer.

- **Write a word problem using the data in the graph you made for Independent Practice Exercise 1.** Check students' answers.

DAILY QUIZ
Tell what graph you would use to show the data.

1. the highest and lowest scores given to three athletes **multiple-bar graph**

2. the number of VCRs in American and Japanese homes over 5 yr **multiple-line graph**

3. the teachers' and students' favorite holidays **multiple-bar graph**

231

EXTENSION

Using Spreadsheets for Data

Use these strategies to extend understanding of ways to display data.

Multiple Intelligences Logical/Mathematical, Visual/Spatial, Bodily/Kinesthetic

Modalities Auditory, Visual, Kinesthetic

Lesson Connection Students will use a spreadsheet to make bar graphs and circle graphs.

Materials Data ToolKit or other spreadsheet software

Program Resources *Intervention and Extension Copying Masters,* p. 18

Challenge

TEACHER NOTES

ACTIVITY 1

Purpose To use a spreadsheet to make a bar graph and a circle graph

Discuss the following questions.

• Why is it important to name a graph and label the axes? to show what information the graph contains

• What are the advantages of using a graph to display data, rather than a table? You can better visualize the relationships among the data.

WHAT STUDENTS DO

The following survey question was asked of 100 students: Which of the following summer activities do you prefer? (Pick one.)

swimming, in-line skating, watching tv, camping, reading

The results are given below:

Activity	Total
Swimming	30
In-line Skating	27
Watching TV	9
Camping	24
Reading	10

1. Enter the data in a spreadsheet like the one shown below.

	A	B	C	D
1	Swimming		30	
2	In-line skating		27	
3	Watching TV		9	
4	Camping		24	
5	Reading		10	

2. Highlight the data you entered. Select the chart function on your spreadsheet, and choose a bar graph. Label the *x*-axis "Activities" and the *y*-axis "Number of Students." Name your graph.

   • Use the graph to write two statements about students' preferences for summer activities. Possible answers: The most popular activity is swimming. The least popular activity is watching TV.

3. Select the data in the spreadsheet again. Select the chart function on your spreadsheet, and choose a circle graph that displays the name of each category and the percent. Make sure to title your graph.

   • What percent chose swimming as their favorite activity? 30%

   • What percent chose watching TV as their favorite activity? 9%

TEACHER NOTES	WHAT STUDENTS DO

ACTIVITY 2

Purpose To use a spreadsheet to make a double bar graph

Compare girls' and boys' favorite activities by using a double bar graph.

1. Enter the data from the following frequency table into a spreadsheet.

Activity	Girls	Boys
Swimming	15	15
In-line skating	13	14
Watching TV	3	6
Camping	14	10
Reading	7	3

2. Select the data you entered. Select the chart function on your spreadsheet, and choose a double bar graph. Label each axis and name the graph.

- Use the graph to write two statements about girls' and boys' favorite summer activities. Possible answer: The most popular summer activity for boys and girls is swimming. More girls than boys prefer reading as a summer activity.

Have students meet in small groups or with partners to discuss the two questions. Have one student report the group's responses to the larger group.

Share Results

- How is the spreadsheet method easier than using pencil and paper? You enter the data. The computer draws the graph for you.

- What does a circle graph show you that a bar graph does not? A circle graph shows you what percent of the whole each category is.

Intervention and Extension Copying Masters, p. 18

ORGANIZER

OBJECTIVE To learn how to make a circle graph from a bar graph

MATERIALS *For each student* $8\frac{1}{2}$-in. × 11-in. or larger paper, markers, scissors, tape

PROGRAM RESOURCES
E-Lab • Activity 12
E-Lab Recording Sheets, p. 12

USING THE PAGE

Begin by asking students:

- **What is a bar graph?** a graph that uses bars to show categorical data

- **What is a circle graph?** a graph in the shape of a circle that shows parts of a whole

Explain to students that in this activity they will discover the relationship between bar graphs and circle graphs.

Explore

A. Observe students as they follow instructions for making a bar graph. Remind them to write a lunch preference directly on each bar. When they tape the bars together, have them place the labeled sides of the bars on the outside of the circle.

B. Demonstrate on the board the steps for making a circle graph. Have a volunteer hold down the circle while you trace around it and mark where each bar begins and ends. Have students trace their circles and mark where the bars begin and end. Before drawing the radii, have students check their drawings and make corrections as needed.

COOPERATIVE LEARNING

GEOMETRY CONNECTION

Exploring Circle Graphs

LAB *Activity*

WHAT YOU'LL EXPLORE
How to make a circle graph from a bar graph

WHAT YOU'LL NEED
$8\frac{1}{2}$-in. × 11-in. or larger paper, markers, scissors, tape

Data can be displayed in more than one kind of graph. In this lab you will show data in a bar graph. Then you will use the bar graph to make a circle graph.

Explore

Work with a partner.

A. Use the following data to make a bar graph.

SCHOOL LUNCHES		
Lunch Preference	Number of Students	Cumulative Frequency
Hot lunch	75	75
Salad	50	125
Pack a lunch	75	200

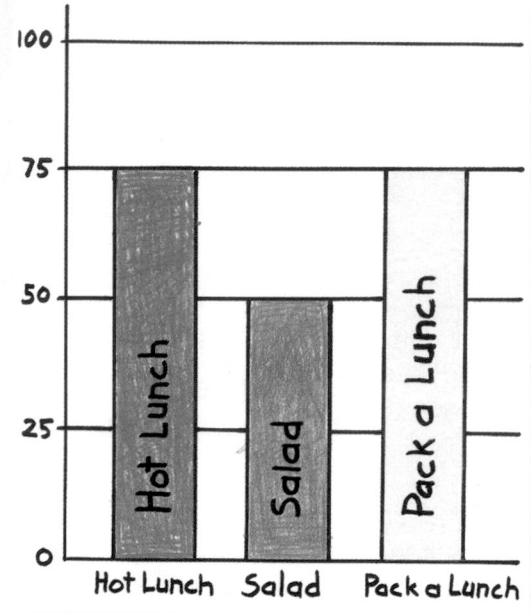

- Make each bar in the graph a different color.

- Write the lunch preference on the corresponding bar in the graph.

- Cut out each bar from your graph.

- Tape the ends of the bars together, without overlapping, to form a circle.

232 Chapter 12

IDEA FILE

SCIENCE CONNECTION

Ask students: What do you think happens to all the old tires that people throw out each year? Tell students that about 78% are buried in landfills or illegally dumped, about 10% are burned for energy, about 7% are recycled, and about 5% are exported.

- Have students make a bar graph and a circle graph to show these data. Check students' graphs.

Modalities Auditory, Kinesthetic, Visual

CONSUMER CONNECTION

Discuss with students the different media used in advertising. Have them make a bar graph and a circle graph to show how much was spent in the United States on the use of each medium in one year.

Money Spent on Advertising (approximate amount in billions of dollars)	
Newspapers	31
TV	11
Radio	10
Direct Mail	27
Other	31

Modalities Auditory, Kinesthetic, Visual

B. You can use your circle of bars to make a circle graph.

- Place your circle on a piece of paper, and trace around it. Mark where each bar begins and ends around the circle.

- Mark the center of the traced circle.

- Draw a radius from each of the lines you marked on the circle.

- Color the sections of the circle to match the colors of the bars. Label each region, and title the graph.

SCHOOL LUNCHES

Pack A Lunch 75 · Hot Lunch 75 · Salad 50

Think and Discuss

- What does the whole circle graph represent? **200 students**

- How is the circle graph different from the bar graph on page 232? **The circle graph shows how the parts relate to the whole.**

- What fraction could you write on your circle graph in each section? $\frac{3}{8}, \frac{1}{4}, \frac{3}{8}$

Try This

- Use the data below to make a bar graph. Then use the bar graph to make a circle graph.

CAROL'S DAY		
Activity	Number of Hours	Cumulative Frequency
School	6	6
Study	3	9
TV	3	12
Sleep	8	20
Other	4	24

Check students' graphs. Graphs should show the following fractional parts: $\frac{1}{4}, \frac{1}{8}, \frac{1}{8}, \frac{1}{3}$, and $\frac{1}{6}$.

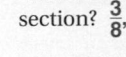 **Link**
E–Lab · Activity 12 Available on CD-ROM and the Internet at http://www.hbschool.com/elab

Note: Students make circle graphs. Use E-Lab p. 12.

233

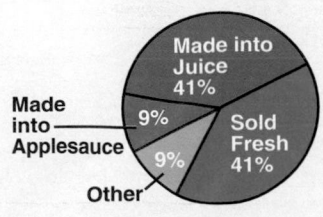
CORRELATION TO OTHER RESOURCES

 Macmillan Visual Almanac, edited by Bruce S. Glassman, 1996, Blackbirch Press.

 CornerStone Mathematics ©1996 SkillsBank Corporation **Level B, Working with Data, Lesson 4**

 Graphers **Sunburst** Students learn about graphs and use graphs to solve problems.

Think and Discuss

Before discussing the questions in the book, have students compare their circle graphs with the bar graph on p. 232. Ask questions such as:

- **How are the categories on the bar graph shown in your circle graph?** Possible answer: in parts or sections of a whole

- **What is the smallest bar on the bar graph?** the bar for *Salad*

- **What is the smallest section on your circle graph?** the section for *Salad*

Discuss the first two questions in the book. Help students with the third question by asking:

- **How many students prefer hot lunches?** 75 students

- **What part of all the students in the survey prefer hot lunches?** $\frac{75}{200}$

- **Write $\frac{75}{200}$ in simplest form.** $\frac{3}{8}$

Repeat the questions above for the other two sections of the circle graph.

Try This

Before students begin making their graphs, ask:

- **How many bars will be in your bar graph?** 5 bars

- **How many sections will be in your circle graph?** 5 sections

Have students refer to the steps described in Explore as they make their graphs.

USING E-LAB

Students use a computer tool to make circle graphs which they then interpret in a variety of ways.

✓ *Assessment Tip*

Use these questions to help students communicate the big ideas of the lesson.

- **What does the whole circle graph for Carol's Day represent?** 24 hr

- **What fraction represents each section of the circle graph for Carol's Day?** School: $\frac{1}{4}$; Study: $\frac{1}{8}$; TV: $\frac{1}{8}$; Sleep: $\frac{1}{3}$; Other: $\frac{1}{6}$

GRADE 5
Make circle graphs.

GRADE 6
Read and make a circle graph.

GRADE 7
Make circle graphs.

ORGANIZER

OBJECTIVE To read and make a circle graph

ASSIGNMENTS Independent Practice
- *Basic* 1–16
- *Average* 1–16
- *Advanced* 1–16

PROGRAM RESOURCES
Reteach 12.4 Practice 12.4 Enrichment 12.4
Cultural Connection • Zimbabwean Export,
    p. 236
Data Tool Kit

PROBLEM OF THE DAY

Transparency 12

Grant didn't have a protractor, so he used a clock face for his circle graph. He drew radii to the 12, the 4, and the 9. How many degrees is each angle? from 12 to 4, a 120° angle; from 4 to 9, a 150° angle; from 9 to 12, a 90° angle

Solution Tab A, p. A12

QUICK CHECK

Find the product.

1. $\frac{1}{10} \times 360$ 36
2. $\frac{1}{9} \times 360$ 40
3. $\frac{1}{5} \times 360$ 72
4. $\frac{1}{3} \times 360$ 120
5. $\frac{1}{8} \times 360$ 45
6. $\frac{1}{6} \times 360$ 60

1 MOTIVATE

MATERIALS: protractor, centimeter ruler

Have students work in pairs as you review how to draw an angle with a protractor.

1. Draw a ray 7 cm long, with point *A* on the left.
2. Place the center mark of the protractor on point *A* with the base of the protractor on the ray.
3. Locate 45° on the bottom scale. Mark a point at this location.
4. Draw another ray from point *A* through the new point.

Repeat this activity for a 90° angle.

12.4

WHAT YOU'LL LEARN
How to read and make a circle graph

WHY LEARN THIS?
To display data that are parts of a whole, such as the different ways an allowance is spent

Science Link

Recycling is one of the easiest ways to conserve resources. Every recycled aluminum can saves 90% of the aluminum ore needed to make a new can. Mr. Veejay's class found that out of 100 items recycled by the cafeteria, 35 were aluminum cans, 28 were steel cans, 24 were glass bottles, and 10 were plastic cups. How many items did not fall into one of the four recycling categories?

3 items

Making Circle Graphs

A circle graph shows parts of a whole. You can use decimals or fractions to divide the circle.

- What is the total weekly allowance? What decimal part of the circle do snacks represent? **$10.00; 0.5**

WEEKLY ALLOWANCE

$5.00 Snacks
$3.00 Games
$1.00 School Stuff
$1.00 Misc.

To make a circle graph, you need to find the number of degrees represented by each part. Since there are 360° in a circle, multiply the fraction or the decimal for each part by 360°.

EXAMPLE Mr. Collins' class earned $400 by recycling different materials. Use the data below to make a circle graph.

aluminum: $200; glass: $100; newspaper: $50; plastic: $50

$200: $\frac{200}{400} = \frac{1}{2}$, $\frac{1}{2} \times 360° = 180°$

$100: $\frac{100}{400} = \frac{1}{4}$, $\frac{1}{4} \times 360° = 90°$

$50: $\frac{50}{400} = \frac{1}{8}$, $\frac{1}{8} \times 360° = 45°$

Write each amount as a part of the whole. Then multiply by 360°.

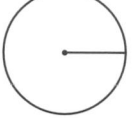

Use a compass to draw a circle. Draw a radius.

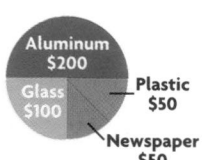

Aluminum $200
Glass $100
Plastic $50
Newspaper $50

Use a protractor to draw a 180° angle. Label it Aluminum $200.

Draw a 90° angle. Label it Glass $100.

Draw two 45° angles. Label them Newspaper $50 *and* Plastic $50.

GUIDED PRACTICE

For Exercises 1–2, use the circle graph in the Example.

1. What decimal and fraction does the glass section represent? 0.25; $\frac{1}{4}$
2. Suppose only aluminum, newspaper, and plastic are recycled. What angle measure would the sections be? 240°, 60°, 60°

IDEA FILE

INTERDISCIPLINARY CONNECTION

One suggestion for an interdisciplinary unit for this chapter is *Our Natural Resources*. In this unit, you can have students

- discuss how people in the United States use fresh water.
- research how we use other natural resources.
- collect data on one of these topics and make a circle graph to show their findings.

Modalities Auditory, Kinesthetic, Visual

RETEACH 12.4

Name _____

LESSON 12.4

Making Circle Graphs

Half of the students in Mrs. Kline's sixth grade class are 12 years old. The rest are either 11 or 13 years old. The circle graph shows the data.

- $\frac{1}{4}$ of the students are 11 years old.
- $\frac{1}{2}$ of the students are 12 years old.
- $\frac{1}{4}$ of the students are 13 years old.

AGE OF STUDENTS IN MRS. KLINE'S SIXTH GRADE CLASS

age 12 $\frac{1}{2}$ $\frac{1}{4}$ age 11 $\frac{1}{4}$ age 13

Each part of the circle graph looks like a piece of pie. So, a circle graph is sometimes called a pie chart. Remember that there are 360° in a complete circle. To find the measure of the angle for a given fraction or decimal, multiply the fraction or decimal by 360°.

The angle for $\frac{1}{4}$ is $\frac{1}{4} \times 360° = 90°$

Find the measure of the angle for each piece in the circle graph at the right.

1. $\frac{1}{3}$ **120°**
2. $\frac{1}{6}$ **60°**
3. $\frac{1}{2}$ **180°**

There are 16 students in Mrs. Albaro's second grade class. Four of the students are 6 years old, and the rest are 7 years old.

MRS. ALBARO'S CLASS

Age 6
Age 7

4. Draw a circle graph to show the ages. **Check students' graphs.**
5. What angle does the piece representing 6-year-olds make?

90° angle

6. Does the circle graph at the top of the page show how many students are in Mrs. Kline's class? Explain.

No, it shows only what fraction of the class is of each age.

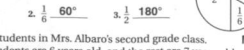

INDEPENDENT PRACTICE

For Exercises 1–2, use the circle graph at the right.

1. How many people were surveyed about their favorite color? **100 people**

2. What fraction of the circle does blue represent? $\frac{20}{100}$, or $\frac{1}{5}$

For Exercises 3–6, use the following data.

FAVORITE COLOR

30 Violet
38 Red
20 Blue
12 Green

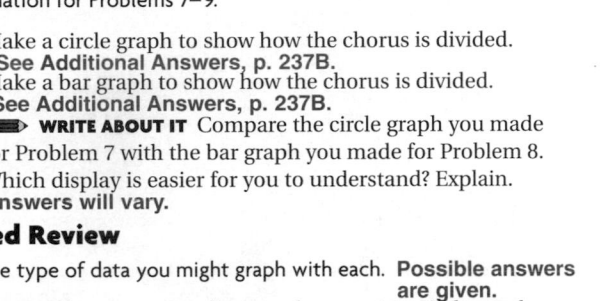

SCOTT'S DAILY ACTIVITIES (IN HOURS)				
Sleep	School	Play	Homework	Other
8	6	3	3	4

3. Into how many sections would you divide a circle graph for the data? **5 sections**

4. For a circle graph, what angle measure would you use for the Sleep section? **120°**

5. Make a circle graph of the data. **See Additional Answers, p. 237B.**

6. What is the total measure of all of the angles in a circle graph of the data? **360°**

Problem-Solving Applications

The chorus at Shoreline Middle School has 60 members. There are 15 sixth graders, 21 seventh graders, and 24 eighth graders. Use this information for Problems 7–9.

7. Make a circle graph to show how the chorus is divided.
See Additional Answers, p. 237B.

8. Make a bar graph to show how the chorus is divided.
See Additional Answers, p. 237B.

9. ✏️ **WRITE ABOUT IT** Compare the circle graph you made for Problem 7 with the bar graph you made for Problem 8. Which display is easier for you to understand? Explain. **Answers will vary.**

Mixed Review

Tell the type of data you might graph with each. **Possible answers are given.**

10. bar graph **categorical**

11. line graph **change over time**

12. circle graph **parts of a whole**

For Exercises 13–14, tell if the sample is biased. Write *yes* or *no*.

13. Ask car makers their favorite car. **yes**

14. Ask people their favorite TV show. **no**

15. Josie jogs 5 times around the football field. The football field is 360 ft long and 160 ft wide. Does she jog 1 mi? Explain. **No. She jogs 5,200 ft, 80 ft short of 1 mi.**

16. **SPORTS** Joe spent $16.50 bowling. It cost $2.50 for shoes and $3.50 for each game. How many games did he bowl? **4 games**

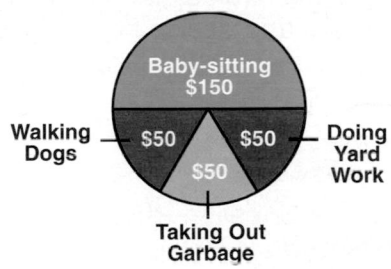

Technology Link

In *Data ToolKit,* you can use data to make a table and then show the data as a circle graph.

Data ToolKit

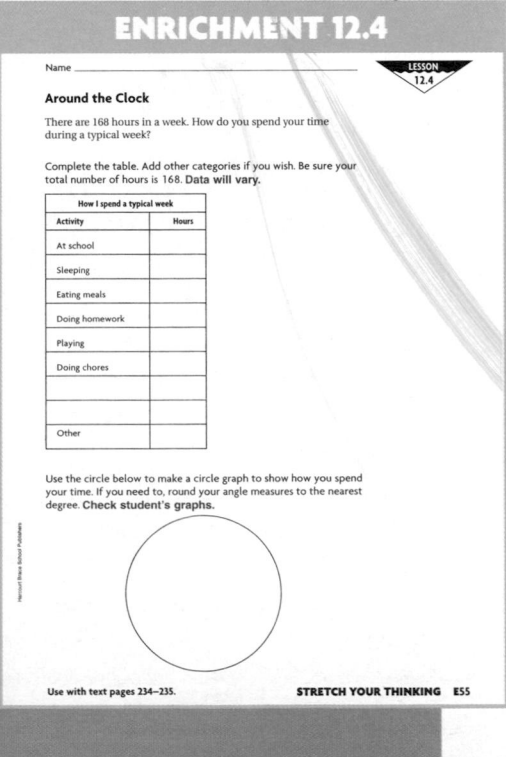

PRACTICE 12.4

Name _____ LESSON 12.4

Making Circle Graphs

For Exercises 1–2, use the circle graph at the right.

1. How many students were surveyed about their favorite subject?
200 students

2. What fraction of the circle does Science represent?
$\frac{60}{200}$, or $\frac{3}{10}$

FAVORITE SUBJECT

History 25
Science 60
English 50
Math 65

For Exercises 3–6, use the following data.
The Video Shack kept track of the categories of videos that were rented last weekend.

VIDEOS RENTED			
Drama	Comedy	Action	Classics
250	250	300	200

3. Into how many sections would you divide a circle graph for the data?
4 sections

4. Make a circle graph of the data.
Check students' graphs.

VIDEOS RENTED

Classics, Drama, Action, Comedy

Mixed Applications

For Exercises 5–6, use the circle graph at the right.

5. The circle graph shows how the Millers plan to spend their money. How many times as much is budgeted for food as for transportation?
4 times as much

6. If the Millers' total income is $30,000, how much do they plan to spend on medical? on housing?
$2,500; $7,500

THE MILLERS' BUDGET

Housing, Food, Medical, Clothing

Use with text pages 234–235. **ON MY OWN** P55

ENRICHMENT 12.4

Name _____ LESSON 12.4

Around the Clock

There are 168 hours in a week. How do you spend your time during a typical week?

Complete the table. Add other categories if you wish. Be sure your total number of hours is 168. **Data will vary.**

How I spend a typical week	
Activity	Hours
At school	
Sleeping	
Eating meals	
Doing homework	
Playing	
Doing chores	
Other	

Use the circle below to make a circle graph to show how you spend your time. If you need to, round your angle measures to the nearest degree. **Check student's graphs.**

Use with text pages 234–235. **STRETCH YOUR THINKING** E55

---

2 TEACH/PRACTICE

Discuss the information about circle graphs on p. 234 as a class. Then work through the example with students.

GUIDED PRACTICE

Complete the exercises as a whole class to make sure students understand how to read a circle graph.

INDEPENDENT PRACTICE

These exercises check students' abilities to determine what fraction each part of a circle graph represents and to decide what angle measure is needed to show a given portion in a circle graph.

ADDITIONAL EXAMPLE

Example, p. 234

Lisa earned $300 doing odd jobs in the summer. She earned $150 baby-sitting, $50 walking dogs, $50 taking out garbage, and $50 doing yard work. Use the data to make a circle graph.

Baby-sitting $150
Walking Dogs $50
$50 Doing Yard Work
$50
Taking Out Garbage

3 WRAP UP

Use these questions to help students focus on the big ideas of the lesson.

- **What is the advantage of displaying data in a circle graph?** Possible answer: You can see how the parts relate to the whole.

✔️ *Assessment Tip*

Have students write the answer.

- **How did you find the answer to Independent Practice Exercise 4?** Possible answer: by writing the number of hours for sleep as part of the whole, $\frac{8}{24} = \frac{1}{3}$, and $\frac{1}{3} \times 360° = 120°$

DAILY QUIZ

Use the circle graph at the top of p. 234.

1. How many degrees are in the whole circle graph? **360°**

2. What angle measure is used for the section *Snacks*? **180°**

3. What fractional part of the circle graph does the section *Games* represent? $\frac{3}{10}$

235

OBJECTIVE To read and make appropriate graphs

USING THE PAGE

Before the Lesson

Have students make a chart showing 10 hours of homework for five days. Use even hours, but vary the time each day spent on homework. Have them show the data in a bar graph and change it to a circle graph.

Cooperative Groups

Have students work together in cooperative groups to answer the question about the graph and to complete Problems 1–4. Remind them to title and label each graph. Ask the groups to study the numbers in each problem and decide how they will mark the intervals on the graph.

> **CULTURAL LINK**
>
> Zimbabwe was called Rhodesia during the years it was a British colony. It declared independence in 1970 and adopted majority rule in 1979.

CULTURAL CONNECTION

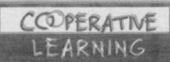

Zimbabwean Exports

Zimbabwe is a country located in Southern Africa. Like many countries, Zimbabwe relies on income earned from exporting goods. Some of the main goods Zimbabwe exports are clothing and textiles, steel, nickel, gold, and cotton lint. These goods help support the economy of Zimbabwe.

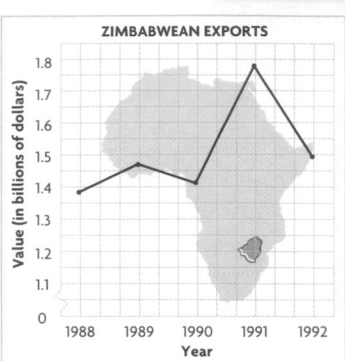

Chuma gathered this information about the growth of exports in Zimbabwe:

1988, $1.39 billion; 1989, $1.47 billion; 1990, $1.41 billion; 1991, $1.77 billion; 1992, $1.54 billion.

Use a line graph to organize and display the data.

ZIMBABWEAN EXPORTS

- For the time shown on the graph, in which year were exports the lowest for Zimbabwe? **1988**

Work Together

1. Use the data about Zimbabwean exports to make a bar graph. Display the changes in values in order from the highest to the lowest. What changes did you have to make to the horizontal axis? **Check students' graphs. The order of the years was changed.**

2. Suppose 25% of the Zimbabwean exports was clothing and textiles, 15% was steel, 18% was nickel, 10% was gold, 12% was cotton lint, and 20% was other products. Make a circle graph to display this data. **Check students' graphs.**

3. Chuma also found this information about the number of livestock in Zimbabwe in 1993: cattle, 4,000,000; pigs, 270,000; sheep, 530,000; goats, 2,500,000. Make an appropriate graph to display this data. **Check students' graphs.**

4. **WRITE ABOUT IT** Make up a country of your own. Tell about the 5 main products your country exports. Make an appropriate graph to display the exports of your country. **Check students' graphs.**

> **CULTURAL LINK**
>
> Gold has been an important part of the Zimbabwean economy since the eighth century. The people of the area used to trade gold to China and India for porcelain and other luxuries.

CHAPTER 12 REVIEW

EXAMPLES

- **Make a bar graph, line graph, and stem-and-leaf plot.** (pages 224–227)

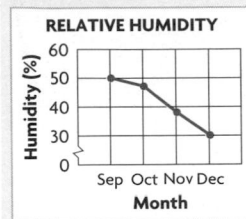

Line graphs show change over time.

- **Make a histogram.** (pages 228–229)

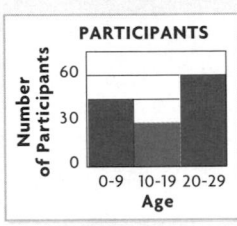

A histogram uses intervals to show groups of data.

3. histogram
5. See Additional Answers, p. 237B.

- **Graph two or more sets of data on one graph.** (pages 230–231)

A multiple-bar graph and a multiple-line graph show two or more sets of data on one graph.

6. multiple-line graph

- **Make a circle graph.** (pages 234–235)

Warm-up: 15 min; training: 45 min

$\frac{15}{60} = \frac{1}{4}, \frac{1}{4} \times 360° = 90°$

$\frac{45}{60} = \frac{3}{4}, \frac{3}{4} \times 360° = 270°$

EXERCISES

Ed found that there were 40 cars, 6 vans, 5 bikes, 1 bus, and 7 trucks on a street.

1. Would a bar graph or a line graph be a better choice for Ed's data? **bar**

2. Use the data below to make a stem-and-leaf plot. **See Additional Answers, p. 237B.**

POINTS SCORED					
33	52	45	47	34	52
34	58	48	52	46	59

3. **VOCABULARY** A(n) _?_ is a bar graph that shows frequencies within intervals.

4. Would a histogram or a bar graph be more appropriate for graphing the numbers of runners in different age groups? **histogram**

5. Use the data to make a histogram.

HEIGHTS OF BUILDINGS (IN FT)						
20	50	80	20	40	45	85
25	30	80	60	10	15	55

6. What type of graph would best show high and low temperatures for a week?

7. Make a multiple-line graph with the data. **See Additional Answers, p. 237B.**

	Sep	Oct	Nov	Dec	Jan
Stock A	$800	$740	$450	$500	$525
Stock B	$500	$525	$525	$500	$450

8. Use the data below to make a circle graph. **See Additional Answers, p. 237B.**

FAVORITE ICE-CREAM FLAVORS	
Flavor	Number of Students
Vanilla	30
Chocolate	50
Strawberry	20

Chapter 12 Review **237**

USING THE PAGE

The Chapter 12 Review reviews making a bar graph, line graph, stem-and-leaf plot, histogram, and circle graph; and graphing two or more sets of data on one graph. Chapter objectives are provided, along with examples and practice exercises for each.

 The Chapter 12 Review can be used as a review or a test. You may wish to place the Review in students' portfolios.

Assessment Checkpoint

For Performance Assessment in this chapter, see page 274, What Did I Learn?, items 2 and 6.

✓ *Assessment Tip*

Student Self-Assessment The How Did I Do? Survey helps students assess both what they have learned and how they learned it. Self-assessment helps students learn more about their own capabilities and develop confidence in themselves.

A self-assessment survey is available as a copying master in the *Teacher's Guide for Assessment*, p. 15.

See *Test Copying Masters*, pp. A45–A47 for the Chapter Test in **multiple-choice** format and pp. B195–B197 for **free-response** format.

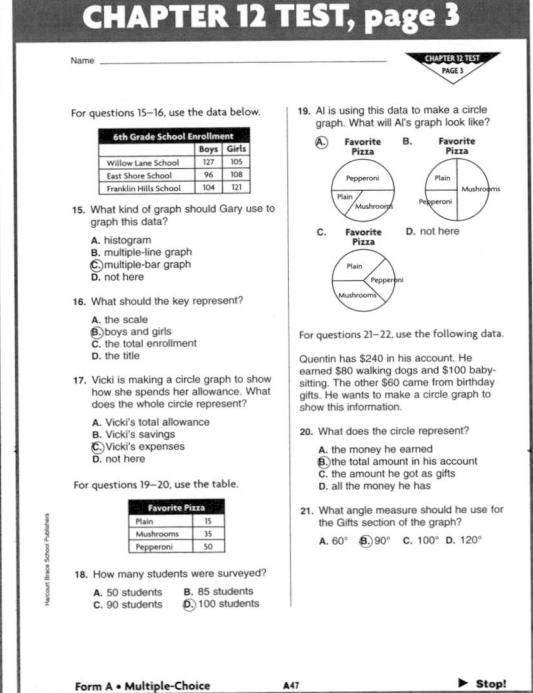

CHAPTER 12 TEST, page 1

Name _____

Choose the letter of the correct answer.

1. For which set of data would a bar graph be the best choice?

   A. changes in temperature
   B. a store's sales from January to June
   C. students' scores on a geography test
   D. the number of magazines sold by 4 different classrooms

2. Which type of graph shows changes over time?

   A. a stem-and-leaf plot
   B. a bar graph
   C. a line graph
   D. not here

For questions 3–4, use the frequency table.

Favorite Sport of Students	
Sport	Frequency
Baseball	12
Basketball	24
Football	11
Soccer	22
Swimming	16
Tennis	14

3. Why is a bar graph the most appropriate graph for this set of data?

   A. The data shows totals for each category.
   B. The data shows changes over time.
   C. The data was gathered in a survey.
   D. The table can be extended to show a cumulative frequency column.

4. If one axis of the graph is labeled Sport, what should the other axis be labeled?

   A. Survey Results
   B. Number of Teams
   C. Number of Students
   D. Favorite

For questions 5–6, use the stem-and-leaf plot of test scores for a class.

Class Test Scores	
Stem	Leaves
5	8 9
6	0 4 7 8 8
7	0 1 3 4 6 7 9 9
8	0 0 1 2 2 4 5 6 6 8 9
9	1 3 5 6 7

5. What score is shown by the fourth stem and its second leaf?

   A. 68   B. 79   C. 81   **D. not here**

6. Where should a score of 90 be placed on this plot?

   A. the fourth stem and its eleventh leaf
   B. the fifth stem and its first leaf
   C. the fifth stem and its third leaf
   D. the fifth stem and its fifth leaf

7. A histogram is a special kind of _?_.

   A. circle graph
   B. stem-and-leaf plot
   C. survey data
   D. bar graph

8. The histogram below shows the ages of people at a movie. What label is missing from the histogram?

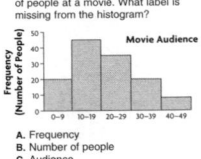

Movie Audience

   A. Frequency
   B. Number of people
   C. Audience
   D. Age

Form A • Multiple-Choice   A45   Go on. ▶

CHAPTER 12 TEST, page 2

Name _____

9. Johanna has data that shows golf scores of twenty players in a tournament. What is the best way to display the data?

   A. a bar graph   **B. a histogram**
   C. a circle graph   D. a line graph

10. What would the histogram for the frequency table below look like?

15-Mile Race	
Minutes	Runners
0 - 59	20
60 - 119	35
120 - 179	10
180 - 239	5

   A. [graph]
   B. [graph]
   C. [graph]
   D. not here

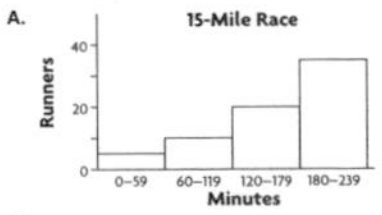

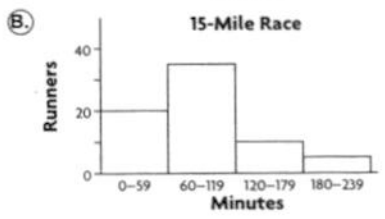

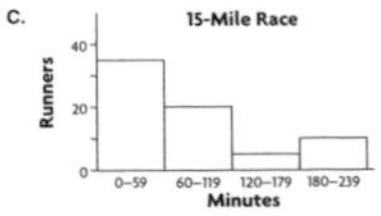

For questions 12–13, use the graph below.

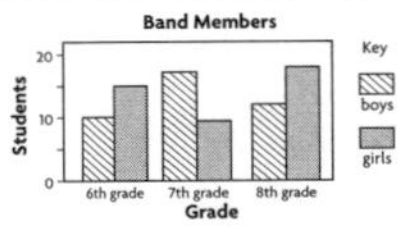

Band Members

12. What tells you which bars represent girls and which bars represent boys?

   A. the data   **B. the key**
   C. the scale   D. the grade

13. Which statement about 8th grade is true?

   A. There are more girls than boys in the band.
   B. There are more boys than girls in the band.
   C. There are fewer 8th graders than 7th graders in the band.
   D. not here

For question 14, use the graph below.

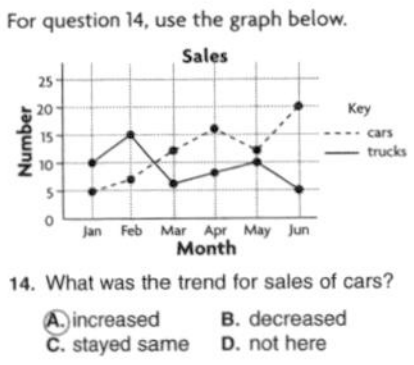

Sales

14. What is the trend for sales of cars?

   A. increased   B. decreased
   C. stayed same   D. not here

Form A • Multiple-Choice   A46   Go on. ▶

CHAPTER 12 TEST, page 3

Name _____

For questions 15–16, use the data below.

6th Grade School Enrollment		
	Boys	Girls
Willow Lane School	127	105
East Shore School	96	108
Franklin Hills School	104	121

15. What kind of graph should Gary use to graph this data?

   A. histogram
   B. multiple-line graph
   C. multiple-bar graph
   D. not here

16. What should the key represent?

   A. the scale
   B. boys and girls
   C. the total enrollment
   D. the title

17. Vicki is making a circle graph to show how she spends her allowance. What does the whole circle represent?

   A. Vicki's total allowance
   B. Vicki's savings
   C. Vicki's expenses
   D. not here

For questions 19–20, use the table.

Favorite Pizza	
Plain	15
Mushrooms	35
Pepperoni	50

18. How many students were surveyed?

   A. 50 students   B. 85 students
   C. 90 students   **D. 100 students**

19. Al is using this data to make a circle graph. What will Al's graph look like?

   A. Favorite Pizza
   B. Favorite Pizza
   C. Favorite Pizza
   D. not here

20. What does the circle represent?

   A. the money he earned
   B. the total amount in his account
   C. the amount he got as gifts
   D. all the money he has

For questions 21–22, use the following data.

Quentin has $240 in his account. He earned $80 walking dogs and $100 baby-sitting. The other $60 came from birthday gifts. He wants to make a circle graph to show this information.

21. What angle measure should he use for the Gifts section of the graph?

   A. 60°   **B. 90°**   C. 100°   D. 120°

Form A • Multiple-Choice   A47   ▶ Stop!

237

ADDITIONAL ANSWERS
Lesson 12.1, page 227

1.
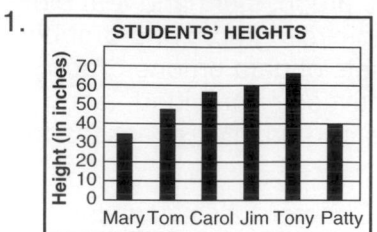

STUDENTS' HEIGHTS

2.
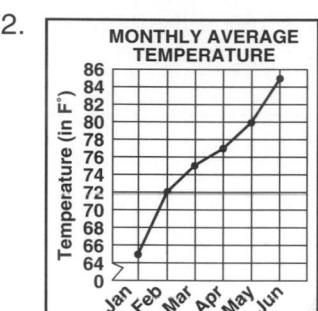

MONTHLY AVERAGE TEMPERATURE

3.

Stem	Leaves
5	9
6	0 5 6
7	0 1 1 5 5 5 7 7
8	0 0 0 1 2 2 4 4
9	2 3 5 8

Math Test Scores

5.
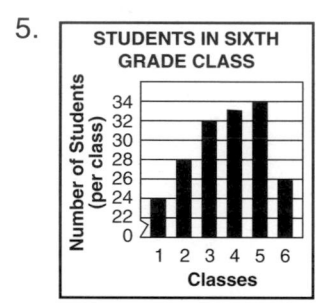

STUDENTS IN SIXTH GRADE CLASS

7.
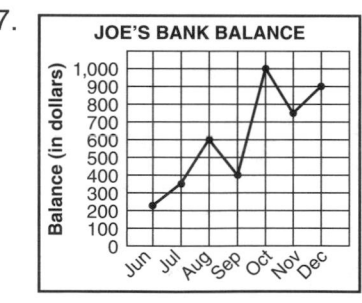

JOE'S BANK BALANCE

8.

Mrs. Green's Golf Scores

Stem	Leaves
7	7
8	0 2 2 2 3 4 5 5 5 8 8 8
9	0 0 2 2 5

10.

Student Ages	
Ages	Frequency
9	1
10	2
11	2
13	1
14	2
15	4
18	2
19	1
20	1

Lesson 12.2, page 229

Guided Practice

4.
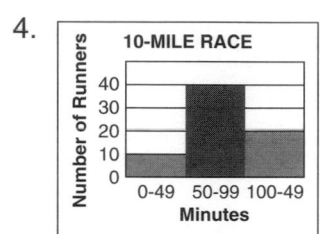

10-MILE RACE

Independent Practice

4.
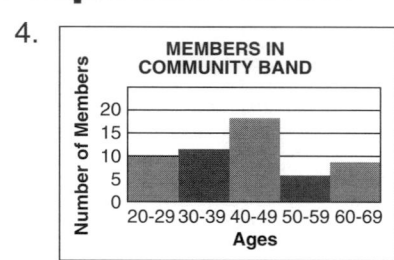

MEMBERS IN COMMUNITY BAND

7.

Buses	
Time	Frequency
7–8:59	18
9–10:59	5
11–12:59	12
1–2:59	6

Lesson 12.3, page 231

Guided Practice

1.
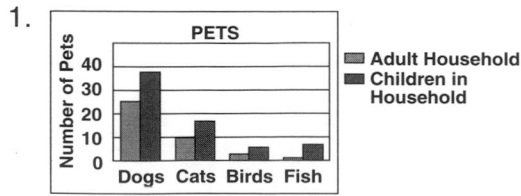

PETS

2.

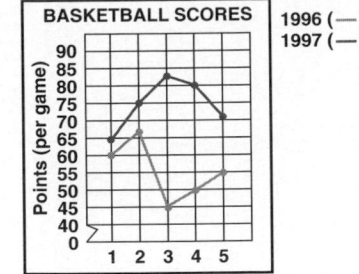

BASKETBALL SCORES

1996 (—)
1997 (—)

Independent Practice

1.

COSTS OF DAIRY PRODUCTS

Regular
Nonfat

2.

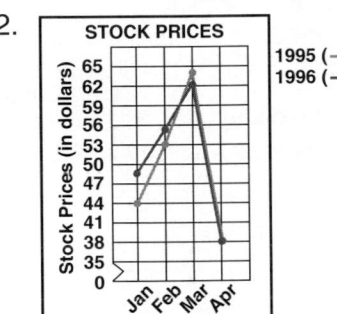

STOCK PRICES

1995 (—)
1996 (—)

4.

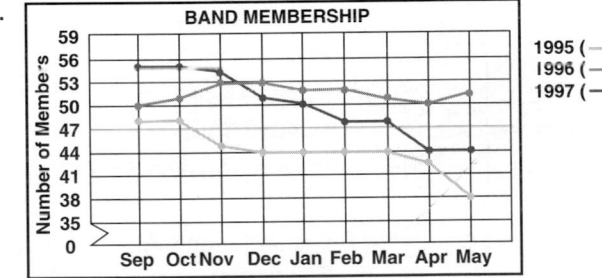

BAND MEMBERSHIP

1995 (—)
1996 (—)
1997 (—)

Lesson 12.4, page 235

Independent Practice

5.

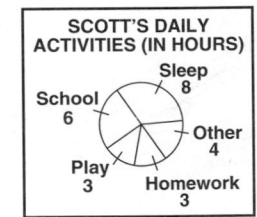

SCOTT'S DAILY ACTIVITIES (IN HOURS)

7.

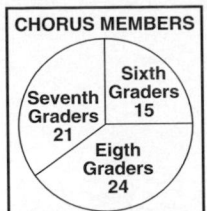

CHORUS MEMBERS

8.

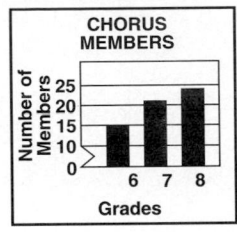

CHORUS MEMBERS

Chapter 12 Review, page 237

2.

Points Scored

Stem	Leaves
3	3 4 4
4	5 6 7 8
5	2 2 2 8 9

5.

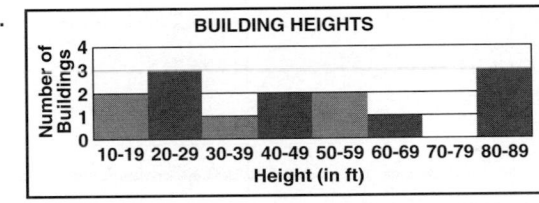

BUILDING HEIGHTS

7.

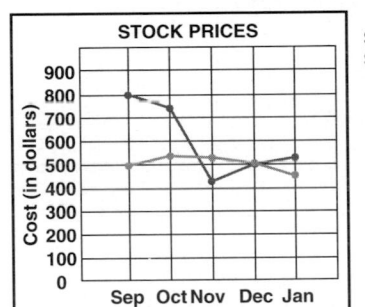

STOCK PRICES

Stock A (—)
Stock B (—)

8.

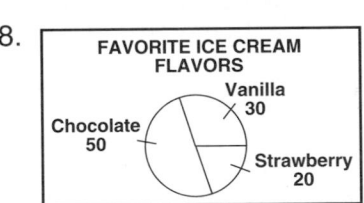

FAVORITE ICE CREAM FLAVORS

Interpreting Data and Predicting

BIG IDEA: Statistics is a tool for solving problems. Analyzing statistical data helps you make predictions and decisions.

Introducing the Chapter p. 238		Team-Up Project • Population 2010 p. 239	
OBJECTIVE	**VOCABULARY**	**MATERIALS**	**RESOURCES**
13.1 Analyzing Graphs pp. 240-241 **Objective:** To analyze data given in a graph **1 DAY**			■ Reteach, ■ Practice, ■ Enrichment 13.1; More Practice, p. H60
13.2 Misleading Graphs pp. 242-243 **Objective:** To identify misleading graphs **1 DAY**			■ Reteach, ■ Practice, ■ Enrichment 13.2; More Practice, p. H60 Math Fun • What's Wrong With This Picture?, p. 254
13.3 Making Predictions pp. 244-245 **Objective:** To make predictions from a graph **1 DAY**	prediction		■ Reteach, ■ Practice, ■ Enrichment 13.3; More Practice, p. H61 Math Fun • Tracking Stock, p. 254
13.4 Mean, Median, and Mode pp. 246-249 **Objective:** To find the mean, median, and mode of a set of data **2 DAYS**	mean median mode		■ Reteach, ■ Practice, ■ Enrichment 13.4; More Practice, p. H61 Math Fun • Name Games, p. 254
EXTENSION • Measures of Central Tendency, Teacher's Edition pp. 250A–250B			
Lab Activity: Exploring Box-and-Whisker Graphs pp. 250-251 **Objective:** To make a box-and-whisker graph and understand its parts	box-and-whisker graph lower extreme upper extreme lower quartile upper quartile	FOR EACH STUDENT: at least eleven 3 in. x 5 in. cards, marker	🖥 **E-Lab • Activity 13** E-Lab Recording Sheets, p. 13
13.5 Box-and-Whisker Graphs pp. 252-253 **Objective:** To analyze a box-and-whisker graph **2 DAYS**			■ Reteach, ■ Practice, ■ Enrichment 13.5; More Practice, p. H61 💾 *Data ToolKit*
CHAPTER ASSESSMENT Chapter 13 Review p. 255			

NCTM CURRICULUM
STANDARDS FOR GRADES 5–8

- ✓ **Problem Solving**
  Lessons 13.1, 13.2, 13.3, 13.4, 13.5
- ✓ **Communication**
  Lessons 13.1, 13.2, 13.3, 13.4, 13.5
- ✓ **Reasoning**
  Lessons 13.1, 13.2, 13.3, 13.4, 13.5
- ✓ **Connections**
  Lessons 13.1, 13.2, 13.3, 13.4, 13.5
- ✓ **Number and Number Relationships**
  Lessons 13.1, 13.4, 13.5
- ✓ **Number Systems and Number Theory**
  Lessons 13.1, 13.2, 13.3, 13.4, 13.5
- ✓ **Computation and Estimation**
  Lessons 13.1, 13.2, 13.3, 13.4, 13.5
- ☐ **Patterns and Functions**
- ☐ **Algebra**
- ✓ **Statistics**
  Lessons 13.1, 13.2, 13.3, 13.4, 13.5
- ☐ **Probability**
- ☐ **Geometry**
- ✓ **Measurement**
  Lessons 13.1, 13.3

ASSESSMENT OPTIONS

CHAPTER LEARNING GOALS

13A.1 To analyze graphs and make predictions from graphs

13A.2 To identify misleading graphs

13A.3 To find the mean, median, and mode

These goals for concepts, skills, and problem solving are assessed in many ways throughout the chapter. See the chart below for a complete listing of both traditional and informal assessment options.

Pretest Options
Pretest for Chapter Content, TCM pp. A48–A50, B198–B200
Assessing Prior Knowledge, TE p. 238; APK p. 15

Daily Assessment
Mixed Review, PE pp. 249, 253
Problem of the Day, TE pp. 240, 242, 244, 246, 252
Assessment Tip and Daily Quiz, TE pp. 241, 243, 245, 249, 251, 253

Formal Assessment
Chapter 13 Review, PE p. 255
Chapter 13 Test, TCM pp. A48–A50, B198–B200
Study Guide & Review, PE pp. 270–271
Test • Chapters 11–14, TCM pp. A53–A54, B203–B204
Cumulative Review, PE p. 273
Cumulative Test, TCM pp. A55–A58, B205–B208

Performance Assessment
Team-Up Project Checklist, TE p. 239
What Did I Learn?, Chapters 11–14, TE p. 274
Interview/Task Test, PA p. 88

Portfolio
Suggested work samples:
Independent Practice, PE pp. 241, 243, 245, 248, 253
Math Fun, PE p. 254

Student Self-Assessment
How Did I Do?, TGA p. 15
Portfolio Evaluation Form, TGA p. 24
Math Journal, TE pp. 238B, 248

Key
Assessing Prior Knowledge Copying Masters: APK
Test Copying Masters: TCM
Performance Assessment: PA

Teacher's Guide for Assessment: TGA
Teacher's Edition: TE
Pupil's Edition: PE

MATH COMMUNICATION

MATH JOURNAL
You may wish to use the following suggestions to have students write about the mathematics they are learning in this chapter:

PE pages 240, 242 • Talk About It

PE pages 241, 243, 245, 249, 253 • Write About It

TE page 248 • Writing in Mathematics

TGA page 15 • How Did I Do?

In every lesson • record written answers to each Wrap-up Assessment Tip from the Teacher's Edition for test preparation.

VOCABULARY DEVELOPMENT
The terms listed at the right will be introduced or reinforced in this chapter. The boldfaced terms in the list are new vocabulary terms in the chapter.

So that students understand the terms rather than just memorize definitions, have them construct their own definitions and pictorial representations and record those in their Math Journals.

prediction, p. 244
mean, pp. 246, 262
median, pp. 246, 252
mode, pp. 246, 252
box-and-whisker graph, pp. 250, 252
lower extreme, pp. 250, 252
upper extreme, pp. 250, 252
lower quartile, p. 250
upper quartile, p. 250

MEETING INDIVIDUAL NEEDS

MEETING INDIVIDUAL NEEDS

RETEACH	PRACTICE
BELOW-LEVEL STUDENTS	**ON-LEVEL STUDENTS**

Practice Activities, TE Tab B
Listing Factors
(Use with Lessons 13.1–13.5.)

Interdisciplinary Connection, TE p. 249

Literature Link, TE p. 238C
What Hearts, Activity 1
(Use with Lessons 13.1–13.5.)

Technology
Data ToolKit
(Use with Lesson 13.5.)

E-Lab • Activity 13
(Use with Lab Activity.)

Practice Game, TE p. 238C
Bull's-eye!
(Use with Lesson 13.4.)

Interdisciplinary Connection, TE p. 249

Literature Link, TE p. 238C
What Hearts, Activity 2
(Use with Lessons 13.1–13.5.)

Technology
Data ToolKit
(Use with Lesson 13.5.)

E-Lab • Activity 13
(Use with Lab Activity.)

CHAPTER IDEA FILE

The activities on these pages provide additional practice and reinforcement of key chapter concepts. The chart above offers suggestions for their use.

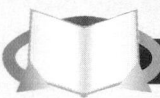

LITERATURE LINK

What Hearts

by Bruce Brooks

This story takes you through Asa's life from first through seventh grade.

Activity 1—As a first grader, Asa was the only student who chose to grow radishes. Have students survey their classmates and make a circle graph showing which vegetables their classmates would choose to grow.

Activity 2—Asa often had to ride his bike to get to friends' houses, which were 5 to 6 mi. away. Have students survey their classmates about the longest bike ride each has ever taken. Then have students make a box-and-whisker graph to show the results.

Activity 3—Like Asa, many students today have to deal with divorce. Have students research divorce statistics for the past ten years and chart their findings. Then ask them to make predictions about divorce in the next five years, based on these findings.

PRACTICE GAME

Bull's-eye!

Purpose To practice finding mean, median, and mode of a set of data

Materials target (TR p. R90), scorecard (TR p. R91), game tokens

About the Game In small groups, each player receives a target, a scorecard, and several game tokens. The player sits in a chair. The target lies on the floor, 2–3 feet in front of the chair. The player tosses a token at the target and records the score on the scorecard. Each player completes a scorecard and finds the mean, median, and mode of his or her scores for the round. A player earns 1 point for the highest mean, for the highest median, and for the lowest mode. The player with the highest total score after 4 rounds wins.

Modalities Visual, Kinesthetic

Multiple Intelligences Logical/Mathematical, Bodily/Kinesthetic

Practice Game, TE p. 238C
Bull's-eye!
(Use with Lesson 13.4)

Interdisciplinary Connection, TE p. 249

 **Literature Link,** TE p. 238C
What Hearts, Activity 3
(Use with Lessons 13.1–13.5.)

 Technology
Data ToolKit
(Use with Lesson 13.5.)

E-Lab • Activity 13
(Use with Lab Activity.)

The purpose of using technology in the chapter is to provide reinforcement of skills and support in concept development.

E-Lab • Activity 13—Students use a computer tool to make box-and-whisker graphs which they then interpret.

Data ToolKit
This program provides a spreadsheet and graphing tool that will help students organize data they collect, choose appropriate graphs for their data, compare graphs they chose, and find measures of central tendency.

INTERDISCIPLINARY SUGGESTIONS

Connecting Data Interpretations and Predictions

Purpose To connect interpretations and predictions of data to other subjects students are studying

You may wish to connect *Interpreting Data and Predicting* to other subjects by using related interdisciplinary units.

Suggested Units

• Cultural Connection: Billionaires Around the World

• Social Studies Connection: State Fairs

• Consumer Connection: Mail Order Shopping

• Science Connection: Insect Population

• Career Connection: College Recruiter

Modalities Auditory, Kinesthetic, Visual

Multiple Intelligences Mathematical/Logical, Bodily/Kinesthetic, Visual/Spatial

Visit our web site for additional ideas and activities.
http://www.hbschool.com

CHAPTER 13

Introducing the Chapter

Encourage students to talk about how they might use graphs to show information. Review the math content students will learn in the chapter.

WHY LEARN THIS?

The question "Why are you learning this?" will be answered in every lesson. Sometimes there is a direct application to everyday experiences. Other lessons help students understand why these skills are needed for future success in mathematics.

PRETEST OPTIONS

Pretest for Chapter Content See the *Test Copying Masters*, pages A48–A50, B198–B200 for a free-response or multiple-choice test that can be used as a pretest.

Assessing Prior Knowledge Use the copying master shown below to assess previously-taught skills that are critical for success in this chapter.

Assessing Prior Knowledge Copying Masters, p. 15

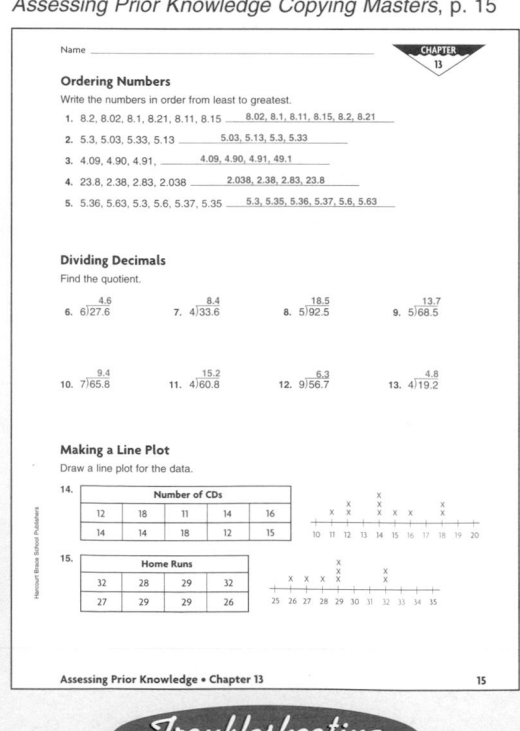

Troubleshooting

If students need help	Then use
• ORDERING NUMBERS	• Key Skills, PE p. H11
• DIVIDING DECIMALS	• Key Skills, PE p. H13
• MAKING A LINE PLOT	• *Data Toolkit* software

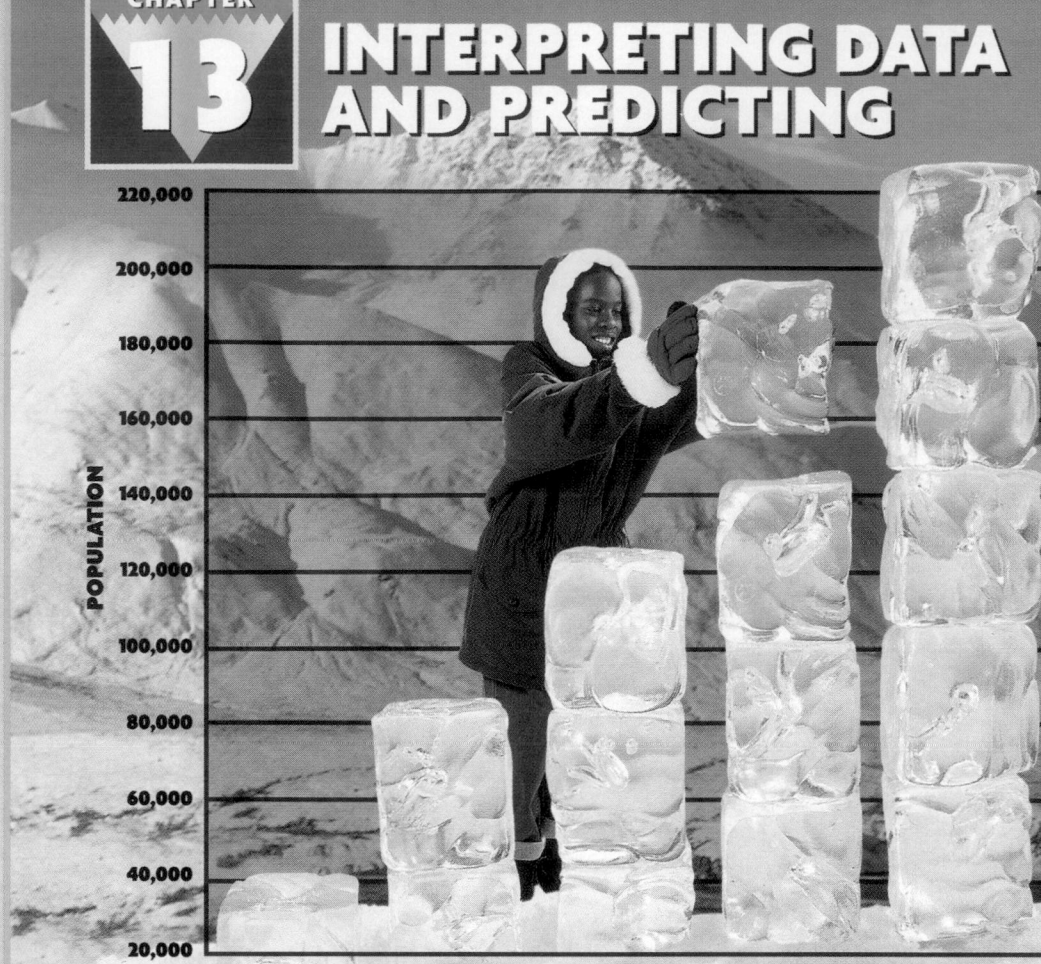

CHAPTER 13 INTERPRETING DATA AND PREDICTING

LOOK AHEAD

In this chapter you will solve problems that involve

• analyzing graphs and making predictions from graphs

• identifying misleading graphs
• finding the mean, median, and mode for a set of data
• displaying and analyzing data in a box-and-whisker graph

238 Chapter 13

 SCHOOL-HOME CONNECTION

You may wish to send to each student's family the *Math at Home* page. It includes an article that stresses the importance of communication about mathematics in the home, as well as helpful hints about homework and taking tests. The activity and extension provide extra practice that develops concepts and skills involving interpreting data.

MATH AT HOME

Interpreting Data and Probability

Activities in this chapter involve analyzing graphs to make predictions; identifying misleading graphs; working with the mean, the median, and the mode; and displaying and analyzing data in a box-and-whisker graph. Look for examples of graphs in newspapers, and discuss these with your child.

Here are an activity and Extension that focus on the mean and the median.

It's All in a Name

Make a list of the last names of at least 25 people. You can list the names of relatives and friends, or if you prefer, you can copy the names of athletes from your favorite teams.
• Find the length (number of letters) of each last name.
• List the names in order of length from least to greatest.
• Find the mean length (sum ÷ 25) of the set of names.
• Find the median length (the middle one) of the set of names.
Which of the measures, mean or median, better describes the set of data you collected?

Extension • The Name Game

To the set of names you collected,
• add 2 names so that the median will not change.
• remove at least 2 names so that the median will not change.
• add at least 2 names and increase the median.
• remove 2 names and decrease the median.
Repeat this process, working with the mean.

For Your Information

Giving children good books to read is a great way to open up a whole new world to them. Books can teach your child about new places and new things, can pose problems, can entertain, and can illustrate ideas and concepts.

Your child might enjoy reading *The Phantom Tollbooth* by Norton Juster (Random House, 1961). In this book Milo travels through Dictionopolis and Digitopolis in a delightful fantasy. During his trip he experiences many language and mathematics concepts. For example, a child says, "Every average family has 2.58 children, so I always have someone to play with." What a wonderful way to read about the concept of average!

VOCABULARY

mean—the arithmetic average of a set of numbers
median—the middle number in a group of numbers arranged in numerical order
mode—the number that occurs most often in a group of numbers
box-and-whisker graph—a graph (an organization of data) that shows the data distributed in four groups
quartile—one of the values that divide a set of ordered data (numbers) into four roughly equal-sized groups

 Math at Home

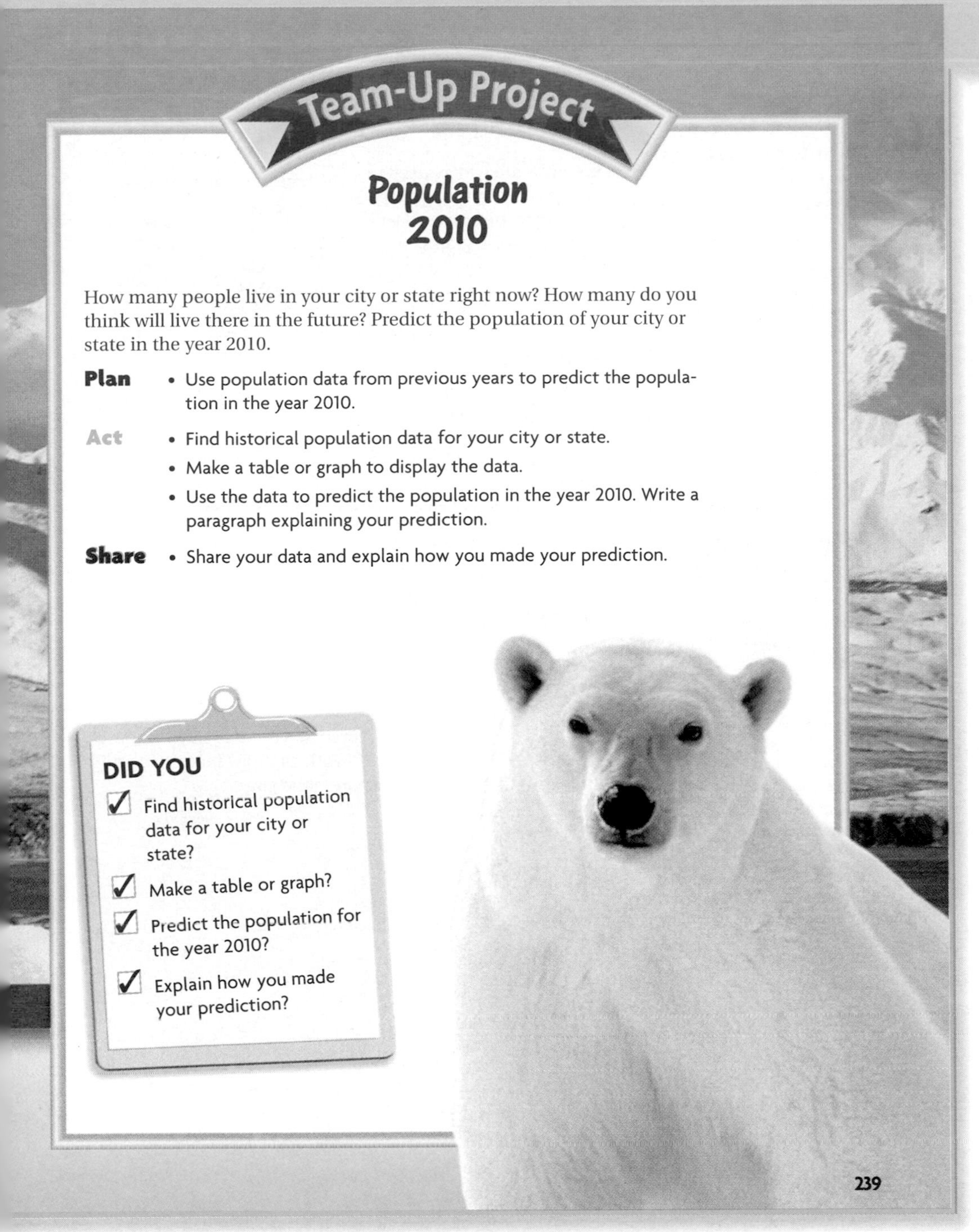

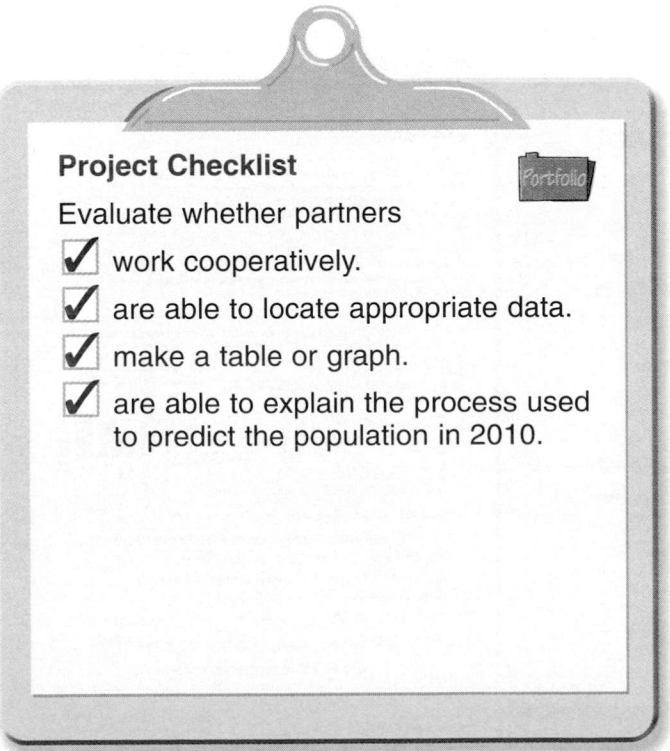

Team-Up Project

Population 2010

How many people live in your city or state right now? How many do you think will live there in the future? Predict the population of your city or state in the year 2010.

Plan
- Use population data from previous years to predict the population in the year 2010.

Act
- Find historical population data for your city or state.
- Make a table or graph to display the data.
- Use the data to predict the population in the year 2010. Write a paragraph explaining your prediction.

Share
- Share your data and explain how you made your prediction.

DID YOU
- ✓ Find historical population data for your city or state?
- ✓ Make a table or graph?
- ✓ Predict the population for the year 2010?
- ✓ Explain how you made your prediction?

239

Project Checklist

Evaluate whether partners
- ✓ work cooperatively.
- ✓ are able to locate appropriate data.
- ✓ make a table or graph.
- ✓ are able to explain the process used to predict the population in 2010.

Using the Project

Accessing Prior Knowledge make a table, make a graph, read and interpret data

Purpose to use population data from previous years to predict the population of a state or city in the year 2010

GROUPING partners
MATERIALS reference materials, rulers, graph paper

Managing the Activity Have students look for both long-term and current trends in their data.

WHEN MINUTES COUNT

A shortened version of this activity can be completed in 15–20 minutes. Instead of researching their own cities and making their own graphs, have students describe the change in the population of Anchorage, Alaska as shown in the text. Then have them predict the population of Anchorage in 2010.

A BIT EACH DAY

Day 1: Students form groups, select a city or state, and decide how to gather population data.

Day 2: Students use reference materials to find historical population data for their city or state.

Day 3: Students make a graph or table to show their data.

Day 4: Students predict the population for the year 2010.

Day 5: Students share their data and explain how they made their prediction for the population in 2010.

HOME HAPPENINGS

Have students ask long-time residents to describe changes that have occurred as the population has grown or declined.

GRADE 5	GRADE 6	GRADE 7
Analyze data given in a graph.	Analyze data given in a graph.	Analyze data given in a graph.

ORGANIZER

OBJECTIVE To analyze data given in a graph
ASSIGNMENTS Independent Practice
- ■ *Basic* 1–10
- ■ *Average* 1–10
- ■ *Advanced* 1–10

PROGRAM RESOURCES
Reteach 13.1 Practice 13.1 Enrichment 13.1

PROBLEM OF THE DAY

Transparency 13

A circle graph shows that half of the ancestors of the 120 students surveyed came to the U.S. from Europe or South America, a quarter came from Africa, and the rest from Asia and Australia. Five times as many came from Europe as from South America. How many students had South American ancestors?

10 students

Solution Tab A, p. A13

QUICK CHECK

Complete.

1. $75 = \underline{\ ?\ } \times 30$ $2\frac{1}{2}$ or 2.5
2. $60 = \underline{\ ?\ } \times 15$ 4
3. $15 = \underline{\ ?\ } \times 20$ $\frac{3}{4}$ or 0.75
4. $100 = \underline{\ ?\ } \times 50$ 2
5. $90 = \underline{\ ?\ } \times 60$ $1\frac{1}{2}$ or 1.5
6. $120 = \underline{\ ?\ } \times 40$ 3

1 MOTIVATE

Review with students the most appropriate graphs for displaying data.

2 TEACH/PRACTICE

Discuss page 240 with students. Help them understand that graphs allow them to see comparisons of data more clearly.

13.1

WHAT YOU'LL LEARN
How to analyze data given in a graph

WHY LEARN THIS?
To use information on graphs to make buying decisions, investments, and other comparisons

Computer Link

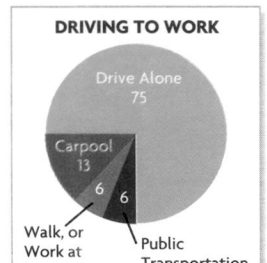

Colorful graphs may be a part of interactive Internet pages. Multimedia authoring tools and computer languages like Java allow attention-grabbing animated graphs and graphs that are updated instantly as new data become available. Make a list of World Wide Web sites that include graphs.

ALGEBRA CONNECTION
Analyzing Graphs

Graphs are widely used in business reports, newspapers, and magazines. Advertisers sometimes use graphs to show how their products compare with competitors' products.

OUR CHIPS ARE CHEAPER

Talk About It

- Look at the graph. Describe the relationship between the height of the bars and the cost of the potato chips. **The greater the height, the greater the cost.**
- How much greater is the cost of brand A than the cost of brand B? **A is 2 times the cost of B in the 10-oz size and is $1\frac{1}{2}$ times B in the 12-oz size or $1 more.**
- Which brand of potato chips is cheaper? How do you know? **B; the bar for B is shorter.**

To analyze a graph, you have to know what relationship to look for. In a bar or line graph, identify the relationships shown by the axes. In a circle graph, compare the parts to the whole. Then, look for greater-than or less-than relationships.

EXAMPLE A survey of 100 people was done to find out how many people drive alone to work and how many use other types of transportation. Analyze the graph.

DRIVING TO WORK

Compare parts with the whole.

75 is $\frac{3}{4}$ of 100.

$13 + 6 + 6 = 25$;

25 is $\frac{1}{4}$ of 100.

Compare parts with other parts.

75 is 3 times as great as 25.

So, the number of people who drive alone to work is 3 times as great as the number of people who use other types of transportation.

IDEA FILE

CORRELATION TO OTHER RESOURCES

📀 **CornerStone Mathematics** ©1996 SkillsBank Corporation **Level C, Working with Data, Lessons 1–3**

📀 *Graphers* **by Sunburst** This math tool helps students learn about graphs and enables them to use graphs to solve problems.

RETEACH 13.1

Name _____

LESSON 13.1

Analyzing Graphs

To analyze a graph, you have to know what relationship to look for. In a bar or line graph, the relationships are shown by the axes. In a circle graph, the parts are compared to the whole.

A survey of 100 yogurt buyers was conducted to find out what brand of yogurt they buy. Analyze the graph.
The greatest number of people buy Brand B.
The least number of people buy Brand D.
Three times as many people buy Brand C as buy Brand D.
As a store owner you could use this survey to decide to stock more Brand B and less Brand D.

Yogurt Purchases

A survey was conducted of 100 people who ate at a local food court. The purpose of the survey was to identify the favorite flavor of ice cream. The bar graph shows the results.

1. Which flavor did the greatest number of people like the best?
   chocolate
2. Which flavor did the least number of people like best?
   mint chip
3. How would you describe the relationship between chocolate and mint chip? **Six times as many people like chocolate best as like mint chip best.**
4. Which two flavors had a combined total equal to that of strawberry? **fudge swirl and mint chip**
5. Suppose you owned an ice cream stand. What effect might this survey have on your orders for next month? **Answers may include: Order six times as much chocolate as mint chip or order half as much strawberry as vanilla.**

FAVORITE FLAVOR ICE CREAM

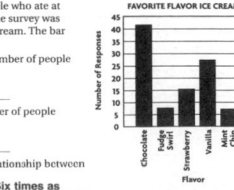

R56 **TAKE ANOTHER LOOK**   Use with text pages 240–241.

GUIDED PRACTICE

The newspaper's movie critic conducted a survey to find out how teens enjoyed the latest teen movie released. The results are displayed in the circle graph.

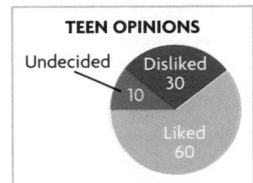

TEEN OPINIONS

Undecided · Disliked 30 · 10 · Liked 60

1. Which opinion forms the largest part of the graph? the smallest? **the number of teens who liked the movie; the number who are undecided**
2. What did the movie critic learn from this survey? **Most of the teens liked the movie.**
3. How many more teens liked the movie than disliked the movie? **30 more, or twice as many**

INDEPENDENT PRACTICE

For Exercises 1–3, use the line graph at the right.

1. Which month had the best sales? Which month had the worst sales? What is the relationship between the two months' sales? **October; July; there were 4.5 times as many sales in October as in July.**
2. What kinds of factors could have affected sales in July? **Possible answers: heat or rain, vacation**
3. What is the overall trend for sales for the seven months? **Sales increased.**

For Exercises 4–7, use the bar graph below at the right.

4. Which flavor was the most popular? the least popular? How do you know? **Coco Delight; Stellar Kiwi; the bar is highest for Coco Delight, lowest for Stellar Kiwi.**
5. How does the number of gallons of Coco Delight sold compare with the number of gallons of Stellar Kiwi sold? **400 more gallons, or 5 times as many gallons**
6. How does the number of gallons of Good n' Fruity sold compare with the number of gallons of Stellar Kiwi sold? **250 more gallons, or 3.5 times as many gallons**
7. For next year's production of the new flavors, what would you recommend? **Possible answer: No longer make Stellar Kiwi; make more Coco Delight.**

Problem-Solving Applications

8. Write a question that can be answered by using a line graph. You may want to use the information in the graph for Exercises 1–3. **Check students' questions.**

9. Find a bar graph, line graph, or circle graph in a magazine or newspaper. Write a question about the graph. **Graphs and questions will vary.**

10. ✏️ **WRITE ABOUT IT** Suppose you want to compare the prices of two products. Would you use a bar graph or a circle graph? Explain. **A bar graph; a circle graph compares the parts with a whole, not two items with each other.**

AUTO SALES — Number of Cars Sold vs. Month (Jun, Jul, Aug, Sep, Oct, Nov, Dec)

NEW ICE-CREAM-FLAVOR SALES — Number of Gallons Sold vs. Flavor (Purple Passion, Stellar Kiwi, Good n'Fruity, Coco Delight)

GUIDED PRACTICE
To make sure students understand how to analyze data in a graph, have them complete Exercises 1–3 orally.

INDEPENDENT PRACTICE
Work through a few problems as a class to make sure students can analyze the data shown in each type of graph.

ADDITIONAL EXAMPLE

Example, p. 240

Justin surveyed 100 people about their favorite type of cookie. Analyze the data.

TYPES OF COOKIES

Cinnamon 5 · Raisins 15 · Peanut Butter 20 · Chocolate Chip 60

The number of people who prefer chocolate chip cookies is $1\frac{1}{2}$ times as great as the number who prefer all other types.

3 WRAP UP

Use these questions to help students focus on the big ideas of the lesson.

- **Is a line graph the most appropriate display of the auto sales data? Explain.** Yes. The data show change over time.

✓ Assessment Tip

Have students write the answer.

- **Can you compare the sales of purple passion ice cream with those of stellar kiwi without using the numbers on the vertical axis? Explain.** Yes. You can compare the heights of the bars.

DAILY QUIZ
Use the Driving to Work circle graph.

1. What type of transportation is most popular among people who don't drive alone? carpool
2. How many people don't go to work in a car? 12
3. What fraction represents the number of people in the survey who walk or work at home? $\frac{3}{50}$
4. What is the relationship between the number of people who use public transportation and the number who walk or work at home? They are the same.

ORGANIZER

GRADE 5	GRADE 6	GRADE 7
Choose the appropriate graph for data.	Identify misleading graphs.	Identify misleading graphs.

OBJECTIVE To identify misleading graphs

ASSIGNMENTS Independent Practice
- ■ *Basic* 1–10
- ■ *Average* 1–10
- ■ *Advanced* 1–10

PROGRAM RESOURCES
Reteach 13.2 Practice 13.2 Enrichment 13.2
Math Fun • What's Wrong With This Picture, p. 254

PROBLEM OF THE DAY

Transparency 13

What is the least number that can be divided evenly by each of the numbers 1–12? 27,720

Solution Tab A, p. A13

QUICK CHECK

Determine an appropriate scale for a graph of the data.

1. 45; 23; 16; 5; 31 0–50 in intervals of 10
2. 398; 244; 103; 189 0–400 in intervals of 100
3. 2.8; 0.8; 2.4; 1.1; 0.3 0–3 in intervals of 0.5

1 MOTIVATE

Have students read Teen Times and discuss with them the importance of analyzing advertisements.

2 TEACH/PRACTICE

Throughout the lesson, emphasize the importance of analyzing scales to determine the validity of graphs.

GUIDED PRACTICE

Discuss the graphs in Exercises 1–3 to make sure students understand how to identify misleading graphs.

13.2

WHAT YOU'LL LEARN
How to identify misleading graphs

WHY LEARN THIS?
To be able to recognize graphs that are made to mislead you

Misleading Graphs

Sometimes advertisements in newspapers or magazines show a graph that is misleading. The data in a misleading graph may be factual, but the presentation of the data is misleading.

OUR SHOES LAST LONGER

Our shoes will last 3 times as long as other brands of shoes!!!

teen times

Teens are part of an important consumer group in the United States. Most CDs, video games, and contemporary clothes are bought by teens and young adults. It is important for teens to become smart consumers. Study all advertisements carefully, especially those with tables and graphs.

Talk About It
- How does the life of Our Brand shoes compare with the life of Brand C? **Our Brand has 2 times the life of Brand C.**
- How is this graph misleading? **The scale makes it appear that Our Brand lasts $2\frac{1}{2}$ times as long as Brand C.**

Sometimes graphs can be misleading when two similar sets of data are compared on graphs that have different scales.

EXAMPLE How are these graphs, taken together, misleading?

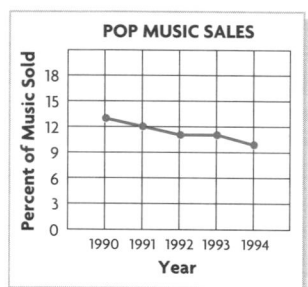
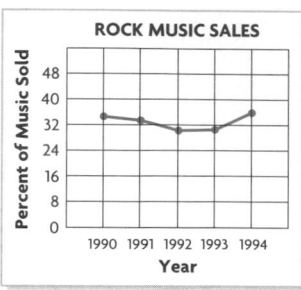

Together, the graphs make it appear as if the sales of pop music and rock music were about the same. However, if you read the scales and find the values of the points, you find that rock music had greater sales.

IDEA FILE

SCIENCE CONNECTION

The graph on the left is realistic. The graph on the right is misleading. What scale was used for the graph on the right?
Possible answer: 0°, 100°, 200°

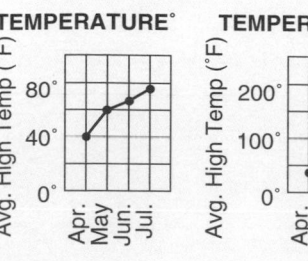

TEMPERATURE°

TEMPERATURE°

Modality Visual

RETEACH 13.2

Name _____

Misleading Graphs

The bar graph at the right shows the amount of school funding over five years.

The bar for 1996 seems to be 4 times as high as the bar for 1992.

Look at the scale. The scale does not have the same interval throughout. The first interval represents $15 million while the other intervals represent $5 million.

The graph is misleading. The funding for 1996 is only 2 times as great as the funding for 1992.

SCHOOL FUNDING

The line graphs below show CD sales for the months of January and July.

CD SALES FOR JANUARY

CD SALES FOR JULY

1. How many CDs were sold during Week 1 in January? in July? ___30; 60___
2. How many CDs were sold during Week 3 in January? in July? ___66; 100___
3. What is the difference in the number of CDs sold during Week 1 and Week 3 in January? in July? ___36; 40___
4. Which graph makes the difference between Week 1 and Week 3 appear greater? Explain why. ___The January graph;___
   ___the intervals in the January graph are 12, while in the___
   ___July graph the intervals are 20.___

Use with text pages 242–243.   TAKE ANOTHER LOOK R57

MEETING INDIVIDUAL NEEDS

GUIDED PRACTICE

Tell whether the graph is misleading. Write *yes* or *no*. If *yes*, explain.

1.
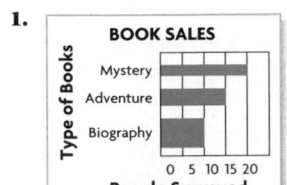

Yes. The bars are different widths.

2.
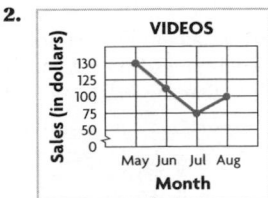

Yes. The distance between 125 and 130 is the same as the distance between 100 and 125.

3.
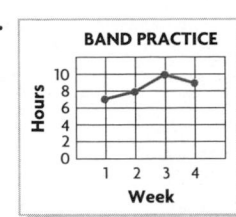

no

INDEPENDENT PRACTICE

For Exercises 1–4, use the bar graph at the right.

1. Who sold more boxes of cookies? **Lucy**

2. About how many times as high is the bar for Lucy's sales as the bar for Terri's sales? **about twice as high**

3. Did Lucy sell twice as many boxes of cookies as Terri? Explain. **No. Lucy sold 5 more boxes of cookies, not 15 more boxes.**

4. Make a new graph, one that is not misleading, to show Lucy's and Terri's cookie sales. **Check students' graphs.**

For Exercises 5–8, use the line graphs at the right.

5. In Graph A, about how many times as great as the sales in Week 4 do the sales in Week 1 appear? **about 4 times**

6. In Graph B, about how many times as great as the sales in Week 4 do the sales in Week 1 appear? **The sales are very close.**

7. What is the actual difference in sales between Week 1 and Week 4? Were the Week 1 sales 4 times as great? **about 40,000; no**

8. Which graph gives a better picture of the data? Explain. **Graph B; sales are not 4 times as great, as shown in Graph A.**

Problem-Solving Applications

9. Share with a classmate graphs you have found in newspapers, magazines, or other sources. Discuss which are misleading and which are not. **Check students' responses.**

10. ✏ **WRITE ABOUT IT** How can you determine whether a graph presents an accurate picture of a set of data? **Possible answers: Check units of measure and scales to make sure they are not exaggerated or incorrect; compare the data in the graph with the source of the data.**

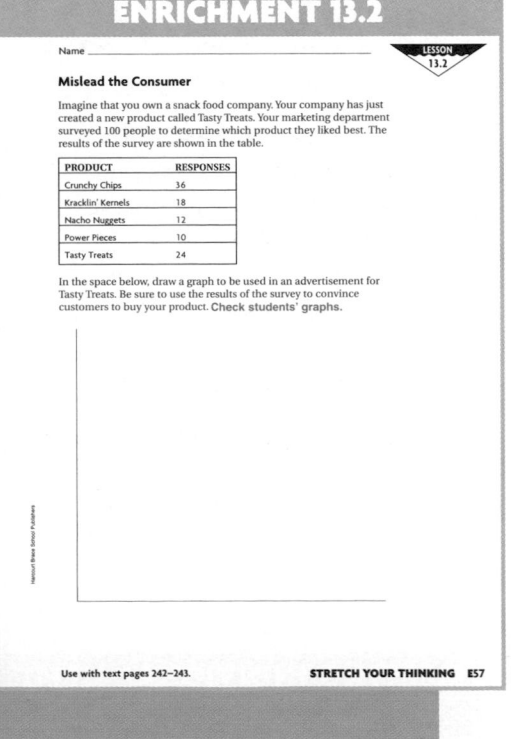

MORE PRACTICE Lesson 13.2, page H60

243

INDEPENDENT PRACTICE

Observe students as they complete the exercises to ensure that they are correctly identifying misleading information.

ADDITIONAL EXAMPLE

Example, p. 242

These bar graphs show the prices of bicycles sold at two different stores. How are these graphs misleading?

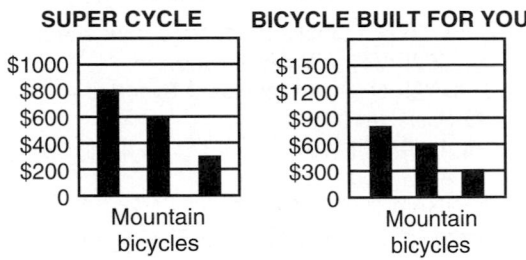

The graphs make it appear that the prices of bicycles at Super Cycle are greater than the prices of bicycles at Bicycle Built For You. However, the prices of bicycles are actually the same at both stores.

3 WRAP UP

Use these questions to help students focus on the big ideas of the lesson.

- **How could you change the scale for the graph in Guided Practice Exercise 2 so that the graph is no longer misleading?** Possible answer: Change 130 to 150 and reposition the first point in the graph.

✓ *Assessment Tip*

Have students write the answer.

- **Why do you think the graph in Guided Practice Exercise 3 is not misleading?** Possible answer: because the scale is reasonable

DAILY QUIZ

For each exercise, use the bar graph.

1. The price of a tie-dyed T-shirt appears to be about how many times as great as that of a striped T-shirt? about three times as great

2. Is the price of a tie-dyed T-shirt three times as great as the price of a striped T-shirt? no

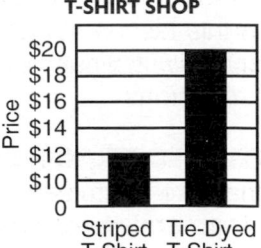

3. What is the actual difference between the price of a tie-dyed T-shirt and that of a striped T-shirt? $8

243

PRACTICE 13.2

Name _____   LESSON 13.2

Misleading Graphs

For Exercises 1–4, use the bar graph at the right.

1. Who sold more magazines? _____ Rico
2. About how many times as high is the bar for Rico's sales as the bar for Jon's sales? ____ about 7 times
3. Did Rico sell seven times as many magazines as Jon? Explain.
   No, the uneven interval between 0 and 6 makes the graph misleading.
4. On the back of this sheet, make a new graph, one that is not misleading, to show Rico's and Jon's magazine sales. Check students' graphs.

For Exercises 5–8, use the line graphs at the right.

5. In Graph A, about how many times as great as the sales in Week 1 do the sales in Week 4 appear? ____ about three times as great
6. In Graph B, about how many times as great as the sales in Week 1 do the sales in Week 4 appear? ____ about twice as great
7. What is the actual difference in sales between Week 1 and Week 4? Were the Week 4 sales three times as great? ____ 60 – 30 = 30; No, they were twice as great.
8. Which graph gives a better picture of the data? Explain. ____ Graph B; the intervals are the same on the scale.

Mixed Applications

9. Is the data in a misleading graph factual? Explain.
   Yes, the data is factual. The presentation is misleading.

10. A battery company claims that its 18-month warranty lasts 3 times as long as any other warranty. According to the company, what is the longest of the other warranties?
    6 months

Use with text pages 242–243.   ON MY OWN   P57

ENRICHMENT 13.2

Name _____   LESSON 13.2

Mislead the Consumer

Imagine that you own a snack food company. Your company has just created a new product called Tasty Treats. Your marketing department surveyed 100 people to determine which product they liked best. The results of the survey are shown in the table.

PRODUCT	RESPONSES
Crunchy Chips	36
Kracklin' Kernels	18
Nacho Nuggets	12
Power Pieces	10
Tasty Treats	24

In the space below, draw a graph to be used in an advertisement for Tasty Treats. Be sure to use the results of the survey to convince customers to buy your product. Check students' graphs.

Use with text pages 242–243.   STRETCH YOUR THINKING   E57

243

ORGANIZER

GRADE 5
Compare graphs.

GRADE 6
Make predictions from a graph.

GRADE 7
Choose an appropriate graph.

OBJECTIVE To make predictions from a graph
VOCABULARY *New* **prediction**
ASSIGNMENTS **Independent Practice**
- *Basic* 1–10
- *Average* 1–10
- *Advanced* 1–10

PROGRAM RESOURCES
Reteach 13.3 Practice 13.3 Enrichment 13.3
Math Fun • Tracking Stock, p. 254

PROBLEM OF THE DAY

Transparency 13

We are a 3-digit and a 2-digit number. None of our digits are greater than 5. Our product is the number of feet in a mile. Who are we? Possible answer: 15 and 352

Solution Tab A, p. A13

QUICK CHECK
Complete the pattern.

1. 80; 85; 90; 95; _?_ ; _?_ ; _?_ 100; 105; 110
2. 1,000; 1,500; 2,000; 2,500; _?_ ; _?_ ; _?_ 3,000; 3,500; 4,000
3. 10; 15; 25; 30; 40 _?_ ; _?_ ; _?_ 45; 55; 60

1 MOTIVATE

Review with students the meaning of the word *trend* as it applies to graphs.

2 TEACH/PRACTICE

In this lesson students use their experiences with analyzing graphs to make predictions.

GUIDED PRACTICE
In these exercises students analyze a graph and make predictions.

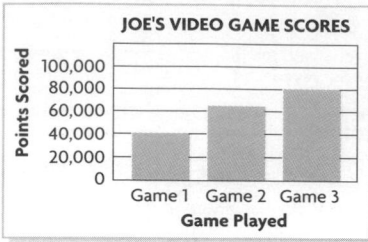

13.3

WHAT YOU'LL LEARN
How to make predictions from a graph

WHY LEARN THIS?
To go beyond the given data, as in predicting what the population will be in the year 2020

WORD POWER
prediction

Business Link
Stock prices for shares traded in most U.S. markets are graphed to show day-to day trends. The Monday-to-Thursday prices for Company X stock are shown in the graph below. What do you predict the price will be on Friday? **$39**

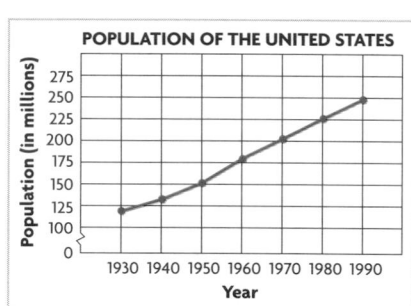

CLOSING PRICES FOR COMPANY X

ALGEBRA CONNECTION

Making Predictions

You can use a bar graph or line graph to make predictions. A **prediction** is an estimate made by looking at a trend over time and then extending that trend to describe a future event.

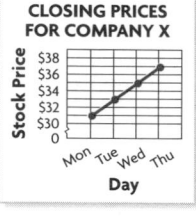

JOE'S VIDEO GAME SCORES

You can look at a graph and observe changes. On this graph, for example, there is a steady growth with no declines. Each increase in the score is about 20,000 points.

- How many points do you think Joe will score in Game 4? Do you think his score will increase indefinitely? Why or why not?
  Possible answer: 100,000; no; he could reach a more difficult ga level

Since a line graph shows change over time, it is used most often to make predictions.

EXAMPLE The line graph below shows the population of the United States every 10 years from 1930 to 1990. Use the graph to predict the population for 2000.

POPULATION OF THE UNITED STATES

The trend is an increase with no declines.

From 1950 to 1990, the population increased about 25 million every 10 years.

Since the population increased about 25 million every 10 years from 1950 to 1990, a good prediction for the year 2000 would be about 250 million + 25 million, or 275 million.

IDEA FILE

CONSUMER CONNECTION

Have students make a line graph using the data in the table. Then have them determine the trend shown in the graph and make a prediction for the years 1997 and 2002.

CATALOGS MAILED IN THE U.S.	
Year	Number of Catalogs Mailed
1982	8.8 billion
1987	11.3 billion
1992	13.6 billion

Modality Visual

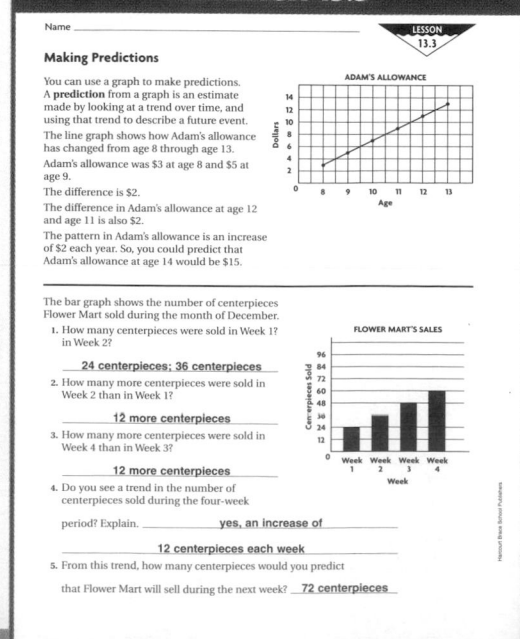

RETEACH 13.3

MEETING INDIVIDUAL NEEDS

GUIDED PRACTICE

Use the graph at the right.

1. What was Regis's height at age 11? at age 12? **55 in.; 57 in.**

2. How much did Regis's height change from age 11 to age 12? **2 in.**

3. What has been the pattern for Regis's growth? **increasing about 3 in. a year**

4. What do you predict Regis's height will be at age 15? **67 in.**

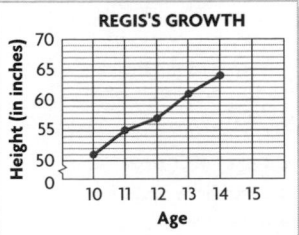

REGIS'S GROWTH

INDEPENDENT PRACTICE

For Exercises 1–4, use the graph at the right.

1. How many CDs did Lucinda have in 1993? in 1994? **10; 15**

2. How many more CDs did Lucinda have in 1996 than in 1995? **5 more**

3. What has been the pattern for the number of CDs Lucinda buys each year? **5 CDs a year**

4. What number of CDs would you predict Lucinda will have in 1998? **35 CDs**

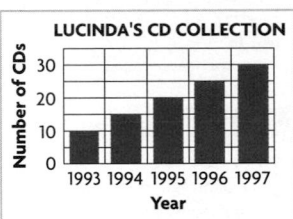

LUCINDA'S CD COLLECTION

Problem-Solving Applications

5. The electric bills for a business over the past four months were $300, $280, $250, and $210. What has the trend been for the electric bills? **decreasing by $20, then $30, and then $40**

6. Use the information in Problem 5 to make a line graph. Look at the graph. What do you think the electric bill will be in the fifth month? **Check students' graphs; $160.**

7. The price of a particular car sold was $14,800. After one year, the car was worth $10,600. After two years, the car was worth $6,200. What has the trend been for the value of the car? **decreasing by about $4,000 each year**

8. Use the information in Problem 7 to make a line graph. Look at the graph. What do you think the car will be worth after 3 years? **about $2,000**

For Problem 9, use the graph at the right.

9. In 1996 Andrew saved $50 each month. He increased his monthly savings each year. From the information in the graph, what do you predict his yearly savings will be in 1999? How much is this per month? **$2,400 a year; $200 a month**

10. ▭▶ **WRITE ABOUT IT** What must a graph display for you to make a prediction? **a pattern or trend**

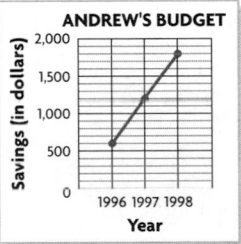

ANDREW'S BUDGET

MORE PRACTICE Lesson 13.3, page H61

245

INDEPENDENT PRACTICE
Exercises check students' ability to identify trends or patterns in graphs.

ADDITIONAL EXAMPLE

Example, p. 244

The line graph shows the amount of money the Bennets spent on snack food. How much will they spend in October?

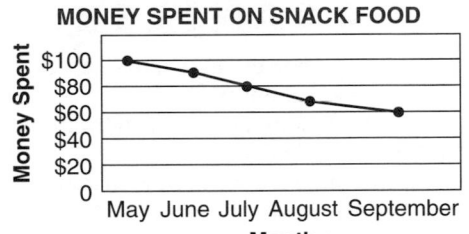

MONEY SPENT ON SNACK FOOD

The trend is a decrease of about $10 each month, so a good prediction for October would be about $50.

3 WRAP UP

Use these questions to help students focus on the big ideas of the lesson.

• **What has been the pattern for Lucinda's CD collection from 1993 to 1997?** increasing by 5 CDs every year

✓ *Assessment Tip*

Have students write the answer.

• **How did you make your prediction for Independent Practice Problem 6?** Possible answer: The trend is a decrease of $10 more each month, so I subtracted $50 from $210.

DAILY QUIZ

For each exercise, use the graph.

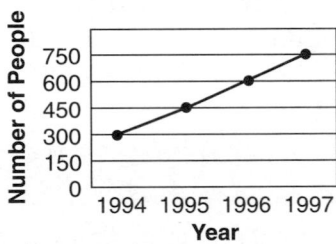

FESTIVAL ATTENDANCE

1. What was the attendance at the festival in 1994? 300 people

2. How many people attended in 1995? 450

3. What has been the trend for attendance at the festival? increasing about 150 every year

4. What do you predict will be the attendance at the festival in 2000? 1,200 people

---

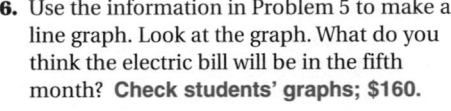

PRACTICE 13.3

Name _____
LESSON 13.3

Making Predictions

For Exercises 1–3, use the graph at the right.
1. How much did Shannan's height change from age 10 to age 11? **2 in.**

SHANNAN'S GROWTH

2. What has been the pattern for Shannan's growth? **increasing about 1 in. per year**

3. What do you predict Shannan's height will probably be at age 16? **45 in.**

For Exercises 4–7, use the graph at the right.
4. How many cards did Scott have in 1992? in 1993? **30; 45**

SCOTT'S BASEBALL CARD COLLECTION

5. How many more cards did Scott have in 1995 than in 1994? **20 more**

6. What relationship exists between the number of cards Scott had in the first and last years shown on the graph? **about 3 times as many cards in 1996 as in 1992**

7. What number of cards would you predict Scott will have in 1997? **120 cards**

Mixed Applications

8. The telephone bills for a pharmacy over the past four months were $220, $209, $197, and $184. What has the trend been for the telephone bills? **decreasing by $11, then $12, then $13**

9. The price a particular car sold at was $15,300. After one year, the car was worth $12,100. After two years, the car was worth $8,900. What is the trend for the value of the car? **decreasing by 3,200 each year**

10. Beth paid for lunch with a twenty dollar bill. She received $15.03 in change. How much was her lunch? **$4.97**

11. Rick paid for school supplies with two five-dollar bills. He received $0.86 change. How much were the school supplies? **$9.14**

P58 **ON MY OWN**
Use with text pages 244–245.

ENRICHMENT 13.3

Name _____
LESSON 13.3

How Many Television Sets?

The pictograph shows the number of households with television sets. Use the pictograph to answer the questions.

1. How many households does each whole symbol represent? **10 million**

UNITED STATES: HOUSEHOLDS WITH TELEVISION SETS

2. How many households had television sets in 1950? **5 million**

3. How many households had television sets in 1980? **75 million**

4. How many more households had television sets in 1960 than in 1950? **40 million**

5. How many more households had television sets in 1980 than in 1970? **15 million**

6. What is the difference in the number of households with television sets in the first and last years shown in the pictograph? **85 million more in 1990 than 1950**

7. Do you see a trend in the data? Explain. **Yes; since 1960 an increase of 15 million sets each decade**

8. Based on this trend, predict the number of households with television sets in 2000. **105 million**

9. How would this amount be represented in the pictograph? **10 whole sets and 1 half set**

10. Why are years prior to 1950 omitted from the pictograph? **TV was not available to consumers.**

E58 **STRETCH YOUR THINKING**
Use with text pages 244–245.

245

ORGANIZER

OBJECTIVE To find the mean, median, and mode of a set of data

VOCABULARY *New* **mean, median, mode**
Review stem, leaf

ASSIGNMENTS Independent Practice
- *Basic* 1–15 odd; 16–18; 19–24
- *Average* 1–18; 19–24
- *Advanced* 8–18; 19–24

PROGRAM RESOURCES
Reteach 13.4 Practice 13.4 Enrichment 13.4
Math Fun • Name Games, p. 254
Extension • Measures of Central Tendency, TE pp. 250A–250B

PROBLEM OF THE DAY

Transparency 13

The mean of these numbers is 14. The greatest number is 21 more than the least. The mode is 18. What are the missing numbers?

3 6 9 ◆ ◆ ◆ ◆

18, 18, 20, 24

Solution Tab A, p. A13

QUICK CHECK

Solve.

1. (4 + 9 + 7 + 3 + 6 + 7) ÷ 6 6
2. (3.7 + 9.2 + 1.9 + 5.6 + 9.6) ÷ 5 6
3. (19 + 45 + 31+ 65) ÷ 4 40
4. (10.7 + 18.6 + 9.2 + 20.9 + 34.6) ÷ 5 18.8

1 MOTIVATE

Use these questions to access students' prior knowledge of mean, median, and mode.

- **Ray bought a $17 shirt from a rack of $15, $12, $22, $17, and $25 shirts.** Median; the median is the middle number in a group of numbers arranged in order.

- **Sue Ellen works out an average of 12 hours every week.** Mean; the mean is the same as average.

13.4

WHAT YOU'LL LEARN
How to find the mean, median, and mode of a set of data

WHY LEARN THIS?
To summarize sets of data, such as test scores or sports scores

WORD POWER
mean
median
mode

Sports Link

In international figure-skating competitions, each skater is scored by 8 judges, but the highest and lowest scores are thrown out. The mean of the remaining 6 scores becomes the skater's final score. In a recent competition, the judges posted the following scores: 5.6, 5.6, 5.4, 5.3, 5.7, 5.7, 5.5, and 5.7. What would the skater's final score be? **5.58**

246 Chapter 13

Mean, Median, and Mode

Suppose a gymnast received these scores:

8.3   7.9   8.3   8.0

There are three measures of central tendency that can be used to describe a set of data as one value: mean, median, and mode.

The **mean**, or average, is the sum of a group of numbers, divided by the number of addends.

The mean of the gymnast's scores is
(8.3 + 7.9 + 8.3 + 8.0) ÷ 4 = 32.5 ÷ 4 = 8.125.

The **median** is the middle number in a group of numbers arranged in numerical order. When there are two middle numbers, the median is the mean of the two middle numbers.

7.9   8.0 ↑ 8.3   8.3   ← numerical order

The median is between 8.0 and 8.3.
(8.0 + 8.3) ÷ 2 = 16.3 ÷ 2 = 8.15

The median of the gymnast's scores is 8.15.

The **mode** of a group of numbers is the number that occurs most often. There may be one mode, more than one mode, or no mode at all.

The mode of the gymnast's scores is 8.3.

EXAMPLE 1   Find the mean, median, and mode for the data.

Basketball Points				
22	18	8	34	18

Mean:
(22 + 18 + 8 + 34 + 18) ÷ 5   *Add the scores, and divide by 5.*
100 ÷ 5 = 20

Median:
8   18   18   22   34   *Order the data. The middle number is 18.*

Mode:
18   *18 occurs twice.*

 Calculator Activities, page H41

IDEA FILE

ALTERNATIVE TEACHING STRATEGY

MATERIALS: connecting cubes

Have students:

1. Model this data set: 8, 10, 5, 2, 5.

2. Find the stacks that have the same number of cubes. 2 stacks of 5 cubes What is 5? mode

3. Place the stacks in order. How many cubes are in the middle stack? 5 cubes What is 5? median

Modalities Kinesthetic, Visual

MEETING INDIVIDUAL NEEDS

RETEACH 13.4

Name _____   LESSON 13.4

Mean, Median, and Mode

The test scores of a math class are shown below. Find the mean, median, and mode for the data.

Test Scores									
78	81	93	91	100	100	81	78	98	100

Mean: Find the sum of the scores and divide by the number of scores.
78 + 81 + 93 + 91 + 100 + 100 + 81 + 78 + 98 + 100 = 900
900 ÷ 10 = 90

Median: Arrange the scores in order from least to greatest. Find the number that is in the middle.
78 78 81 81 91 93 98 100 100 100
92←The middle number is between 91 and 93.

Mode: Find the number or numbers that appear most often.
100

So, the mean test score is 90. The median score is 92. The mode score is 100.

Find the mean, median, and mode for each set of data.

1.
Test Scores									
85	77	91	97	95	95	79	83	95	93

mean: ____89____
median: ____92____
mode: ____95____

2.
Golf Scores					
88	76	92	91	69	76

mean: ____82____
median: ____82____
mode: ____76____

3.
Number of School Days								
Japan	England	Israel	Germany	Netherlands	USA	Thailand	Sweden	Canada
243	200	216	210	200	180	200	180	180

mean: ____201____   median: ____200____   mode: ____180 and 200____

Use with text pages 246–249.   **TAKE ANOTHER LOOK** R59

For some sets of data, you can use a line plot or a stem-and-leaf plot to help you find the median and the mode.

EXAMPLE 2 Use Ms. Jones's class test-scores data below to make a line plot. Find the mode and the median.

Ms. Jones's Class Test Scores										
72	82	83	78	81	78	73	74	75	73	76
71	75	80	83	72	72	78	81	79	82	76

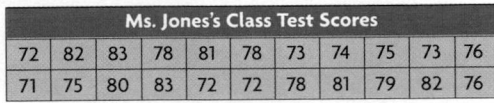

```
                    X              X
      X     X    X  X       X  X  X
X  X  X  X  X    X  X  X  X  X  X
+--+--+--+--+--+--+--+--+--+--+--+--+--+--+--+
70 71 72 73 74 75 76 77 78 79 80 81 82 83 84
```

72 and 78 each occur three times.

There are 22 scores. The median is between the 11th and 12th scores.

Modes: 72 and 78
Median: (76 + 78) ÷ 2 = 77

EXAMPLE 3 Use the fitness test data to make a stem-and-leaf plot. Find the mode and the median.

Fitness Test Scores									
97	88	74	96	98	58	68	90	80	90
72	86	69	78	93	84	99	92	85	

Stem	Leaves
5	8
6	8 9
7	2 4 8
8	0 4 5 6 8
9	0 0 2 3 6 7 8 9

90 occurs more than any other number.

There are 19 scores. The median is the 10th.

Mode: 90
Median: 86

> **REMEMBER:**
> A stem-and-leaf plot is a way to organize data.
>
> | Stem | Leaves |
> |---|---|
> | 2 | 5 7 9 |
> | 3 | 2 2 4 6 8 |
> | 4 | 0 1 3 |
>
> In this example, the tens digits are the *stems*. The ones digits are the *leaves*. See page 226.

GUIDED PRACTICE

For Exercises 1–3, use the table below.

Day	Sun	Mon	Tue	Wed	Thu	Fri	Sat
Miles Jogged	4	4	5	5	6	7	4

1. Find the mean. **5**
2. Find the mode. **4**
3. Find the median. **5**

PRACTICE 13.4

Name _____

LESSON 13.4

Mean, Median, and Mode

Vocabulary

Write the correct letter from Column 2.

Column 1	Column 2
b 1. mean	a. number that appears most often in a group of numbers
c 2. median	b. sum of a group of numbers divided by the number of addends
a 3. mode	c. middle number in a group of numbers arranged in numerical order

Complete the table.

Data	Mean	Median	Mode
4. 12, 15, 11, 15, 13, 10, 15	13	13	15
5. 68, 74, 71, 69, 74, 78, 70	72	71	74
6. 7.6, 6.2, 6.0, 6.2, 8.1, 6.7	6.8	6.45	6.2

Write *true* or *false* for each.
7. The mode is always one of the numbers in a set of data. **true**
8. Every set of data has a mean. **true**

For Exercises 9–10, use the table below.

Test Scores									
98	88	82	91	83	76	98	100	84	90

9. Make a line plot and use it to find the median and the mode.
Check students' plots.

10. Make a stem-and-leaf plot and use it to find the median and the mode.
Check students' plots.

Mixed Applications
11. Five books have the following prices: $12.00, $14.95, $13.50, $13.75, and $13.80. What are the mean and median?
mean: $13.60; median: $13.75

12. A baseball league had 576 members. There are 16 members on each team. How many teams are in the league?
36 teams

Use with text pages 246–249. **ON MY OWN P59**

ENRICHMENT 13.4

Name _____

LESSON 13.4

Number Puzzles

Use the information to find the numbers in the group. The first one is done for you.

1. There are 7 whole numbers in a group. The least number in the group is 6. The greatest number in the group is 16. The mode of the group is 15. The median is 10 and the mean is 11.
6, 7, 8, 10, 15, 15, 16

2. There are 5 whole numbers in a group. The least number is 7 and the greatest is 14. The mode is 9 and the median is 9. The mean is 11.
7, 9, 9, 11, 14

3. There are 7 whole numbers in a group. The greatest number is 20 and the least is 8. The median is 12 and the mode is 12. The mean is 13.
8, 10, 12, 12, 14, 15, 20

4. There are 7 whole numbers in a group. The least number is 5 and the greatest is 15. The mean and the median are 11. The mode is 15.
5, 7, 9, 11, 15, 15, 15

5. There are 7 whole numbers in a group. The least number is 11 and the greatest is 17. The mean and the median are 14. There is no mode.
11, 12, 13, 14, 15, 16, 17

6. There are 7 whole numbers in a group. The least number is 15 and the greatest is 33. The mean is 23. The median is 22. The mode is 19.
15, 19, 19, 22, 25, 28, 33

7. There are 7 whole numbers in a group. The greatest number is 37 and the least is 21. The median is 29. The mean is 28. The mode is 31.
21, 23, 24, 29, 31, 31, 37

Use with text pages 246–249. **STRETCH YOUR THINKING E59**

• **Most students scored 100% on their history test.** Mode; the mode is the number that occurs most often in a group of numbers.

2 TEACH/PRACTICE

Read and discuss the definitions of the measures of central tendencies and have students calculate each measure for the gymnast's scores. As you discuss the examples, point out to students that a set of data always has only one mean and one median.

ADDITIONAL EXAMPLES

Example 1, p. 246

Find the mean, median, and mode of the data set below.

Ages of Students in Math Club						
11	12	12	13	10	14	12

Mean: Add the ages and divide by 7.

(11 + 12 + 12 + 13 + 10 + 14 + 12) ÷ 7

84 ÷ 7 = 12

Median: Arrange the data in increasing order. Identify the number that is in the middle.

10 11 12 12 12 13 14

Mode: 12 occurs three times, more often than any other age. The mean, median, and mode are 12.

Example 2, p. 247

Use the scores below to make a line plot. Find the mode and the median.

Score of Hamilton High School Golf Team Members						
84	82	77	78	79	77	81
79	80	83	77	84	85	82

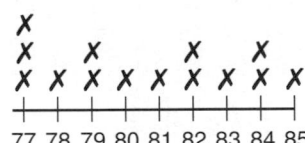

```
X
X     X     X     X
X  X  X  X  X  X  X  X
+--+--+--+--+--+--+--+--+--+
77 78 79 80 81 82 83 84 85
```

77 occurs three times.

Mode: 77

There are 14 scores. The median is between the 7th and 8th scores.

Median: (80 + 81) ÷ 2 = 80.5 or 81

GUIDED PRACTICE

Complete Exercises 1–3 orally as a class to make sure students understand how to find the mean, median, and mode.

INDEPENDENT PRACTICE

To ensure student success, have students work in pairs to complete Exercises 1–9 before you assign the remaining exercises.

ADDITIONAL EXAMPLES

Example 3, p. 247

Use the data from the table below to make a stem-and-leaf plot. Find the mode and the median.

Number of Cans Collected by Each Class in the Canned-Food Drive

48	56	77	63	69	71	40
51	53	78	77	76	42	51

Stem	Leaves
4	028
5	1136
6	39
7	16778

51 and 77 occur more than any other number.

Modes: 51 and 77

There are 14 numbers. The median is between the 7th and 8th numbers.

Median: (56 + 63) ÷ 2 = 59.5 or 60

Example 4, p. 248

A CD producer keeps track of the number of songs recorded on each CD her company makes. Find the mean, median, and mode of the data in the graph. Then tell which measure of central tendency best represents the data.

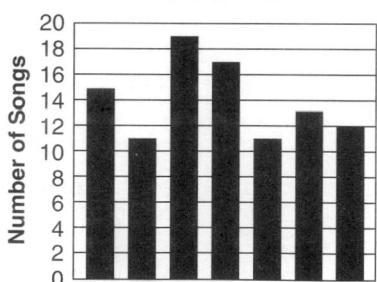
SONGS ON CD'S

Mean: (15 + 11 + 19 + 17 + 11 + 13 + 12) ÷ 7   98 ÷ 7 = 14

The mean is 14.

Median: 11 11 12 13 15 17 19

The median is 13.

Mode: 11 occurs twice, more than any other number.

The mode is 11.

When you want to summarize a set of data as one value, you can use one of the three measures of central tendency.

EXAMPLE 4 As a T-ball coach, you kept track of the number of runs each of your players scored for the season. Use the data in the graph to find the mean, median, and mode. Then tell which measure of central tendency best represents the data.

Mean:
(16 + 15 + 15 + 15 + 15 + 18 + 16 + 17 + 30 + 30) ÷ 10 = 187 ÷ 10 = 18.7

So, the mean is 18.7.

Median:
15	15	15	15	16
16	17	18	30	30

So, the median is 16.

Mode:
The mode is 15. It occurs four times in the data.

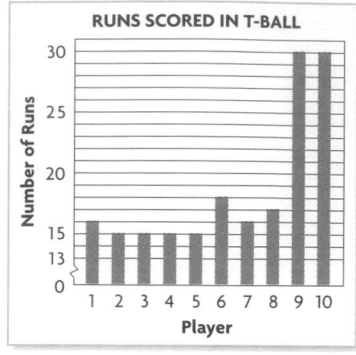
RUNS SCORED IN T-BALL

Because of the two high scores of 30, the mean is much larger than the mode and the median. So, the mean is not a good measure of the data. Either the median or the mode represents the data better.

- **CRITICAL THINKING** Which part of the data would have to be different for the mean to be a good measure of central tendency? **the two scores of 30 runs**

Business Link

Television advertisers often use the median income of an audience as the basis for deciding what products to advertise during a particular show. Why would knowing the median income of the audience be better than knowing the mean income of the same group?

The median is not affected by a few people with very high incomes, as the group's mean would be.

INDEPENDENT PRACTICE

Copy and complete the table.

	Data	Mean	Median	Mode			
1.	7, 6, 11, 7, 9	?	?	?	8	7	7
2.	84, 73, 92, 77, 89	?	?	?	83	84	none
3.	16, 32, 24, 10, 48, 32	?	?	?	27	28	32
4.	9.5, 8.2, 7.1, 8.0	?	?	?	8.2	8.1	none
5.	45.7, 45.9, 45.7, 48.3	?	?	?	46.4	45.8	45.7
6.	27, 48, 83, 76, 48, 27	?	?	?	51.5	48	27 and 48
7.	1.3, 7.8, 0.7, 7.8, 3.4	?	?	?	4.2	3.4	7.8

Technology

In *Data ToolKit*, you can make a stem-and-leaf plot.

HARCOURT BRACE

Data Too

IDEA FILE

EXTENSION

Have students find the value of each item in data sets with the measures shown in the table.

	Mean	Mode	Median	# of Items
1.	4	4	4	4
2.	31	37	37	5
3.	47	42	42	5
4.	79	77	78	6

Possible answers:
1. 3, 4, 4, 5. **2.** 15, 24, 37, 37, 42.
3. 28, 42, 42, 59, 64. **4.** 66, 77, 77, 79, 80, 95.

Modality Visual

WRITING IN MATHEMATICS

Math Journal

Have students describe situations that represent mean, mode, or median and have their classmates determine which measure of central tendency each situation represents. Provide this example:

- Several of the cereal brands tested contain 2 grams of sugar per serving. **mode**

Modalities Auditory, Visual

Write *true* or *false* for each.

8. The mean is always one of the numbers in a set of data. **false**

9. The median may be the average of the two middle numbers in the data. **true**

For Exercises 10–13, use the table below.

Test Scores									
78	82	96	65	100	100	94	61	70	88

10. Make a line plot. **See Additional Answers, p. 255A.**

11. Use your line plot to find the median and the mode. **85; 100**

12. Make a stem-and-leaf plot. **See Additional Answers, p. 255A.**

13. Use your stem-and-leaf plot to find the median and the mode. **85; 100**

14. In Exercises 10 and 12, which display made it easier to find the median and the mode? Explain. **Possible answer: Stem-and-leaf plot; it took less time to make.**

15. Can you tell what the mean is for this set of data by looking at the stem-and-leaf plot? Explain. **No. You have to find the sum of the data and divide by 10.**

Problem-Solving Applications

16. Use the data shown in the graph to find the mean, median, and mode. Then tell which measure of central tendency best represents the data. **20.6; 18; 18; median or mode; mean is distorted by extreme score, 30.**

BASKETBALL POINTS

Player / Points Scored / 0 18 20 25 30

17. Eight joggers ran the following number of miles: 8, 5, 6, 4, 8, 8, 7, and 10. Determine the mean, median, and mode of the miles jogged. **7; 7.5; 8**

18. ▶ **WRITE ABOUT IT** Explain the differences between the mean, median, and mode of a group of numbers. **mean: sum divided by count; median: middle number in ordered scores; mode: score that occurs most often.**

Mixed Review

For Exercises 19–21 use the data 10, 12, 28, 9, and 17.

19. Find the lowest value. **9**

20. Find the highest value. **28**

21. Find the mean. **15.2**

22. Use the data in the table to the right to make a bar graph. **See Additional Answers, p. 255A.**

NUMBER OF BOOKS READ		
Grade 6	Grade 7	Grade 8
59	70	65

23. **NUMBER SENSE** There are 180 players in a baseball league. Each team has 12 players. How many teams are in the league? **15 teams**

24. **CONSUMER** A company claims that its 8-yr warranty lasts 4 times as long as any other warranty. What is the longest of the other warranties? **2 yr**

MORE PRACTICE Lesson 13.4, page H61

249

INTERDISCIPLINARY CONNECTION

One interdisciplinary unit you can use to integrate interpreting data and predicting is Cultural Connection: Billionaires Around the World. In this unit you can have students

- make and use a bar graph to show the countries with the greatest number of billionaires in the world.

- determine what measure of central tendency best represents the data. median

Country	Number of Billionaires
United States	129
Germany	48
Japan	34
Hong Kong	12
Thailand	12

Modality Visual

MEETING INDIVIDUAL NEEDS

TEACHING TIP

When discussing Example 4, have students give additional examples of situations in which mean would not be a good representation of the data.

CRITICAL THINKING **Can the mean, median, and mode of a set of data all be the same number? Give an example.** Yes. Possible example: 10, 11, 11, 12, 12, 12, 13, 13, 14; the mean, median, and mode are 12. **Is it easier to find mode and median from a stem-and-leaf plot or from a table? Explain.** Most students will say stem-and-leaf plot because the data are organized in numerical order.

3 WRAP UP

Use these questions to help students focus on the big ideas of the lesson.

- **How do you find the mean of a set of data?** Possible answer: Find the sum of the numbers and divide the sum by the number of addends.

✓ *Assessment Tip*

Have students write the answer.

- **Can you tell from a line plot what the mode of a data set is? Explain.** Yes. It is the number with the greatest number of x's above it.

DAILY QUIZ

The table below shows the number of hours Jenny's cat, Sylvester, slept every day for two weeks. Use the table to complete exercises 1–4.

Number of Hours Sylvester Slept Each Day	
Week 1	17 18 20 17 18 19 21
Week 2	12 12 22 18 21 18 19

1. Make a stem-and-leaf plot for the data.

Stem	Leaves
1	2 2 7 7 8 8 8 8 9 9
2	0 1 1 2

2. What is the mean number of hours Sylvester slept? 18 hr

3. What is the median number of hours Sylvester slept? 18 hr

4. What is the mode? 18 hr

249

EXTENSION

MEETING INDIVIDUAL NEEDS

Measures of Central Tendency

Use these strategies to extend understanding of measures of central tendency.

Multiple Intelligences Logical/Mathematical, Visual/Spatial, Bodily/Kinesthetic

Modalities Kinesthetic, Visual

Lesson Connection Students will collect data from their classmates and find the mean, median, and mode of the data.

Materials Calculator

Program Resources *Intervention and Extension Copying Masters,* p. 19

Challenge

TEACHER NOTES

ACTIVITY 1

Purpose To record the heights of students and find the mean, median, and mode of the data

Vocabulary Review

- **mean**—the sum of a set of numbers divided by the number of addends

- **median**—the middle number or mean of the middle two numbers in an ordered set of data

- **mode**—the number that occurs most frequently in a set of data

Discuss the following questions.

- Give an example of a set of data that has no mode. Answers will vary.

- Give an example of a set of data that has more than one mode. Answers will vary.

- How do you find the median when the data set has an even number of items? Find the mean of the middle two numbers.

WHAT STUDENTS DO

1. Measure your height to the nearest centimeter. Check students' measurements.

2. Record your height and the heights of your classmates.

3. Use a line plot to organize the data. Check students' line plots.

   - What do you notice about the shape of the line plot? Possible answer: It resembles a bell-shaped curve.

4. Find the mean, median, and mode of the data. Answers will vary.

   - How do the measures of central tendency compare? Possible answer: Their values are about the same.

5. Which measures of central tendency best represent the data? Possible answer: Any of the measures can be used to represent the data because their values are similar.

TEACHER NOTES

ACTIVITY 2

Purpose To record the number of years students have lived in the area and find the mean, median, and mode of the data

Have students meet in small groups or with a partner to discuss these two questions. Have one student from each group report the group's conclusions to the class.

WHAT STUDENTS DO

1. How many years have you lived in this area? Answers will vary.

2. Record your response and the responses of your classmates.

3. Use a line plot to organize the data. Check students' line plots.

   • What do you notice about the shape of the line plot? Possible answer: The data is clustered around 10, 11, and 12.

4. Find the mean, median, and mode of the data. Answers will vary.

   • How do the measures of central tendency compare? Answers will vary.

5. Which measures of central tendency best represent the data? Possible answer: The mode or median best represents the data since most students have lived in the area all their lives.

6. Why might the mean not be a good representation of the data? Possible answer: The mean is most affected by the responses of students who recently moved to the area.

Share Results

• Describe a set of data where the mean, median, and mode are the same or close to the same. Describe the shape of the line plot for the data. Answers will vary.

• Describe a set of data where the mean, median, and mode are not close to the same. Describe the shape of the line plot for the data. Answers will vary.

Intervention and Extension Copying Masters, p. 19

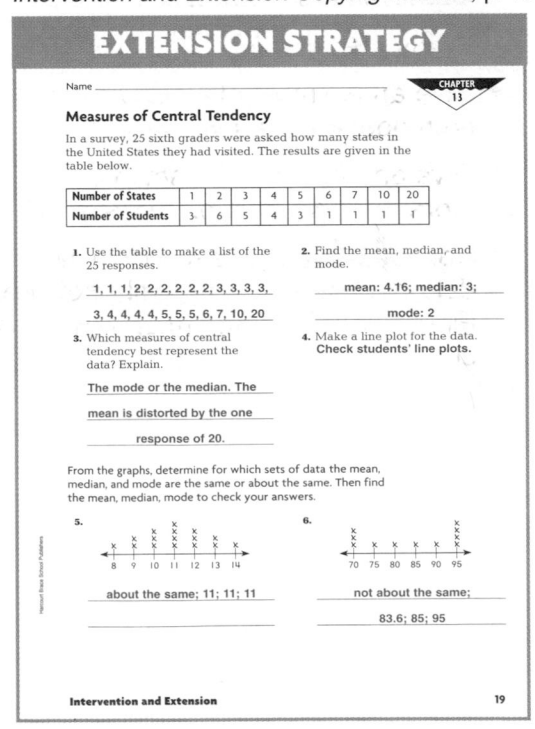

ORGANIZER

OBJECTIVE To make a box-and-whisker graph and understand its parts

VOCABULARY *New* **box-and-whisker graph, lower extreme, upper extreme, lower quartile, upper quartile**

MATERIALS *For each student* at least eleven 3 × 5 cards, marker

PROGRAM RESOURCES
E-Lab • Activity 13
E-Lab Recording Sheets, p. 13

USING THE PAGE

Review with students the advantages of using different graphs to display data. Ask:

- **Why would you display data in a stem-and-leaf plot?** to show each item in the data

- **Why would you display data in a line graph?** to show change over time

- **Why would you display data in a histogram?** to show the frequency data occur in intervals

- **Why would you display data in a bar graph?** to show categorical data

- **Why would you display data in a circle graph?** to show parts of a whole

Explain to students they can display data in a box-and-whisker graph if they want to show how data are distributed.

Activity 1

Explore

If necessary, assist students as they organize data into the parts that make up a box-and-whisker graph. Record each step on the board as students work with their partners. If some students need help with the vocabulary, use the ESL Tip in the Idea File.

Think and Discuss

Have students check to be sure their data are separated into the parts described in the first question. Explain that the three remaining questions deal with the four *parts* of the data, not the *number* of data items. Discuss the box-and-whisker graph labeled with the parts. Have students associate the parts with the numbers in the graph.

CO-OPERATIVE LEARNING

LAB *Activity*

WHAT YOU'LL EXPLORE
How to make a box-and-whisker graph and understand its parts

WHAT YOU'LL NEED
at least eleven 3-in × 5-in cards, marker

WORD POWER
box-and-whisker graph
lower extreme
upper extreme
lower quartile
upper quartile

250   Chapter 13

Exploring Box-and-Whisker Graphs

A **box-and-whisker graph** shows how far apart and how evenly data are distributed.

Explore

Work with a partner.

- Write each of the hours worked, shown in the table below, on a separate card.

Number of Hours Worked									
30	16	19	27	31	15	19	24	22	23

- Order the data from least to greatest.

- What is the least number of hours worked? What is the greatest number of hours worked? **15; 31**

- The least number is called the **lower extreme**. The greatest number is called the **upper extreme**. Draw a star on the cards with the lower extreme and the upper extreme.

- What is the median of the data? If the median is not one of the numbers already written, write it on a card, put it in the middle of the data, and circle it. **22.5**

- The lower half of the data is to the left of the median. What is the median of the lower half? Circle it. This median is called the **lower quartile**. Separate the data to the left of the lower quartile from the rest of the data. **19**

- The upper half of the data is to the right of the overall median. What is the median of the upper half? Circle it. This median is called the **upper quartile**. Separate the data to the right of the upper quartile from the rest of the data. **27**

IDEA FILE

ENGLISH AS A SECOND LANGUAGE

Tip • Have students use these tips to help them understand vocabulary related to box-and-whisker graphs.

box-and-whisker graph—Look at the graph and think of a cat's face.

extreme—Think of extreme temperatures: very hot and very cold.

quartile—Think of a quarter, $\frac{1}{4}$ of a dollar; and a quart, $\frac{1}{4}$ of a gallon.

Modality Visual

HOME CONNECTION

HOME NOTE Have students conduct a survey of their classmates who have pets to estimate how much they spend on pet food each week. Have them make a large box-and-whisker graph of the data and display the graph in the classroom.

Modalities Auditory, Kinesthetic, Visual

Think and Discuss

- Into how many parts do the lower quartile, the median, and the upper quartile separate the data? **4 parts**

- What fraction of the data are to the left of the lower quartile? $\frac{1}{4}$

- What fraction of the data are to the right of the upper quartile? $\frac{1}{4}$

- What fraction of the data are between the lower quartile and the upper quartile? $\frac{1}{2}$

You have found all you need to make a box-and-whisker graph. Below is the box-and-whisker graph of the data, labeled with the parts.

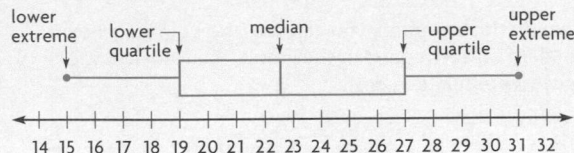

14 15 16 17 18 19 20 21 22 23 24 25 26 27 28 29 30 31 32

- How do you know where to start and end the box in the graph? **The box starts at the lower quartile and stops at the upper quartile.**

Try This

- For the data below, find the lower and upper extremes, the median, and the lower and upper quartiles.
  2; 11; 6.5; 4.5; 8

2	5	11	7	9	8
8	3	4	8	5	6

- Make a box-and-whisker graph of the data.
  See Additional Answers, p. 255A.

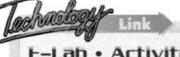

E-Lab • Activity 13 Available on CD-ROM and the Internet at http://www.hbschool.com/elab

Note: Students make a box-and-whisker graph. Use E-Lab p. 13.

Try This

After finding the parts of the data, students should follow these steps for making a box-and-whisker graph.

1. Draw and label a number line with consecutive numbers from one less than the lower extreme to one more than the upper extreme.
2. Label the lower and upper extremes, the median, and the lower and upper quartiles.
3. Make a box that extends from the lower quartile to the upper quartile and includes the median.
4. Connect the ends of the box to the lower and upper extremes with whiskers.

Have students share their graphs with the class and describe the parts that make up the graph. Repeat the questions from Think and Discuss.

USING E-LAB

Students use a computer tool to make box-and-whisker graphs which they then interpret.

✓ Assessment Tip

Use these questions to help students communicate the big ideas of the lesson.

- **What are lower and upper quartiles?** lower quartile: the median of the data to the left of the overall median; upper quartile: the median of the data to the right of the overall median

- **What are upper and lower extremes?** the greatest number and the least number, respectively, in a data set

REVIEW

Have students research the prices of ten different calculators and find the mean, median, and mode of the data. Then have them determine which measure of central tendency best represents the data.

Modality Visual

SCIENCE CONNECTION

Present this problem to students:

- Elaine heard there are more than 1,000,000 insects for every one of the 5,292,000,000 people on earth. She contacted seven of her pen pals from around the world to ask them how many insects they could find in one day. Here are the results of their search. Make a box-and-whisker graph of the data.

  29 53 7 40 33 21 19

Modality Visual

ORGANIZER

OBJECTIVE To analyze a box-and-whisker graph

ASSIGNMENTS Independent Practice
- *Basic* 1–8 even; 9–14
- *Average* 1–8; 9–14
- *Advanced* 5–8; 9–14

PROGRAM RESOURCES
Reteach 13.5 Practice 13.5 Enrichment 13.5
Data Toolkit

PROBLEM OF THE DAY

  Transparency 13

Unscramble these letters to form math words used in this chapter.

1. **ARB** BAR
2. **NAME** MEAN
3. **LICCER** CIRCLE
4. **DAMIEN** MEDIAN
5. **MEST-DAN-FALE** STEM-AND-LEAF
6. **DOME** MODE
7. **NILE** LINE
8. **XOB-NAD-SHEWIRK** BOX-AND-WHISKER
9. **CROPINTIDE** PREDICTION

Solution Tab A, p. A13

QUICK CHECK
Find the median.

1. 19, 26, 18, 33, 51 26
2. 2.7, 4.3, 1.2, 0.5, 1 1.2
3. 67, 55, 74, 45 61
4. 789, 555, 483, 768, 601, 443 578

1 MOTIVATE

Use the first bar-and-whisker graph to show students how these graphs can be used. Read and discuss Teen Times.

2 TEACH/PRACTICE

Read and discuss page 252. Focus attention on how data are distributed in both graphs.

13.5

WHAT YOU'LL LEARN
How to analyze a box-and-whisker graph

WHY LEARN THIS?
To understand how data such as test scores, sports statistics, and survey responses are grouped

Box-and-Whisker Graphs

Angela made a box-and-whisker graph to represent the number of cookies she sold each day for one week.

- What is the least number of cookies Angela sold in one day? What is the greatest number? **4; 12**

- Look at the graph. Which of the following can you determine: mean, median, mode, or range? **median and range**

In a box-and-whisker graph, you can see how your data are distributed. In different parts of the graph, the values may be closer together or farther apart.

EXAMPLE A class of 20 students took a math test. Their scores were used to make the graph below. What does the graph show about how the scores are distributed?

The scores in the lowest $\frac{1}{4}$ of the data are very close together. The scores in each part of the middle $\frac{1}{2}$ are farther apart. These scores are closer to the lower extreme than to the upper extreme. The scores in the highest $\frac{1}{4}$ of the data are even farther apart than those in the middle. At least one student scored 100.

- **CRITICAL THINKING** What fraction of the class had scores between 90 and 100? What fraction had scores between 80 and 90? If 90 was needed for an A, how many students got an A? $\frac{1}{4}$; $\frac{3}{4}$; **about 5**

In a box-and-whisker graph, the only actual values you can identify from the data set are the extremes, the highest and lowest values.

- What are the two actual scores you can identify in the box-and-whisker graph for the example? **82, 100**

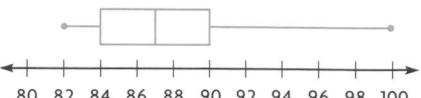

teen times

If you have ever taken a standardized test, you may have noticed that quartiles were reported. These can tell you how your performance compared with that of others who took the test.

IDEA FILE

CAREER CONNECTION

College Recruiter One of the factors a college recruiter looks at when considering applicants for a school is the quartile the applicants' standardized test scores are in.

You may want to have a local recruiter speak with students about how careful analysis of data is important to the job of a recruiter.

Modality Auditory

RETEACH 13.5

Name _____

LESSON 13.5

Box-and-Whisker Graphs

The owner of a gift shop made a box-and-whisker graph to represent the number of customers who came into the shop each day.

The least number in a box-and-whisker graph is called the **lower extreme**. It is the black dot at the end of the left whisker. In this graph, the lower extreme represents the least number of customers to enter the shop in a single day. The lower extreme is 12.

The greatest number in a box-and-whisker graph is called the **upper extreme**. It is the black dot at the end of the right whisker. In this graph, the upper extreme represents the greatest number of customers to enter the shop in one day. The upper extreme is 28.

The range is the difference between the upper and lower extremes. The range of this set of data is 16.

The median of a set of data is the middle number when all numbers are arranged in numerical order. In a box-and-whisker graph, the median is represented by a dotted line. The median of this set is 21.

Another store owner also made a box-and-whisker graph of the customers who entered her store daily.

1. What is the lower extreme? __30__
2. What is the upper extreme? __54__
3. What is the lower quartile? __40__
4. What is the upper quartile? __50__
5. What is the range of this set of data? __24__
6. What is the median for this set of data? __44__

R60 **TAKE ANOTHER LOOK** Use with text pages 252–253.

GUIDED PRACTICE

Use the box-and-whisker graph at the right.

1. What is the median? **28**

2. What are the lower and upper quartiles? **27; 31**

3. What are the lower and upper extremes? **25; 32**

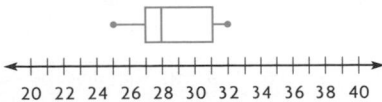

20 22 24 26 28 30 32 34 36 38 40

INDEPENDENT PRACTICE

For Exercises 1–5, use the table at the right.

1. What is the median? **22**

2. What are the lower and upper quartiles?
   16.5; 27
3. What are the lower and upper extremes?
   15; 32
4. What is the range? **32 − 15 = 17**

Lengths of Phone Calls (in min)					
20	24	21	16	15	26
17	32	30	28	16	23

5. Make a box-and-whisker graph. **See Additional Answers, p. 255A.**

Problem-Solving Applications

This box-and-whisker graph shows the number of runs scored by a baseball team in one season's games.

6. What was the least number of runs? What was the greatest number of runs? **1; 9**

0 1 2 3 4 5 6 7 8 9 10

7. Describe how the data are distributed. **The numbers in the lowest and highest $\frac{1}{4}$ of the data are close together; the data in the middle are more spread out.**

8. ✏ **WRITE ABOUT IT** Explain how a box-and-whisker graph is divided into four parts. **See Additional Answers, p. 255A.**

Mixed Review

Write the ratio as a fraction.

9. 3 out of 4 $\frac{3}{4}$ 10. 7 out of 8 $\frac{7}{8}$ 11. 1 out of 2 $\frac{1}{2}$

12. Use the data in the table to make a multiple-bar graph.

Club Name	Grade 6	Grade 7	Grade 8
Computer	19	9	7
Math	10	10	9
Yearbook	4	7	10

See Additional Answers, p. 255A.

13. **NUMBER SENSE** Yolonda gave a clerk $10.03 for frozen yogurt and received $7.00 in change. How much was the yogurt? **$3.03**

14. **CRITICAL THINKING** The mean of 4 numbers is 13.5. What is the sum of the numbers? **54**

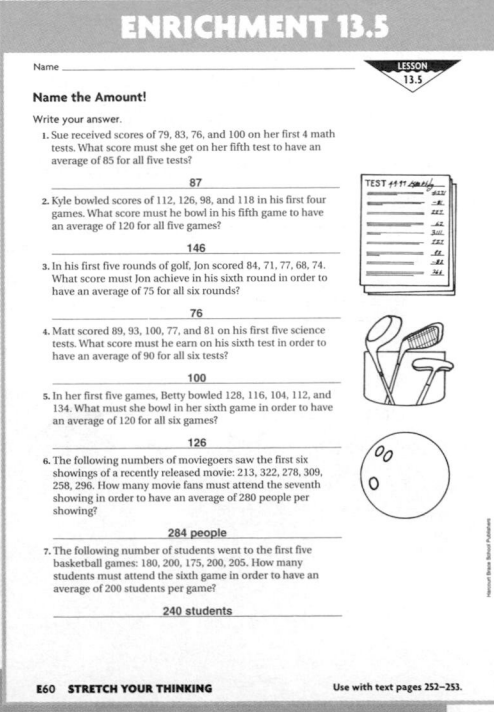

Technology **Link**

In **Data ToolKit**, you can use data to make a table and then show these data as a box-and-whisker graph.

MORE PRACTICE Lesson 13.5, page H61

253

GUIDED PRACTICE

Discuss these exercises to make sure students understand how to analyze a box-and-whisker graph.

INDEPENDENT PRACTICE

Students may need to be reminded that a box-and-whisker graph divides a set of data into four quartiles. If necessary, help them identify the quartiles in Exercises 2 and 7.

ADDITIONAL EXAMPLE

Example, p. 252

Jason surveyed 20 restaurants for the price of a large pizza with two toppings. The results were used to make the graph below. What does the graph show about how the prices are distributed?

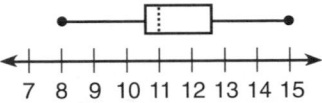

7 8 9 10 11 12 13 14 15

The prices in the lowest $\frac{1}{4}$ of the data are far apart. The prices in each part of the middle $\frac{1}{2}$ are closer together. These prices are closer to the lower extreme than to the upper extreme. At least one price is $15 and another is $8.

3 WRAP UP

Use these questions to help students focus on the big ideas of the lesson.

- **What three medians divide the box-and-whisker graph at the top of page 252?** 5, 6, 9

✔ *Assessment Tip*

Have students write the answer.

- **Why can't you use a box-and-whisker graph to find mean?** Possible answer: because a box-and-whisker graph does not show every data item

DAILY QUIZ

For each exercise, use the box-and-whisker graph.

1. What is the median? 2.5
2. What are the lower and upper quartiles? 2; 4
3. What are the lower and upper extremes? 1; 7

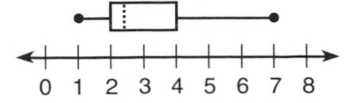

0 1 2 3 4 5 6 7 8

253

USING THE PAGE

To provide additional practice activities related to Chapter 13: *Interpreting Data and Predicting*

WHAT'S WRONG WITH THIS PICTURE?

Have students work individually to complete this activity.

In this activity, students identify misleading graphs.

Multiple Intelligences Visual/Spatial, Logical/Mathematical, Verbal/Linguistic, Interpersonal/Social, Intrapersonal/Introspective
Modalities Auditory, Visual

TRACKING STOCK

MATERIALS: *For each student* daily newspaper's stock section for a week

Have students work individually to complete this activity.

In this activity, students make a line graph based on one week's worth of a stock's prices. Students then use the line graph to predict future performance of the stock.

Multiple Intelligences Visual/Spatial, Logical/Mathematical, Intrapersonal/Introspective
Modality Visual

NAME GAMES

MATERIALS: *For each student* 10 index cards

Have students work individually to complete this activity.

In this activity, students find the mean, median, and mode from a set of data. Students then try to add or delete names to make the requested changes to the data.

Multiple Intelligences Logical/Mathematical, Bodily/Kinesthetic, Intrapersonal/Introspective
Modalities Kinesthetic, Visual

MATH FUN!

WHAT'S WRONG WITH THIS PICTURE?

PURPOSE To practice identifying misleading graphs (pages 242–243)
YOU WILL NEED graph paper

Lynn and Linda are amazingly alike. They were born on the same day in the same year by the same parents. You think they are twins but they are not. How can you explain that?

Like words, graphs can be designed to mislead you. Look at the graph. Why is it misleading? **The population does not change.** **They are 2 of triplets, quadruplets, or quintuplets; the others are not mentioned.**

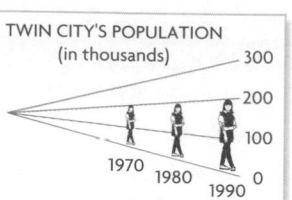

TWIN CITY'S POPULATION (in thousands)

HOME NOTE How many graphs in newspapers or magazines can you find that present data in misleading ways?

Portfolio Tracking Stock

PURPOSE To practice predicting trends (pages 244–245)
YOU WILL NEED daily newspaper's stock section for a week

Look through a daily newspaper that you can get each day for a week. Choose a stock. Record the closing prices of your chosen stock each day for a week.

Make a line graph using your recordings. **Check students' graphs.**
What was the lowest closing price? the highest closing price? **Answers will vary.**

NAME GAMES

PURPOSE To practice finding the mean, median, and mode from a set of data (pages 246–253)
YOU WILL NEED 11 index cards

Write each of the names on an index card. On the back of the card, write the number of letters in the name. Order the data. Find the mean, median, and mode of the numbers of letters. **11.7; 12; 11,12, and 14**

1. Add 2 names without changing the median.
2. Add 2 names so the median decreases.

Names
Shawna Smith
Jonathan Newton
Roxanne Chang
Rebecca Brown
David Miller
Lois Lotz
Sharon Robinson

254 Math Fun!

ADDITIONAL ANSWERS

Name lengths may vary.
1. Check that students added one name that has a length less than the median and one name that has a length greater than the median.

2. Check that students added two names whose lengths are less than the median.

CHAPTER 13 REVIEW

EXAMPLES

• **Analyze graphs.**
(pages 240–241)
The number of fans in Grade 6 is 2.5 times the number in Grade 8.

• **Identify misleading graphs.**
(pages 242–243)
The broken scale makes Bob's hours look several times the others' hours.

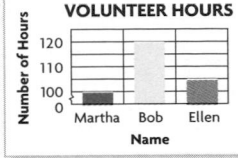

• **Make predictions from graphs**
(pages 244–245)
The trend is a decrease.

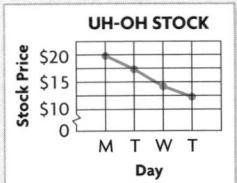

5. about $2.50
6. about $2.50

• **Find the mean, median, and mode.**
(pages 246–249)

9. 86.8; 87; none

data: 10, 20, 5, 15, 5   10. 18.8; 19; 19 and 20
mean: $(10 + 20 + 5 + 15 + 5) \div 5 = 11$
median: 10   mode: 5

• **Analyze box-and-whisker graphs.**
(pages 252–253)

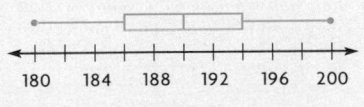

EXERCISES

For Exercises 1–2, use the graph at the left.
1. Describe the relationship between the height of the bars and the number of students who like comedies.
2. Which grade likes comedies the least? the most? **eighth; sixth**

1. **The greater the height, the greater the number of students who like comedy.**

For Exercises 3–4, use the graph at the left.
3. The bar for Bob is how many times as long as the bar for Martha? **5 times as long**
4. How can you change the graph so that it is not misleading? **by showing a scale from 0 to 120 with no break**

For Exercises 5–7, use the graph at the left.
5. About how much did the stock price decrease from Monday to Tuesday?
6. About how much did the stock price decrease from Tuesday to Wednesday?
7. What do you think the stock price will be on Friday? **about $10**

8. **VOCABULARY** The __?__ is the sum of a group of numbers, divided by the number of addends. **mean**

Find the mean, median, and mode.
9. 81, 76, 92, 98, 87   10. 19, 20, 20, 19, 16

For Exercises 11–13, use the graph at the left.
11. What is the median? **190**
12. What is the lower extreme? the upper extreme? **180; 200**
13. What is the lower quartile? the upper quartile? **186; 194**

Chapter 13 Review   255

USING THE PAGE

The Chapter 13 Review reviews analyzing graphs; identifying misleading graphs; making predictions from graphs; finding the mean, median, and mode; analyzing box-and-whisker graphs. Chapter objectives are provided, along with examples and practice exercises for each.

The Chapter 13 Review can be used as a review or a test. You may wish to place the Review in students' portfolios.

Assessment Checkpoint

For Performance Assessment in this chapter, see page 274, What Did I Learn?, items 3 and 7.

✓ *Assessment Tip*

Student Self-Assessment The How Did I Do? Survey helps students assess both what they have learned and how they learned it. Self-assessment helps students learn more about their own capabilities and develop confidence in themselves.

A self-assessment survey is available as a copying master in the *Teacher's Guide for Assessment*, p. 15.

See *Test Copying Masters*, pp. A48–A50 for the Chapter Test in **multiple-choice** format and pp. B198–B200 for **free-response** format.

CHAPTER 13 TEST, page 1

Choose the letter of the correct answer.

For questions 1–2, use the graph.

The schools in Northwest County conducted a survey of 100 students to find out if they liked the idea of changing the school schedule.

1. Which opinion forms the largest part of the graph?
A. Not Sure   B. Strongly Dislike
C. Dislike   D. Strongly Like

2. What did the schools learn from the survey?
A. Of the students surveyed, 25% liked the idea.
B. Of the students surveyed, 60% disliked the idea.
C. Opinions were evenly divided.
D. not here

For questions 3–5, use the graph below.

The graph shows how many of each type of candy bar were sold.

3. Which kind of candy bar was the most popular?
A. Chocolate   B. Nutty
C. Crunchy   D. Mixed

4. How does the number of Mixed candy bars sold compare with the number of Crunchy candy bars sold?
A. The same number of each was sold.
B. Twice as many Crunchy candy bars were sold.
C. Twice as many Mixed candy bars were sold.
D. not here

5. The students plan to sell Fruit-and-Nut candy bars for their next sale. They still want only four kinds of candy bars. Which candy bar should they stop selling?
A. Chocolate   B. Nutty
C. Crunchy   D. Mixed

6. Decide whether the graph below is misleading. If so, explain why.

A. No.
B. No; the bars are same widths.
C. Yes; the scale makes the price double that of Our Store.
D. Yes; the graph doesn't show which store has more records.

Form A • Multiple-Choice   A48   Go on. ▶

CHAPTER 13 TEST, page 2

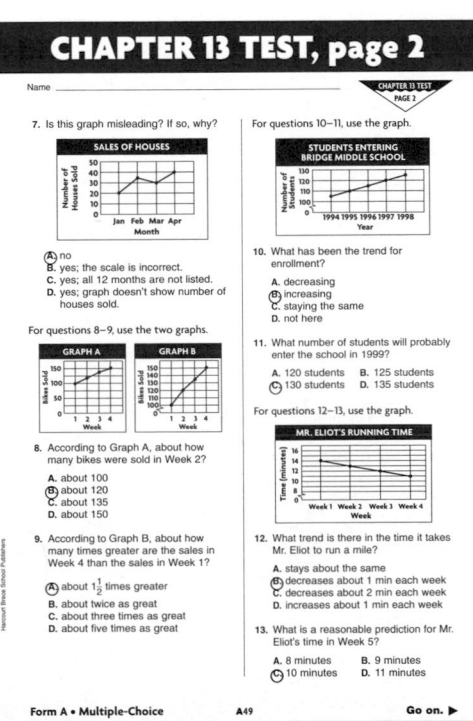

7. Is this graph misleading? If so, why?

A. no
B. yes; the scale is incorrect.
C. yes; all 12 months are not listed.
D. yes; graph doesn't show number of houses sold.

For questions 8–9, use the two graphs.

8. According to Graph A, about how many bikes were sold in Week 2?
A. about 100
B. about 120
C. about 135
D. about 150

9. According to Graph B, about how many times greater are the sales in Week 4 than the sales in Week 1?
A. about 1½ times greater
B. about twice as great
C. about three times as great
D. about five times as great

For questions 10–11, use the graph.

10. What has been the trend for enrollment?
A. decreasing
B. increasing
C. staying the same
D. not here

11. What number of students will probably enter the school in 1999?
A. 120 students   B. 125 students
C. 130 students   D. 135 students

For questions 12–13, use the graph.

12. What trend is there in the time it takes Mr. Eliot to run a mile?
A. stays about the same
B. decreases about 1 min each week
C. decreases about 2 min each week
D. increases about 1 min each week

13. What is a reasonable prediction for Mr. Eliot's time in Week 5?
A. 8 minutes   B. 9 minutes
C. 10 minutes   D. 11 minutes

Form A • Multiple-Choice   A49   Go on. ▶

CHAPTER 13 TEST, page 3

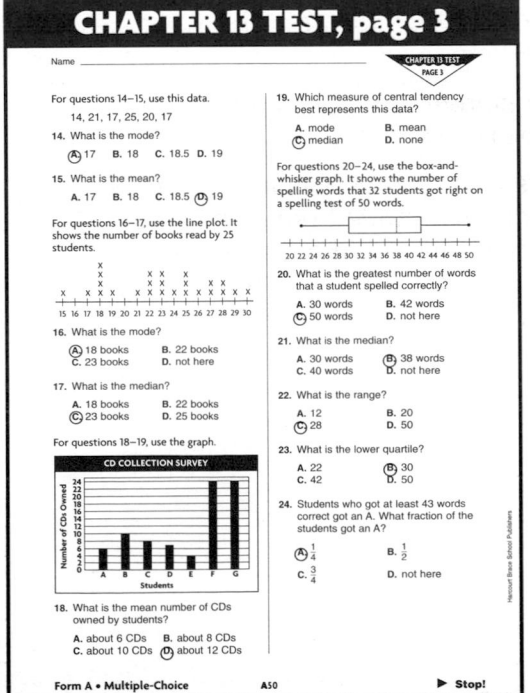

For questions 14–15, use this data.
14, 21, 17, 25, 20, 17

14. What is the mode?
A. 17   B. 18   C. 18.5   D. 19

15. What is the mean?
A. 17   B. 18   C. 18.5   D. 19

For questions 16–17, use the line plot. It shows the number of books read by 25 students.

16. What is the mode?
A. 18 books   B. 22 books
C. 23 books   D. not here

17. What is the median?
A. 18 books   B. 22 books
C. 23 books   D. 25 books

For questions 18–19, use the graph.

18. What is the mean number of CDs owned by students?
A. about 6 CDs   B. about 8 CDs
C. about 10 CDs   D. about 12 CDs

19. Which measure of central tendency best represents this data?
A. mode   B. mean
C. median   D. none

For questions 20–24, use the box-and-whisker graph. It shows the number of spelling words that 32 students got right on a spelling test of 50 words.

20. What is the greatest number of words that a student spelled correctly?
A. 30 words   B. 42 words
C. 50 words   D. not here

21. What is the median?
A. 30 words   B. 38 words
C. 40 words   D. not here

22. What is the range?
A. 12   B. 20
C. 28   D. 50

23. What is the lower quartile?
A. 22   B. 30
C. 42   D. 50

24. Students who got at least 43 words correct got an A. What fraction of the students got an A?
A. ¼   B. ½
C. ¾   D. not here

Form A • Multiple-Choice   A50   ▶ Stop!

255

ADDITIONAL ANSWERS
Lesson 13.4, page 249

10.

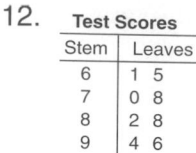

12.

Test Scores	
Stem	Leaves
6	1 5
7	0 8
8	2 8
9	4 6
10	0 0

22.

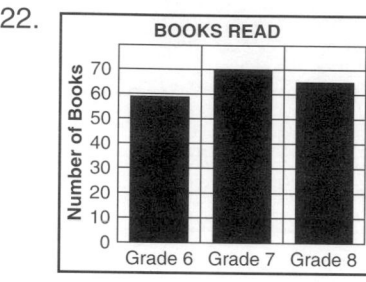

Chapter 13, Lab, page 251

Try This

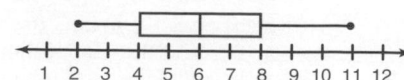

Lesson 13.5, page 253

Independent Practice

5.

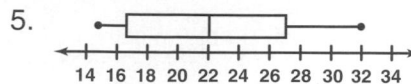

8. From the lower extreme to the lower quartile is 1 part; from the lower quartile to the median is 1 part; from the median to the upper quartile is 1 part; and from the upper quartile to the upper extreme is 1 part.

12.

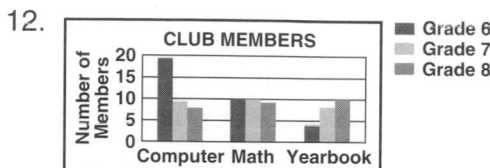

14

PLANNING GUIDE

Probability

BIG IDEA: The likelihood of an event occurring is measured by its probability, usually expressed as a fraction, decimal, or percent.

Introducing the Chapter p. 256		Team-Up Project • Which Item? p. 257	
OBJECTIVE	**VOCABULARY**	**MATERIALS**	**RESOURCES**
14.1 Problem Solving: Account for All Possibilities pp. 258-259 **Objective:** To use the strategy account for all possibilities to solve problems **1 DAY**			■ Reteach, ■ Practice, ■ Enrichment 14.1; More Practice, p. H62 Cultural Connection • Navajo Blanket Design, p. 270 Problem Solving Think Along, TR p.R1
14.2 Probability pp. 260-263 **Objective:** To find the probability of equally likely events **2 DAYS**	probability		■ Reteach, ■ Practice, ■ Enrichment 14.2; More Practice, p. H62 Cultural Connection • Navajo Blanket Design, p. 270
14.3 More on Probability pp. 264-265 **Objective:** To find the probability of events that are not equally likely **1 DAY**			■ Reteach, ■ Practice, ■ Enrichment 14.3; More Practice, p. H62 *Number Heroes • Probability*
Lab Activity: Simulations pp. 266-267 **Objective:** To use a simulation to model an experiment		FOR EACH PAIR: number cube numbered 1 to 6, calculator	**E-Lab • Activity 14** E-Lab Recording Sheets, p. 14
14.4 Experimental Probability pp. 268-269 **Objective:** To find the experimental probability of an event **2 DAYS**	experimental probability		■ Reteach, ■ Practice, ■ Enrichment 14.4; More Practice, p. H63

EXTENSION • Independent and Dependent Events, Teacher's Edition pp. 270A–270B

CHAPTER ASSESSMENT   Chapter 14 Review   p. 271

ASSESSMENT CHECKPOINT   Review • Chapter 11–14   pp. 272–273
Performance Assessment • Chapter 11–14   p. 274
Cumulative Review • Chapter 11–14   p. 275

NCTM CURRICULUM
STANDARDS FOR GRADES 5–8

- ✓ **Problem Solving**
  Lessons 14.1, 14.2, 14.3, 14.4
- ✓ **Communication**
  Lessons 14.1, 14.2, 14.3, 14.4
- ✓ **Reasoning**
  Lessons 14.1, 14.2, 14.3, 14.4
- ✓ **Connections**
  Lessons 14.1, 14.2, 14.3, 14.4
- ✓ **Number and Number Relationships**
  Lessons 14.1, 14.2, 14.3
- ✓ **Number Systems and Number Theory**
  Lessons 14.1, 14.2, 14.3, 14.4
- ✓ **Computation and Estimation**
  Lessons 14.1, 14.2, 14.3, 14.4
- ✓ **Patterns and Functions**
  Lesson 14.1
- ✓ **Algebra**
  Lesson 14.1
- ✓ **Statistics**
  Lesson 14.4, Lab Activity
- ✓ **Probability**
  Lessons 14.1, 14.2, 14.3, 14.4
- ✓ **Geometry**
  Lesson 14.2
- ✓ **Measurement**
  Lesson 14.2

ASSESSMENT OPTIONS

CHAPTER LEARNING GOALS

14A.1 To use the strategy account for all possibilities to solve problems

14A.2 To find the probability of events

14A.3 To find the experimental probability of an event

These goals for concepts, skills, and problem solving are assessed in many ways throughout the chapter. See the chart below for a complete listing of both traditional and informal assessment options.

Pretest Options
Pretest for Chapter Content, TCM pp. A51–A52, B201–B202
Assessing Prior Knowledge, TE p. 256; APK p. 16

Daily Assessment
Mixed Review, PE pp. 263, 269
Problem of the Day, TE pp. 258, 260, 264, 268
Assessment Tip and Daily Quiz, TE pp. 259, 263, 265, 267, 269

Formal Assessment
Chapter 14 Review, PE p. 271
Chapter 14 Test, TCM pp. A51–A52, B201–B202
Study Guide & Review, PE pp. 272–273
Test • Chapters 11–14, TCM pp. A53–A54, B203–B204
Cumulative Review, PE p. 275
Cumulative Test, TCM pp. A55–A58, B205–B208

Performance Assessment
Team-Up Project Checklist, TE p. 257
What Did I Learn?, Chapters 11–14, TE p. 274
Interview/Task Test, PA p. 89

Portfolio
Suggested work samples:
Independent Practice, PE pp. 262, 265, 269
Cultural Connection, PE p. 270

Student Self-Assessment
How Did I Do?, TGA p. 15
Portfolio Evaluation Form, TGA p. 24
Math Journal, TE pp. 256B, 266

Key
Assessing Prior Knowledge Copying Masters: APK
Test Copying Masters: TCM
Performance Assessment: PA
Teacher's Guide for Assessment: TGA
Teacher's Edition: TE
Pupil's Edition: PE

MATH COMMUNICATION

MATH JOURNAL
You may wish to use the following suggestions to have students write about the mathematics they are learning in this chapter:

PE pages 262, 268 • Talk About It

PE pages 259, 263, 265, 269 • Write About It

TE page 266 • Writing in Mathematics

TGA page 15 • How Did I Do?

In every lesson • record written answers to each Wrap-up Assessment Tip from the Teacher's Edition for test preparation.

VOCABULARY DEVELOPMENT
The terms listed at the right will be introduced or reinforced in this chapter. The boldfaced terms in the list are new vocabulary terms in the chapter.

So that students understand the terms rather than just memorize definitions, have them construct their own definitions and pictorial representations and record those in their Math Journals.

probability, pp. 260, 261, 264

experimental probability, p. 268

MEETING INDIVIDUAL NEEDS

RETEACH
BELOW-LEVEL STUDENTS

Practice Activities, TE Tab B
Does It Divide Evenly?
(Use with Lessons 14.1–14.4.)

Interdisciplinary Connection, TE p. 262

Literature Link, TE p. 256C
Skinny Bones, Activity 1
(Use with Lessons 14.1–14.3.)

Technology
Number Heroes • *Probability*
(Use with Lesson 14.3.)

E-Lab • Activity 14
(Use with Lab Activity.)

PRACTICE
ON-LEVEL STUDENTS

Practice Game, TE p. 256C
Celebrity Challenge
(Use with Lessons 14.1–14.4.)

Interdisciplinary Connection, TE p. 262

Literature Link, TE p. 256C
Skinny Bones, Activity 2
(Use with Lesson 14.4.)

Technology
Number Heroes • *Probability*
(Use with Lesson 14.3.)

E-Lab • Activity 14
(Use with Lab Activity.)

CHAPTER IDEA FILE

The activities on these pages provide additional practice and reinforcement of key chapter concepts. The chart above offers suggestions for their use.

LITERATURE LINK

Skinny Bones
by Barbara Parks

Alex, nicknamed Skinny Bones because of his size, is constantly ridiculed and has a hard time finding his niche with friends.

Activity 1—Alex was the only player on his team to need a size small uniform. Have students poll classmates to see what size shirt each would order, and then write the probability that a student chosen at random would need a size small.

Activity 2—Have students research the scores of the last 20 games of their favorite baseball teams and compute the experimental probability for each number of runs. Discuss with students whether the random number function on their calculator would be a good way to predict the next score for the team.

Activity 3—There are 9 players on a baseball team. Have students figure out the number of possible lineups. Discuss with students how this number would change if one of the players could play only a certain position, such as first base.

PRACTICE GAME

Celebrity Challenge

Purpose To practice finding probability

Materials shoe box, index cards, frequency tables (TR p. R92)

About the Game In groups of 4, each player writes his or her favorite celebrity's name on 2, 3, 4, or 5 index cards. Players predict how many times each name will be picked in 20 turns and record their predictions in the frequency table. Each player, in turn, draws a card, tallies the name in the frequency table, and replaces the card. After 20 draws, players score 1 point for each correct prediction. After 3 rounds the player with the most points wins.

Modalities Visual, Kinesthetic

Multiple Intelligences
Logical/Mathematical,
Visual/Spatial,
Bodily/Kinesthetic

Points Scored on Target				
Round 1				
Round 2				
Round 3				
Round 4	—	—	—	—

Practice Game, TE p. 256C
Celebrity Challenge
(Use with Lessons 14.1–14.4.)

Interdisciplinary Connection, TE p. 262

 **Literature Link,** TE p. 256C
Skinny Bones, Activity 3
(Use with Lessons 14.1–14.4.)

 Technology
Number Heroes • Probability
(Use with Lesson 14.3.)

E-Lab • Activity 14
(Use with Lab Activity.)

INTERDISCIPLINARY SUGGESTIONS

Connecting Probability

Purpose To connect probability to other subjects students are studying

You may wish to connect ***Probability*** to other subjects by using related interdisciplinary units.

Suggested Units

- Language Connection: The Language of Probability

- Science Connection: Genetics and Probability

- Sports Connection: Heads or Tails, Probability in Sports

- Sports Connection: Hitting the Target, Geometric Probability in Sharpshooting, Archery, and Darts

- Health Connection: Vaccinations Reduce the Probability of Contracting Disease

Modalities Auditory, Kinesthetic, Visual

Multiple Intelligences Mathematical/Logical, Bodily/Kinesthetic, Visual/Spatial

---

Technology

The purpose of using technology in the chapter is to provide reinforcement of skills and support in concept development.

E-Lab • Activity 14—Students use computer simulation software to determine the area of a plane figure through randomness and ratios.

Number Heroes • Probability—The activities on this CD-ROM provide students with opportunities to practice interpreting unequal probability.

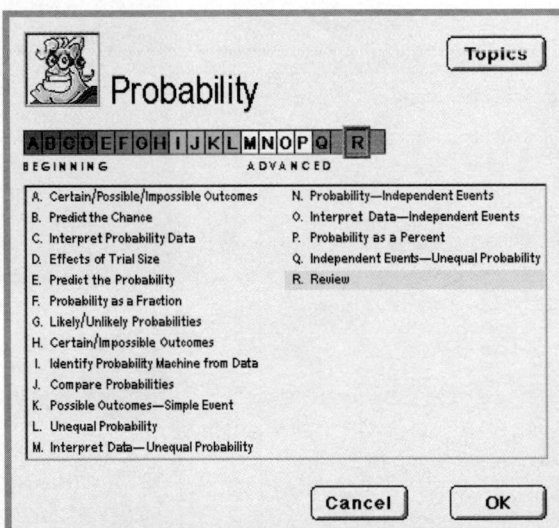

Each Number Heroes activity can be assigned by degree of difficulty. See Teacher's Edition page 265 for Grow Slide information.

Visit our web site for additional ideas and activities.
http://www.hbschool.com

Introducing the Chapter

Encourage students to talk about the probabilities of various events. Review the math content students will learn in the chapter.

WHY LEARN THIS?

The question "Why are you learning this?" will be answered in every lesson. Sometimes there is a direct application to everyday experiences. Other lessons help students understand why these skills are needed for future success in mathematics.

PRETEST OPTIONS

Pretest for Chapter Content See the *Test Copying Masters*, pages A51–A52, B201–B202 for a free-response or multiple-choice test that can be used as a pretest.

Assessing Prior Knowledge Use the copying master shown below to assess previously taught skills that are critical for success in this chapter.

Assessing Prior Knowledge Copying Masters, p. 16

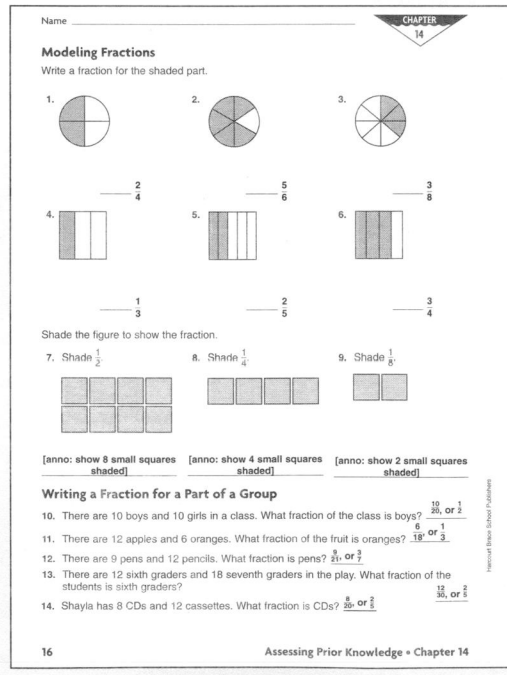

 *Troubleshooting*

If students need help	Then use
• MODELING FRACTIONS	• Key Skills, PE p. H14
• WRITING A FRACTION FOR PART OF A GROUP	• Key Skills, PE p. H14

CHAPTER 14 — PROBABILITY

LOOK AHEAD

In this chapter you will solve problems that involve

• the problem-solving strategy *account for all possibilities*

• using mathematical and experimental probability

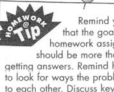 **SCHOOL-HOME CONNECTION**

You may wish to send to each student's family the *Math at Home* page. It includes an article that stresses the importance of communication about mathematics in the home, as well as helpful hints about homework and taking tests. The activity and extension provide extra practice that develops concepts and skills involving probability.

Probability: Chances Are

Your child is beginning a chapter that explores the idea of probability. Problems include finding probabilities by experimenting and finding probabilities by using a mathematical formula. Probability is an exciting topic to study. Invite your child to share with you the probability activities he or she does in school.

Here is an experiment in probability for you and your child to do. The Extension asks you to use your problem-solving skills to explain the outcomes in a given situation.

A Pointed Discovery

Place ten thumb tacks in a paper cup. Shake the cup. Spill the tacks onto a table. Record how many of the tacks landed point up and how many landed point down. Repeat the experiment 10 times. This means you will have results for 100 tacks. How many tacks in all landed point up? Write this as a fraction over 100, the total number of tacks. This is the experimental probability for a tack landing point up. Write the experimental probability for a tack landing point down. Compare the two fractions. Which event has the greater probability? Repeat the experiment and see if you get the same result. What might affect the way the tack lands?

Extension • Digit Dilemma

Using the digits 3, 0, 7, 4 only once in each number, how many three-digit numbers can you write? What is the probability that any three-digit numbers you pick can be divided by 5?

For Your Information

Did you know that the chance of being struck by lightning is 1 in 600,000? Does that mean it's likely to happen or not likely to happen?

Chance enters our lives in many ways—in sports, lotteries, politics, weather predictions, and insurance rates. An understanding of chance helps a person deal with these everyday situations.

Probability is the study of chance. Your child learns about probability in mathematics. As a result, he or she will be able to make useful predictions about many things in life. Probability problems provide many chances for students to use what they have learned about fractions, decimals, percents, and ratios.

VOCABULARY

probability—a comparison of the number of desired outcomes with the total number of outcomes
experimental probability—the number of times the desired event happens compared with the total number of times the activity is done

14

Math at Home

Team-Up Project

Which Item?

If you randomly select one item from a mixture, does each item have the same chance of being selected?

YOU WILL NEED: animal crackers or colored candy or marbles

Plan
- Work with a partner to count the number of each type of item in a mixture. Find the likelihood of selecting each item.

Act
- Find a mixture. Examples might be a box of animal crackers, a bag of different colored candy, or a package of marbles.
- List each item in the mixture.
- Make a tally table to show the number of each item in the mixture.
- Decide if each item has the same chance of being randomly selected.

Share
- Show your mixture and tally table to the class and explain the likelihood of randomly selecting each item.

Gumballs

blue	ᚸᚸ ᚸᚸ ᚸᚸ III	= 23
green	ᚸᚸ ᚸᚸ ᚸᚸ ᚸᚸ ᚸᚸ	= 25
orange	ᚸᚸ ᚸᚸ ᚸᚸ ᚸᚸ II / ᚸᚸ ᚸᚸ ᚸᚸ II	= 42
red	ᚸᚸ I	= 11
purple	ᚸᚸ ᚸᚸ ᚸᚸ IIII	= 19

DID YOU
- ✔ List each item in the mixture?
- ✔ Make a tally table?
- ✔ Decide if each possibility was equally likely?

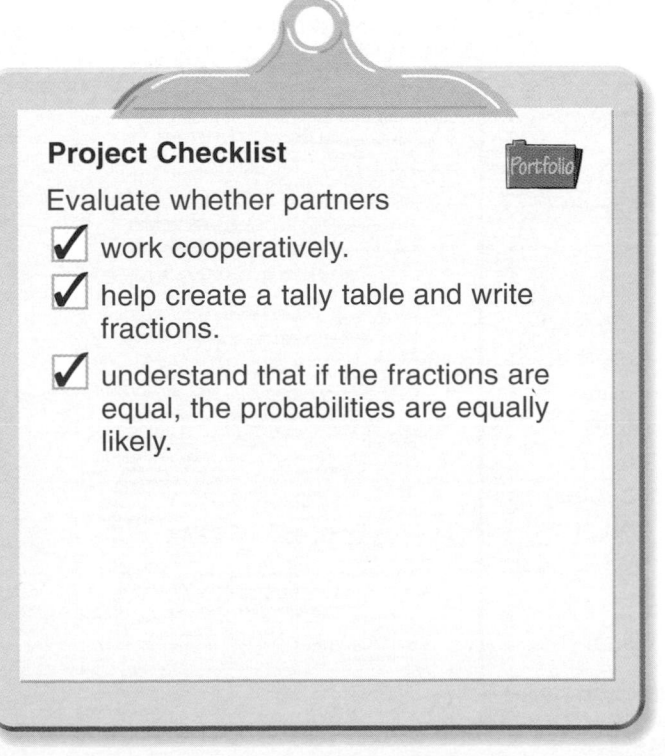

Project Checklist

Portfolio

Evaluate whether partners
- ✔ work cooperatively.
- ✔ help create a tally table and write fractions.
- ✔ understand that if the fractions are equal, the probabilities are equally likely.

Using the Project

Accessing Prior Knowledge making tally tables, writing fractions

Purpose to determine whether each item in a mixture is equally likely to be selected if one item is chosen from the mixture

GROUPING partners
MATERIALS mixtures

Managing the Activity You may want to make up mixtures or provide photos of mixtures like the one on the student page.

WHEN MINUTES COUNT

A shortened version of this activity can be completed in 15–20 minutes. The student page shows a tally table for selecting a colored gumball. Ask students to write fractions and decide on the probability of selecting each color of gumball.

A BIT EACH DAY

Day 1: Students select partners and choose a mixture to examine.

Day 2: Students set up tally sheets and figure out all of the possible items in the mixture.

Day 3: Students tally the number of each item in the mixture.

Day 4: Students write fractions and determine if each possibility is equally likely.

Day 5: Students share their work.

HOME HAPPENINGS

Ask students to watch for examples of mixtures in their everyday lives. Examples might include letters in alphabet soup, flowers in a vase, pencils and pens in a cup, and so on.

ORGANIZER

OBJECTIVE To use the strategy *account for all possibilities* to solve problems

ASSIGNMENTS Practice/Mixed Applications
- Basic 1–12
- Average 1–12
- Advanced 1–12

PROGRAM RESOURCES
Reteach 14.1 Practice 14.1 Enrichment 14.1
Problem-Solving Think Along, TR p. R1

PROBLEM OF THE DAY

Transparency 14

How many different four-digit numbers can you write using the digits 2, 8, 5 and 4 in each number?
24 four-digit numbers

Solution Tab A, p. A14

QUICK CHECK

Regroup factors to use mental math to find the product.

1. $25 \times 7 \times 4$ $4 \times 25 \times 7 = 700$
2. $5 \times 34 \times 2$ $2 \times 5 \times 34 = 340$
3. $4 \times 6 \times 10$ $6 \times 10 \times 4 = 240$

1 MOTIVATE

Have three students come to the front of the class. Ask:

- **How many different ways can we line up these three students?**

Have students model each possibility while the rest of the class records the outcomes. Six ways: *ABC, ACB, BCA, BAC, CAB, CBA*

2 TEACH/PRACTICE

Have a volunteer read the problem presented at the top of p. 258. Then discuss the tree diagram with the class.

14.1

WHAT YOU'LL LEARN
How to use the strategy *account for all possibilities* to solve problems

WHY LEARN THIS?
To find the number of possible choices, such as the number of choices for a yogurt cone

Problem Solving
- Understand
- Plan
- Solve
- Look Back

Business Link

When Ben & Jerry's® Ice Cream Company began making and selling its own Low Fat Frozen Yogurt, company sales increased from $77,000,000 to $97,000,000 in one year. How much did the company's sales increase?

$20,000,000

PROBLEM-SOLVING STRATEGY
Account for All Possibilities

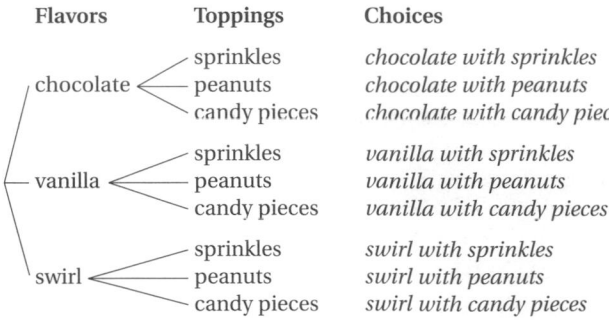

At the Frozen Treat yogurt shop, Jorge wants to buy a yogurt cone. He can have chocolate, vanilla, or swirl yogurt, and sprinkles, peanuts, or candy pieces for the topping. How many different yogurt cones are possible?

☑ **UNDERSTAND**   What are you asked to find?
the number of different cones possible
What information is given?
There are three flavors of yogurt and
☑ **PLAN**   What strategy will you use?   three toppings.

You can use the strategy *account for all possibilities* to find the number of possible choices.

☑ **SOLVE**   How will you solve the problem?

One way to find the number of possible choices is to make a tree diagram.

Flavors	Toppings	Choices
chocolate	sprinkles	*chocolate with sprinkles*
	peanuts	*chocolate with peanuts*
	candy pieces	*chocolate with candy pieces*
vanilla	sprinkles	*vanilla with sprinkles*
	peanuts	*vanilla with peanuts*
	candy pieces	*vanilla with candy pieces*
swirl	sprinkles	*swirl with sprinkles*
	peanuts	*swirl with peanuts*
	candy pieces	*swirl with candy pieces*

There are 9 choices.

Another way to find the number of choices Jorge has is to multiply the flavors and the toppings.

Flavors		Toppings		
3	×	3	=	9 choices

☑ **LOOK BACK**   What other strategy could you use to solve the problem? **act it out**

What if . . . Jorge could have one of 4 toppings with any of the 3 flavors? How many different cones would be possible?
12 cones

IDEA FILE

ALTERNATIVE TEACHING STRATEGY

MATERIALS: 6 blocks of different colors

Have students work in groups and use the strategy *act it out* to solve the problem presented at the top of p. 258. Ask students to designate a color to represent each flavor and each topping. Students should continue placing a topping block on a flavor block until they have shown all possible choices.

Modalities Kinesthetic, Visual

RETEACH 14.1

Name _____

LESSON 14.1

Problem-Solving Strategy

Account for All Possibilities

Jill is choosing an outfit to wear. She can wear tan pants, blue pants, black pants, or a skirt, with either a white sweater or a red sweater. How many choices does she have?

UNDERSTAND

1. What are you asked to find? ___ total number of outfits

2. What information is given? ___ She has 3 pairs of pants
   and 1 skirt that go with 2 different tops.

PLAN

3. What strategy will you use? The strategy *account for all possibilities* can be used to find the total number of outfit choices.

SOLVE

4. How can you use a tree diagram to solve the problem?
   List the various pants and skirts, and match each with
   both tops. Then count the total number of choices.

5. Make a tree diagram to find the number of choices Jill has.

BOTTOMS	TOPS	CHOICES
Tan pants	White sweater	Tan pants, white sweater
	Red sweater	Tan pants, red sweater
Blue pants	White sweater	Blue pants, white sweater
	Red sweater	Blue pants, red sweater
Black pants	White sweater	Black pants, white sweater
	Red sweater	Black pants, red sweater
Skirt	White sweater	Skirt, white sweater
	Red sweater	Skirt, red sweater

6. What is another way to find her total choices? Multiply.

7. Use this second strategy to find her total number of choices.
   4 bottoms × 2 tops = 8 choices

LOOK BACK

8. What if Jill had 4 different tops to choose from? How many choices would she have?
   16 choices

Use with text pages 258–259.   TAKE ANOTHER LOOK R61

PRACTICE

Use the strategy *account for all possibilities* to solve.

1. Matthew is buying a new car. He can have a white or blue exterior, with a red, white, or gray interior. How many choices does he have in all? **6 choices**

2. Ms. Clark is making a dental appointment. It can be on Monday, Tuesday, or Wednesday, at 10:00 A.M., 10:30 A.M., or 2:00 P.M. Find the total number of choices. **9 choices**

3. Michelle is buying a new outfit. She can buy a blouse that is a solid color, a stripe, or a print, and a skirt that is short, long, or wraparound. How many choices does she have? **9 choices**

4. The sixth-grade debate team is having lunch. The students can have pizza, a garden salad, macaroni, a tuna sub, or a taco, with milk, fruit juice, or soda. Find the total number of choices. **15 choices**

MIXED APPLICATIONS   CHOOSE

Choose a strategy and solve. **Choices of strategies will vary.**

Problem-Solving Strategies
- Make a Table
- Draw a Diagram
- Account for All Possibilities
- Write a Proportion
- Find a Pattern
- Guess and Check

5. Julia has been working on a jigsaw puzzle for an hour and has put 90 of the 360 pieces together. How much longer will it take her to finish the puzzle if she continues to put 90 pieces together every hour? **3 hr**

6. Amy is going to buy a music CD. She can buy a single or an album of rock-and-roll or country music. How many choices does she have? **4 choices**

7. The dance company performs in groups of 8 and 10. There are 64 dancers in all. There are more groups of 10 than groups of 8. How many groups of each size are there? **4 groups of 10, 3 groups of 8**

8. On Saturday the museum sold 100 more than twice the number of tickets it sold on Friday. On Saturday 500 tickets were sold. How many tickets were sold on Friday? **200 tickets**

9. Melissa will take 1-hr swimming lessons 3 times a week. She can pay $12 in advance each week, or she can pay $5 an hour. Which payment plan will cost her less? **$12 each week**

10. Elena is planting a garden. She knows that 50 plants cost $20. She is waiting for a sale so she can pay only $\frac{3}{4}$ of that price. How much is she willing to pay? **$15**

11. For 30 days Leon works every third day. He starts on Saturday. What day of the week is his last day? **Friday**

12. **WRITE ABOUT IT** Write a problem using the strategy *account for all possibilities*. Explain how to solve the problem. **Answers will vary.**

Portfolio

MORE PRACTICE Lesson 14.1, page H62

259

CRITICAL THINKING **Why can we use multiplication to find the number of branches in the tree diagram?** Each of the 3 flavors has 3 branches.

PRACTICE

You may want to use the first problem of the Practice as a whole-class activity. Ask partners to work through the problem and to complete the Problem-Solving Think Along, TR p. R1.

MIXED APPLICATIONS

These problems provide students an opportunity to use a variety of problem-solving strategies.

3 WRAP UP

Use these questions to help students focus on the big ideas of the lesson.

- **Why is a tree diagram a good way to account for all possibilities?** It lets you see all possible combinations. By counting all the branches, you can find how many possibilities there are.

- **If you know the number of ways that each choice can occur, how can you find the total number of choices *without* constructing a tree diagram?** Multiply the number of ways the choices can occur by the number of choices.

✔ *Assessment Tip*

Have students write the answer.

- **Use a tree diagram to show all possible three-letter combinations using the letters *A, E,* and *T.***

- **How many three-letter combinations are there?** 6 combinations

- **Which combinations form actual words?** *ATE, EAT, TEA*

DAILY QUIZ

Solve each problem by using a tree diagram. Check students' diagrams.

1. For dessert, Sam has a choice of two kinds of pie (apple and peach) and three toppings (whipped cream, vanilla ice cream, and chocolate ice cream). How many different desserts are possible? 6 desserts

2. After school, Georgia has a choice of three foreign-language clubs and four sports clubs. Georgia wants to join two clubs. How many possible choices does she have? 12 choices

---

PRACTICE 14.1

Name _____   LESSON 14.1

Problem-Solving Strategy

Account for All Possibilities

Use the strategy *account for all possibilities* to solve.

1. Tina is making an appointment for a haircut. It can be on Tuesday, Wednesday, or Friday at 11:00 A.M., 1:00 P.M., 2:30 P.M., or 4:00 P.M. Find the total number of choices.

_____ **12 choices**

2. Rod is choosing an outfit. He can wear tan, blue, or black pants with a white, green, or print shirt. Find the total number of choices.

_____ **9 choices**

3. Larry is buying a new car. He can have a silver, black, white, or blue exterior, with a tan, gray, white, or red interior. How many choices does he have in all?

_____ **16 choices**

4. The cafeteria is offering a hamburger, garden salad, turkey sandwich, tuna sub, or pizza, with either milk or juice. How many choices do the students have?

_____ **10 choices**

Mixed Applications

Solve.   CHOOSE A STRATEGY

- Account for All Possibilities
- Find a Pattern
- Guess and Check

Choices of strategies will vary.

5. For three weeks, Ted works every fourth day. He starts on Tuesday. What day of the week is his last day?

_____ **Monday**

6. On Friday, the theater sold three times as many tickets as it sold on Wednesday. On Friday, 225 tickets were sold. How many tickets were sold on Wednesday?

_____ **75 tickets**

7. Ann is choosing an outfit. She can wear shorts, a skirt, blue pants, or tan pants with a white sweater, red sweater, or striped blouse. How many choices does she have?

_____ **12 choices**

8. Five students are standing in a line. Jon is in front of Terri. Joan is behind Carol. Zack is in front of Jon. Carol is fourth in line. In what order are the students standing?

_____ **Zack, Jon, Terri, Carol, Joan**

Use with text pages 258–259.   **ON MY OWN** P61

ENRICHMENT 14.1

Name _____   LESSON 14.1

Polygon Percents

Express each area as a percent.

	SHADED AREA	UNSHADED AREA
1. Square *ACEG*	25%	75%
2. Rectangle *ACDH*	50%	50%
3. Trapezoid *BCDH*	$66\frac{2}{3}$%	$33\frac{1}{3}$%
4. Pentagon *BCDFH*	40%	60%
5. Rectangle *HDEG*	0%	100%
6. Pentagon *BCEFH*	$33\frac{1}{3}$%	$66\frac{2}{3}$%
7. Triangle *BHD*	100%	0%
8. Triangle *FHB*	50%	50%
9. Rectangle *CEFB*	25%	75%
10. Square *HIFG*	0%	100%
11. Square *ABIH*	50%	50%
12. Triangle *BID*	100%	0%

Use with text pages 258–259.   **STRETCH YOUR THINKING** E61

259

14.2

GRADE 5	GRADE 6	GRADE 7
Compare probabilities of simple events.	Find the probability of equally likely events.	Find the probability of an event.

ORGANIZER

OBJECTIVE To find the probability of equally likely events

VOCABULARY *New* **probability**

ASSIGNMENTS **Independent Practice**
- Basic 1–21, even; 22–27; 28–36
- Average 1–27; 28–36
- Advanced 1–14; 22–27; 28–36

PROGRAM RESOURCES
Reteach 14.2 Practice 14.2 Enrichment 14.2
Cultural Connection • Navajo Blanket Design, p. 270

PROBLEM OF THE DAY

Transparency 14

Betty's dentist does not work on weekends. What is the probability that Betty's next appointment is on Monday? $\frac{1}{5}$ on Saturday? $\frac{0}{5}$ on Tuesday, Wednesday, or Thursday? $\frac{3}{5}$

Solution Tab A, p. A14

QUICK CHECK
Tell whether the fraction is less than $\frac{1}{2}$, equal to $\frac{1}{2}$, or greater than $\frac{1}{2}$.

1. $\frac{3}{6}$ equal to $\frac{1}{2}$
2. $\frac{3}{7}$ less than $\frac{1}{2}$
3. $\frac{2}{3}$ greater than $\frac{1}{2}$
4. $\frac{5}{8}$ greater than $\frac{1}{2}$

1 MOTIVATE

Ask students to discuss the meaning of the expression *chances are 50-50.*
Possible answer: The outcomes are equally likely or equally probable events.

Have volunteers complete the sentence starter, "It is equally likely that. . . ."
Students may fill in with such clauses as, "it will rain today" or "I will go to the beach" or "our team will win the pennant."

260

14.2

WHAT YOU'LL LEARN
How to find the probability of equally likely events

WHY LEARN THIS?
To understand your chances of winning when you play games with numbers or spinners

WORD POWER
probability

teen times

Many of the things you do have a high probability of success. Every time you get on a bus, the probability is high that you will arrive safely at your destination. The probability is high that you will graduate from high school. In fact, most of the routine things you do have high probabilities of success.

260 Chapter 14

Probability

You use the ideas of probability in everyday life. When you have a lengthy chore to do, you may think, "I probably won't finish this today." While playing basketball, you may say, "She has a 50-50 chance of making that basket."

These expressions indicate your understanding of probability. **Probability**, P, is a comparison of the number of favorable outcomes to the number of possible outcomes.

The mathematical probability of an event can be written as a fraction.
$$P = \frac{\text{number of favorable outcomes}}{\text{number of possible outcomes}}$$

EXAMPLE 1 Each letter of the word *MATH* is written on a card and placed in a bag. Find P(A), the probability of choosing an A.

M A T H

1 favorable outcome: *A* — *List the favorable outcomes.*

4 possible outcomes: *M, A, T, H* — *List all the possible outcomes.*

$P(A) = \frac{\text{number of favorable outcomes}}{\text{number of possible outcomes}} = \frac{1}{4}$ — *Write the probability as a fraction.*

So, the probability of choosing an A is $\frac{1}{4}$.

- How many favorable outcomes are there for choosing *T* or *H*? **2**

You can use a number cube to find favorable outcomes.

EXAMPLE 2 You roll a number cube numbered 1 to 6. Find P(1), P(4), and P(1 or 4).

$P(1) = \frac{1}{6}$ ← 1 choice out of 6

$P(4) = \frac{1}{6}$ ← 1 choice out of 6

$P(1 \text{ or } 4) = \frac{1+1}{6}$ ← 2 choices out of 6

$= \frac{2}{6} = \frac{1}{3}$ *Write in simplest form.*

- What is the probability of rolling a number greater than 2?
$\frac{4}{6}$, or $\frac{2}{3}$

RETEACH 14.2

Name _____
LESSON 14.2

Probability

To find the probability, make a fraction.

Probability: $\frac{\text{numerator}}{\text{denominator}}$ ← number of favorable outcomes / ← number of possible outcomes

1. You have a spinner numbered 1 to 5. What is the probability of spinning a 2?

   a. How many possible outcomes are there? List them.
   _____ 5; 1, 2, 3, 4, 5 _____

   b. How many favorable outcomes are there? ___1___

   c. You can write as a fraction the probability of spinning a 2.
   $P(2) = \frac{\text{number of favorable outcomes}}{\text{number of possible outcomes}}$
   Write the probability as a fraction. $\frac{1}{5}$

2. Each letter of the word *FRACTION* is written on a card and placed in a bag. What is the probability of choosing a *C*?
   a. How many possible outcomes are there? List them.
   _____ 8; F, R, A, C, T, I, O, N _____

   b. How many favorable outcomes are there? List them. 1; C

   c. You can write the probability of choosing a C as a fraction.
   $P(C) = \frac{\text{number of favorable outcomes}}{\text{number of possible outcomes}}$
   Write the probability as a fraction. $\frac{1}{8}$

3. Each letter of the word *NUMBER* is written on a card and placed in a bag. What is the probability of choosing a vowel?
   a. How many possible outcomes are there? List them.
   _____ 6; N, U, M, B, E, R _____

   b. How many favorable outcomes are there? List them.
   _____ 2; U, E _____

   c. You can write the probability of choosing a vowel as a fraction.
   $P(\text{vowel}) = \frac{\text{number of favorable outcomes}}{\text{number of possible outcomes}}$
   Write the probability as a fraction in simplest form. $\frac{1}{3}$

R62 **TAKE ANOTHER LOOK**
Use with text pages 260–263.

PRACTICE 14.2

Name _____
LESSON 14.2

Probability

Vocabulary

1. What is probability? _____ Probability is a comparison of the number of favorable outcomes and the number of possible outcomes.

For Exercises 2–5, use the spinner at the right.

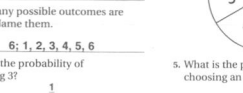

2. How many favorable outcomes are there for choosing 3? ___1___

3. How many possible outcomes are there? Name them.
   6; 1, 2, 3, 4, 5, 6

4. What is the probability of choosing 3? $\frac{1}{6}$

5. What is the probability of choosing an even number? $\frac{3}{6}$, or $\frac{1}{2}$

The letters H, O, L, I, D, A, and Y are put in a bag. Find each probability.

6. P(Y) $\frac{1}{7}$
7. P(H or L) $\frac{2}{7}$
8. P(A, I, or O) $\frac{3}{7}$

For Exercises 9–12, use the rectangle at the right. Find each probability.

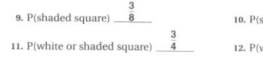

9. P(shaded square) $\frac{3}{8}$
10. P(striped square) $\frac{1}{4}$
11. P(white or shaded square) $\frac{3}{4}$
12. P(white square) $\frac{3}{8}$

Mixed Applications

13. If you have a cube numbered 6 through 11, what is the probability you will roll an odd number? $\frac{1}{2}$

14. There are 57 students in an after-school club. There are twice as many boys as girls in the club. How many girls are in the club? 19 girls

P62 **ON MY OWN**
Use with text pages 260–263.

Sometimes outcomes, or events, cannot occur. For example, what is the probability there will be eight days in the week to come? **0**

The diagram shows that the probability of an event ranges from 0, or impossible, to 1, or certain. A probability is always 0, 1, or a fraction between 0 and 1.

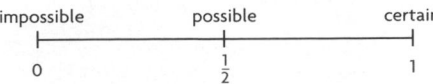

impossible   possible   certain

0        $\frac{1}{2}$ 1

EXAMPLE 3   Look at the spinner at the right to find each probability.

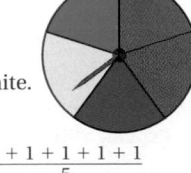

A. $P(\text{white}) = \frac{0}{5}$. None of the sections are white.

B. $P(\text{blue or green}) = \frac{1+1}{5} = \frac{2}{5}$

C. $P(\text{green, pink, brown, blue, or yellow}) = \frac{1+1+1+1+1}{5}$
$= \frac{5}{5} = 1$

GUIDED PRACTICE

For Exercises 1–4, use the numbered cards.

| 4 | 6 | 1 | 7 | 2 | 8 | 3 | 9 | 0 |

1. List the favorable outcomes for choosing an odd number. **1, 7, 3, 9**

2. How many possible outcomes are there? **9**

3. Find P(odd). $\frac{4}{9}$

4. Find P(6 or 0). $\frac{2}{9}$

5. You are given a choice of three answers to a test question. If you don't know the answer, what is the probability of guessing the correct answer? $\frac{1}{3}$

Geometric Probability

You can use the areas of geometric figures to find probabilities.

In a certain dart game, the target looks like the square at the right. The total area of the target is 144 in.² To find the probability of hitting the blue section with a dart, compare the areas. Assume the dart hits the target.

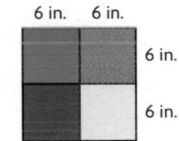
6 in.   6 in.
6 in.
6 in.

$P(\text{blue}) = \dfrac{\text{area of blue section}}{\text{area of target}} = \dfrac{6 \times 6}{144} = \dfrac{36}{144} = \dfrac{1}{4}$

261

REMEMBER:

The area is the number of square units needed to cover a surface.
See page H25.

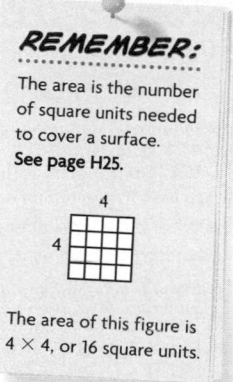

4
4

The area of this figure is 4×4, or 16 square units.

Discuss the information about how probability is used in everyday life. Ask students to give examples of times when they have used probability. As you discuss Examples 1–3, explain that an event can be one outcome or a set of outcomes and that *favorable outcome* and *success* have the same meaning.

Discuss the probability number scale shown on p. 261. Explain that this scale shows all probabilities from 0 for an impossible event (an event that cannot occur) to 1 for a certain event (an event that will definitely happen).

Discuss the concept of and model for geometric probability on pp. 261–262. Review with students the formulas for finding the area of a square and the area of a rectangle: $A = s^2$ and $A = lw$, respectively.

The Talk About It questions on p. 262 are intended to help students clarify the relationship between probability and area.

CRITICAL THINKING   Which of these probabilities indicates a highly-probable event? Which indicates an unlikely event?

a. $\frac{1}{32}$ unlikely   **b.** $\frac{63}{64}$ highly probable

c. $\frac{7}{8}$ highly probable   **d.** $\frac{2}{72}$ unlikely

GUIDED PRACTICE

Discuss the exercises as a class to check students' conceptual understanding of probability.

ADDITIONAL EXAMPLES

Example 1, p. 260

Each letter of the word *HOLIDAY* is written on a card and placed in a bag. What is the probability of choosing a vowel? $\frac{3}{7}$

Example 2, p. 260

You roll a number cube labeled 1, 3, 5, 7, 8, 9.

Find P(odd number), P(even number), and P(5). $P(\text{odd number}) = \frac{5}{6}$, $P(\text{even number}) = \frac{1}{6}$, $P(5) = \frac{1}{6}$

Example 3, p. 261

You roll a number cube labeled 3, 6, 9, 12, 15, 18. What is the probability of rolling a number that is divisible by 3? 1

What is the probability of rolling a number that is divisible by 10? 0

What is the probability of rolling a number less than 10? $\frac{1}{2}$

261

ENRICHMENT 14.2

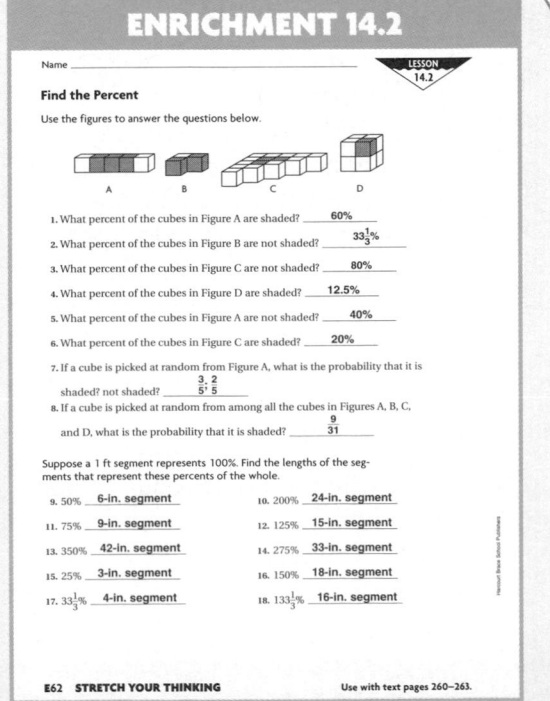

Name _____

LESSON
14.2

Find the Percent

Use the figures to answer the questions below.

A   B   C   D

1. What percent of the cubes in Figure A are shaded? **60%**
2. What percent of the cubes in Figure B are not shaded? **$33\frac{1}{3}$%**
3. What percent of the cubes in Figure C are not shaded? **80%**
4. What percent of the cubes in Figure D are shaded? **12.5%**
5. What percent of the cubes in Figure A are not shaded? **40%**
6. What percent of the cubes in Figure C are shaded? **20%**
7. If a cube is picked at random from Figure A, what is the probability that it is shaded? not shaded? **$\frac{3}{5}$, $\frac{2}{5}$**
8. If a cube is picked at random from among all the cubes in Figures A, B, C, and D, what is the probability that it is shaded? **$\frac{9}{31}$**

Suppose a 1 ft segment represents 100%. Find the lengths of the segments that represent these percents of the whole.

9. 50% **6-in. segment**
10. 200% **24-in. segment**
11. 75% **9-in. segment**
12. 125% **15-in. segment**
13. 350% **42-in. segment**
14. 275% **33-in. segment**
15. 25% **3-in. segment**
16. 150% **18-in. segment**
17. $33\frac{1}{3}$% **4-in. segment**
18. $133\frac{1}{3}$% **16-in. segment**

E62 STRETCH YOUR THINKING   Use with text pages 260–263.

INDEPENDENT PRACTICE

These exercises check students' ability to identify the number of favorable outcomes, the number of possible outcomes, and the probability of a certain outcome occurring.

What is the probability that a dart will hit a blue section of this dart target?
P(blue) = $\frac{5}{9}$

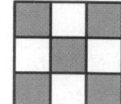

EXAMPLE 4 If a randomly thrown dart hits the target at the right, what is the probability of it hitting the red section?

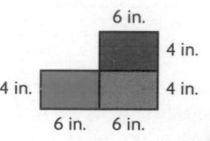

6 in.
4 in.
4 in.
4 in.
6 in.   6 in.

$4 \times 6 = 24$
area of red section: 24 in.^{2} *Find the areas of the red section and the target.*
$3 \times (4 \times 6) = 72$
area of target: 72 in.^{2}

P(red) = $\frac{24}{72} = \frac{1}{3}$ *Find the probability of hitting the red section.*

So, the probability of hitting the red section is $\frac{1}{3}$.

Talk About It

• Look at the target in Example 4. If you threw a dart and it hit the target, would you have a greater chance of hitting the red section than the green section? Explain. **No. The areas are the same.**
• Suppose you wanted the probability of a dart hitting any section to be $\frac{1}{5}$. How many sections would you have on the target? **5**

• 🔍 **CRITICAL THINKING** What if the probability of a randomly thrown dart hitting red is $\frac{1}{2}$ and the probability of hitting green is $\frac{1}{4}$? What do you know about the areas of those two sections on the target? **The red area is twice as great as the green area.**

INDEPENDENT PRACTICE

For Exercises 1–3, use the spinner at the right.

1. How many favorable outcomes are there for choosing 1? **1**
2. How many possible outcomes are there? Name them. **4; 1, 2, 3, and 4**
3. What is the probability of choosing 1? $\frac{1}{4}$

A number cube is numbered 2, 4, 6, 8, 10, and 12. Find each probability.

4. P(2) $\frac{1}{6}$ 5. P(8) $\frac{1}{6}$ 6. P(4 or 10) $\frac{1}{3}$ 7. P(a number less than 8) $\frac{1}{2}$

The letters *E, D, U, C, A, T, I, O,* and *N* are put in a bag. Find each probability.

8. P(*A*) $\frac{1}{9}$ 9. P(*E* or *T*) $\frac{2}{9}$ 10. P(*S*) **0**   11. P(*A, C, D, E, O,* or *U*) $\frac{2}{3}$

IDEA FILE

INTERDISCIPLINARY CONNECTION

One unit you can use to relate probability to other areas of study is The Language of Probability. In this unit, have students

• make a list of vocabulary words related to probability.

• use each term in a sentence to illustrate its meaning.

Modality Visual

SCIENCE CONNECTION

Gregor Mendel (1822–1884), an Austrian monk, applied concepts of probability to the science of genetics. When he crossbred garden pea plants having round-pea genes with plants having wrinkled-pea genes, the second-generation offspring all had round peas. When these plants were bred with each other, $\frac{3}{4}$ of the third-generation offspring had round peas and $\frac{1}{4}$ had wrinkled peas.

• According to these results, what is the probability of getting a plant with wrinkled peas in the third generation? $\frac{1}{4}$

Modality Auditory

MEETING INDIVIDUAL NEEDS

For Exercises 12–14, use the rectangle at the right. A point on the figure is chosen at random. What is the probability that it has this color?

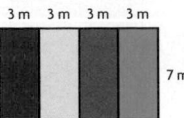

3 m 3 m 3 m 3 m
7 m

12. red $\frac{1}{4}$ **13.** green $\frac{1}{4}$ **14.** blue or green $\frac{1}{2}$

For Exercises 15–18, use the figure at the right. Find each probability.

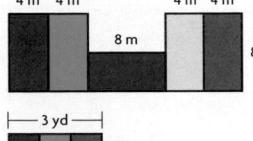

4 m 4 m   4 m 4 m
8 m
8 m

15. P(yellow) $\frac{1}{5}$ **16.** P(white) **0**   **17.** P(red or green) $\frac{2}{5}$

18. P(blue, green, red, purple, or yellow) **1**

For Exercises 19–21, use the figure at the right. Find each probability.

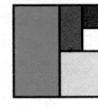

3 yd
2 yd

19. P(red) $\frac{1}{6}$ **20.** P(blue or green) $\frac{1}{3}$ **21.** P(yellow, black, or green) $\frac{1}{2}$

Problem-Solving Applications

22. There are 3 desserts on the menu, but one is not available. What is the probability that in making a random choice you choose that dessert? $\frac{1}{3}$

23. **CRITICAL THINKING** In a new board game, the probability of landing on each section of the board is $\frac{1}{12}$. How many sections are on the board? **12 sections**

24. Katie has a choice of 6 chores. She draws to see which chore she has for this week. What is the probability she will draw the same chore as last week? $\frac{1}{6}$

25. If you have a cube numbered 1 to 6, what is the probability you will roll a number less than 7? **1**

26. Georgia is coloring a poster. She has 1 blue, 1 green, 1 purple, and 1 yellow marker in a box. She chooses a marker without looking. What is the probability of choosing a green marker? $\frac{1}{4}$

27. **WRITE ABOUT IT** Explain the difference between a favorable outcome and a possible outcome. Give an example. **Favorable: a selected event; possible: any event; examples will vary.**

Mixed Review

Write a fraction that represents the part of the figure that is

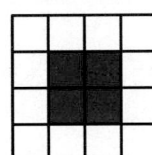

28. green $\frac{1}{2}$ **29.** yellow $\frac{1}{4}$ **30.** red $\frac{1}{16}$ **31.** blue $\frac{1}{8}$

Determine the median, mode, or mean.

32. median: 1, 2, 3, 4, 5, 6 **3.5**   **33.** mode: 4, 5, 10, 4, 2, 11 **4**   **34.** mean: 2, 4, 6, 8, 10 **6**

35. **NUMBER SENSE** There are 24 boys and girls in the concert band. There are twice as many girls as boys. How many girls are there? **16 girls**

36. **FRACTIONS** Margaret spent $\frac{1}{3}$ of the dimes she saved. She gave her brother $\frac{1}{4}$ of the remaining ones. Then she had 45 dimes left. How many dimes had she saved? **90 dimes**

MORE PRACTICE Lesson 14.2, page H62

263

Use these questions to help students focus on the big ideas of the lesson.

• **How do you find the probability of getting an odd number when rolling a number cube labeled I–6?** Find the number of favorable outcomes and divide by the number of possible outcomes, so P(odd number) = $\frac{3}{6}$, or $\frac{1}{2}$.

✔ Assessment Tip

Have students write the answer.

• **A 4-in. × 4-in. square is divided into 16 square units. Explain how to find the probability that a randomly placed point lies inside the red area.** The area of the red square is 4 in.^{2}, and the area of the large square is 16 in.^{2} The ratio of the areas is the probability $\frac{4}{16}$, or $\frac{1}{4}$.

DAILY QUIZ

A spinner is divided into 8 congruent sections: 4 red, 3 blue, and 1 yellow. Find the probability.

1. P(red) $\frac{1}{2}$ **2.** P(blue) $\frac{3}{8}$
3. P(green) 0   **4.** P (blue or yellow) $\frac{1}{2}$

14.3

GRADE 5
Compare probabilities of simple events.

GRADE 6
Find the probability of events that are not equally likely.

GRADE 7
Find probability.

ORGANIZER

OBJECTIVE To find the probability of events that are not equally likely

ASSIGNMENTS Independent Practice
- Basic 1–23, odd; 24–27
- Average 1–27
- Advanced 7–27

PROGRAM RESOURCES
Reteach 14.3 Practice 14.3 Enrichment 14.3
Cultural Connection • Navajo Blanket
   Design, p. 270

PROBLEM OF THE DAY

Transparency 14

In her coin purse, Rachel has pennies with the following dates: 1956, 1981, 1990, 1983, 1994, and 1985. What is the probability that she will choose a coin minted in the 1990's? 2 out of 6, or $\frac{1}{3}$ **in an odd-numbered year?** 3 out of 6 or $\frac{1}{2}$ **in the twentieth century?** 6 out of 6, or $\frac{6}{6}$, or 1

Solution Tab A, p. A14

QUICK CHECK

A spinner is divided into 5 congruent sections, labeled *A, E, I, O, S.* Find the probability.

1. P(*A*) $\frac{1}{5}$
2. P(*E* or *S*) $\frac{2}{5}$
3. P(vowel) $\frac{4}{5}$
4. P(not *E*) $\frac{4}{5}$

1 MOTIVATE

Help students access their prior knowledge by asking:

- **If you toss a cube, what is the probability that any particular face will come up on top?** $\frac{1}{6}$

2 TEACH/PRACTICE

To help students understand that all outcomes don't have the same probability, discuss Examples 1 and 2 with the class.

WHAT YOU'LL LEARN
How to find the probability of events that are not equally likely

WHY LEARN THIS?
To understand that all events do not have the same probability of happening

Consumer Link

The probability of a pepper plant growing red peppers is the same as the probability of a pepper plant growing green peppers. However, consumers believe red peppers are less common, so farmers grow fewer red peppers and charge more for them. If green peppers are $0.29 per pound and red peppers are $0.89 per pound, how much more would 11 lb of red peppers cost than 11 lb of green peppers?

$6.60 more

264 Chapter 14

More on Probability

You have learned that probability is a comparison of the number of favorable outcomes and the number of possible total outcomes. Sometimes there is a greater chance of one event occurring than another.

On this spinner, there are twice as many orange sections as yellow or green sections.

EXAMPLE 1 Find the probability of the pointer stopping on each color. Which color do you think the pointer will stop on most often?

P(yellow) $= \frac{1}{4}$ *Find each probability.*

P(green) $= \frac{1}{4}$

P(orange) $= \frac{2}{4} = \frac{1}{2}$

$\frac{1}{2} > \frac{1}{4}$ *Compare the probabilities.*

So, the pointer will stop on orange most often.

EXAMPLE 2 Tamara has a bag with 8 cubes all the same size: 1 red, 4 yellow, and 3 blue. Without looking, she chooses a cube from the bag. After recording the color, she places the cube back in the bag and chooses again. What is the probability of Tamara choosing a yellow cube each time?

P(yellow) $= \frac{4}{8} = \frac{1}{2}$ *4 yellow cubes; 8 cubes in all*

So, the probability of a yellow cube is $\frac{1}{2}$ each time.

- **CRITICAL THINKING** Tamara chose a blue cube and did not return it to the bag. What is the probability of choosing a yellow cube once a blue cube has been removed? $\frac{4}{7}$

GUIDED PRACTICE

Use the spinner at the right to find each probability.

1. P(red) $\frac{1}{4}$
2. P(blue) $\frac{1}{2}$
3. P(yellow) $\frac{1}{4}$
4. P(blue or yellow) $\frac{3}{4}$
5. P(white) 0

IDEA FILE

CORRELATION TO OTHER RESOURCES

 Probability Toolkit
Ventura Math Power Series.
Students can simulate chance events in an easy-to-use computer-based environment.

MEETING INDIVIDUAL NEEDS

RETEACH 14.3

Name _____

More on Probability

Probabilities can change.
Suppose a box contains 8 crayons, and 2 of them are red. Then the probability of drawing a red crayon is $\frac{2}{8}$, or $\frac{1}{4}$. Now, suppose you remove a red crayon, leaving just 7 crayons. Since just 1 of the 7 is red, the probability of drawing a red crayon now becomes $\frac{1}{7}$.

A box contains 8 crayons: 2 are blue, 4 are green, 1 is red, and 1 is yellow.

1. You choose 1 crayon without looking. What is the probability of the crayon being either blue or green?
   a. How many favorable outcomes are there? List them.
      6: blue, blue, green, green, green, green
   b. How many possible outcomes are there? List them.
      8: blue, blue, green, green, green, green, red, yellow
   c. Write the probability of choosing a blue or green crayon as a fraction.
      $\frac{6}{8}$, or $\frac{3}{4}$

2. Suppose you remove 1 green crayon. What is the probability of choosing a yellow crayon now?
   a. How many favorable outcomes are there? List them.
      1: yellow
   b. How many possible outcomes are there? List them.
      7: blue, blue, green, green, green, yellow, red
   c. Write the probability of choosing a yellow crayon as a fraction.
      $\frac{1}{7}$

3. Suppose you now remove 1 yellow crayon, leaving a total of 6 crayons. What is the probability of now choosing a blue crayon?
   a. How many favorable outcomes are there? **2**
   b. How many possible outcomes are there? **6**
   c. Write the probability of choosing a blue crayon. $\frac{2}{6}$, or $\frac{1}{3}$

Use with text pages 264–265.   **TAKE ANOTHER LOOK** R63

INDEPENDENT PRACTICE

For Exercises 1–6, use the spinner at the right.
Find each probability on a single spin.

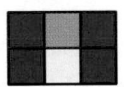

1. P(F) $\frac{1}{4}$ **2.** P(T) $\frac{1}{8}$ **3.** P(B) 0 **4.** P(T, D, or O) $\frac{1}{2}$

5. P(O or D) $\frac{3}{8}$ **6.** P(F, A, S, or T) $\frac{5}{8}$

A bag contains 4 red, 2 blue, 1 yellow, and 3 green pencils. You choose one pencil without looking. Find each probability.

7. P(yellow) $\frac{1}{10}$ **8.** P(red) $\frac{2}{5}$ **9.** P(blue or green) $\frac{1}{2}$ **10.** P(yellow, green, or red) $\frac{4}{5}$

Cards numbered 1, 1, 2, 2, 3, 4, 5, and 6 are placed in a hat. You choose one card without looking. Find each probability.

11. P(1) $\frac{1}{4}$ **12.** P(1 or 2) $\frac{1}{2}$ **13.** P(1, 2, or 3) $\frac{5}{8}$ **14.** P(even numbers) $\frac{1}{2}$

15. P(1, 2, 3, or 4) $\frac{3}{4}$ **16.** P(1, 2, 3, or 6) $\frac{3}{4}$ **17.** P(8) 0

For Exercises 18–23, use the figure at the right. A point in the figure is chosen at random. Find each probability. Each measurement is in feet.

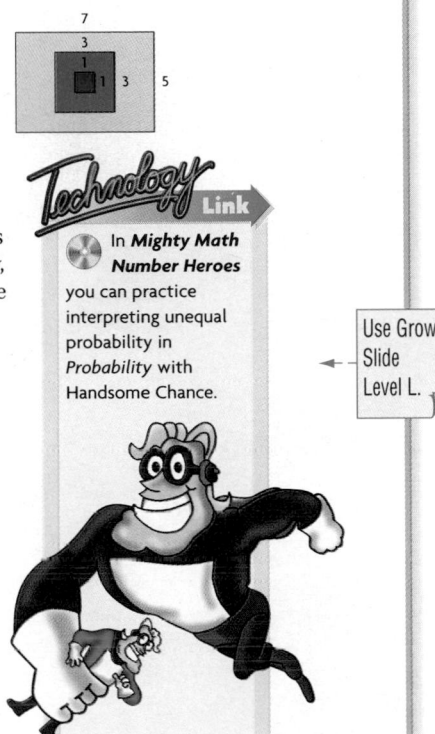

18. P(red) $\frac{1}{35}$ **19.** P(blue) $\frac{8}{35}$ **20.** P(green) $\frac{26}{35}$

21. P(blue or red) $\frac{9}{35}$ **22.** P(green or blue) $\frac{34}{35}$ **23.** P(blue, red, or green) 1

Problem-Solving Applications

24. Tasha wants to borrow a CD from her brother, Carlos. Carlos has 6 country music CDs and 2 rock CDs. Tasha is in a hurry, so she picks one without looking. What is the probability she chose a country music CD? $\frac{3}{4}$

25. Brenda can't decide which movie to pick for her party. She writes the titles on separate pieces of paper and puts them in a bag. There are 2 adventure movies, 3 comedies, and 1 mystery. Brenda will choose one piece of paper at random. What is the probability that she will choose a comedy? $\frac{1}{2}$

26. Rafael is going to his grandmother's house for dinner and must wear a tie. He has 4 ties in his closet: 1 blue, 2 print, and 1 striped. He takes one at random and puts it on. What is the probability that he chose the striped tie? $\frac{1}{4}$

27. 📣 **WRITE ABOUT IT** Write a problem that involves finding the probability of events that are not equally likely to happen. **Problems will vary.**

Technology **Link**

In *Mighty Math Number Heroes* you can practice interpreting unequal probability in *Probability* with Handsome Chance.

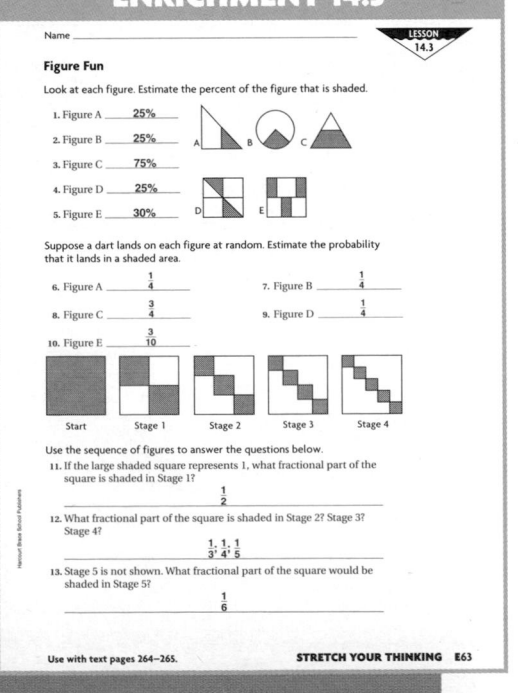

MORE PRACTICE Lesson 14.3, page H62

265

Use Grow Slide Level L.

GUIDED PRACTICE
Complete the exercises as a whole group to check students' understanding of the concepts.

INDEPENDENT PRACTICE
These exercises check students' ability to determine the probability of events that are not equally likely.

ADDITIONAL EXAMPLES

Example 1, p. 264

Find the probability that a dart will land on each color. Which color will the dart land on most often?

P(red) = $\frac{4}{6}$ = $\frac{2}{3}$; P(blue) = $\frac{1}{6}$; P(yellow) = $\frac{1}{6}$; since $\frac{2}{3} > \frac{1}{6}$, the dart will land on red most often.

Example 2, p. 264

Joe puts 8 colored balls in a box: 2 green, 3 purple, and 3 pink. He chooses a ball, records the color, and puts the ball back in the box. What is the probability that Joe will choose a purple ball each time? $\frac{3}{8}$

3 WRAP UP

Use these questions to help students focus on the big ideas of the lesson.

- **Five identical cards are placed in a box. Three are marked with an *X* and two with a *Y*. What is the probability that you will choose a card with an *X*?** $\frac{3}{5}$

✔ *Assessment Tip*
Have students write the answer.

- **There are red, blue, green, and yellow cubes in a bag. Can you find the probability of choosing a red cube? Why or why not?** No. You don't know the number of cubes of each color or the total number of cubes in the bag.

DAILY QUIZ

Of the faces of a cube, 3 are red, 2 are blue, and 1 is green. Find the probability of each.

1. P(red) $\frac{3}{6}$, or $\frac{1}{2}$
2. P(green or blue) $\frac{3}{6}$, or $\frac{1}{2}$
3. P(red or blue or green) 1

265

ORGANIZER

OBJECTIVE To use a simulation to model an experiment
MATERIALS *For each pair* number cube, calculator
PROGRAM RESOURCES
E-Lab • Activity 14
E-Lab Recording Sheets, p. 14

USING THE PAGE

Ask students to describe times when they tried to guess how many times something would occur before the desired result was achieved. For example, they may have tried to win the best of seven games in basketball. To do this, a total of 4, 5, 6, or 7 games would have to be played.

Activity 1

Explore

Have students predict the number of rolls that will be needed to get all the numbers at least once. Explain that the numbers are *random* numbers because the probability of getting each number is the same. Ask:

- **Do you expect to get all the numbers on the first six rolls? Explain.** Possible answer: No. It is equally likely that any number will occur on any roll.

Think and Discuss

As students answer the questions, they should come to realize that the number they obtain for the average is an *estimate* of the number of bottles they would have to buy. They would still *not* be sure to win.

Try This

Remind students that the first average cannot be used to find the second one. They must actually average the number of rolls in all five experiments. Help students see that the second average is a better estimate since they conducted more experiments. You may also wish to have two or more pairs of students average the number of rolls it took in all of their experiments in order to get an even better estimate.

COOPERATIVE LEARNING

Simulations

LAB *Activity*

WHAT YOU'LL EXPLORE
How to use a simulation to model an experiment

WHAT YOU'LL NEED
number cube numbered 1 to 6, calculator

A juice company is having a contest. To win a prize, you have to collect six bottle caps that spell out *ORANGE*. One of the six letters is put under each bottle cap when the cap is produced. The letters are divided equally among the juice bottles.

You can conduct an experiment to simulate how many bottles of juice you have to buy to get all six letters.

ACTIVITY 1

Explore

- Work with a partner. Use a number cube to generate random numbers. Each of the numbers 1 to 6 will represent one of the letters in the word *ORANGE*.

O	R	A	N	G	E
1	2	3	4	5	6

- Roll the number cube, and tally the numbers you get.

- Continue to roll the number cube until you get all the numbers at least once.

- Repeat the experiment.

Number	Times Rolled
1	II
2	I
3	IIII
4	III
5	I
6	I

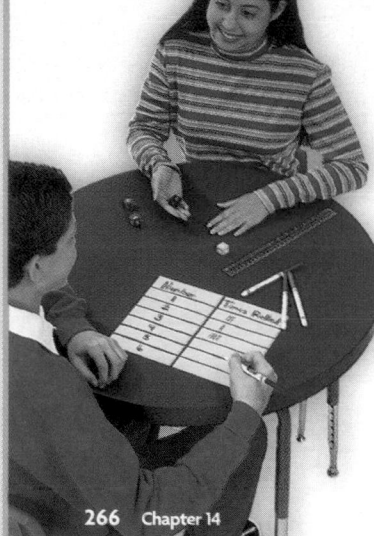

Think and Discuss Answers will vary.

- How many rolls did it take in the first experiment to get all six numbers?

- How many rolls did it take in the second experiment?

- What is the average of the rolls in your two experiments?

- How many bottles of juice do you expect you will have to buy? If you bought this many bottles, would you be sure to win? **no**

Try This

- Repeat the experiment three more times.

- Guess how many times each number will be rolled. Record your guess and compare it with the results.

- Find the average of the rolls in all five experiments. Compare your results with your classmates.
  Answers will vary.

 Calculator Activities, page H39

266 Chapter 14

IDEA FILE

WRITING IN MATHEMATICS

Have students write a letter to a friend to explain at least three ideas about the concept of probability. Students might explain what they have done in class, what they found most interesting, and what they consider easy or difficult to understand about probability.

Modality Auditory

CAREER CONNECTION

Meteorologist Computer simulations have made it possible to study weather patterns and to make predictions about the paths of storms. Advance warnings can save lives and property.

- Ask students to watch local weather reports on TV or to listen to them on the radio for one week and to record all predictions that mention the words *probability* or *chance*. Have students bring their lists to class for discussion.

Modalities Auditory, Visual

Instead of rolling a number cube, you can produce random numbers using a calculator. Use a calculator to do the following activity with your partner.

ACTIVITY 2

Explore

- Use this key sequence to produce random numbers from 1 to 10.

  1 [2nd] [RAND] 10 [=]

- Record the number you get each time.

- Repeat the key sequence until you have gotten each of the numbers 1 to 10.

- Make a bar graph to show the results.

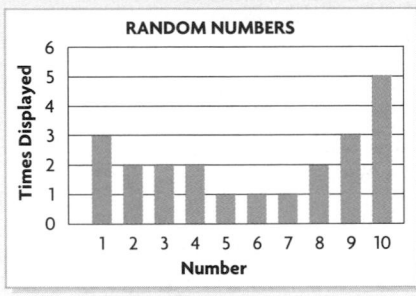

RANDOM NUMBERS

Think and Discuss

- How many times did you have to use the key sequence to get all ten numbers? **Answers will vary.**

- How do your results compare with those of your classmates? **Answers will vary.**
- Look at the results in the graph. What was the last random number produced? **5, 6, or 7**

Try This

- Use the calculator to produce random numbers from 1 to 20. Record the number of times you get each number. Continue until you get each number.

Technology Link
E-Lab • Activity 14 Available on CD-ROM and the Internet at http://www.hbschool.com/elab

267

Note: Students use simulations. Use E-Lab p. 14.

Activity 2
Explore
Explain to students that they can use their calculators to generate random numbers. When they key the sequence for random numbers, the calculator is programmed so that each outcome is equally likely.

Think and Discuss
After discussing the questions, have students combine their experimental data. Ask:

What is the mathematical probability of generating a 3? $\frac{1}{10}$

Have students compare the mathematical probability with the experimental probability of individual student's results and then of the combined class results. The experimental probability should be closer to the mathematical probability in the pooled data than in any individual data set.

Try This
Have students predict the experimental probability of randomly generating a 10 and record the prediction before they begin the experiment. After students have gathered data experimentally, have them compare their predictions with results.

USING E-LAB
Students use computer-simulation software to determine the area of a plane figure through randomness and ratios.

✔ *Assessment Tip*
Have students write the answer.

- **How is finding experimental outcomes by using a calculator similar to finding experimental outcomes by using a number cube?** In each case the mathematical probability of each outcome is equal; the experimental probabilities may or may not be close to the mathematical probabilities, depending on the number of trials.

EXTENSION

Have students work in groups. Present this situation to groups:

- A student has decided to guess the answers to a true-false test of 5 questions. What is the probability that he will get all 5 questions correct? Use a coin to simulate a guess: Heads = R (right) and Tails = W (wrong). Flip the coin 5 times (for 5 guesses) and record the results. This is one trial. Perform 25 trials. Five R's (5 correct guesses) is success. Have students express the probability as the ratio of the number of successes to the total number of trials.

Modalities Kinesthetic, Visual

ENGLISH AS A SECOND LANGUAGE

Tip • Help students acquiring English express important ideas about probability by completing a sentence starter: "One thing I know about *probability* is ___."
Replace the underlined vocabulary term with such other terms as *event, chance, possible outcome, impossible outcome, likely event, unlikely event.*

Modality Auditory

ORGANIZER

GRADE 5	GRADE 6	GRADE 7
Identify possible outcomes of an experiment.	Find the experimental probability of an event.	Find the experimental probability of an event.

OBJECTIVE To find the experimental probability of an event

VOCABULARY New **experimental probability**

ASSIGNMENTS Independent Practice
- ■ *Basic* 1–9; 10–17
- ■ *Average* 1–9; 10–17
- ■ *Advanced* 1–9; 10–17

PROGRAM RESOURCES
Reteach 14.4 Practice 14.4 Enrichment 14.4
Extension • Independent and Dependent Events, TE pp. 270A–270B

PROBLEM OF THE DAY

Transparency 14

For their probability experiments, three students used a spinner divided into 3 congruent sections, a number cube, and a coin. Carlo's and Judy's probabilities could be $\frac{1}{3}$. Martha's and Carlo's probabilities could be $\frac{1}{2}$. Which object did each student use? Carlo used the number cube; Judy used the spinner; and Martha used the coin.

Solution Tab A, p. A14

QUICK CHECK

1. If you toss a coin, the probability of getting heads is $\frac{1}{2}$. About how many heads do you expect to get in 50 tosses of the coin? about $\frac{1}{2} \times 50$, or 25
2. Could you get 40 heads? yes
3. Could you get 0 heads? yes

1 MOTIVATE

Access students' prior knowledge by asking them the purpose of a simulation.

2 TEACH/PRACTICE

Discuss the examples on p. 268. Tell students that each time they do an activity in an experiment, they do a *trial*.

WHAT YOU'LL LEARN
How to find the experimental probability of an event

WHY LEARN THIS?
To understand that real-life probabilities of events may be different than the mathematical probabilities

WORD POWER
experimental probability

Science Link

The inheritance of many characteristics is based on probabilities. In a plant called the four o'clock, the flowers may be red, white, or pink. Pink flowers are twice as common as either red or white flowers. If a four o'clock has 140 flowers, how many should be red? white? pink?

35 red, 35 white, 70 pink

Experimental Probability

With a number cube numbered 1 to 6, the mathematical probability of rolling each number is $\frac{1}{6}$. By performing an experiment, you can find the experimental probability of rolling each number. The **experimental probability** of an event is the number of times a success occurs compared with the total number of times you do the activity.

Bailey rolled a cube numbered 1 to 6. The table below shows the results of rolling the cube 50 times.

Number	1	2	3	4	5	6
Times rolled	4	9	11	6	10	10

$$\text{experimental probability} = \frac{\text{number of times success occurs}}{\text{total number of trials}}$$

EXAMPLE 1 Use Bailey's results to find the experimental probability of rolling each number. Write each fraction in simplest form.

$P(1) = \frac{4}{50} = \frac{2}{25}$ $P(2) = \frac{9}{50}$ $P(3) = \frac{11}{50}$

$P(4) = \frac{6}{50} = \frac{3}{25}$ $P(5) = \frac{10}{50} = \frac{1}{5}$ $P(6) = \frac{10}{50} = \frac{1}{5}$

Talk About It

- What if Bailey rolled the number cube 50 more times? Do you think the results would be the same as in Example 1? Explain. **No. Each outcome is random and cannot be predicted with certainty.**
- How does the experimental probability of each number compare with the mathematical probability of $\frac{1}{6}$? **P(1), P(4): less than; P(2), P(3), P(5), P(6): greater than**

You can use the experimental probability to predict future events.

EXAMPLE 2 Based upon his experimental results, how many times can Bailey expect to roll a 6 in his next 10 rolls?

$P(6) = \frac{1}{5}$ *Use the experimental probability.*

$\frac{1}{5} \times 10 = 2$ *Multiply 10 rolls by $\frac{1}{5}$.*

So, he can expect to roll a 6 two times in the next 10 rolls, based on his past rolls.

IDEA FILE

EXTENSION

MATERIALS: 2 number cubes

Have students toss the cubes 36 times and record the sum of the numbers that face up. Ask:

- Which sum appeared most often?
- Which sum appeared least often?
- What is the probability of getting each sum when you toss two number cubes?

Modalities Kinesthetic, Visual

RETEACH 14.4

Name _____

LESSON 14.4

Experimental Probability

You have a box containing thousands of radish seeds. How many of them will sprout? You can experiment to find out.
Suppose you plant 20 of the seeds, and 15 of them sprout.

$$\frac{15 \text{ seeds sprout}}{20 \text{ seeds in all}} \rightarrow \frac{15}{20} = \frac{3}{4}$$

You might conclude, based on your experiment, that about $\frac{3}{4}$ of the seeds will sprout. So, if you select a seed at random from the box, the **experimental probability** that it will sprout is $\frac{3}{4}$.

Mel spins a spinner numbered 1 to 5. The table below shows the results of 25 spins.

Number	1	2	3	4	5
Frequency	4	2	6	8	5

1. What is the experimental probability of spinning a 4?
   a. How many times did Mel spin a 4? __8__
   b. How many times did Mel spin in all? __25__
   c. You can write the experimental probability as a fraction.
      $$\text{Experimental probability} = \frac{\text{number of times the event occurs}}{\text{total number of trials}}$$
      Use the formula to write the experimental probability of spinning 4 as a fraction in simplest form. __$\frac{8}{25}$__
2. Suppose Mel spins the spinner 50 more times. How often can he expect to spin a 3?
   a. Begin by finding the experimental probability of spinning a 3.
      How many times did Mel spin a 3? __6__
   b. How many times did Mel spin? __25__
   c. Use the formula to write the experimental probability as a fraction in simplest form. __$\frac{6}{25}$__
   d. Multiply 50 spins by the experimental probability.
      $$50 \times \frac{6}{25} = 12$$
   e. How many times can Mel expect to spin a 3? __12 times__

R64 TAKE ANOTHER LOOK   Use with text pages 268–269.

GUIDED PRACTICE

Rickie tossed a coin 40 times and got these results. For Exercises 1–2, use the table at the right to find the experimental probability.

Coin	Heads	Tails
Toss	28	12

1. P(heads) $\frac{7}{10}$ 2. P(tails) $\frac{3}{10}$ 3. What is the mathematical probability of getting tails? $\frac{1}{2}$

4. Is the experimental probability of getting tails, based on these results, less than the mathematical probability? Explain.
   Yes; $\frac{3}{10} < \frac{1}{2}$

INDEPENDENT PRACTICE

Tania spins the pointer of the spinner 50 times. For Exercises 1–4, use her results in the table below to find the experimental probability.

Color	Blue	Green	Red	Gray
Times lands on	20	10	12	8

1. P(blue) $\frac{2}{5}$ 2. P(green) $\frac{1}{5}$ 3. P(red) $\frac{6}{25}$ 4. P(gray) $\frac{4}{25}$

5. What is the mathematical probability for each color? $\frac{1}{4}$

6. Based on her past experimental results, how many times can Tania expect the pointer to land on blue in the next 20 spins? **8 times**

7. Based on her past experimental results, how many times can Tania expect the pointer to land on red in the next 100 spins? **24 times**

Problem-Solving Applications

8. Fran was in a batting cage. Out of 100 balls, she hit 30 of them. How many balls can she expect to hit in her next 30 tries? **9 balls**

9. ✏ **WRITE ABOUT IT** Explain how you get the experimental probability of the number 4 when rolling a number cube labeled 1 to 6. **number of times 4 lands ÷ total rolls**

Mixed Review

Find the value.

10. 6 + 8 + 4 **18**   11. 6 × 3 + 5 **23**   12. 18 − 3 × 2 **12**   13. 16 ÷ 4 + 5 **9**

A bar graph showed that more children than adults have pets.

14. Would the taller bar represent children or adults? **children**

15. What does a taller bar show? **a larger amount**

16. **AVERAGES** Emily wants to get a 90 average in math. She will have a total of 8 tests. How many total points will she need for all 8 tests? **720 points**

17. **NUMBER SENSE** A giant tortoise travels at a speed of 0.2 km per hour. How far does it travel in $2\frac{1}{2}$ hours? **0.5 km**

MORE PRACTICE Lesson 14.4, page H63

269

INDEPENDENT PRACTICE
These exercises check students' ability to use experimental and mathematical probabilities.

ADDITIONAL EXAMPLES

Example 1, p. 268

Juan tosses a thumbtack 25 times and finds that it lands point down 11 times. What is the experimental probability that the thumbtack will land point down? $\frac{11}{25}$

Example 2, p. 268

Use the experimental probability to determine how many times Bailey can expect to roll a 1 in his next 10 rolls. $\frac{2}{25} \times 10$ is less than 1 time in 10 rolls.

3 WRAP UP

Use these questions to help students focus on the big ideas of the lesson.

- **There are pink, green, and yellow cards in a box. How can you estimate the probability of choosing a pink card?** Perform an experiment: Choose a card, record the color, and put the card back in the box. Do this 25 times. Use the experimental probability formula.

✓ *Assessment Tip*

Have students write the answer.

- **There are 90 colored blocks in a bag. Some are red, some are blue, and some are green. The experimental probability of picking a red block is $\frac{1}{6}$. About how many red blocks might be in the bag?** about $\frac{1}{6} \times 90$, or 15 red blocks

DAILY QUIZ

Find the expected values, using the mathematical probability and the experimental probability.

Experimental	Mathematical	Number of Trials
$\frac{1}{5}$ 6	$\frac{1}{6}$ 5	30
$\frac{1}{2}$ 25	$\frac{11}{25}$ 22	50
$\frac{3}{8}$ 24	$\frac{11}{32}$ 22	64

PRACTICE 14.4

Name _____

LESSON 14.4

Experimental Probability

Vocabulary

1. What is experimental probability?

   The experimental probability of an event is the number of times the event occurs compared with the total number of times you do the activity.

Adam tossed a coin 50 times. For Exercises 2–3, use the table at the right to find the experimental probability.

Coin	Heads	Tails
Toss	22	28

2. P(Heads) $\frac{11}{25}$ 3. P(Tails) $\frac{14}{25}$

4. What is the mathematical probability of getting heads? $\frac{1}{2}$

Sarah rolled a number cube numbered 1 to 6. The table below shows the results of rolling the cube 50 times. Use the results in the table to find the experimental probability.

Number	1	2	3	4	5	6
Times rolled	6	11	5	10	16	2

5. P(3) $\frac{1}{10}$ 6. P(4 or 5) $\frac{13}{25}$ 7. P(1 or 2) $\frac{17}{50}$

8. What is the mathematical probability for each number? $\frac{1}{6}$

Mixed Applications

9. Mitch spins the pointer of a spinner 20 times. It lands on blue 15 times. What is the experimental probability of landing on blue? $\frac{3}{4}$

10. Ryan was hitting in the batting cage. Out of 100 balls, he hit 42 of them. How many balls can he expect to hit in his next 50 tries? **21**

11. A car is traveling at a speed of 52 mph. How far does it travel in $3\frac{1}{2}$ hr? **182 mi**

12. Scott wants to get a 94 average in science. He will have a total of 9 tests. How many total points will he need for all 9 tests? **846 points**

P64 **ON MY OWN**   Use with text pages 268–269.

ENRICHMENT 14.4

Name _____

LESSON 14.4

Comparing Coins

For Problems 1–12, write a ratio that compares the dollar values of each set of coins.

1. 5 dimes to 6 quarters   1:3
2. 16 nickels to 2 dimes   4:1
3. 6 quarters to 6 nickels   5:1
4. 10 dimes to 20 quarters   1:5
5. 3 nickels to 3 quarters   1:5
6. 1 quarter to 20 dimes   1:8
7. 9 quarters to 15 nickels   3:1
8. 48 dimes to 12 nickels   8:1
9. 5 quarters to 1 nickel   25:1
10. 11 pennies to 11 quarters   1:25
11. 7 nickels to 21 dimes   1:6
12. 30 nickels to 3 dimes   5:1

Solve.

13. You have $3 in dimes and $3 in quarters. Find the probability that a coin picked at random is a dime. $\frac{5}{7}$

14. You have $1 in pennies, $1 in nickels, and $1 in dimes. Find the probability that a coin picked at random is worth more than 2 cents. $\frac{3}{13}$

15. You have $2 in nickels and $2 in quarters. Find the probability that a coin picked at random is a dime. 0

E64 **STRETCH YOUR THINKING**   Use with text pages 268–269.

269

EXTENSION

Independent and Dependent Events

Use these strategies to extend understanding of probability.

Multiple Intelligences Logical/Mathematical

Modalities Auditory, Kinesthetic, Visual

Lesson Connection Students will determine whether two events are dependent or independent and will find the probability of two independent events.

Program Resources *Intervention and Extension Copying Masters,* p. 20

Challenge

TEACHER NOTES

WHAT STUDENTS DO

ACTIVITY 1

Purpose To determine whether two events are dependent or independent

Discuss the following question.

- What are the definitions of the words *dependent* and *independent*? How do these meanings relate to the use of the words in probability? *Dependent* means "determined or conditioned by another." *Independent* means "not requiring or relying on something else." In probability, dependent events are events of which one is conditioned by the other. Independent events are events of which neither relies on the other.

1. You toss a coin. Then you roll a number cube. Does the outcome of tossing the coin affect the outcome of rolling the number cube? Explain. No. They are two separate events.

2. You choose a card from a deck and put the card aside. Then you choose another card from the remaining deck. Does the outcome of the first card choice affect the outcome of the second card choice? Explain. Yes. There is one less card to choose from for your second choice.

   When the outcome of one event affects the outcome of another event, the events are **dependent**. When the outcome of one event does not affect the outcome of the second event, the events are **independent**.

3. Decide if the following pairs of events are dependent or independent.

   a. Toss a coin. Spin a spinner. independent

   b. Draw a card from a deck. Replace it. Draw another card. independent

   c. Draw a marble from a bag. Draw another marble from those remaining in the bag. dependent

   d. Roll a number cube. Then roll it again. independent

   e. Choose a student from the class. Choose another student from the remaining students in the class. dependent

4. Describe two events that are dependent. Answers will vary.

5. Describe two events that are independent. Answers will vary.

TEACHER NOTES

ACTIVITY 2

Purpose To determine the probability of two independent events

Have students meet in small groups or with partners to discuss the given question. Have one student report the group's responses to the larger group.

WHAT STUDENTS DO

To find the probability of two independent events, find the probability of each event. Then find the product of the two probabilities.

1. You roll a number cube and toss a coin. Complete the following steps to find the probability of rolling a 3 and getting heads.

    Make sure the events are independent.

    Find P(tossing 3): P(tossing 3) $= \frac{1}{6}$

    Find P(heads): P(heads) $= \frac{1}{2}$

    Find the product: P(tossing 3 and heads) $= \frac{1}{6} \times \frac{1}{2} = \frac{1}{12}$

2. You toss two coins. What is the probability of getting the following?

    a. two heads $\frac{1}{2} \times \frac{1}{2} = \frac{1}{4}$

    b. one head and one tails $\frac{1}{2} \times \frac{1}{2} = \frac{1}{4}$

    c. two tails $\frac{1}{2} \times \frac{1}{2} = \frac{1}{4}$

    • What do you notice? The probabilities are the same.

3. You spin the spinner shown and roll a number cube. What is the probability of each of the following?

    a. red and 6 $\frac{1}{4} \times \frac{1}{6} = \frac{1}{24}$

    b. blue and an odd number $\frac{1}{4} \times \frac{3}{6} = \frac{3}{24} = \frac{1}{8}$

    c. green and a prime number $\frac{1}{4} \times \frac{3}{6} = \frac{3}{24} = \frac{1}{8}$

Share Results

• Explain the steps for finding the probability of two independent events. Make sure the events are independent; find the probability of each event; find the product of the probabilities.

Intervention and Extension Copying Masters, p. 20

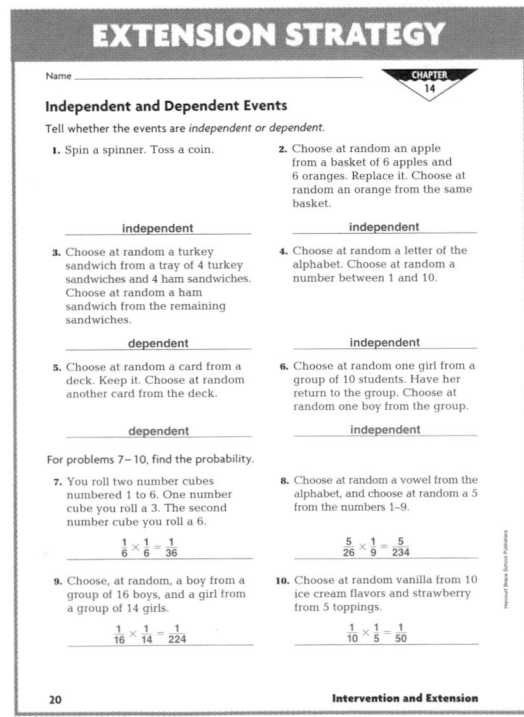

OBJECTIVE To solve geometric probability problems

USING THE PAGE

Before the Lesson

Number two sets of six cards from 1–6. Shuffle the two sets together, and have a student volunteer choose a number from 1–6. Ask the class to predict how many cards will have to be drawn before the number appears. Compare the prediction and the result. Discuss whether the probability is greater, less, or the same if the number appears once in 6 cards or twice in 12 cards. the same, because $\frac{1}{6} = \frac{2}{12}$

Cooperative Groups

Have the groups work together to answer the question and solve Problems 1–7. Encourage groups to be creative with their designs for Problem 7.

> **CULTURAL LINK**
>
> Members of the Navajo Nation are well known for many artistic endeavors, including jewelry and pottery. One Navajo potter, Maria Martinez, rediscovered an ancient method for producing black designs on black pottery. Today, the technique is being used by several Navajo potters.

CULTURAL CONNECTION

Navajo Blanket Design

The Navajo are the largest group of Native American people in the United States. One of the most popular items produced by the Navajo are handwoven blankets.

Handweaving is a time-honored activity for the Navajo. They rarely sketch their plans, since children become very good at blanket weaving at a young age.

The Navajo are noted for their use of various geometric patterns on their blankets and rugs. The blanket shown at the right makes use of line segments, triangles, rectangles, and rhombi.

Suppose the blanket at the right has an area of 3,000 in.^{2}, with red covering 1,200 in.^{2}, beige covering 1,000 in.^{2}, and blue covering 800 in.^{2} What is the geometric probability that a marble tossed randomly onto the blanket will land on a blue area?

$$P(\text{blue}) = \frac{\text{area of blue section}}{\text{area of blanket}} = \frac{800}{3000} = \frac{4}{15}$$

• What is the geometric probability that a marble tossed randomly onto the blanket will land on a red or beige area? $\frac{11}{15}$

 Work Together

Look at the blanket design on the grid. Find the geometric probability that a marble randomly tossed onto the blanket will land on an area of the given color.

1. P(blue) $\frac{8}{35}$ 2. P(red) $\frac{1}{7}$

3. P(yellow) $\frac{1}{5}$ 4. P (white) $\frac{3}{7}$

5. P(blue or white) $\frac{23}{35}$ 6. P(yellow or red) $\frac{12}{35}$

7. On a sheet of grid paper, make a Navajo blanket design using at least three different colors. Find the geometric probability that a randomly tossed marble will land on each of the colors. **Check students' drawings.**

CHAPTER 14 REVIEW

EXAMPLES

- **Use the strategy *account for all possibilities* to solve problems.** (pages 258–259)

PROBLEM-SOLVING TIP: For help in solving problems by accounting for all possibilities, pages 7 and 258.

- **Find the probability of equally likely events.** (pages 260–263)

apple
banana
lemon
grapefruit

$P(\text{lemon}) = \frac{1}{4}$

$P(\text{apple or grapefruit}) = \frac{1}{2}$

- **Find the probability of events that are not equally likely.** (pages 264–265)

$P(\text{green}) = \frac{1}{5}$

$P(\text{red or yellow}) = \frac{4}{5}$

- **Find the experimental probability of an event.** (pages 268–269)

The table shows the results of spinning the pointer of a spinner 20 times.

Color	Blue	White	Red
Number of spins	8	3	9

$P(\text{blue}) = \frac{8}{20} = \frac{2}{5}$ $P(\text{white}) = \frac{3}{20}$

$P(\text{red}) = \frac{9}{20}$

EXERCISES

1. The A.M. Cafe serves a choice of cereal, a bagel, or pancakes, and a choice of milk or juice for breakfast. Find the number of possible food and drink choices. **6 choices**

2. For dessert, Cary has a choice of 6 different fruits and 4 different pies. He will choose one of each. Find the number of possible choices. **6 × 4 = 24 choices**

3. **VOCABULARY** A comparison of the number of favorable outcomes to the number of possible outcomes is __?__.

A bag has slips of paper numbered 2 to 10. Find the probability of randomly drawing the number from the bag.

4. P(5) $\frac{1}{9}$ 5. P(7 or 9) $\frac{2}{9}$ 6. P(1) **0**

7. P(2, 5, or 9) $\frac{1}{3}$ 8. P(even numbers) $\frac{5}{9}$

A bag has 10 new pencils: 3 red, 4 yellow, 1 blue, and 2 green. Find the probability of randomly choosing the color of pencil from the bag.

9. P(green) $\frac{1}{5}$ 10. P(yellow) $\frac{2}{5}$

11. P(brown) **0**   12. P(blue or red) $\frac{2}{5}$

13. P(red or green) $\frac{1}{2}$

14. P(red, yellow, or blue) $\frac{4}{5}$

15. **VOCABULARY** The __?__ compares the number of times a success occurs to the total number of times the event was done. **experimental probability**

The table shows the results of rolling a cube 100 times. Use the table to answer Exercises 16–20.

Number	1	2	3	4	5	6
Times rolled	8	12	10	21	15	34

16. P(3) $\frac{1}{10}$ 17. P(6) $\frac{17}{50}$ 18. P(4) $\frac{21}{100}$

19. P(1) $\frac{2}{25}$ 20. P(2 or 5) $\frac{27}{100}$

USING THE PAGE

The Chapter 14 Review reviews using the strategy *account for all possibilities* to solve problems; finding the probability of equally likely events and of events that are not equally likely; and finding the experimental probability of an event. Chapter objectives are provided, along with examples and practice exercises for each.

 The Chapter 14 Review can be used as a review or a test. You may wish to place the Review in students' portfolios.

Assessment Checkpoint

For Performance Assessment in this chapter, see page 274, What Did I Learn?, items 4 and 8.

✔ Assessment Tip

Student Self-Assessment The How Did I Do? survey helps students assess both what they have learned and how they learned it. Self-assessment helps students learn more about their own capabilities and develop confidence in themselves.

A self-assessment survey is available as a copying master in the *Teacher's Guide for Assessment,* p. 15.

CHAPTER 14 TEST, page 1

Name _____

Choose the letter of the correct answer.

1. Vinnie has signed up for trumpet lessons. He can take lessons on Monday, Tuesday, Thursday, or Friday, at 3:30 P.M. or at 4:30 P.M. How many choices does he have?

    A. 4 choices   B. 6 choices
    C. 8 choices   D. 10 choices

2. Notebooks come in green, red, blue, black, or yellow. They come with or without an inside pocket for papers. How many choices are there?

    A. 7 choices   **B. 10 choices**
    C. 12 choices   D. 15 choices

3. A restaurant offers a tossed salad or a Caesar salad. There is a choice of French, Italian, or Ranch dressing. How many choices are there?

    A. 3 choices   **B. 4 choices**
    C. 5 choices   D. 6 choices

4. The bagel store has plain, onion, sesame, raisin, rye, or spinach bagels. Bagels can be served with plain cream cheese, herb cream cheese, or honey cream cheese. How many choices of a bagel with cream cheese are there?

    A. 9 choices   **B. 15 choices**
    C. 16 choices   D. 18 choices

5. When finding the mathematical probability of an event, what fraction should you use?

    A. number of possible outcomes
    B. number of unfavorable outcomes
    C. number of favorable outcomes
    D. number 6

6. You toss a cube numbered 1, 3, 5, 7, 9, 11. What is P(3)?

    A. $\frac{1}{6}$ B. $\frac{1}{4}$ C. $\frac{1}{3}$ D. $\frac{1}{2}$

7. Use the spinner. What is P(2 or 4)?

    A. $\frac{1}{4}$ **B.** $\frac{1}{2}$ C. $\frac{2}{3}$ D. $\frac{3}{4}$

8. What is the probability of hitting the shaded section of the dart target?

    A. $\frac{1}{9}$ B. $\frac{1}{4}$ C. $\frac{1}{6}$ **D.** $\frac{1}{3}$

9. There are 4 answer choices for a test question. What is the probability of guessing the correct answer?

    A. $\frac{1}{10}$ B. $\frac{1}{6}$ C. $\frac{1}{5}$ **D.** $\frac{1}{4}$

10. Ron tosses a number 4 on a cube numbered 1–6. What are Sue's chances of rolling a 5 or a 6 to go first?

    A. $\frac{1}{4}$ **B.** $\frac{1}{3}$ C. $\frac{1}{2}$ D. $\frac{5}{6}$

11. There are 10 sections on a spinner. There are 3 green sections, 1 gold section, 2 blue sections, and 4 red sections. Which color will the pointer probably stop on most often?

    ☐ green
    ☒ blue
    ☐ gold
    ☐ red

    A. green   B. gold
    C. red   D. blue

Form A • Multiple-Choice   A51   Go on. ▶

CHAPTER 14 TEST, page 2

Name _____

For questions 12–15, use the spinner to find each probability.

12. What is P(Z)?

    A. $\frac{1}{8}$ **B.** $\frac{1}{4}$ C. $\frac{3}{8}$ D. $\frac{5}{8}$

13. What is P(N)?

    A. 0   B. $\frac{1}{8}$ C. $\frac{1}{4}$ D. not here

14. What is P(A, E, or I)?

    A. $\frac{1}{8}$ B. $\frac{2}{8}$ **C.** $\frac{1}{2}$ D. $\frac{3}{8}$

15. What is P(A, E, I, P, or Z)?

    A. 0   B. $\frac{1}{2}$ C. $\frac{5}{8}$ **D.** not here

16. Karla has 3 red, 1 black, 2 green, and 2 blue T-shirts. She chooses a shirt without looking. What is the probability of choosing a red T-shirt?

    A. $\frac{1}{8}$ B. $\frac{1}{4}$ **C.** $\frac{3}{8}$ D. $\frac{3}{4}$

17. There are 12 girls and 18 boys in the school band. The band director puts everyone's name in a bag and then draws one name without looking. What is the probability of choosing a boy's name?

    A. $\frac{1}{3}$ B. $\frac{1}{2}$ C. $\frac{2}{3}$ **D.** $\frac{3}{5}$

18. What is the experimental probability of getting tails?

    Results of Coin Tossed 30 Times

Coin	heads	tails
Toss	10	20

    A. $\frac{3}{10}$ B. $\frac{1}{3}$ C. $\frac{1}{2}$ **D.** $\frac{2}{3}$

For questions 19–20, use the data in the table which shows the results of Ramon spinning a pointer of the spinner 60 times.

Number	1	2	3
Times lands on	22	12	26

19. What was the experimental probability of getting a 2?

    A. $\frac{1}{6}$ **B.** $\frac{1}{5}$ C. $\frac{1}{3}$ D. $\frac{5}{12}$

20. How many times can Ramon expect the pointer to land on 3 in the next 30 spins?

    A. 11 times   B. 12 times
    C. 13 times   D. not here

21. Francine tosses a number cube 100 times. The number 4 comes up 20 times. What is the experimental probability of tossing a 4?

    A. $\frac{1}{6}$ **B.** $\frac{1}{5}$ C. $\frac{1}{4}$ D. $\frac{1}{3}$

22. Ahmed has been successful in 32 of 100 foul shots. How many foul shots can he expect to hit in his next 50 tries?

    A. 3 foul shots   B. 8 foul shots
    C. 16 foul shots   D. 24 foul shots

Form A • Multiple-Choice   A52   ▶ Stop!

See *Test Copying Masters,* pp. A51–A52 for the Chapter Test in **multiple-choice** format and pp. B201–B202 for **free-response** format.

✓ *Assessment Checkpoint*

USING THE PAGES
This Study Guide reviews the following content:

Chapter 11 Organizing Data
pp. 206–221

Chapter 12 Displaying Data,
pp. 222–237

Chapter 13 Interpreting Data and
Predicting, pp. 238–255

Chapter 14 Probability, pp. 256–271

CHAPTERS 11–14 STUDY GUIDE & REVIEW

VOCABULARY CHECK

1. The difference between the greatest number and the least number in a set of numbers is the __?__. (page 218) **range**

2. The sum of a group of numbers, divided by the number of addends, is the __?__, or average. (page 246) **mean**

3. A comparison of the number of favorable outcomes and the number of possible outcomes is called __?__. (page 260) **probability**

EXAMPLES

• **Determine whether a sample or a question is biased.** (pages 212–213)

Bill surveys 5 out of 15 boys in his class about class elections. Is the sample biased? Explain. Yes. His sample should include girls.

• **Collect and organize data.** (pages 216–219)

A cumulative frequency keeps a running total of the number of people surveyed.

Ages of Zoo Visitors

Age Group	Frequency	Cumulative Frequency
1–20	10	10
21–40	3	13
41–60	2	15
61–80	1	16

• **Graph two or more sets of data on one graph.** (pages 230–231)

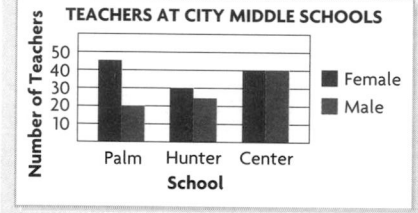

TEACHERS AT CITY MIDDLE SCHOOLS
■ Female
■ Male

EXERCISES

4. Connie surveys 10 out of 30 sixth-grade students on what the voting age should be in city elections. Is this sample biased or not biased? Explain. **biased; excludes people of other ages**

For Exercises 5–6, use the following data.

Ages of Neighborhood Cats

3	9	1	1	4	12	7	1
6	5	4	2	10	6	3	4

See Additional Answers, p. 273A.

5. Make a line plot.

6. Make a cumulative frequency table with intervals.

7. Make a multiple-line graph with the data. See Additional Answers, p. 273A.

High and Low Temperatures

	Mon	Tue	Wed	Thu	Fri
Highs	75°	80°	75°	85°	80°
Low	60°	70°	65°	70°	65°

See *Test Copying Masters*, pp. A53–A54 for copying masters of the Test • Chapters 11–14 in **multiple-choice** format and pp. B203–B204 for **free-response** format.

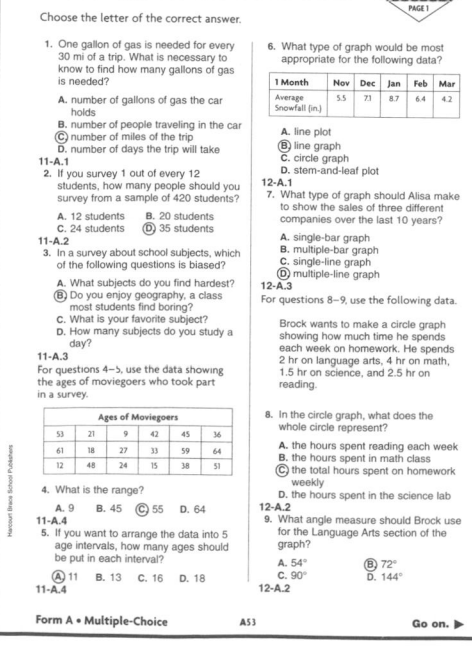

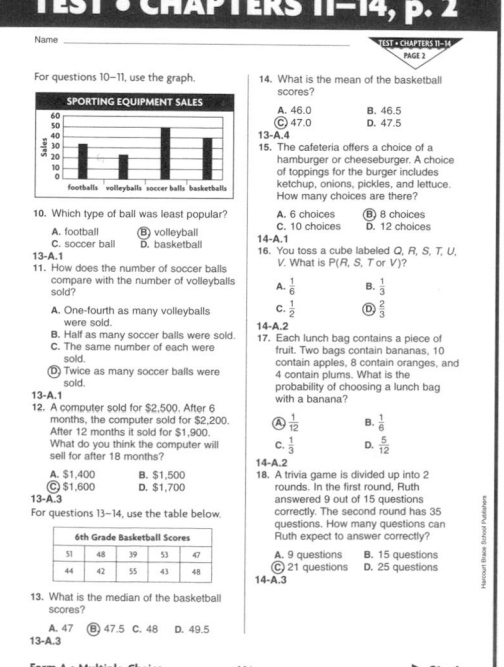

- **Make predictions from graphs.** (pages 244–245)

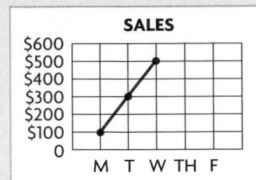

SALES

For Exercises 8–10, use the graph at the left.

8. How much did sales rise from Monday to Tuesday? **$200**

9. How much did sales rise from Tuesday to Wednesday? **$200**

10. What do you predict sales will be on Friday? **$900**

- **Find the mean, median, and mode.** (pages 246–249)

data: 15, 15, 20, 25, 30
mean: $(15 + 15 + 20 + 25 + 30) \div 5 = 21$
median: 20 mode: 15

Find the mean, median, and mode.

11. 4, 6, 7, 5, 8, 9, 3 **6; 6; none**

12. 4.2, 5.5, 8.3, 4.2, 7.8 **6; 5.5; 4.2**

13. 90, 100, 84, 75, 110, 90, 95 **92; 90; 90**

- **Find the probability of events that are not equally likely.** (pages 264–265)

A bag has 10 marbles in it: 4 blue, 3 red, 2 green, and 1 yellow. One marble is drawn.

Find P(blue). $\frac{4}{10} = \frac{2}{5}$

For Exercises 14–17, use the description of the marbles at the left. Find each probability.

14. P(purple) **0**

15. P(yellow) $\frac{1}{10}$

16. P(red or green) $\frac{5}{10} = \frac{1}{2}$

17. P(blue, red, or green) $\frac{9}{10}$

- **Find the experimental probability of an event.** (pages 268–269)

The table shows the results of rolling a number cube 30 times.

Number	1	2	3	4	5	6
Times Rolled	2	3	10	6	5	4

experimental probability of rolling 1 $= \frac{2}{30} = \frac{1}{15}$

Use the table at the left to find each experimental probability.

18. P(2) $\frac{3}{30} = \frac{1}{10}$

19. P(3 or 4) $\frac{16}{30} = \frac{8}{15}$

20. P(even number) $\frac{13}{30}$

21. P(2, 3, or 5) $\frac{18}{30} = \frac{3}{5}$

PROBLEM-SOLVING APPLICATIONS

Solve. Explain your method.
Explanations will vary.

22. The school band has 60 members. There are 36 girls and 24 boys. If you used a circle graph to display this data, what would be the measurements of the angles? (pages 234–235) **216° and 144°**

23. The mean of eight numbers is 100. What is the sum of the numbers? (pages 246–249) **800**

Portfolio

PORTFOLIO SUGGESTIONS
The portfolio illustrates the growth, talents, achievements, and reflections of the mathematics learner. You may wish to have students spend a short time selecting work samples for their portfolios and complete the Portfolio Summary Guide in the *Teacher's Guide for Assessment*, page 23. Students may wish to address the following questions:

- **What new understanding of math have I developed in the past several weeks?**

- **What growth in understanding or skills can I see in my work?**

- **What can I do to improve my understanding of math ideas?**

- **What would I like to learn more about?**

For information about how to organize, share, and evaluate portfolios, see the *Teacher's Guide for Assessment*, pages 20–22.

ADDITIONAL ANSWERS
Grade 6, Chapters 11–14

PE Assessment, Study Guide & Review

5.
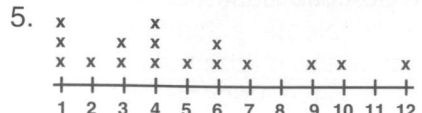

6.

Age Group	Frequency	Cumulative Frequency
1–4	9	9
5–8	4	13
9–12	3	16

7.
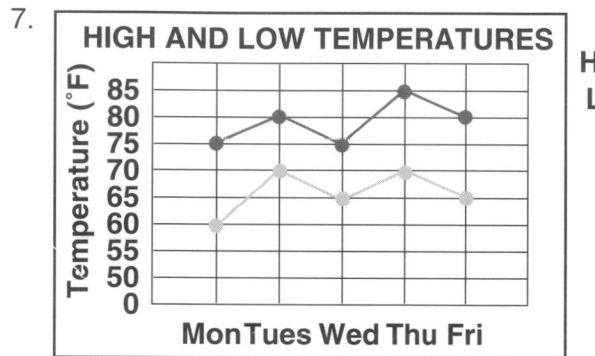

High (—)
Low (—)

✓ *Assessment Checkpoint*

USING PERFORMANCE ASSESSMENT

With the *Performance Assessment Teacher's Guide,*

- have students work individually or in pairs as an alternative to formal assessment.
- use the evaluation checklist to evaluate items 1-4.
- use the rubric and model papers below to evaluate items 5-8.
- use the copying masters to evaluate interview/task tests on pp. 86-89.

SCORING RUBRIC FOR PROBLEM SOLVING TASKS, ITEMS 5–8

Levels of Performance

Level 3   Accomplishes the purposes of the task. Student gives clear explanations, shows understanding of mathematical ideas and processes, and computes accurately.

Level 2   Purposes of the task not fully achieved. Student demonstrates satisfactory but limited understanding of the mathematical ideas and processes.

Level 1   Purposes of the task not accomplished. Student shows little evidence of understanding the mathematical ideas and processes and makes computational and/or procedural errors.

STUDENT WORK SAMPLES for Item 6

CHAPTERS
11–14
WHAT DID I LEARN?

✏ Write About It

For items 1–4, see Evaluation Checklist in the *Performance Assessment Guide.* Explanations will vary.

1. There are 750 students in your school. A dance is planned. The dance committee wants to find out about how many people will attend. How will you survey the students? (pages 210–211) **Checks students' explanations; sample size should be 75 students.**

2. The table to the right shows the number of home runs hit by a little league team. Use the data to make a histogram. (pages 228–229) **Checks students' histograms.**

Runs					
24	3	8	47	19	28
21	40	36	15	2	16

3. Use the information in the table to find the mean, median, and mode. Explain your method. (pages 246–249) **Check students' explanations; mean: 95, median: 96.5, mode: 97**

David's Test Scores				
97	88	96	93	93
98	92	97	99	97

4. You roll 1 number cube numbered 1 to 6. Find P(even number). (pages 260–265) $\frac{3}{6}$ or $\frac{1}{2}$

✓ Performance Assessment

Choose a strategy and solve. Explain your method.

> **Problem-Solving Strategies**
> - Find a Pattern   • Draw a Diagram   • Make a Model
> - Make a Table   • Write an Equation   • Make a Graph

Choice of strategies will vary. Explanations will vary.

5. The scores on a science test were 100, 83, 88, 61, 95, 85, 51, 89, 51, 75, 63, 58, 99, 67, 74, 72, 94, 88, 72, 70, 83, 89, 80, 85, and 87. Find the range. Then make a tally table and a frequency table using intervals. (pages 216–219) **Check students' charts and tables; see additional answers, page 275A.**

6. Use the data below to make a circle graph. (pages 234–235)

Activities			
Reading	Math	Science	Homeroom
90 min	36 min	36 min	18 min

Check students' graphs; 180°; 72°; 72°; 36°

7. Jen bought a car for $5,100. It was worth $4,650 after 6 months, $4,200 after 12 months, and $3,750 after 18 months. What will the car be worth after 3 years? (pages 244–247) **Check students' explanations; $2,400**

8. Pizzas come in small, medium, and large. Toppings include pepperoni, mushrooms, onions, and green peppers. How many different types of one-topping pizzas are possible? (pages 258–259) **12 types of pizza**

Level 3

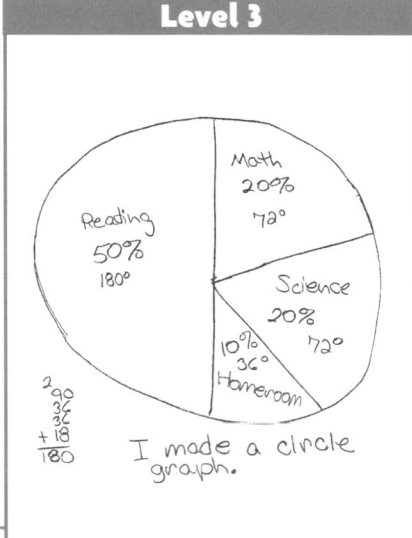

Level 2

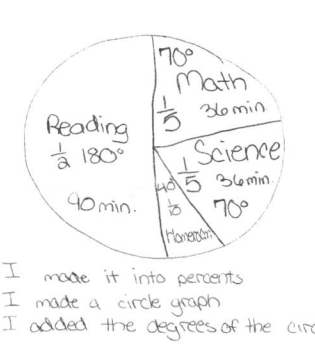

Level 1

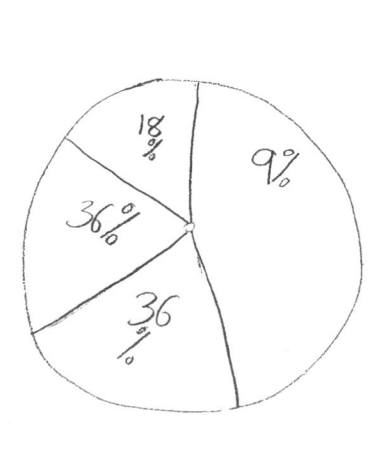

The student added the number of minutes to determine the total minutes. Then he drew an accurate and well labeled circle graph.

The student drew and labeled a circle graph. The fractions she used were correct, but not all of the measures of the central angles are correct.

The student tried to draw a circle graph. He did not understand what the numbers in the table represented and used them as percents.

CUMULATIVE REVIEW

Solve the problem. Then write the letter of the correct answer.

1. Compare. ⁻70 ● ⁺1 (pages 28–29)

  A. <   **B.** >
  C. =   **D.** − **1A.4**

2. Find the quotient. 8)818 (pages 40–43)

  A. 12 r2   **B.** 101
  C. 102   **D.** 102 r2 **2A.3**

3. Find the LCM for 8 and 10. (pages 80–83)

  A. 2   **B.** 10
  C. 40   **D.** 80 **4A.3**

4. Estimate the sum. $2\frac{3}{4} + 2\frac{1}{3}$ (pages 122–123)

  A. ⁻3   **B.** 0
  C. 5   **D.** 6 **6A.2**

For Exercises 5–6, use the diagram below. (pages 152–153) **8A.2**

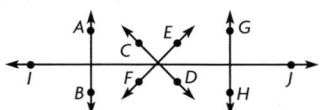

5. $\overleftrightarrow{AB} \parallel$ __?__

  A. $\overleftrightarrow{CD}$ **B.** $\overleftrightarrow{EF}$
  C. $\overleftrightarrow{GH}$ **D.** $\overleftrightarrow{IJ}$ **8A.2**

6. $\overleftrightarrow{GH} \perp$ __?__

  A. $\overleftrightarrow{AB}$ **B.** $\overleftrightarrow{CD}$
  C. $\overleftrightarrow{EF}$ **D.** $\overleftrightarrow{IJ}$ **8A.2**

7. If you survey 1 out of every 10 students, how many would you survey out of a group of 350 students? (pages 210–211)

  A. 10   **B.** 35
  C. 70   **D.** 3,500 **11A.2**

8. You want to graph some data that show favorite subjects of a sample of students. Which type of graph would be the best choice? (pages 224–229) **12A.1**

  A. bar graph   **B.** histogram
  C. line graph   **D.** stem-and-leaf plot

9. You want to make a circle graph. A period of 15 min out of a 60-min workout is spent in warm-up exercises. What angle measurement should you use to represent 15 min? (pages 234–235)

  A. 15°   **B.** 25°
  C. 90°   **D.** 180° **12A.2**

For Exercises 10–11, use the graph below. (pages 240–241) **13A.1**

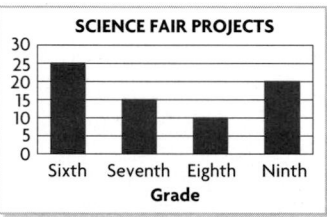

SCIENCE FAIR PROJECTS

Grade

10. Which grade had the most projects?

  A. sixth   **B.** seventh
  C. eighth   **D.** ninth

11. How many more projects did the ninth grade have than the eighth grade?

  A. 0   **B.** 5
  C. 10   **D.** 20

12. For dessert, there are choices of four different pies and three flavors of ice cream. How many possible choices of pie with ice cream are there? (pages 258–259)

  A. 3   **B.** 4
  C. 7   **D.** 12 **14A.1**

Cumulative Review 275

✓ Assessment Checkpoint

USING THE PAGE

This review is in multiple-choice format to provide the students with practice for standardized testing. More in-depth Cumulative Tests are provided in free-response and multiple-choice formats. See *Test Copying Masters*, pages A55–A58 for the multiple-choice test and pages B205–B208 for the free-response test.

CUMULATIVE TEST FREE-RESPONSE FORMAT CHAPTERS 1-14, PAGES 1-4

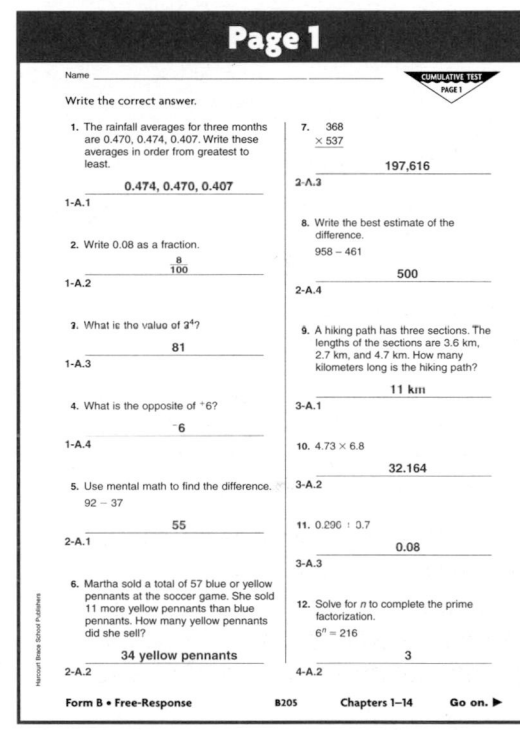

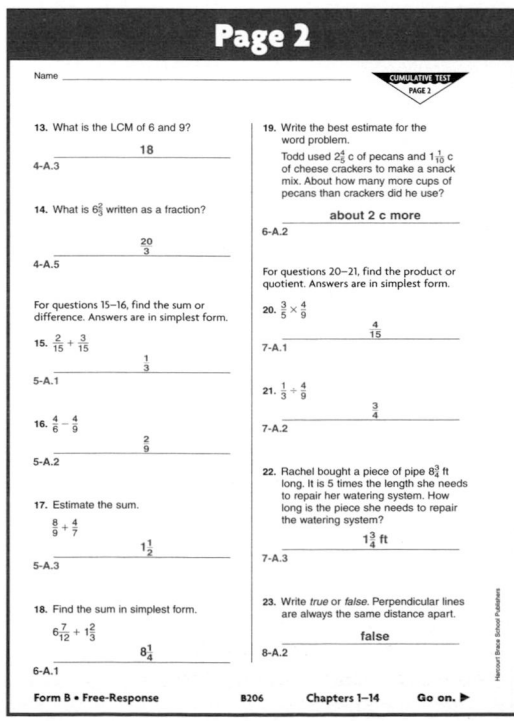

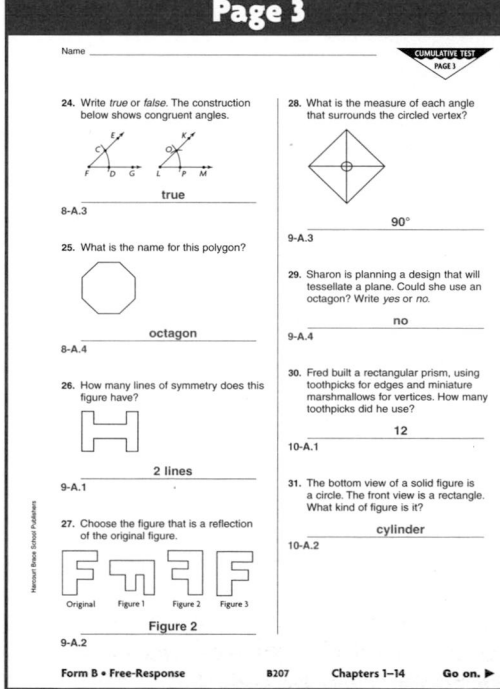

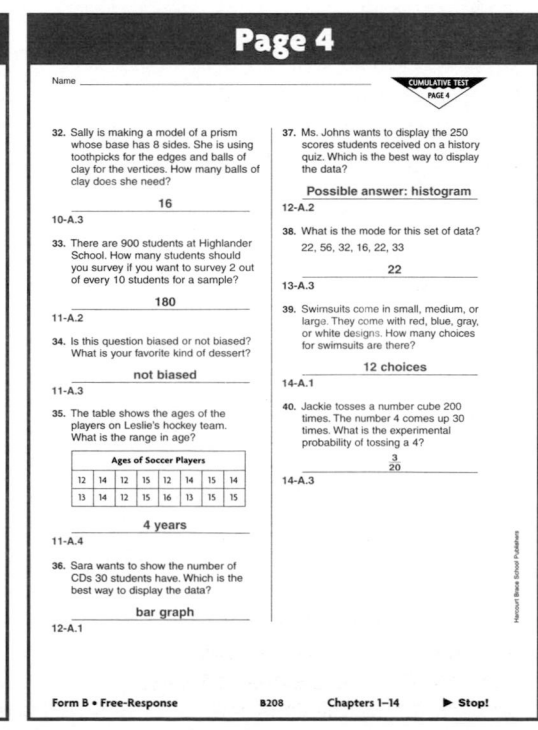

ADDITIONAL ANSWERS
Grade 6, Chapters 11–14

PE Assessment Cumulative Review

What Did I Learn?

5.

Science Test Scores		
Score	Tally	Frequency
90–100	////	4
80–89	//// ////	10
70–79	/////	5
60–69	///	3
50–59	///	3

STUDENT HANDBOOK

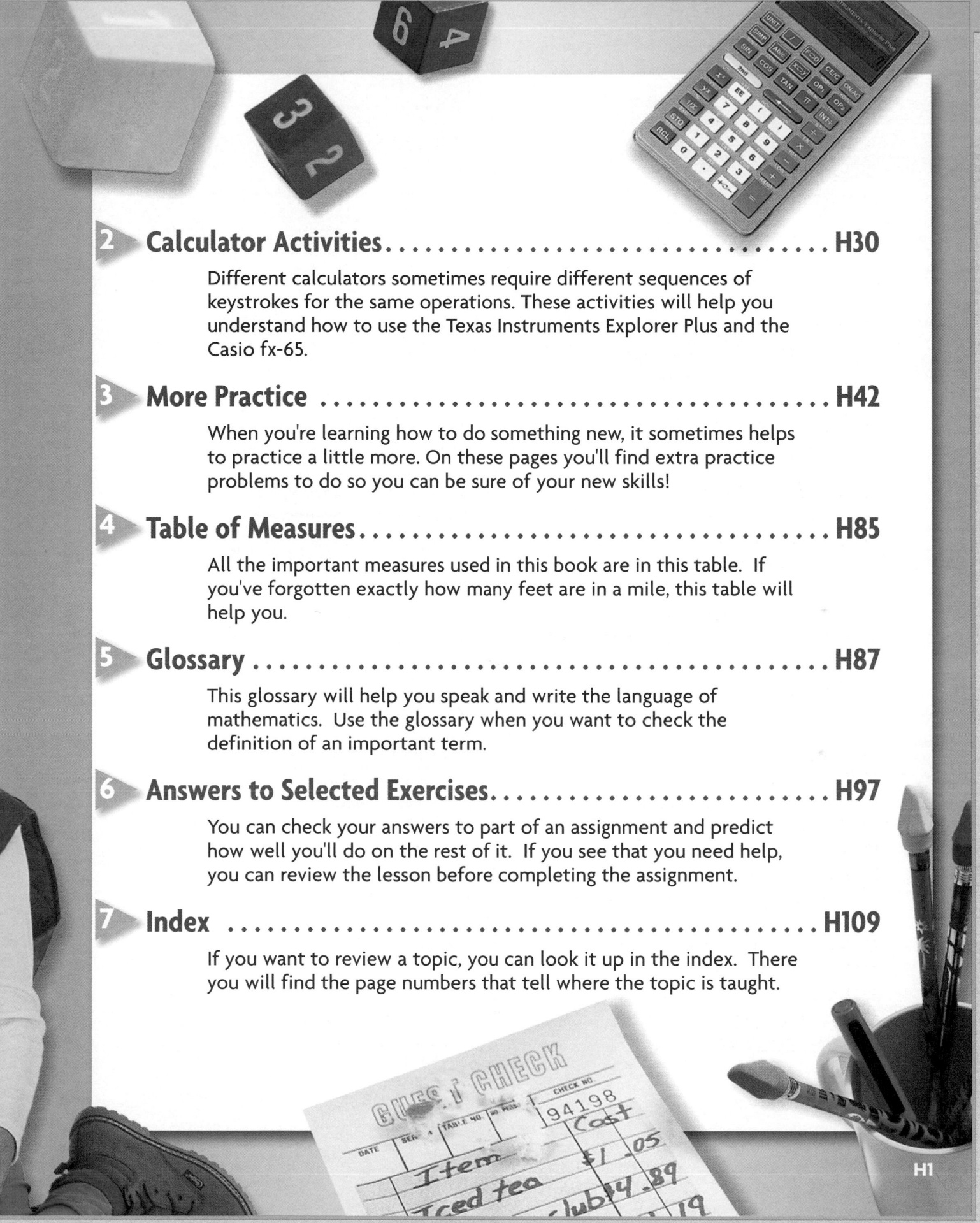

1 PRIME NUMBERS

Numbers which have exactly two different factors, one and the number itself, are **prime numbers**.

Prime	Reason	Not Prime	Reason
2	2 and 1 are the only factors of 2.	6	6 has other factors like 2 and 3.
7	7 and 1 are the only factors of 7.	8	8 has other factors like 2 and 4.
23	23 and 1 are the only factors of 23.	24	24 has other factors like 4 and 6.
47	47 and 1 are the only factors of 47.	36	36 has other factors like 3 and 12.

Examples Determine whether the number is prime or not prime.

A. 15
$3 \times 5 = 15$
15 *is not prime since it has factors other than 1 and 15.*

B. 5
$1 \times 5 = 5$
5 *is prime since it has only two factors, 1 and 5 itself.*

C. 29
$1 \times 29 = 29$
29 *is prime since it has only two factors, 1 and 29 itself.*

PRACTICE Determine if the numbers are prime or not prime. Write *yes* or *no*.

1. 3 yes **2.** 11 yes **3.** 12 no **4.** 19 yes **5.** 14 no **6.** 17 yes

2 COMPOSITE NUMBERS

Numbers which have more than two factors are **composite numbers**.

Composite	Reason	Not Composite	Reason
4	1, 2, and 4 are factors of 4.	5	1 and 5 are only factors of 5.
9	1, 2, and 3 are factors of 9.	7	1 and 7 are only factors of 7.
10	1, 2, 5, and 10 are factors of 10.	17	1 and 17 are only factors of 17.
27	1, 3, 9, and 27 are factors of 27.	29	1 and 29 are only factors of 29.

Examples Determine whether the number is composite or not composite.

A. 15
factors: 1, 3, 5, 15
15 *is composite since it has more than 2 factors.*

B. 13
factors: 1, 13
13 *is not composite since it has only two factors.*

C. 36
factors: 1, 2, 3, 4, 6, 9, 12, 18, 36
36 *is composite since it has more than 2 factors.*

PRACTICE Determine if the numbers are composite or not. Write *yes* or *no*.

1. 6 yes **2.** 21 yes **3.** 2 no **4.** 18 yes **5.** 11 no **6.** 26 yes

3 MULTIPLES

Multiples of a number can be found by multiplying the number by 1, 2, 3, 4, and so on.

Example Find the first five multiples of 3.

$3 \times 1 = 3$
$3 \times 2 = 6$
$3 \times 3 = 9$ → *The numbers 3, 6, 9, 12, and 15 are the first five multiples of 3.*
$3 \times 4 = 12$
$3 \times 5 = 15$

PRACTICE Write the first five multiples of each number.

1. 4 4, 8, 12, 16, 20 **2.** 2 2, 4, 6, 8, 10 **3.** 5 5, 10, 15, 20, 25 **4.** 7 7, 14, 21, 28, 35 **5.** 8 8, 16, 24, 32, 40 **6.** 12 12, 24, 36, 48, 60
7. 9 9, 18, 27, 36, 45 **8.** 10 10, 20, 30, 40, 50 **9.** 20 20, 40, 60, 80, 100 **10.** 15 15, 30, 45, 60, 75 **11.** 50 50, 100, 150, 200, 250 **12.** 18 18, 36, 54, 72, 90

4 FACTORS

When two numbers are multiplied to form a third, the two numbers are said to be **factors** of the third number.

$$4 \times 8 = 32$$
$$\uparrow \quad \uparrow$$
$$factors$$

Example List all the factors of 32.

Think: The only possibilities for factors of 32 are whole numbers from 1 to 32. Begin with 1, since every number has 1 and itself as factors.

$1 \times 32 = 32$ ← *The numbers 1 and 32 are factors of 32.*
$2 \times 16 = 32$ ← *The number 2 and 16 are factors of 32.*
$3 \times \ ? = 32$ ← *No whole number times 3 equals 32, so 3 is not a factor of 32.*
$4 \times \ 8 = 32$ ← *The numbers 4 and 8 are factors of 32.*

Think: The only remaining possibilities must be between 4 and 8. Since no whole number multiplied by 5, 6, or 7 equals 32, they are not factors of 32.

The factors of 32 are 1, 2, 4, 8, 16, and 32.

PRACTICE List all the factors of each number.

1. 8 1, 2, 4, 8 **2.** 20 1, 2, 4, 5, 10, 20 **3.** 9 1, 3, 9 **4.** 51 1, 3, 17, 51 **5.** 16 1, 2, 4, 8, 16 **6.** 27 1, 3, 9, 27
7. 18 1, 2, 3, 6, 9, 18 **8.** 63 1, 3, 7, 9, 21, 63 **9.** 50 1, 2, 5, 10, 25, 50 **10.** 17 1, 17 **11.** 76 1, 2, 4, 19, 38, 76 **12.** 23 1, 23

5 DIVISIBILITY RULES

A number is divisible by another number when the division results in a remainder of 0. You can determine divisibility by some numbers with divisibility rules.

A number is divisible by	Divisible	Not Divisible
2 if the last digit is an even number.	11,994	2,175
3 if the sum of the digits is divisible by 3.	216	79
4 if the last two digits form a number divisible by 4.	1,024	621
5 if the last digit is 0 or 5.	15,195	10,007
6 if the number is divisible by 2 and 3.	1,332	44
9 if the sum of the digits is divisible by 9.	144	33
10 if the last digit is 0.	2,790	9,325

PRACTICE Determine whether each number is divisible by 2, 3, 4, 5, 6, 9, or 10.

1. 56 2, 4 **2.** 200 2, 4, 5, 10 **3.** 75 3, 5 **4.** 324 2, 3, 4, 6, 9 **5.** 42 2, 3, 6 **6.** 812 2, 4
7. 784 2, 4 **8.** 501 3 **9.** 2,345 5 **10.** 555,555 3, 5 **11.** 3,009 3 **12.** 2,001 3

6 PRIME FACTORIZATION

A composite number can be expressed as a product of prime numbers. This is the **prime factorization** of the number. To find the prime factorization of a number, you can use a factor tree.

Example Find the prime factorization of 24 by using a factor tree.

Think: Use 2×12, 3×8, or 4×6.

24	24	24
2×12	3×8	4×6
$2 \times 3 \times 4$	$3 \times 2 \times 4$	$2 \times 2 \times 2 \times 3$
$2 \times 3 \times 2 \times 2$	$3 \times 2 \times 2 \times 2$	
all primes	all primes	all primes

The prime factorization of 24 is $2 \times 2 \times 2 \times 3$, or $2^3 \times 3$.

PRACTICE Find the prime factorization by using a factor tree.

1. 25 5×5 or 5^2 **2.** 16 $2 \times 2 \times 2 \times 2$ or 2^4 **3.** 56 $2 \times 2 \times 2 \times 7$ or $2^3 \times 7$ **4.** 18 $2 \times 3 \times 3$ or 2×3^2 **5.** 72 $2 \times 2 \times 2 \times 3 \times 3$ or $2^3 \times 3^2$ **6.** 40 $2 \times 2 \times 2 \times 5$ or $2^3 \times 5$

7 GREATEST COMMON FACTOR

The **greatest common factor** (*GCF*) of two whole numbers is the greatest factor the numbers have in common.

Example Find the GCF of 24 and 32.

Method 1
List all the factors for both numbers. Find all the common factors.

24: 1, 2, 3, 4, 6, 8, 12, 24
32: 1, 2, 4, 8, 16, 32

The common factors are 1, 2, 4, and 8.

So, the *GCF* is 8.

Method 2
Find the prime factorizations. Then find the common prime factors.

24: $2 \times 2 \times 2 \times 3$
32: $2 \times 2 \times 2 \times 2 \times 2$

The common prime factors are 2, 2, and 2. The product of these is the *GCF*.

So, the *GCF* is $2 \times 2 \times 2 = 8$.

PRACTICE Find the GCF of each pair of numbers by either method.

1. 9, 15 3 **2.** 25, 75 25 **3.** 18, 30 6 **4.** 4, 10 2 **5.** 12, 17 1 **6.** 30, 96 6
7. 54, 72 18 **8.** 15, 20 5 **9.** 40, 60 20 **10.** 40, 50 10 **11.** 14, 21 7 **12.** 14, 28 14

8 LEAST COMMON MULTIPLE

The **least common multiple** (*LCM*) of two numbers is the smallest common multiple the numbers share.

Example Find the least common multiple of 8 and 10.

Method 1
List multiples of both numbers.

8: 8, 16, 24, 32, 40, 48, 56, 64, 72, 80
10: 10, 20, 30, 40, 50, 60, 70, 80, 90

The smallest common multiple is 40.

So, the *LCM* is 40.

Method 2
Find the prime factorizations.

8: $2 \times 2 \times 2$
10: 2×5

The *LCM* is found by finding a product of factors.

$2 \times 2 \times 2 \times 5 = 40$. So, the *LCM* is 40.

PRACTICE Find the LCM of each pair of numbers by either method.

1. 2, 4 4 **2.** 3, 15 15 **3.** 10, 25 50 **4.** 10, 15 30 **5.** 3, 7 21 **6.** 18, 27 54

9 PROPERTIES

The following are basic properties of addition and multiplication.

ADDITION	MULTIPLICATION
Commutative: $a + b = b + a$	**Commutative:** $a \times b = b \times a$
Associative: $(a + b) + c = a + (b + c)$	**Associative:** $(a \times b) \times c = a \times (b \times c)$
Identity Property of Zero:	**Identity Property of One:**
$a + 0 = a$ and $0 + a = a$	$a \times 1 = a$ and $1 \times a = a$
	Property of Zero: $a \times 0 = 0$ and $0 \times a = 0$
	Distributive: $a \times (b + c) = a \times b + a \times c$

PRACTICE Name the property illustrated.

1. $4 + 0 = 4$
  Identity of Addition
2. $(6 + 3) + 1 = 6 + (3 + 1)$
  Associative of Addition
3. $7 \times 51 = 51 \times 7$
  Commutative of Mult.
4. $5 \times 456 = 456 \times 5$
  Commutative of Mult.
5. $17 \times (1 + 3) = 17 \times 1 + 17 \times 3$
  Distributive
6. $1 \times 5 = 5$
  Identity of Mult.
7. $(8 \times 2) \times 5 = 8 \times (2 \times 5)$
  Associative of Mult.
8. $72 + 1,234 = 1,234 + 72$
  Commutative of Addition
9. $0 \times 12 = 0$
  Property of Zero

10 WHOLE-NUMBER PLACE VALUE

The place value chart below can help you read and write whole numbers. Each section is called a period. A period is made up of the hundreds, tens, and ones place. The number 527,019,346,912 is read as "five hundred twenty-seven billion, nineteen million, three hundred forty-six thousand nine hundred twelve.

BILLIONS			MILLIONS			THOUSANDS			ONES		
Hundreds	Tens	Ones	Hundreds	Tens	Ones	Hundreds	Tens	Ones	Hundreds	Tens	Ones
5	2	7,	0	1	9,	3	4	6,	9	1	2

Examples Name the place value of the digit.

A. 5 in the billions period
  5 → hundred billions place
B. 4 in the thousands period
  4 → ten thousands place
C. 9 in the ones period
  9 → hundreds place

PRACTICE Name the place value of the underlined digit.

1. 234,765,890,173
  ones
2. 234,765,890,173
  hundred millions
3. 234,765,890,173
  ten billions
4. 234,765,890,173
  thousands
5. 234,765,890,173
  ten thousands
6. 234,765,890,173
  billions
7. 234,765,890,173
  tens
8. 234,765,890,173
  ten millions

11 COMPARING AND ORDERING WHOLE NUMBERS

The symbols $<, \leq, =, \geq, >, \neq$ can be used to compare whole numbers. The symbol $<$ means "is less than," and $>$ means "is greater than."

Example

A. Use $<$, $>$, or $=$ to compare.

A. 345 ● 350 → 345 $<$ 350
B. 345 ● 340 → 345 $>$ 340
C. 345 ● 345 → 345 $=$ 345
D. 829 ● 828 → 829 $>$ 828

B. Write the numbers in order from least to greatest. Use $<$.

A. 73; 68; 84
  68 $<$ 73 $<$ 84
B. 4,687; 4,874; 4,784
  4,687 $<$ 4,784 $<$ 4,874
C. 123; 119; 175
  119 $<$ 123 $<$ 175

PRACTICE Use $<$, $>$, or $=$ to compare.

1. 899 ● 895 $>$
2. 35 ● 30 $>$
3. 9,876 ● 9,880 $<$
4. 237 ● 237 $=$

Write the numbers in order from least to greatest. Use $<$.

5. 12; 8; 17
  8 $<$ 12 $<$ 17
6. 713; 698; 624
  624 $<$ 698 $<$ 713
7. 1,273; 1,256; 1,114
  1,114 $<$ 1,256 $<$ 1,273
8. 46; 44; 49
  44 $<$ 46 $<$ 49

12 ADDING AND SUBTRACTING WHOLE NUMBERS

When adding or subtracting whole numbers, align corresponding digits.

Examples

Find the sum. 247 + 1,496 + 89

Carry extra digits to the next column.

$$\begin{array}{r} 22 \\ 247 \\ 1,496 \\ +\ 89 \\ \hline 1,832 \end{array}$$ *Think: 7 + 6 + 9 = 22*

So, 247 + 1,496 + 89 is 1,832.

Find the difference. 3,000 − 1,650

Borrow from the next column when necessary.

$$\begin{array}{r} 9 \\ 2\ \cancel{10}\ 10 \\ 3,\cancel{0}\ \cancel{0}\ \cancel{0} \\ -\ 1,\ 6\ 5\ 0 \\ \hline 1,\ 3\ 5\ 0 \end{array}$$

So, 3,000 − 1,650 is 1,350.

PRACTICE Find the sum or difference.

1. 84 + 132 + 1,512
  1,728
2. 79 + 56 + 99
  234
3. 3,492 − 2,270
  1,222
4. 895 − 689
  206
5. 8,906 + 476 + 98
  9,480
6. 340 + 199 + 637
  1,176
7. 9,003 − 5,674
  3,329
8. 1,148 − 739
  409

13 MULTIPLYING WHOLE NUMBERS

When multiplying whole numbers, align corresponding digits.

Example Find the product. 27 × 86

STEP 1 *Multiply by the 6 ones.*

$$\begin{array}{r} 27 \\ \times\ 86 \\ \hline 162 \\ +\ 2160 \\ \hline 2,322 \end{array}$$

STEP 2 *Multiply by the 8 tens.*

$$\begin{array}{r} 27 \\ \times\ 86 \\ \hline 162 \\ +\ 2160 \\ \hline 2,322 \end{array}$$

STEP 3 *Add the partial products.*

$$\begin{array}{r} 27 \\ \times\ 86 \\ \hline 162 \\ +\ 2160 \\ \hline 2,322 \end{array}$$

PRACTICE Find the product.

1. 25 × 15
  375
2. 84 × 37
  3,108
3. 45 × 23
  1,035
4. 67 × 76
  5,092
5. 417 × 12
  5,004
6. 895 × 84
  75,180
7. 365 × 30
  10,950
8. 642 × 205
  131,610

14 DIVIDING WHOLE NUMBERS

In the division problem 146 ÷ 9, the **dividend** is 146 and the **divisor** is 9. In long-division the divisor is outside the long-division symbol and the dividend is inside the long-division symbol. The answer to a division problem is the **quotient**.

divisor
↓
$9\overline{)146}$
↑
dividend

Example Find the quotient. 146 ÷ 9

STEP 1
Divide the hundreds.

$9\overline{)146}$

There are not enough hundreds.

STEP 2
Divide the 14 tens.

$$9\overline{)146} \begin{array}{r} 1 \\ -\ 9\downarrow \\ \hline 56 \end{array}$$

Multiply 1 and 9 and subtract. Bring down the 6.

STEP 3
Divide the 56 ones.

$$9\overline{)146} \begin{array}{r} 16 \\ -\ 9 \\ \hline 56 \\ -\ 54 \\ \hline 2 \end{array}$$

Multiply 6 and 9 and subtract.

STEP 4
Write the remainder.

$$9\overline{)146} \begin{array}{r} 16\ \text{r2} \\ -\ 9 \\ \hline 56 \\ -\ 54 \\ \hline 2 \end{array}$$

The quotient is 16. The remainder is 2.

PRACTICE Find the quotient.

1. 167 ÷ 7 **23 r6**
2. 296 ÷ 3 **98 r2**
3. 510 ÷ 8 **63 r6**
4. 909 ÷ 9 **101**
5. 375 ÷ 15 **25**
6. 780 ÷ 30 **26**
7. 250 ÷ 20 **12 r10**
8. 896 ÷ 45 **19 r41**

15 INVERSE OPERATIONS

Addition and subtraction are inverse operations. Multiplication and division are inverse operations. When performing an operation, you can use its inverse to check your answer.

Examples Use the inverse operation to check the answer.

A.
$$\begin{array}{r} 3,465 \\ -\ 2,197 \\ \hline 1,268 \end{array}$$
Check:
$$\begin{array}{r} 1,268 \\ +\ 2,197 \\ \hline 3,465 \end{array}$$

The inverse operation is addition. Since 1,268 + 2,197 = 3,465, the answer checks.

B.
$$\begin{array}{r} 72 \\ \times\ 9 \\ \hline 648 \end{array}$$
Check: $9\overline{)648}$ 72

The inverse operation is division. Since 648 ÷ 9 = 72, the answer checks.

C.
$$\begin{array}{r} 155 \\ +\ 268 \\ \hline 423 \end{array}$$
Check:
$$\begin{array}{r} 423 \\ -\ 268 \\ \hline 155 \end{array}$$

The inverse operation is subtraction. Since 423 − 268 = 155, the answer checks.

D. $8\overline{)584}$ 73
Check:
$$\begin{array}{r} 73 \\ \times\ 8 \\ \hline 584 \end{array}$$

The inverse operation is multiplication. Since 73 × 8 = 584, the answer checks.

PRACTICE Perform the operation. Check the answer.

1. 36 × 9
  324
2. 425 − 364
  61
3. 984 + 87
  1,071
4. 256 ÷ 8
  32
5. 872 × 20
  17,440
6. 967 − 92
  875

16 RULES FOR ROUNDING

To round to a certain place, follow these steps.
1. Locate the digit in that place, and consider the next digit to the right.
2. If the digit to the right is 5 or greater, round up. If the digit to the right is 4 or less, round down.
3. Change each digit to the right of the rounding place to zero.

Examples

A. Round 125,439.378 to the nearest thousand.

Locate digit.
↓
125,439.378
↑

The digit to the right is less than 5, so the digit in the rounding location stays the same.
↓
125,000 ← *Each digit to the right becomes zero.*

B. Round 125,439.378 to the nearest tenth.

Locate digit.
↓
125,439.378
↑

The digit to the right is greater than 5, so the digit in the rounding location increases by 1.
↓
125,439.4 ← *Each digit to the right is dropped.*

PRACTICE Round 259,345.278 to the place indicated.

1. hundred thousand
  300,000.000
2. ten thousand
  260,000.000
3. thousand
  259,000.000
4. hundred
  259,300.000

17 COMPATIBLE NUMBERS

Compatible numbers divide without a remainder, are close to the actual numbers, and are easy to compute mentally. Use compatible numbers to estimate quotients.

Examples

A. Use compatible numbers to estimate the quotient. $6,134 \div 35$.

$6,134 \div 35$
$6,000 \div 30 = 200 \leftarrow estimate$
↑   ↑
compatible numbers

B. Use compatible numbers to estimate the quotient. $647 \div 7$.

$647 \div 7$
$630 \div 7 = 90 \leftarrow estimate$
↑   ↑
compatible numbers

PRACTICE Estimate the quotient by using compatible numbers. Possible answers are given.
1. $345 \div 5$ 70
2. $5,474 \div 23$ 200
3. $46,170 \div 18$ 2,500
4. $749 \div 7$ 100
5. $861 \div 41$ 20
6. $1,225 \div 2$ 600
7. $968 \div 47$ 20
8. $3,456 \div 432$ 8
9. $5,765 \div 26$ 200
10. $25,012 \div 64$ 500

18 DECIMALS AND PLACE VALUE

A place-value chart can help you read and write decimal numbers. The number 912.5784 (nine hundred twelve and five thousand seven hundred eighty-four ten-thousandths) is shown. Remember that the decimal point is read as "and."

Hundreds	Tens	Ones	Tenths	Hundredths	Thousandths	Ten-Thousandths
9	1	2	5	7	8	4

Examples Name the place value of the given digit from 912.5784.

A. 7 → hundredths
B. 8 → thousandths
C. 5 → tenths
D. 4 → ten-thousandths

PRACTICE Name the place value of the underlined digit.

1. 3.0594   hundredths
2. 3.0594   ten-thousandths
3. 3.0594   thousandths
4. 3.0594   tenths

Write each number in words.

5. 2.345   See Additional Answers, p. H29A.
6. 0.43
7. 10.2062
8. 105.0007

19 COMPARING AND ORDERING DECIMALS

The symbols <, ≤, =, ≥, >, ≠ can be used to compare decimals. To find how decimals compare, align the decimal points and compare the corresponding digits.

Examples Use <, >, or = to compare.

A. 4.135 ● 4.137
4.135
4.137
5 < 7
So, 4.135 < 4.137.

B. 19.010 ● 19.005
19.010
19.005
1 > 0
So, 19.010 > 19.005.

Example Write the numbers in order from least to greatest. Use <.

2.046, 2.039, 2.064
2.046
2.039 → 3 < 4 < 6
2.064
So, 2.039 < 2.046 < 2.064.

PRACTICE Use <, >, or = to compare.

1. 7.456 ● 7.476   <
2. 19.072 ● 19.100   <
3. 0.029 ● 0.029   =
4. 8.103 ● 8.099   >

Write the numbers in order from least to greatest. Use <.

5. 2.15, 2.95, 2.45   2.15 < 2.45 < 2.95
6. 4.175, 4.173, 4.17   4.17 < 4.173 < 4.175
7. 0.043, 0.03, 0.072   0.03 < 0.043 < 0.072
8. 7.29, 7.07, 7.33   7.07 < 7.29 < 7.33

20 MODELING DECIMALS

Decimals can be modeled with decimal squares.

Examples

A. Model 0.7 with a decimal square.
Shade 7 of 10 columns since 0.7 is seven tenths.

B. Model 1.2 with decimal squares.
Shade one entire decimal square and 2 of 10 columns since 1.2 is one and two tenths.

C. Model 0.24 with a decimal square.
Shade 24 of 100 squares since 0.24 is twenty-four hundredths.

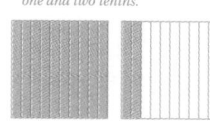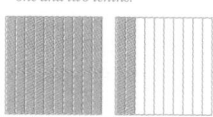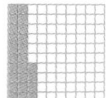

PRACTICE Model with decimal squares. See Additional Answers, p. H29A.
1. 0.4
2. 0.9
3. 0.1
4. 1.7
5. 1.3
6. 0.94
7. 0.47
8. 0.02

21 ADDING AND SUBTRACTING DECIMALS

When adding and subtracting decimals, you must remember to line up the decimal points vertically. You may add zeros to the right of the decimal point as place holders. Adding zeros to the right of the last digit after the decimal point doesn't change the value of the number.

Examples

A. Find the sum. $3.54 + 1.7 + 22 + 13.409$

  3.540
  1.700
22.000
+ 13.409
40.649

B. Find the difference. $636.2 - 28.538$

636.200
− 28.538
607.662

PRACTICE Find the sum or difference.

1. $0.687 + 0.9 + 27.25$ 28.837
2. $87.34 - 6.8$ 80.54
3. $65 + 0.0004 + 2.57$ 67.5704
4. $17 - 0.095$ 16.905
5. $263.7 - 102.08$ 161.62
6. $27 + 3.24 + 0.256 + 0.3689$ 30.8649

22 MULTIPLYING AND DIVIDING DECIMALS BY POWERS OF 10

Notice the pattern below.

$0.24 \times 10 = 2.4$
$0.24 \times 100 = 24$
$0.24 \times 1,000 = 240$
$0.24 \times 10,000 = 2,400$

$10 = 10^1$
$100 = 10^2$
$1,000 = 10^3$
$10,000 = 10^4$

Think: *When multiplying decimals by powers of 10, move the decimal point one place to the right for each power of 10, or for each zero.*

Notice the pattern below.

$0.24 \div 10 = 0.024$
$0.24 \div 100 = 0.0024$
$0.24 \div 1,000 = 0.00024$
$0.24 \div 10,000 = 0.000024$

Think: *When dividing decimals by powers of 10, move the decimal point one place to the left for each power of 10, or for each zero.*

PRACTICE Find the product or quotient.

1. 10×9.26 92.6
2. 0.642×100 64.2
3. $10^3 \times 84.2$ 84,200
4. 0.44×10^4 4,400
5. $69.7 \times 1,000$ 69,700
6. $11.32 \div 10$ 1.132
7. $1.276 \div 1,000$ 0.001276
8. $536.5 \div 10^2$ 5.365
9. $5.92 \div 10^3$ 0.00592
10. $25 \div 10,000$ 0.0025

23 MULTIPLYING DECIMALS

When multiplying decimals, multiply as you would with whole numbers. The sum of the decimal places in the factors equals the number of decimal places in the product.

Examples Find the product.

A. 81.2×6.547

    6.547 ← *3 decimal places*
× 81.2 ← *1 decimal place*
  13094
 65470
+ 5237600
531.6164 ← *4 decimal places*

B. 0.376×0.12

   0.376 ← *3 decimal places*
× 0.12 ← *2 decimal places*
   752
+ 3760
0.04512 ← *5 decimal places*

PRACTICE Find the product.

1. 6.8×3.4 23.12
2. 2.56×4.6 11.776
3. 6.787×7.6 51.5812
4. 0.98×4.6 4.508
5. 0.97×0.76 0.7372
6. 0.5×3.761 1.8805
7. 42×17.654 741.468
8. 7.005×32.1 224.8605
9. 9.76×16.254 158.63904
10. 296.5×2.4 711.60, or 711.6

24 DIVIDING DECIMALS

When dividing with decimals, set up the division as you would with whole numbers. Pay attention to the decimal places, as shown below.

Examples

A. Find the quotient. $89.6 \div 16$

Place decimal point.
↓
     5.6
16)89.6
− 80
  96
− 96
   0

B. Find the quotient. $3.4 \div 4$

Place decimal point.
↓
   0.85
4)3.40 ← *Insert zeros if necessary.*
−32
  20
−20
   0

PRACTICE Find the quotient.

1. $242.76 \div 68$ 3.57
2. $40.5 \div 18$ 2.25
3. $121.03 \div 98$ 1.235
4. $3.6 \div 4$ 0.9
5. $1.58 \div 5$ 0.316
6. $0.2835 \div 2.7$ 0.105
7. $8.1 \div 0.09$ 90
8. $0.42 \div 0.28$ 1.5
9. $15.12 \div 0.063$ 240
10. $480.48 \div 7.7$ 62.4

25 UNDERSTANDING FRACTIONS

You can use a fraction to describe part of a whole. You can also use a fraction to describe part of a group.

Examples Write the fraction for the part that is shaded. Tell whether it represents part of a whole or part of a group.

A.
$\frac{5}{8}$

B. (circles)
$\frac{7}{10}$

Since 5 of 8 parts of one square are shaded, the fraction represents part of a whole.

Since 7 of 10 parts of a group of circles are shaded, the fraction represents part of a group.

PRACTICE Write the fraction for the part that is shaded. Tell whether it represents part of a whole or part of a group.

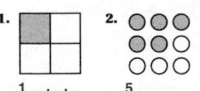 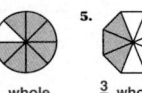

1. $\frac{1}{4}$, whole 2. $\frac{5}{9}$, group 3. $\frac{5}{6}$, group 4. $\frac{7}{8}$, whole 5. $\frac{3}{8}$, whole 6. $\frac{2}{5}$, group

26 EQUIVALENT FRACTIONS

Equivalent fractions are fractions that name the same amount.

Examples

A. Write two equivalent fractions for $\frac{15}{30}$.

Method 1 *Multiply both the numerator and denominator by a whole number.*

$$\frac{15 \times 2}{30 \times 2} = \frac{30}{60}$$

Method 2 *Divide by a common factor of the numerator and denominator.*

$$\frac{15 \div 15}{30 \div 15} = \frac{1}{2}$$

The fractions $\frac{30}{60}$ and $\frac{1}{2}$ are equivalent to $\frac{15}{30}$.

B. Are $\frac{4}{6}$ and $\frac{12}{18}$ equivalent?

Method 1 *Write the fractions in simplest form and compare.*

$$\frac{4 \div 2}{6 \div 2} = \frac{2}{3} \qquad \frac{12 \div 6}{18 \div 6} = \frac{2}{3}$$

The fractions are equivalent.

Method 2 *Cross multiply and compare.*

$4 \times 18 = 72 \qquad \frac{4}{6} = \frac{12}{18} \qquad 12 \times 6 = 72$

The fractions are equivalent since the cross products both equal 72.

PRACTICE Tell whether the fractions are equivalent. Write *yes* or *no*. Then, write two equivalent fractions for each. Possible answers are given.

1. $\frac{15}{30}, \frac{1}{2}$ yes; $\frac{3}{6}, \frac{2}{4}, \frac{5}{10}$
2. $\frac{3}{18}, \frac{1}{3}$ no; $\frac{1}{6}, \frac{2}{12}, \frac{3}{18}$
3. $\frac{3}{6}, \frac{9}{12}$ yes; $\frac{6}{8}, \frac{12}{16}, \frac{15}{20}$
4. $\frac{1}{2}, \frac{5}{10}$ yes; $\frac{2}{4}, \frac{3}{6}, \frac{4}{8}$
5. $\frac{4}{16}, \frac{12}{18}$ no; $\frac{1}{4}, \frac{2}{8}, \frac{6}{24}$
6. $\frac{12}{21}, \frac{4}{7}$ yes; $\frac{16}{28}, \frac{24}{42}, \frac{8}{14}$

27 SIMPLEST FORM OF FRACTIONS

A fraction is in **simplest form** when the numerator and denominator have no common factor other than 1.

Example Find the simplest form of $\frac{32}{40}$.

METHOD 1 *Divide the numerator and denominator by common factors until the only common factor is 1.*

$$\frac{32 \div 2}{40 \div 2} = \frac{16}{20} \qquad \frac{16 \div 4}{20 \div 4} = \frac{4}{5}$$

The simplest form of $\frac{32}{40}$ is $\frac{4}{5}$.

Method 2 *Find the GCF of 32 and 40. Divide both the numerator and the denominator by the GCF.*

$$\frac{32 \div 8}{40 \div 8} = \frac{4}{5} \leftarrow \text{GCF: 8}$$

The simplest form of $\frac{32}{40}$ is $\frac{4}{5}$.

PRACTICE Write in simplest form.

1. $\frac{20}{24}$ $\frac{5}{6}$ 2. $\frac{4}{12}$ $\frac{1}{3}$ 3. $\frac{14}{49}$ $\frac{2}{7}$ 4. $\frac{60}{72}$ $\frac{5}{6}$ 5. $\frac{40}{75}$ $\frac{8}{15}$ 6. $\frac{12}{12}$ 1

7. $\frac{18}{24}$ $\frac{3}{4}$ 8. $\frac{5}{10}$ $\frac{1}{2}$ 9. $\frac{15}{45}$ $\frac{1}{3}$ 10. $\frac{17}{51}$ $\frac{1}{3}$ 11. $\frac{6}{32}$ $\frac{3}{16}$ 12. $\frac{26}{39}$ $\frac{2}{3}$

28 COMPARING FRACTIONS

The symbols $<, \le, =, \ge, >, \ne$ can be used to compare fractions. To find how fractions compare, rename the fractions so they have a common denominator. Then compare the numerators.

Examples Use $<, >,$ or $=$ to compare.

A. $\frac{3}{4} \bullet \frac{4}{5}$ *Rename the fractions.* $\frac{15}{20} \quad \frac{16}{20}$ $15 < 16$ So, $\frac{3}{4} < \frac{4}{5}$.

B. $\frac{2}{3} \bullet \frac{5}{8}$ *Rename the fractions.* $\frac{16}{24} \quad \frac{15}{24}$ $16 > 15$ So, $\frac{2}{3} > \frac{5}{8}$.

C. $\frac{14}{21} \bullet \frac{4}{6}$ *Rename the fractions.* $\frac{2}{3} \quad \frac{2}{3}$ $2 = 2$ So, $\frac{14}{21} = \frac{4}{6}$.

PRACTICE Use $<, >,$ or $=$ to compare.

1. $\frac{2}{3} < \frac{3}{4}$ 2. $\frac{7}{8} > \frac{5}{6}$ 3. $\frac{1}{4} > \frac{1}{8}$ 4. $\frac{2}{4} < \frac{3}{4}$ 5. $\frac{5}{6} > \frac{3}{4}$ 6. $\frac{1}{2} = \frac{3}{6}$

7. $\frac{39}{40} > \frac{7}{8}$ 8. $\frac{1}{3} = \frac{15}{45}$ 9. $\frac{4}{5} > \frac{2}{3}$ 10. $\frac{24}{48} < \frac{27}{36}$ 11. $\frac{4}{6} > \frac{7}{14}$ 12. $\frac{9}{27} = \frac{11}{33}$

29 ORDERING FRACTIONS

To order fractions, rename the fractions so they have a common denominator. Then compare the numerators.

Example Write the fractions $\frac{5}{6}, \frac{1}{2},$ and $\frac{3}{4}$ in order from least to greatest. Use $<$.

STEP 1 *Rename the fractions.*

$$\frac{5}{6} \qquad \frac{1}{2} \qquad \frac{3}{4}$$
$$\downarrow \qquad \downarrow \qquad \downarrow$$
$$\frac{10}{12} \qquad \frac{6}{12} \qquad \frac{9}{12}$$

STEP 2 *Compare the numerators.*

$$6 < 9 < 10$$

STEP 3 *Order the fractions.*

$$\frac{6}{12} < \frac{9}{12} < \frac{10}{12}$$
So, $\frac{1}{2} < \frac{3}{4} < \frac{5}{6}$.

PRACTICE Write the fractions in order from least to greatest. Use $<$.

1. $\frac{4}{5}, \frac{7}{10}, \frac{3}{5}$ $\frac{3}{5} < \frac{7}{10} < \frac{4}{5}$
2. $\frac{7}{2}, \frac{1}{3}, \frac{2}{4}$ $\frac{1}{3} < \frac{2}{4} < \frac{7}{2}$
3. $\frac{1}{3}, \frac{1}{2}, \frac{3}{4}$ $\frac{1}{3} < \frac{1}{2} < \frac{3}{4}$
4. $\frac{5}{6}, \frac{1}{14}, \frac{3}{7}$ $\frac{1}{2} < \frac{3}{4} < \frac{5}{6}$
5. $\frac{4}{14}, \frac{7}{6}, \frac{5}{7}$ $\frac{7}{14} < \frac{4}{6} < \frac{5}{7}$
6. $\frac{4}{9}, \frac{2}{3}, \frac{3}{4}$ $\frac{4}{9} < \frac{2}{3} < \frac{3}{4}$

30 MIXED NUMBERS AND FRACTIONS

Mixed numbers can be written as fractions greater than 1, and fractions greater than 1 can be written as mixed numbers.

Examples

A. Write $\frac{23}{5}$ as a mixed number.

$\frac{23}{5} \rightarrow$ *Divide the numerator by the denominator.*

$5\overline{)23} \rightarrow 4\frac{3}{5} \leftarrow$ *Write the remainder as the numerator of a fraction.*
$\underline{-20}$
3

B. Write $6\frac{2}{7}$ as a fraction.

Multiply the denominator by the whole number. *Add the product to the numerator.*

$6\frac{2}{7} \rightarrow 7 \times 6 = 42 \rightarrow 42 + 2 = 44$

Write the sum over the denominator. $\rightarrow \frac{44}{7}$

PRACTICE Write each mixed number as a fraction. Write each fraction as a mixed number.

1. $\frac{22}{5}$ $4\frac{2}{5}$ 2. $9\frac{1}{7}$ $\frac{64}{7}$ 3. $\frac{41}{8}$ $5\frac{1}{8}$ 4. $5\frac{7}{9}$ $\frac{52}{9}$ 5. $\frac{7}{3}$ $2\frac{1}{3}$ 6. $4\frac{9}{11}$ $\frac{53}{11}$

7. $\frac{47}{16}$ $2\frac{15}{16}$ 8. $3\frac{3}{8}$ $\frac{27}{8}$ 9. $\frac{31}{9}$ $3\frac{4}{9}$ 10. $8\frac{2}{3}$ $\frac{26}{3}$ 11. $\frac{33}{5}$ $6\frac{3}{5}$ 12. $12\frac{1}{9}$ $\frac{109}{9}$

31 ADDING LIKE FRACTIONS

When adding like fractions, add the numerators. Simplify if necessary.

Example Add. Write the sum in simplest form. $\frac{5}{12} + \frac{3}{12}$

Since the denominators are the same, add the numerators.

$$\frac{5}{12} + \frac{3}{12} = \frac{5+3}{12} = \frac{8}{12}$$

Write in simplest form.

$$\frac{8}{12} = \frac{8 \div 4}{12 \div 4} = \frac{2}{3}$$

PRACTICE Add. Write the sum in simplest form.

1. $\frac{1}{4} + \frac{2}{4}$ $\frac{3}{4}$ 2. $\frac{3}{10} + \frac{2}{10}$ $\frac{1}{2}$ 3. $\frac{12}{21} + \frac{2}{21}$ $\frac{2}{3}$ 4. $\frac{3}{8} + \frac{1}{8}$ $\frac{1}{2}$ 5. $\frac{2}{5} + \frac{3}{5}$ 1 6. $\frac{2}{10} + \frac{4}{10}$ $\frac{3}{5}$

7. $\frac{4}{15} + \frac{6}{15}$ $\frac{2}{3}$ 8. $\frac{5}{9} + \frac{2}{9}$ $\frac{7}{9}$ 9. $\frac{7}{12} + \frac{2}{12}$ $\frac{3}{4}$ 10. $\frac{7}{10} + \frac{1}{10}$ $\frac{4}{5}$ 11. $\frac{15}{24} + \frac{5}{24}$ $\frac{5}{6}$ 12. $\frac{3}{48} + \frac{13}{48}$ $\frac{1}{3}$

13. $\frac{2}{7} + \frac{4}{7}$ $\frac{6}{7}$ 14. $\frac{6}{8} + \frac{2}{8}$ 1 15. $\frac{8}{20} + \frac{5}{20}$ $\frac{13}{20}$ 16. $\frac{5}{16} + \frac{4}{16}$ $\frac{9}{16}$ 17. $\frac{12}{14} + \frac{1}{14}$ $\frac{13}{14}$ 18. $\frac{10}{28} + \frac{10}{28}$ $\frac{5}{7}$

32 SUBTRACTING LIKE FRACTIONS

When subtracting like fractions, subtract the numerators. Simplify if necessary.

Example Subtract. Write the sum in simplest form. $\frac{5}{12} - \frac{3}{12}$

Since the denominators are the same, subtract the numerators.

$$\frac{5}{12} - \frac{3}{12} = \frac{5-3}{12} = \frac{2}{12}$$

Write in simplest form.

$$\frac{2}{12} = \frac{2 \div 2}{12 \div 2} = \frac{1}{6}$$

PRACTICE Subtract. Write the difference in simplest form.

1. $\frac{9}{10} - \frac{8}{10}$ $\frac{1}{10}$ 2. $\frac{19}{20} - \frac{3}{20}$ $\frac{4}{5}$ 3. $\frac{11}{12} - \frac{8}{12}$ $\frac{1}{4}$ 4. $\frac{5}{8} - \frac{1}{8}$ $\frac{1}{2}$ 5. $\frac{3}{5} - \frac{1}{5}$ $\frac{2}{5}$ 6. $\frac{7}{9} - \frac{1}{9}$ $\frac{2}{3}$

7. $\frac{17}{18} - \frac{11}{18}$ $\frac{1}{3}$ 8. $\frac{7}{8} - \frac{5}{8}$ $\frac{1}{4}$ 9. $\frac{13}{16} - \frac{1}{16}$ $\frac{3}{4}$ 10. $\frac{13}{14} - \frac{3}{14}$ $\frac{5}{7}$ 11. $\frac{7}{16} - \frac{5}{16}$ $\frac{1}{8}$ 12. $\frac{19}{21} - \frac{5}{21}$ $\frac{2}{3}$

13. $\frac{8}{9} - \frac{4}{9}$ $\frac{4}{9}$ 14. $\frac{12}{15} - \frac{5}{15}$ $\frac{7}{15}$ 15. $\frac{14}{15} - \frac{7}{15}$ $\frac{7}{15}$ 16. $\frac{5}{24} - \frac{4}{24}$ $\frac{1}{24}$ 17. $\frac{2}{4} - \frac{1}{4}$ $\frac{1}{4}$ 18. $\frac{1}{12} - \frac{1}{12}$ 0

33 RENAMING MIXED NUMBERS

A mixed number is the sum of a whole number and a fraction. Sometimes you need to rename a mixed number in order to add or subtract.

Examples

A. Rename $5\frac{5}{4}$ so the fraction is not greater than 1.

$\frac{5}{4} = 1\frac{1}{4}$ *Rename the fraction.*

$5\frac{5}{4} = 5 + \frac{5}{4} = 5 + 1\frac{1}{4} = 6\frac{1}{4}$ *Rename the mixed number.*

So, $5\frac{5}{4} = 6\frac{1}{4}$.

B. Rename $3\frac{1}{4}$ so the whole number is 2.

$3 = 2 + \frac{4}{4}$ *Rename 3 using 2 and a fraction.*

$3\frac{1}{4} = 2 + \frac{4}{4} + \frac{1}{4} = 2\frac{5}{4}$ *Rename the mixed number.*

So, $3\frac{1}{4} = 2\frac{5}{4}$.

PRACTICE Rename the mixed number so the fraction is not greater than 1.

1. $3\frac{6}{5}$ $4\frac{1}{5}$ **2.** $5\frac{7}{4}$ $6\frac{3}{4}$ **3.** $8\frac{5}{8}$ $9\frac{1}{8}$ **4.** $1\frac{13}{8}$ $2\frac{5}{8}$ **5.** $4\frac{10}{7}$ $5\frac{3}{7}$ **6.** $9\frac{3}{2}$ $10\frac{1}{2}$

Rename the mixed number so the whole number is 1 less.

7. $4\frac{1}{2}$ $3\frac{3}{2}$ **8.** $6\frac{2}{5}$ $5\frac{7}{5}$ **9.** $3\frac{2}{7}$ $2\frac{9}{7}$ **10.** $8\frac{1}{4}$ $7\frac{5}{4}$ **11.** $2\frac{5}{8}$ $1\frac{13}{8}$ **12.** $7\frac{6}{11}$ $6\frac{17}{11}$

34 ADDING MIXED NUMBERS

To add mixed numbers, add the fractions and the whole numbers. If the fraction in the sum is greater than 1, rename it.

Example Add. $7\frac{4}{5} + 1\frac{3}{5}$

Add the fractions.   *Add the whole numbers.*   *Rename so the fraction is less than 1.*

$\begin{array}{r} 7\frac{4}{5} \\ + 1\frac{3}{5} \\ \hline \frac{7}{5} \end{array}$

$\begin{array}{r} 7\frac{4}{5} \\ + 1\frac{3}{5} \\ \hline 8\frac{7}{5} \end{array}$

$8\frac{7}{5} = 8 + \frac{7}{5} = 8 + 1\frac{2}{5} = 9\frac{2}{5}$

PRACTICE Add. Rename the sum if the fraction in the sum is greater than 1.

1. $2\frac{1}{3} + 1\frac{1}{3}$ $3\frac{2}{3}$ **2.** $7\frac{3}{5} + 2\frac{1}{5}$ $9\frac{4}{5}$ **3.** $3\frac{3}{7} + 5\frac{6}{7}$ $9\frac{2}{7}$ **4.** $4\frac{5}{8} + 4\frac{4}{8}$ $9\frac{1}{8}$ **5.** $9\frac{7}{12} + 5\frac{9}{12}$ $15\frac{1}{3}$

35 SUBTRACTING MIXED NUMBERS

To subtract mixed numbers, subtract the fractions first. Borrow from the ones if necessary. Then subtract the whole numbers.

Example Subtract. $6\frac{1}{5} - 2\frac{3}{5}$

Since $\frac{1}{5} < \frac{3}{5}$, borrow from the 6.   *To borrow from the 6, rename $6\frac{1}{5}$.*   *Subtract the fractions. Then subtract the whole numbers.*

$\begin{array}{r} 6\frac{1}{5} \\ - 2\frac{3}{5} \\ \hline \end{array}$

$6\frac{1}{5} = 5 + \frac{5}{5} + \frac{1}{5} = 5\frac{6}{5}$

$\begin{array}{r} 6\frac{1}{5} = \\ - 2\frac{3}{5} = \\ \hline \end{array} \begin{array}{r} 5\frac{6}{5} \\ - 2\frac{3}{5} \\ \hline 3\frac{3}{5} \end{array}$

So, $6\frac{1}{5} - 2\frac{3}{5} = 3\frac{3}{5}$.

PRACTICE Subtract.

1. $9\frac{5}{8} - 3\frac{4}{8}$ $6\frac{1}{8}$ **2.** $3\frac{7}{9} - 1\frac{5}{9}$ $2\frac{2}{9}$ **3.** $8\frac{2}{5} - 4\frac{4}{5}$ $3\frac{3}{5}$ **4.** $6\frac{4}{7} - 4\frac{6}{7}$ $1\frac{5}{7}$ **5.** $7\frac{1}{3} - 2\frac{2}{3}$ $4\frac{2}{3}$

6. $4\frac{3}{4} - 2\frac{1}{4}$ $2\frac{2}{4}$ **7.** $11\frac{5}{7} - 5\frac{1}{7}$ $6\frac{4}{7}$ **8.** $3\frac{8}{9} - 3\frac{1}{9}$ $\frac{7}{9}$ **9.** $4\frac{3}{8} - 2\frac{7}{8}$ $1\frac{4}{8}$ **10.** $17\frac{1}{10} - 14\frac{7}{10}$ $2\frac{4}{10}$

36 WRITING A FRACTION AS A DECIMAL

A fraction can be thought of as a division problem. The fraction $\frac{3}{4}$ can be read as "3 divided by 4." To write a fraction as a decimal, you can divide the numerator by the denominator.

Example Write $\frac{3}{4}$ as a decimal.

$\begin{array}{r} 0.75 \\ 4\overline{)3.00} \\ -2\,8 \\ \hline 20 \\ -20 \\ \hline 0 \end{array}$ *Think of $\frac{3}{4}$ as a division problem. Divide the numerator, 3, by the denominator, 4.*   So, $\frac{3}{4} = 0.75$.

PRACTICE Write the fraction as a decimal.

1. $\frac{1}{2}$ 0.5   **2.** $\frac{1}{4}$ 0.25   **3.** $\frac{3}{5}$ 0.6   **4.** $\frac{9}{20}$ 0.45   **5.** $\frac{1}{5}$ 0.2   **6.** $\frac{7}{8}$ 0.875

7. $\frac{3}{4}$ 0.75   **8.** $\frac{1}{8}$ 0.125   **9.** $\frac{8}{9}$ 0.889   **10.** $\frac{3}{8}$ 0.375   **11.** $\frac{1}{10}$ 0.1   **12.** $\frac{7}{10}$ 0.7

37 PARTS OF A CIRCLE

A circle is the set of all points a given distance from a point called the **center**. A **radius** is a line segment with one endpoint at the center of a circle and one endpoint on the circle. A **diameter** is a line segment that passes through the center of the circle and has both endpoints on the circle. A circle is named by its center.

Example Name the circle, the center, a radius, and a diameter of the circle.

name: circle O radius: $\overline{OA}$, $\overline{OB}$, or $\overline{OC}$

center: O diameter: $\overline{BC}$

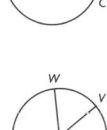

PRACTICE Name each.

1. the circle   circle A **2.** the center A
3. two radii   Possible answers: $\overline{AY}$, $\overline{AW}$ **4.** two diameters. $\overline{WZ}$, $\overline{YV}$

38 TRANSFORMATIONS

Three types of transformations are shown below.

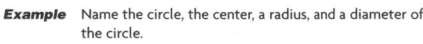

A **translation** occurs when a figure slides along a straight line.   A **reflection** occurs when a figure is flipped over a line.   A **rotation** occurs when a figure is turned around a point.

PRACTICE Identify the transformation as a *translation*, *reflection*, or *rotation*.

1. 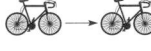   **2.**   **3.**
rotation   reflection   translation

4.    **5.**   **6.**
translation   rotation   reflection

39 TYPES OF POLYGONS

A **polygon** is a closed plane figure with at least three sides. Polygons are classified by number of sides and angles.

triangle	quadrilateral	pentagon	hexagon	octagon

3 sides, 3 angles	4 sides, 4 angles	5 sides, 5 angles	6 sides, 6 angles	8 sides, 8 angles

A **regular polygon** has all sides congruent and all angles congruent.

PRACTICE Name the polygon.

1.    **2.**   **3.**   **4.**   **5.**
quadrilateral   pentagon   triangle   octagon   hexagon

40 TYPES OF TRIANGLES

Triangles are classified by the lengths of their sides and the measures of their angles.

acute triangle	obtuse triangle	right triangle
	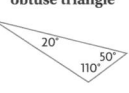	
all angles < 90°	one angle > 90°	one angle = 90°

scalene triangle	equilateral triangle	isosceles triangle
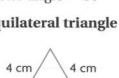		
All sides have different lengths.	All sides have the same length.	Two sides have the same length.

PRACTICE Classify the triangles by using the given information.

1.    **2.**   **3.**   **4.** 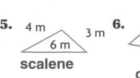   **5.** 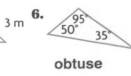   **6.**
equilateral   right   acute   isosceles   scalene   obtuse

41 TYPES OF QUADRILATERALS

A **quadrilateral** is a polygon with 4 sides and 4 angles. There are many different kinds of quadrilaterals with special properties. Some are shown below.

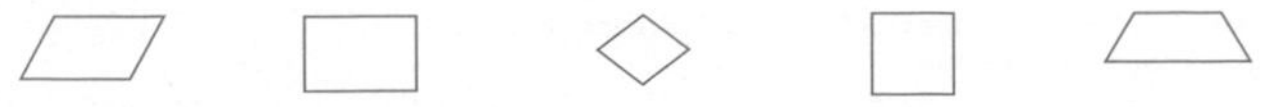

parallelogram	rectangle	rhombus	square	trapezoid
opposite sides parallel and congruent	parallelogram with 4 right angles	parallelogram with 4 congruent sides	rectangle with 4 congruent sides	quadrilateral with exactly 2 parallel sides

PRACTICE Give the most exact name for the figure.

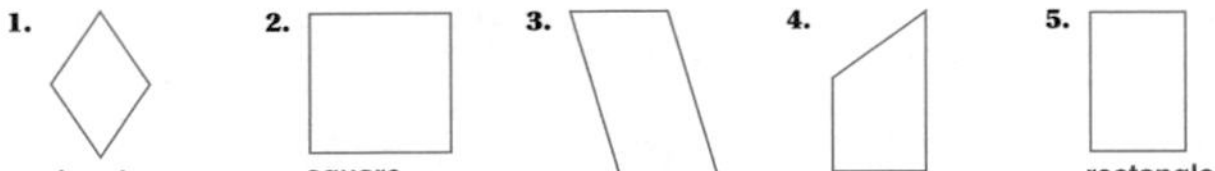

1. rhombus
2. square
3. parallelogram
4. trapezoid
5. rectangle

42 SOLID FIGURES

Five basic types of solid figures are shown below.

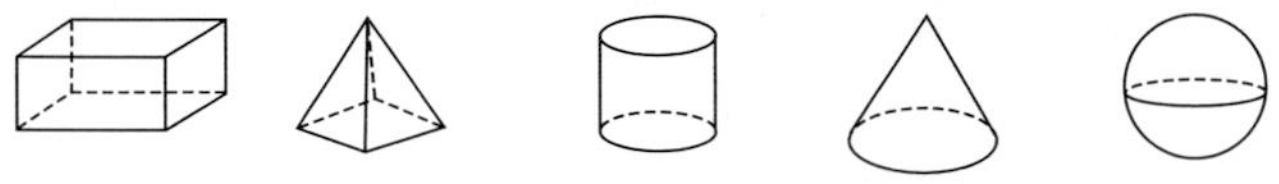

rectangular prism	square pyramid	cylinder	cone	sphere
The base is a rectangle.	The base is a square.	The base is a circle.	The base is a circle.	

PRACTICE Name the figure.

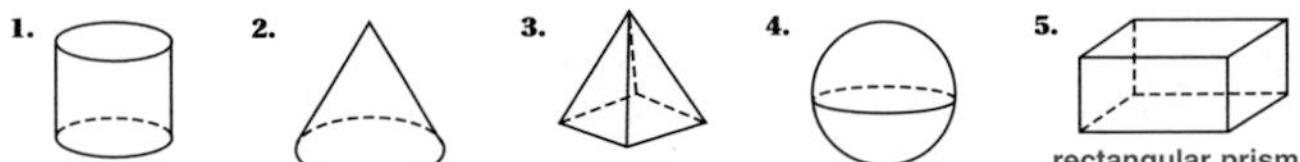

1. cylinder
2. cone
3. square pyramid
4. sphere
5. rectangular prism

43 FACES, EDGES, AND VERTICES

Solid figures made from polygons have faces, edges, and vertices. Each polygon is a **face**. The faces meet to form line segments called **edges**. The edges meet to form points called **vertices**.

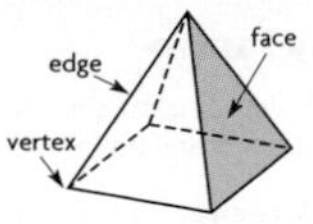

Examples Tell the number of faces, edges, and vertices for the figure.

A.

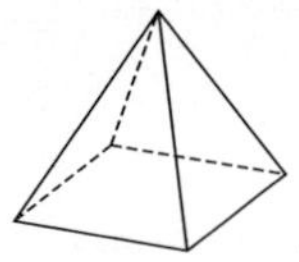

faces: 5
edges: 8
vertices: 5

B.

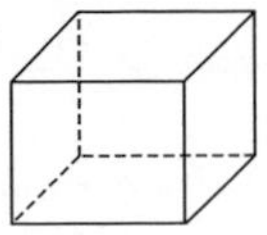

faces: 6
edges: 12
vertices: 8

PRACTICE Tell the number of faces, edges, and vertices for the figure.

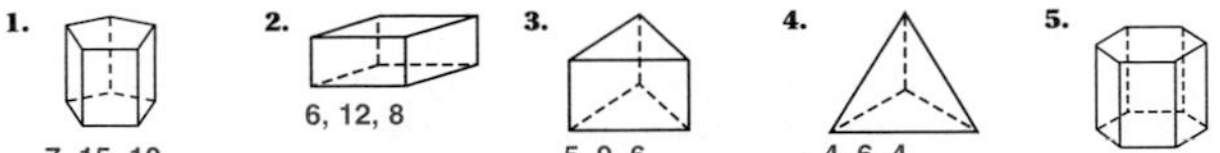

1. 7, 15, 10
2. 6, 12, 8
3. 5, 9, 6
4. 4, 6, 4
5. 8, 18, 12

44 CONGRUENT FIGURES

Congruent figures are figures that have the same size and shape.

Examples Tell if the figures are congruent or not.

A.

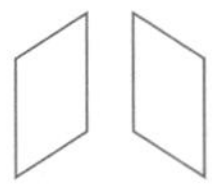

The figures are congruent since they are the same size and shape.

B.

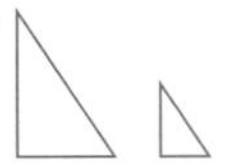

The figures are not congruent since they are not the same size.

PRACTICE Tell if the figures are congruent or not. Write *yes* or *no*.

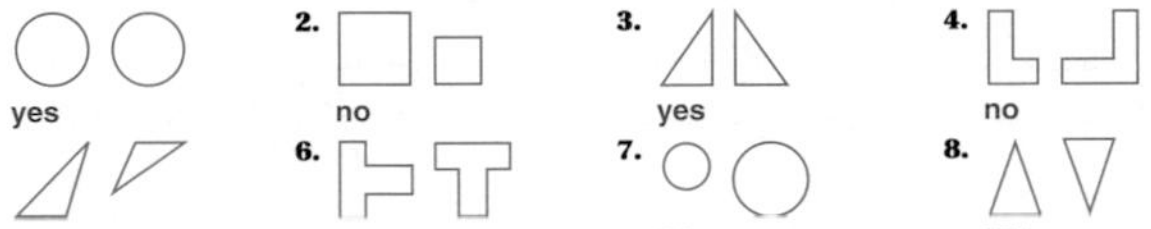

1. yes
2. no
3. yes
4. no
5. no
6. yes
7. no
8. yes

45 CUSTOMARY MEASURES

To convert a measurement from one unit to another, you multiply or divide.

Customary Units of Length

12 inches (in.) = 1 foot (ft)
36 inches (in.) = 1 yard (yd)
3 feet (ft) = 1 yard (yd)
5,280 feet (ft) = 1 mile (mi)
1,760 yards (yd) = 1 mile (mi)

Customary Units of Capacity and Weight

8 fluid ounces (fl oz) = 1 cup (c)
2 cups (c) = 1 pint (pt)
2 pints (pt) = 1 quart (qt)
4 quarts (qt) = 1 gallon (gal)
16 ounces (oz) = 1 pound (lb)
2,000 pounds (lb) = 1 ton (t)

Examples

A. 4 feet = ? inches

4 ft × 12 in. per ft = 48 in.

B. 78 fluid ounces = ? cups

78 fl oz ÷ 8 fl oz per c = 9.75 c

PRACTICE Convert the measure to the given unit.

1. 3 ft = ? in. **36 in.**
2. 3 mi = ? ft **15,840 ft**
3. 72 ft = ? yd **24 yd**
4. 18 yd = ? ft **54 ft**
5. 55 gal = ? qt **220 qt**
6. 19 c = ? fl oz **152 fl oz**
7. 54 qt = ? gal **13.5 gal**
8. 64 oz = ? lb **4 lb**

46 METRIC MEASURES

The metric system is based on the decimal system. When you move from left to right on the chart below, you multiply by powers of 10. When you move from right to left, you divide by powers of 10.

kilometer	hectometer	dekameter	meter	decimeter	centimeter	millimeter
1 km	1 hm	1 dam	1 m	1 dm	1 cm	1 mm
=	=	=	=	=	=	=
1,000 m	100 m	10 m	1 m	0.1 m	0.01 m	0.001 m

Examples

A. 4.5 m = ? mm

Think: *You move from left to right 3 places, so multiply by 1,000 or move the decimal point 3 places to the right.*

4.5 m = 4,500 mm

B. 4.5 m = ? hm

Think: *You move from right to left 2 places, so divide by 100 or move the decimal point 2 places to the left.*

4.5 m = 0.045 hm

PRACTICE Convert the measure to the given unit.

1. 4.5 m = ? cm **450 cm**
2. 7.9 m = ? km **0.0079 km**
3. 0.09 dm = ? m **0.009 m**
4. 0.15 m = ? hm **0.0015 hm**
5. 12 m = ? mm **12,000 mm**
6. 86 dam = ? m **860 m**
7. 0.34 km = ? m **340 m**
8. 480 cm = ? m **4.8 m**

47 PERIMETER

Perimeter is the distance around a figure.

Examples Find the perimeter of the figure.

A.

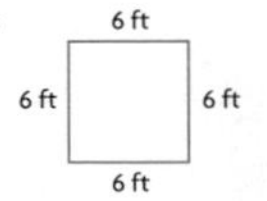

$P = 4 \times s$
$P = 4 \times 6$
$P = 24$ ft

B.

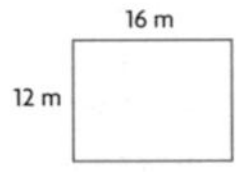

$P = 2(l + w)$
$P = 2(16 + 12)$
$P = 2 \times 28$
$P = 56$ m

C.

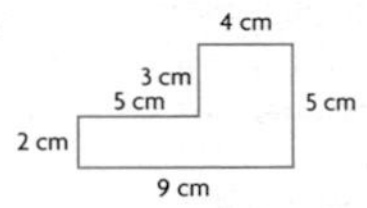

$P = a + b + c + d + e + f$
$P = 9 + 2 + 5 + 3 + 4 + 5$
$P = 28$ cm

PRACTICE Find the perimeter of the figure.

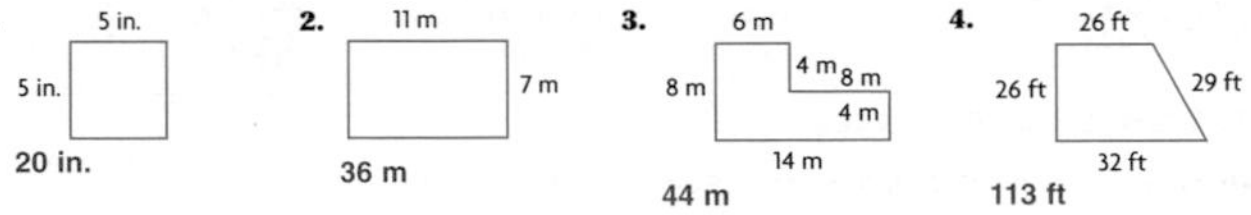

1. 20 in.
2. 36 m
3. 44 m
4. 113 ft

48 AREA

To find the area of a figure, you can count the number of square units covered by the figure.

Examples Find the area of the shaded region.

A.

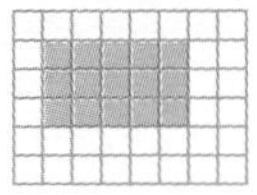

15 full squares

$A = 15 \text{ units}^2$

B.

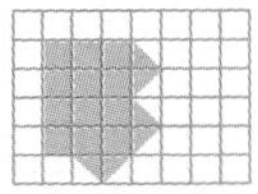

12 full squares
6 half squares

$A = \text{no. of full squares} + \frac{1}{2}(\text{no. of half squares})$
$A = 12 + \frac{1}{2}(6)$
$A = 12 + 3 = 15$

15 units^2

PRACTICE Find the area of the shaded region.

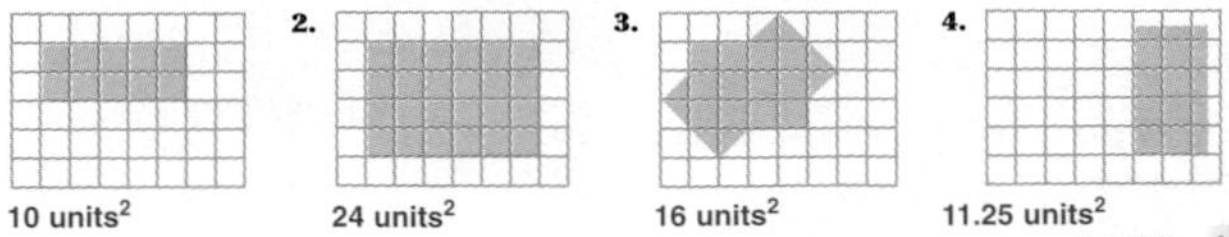

1. 10 units^2
2. 24 units^2
3. 16 units^2
4. 11.25 units^2

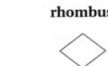

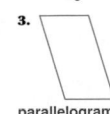

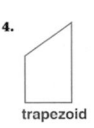

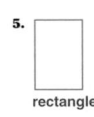

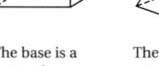

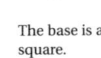

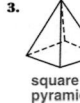

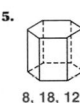

49 MEAN

The **mean** is the average of a set of numbers. To find the mean, add the numbers and divide the sum by the number of addends.

Examples Find the mean of the set of numbers. 36, 74, 43, 36, 41

Find the sum of the numbers.

$36 + 74 + 43 + 36 + 41 = 230$

Divide the sum by the number of addends.

$230 \div 5 = 46$

So, the mean is 46.

PRACTICE Find the mean of the set of numbers.

1. 32, 87, 45, 63, 73
60

2. 17, 44, 33, 10
26

3. 6, 52, 41, 21, 36, 48
34

4. 126, 99, 234.3
153.1

50 MEDIAN AND MODE

The **median** is the middle number of a set of numbers in numerical order. If there are two middle numbers, the median is the mean of those two numbers. The **mode** of a set of numbers is the most commonly occurring number in the set. There may be more than one mode, or there may be no mode at all.

Examples Find the median and mode of the set of numbers.

A. 4, 25, 72, 36, 41, 2, 98

Median

Put the numbers in numerical order.

2, 4, 25, 36, 41, 72, 98

So, the median is 36.

Mode

Since no number in the set occurs more than once, there is no mode.

B. 4, 25, 72, 4, 36, 4, 2, 25, 25, 98

Median

Put the numbers in numerical order.

2, 4, 4, 4, 25, 25, 25, 36, 72, 98

$\frac{25 + 25}{2} = \frac{50}{2} = 25$ So, the median is 25.

Mode

Since both 4 and 25 occur three times, 4 and 25 are the modes.

PRACTICE Find the median and the mode of the set of numbers.

1. 2, 1, 2, 5, 7, 9
3.5; 2

2. 42, 8, 54, 192, 8, 0, 44, 16
29; 8

3. 12, 44, 324, 17, 41
41; none

4. 0, 77, 125, 77, 2, 2, 3, 5, 3
3; 2, 3, 77

5. 5, 0, 15, 25, 5, 23, 1
5; 5

6. 297, 7, 12, 18, 5, 21, 17
17; none

51 RANGE

The **range** is the distance between extremes in a set of numbers. You can find the range by finding the difference between the greatest number and the least number. You can find the size of intervals for the data by dividing the range by the number of intervals needed.

Example Find the range of the set of numbers. Divide the range into 5 intervals for the data.

range: $44 - 2 = 42$

So, the range is 42.

2	8	12	15
15	20	25	30
35	40	42	44

5 intervals: $42 \div 5 \approx 9$

So, use 5 intervals of 9 or 10 values. Intervals of 10 are:
0–9, 10–19, 20–29, 30–39, 40–49.

PRACTICE Find the range of the set of numbers. Divide the range into 5 intervals for the data. Intervals may vary. Possible sizes are given.

1. 8, 24, 32, 44, 26,
47, 17, 14, 19
39; 8 or 9

2. 115, 42, 120, 99,
80, 56, 79, 84
78; 16 or 17

3. 77, 72, 70, 88, 94,
99, 83, 78, 99
29; 5 or 6

4. 346, 250, 225, 194,
355, 290, 323
161; 33 or 34

52 RATIOS

A **ratio** is a comparison of two numbers. A ratio, like a fraction, can be written in simplest form. There are three ways to write a ratio.

Example In a group of sixth graders, there are 20 boys and 25 girls. Write the ratio of boys to girls in three different ways. Write the ratios in simplest form.

$\frac{20}{25}$, or $\frac{4}{5}$

20:25, or 4:5

20 to 25, or 4 to 5

The ratio in simplest form is read "4 to 5," no matter how it is written. For every 4 boys, there are 5 girls.

PRACTICE In a bag of marbles, 7 are green, 3 are red, 10 are blue, 14 are yellow, and 2 are orange. Write the ratio in simplest form in three different ways.

1. green to red
7 to 3, 7:3, $\frac{7}{3}$

2. blue to yellow
5 to 7, 5:7, $\frac{5}{7}$

3. orange to green
2 to 7, 2:7, $\frac{2}{7}$

4. green to orange
7 to 2, 7:2, $\frac{7}{2}$

53 PERCENTS

Percent means "per hundred". Percents can be modeled with 10×10 decimal squares.

Examples Use the decimal square to answer the questions.

What percent of the decimal square is shaded blue?

Since 42 of 100 squares are shaded blue, 42% of the decimal square is shaded blue.

What percent of the decimal square is shaded red?

Since 39 of 100 squares are shaded red, 39% of the decimal square is shaded red.

PRACTICE Use the decimal square at the right.

1. What percent is shaded red? **29%**

2. What percent is shaded blue? **24%**

3. What percent is shaded green? **14%**

4. What percent is shaded yellow? **10%**

5. What percent is shaded red or blue? **53%**

6. What percent is not shaded? **23%**

54 PERCENTS AND DECIMALS

You can use a 10×10 decimal square to show the relationship between decimals and percents.

Examples Write a decimal and a percent for the amount shaded.

A. 0.57, 57%

B. 0.07, 7%

To change from a decimal to a percent, move the decimal point two places to the right and add a percent symbol. To change from a percent to a decimal, move the decimal point two places to the left and drop the percent symbol.

PRACTICE Write a decimal and a percent for the amount shaded.

1. 0.67, 67% **2.** 0.13, 13% **3.** 0.94, 94% **4.** 0.03, 3%

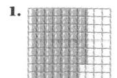

Change the percent to a decimal or the decimal to a percent.

5. 72% **0.72** **6.** 8% **0.08** **7.** 0.06 **6%** **8.** 0.72 **72%**

55 PERCENTS AND RATIOS

A percent can be written as a ratio, and a ratio can be written as a percent.

Examples

A. Write 39% as a ratio.

Percent means "per hundred."

$39\% = \frac{39}{100}$

B. Write $\frac{47}{100}$ as a percent.

Percent means "per hundred."

$\frac{47}{100} = 47\%$

C. Write $\frac{3}{5}$ as a percent.

Change $\frac{3}{5}$ to a decimal. Then change the decimal to a percent.

$\frac{3}{5} = 3 \div 5 = 0.6 = 60\%$

So, $\frac{3}{5} = 60\%$.

PRACTICE Write the percent as a ratio.

1. 31% $\frac{31}{100}$ **2.** 79% $\frac{79}{100}$ **3.** 19% $\frac{19}{100}$ **4.** 27% $\frac{27}{100}$ **5.** 61% $\frac{61}{100}$

Write the ratio as a percent.

6. $\frac{53}{100}$ 53% **7.** $\frac{7}{100}$ 7% **8.** $\frac{91}{100}$ 91% **9.** $\frac{4}{5}$ 80% **10.** $\frac{13}{20}$ 65%

KEY SKILLS
ADDITIONAL ANSWERS
Key Skill #18, page H10

Decimals and Place Value

Two and three hundred forty-five thousandths

Forty-three hundredths

Ten and two thousand sixty-two ten-thousandths

One hundred five and seven ten-thousandths

Key Skill #20, page H11

Modeling Decimals

1.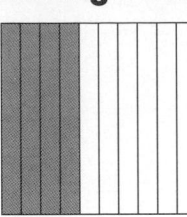

2.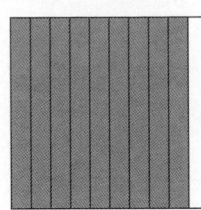

3.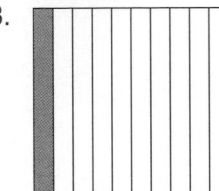

4.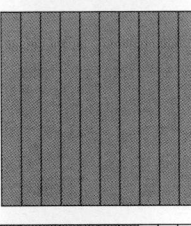

5.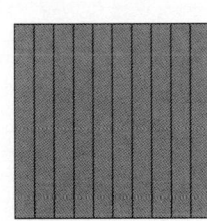

6.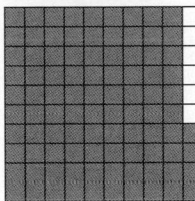

7.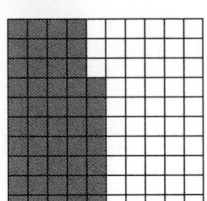

8.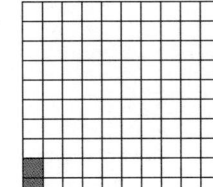

- Lesson 1.4 Exponents, pp. 24–25
- Lesson 2.4 Multiplication and Division, pp. 40–41
- Lesson 4.5 Mixed Numbers and Fractions, pp. 88–89

CALCULATOR Activities

LOTS AND LOTS OF KIDS
Scientific Notation

Michali had a difficult time imagining how many young people the table represented. Linell helped her understand. Follow along as Linell explains how to use the *TI Explorer Plus* to work with the numbers.

Total Population Under 15 Years Old	
China	3.159×10^8
India	3.044×10^8
Indonesia	6.574×10^7
United States	5.384×10^7
Brazil	5.165×10^7
Pakistan	5.163×10^7
Nigeria	5.141×10^7
Bangladesh	4.707×10^7
Mexico	3.211×10^7
Iran	2.676×10^7

Using the Calculator

"How many more children are there in Indonesia than in the United States?" Michali asked. Linell entered the following key sequence to answer her question.

"OK," sighed Michali. "But what number is that, really?" Linell demonstrated to Michali how to show the number in decimal notation.

3. 1.18049×10^1, or 11.80493274

PRACTICE

Use your calculator and the population data.

1. Calculate the total number of children in the North and South American countries shown. **1.3759 × 10⁸**

2. How many more children live in Nigeria than in Mexico? **1.930×10^7**

3. For every child in Iran, how many children live in China?

4. The total population of China is one billion, one hundred fifty three million. The total population of the United States is two hundred fifty million. How many more people live in China than in the United States? Use scientific notation to express the answer. **9.03×10^8**

> **REMEMBER:**
> Scientific notation always includes a decimal number with one digit in the ones place and the remaining digits to the right of the decimal point. The exponent of 10 indicates the true size of the number.

ALL AT ONCE
Division with Remainders

Hanako needs to freeze 69 dumplings she made. Each of her pans can hold 20 dumplings at a time. How many pans does she need for all of the dumplings?

$69 \div 20 = 3 \text{ r}9$

Hanako needs 4 pans to hold all the dumplings.

Using the Calculator

There are 4 people in Hanako's family. She wants to separate the dumplings into packages of 8 dumplings so each family member can have 2 dumplings as a side dish at a meal. How many packages will she be able to make? How many dumplings will be left over?

Use the following keys on the *Casio fx-65* to find the answer.

Use the following keys on the *TI Explorer Plus* to find the answer.

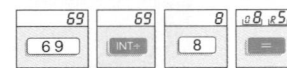

> **REMEMBER:**
> Integers are all the whole numbers, all the negative whole numbers, and zero.
> These are integers:
> 2   ⁻56   0
> ⁻987,900   12
> 390,218,900
> These are not integers:
> ⁻2.3 0.009 $\frac{4}{5}$

PRACTICE

Use your calculator to find the quotient as an integer and remainder.

1. $789 \div 12$ **65 r9**
2. $59 \div 5$ **11 r4**
3. $85 \div 8$ **10 r5**
4. $3,258 \div 25$ **130 r8**
5. $359,523 \div 40$ **8,988 r3**
6. $70 \div 3$ **23 r1**
7. $82 \div 7$ **11 r5**
8. $29 \div 4$ **7 r1**
9. $102 \div 18$ **5 r12**

10. Tyesha waited in the line for the carousel with William, her younger brother. There were 15 horses on the carousel. Tyesha and William counted 64 people in line ahead of them. How many more times would the carousel ride begin before Tyesha and William could get on? **4 times; Tyesha and William could ride on the fifth turn.**

11. Tyesha attends Division Middle School. The principal at Division Middle School has ordered 1,190 desks. He plans to divide them equally among 45 classrooms. How many desks will be in each classroom? Will there be any extra desks? If so, how many? **26; there will be 20 extra desks.**

FIRST THIS ONE, THEN THE NEXT
Solve Multistep Problems

Alison's class saved their change for several months. They had 692 pennies, 465 dimes, 274 nickels, and 178 quarters. How much money did they save?

Using the Calculator

Use the *Casio fx-65* to calculate the class savings.

Use the *TI Explorer Plus* to calculate the class savings.

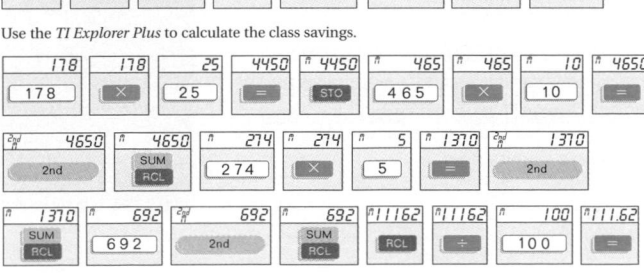

The class saved $111.62.

> **REMEMBER:**
> Money can be counted as cents or as dollars. One cent is one hundredth of one dollar; one dollar is one hundred times one cent.
> 0.01 × 100 = 1.00
> 1¢ × 100 = $1.00
> 1.00 ÷ 100 = 0.01
> $1.00 ÷ 100 = 1¢

PRACTICE

Use your calculator.

1. Mrs. Saloman counted the number of students in each school group that visited the museum where she worked. During one week the following numbers of students visited: 4 groups of 42; 5 groups of 23; 8 groups of 31; 2 groups of 57; and 3 groups of 27. How many students visited the museum as members of school groups? **726 students**

FIRST THINGS FIRST
Order of Operations

Eton and Luba both found the value of $9 + 132 \div 3 - 18$. Can you explain Eton's mistake?

Eton	Luba
$9 + 132 \div 3 - 18$	$9 + 132 \div 3 - 18$
$141 \div ⁻15$	$9 + 44 - 18$
$⁻9.4$	$53 - 18$
	35

Using the Calculator

You can use the *Casio fx-65* to solve the problem.

You can use the *TI Explorer Plus* to solve the problem.

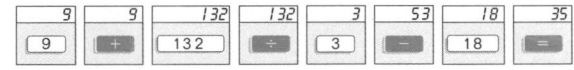

If Eton added parentheses to his work, his answer would be correct. Here's how to include parentheses on the *Casio fx-65*.

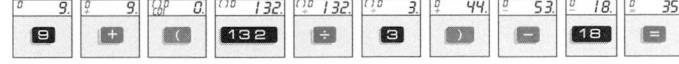

Here's how to include parentheses on the *TI Explorer Plus*.

> **REMEMBER:**
> Use **PEMDAS** to remember the order of operations. The letters stand for the following:
> **P**arentheses
> **E**xponents
> **M**ultiplication and **D**ivision
> **A**ddition and **S**ubtraction

PRACTICE

Use your calculator.

1. $2 \times 11 + 3 \times 4$ **34**
2. $(2 \times 11 + 3) \times 4$ **100**
3. $2 \times (11 + 3 \times 4)$ **46**
4. $(270 - 140) \div 2 + 3 \times 6$ **83**
5. $270 - 140 \div (2 + 3 \times 6)$ **263**
6. $270 - 140 \div 2 + 3 \times 6$ **218**

- Lesson 5.3 Adding Unlike Fractions, pp.100–101
- Lesson 6.3 Adding and Subtracting Mixed Numbers, pp. 118–121

- Lesson 1.3 Decimals and Fractions, pp. 20–23
- Lesson 17.3 Percents, pp. 320–323

THUMBS ACROSS THE UNITED STATES
Operations with Fractions, Simplify

Mr. Jones challenged his class to measure something in a way that it had not been measured before. Brian and Joon thought and thought. Then they decided to use the widths of their thumbs to measure across their desk map of the United States.

Brian's thumb is $\frac{7}{16}$ in. wide, and Joon's is $\frac{3}{4}$ in. wide. They worked together and each used 14 thumb widths to cover the distance from San Francisco to New York. How many inches did they cover?

REMEMBER:
When you need to simplify a fraction, you don't want to change the relationship between the numerator and denominator. To keep the relationship the same, you can divide the fraction only by a number that equals 1.
$$\frac{112}{1,016} \div \frac{8}{8} = \frac{14}{127} \qquad \frac{8}{8} = 1$$

Using the Calculator

You can use the *Casio fx-65* to calculate the measurements with these keystrokes.

You can use the *TI Explorer Plus* to calculate the measurements with these keystrokes.

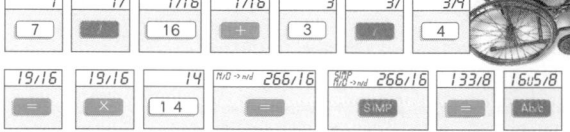

So, they covered $16\frac{5}{8}$ in.

PRACTICE
Use your calculator. Simplify your answers.

1. $\frac{6}{7} + \frac{9}{11}$ $1\frac{52}{77}$
2. $\frac{67}{16} - 3\frac{3}{4}$ $3\frac{7}{16}$
3. $\frac{484}{28} - 13$ $4\frac{2}{7}$
4. $\frac{107}{12} \times \frac{19}{20}$ $8\frac{113}{240}$
5. $\frac{4}{5} \times \frac{139}{8}$ $13\frac{9}{10}$

IT'S THE SAME, ONLY DIFFERENT
Change Decimals to Fractions and Fractions to Decimals

Robert and Keisha are changing fractions to decimals. Keisha wrote the following:

$$\frac{2}{5} = 0.4$$

"But that doesn't make any sense!" Robert exclaimed. "How can a fraction and a decimal be the same thing? Two fifths doesn't even have a four in it!"

"Watch," Keisha explained. "I'll show you how to do the conversion with your calculator."

REMEMBER:
Fractions can be thought of as division problems. When you do the division, the quotient is in decimal form.
$$\frac{3}{4} = 4\overline{)3} = 0.75$$
$$\frac{5}{16} = 16\overline{)5} = 0.3125$$

Using the Calculator

Follow along as Keisha shows Robert how to change fractions to decimals and decimals to fractions on the *Casio fx-65*.

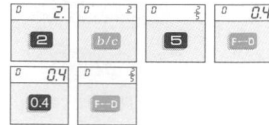

You can use the *TI Explorer Plus* to change fractions to decimals and decimals to fractions.

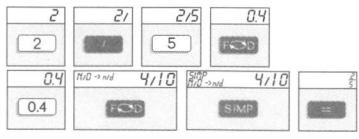

PRACTICE
Use your calculator. Change the fractions to decimals. Change the decimals to fractions. Write the fractions in simplest form.

1. 0.625 $\frac{5}{8}$
2. $\frac{4}{5}$ 0.8
3. 0.55 $\frac{11}{20}$
4. $\frac{7}{4}$ 1.75
5. $\frac{1}{6}$ 0.166666667
6. 0.12 $\frac{3}{25}$
7. $\frac{16}{3}$ 5.333333333
8. 2.125 $2\frac{1}{8}$

- Lesson 18.2 Percent of a Number, pp. 336–339

PART OF THE WHOLE
Solve Percent Problems

The Hamilton-Martin School's 56 sixth graders plan to visit the Native American Art Museum. When Yona and Letitia telephoned to make the arrangements, they were told that the normal admission price is $4.50 per person.

However, the museum also told the girls they could have a 15% discount. What is the total admission cost for the sixth grade?

Using the Calculator

You can use the following key sequence to calculate the total cost of admission on a *Casio fx-65*.

You can use the following key sequence to calculate the total cost of admission on a *TI Explorer Plus*.

So, the total admission is $214.20.

PRACTICE
Use your calculator.

1. What is 5% of 175? 8.75
2. What is 125% of 36? 45
3. What is 20% of 45? 9
4. What is 150% of 70? 105
5. What is 70% of 85? 59.5
6. What is 140% of 90? 126
7. At the museum gift shop, each of the Hamilton-Martin students bought a souvenir for $2.50. The students each paid sales tax of 4%. What was the total amount of money the students spent at the gift shop? $145.60
8. Yona bought another souvenir at the museum gift shop. She bought a small basket that had a Native American design on it. The price of the basket was $12.00, but the shop gave Yona a 20% discount. How much did Yona pay for the basket? $9.60

- Lab Activity Circumference, pp. 398–399
- Lesson 22.5, Finding the Area of a Circle, pp. 416–417

ROUND AND ROUND HE GOES
Area and Circumference of a Circle

Reena and Susannah tied the end of their dog's leash to a stake in their backyard. The dog's leash is 15 feet long. How large is the area that the dog can roam? Round your answer to the nearest thousandth.

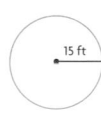
15 ft

REMEMBER:
The formulas to find the area and circumference of a circle are easy to confuse.
$$A = \pi r^2 \qquad C = \pi d$$
$$\text{or } C = 2\pi r$$
To find each for a circle with a radius of 6 cm:
$$A = \pi \times 6^2 \qquad C = \pi \times 12$$

Using the Calculator

To find the area, press these keys on a *Casio fx-65*:

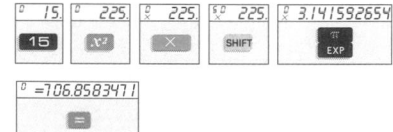

To find the area, press these keys on a *TI Explorer Plus*:

So, the area is 706.858 ft².

PRACTICE
Use your calculator to find the area and circumference of each circle. Round each answer to the nearest thousandth.

1. radius = 2.5 mi
2. radius = 0.32 mm
3. diameter = 52 yd
4. Natalie plans to make a wreath out of six strands of bittersweet, a vine that grows near her house. She would like the wreath to have a diameter of 16 inches. About how much bittersweet does Natalie need to cut? about 300 in., or 25 ft
5. Jason is installing a sprinkler that sprays in a circle with a radius of 12 ft. How much lawn is covered by this sprinkler? 452.389 ft²

1. $A = 19.635$ mi²
   $C = 15.708$ mi
2. $A = 0.322$ mm²
   $C = 2.011$ mm
3. $A = 2,123.717$ yd²
   $C = 163.363$ yd

SECOND DIMENSIONS
Square and Square Root

Tiernan is saving his money to buy a Japanese fighting fish. He wants to put a small square fish tank on a shelf in his room. The shelf is only 5 in. wide. If the tank is the same width as the shelf, what is the area the tank will cover?

Using the Calculator

You can calculate the answer by using the *Casio fx-65.*

You can calculate the answer by using the *TI Explorer Plus.*

So, the area the tank will cover is 25 in.²

Tiernan noticed that he had 64 in.² of space on the corner of his desk. What is the greatest possible length a square fish tank could be to fit on the desk?

You can find the answer by using the *Casio fx-65.*

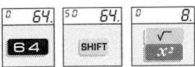

You can find the answer by using the *TI Explorer Plus.*

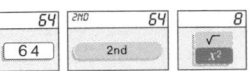

So, the greatest length is 8 in.

PRACTICE

Use your calculator.
1. 11^2 121   2. 20^2 400   3. 15^2 225   4. 46^2 2,116

Round to the nearest hundredth.
5. $\sqrt{1,225}$ 35   6. $\sqrt{9,801}$ 99   7. $\sqrt{5,869}$ 76.61   8. $\sqrt{6,149}$ 78.42

> **REMEMBER:**
> A number squared is that number multiplied by itself. You can see the multiplication pattern by making squares of blocks. The number of blocks in the square is the square of the number of blocks on each side.
>
> $4^2 = 16$ $3^2 = 9$

RANDOMLY CHOSEN
Random Number Generator

Ms. Salvadori, the principal, had to choose 3 of 1,000 people at Benson Corners Middle School to represent the school at the town Founders' Parade. Each deserved an equal chance. Ms. Salvadori gave each person a number between 0 and 0.999. But she didn't really know what to do next. How could she choose?

Using the Calculator

You can help Ms. Salvadori by using a *Casio fx-65.* Enter these keystrokes three times to get three different numbers.

You can help Ms. Salvadori by using a *TI Explorer Plus.* Enter these keystrokes three times to get three different numbers.

Valentin wants to survey a random sample of 10 of the 100 sixth graders in the Benson Corners Middle School. He assigns each student a number between 1 and 100. He repeats the keystrokes below ten times using his *TI Explorer Plus* to choose ten numbers at random.

PRACTICE

Use your calculator.

1. Sarah and April played a game with the spinner at the right. If the number was odd, April got one point. If the number was even, Sarah got one point. Simulate their game by using your calculator. Assign numbers ending in 1, 3, 5, 7, and 9 to April and the rest to Sarah. Who won in your simulation? **Answers will vary.**

2. International Pizzerias would like to survey 75 people in your school about their favorite pizza. Write a letter to the company's president, Marshall Spangler, explaining how to use a calculator to choose the 75 people randomly. Include information about random numbers. **Check students' answers.**

> **REMEMBER:**
> Numbers that were chosen randomly each had the exact same mathematical probability. They each had the same chance of being chosen.

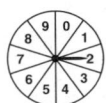

UP WITH THE POSITIVE, DOWN WITH THE NEGATIVE
Operations with Integers

The Brooklyn Bridge was opened on May 24, 1883, and still carries pedestrian and motor vehicle traffic across the East River between Manhattan and Brooklyn. Each of the towers is sunk deep into the riverbed, 79 ft below the surface of the water, and soars 266.5 ft high above the surface of the water.

We can say that the height of the underwater part of the tower is ⁻79 ft in relation to the surface of the water at 0 ft. To find the total distance from the top of the tower to the bottom, we have to subtract ⁻79 from 266.5. How do you do that?

Using the Calculator

You can find the total distance from the top of the tower to the bottom by using the *Casio fx-65.*

You can find the total distance from the top of the tower to the bottom by using the *TI Explorer Plus.*

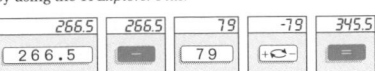

PRACTICE

Use your calculator.
1. ⁻18 ÷ 6 × 3 ⁻9   2. ⁻2 − (⁻6) + 12 16   3. 7 × (⁻4) + (⁻20) ⁻48
4. 369 ÷ ⁻3 × ⁻2 246   5. 2,002 × ⁻4,100 6102   6. 3 − ⁻2 + 1 − ⁻19 + ⁻9 + 6 22

> **REMEMBER:**
> *Negative* can be thought of as "the opposite of." *The opposite* means "the same distance from zero on a number line."
>
> These numbers are opposites:
> 3 and ⁻3
> 45 and ⁻45
> 187.4 and ⁻187.4

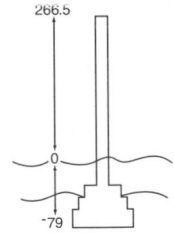

CRANKY NUMBERS, MEAN DATA
Find the Mean of a Set of Data

Isabel and Felipe collected data on the number of centimeters a basketball bounced when dropped from the same height with differing amounts of air in it. Here is their data for a regulation ball:

80 cm, 75 cm, 82 cm, 74 cm, 74 cm

What is the average bounce?

Using the Calculator

You can use these keystrokes to find the sum and the mean of a data set by using the *TI Explorer Plus.*

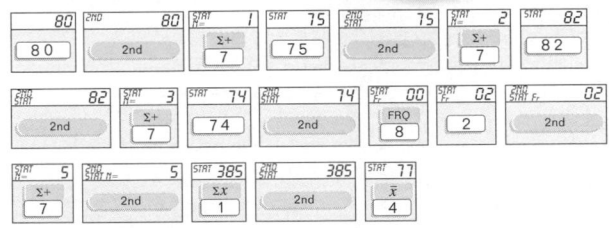

So, the sum of the data set is 385. The mean is 77.

PRACTICE

Use your calculator. Find the sum and mean of the data sets. Round the mean to the nearest hundredth.
1. Men's Long Jump: First Place, Olympics, 1960–1992 76.15 m; 8.46 m
   8.67 m, 8.72 m, 8.54 m, 8.54 m, 8.35 m, 8.24 m, 8.9 m, 8.07 m, 8.12 m
2. Number of Push-ups Niko Did Each Day
   16, 18, 21, 17, 15, 15, 21, 15, 21, 14, 19 192, 17.45
3. Number of Lawns Mowed by Kurt's Lawn Service Last Week
   3, 5, 4, 6, 5, 7, 8 38; 5.43
4. Jared scored 80, 83, 96, 91, 78, 88, 90, 92, and 94 on his social studies tests. His goal was to have an average of 85 on his tests. Use your calculator to find Jared's mean test score. Did he meet his goal?
   88; Yes

> **REMEMBER:**
> The mean of a set of data is calculated by finding the sum of the data, and then dividing by the number of items in the set of data.

More Practice

CHAPTER 1

Lesson 1.1
Write the value of the digit 3.

1. 301
3 hundreds

2. 0.0231
3 thousandths

3. 8.3
3 tenths

Write the number in words.

4. 18 eighteen

5. 0.05 five hundredths

6. 2,000,500
two million, five hundred

7. 200.05
two hundred and five hundredths

Write the number in standard form.

8. thirteen hundredths
0.13

9. two hundred, five and five tenths
205.5

10. two hundred fifty thousand, four hundred and two hundredths
250,400.02

11. three hundred sixteen 316 **12.** one and thirty-seven ten thousandths 1.0037

Lesson 1.2
Compare the numbers. Write $<$, $>$ or $=$.

1. 5.099 ? 5.999 $<$

2. 226.5 ? 226.4 $>$

3. 251.36 ? 241.36 $>$

4. 18.3 ? 18.30 $=$

5. 418 ? 428 $<$

6. 49.089 ? 49.098 $<$

Write the numbers in order from least to greatest. Use $<$.

7. 82.16, 82, 82.15
82 $<$ 82.15 $<$ 82.16

8. 141.14, 114.41, 141.41
114.41 $<$ 141.14 $<$ 141.41

Write the numbers in order from greatest to least. Use $>$.

9. 14.63, 14.36, 13.46, 13.64
14.63 $>$ 14.36 $>$ 13.64 $>$ 13.46

10. 0.300, 3.000, 0.030, 0.303
3.000 $>$ 0.303 $>$ 0.300 $>$ 0.030

Lesson 1.3
Use place value to change the decimal to a fraction.

1. 0.16 $\frac{16}{100}$ **2.** 0.021 $\frac{21}{1000}$ **3.** 0.02 $\frac{2}{100}$ **4.** 0.7 $\frac{7}{10}$

Write as a decimal.

5. $\frac{5}{8}$ 0.625 **6.** $\frac{14}{16}$ 0.875 **7.** $\frac{9}{100}$ 0.09 **8.** $\frac{2}{8}$ 0.25

Write $<$ or $>$.

9. $\frac{1}{7}$? $\frac{2}{5}$ $<$ **10.** $\frac{2}{7}$? $\frac{1}{3}$ $<$ **11.** $\frac{5}{9}$? 0.56 $<$ **12.** 0.6 ? $\frac{62}{60}$ $<$

Lesson 1.4
Write in exponent form.

1. $3 \times 3 \times 3 \times 3$
3^4

2. $1 \times 1 \times 1 \times 1 \times 1 \times 1$
1^6

3. 10×10
10^2

4. $12 \times 12 \times 12$
12^3

5. $21 \times 21 \times 21 \times 21 \times 21$
21^5

6. $1 \times 1 \times 1 \times 1$
1^4

Find the value.

7. 10^4 10,000 **8.** 5^6 15,625 **9.** 2^3 8

10. 7^4 2,401 **11.** 15^1 15 **12.** 27^2 729

Lesson 1.5
Compare the integers. Use $<$ or $>$.

1. $^-2$? $^-6$ $>$ **2.** $^-6$? $^+6$ $<$ **3.** $^+3$? $^+1$ $>$ **4.** 0 ? $^-3$ $>$

5. $^+1$? $^-8$ $>$ **6.** $^-5$? 0 $<$ **7.** $^+7$? $^-2$ $>$ **8.** $^-3$? $^+3$ $<$

9. $^-2$? $^+3$ $<$ **10.** 0 ? $^+4$ $<$ **11.** $^-8$? $^-3$ $<$ **12.** $^-7$? $^+1$ $<$

Order the integers from least to greatest. Use $<$.

13. $^+3$, $^-2$, $^+6$, 0 $^-2$, 0, $^+3$, $^+6$

14. 0, $^-6$, $^-8$, $^-2$ $^-8$, $^-6$, $^-2$, 0

15. $^+2$, $^-2$, $^+6$, $^-6$ $^-6$, $^-2$, $^+2$, $^+6$

16. $^+3$, $^-5$, 0 $^+2$ $^-5$, 0, $^+2$, $^+3$

Name the opposite of the given integer.

17. $^-3$ $^+3$ **18.** $^+2$ $^-2$ **19.** 0 0 **20.** $^-17$ $^+17$

CHAPTER 2

Lesson 2.1
Use mental math to add.

1. $15 + 17 + 5$ 37 **2.** $13 + 9 + 7$ 29 **3.** $37 + 3 + 19$ 59

4. $51 + 7 + 29$ 87 **5.** $15 + 1 + 15$ 31 **6.** $19 + 52 + 21$ 92

Use compensation to add.

7. $26 + 18$ 44 **8.** $14 + 15$ 29 **9.** $28 + 41$ 69

10. $19 + 26$ 45 **11.** $38 + 15$ 53 **12.** $19 + 31$ 50

Use compensation to subtract.

13. $51 - 16$ 35 **14.** $54 - 42$ 12 **15.** $47 - 29$ 18

16. $35 - 14$ 21 **17.** $52 - 21$ 31 **18.** $38 - 22$ 16

Lesson 2.2
Guess and check to solve.

1. Harry ate three meals yesterday. The total calories were 2,300. Dinner included 500 calories more than breakfast and lunch together. How many calories did dinner include? 1,400 calories

2. Rebecca has 36 feet of fencing to create the border for her new garden. She wants the garden to be square. How long is each side? 9 feet

3. Tom took his new convertible car for a drive to his friend's farm. He drove a total of 84 miles. He drove 12 miles longer to get there than he drove to get home. How long was the ride home? 36 miles

4. Sean's model rocket flew for a total of 42 seconds on its first flight. The time it took to reach its highest point was 22 seconds shorter than the time it took to return to the ground. How long did it take to reach its highest point? 10 seconds

Lesson 2.3
Find the missing factor(s).

1. $36 \times 5 = (30 \times 5) + (\underline{?} \times 5)$ 6

2. $51 \times 8 = (50 \times 8) + (\underline{?} \times 8)$ 1

3. $43 \times 7 = (\underline{?} \times 7) + (3 \times 7)$ 40

4. $26 \times 3 = (\underline{?} \times 3) + (6 \times 3)$ 20

5. $24 \times \underline{?} = (24 \times 30) + (24 \times 1)$ 31

6. $63 \times 17 = (63 \times \underline{?}) + (63 \times 7)$ 10

7. $42 \times 15 = (42 \times 10) + (42 \times \underline{?})$ 5

8. $12 \times \underline{?} = (12 \times 30) + (12 \times 7)$ 37

9. $36 \times 12 = (36 \times \underline{?}) + (36 \times \underline{?})$ 10,2

10. $3 \times 21 = (3 \times \underline{?}) + (3 \times 1)$ 20

11. $27 \times 21 = (27 \times \underline{?}) + (27 \times \underline{?})$ 20,1

12. $16 \times \underline{?} = (16 \times 20) + (16 \times 8)$ 28

13. $36 \times 8 = (\underline{?} \times 8) + (6 \times 8)$ 30

14. $41 \times 36 = (\underline{?} \times 36) + (1 \times 36)$ 40

Lesson 2.4
Find the product.

1. 23×12 276 **2.** 28×18 504 **3.** 127×19 2,413

4. 156×83 12,948 **5.** 540×81 43,740 **6.** $1,206 \times 15$ 18,090

7. $2,150 \times 93$ 199,950 **8.** $4,301 \times 179$ 769,879 **9.** $5,220 \times 860$ 4,489,200

10. $5,825 \times 703$ 4,094,975 **11.** $8,710 \times 50$ 435,500 **12.** $7,007 \times 598$ 4,190,186

Find the quotient.

13. $5\overline{)125}$ 5 **14.** $7\overline{)630}$ 90 **15.** $8\overline{)96}$ 12

16. $11\overline{)2,541}$ 231 **17.** $9\overline{)153}$ 17 **18.** $13\overline{)169}$ 13

19. $5\overline{)29}$ 5 r 4 **20.** $12\overline{)53}$ 4 r 5 **21.** $25\overline{)612}$ 24 r 12

22. $21\overline{)85}$ 4 r 1 **23.** $77\overline{)426}$ 5 r 41 **24.** $12\overline{)4,285}$ 357 r 1

Lesson 2.5
Estimate the sum or difference. Possible estimates given.

1. $1,200 + 4,006 + 3,210$ 8,000 **2.** $378 + 98 + 682$ 1,200

3. $5,403 + 4,320$ 9,700 **4.** $482 - 71$ 410

5. $340 - 208$ 130 **6.** $8,710 - 4,203$ 4,400

Estimate the product. Possible estimates given.

7. 38×4 160 **8.** 74×3 210 **9.** 69×42 2,800

10. 620×32 18,000 **11.** 71×29 2,100 **12.** 487×311 150,000

Estimate the quotient. Possible estimates given.

13. $481 \div 59$ 8 **14.** $448 \div 88$ 5 **15.** $2,844 \div 67$ 40 **16.** $6,340 \div 71$ 90

17. $3,480 \div 48$ 70 **18.** $6,319 \div 81$ 80 **19.** $4,233 \div 608$ 7 **20.** $7,197 \div 810$ 9

CHAPTER 3

Lesson 3.1
Estimate the sum. Possible estimates given.

1. $0.4 + 0.9$ 1 **2.** $2.6 + 8.1$ 11 **3.** $0.31 + 2.04$ 2

4. $5.92 + 3.15$ 9 **5.** $0.218 + 2.143$ 2 **6.** $5.816 + 3.215$ 9

Find the sum.

7. $2.14 + 6.08$
8.22

8. $7.04 + 0.13$
7.17

9. $0.126 + 0.408$
0.534

10. $8.360 + 5.216$
13.576

11. $17.31 + 12.06$
29.37

12. $0.08 + 7.42$
7.5

13. $\$18.56 + \$3.12$
$21.68

14. $\$8.26 + \$26.15$
$34.41

15. $\$31.18 + \$125.50$
$156.68

Lesson 3.2
Find the difference.

1. $11.4 - 6.2$
5.2

2. $12.8 - 4.1$
8.7

3. $17.3 - 16.5$
0.8

4. $\$13.20 - \$8.17$
$5.03

5. $\$21.80 - \$17.40$
$4.40

6. $\$35.51 - \$23.17$
$12.34

7. $8.026 - 7.317$
0.709

8. $19.408 - 1.582$
17.826

9. $5.888 - 5.261$
0.627

10. $13.601 - 10.311$
3.29

11. $5 - 3.021$
1.979

12. $628.71 - 527.21$
101.5

Find the missing number.

13. $2.56 + \underline{?} = 8.21$ 5.65 **14.** $12.18 + \underline{?} = 17.81$ 5.63

15. $3.8 - \underline{?} = 2.6$ 1.2 **16.** $721 - \underline{?} = 385.51$ 335.49

Lesson 3.3

Estimate the product. **Possible estimates given.**

1. 2.9×3.7 **12**   2. 18.8×8.3 **200**   3. 5.8×11.4 **50**

4. 9.8×21.513 **200**   5. 788.3×9.2 **8,000**   6. 308.9×58.3 **18,000**

Copy the problem. Place the decimal point in the product.

7. $3.12 \times 6.4 = 19968$ **19.968**
8. $0.48 \times 6.2 = 2976$ **2.976**
9. $21.3 \times 18.4 = 39192$ **391.92**
10. $7.03 \times 7.05 = 495615$ **49.5615**
11. $12.31 \times 17.42 = 2144402$ **214.4402**
12. $212.3 \times 3.26 = 692098$ **692.098**
13. $360.05 \times 12.6 = 453663$ **4536.63**
14. $762 \times 3.285 = 2503170$ **2503.170**

Find the product.

15. 3×0.4 **1.2**   16. 7×2.1 **14.7**   17. 8×1.56 **12.48**
18. 8.12×5.4 **43.848**   19. 22.35×2.04 **45.594**   20. 212.021×0.24 **50.88504**

Lesson 3.4

Complete.

1. $33.6 \div 11.2 = 336 \div$? **112**   2. $31.5 \div 5 = 3.15 \div$? **0.5**

3. $16.92 \div 12 =$? $\div 1.2$ **1.692**   4. $661.44 \div$? $= 66{,}144 \div 312$ **3.12**

Find the quotient.

5. $16.4 \div 4$ **4.1**   6. $36.03 \div 3$ **12.01**   7. $2.88 \div 0.4$ **7.2**

8. $9.72 \div 1.2$ **8.1**   9. $25.0224 \div 3.12$ **8.02**   10. $20.801 \div 6.1$ **3.41**

CHAPTER 4

Lesson 4.1

Name the first four multiples.

1. 19 **19, 38, 57, 76**   2. 12 **12, 24, 36, 48**   3. 22 **22, 44, 66, 88**

Find the missing multiple or multiples.

4. ?, 14, 21, ?, 35 **7, 28**   5. 12, ?, ?, 48, 60 **24, 36**

Write the factors.

6. 26 **1, 2, 13, 26**   7. 32 **1, 2, 4, 8, 16, 32**   8. 40 **1, 2, 4, 5, 8, 10, 20, 40**

Write P for prime or C for composite.

9. 32 **C**   10. 63 **C**   11. 59 **P**

Lesson 4.2

Write the prime factorization in exponent form.

1. 18 **2×3^2**   2. 40 **$2^3 \times 5$**   3. 36 **$2^2 \times 3^2$**
4. 100 **$2^2 \times 5^2$**   5. 150 **$2 \times 3 \times 5^2$**   6. 280 **$2^3 \times 5 \times 7$**
7. 28 **$2^2 \times 7$**   8. 420 **$2^2 \times 3 \times 5 \times 7$**   9. 48 **$2^4 \times 3$**
10. 72 **$2^3 \times 3^2$**   11. 45 **$3^2 \times 5$**   12. 147 **3×7^2**
13. 98 **2×7^2**   14. 180 **$2^2 \times 3^2 \times 5$**   15. 80 **$2^4 \times 5$**

Solve for n to complete the prime factorization.

16. $2 \times 5 \times n = 50$ **5**   17. $2 \times n \times 5 = 30$ **3**   18. $2 \times 3 \times n = 18$ **3**
19. $n \times 3 \times 3 = 27$ **3**   20. $2 \times 2 \times 2 \times n = 16$ **2**   21. $n \times 2 \times 3 = 42$ **7**

Lesson 4.3

Find the LCM for each set of numbers.

1. 3, 8 **24**   2. 4, 15 **30**   3. 3, 20 **60**   4. 4, 16 **16**
5. 10, 35 **70**   6. 12, 20 **60**   7. 8, 15 **120**   8. 10, 16 **80**

Find the GCF for each set of numbers.

9. 30, 18 **6**   10. 20, 35 **5**   11. 30, 50 **10**   12. 36, 48 **12**
13. 48, 84 **12**   14. 52, 39 **13**   15. 75, 105 **15**   16. 45, 108 **9**

Lesson 4.4

Write the fraction in simplest form.

1. $\frac{15}{20}$ **$\frac{3}{4}$**   2. $\frac{16}{24}$ **$\frac{2}{3}$**   3. $\frac{9}{27}$ **$\frac{1}{3}$**   4. $\frac{21}{27}$ **$\frac{7}{9}$**   5. $\frac{8}{56}$ **$\frac{1}{7}$**   6. $\frac{10}{95}$ **$\frac{2}{19}$**
7. $\frac{24}{80}$ **$\frac{3}{10}$**   8. $\frac{14}{63}$ **$\frac{2}{9}$**   9. $\frac{24}{56}$ **$\frac{3}{7}$**   10. $\frac{36}{117}$ **$\frac{4}{13}$**   11. $\frac{16}{40}$ **$\frac{2}{5}$**   12. $\frac{33}{187}$ **$\frac{3}{17}$**
13. $\frac{7}{154}$ **$\frac{1}{22}$**   14. $\frac{27}{81}$ **$\frac{1}{3}$**   15. $\frac{9}{120}$ **$\frac{3}{40}$**   16. $\frac{15}{60}$ **$\frac{1}{4}$**   17. $\frac{32}{60}$ **$\frac{8}{15}$**   18. $\frac{50}{90}$ **$\frac{5}{9}$**

Write the missing number.

19. $\frac{9}{12} = \frac{\square}{4}$ **3**   20. $\frac{7}{42} = \frac{\square}{6}$ **1**   21. $\frac{25}{\square} = \frac{5}{10}$ **50**
22. $\frac{1}{12} = \frac{\square}{144}$ **12**   23. $\frac{9}{20} = \frac{\square}{60}$ **27**   24. $\frac{5}{\square} = \frac{15}{27}$ **9**

Lesson 4.5

Write the fraction as a mixed number or a whole number.

1. $\frac{7}{3}$ **$2\frac{1}{3}$**   2. $\frac{9}{2}$ **$4\frac{1}{2}$**   3. $\frac{36}{5}$ **$7\frac{1}{5}$**
4. $\frac{10}{2}$ **5**   5. $\frac{13}{4}$ **$3\frac{1}{4}$**   6. $\frac{36}{6}$ **6**
7. $\frac{25}{12}$ **$2\frac{1}{12}$**   8. $\frac{54}{13}$ **$4\frac{2}{13}$**   9. $\frac{33}{4}$ **$8\frac{1}{4}$**
10. $\frac{52}{7}$ **$7\frac{3}{7}$**   11. $\frac{14}{9}$ **$1\frac{5}{9}$**   12. $\frac{44}{6}$ **$7\frac{1}{3}$**
13. $\frac{17}{5}$ **$3\frac{2}{5}$**   14. $\frac{27}{11}$ **$2\frac{5}{11}$**   15. $\frac{26}{4}$ **$6\frac{1}{2}$**

Write the mixed number as a fraction.

16. $4\frac{1}{2}$ **$\frac{9}{2}$**   17. $5\frac{2}{3}$ **$\frac{17}{3}$**   18. $2\frac{1}{8}$ **$\frac{17}{8}$**
19. $7\frac{4}{7}$ **$\frac{53}{7}$**   20. $1\frac{6}{11}$ **$\frac{17}{11}$**   21. $4\frac{3}{8}$ **$\frac{35}{8}$**
22. $9\frac{2}{3}$ **$\frac{29}{3}$**   23. $3\frac{4}{7}$ **$\frac{25}{7}$**   24. $8\frac{1}{3}$ **$\frac{25}{3}$**
25. $6\frac{4}{9}$ **$\frac{58}{9}$**   26. $3\frac{3}{8}$ **$\frac{27}{8}$**   27. $2\frac{3}{11}$ **$\frac{25}{11}$**
28. $10\frac{1}{4}$ **$\frac{41}{4}$**   29. $2\frac{5}{12}$ **$\frac{29}{12}$**   30. $8\frac{4}{5}$ **$\frac{44}{5}$**

CHAPTER 5

Lesson 5.1

Find the sum or difference. Write the answer in simplest form.

1. $\frac{8}{10} + \frac{1}{10}$ **$\frac{9}{10}$**   2. $\frac{2}{7} + \frac{4}{7}$ **$\frac{6}{7}$**   3. $\frac{1}{8} + \frac{3}{8}$ **$\frac{1}{2}$**   4. $\frac{8}{11} + \frac{4}{11}$ **$1\frac{1}{11}$**
5. $\frac{2}{5} + \frac{4}{5}$ **$1\frac{1}{5}$**   6. $\frac{2}{12} + \frac{5}{12}$ **$\frac{7}{12}$**   7. $\frac{9}{11} - \frac{3}{11}$ **$\frac{6}{11}$**   8. $\frac{4}{7} - \frac{3}{7}$ **$\frac{1}{7}$**
9. $\frac{11}{14} - \frac{5}{14}$ **$\frac{3}{7}$**   10. $\frac{9}{12} - \frac{5}{12}$ **$\frac{1}{3}$**   11. $\frac{7}{11} - \frac{3}{11}$ **$\frac{4}{11}$**   12. $\frac{13}{20} - \frac{6}{20}$ **$\frac{7}{20}$**

Lesson 5.2

Draw a diagram to find each sum or difference.

1. $\frac{1}{2} + \frac{1}{3}$ **$\frac{5}{6}$**   2. $\frac{3}{4} - \frac{1}{2}$ **$\frac{1}{4}$**   3. $\frac{7}{12} + \frac{1}{3}$ **$\frac{11}{12}$**
4. $\frac{1}{8} + \frac{3}{4}$ **$\frac{7}{8}$**   5. $\frac{1}{4} + \frac{2}{3}$ **$\frac{11}{12}$**   6. $\frac{1}{8} + \frac{1}{4}$ **$\frac{3}{8}$**
7. $\frac{7}{10} - \frac{3}{5}$ **$\frac{1}{10}$**   8. $\frac{1}{4} - \frac{1}{8}$ **$\frac{1}{8}$**   9. $\frac{5}{6} - \frac{1}{2}$ **$\frac{1}{3}$**
10. $\frac{4}{6} + \frac{1}{4}$ **$\frac{11}{12}$**   11. $\frac{3}{8} + \frac{1}{4}$ **$\frac{5}{8}$**   12. $\frac{1}{2} - \frac{1}{6}$ **$\frac{2}{6}$, or $\frac{1}{3}$**

Lesson 5.3

Write equivalent fractions with the LCD.

1. $\frac{3}{4} + \frac{1}{2}$ **$\frac{3}{4} + \frac{2}{4}$**   2. $\frac{2}{5} + \frac{8}{15}$ **$\frac{6}{15} + \frac{8}{15}$**   3. $\frac{1}{4} + \frac{3}{5}$ **$\frac{5}{20} + \frac{12}{20}$**
4. $\frac{1}{3} + \frac{5}{8}$ **$\frac{8}{24} + \frac{15}{24}$**   5. $\frac{1}{4} + \frac{1}{8}$ **$\frac{2}{8} + \frac{1}{8}$**   6. $\frac{1}{6} + \frac{2}{3}$ **$\frac{1}{6} + \frac{4}{6}$**

Find the sum. Write your answer in simplest form.

7. $\frac{1}{6} + \frac{3}{4}$ **$\frac{11}{12}$**   8. $\frac{5}{6} + \frac{1}{4}$ **$1\frac{1}{12}$**   9. $\frac{2}{5} + \frac{1}{3}$ **$\frac{11}{15}$**
10. $\frac{1}{6} + \frac{1}{12}$ **$\frac{1}{4}$**   11. $\frac{5}{8} + \frac{1}{20}$ **$\frac{27}{40}$**   12. $\frac{5}{12} + \frac{1}{3}$ **$\frac{3}{4}$**
13. $\frac{3}{8} + \frac{1}{5}$ **$\frac{23}{40}$**   14. $\frac{4}{9} + \frac{2}{5}$ **$\frac{38}{45}$**   15. $\frac{1}{2} + \frac{1}{8}$ **$\frac{5}{8}$**
16. $\frac{1}{10} + \frac{3}{8}$ **$\frac{19}{40}$**   17. $\frac{6}{20} + \frac{3}{5}$ **$\frac{9}{10}$**   18. $\frac{4}{16} + \frac{1}{9}$ **$\frac{13}{36}$**

Lesson 5.4

Subtract. Write the answer in simplest form.

1. $\frac{5}{8} - \frac{1}{4}$ **$\frac{3}{8}$**   2. $\frac{2}{5} - \frac{1}{4}$ **$\frac{3}{20}$**   3. $\frac{5}{9} - \frac{1}{2}$ **$\frac{1}{18}$**
4. $\frac{5}{6} - \frac{3}{4}$ **$\frac{1}{12}$**   5. $\frac{4}{7} - \frac{1}{3}$ **$\frac{5}{21}$**   6. $\frac{9}{10} - \frac{2}{5}$ **$\frac{1}{2}$**
7. $\frac{7}{9} - \frac{2}{3}$ **$\frac{1}{9}$**   8. $\frac{5}{7} - \frac{2}{3}$ **$\frac{1}{21}$**   9. $\frac{2}{3} - \frac{3}{5}$ **$\frac{1}{15}$**
10. $\frac{4}{9} - \frac{1}{3}$ **$\frac{1}{9}$**   11. $\frac{5}{8} - \frac{1}{6}$ **$\frac{11}{24}$**   12. $\frac{11}{18} - \frac{1}{4}$ **$\frac{13}{36}$**
13. $\frac{3}{5} - \frac{1}{9}$ **$\frac{22}{45}$**   14. $\frac{10}{12} - \frac{1}{3}$ **$\frac{1}{2}$**   15. $\frac{7}{8} - \frac{1}{16}$ **$\frac{13}{16}$**

Lesson 5.5

Round each fraction. Write about 0, about $\frac{1}{2}$, or about 1.

1. $\frac{3}{5}$ **$\frac{1}{2}$**   2. $\frac{4}{7}$ **$\frac{1}{2}$**   3. $\frac{11}{12}$ **1**
4. $\frac{1}{5}$ **0**   5. $\frac{3}{11}$ **0**   6. $\frac{6}{7}$ **1**

Estimate the sum or difference. **Possible estimates given.**

7. $\frac{4}{7} - \frac{3}{8}$ **0**   8. $\frac{8}{9} + \frac{3}{5}$ **$1\frac{1}{2}$**   9. $\frac{5}{9} - \frac{3}{7}$ **0**
10. $\frac{3}{7} + \frac{1}{9}$ **$\frac{1}{2}$**   11. $\frac{4}{9} + \frac{2}{3}$ **1**   12. $\frac{9}{10} - \frac{2}{5}$ **$\frac{1}{2}$**
13. $\frac{4}{5} - \frac{1}{3}$ **$\frac{1}{2}$**   14. $\frac{8}{9} - \frac{5}{7}$ **0**   15. $\frac{5}{9} + \frac{2}{6}$ **1**

CHAPTER 6

Lesson 6.1

Draw a diagram to find each sum. Write the answer in simplest form.

1. $1\frac{3}{8} + 1\frac{5}{8}$ 3   **2.** $2\frac{3}{8} + 1\frac{1}{4}$ $3\frac{5}{8}$

3. $1\frac{3}{8} + 2\frac{1}{8}$ $3\frac{1}{2}$ **4.** $1\frac{2}{3} + 2\frac{1}{6}$ $3\frac{5}{6}$

Lesson 6.2

Write each mixed number by renaming one whole.

1. $2\frac{1}{3}$ $1\frac{4}{3}$ **2.** $3\frac{2}{7}$ $2\frac{9}{7}$ **3.** $4\frac{1}{12}$ $3\frac{13}{12}$

4. $3\frac{4}{9}$ $2\frac{13}{9}$ **5.** $2\frac{5}{7}$ $1\frac{12}{7}$ **6.** $4\frac{5}{6}$ $3\frac{11}{6}$

Draw a diagram to find the difference. Write the answer in simplest form.

7. $2\frac{2}{7} - 1\frac{3}{7}$ $\frac{6}{7}$ **8.** $3\frac{3}{5} - 2\frac{1}{5}$ $1\frac{2}{5}$ **9.** $2\frac{1}{3} - 1\frac{1}{6}$ $1\frac{1}{6}$

10. $3\frac{1}{3} - 1\frac{3}{4}$ $1\frac{7}{12}$ **11.** $4\frac{5}{12} - 2\frac{1}{6}$ $2\frac{1}{4}$ **12.** $5\frac{2}{5} - 3\frac{3}{10}$ $2\frac{1}{10}$

Lesson 6.3

Rename the fractions using the LCD. Rewrite the problem.

1. $1\frac{1}{2} + 2\frac{3}{5}$ $1\frac{5}{10} + 2\frac{6}{10}$ **2.** $2\frac{5}{6} + 2\frac{2}{9}$ $2\frac{15}{18} + 2\frac{4}{18}$

3. $3\frac{3}{4} + 1\frac{1}{5}$ $3\frac{15}{20} + 1\frac{4}{20}$

Rename the fraction as a mixed number. Write the new mixed number.

4. $2\frac{5}{4}$ $1\frac{1}{4}$, $3\frac{1}{4}$ **5.** $3\frac{9}{7}$ $1\frac{2}{7}$, $4\frac{2}{7}$ **6.** $4\frac{10}{6}$ $1\frac{2}{3}$, $5\frac{2}{3}$

Tell whether you must rename to subtract. Write *yes* or *no*.

7. $3\frac{2}{7} - 1\frac{3}{5}$ yes   **8.** $3\frac{5}{7} - 1\frac{1}{7}$ no   **9.** $2\frac{1}{6} - 1\frac{3}{5}$ yes

10. $8\frac{1}{7} - 6\frac{2}{19}$ no   **11.** $4\frac{1}{4} - 3\frac{2}{7}$ yes   **12.** $3\frac{2}{3} - 2\frac{1}{6}$ no

Find the sum. Write the answer in simplest form.

13. $2\frac{1}{4} + 3\frac{7}{12}$ $5\frac{5}{6}$ **14.** $1\frac{1}{7} + 3\frac{2}{3}$ $4\frac{17}{21}$ **15.** $3\frac{5}{6} + 4\frac{5}{9}$ $8\frac{7}{18}$

16. $1\frac{3}{4} + 5\frac{4}{5}$ $7\frac{11}{20}$ **17.** $5\frac{1}{3} + 4\frac{1}{9}$ $9\frac{4}{9}$ **18.** $5\frac{1}{4} + 5\frac{2}{3}$ $10\frac{11}{12}$

Find the difference. Write the answer in simplest form.

19. $4\frac{5}{9} - 3\frac{5}{6}$ $\frac{13}{18}$ **20.** $5\frac{1}{3} - 4\frac{1}{9}$ $1\frac{2}{9}$ **21.** $6\frac{2}{3} - 3\frac{1}{4}$ $3\frac{5}{12}$

Lesson 6.4

Estimate the sum or difference. Possible estimates given.

1. $8\frac{7}{8} + 5\frac{1}{2}$ $14\frac{1}{2}$ **2.** $3\frac{10}{13} + 4\frac{1}{9}$ 8   **3.** $2\frac{8}{9} - 1\frac{3}{5}$ $1\frac{1}{2}$

4. $7\frac{4}{7} - 5\frac{3}{7}$ 2   **5.** $8\frac{5}{6} - 6\frac{2}{4}$ $2\frac{1}{2}$ **6.** $3\frac{3}{4} + 2\frac{7}{8}$ 7

7. $5\frac{4}{5} - 3\frac{1}{2}$ $2\frac{1}{2}$ **8.** $3\frac{7}{9} + 1\frac{10}{11}$ 6   **9.** $2\frac{5}{9} + 3\frac{3}{7}$ 6

10. $7\frac{5}{9} - 5\frac{7}{8}$ 1   **11.** $3\frac{1}{2} + 4\frac{1}{5}$ $7\frac{1}{2}$ **12.** $15\frac{1}{4} - 14\frac{1}{3}$ 1

13. $7\frac{9}{9} - 5\frac{1}{6}$ $2\frac{1}{2}$ **14.** $8\frac{9}{11} + 2\frac{3}{8}$ $11\frac{1}{2}$ **15.** $5\frac{8}{9} - 4\frac{1}{6}$ 2

CHAPTER 7

Lesson 7.1

Find the product. Write it in simplest form.

1. $\frac{2}{4} \times \frac{3}{3}$ $\frac{3}{8}$ **2.** $\frac{2}{5} \times \frac{1}{4}$ $\frac{1}{10}$ **3.** $\frac{3}{8} \times \frac{3}{3}$ $\frac{9}{40}$

4. $\frac{2}{7} \times 8$ $2\frac{2}{7}$ **5.** $\frac{5}{6} \times \frac{3}{8}$ $\frac{5}{16}$ **6.** $\frac{5}{9} \times \frac{1}{5}$ $\frac{1}{9}$

7. $\frac{1}{9} \times \frac{2}{3}$ $\frac{2}{27}$ **8.** $\frac{2}{7} \times 5$ $1\frac{3}{7}$ **9.** $8 \times \frac{2}{3}$ $5\frac{1}{3}$

10. $\frac{1}{3} \times \frac{3}{7}$ $\frac{1}{7}$ **11.** $\frac{5}{9} \times \frac{3}{5}$ $\frac{1}{3}$ **12.** $\frac{3}{8} \times \frac{1}{6}$ $\frac{1}{16}$

13. $\frac{4}{9} \times \frac{3}{8}$ $\frac{1}{6}$ **14.** $\frac{1}{4} \times 8$ $\frac{2}{9}$ **15.** $\frac{4}{5} \times \frac{3}{7}$ $\frac{12}{35}$

Lesson 7.2

Tell the GCF you would use to simplify the fractions.

1. $\frac{2}{5}, \frac{5}{7}$ 5   **2.** $\frac{3}{9}, \frac{9}{27}$ 3,9   **3.** $\frac{3}{4}, \frac{2}{3}$ 3,2

4. $\frac{3}{20}, \frac{5}{6}$ 3,5   **5.** $\frac{7}{27}, \frac{6}{9}$ 3   **6.** $\frac{1}{6}, \frac{3}{10}$ 3

7. $\frac{6}{8}, \frac{3}{12}$ 6   **8.** $\frac{6}{12}, \frac{5}{18}$ 6   **9.** $\frac{10}{25}, \frac{5}{20}$ 2,25

10. $\frac{1}{2}, \frac{4}{9}$ 2   **11.** $\frac{9}{10}, \frac{7}{1}$ 9   **12.** $\frac{2}{5}, \frac{1}{4}$ 2

13. $\frac{3}{26}, \frac{13}{15}$ 3,13   **14.** $\frac{3}{7}, \frac{14}{27}$ 3,7   **15.** $\frac{5}{12}, \frac{24}{25}$ 5,12

Use GCF's to simplify the factors so that the answer is in simplest form.

16. $\frac{3}{8} \times \frac{10}{15}$ $\frac{1}{4}$ **17.** $\frac{9}{12} \times \frac{6}{27}$ $\frac{1}{6}$ **18.** $\frac{2}{3} \times \frac{6}{10}$ $\frac{2}{5}$

19. $\frac{6}{10} \times \frac{5}{7}$ $\frac{3}{7}$ **20.** $\frac{6}{7} \times \frac{7}{9}$ $\frac{2}{3}$ **21.** $\frac{15}{20} \times \frac{20}{27}$ $\frac{5}{9}$

22. $\frac{3}{4} \times 16$ 12   **23.** $8 \times \frac{3}{4}$ 6   **24.** $27 \times \frac{5}{6}$ $22\frac{1}{2}$

Lesson 7.3

Rewrite the problem by changing each mixed number to a fraction.

1. $\frac{3}{8} \times 5\frac{1}{3}$ $\frac{3}{8} \times \frac{16}{3}$ **2.** $\frac{5}{7} \times 3\frac{2}{3}$ $\frac{5}{7} \times \frac{11}{3}$ **3.** $2\frac{4}{9} \times 1\frac{1}{5}$ $\frac{22}{9} \times \frac{6}{5}$

4. $5\frac{2}{3} \times 4\frac{1}{7}$ $\frac{17}{3} \times \frac{29}{7}$ **5.** $1\frac{5}{8} \times 3\frac{4}{8}$ $\frac{13}{8} \times \frac{19}{5}$ **6.** $2\frac{2}{5} \times 3\frac{2}{5}$ $\frac{12}{5} \times \frac{17}{5}$

7. $\frac{5}{7} \times 3\frac{1}{3}$ $\frac{5}{7} \times \frac{10}{3}$ **8.** $6\frac{5}{6} \times 2\frac{9}{11}$ $\frac{41}{6} \times \frac{31}{11}$ **9.** $3\frac{2}{3} \times 1\frac{6}{6}$ $\frac{11}{3} \times \frac{1}{6}$

10. $2\frac{2}{3} \times 3\frac{2}{7}$ $\frac{8}{3} \times \frac{23}{7}$ **11.** $2\frac{1}{5} \times 3\frac{1}{7}$ $\frac{11}{5} \times \frac{22}{7}$ **12.** $3\frac{2}{3} \times 4\frac{3}{5}$ $\frac{11}{3} \times \frac{23}{5}$

Find the product. Write it in simplest form.

13. $2\frac{2}{3} \times 2\frac{1}{4}$ 6   **14.** $2\frac{1}{2} \times \frac{1}{5}$ $\frac{1}{2}$ **15.** $\frac{1}{5} \times 2\frac{2}{3}$ $\frac{8}{15}$

16. $2\frac{3}{8} \times \frac{5}{6}$ $2\frac{7}{24}$ **17.** $2\frac{2}{7} \times \frac{2}{5}$ $\frac{32}{35}$ **18.** $3\frac{2}{5} \times 2\frac{9}{11}$ $9\frac{8}{11}$

19. $3\frac{4}{9} \times 1\frac{4}{8}$ $5\frac{1}{6}$ **20.** $7\frac{1}{6} \times \frac{3}{4}$ $5\frac{3}{8}$ **21.** $2\frac{1}{3} \times 2\frac{3}{7}$ $5\frac{2}{3}$

22. $8\frac{5}{8} \times 5\frac{1}{8}$ $44\frac{13}{64}$ **23.** $2\frac{1}{12} \times 9\frac{3}{5}$ 20   **24.** $6\frac{1}{10} \times 4\frac{2}{5}$ $26\frac{21}{25}$

Lesson 7.4

Find the quotient. Write it in simplest form.

1. $\frac{2}{7} \div \frac{4}{7}$ $\frac{1}{2}$ **2.** $\frac{3}{10} \div \frac{6}{5}$ $\frac{1}{4}$ **3.** $\frac{1}{7} \div \frac{25}{49}$ $\frac{7}{25}$

4. $\frac{2}{9} \div \frac{8}{18}$ $\frac{1}{2}$ **5.** $\frac{3}{8} \div \frac{9}{4}$ $\frac{1}{6}$ **6.** $2 \div \frac{8}{3}$ $\frac{3}{4}$

7. $\frac{3}{4} \div \frac{1}{8}$ 6   **8.** $\frac{5}{6} \div \frac{5}{3}$ $\frac{1}{2}$ **9.** $\frac{7}{8} \div \frac{7}{2}$ $\frac{1}{4}$

10. $\frac{7}{8} \div \frac{3}{4}$ $1\frac{1}{6}$ **11.** $\frac{12}{5} \div \frac{6}{10}$ 4   **12.** $\frac{3}{7} \div \frac{18}{21}$ $\frac{1}{2}$

13. $\frac{8}{3} \div \frac{24}{36}$ 4   **14.** $\frac{4}{5} \div \frac{12}{15}$ 1   **15.** $\frac{4}{9} \div \frac{20}{18}$ $\frac{2}{5}$

Lesson 7.5

Work Backward to solve.

1. Sandy is designing a pool for her garden. It is to have an area of $24\frac{3}{4}$ square feet. The length of the pool is to be $7\frac{1}{2}$ feet. Find its width. **$3\frac{11}{45}$ feet**

2. A spacecraft has left Earth for a star that is $4\frac{1}{3}$ light years away. One-fourth of the way there it will stop at a space station. How far from Earth is the space station? **$1\frac{1}{12}$ light years**

3. The tallest tree in Maria's yard is $27\frac{1}{2}$ feet tall. The shortest tree is one-third the height of the tallest one. How tall is the shortest one? **$9\frac{1}{6}$ feet**

4. A window in Susan's family room is $5\frac{1}{3}$ feet wide with an area of $40\frac{1}{2}$ square feet. How tall is it? **$7\frac{19}{32}$ feet**

CHAPTER 8

Lesson 8.1

For Exercises 1–4, use the figure below.

All possible answers are given.

1. Name four points. **point R, point S, point T, and point W**

2. Name three line segments. $\overline{RS}$, $\overline{SR}$, $\overline{ST}$, $\overline{TS}$, $\overline{TW}$, $\overline{WT}$, $\overline{RT}$, $\overline{TR}$, $\overline{RW}$, $\overline{WR}$, $\overline{SW}$, $\overline{WS}$

3. Name two line rays. $\overrightarrow{RS}$, $\overrightarrow{ST}$, $\overrightarrow{TS}$, $\overrightarrow{TW}$, $\overrightarrow{WT}$, $\overrightarrow{RT}$, $\overrightarrow{TR}$, $\overrightarrow{RW}$, $\overrightarrow{WR}$, $\overrightarrow{SW}$, $\overrightarrow{WS}$

4. Name one line. $\overleftrightarrow{RS}$, $\overleftrightarrow{SR}$, $\overleftrightarrow{ST}$, $\overleftrightarrow{TS}$, $\overleftrightarrow{TW}$, $\overleftrightarrow{WT}$, $\overleftrightarrow{RT}$, $\overleftrightarrow{TR}$, $\overleftrightarrow{RW}$, $\overleftrightarrow{WR}$, $\overleftrightarrow{SW}$, $\overleftrightarrow{WS}$

Name the geometric figure that is suggested in Exercises 5–8.

5. a period at the end of a sentence **point**

6. a teacher pointing the way to the cafeteria with one arm **ray**

7. main street passing through your town **a line**

8. the line that separates the top and bottom numbers in a fraction **a line segment**

Lesson 8.2

Use the figure at the right for Exercises 1–2. **All possible answers are given.**

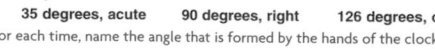

1. Name a line that is parallel to *BE*. **AG, CD**

2. Name a line that is perpendicular to and intersects *BE*. **AF, FD or AD**

Write *true* or *false*. Change any false statement into a true statement.

3. Parallel lines can be perpendicular. **False, parallel lines never intersect; perpendiculars do.**

4. If two lines are perpendicular to the same line, then those two lines are parallel. **True**

Lesson 8.3

Measure the angle. Write the measure of the angle, and also the word *acute*, *right*, or *obtuse*.

1.   **2.**   **3.**

35 degrees, acute   **90 degrees, right**   **126 degrees, obtuse**

For each time, name the angle that is formed by the hands of the clock.

4. 5:00 **obtuse**   **5.** 4:15 **acute**   **6.** 6:00 **straight**

Lesson 8.4

Use a compass to determine if the line segments in each pair are congruent. Write *yes* or *no*.

1. **yes**   2. **no**   3. **no**

Use a protractor to measure the angles in each pair. Tell whether they are congruent. Write each measure, and then *yes* or *no*.

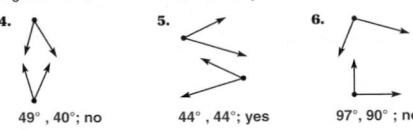

4. **49° , 40°; no**   5. **44° , 44°; yes**   6. **97°, 90° ; no**

Lesson 8.5

Name the quadrilateral. Tell whether both pairs of opposite sides are parallel. Write *yes* or *no*. **Check students' descriptions.**

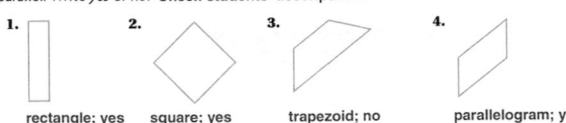

1. **rectangle; yes**   2. **square; yes**   3. **trapezoid; no**   4. **parallelogram; yes**

CHAPTER 9

Lesson 9.1

Trace the given figures. Draw all lines of symmetry. **Check students' drawings.**

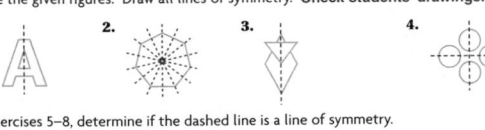

In Exercises 5–8, determine if the dashed line is a line of symmetry. Write *yes* or *no*.

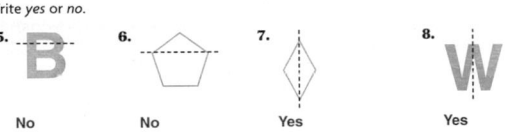

5. **No**   6. **No**   7. **Yes**   8. **Yes**

Lesson 9.2

Tell whether the second figure in each set is a *translation*, or a *rotation* of the first figure.

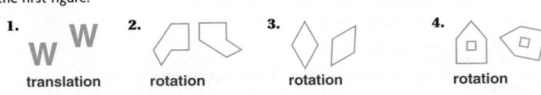

1. **translation**   2. **rotation**   3. **rotation**   4. **rotation**

Trace the figure, the line, and point. Draw a reflection about the line and then rotate that figure 90° clockwise. **Check students' drawings.**

5.   6.   7.   8.

Lesson 9.3

Trace each polygon several times, and cut out your tracings. Tell whether the polygon forms a tessellation. Write *yes* or *no*.

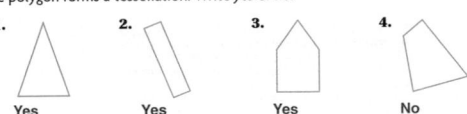

1. **Yes**   2. **Yes**   3. **Yes**   4. **No**

Find the measure of the angles that surround the circled vertex. Then find the sum of the measures.

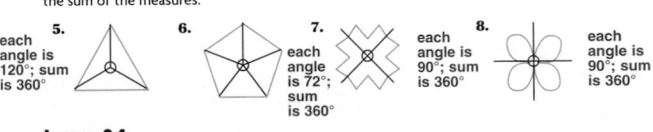

5. **each angle is 120°; sum is 360°**   6. **each angle is 72°; sum is 360°**   7. **each angle is 90°; sum is 360°**   8. **each angle is 90°; sum is 360°**

Lesson 9.4

Make a model to solve.

1. Jamal is making a design from the shape shown at the right. He wants the design to tessellate a plane. Can he use this figure to tessellate a plane? **Yes, trace the figure several times; cut them out and try!**
2. Draw your own design that will tessellate a plane. **Check students' drawings.**
3. Draw your own design that would not tessellate a plane. **Check students' drawings.**

CHAPTER 10

Lesson 10.1

Write *base* or *lateral face* to identify the shaded face.

1. **base**   2. **lateral face**   3. **lateral face**   4. **lateral face**

For exercises 5–8 name each figure shown in Exercises 1–4 above.
5. **pentagonal prism** 6. **rectangular prism** 7. **cylinder** 8. **hexagonal pyramid**

Write *true* or *false* for each statement. Change each false statement into a true statement.

9. If a cone is placed on its lateral surface, it can be rolled in a straight line. **False, try it!**
10. In a rectangular prism, the bases are congruent. **True**
11. A cylinder may have a polygon as its base. **False, the base will be a circle.**

Lesson 10.2

Use the figure at the right for Exercises 1–5.

1. Name the vertices. **A, B, C, D, E, F, G, H, I, and J**
2. Name the edges. **AB, BC, CD, DE, AE, FG, GH, HI, IJ, FJ, AF, BG, CH, DI, and EJ**
3. Name the faces. **ABCDE, FGHIJ, AEJF, ABGF, BCHG, CDIH, and DEJI**
4. Name the figure. **a pentagonal prism**
5. Write *polyhedron* or *not a polyhedron* to describe the figure. **polyhedron**

Lesson 10.3

Use the rectangular prism at the right for Exercises 1–3.

1. What are the dimensions of the faces? **two faces are 2 in. × 3 in.; two faces are 3 in. × 4 in.; and two faces are 2 in. × 4 in.**
2. Draw the faces, with the correct measurements, on an index card. **Check students' drawings.**
3. Cut out the faces, and arrange them to form a net for the prism. Tape the pieces together to form a prism. **Check students' prisms.**

Will the arrangement of squares fold to form a cube? Write *yes* or *no*.

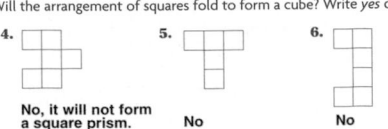

4. **No, it will not form a square prism.**   5. **No**   6. **No**

Lesson 10.4

Name the solid figure that has the given views.

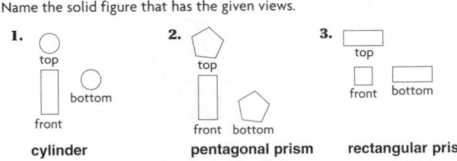

1. **cylinder**   2. **pentagonal prism**   3. **rectangular prism**

Name the solids that have the following features.

4. The figure contains circles. **either a cone, or a cylinder**
5. The figure has one rectangle and several triangles. **A rectangular pyramid**
6. The figure has six rectangles. **A rectangular prism**
7. The figure has two triangles and three rectangles. **A triangular prism**

Lesson 10.5

Solve by first solving a simpler problem.

1. Jeremy wants to make a model of a prism whose base has 12 sides. He will use balls of clay for the vertices and straws for the edges. How many balls of clay will he need? **24 balls of clay**
2. Look back at Exercise 1. How many straws will Jeremy need? How many faces will his prism have? **36 straws ; 14 faces**
3. Mara wants to make a model of a pyramid whose base has 12 sides. She will use balls of clay for the vertices and straws for the edges. How many balls of clay will she need? **13 balls of clay**
4. Look back at Exercise 3. How many straws will Mara need? How many faces will her pyramid have? **24 straws ; 13 faces**

CHAPTER 11

Lesson 11.1

The Band Boosters are planning a spaghetti dinner fund raiser to purchase new uniforms. Determine whether each is a decision they will have to make. Write *yes* or *no*.

1. The location of the dinner **yes**
2. The price to charge per person **yes**
3. Whether the cafeteria is available **yes**
4. The availability of the auditorium for the choir **no**
5. The color of the new football uniforms **no**
6. How many people might attend the dinner **yes**

Lesson 11.2

Manuel and his committee are in charge of gathering information to determine a school mascot for the new middle school. There are 680 sixth, seventh and eighth graders in the new school this September.

1. How many people should the committee survey? **68 people (recommended is 1 per 10)**
2. Whom should Manuel and his committee survey? **some from each of the three grades**
3. How can Manuel get a random sample? **Possible response: by randomly choosing students as they enter, or leave school for several days.**
4. Is a random sample of the band members fair? Explain. **No; not all students are in the band. The sample should include some students of every possible type.**

Lesson 11.3

Tell whether the sampling method is *biased* or *not biased*. Explain.

The 680 students want good school lunches, and a variety of foods from which to select. A survey was conducted to find out the favorite foods.

1. Randomly survey the teachers. **Biased; excludes all the students.**

2. Randomly survey 1 out every 10 students in sixth, seventh, and eighth grades. **Not biased.**

3. Ask all your friends in the sixth grade. **Biased; you are only asking your friends and also not seventh graders or eighth graders are represented.**
4. Randomly survey 10 people who get off the school bus one morning. **Biased; not all people ride the school bus; also the sample is too small.**
5. Randomly ask people in each of the grades whether their favorite fast food is hot dogs or hamburgers. **Biased; does not include other types of fast food, such as pizza; also, does not state how many should be surveyed.**
6. Randomly survey several people from each sports team. **Biased; not all people ride the school bus; also the sample is too small.**

Lesson 11.4

For Exercises 1–3, use the table at the right.

1. Copy and complete the table, by first making a frequency table, and then also including the values for a cumulative frequency column.

2. How large was the sample size? **63**

3. Could you use the data in the table to make a line plot? Explain. **Possible answer: No. A line plot shows the frequency of numerical data; the data in the table is categorical.**

Favorite Vegetable	Tally	Frequency	Cumulative Frequency
Peas	1111 1	**?** 6	**?** 6
Green Beans	1111 1111	**?** 9	**?** 15
Potatoes	1111 1111 1111 1111 11	**?** 22	**?** 37
Carrots	1111 1111 1	**?** 11	**?** 48
Corn	1111 1111 1111	**?** 15	**?** 63

CHAPTER 12

Lesson 12.1

Choose the appropriate graph for each set of data. Then make the graph. **Check students' graphs.**

1.

WEIGHTS (in lbs.) OF CLASSMATES					
104	115	156	175	121	142
112	102	133	134	143	138
114	134	132	132	132	113
125	136	127	133	122	105
134	168	167	145	113	152

Weights of Classmates Use a Stem and Leaf

2.

TEAM'S PERCENTAGE OF WINS					
92-93	93-94	94-95	95-96	96-97	97-98
0.188	0.500	0.389	0.245	0.875	0.935

Team's Percentage of Wins Use a Line Graph

Lesson 12.2

Tell whether a bar graph or a histogram is more appropriate.

1. Heights to the nearest inch of all students in seventh grade. **histogram (intervals)**

2. Points for each starting player on the junior high basketball team. **bar graph**

3. Ages of all the cars in the school parking lot. **histogram**

For Exercises 4–5, use the table below.

NOON TEMPERATURES FOR THIRTY DAYS IN JUNE						
Interval	75-79	80-84	85-89	90-94	95-99	100-104
Number of days	2	4	7	10	6	1

4. Make a histogram. **Check students' graphs.**

5. How would the number of days per interval change if instead of six intervals, only three intervals were used? **The number of days per interval would increase.**

Lesson 12.3

1. Make a multiple-bar graph showing the following before weights of these five students, and the weights after a six-week physical fitness program. **Check students' graphs.**

WEIGHTS IN POUNDS BEFORE AND AFTER					
	Jamie	Todd	Marquis	Eddie	Sammy
Before	124	132	118	148	142
After	118	118	116	128	126

2. Make a multiple-line graph showing the following population changes from 1960 to 1990 for Phoenix and San Antonio. **Check students' graphs.**

POPULATION IN PHOENIX AND SAN ANTONIO				
	1960	1970	1980	1990
Phoenix	439,170	584,303	789,704	983,403
San Antonio	587,718	654,153	785,940	935,393

Lesson 12.4

For Exercises 1–2, use the circle graph at the right.

1. How many students were surveyed about their favorite sport to watch on TV? **120 students**

2. What fraction of the circle does watching auto racing represent? $\frac{12}{120}$ **or** $\frac{1}{10}$ **or 10%**

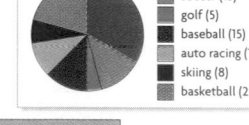

- football (40)
- soccer (15)
- golf (5)
- baseball (15)
- auto racing (12)
- skiing (8)
- basketball (25)

For Exercises 3–5, use the following data.

MONTHLY LIVING EXPENSES					
Rent	Car	Entertainment	Laundry	Food	Miscellaneous
$350.00	$120.00	$105.00	$60.00	$275.00	$90.00

3. Into how many sections would you divide a circle graph for the data? **Six sections**
4. For a circle graph, what angle measure would you use to represent rent? $\frac{350}{1000}$ **or 35% or** $\frac{350}{1000}$ **times 360° = 126 degrees**
5. Make a circle graph for the data. **Check students' graphs.**

CHAPTER 13

Lesson 13.1

For Exercise 1–2, use the bar graph.

1. Which activity had the most participants? Which had the least? **Intramurals had the most, around 280; Basketball had the least, around 20.**
2. Do intramurals have more or less participants than the total participating on the baseball, basketball and football teams combined? **Intramurals (about 280) have more participants than baseball (about 25), basketball (about 20) plus football (about 55)**

Lesson 13.2

For Exercises 1–3, use the line graph.

1. Looking at the graph only, how many times as high is the 9th grade bar, compared to the 7th grade bar? **It appears the 9th graders spent twice as many hours.**
2. Actually, how many more hours did the 9th grade class volunteer than the 7th grade class? **Only 1**
3. Explain why the graph is misleading. Then make a graph that is not misleading. **The graph is misleading because the intervals on the "hours axis" are not equal; nor do they start at zero; nor is a proper break shown on this scale. Check students' graphs for better representations.**

Lesson 13.3

For Exercises 1–4, use the graph at the right, which shows the average cost of a textbook over the time period from 1990 to 1997.

1. How much did an average textbook cost in 1990? in 1995? **$30; $55**

2. How much did the average price increase from 1990 to 1997? **about $35**

3. What has been the pattern for the cost of a textbook from one year to the next year? **about a $5 increase a year**
4. Based on the data given, what would you predict the average price of a textbook to be in the year 1998? **The average increase per year is about $4 or $5; so in 1998 a predicted price might be around $69 or $70.**

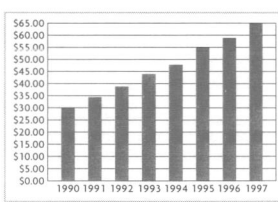

Lesson 13.4

Copy and complete the table for Exercises 1–4. **Check students' work.**

Data	Mean	Median	Mode
1. 4, 5, 7, 8, 8	**6.4**	**7**	**8**
2. 6.1, 8.1, 7.4, 7.2	**7.2**	**7.3**	**none**
3. 44, 55, 77, 88, 88	**70.4**	**77**	**88**
4. 95, 95, 95, 95, 95	**95**	**95**	**95**

Write *true* or *false* for each.

5. Some sets of data have no mean. **False**

6. The median is always one of the numbers in the set of data. **False**

7. The mode is always one of the numbers in the set of data. **True**

8. The mean, median, and mode can never all be the same number. **False**

Lesson 13.5

For Exercises 1–7, use the table at the right.

1. What is the median? **36**

2. What are the lower and upper quartiles? **31; 43**

3. What are the lower and upper extremes? **28; 44**

Weekly Sales of Different Pants Sizes					
28	44	28	44	38	42
34	40	34	32	30	44

4. What is the range? **44 − 28 = 16**

5. Make a box-and-whisker graph. **Check students' graphs.**

6. What fractional part of the data is less than 31? greater than 43? $\frac{1}{4}$; $\frac{1}{4}$

7. How are the data distributed? **The sizes in the lowest $\frac{1}{4}$ are close together; the sizes in the highest $\frac{1}{4}$ are all within one size of the upper quartile. The sizes in each part of the middle $\frac{1}{2}$ are distributed differently, with the sizes from the median to the upper quartile more spread out. Quite a few pants sizes are beyond the lower and upper quartiles, with three people buying pant sizes at the upper extreme (44).**

CHAPTER 14

Lesson 14.1

Use the strategy *account for all possibilities* to solve.

1. Sarah's clothes are color coordinated. If she has three pairs of slacks (light green, light gray, and dark green), three blouses (white, tan, yellow), and two sweater vests (beige and cream), how many different choices does she have in all, if she wears one each (slacks, blouse, and vest)? By making a tree diagram, there are $3 \times 3 \times 2 = 18$ choices.

2. On your hamburger deluxe, you may have one beef patty *or* two, provolone cheese *or* swiss cheese, *one* of these condiments (ketchup, mustard, or mayonnaise), and *one* of these health items (lettuce, tomato, or onion). How many choices do you have? $2 \times 2 \times 3 \times 3 = 36$ choices; making a tree diagram will show all these possibilities

Lesson 14.2

For Exercises 1–4, use the spinner at the right.

1. How many outcomes are there for choosing 5? 1

2. How many outcomes are there? Name them. Eight; A, red, 5, *, 3, green, #, Z

3. How many favorable outcomes are there for choosing a letter? 2

4. What is the probability of choosing a color? $\frac{2}{8}$ or $\frac{1}{4}$

A number cube is numbered 5, 10, 25, 50, 100, 2000. Find each probability.

5. $P(25)$ $\frac{1}{6}$

6. $P(5 \text{ or } 25)$ $\frac{1}{3}$

7. $P(\text{a number ending in zero })$ $\frac{2}{3}$

For Exercises 8–11, use the rectangle grid at the right. Find each probability.

8. $P(\text{yellow})$ $\frac{1}{24}$

9. $P(\text{green or blue})$ $\frac{11}{24}$

10. $P(\text{red})$ $\frac{1}{2}$

11. $P(\text{red or blue or green})$ $\frac{23}{24}$

	A	B	C	D
1	BLUE	GREEN	GREEN	YELLOW
2	BLUE	GREEN	GREEN	GREEN
3	BLUE	GREEN	GREEN	GREEN
4	RED	RED	RED	RED
5	RED	RED	RED	RED
6	RED	RED	RED	RED

Lesson 14.3

For Exercises 1–6, use the spinner at the right. Find each probability.

1. $P(\text{Al})$ $\frac{1}{8}$

2. $P(\text{Ed or Dee})$ $\frac{1}{4}$

3. $P(\text{not Al or Ed})$ $\frac{3}{4}$

4. $P(\text{Garth, Al, or Cara})$ $\frac{3}{8}$

5. $P(\text{Jim})$ $\frac{0}{8}$

6. $P(\text{not Garth})$ $\frac{7}{8}$

Cards with these team mascots (4 lions, 3 bears, 5 tigers, 8 cheetahs) are placed in a hat. You choose one card without looking. Find each probability.

7. $P(\text{lion})$ $\frac{4}{20}$ or $\frac{1}{5}$

8. $P(\text{tiger or lion})$ $\frac{9}{20}$

9. $P(\text{member of cat family})$ $\frac{17}{20}$

A bag contains some pencils: 8 blue, 12 brown, 10 red, 4 green, and 6 yellow. You choose one piece without looking. Find each probability.

10. $P(\text{brown})$ $\frac{12}{40}$ or $\frac{3}{10}$

11. $P(\text{brown or blue})$ $\frac{1}{2}$

12. $P(\text{orange})$ $\frac{0}{40}$ or 0

Lesson 14.4

1. Carlos spins a pointer on a circular disc with the following letters (A, E, I, O, U), one letter in each of the five equally spaced regions. He spins the pointer 100 times and records his results in the table below. Find the experimental probability of each letter.

Letter	A	E	I	O	U
Times Landed On	10	30	15	5	40

2. What is the mathematical probability for each letter? $\frac{1}{5}$

3. How many times can Carlos expect the pointer to land on E in the next twenty times? Experimentally, E's probability was $\frac{30}{100}$, so $\frac{3}{10} = \frac{6}{20}$; 6.

4. How many times can Carlos expect the pointer to land on U if he spins 2,000 times? 800 times

CHAPTER 15

Lesson 15.1

Write a numerical or algebraic expression for the word expression.

1. 9 less than 12 $12 - 9$

2. 13 more than a number, x $x + 13$

3. 22 multiplied by 3 22×3

4. 52 divided by 2 $52 \div 2$

5. the product of one half and three fourths $\frac{1}{2} \times \frac{3}{4}$

6. the sum of 9 squared and 12 $9^2 + 12$

7. 24 decreased by $\frac{1}{5}$ $24 - \frac{1}{5}$

8. x more than y $y + x$

9. 42 times a number, y $42 \times y$

10. 22 divided by 2 $22 \div 2$

Lesson 15.2

Evaluate the numerical expression. Remember the order of operations.

1. $8 - 6 \times 5$ ⁻22

2. $4.8 + 6.2 - 8$ 3

3. $9 \div 3 \times 2$ 6

4. $(4.1 - 3.2) + 2.1$ 3

5. $42 - 0.5$ 41.5

6. $(28 - 12) + 4$ 20

7. $32 \div 8 + 4$ 8

8. $(16 \div 2)^2$ 64

Evaluate the algebraic expressions for the given value of the variable.

9. $x - 6$, for $x = 12$ 6

10. $y + 13$, for $y = 25$ 38

11. $42 - k$, for $k = 30$ 12

12. $38 + p$, for $p = 22$ 60

13. $a^2 - 6$, for $a = 3$ 3

14. $x^3 - 6$, for $x = 2$ 2

15. $x^3 - 9 + 12$, for $x = 3$ 30

16. $32 - a^4$, for $a = 2$ 16

17. $4 \times p$, for $p = 13.5$ 54

18. $7z + 12$, for $z = 3$ 33

19. $(19 + 5) \div x$, for $x = 8$ 3

20. $12 \times 3 - p$, for $p = 14$ 22

21. $x^2 - x + 10$, for $x = 3$ 16

22. $(12 \div y) \times 9$, for $y = 6$ 18

23. $(7 \times p) \div 2$, for $p = 8$ 28

Lesson 15.3

Compute the output for each input.

1. $x + 9$; for $x = 4, 5, 6$ 13, 14, 15

2. $x + 6.4$; for $x = 9.6, 10.1, 11.6$ 16, 16.5, 18

3. $y \times 14$; for $y = 8, 12, 10.5$ 112, 168, 147

4. $y^3 - 8$; for $y = 2, 3, 4$ 0, 19, 56

Make an input-output table for the algebraic expression. Evaluate the expression for 2, 3, 4, and 5.

5. $x + 12$ 14, 15, 16, 17

6. $36 - y$ 34, 33, 32, 31

7. $30 \div z$ 15, 10, 7.5, 6

8. $6.5 \times y$ 13, 19.5, 26, 32.5

Determine the input for the given output. Make a table if necessary.

9. $x + 9$ 4 output = 13

10. $x - 14$ 23 output = 9

11. $x \div 4$ 36 output = 9

12. $y \times 8$ 4 output = 32

Lesson 15.4

Determine whether the given value is a solution of the equation. Write *yes* or *no*.

1. $x + 8 = 12, x = 3$ no

2. $x \div 8 = 3, x = 24$ yes

3. $12 \times y = 36, y = 3$ yes

4. $8 - x = 2, x = 7$ no

5. $2.5 \times z = 10, z = 4$ yes

6. $8.9 - x = 4.4, x = 3.5$ no

7. $x - 12.1 = 8.2, x = 20.3$ yes

8. $9.25 \times p = 37, p = 4$ yes

9. $x + 13 - 4 = 16, x = 6$ no

Use inverse operations to solve. Check your solution.

10. $x + 6 = 12$ $x = 6$

11. $16 - r = 8$ $r = 8$

12. $p - 12 = 4$ $p = 16$

13. $14 = k + 8$ $k = 6$

14. $x^2 - 9 = 7$ $x = 4$

15. $21 = p \times 3$ $p = 7$

16. $48 \div x = 3$ $x = 16$

17. $x + 450 = 1120$ $x = 670$

18. $4.7 + x = 10$ $x = 5.3$

19. $64 = x - 28$ $x = 92$

20. $41.7 = x + 30.3$ $x = 11.4$

21. $y + 9 = 26$ $y = 17$

CHAPTER 16

Lesson 16.1

Determine whether the given value is a solution of the equation. Write *yes* or *no*.

1. $3x = 12; x = 4$ yes

2. $7k = 56; k = 7$ no

3. $\frac{p}{16} = 3; p = 48$ yes

4. $\frac{a}{14} = 2; a = 26$ no

5. $27 = 3x; x = 8$ no

6. $88 = 22n; n = 6$ no

Use inverse operations to solve. Check your solution.

7. $4x = 24$ $x = 6$

8. $8x = 32$ $x = 4$

9. $9 = \frac{p}{3}$ $p = 27$

10. $56 = 7p$ $p = 8$

11. $21 = \frac{s}{3}$ $s = 63$

12. $45 = 9n$ $n = 5$

13. $180 = 3d$ $d = 60$

14. $462 = \frac{a}{3}$ $a = 1386$

15. $1486 = \frac{a}{2}$ $a = 2972$

Lesson 16.2

Find the numbers of quarters, dimes, nickels, and pennies in the given dollar amount.

1. $13.00 52; 130; 260; 1300

2. $7.00 28; 70; 140; 700

3. $45.00 180; 450; 900; 4500

4. $17.00 68; 170; 340; 1700

5. $3.75 15; 37; 75; 375

6. $21.50 86; 215; 430; 2150

7. $13.25 53; 132; 265; 1325

8. $52.75 211; 527; 1055; 5275

Use the formulas on page 300 to convert the English pounds and German marks to U.S. dollars. Round to the nearest penny.

9. 35 pounds $56.91

10. 235 pounds $382.11

11. 80 marks $49.60

12. 283 marks $175.46

Lesson 16.3

Convert the temperatures from degrees Celsius to degrees Fahrenheit. Round the answer to the nearest degree.

1. 40°C 104°F

2. 2.3°C 37.4°F

3. 14°C 57.2°F

4. 35°C 95°F

5. 85°C 185°F

6. 60°C 140°F

Convert the temperatures from degrees Fahrenheit to degrees Celsius. Round the answer to the nearest degree.

7. 32°F 0°C

8. 47°F 8°C

9. 79°F 26°C

10. 105°F 40°C

11. 72°F 22°C

12. 98°F 36°C

Lesson 16.4

Use the formula $d = r \times t$ to complete.

1. $d = ?$ 140 mi
$r = 35$ mi per hr
$t = 4$ hr

2. $d = ?$ 1100 ft
$r = 22$ ft per sec
$t = 50$ sec

3. $d = ?$ 93.33 km
$r = 18.3$ km per hr
$t = 5.1$ hr

4. $d = 200$ mi
$r = ?$ 40 mi per hr
$t = 5$ hr

5. $d = 1600$ km
$r = ?$ 4 km per min
$t = 400$ min

6. $d = 180$ ft
$r = ?$ 3 ft per sec
$t = 60$ sec

7. $d = 1300$ mi
$r = 65$ mi per hr
$t = ?$ 20 hr

8. $d = 4880$ ft
$r = 20$ ft per sec
$t = ?$ 244 sec

9. $d = 2100$ km
$r = 70$ km per sec
$t = ?$ 30 sec

10. $d = ?$ 495 mi
$r = 55$ mi per hr
$t = 9$ hr

11. $d = 3500$ mi
$r = ?$ 70 mi per hr
$t = 50$ hr

12. $d = 3700$ ft
$r = 740$ ft per sec
$t = ?$ 5 sec

Lesson 16.5

Solve by making a table.

1. Jason contributes to a charity that helps children with serious medical problems. For every dollar he contributes, a local radio station contributes $4. He wants to make sure that a total of $50 goes to the charity. How much does he have to contribute?
$10

2. Shawn runs 5 miles every day. His coach wants him to run 10 km. Is Shawn running far enough? Explain. (HINT: 1 mi = 1.609 km.)
No; Shawn is running 8.045 km.

CHAPTER 17

Lesson 17.1

Write the ratio in three ways.

1. four to nine
4 to 9, 4:9, $\frac{4}{9}$

2. seven to thirteen
7 to 13, 7:13, $\frac{7}{13}$

3. eight to three
8 to 3, 8:3, $\frac{8}{3}$

4. seven to three
7 to 3, 7:3, $\frac{7}{3}$

5. seven to eighteen
7 to 18, 7:18, $\frac{7}{18}$

6. eleven to thirty-one
11 to 31, 11:31, $\frac{11}{31}$

7. nineteen to eight
19 to 8, 19:8, $\frac{19}{8}$

8. thirty-seven to four
37 to 4, 37:4, $\frac{37}{4}$

For Exercises 9–10, use the figure at right.

9. Find the ratio of blue sections to red sections. Then write three equivalent ratios. 4:2, possible answers: 2:1, 8:4, 12:6

10. Find the ratio of yellow sections to all sections. Then write three equivalent ratios. 2:8, possible answers: 1:4, 4:16, 6:24

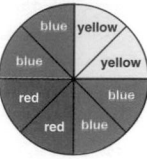

Find the missing term that makes the ratios equivalent.

11. $\frac{2}{7}, \frac{\square}{21}$
6

12. 7 to 3, $\square$ to 6
14

13. 9 : $\square$, 18:8
4

14. $\frac{3}{7}, \frac{15}{\square}$
35

Lesson 17.2

Write a ratio in fraction form for each rate.

1. 247 points in 8 games
$\frac{247 \text{ points}}{8 \text{ games}}$

2. $8 for 2 lb
$\frac{\$8}{2 \text{ lb}}$

3. 150 minutes in 3 hr
$\frac{150 \text{ minutes}}{3 \text{ hr}}$

4. 15¢ per page
$\frac{15¢}{1 \text{ page}}$

5. 75 words in 2 minutes
$\frac{75 \text{ words}}{2 \text{ minutes}}$

6. 12 pages in 3 hr
$\frac{12 \text{ pages}}{3 \text{ hr}}$

Write the unit rate in fraction form.

7. $2.50 for 10
$\frac{\$0.25}{1}$

8. $2.70 for 12
$\frac{\$0.23}{1}$

9. 360 people for 3 sq mi
$\frac{120}{1 \text{ sq mi}}$

10. 240 miles in 6 hr
$\frac{40 \text{ miles}}{1 \text{ hr}}$

11. $2.75 for 25
$\frac{11¢}{1}$

12. 500 miles per 20 gallons
$\frac{25 \text{ miles}}{1 \text{ gal}}$

Lesson 17.3

Write the ratio as a percent.

1. $\frac{27}{100}$ 27%
2. $\frac{42}{100}$ 42%
3. 74:1 74%
4. 19:100 19%

Write the decimal as a percent.

5. 0.3 30%
6. 0.09 9%
7. 0.43 43%
8. 0.7 70%

Write the ratio as a percent.

9. $\frac{3}{5}$ 60%
10. $\frac{1}{3}$ 33%
11. $\frac{5}{25}$ 20%
12. $\frac{3}{8}$ 37.5%

Write the percent as a decimal and as a ratio in simplest form.

13. 30% 0.3, $\frac{3}{10}$
14. 75% 0.75, $\frac{3}{4}$
15. 96% 0.96; $\frac{24}{25}$

Complete by using <, >, or =.

16. $\frac{3}{50} \square$ 5% >
17. 40% $\square \frac{7}{20}$ >
18. 50% $\square \frac{13}{20}$ <
19. 0.015 $\square$ 1.5% =
20. $\frac{5}{16} \square$ 6.25% >
21. 25% $\square \frac{1}{2}$ <

Lesson 17.4

Tell what percent of the figure is shaded.

1. 36%
2. 50%
3. 25%

4. 44%
5. 18.75%
6. 37.5%

Lesson 17.5

Write a proportion to solve.

1. 7 bags of potatoes weigh 28 lbs. How much do 3 bags weigh? 12 lbs

2. Theresa's car uses 5 gallons of gasoline on a 250 mile trip. How much gasoline does she need to travel 800 miles? 16 gals

3. The ratio of rolls of film to exposures is 1 to 36. How many exposures are in 5 rolls of film? 180 exposures

CHAPTER 18

Lesson 18.1

Solve the problem by acting it out.

1. Bill reviewed his spending for the past year. His expenditures totaled $15,000. Of these expenditures, 60% went to necessities. How much did he spend on necessities? $9,000

2. Samantha and her husband paid an interest rate of 9% last year on their $200,000 home loan. How much did they spend in interest for the year? $18,000

Lesson 18.2

Use a ratio in simplest form to find the percent of the number.

1. 20% of 8 $1\frac{3}{5}$
2. 30% of 90 27
3. 45% of 75 $33\frac{3}{4}$
4. 25% of 60 15
5. 85% of 220 187

Use a decimal to find the percent of the number.

6. 14% of 20 2.8
7. 35% of 90 31.5
8. 47% of 71 33.37
9. 91% of 37 33.67
10. 38% of 42 15.96

Use the method of your choice to find the percent of the number.

11. 20% of 15 3
12. 150% of 300 450
13. 60% of 120 72
14. 75% of 90 67.5
15. 300% of 85 255
16. 6.5% of 70 4.55
17. 53% of 106 56.18
18. 82% of 82 67.24
19. 3% of 24 0.72

Find the sales tax. Round to the nearest cent when necessary.

20. price: $30 $1.80
tax rate: 6%

21. price: $29.99 $2.25
tax rate: 7.5%

22. price: $9.28 $0.74
tax rate: 8%

Lesson 18.3

1. Find the angles you would use to make a circle graph.

2. Make a circle graph of the data in Exercise 1.
Check students' graphs.

ICE CREAM		
Item	Percent	Angle
Mint	10%	36
Vanilla	35%	126
Chocolate	37.5%	135
Strawberry	17.5%	63

Lesson 18.4

Find the amount of discount.

1. regular price: $31.00
25% off $7.75

2. regular price: $65.00
50% off $32.50

3. regular price: $42.00
75% off $31.50

4. regular price: $116.50
30% off $34.95

5. regular price: $87.00
50% off $43.50

6. regular price: $97.00
15% off $14.55

7. regular price: $38.20
35% off $13.37

8. regular price: $137.50
20% off $27.50

9. regular price: $95.00
40% off $38.00

10. regular price: $325.00
85% off $276.25

11. regular price: $180.00
60% off $108.00

12. regular price: $225.20
25% off $56.30

13. regular price: $37.00
75% off $27.75

14. regular price: $223.60
45% off $100.62

15. regular price: $14.40
50% off $7.20

Lesson 18.5

Find the interest.

	Principal	Yearly Rate	Interest for 1 year	Interest for 2 years
1.	$65.00	4%	? $2.60	? $5.20
2.	$210.00	2.1%	? $4.41	? $8.82
3.	$735.00	7%	? $51.45	? $102.90
4.	$1300.00	3.9%	? $50.70	? $101.40
5.	$2250.00	2.3%	? $51.75	? $103.50
6.	$7200.00	9%	? $648	? $1,296

Chapter 19

Lesson 19.1

Look at each figure. Tell whether each pair of shapes appear to be *similar, congruent, both,* or *neither.*

1. both
2. similar
3. neither
4. both
5. neither
6. both
7. both
8. similar

9. Draw two figures that are congruent. Then draw two figures that are not congruent but are similar. Check students' drawings.

Lesson 19.2

Name the corresponding sides and angles. Write the ratio of the corresponding sides in simplest form.

1. 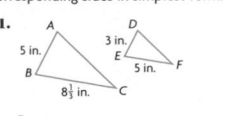   **2.**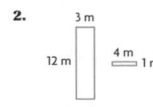

$\frac{5}{3}$ 3

Tell whether the figures in each pair are similar. Write *yes* or *no*.

3.   **4.**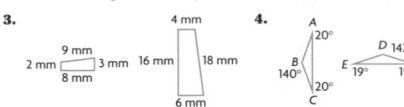

yes   no

Lesson 19.3

The figures in each pair are similar. Find *n*.

1.   **2.**   **3.**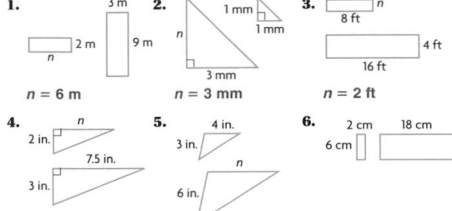

$n = 6$ m $n = 3$ mm $n = 2$ ft

4.   **5.**   **6.**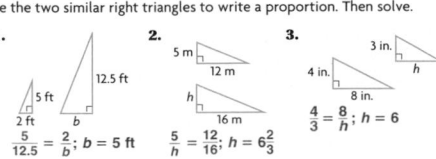

$n = 5$ in. $n = 8$ in. $n = 6$ cm

Lesson 19.4

Use the two similar right triangles to write a proportion. Then solve.

1.   **2.**   **3.**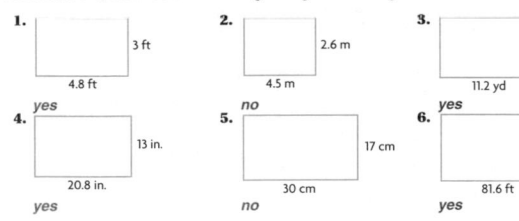

$\frac{5}{12.5} = \frac{2}{b}$; $b = 5$ ft $\frac{5}{h} = \frac{12}{16}$; $h = 6\frac{2}{3}$ $\frac{4}{3} = \frac{8}{h}$; $h = 6$

CHAPTER 20

Lesson 20.1

Find the ratio of length to width.

1. $l = 6$, $w = 4\frac{3}{2}$ **2.** $l = 5$, $w = 15\frac{1}{3}$ **3.** $l = 2.7$, $w = 0.9\frac{3}{1}$

4. $l = 12$, $w = 2\frac{6}{1}$ **5.** $l = 28$, $w = 27\frac{28}{27}$ **6.** $l = 8.1$, $w = 0.9\frac{9}{1}$

7. $l = 72$, $w = 9\frac{8}{1}$ **8.** $l = 17$, $w = 13\frac{17}{13}$ **9.** $l = 27$, $w = 5\frac{27}{5}$

Find the missing dimension.

10. scale: 1 in. : 5 ft
drawing length : 7 in.
actual length : ☐ ft
35

11. scale: 1 in. : 5 ft
drawing length : 4 in.
actual length : ☐ ft
20

12. scale: 1 in. : 5 ft
drawing length : ☐ in.
actual length : 15 ft
3

13. scale: 1 in. : 8 ft
drawing length : 2 in.
actual length : ☐ ft
16

14. scale: 1 in. : 8 ft
drawing length : ☐ in.
actual length : 72 ft
9

15. scale: 5 cm : 1 mm
drawing length : ☐ cm
actual length : 8 mm
40

16. scale: 5 cm : 1 mm
drawing length : 15 cm
actual length : ☐ mm
3

17. scale : 5 m : 1 km
drawing length : 20 m
actual length : ☐ km
4

18. scale: 5 m : 1 km
drawing length : ☐ m
actual length : 10 km
50

Lesson 20.2

Write and solve a proportion to find the actual miles. Use a map scale of 1 inch = 25 miles.

1. map distance: $1\frac{1}{2}$ in. $\frac{1}{25} = \frac{1.5}{n}$, **37.5 mi**

2. map distance: 3 in. $\frac{1}{25} = \frac{3}{n}$, **75 mi**

3. map distance: $5\frac{1}{2}$ in. $\frac{1}{25} = \frac{5.5}{n}$, **137.5 mi**

4. map distance: 7 in. $\frac{1}{25} = \frac{7}{n}$, **175 mi**

5. map distance: 9 in. $\frac{1}{25} = \frac{9}{n}$, **225 mi**

6. map distance: 18 in. $\frac{1}{25} = \frac{18}{n}$, **450 mi**

Lesson 20.3

Draw a diagram to solve.

1. Enya is on a 5 day hiking trip. The first day she hikes 4 miles north. The second day she hikes 3 miles east. The third day she hikes 7 miles south. The fourth day she hikes 3 miles west. In what direction and how many miles does she go on a fifth day to return to the starting point?
north 3 miles

2. Tom sails 200 yards north, and then goes 300 yards east. Next he goes north again 500 yards and west for 300 yards. In what direction and how far must he go from that point to meet the first part of his path? How far has it been to his starting point?
south 500 yards; south 700 yards

3. To get to the library, Michele walks 3 blocks north, and then 1 block west. Draw a map of her walk, using the scale 1 in. = 1 block.
Check Students' maps.

4. To get to a gas station, Dave walked 4 blocks south, and 3 blocks east. In what directions and what distances must Dave walk to get back to his car?
3 blocks west, and 4 blocks north

Lesson 20.4

In Exercises 1–6, tell whether the rectangle is a golden rectangle.

1.   **2.**   **3.**

yes   no   yes

4.   **5.**   **6.**

yes   no   yes

CHAPTER 21

Lesson 21.1

Change to the given unit.

1. 40 ft = _?_ in. **480**   **2.** 50 yd = _?_ ft **150**   **3.** 20 c = _?_ fl oz **160**

4. 16 lbs = _?_ oz **256**   **5.** 13 weeks = _?_ days **91**   **6.** 600 ft = _?_ yds **200**

Use a proportion to change to the given unit.

7. 6 years = _?_ mo **72**   **8.** 6 tons = _?_ lbs **12,000**

9. 3 yds = _?_ in. **108**   **10.** 10 gal = _?_ qt **40**

11. 72 pt = _?_ gal **9**   **12.** 4 mi = _?_ ft **21,120**

13. 10,560 ft = _?_ mi **2**   **14.** 272 ft = _?_ yd _?_ ft **90;2**

15. 248 min = _?_ hours _?_ min **4;8**   **16.** $7\frac{1}{2}$ ft = _?_ yd _?_ ft _?_ in. **2;1;6**

Lesson 21.2

Change to the given unit.

1. 300 kL = _?_ L **300,000**   **2.** 50 dm = _?_ cm **500**   **3.** 600 m = _?_ km **0.6**

4. 20 g = _?_ kg **0.02**   **5.** 4,000 mm = _?_ km **0.004**   **6.** 0.004 mm = _?_ m **0.000004**

Use a proportion to change to the given units.

7. 12 L = _?_ mL **12,000**   **8.** 22 kL = _?_ L **22,000**   **9.** 0.22 m = _?_ km **0.00022**

10. 650,000 mg = _?_ g **650**   **11.** 450 mm = _?_ cm **45**   **12.** 0.042 m = _?_ mm **42**

13. 0.057 L = _?_ mL **57**   **14.** 200 g = _?_ kg **0.2**   **15.** 0.0030 kL = _?_ L **3**

Lesson 21.3

Tell which measurement is more precise.

1. 8 ft or 97 in. **97 in.**   **2.** 58 mm or 5 cm **58 mm**   **3.** 95 mm or 9 cm **95 mm**

4. 73 in. or 2 yds **73 in.**   **5.** 28 in. or 2 ft **28 in.**   **6.** 22 mm or 2 cm **22 mm**

7. 482 in. or 40 ft **482 in.**   **8.** 9.3 cm or 93.3 mm **93.3 mm**   **9.** 600 m or 0.6 km **600 m**

10. 29 yd or 89 ft **89 ft**   **11.** 5,282 ft or 1 mile **5,282 ft**   **12.** 50 cm or 500 mm **500 mm**

13. 85 mm or 8 cm **85 mm**   **14.** 37 ft or 12 yds **37 ft**   **15.** 9.1 cm or 9 cm **9.1 cm**

16. 4 m or 403 cm **403 cm**   **17.** 532 mm or 53 cm **532 mm**   **18.** 19 yds or 57 ft **57 ft**

Lesson 21.4

Use the network to find the shorter route. Give the distance. Distances are in kilometers.

1. BDCE or BFEC **BFEC**, 57   **2.** ECBF or DBCF **ECBF**, 59

3. CFBC or CBFE **CBFE**, 37   **4.** EFCD or DEFC **DEFC**, 42

5. FEDB or CFBD **CFBD**, 55   **6.** DCEF or EDCB **DCEF**, 72

7. BFCD or CEDB **BFCD**, 50   **8.** CBDE or BDCF **CBDE**, 56

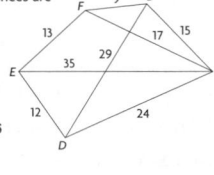

Lesson 21.5

Find the perimeter.

1.   **2.**   **3.**

15 cm   117 yd   22 in.

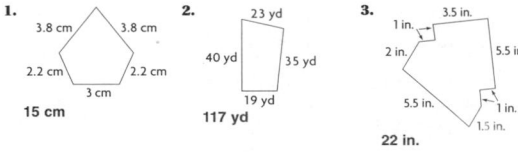

The perimeter is given. Find the missing length.

4. $P = 230$ $x = 35$ **5.** $P = 124$ yd $y = 17$ yd   **6.** $P = 401$ cm $d = 53.7$ cm

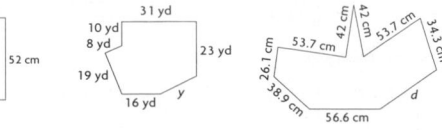

CHAPTER 22

Lesson 22.1

Estimate the area of the figure. Each square is 1 in².

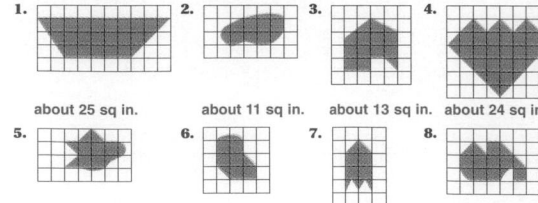

1. about 25 sq in. 2. about 11 sq in. 3. about 13 sq in. 4. about 24 sq in.

5. about 8 in.² 6. about 7 in.² 7. about 6 in.² 8. about 11 in.²

Lesson 22.2

Use a formula to solve.

1. A garden measures 18 ft on each side. What is the area of the garden? **324 sq ft**

2. A patio is built out of poured blocks. Each block is 15 in. on each side. 244 blocks were used to build the patio. At a cost of $3.25 for each block, find the area of the patio and its cost. **54,900 sq in., $793**

3. Evan marked a 215 ft by 170 ft rectangular section of a parking lot to make an unloading zone. What is the area of the unloading zone? **36,550 ft²**

4. Dean has an oriental rug that is 6 ft long and 12 ft wide. What is the area of the rug? **72 ft²**

Lesson 22.3

Use the formula to find the area of the parallelogram.

1. $b = 9$ in. $h = 5$ in. **45 sq in.**
2. $b = 6.2$ m $h = 3.1$ m **19.22 sq m**
3. $b = 13\frac{1}{2}$ ft $h = 10$ ft **135 sq ft**

4. $b = 7$ yd $h = 4$ yd **28 sq yd**
5. $b = 5$ ft $h = 2$ ft **10 sq ft**
6. $b = 12$ yd $h = 6$ yd **72 sq yd**

Use the formula to find the area of the triangle.

7. $b = 40$cm $h = 20$ cm **400 sq cm**
8. $b = 4$ ft $h = 1\frac{1}{2}$ ft **3 sq ft**
9. $b = 21$ ft $h = 4$ ft **42 sq ft**

10. $b = 6$ cm $h = 0.6$ cm **1.8 sq cm**
11. $b = 16$ m $h = 1.6$ m **12.8 sq m**
12. $b = 32$ in. $h = 0.8$ in. **12.8 sq in.**

Student Handbook

Lesson 22.4

Find the perimeter and area of each figure. Then double the dimensions and find the new perimeter and area.

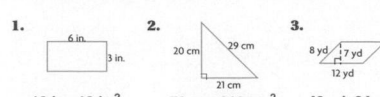

1. **18 in., 18 in².; 36 in., 72 in².**
2. **70 cm, 210 cm².; 140 cm, 840 cm²**
3. **40 yd, 84 yd².; 80 yd, 336 yd²**

Find the perimeter and area of each figure. Then halve the dimensions and find the new perimeter and area.

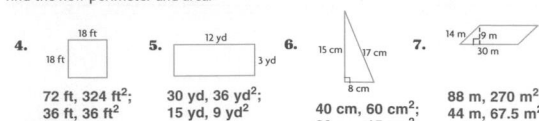

4. **72 ft, 324 ft².; 36 ft, 36 ft²**
5. **30 yd, 36 yd².; 15 yd, 9 yd²**
6. **40 cm, 60 cm².; 20 cm, 15 cm²**
7. **88 m, 270 m².; 44 m, 67.5 m²**

Lesson 22.5

Find the area of each circle. Round to the nearest tenth.

1. $r = 9$ yd **254.5 sq yd**
2. $d = 12$ mm **113.1 sq mm**
3. $r = 7$ ft **153.9 sq ft**
4. $d = 24$ cm **452.4 sq cm**
5. $r = 9.2$ mm **265.9 sq mm**
6. $d = 8$ yd **50.3 sq yd**
7. $r = 45$ mm **6,358.5 sq mm**
8. $r = 16$ in. **803.8 sq in.**
9. $r = 27$ cm **2,289.1 sq cm**

CHAPTER 23

Lesson 23.1

Find the volume of each rectangular prism.

1. length = 6 in. width = 3 in. height = 4 in. **72 in.³**
2. length = 8 ft width = 5 ft height = 10 ft **400 ft³**
3. length = 10 m width = 2 m height = 5 m **100 m³**

Find the volume of each triangular prism.

4. length = 12 cm width = 2 cm height = 1 cm **12 cm³**
5. length = 10 in. width = 3 in. height = 2 in. **30 in.³**
6. length = 10 ft width = 2 ft height = 2 ft **20 ft³**

7. length = 8 ft width = 2 ft height = 2 ft **16 ft³**
8. length = 20 m width = 10 m height = 10 m **1000 m³**
9. length = 20 in. width = 5 in. height = 10 in. **500 in.³**

More Practice

Lesson 23.2

Find the volume of each prism. Then double the dimensions and find the new volume.

1. length = 4 cm width = 2 cm height = 1 cm **8 cm³, 64 cm³**
2. length = 8 in width = 2 in height = 2 in **32 in³, 256 in³**
3. length = 10 ft width = 5 ft height = 2 ft **100 ft³, 800 ft³**
4. length = 8 ft width = 8 ft height = 8 ft **512 ft³, 4096 ft³**
5. length = 4 m width = 2 m height = 2 m **16 m³, 128 m³**
6. length = 6 in width = 6 in height = 3 in **108 in³, 864 in³**

Find the volume of each prism. Then halve the dimensions and find the new volume.

7. length = 8 cm width = 4 cm height = 2 cm **64 cm³, 8 cm³**
8. length = 8 in width = 8 in height = 1 in **64 in³, 8 in³**
9. length = 16 ft width = 3 ft height = 2 ft **96 ft³, 12 ft³**
10. length = 12 ft width = 4 ft height = 2 ft **96 ft³, 12 ft³**
11. length = 5 m width = 4 m height = 4 m **80 m³, 10 m³**
12. length = 7 in width = 4 in height = 2 in **56 in³, 7 in³**

Lesson 23.3

Find the volume of each cylinder. Round to the nearest whole number.

1. height = 6 in radius = 3 in **170 in³**
2. height = 5 ft radius = 2 ft **63 ft³**
3. height = 9 m radius = 1 m **28 m³**
4. height = 5 in radius = 7 in **770 in³**
5. height = 6 cm radius = 9 cm **1,527 cm³**
6. height = 2 in radius = 12 in **905 in³**
7. height = 7 ft radius = 7 ft **1,078 ft³**
8. height = 6 in radius = 8 in **1,206 in³**
9. height = 3 ft radius = 4 ft **151 ft³**
10. height = 7 m radius = 2 m **88 m³**
11. height = 10 in radius = 14 in **6,158 in³**
12. height = 2 ft radius = 6 ft **226 ft³**
13. height = 6 m radius = 16 m **4,825 m³**
14. height = 7 in radius = 14 in **4,310 in³**
15. height = 5 cm radius = 8 cm **1,005 cm³**

Lesson 23.4

Find the surface area of each prism.

1. **132 in²**
2. **276 cm²**
3. **122 m²**

Student Handbook

CHAPTER 24

Lesson 24.1

Write an integer to represent each situation. Then describe the opposite situation, and write an integer to represent it.

1. a temperature increase of 5 degrees **+5, a temperature decrease of 5 degrees, −5**
2. the wind speed decreases by 12 mph **−12, the wind speed increases by 12 mph, +12**
3. depositing $510 into a savings account **+510, withdrawing $510 from a savings account, −510**
4. a temperature decrease of 7 degrees **−7, a temperature increase of 7 degrees, +7**

Write the absolute value.

5. $|-3|$ **3**
6. $|-12|$ **12**
7. $|+36|$ **36**
8. $|-160|$ **160**
9. $|+62|$ **62**
10. $|+4|$ **4**
11. $|+400|$ **400**
12. $|-4|$ **4**
13. $|+1|$ **1**
14. $|-7|$ **7**
15. $|+27|$ **27**
16. $|-271|$ **271**

Lesson 24.2

Write each rational number in the form $\frac{a}{b}$.

1. $3\frac{1}{6}$ **$\frac{19}{6}$**
2. 0.5 **$\frac{5}{10}$**
3. 0.27 **$\frac{27}{100}$**
4. 13.4 **$13\frac{4}{10}$**
5. $2\frac{2}{5}$ **$\frac{12}{5}$**
6. 3.18 **$3\frac{18}{100}$**
7. 10.02 **$\frac{1002}{100}$**
8. 300 **$\frac{300}{1}$**
9. 13 **$\frac{13}{1}$**
10. 0.04 **$\frac{4}{100}$**
11. $5\frac{2}{5}$ **$\frac{27}{5}$**
12. $7\frac{2}{3}$ **$\frac{23}{3}$**
13. 0.36 **$\frac{36}{100}$**
14. $5\frac{1}{3}$ **$\frac{16}{3}$**
15. 312 **$\frac{312}{1}$**
16. $4\frac{1}{4}$ **$\frac{17}{4}$**

Lesson 24.3

Find the terminating decimal for the fraction.

1. $\frac{3}{10}$ **0.3**
2. $\frac{2}{5}$ **0.4**
3. $\frac{3}{8}$ **0.375**
4. $\frac{9}{25}$ **0.36**
5. $\frac{3}{20}$ **0.15**
6. $\frac{3}{16}$ **0.1875**
7. $\frac{3}{25}$ **0.12**
8. $\frac{17}{200}$ **0.085**
9. $\frac{4}{5}$ **0.8**
10. $\frac{19}{25}$ **0.76**
11. $\frac{9}{16}$ **0.5625**
12. $\frac{4}{8}$ **0.5**

Find the repeating decimal for the fraction.

13. $\frac{3}{11}$ **$0.\overline{27}$**
14. $\frac{7}{3}$ **$2.\overline{3}$**
15. $\frac{2}{9}$ **$0.\overline{2}$**
16. $\frac{4}{45}$ **$0.0\overline{8}$**
17. $\frac{17}{9}$ **$1.\overline{8}$**
18. $\frac{7}{60}$ **$0.11\overline{6}$**
19. $\frac{4}{15}$ **$0.2\overline{6}$**
20. $\frac{7}{11}$ **$0.\overline{63}$**
21. $\frac{8}{9}$ **$0.\overline{8}$**

Write the fraction as a decimal.

22. $\frac{7}{50}$ **0.14**
23. $\frac{19}{20}$ **0.95**
24. $\frac{12}{15}$ **0.8**
25. $\frac{13}{3}$ **$4.\overline{3}$**
26. $\frac{3}{40}$ **0.075**
27. $\frac{7}{4}$ **1.75**

More Practice

Lesson 24.4

Find a rational number between the two given numbers.

1. $\frac{3}{8}$ and $\frac{5}{6}$ $\frac{5}{12}$
2. $\frac{1}{8}$ and $\frac{1}{4}$ $\frac{3}{16}$
3. $1\frac{1}{8}$ and $1\frac{1}{3}$ $1\frac{1}{4}$
4. $-\frac{1}{4}$ and $-\frac{1}{8}$ $-\frac{3}{16}$
5. 1.8 and 1.9 1.85
6. -1.5 and -1.3 -1.4
7. -3.55 and -3.5 -3.52
8. 3.02 and 3.03 3.025
9. -5.16 and -5.15 -5.155
10. $\frac{1}{3}$ and 0.4 0.35
11. $\frac{1}{4}$ and 0.28 0.26
12. $\frac{2}{3}$ and 0.8 0.7
13. -6.01 and -6 -6.008
14. 1.4 and 1.41 1.407
15. -7.08 and -7.07 -7.077
16. $\frac{1}{7}$ and $\frac{3}{14}$
17. $\frac{1}{8}$ and $\frac{2}{6}$ $\frac{1}{4}$
18. 2.5 and 2.6 2.52
19. 3.8 and 3.82 3.81
20. -5 and -4.9 -4.94

Lesson 24.5

Compare. Write <, > or =.

1. $0.4 \,\square\, 0.38$ >
2. $\frac{2}{7} \,\square\, 0.25$ >
3. $-0.6 \,\square\, -\frac{2}{5}$ <
4. $0.28 \,\square\, \frac{2}{7}$ <
5. $\frac{3}{13} \,\square\, 0.23$ >
6. $-\frac{4}{9} \,\square\, -\frac{2}{5}$ <
7. $2\frac{1}{5} \,\square\, 2\frac{4}{13}$ <
8. $0.87 \,\square\, 0.868$ >

Compare the rational numbers and order them from least to greatest.

9. $\frac{1}{3}, \frac{2}{5}, 0.38, \frac{1}{2}$ $\frac{1}{3}, 0.38, \frac{2}{5}, \frac{1}{2}$
10. $\frac{2}{7}, \frac{1}{3}, \frac{1}{4}, 0.26$ $\frac{1}{4}, 0.26, \frac{2}{7}, \frac{1}{3}$
11. $0.23, \frac{2}{9}, \frac{1}{4}, \frac{1}{5}$ $\frac{1}{5}, 0.23, \frac{2}{9}, \frac{1}{4}$
12. $-\frac{2}{5}, -\frac{1}{3}, -\frac{3}{8}, -\frac{1}{2}$ $-\frac{1}{2}, -\frac{2}{5}, -\frac{3}{8}, -\frac{1}{3}$

CHAPTER 25

Lesson 25.1

Find the sum.

1. +2 + -6 -4
2. -5 + +5 0
3. -82 + -8 -90
4. -4 + +8 +4
5. -33 + -18 -51
6. -9 + -4 -13
7. +4 + +12 +16
8. +64 + -31 +33
9. -82 + +71 -11
10. -21 + -7 -28
11. +120 + -42 +78
12. -19 + +19 0
13. -67 + +58 -9
14. -14 + -57 -71
15. +45 + -33 +12
16. -90 + -111 -201
17. +47 + -12 +35
18. +83 + -15 +68

Lesson 25.2

Find the difference.

1. -2 - +6 -8
2. -3 - -5 +2
3. +5 - +3 +2
4. +5 - -3 +8
5. -8 - -3 -5
6. -8 - +3 -11
7. -7 - -7 0
8. -12 - -10 -2
9. +3 - +6 -3
10. +7 - -3 +10
11. -13 - +6 -19
12. -2 - -2 0
13. -7 - +7 -14
14. +23 - +31 -8
15. 0 - -4 +4

Lesson 25.3

Complete the pattern.

1.
-4 × +2 = -8
-4 × +1 = -4
-4 × 0 = 0
-4 × -1 = □
-4 × -2 = □
-4 × -3 = □
+4, +8, +12

2.
-5 × +3 = -15
-5 × +2 = -10
-5 × +1 = -5
-5 × 0 = □
-5 × -1 = □
-5 × -2 = □
-5 × -3 = □
0, +5, +10, +15

3.
-8 × +3 = -24
-8 × +2 = -16
-8 × +1 = -8
-8 × 0 = 0
-8 × -1 = □
-8 × -2 = □
-8 × -3 = □
+8, +16, +24

Find the product.

4. -3 × -2 +6
5. -8 × +6 -48
6. -3 × -12 +36
7. -8 × +9 -72
8. +7 × -12 -84
9. -30 × -20 +600
10. -31 × +6 -186
11. -12 × -5 +60
12. -7 × -3 +21
13. -10 × -8 +80
14. +17 × -9 -153
15. -23 × -5 +115

Lesson 25.4

Find the quotient.

1. -48 ÷ -8 +6
2. -72 ÷ +9 -8
3. +115 ÷ -23 -5
4. -126 ÷ +7 -18
5. -36 ÷ +3 -12
6. -36 ÷ -3 +12
7. -60 ÷ -6 +10
8. -84 ÷ -12 +7
9. +21 ÷ -7 -3
10. +6 ÷ -2 -3
11. +186 ÷ -6 -31
12. +126 ÷ -7 -18
13. -260 ÷ -65 +4
14. +84 ÷ -6 -14
15. +176 ÷ -11 -16

CHAPTER 26

Lesson 26.1

Evaluate the numerical expression. Remember the order of operations.

1. 8 + 1 × 2 10
2. 9 - 6 × 4 -15
3. -4 × -2 + 2 10
4. (-20 - 14) + 7 -27
5. 62 × -2 -124
6. -8 × (-28 + 26) 16
7. $-4 \times \frac{6}{4}$ -6
8. 9 × -2 + 10 -8

Evaluate the algebraic expression for the given value of the variable.

9. x - 6, for x = 12 6
10. x - 32, for x = 11 -21
11. 42 + k, for k = 18 60
12. 21 + a, for a = -14 7
13. $k^2 - 8$, for k = -4 24
14. $x^3 - 3$, for x = 3 30
15. 5 + k - 16, for k = -4 -15
16. $-21 + a^3$, for a = 3 6
17. 8z, for z = -3.5 -28
18. y - -18, for y = -4 14
19. $\frac{8}{j}$, for j = -64 $-\frac{1}{8}$
20. $-\frac{72}{y}$, for y = -8 9

Lesson 26.2

Solve and check.

1. x + 8 = 4   x = -4
2. 13 = y + 20   y = -7
3. k + 6 = -2   k = -8
4. -9 = n + 8   n = -17
5. c - 10 = -4   c = 6
6. -9 = w - 13   w = 4
7. r - 9 = -6   r = 3
8. -21 = x - 13   x = -8
9. x + 120 = 14   x = -106
10. -32 = y + 147   y = -179
11. p - 243 = -182   p = 61
12. -847 = y - 391   y = -456
13. -4y = 36   y = -9
14. -6x = 72   x = -12
15. -7x = 42   x = -6
16. -3m = -69   m = 23
17. 10y = -40   y = -4
18. 8d = -56   d = -7

Lesson 26.3

Find the whole-number solutions to the inequality.

1. x < 4   0, 1, 2, 3
2. x < 9   0, 1, 2, 3, 4, 5, 6, 7, 8
3. x ≤ 3   0, 1, 2, 3
4. x ≤ 6   0, 1, 2, 3, 4, 5, 6

Write the algebraic inequality represented by the number line.

5.
x ≥ 1

6.
x < -3

Lesson 26.4

Write the ordered pair for the point on the coordinate plane.

1. point A (3, 4)
2. point B (1, 0)
3. point C (-2, 2)
4. point D (-8, 6)
5. point E (-8, -5)
6. point F (-3, -3)
7. point G (2, -5)
8. point H (5, -3)
9. point I (8, 4)
10. point J (6, 6)
11. point K (-5, 3)
12. point L (-5, -4)
13. point M (-7, 0)
14. point N (-2, 5)
15. point P (5, 2)
16. point R (5, -6)

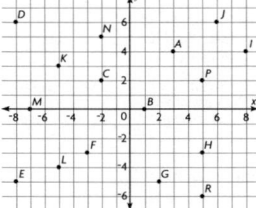

Lesson 26.5

Use the first three values of x and y to complete the table.

1. 6;7

input (x)	1	2	3	4	5
output (y)	3	4	5	?	?

2. 24;30

input (x)	1	2	3	4	5
output (y)	6	12	18	?	?

3. 4;5

input (x)	3	4	5	6	7
output (y)	1	2	3	?	?

4. -4; -8

input (x)	2	1	0	-1	-2
output (y)	8	4	0	?	?

5. What expression can you write using x to get the value of y in Exercise 1? x + 2

6. What expression can you write using x to get the value of y in Exercise 2? 6x

CHAPTER 27

Lesson 27.1

Copy each figure onto a coordinate grid. Perform the indicated transformation. Give the new coordinates of the vertices.

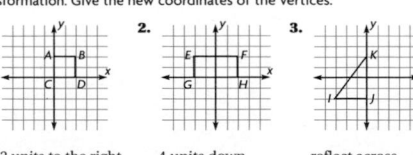

1. 2 units to the right
A'(2,2), B'(4,2), C'(2,0), D' (4,0)
2. 4 units down
E'(-2,-2), F'(2,-2), G'(-2,-4), H'(2,-4)
3. reflect across the y-axis
I'(3,-2), J'(0,-2), K'(0,2)

H78 Student Handbook

More Practice H79

H80 Student Handbook

More Practice H81

More Practice

Student Handbook H78-H81

Lesson 27.2

Find a pattern and solve.

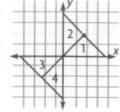

1. Sandra made the design at right.
   What pattern of transformations did she use?
   rotation, reflection, rotation

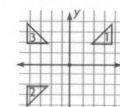

2. Sherry made the design at right.
   What pattern of transformations did she use? **rotation, reflection, rotation**

Lesson 27.3

Draw the next two figures in each geometric pattern.

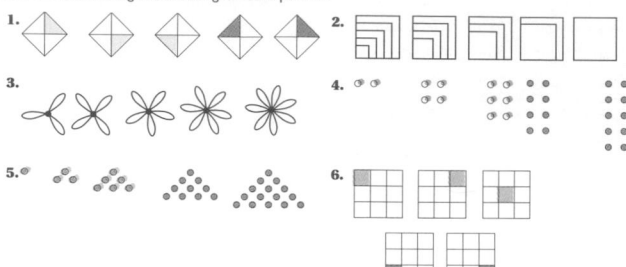

1.

2.

3.

4.

5.

 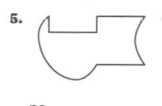

6.

Lesson 27.4

Make the tessellation shape described by each pattern. Then form two rows of a tessellation. **See students' drawings.**

1.

2.

3.

Write *yes* or *no* to tell whether the figure forms a tessellation.

4.

yes

5.

no

6.

no

CHAPTER 28

Lesson 28.1

Find the next three terms in the sequence.

1. 7, 10, 13, 16, ...
   19, 22, 25
2. 23, 230, 2300, 23000, ...
   230000, 2300000, 23000000
3. 11, 23, 35, 47, ..
   59, 71, 83
4. 2, 6, 18, ...
   54, 162, 486
5. 7, 17, 37, 67, ...
   107, 157, 217
6. 34, 35, 37, 40, ...
   44, 49, 55
7. 95, 89, 83, 77, ...
   71, 65, 59
8. 0.76, 0.73, 0.70, 0.67, ...
   0.64, 0.61, 0.58
9. 440, 44, 4.4, 0.44, ...
   0.044, 0.0044, 0.00044
10. 8000, 4000, 2000, 1000, ...
    500, 250, 125
11. 2.3, 3.9, 5.5, ...
    7.1, 8.7, 10.3
12. 62.4, 62.2, 62.0, 61.8, ...
    61.6, 61.4, 61.2

Lesson 28.2

Find the next three terms in the sequence.

1. $\frac{2}{5}, \frac{4}{5}, 1\frac{1}{5}, ...$ **$1\frac{3}{5}, 2, 2\frac{2}{5}$**
2. $1\frac{3}{5}, 2\frac{1}{5}, 2\frac{4}{5}, ...$ **$3\frac{2}{5}, 4, 4\frac{3}{5}$**
3. $\frac{1}{10}, \frac{1}{5}, \frac{3}{10}, ...$ **$\frac{2}{5}, \frac{1}{2}, \frac{3}{5}$**
4. $\frac{3}{14}, \frac{3}{7}, \frac{9}{14}, ...$ **$1\frac{1}{14}, 1\frac{2}{7}, 1\frac{1}{2}$**
5. $2, 3\frac{1}{2}, 5, ...$ **$6\frac{1}{2}, 8, 9\frac{1}{2}$**
6. $20, 18\frac{2}{3}, 17\frac{1}{3}, ...$ **$16, 14\frac{2}{3}, 13\frac{1}{3}$**
7. $\frac{3}{8}, \frac{7}{8}, \frac{11}{8}, ...$ **$\frac{15}{8}, \frac{19}{8}, \frac{23}{8}$**
8. $5, 4\frac{11}{12}, 4\frac{5}{6}, ...$ **$4\frac{3}{4}, 4\frac{2}{3}, 4\frac{7}{12}$**

Lesson 28.3

Find the next three terms in the sequence.

1. $\frac{1}{2}, \frac{3}{2}, \frac{9}{2}, \frac{27}{2}, ...$ **$\frac{81}{2}, \frac{243}{2}, \frac{729}{2}$**
2. $\frac{1}{4}, \frac{1}{0}, \frac{1}{16}, \frac{1}{32}, ...$ **$\frac{1}{64}, \frac{1}{128}, \frac{1}{256}$**
3. $\frac{1}{27}, \frac{1}{9}, \frac{1}{3}, 1, ...$ **3, 9, 27**

Write the first four terms of the sequence.

4. pattern: multiply by 4
   first term: $\frac{1}{2}$
   $\frac{1}{2}$, 2, 8, 32
5. pattern: multiply by $\frac{1}{3}$
   first term: $\frac{1}{6}$
   $\frac{1}{6}, \frac{1}{18}, \frac{1}{54}, \frac{1}{162}$
6. pattern: multiply by $\frac{1}{2}$
   first term: $\frac{1}{2}$
   $\frac{1}{2}, \frac{1}{4}, \frac{1}{8}, \frac{1}{16}$
7. pattern: multiply by $\frac{2}{5}$
   first term: $\frac{1}{2}$
   $\frac{1}{2}, \frac{2}{10}, \frac{4}{50}, \frac{8}{250}$
8. pattern: multiply by $\frac{3}{8}$
   first term: $\frac{1}{2}$
   $\frac{1}{2}, \frac{3}{16}, \frac{9}{128}, \frac{27}{1,024}$
9. pattern: multiply by $\frac{1}{12}$
   first term: $\frac{1}{3}$
   $\frac{1}{3}, \frac{1}{36}, \frac{1}{432}, \frac{1}{5,184}$

Lesson 28.4

Find the next three terms in the sequence.

1. −13, −9, −5, −1, ...
   +3, +7, +11
2. 4, 1, −2, −5, ...
   −8, −11, −14
3. 3, −6, 12, −24, ...
   48, −96, 192
4. 3000, −300, 30, −3, ...
   0.3, −0.03, 0.003
5. 14, 7, 0, −7, ...
   −14, −21, −28
6. −33, −22, −11, 0, ...
   11, 22, 33
7. 4, −8, 16, −32, ...
   64, −128, 256
8. 27, −9, 3, −1, ...
   $\frac{1}{3}, -\frac{1}{9}, \frac{1}{27}$
9. −35, −20, −5, 10, ...
   25, 40, 55
10. −64, −53, −42, −31, ...
    −20, −9, 2
11. 2, −10, 50, −250, ...
    1,250; −6,250; 31,250
12. −5000, 1000, −200, 40, ...
    $-8, \frac{8}{5}, -\frac{8}{25}$
13. 20, −2, 0.2, −0.02, ...
    0.002, −0.0002, 0.00002
14. 0.4, −1.2, 3.6, −7.2, ...
    21.6, −64.6, 194.4
15. $\frac{250}{2}, -\frac{50}{2}, 10, -2, ...$
    $\frac{2}{5}, -\frac{2}{25}, \frac{2}{125}$
16. 1, −2, 4, −8, ...
    16, −32, 64

TABLE OF MEASURES

METRIC UNITS | CUSTOMARY UNITS

Length

METRIC UNITS	CUSTOMARY UNITS
1 millimeter (mm) = 0.001 meter (m)	1 foot (ft) = 12 inches (in.)
1 centimeter (cm) = 0.01 meter	1 yard (yd) = 36 inches
1 decimeter (dm) = 0.1 meter	1 yard = 3 feet
1 kilometer (km) = 1,000 meters	1 mile (mi) = 5,280 feet
	1 mile = 1,760 yards
	1 nautical mile = 6,076.115 feet

Capacity

METRIC UNITS	CUSTOMARY UNITS
1 milliliter (mL) = 0.001 liter (L)	1 teaspoon (tsp) = $\frac{1}{6}$ fluid ounce (fl oz)
1 centiliter (cL) = 0.01 liter	1 tablespoon (tbsp) = $\frac{1}{2}$ fluid ounce
1 deciliter (dL) = 0.1 liter	1 cup (c) = 8 fluid ounces
1 kiloliter (kL) = 1,000 liters	1 pint (pt) = 2 cups
	1 quart (qt) = 2 pints
	1 quart = 4 cups
	1 gallon (gal) = 4 quarts

Mass/Weight

METRIC UNITS	CUSTOMARY UNITS
1 milligram (mg) = 0.001 gram (g)	1 pound (lb) = 16 ounces (oz)
1 centigram (cg) = 0.01 gram	1 ton (T) = 2,000 pounds
1 decigram (dg) = 0.1 gram	
1 kilogram (kg) = 1,000 grams	
1 metric ton (t) = 1,000 kilograms	

Volume/Capacity/Mass for Water

1 cubic centimeter (cm^3) → 1 milliliter → 1 gram
1,000 cubic centimeters → 1 liter → 1 kilogram

TIME

1 minute (min) = 60 seconds (sec)	1 year (yr) = 12 months (mo)
1 hour (hr) = 60 minutes	1 year = 52 weeks
1 day = 24 hours	1 year = 365 days
1 week (wk) = 7 days	

FORMULAS

Perimeter

Polygon	P = sum of the lengths of the sides
Rectangle	$P = 2(l + w)$
Square	$P = 4s$

Circumference

Circle	$C = 2\pi r$, or $C = \pi d$

Area

Circle	$A = \pi r^2$
Parallelogram	$A = bh$
Rectangle	$A = lw$
Square	$A = s^2$
Trapezoid	$A = \frac{1}{2}h(b_1 + b_2)$
Triangle	$A = \frac{1}{2}bh$

Surface Area

Rectangular Prism	$S = 2(lh + lw + wh)$

Volume

Cylinder	$V = \pi r^2 h$
Pyramid	$V = \frac{1}{3}Bh$
Rectangular Prism	$V = lwh$
Triangular Prism	$V = \frac{1}{2}lwh$

Other

Celsius (°C)	$C = \frac{5}{9} \times (F - 32)$
Diameter	$d = 2r$
Fahrenheit (°F)	$F = (\frac{9}{5} \times C) + 32$
Pythagorean Property	$c^2 = a^2 + b^2$

Consumer

Distance traveled	$d = rt$
Interest (simple)	$I = prt$

SYMBOLS

Symbol	Meaning	Symbol	Meaning
$<$	is less than	1:2	ratio of 1 to 2
$>$	is greater than	%	percent
$\leq$	is less than or equal to	$\cong$	is congruent to
$\geq$	is greater than or equal to	$\approx$	is approximately equal to
$=$	is equal to	$\perp$	is perpendicular to
$\neq$	is not equal to	$\parallel$	is parallel to
10^2	ten squared	$\overleftrightarrow{AB}$	line AB
10^3	ten cubed	$\overrightarrow{AB}$	ray AB
10^4	the fourth power of 10	$\overline{AB}$	line segment AB
2^3	the third power of 2	$\angle ABC$	angle ABC
$2.\overline{6}$	repeating decimal 2.666 . . .	$m\angle A$	measure of $\angle A$
$^{+}7$	positive 7	$\triangle ABC$	triangle ABC
$^{-}7$	negative 7	°	degree (angle or temperature)
(4,7)	the ordered pair 4,7	π	pi (about 3.14)
$5/hr	the rate $5 per hour	P(4)	the probability of the outcome 4

GLOSSARY

absolute value The distance from a point on the number line to zero *(page 446)*

acute angle An angle whose measure is greater than 0° and less than 90° *(page 154)*
Example:

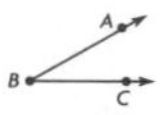

acute triangle A triangle whose three angles are acute *(page H21)*
Example:

algebraic expression An expression that is written using one or more variables *(page 278)*
Examples: $x - 4$; $2a + 5$

area The number of square units needed to cover a given surface *(pages 404, H25)*

Associative Property Addends can be grouped differently; the sum is always the same. Factors can be grouped differently; the product is always the same. *(pages 34, 38)*
Examples: $(8 + 7) + 4 = 8 + (7 + 4)$
$(5 \times 2) \times 6 = 5 \times (2 \times 6)$

average The number obtained by dividing the sum of a set of numbers by the number of addends *(page 246)*

axes The horizontal line (x-axis) and the vertical line (y-axis) on the coordinate plane *(page 487)*

bar graph A graph that uses separate bars (rectangles) of different heights (lengths) to show and compare data *(page 224)*

base A number used as a repeated factor *(page 24)*
Example: $8^3 = 8 \times 8 \times 8$
The base is 8. It is used as factor three times.

base A side of a polygon or a face of a solid figure by which the figure is measured or named *(page 186)*
Examples:

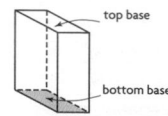
top base
bottom base

biased sample A sample that is not representative of the population *(page 212)*

bisect To divide into two equal parts *(page 160)*

box-and-whisker graph A graph that shows how far apart and how evenly data are distributed *(page 250)*
Example:

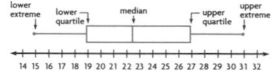
lower extreme · lower quartile · median · upper quartile · upper extreme
14 15 16 17 18 19 20 21 22 23 24 25 26 27 28 29 30 31 32

center of a circle The point inside a circle that is the same distance from each point on the circle *(page H20)*

circle The set of points in a plane that are the same distance from a given point called the center of the circle *(page H20)*

circle graph A graph using a circle that is divided into pie-shaped sections showing percents or parts of the whole *(pages 232, 234, 340)*

circumference The distance around a circle *(page 398)*

clustering A method used in estimation when all addends are about the same *(page 46)*

Commutative Property The property which states that numbers can be added in any order or can be multiplied in any order without changing the sum or the product. *(pages 34, 38)*
Examples: $9 + 4 = 4 + 9$
$6 \times 3 = 3 \times 6$

compatible numbers Pairs of numbers that are easy to compute mentally *(page 48)*

compensation An estimation strategy in which you change one addend to a multiple of ten and then adjust the other addend to keep the balance *(page 34)*
Example: $16 + 9$
$(16 - 1) + (9 + 1)$
$15 + 10 = 25$

composite number A whole number greater than 1 with more than two whole-number factors *(page 76)*

cone A solid figure with a circular base and one vertex *(page 187)*
Example:

congruent Having the same size and shape *(pages 156, 157)*

congruent figures Figures that have the same size and shape *(page H23)*
Example:

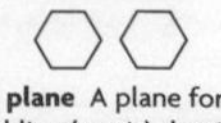

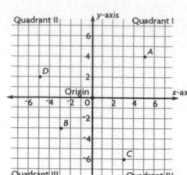

coordinate plane A plane formed by a horizontal line (x-axis) that intersects a vertical line (y-axis) at a point called the origin *(page 487)*
Example:

(coordinate plane diagram with Quadrants I–IV, x-axis, y-axis, Origin)

cross products Two equal products obtained by multiplying the second term of each ratio by the first term of the other ratio in a proportion *(page 327)*

cube A rectangular solid figure with six congruent faces *(page 193)*
Example:

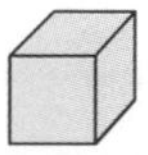

(cube diagram)

cumulative frequency A column in a table that keeps a running total *(page 217)*

customary measurement system A measurement system that measures length in inches, feet, yards, and miles; capacity in cups, pints, quarts, and gallons; weight in ounces, pounds, and tons; and temperature in degrees Fahrenheit *(pages 388, H24, H85)*

cylinder A solid figure with two parallel bases that are congruent circles *(page 187)*

data A set of information *(page 216)*

decimal system A numeration system based on grouping by tens *(page 16)*

degree A unit for measuring angles *(page H86)*

degree Celsius (°C) A metric unit for measuring temperature *(page H86)*

degree Fahrenheit (°F) A customary unit for measuring temperature *(page H86)*

diameter A line segment that passes through the center of a circle *(page H20)*

difference The answer in a subtraction problem *(page H7)*

discount The amount by which the original price is reduced *(page 342)*

Distributive Property Multiplying the sum by a number is the same as multiplying each addend by the number and then adding the products. *(page 38)*
Example: $4 \times (3 + 5) = 32$
$(4 \times 3) + (4 \times 5) = 32$

dividend The number to be divided in a division problem *(page H8)*

divisible A number is divisible by another number if the quotient is a whole number and the remainder is zero. *(page H4)*

divisor The number by which a dividend is divided in a division problem *(page H8)*

edge The line segment along which two faces of a solid figure meet *(page 190)*
Example:

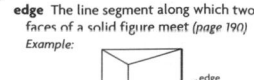
edge

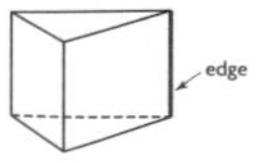

equation An algebraic or numerical sentence that shows two quantities are equal *(page 284)*

equilateral triangle A triangle with three congruent sides *(page H21)*
Example:

4 cm · 4 cm · 4 cm

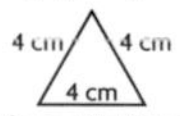

equivalent fractions Fractions that name the same amount or part *(page 84)*

equivalent ratios Ratios that make the same comparisons *(page 316)*

estimate An answer that is close to the exact answer and is found by rounding, by clustering, or by using compatible numbers *(pages 46, 104, 122, H9, H10)*

evaluate In a numerical expression, perform the operations and write the expression as one number. In an algebraic expression, replace the variable with a number and perform the operation in the expression. *(page 280)*

experimental probability The ratio of the number of times the event occurs to the total number of trials or times the activity is performed *(page 268)*

exponent A number that tells how many times a base is to be used as a factor *(page 24)*
Example: $2^3 = 2 \times 2 \times 2 = 8$
The exponent is 3, indicating that 2 is multiplied by itself 3 times

face One of the polygons of a solid figure *(pages 190, H23)*

factor A number that is multiplied by another number to find a product *(page H3)*

factor tree A diagram that shows the prime factors of a number *(pages 78, H4)*

fractal-like A mathematical figure that appears to have self-similarity *(page 511)*

fraction A number that names part of a group or part of a whole *(page H14)*

frequency table A table that organizes the total for each category or group *(page 217)*

Golden Ratio A ratio equivalent to the value of about 1.6 *(page 378)*

Golden Rectangle A rectangle with a length-to-width ratio of about 1.6 to 1 *(page 378)*
Example:

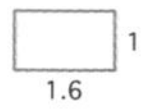

1
1.6

greater than (>) More than in size, quantity, or amount; the symbol > stands for *is greater than.* *(pages H7, H86)*
Example:
Read $7 > 5$ as seven is greater than five.

greatest common factor (GCF) The largest number that is a factor of two or more numbers *(page 81)*

hexagon A six-sided polygon *(page H21)*
Example:

histogram A bar graph that shows the number of times data occur within certain ranges or intervals *(page 228)*

Identity Property of One The property which states that the product of 1 and any factor is the factor *(page H6)*

Identity Property of Zero The property which states that the sum of any number and zero is that number *(page H6)*

indirect measurement The technique of using similar figures and proportions to find a measure *(page 362)*

inequality A mathematical sentence containing <, >, ≤, or ≥ to show that two expressions do not represent the same quantity *(page 484)*
Examples:
$2 \times 3 < 8$; $6 + 5 > 9$

integers The set of whole numbers, their opposites, and zero *(pages 28, 446)*
Example:

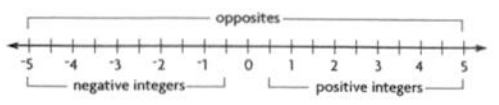

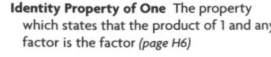
opposites
-5 -4 -3 -2 -1 0 1 2 3 4 5
negative integers · positive integers

intersecting lines Lines that cross at exactly one point *(page 152)*
Example:

E · C · D · H · F

inverse operations Operations that undo each other; addition and subtraction are inverse operations; multiplication and division are inverse operations. *(page 286)*

isosceles triangle A triangle with two congruent sides and two congruent angles *(page H21)*

lateral face In a prism or a pyramid, a face that is not a base *(page 186)*

least common denominator (LCD) The smallest number, other than zero, that is a multiple of two or more denominators *(page 100)*

least common multiple (LCM) The smallest number, other than zero, that is a multiple of two or more given numbers *(page 80)*

less than (<) Smaller in size, quantity, or amount; the symbol < stands for *is less than. (page 86)*
Example: Read 5 < 7 as five is less than seven.

line A straight path that goes on forever in opposite directions *(page 150)*

line graph A graph in which line segments are used to show changes over time *(page 225)*

line of symmetry A line that divides a figure into two congruent parts *(page 168)*
Example:

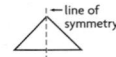
line of symmetry

line segment Part of a line; it has two endpoints. *(page 150)*

line symmetry A figure has line symmetry if a line can separate the figure into two congruent parts *(page 168)*

liter A metric unit for measuring capacity *(pages 390, H85)*

lower extreme The least number in a set of data *(page 250)*

lower quartile The median of the lower half of a set of data *(page 250)*

mean The average of a group of numbers *(page 246)*

measure of central tendency A measure used to describe data; the mean, median, and mode are measures of central tendency. *(page 246)*

median The middle number or the average of the two middle numbers in an ordered set of data *(page 246)*

meter A metric unit for measuring length *(pages 390, H24, H85)*

metric system A measurement system that measures length in millimeters, centimeters, meters, and kilometers; capacity in liters and milliliters; mass in grams and kilograms; and temperature in degrees Celsius *(pages 390, H24, H85)*

mixed number A number that is made up of a whole number and a fraction *(page 88)*
Examples:
$3\frac{3}{4}; 1\frac{7}{8}; 2\frac{1}{6}$

mode The number or numbers that occur most often in a collection of data; there can be more than one mode or none at all. *(page 246)*

multiple The product of a given number and a whole number *(pages 76, H3)*

negative integers Integers less than 0 *(pages 28, 446)*

net An arrangement of two-dimensional figures that folds to form a solid figure *(page 192)*

network A graph with vertices and edges *(page 394)*
Example:

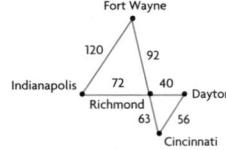
Fort Wayne
120   92
Indianapolis   72   40   Dayton
Richmond   63   56
Cincinnati

numerical expression A mathematical phrase that includes only numbers and operation symbols *(page 278)*

obtuse angle An angle whose measure is greater than 90° and less than 180° *(page 154)*

octagon An eight-sided polygon *(page 162)*
Example:

opposites Two numbers that are an equal distance from 0 on the number line *(page 28)*

ordered pair A pair of numbers used to locate a point on a coordinate plane *(page 486)*

order of operations The order in which operations are done; first, do the operations within parentheses; next, clear exponents; then, multiply and divide from left to right; and last, add and subtract from left to right. *(page 44)*
Example:
$3 \times (8 - 3) + 2^2$
$3 \times (5) + 2^2$
$3 \times (5) + 4$
$15 + 4 = 19$

origin The point on the coordinate plane where the *x*-axis and the *y*-axis intersect, (0,0) *(page 487)*

outcome A possible result in a probability experiment *(page 260)*

parallel lines Lines in a plane that do not intersect *(page 152)*
Example:

Read: Line *AB* || to line *CD*.

parallelogram A quadrilateral whose opposite sides are parallel and congruent *(page H22)*

pentagon A five-sided polygon *(page 162)*

percent The ratio of a number to 100; *percent* means "per hundred." *(page 320)*
Example:
$25\% = \frac{25}{100}$

perimeter The distance around a polygon *(page 396)*

perpendicular lines Lines that intersect to form 90°, or right, angles *(page 152)*
Example:

Read: Line *RS* is perpendicular to line *MN*.

pi (π) The ratio of the circumference of a circle to its diameter; $\pi \approx 3.14$ or $\frac{22}{7}$ *(page 399)*

place value The value of a digit as determined by its position in a number *(page 16)*

plane A flat surface that goes on forever in all directions *(page 150)*

point An exact location in space, usually represented by a dot *(page 150)*

polygon A closed plane figure formed by three or more line segments *(page 162)*

polyhedron A solid figure whose faces are polygons *(page 186)*

population A particular group of people, such as sixth graders *(page 210)*

positive integer A whole number greater than zero *(page 28)*

precision A property of measurement that is related to the unit of measure used; the smaller the unit of measure used, the more precise the measurement is. *(page 392)*

prediction An estimate made by looking at a trend over time and then extending that trend to describe a future event *(page 244)*

prime factorization A number written as the product of all its prime factors *(page 78)*
Example:

$24 = 2 \times 2 \times 2 \times 3 \text{ or } 2^3 \times 3$

prime number A whole number greater than 1 whose only factors are itself and 1 *(page 76)*

principal The amount of money borrowed or saved *(page 344)*

prism A solid figure whose bases are congruent, parallel polygons and whose other faces are parallelograms *(page 186)*
Examples:

rectangular prism   triangular prism

probability (P) The chance that an event will occur expressed as the ratio of the number of favorable outcomes to the number of possible outcomes *(page 260)*

product The answer in a multiplication problem *(page H8)*

Property of Zero for Multiplication The property which states that the product of 0 and any number is 0 *(page H6)*

proportion A number sentence or an equation that states that two ratios are equivalent *(page 326)*
Example: $\frac{2}{3} = \frac{4}{6}$

pyramid A solid figure whose base is a polygon and whose other faces are triangles with a common vertex *(page 188)*
Examples:

rectangular pyramid   triangular pyramid

quadrant One of the four regions of the coordinate plane *(page 487)*

quadrilateral A four-sided polygon *(page 162)*
Example:

quotient The answer in a division operation *(page H8)*

radius A line segment with one endpoint at the center of a circle and the other endpoint on the circle (page H20)

random sample A group chosen by chance so that each item has an equal chance of being selected *(page 210)*

range The difference between the greatest and least numbers in a set of numbers *(page 218)*

rate A ratio that compares two quantities having different units of measure *(page 318)*

ratio A comparison of two numbers *(page 316)*
Example: 3 to 5, or 3:5, or $\frac{3}{5}$

rational number Any number that can be expressed as a ratio in the form of $\frac{a}{b}$ where *a* and *b* are integers and $b \neq 0$ *(page 448)*

ray A part of a line that has one endpoint and goes on forever in only one direction *(page 150)*

reciprocal One of two numbers whose product is 1 *(page 138)*

rectangle A parallelogram with four right angles *(page H22)*

reflection The figure formed by flipping a geometric figure across a line of symmetry to obtain a mirror image *(page 174)*
Example:

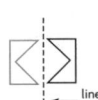
line of reflection

regular polygon A polygon in which all sides and all angles are congruent *(page 162)*

relation A set of ordered pairs *(page 490)*

repeating decimal A decimal in which one or more digits repeat endlessly *(page 451)*
Example: 0.333 . . . , or 0.3

rhombus A parallelogram whose four sides are congruent and whose opposite angles are congruent *(page H22)*
Example:

right angle An angle whose measure is 90° *(page 154)*

right triangle A triangle with exactly one right angle *(page H21)*

rotation A turning of a figure about a fixed point without reflection *(page 174)*
Example:

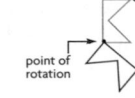
point of rotation

rotational symmetry A figure has rotational symmetry when it can be rotated less than 360° around a central point, or point of rotation. *(page 169)*

sample A group of people or objects chosen from a larger group to provide data to make predictions about the larger group *(page 210)*

scale The ratio between two sets of measurements *(page 369)*

scale drawing A reduced or enlarged drawing whose shape is the same as an actual object and whose size is determined by the scale *(page 368)*

scalene triangle A triangle with no congruent sides *(page H21)*
Example:

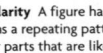

self-similarity A figure has self-similarity if it contains a repeating pattern of smaller and smaller parts that are like the whole, but different in size *(page 510)*
Example:

sequence An ordered set of numbers *(page 516)*

similar figures Geometric figures that have the same shape and angles of the same size *(page 352)*

simple interest The amount obtained by multiplying the principal by the rate by the time; I = prt *(page 344)*

simplest form A fraction is in simplest form when the numerator and denominator have no common factors other than 1. *(page 86)*

solid figure A geometric figure that exists in three or more planes *(pages 186, H22)*

solution The value of the variable that makes an equation true *(page 285)*

solve To find the value of a variable that makes an equation true *(page 285)*

square To square a number means to multiply it by itself *(page 26)*

square A rectangle with four congruent sides *(page H22)*
Example: $3^2 = 3 \times 3 = 9$

square root One of the two equal factors of a number *(page 27)*
Example: 5 is the square root of 25 because $5^2 = 25$.

standard form The form in which numerals are usually written, with digits 0 through 9, separated into periods by commas *(page 16)*
Example: 634,578,910

stem-and-leaf plot A method of organizing data in order to make comparisons; the ones digits appear horizontally as leaves, and the tens digits and greater appear vertically as stems. *(page 226)*
Example:

Card-Stacking Competition

Stem	Leaves
1	1 2 6 8 8 9
2	1 1 1 9
3	1 2 3 3
4	2 7
5	0 4

Key: 3 | 2 = 32

straight angle An angle whose measure is 180° *(page 154)*

sum The answer in an addition problem *(page 46)*

surface area The sum of the areas of all the faces of a solid figure *(page 434)*

T

tally table A table with categories for recording each piece of data as it is collected *(page 216)*

term Each of the numbers in a sequence *(page 516)*

terminating decimal A decimal that ends; a decimal for which the division operation results in a remainder of zero *(page 450)*
Examples: $\frac{1}{2} = 0.5$
$\frac{5}{8} = 0.625$

tessellation An arrangement of shapes that completely covers a plane, with no gaps and no overlaps *(page 178)*

transformation A movement that doesn't change the size or shape of a figure *(page 174)*

translation A movement of a geometric figure to a new position without turning or flipping it *(page 174)*
Example:

trapezoid A quadrilateral with only two parallel sides *(page H22)*

tree diagram A diagram that shows all the possible outcomes of an event *(page 258)*

triangle A three-sided polygon *(page 162)*

triangular number A geometric pattern of triangular arrays *(page 516)*
Example:

U

unit rate A rate in which the second term is 1 *(page 318)*

unlike fractions Fractions whose denominators are not the same *(page 96)*

upper extreme The greatest number in a set of data *(page 250)*

upper quartile The median of the upper half of a set of data *(page 250)*

V

variable A letter or symbol that stands for one or more numbers *(page 278)*

vertex The point where two or more rays meet; the point of intersection of two sides of a polygon; the point of intersection of three edges of a solid figure *(page 154)*

volume The number of cubic units needed to occupy a given space *(page 422)*

X

x-axis The horizontal axis on the coordinate plane *(page 487)*

Y

y-axis The vertical axis on the coordinate plane *(page 487)*

Glossary

Answers to Selected Exercises

Chapter 1

Page 17

1. 4 ones   **3.** 4 thousandths   **5.** two tens
7. eight hundred thousands   **9.** forty six
11. one thousand, five hundred, and one
tenth   **13.** one million, thirty-seven
thousand, eight hundred four   **15.** 0.15
17. 850,247.56   **19.** 9,456,302

Page 19

1. <   **3.** >   **5.** >   **7.** =   **9.** >   **11.** >
13. 4,555 < 4,556 < 4,566   **15.** 5.004 <
5.040 < 5.4   **17.** 162.32 > 126.33 > 126.3
19. *a.* 0.7 < 1.3 < 1.4 < 1.6 < 2.0 < 2.4 <
3.4 < 3.6 < 4.2 < 5.6 < 6.0 < 6.3
b. July; because it has the least rainfall
21. Compare place values: 6 tens > 1 ten,
so 5,361 > 5,316

Pages 22–23

1. $\frac{2}{10}$ **3.** $\frac{17}{100}$ **5.** $\frac{99}{100}$ **7.** $\frac{25}{100}$ **9.** $\frac{325}{1,000}$
11. $\frac{3,525}{10,000}$ **13.** 0.2   **15.** 0.6   **17.** 0.9375
19. 0.0087   **21.** 0.25   **23.** 0.4375   **25.** <
27. >   **29.** >   **31.** <   **33.** >   **35.** 0.125
37. 0.75 lb   **39.** The 3 keeps repeating; the
remainder is not 0.   **41.** 16   **43.** 216
45. 5 in.^{2} **47.** 1.25, 1.50

Page 25

1. 12^3 **3.** 4^4 **5.** 7^2 **7.** 2^5 **9.** 1,024   **11.** 1
13. 8   **15.** 14   **17.** 100,000,000   **19.** 1,024
21. 8^2 **23.** 10^3 **25.** 10^3 **27.** Take 2 steps
backward.   **29.** Jump down.   **31.** 200,024
33. $\frac{3}{5}$ of a pizza

Page 29

1. <   **3.** >   **5.** >   **7.** <   **9.** >   **11.** <
13. >   **15.** >   **17.** $^-6 < ^-4 < ^-3 < ^+6$
19. $^-10 < ^-3 < 0 < ^+2$ **21.** $^-8 < ^-6 < ^-2 < ^+5$
23. $^-11 < ^-9 < ^+2 < ^+20$ **25.** $^-9$ **27.** $^+11$
29. $^-14$ **31.** $^+12$ **33.** $^+24$ **35.** $^+98$

37. the Jackson's house
39. communications check

Page 31

1. 255   **3.** 821,202.4   **5.** <   **7.** >   **9.** <
11. $\frac{1}{10}$ **13.** 0.03   **15.** Jean   **17.** 10,000
19. 3,375   **21.** opposites   **23.** >   **25.** >

Chapter 2

Page 35

1. 36   **3.** 45   **5.** 60   **7.** 70   **9.** 80   **11.** 90
13. 41   **15.** 41   **17.** 44   **19.** 80   **21.** 63
23. 23   **25.** 21   **27.** 14   **29.** 23   **31.** 51
33. 70 CDs   **35.** 11 CDs

Page 39

1. 8   **3.** 55   **5.** 153   **7.** 192   **9.** 90   **11.** 63
13. 210   **15.** 72   **17.** 52   **19.** 224   **21.** 96
23. 440   **25.** 240 items   **27.** Possible
answer: $24 \times 14 = (24 \times 10) + (24 \times 4)$,
$26 \times 14 = (26 \times 10) + (26 \times 4)$, $28 \times 14 =$
$(28 \times 10) + (28 \times 4)$

Pages 42–43

1. 297   **3.** 10,920   **5.** 26,522   **7.** 1,245,454
9. 1,312,800   **11.** 14   **13.** 12   **15.** 26 r2
17. 503 r2   **19.** 301   **21.** 125   **23.** 89
25. 14   **27.** $20\frac{5}{12}$ **29.** $241\frac{11}{14}$ **31.** $25\frac{11}{15}$
33. $317\frac{7}{9}$ **35.** $50\frac{4}{9}$ **37.** \$84   **39.** $49\frac{4}{6}$
43. 13,000   **45.** $^-3 < ^-2 < 4 < 5$
47. $^-6 < ^-3 < < 6 < 7$ **49.** 21

Pages 48–49

1. 6,000   **3.** 8,000   **5.** 14,000   **7.** 10
9. 2,000   **11.** 360   **13.** 1,800   **15.** 63,000
17. 2,000   **19.** 150,000   **21.** 54   **23.** 50
25. 200   **27.** 60   **29.** 12   **31.** 920   **33.** 300
35. 15   **37.** about 130 plants   **39.** about
12,000 people   **41.** It is easier to use
compatible numbers, because they divide

evenly and are easy to compute mentally.
43. 1.45, 1.450   **45.** 0.5   **47.** 0.75
49. 0.625   **51.** $60

Page 51
1. compensation   **3.** 113   **5.** 30 red
marbles, 18 yellow marbles   **7.** 4,590   **9.** 120
11. 180   **13.** 1,472   **15.** 43,824   **17.** 65 r16
19. 68 r18   **21.** 16,000   **23.** 300   **25.** 70

Chapter 3
Page 55
1. 1   **3.** 12   **5.** 10   **7.** 1   **9.** 7.8   **11.** 6.32
13. 3.997   **15.** 93.7646   **17.** 2.177   **19.**
$456.00   **21.** 62.814   **23.** 1,221.874   **25.**
1,274.5429   **27.** about $10

Page 57
1. 3.2   **3.** 2.69   **5.** 1.746   **7.** 3.7586
9. 5.384   **11.** 13.243   **13.** 0.106
15. 91.7828   **17.** 26.953   **19.** 38.692
21. 99.43   **23.** 1.7 lb   **25.** $8   **29.** 7,897
31. 496,976   **33.** $(30 \times 6) + (5 \times 6) = 210$
35. $(80 \times 4) + (9 \times 4) = 356$ **37.** 42
newspapers

Pages 60–61
1. 4   **3.** 400   **5.** 100   **7.** 5,400   **9.** 22.456
11. 194.208   **13.** 12   **15.** $12.80   **17.** 0.88
19. 10.48   **21.** $16.92   **23.** 48.32   **25.** 5.67
27. 14.445   **29.** 5.5080   **31.** 82.5286
33. 163.5147   **35.** 4.51968   **37.** 10.8934
39. 11,379.8938   **41.** $6   **43.** 45 cups
45. $3.60   **49.** 4   **51.** 40   **53.** 85   **55.** He
should continue saving; the cost will be less.

Page 67
1. $96 \div 16$ **3.** $630 \div 18$ **5.** 4   **7.** 21
9. 239,000   **11.** 7.2   **13.** 8.9   **15.** 9.91
17. 0.12   **19.** 5.5   **21.** 220   **23.** 0.7   **25.** 0.9
27. 35   **29.** 21.2   **31.** 4.08   **33.** 6.4
35. $0.60   **37.** $5.75   **39.** 4 floors   **41.** He
didn't multiply the dividend by 100; 17.

Page 69
1. 3   **3.** 23   **5.** 11.1   **7.** 32.35   **9.** $85.49
11. 9.11   **13.** 203.794   **15.** 10.776   **17.** 0.75
19. 17.64   **21.** 0.0364   **23.** $84.00   **25.** 4.85

27. 16.9   **29.** $0.45

Chapter 4
Page 77
1. 25, 50, 75, 100   **3.** 21, 42, 63, 84   **5.** 11, 22,
33, 44   **7.** 32, 40   **9.** 7, 35   **11.** 1, 2, 4, 8, 16
13. 1, 37   **15.** 1, 2, 3, 6, 9, 18   **17.** 1, 2, 3, 6,
7, 14, 21, 42   **19.** 1, 2, 3, 4, 6, 7, 12, 14, 21, 28,
42, 84   **21.** 1, 2, 3, 6, 9, 18, 27, 54   **23.** 1, 7,
11, 77   **25.** P   **27.** C   **29.** C   **31.** C   **33.** C
35. C   **37.** It has only two factors, 1 and itself.
39. 37, 41, 43   **41.** 3 times; 8, 16, 24   **43.**
Yes. Composite numbers can have 2, 3, 5, 7,
and so on as factors.

Page 79
1. $2 \times 2 \times 2 \times 3$ **3.** 3×5 **5.** $2 \times 2 \times 2 \times 5$
7. 2×13 **9.** $3 \times 5 \times 5$ **11.** $3 \times 3; 3^2$
13. $2 \times 2 \times 2 \times 2 \times 2; 2^5$ **15.** $3 \times 3 \times 3; 3^3$
17. $2 \times 2 \times 13; 2^2 \times 13$ **19.** $2 \times 2 \times 19;$
$2^2 \times 19$ **21.** $2 \times 2 \times 2 \times 7; 2^3 \times 7$ **23.** $2 \times$
$2 \times 2 \times 3 \times 3; 2^3 \times 3^2$ **25.** $2 \times 2 \times 3 \times 7;$
$2^2 \times 3 \times 7$ **27.** 11   **29.** 5   **31.** 5   **33.** 13
35. 3×5^2 **37.** $a = 3$, or 2   **39.** No. The
order of factors does not affect the value of
their product.

Page 83
1. 12   **3.** 6   **5.** 20   **7.** 28   **9.** 24   **11.** 60
13. $2 \times 2 \times 2; 8$ **15.** $2 \times 2; 4$ **17.** $2 \times 3; 6$
19. 4   **21.** 4   **23.** 1   **25.** 8   **27.** 5   **29.** 12
31. 4 packages of toothpaste samples, 3
packages of pamphlets   **33.** It means that
two or more numbers have an identical
multiple or factor. For 18 and 24, the least
common multiple is 72 and the greatest
common factor is 6.   **35.** 4   **37.** 5   **39.** 4.01

Page 87
1. 1;1   **3.** 1, 3; 3   **5.** 1, 5; 5   **7.** $\frac{1}{6}$ **9.** $\frac{1}{8}$
11. $\frac{5}{9}$ **13.** $\frac{1}{5}$ **15.** $\frac{7}{11}$ **17.** $\frac{7}{10}$ **19.** b   **21.** 6
23. 8   **25.** 12   **27.** any multiple of 4 or 12
29. when 1 is the GCF of the numerator and
the denominator   **31.** 3   **33.** 0.035   **35.** 0.6
37. $138.00

Page 89

1. $1\frac{3}{4}$ 3. $5\frac{1}{2}$ 5. 9 7. $5\frac{1}{6}$ 9. $4\frac{1}{7}$

11. $4\frac{11}{16}$ 13. $3\frac{4}{9}$ 15. $1\frac{1}{5}$ 17. $5\frac{1}{2}$ 19. $\frac{11}{3}$

21. $\frac{16}{3}$ 23. $\frac{37}{9}$ 25. $\frac{37}{4}$ 27. $\frac{53}{11}$ 29. $\frac{33}{5}$

31. $\frac{29}{4}$ 33. $\frac{15}{2}$ 35. $\frac{25}{12}$ 37. $9\frac{1}{2}$ and $\frac{19}{2}$

39. $\frac{28}{3}$ and $9\frac{1}{3}$ 41. 11 43. 5 45. yes; $\frac{5}{3} = 1\frac{2}{3}$ 47. No. The numerator must be greater than the denominator.

Page 91

1. prime numbers 3. composite 5. composite 7. prime 9. prime 11. 3^2 13. 2×7 15. $2^2 \times 3$ 17. 2×5^2 19. greatest common factor 21. 6, 2 23. 30, 5 25. 27, 9 27. 60, 3 29. $\frac{1}{3}$ 31. $\frac{3}{8}$ 33. $\frac{3}{4}$ 35. $1\frac{4}{7}$ 37. $\frac{17}{7}$ 39. $\frac{13}{6}$ 41. $2\frac{1}{3}$

Chapter 5

Page 95

1. $\frac{3}{5}$ 3. $\frac{3}{4}$ 5. $\frac{5}{6}$ 7. $\frac{1}{5}$ 9. $\frac{2}{7}$ 11. $\frac{11}{12}$

13. $\frac{11}{20}$ 15. $1\frac{2}{9}$ 17. $\frac{2}{7}$ 19. $\frac{1}{4}$ 21. $\frac{3}{4}$ 23. $\frac{1}{3}$

25. $1\frac{3}{5}$ 27. $\frac{3}{10}$ 29. 1 31. $\frac{3}{12}$, or $\frac{1}{4}$ ft 33. $\frac{4}{5}$ of the lumber 35. $\frac{2}{4}$

Page 99

1. $\frac{1}{4} + \frac{1}{12} = \frac{4}{12}$; or $\frac{1}{3}$ 3. $\frac{2}{5} + \frac{1}{10} = \frac{5}{10}$, or $\frac{1}{2}$

5. $\frac{4}{6}$, or $\frac{2}{3}$ 7. $\frac{11}{12}$ 9. $\frac{9}{12}$, or $\frac{3}{4}$ 11. $\frac{3}{10}$ 13. $\frac{3}{6}$, or $\frac{1}{2}$ 15. $\frac{7}{12}$ 17. $\frac{11}{12}$ yd 19. $\frac{2}{10}$ mi, or $\frac{1}{5}$ mi farther

Page 101

1. $\frac{5}{6} + \frac{4}{6}$ 3. $\frac{3}{15} + \frac{7}{15}$ 5. $\frac{2}{8} + \frac{5}{8}$ 7. $\frac{15}{20} + \frac{16}{20}$

9. $\frac{3}{8} + \frac{4}{8}$ 11. $\frac{9}{12} + \frac{2}{12}$ 13. $\frac{5}{6}$ 15. $\frac{4}{5}$

17. $\frac{11}{8}$, or $1\frac{3}{8}$ 19. $\frac{7}{12}$ 21. $\frac{13}{12}$, or $1\frac{1}{12}$

23. $\frac{9}{10}$ 25. $\frac{9}{10}$ 27. $\frac{14}{15}$ 29. $\frac{5}{6}$ can

31. $\frac{17}{12}$ hr, or $1\frac{5}{12}$ hr

Page 103

1. $\frac{1}{12}$ 3. $\frac{7}{12}$ 5. $\frac{5}{18}$ 7. $\frac{1}{8}$ 9. $\frac{1}{12}$ 11. $\frac{2}{9}$

13. $\frac{17}{24}$ 15. $\frac{7}{20}$ 17. $\frac{1}{2}$ 19. $\frac{1}{20}$ 21. $\frac{2}{12}$ mi, or $\frac{1}{6}$ mi farther 23. $\frac{7}{8}$ tank 25. much less than 27. about $\frac{1}{2}$ 29. $\frac{1}{2}$ 31. $\frac{3}{4}$ 33. 4 of each kind

Pages 106–107

1. about 1 3. about $\frac{1}{2}$ 5. about 0 7. 1 9. $\frac{1}{2}$ 11. 1 13. $1\frac{1}{2}$ 15. 2 17. 1 19. $1\frac{1}{2}$

21. $\frac{1}{2}$ 23. 0 25. $\frac{1}{2}$ 27. $\frac{1}{2}$ 29. 2 31. $\frac{1}{2}$

33. About $\frac{1}{2}$ more of the flowers bloomed in May. 35. No; he has about $2\frac{1}{2}$ or 3 yd of fabric 37. Yes; $\frac{1}{2} + 1 + 0 \approx 1\frac{1}{2}$ 39. $\frac{7}{4}$

41. $\frac{21}{8}$ 43. $\frac{25}{6}$ 45. 3 47. 5 49. 4

51. $1,725

Page 109

1. $\frac{5}{4}$, or $1\frac{1}{4}$ 3. $\frac{5}{7}$ 5. $\frac{1}{2}$ c 7. $\frac{5}{6}$ 9. $\frac{9}{10}$

11. $\frac{3}{2}$, or $1\frac{1}{2}$ 13. $\frac{19}{12}$ yd, or $1\frac{7}{12}$ yd 15. $\frac{5}{8}$

17. $\frac{13}{18}$ 19. $\frac{1}{3}$ 21. $\frac{1}{2}$ 23. 0 25. 1 or $\frac{1}{2}$

27. about $1\frac{1}{2}$ dollars

Chapter 6

Page 113

1. $1\frac{1}{5} + 1\frac{3}{5} = 2\frac{4}{5}$ 3. $2\frac{6}{8}$, or $2\frac{3}{4}$ 5. $3\frac{4}{8}$, or $3\frac{1}{2}$

7. $3\frac{1}{2}$ hr 9. $6\frac{5}{8}$

Page 117

1. $1\frac{3}{2}$ 3. $2\frac{5}{3}$ 5. $4\frac{13}{8}$ 7. $1\frac{2}{5}$ 9. $1\frac{1}{2}$

11. $\frac{1}{6}$ 13. $2\frac{2}{3}$ 15. $1\frac{1}{2}$ qt 17. $1\frac{11}{12}$ mi

Pages 120–121

1. $1\frac{2}{8} + 1\frac{3}{8}$ 3. $2\frac{5}{10} + 3\frac{4}{10}$ 5. $2\frac{5}{20} + 3\frac{8}{20}$

7. $1\frac{3}{4}; 2\frac{3}{4}$ 9. $1\frac{5}{6}; 5\frac{5}{6}$ 11. $1\frac{3}{10}; 6\frac{3}{10}$

13. yes 15. no 17. $2\frac{3}{8}$ 19. $8\frac{3}{10}$

21. $11\frac{3}{20}$ **23.** $13\frac{4}{9}$ **25.** $10\frac{1}{18}$ **27.** $3\frac{3}{10}$
29. $2\frac{1}{12}$ **31.** $2\frac{13}{20}$ **33.** $1\frac{4}{5}$ **35.** $1\frac{3}{8}$
37. $28\frac{1}{8}$ dollars **39.** $41\frac{7}{9}$ mi **41.** You must write unlike fractions in mixed numbers as like fractions before you can add or subtract.
43. 1 **45.** 0 **47.** $\frac{1}{2}$ **49.** $\frac{2}{3}$ **51.** $\frac{5}{6}$
53. 90,000; 5,400,000; 129,600,000

Page 123
1. 6 **3.** $3\frac{1}{2}$ **5.** $2\frac{1}{2}$ **7.** 13 **9.** 17 **11.** 4
13. $5\frac{1}{2}$ **15.** 2 **17.** about 20 gal **19.** Round the mixed number to the nearest whole number. Then add or subtract. **21.** 0.15
23. 0.025 **25.** $1\frac{4}{15}$ **27.** $\frac{13}{24}$ **29.** $12.99

Page 125
1. $5\frac{7}{8}$ **3.** $3\frac{7}{8}$ **5.** $4\frac{5}{6}$ **7.** $4\frac{7}{10}$ **9.** $\frac{9}{10}$ mi
11. $2\frac{5}{12}$ **13.** $5\frac{1}{6}$ **15.** $8\frac{14}{15}$ **17.** $7\frac{1}{8}$ **19.** 7
21. $6\frac{1}{2}$ **23.** 4 **25.** about $1\frac{1}{2}$ ft taller

Chapter 7
Page 131
1. $\frac{3}{8}$ **3.** $\frac{3}{10}$ **5.** $\frac{7}{2}$, or $3\frac{1}{2}$ **7.** $\frac{1}{4}$ **9.** $\frac{2}{15}$
11. $\frac{7}{10}$ **13.** $\frac{2}{7}$ **15.** $\frac{1}{6}$ **17.** $\frac{3}{4}$ **19.** 2 **21.** $\frac{25}{6}$, or $4\frac{1}{6}$ **23.** $\frac{5}{9}$ **25.** $\frac{28}{5}$, or $5\frac{3}{5}$ **27.** 3
29. 2, 3, or 3, 2 **31.** $93\frac{3}{4}$, or $93.75 **33.** Use a model; multiply the numerators and the denominators, and write the answer in simplest form. **35.** 4 **37.** 5 **39.** $\frac{11}{12}$
41. $3\frac{9}{10}$ **43.** 7:40 A.M.

Page 133
1. 2 **3.** 7, 3 **5.** 4 **7.** 2, 9 **9.** $\frac{3}{1} \times \frac{1}{7}$
11. $1 \times \frac{1}{2}$ **13.** $\frac{1}{2} \times \frac{3}{5}$ **15.** $\frac{1}{5} \times \frac{1}{2}$ **17.** $\frac{1}{9}$
19. $\frac{1}{2}$ **21.** $\frac{2}{3}$ **23.** $\frac{3}{7}$ **25.** $\frac{1}{6}$ **27.** $\frac{7}{12}$ **29.** 7
31. 20 **33.** 13 teens **35.** 245 apartments

37. You do not have to simplify the product; multiplying the factors is easier.

Page 135
1. $\frac{19}{4} \times \frac{1}{8}$ **3.** $\frac{4}{3} \times \frac{15}{4}$ **5.** $\frac{14}{5}$, or $2\frac{4}{5}$ **7.** $\frac{143}{24}$, or $5\frac{23}{24}$ **9.** $\frac{49}{6}$, or $8\frac{1}{6}$ **11.** $\frac{11}{12}$ **13.** 15
15. 85 **17.** $\frac{50}{7}$, or $7\frac{1}{7}$ **19.** $\frac{13}{2}$, or $6\frac{1}{2}$
21. 4 mi **23.** $40\frac{1}{4}$ mi **25.** $\frac{14}{12}$

Page 139
1. $\frac{3}{4}$ **3.** $2\frac{5}{8}$ **5.** 20 **7.** $\frac{20}{27}$ **9.** 8 **11.** $\frac{5}{32}$
13. $\frac{1}{40}$ **15.** $\frac{7}{8}$ **17.** 48 hamburgers
19. 80 times **21.** $\frac{7}{3}$ **23.** $\frac{9}{2}$ **25.** $\frac{14}{5}$ **27.** $\frac{4}{9}$
29. $4\frac{7}{18}$ **31.** two 6-in. sandwiches

Page 143
1. $\frac{1}{10}$ **3.** $6\frac{2}{3}$ **5.** $\frac{1}{9}$ **7.** $\frac{15}{28}$ **9.** 4 **11.** $\frac{5}{4}$, or $1\frac{1}{4}$ **13.** $\frac{5}{1}$, or 5 **15.** $\frac{9}{8}$, or $1\frac{1}{8}$ **17.** reciprocal
19. $\frac{9}{4}$, or $2\frac{1}{4}$ **21.** 24 **23.** $\frac{1}{5}$ **25.** $3\frac{1}{2}$

Chapter 8
Page 151
1. $\overline{PQ}, \overline{QR}, \overline{PR}$ **3.** $\overleftrightarrow{PQ}, \overleftrightarrow{QR}, \overleftrightarrow{QP}, \overleftrightarrow{RP}$ **5.** line
7. point **9.** point **11.** line segment

Page 153
1. $\overleftrightarrow{BD}, \overleftrightarrow{EG}, \overleftrightarrow{FH}$ **3.** $\overleftrightarrow{AB}, \overleftrightarrow{CD}, \overleftrightarrow{BF}, \overrightarrow{DH}$ **7.** less than **9.** greater than **11.** $\frac{25}{3}$, or $8\frac{1}{3}$ **13.** 15
15. 3^2 coins

Page 155
1. 77°, acute **3.** 116°, obtuse **5.** 162°, obtuse **7.** acute **9.** straight **11.** obtuse
13. right **15.** 180° **17.** acute angle: $< 90°$; obtuse angle: $> 90°$; obtuse angle: $> 90°$ and $< 180°$

Pages 158–159
1. no **3.** yes **5.** 114°, 145°; no **13.** 3; triangle **15.** 6; hexagon **17.** $4\frac{2}{3}$
19. $1\frac{1}{9}$ **21.** 24 gal

Page 163
1. parallelogram; yes   3. trapezoid; no
5. Figure 4   7. scalene   9. 12 triangles; 11 quadrilaterals   11. hexagon

Page 165
1. line segment   3. line segment, $\overline{CD}$
5. $\overleftrightarrow{JK}$, or $\overleftrightarrow{HI}$ 7. vertex   9. 130°, obtuse
13. regular polygon   15. octagon

Chapter 9
Page 171
5. yes   7. no   9. no   11. yes   13. yes
15. no   17. $\frac{1}{2}$, 180°   19. $\frac{1}{2}$, 180°
23. 8 pieces   29. parallel   31. perpendicular and intersecting 33. 10 qt

Pages 176–177
1. translation   3. rotation   9. false; rotation
11. true   13. rotation, translation, reflection or rotation   15. reflection, rotation, rotation, reflection   17. right   19. pop   21. 90°   23. 45°   25. obtuse   27. right   29. 5 tries

Page 179
1. yes   3. no   5. each angle; 90°; sum: 360°
7. each angle: 90°; sum: 360°   9. 360°
11. two squares, a triangle, and a hexagon
13. yes   17. They cover a plane with no gaps or overlaps, and the angles at each vertex have a sum of 360°

Page 183
1. line symmetry   5. $\frac{1}{4}$, or 90°   7. reflection
9. reflection   11. tessellataion   13. no
15. yes   17. Yes

Chapter 10
Pages 188–189
1. base   3. lateral face   5. rectangular prism   7. triangular prism   9. cone
11. true   13. true   15. true   17. False; a cylinder is not a polyhedron   19. six sides
23. 6, 6; hexagon   25. 4, 4; rectangle

27. yes   29. no   31. 11
Page 191
1. N, M, R, P, S   3. *MRS, MSP, MPN, MNR, RSPN*   5. polyhedron   7. 4, 6, 5, 8, 6, 10
9. 4 colors   11. pentagonal prism, hexagonal pyramid

Page 193
3. yes   5. yes   7. 5   17. 2 and 4

Page 195
1. cone   3. hexagonal prism   5. triangular pyramid   7. cone and cylinder   13. Possible answer: cereal box

Page 201
1. triangular prism; polyhedron
3. rectangular pyramid; polyhedron   7. yes
9. 2 faces: 2 in. × 1 in.; 2 faces: 1 in. × 3 in.; 2 faces: 2 in. × 3 in.   13. cylinder or cone
15. 8 sides

Chapter 11
Page 209
1. a and b   5. 3   7. 32   9. cone
11. cylinder   13. 120 national stamps; 40 international stamps

Page 211
1. 57 people   3. Possible response: by randomly choosing students at lunch   5. Yes; all voters had equal chances of selection.
7. 100 students   9. about 56 students

Page 213
1. biased; excludes female teachers   3. biased, sample not large enough   5. biased, sample not large enough   7. biased, excludes men   9. biased   11. not biased   13. No; boys and girls are randomly selected

Pages 218–219
1. 20; 46; 80   7. 9 heights   13. line graph
15. 6 faces   17. 4 hr

Page 221

1. a   3. 35 students   5. No; his sample should include other classes.   7. not biased

Chapter 12

Page 227

1. bar graph   3. stem−and−leaf plot
9. Answers will vary.   11. 10   13. 62
15. 44 yd

Page 229

1. histogram   3. bar graph   5. The number of members in the different age groups would increase.   9. Answers will vary.

Page 231

3. multiple−line graph; changes over time
5. 3 lines; 1 line for each year's data
7. multiple−line graph

Page 235

1. 100 people   3. 5 sections   11. change over time   13. yes   15. No. She jogs 5,200 ft, 80 ft short of 1 mi.

Page 237

1. bar   3. histogram

Chapter 13

Page 241

1. October; July; there were 4.5 times as many sales in October as in July.   3. Sales increased.   5. 400 more gallons, or 5 times as many gallons   7. No longer make Stellar Kiwi; make more CocoDelight.   9. Graphs and questions will vary.

Page 243

1. Lucy   3. No. Lucy sold 5 more boxes of cookies, not 15 more boxes.   5. about 5 times   7. about 40,000; no

Page 245

1. 10; 15   3. 5 CDs a yr   5. decreasing by $20, then $30, and then $40   7. decreasing by about $4,000 each year   9. $2,400 a year; $200 a month

Pages 248– 248

1. 8, 7, 7   3. 27, 28, 32   5. 46.4, 45.8, 45.7
7. 4.2, 3.4, 7.8   9. true   11. 85; 100   13. 85; 100   15. No. You have to find the sum of the data and divide by 10.   17. 7; 7.5; 8   19. 9
21. 15.2   23. 15 teams

Page 253

1. 22   3. 15; 32   7. The numbers in the lowest and highest $\frac{1}{4}$ of the data are close together; the data in the middle are more spread out.   9. $\frac{3}{4}$ 11. $\frac{1}{2}$ 13. $3.03

Page 255

1. The greater the height, the greater the number of students who like comedy.   3. 5 times as long   5. about $2.50   7. about $10
9. 86.8; 87; none   11. 190   13. 186; 194

Chapter 14

Pages 262–263

1. 1   3. $\frac{1}{4}$ 5. $\frac{1}{6}$ 7. $\frac{1}{2}$ 9. $\frac{2}{9}$ 11. $\frac{2}{3}$
13. $\frac{1}{4}$ 15. $\frac{1}{5}$ 17. $\frac{2}{5}$ 19. $\frac{1}{6}$ 21. $\frac{1}{2}$ 23. 12
25. 1   27. Favorable; a selected event; possible: all possible events; examples will vary.   29. $\frac{1}{4}$ 31. $\frac{1}{8}$ 33. 4   35. 16 girls

Page 265

1. $\frac{1}{4}$ 3. 0   5. $\frac{3}{8}$ 7. $\frac{1}{10}$ 9. $\frac{1}{2}$ 11. $\frac{1}{4}$
13. $\frac{5}{8}$ 15. $\frac{3}{4}$ 17. 0   19. $\frac{8}{35}$ 21. $\frac{9}{35}$
23. 1   25. $\frac{1}{2}$

Page 269

1. $\frac{2}{5}$ 3. $\frac{6}{25}$ 5. $\frac{1}{4}$ 7. 24 times   9. number of times 4 lands ÷ total rolls   11. 23   13. 9
15. a larger amount   17. 0.5 km

Page 271

1. 6 choices   3. probability   5. $\frac{2}{9}$ 7. $\frac{1}{3}$
9. $\frac{1}{5}$ 11. 0   13. $\frac{1}{2}$ 15. experimental probability   17. $\frac{17}{50}$ 19. $\frac{2}{25}$

Chapter 15

Page 279

1. $16 \times \frac{1}{4}$ **3.** $\frac{1}{2} \times \frac{2}{5}$ **5.** $87 - k$
7. $32.7 \times p$ **9.** the sum of 42 and 2.5
11. twelve less than c **13.** $s - 3$ **15.** 26
17. 2.6   **19.** $\frac{1}{6}$ **21.** $\frac{1}{3}$ **23.** $\frac{1}{2}$ **25.** 23 points

Page 281

1. 17 **3.** 40 **5.** 6.1 **7.** $\frac{3}{4}$ **9.** 8 **11.** 5 **13.** 52
15. 17.5   **17.** 14   **19.** 19   **21.** $30.00 -
$12.00; $18.00   **23.** $45 \div$ f; 4.5 pieces

Page 283

1. $6 - 6, 0; 7 - 6, 1; 8 - 6, 2$ **3.** $100 \div 4, 25;$
$102 \div 4, 25.5; 104 \div 4, 26$ **5.** $25 + 3^2, 34;$
$25 + 4^2, 41; 25 + 5^2, 50$ **7.** 12, 13, 14, 15
9. $10, 6\frac{2}{3}, 5, 4$ **11.** 5   **13.** 8   **15.** $5.25 \times h;$
23 hr.

Pages 288–289

1. yes   **3.** no   **5.** no   **7.** no   **9.** yes
11. $a = 2$ **13.** $r = 35$ **15.** $b = 55$
17. $c = 54$ **19.** $m = 902$ **21.** $y = 1,574$
23. $x = 14.6$ **25.** $s = 103\frac{2}{5}$ **27.** $y - 9 = 17$
29. $m =$ Mary's height; $m - 60 = 6$
31. $x =$ amount in savings; $x + $125 = $324;$
$x = 199 **33.** 42   **35.** 7   **37.** $\frac{3}{7}$ **39.** $\frac{2}{7}$
41. $\frac{6}{7}$ **43.** 32 hr and 25 min

Page 291

1. numerical expression   **3.** algebraic   **5.** 17
7. 10   **9.** 137   **11.** $12 \div 3, 4$ **13.** $18 \div 3, 6$
15. 0.75, 1.25, 1.75   **17.** 14, 15, 16   **19.** yes
21. $x = 3$ **23.** $x = 9$

Chapter 16

Page 299

1. yes   **3.** no   **5.** yes   **7.** no   **9.** $x = 2$
11. $y = 35$ **13.** $m = 12$ **15.** $s = 75$
17. $d = 10$ **19.** $b = 80$ **21.** $11x = 66$
23. $n =$ number of cookies sold; $2n = 112$
25. $g =$ games won; $5g = 120$; $g = 24$; 24 games
27. Uses the inverse operation.   **29.** 60; 150;
300; 1,500   **31.** 170; 425; 850; 4,250

33. $y = 22$ **35.** $d = 20$ **37.** 26; $26^2 = 676$;
$27^2 = 729$

Page 301

1. 32; 80; 160; 800   **3.** 108; 270; 540; 2,700
5. 37; 92; 185; 925   **7.** 173; 432; 865; 4,325
9. $40.65   **11.** $37.20   **13.** 40 quarters and
40 nickels   **15.** Jenny; she has 157 nickels.
Reggie has 107 quarters.   **17.** 15   **19.** 72
21. 76   **23.** 9   **25.** 3   **27.** 108; 324; 972;
2,916

Page 303

1. 68°F   **3.** 140°F   **5.** 36°F   **7.** 25°C
9. 70°C   **11.** 31°C   **13.** 32°F   **15.** 752°F

Page 305

1. 75 mi   **3.** 8 km per min   **5.** 5 km per min
7. 12 hr   **9.** 20 min   **11.** 53.34 km   **13.** 80
km per hr   **15.** about 4 hr   **17.** 59 sec

Page 309

1. $x = 4$ **3.** $a = 7$ **5.** $y = 77$ **7.** 53 pounds
9. 100; 500   **11.** $4.96   **13.** $9.30   **15.** 32°C
17. 68°C   **19.** 199°F   **21.** 382 sec   **23.** 64 mi
per hr

Chapter 17

Page 317

1. 5 to 9, 5:9, $\frac{5}{9}$ **3.** 10 to 1, 10:1, $\frac{10}{1}$ **5.** 9 to
20, 9:20, $\frac{9}{20}$ **7.** 21 to 10, 21:10, $\frac{21}{10}$ **9.** 4:2;
possible answers: 2:1, 8:4, 12:6   **11.** 6
13. 25   **19.** 12:4, or 3:1   **21.** 18 dimes

Page 319

1. $\frac{12 \text{ eggs}}{$1.10}$ **3.** $\frac{90 \text{ words}}{2 \text{ mi}}$ **5.** $\frac{60 \text{ mi}}{3 \text{ gal}}$ **7.** $\frac{$1.89}{3 \text{ pens}}$
9. $\frac{$15}{5 \text{ tapes}}$ **11.** $\frac{$0.15}{1}$ **13.** $\frac{50 \text{ mi}}{1 \text{ hr}}$ **15.** $\frac{20 \text{ mi}}{1 \text{ gal}}$
17. $\frac{$52}{1 \text{ tire}}$ **19.** $0.35   **21.** $0.37   **23.** P.O.;
$0.32; drugstore: $0.35   **25.** the rate when
the second term is 1

Pages 322–323

1. 35%   **3.** 84%   **5.** 40%   **7.** 10%   **9.** 38%
11. 80%   **13.** 80%   **15.** 24%   **17.** $66\frac{2}{3}\%$

19. $0.2; \frac{1}{5}$ **21.** $0.9; \frac{9}{10}$ **23.** $0.48; \frac{12}{25}$

25. 100 **27.** 15 **29.** 29 **31.** $<; \frac{2}{50} = 4\%$

33. $<; 0.0018 = 0.18\%$ **35.** 88% **37.** 36% want supreme **39.** 20%; 20% of 35 = 7 and

$\frac{1}{5}$ of 35 = 7 **41.** $\frac{1}{4}$ **43.** $\frac{1}{2}$ **45.** 5 **47.** 9

49. 45 cookies

Page 325
1. 40% **3.** 25% **5.** 60% **7.** 12.5%
9. 20%; 40% **11.** 50%

Page 331
1. ratio **3.** 10:1, 10 to 1, $\frac{10}{1}$

5. $\frac{2}{3}; \frac{12}{18}, \frac{18}{27}$ **7.** unit rate **9.** $\frac{210 \text{ km}}{3 \text{ hr}}, \frac{70 \text{ km}}{1 \text{ hr}}$

11. 25% **13.** $0.20, 0.2; \frac{1}{5}$ **15.** $0.35; \frac{7}{20}$

17. 50% **19.** proportion

Chapter 18
Page 339
1. $\frac{3}{5}$ **3.** $4\frac{1}{2}$ **5.** 108 **7.** 5.6 **9.** 59.64 **11.** 3
13. 32.2 **15.** 0.24 **17.** 98.01 **19.** 32
21. 3.85 **23.** 200 **25.** 225 **27.** $1.30
29. $6.25 **31.** $24.80 **33.** Change the percent to a ratio or decimal and multiply by

the number. **35.** 36 **37.** 45 **39.** $\frac{1}{10}, 0.1$

41. $\frac{9}{20}, 0.45$ **43.** $\frac{8}{25}, 0.32$ **45.** Maria

Page 341
1. 30% = 108°, 25% = 90°, 12.5% = 45°, 32.5% = 117° **3.** 175 people.

Page 343
1. $42.00 **3.** $3.60 **5.** $8.75 **7.** $37.50
9. $15 **11.** $112.50; $337.50 **13.** $3.60; $4

Page 345
1. 1.00, 2.00 **3.** 11.50, 23.00 **5.** 82.80, 165.60 **7.** $109 **9.** 1 **11.** 3.6 **13.** 25%
15. 33.3% **17.** 2 hr

Page 349
1. 980 students **3.** 120 **5.** 63 **7.** $2.49
9. $800 **11.** $500 **13.** $264 **15.** $5.40
17. $7.94 **19.** Simple interest **21.** $560

Chapter 19
Page 355
1. both **3.** neither **5.** neither **7.** both
9. 15 **11.** 15 **13.** yes; because they are the same shape

Page 359
3. *No.* Angles are not congruent, and ratios are not equivalent. **5.** yes **7.** Check that the corresponding angles are congruent and that the ratios of the lengths of the corresponding sides are equivalent.
9. $x = 24$ **11.** $9 **13.** $44.25 **15.** $550

Page 361
1. $n = 12$ ft **3.** $n = 2.5$ cm **5.** $n = 3$ mi
7. No. Their ratios are not equivalent.
9. 2 in. wide

Page 363
1. $\frac{2}{h} = \frac{4}{10}; h = 5$ yd **3.** 8 m **7.** 6 **9.** 6
11. $0.50 **13.** $2.19 **15.** 63 cm

Page 365
1. similar **3.** corresponding **5.** $y = 18$ cm
7. $y = 6$ m **9.** $\frac{6}{3} = \frac{a}{5}; a = 10$ ft

Chapter 20
Page 371
1. $\frac{4}{3}$ **3.** $\frac{2.4}{1.2}$, or $\frac{2}{1}$ **5.** 15 **7.** 36 **9.** 48

11. $\frac{1}{2} = \frac{3}{n}; n = 6$ ft **13.** $x = 10$ **15.** $x = 20$
17. yes **19.** 10,000 pesos

Page 375
1. $\frac{1}{50} = \frac{1.5}{n}$; 75 mi **3.** $\frac{1}{50} = \frac{4.5}{n}$; 225 mi

5. $\frac{1}{50} = \frac{12}{n}$; 600 mi **7.** $\frac{3}{4}$ in; 69 mi **9.** 2 in.; 184 mi; no; there is no road. **11.** about 138 mi; about 2 hr and 50 min **13.** Use the map distance and scale to write a proportion. Solve for the unknown actual distance.

Page 379
1. yes; $\frac{4}{2.5} = 1.6$ **3.** no; $\frac{12}{4.5} \approx 2.67$ **5.** no;

$\frac{12}{10} = 1.2$ **7.** yes; $\frac{101}{60} \approx 1.6$ **9.** If its $\frac{l}{w} \approx 1.6$, then it is a Golden Rectangle.   **11.** 3
13. $x = 12\frac{1}{2}$ ft   **15.** $0.25 more per hr; 4% is $0.20 more per hr.

Page 381
1. scale   **3.** $28\frac{1}{8}$ ft   **5.** 22 cm   **7.** 1,800 mi
9. $1,012\frac{1}{2}$ mi   **11.** $8\frac{7}{10}$ in.   **13.** Golden
Rectangle   **15.** $\frac{6.4}{4}$; 1.6; yes   **17.** $\frac{3.68}{2.3}$; 1.6; yes

Chapter 21
Page 389
1. 240   **3.** 90   **5.** 12   **7.** 60   **9.** 4   **11.** 20
13. 15 yd   **15.** 840ft   **17.** centimeter
19. liter   **21.** 6   **23.** 42.5   **25.** 60°

Page 391
1. divide by 1,000   **3.** divide by 1,000
5. divide by 1,000   **7.** multiply by 1,000
9. 10; 1   **11.** 0.1; 0.01; 0.001   **13.** 4,000
15. 0.01   **17.** 0.001   **19.** 9,000   **21.** 0.00034
23. 32   **25.** 2.512 km   **27.** 2.5 L

Page 393
1. 1 in.; $1\frac{1}{2}$ in.   **3.** 85 in.   **5.** 71 mm
7. 111 in.   **9.** 395 in.   **11.** 400 m
13. 10,520 ft   **15.** 65 mm   **17.** 62 mm
19. 211 mm   **21.** 4,000 m   **23.** 2,650 yd
25. 30 ft

Page 395
1. BCDE, 52 km   **3.** FCBE, 101 km
5. EFDC, 96 km   **7.** CBFE, 73 km   **9.** HBS,
29 mi   **11.** by listing the different routes,
finding the distance for each route, and
identifying the shortest route   **13.** 208.4
15. 6   **17.** 6   **19.** $36

Page 397
1. $x = 12$ m; 43 m   **3.** $x = 27$ ft, y = 58 ft;
610 ft   **5.** $y = 34$ yd   **7.** 20km   **9.** Possible
answers: $P = s + s + s + s$, or $P = 4s$; 460 m

Page 401
1. divide by 12   **3.** multiply by 2   **5.** 4

7. 2.5   **9.** 32   **11.** 0.018   **13.** 143 mm
15. 14 in.   **17.** network   **19.** 21.6   **21.** 23.8
23. x = 3.4; 17 km

Chapter 22
Page 405
1. about 27 in.^{2} **3.** about 7 in.^{2} **5.** about
30 in.^{2} **7.** about 21 in.^{2} **9.** about 18 ft^2
11. Count full squares, almost-full squares,
and half-full squares   **13.** 12   **15.** 5
17. 30   **19.** 4 girls, 3 boys

Pages 410–411
1. 10 ft^2 **3.** 102.3 m^2 **5.** 112 in.^{2} **7.** 92 ft^2
9. 0.2 m^2 **11.** 14.175 m^2 **13.** 100 cm^2
15. 45 in.^{2} **17.** 9 cm^2; 4.5 cm^2 **19.** 320 ft^2
21. 60 ft^2 **23.** 32   **25.** 6   **27.** 0.01
29. $56.25

Page 413
1. 20 in., 16 in.^{2}; 40 in., 64 in.^{2} **3.** 36 yd, 48
yd^2; 72 yd, 192 yd^2 **5.** 80 ft, 400 ft^2; 40 ft, 100
ft^2 **7.** 56 cm, 84 cm^2; 28 cm, 21 cm^2 **9.** 44 ft
11. No. The area only doubled, because the
width didn't change.

Page 417
1. 201 yd^2 **3.** 260 cm^2 **5.** 113 yd^2
7. 7.1 m^2 **9.** 88.2 cm^2 **11.** 1,962.5 yd^2
13. 38.5 m^2 **15.** 60.8 cm^2 **17.** about 7,850
mi^2 **19.** The one with the 9 in. radius; its
area is about 254 in.^{2} **21.** about 3.44 cm^2
23. Area is the number of square units
needed to cover the circle; circumference is
the distance around.

Page 419
1. about 9 ft^2 **3.** about 10 mi^2 **5.** 180 ft^2
7. 97.5 yd^2 **9.** 9 m^2 **11.** $P = 15$ m;
$A = 12.5$ m^2 **13.** $P = 4.5$ in.; $A = 1.5$ in.^{2}
15. about 50 m^2

Chapter 23
Page 425
1. 24 in.^{3} **3.** 75 cm^3 **5.** 324 ft^3 **7.** 900 cm^3
9. 162 cm^3 **11.** $12   **13.** It has three
dimensions.   **15.** 1,280   **17.** 720   **19.** 90 ft^2
21. 14.4 cm^2 **23.** $515

Page 429

1. 24 cm^3; 192 cm^3 3. 60 m^3; 480 m^3 5. 60 m^3, 7.5 m^3 7. 840 ft^3, 105 ft^3 9. 120 in.^{3}
11. 13 ft^2 13. 79 m^2 15. 3 in.^{2} 17. 40 m^2
19. $14.11

Page 433

1. about 50 m^3 3. about 1,766 ft^3 5. about 1,964 cm^3 7. about 603 cm^3 9. about 16 hr
11. πr^2 is the area of the base, and h is the height.

Page 435

1. 198 in.^{2} 3. 148 m^2 5. 416 ft^2 7. 440 ft^2
9. 142 cm^2

Page 437

1. volume   3. 682.5 cm^3 5. $V = 2,016$ cm^3
7. $V = 5$ yd^3 9. about 126 cm^3 11. about 57 ft^3 13. 190 yd^2

Chapter 24
Page 447

1. $^+2$; losing 2 pounds; $^-2$ 3. $^+3$; going down 3 flights; $^-3$ 5. 1   7. 7   9. 4   11. 28
13. 35   15. 150   17. $^-1,310$ 19. $^+5$ and $^-5$
21. $\frac{1}{4}$ 23. $\frac{5}{2}$ 25. $\frac{4}{5}$ 27. 30 m^3
29. 15 games

Page 449

1. $\frac{21}{4}$ 3. $\frac{32}{100}$, or $\frac{8}{25}$ 5. $\frac{10}{7}$ 7. $\frac{100}{1}$ 9. $\frac{5}{2}$
11. $\frac{260}{1}$ 13. R   15. I and R   17. R   19. R
21. R   23. R   25. all   27. R   29. $\frac{^-10}{1} = ^-10$
31. $1\frac{1}{2}$, or 1.5   33. $\frac{20}{100}$ or $\frac{1}{5}$ 35. $\frac{123}{2}$

Pages 452–453

1. 0.1   3. 0.625   5. 0.7   7. $0.\overline{54}$ 9. $1.\overline{3}$
11. $0.\overline{1}$ 13. $0.\overline{4}$ 15. $0.\overline{6}$ 17. $0.\overline{8}$ 19. 0.06
21. 0.875   23. 0.3   25. $0.91\overline{6}$ 27. 0.82
29. 0.025   31. 0.5 in.; yes; he recorded 0.15 in. more   33. Yes. $9\frac{3}{4}$ is equal to 9.75.
35. $\frac{1}{5}$; 0.2; 20%   37. 4   39. 24
41. 301.44 cm^3 43. 72 ft^2

Page 455

1. $^-1\frac{3}{4}$ 3. $^-1\frac{3}{10}$ 5. $\frac{5}{24}$ 7. $1\frac{5}{16}$ 9. 2.38
11. $^-7.44$ 13. $1\frac{5}{8}$ 15. $\frac{3}{20}$, or 0.15   17. Yes; $\frac{5}{8}$ is between $\frac{1}{2}$ and $\frac{3}{4}$.   21. Possible answer: Between 0.25 and 0.50; you don't have to find a common denominator.

Page 457

1. 2.5   3. >   5. <   7. $-1.6 < \frac{4}{5} < \frac{6}{3} < 3.8$
9. $^-1.25 < \frac{^-1}{2} < 0 < \frac{1}{10} < 0.3$ 11. $0.64 < 6.02 < 6\frac{1}{8} < 6\frac{3}{4} < 6.8$ 13. $3.3 > 2.4 > 1.5 > ^-1.9 > ^-2$ 15. John   17. Sheila

Page 459

1. integers   3. $^-12$ 5. 12   7. 17   9. all
11. I and R   13. terminating   15. 0.6
17. $0.7\overline{3}$ 19. 1.31   21. -4.36 23. <
25. >

Chapter 25
Page 465

1. $^+3 + ^+2 = ^+5$ 3. $^+12$ 5. 0   7. $^-1$
9. $^-60$ 11. $^-116$ 13. $^+2$ 15. $^-17$
17. $^+51$ 19. $^+64$ 21. $^+5 + ^-2 = ^+3$; 3rd floor   23. $^+4°F$

Page 469

1. $^-3 - ^+5$; $^-3 + ^-5$; $^-8$ 3. $^-5$ 5. $^+9$
7. $^-15$ 9. 0   11. $^+20$ 13. $^-16$ 15. $^+5$
17. $^-54$ 19. $^-25°$ 21. Change subtraction to addition, and use the opposite of the number to be subtracted.   23. 60   25. 504
27. $0.\overline{3}$ 29. $0.\overline{2}$ 31. 0.3   33. $65

Page 471

1. $^+4$; $^+6$ 3. $^-9$; $^-6$ 5. $^-48$ 7. $^-20$
9. $^-21$ 11. $^-96$ 13. $^-60$ 15. $^-30$
17. $^+75$ 19. $^-180$ 21. $^-480$ 23. $^+1,000$
25. $^-8$ in.   27. negative; positive

Page 473

1. $^+3$ 3. $^+6$ 5. $^+2$ 7. $^-16$ 9. $^+5$
11. $^+10$ 13. $^+11$ 15. $^-20$ 17. $^-30$
19. $^-27$ 21. $^-4°$ 23. $^-40 \div ^-10 = ^+4$; a positive number is greater than a negative

number.   **25.** 8   **27.** 5   **29.** $^-6.5 < {}^-5.1 <$ $6.2 < 7.1 < 7.7 < 8.0$ **31.** Larger; the square is 64 ft^2; the circle is 50.24 ft^2.

Page 475
1. $^+4 + {}^-6 = {}^-2$ **3.** $^-8$ **5.** $^-3$ **7.** $^-10$
9. $^+8$ **11.** $^-4$ **13.** $^-4 + {}^-4 = {}^-8; 2 \times {}^-4 =$ $^-8$ **15.** $^+45$ **17.** $^-20$ **19.** $^-9$ **21.** $^+6$

Chapter 26

Page 479
1. 13   **3.** 21   **5.** $^-192$ **7.** $^-12$ **9.** 7
11. 47   **13.** 29   **15.** $^-20$ **17.** $^-40.5$
19. $^-7$ **21.** $^-4$ **23.** $^-147$ **25.** $^-\$9$
29. $a = 40$ **31.** $x = 15$ **33.** $^-11$ **35.** $^-14$
37. 22.5 mi per gallon

Page 481
1. $x = {}^-3$ **3.** $k = {}^-6$ **5.** $c = 5$ **7.** $r = {}^-3$
9. $x = {}^-113$ **11.** $k = 55$ **13.** $v = {}^-8$
15. $p = 8$ **17.** $y = {}^-5$ **19.** $y = {}^-24$
21. $y = {}^-30$ **23.** $m = 15$ **25.** $p = {}^-72$
27. $x = 19{,}270$ **29.** $^-7x = 63$ **31.** $x + 7 =$ $^-34$ **33.** $d - 30 = {}^-67; {}^-37$ ft is her original depth.   **35.** Multiply each side by $^-7$. Replace x with the solution and divide by $^-7$.

Page 485
1. 0, 1   **3.** 0, 1, 2, 3, 4, 5, 6, 7   **5.** 0, 1, 2, 3, 4, 5
17. $x \leq 3$, or $x < 4$ **19.** $x < {}^-2$, or $x \leq {}^-3$
21. $y \geq 12$ **23.** $y > 9$ **25.** $g > j$ or $j < g$

Page 489
1. (2, 6)   **3.** (2, $^-$2)   **5.** (0, 7)   **7.** (3, 0)
9. (6, $^-$4)   **11.** ($^-$7, $^-$6)   **13.** (4, $^-$7)
27. triangle   **29.** They are located in the first quadrant. They are located in the third quadrant.   **31.** $y = 15$ **33.** $y = {}^-7$
35. $^-72$ **37.** 0

Page 491
1. 5; 6   **3.** 6; 7   **7.** $x + 1$ **9.** (0, 5), (1, 6), (2, 7), (3, 8), (4, 9)   **11.** (7, 35), (8, 40), (9, 45), (10, 50)

Page 493
1. 4   **3.** 17   **5.** 36   **7.** $x = {}^-3$ **9.** $k = {}^-5$
11. $p = {}^-70$ **19.** axes   **21.** 2 right, 4 up
23. 1 left, 6 down

Chapter 27
Page 503
1. E' (0, $^-$1), F' (3, $^-$1), G' (3, $^-$4), H' (0, $^-$4)
3. S'(2, 6), T' (2, 2), V' (4, 2)   **5.** A' (0, 0), B' ($^-$6, 0), C' ($^-$4, 2), D' ($^-$2, 2)   **7.** E' (1, 3), F' (4, 3), G' (4, 1), H' (1, 1)   **9.** A' (0, 0), B' (0, $^-$2), C' ($^-$3, 0)   **11.** 10   **13.** $^-5$
15. $y = 3$ **17.** $c = 6$ **19.** 80°F

Page 507
1. right, row 2; left, row 3; right, row 3   **3.** 4^2, 5^2, 6^2 **5.** Add 2 more cubes in each row, and start a new row of 2 cubes on top.   **7.** 25 rectangular prisms; 36 rectangular prisms
9. yes   **11.** no   **13.** Quadrant I
15. Quadrant IV   **17.** Quadrant I   **19.** 6 orders

Page 509

5. yes   **7.** no

Page 513
1. D' (1, 3), E' (2, 3), F' (3, 1), G' (0, 1)
3. A' (5, $^-$1), B' (2, $^-$3), C' (2, $^-$1)
5. reflection

Chapter 28
Page 519
1. add four to each term   **3.** subtract 12 from each term   **5.** add 0.89 to each term
7. 28, 32, 36   **9.** 24, 30, 37   **11.** 335, 485, 665
13. 16, 8, 0   **15.** 0.03, 0.003, 0.0003   **17.** 7.2, 7.95, 8.7   **19.** $200, $220, $240, . . .; $300
21. $55, $59, $63, $67, . . .; 12 weeks   **23.** 6
25. 15   **27.** 9   **31.** Possible answers: 2 dimes, 6 nickels or 1 quarter, 2 dimes, 5 pennies

Page 523
1. add $\frac{2}{32}$ to each term   **3.** add $\frac{4}{27}$ to each term   **5.** subtract $\frac{1}{5}$ from each term
7. subtract $\frac{1}{12}$ from each term   **9.** add $\frac{12}{10}$ to each term   **11.** $\frac{1}{4}, \frac{5}{16}, \frac{3}{8}$ **13.** 17, $18\frac{1}{2}$, 20
15. $8\frac{1}{9}, 9\frac{1}{3}, 10\frac{5}{9}$ **17.** $17\frac{1}{2}, 16\frac{1}{4}$, 15

19. $7\frac{1}{3}, 6\frac{1}{6}, 5$ **21.** $7\frac{1}{2}, 6\frac{4}{9}, 5\frac{7}{18}$

23. $\frac{1}{2}, 1\frac{1}{4}, 2, \ldots; 2\frac{3}{4}$ hr **25.** Possible
answer: Find the common denominator, and
then find the pattern.

Page 525

1. $\frac{16}{3}, \frac{32}{3}, \frac{64}{3}$ **3.** 2, 4, 8 **5.** $\frac{1}{8}, \frac{5}{8}, \frac{25}{8}$,
$\frac{125}{8}$ **7.** $\frac{7}{2}, \frac{21}{20}, \frac{63}{200}, \frac{189}{2,000}$ **9.** $\frac{1}{16}$ **13.** $^+10$

15. $^-21$ **17.** $^+48$ **19.** no **21.** yes

23. 4 trips

Page 527

1. $^-4, 0, ^+4$ **3.** $^-512; ^+2,048; ^-8,192$
5. $^-20, ^-30, ^-40$ **7.** $^-72, ^+144, ^-288$
9. $^-5, 0, ^+5$ **11.** $^-375; ^+1,875; ^-9,375$
13. $^-0.025, ^+0.0125, ^-0.00625$ **15.** $^+4, ^-2, ^+1$
17. 68°F **19.** $^-120$ ft **21.** 13 **23.** In all,
you find a pattern and repeat it.

Page 529

1. sequence **3.** 41, 53, 65 **5.** 4, 2, 1
7. $\frac{13}{8}, \frac{17}{8}, \frac{21}{8}$ **9.** $3\frac{1}{4}, 2, \frac{3}{4}$ **11.** $3, 2\frac{2}{3}, 2\frac{1}{3}$
13. 6, 12, 24 **15.** $\frac{28}{3}, \frac{56}{3}, \frac{112}{3}$ **17.** $\frac{5}{16}, \frac{5}{32}, \frac{5}{64}$
19. $^-4, ^-5, ^-6$ **21.** $^+81, ^-243, ^+729$

23. $^+32, ^-64, ^+128$ **25.** 5°C

A Problem of the Day

This section provides complete solutions for the Problems of the Day in each lesson plan. The problems include all types—one-step, multi-step, applied, process, nonroutine, open-ended, and puzzle problems—and provide options for students to develop their ability to use logical reasoning, to choose and apply problem-solving strategies, and to apply the process—Understand, Plan, Solve, Look Back—to varied and interesting situations.

The problems are also available on transparencies. Each transparency includes all of the Problem of the Day problems for a chapter.

CHAPTER 1 • *Looking at Numbers*

Answer Key

Lesson 1.1
PROBLEM
Michael's number has six twos and five zeros. Each 2, except the one in the ten-thousandths place, is 100 times greater than the 2 to its right. Write his number in words.

SOLUTION
Strategy: Use Logical Reasoning
Write a 2 in the ten-thousandths place.
A digit one place-value position to the left of another is 10 times greater than that digit. So a digit two place-value positions to the left of another is 100 times greater than that digit. Enter the other five twos. Fill in with zeros.

PLACE VALUE											
M	HT	TT	T	H	T	O	T	H	T	TT
2	0	2	0	2	0	2	0	2	0	2

The number is 2,020,202.0202.
It is read as two million, twenty thousand, two hundred two and two hundred two ten-thousandths.

Lesson 1.2
PROBLEM
Two numbers have the same seven digits in the same order. The leftmost place is the millions place. The rightmost place is the ten-thousandths place. The difference between the numbers is 1,975,427.4375. What are the two numbers?

SOLUTION
Strategy: Make a Table
You know the first 3 digits in each number must be 1, 9, and 7. Then, write zeros in the tenths place and following for the first number. Write the difference in the last row. Compute to find the remaining digits.

PLACE VALUE											
M	HT	TT	T	H	T	O	T	H	T	TT
1	9	7	?	?	?	?	0	0	0	0
				1	9	7	?	?	?	?
1	9	7	5	4	2	7	4	3	7	5

PLACE VALUE											
M	HT	TT	T	H	T	O	T	H	T	TT
1	9	7	5	6	2	5	0	0	0	0
				1	9	7	5	6	2	5
1	9	7	5	4	2	7	4	3	7	5

The numbers are: 1,975,625 and 197.5625.

Lesson 1.3
PROBLEM
From 4 o'clock to 5 o'clock, Jacob, Lisa, and Chelsea took turns at the same computer game. Jacob played $\frac{1}{3}$ hour and Lisa played 0.25 hour. For how many minutes did Chelsea play the game?

SOLUTION
Strategy: Write a Number Sentence
Jacob played $\frac{1}{3}$ hour or 20 minutes.
Lisa played 0.25 hour or $(\frac{1}{4})$ hour or 15 minutes.
They played a total of 35 minutes, so Chelsea played the game for 25 minutes (60 − 35).
Some students may need to draw a picture to find the answer.

Lesson 1.4
PROBLEM
Replace the letters *a, b, c, and d* with the numbers 2, 3, 4, and 5 to make a true sentence.
$a^b + a^b = c^d$

SOLUTION
Strategy: Guess and Check
$$2^5 + 2^5 = 4^3$$
$$(2 \times 2 \times 2 \times 2 \times 2) + (2 \times 2 \times 2 \times 2 \times 2) = (4 \times 4 \times 4)$$
$$32 \quad + \quad 32 \quad = 64$$

Lesson 1.5
PROBLEM
One side of Jessica's square array is 2 tiles longer than a side of Dave's square array. Together, they used a total of 100 tiles. How many tiles are on each side of Dave's array?

SOLUTION
Strategy: Guess and Check
Dave's array	6×6	=	36
Jessica's array	8×8	=	64
			100

There are 6 tiles on each side of Dave's array.

CHAPTER 2 • *Using Whole Numbers*

PROBLEM OF THE DAY

Answer Key

Lesson 2.1

PROBLEM

Subtract vertically and horizontally.

93	38	?
68	29	?
?	?	?

SOLUTION

Strategy: Write a Number Sentence

$$\begin{array}{ccc} 93 & 38 & 55 \\ -68 & -29 & -39 \\ \hline 25 & 9 & 16 \end{array}$$

$$25 - 9 = 16$$

Lesson 2.2

PROBLEM

At noon on Monday, Latisha sets her watch to the correct time. If her watch loses one minute each hour, and she does not correct it, what time will her watch show at noon on Wednesday? At noon on what day will her watch show 10:00?

SOLUTION

Strategy: Make a Table
At noon on Wednesday her watch will show 11:12.
At noon on Saturday her watch will show 10:00.
OR
Strategy: Write a Number Sentence
$24 \times 1 = 24$ minutes are lost each day.
Subtract 24 minutes each day.
Results:

Monday noon:	12:00
Tuesday noon:	11:36
Wednesday noon:	11:12
Thursday noon:	10:48
Friday noon:	10:24
Saturday noon:	10:00

Lesson 2.3

PROBLEM

Replace the dots with the digits 0–9 to make a correct number sentence. Use each digit only once.

- ● × ● = 18
- ● × ● = 24
- ● × ● = 0
- ● × ● = 28
- ● × ● = 6

SOLUTION

Strategy: Use Logical Reasoning (or Guess and Check)
Students can write the digits 0–9, and then cross them off as they use them. Students look for a product which can have only one correct set of factors, i.e. 28.
$4 \times 7 = 28$ or $7 \times 4 = 28$. (Digits may be in any order.)
Since 4 is already used,   only $3 \times 8 = 24$.
Since 3 is used,   only $2 \times 9 = 18$.
Since 2 and 3 are used,   only $6 \times 1 = 6$.
The only digits left are 5 and 0, so $5 \times 0 = 0$.

$$\begin{array}{ccccc} 2 & \times & 9 & = & 18 \\ 3 & \times & 8 & = & 24 \\ 5 & \times & 0 & = & 0 \\ 4 & \times & 7 & = & 28 \\ 1 & \times & 6 & = & 6 \end{array}$$

Lesson 2.4

PROBLEM

Find the product. Compare the product with the first factor.

SOLUTION

Strategy: Find a Pattern
1. $13 \times 11 = 143$
2. $72 \times 11 = 792$
3. $326 \times 11 = 3{,}586$
4. $6{,}045 \times 11 = 66{,}495$

A shortcut for multiplying a 2-digit factor by 11: $25 \times 11=$
1. Write the units digit of the other factor. <u>5</u>
2. Write the sum of the units and tens digits. <u>75</u>
3. Write the tens digit of the other factor. <u>275</u>
For multiplying greater factors by 11:
1. Write the units digit of the other factor.
2. Write the sum of the units and tens digits (the sum of the tens and hundreds, etc.).
3. Write the digit in the greatest place-value position. Some students may try problems in which the sum of 2 digits is greater than 9. The pattern remains the same, but the 1 which is "carried" is added to the next pair of digits.

Lesson 2.5

PROBLEM

Although not all the digits are shown, how do you know the quotients cannot be correct?

1. $\overset{78}{\diamond 7)\overline{3,\diamond\diamond 5}}$ 2. $\overset{306}{58)\overline{\diamond,\diamond\diamond\diamond}}$

SOLUTION

Strategy: Use Logical Reasoning
1. $8 \times 7 = 56$, but there is a 5 in the units place in the dividend.
2. $306 \times 58 = 17{,}748$, but there are only four places in the dividend.

CHAPTER 3 • *Using Decimals*

Answer Key

Lesson 3.1

PROBLEM

Replace each [♥] with a digit from 0–9 to make a true number sentence.

[♥].[♥] [♥] [♥] + [♥] [♥].[♥] [♥] + [♥].[♥] = 22.815

SOLUTION

Strategy: Guess and Check
Possible answer:
0.725 + 13.69 + 8.4 = 22.815

Lesson 3.2

PROBLEM

Carrie and her four friends noticed that the change each got from a $10.00 bill had the same digits as the amount spent. Carrie spent $4.55 and got $5.45 in change. Each of her friends spent different amounts. How much did each spend? What was the change?

SOLUTION

Strategy: Guess and Check

1.	$9.05	$0.95
2.	$1.85	$8.15
3.	$2.75	$7.25
4.	$3.65	$6.35

Lesson 3.3

PROBLEM

The sum of two decimal numbers is 9.3.
Their difference is 4.3, and their product is 17.00.
What are the numbers?

SOLUTION

Strategy: Use Logical Reasoning

There is a 3 in the tenths place for the sum <u>and</u> the difference, so one of the digits in the tenths place must be a zero or a 5. If one digit is 0, the other tenths digit is 3; if one digit is 5, the other tenths digit is 8. 17.00 is divisible by 5 and 8, but not by 3. So the digits in the tenths place are 5 and 8.

The digit 1 may not be great enough for the ones place.

THINK: One number is 4.3 greater than the other.

Try 2.8 + 4.3 = 7.1 (Not correct. There should be a 5 in the tenths place.)

Try 2.5 + 4.3 = 6.8 (One number has a 5 and the other has an 8 in the tenths place.)

Check: 2.5 + 6.8 = 9.3
          6.8 − 2.5 = 4.3
          2.5 × 6.8 = 17.00

The two numbers are 2.5 and 6.8.
Some students may need the strategy *Guess and Check.*

Lesson 3.4

PROBLEM

Lashonda and Mark each have the same number of coins. Lashonda has $8.25 in quarters. Mark has all dimes. How much more money does Lashonda have than Mark?

SOLUTION

Strategy: Write a Number Sentence

Lashonda: $8.25 ÷ 0.25 = 33 quarters
Mark: 33 dimes × $0.10 = $3.30
         $8.25 − $3.30 = $4.95

Lashonda has $4.95 more than Mark.

Answer Key

Lesson 4.1

PROBLEM

The sum of the ages of Mr. and Mrs. Olsen and their two children is 108. Their ages are sets of twin prime numbers. What are their ages? NOTE: Twin primes are two prime numbers whose difference is 2.

SOLUTION

Strategy: Guess and Check
Students can list prime numbers and circle twin primes.
They guess and check to see which sets total 68.
3, 5, 7, 11, 13, 17, 19, 29, 31, 41, 43, 59, 61,...
Ages: $11 + 13 + 41 + 43 = 108$.

Lesson 4.2

PROBLEM

Which product is greater?
$95 \times 21 = ?$ or $57 \times 35 = ?$

SOLUTION

Strategy: Use Logical Thinking
$95 \times 21 = 1,995$ or $57 \times 35 = 1,995$
The products are the same.

Lesson 4.3

PROBLEM

Which number in each group does not belong with the rest? Why?

1.	19	15	20	18	21
2.	426	792	158	345	620
3.	36	56	63	84	49
4.	24	39	51	16	67

SOLUTION

Strategy: Look for a Pattern
1. 15 → The rest are consecutive counting numbers.
2. 345 → The rest are even numbers.
3. 36 → The rest are multiples of 7.
4. 67 → The rest are composite numbers.

Lesson 4.4

PROBLEM

Write these everyday items in "simplest form."
1. 50 pennies
2. 24 eggs
3. 8 ounces of butter
4. 2 quarts of ice cream
5. 36 inches of material

SOLUTION

Strategy: Use Logical Thinking
Students should be able to answer these from everyday experience.
1. 1 half dollar
2. 2 dozen eggs
3. $\frac{1}{2}$ pound of butter
4. 1 half-gallon of ice cream
5. one yard of material
Some may be able to write the answers as fractions in simplest form.

1. $\frac{50}{100} = \frac{1}{2}$
2. $\frac{24}{12} = 2$
3. $\frac{8}{16} = \frac{1}{2}$
4. $\frac{2}{4} = \frac{1}{2}$
5. $\frac{36}{36} = 1$

Lesson 4.5

PROBLEM

Brad had 100 coins, all of which were different from Chelsea's 50 coins. Brad and Chelsea took their coins to the bank and exchanged them for bills. Each got the same three bills, with no coins left over. What bills did they receive? What coins did each start with?

SOLUTION

Strategy: Guess and Check

Brad:	10 quarters	→	$2.50
	40 dimes	→	4.00
	50 pennies	→	0.50
Chelsea:	10 half-dollars	→	$5.00
	40 nickels	→	2.00

Each had $7.00 worth of coins.
Each received 1 five-dollar bill and 2 one-dollar bills.

CHAPTER 5 • *Adding and Subtracting Fractions*

PROBLEM OF THE DAY

Answer Key

Lesson 5.1

PROBLEM

Write the next 2 terms of the pattern. Explain.

1.

$$\frac{1}{6} = \frac{1}{6}$$

$$\frac{2}{6} = \frac{1}{3}$$

$$\frac{3}{6} = \frac{1}{2}$$

2.

$$\frac{1}{8} = \frac{1}{8}$$

$$\frac{2}{8} = \frac{1}{4}$$

$$\frac{3}{8} = \frac{3}{8}$$

3.

$$\frac{1}{10} = \frac{1}{10}$$

$$\frac{2}{10} = \frac{1}{5}$$

$$\frac{3}{10} = \frac{3}{10}$$

SOLUTION

Strategy: Find a Pattern

1.

$$\frac{4}{6} = \frac{2}{3}$$

$$\frac{5}{6} = \frac{5}{6}$$

Count (or add) by $\frac{1}{6}$ in both columns, but write in simplest form in second column.

2.

$$\frac{4}{8} = \frac{1}{2}$$

$$\frac{5}{8} = \frac{5}{8}$$

Count (or add) by $\frac{1}{8}$ in both columns, but write in simplest form in second column.

3.

$$\frac{4}{10} = \frac{2}{5}$$

$$\frac{5}{10} = \frac{1}{2}$$

Count (or add) by $\frac{1}{10}$ in both columns, but write in simplest form in second column.

Lesson 5.2

PROBLEM

How many minutes are in $\frac{1}{2}$ hr? in $\frac{1}{3}$ hr? in $\frac{1}{4}$ hr? in $\frac{1}{5}$ hr? in $\frac{1}{6}$ hr?

SOLUTION

Strategy: Draw a Picture

Students draw pictures of clock faces, divide them into the corresponding number of sections, and color one section.

$\frac{1}{2}$ hr = 30 min $\frac{1}{3}$ hr = 20 min $\frac{1}{4}$ hr = 15 min

$\frac{1}{5}$ hr = 12 min $\frac{1}{6}$ hr = 10 min

Lesson 5.3

PROBLEM

Add horizontally and vertically.

SOLUTION

Strategy: Write a Number Sentence

$\frac{1}{4}$	$\frac{1}{12}$	? $\frac{1}{3}$
$\frac{1}{3}$	$\frac{1}{6}$	? $\frac{1}{2}$
? $\frac{7}{12}$	? $\frac{1}{4}$	? $\frac{5}{6}$

Lesson 5.4

PROBLEM

Subtract horizontally and vertically.

SOLUTION

Strategy: Write a Number Sentence

$\frac{8}{9}$	$\frac{1}{3}$	? $\frac{5}{9}$
$\frac{1}{2}$	$\frac{1}{6}$	? $\frac{1}{3}$
? $\frac{7}{18}$	? $\frac{1}{6}$	? $\frac{2}{9}$

Lesson 5.5

PROBLEM

Mike's and Kay's numbers are both less than 1. The digit in Mike's numerator is the same as the digit in Kay's denominator. Kay's number is $\frac{1}{10}$ greater than Mike's. What are Mike's and Kay's numbers?

SOLUTION

Strategy: Use Logical Reasoning, Guess and Check

Mike's number is less than 1 and $\frac{1}{10}$ less than Kay's.

Try $\frac{1}{2}$ for Kay's number. $\frac{1}{2} = \frac{5}{10}$, $\frac{5}{10} - \frac{1}{10} = \frac{4}{10}$. So Mike's number is $\frac{4}{10}$.

Simplest form:→ Mike's number is $\frac{2}{5}$ and Kay's is $\frac{1}{2}$.

Check: Yes, the 2 in Mike's numerator is the same digit as the 2 in Kay's denominator.

CHAPTER 6 • *Adding and Subtracting Mixed Numbers*

Answer Key

Lesson 6.1

PROBLEM

Melissa rides the bus $1\frac{2}{3}$ mi north and $3\frac{1}{4}$ mi east to get to school. Brandon rides his bike $2\frac{3}{4}$ mi south and $2\frac{1}{6}$ mi west to go to the same school. Who travels a longer distance to school?

SOLUTION

Strategy: Write a Number Sentence

Melissa: $1\frac{2}{3} + 3\frac{1}{4} = 1\frac{8}{12} + 3\frac{3}{12} = 4\frac{11}{12}$ mi

Brandon: $2\frac{3}{4} + 2\frac{1}{6} = 2\frac{9}{12} + 2\frac{2}{12} = 4\frac{11}{12}$ mi

They both travel the same distance.

Lesson 6.2

PROBLEM

Write the next 4 numbers.

HINT: What is being subtracted in each sequence?

1. 10, $8\frac{3}{4}$, $7\frac{1}{2}$, $6\frac{1}{4}$, __, __, __, __

2. 9, $7\frac{7}{8}$, $6\frac{3}{4}$, $5\frac{5}{8}$, __, __, __, __

3. $11\frac{7}{10}$, $10\frac{2}{5}$, $9\frac{1}{10}$, $7\frac{4}{5}$, __, __, __, __

SOLUTION

Strategy: Find a Pattern

1. 10, $8\frac{3}{4}$, $7\frac{1}{2}$, $6\frac{1}{4}$, 5, $3\frac{3}{4}$, $2\frac{1}{2}$, $1\frac{1}{4}$
    (Subtract $1\frac{1}{4}$)

2. 9, $7\frac{7}{8}$, $6\frac{3}{4}$, $5\frac{5}{8}$, $4\frac{1}{2}$, $3\frac{3}{8}$, $2\frac{1}{4}$, $1\frac{1}{8}$
    (Subtract $1\frac{1}{8}$)

3. $11\frac{7}{10}$, $10\frac{2}{5}$, $9\frac{1}{10}$, $7\frac{4}{5}$, $6\frac{1}{2}$, $5\frac{1}{5}$, $3\frac{9}{10}$, $2\frac{3}{5}$
    (Subtract $1\frac{3}{10}$)

Lesson 6.3

PROBLEM

Add vertically and horizontally.

SOLUTION

Strategy: Write a Number Sentence

$1\frac{2}{3}$	$2\frac{1}{4}$	? $3\frac{11}{12}$
$3\frac{1}{6}$	$4\frac{1}{2}$	? $7\frac{2}{3}$
? $4\frac{5}{6}$	? $6\frac{3}{4}$	? $11\frac{7}{12}$

If needed:

$$1\frac{2}{3} + 2\frac{1}{4} = 1\frac{8}{12} + 2\frac{3}{12} = 3\frac{11}{12}$$
$$+ \quad 3\frac{1}{6} + 4\frac{1}{2} = 3\frac{2}{12} + 4\frac{6}{12} = 7\frac{8}{12}$$
$$\overline{4\frac{10}{12} + 6\frac{9}{12} = 10\frac{19}{12} \qquad = 11\frac{7}{12}}$$

Lesson 6.4

PROBLEM

Complete the Magic Square.
The magic sum is $1\frac{1}{4}$.

$\frac{1}{2}$		$\frac{1}{6}$
	$\frac{5}{12}$	

SOLUTION

Strategy: Write a Number Sentence

$\frac{1}{2}$	$\frac{7}{12}$	$\frac{1}{6}$
$\frac{1}{12}$	$\frac{5}{12}$	$\frac{3}{4}$
$\frac{2}{3}$	$\frac{1}{4}$	$\frac{1}{3}$

If necessary:

$\frac{6}{12}$	$\frac{7}{12}$	$\frac{2}{12}$
$\frac{1}{12}$	$\frac{5}{12}$	$\frac{9}{12}$
$\frac{8}{12}$	$\frac{3}{12}$	$\frac{4}{12}$

CHAPTER 7 • *Multiplying and Dividing Fractions*

Answer Key

Lesson 7.1

PROBLEM

In a jump-rope marathon, Cara will earn and donate to charity $5 for each half hour or fraction of a half hour that she jumps rope. How much money will Cara earn if she jumps rope for 175 min?

SOLUTION

Strategy: Write a Number Sentence

$\frac{1}{2}$ hour $\quad = \quad$ 30 minutes

$175 \div 30 \quad = \quad$ 5 r25 or 6 half hours

$6 \times \$5.00 \quad = \quad \$30.00$

Lesson 7.2

PROBLEM

Use the digits 1–9 to make as many pairs of equivalent fractions as you can. You may use each number only once in each pair.

SOLUTION

Strategy: Make an Organized List

$\frac{1}{2}=\frac{2}{4}$ $\frac{1}{2}=\frac{3}{6}$ $\frac{1}{2}=\frac{4}{8}$	$\frac{2}{4}=\frac{3}{6}$ $\frac{2}{4}=\frac{4}{8}$	$\frac{3}{6}=\frac{4}{8}$
$\frac{2}{1}=\frac{4}{2}$ $\frac{2}{1}=\frac{6}{3}$ $\frac{2}{1}=\frac{8}{4}$	$\frac{4}{2}=\frac{6}{3}$ $\frac{4}{2}=\frac{8}{4}$	$\frac{6}{3}=\frac{8}{4}$
$\frac{1}{3}=\frac{2}{6}$ $\frac{1}{3}=\frac{3}{9}$	$\frac{3}{1}=\frac{6}{2}$ $\frac{3}{1}=\frac{9}{3}$	
$\frac{2}{3}=\frac{4}{6}$ $\frac{2}{3}=\frac{6}{9}$	$\frac{3}{2}=\frac{6}{4}$ $\frac{3}{2}=\frac{9}{6}$	
$\frac{1}{4}=\frac{2}{8}$	$\frac{4}{1}=\frac{8}{2}$	
$\frac{3}{4}=\frac{6}{8}$	$\frac{4}{3}=\frac{8}{6}$	

Lesson 7.3

PROBLEM

It took André 1 min to fill his aquarium $\frac{1}{3}$ full. How long will it take him to fill the aquarium $\frac{3}{4}$ full?

SOLUTION

Strategy: Draw a Picture

It takes André 1 minute to fill the aquarium one-third full. Shade one third of the picture. To fill it $\frac{3}{4}$ full, divide the aquarium into twelfths. Now each third is divided into fourths. Each section takes one-fourth minute or 15 seconds to fill.

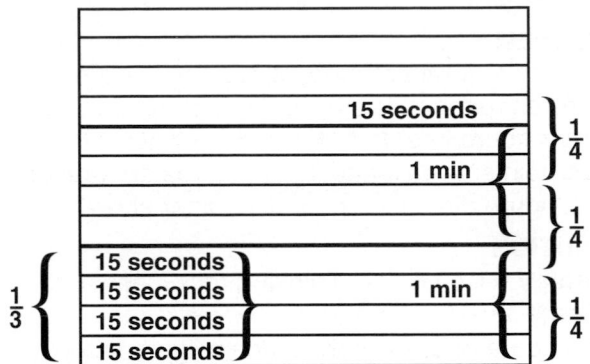

$\frac{3}{4}=\frac{9}{12}$ so if he fills the aquarium $\frac{3}{4}$ full, he fills 9 sections.

4 sections $\quad = \quad$ 1 min

1 section $\quad = \quad$ 15 sec

9 sections $\quad = \quad$ 2 min and 15 sec or $2\frac{1}{4}$ min

Lesson 7.4

PROBLEM

A hawk flies $\frac{1}{3}$ mi in 30 sec. How far can the hawk fly in 1 min? How fast does it fly, in miles per hour?

SOLUTION

Strategy: Write a Number Sentence

A hawk flies $\frac{1}{3}$ mi in 30 sec so it flies

$2 \times \frac{1}{3}$ mi in 2×30 sec or

$\frac{2}{3}$ mi in 60 sec, or 1 min.

If the hawk flies $\frac{2}{3}$ mi in 1 min, it flies

$60 \times \frac{2}{3}$ mi in 60×1 min, or 40 mi in

60 min, or 1 hr.

So the hawk's speed is 40 miles per hour.

Lesson 7.5

PROBLEM

Ming Li ran 90 ft from first base to second base. Each stride was $3\frac{4}{5}$ ft long. About how many strides did Ming Li take?

SOLUTION

Strategy: Write a Number Sentence

Using fractions: $\qquad$ or using decimals:

$90 \div 3\frac{4}{5}$ $\qquad$ $3\frac{4}{5} \quad = \quad 3.8$

$\frac{90}{1} \div \frac{19}{5} \quad = \quad 90 \quad \div \quad 3.8 = 23.684$ or about 24

$\frac{90}{1} \times \frac{5}{19} \quad = \quad 23\frac{13}{19}$ $\qquad$ or about 24

Ming Li took about 24 strides.

Answer Key

Lesson 8.1

PROBLEM
Kate and four other students each made a poster of a different geometric figure. Tim did not draw the ray. Marty's figure has no symbol. Joe's figure needs three points. Lyda's and Tim's figures go forever. What figure did each student's poster show?

SOLUTION
Strategy: Use Logical Thinking (or Guess and Check)
Joe's figure needs 3 points. Joe's figure is a plane. Marty's figure has no symbol. Neither points nor planes have symbols. Joe's figure is a plane, so Marty's figure is a point. Lyda's and Tim's figures go on forever. Lines and rays go on forever. Tim did not draw the ray, so Lyda drew the ray, and Tim drew a line.The figure left over is the line segment, so Kate drew the line segment.

Lesson 8.2

PROBLEM
Thad describes his drawing with these symbols: $\overleftrightarrow{AB} \parallel \overleftrightarrow{CD}$, $\overleftrightarrow{AC} \perp \overleftrightarrow{AB}$, $\overleftrightarrow{AB} \perp \overleftrightarrow{BD}$. What symbols can he use to describe $\overleftrightarrow{AC}$ and $\overleftrightarrow{BD}$?

SOLUTION
Strategy: Draw a Picture
Students draw $\overleftrightarrow{AB}$ parallel to $\overleftrightarrow{CD}$.
Then they draw $\overleftrightarrow{AC}$ perpendicular to $\overleftrightarrow{AB}$; and $\overleftrightarrow{AB}$ perpendicular to $\overleftrightarrow{BD}$.
$\overleftrightarrow{AC}$ and $\overleftrightarrow{BD}$ are parallel to each other. $\overleftrightarrow{AC} \parallel \overleftrightarrow{BD}$.

Lesson 8.3

PROBLEM
The measure of Marlon's angle is twice that of Amee's and half that of Nate's. The sum of the measures of Amee's and Trisha's angles is equal to the sum of Marlon's and Nate's angles. The sum of the measures of all the angles is equal to the measure of a straight angle. What is the measure of each student's angle?

SOLUTION
Strategy: Guess and Check

Amee	+	Trisha	=	Marlon	+	Nate
15°	+	75°	=	30°	+	60°

Or

Another Strategy: Use Logical Reasoning
The sum of the measures of all the angles is 180°.
measures of girls' angles = measures of boys' angles
measures of girls' angles + measures of boys' angles = 180°

$$90° + 90° = 180°$$

Marlon	+	Nate	=	90°		90°	−	Amee	=	Trisha
30°	+	60°	=	90°		90°	−	15°	=	75°

Lesson 8.4

PROBLEM
How many squares are there in all?

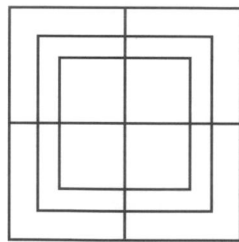

SOLUTION
Strategy: Make an Organized List
Label the vertices.

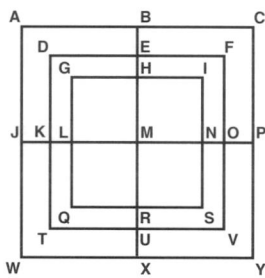

15 squares in all

ACYW	DFVT	GISQ
ABMJ	DEMK	GHML
BCPM	EFOM	HINM
JMXW	KMUT	LMRQ
MPYX	MOVU	MNSR

Lesson 8.5

PROBLEM
Jason used 4 clock faces to draw regular polygons with 3 to 6 sides. One vertex of each polygon was on the dot for 12:00. The other vertices of his triangle were at 20 minutes past and 20 minutes to the hour. Where were the vertices of his other polygons?

SOLUTION
Strategy: Draw a Picture
Students use clock faces to draw a triangle, a square, a pentagon and a hexagon.
Check students' drawings.
Square: 12:00, 15 past, 30 past, 45 past
Pentagon: 12:00, 12 past, 24 past, 36 past, 48 past
Hexagon: 12:00, 10 past, 20 past, 30 past, 40 past, 50 past

PROBLEM OF THE DAY
Answer Key

Lesson 9.1

PROBLEM
Not only letters but also whole words can have lines of symmetry. Write at least five words and draw their lines of symmetry.

DECIDED

TOT MAAM EXCEED

SOLUTION
Strategy: Guess and Check
Some solutions are:
HAH, HUH, TAT, TUT, MUM, DICE, DICED, CODE, CODED, DECODE, DECODED, DID, DEED, DIE, DIED, DODO, COO, COOED, BID, BODE, BODED, BOX, BOXED, BOO, BOOED, COB, HOE, HOED, HID, HIDE, CHIDE, HOBO, HOOD, EXCEEDED, BEECH

Lesson 9.2

PROBLEM
Rashan and two classmates each moved one of these letters. Louis did not flip or slide his letter. Meg did not use reflection to move her letter. Which letter did each move and what transformation did each use?

G F P
G Ŧ Ƿ

SOLUTION
Strategy: Use Logical Reasoning
Louis did not flip or slide his letter, so he turned the letter P (rotation). Meg did not use reflection, so she slid the letter G (translation). So Rashan flipped the letter F (reflection).

Lesson 9.3

PROBLEM
Which term does not belong with the others?
1. intersecting   perpendicular   protractor   parallel
2. rotation   translation   reflection   tessellation
3. acute   vertex   obtuse   right
4. straight   scalene   equilateral   isosceles

SOLUTION
Strategy: Find a Pattern
1. protractor → The others tell how two lines relate to each other.
2. tessellation → The others are transformations.
3. vertex → The others are kinds of angles.
4. straight → The others are kinds of triangles.

Lesson 9.4

PROBLEM
Make each statement true by writing *All* or *Some.*
1. _____?_____ equilateral triangles are isosceles triangles.
2. _____?_____ isosceles triangles have a right angle.
3. _____?_____ scalene triangles have an obtuse angle.
4. _____?_____ triangles with all acute angles are equilateral triangles.
5. _____?_____ triangles have at least two acute angles.

SOLUTION
Strategy: Use Logical Reasoning
1. All
2. Some
3. All
4. Some
5. All

CHAPTER 10 • Solid Figures

PROBLEM OF THE DAY
Answer Key

Lesson 10.1

PROBLEM

Complete the comparisons.

1. Sphere is to no flat faces as _____?_____ is to all flat faces.
2. _____?_____ is to all triangular faces as rectangular prism is to all rectangular faces.
3. Square is to _____?_____ as circle is to cone.
4. Circle is to _____?_____ as triangle is to triangular prism.
5. Circle is to square as sphere is to _____?_____.

SOLUTION

Strategy: Find a Pattern

1. polyhedron
2. Triangular pyramid
3. square pyramid
4. cylinder
5. cube

Lesson 10.2

PROBLEM

Fred, Lee, and Ann have drawn different figures. Fred's figure has as many vertices as Lee's has faces. Lee's figure has twice as many vertices and twice as many edges as Ann's. Ann's figure has as many faces as it has vertices. Fred's figure has one more face than Ann's. What figure has each drawn?

SOLUTION

Strategy: Guess and Check

Ann → triangular pyramid
Fred → triangular prism
Lee → rectangular prism

Lesson 10.3

PROBLEM

How many arrangements of five squares can you make in which at least one side of each square is touching another? How many of these arrangements can you fold into a five-sided box?

SOLUTION

Strategy: Guess and Check, Make a Model

12 arrangements; 8 arrangements (unshaded below) can be folded into a 5-sided box.

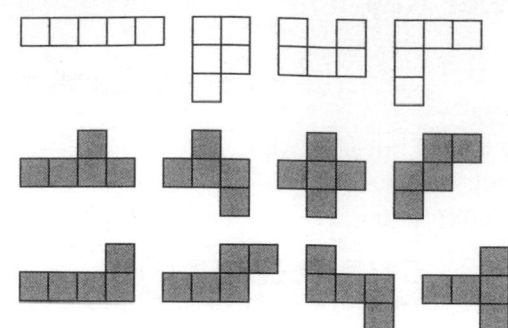

Lesson 10.4

PROBLEM

What is the least number of toothpicks you can move to change 629 to 538?

SOLUTION

Strategy: Guess and Check

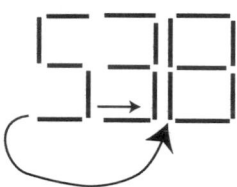

2 toothpicks

Lesson 10.5

PROBLEM

Ms. Courtney told nine students to make three straight lines with four students in each line. How did they do it?

SOLUTION

Strategy: Use a Model

Students may use counters to represent students, and guess and check to find the answer.
They stood in the shape of a triangle.

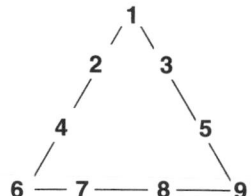

CHAPTER 11 • *Organizing Data*

PROBLEM OF THE DAY
Answer Key

Lesson 11.1

PROBLEM
Frank used centimeter cubes to make a solid 3 cubes wide, 5 cubes long and 4 cubes high. How many more cubes does he need if he wants to make the solid

  a. twice as wide, half as high, but with the same length?
  b. twice as wide, twice as long, and twice as high?

SOLUTION
Strategy: Write a Number Sentence
Frank's solid: $3 \times 4 \times 5 = 60$ cubes

  a. $6 \times 5 \times 2 = 60 \rightarrow$ He will not need any more cubes.

  b. $6 \times 10 \times 8 = 480$ cubes
     $480 - 60 = 420 \rightarrow$ He will need 420 more cubes.

Lesson 11.2

PROBLEM
Dana's survey showed that 3 out of 8 students preferred vegetarian pizza and 1 out of 6 students preferred cheese pizza. How many more of the 72 students surveyed by Dana liked vegetarian pizza than liked cheese pizza?

SOLUTION
Strategy: Write a Number Sentence

$\frac{3}{8}$ of 72 = 27 students

$\frac{1}{6}$ of 72 = 12 students

$27 - 12 =$ 15 more students liked vegetarian pizza.

Lesson 11.3

PROBLEM
In a survey, 90 students were in favor of a field trip to a nearby factory, 48 were against it, and 12 couldn't make up their minds. What fraction of the students were in favor of the trip?

SOLUTION
Strategy: Write a Number Sentence
$90 + 48 + 12 = 150$ students
$\frac{90}{150}$, or $\frac{3}{5}$, of the students were in favor of the trip.

Lesson 11.4

PROBLEM
Anne's line plot of ages of students has a range of 5. Each age has twice as many x's as the previous one. If the last age has 16 x's, how many students did Anne include in her data?

SOLUTION
Strategy: Use Logical Reasoning
It does not matter what the ages were, but there are five sets of ages. So work backward, dividing by 2 each time. Find the sum.
$16 + 8 + 4 + 2 + 1 = 31$ students

CHAPTER 12 • *Displaying Data*

Answer Key

Lesson 12.1

PROBLEM

Maurie made a stem-and-leaf plot of his relatives' ages. He is the second-youngest of 4 teenagers, all 2 years apart in age. His mother is 3 times as old as he is and 24 years younger than her father. How does Maurie show his grandfather's age on the plot?

SOLUTION

Strategy: Use Logical Reasoning

The teenagers must be 13, 15, 17, and 19 years old. Maurie must be 15 years old.

His mother is 45 years old.   (3×15)

His grandfather is 69 years old. ($45 + 24$)

Key: $6 \mid 9 = 69$ years old

or

Strategy: Make a Stem-and-Leaf Plot

Ages of Relatives

Stem	Leaves
1	3 5 7 9
4	5
6	9 (grandfather's age)

Lesson 12.2

PROBLEM

Draw the circles and replace the ?'s with the numbers 1–10. The sum of the numbers in the outer circle is 4 times the sum of the 3 odd numbers in the inner circle. The sum of the inner-circle numbers is 11.

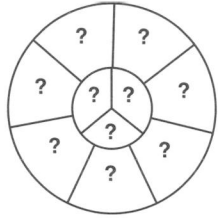 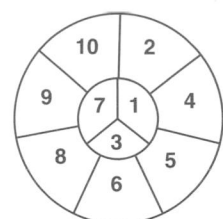

SOLUTION

Strategy: Guess and Check

The sum of the numbers in the inner circle must be 11. Find 3 odd numbers that sum to 11. (1, 3, 7)

Check: $1 + 3 + 7 = 11$

$\qquad 2 + 4 + 5 + 6 + 8 + 9 + 10 = 44$

$\qquad$ 44 is 4 times 11.

Lesson 12.3

PROBLEM

On the first day of each month, Irene records on a multiple-line graph the times of the sunrise and sunset. In January, sunrise is 7:54 A.M. and sunset is 5:17 P.M.; in February, the times are 7:40 A.M. and 5:50 P.M.; and in March, 7:05 A.M. and 6:23 P.M. Compare the 2 lines.

SOLUTION

Strategy: Use Logical Reasoning

If the morning hours are at the top of the graph, the sunrise line will be going up, and the sunset line will be going down. There will be more space between the lines because the days are getting longer where Irene lives.

Lesson 12.4

PROBLEM

Grant didn't have a protractor, so he used a clock face for his circle graph. He drew radii to the 12, the 4, and the 9. How many degrees is each angle?

SOLUTION

Strategy: Write a Number Sentence

Degrees in a circle ÷ Minutes in 1 hour = Degrees in 1 minute

$\qquad 360 \qquad ÷ \qquad 60 \qquad = \qquad 6$

From 12 to 4 is 20 minutes. $20 \times 6 = 120$ degree angle

From 4 to 9 is 25 minutes. $25 \times 6 = 150$ degree angle

From 9 to 12 is 15 minutes. $15 \times 6 = 90$ degree angle

or

Strategy: Make a Model

Students may draw a clock face, draw the radii, and use a protractor to measure the degrees of each angle.

From 12 to 4 is a 120 degree angle.

From 4 to 9 is a 150 degree angle.

From 9 to 12 is a 90 degree angle.

CHAPTER 13 • *Interpreting Data and Predicting*

Answer Key

Lesson 13.1

PROBLEM
A circle graph shows that half of the ancestors of the 120 students surveyed came to the U.S. from Europe or South America, a quarter came from Africa, and the rest from Asia and Australia. Five times as many came from Europe as from South America. How many students had South American ancestors?

SOLUTION
Strategy: Write an Equation
$120 \div 2 = 60$ students had ancestors from Europe or South America.

$$
\begin{aligned}
x + 5x &= 60 \\
6x &= 60 \\
x &= 60
\end{aligned}
$$

Ten students had South American ancestors.

Lesson 13.2

PROBLEM
What is the least number that can be divided evenly by each of the numbers 1–12?

SOLUTION
Strategy: Guess and Check
Students can multiply some of the numbers together and see if the product is divisible by the other numbers.
The number is 27,720.
or
Strategy: Use Logical Reasoning
Numbers 1, 2, 3, 4, and 6 are factors of 12, so if the number is divisible by 12 it is divisible by all these also. Remember, if a number is divisible by 5 and 2, it is divisible by 10. Multiply the odd numbers:
$5 \times 7 \times 9 \times 11 = 3,465$
3,465 is not divisible by 8 so multiply it by 8.
$3,465 \times 8 = 27,720$
27,720 is divisible by 12.
Check by dividing 27,720 by each of the numbers 1–12.

Lesson 13.3

PROBLEM
We are a 3-digit and a 2-digit number.
None of our digits are greater than 5.
Our product is the number of feet in a mile.
Who are we?

SOLUTION
Strategy: Guess and Check
One answer is: You are 15 and 352.

Lesson 13.4

PROBLEM
The mean of these numbers is 14.
The greatest number is 21 more than the least.
The mode is 18. What are the missing numbers?
3   6   9   ♦   ♦   ♦   ♦

SOLUTION
Strategy: Use Logical Reasoning
3   6   9   ♦   ♦   ♦   24
$(3 + 21 = 24)$
There are 7 numbers with a mean of 14, so $7 \times 14 = 98$, the total. The mode is 18. Two or three of the missing numbers must be 18. Try three 18's and find the total.
3   6   9   18   18   18   24
(Sum is 96, which is 2 less than 98.)
So try two 18's and 20 (18 + 2).

3   6   9   18   18   20   24
(Yes, the sum is 98.)

Lesson 13.5

PROBLEM
Unscramble these letters to form math words used in this chapter.
1. ARB   **2.** NAME   **3.** LICCER
4. DAMIEN **5.** MEST-DAN-FALE   **6.** DOME
7. NILE   **8.** XOB-NAD-SHEWIRK
9. CROPINTIDE

SOLUTION
Strategy: Guess and Check
1. BAR   2. MEAN   3. CIRCLE
4. MEDIAN   5. STEM-AND-LEAF   6. MODE
7. LINE   8. BOX-AND-WHISKER
9. PREDICTION

CHAPTER 14 • Probability

PROBLEM OF THE DAY
Answer Key

Lesson 14.1

PROBLEM
How many different four-digit numbers can you write using the digits 2, 8, 5 and 4 in each number?

SOLUTION
Strategy: Make an Organized List

2458	4258	5248	8245
2485	4285	5284	8254
2548	4528	5428	8425
2584	4582	5482	8452
2845	4825	5824	8524
2854	4852	5842	8542

24 numbers

Lesson 14.2

PROBLEM
Betty's dentist does not work on weekends. What is the probability that Betty's next appointment is on Monday? on Saturday? on Tuesday, Wednesday or Thursday?

SOLUTION
Strategy: Use Logical Reasoning
The dentist works only 5 days.

Monday: $\frac{1}{5}$

Saturday: $\frac{0}{5}$

Tuesday, Wednesday or Thursday: $\frac{3}{5}$

Lesson 14.3

PROBLEM
In her coin purse, Rachel has pennies with the following dates: 1956, 1981, 1990, 1983, 1994, and 1985. What is the probability that she will choose a coin minted in the 1990's? in an odd-numbered year? in the 20th century?

SOLUTION
Strategy: Use Logical Reasoning
There are 6 possibilities.

1990's: 2 out of 6, or $\frac{1}{3}$

odd-numbered year: 3 out of 6, or $\frac{1}{2}$

20th century: 6 out of 6, or $\frac{6}{6}$

Lesson 14.4

PROBLEM
For their probability experiments, three students used a spinner divided into 3 congruent sections, a number cube, and a coin. Carlo's and Judy's probabilities could be $\frac{1}{3}$. Martha's and Carlo's probabilities could be $\frac{1}{2}$. Which object did each use?

SOLUTION
Strategy: Use Logical Reasoning
The spinner could only have a probability of $\frac{1}{3}$.

The coin could only have a probability of $\frac{1}{2}$.

The number cube could have a probability of either $\frac{1}{3}$, or $\frac{1}{2}$.

So Carlo used the number cube; Judy used the spinner; and Martha used the coin.

B Practice Activities

These engaging activities prepare students for this year's mathematics. They review and reinforce skills and concepts taught previously. Photocopy or laminate these games and projects to place in your learning center, use as seatwork, or to send home for families to enjoy. Assign Practice Activities when you see references in "Meeting Individual Needs" and "Assessing Prior Knowledge," or use your grade's list of review topics at the right as a guide.

Grade 1
Number Concepts
Basic Facts
Place Value
Money

Grade 2
Basic Facts
Place Value
Money
Time
Computation

Grade 3
Basic Facts
Computation
Time
Money
Place Value

Grade 4
Basic Facts
Place Value
Time
Money
Computation
Fraction Concepts

Grade 5
Place Value
Basic Facts
Computation
Decimal Concepts
Decimal Operations
Fraction Concepts

Grade 6
Place Value
Computation
Decimal Operations
Number Theory
Fraction Concepts
Fraction Operations
Concepts of Ratio
   and Percent

Grade 7
Concepts of Ratio
   and Percent
Whole Number Operations
Decimal Operations
Number Theory
Fraction Operations
Integer Concepts
Integer Operations
Algebra

Grade 8
Number Theory
Concepts of Ratio
   and Percent
Integer Concepts
Integer Operations
Computation with
   Rational Numbers
Algebra

CONTENTS

Activity 1A

SMALL GROUPS

What You Will Learn **A** To build, write, and read numbers to hundred millions **B** To model numbers to hundred millions

What You Need **A** ten-section spinner (TR p. R69), marker, paper, pencils **B** place-value chart to hundred millions (TR p. R3)

Getting Ready **A** Label the sections of the spinner 0–9. **B** Make a blank place-value recording sheet as shown below.

```
                    _ 4 3
                  _ , _ 4 3
                _ _ , _ 4 3
              _ _ _ , _ 4 3
            _ _ , _ _ _ , _ 4 3
          _ _ _ , _ _ _ , _ 4 3
```

Standard Form:
_____
_____

Building a Number

What You Will Do

• The first player spins the pointer and writes the digit in the ones column of each row on the recording sheet.

• The second player spins the pointer, writes the digit in the tens column of each row on the recording sheet, and reads aloud the new number.

• Each player in turn spins the pointer, writes the digit in the next column of each row on the recording sheet, and reads aloud the new number.

• The player who writes the ninth digit also writes the number in standard form.

Activity 1B

• After each player writes the digit spun on the recording sheet, he or she places the same number on the place-value chart.

Activity 2A

PARTNERS

What You Will Learn **A** and **B** To round whole numbers to the nearest thousand, ten thousand, hundred thousand, or million; to add rounded numbers

What You Need **A** and **B** sixteen index cards, paper, pencils, calculator

Getting Ready **A** Each partner prepares four cards—one that says *Round to the nearest thousand,* one that says *Round to the nearest ten thousand,* one that says *Round to the nearest hundred thousand,* and one that says *Round to the nearest million.* Write four numbers on each card—two that will have to be rounded down and two that will have to be rounded up. **B** Each partner prepares four cards—one that says *Round to the nearest thousand,* one that says *Round to the nearest ten thousand,* one that says *Round to the nearest hundred thousand,* and one that says *Round to the nearest million.* Write six numbers on each card—three that will have to be rounded down and three that will have to be rounded up.

Round and Round We Go

What You Will Do

• Shuffle the cards, and place them face down in a stack.

• Take the top card, round each number to the place indicated, read aloud the rounded numbers, write them on a sheet of paper, and find their sum.

• Your partner uses the calculator to find the sum of the numbers on the card. He or she then rounds that sum to the place indicated on the card.

• Compare your rounded numbers. If the numbers do not match, change roles and do the activity again.

• Alternate roles, and play until you have used all the cards.

Activity 2B

• Play the game with the new set of cards.

Round to the nearest thousand
3,648
12,096
5,431
41,725

Round to the nearest ten thousand
265,432
50,756
39,802
123,456

Activity 3A

SMALL GROUPS

What You Will Learn A and B To compare numbers

What You Need A and B number line by hundreds (TR p. R10), index card, marker, paper, pencils

Getting Ready Draw an arrow on the index card.

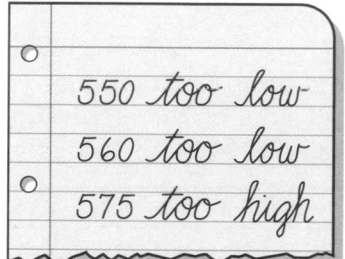

550 too low
560 too low
575 too high

Name the Mystery Number

What You Will Do
• One player selects a "mystery number" (between 0 and 1,000) and writes it down so that the others can't see it. The player then places the index card with the arrow on the number line to find the approximate location of the number.

• In turn, each player guesses the number and writes his or her guess on a piece of paper. The player who chose the number writes *too low* or *too high* next to each guess, then passes the paper to next player. Play continues until the mystery number is found. The player who names the mystery number scores 1 point.

• Take turns selecting the mystery number. Play until one player has scored 5 points.

Activity 3B

• Play the game without using the arrow card.

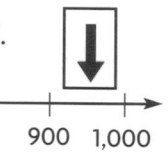

← 0   100   200   300   400   500   600   700   800   900   1,000 →

Activity 4A

PARTNERS

What You Will Learn A and B To read, write, and compare place value for numbers to hundred millions

What You Need A and B ten-section spinner (TR p. R69), marker, nine index cards, paper, pencils

Getting Ready A and B Write the numbers 0–9 on the spinner. On each index card, write a place-value name from ones to hundred millions.

364,291,450
364,921,450

Switching Digits

What You Will Do
• Shuffle the cards, and place them face down in a stack. Spin the pointer nine times, and record your 9-digit number.

• Your partner takes the top card, reads aloud the name of the place value, and switches the digit in that place with any other digit in the number to make the greatest possible number. He or she writes the new number under the first number and reads aloud the new number.

• Take turns picking cards and switching digits to make the greatest possible number until all the place-value cards have been used.

Activity 4B

• Switch the digits to make the least possible number.

hundred thousands

Activity 5A

PARTNERS

What You Will Learn **A** To estimate products **B** To estimate sums or differences

What You Need **A** and **B** four blank number cubes (TR p. R71), markers, paper, pencils, calculator

Getting Ready **A** and **B** Label two number cubes 0–5 and two number cubes 4–9.

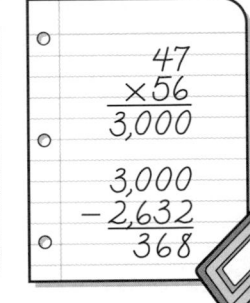

Getting Close to the Answer

What You Will Do

- The first player tosses the four number cubes and uses them to form a 2-digit × 2-digit multiplication problem. Each player writes an estimate of the product.

- The player who formed the problem uses a calculator to find the exact product. Players subtract to find the difference between their estimate and the exact product. The player whose estimated product is closer to the exact product scores 1 point.

- Players take turns tossing the cubes for each round. Play continues until a player has scored 3 points.

Activity 5B

- Use the number cubes to estimate sums or differences.

Activity 6A

SMALL GROUPS

What You Will Learn **A** To divide by 1-digit and 2-digit numbers **B** To multiply a 3-digit number by a 2-digit number

What You Need **A** and **B** ten-section spinner (TR p. R69), marker, paper, pencils, calculator

Getting Ready **A** Label the sections of the spinner 3, 7, 9, 16, 20, 25, 28, 36, 41, and 52. **B** Label the sections of the spinner with only 2-digit numbers.

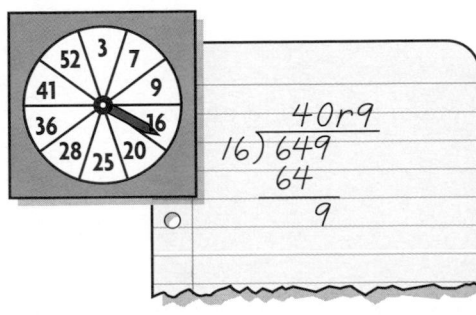

Spinning for a Divisor

What You Will Do

- The first player names a 3-digit number that will be the dividend for the round. Each player writes that dividend, then spins the pointer to find the number to use as his or her divisor.

- Players find the quotients for their problems and use a calculator to check the quotient of the player to their right. For each round, players score 1 point for a correct quotient. The player or players with the greatest remainder score 1 additional point.

- Players take turns naming the 3-digit number for each round. Play continues until one player has scored 6 points.

Activity 6B

- Each player multiplies the 3-digit number chosen for the round by the number he or she spins.

Activity 7A

SMALL GROUPS

What You Will Learn **A** To estimate and calculate an average of three 3-digit numbers **B** To estimate and calculate an average of three 4-digit numbers

What You Need **A** and **B** ten index cards, calculator

Getting Ready **A** and **B** Write the numbers 0–9 on the cards.

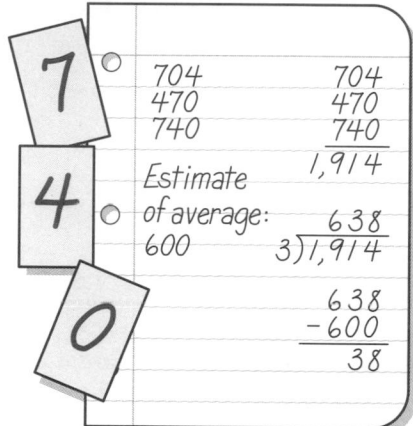

Finding Averages

What You Will Do

• The first player shuffles the cards, places them face down in a stack, and turns over three cards. Each player uses the numbers on the cards to write three different 3-digit numbers.

• Each player then writes an estimate of the average of the three numbers, and passes his or her paper to the player on the right.

• The player on the right uses a calculator to find the average and returns the paper to the player who made the estimate.

• Players then subtract to find the difference between their estimated average and the exact average. The player whose estimated average is closest to the exact average scores 1 point.

• Players take turns turning over the cards for each round. Play continues until a player has scored 3 points.

Activity 7B

• Use over four cards. Rearrange them, and estimate their average.

Activity 8A

SMALL GROUPS

What You Will Learn **A** and **B** To add, subtract, multiply, and divide with whole numbers

What You Need **A** and **B** ten-section spinner (TR p. R69), eight index cards, calculators

Getting Ready **A** and **B** Label the sections of a spinner 0–9. Label the index cards *greatest possible sum, least possible sum, greatest possible difference, least possible difference, greatest possible product, least possible product, greatest possible quotient,* and *least possible quotient.*

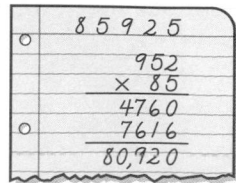

greatest possible product

Greatest Answer–Least Answer

What You Will Do

• Place the cards face down in a stack. Turn over the top card, and spin the pointer five times. Each player writes the five digits on his or her paper, then uses them to write a 2-digit number and a 3-digit number for a problem with an answer described on the card.

• Each player uses a calculator to check the answer of the player on his or her right. Each player whose answer is correct and described on the card gets 1 point.

• Take turns until all the cards have been used. The player with the most points wins.

Activity 8B

• Spin the pointer four times, and make two 2-digit numbers.

PARTNERS

What You Will Learn **A** To use a model to check the product of two decimals (tenths) **B** To match a model with a written problem

What You Need **A** and **B** ten-section spinner (TR p. R69), red and blue crayons or markers, ten 10 × 10 decimal squares (TR p. R104), 20 index cards, glue, poster board

Getting Ready **A** and **B** Label the sections of the spinner 0.1–1.0. Glue the decimal squares to poster board and cut out.

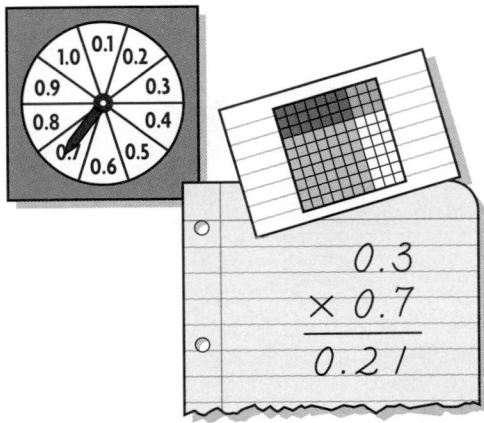

Multiplying Tenths

What You Will Do

• Spin the pointer. On a blank index card, write the number as the first factor in a multiplication problem. Shade that part of the rows of the decimal square with a red crayon or marker.

• Your partner spins the pointer, and uses a blue crayon or marker to shade that part of the columns in the same decimal square. He or she records that number as the second factor in the problem and writes the product on the index card.

• You check the product by counting the number of squares that are shaded by both colors.

• Alternate roles until you have made ten different models.

• Shuffle all 20 cards, and lay them face down in an array.

• Turn over two cards at a time. If the model and the problem match, keep the cards and take another turn. If the cards do not match, turn them face down again.

• Alternate turns until all the cards have been matched. The player with more matching pairs of cards is the winner.

SMALL GROUPS

What You Will Learn **A** and **B** To multiply with decimals (tenths and hundredths)

What You Need **A** and **B** 16 index cards, markers, 3-minute timer (or stopwatch), paper, pencils, calculators

Getting Ready **A** Write these products on the cards 1.2, 0.12, 2.4, 0.24, 3.6, 0.36, 4.8, and 0.48. **B** Write these products on the cards 0.4, 0.04, 0.6, 0.06, 6.4, 0.64, 0.8, and 0.08.

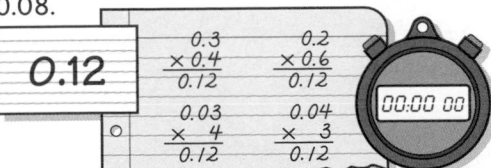

Problems for Products

What You Will Do

• Shuffle the product cards, and place them face down in a stack. Turn over the top card, and start the 3-minute timer.

• Each player writes multiplication problems that have products that match the product on the card.

• When 3 minutes is up, each player uses a calculator to check the problems of the player to the right. Score 1 point for each correct problem.

• Take turns until all the cards have been played. The player with the most points is the winner.

• Play the game with the new set of cards.

Activity 11A

SMALL GROUPS

What You Will Learn **A** To add, subtract, multiply, and divide with decimals **B** To add and subtract with decimals, or to multiply and divide with decimals

What You Need **A** and **B** blank number cube (TR p. R71), markers, gameboard (TR p. R105), oaktag, glue, counters, paper, pencils, calculator

Getting Ready **A** and **B** Write the numbers 1, 1, 2, 2, 3, and 3 on the faces of the cube. **A** Write a decimal problem in each of the spaces on the gameboard. Glue the gameboard on to the oaktag. **B** Write problems for addition and subtraction of decimals, or problems for multiplication and division of decimals.

Racing to the Finish Line

What You Will Do

• Take turns tossing the number cube, moving that number of spaces, and writing the problem and your answer. The player on the right uses a calculator to check your answer. If the answer is correct, stay on the space. If the answer is incorrect, move back two spaces.

• Take turns. The first player to land on Finish is the winner.

Activity 11B

• Play the game with the new gameboard.

	Start								
	13.5 − 2.7	8.09 × 3.5	12.74 +9.20	1.78 ×0.34	8.342 −1.651	6)49.2	0.625 ×0.24	7.63 +2.79	0.6 × 0.3
									9.0 − 3.6
0.08 × 0.5	2)19.62	9.72 +4.84	3)0.456	37 −14.3	2.62 +4.79	6.5 × 3.2	4.95 −2.007	7.42 ×0.28	
14.3 − 6.9									
2.7 × 4.6	14.36 +2.07	6.40 − 2.19	4.52 × 3.2	3.69 +2.05	5)2.45	1.35 −0.79	2.35 −0.98	4.68 × 3.5	
									Finish

Activity 12A

PARTNERS

What You Will Learn **A** To find multiples of 2, 3, 5, 7, 8, and 12 **B** To find numbers between 1 and 100 that are not multiples of 2, 3, 5, 7, 8, or 12

What You Need **A** and **B** blank number cube (TR p. R71), marker, hundred number table (TR p. R5)

Getting Ready **A** and **B** Label the faces of the cube 2, 3, 5, 7, 8, and 12.

Multiples Game

What You Will Do

• The first player rolls the number cube and marks an X over four multiples of the number rolled.

• The second player rolls the number cube and marks an O over four multiples of the number rolled that have not already been marked.

• Play continues until each player goes for two rounds without being able to mark a multiple. The player with more numbers marked is the winner.

Activity 12B

• Play until there are exactly 22 numbers unmarked.

1	2	3	4	5	6	7	8	9	10
11	12	13	14	15	16	17	18	19	20
21	22	23	24	25	26	27	28	29	30
31	32	33	34	35	36	37	38	39	40
41	42	43	44	45	46	47	48	49	50
51	52	53	54	55	56	57	58	59	60
61	62	63	64	65	66	67	68	69	70
71	72	73	74	75	76	77	78	79	80
81	82	83	84	85	86	87	88	89	90
91	92	93	94	95	96	97	98	99	100

Activity 13A

PARTNERS

What You Will Learn **A** To find factors of a number **B** To find the greatest common factor of two numbers

What You Need **A** and **B** two blank number cubes (TR p. R71), markers, centimeter cubes, paper, pencils

Getting Ready **A** and **B** Label the faces of both number cubes 1–6.

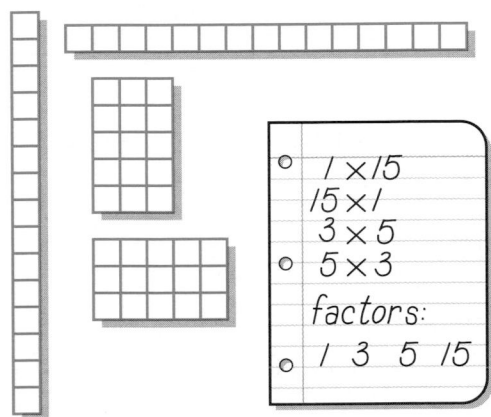

Listing Factors

What You Will Do

- Toss both cubes, and use the numbers to make the 2-digit number you think has more factors. Use the cubes to make rectangular arrangements that show all the ways to factor the number. Write the related multiplication facts for the arrangements, and list all the factors of your number.

- Your partner checks your work. You score 1 point for each correct factor. Your partner scores 2 points for any additional factors he or she finds. Alternate roles. The player with more points after five rounds is the winner.

Activity 13B

- At the end of each round, compare the arrangements you have both made, and find the greatest common factor of the two numbers used in the round.

Activity 14A

PARTNERS

What You Will Learn **A** To use divisibility rules for 2, 3, 5, and 10 **B** To identify divisibility rules for 2, 3, 5, and 10

What You Need **A** and **B** eight index cards, scissors, markers, four-section spinner (TR p. R66), counters

Getting Ready **A** and **B** Cut each card in half. Write these numbers on the cards 6, 10, 12, 15, 18, 20, 24, 30, 36, 40, 42, 45, 48, 50, 56, and 60. Label the sections of the spinner 2, 3, 5, and 10.

Does It Divide Evenly?

What You Will Do

- Place the cards face down in a stack. Spin the pointer, take the top card, and tell whether the number on the card is divisible by the spinner number.

- Your partner checks the answer by taking counters for the number on the card and dividing the counters into groups. Each group must have the number of counters shown on the spinner. If the number on the card is divisible by the number on the spinner, your partner will have no counters left over. Score 1 point for each correct answer. Alternate roles, and play until both players have scored at least 5 points.

Activity 14B

- Name the divisibility rule you used to decide whether the number was divisible.

Activity 15A

SMALL GROUPS

What You Will Learn **A** and **B** To match equivalent fractions

What You Need **A** and **B** 18 index cards, scissors, markers

Getting Ready **A** and **B** Prepare a set of domino cards by cutting each card in half. Draw a line that divides each card in half, then write one fraction on each half.

$\frac{3}{6}, \frac{4}{10}, \frac{10}{16}, \frac{4}{6}, \frac{10}{12}, \frac{2}{6}, \frac{3}{4}, \frac{1}{3}, \frac{15}{18}, \frac{15}{24}, \frac{3}{9}, \frac{20}{24},$
$\frac{5}{6}, \frac{2}{3}, \frac{6}{9}, \frac{30}{36}, \frac{5}{8}, \frac{10}{25}, \frac{25}{30}, \frac{6}{8}, \frac{9}{12}, \frac{10}{15}, \frac{4}{8}, \frac{8}{20},$
$\frac{4}{12}, \frac{8}{12}, \frac{12}{18}, \frac{5}{15}, \frac{25}{40}, \frac{6}{12}, \frac{5}{10}, \frac{15}{20}, \frac{12}{30}, \frac{6}{18},$
$\frac{6}{15};$ and $\frac{2}{5}.$

Fraction Dominoes

What You Will Do

- Place the cards face down in a stack. Each player in turn takes four cards. Turn the top card of the stack face up, and place it in the center of the table.

- The first player matches an equivalent fraction to either side of the domino that is face up. If no match can be made, the player draws cards until a match can be made.

- Take turns. Continue playing until one player has no more cards. He or she is the winner.

Activity 15B

- Before you lay down your domino card, name the simplest form of the fractions you are matching.

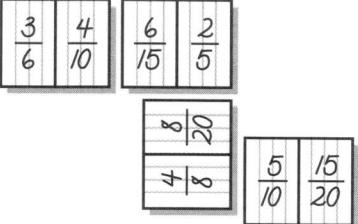

Activity 16A

SMALL GROUPS

What You Will Learn **A** and **B** To find equivalent fractions

What You Need **A** and **B** 48 index cards, markers **B** fraction circles (TR pp. R19–20), scissors, glue

Getting Ready **A** Write these fractions on 16 of the cards $\frac{1}{2}, \frac{1}{3}, \frac{2}{3}, \frac{1}{4}, \frac{3}{4}, \frac{1}{5}, \frac{3}{5}, \frac{1}{6}, \frac{5}{6}, \frac{1}{8}, \frac{3}{8},$ $\frac{5}{8}, \frac{7}{8}, \frac{1}{9}, \frac{5}{9},$ and $\frac{8}{9}.$ For each fraction, write two equivalent fractions on other cards. **B** Color a fraction circle for each of the 16 fractions listed. Glue the circles to the index cards.

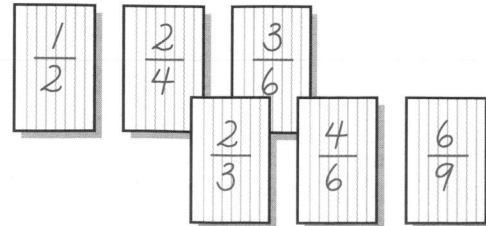

Fraction Match

What You Will Do

- Shuffle all the cards together. Deal six cards to each player, and put the remaining cards face down in a stack. The object of the game is to get sets of three equivalent fractions.

- The first player takes a card from the stack and chooses whether to discard it or keep it and discard one of the other cards in his or her hand. A discarded card is placed face up next to the stack.

- The next player can either pick up the discarded card or choose a card from the stack.

- Take turns. The first player to collect two sets of three matching cards calls "Fraction match," displays the matching sets, and scores 1 point.

- The first player to score 5 points wins the game.

Activity 16B

- Make sets of three cards—one showing a fraction circle and the other two showing fractions equivalent to the model.

Activity 17A

PARTNERS

What You Will Learn **A** To compare fractions **B** To order fractions

What You Need **A** and **B** forty 3 × 5 index cards, markers, paper, pencils

Getting Ready **A** and **B** Make four sets of number cards for 1–10.

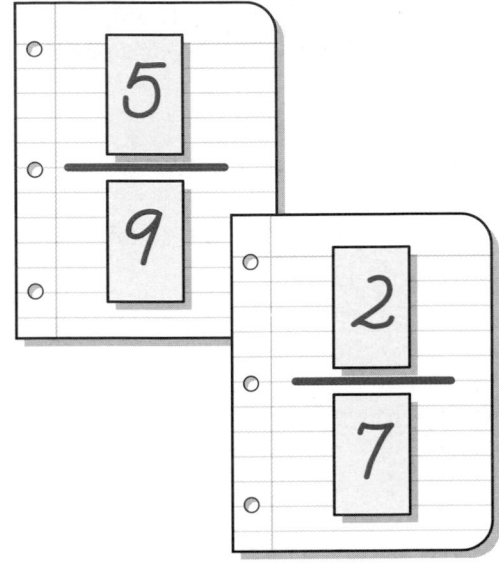

Dare to Compare

What You Will Do

• Shuffle all the cards together, and deal 20 cards to each player. Each player places his or her cards face down in a stack. Each player then draws a fraction line on his or her paper.

• At the same time, the players turn over two cards from their stack and arrange them on the paper with the fraction line to form a fraction, with the greater number as the denominator.

• The player with the greater fraction takes all four cards and adds them to the bottom of his or her stack. If players need help in comparing the two fractions, they change them to fractions with like denominators.

• If the fractions are equivalent, each player turns over two more cards and builds another fraction. The player with the greater fraction wins all the cards showing.

• Play ten rounds or until one player has no cards. The player who collects more cards wins.

Activity 17B

• Turn over six cards, and work together to form three fractions. Order the fractions from least to greatest.

Activity 18A

SMALL GROUPS

What You Will Learn **A** To rename improper fractions **B** To order mixed numbers

What You Need **A** and **B** nine-section spinner (TR p. R69)

Getting Ready **A** and **B** Label the sections of the spinner 1–9.

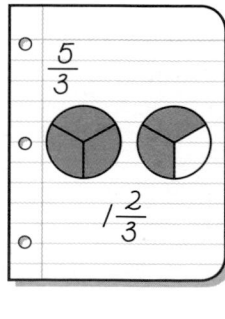

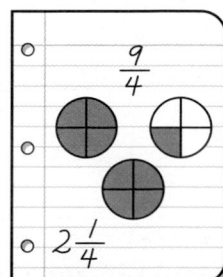

Make the Most of It

What You Will Do

• Each player spins the pointer twice, uses the numbers to write an improper fraction, and then draws fraction circles to model the fraction. The player then writes a mixed number equivalent to his or her fraction.

• Each player checks the fraction and mixed number of the player to his or her right.

• The player with the greatest mixed number scores 1 point for the round.

• Continue playing until one player has scored 5 points.

Activity 18B

• After each round, order from least to greatest all the mixed numbers you have written.

Activity 19A

PARTNERS

What You Will Learn A and B To model the multiplication of fractions

What You Need A and B four blank number cubes (TR p. R71), markers, recording sheet of squares (TR p. R106), different colored pencils

Getting Ready A Label the number cubes 1–6. B Label one number cube 1–6 and the other number cube 5–10.

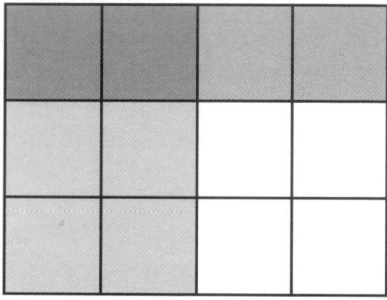

$$\frac{2}{4} \times \frac{1}{3} = \frac{2}{12}$$

Picturing Fraction Products

What You Will Do

• Toss the two cubes. Use the numbers to write a proper fraction on the line below a square on the recording sheet. That fraction will be the first factor in a multiplication problem. Draw vertical lines and shade the square to model the first factor.

• Your partner tosses the two cubes and writes a proper fraction that becomes the second factor in the problem. He or she then draws horizontal lines in the square and shades it to model the second factor.

• Use the model to find the product of the factors.

• Alternate roles, and continue the activity.

Activity 19B

• Use the new number cubes to play the game.

Activity 20A

PARTNERS

What You Will Learn A and B To add, subtract, and multiply fractions

What You Need A and B blank number cubes (TR p. R71), markers, 24 index cards, ladder gameboard (TR p. R107), 9 × 12 oaktag, counters, calculator with fraction function

Getting Ready A and B Write the numbers 1, 1, 2, 2, 3, and 3 on the faces of the cube. Glue the gameboard on to the oaktag. A Write an addition, subtraction, or multiplication of fractions problem on 12 cards. B Make a set of cards for only one fraction operation.

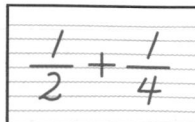

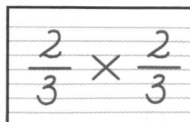

Climb the Ladder

What You Will Do

• Shuffle the cards, and place them face down in a stack.

• Toss the number cube, and move that number of rungs up the ladder. Take the top card from the stack, and solve the problem on it.

• Your partner checks the answer with the calculator. If the answer is correct, stay on the rung. If the answer is incorrect, move down the ladder two rungs.

• Alternate roles. The first player to reach the top wins.

Activity 20B

• Play the game with the new set of cards.

Activity 21A

What You Will Learn **A** To subtract mixed numbers with like and unlike denominators **B** To add mixed numbers with like and unlike denominators

What You Need **A** and **B** blank number cube (TR p. R71), target gameboard (TR p. R109), markers of two different colors, calculator

Getting Ready **A** and **B** Write the following mixed numbers on the faces of the number cube $4\frac{2}{5}$, $4\frac{1}{2}$, $5\frac{1}{4}$, $5\frac{1}{3}$, $6\frac{2}{3}$, and $6\frac{3}{4}$. Write these numbers in the middle ring of the target gameboard $1\frac{3}{3}$, $1\frac{2}{2}$, $1\frac{8}{8}$, $2\frac{4}{4}$, $2\frac{5}{5}$, $2\frac{6}{6}$, $3\frac{6}{6}$, and $3\frac{4}{4}$.

Hit the Target Number

What You Will Do
- Toss the cube, and write the number in pencil in the center of the target. This number will be the sum of the numbers in the middle and outer rings.

- Write a missing addend, in simplest form, in one section of the target, and pass the target to your partner. Your partner checks the answer, using the calculator if necessary.

- Alternate roles, each player using a different-colored marker. Score 1 point for each correct answer. When the target is completed, the player with more points wins the round.

Activity 21B

- Use the center and middle ring numbers as addends, and write the sums in the outer ring.

Activity 22A

What You Will Learn **A** and **B** To define and write ratios

What You Need **A** and **B** ten colored cubes, ten colored cubes of a second color, paper bag, two blank number cubes (TR p. R71) **B** ten colored cubes of a third color

Getting Ready **A** and **B** Label each number cube 1–6.

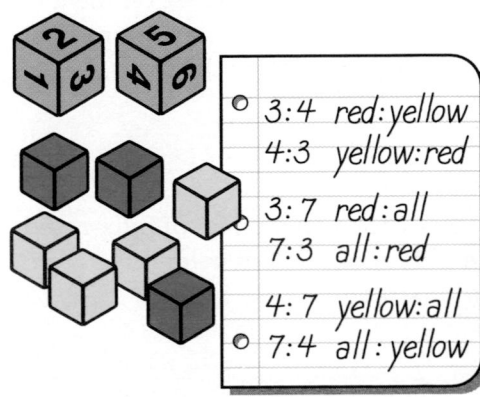

3:4	red:yellow
4:3	yellow:red
3:7	red:all
7:3	all:red
4:7	yellow:all
7:4	all:yellow

Cube Ratios

What You Will Do
- Put the colored cubes in the bag.

- Toss the number cubes, name the sum, and take that number of colored cubes from the bag. Write a ratio comparing the colored cubes, and pass the paper to your partner.

- Your partner writes a description of the ratio, writes a different ratio comparing the colored cubes, and passes the paper back to you.

- You write a description of that ratio and pass the paper back to your partner.

- Continue until neither of you can write another ratio.

Activity 22B

- Add ten cubes of a third color to the bag, and write as many ratios as you can to compare the three sets.

Activity 23A

SMALL GROUPS

What You Will Learn A and **B** To write equivalent ratios

What You Need A and **B**
4-in. × 4-in. squares of paper, three different-colored markers, paper, pencils

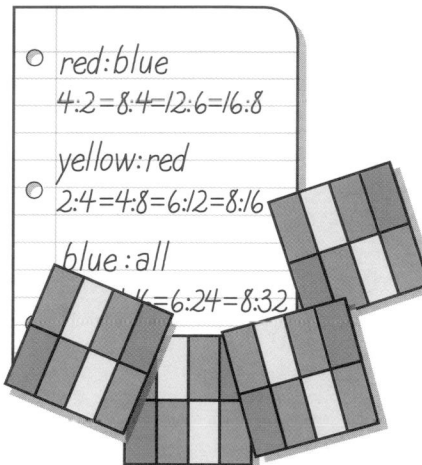

red:blue
4:2=8:4=12:6=16:8

yellow:red
2:4=4:8=6:12=8:16

blue:all
...6=6:24=8:32

Pattern It!

What You Will Do

• Each player folds a square of paper in half three times to form eight equal sections. The first player colors his or her square using one of the three different-colored markers for each section. The other players color their squares to match the square of the first player. (If there are only two players, each one completes two squares.)

• The first player writes three different ratios to compare the colors in his or her square, then passes his or her square and the paper to the right. The second player writes equivalent ratios represented by the two squares, then passes both squares and the paper to the right. Continue passing squares and writing equivalent ratios until at least four equivalent ratios have been written for each original ratio.

Activity 23B

• The first player writes as many different ratios as possible for his or her square. The other players write equivalent ratios for all the original ratios.

Activity 24A

SMALL GROUPS

What You Will Learn A
To relate ratio and percent
B To model ratios and percents

What You Need A and **B**
decimal squares in hundredths (TR p. R7), three different-colored markers, paper, pencils

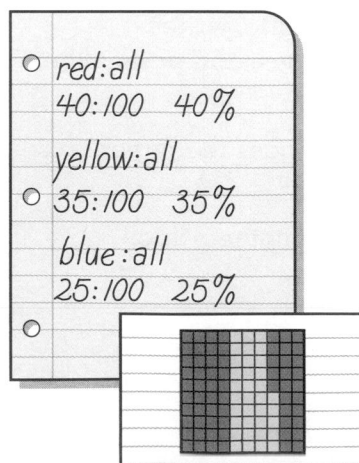

red:all
40:100   40%

yellow:all
35:100   35%

blue:all
25:100   25%

Modeling Percents

What You Will Do

• Each player takes a decimal square and colors some squares each of three different colors. On the back of the decimal square, each player writes the following ratios for the grid: squares colored in first color to all squares, squares colored in second color to all squares, squares colored in third color to all squares.

• Next to the ratios, they write the percent of the grid that is that color. Pass the grids, color side up, to the player on the left.

• Each player writes the ratios and percents for the new grid on a sheet of paper, then compares them to the numbers on the back of the grid. Discuss any disagreements and make sure that the numbers on the backs of the grids are correct. Continue to pass grids to the left until each player has written the ratios and percents of all of the grids.

Activity 24B

• Each player writes the three ratios listed above on a sheet of paper and passes it to the player on the left. The player on the left colors a decimal square to model the ratios and percents.

Activity 25A

PARTNERS

What You Will Learn **A** To name the ratio and percent for a model; to write equivalent ratios **B** To order percent models

What You Need **A** and **B**
16 decimal squares in hundredths (TR p. R7), markers, poster board, glue, paper, pencils

Getting Ready **A** and **B** Color the decimal squares to show these percents 0%, 8%, 10%, 15%, 20%, 25%, 30%, 38%, 42%, 48%, 56%, 68%, 75%, 80%, 94%, and 100%. Glue each grid on to poster board and cut out.

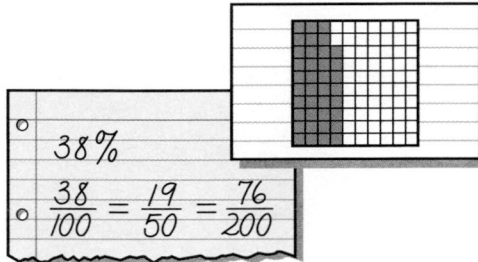

Name the Grid

What You Will Do

• Shuffle the cards, and place them face down in a stack. Turn over the top card.

• Write the percent of the model that is colored. Your partner writes the ratio of the colored squares to the total number of squares of the model.

• Together you and your partner write two more ratios equivalent to the first ratio.

• Alternate turning over cards until all the cards have been used.

Activity 25B

• Turn over four cards, and order the percents from least to greatest.

C Scope & Sequence Correlations

This section contains the following:

- **Scope and Sequence**
  The scope and sequence shows the development of all strands of math across the grades—Kindergarten through Grade 8. In addition, each grade level has a detailed scope and sequence for that grade level.

- **Correlation to Standardized Tests**
  The standardized test correlations will assist you as you prepare children for the following standardized tests:
  - CAT - California Achievement Test
  - CTBS - Comprehensive Tests of Basic Skills
  - ITBS - Iowa Tests of Basic Skills
  - MAT - Metropolitan Achievement Test
  - SAT - Stanford Achievement Test

- **Correlation to Commercial Manipulatives**
  The manipulative correlation shows you the grade levels at which commercial manipulatives are used.

- **Correlation to Readily Available Materials**
  This correlation provides information on how you can use readily available materials should you not have all of the manipulatives used in the lessons.

- **Correlation to Technology**
  The technology correlation shows the chapters in which the technology developed to link directly to the content in Math Advantage is used.

Scope and Sequence

PROBLEM SOLVING, REASONING, AND PROCESSES

PROBLEM SOLVING	K	1	2	3	4	5	6	7	8
Heuristic		■	●	●	●	●	●	●	●
Strategies									
Write a number sentence		●	●	●	●	●			
Make a model		●	●	●	●	●	●	●	●
Draw a picture/diagram		●	●	●	●	●	●	●	●
Act it out	■	●	●	●	●	●	●	●	●
Use or make a table/chart/graph	■	●	●	●	●	●	●	●	●
Work backward				●	●	●	●	●	●
Use guess and check		●	●	●	●	●	●	●	●
Find a pattern	■	●	●	●	●	●	●	●	●
Account for all possibilities							●	●	●
Solve a simpler problem							●	●	●
Write an equation/proportion							●	●	●
Use a formula						●	●	●	●
Skills									
Too much information			●	●	●	●	▲	▲	▲
Choose the operation		●	●	●	●	●	▲	▲	▲
Choose a strategy			●	▲	▲	▲	▲	▲	▲
Choose a model		●	●	▲	▲	▲	▲	▲	▲
Make an organized list					●	●	▲	▲	▲
Interpret the remainder					●	●	▲	▲	▲
Formulate problems and questions		●	●	●	●	●	●	●	●
Solve multistep problems			■	●	●	●	▲	▲	▲
Solve nonroutine problems		●	●	●	●	●	●	●	●
Applications									
Number and quantitative reasoning	■	●	●	●	●	●	●	●	●
Operation and quantitative reasoning	■	●	●	●	●	●	●	●	●
Algebra and quantitative reasoning	■	●	●	●	●	●	●	●	●
Patterns, relationships, and algebraic thinking	■	●	●	●	●	●	●	●	●
Geometry and spatial reasoning	■	●	●	●	●	●	●	●	●
Measurement and time	■	●	●	●	●	●	●	●	●
Statistics and probability	■	●	●	●	●	●	●	●	●

■ **Introduce Concept Only** ● **Teach and Test** ▲ **Reinforce**

PROBLEM SOLVING, REASONING, AND PROCESSES

PROBLEM SOLVING

Heuristic 1–13, 36–37, 140–141, 180–181, 198–199, 258–259, 306–307, 328–329, 334–335, 376–377, 406–407, 504–505

Strategies

Make a model 4, 180–181, 406–407

Draw a picture/diagram 2, 141, 198–199, 258–259, 335, 376–377, 407, 504–505

Act it out 3, 37, 181, 258–259, 334–335, 377, 504–505

Use or make a table/chart/graph 9, 36–37, 141, 198–199, 258–259, 282–283, 306–307, 335, 377, 407, 505

Work backward 6, 140–141, 181, 199, 307, 329, 335, 407

Use *guess and check* 5, 36–37, 141, 181, 199, 259, 307, 329, 335, 377, 505

Find a pattern 8, 37, 198–199, 259, 307, 329, 407, 504–505

Account for all possibilities 7, 258–259, 377

Solve a simpler problem 10, 37, 198–199

Write an equation/proportion 12, 37, 141, 181, 199, 259, 307, 328–329, 335, 377, 407, 505

Use a formula 11, 141, 199, 302–305, 329, 406–407, 505

Skills

Too much information 208–209, 210–211

Choose the operation 1–13, 36–37, 140–141, 180–181, 198–199, 258–259, 306–307, 328–329, 334–335, 376–377, 406–407, 504–505

Choose a strategy 13, 37, 141, 181, 199, 259, 307, 329, 335, 377, 407, 505

Choose a model 4, 180–181, 406–407

Make an organized list 7, 9, 37, 258–259, 306–307

Interpret the remainder 41–43, 61

Formulate problems and questions 37, 50, 207–211, 259, 307, 335, 407

Solve multistep problems 5, 11, 12, 13, 36, 44, 45, 53, 123, 127, 139, 140, 141, 181, 199, 254, 259, 281, 283, 302, 303, 307, 328, 329, 330, 335, 337, 339, 342, 343, 347, 361, 363, 364, 369, 370, 371, 374, 375, 377, 389, 391, 407, 435, 479, 483

Solve nonroutine problems 3, 6, 7, 9, 11, 14, 36–37, 87, 140–41, 153, 198–199, 235, 279, 334–335, 345, 362–363, 519

Applications

Number and quantitative reasoning 36–37, 140–141, 306–307, 334–335

Operation and quantitative reasoning 1–13, 36–37, 140–141, 180–181, 198–199, 258–259, 306–307, 328–329, 334–335, 376–377, 406–407, 504–505

Algebra and quantitative reasoning 306–307, 328–329, 406–407

Patterns, relationships, and algebraic thinking 180–181, 198–199, 504–505

Geometry and spatial reasoning 180–181, 198–199, 406–407

Measurement and time 304–305, 306–307, 376–377, 406–407

Statistics and probability 258–259

▶ Type printed in red indicates that a topic is being introduced for the first time.

Scope & Sequence/Correlations

C1

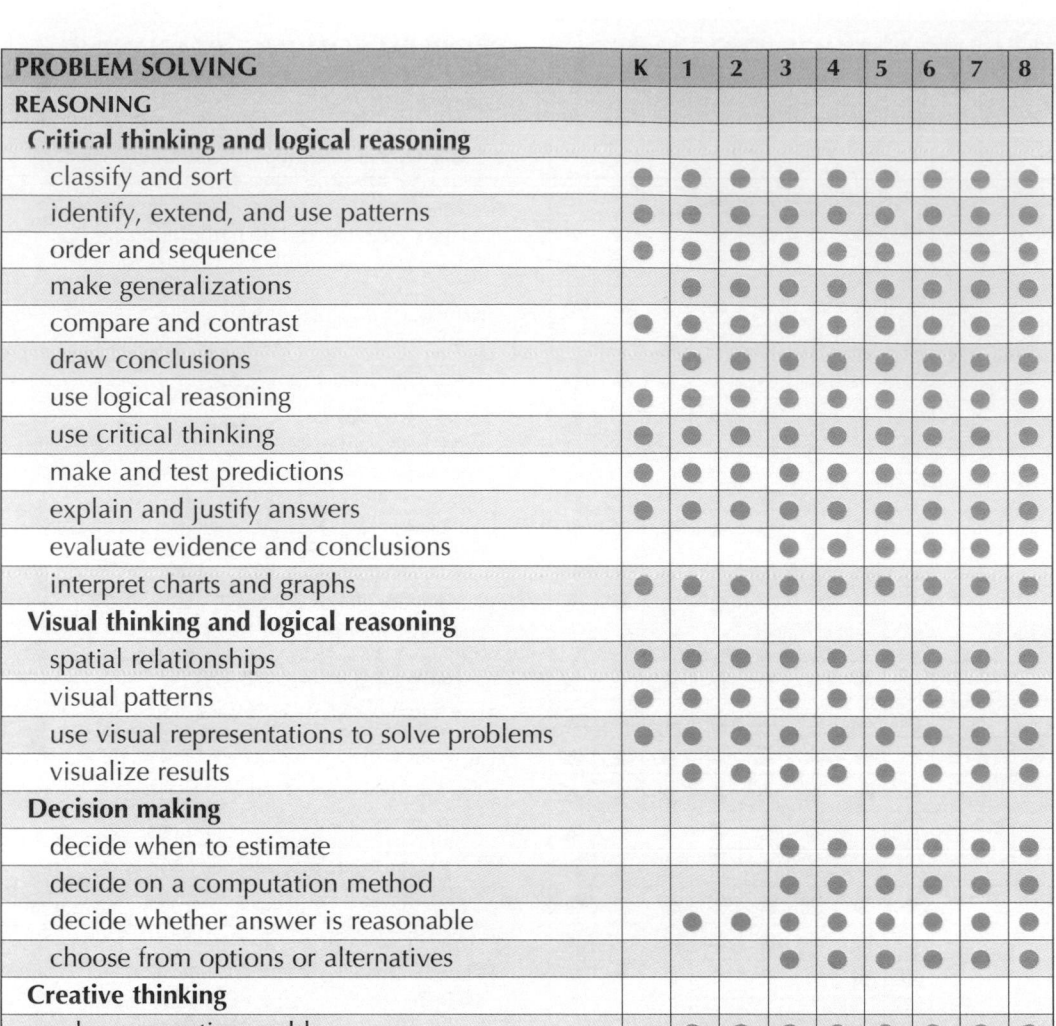

PROBLEM SOLVING	K	1	2	3	4	5	6	7	8
REASONING									
Critical thinking and logical reasoning									
classify and sort	●	●	●	●	●	●	●	●	●
identify, extend, and use patterns	●	●	●	●	●	●	●	●	●
order and sequence	●	●	●	●	●	●	●	●	●
make generalizations		●	●	●	●	●	●	●	●
compare and contrast	●	●	●	●	●	●	●	●	●
draw conclusions		●	●	●	●	●	●	●	●
use logical reasoning	●	●	●	●	●	●	●	●	●
use critical thinking	●	●	●	●	●	●	●	●	●
make and test predictions	●	●	●	●	●	●	●	●	●
explain and justify answers	●	●	●	●	●	●	●	●	●
evaluate evidence and conclusions				●	●	●	●	●	●
interpret charts and graphs	●	●	●	●	●	●	●	●	●
Visual thinking and logical reasoning									
spatial relationships	●	●	●	●	●	●	●	●	●
visual patterns	●	●	●	●	●	●	●	●	●
use visual representations to solve problems	●	●	●	●	●	●	●	●	●
visualize results		●	●	●	●	●	●	●	●
Decision making									
decide when to estimate				●	●	●	●	●	●
decide on a computation method				●	●	●	●	●	●
decide whether answer is reasonable		●	●	●	●	●	●	●	●
choose from options or alternatives				●	●	●	●	●	●
Creative thinking									
solve nonroutine problems		●	●	●	●	●	●	●	●
generate problems	●	●	●	●	●	●	●	●	●
choose alternative ways to solve problems			●	●	●	●	●	●	●

■ **Introduce Concept Only**　　● **Teach and Test**　　▲ **Reinforce**

PROBLEM SOLVING
REASONING
Critical thinking and logical reasoning
　classify and sort 80, 162, 448
　identify, extend, and use patterns 23, 171, 320, 347, 470, 518, 524, 527
　order and sequence 16, 22, 454
　make generalizations 80, 100, 104, 175, 318, 347, 396, 398, 471
　compare and contrast 22, 65, 129, 187, 286, 357, 369, 487, 501
　draw conclusions 42, 47, 58, 100, 244–245, 490
　use logical reasoning 23, 27, 175, 177, 190, 230, 263, 289, 304, 337, 398, 424, 425, 489, 502
　use critical thinking 16, 22, 23, 26, 27, 42, 47, 58, 65, 67, 79, 80, 82, 100, 104, 119, 120, 129, 132, 155, 161, 162, 169, 171, 175, 179, 187, 190, 217, 218, 225, 230, 248, 252, 253, 262, 263, 264, 285, 286, 300, 304, 318, 320, 323, 337, 342, 347, 357, 362, 369, 370, 396, 398, 408, 411, 416, 424, 428, 448, 454, 455, 469, 470, 471, 472, 487, 490, 501, 502, 518, 524, 527
　make and test predictions 244–245, 262, 264, 428
　explain and justify answers 22, 27, 47, 67, 80, 132, 155, 179, 218, 286, 323, 357, 362, 396, 398, 408, 469, 472, 490, 527
　evaluate evidence and conclusions 217, 218, 262
　interpret charts and graphs 217, 218, 225, 252, 490
Visual thinking and logical reasoning
　spatial relationships 174–177, 261–262, 320–321, 324–325, 346–347, 369, 370, 372–373, 500–503
　visual patterns 178–181, 352–353, 504–505, 506–507, 508–509, 510–511
　use visual representations to solve problems 26–27, 58–59, 62–63, 84–85, 96–99, 112–117, 128–129, 136, 172–173, 192–193, 284–285, 294–295, 326–327, 376–377, 422–423, 426, 428, 430–431, 464–465, 466–467, 468–469, 470–471, 482–483
　visualize results 194–195, 196–197, 404–405
Decision making
　decide when to estimate 46–49, 58–59, 104–107, 122–123, 367, 404–405
　decide on a computation method 13, 34–35, 37, 79, 85, 101, 131, 141, 181, 199, 259, 307, 329, 335, 377, 407, 505
　decide whether answer is reasonable 47, 58, 106, 107, 242–243
　choose from options or alternatives 23, 34–35, 38–39, 48, 82, 88, 177, 225–227, 296–298, 320–321, 327
Creative thinking
　solve nonroutine problems 3, 6, 7, 9, 11, 14, 36–37, 87, 140–141, 153, 198–199, 235, 279, 334–335, 345, 362–363, 519
　generate problems 37, 50, 207–211, 259, 307, 335, 407, 519
　choose alternative ways to solve problems 42, 48, 78, 82, 88, 208–209, 297, 320–322, 327

Type printed in red indicates that a topic is being introduced for the first time.

PROCESSES	K	1	2	3	4	5	6	7	8
COMMUNICATION									
Drawing	●	●	●	●	●	●	●	●	●
Writing		●	●	●	●	●	●	●	●
Talking	●	●	●	●	●	●	●	●	●
CONNECTIONS									
Mathematical	●	●	●	●	●	●	●	●	●
Cross-curricular	●	●	●	●	●	●	●	●	●
Everyday	●	●	●	●	●	●	●	●	●
Cultural	●	●	●	●	●	●	●	●	●
MULTIPLE REPRESENTATIONS									
Different manipulatives	●	●	●	●	●	●	●	●	●
Different models	●	●	●	●	●	●	●	●	●
Manipulatives and models	●	●	●	●	●	●	●	●	●
Manipulatives, words, and symbols	●	●	●	●	●	●	●	●	●
Models, words, and symbols	●	●	●	●	●	●	●	●	●
TOOLS									
Calculator	■	●	●	●	●	●	●	●	●
Software	●	●	●	●	●	●	●	●	●
Manipulatives	●	●	●	●	●	●	●	●	●
Measuring tools	●	●	●	●	●	●	●	●	●
Compass						■	●	●	●

■ Introduce Concept Only   ● Teach and Test   ▲ Reinforce

PROCESSES

COMMUNICATION

Drawing 58–59, 62–63, 98, 112, 116, 151, 156–159, 160–161, 172, 174–177, 178–179, 180–181, 194–195, 352, 368, 372–373, 376–377, 378–379, 500–503, 506–507, 508–509

Writing 35, 37, 39, 43, 49, 55, 57, 61, 77, 79, 87, 121, 135, 139, 163, 171, 189, 195, 227, 231, 241, 259, 265, 279, 289, 299, 305, 323, 325, 339, 359, 389, 397, 405, 407, 425, 435, 447, 453, 469, 471, 485, 503, 507, 523, 525

Talking 22, 26, 38, 42, 47, 59, 65, 80, 106, 119, 132, 152, 175, 178, 186, 190, 208, 215, 226, 233, 240, 242, 251, 262, 268, 286, 297, 300, 318, 336, 337, 354, 357, 362, 369, 370, 396, 408, 412, 424, 426, 448, 450, 468, 472, 478, 480, 486, 501

CONNECTIONS

Mathematical 23, 43, 67, 87, 103, 135, 155, 179, 191, 219, 235, 253, 265, 281, 303, 323, 345, 359, 371, 397, 411, 433, 453, 465, 481, 509, 527

Cross-curricular 20, 28, 34, 47, 60, 64, 78, 98, 100, 105, 122, 129, 130, 132, 150, 152, 154, 178, 187, 192, 217, 230, 234, 240, 244, 246, 248, 258, 261, 268, 278, 286, 300, 302, 304, 318, 320, 337, 357, 358, 360, 362, 370, 378, 388, 390, 396, 409, 412, 424, 446, 448, 451, 464, 468, 470, 472, 478, 487, 502, 506, 508, 516, 526

Everyday 18, 56, 66, 80, 86, 88, 102, 116, 138, 156, 170, 174, 176, 190, 194, 208, 210, 212, 225, 226, 242, 252, 260, 264, 282, 297, 298, 316, 322, 324, 342, 344, 368, 369, 370, 392, 406, 416, 427, 428, 432, 450, 456, 486, 517, 524

Cultural 40, 50, 58, 90, 124, 157, 164, 200, 236, 270, 308, 336, 348, 380, 408, 418, 458, 492, 528

MULTIPLE REPRESENTATIONS

Different manipulatives 84–85, 96–97, 114–115, 136–137

Different models 264–265, 352–353

Manipulatives and models 96–99, 196–197, 352–353, 462–465, 466–469

Manipulatives, words, and symbols 84–85, 96–97, 114–115, 136–137, 284–285, 294–295, 326–327, 336, 462–463, 466–467, 482–483

Models, words, and symbols 26–27, 58–59, 62–63, 128–129, 346–347, 422–423, 426–428

TOOLS

Calculator 20–21, 26–27, 42, 44–45, 65, 87, 88, 134–135, 322, 338, 399, 416, 450, 452, 470, 520–521, H30–H41

Software 23, 27, 45, 63, 85, 97, 115, 137, 161, 173, 197, 215, 219, 233, 235, 251, 253, 267, 285, 295, 327, 347, 353, 373, 399, 415, 431, 445, 467, 483, 511, 521

Manipulatives 84–85, 96–97, 114–115, 136–137, 162, 178, 194, 196–197, 260–262, 284–285, 294–295, 336, 338, 352–353, 422–423, 426–428, 462–463, 466–467, 474, 482–483, 512, 516

Measuring tools 154–155, 160–161, 177, 234–235, 368–369, 374–375, 376–377, 392–393, 396, 398–399, 508

Compass 156–159, 160–161, 234–235, 398–399, 414–415

Type printed in red indicates that a topic is being introduced for the first time.

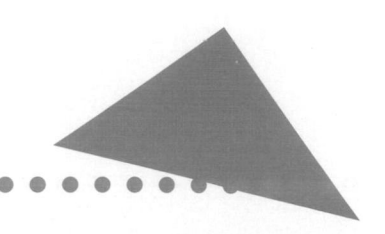

NUMBER, OPERATION, QUANTITATIVE REASONING, AND ALGEBRA

NUMBER AND QUANTITATIVE REASONING	K	1	2	3	4	5	6	7	8
WHOLE NUMBERS									
Meaning of numbers	●	●	●	●	●	▲	▲		
Read and write numbers	●	●	●	●	●	●	▲		
Count	●	●	●	●					
Ordinal numbers	●	●	●	●	●				
Compare and order	●	●	●	●	●	●	▲	▲	▲
Even/odd		●	●	●					
Place value	■	●	●	●	●	●	▲	▲	▲
Estimate quantities		●	●	●	●	●	▲	▲	▲
Rounding			■	●	▲	▲	▲	▲	▲
Multiples				■	●	●	●	▲	▲
Divisibility						■	●	●	▲
Prime and composite						■	▲	▲	▲
Least common multiple							●	●	▲
Common factors							●	●	▲
Greatest common factor							●	●	▲
Powers and exponents						■	●	●	●
Prime factors							●	●	●
Prime factorization							●	●	●
Square numbers and square roots							●	●	●
MONEY									
Identify coins	●	●	●	▲					
Value of coins	●	●	●	▲	▲	▲			
count and trade amounts		●	●	●	▲	▲			
make change				●	▲	▲			
Value of collection of coins and bills			●	▲	▲	▲			
count and trade amounts				●	▲	▲			
make change				●	▲	▲			
Compare amounts and prices			●	●	●	●	▲	▲	▲
Equivalent amounts			●	●	●	▲	▲	▲	▲
Relate to decimals				●	●	●			
DECIMALS									
Meaning of decimals				●	●	●	●	▲	▲
Read and write decimals				●	●	●	●	▲	▲
Decimals as fractions				●	●	●	●	●	●
Decimal place value				■	●	●	●	▲	▲
Compare and order				■	●	●	●	●	▲
Equivalent decimals					■	●	▲	▲	▲
Decimals and percent							●	●	●
Terminating/repeating decimals							●	●	●
Nonrepeating decimals							■	●	●
Scientific notation								●	●
FRACTIONS									
Part of a whole	●	●	●	●	●	▲	▲		
Part of a group	■	●	●	●	●	▲	▲		
Read and write fractions		●	●	●	●	▲	▲		

■ **Introduce Concept Only**   ● **Teach and Test**   ▲ **Reinforce**

NUMBER, OPERATION, QUANTITATIVE REASONING, AND ALGEBRA

NUMBER AND QUANTITATIVE REASONING

WHOLE NUMBERS

Meaning of numbers
  through millions 16–17
Read and write numbers 16–17
Compare and order 18–19
Place value
  through millions 16–17
Estimate quantities 46–49
Rounding 46–49, H9
Multiples 76–77, 80–83, 96, H5
Divisibility H4
Prime and composite 76–77
Least common multiple
  use a number line to find 80–83, 96, H5
Common factors 80–83
Greatest common factor
  use a number line to find 81–83, H5
  use prime factors to find 82–83
Powers and exponents
  powers of ten 24–25, 65, H12
  exponents 24–25
Prime factors
  identify 76–77, H2
  write with exponents 78–79, H4
  divide to find 78–79
  use a factor tree 78–79, H4
Prime factorization 78–79, H4
Square numbers and square roots
  squares 26
  square roots 27

MONEY

Compare amounts and prices
  discounts 342–343
  sales tax 338–339
Equivalent amounts
  relationships 300–301

DECIMALS

Meaning of decimals
  through ten thousandths 16–17
Read and write decimals 16–17
Decimals as fractions 20–21
Decimal place value
  through ten thousandths 16–17
Compare and order 18–19, 22–23, 30, H11
Equivalent decimals 18
Decimals and percent 322–323, H28
Terminating/repeating decimals 450–451
Nonrepeating decimals 450

FRACTIONS

Part of a whole H14
Part of a group H14
Read and write fractions H14

Type printed in red indicates that a topic is being introduced for the first time.

NUMBER AND QUANTITATIVE REASONING	K	1	2	3	4	5	6	7	8
Fractions as decimals				●	●	●	●	▲	▲
Compare, like denominators				●	●	●	▲		
Compare, unlike denominators				■	■	●	▲		
Fractions equal to 1 or greater				■	●	●	▲	▲	▲
Equivalent fractions				■	●	●	●	▲	▲
Order, like denominators						●	●	▲	
Order, unlike denominators						■	●	▲	
Simplest form					■	●	●	▲	▲
Least common denominator						●	●	▲	▲
Fractions and percent						●	●	●	●
Reciprocals							■	●	●
Rational numbers							■	●	●
INTEGERS									
Negative numbers					●	●	●	●	●
Meaning of integers							■	●	●
Integers on the number line							■	●	●
Absolute value							■	●	●
Compare and order							■	●	●
RATIONAL NUMBERS									
Meaning of rational numbers							■	●	●
Compare and order							■	●	●
Scientific notation								●	●
REAL NUMBERS									
Square and square root							●	●	●
Meaning of irrational numbers								■	●
Negative exponents								■	●
Density Property								■	●
NUMBER SENSE									
Meaning of whole numbers	●	●	●	●	●	●	●	▲	
Number relationships	●	●	●	●	●	●	●	●	●
Meaning of fractions			●	●	●	●	●	▲	▲
Meaning of decimals				●	●	●	▲	▲	▲
Equivalent forms of numbers				●	●	●	●	●	●
Effects of operations						●	●	●	●
Meaning of percent						●	●	●	●
Meaning of integers							■	●	●
Meaning of rational numbers							■	●	●
Meaning of real numbers									●
MENTAL MATH									
Skip-counting		●	●	▲	▲				
Use properties		●	●	●	●	●	▲	▲	▲
Patterns, multiples, and powers of 10				●	●	●	▲	▲	▲
Compatible numbers					●	●	●	●	▲
ESTIMATE QUANTITIES									
Benchmarks		●	●	●	●	●	▲	▲	▲
Rounding			■	●	▲	▲	▲	▲	▲

■ Introduce Concept Only   ● Teach and Test   ▲ Reinforce

Scope & Sequence/Correlations

Type printed in red indicates that a topic is being introduced for the first time.

OPERATION AND QUANTITATIVE REASONING	K	1	2	3	4	5	6	7	8
WHOLE NUMBERS									
Addition									
Meaning of addition	●	●	●	▲	▲	▲			
Addition facts		●	●	●	▲				
Basic-fact strategies									
counting on		●	●	●					
doubles, doubles plus one/minus one		●	●	●					
make a ten		●	●	●					
Fact families		●	●	●					
Properties		●	●	●	▲	▲	▲	▲	▲
3 or more addends		●	●	●	▲	▲	▲		
2-digit numbers, with/without regrouping		■	●	●	▲	▲	▲		
3-digit numbers			●	●	▲	▲	▲		
4-digit and greater numbers				■	●	●	▲		
Estimate sums				●	●	●	▲		
Subtraction									
Meaning of subtraction	●	●	●	●	●	▲	▲		
Subtraction facts		●	●	●	▲				
Basic-fact strategies									
counting back		●	●	●					
Missing addends		●	●	●	▲	▲	▲		
2-digit numbers, with/without regrouping		■	●	●	▲	▲	▲		
Checking subtraction			●	●	●	▲	▲		
With zeros				●	●	●	▲		
3-digit numbers			●	●	▲	▲	▲		
4-digit and greater numbers				■	●	●	▲		
Estimate differences				●	●	●	▲		
Multiplication									
Meaning of multiplication		●	●	●	●	▲			
Multiplication facts			■	●	●	▲			
Basic-fact strategies									
skip-counting				●	▲				
break-apart numbers				■	●	●			
pattern of nines				●	●				
Fact families				●	▲	▲			
1-digit factor, with/without regrouping				●	●	●	▲		
Estimate products					●	●	●		
Multiples of 10					●	●	▲		
2-digit factor					●	●	▲		
3-digit and greater factor						●	▲		
Exponents						■	●	●	●
Properties				●	●	●	▲	▲	▲
Division									
Meaning of division		●	●	●	●	▲			
Division facts				●	●	▲			
Missing factors				●	●	▲	▲		

■ **Introduce Concept Only** ● **Teach and Test** ▲ **Reinforce**

OPERATION AND QUANTITATIVE REASONING

WHOLE NUMBERS

Addition
  Properties
    Commutative 34–35, H6
    Associative 34–35, H6
  3 or more addends 34–35, 36–37, H7
  2-digit numbers, with/without regrouping 34–35, 36–37, H7
  3-digit numbers 36 37, H7
  4-digit and greater numbers 36–37, H7
  Estimate sums
    rounding 46–47
    clustering 46–47

Subtraction
  Meaning of subtraction H7
  Missing addends 35
  2-digit numbers, with/without regrouping 34–35, 36–37
  Checking subtraction 36–37
  With zeros 36–37, H7
  3-digit numbers 36–37, H7
  4-digit and greater numbers H7
  Estimate differences
    rounding 46–47

Multiplication
  1-digit factor, with/without regrouping 43
  Estimate products
    rounding 47–49
  Multiples of 10 H12
  2-digit factor
    symbolic 40–41, 42, H8
  3-digit and greater factor
    symbolic 41, 42, H8
  Exponents H12
  Properties
    Commutative 38–39, H6
    Associative 38–39, H6
    Distributive 38–39, H6
    Property of Zero H6
    Identity Property of One H6

Division
  Missing factors 39, 79, 83, 281, 282–283, 294–295, 296–299

Type printed in red indicates that a topic is being introduced for the first time.

OPERATION AND QUANTITATIVE REASONING	K	1	2	3	4	5	6	7	8
WHOLE NUMBERS									
Basic-fact strategies									
inverse operations				●	●	▲	▲		
Fact families				●	●	▲			
Remainders				●	●	●	▲		
Interpret remainders				●	●	●	▲		
1-digit divisor				●	●	●	▲		
2-digit divisor					●	●	▲		
Zeros in the quotient					●	●	▲		
Estimate quotients					●	●	●		
Divisibility						●	●	●	▲
MONEY									
Add		●	●	●	●	●	▲		
Subtract		●	●	●	●	●	▲		
Multiply					●	●	▲		
Divide					■	●	▲		
Estimate sums, differences, products, quotients						●	▲		
DECIMALS									
Addition									
Model and compute sums				●	●	●			
Sums of tenths and hundredths				●	●	●	●	▲	▲
Sums of tenths, hundredths, thousandths						●	●	▲	▲
Sums of ten-thousandths and beyond							●	▲	▲
Subtraction									
Model and compute differences				■	●	●			
Differences of tenths and hundredths				■	●	●	▲	▲	▲
Differences of tenths, hundredths, thousandths						●	●	▲	▲
Differences of ten-thousandths and beyond							●	▲	▲
Estimate sums and differences					●	●	▲	▲	▲
Multiplication									
Model and compute products						●	●		
Decimal by a whole number						●	●	●	▲
Decimal by a decimal						●	●	●	▲
By powers of 10						●	●	●	▲
Place the decimal point						●	●	●	▲
Place zeros in the product						●	●	●	▲
Estimate products						●	●	●	▲
Division									
Model and compute quotients						●	●		
Decimal by a whole number						●	●	●	▲
Decimal by a decimal							●	●	▲
By powers of 10						●	●	●	▲
Place the decimal point						●	●	●	▲
Place zeros in the quotient						●	●	●	▲
Estimate quotients (compatible numbers)						■	●	●	▲
Repeating/terminating quotients							●	●	▲

■ **Introduce Concept Only**   ● **Teach and Test**   ▲ **Reinforce**

OPERATION AND QUANTITATIVE REASONING

WHOLE NUMBERS
Inverse operations 296–299, H9
Remainders 42–43
Interpret remainders 42
1-digit divisor 43
2-digit divisor
  symbolic 42–43
Zeros in the quotient 42–43
Estimate quotients
  using rounding 48–49
  using compatible numbers 48–49
Divisibility H4
MONEY
Add
  dollars and cents 54–55
Subtract
  dollars and cents 56–57
Multiply
  multiplication 58–61
Divide
  dollars and cents 64–65
Estimate sums, differences, products, quotients
  estimate sums 54–55
DECIMALS
Addition
  Sums of tenths and hundredths 54–55
  Sums of tenths, hundredths, thousandths 55
  Sums of ten-thousandths and beyond 55
Subtraction
  Differences of tenths and hundredths
    symbolic 56–57
  Differences of tenths, hundredths, thousandths
    56–57
  Differences of ten-thousandths and beyond 56–57
  Estimate sums and differences
    estimate sums 54–55
Multiplication
  Model and compute products 58–61
  Decimal by a whole number
    model 58–59
    symbolic 58–59, 61
  Decimal by a decimal
    model 59
    symbolic 60–61, H13
  By powers of 10 H12
  Place the decimal point
    by estimating 58–59
    by counting places 59–60, H13
    zeros as placeholders 60
  Place zeros in the product 58–61, H13
  Estimate products 59–60
Division
  Model and compute quotients 62–63
  Decimal by a whole number 64–65, 67, H13
  Decimal by a decimal 65–67
  By powers of 10 66, H12
  Place the decimal point 64–67, H13
  Place zeros in the quotient 64–67, H13
  Estimate quotients (compatible numbers) 48–49
  Repeating/terminating quotients 450–453

Type printed in red indicates that a topic is being introduced for the first time.

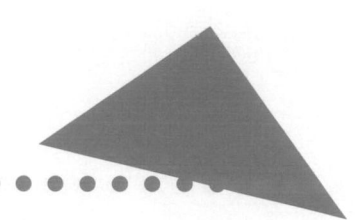

OPERATION AND QUANTITATIVE REASONING	K	1	2	3	4	5	6	7	8
FRACTIONS									
Addition									
Model and compute sums					●	●	●		
Like denominators					●	●	●	▲	▲
Unlike denominators					■	●	●	●	▲
Estimate sums					■	●	●	●	
Mixed numbers					●	●	●	●	▲
Subtraction									
Model and compute differences					●	●	●		
Like denominators					●	●	●	▲	▲
Unlike denominators					■	●	●	●	▲
Estimate differences					■	●	●	●	
Mixed numbers					●	●	●	●	▲
Multiplication									
Model and compute products						●	●		
Fraction of a whole number						●	●	▲	▲
Fraction of a fraction						●	●	●	▲
Estimate products						●	●		▲
Fraction and mixed number						●	●	●	▲
Mixed numbers							●	●	▲
Division									
Model and compute quotients						■	●		
Fraction by a whole number							■	●	▲
Fraction by a fraction							■	●	▲
Mixed numbers							■	●	▲
Estimate quotients								●	▲
INTEGERS									
Model addition and subtraction							●	●	●
Use a number line to compute							●	●	●
Sums and differences							●	●	●
Products and quotients							●	●	●
RATIONAL NUMBERS									
Add, subtract, multiply, and divide								●	●
Products and quotients of powers									●
Inverse operations								●	●
solve equations								●	●
Scientific notation								●	●
REAL NUMBERS									
Squares and square roots							●	●	●
Negative exponents								●	●

■ Introduce Concept Only ● Teach and Test ▲ Reinforce

OPERATION AND QUANTITATIVE REASONING

FRACTIONS

Addition
Model and compute sums 94–95, 96, 98–99, 112–113
Like denominators
model 94–95, H17
Unlike denominators
model 96, 98–99
symbolic 100–101
use least common denominator 100–101
Estimate sums 104–107
Mixed numbers
like denominators H18
unlike denominators, using models 112–113
symbolic 118–121
estimate sums 122–123

Subtraction
Model and compute differences 94–95, 96–97, 98–99
Like denominators
model 94–95, H17
Unlike denominators
model 97, 98–99
using least common denominator 102–103
Estimate differences 104–107
Mixed numbers
like denominators H19
unlike denominators, using models 114–115, 116–117
symbolic 119–121
estimate differences 122–123

Multiplication
Model and compute products 128–129
Fraction of a whole number 130–131
Fraction of a fraction 128–129, 131, 132–133
Estimate products 130
Fraction and mixed number 134–135
Mixed numbers 134–135

Division
Model and compute quotients 136–137
Fraction by a whole number 136–137, 138–139
Fraction by a fraction 137, 138–139, 142
Mixed numbers 140–141

INTEGERS
Model addition and subtraction
modeling addition 462–463, 464–465
modeling subtraction 466–467, 468–469
modeling multiplication 470–471
Use a number line to compute
addition 464–465, 475
subtraction 468–469, 475
multiplication 470–471, 475
Sums and differences
addition 462–463, 464–465, 475
subtraction 466–467, 468–469, 475
Products and quotients
multiplication 470–471, 475
division 472–473, 475

REAL NUMBERS
Squares and square roots
squares 26
square roots 27

Type printed in red indicates that a topic is being introduced for the first time.

OPERATION AND QUANTITATIVE REASONING	K	1	2	3	4	5	6	7	8
ESTIMATE ANSWERS									
Rounding				●	●	●	▲	▲	▲
Strategy									
Benchmarks				●	●	●	▲	▲	▲
Compatible numbers						●	●	▲	▲
ALGEBRA									
PROPERTIES									
Of whole numbers			●	●	●	●	●	▲	▲
Of integers								●	●
Of rational numbers								●	●
EQUATIONS AND EXPRESSIONS									
Number sentences	●	●	●	●	●	●	●	●	●
Missing addend			●	●	●	●	▲	▲	▲
Missing factor			●	●	●	●	▲	▲	▲
Equivalent fractions and decimals				●	●	●	●	●	●
Input/output tables				●	●	●	●	●	●
Evaluate and write numerical expressions		●	●	●	●	●	●	●	●
Formulas				■	■	●	●	●	●
Order of operations						■	●	●	●
Variables						●	●	●	●
Evaluate and write algebraic expressions						■	●	●	●
Solve equations									
modeling like terms								●	●
1-step addition and subtraction equations							●	●	●
1-step multiplication and division equations							●	●	●
modeling 2-step equations							■	●	●
2-step equations								●	●
like terms								●	●
combine like terms								●	●
equations with two variables								●	●
solve equations with integers							●	●	●
solve equations with rational numbers								●	●
solve linear equations								●	●
slope-intercept form									●
Graph equations								●	●
Polynomials									●
Solve systems of equations									●
Functions and relations							■	●	●

■ Introduce Concept Only   ● Teach and Test   ▲ Reinforce

OPERATION AND QUANTITATIVE REASONING

ESTIMATE ANSWERS

Rounding
  Sums
    whole number 46–47
    column addition 46–47
    fraction 103–105
    mixed number 122–123
    decimal 54–55
    clustering 46–47
  Differences
    whole number 46–47
    fraction 106–107
    mixed number 122–123
  Products
    whole numbers 47–49
    decimals 58–59
  Quotients
    rounding 48–49
    compatible numbers 48–49

Strategy
  Benchmarks 422–425, 436
  Compatible numbers 48–49

ALGEBRA

PROPERTIES

Of whole numbers 34–35, 38–39, H6

EQUATIONS AND EXPRESSIONS

Number sentences 284–285, 286–289, 294–295, 300–301

Missing addend 35, 57, 284–285, 286–289

Missing factor 39, 67, 79, 137, 294–295, 296–299

Equivalent fractions and decimals
  fractions 84–85, 108, H14
  decimals 19

Input/output tables 282–283, 490–491, 493

Evaluate and write numerical expressions 278–279, 280–281, 478–479

Formulas 300–301, 302–303, 304–305, 396–397, 398–399, 406–407, 408–411, 412–413, 414–415, 416–417, 424–425, 426–429, 430–431, 432–433, 434–435
  temperature 302–303
  time and distance 304–305
  perimeter 396–397
  circumference 398–399
  area 406–407, 408–411, 412–413, 414–415, 416–417
  volume 424–425, 426–429, 430–431, 432–433
  surface area 434–435

Order of operations 44–45

Variables 278–279, 280–289, 294–305

Evaluate and write algebraic expressions 278–279, 280–281, 282–283, 478–479

Solve equations
  1-step addition and subtraction equations 284–285, 286–289
  1-step multiplication and division equations 294–295, 296–299, 300–301, 304–305, 326–327
  modeling 2-step equations 482–483
  solve equations with integers
    addition 480–481, 482–483

Functions and relations
  relations 490–491

Type printed in red indicates that a topic is being introduced for the first time.

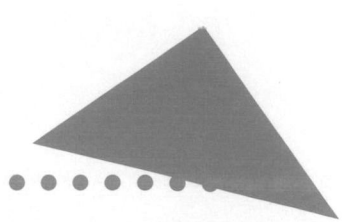

ALGEBRA	K	1	2	3	4	5	6	7	8
INEQUALITIES									
Compare numbers	●	●	●	●	●	●	●	●	●
Use inequality symbols			●	●	●	●	●	●	●
Algebraic inequality							●	●	●
Solve inequalities							●	●	●
Inequality with two variables									●
GRAPHING ON A NUMBER LINE									
Graph inequalities on a number line							●	●	●
Graph rational numbers								●	●
COORDINATE GRAPHING									
Ordered pairs		■	■	●	●	●	●	●	●
Relations							■	●	●
Functions								●	●
Linear equations								●	●
Nonlinear equations									●
Slope									●
Intercepts									●
Systems of equations									●
Graph inequalities									●
RELATIONS AND FUNCTIONS									
Input, output tables				●	●	●	●	●	●
Function graphs								●	●
Graph from a rule								●	●
Linear functions								●	●
Nonlinear functions									●

■ **Introduce Concept Only**   ● **Teach and Test**   ▲ **Reinforce**

ALGEBRA

INEQUALITIES

Compare numbers 18–19, 22–23, 28–29, 456–457, 484–485, H7, H15
Use inequality symbols 18–19, 22–23, 28–29, 456–457, 484–485, H7, H15
Algebraic inequality 484–485
Solve inequalities 484–485

GRAPHING ON A NUMBER LINE

Graph inequalities on a number line 484–485

COORDINATE GRAPHING

Ordered pairs 486–489, 490–491, 500–503
Relations 490–491

RELATIONS AND FUNCTIONS

Input, output tables 282–283, 490–491, 493

Type printed in red indicates that a topic is being introduced for the first time.

PATTERNS, RELATIONSHIPS, AND ALGEBRAIC THINKING

PATTERNS	K	1	2	3	4	5	6	7	8
GEOMETRIC PATTERNS									
Copy geometric patterns	●	●	●	●	●	●			
Continue geometric patterns	●	●	●	●	●	●	●	●	●
Describe by color, size, and shape		●	●	●	●	●	●	●	●
Generate geometric patterns		●	●	●	●	●	●	●	●
Tessellations				■	●	●	●	●	●
Self-similar							■	●	●
Fractals							■	●	●
NUMBER									
Counting patterns	■	●	●	●	●	●	▲		
Continue number patterns		●	●	●	●	●	●	●	●
Generate number patterns			●	●	●	●	●	●	●
Odd and even		●	●	●					
Place-value patterns			●	●	●				
OPERATION									
Fact strategies		●	●	●	●	▲			
count on		●	●	●					
count back		●	●	●					
Figurate numbers							●	●	▲
Multiply, divide by multiples/powers of 10				●	●	●	▲	▲	▲
Arithmetic sequences				●	●	●	●	●	●
Geometric sequences				●	●	●	●	●	●
Pascal's triangle								●	●
Fibonacci sequence								●	●
RELATIONSHIPS									
NUMBER									
Money and decimals				●	●	●			
Coin value relationships	■	●	●	●	●	●			
Fractions and decimals				●	●	●	●	●	●
Ratios, rates						●	●	●	●
Convert between percents, decimals, fractions						●	●	●	●
Scatterplots								●	●
OPERATION									
Fact families		●	●	●	▲				
Multiplication and addition			●	●	▲	▲			
Division and subtraction				●	●	▲			
Inverse operations		●	●	●	●	●	●	●	●
PROPORTIONAL REASONING									
Ratio						■	●	●	●
concept							●	●	●
read and write							●	●	●
equivalent ratios							●	●	●
cross products							●	●	●
rates, unit rates							●	●	●
slope									●
tangent, sine, and cosine									●

■ **Introduce Concept Only**   ● **Teach and Test**   ▲ **Reinforce**

PATTERNS, RELATIONSHIPS, AND ALGEBRAIC THINKING

PATTERNS

GEOMETRIC PATTERNS

Continue geometric patterns 504–505, 506–507, 508, 511–512, 516

Describe by color, size, and shape 506–507

Generate geometric patterns 162–163, 508–509, 516, 519

Tessellations 178–181, 508–509

Self-similar 510–511

Fractals 510–511

NUMBER

Counting patterns 517–518

Continue number patterns 516–519, 520, 522–523, 524–525, 527

Generate number patterns 521, 526

OPERATION

Figurate numbers 516–519

Multiply, divide by multiples/powers of 10, 65–66, 321, 322, H12

Arithmetic sequences 516–519, 520–521, 522–523, 526–527

Geometric sequences 516–519, 524–525

RELATIONSHIPS

NUMBER

Fractions and decimals 20–21, 22–23, H19

Ratios, rates 318–319

Convert between percents, decimals, fractions 320–324

OPERATION

Inverse operations 296–299, H9

PROPORTIONAL REASONING

Ratio

   concept 316–317

   read and write 316–317

   equivalent ratios 316–317, 326–327

   cross products 327, 328–329

   rates, unit rates 318–319

Type printed in red indicates that a topic is being introduced for the first time.

Scope & Sequence/Correlations

RELATIONSHIPS	K	1	2	3	4	5	6	7	8
PROPORTIONAL REASONING									
Proportion									
meaning of proportion							●	●	●
solve proportions							●	●	●
applications									
indirect measurement							●	●	●
scale drawings						■	●	●	●
similar figures						■	●	●	●
tangent, sine, and cosine ratios									●
Percent							●	●	●
meaning of percent							●	●	●
percent and decimals							●	●	●
percent and fractions							●	●	●
percents greater than 100%, less than 1%							●	●	●
find percent of a number						■	●	●	●
find percent one number is of another							■	●	●
find number when percent is known								●	●
estimate percents							●	●	●
applications									
circle graph						■	●	●	●
sales tax							●	●	●
simple interest							●	●	●
discount							●	●	▲
markup								●	▲
commission									●
percent of increase or decrease									●
GEOMETRY									
Solid and plane figures	●	●	●	●	●	●	●	●	●
Plane figures	●	●	●	●	●	●	●	●	●
Solid figures				●	●	●	●	●	●
number of faces, edges, and vertices				■	■	■	●	●	●
Similarity					■	■	●	●	●
Tangent, sine, and cosine									●
MEASUREMENT AND TIME									
Perimeter and area				●	●	●	●	●	●
Circumference and diameter						●	●	●	▲
Proportional reasoning					■	●	●	●	▲
Customary units				●	●	●	●	▲	▲
Metric units				●	●	●	▲	▲	▲
Time units				●	●	●	▲		
Computations with denominate numbers						●	▲	▲	▲
Pythagorean Property								●	●
PROBABILITY AND STATISTICS									
Probability as a ratio							●	●	●
Compare/choose appropriate graph					●	●	●	●	●
Measures of central tendency					■	●	●	●	●

■ **Introduce Concept Only**   ● **Teach and Test**   ▲ **Reinforce**

Type printed in red indicates that a topic is being introduced for the first time.

ALGEBRAIC THINKING	K	1	2	3	4	5	6	7	8
PROPERTIES									
Addition of whole numbers		●	●	●	●	●	▲		
integers and rational numbers								●	▲
Multiplication of whole numbers				●	●	●	▲		
integers and rational numbers								●	▲
EQUATIONS AND EXPRESSIONS									
Number sentences	●	●	●	●	●	●	●	●	●
Missing addend			●	●	●	●	▲	▲	▲
Missing factor				●	●	●	▲	▲	▲
Equivalent fractions and decimals				●	●	●	●	●	●
Evaluate and write numerical expressions		●	●	●	●	●	●	●	●
Order of operations						■	●	●	●
Variables					●	●	●	●	●
Evaluate and write algebraic expressions						■	●	●	●
Identify and combine like terms								■	●
Solve equations									
1-step addition and subtraction equations							●	●	●
1-step multiplication and division equations							●	●	●
2-step equations							■	●	●
with like terms								●	●
equations with two variables								●	●
solve equations with integers							●	●	●
solve equations with rational numbers								●	●
Polynomials									●
meaning of polynomials									●
simplifying									●
adding and subtracting									●
Relations and functions							■	●	●
Formulas				■	■	●	●	●	●
INEQUALITIES									
Compare numbers	●	●	●	●	●	●	●	●	●
Use inequality symbols		●	●	●	●	●	●	●	●
Algebraic inequality							●	●	●
Solve inequalities							●	●	●
Inequality with two variables									●
GRAPHING ON A NUMBER LINE									
Graph inequalities on a number line							●	●	●
Graph rational numbers								●	●
COORDINATE GRAPHING									
Ordered pairs		■	■	●	●	●	●	●	●
Coordinate plane							■	●	●
origin							●	●	●
1 quadrant				●	●	●	●	●	●
4 quadrants							●	●	●
axes							●	●	●
Linear equations								●	●
Nonlinear equations								■	●
Slope									●

■ **Introduce Concept Only**   ● **Teach and Test**   ▲ **Reinforce**

ALGEBRAIC THINKING

PROPERTIES
Addition of whole numbers
  Commutative 34–35, H6
  Associative 34–35, H6
Multiplication of whole numbers
  Commutative 38–39, H6
  Associative 38–39, H6
  Distributive 38–39, H6
  Property of Zero H6
  Identity Property of One H6
EQUATIONS AND EXPRESSIONS
Number sentences 284–285, 286–289, 294–295, 300–301
Missing addend 35, 57, 284–285, 286–289
Missing factor 39, 67, 79, 137, 294–295, 296–299
Equivalent fractions and decimals
  fractions 18, 84–85, 108, H14
  decimals 18
Evaluate and write numerical expressions 278–279, 280–281, 478–479
Order of operations 44–45
Variables 278–279, 280–289, 294–305
Evaluate and write algebraic expressions 278–279, 280–281, 282–283, 478–479
Solve equations
  1-step addition and subtraction equations 284–285, 286–289
  1-step multiplication and division equations 294–295, 296–299, 300–301, 304–305, 326–327
  2-step equations 482–483
  solve equations with integers 480–481, 482–483
Relations and functions 282–283, 490–491, 493
Formulas 300–301, 302–303, 304–305, 396–397, 398–399, 406–407, 408–411, 412–413, 414–415, 416–417, 424–425, 426–429, 430–431, 432–433, 434–435
INEQUALITIES
Compare numbers 18–19, 22–23, 28–29, 456–457, 484–485, H7, H15
Use inequality symbols 18–19, 22–23, 28–29, 456–457, 484–485, H7, H15
Algebraic inequality 484–485
Solve inequalities 484–485
GRAPHING ON A NUMBER LINE
Graph inequalities on a number line 484–485
COORDINATE GRAPHING
Ordered pairs 486–489, 490–491, 500–503
Coordinate plane 486–489, 490–491, 500–503, 504–505
  origin 487–489
  1 quadrant 504
  4 quadrants 487–489, 500–503, 504–505
  axes 487–489, 500–501

Type printed in red indicates that a topic is being introduced for the first time.

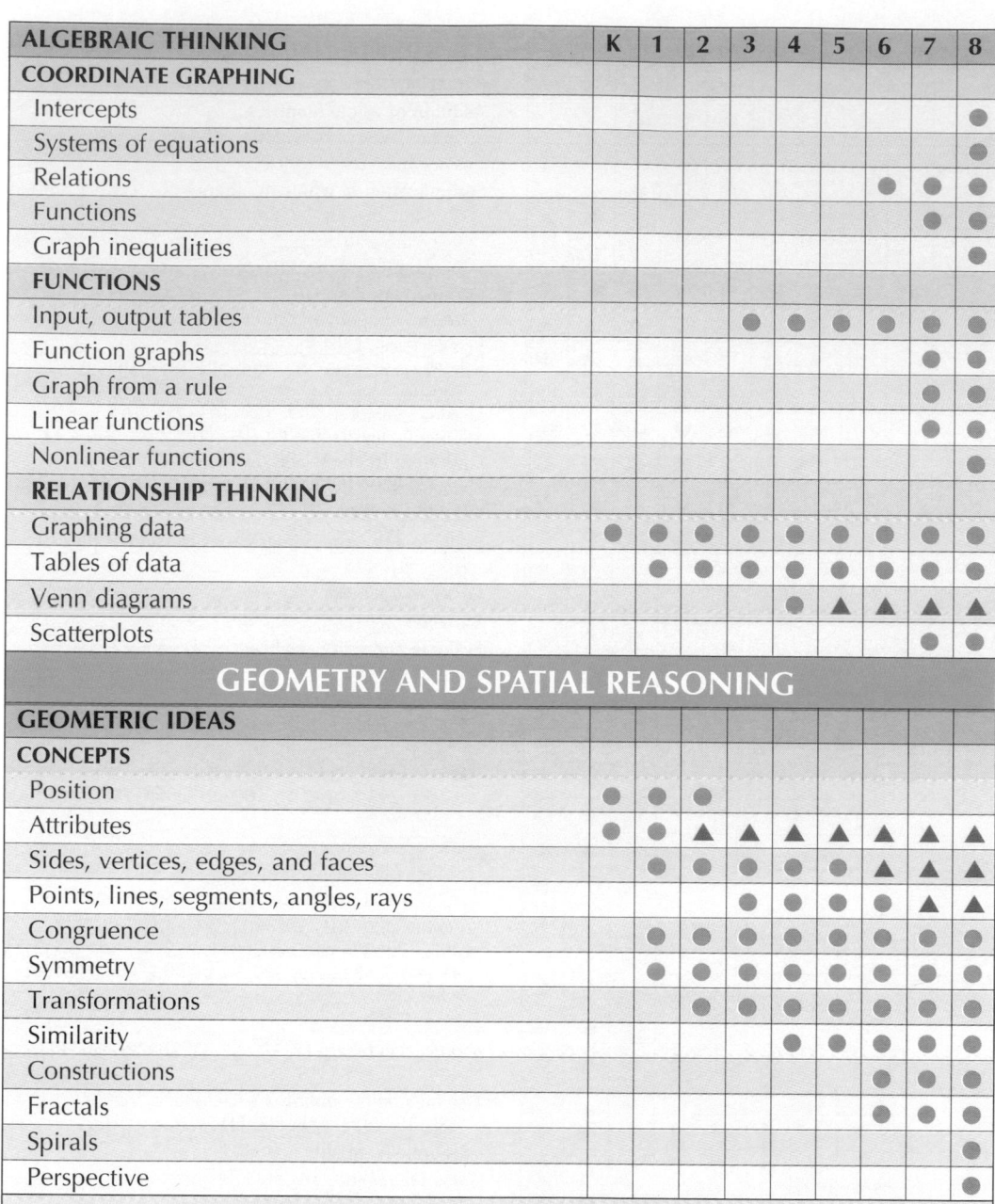

ALGEBRAIC THINKING	K	1	2	3	4	5	6	7	8
COORDINATE GRAPHING									
Intercepts									●
Systems of equations									●
Relations							●	●	●
Functions								●	●
Graph inequalities									●
FUNCTIONS									
Input, output tables				●	●	●	●	●	●
Function graphs								●	●
Graph from a rule								●	●
Linear functions								●	●
Nonlinear functions									●
RELATIONSHIP THINKING									
Graphing data	●	●	●	●	●	●	●	●	●
Tables of data		●	●	●	●	●	●	●	●
Venn diagrams					●	▲	▲	▲	▲
Scatterplots								●	●

GEOMETRY AND SPATIAL REASONING

GEOMETRIC IDEAS	K	1	2	3	4	5	6	7	8
CONCEPTS									
Position	●	●	●						
Attributes	●	●	▲	▲	▲	▲	▲	▲	▲
Sides, vertices, edges, and faces		●	●	●	●	●	▲	▲	▲
Points, lines, segments, angles, rays				●	●	●	●	▲	▲
Congruence		●	●	●	●	●	●	●	●
Symmetry		●	●	●	●	●	●	●	●
Transformations			●	●	●	●	●	●	●
Similarity					●	●	●	●	●
Constructions							●	●	●
Fractals							●	●	●
Spirals									●
Perspective									●

■ **Introduce Concept Only**   ● **Teach and Test**   ▲ **Reinforce**

Type printed in red indicates that a topic is being introduced for the first time.

GEOMETRIC IDEAS	K	1	2	3	4	5	6	7	8
SOLID FIGURES									
Attributes and properties	■	●	●	●	●	●	●	●	●
Identify	■	●	●	●	●	●	▲	▲	▲
Sort and classify	■	●	●	●					
polyhedrons						■	●	●	●
not polyhedrons						■	●	●	●
Represent and visualize	●	●	●	●	●	●	●	●	●
Build/take apart				●	●	●	●	●	●
Identify or draw different views						●	●	●	●
Identify cross sections								●	●
Make nets				■	●	●	●	●	●
Measure									
volume						■	●	●	●
surface area							●	●	●
PLANE FIGURES									
Attributes and properties	■	●	●	●	●	●	▲	▲	▲
Identify	●	●	●	●	●	●	●	●	●
Sort and classify	●	●	●	●					
plane figures				●	●	●	▲	▲	▲
polygons					●	●	▲	▲	▲

■ **Introduce Concept Only**   ● **Teach and Test**   ▲ **Reinforce**

GEOMETRIC IDEAS

SOLID FIGURES

Attributes and properties 186–189

Identify
  cones 187
  polyhedrons 186–189
  prisms
    square 187
    rectangular 187–189, 190–191
    triangular 187–189, 190–191
    pentagonal 187–189, 190–191
    hexagonal 187–189
  cylinders 187–189
  pyramids
    square 188–189
    triangular 188–189, 190–191
    pentagonal 189, 190–191
    hexagonal 188–189
    rectangular 190–191

Sort and classify
  polyhedrons 187–188
  not polyhedrons 187–188

Represent and visualize
  describe, identify, and draw different views 194–195
  build solids, draw different views 196–197

Build/take apart
  construct to match a given solid using cubes 196–197
  net 192–193, 422

Identify or draw different views
  describe, identify, and draw different perspectives 194–195
  build solids, draw different views 196–197

Make nets 192–193, 422

Measure
  volume
    rectangular prism 422–425
    triangular prism 422–425
    cylinder 430–431, 432–433
    volume/dimensions relationships 426–429
  surface area
    rectangular prism 434–435

PLANE FIGURES

Attributes and properties 162–163

Identify
  circles 398–399, 414–415
  triangles 162–163
  hexagons 162–163
  parallelograms 163
  trapezoids 163
  octagons 162–163
  quadrilaterals 162–163
  pentagons 162–163
  polygons 162–163
  regular polygons 162–163, 168
  Golden Rectangles 378–379

Sort and classify
  plane figures 154–155
  polygons 162–163
    regular/not regular 162
    triangles
      by sides
        isosceles 162
        scalene 162
        equilateral 162
      by angles
        right 162
        acute 162
        obtuse 162
    quadrilaterals 162–163

Type printed in red indicates that a topic is being introduced for the first time.

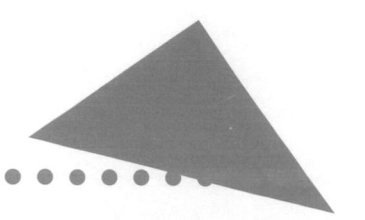

GEOMETRIC IDEAS	K	1	2	3	4	5	6	7	8
PLANE FIGURES									
Represent and visualize	●	●	●	●	●	●	●	●	●
Build/take apart			●	●	●	●	●	●	●
Measure									
angles							●	▲	▲
perimeter/circumference				●	●	●	●	●	●
area			■	●	●	●	●	●	●
ALGEBRAIC THINKING									
Formulas				■	■	●	●	●	●
Pythagorean Property								●	●
Graphing figures							●	●	●
Scatterplots								●	●
Correlations								●	●
Line of best fit									●
Similar figure applications									
indirect measurement							●	●	●
Tangent, Sine, and Cosine									●
SPATIAL SENSE									
VISUAL THINKING									
Patterns	●	●	●	●	●	●	●	●	●
tessellations				■	●	●	●	●	●
nets				■	●	●	●	●	●
fractals							●	●	●
spirals									●
Congruence		●	●	●	●	●	●	●	●
Symmetry		●	●	●	●	●	●	●	●
line		●	●	●	●	●	●	●	●
point (rotational)				■	●	●	●	●	●
Similarity					■	●	●	●	●
Transformations									
translations, reflections, and rotations				●	●	●	●	●	●
dilations							■	●	●
Tessellations					●	●	●	●	●
Fractals							●	●	●
Spirals									●
Representing									
building, drawing 3-D figures							■	●	●
different views							●	▲	▲
Perspective									●
Networks							●	●	●
Geometric probability							●	●	●
COORDINATE GEOMETRY									
COORDINATE PLANE									
Ordered pairs			■	■	●	●	●	●	●
Graph points and figures							●	●	●
Graph equations								●	●
with fractions and decimals								●	●
with integers								●	●
with rational numbers								●	●

■ **Introduce Concept Only**   ● **Teach and Test**   ▲ **Reinforce**

GEOMETRIC IDEAS

PLANE FIGURES

Represent and visualize 172–173, 352–353
Build/take apart
  tangrams 172–173
  circles 398–399, 414–415
  tessellations 178–181, 508–509
  construct and bisect segments 156–157, 159, 160–161
  construct angles 157–159
  Golden Rectangles 378
  stretch and shrink 372–373
Measure
  angles 154–155, 159
  perimeter/circumference 396–397, 398–399
  area
    of rectangles 404–405, 406–407
    of triangles 408–409
    of parallelograms 408–409
    of a circle 414–415, 416–417

ALGEBRAIC THINKING

Formulas 406–407, 408–411, 416–417, 424–425, 426–429, 432–433, 434–435
Graphing figures 500–503, 504–505
Similar figure applications
  indirect measurement 362–363
  scale drawings 368–371
  Golden Rectangles 378–379
  map scales 374–375, 376–377

SPATIAL SENSE

VISUAL THINKING

Patterns
  tessellations 178–181, 508–509
  nets 192–193, 422, 434–435
  fractals 510–511
Congruence 156–159, 352, 354–355
Symmetry
  line 168–171, 172–173
  point (rotational) 169–171, 173
Similarity 352–353, 354–355, 356–359, 360–361, 362–363, 368–371
Transformations
  translations, reflections, and rotations 174–177, 500–503, 506–507
  dilations 372–373
Tessellations 178–181, 508–509
Fractals 510–511
Representing
  building, drawing 3-D figures 196–197
  different views 194–195
Networks 394–395
Geometric probability 261–263

COORDINATE GEOMETRY

COORDINATE PLANE

Ordered pairs
  4 quadrants 486–488
Graph points and figures 486–488, 490–491, 500–503, 504–505

Type printed in red indicates that a topic is being introduced for the first time.

COORDINATE GEOMETRY	K	1	2	3	4	5	6	7	8
COORDINATE PLANE									
Slope									●
Intercepts									●
Systems of equations									●
Graph inequalities									●
Relations and functions							●	●	●
Identify functions								●	●
linear functions								●	●
nonlinear functions									●
Translations, reflections, rotations						●	●	●	●
Dilations								■	●

MEASUREMENT AND TIME	K	1	2	3	4	5	6	7	8
MEASURING OBJECTS									
CONCEPTS									
Choose appropriate tools/units to measure			●	●	●	●	▲	▲	▲
Precision/Accuracy						●	●	●	●
significant digits									●
Networks							●	●	●
LENGTH									
Meaning of linear measurement	●	●	●	●	●	●	▲		
Nonstandard units	●	●	●	●	▲				
Compare and order	●	●	●	●	●	▲			
Estimate	■	●	●	●	●	●			
Customary units		●	●	●	●	●	●	▲	▲
Metric units		●	●	●	●	●	●	▲	▲
Relate units				●	●	●	●	●	●
choose appropriate units				●	●	●	▲		
Change units within systems				■	●	●	●	▲	▲
Apply distance formula							●	●	●
CAPACITY									
Meaning of capacity	■	●	●	●	●	●			
Nonstandard units		●	●	●	●				
Compare and order		●	●	●	●	●			
Estimate		●	●	●	●	●			
Customary units			●	●	●	●	●	▲	▲
Metric units				●	●	●	●	▲	▲
Relate units				●	●	●	●	▲	▲
choose appropriate units				●	●	●			
Change units within systems				●	●	●	●	▲	▲
WEIGHT/MASS									
Meaning of weight/mass	■	●	●	●	●	●			
Nonstandard units		●	●	●	●	▲			
Compare and order		●	●	●	●	●			
Estimate		●	●	●	●	●			
Customary units			●	●	●	●	●	▲	▲
Metric units				●	●	●	●	▲	▲
Relate units				●	●	●	●	▲	▲
choose appropriate units				●	●	●			
Change units within systems				●	●	●	●	▲	▲

■ Introduce Concept Only   ● Teach and Test   ▲ Reinforce

COORDINATE GEOMETRY
COORDINATE PLANE
Relations and functions 490–491
Translations, reflections, rotations 500–503

MEASUREMENT AND TIME
MEASURING OBJECTS
CONCEPTS
Choose appropriate tools/units to measure 393
Precision/Accuracy 392–393
Networks 394–395
LENGTH
Meaning of linear measurement 392
Customary units 388–389, 392–393, 394–395
Metric units 390–391, 392–393, 394–395
Relate units
   choose appropriate units 392–393
Change units within systems
   customary units 388–389
   metric units 390–391
Apply distance formula 304–305
CAPACITY
Customary units 388–389
Metric units 390–391
Relate units 388–389, 390–391
Change units within systems
   customary units 388–389
   metric units 390–391
WEIGHT/MASS
Customary units 388–389
Metric units 390–391
Relate units 306–307, 388–389, 390 391
Change units within systems
   customary units 388–389
   metric units 390–391

▶ Type printed in red indicates that a topic is being introduced for the first time.

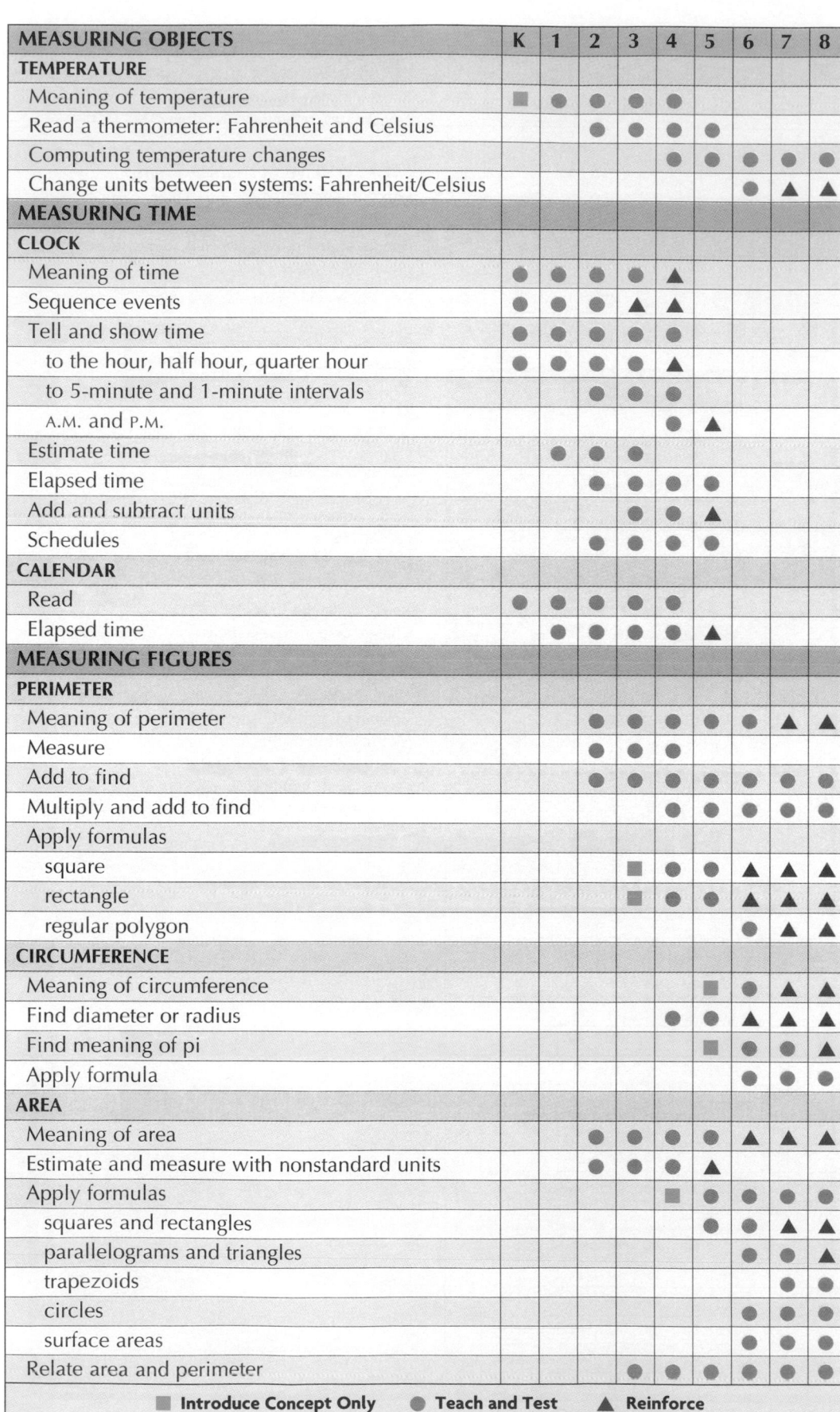

MEASURING OBJECTS	K	1	2	3	4	5	6	7	8
TEMPERATURE									
Meaning of temperature	■	●	●	●	●				
Read a thermometer: Fahrenheit and Celsius			●	●	●	●			
Computing temperature changes					●	●	●	●	●
Change units between systems: Fahrenheit/Celsius							●	▲	▲
MEASURING TIME									
CLOCK									
Meaning of time	●	●	●	●	▲				
Sequence events	●	●	●	▲	▲				
Tell and show time	●	●	●	●	●				
to the hour, half hour, quarter hour	●	●	●	●	▲				
to 5-minute and 1-minute intervals			●	●	●				
A.M. and P.M.					●	▲			
Estimate time			●	●					
Elapsed time			●	●	●	●			
Add and subtract units				●	●	▲			
Schedules			●	●	●				
CALENDAR									
Read	●	●	●	●	●				
Elapsed time		●	●	●	●	▲			
MEASURING FIGURES									
PERIMETER									
Meaning of perimeter			●	●	●	●	●	▲	▲
Measure			●	●	●				
Add to find			●	●	●	●	●	●	●
Multiply and add to find					●	●	●	●	●
Apply formulas									
square				■	●	●	▲	▲	▲
rectangle				■	●	●	▲	▲	▲
regular polygon							●	▲	▲
CIRCUMFERENCE									
Meaning of circumference						■	●	▲	▲
Find diameter or radius					●	●	▲	▲	▲
Find meaning of pi						■	●	●	▲
Apply formula							●	●	●
AREA									
Meaning of area			●	●	●	●	▲	▲	▲
Estimate and measure with nonstandard units			●	●	●	▲			
Apply formulas					■	●	●	●	●
squares and rectangles						●	●	▲	▲
parallelograms and triangles							●	●	▲
trapezoids								●	●
circles							●	●	●
surface areas							●	●	●
Relate area and perimeter				●	●	●	●	●	●

■ **Introduce Concept Only** ● **Teach and Test** ▲ **Reinforce**

MEASURING OBJECTS

TEMPERATURE

Computing temperature changes 465, 468–469, 471, 473

Change units between systems: Fahrenheit/Celsius 302–303

MEASURING FIGURES

PERIMETER

Meaning of perimeter 396–397, 398–399
Add to find 396
Multiply and add to find 396
Apply formulas
  square 396–397
  rectangle 396–397
  regular polygon 396

CIRCUMFERENCE

Meaning of circumference 398–399
Find diameter or radius 398
Meaning of pi 398–399
Apply formula 399

AREA

Meaning of area 404
Apply formulas
  squares and rectangles 406–407
  parallelograms and triangles 408–411
  circles 414–415, 416–417
  surface areas 434–435
Relate area and perimeter 412–413

Type printed in red indicates that a topic is being introduced for the first time.

MEASURING FIGURES	K	1	2	3	4	5	6	7	8
SURFACE AREA									
Meaning of surface area							●	●	●
Apply formulas									
prism							●	●	●
pyramid								●	●
cylinder								●	●
cone									●
VOLUME									
Meaning of volume					■	●	●	●	▲
Estimate and measure					■	●			
prisms and pyramids							●	●	●
cylinders and cones							●	●	●
spheres									●
Apply formula						●	●	●	●
Relate volume of prism to pyramid								●	●
Relate volume of cylinder to cone								●	●
Relate perimeter, area, and volume						●	●	●	●
ANGLES									
Compare to right angle				■	●	●	●	▲	▲
Identify and classify					■	●	●	▲	▲
Measure, draw, and construct						■	●	●	●
ALGEBRAIC THINKING									
Networks							●	●	●
apply formulas									●
Indirect measurement						■	●	●	●
similar figure applications							●	●	●
Pythagorean Property								●	●
tangent, sine, cosine									●

■ Introduce Concept Only   ● Teach and Test   ▲ Reinforce

MEASURING FIGURES

SURFACE AREA
Meaning of surface area 434–435
Apply formulas
  prism 434–435
VOLUME
Meaning of volume 422–424
Estimate and measure
  prisms and pyramids 422–425, 426–429
  cylinders 430–431, 432–433
Apply formula 424–425, 431, 432–433
Relate perimeter, area, and volume 426–429, 431
ANGLES
Compare to right angle 154
Identify and classify 154–155
Measure, draw, and construct 154–155, 156–159
ALGEBRAIC THINKING
Networks 394–395
Indirect measurement
  similar figure applications 362–363, 368–371, 374–375, 376–377, 378–379

Type printed in red indicates that a topic is being introduced for the first time.

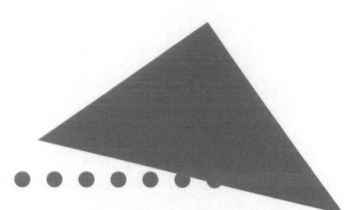

STATISTICS, GRAPHING, AND PROBABILITY

STATISTICS AND GRAPHING	K	1	2	3	4	5	6	7	8
COLLECTING DATA									
Conduct survey			●	●	●	●	●	●	▲
Sampling							●	●	●
bias							●	●	▲
ORGANIZING DATA									
Tally	■	●	●	●	●	●	▲	▲	▲
Frequency table				●	●	▲	▲	▲	
cumulative frequency					●	▲	▲	▲	
List									
account for all possibilities					●	●	●	●	●
Stem-and-leaf plot					●	●	●	●	▲
Line plot					●	●	●	●	▲
Spreadsheets								●	●
DISPLAYING DATA									
Picture graph	●	●	●						
Pictograph			●	●	▲				
Bar graph	■	●	●	●	●	●	▲	▲	▲
Line graph					■	●	▲	▲	▲
Circle graph					■	●	●	●	●
Histograms						■	●	▲	▲
Box-and-whisker graph							●	●	●
Stacked bar graph								●	●
Scatterplots								●	●
ANALYZING DATA									
Interpret graphs	■	●	●	●	●	●	●	●	●
Interpret tables	■	●	●	●	●	●	●	●	●
Compare data			●	●	●	●	●	●	●
Compare/choose representations						●	●	●	●
Identify misleading graphs						●	●	●	●
Choose scale						●	▲	▲	▲
Find mean (average)					■	●	●	▲	▲
Find median						●	●	▲	▲
Find mode						●	●	▲	▲
Find range					■	●	●	▲	▲
Determine best measure of central tendency							●	●	●
Interpolate, Extrapolate									●

■ **Introduce Concept Only** ● **Teach and Test** ▲ **Reinforce**

STATISTICS, GRAPHING, AND PROBABILITY

STATISTICS AND GRAPHING

COLLECTING DATA

Conduct survey
analyze survey questions 214–215
define the problem 208–209

Sampling
random samples 210–211
bias in surveys 212–213

ORGANIZING DATA

Tally 216–219

Frequency table 217–219, 228
cumulative frequency 217–219

List
account for all possibilities 258–259

Stem-and-leaf plot 226–227, 247

Line plot 216–219

DISPLAYING DATA

Bar graph 224, 227, 232–233, 235
multiple-bar graph 230–231

Line graph 225, 227
multiple-line graph 231

Circle graph
sections as fractions 233, 234
with decimals 234–235
with degrees 234–235
with percents 340–341

Histogram 228–229

Box-and-whisker graph 250–251

ANALYZING DATA

Interpret graphs 224–225, 228–229, 230, 240–241, 242–243, 244–245, 247, 248–249, 252–253
describe relationships within graphs 240–241
make predictions based on graphs 244–245

Interpret tables 217–219, 228

Compare data 228, 240–241, 242–243, 244–245

Compare/choose representations 225, 227, 228, 229, 231, 235

Identify misleading graphs 242–243

Choose scale 225, 230

Find mean (average) 246–249

Find median 246–249, 250–251, 252–253

Find mode 246–249

Find range 218–219, 252–253

Determine best measure of central tendency 246, 249

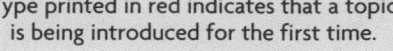

Type printed in red indicates that a topic is being introduced for the first time.

PROBABILITY	K	1	2	3	4	5	6	7	8
CONCEPTS									
Likelihood of events		●	●	●	●	●	▲	▲	▲
certain, possible, impossible		●	●	●	●	●	▲	▲	
likely, more likely, less likely		●	●	●	●	●	▲	▲	
Fairness					●	●	▲	▲	
Randomness							●	●	●
FINDING OUTCOMES									
Outcomes			●	●	●	●	●	●	●
Tree diagram							●	●	●
Sample spaces							●	●	▲
Combinations					●	●	●	●	●
Fundamental counting principle							■	●	●
Permutations								●	●
MATHEMATICAL PROBABILITY									
Meaning of mathematical probability						■	●	●	●
Simple events				●	●	●	●	▲	▲
Compound events						■	●	▲	▲
Geometric probability							●	●	●
Odds									●
Independent/dependent events									●
EXPERIMENTAL PROBABILITY									
Predict based on experiment					■	●	●	●	●
Simulations						■	●	●	●
Random numbers							●	●	●

■ **Introduce Concept Only**   ● **Teach and Test**   ▲ **Reinforce**

PROBABILITY

CONCEPTS

Likelihood of events
  certain, possible, impossible 261
  likely, more likely, less likely 260, 264
Fairness 264–265
Randomness 266–267, 268

FINDING OUTCOMES

Outcomes 260–263
  list all possible 260
  write as a fraction 260–263, 264–265
  compare probabilities 264
Tree diagram 258–259
Sample spaces 260–262
Combinations
  use a tree diagram 258–259
  use multiplication 258–259
Fundamental counting principle 258–259

MATHEMATICAL PROBABILITY

Meaning of mathematical probability 260–261
Simple events 260–263, 264–265, 268
Compound events 264–265
Geometric probability 261–263

EXPERIMENTAL PROBABILITY

Predict based on experiment 266, 268
Simulations 266–267, 268–269
Random numbers 266–267

▶ Type printed in red indicates that a topic is being introduced for the first time.

Standardized Test Correlations

Learning Goals for GRADE 6		CAT	CTBS	ITBS	MAT	SAT
Chapter 1	**Looking at Numbers**					
1-A.1	To use place value to express and compare whole numbers and decimals	CAT	CTBS	ITBS	MAT	SAT
1-A.2	To write a decimal as a fraction and a fraction as a decimal			ITBS		SAT
1-A.3	To use exponents to represent numbers	CAT	CTBS	ITBS	MAT	SAT
1-A.4	To order integers and identify opposite integers	CAT	CTBS	ITBS		SAT
Chapter 2	**Using Whole Numbers**					
2-A.1	To use properties and mental math to find sums and differences		CTBS	ITBS	MAT	SAT
2-A.2	To solve problems by using the *guess and check* strategy				MAT	
2-A.3	To multiply and divide whole numbers	CAT	CTBS	ITBS	MAT	SAT
2-A.4	To use estimation to find sums, differences, products, and quotients		CTBS	ITBS	MAT	SAT
Chapter 3	**Using Decimals**					
3-A.1	To add and subtract decimals	CAT	CTBS	ITBS	MAT	SAT
3-A.2	To multiply decimals	CAT	CTBS	ITBS	MAT	SAT
3-A.3	To divide decimals	CAT	CTBS	ITBS	MAT	SAT
Chapter 4	**Number Theory and Fractions**					
4-A.1	To identify factors and multiples of a number and tell whether a number is prime or composite	CAT	CTBS	ITBS	MAT	SAT
4-A.2	To write a composite number as the product of prime numbers		CTBS		MAT	SAT
4-A.3	To find the least common multiple and greatest common factor	CAT	CTBS	ITBS	MAT	SAT
4-A.4	To write fractions in simplest form	CAT	CTBS		MAT	
4-A.5	To write fractions as mixed numbers and mixed numbers as fractions	CAT	CTBS	ITBS		SAT
Chapter 5	**Adding and Subtracting Fractions**					
5-A.1	To add and subtract like fractions	CAT	CTBS	ITBS	MAT	SAT
5-A.2	To add and subtract unlike fractions	CAT	CTBS	ITBS	MAT	SAT
5-A.3	To estimate sums and differences of fractions		CTBS	ITBS		
Chapter 6	**Adding and Subtracting Mixed Numbers**					
6-A.1	To add and subtract mixed numbers	CAT	CTBS	ITBS	MAT	SAT
6-A.2	To estimate sums and differences of mixed numbers		CTBS	ITBS		SAT
Chapter 7	**Multiplying and Dividing Fractions**					
7-A.1	To simplify factors and multiply fractions and mixed numbers	CAT	CTBS	ITBS	MAT	SAT
7-A.2	To divide fractions	CAT	CTBS	ITBS	MAT	SAT
7-A.3	To solve problems by *working backward*				MAT	
Chapter 8	**Geometric Figures**					
8-A.1	To identify and describe points, lines, and planes	CAT	CTBS	ITBS	MAT	
8-A.2	To classify lines and angles and measure angles	CAT	CTBS	ITBS	MAT	SAT
8-A.3	To construct congruent line segments and angles					
8-A.4	To identify polygons by the number of sides and angles		CTBS	ITBS	MAT	SAT
Chapter 9	**Symmetry and Transformations**					
9-A.1	To identify line symmetry and rotational symmetry		CTBS			SAT
9-A.2	To identify and use transformations of geometric shapes		CTBS		MAT	SAT
9-A.3	To use polygons to make tessellations					SAT
9-A.4	To use a model to solve a problem					
Chapter 10	**Solid Figures**					
10-A.1	To identify solid figures and their parts	CAT	CTBS	ITBS	MAT	SAT
10-A.2	To identify nets for solid figures and different points of view					
10-A.3	To solve problems by using the strategy *solving a simpler problem*					
Chapter 11	**Organizing Data**					
11-A.1	To identify information needed to make decisions					
11-A.2	To identify sample size and types of samples when conducting a survey					
11-A.3	To determine whether a sample or question in a survey is biased					
11-A.4	To use and and organize data from a survey		CTBS			
Chapter 12	**Displaying Data**					
12-A.1	To display data in a bar graph, a line graph, and a stem-and-leaf plot	CAT	CTBS	ITBS	MAT	SAT
12-A.2	To make histograms and circle graphs	CAT	CTBS	ITBS		SAT
12-A.3	To graph two or more sets of data					
Chapter 13	**Interpreting Data and Predicting**					
13-A.1	To analyze graphs and make predictions from graphs	CAT	CTBS	ITBS	MAT	SAT
13-A.2	To identify misleading graphs					
13-A.3	To find the mean, median, and mode	CAT	CTBS	ITBS	MAT	SAT
Chapter 14	**Probability**					
14-A.1	To use the strategy *account for all possibilities* to solve problems					
14-A.2	To find the probability of events	CAT	CTBS	ITBS	MAT	SAT
14-A.3	To find the experimental probability of an event		CTBS			SAT
Chapter 15	**Algebra: Expressions and Equations**					
15-A.1	To write, interpret, and evaluate numerical and algebraic expressions	CAT	CTBS	ITBS	MAT	SAT
15-A.2	To use input/output tables to evaluate algebraic expressions		CTBS			SAT
15-A.3	To solve addition and subtraction equations	CAT	CTBS	ITBS	MAT	SAT

Learning Goals for GRADE 6		CAT	CTBS	ITBS	MAT	SAT
Chapter 16	**Algebra: Real-life Relationships**					
16-A.1	To solve multiplication and division equations	CAT	CTBS	ITBS		SAT
16-A.2	To use equations to show money relationships	CAT				
16-A.3	To convert temperature between the Fahrenheit and Celsius scales					
16-A.4	To calculate distance, rate, or time by solving an equation		CTBS			SAT
16-A.5	To use the strategy *make a table* to solve problems					
Chapter 17	**Ratio, Proportions, and Percents**					
17-A.1	To write a ratio to compare two objects		CTBS	ITBS	MAT	
17-A.2	To find rates and unit rates			ITBS		SAT
17-A.3	To write ratios and decimals as percents and write percents as decimals and ratios	CAT	CTBS			
17-A.4	To express parts of a whole as percents	CAT	CTBS	ITBS	MAT	
17-A.5	To write a proportion to solve a problem		CTBS	ITBS	MAT	SAT
Chapter 18	**Percent and Change**					
18-A.1	To find the percent of a number	CAT	CTBS	ITBS	MAT	SAT
18-A.2	To use percents to make a circle graph and interpret circle graphs that use percent		CTBS	ITBS		
18-A.3	To find the amount of discount and the sale price of an item if given the price and the discount rate	CAT		ITBS	MAT	SAT
18-A.4	To find simple interest	CAT	CTBS			
Chapter 19	**Ratio, Proportion, and Similar Figures**					
19-A.1	To identify similar and congruent figures	CAT	CTBS	ITBS		
19-A.2	To use ratios to identify similar figures		CTBS			SAT
19-A.3	To use similar figures to find the unknown length of a side		CTBS			SAT
19-A.4	To use proportions to measure indirectly		CTBS			SAT
Chapter 20	**Applications of Ratio and Proportion**					
20-A.1	To use scale to find dimensions of drawing or actual length	CAT	CTBS			SAT
20-A.2	To read and use scales on a map		CTBS	ITBS		SAT
20-A.3	To draw a diagram to show directions					
20-A.4	To use ratios to determine if similar rectangles are Golden Rectangles					
Chapter 21	**Measurement**					
21-A.1	To change one customary unit of measurement to another		CTBS			SAT
21-A.2	To change one metric unit of measurement to another					SAT
21-A.3	To measure length by using precise measurements	CAT	CTBS			
21-A.4	To use a network to find the distance from one place to another					
21-A.5	To find the perimeters of polygons		CTBS	ITBS		SAT
Chapter 22	**Measuring Area**					
22-A.1	To estimate the area of irregular figures			ITBS	MAT	SAT
22-A.2	To *use a formula* to solve problems	CAT	CTBS	ITBS		SAT
22-A.3	To find the area of triangles and parallelograms	CAT	CTBS			SAT
22-A.4	To double the dimensions of a polygon and determine the effects on its perimeter and area					
22-A.5	To find the area of a circle				MAT	SAT
Chapter 23	**Measuring Solids**					
23-A.1	To estimate and find the volume of rectangular and triangular prisms				MAT	SAT
23-A.2	To determine how the volume of a rectangular prism changes when the dimensions change					
23-A.3	To find the volume of a cylinder					
23-A.4	To find the surface area of a rectangular prism				MAT	
Chapter 24	**Algebra: Number Relationships**					
24-A.1	To classify and compare sets of numbers					
24-A.2	To write fractions as terminating and repeating decimals					
24-A.3	To find a rational number between two rational numbers				MAT	
24-A.4	To compare and order rational numbers	CAT				
Chapter 25	**Operations with Integers**					
25-A.1	To add and subtract integers	CAT	CTBS			SAT
25-A.2	To multiply and divide integers	CAT	CTBS			SAT
Chapter 26	**Algebra: Equations and Relationships**					
26-A.1	To evaluate numerical and algebraic expressions involving integers	CAT	CTBS	ITBS		
26-A.2	To solve one-step equations and inequalities involving integers	CAT	CTBS			
26-A.3	To locate points and graph relations on a coordinate plane	CAT	CTBS	ITBS		SAT
Chapter 27	**Geometric Patterns**					
27-A.1	To transform figures on a coordinate plane				MAT	SAT
27-A.2	To find a pattern to solve problems that involve transformations on the coordinate plane					
27-A.3	To identify the next two- or three-dimensional figures in a geometric pattern	CAT	CTBS	ITBS	MAT	SAT
27-A.4	To make shapes for tessellations					
Chapter 28	**Patterns and Operations**					
28-A.1	To identify, extend, and make number patterns with whole numbers and decimals		CTBS	ITBS	MAT	SAT
28-A.2	To identify, extend, and make patterns with fractions	CAT	CTBS			
28-A.3	To identify, extend, and make sequences of integers	CAT	CTBS			SAT

Manipulatives Chart

Manipulatives	Kit Sources	Pupil's Edition Pages	Teacher's Edition Pages
11 x 11 Geoboards	Core Kit Classroom Kit Teacher Modeling Kit Build-A-Kit® Module Q	162, 352, 353	26, 163, 352, 353
Two-Color Counters	Core Kit Classroom Kit Teacher Modeling Kit Build-A-Kit® Module C	326, 327, 336, 338, 462, 463, 466, 467, 474, 512, 516	326, 327, 462, 463, 466, 467, 474, 512, 516, 517
Fraction Bars	Core Kit Classroom Kit Teacher Modeling Kit Build-A-Kit® Module I	84, 85, 96, 97, 114, 115	84, 85, 96, 97, 114, 115, 130
Fraction Circles	Core Kit Classroom Kit Teacher Modeling Kit Build-A-Kit® Module J	136, 137	107, 136, 137
Pattern Blocks	Core Kit Classroom Kit Teacher Modeling Kit Build-A-Kit® Module L	178	354
Algebra Tiles	Core Kit Classroom Kit Teacher Modeling Kit Build-A-Kit® Module N	284, 285, 294, 295, 482, 483	284, 285, 294, 295, 482, 483
Spinners	Core Kit Classroom Kit Teacher Modeling Kit Build-A-Kit® Module O	68, 261, 262, 264, 265, 269, 290	68, 263, 264, 290
Polyhedron Number Cubes	Core Kit Classroom Kit Teacher Modeling Kit	108, 142, 260, 262, 266, 268, 330, 474	108, 142, 261, 264, 266, 268, 284, 330, 466, 474
Power Solids	Core Kit Classroom Kit Teacher Modeling Kit	186, 187, 188, 189, 190, 191, 192, 194, 195, 198, 422, 423, 424, 425, 426, 427, 428, 429, 430, 431, 432, 433, 434, 435	186, 187, 188, 189, 190, 191, 192, 194, 195, 198, 422, 423, 426, 427, 428, 429, 430, 431, 432, 433, 434, 435
Dot Dice	Core Kit Classroom Kit Teacher Modeling Kit	142, 260, 266, 268, 330, 474	142, 264, 266, 268, 284, 330, 466, 474

Manipulatives: Commercial and Noncommercial

The following chart shows the variety of manipulatives that can be used to develop understanding of a particular topic or concept.

Manipulatives	Problem Solving	Whole Number Concepts & Operations	Fraction Concepts & Operations	Decimal Concepts & Operations	Algebra	Patterns, Relations, & Functions	Statistics, Graphing, & Probability	Ratio, Proportion, and Percent	Measurement, Time, and Money	Geometry
Attrilinks™	■					■				■
Balance	■	■			■				■	
Base-Ten Blocks	■	■		■		■		■	■	
Bills	■	■		■		■		■	■	
Buttons	■	■					■			
Clock	■		■			■				■
Clock Dials	■		■			■				■
Coins (plastic)	■	■		■		■	■	■	■	
Compass	■		■							■
Connecting Cubes	■	■	■		■	■	■	■	■	■
Counters (two-color)	■	■	■		■	■	■	■	■	
Fraction Bars	■		■	■		■		■		
Fraction Circles	■		■			■		■		
Geoboards	■					■				■
Geometric Solids	■					■			■	■
Number Cubes	■	■	■		■	■	■			
Pattern Blocks	■		■			■				■
Plane Figures	■					■		■	■	■
Protractor	■							■	■	■
Spinners	■	■				■	■			
Tangrams	■					■				■

The following chart gives a list of common materials that can be used to develop understanding of a particular topic or concept.

Common Materials	Problem Solving	Whole Number Concepts & Operations	Fraction Concepts & Operations	Decimal Concepts & Operations	Algebra	Patterns, Relations, & Functions	Statistics, Graphing, & Probability	Ratio, Proportion, and Percent	Measurement, Time, and Money	Geometry
Abacus	■	■		■						
Beads	■	■	■			■	■		■	
Bottle Caps	■	■				■				
Clay	■								■	■
Craft Sticks	■	■							■	
Egg Cartons	■	■	■							
Keys	■		■			■	■			
Lids	■								■	■
Noodles; Noodle Stick	■	■							■	
Pipe Cleaners	■								■	■
Plastic Containers	■							■	■	■
Popcorn	■						■		■	
Rice	■								■	
Scales	■	■			■				■	
Shells	■	■	■			■	■			
Stirrers	■								■	
Straws	■	■								■
Tiles	■		■			■	■	■	■	■
Timers	■			■					■	
Toothpicks	■								■	■
Towel Rolls	■							■	■	■

Technology Correlation

Integrated Software	Publisher	CD	Disk	IBM	Mac	Pupil's Edition Pages
Data Toolkit	Harcourt Brace		■	■	■	9, 219, 225, 235, 248, 253, 427
E-Lab	Harcourt Brace	■		■	■	45, 63, 85, 97, 115, 137, 161, 173, 197, 215, 233, 251, 267, 285, 295, 327, 347, 353, 373, 399, 415, 431, 445, 467, 483, 511, 521
Mighty Math Astro Algebra™	Edmark / Harcourt Brace	■		■	■	281, 303, 323, 345, 453, 465, 481, 527
Mighty Math Calculating Crew™	Edmark / Harcourt Brace	■		■	■	23, 43, 113, 191
Mighty Math Cosmic Geometry™	Edmark / Harcourt Brace	■		■	■	155, 179, 359, 371, 397, 411, 433, 509
Mighty Math Number Heroes™	Edmark / Harcourt Brace	■		■	■	67, 87, 103, 135, 265

Correlation to Other Software	Publisher	CD	Disk	IBM	Mac	Teacher's Edition Pages
Alge-Blaster 3	Davidson	■	■	■	■	289
Balancing Act	Ventura Math Power Series	■	■	■	■	489
CornerStone Mathematics	SkillsBank	■		■	■	43, 66, 86, 134, 233, 240, 289, 302, 323, 340, 352, 390, 414, 454, 466, 489, 511
Counting on Frank: A Real Math Adventure	Creative Wonders	■		■	■	302
Fraction Attraction	Sunburst	■	■	■	■	134
Graphers	Sunburst	■			■	233, 240
Hot Dog Stand: The Works	Sunburst	■		■	■	340
How the West Was -1	Sunburst			■	■	454, 466
Math Heads™	Theatrix Interactive	■		■	■	323
Mystery Math Island	Lawrence Productions	■		■	■	390
Shape Up!	Sunburst			■	■	352, 414
TesselMania!	MECC	■		■	■	511

Integrated Calculators	Pupil's Edition Pages
Casio fx-65	42, 43, 45, 88, 450, 452, 470, 520, H30-H41
TI-Explorer Plus	20, 21, 26, 27, 45, 134, 267, 322, 338, 399, 416, H30-H41
	Teacher's Edition Pages
Calculator Connections	22, 62, 66, 338, 464

Professional Handbook

following articles by the authors of
h Advantage describe the philosophy
the research base that guided
elopment of the program.

NCTM Standards and International Studies

by Dr. Karen A. Schultz

"Central to the Curriculum and Evaluation Standards is the development of mathematical power for all students. Mathematical power includes the ability to explore, conjecture, and reason logically; to solve non-routine problems; to communicate about and through mathematics; and to connect ideas within mathematics and between mathematics and other intellectual activity. Mathematical power also involves the development of personal self-confidence and a disposition to seek, evaluate, and use quantitative and spatial information in solving problems and making decisions." NCTM, 1991.

The *Curriculum and Evaluation Standards for School Mathematics* (1989) describes a vision for mathematics designed to develop mathematically literate citizens. The Content Standards emphasize problem solving, written and oral communication, mathematical reasoning, and connections between ideas in mathematics, between mathematics and other disciplines, and between mathematics and everyday activities. The Standards also emphasize a balance between conceptual understanding and computational proficiency and recommend increased attention to data analysis, statistics, number sense, estimation, and algebraic reasoning. This content focus is clearly linked to the increased use of technology and the importance of understanding mathematical relationships as a prerequisite for success in algebra.

The *Professional Standards for Teaching Mathematics* (1991) describes the teaching and learning methodologies that will help students

- work together to make sense of mathematics.
- reason mathematically.
- depend on their own abilities to determine whether something is mathematically correct.
- become good problem solvers.
- connect mathematics, its ideas, and its applications.

NCTM Standards—Their Application in the Classroom

The authors, advisors, editors, field-test teachers, and reviewers designed *Math Advantage* to implement the content and the methodologies described in the two NCTM Standards publications and in curriculum guidelines from states and districts across the country.

Throughout the program you will find examples of this careful attention to both national and local curriculum guidelines. The Chapter Planning Guide for every chapter in *Math Advantage* includes a chart that shows which Standards are developed in that chapter. The teaching suggestions in each lesson plan reflect the recommendations of the Teaching Standards. Critical thinking questions and assessment suggestions focus on stimulating students to synthesize and express the mathematical understanding developed in the lesson. Practice items are carefully crafted to build both conceptual understanding and skill proficiency. Talk About It and Write About It opportunities in the Pupil's Editions help students reason mathematically and communicate their thinking effectively. Opportunities for students to work together cooperatively as they do projects and activities, solve problems, and engage in hands-on activities will help you build a classroom community of learners as well as a collection of individuals.

You can be assured that this approach to mathematics education is based on solid research. In the *Journal for Research in Mathematics Education*, NCTM president Jack Price (1996) cites concrete evidence of the positive effects of the Standards. Studies, particularly those funded by the National Science Foundation, show that students like to solve new problems rather than do repetitious exercises and that solving new problems helps students learn more mathematics in less time. According to Price, "We no longer have to ask, 'Are the Standards working?' We now have evidence that they are" (p. 607).

Among the more pervasive curriculum changes advocated in the Standards is an increased focus on algebra throughout the elementary and middle school grades. Usiskin (1997) urges teachers of kindergarten through Grade 4 to teach algebra's major themes: unknowns, formulas, generalized patterns, placeholders, and relationships. He says that when any of these ideas are taught beginning in kindergarten, the instruction serves as an introduction to the language of algebra, giving children an unsurpassable advantage. Research shows that if the K–6 curriculum focuses on developing algebraic thinking, a significant number of the students will be ready for algebra by Grade 7 (Usiskin, 1997, p. 356).

In *Math Advantage*, the topics of algebra are systematically introduced and developed throughout the elementary grades. In many lesson plans, under the heading "Developing Algebraic Thinking," you will find information about how algebra topics are developed. A major focus of the curriculum design in *Math Advantage* is to carefully and systematically develop over time the critical topics that are prerequisites for success in algebra.

TIMSS—Impact on Classroom Practice

The Third International Mathematics and Science Study (TIMSS), the most comprehensive international study of mathematics and science ever conducted, tested a half million students, of whom 8% were from the United States. The TIMSS report showed that U.S. eighth graders were below average in mathematics, compared with their peers in the United Kingdom, Canada, France, Germany, and Japan. However, U.S. students were found to spend as much or more time in class discussing mathematics and the same amount of time studying mathematics outside of class, compared with their peers in Germany and Japan. Among the recommendations made by the researchers is that students in the United States spend more time focusing on fewer topics at each grade level. This concentration on a few key topics at a grade level ensures a deeper understanding and a higher degree of skill proficiency, alleviating the need to review and reteach skills year after year.

The organization of *Math Advantage* helps you focus on the topics that students must master at your grade level. Shorter chapters, which focus on key concepts and skills, allow you to sequence the program to align with your local curriculum guidelines and the needs of your students. The Assessing Prior Knowledge checkups and frequent assessments allow you to quickly diagnose and remediate students' weaknesses. Practice Activities found in Tab B of the Teacher's Editions provide activities that you can use to improve your students' skill proficiency in critical areas.

The solid research base of *Math Advantage* and the careful attention to core content at every grade level ensure that you can meet the individual needs of your students. *Math Advantage* enables you to develop a classroom community of students who can use mathematics to solve problems each day and to prepare for adult life in the next century.

References

National Council of Teachers of Mathematics (1989). *Curriculum and evaluation standards for school mathematics*. Reston, Virginia.

National Council of Teachers of Mathematics (1991). *Professional standards for teaching mathematics*. Reston, Virginia.

Price, J. (1996). President's report: Building bridges of mathematical understanding for all children. *Journal for research in mathematics education, 27,* 603–608.

Usiskin, Z. (1997). Doing algebra in Grades K–4. *Teaching children mathematics, 3,* 346–356.

Building Mathematical Concepts and Understanding

by Dr. Evan M. Maletsky

"Representations are crucial to the development of mathematical thinking, and through their use, mathematical ideas can be modeled, important relationships identified and clarified, and understandings fostered. Physical models, materials, calculators, and computers help provide the array of rich and substantive experiences needed to build a deep and comprehensive knowledge of mathematical concepts and procedures." NCTM, 1991.

In elementary school and middle school, students begin to develop their own abilities to think critically and solve problems in mathematics. The support for this thinking comes from their level of understanding and skill in dealing with the mathematical concepts they have been taught. What is it that establishes this necessary foundation?

Manipulatives—Powerful Tools for Building Understanding

Many abstract concepts have their beginnings in concrete, hands-on experiences. This is especially true when it comes to the learning process that takes place in the mathematics classroom. Manipulatives offer powerful tools for developing solid understanding of mathematical concepts.

For example, put some 2-in. x 8-in. strips of paper in the hands of your students when they are studying the concept of volume. Although these strips start out as flat, two-dimensional plane figures, they can easily be folded into models that are three-dimensional. Provide 1-in. wooden cubes as units of volume, and have students fill paper-strip models of a 2-in. cube and a 1-in. x 3-in. x 2-in. rectangular prism. This activity sets in motion an understanding of how linear dimensions relate to volume. By changing the folding pattern, your students can easily construct and explore prisms of other shapes.

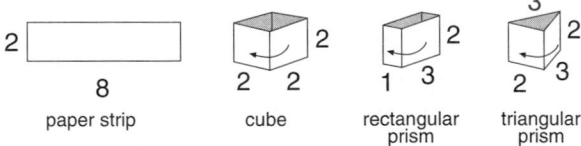

paper strip   cube   rectangular prism   triangular prism

Sometimes the best manipulatives for the classroom are the simplest ones. A 2-in. x 8-in. strip of paper may not look like much to work with, at first. But it may, in fact, be just the needed visual model to give meaning to an otherwise abstract concept.

Have students fold a strip in half and in half again, open it up, and look at the unfolded strip. What mathematical ideas would you want your students to see when they look at the unfolded strip?

Visual Models—A Way of "Seeing" Mathematics

Many arithmetic and algebraic concepts and processes are abstractions to students because they don't see any reality in these concepts and processes. Here is where manipulatives, such as folded paper strips, can offer visual support. They become concrete models that can be used, as needed, to solve problems.

Consider for a moment the concept of percent. Many students in the middle grades struggle with percent. They've been taught the computational algorithms, but what do they see? Even many of those who get the correct numerical answers to percent problems have little if anything to say about what those numbers mean.

Think again about our folded strip. If the whole strip represents 100%, then each of the four squares is a visual model for 25%. The squares can be combined with the eye to show 50% and 75% as well.

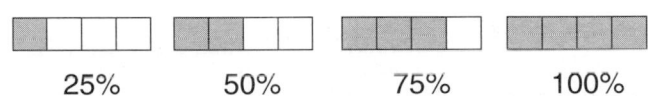

25%   50%   75%   100%

Of course, if a small square represents a whole, then from the very same model, students can see 100%, 200%, 300%, and 400%. But then, you can assign any value at all to the squares or to the strip. Suppose you assign the value of 12 to the whole strip. Then, using the concept of area, students quickly see the numbers 3, 6, and 9 as well.

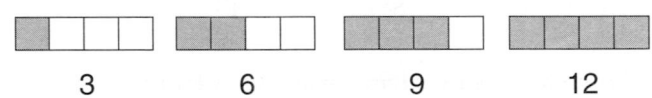

3   6   9   12

Maybe your students see even more. The same numbers of squares are shaded to show 9 and 75%. Does this mean that 75% of 12 is 9? Clearly, it does.

On one of the past National Assessment of Educational Progress tests, eighth graders were asked to find 75% of 12. Sadly, less than half answered correctly. One wonders whether those who did not could "see" a visual model to represent this problem.

Examples—Models for Thinking

Students begin to develop their own mathematical skills and problem-solving processes by following the examples shown in their textbooks and demonstrated by their teachers and peers. When these examples can involve hands-on experiences or visual models, they are all the more powerful. Where possible, make your classroom examples dynamic and motivating by involving action and change, and embed them in mathematical thinking and problem-solving situations.

Think again about the folded strip. This time, number the squares 1, 2, 3, and 4 on one side and 5, 6, 7, and 8 on the other side, with the 8 behind the 1. Now cut the strip apart into four squares. Tell students to think of the numbers as digits, and ask the following questions about arrangements that can be explored hands-on:

- How many two-digit numbers can be formed with the digit 2 in the tens place?

21   23   24   25   26   28

- Do you see why the number 27 cannot be formed?
- What are the 24 different three-digit numbers possible with 2 in the tens place?
- What are the 48 different four-digit numbers that have 2 in the tens place?

You can adapt the example to any grade level. One adaptation is to simplify the problem by using only the digits 1, 2, 3, and 4 on one side. At a more challenging level, make the problem a cooperative learning activity in which students count all possible arrangements with one, two, three, or four digits.

$$8 + \quad 8 \times 6 + \quad 8 \times 6 \times 4 + \quad 8 \times 6 \times 4 \times 2$$
one-   two-   three-   four-digit choices

$$= 8 + 48 + 192 + 384 = 632$$

Repeated digits are possible if students work in groups of four with digits chosen from the combined 16 squares of the four students. Now the count for all possible arrangements with one, two, three, and four digits is 4,680.

$$8^1 + \quad 8^2 + \quad 8^3 + \quad 8^4$$
one-   two-   three-   four-digit choices

$$= 8 + 64 + 512 + 4,096 = 4,680$$

Examples linked to hands-on experiences or visual models are powerful models for thinking. Careful development of concepts and procedures through the use of examples ensures that practice exercises will deepen understanding and build proficiency.

Reasoning—A Key to Unlocking Mathematics as a Way of Thinking

In its *Curriculum and Evaluation Standards for School Mathematics*, the National Council of Teachers of Mathematics highlighted reasoning as one of its key

standards, and this emphasis is being widely maintained in the standards of individual states and districts. Reasoning is at the heart of mathematical thinking as we know it today, and it has been so for some 2,500 years, since the time of the ancient Greek philosophers such as Plato and Socrates. We need to bring this same emphasis on reasoning into mathematics classrooms every day. The following problem provides an example of reasoning that uses the four numbered paper squares described earlier:

- What arrangement of four digits will form two numbers with the greatest possible sum? the greatest possible product?

Trial and error is one approach to this problem, but careful reasoning is another. Here are two promising choices:

$$\begin{array}{r} 8\ 6 \\ +\ 7\ 5 \end{array} \qquad \begin{array}{r} 8\ 5 \\ +\ 7\ 6 \end{array}$$

These may look like winners, but they are not: 881—a much greater sum—is also possible with the same four digits. Remember, nothing was said about both numbers having two digits.

As for the greatest possible product, consider the choices shown below. Which product do you think is the greatest? Why? Is there a better choice?

$$\begin{array}{r} 8\ 7 \\ \times\ 6\ 5 \end{array} \qquad \begin{array}{r} 8\ 6 \\ \times\ 7\ 5 \end{array} \qquad \begin{array}{r} 8\ 5 \\ \times\ 7\ 6 \end{array} \qquad \begin{array}{r} 6\ 8 \\ \times\ 5\ 7 \end{array}$$

George Polya, in his delightful book titled *How to Solve It* (1971), presents the following four key steps in solving a problem:

- Understand the problem.
- Devise a plan.
- Carry out the plan.
- Look back.

Although they have been stated in many different ways, these components of problem solving remain the grand design. These must be the focus of teaching as you and your students talk through and then work through examples. In the arrangement example above, at one level the *plan* might be to physically arrange the paper squares to form the different numbers, making a list as you go. At another level the physical arranging may be used essentially to *understand* the problem, while the planning stage is used to seek out the multiplication property of counting as the best process.

In *Math Advantage* the approach to building mathematical concepts is to provide learning experiences based on reasoning before presenting practice. This approach develops mathematics as a way of thinking, not an isolated group of skills to be memorized and practiced for automaticity.

Writing—A Record of Thinking and Learning

With today's emphasis on objective testing, teachers often miss powerful alternative methods of assessing students' knowledge and thinking. Having students write about their thoughts and methods when solving problems can be very revealing.

Think again about our folded strip. This time, have students keep the strip folded into a square and visualize what it will look like when it is unfolded. Ask them to write about their mental model by answering the following question:

- How many rectangles of all sizes will be in the strip when it is unfolded?

Some students will see only the four squares. Others will just randomly count different rectangles as they mentally see them. Still others will do some organized, systematic counting and find that there are ten different rectangle possibilities. Having students write about their methods, as well as give their answers, brings another dimension into the assessment process.

Some students will tackle this problem by first using the strategy of simplification. Some may start with a simpler situation, such as a strip with zero, one, or two creases. Some may list their results in a table such as this one and see some interesting number patterns emerge.

Number of creases	0	1	2	3	4	5	6
Number of parts	1	2	3	4	5	6	7
Number of rectangles	1	3	6	10	15	21	28

The entries in the last row are the triangular numbers. You may be surprised to see them embedded in this problem.

Writing helps students organize their thoughts, use the mathematical language they have developed, and build a record of their thinking and learning. Throughout *Math Advantage*, students are given many opportunities to write about the mathematics they are learning. Students can keep this writing in their Math Journals and include samples of it in their portfolios, providing an opportunity for you to assess students' growth in reasoning and in written representations of that reasoning.

Vocabulary—Building the Language of Mathematics

Mathematics has many faces. In certain respects mathematics is a language—a universal language. The words *triangular numbers* may not be meaningful to all those who speak the 3,000 different languages of the world. But when the ideas that those words stand for are displayed in arithmetic, geometric, or algebraic form, they can be read by millions upon millions who speak the universal language of mathematics.

In this arithmetic view of the triangular numbers, we see them as sums of successive sets of counting numbers. The numerical pattern can be easily recognized and readily extended.

$$
\begin{aligned}
1 &= 1 \\
3 &= 1 + 2 \\
6 &= 1 + 2 + 3 \\
10 &= 1 + 2 + 3 + 4 \\
15 &= 1 + 2 + 3 + 4 + 5 \\
21 &= 1 + 2 + 3 + 4 + 5 + 6
\end{aligned}
$$

In this geometric view of the triangular numbers, we see them as triangular arrangements. Here the geometric pattern is apparent.

| 1 | 3 | 6 | 10 | 15 |

In this algebraic view, a variable, n, is used in a general formula to describe the character of the nth triangular number, $T(n)$.

$$T(n) = \frac{n(n + 1)}{2}$$

To find the tenth triangular number, substitute 10 for n. The tenth triangular number is 55.

$$T(10) = \frac{10(10 + 1)}{2} = \frac{10(11)}{2} = 55$$

The tenth triangular number, whether modeled in arithmetic, geometric, or algebraic form, is 55.

Throughout the years, mathematics has been the model for logic and reasoning because of the precision of its language, its vocabulary. In fact, it is the abstract, deductive nature of mathematics—its emphasis on definitions, reasoning, and proof—that makes mathematics unique among the disciplines. The foundation for future success in mathematics is built on the basic concepts and skills developed in the elementary and middle school years and on the mathematical thinking inherent in problem-solving experiences at these levels. You can promote students' success in large measure by helping them use mathematical language to explain and clarify their thinking.

At every level, a mathematics program must offer a rich blend of concept development, skill-oriented activities, and problem-solving opportunities. Attention must be given throughout to reasoning, writing, and visualization, with special emphasis on manipulatives and visual models as powerful tools for exploring, enriching, and extending mathematical ideas. *Math Advantage* captures these ideas in a new and refreshing way. It offers the best way to reach and teach your students.

Mathematics and the mathematical experience must tickle the senses as well as sharpen and stretch the mind. It is through handling, seeing, and thinking that students can sense the excitement, appreciate the beauty, and share in the creativity of the subject.

References

National Council of Teachers of Mathematics (1989). *Curriculum and evaluation standards for school mathematics*. Reston, Virginia.

National Council of Teachers of Mathematics (1991). *Professional standards for teaching mathematics*. Reston, Virginia.

Polya, G. (1971). *How to solve it*. Princeton, New Jersey: Princeton University Press.

Strategies for Teaching Mathematics

by Dr. Grace M. Burton

"The final success for any teacher is the integration of theory and practice." NCTM, 1991.

Teachers intuitively know that students learn best when they are challenged to go beyond their present understandings and competencies, when they are helped to form questions that will enrich and extend their knowledge, and when they are given opportunities to apply their knowledge in both familiar and novel situations. The key to a teacher's success is his or her ability to plan and deliver instruction that provides this rich environment and that meets the varied needs of all the students in the classroom.

The research and writings of classical theorists such as John Dewey, Jean Piaget, Lev Vygotsky, Jerome Bruner, and others have suggested theories of how children learn and how teachers can facilitate the learning process. More recently, educators such as David and Roger Johnson, in publications on cooperative learning; Howard Gardner, in publications on multiple intelligences; and Mary Baratta-Lorton, in instructional materials for primary grades, have substantially added to our understanding of the teaching-learning process. Their work has provided practical guides for planning and delivering instruction that is rich, challenging, and appropriate to children's developmental levels.

Developmental Learning–From Concrete to Pictorial to Abstract

The work of Jean Piaget and the later work of Mary Baratta-Lorton have convinced many teachers that the most effective way to develop mathematical concepts is to have children model mathematical situations with concrete materials. These *concrete experiences* allow children to use appropriate manipulatives to build models of their understanding. Once children have had the experience of manipulating objects, they can then describe their actions and their models in informal oral and written language–a necessary first step along the path toward symbolic representation and the development of formal math vocabulary.

Children can then transfer the initial understandings, developed through manipulation and communication, by representing the models in pictures and drawings. The *pictorial representation* stage of development helps children build a reference bank of mental images that they can access to interpret mathematical symbols. Drawing pictures is a powerful form of communication that can in turn help children develop verbal communication skills as they describe their representations.

Finally, children link their models and their pictures representing those models to the formal symbolic language of mathematics. Children are now able to understand mathematics in its *abstract* presentation because they can link the abstractions back to their own mental models and pictures.

This developmental sequence forms the basis for the instructional approach in **Math Advantage**. Students build a rich repertoire of models and drawings, mental images, and symbolic mathematical language that leads to proficiency in both skill development and problem solving and that facilitates algebraic reasoning.

Cooperative Learning–Preparation for Real-Life Experiences

Adults often find it helpful to work with others. This approach predominates in many work settings. Johnson and Johnson (1994) have described an instructional method–working in cooperative groups–that provides opportunities for students to work together to solve problems and communicate their thinking. This instructional strategy develops social skills as well as academic skills and prepares students for the world of work.

Teachers can promote active involvement in mathematics by assigning students to mixed-ability groups of four to six students and giving each group a problem to solve. The key to cooperation within the group is the development of the social skills that enable students to work with others on complex tasks. In addition, students need defined roles within the group to ensure that each group member is actively involved in group management as well as in the learning activity.

In *Math Advantage* students have many opportunities to work in cooperative groups on hands-on, skill development, and problem-solving activities. These opportunities include the following:

- *Learning Centers* in every chapter in Kindergarten through Grade 5 provide concept development and skill practice activities.

- *Activity Options* in every lesson in Kindergarten through Grade 2 provide cooperative learning activities directly tied to the lesson objective.

- *Problem of the Day* problems in every lesson in Kindergarten through Grade 8 encourage students to choose strategies, try multiple strategies, discuss their thinking, and present their solutions to application, non-routine, and open-ended problems.

- *Team-Up Time* (Grades 3-5) and *Team-Up Project* (Grades 6-8) provide activities in which students work together to plan their approach, complete their task, and share their results with classmates. A checklist allows students to assess their own work before sharing it.

- *Problem-Solving Strategy* lessons in Grades 1-8 provide opportunities for students to use the Problem-Solving Think Along®, a guide for using the four-step approach to thinking through a problem, in practice activities designed for cooperative learning groups.

- *Practice Activities*, found in Tab B in the Teacher's Editions for Grades 1-8, provide opportunities for cooperative group activities that help students strengthen key skills.

Multimodal Strategies–A Way to Reach Every Child

Every learner is unique in the ways in which he or she processes new information and integrates it with existing mental structures. Some students learn best when they have visual input to process. For these students the saying "A picture is worth a thousand words" is an apt description of their preferred learning modality. Other students learn and remember best when lessons are presented in an auditory modality. Hearing is the key to these students' new learning. Still other students learn best when they manipulate objects and use their whole body in learning experiences. For these students learning occurs when they can involve their sense of touch and use body movement as their preferred modality.

Math Advantage offers many opportunities to meet the needs of every student in your classroom by providing activities that are built on the use of all modalities. You will find each activity clearly labeled with its predominant modality:

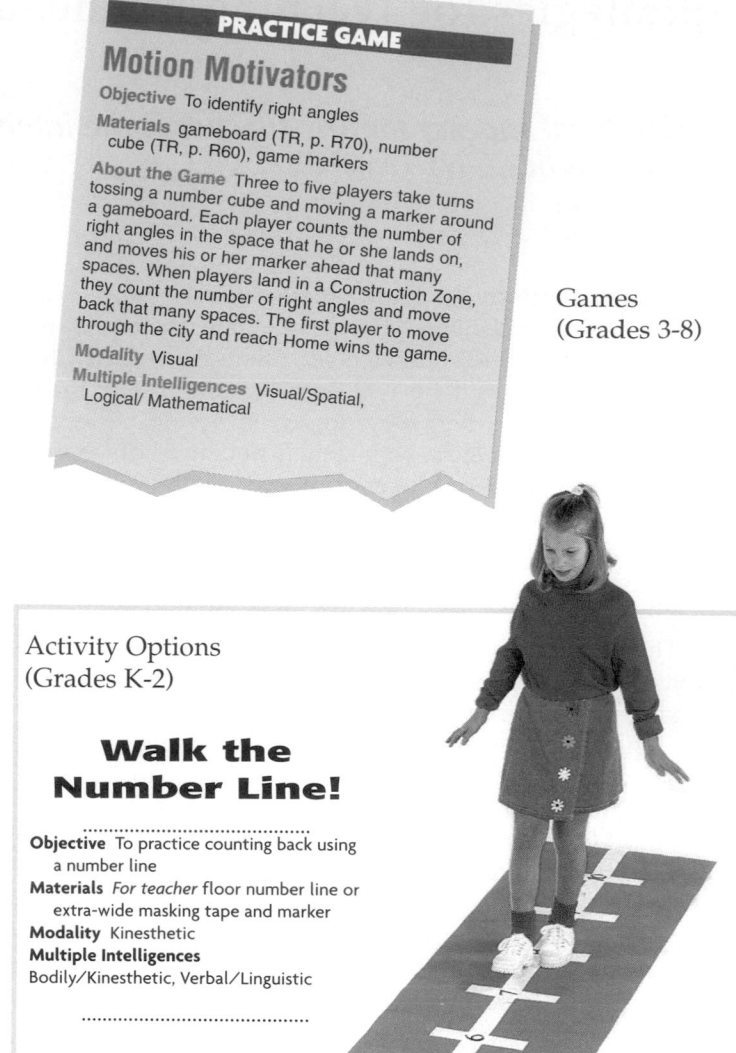

Games
(Grades 3-8)

Activity Options
(Grades K-2)

Idea File Activities
(Grades K-8)

Multiple Intelligences–A New Dimension for Mathematics

In his book *Frames of Mind* (1983), Howard Gardner introduced his theory of multiple intelligences. He defined seven intelligences that each of us has–logical/mathematical, verbal/linguistic, visual/spatial, musical/rhythmic, bodily/kinesthetic, interpersonal/social, and intrapersonal/introspective. The key to applying Gardner's theory to classroom practice is to evaluate how each student demonstrates intelligences and how the math topics you are teaching can be approached through the various intelligences. Certainly it is not possible to approach every topic from the perspective of all seven intelligences. However, you can use Gardner's theory to choose a variety of instructional strategies that encompass multiple intelligences. To assist you in making these instructional decisions, activities in *Math Advantage* are labeled to show the predominant intelligences developed in each activity:

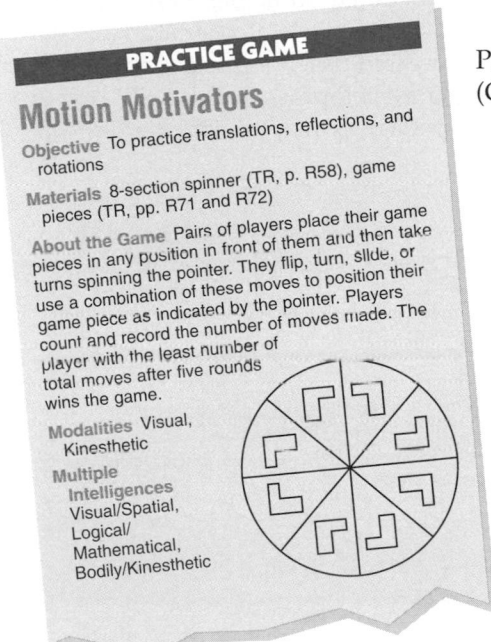

Practice Game
(Grades K-8)

Follow-Up Strategies and Activities
(Grades K-2)

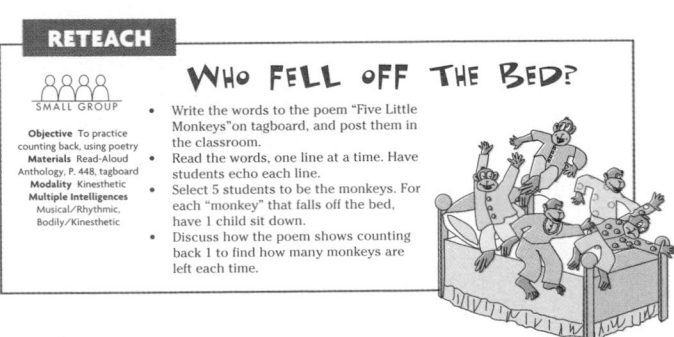

Interdisciplinary Suggestions
(Grades 6-8)

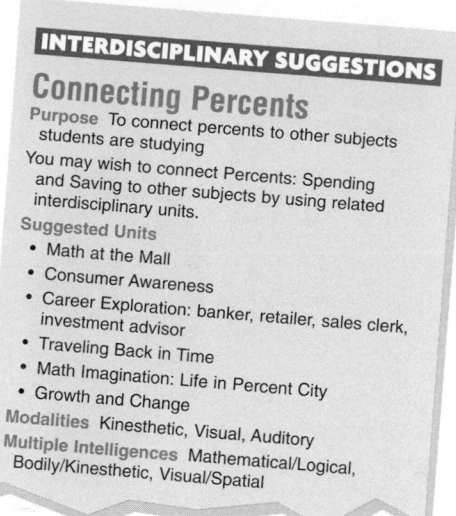

Learning Centers

Among the most important decisions that teachers make is how to provide time and space for remedial work and for extension and enrichment opportunities. An effective way to do this is by making learning centers a vital part of the mathematics curriculum. Center activities offer unique opportunities for concept and skill development, problem-solving practice, and the development of multiple intelligences.

Learning Center Cards
(Grades K-5)

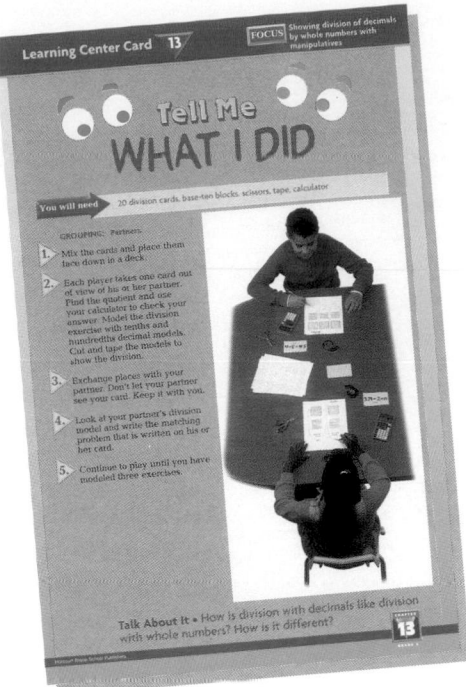

The Learning Centers in *Math Advantage* encourage cooperative learning and the development of interpersonal intelligence. They provide skill practice through the use of games and tasks that differ from the paper-and-pencil practice in the Pupil's Editions. The Practice Activities found in Tab B of the Teacher's Editions are appropriate for learning centers and can provide review and remediation in a motivating context. Many of the ideas found in the Idea File activities are appropriate for both review and extension centers.

Using learning centers as a part of your instructional program allows you to meet the needs of individual students, to group students according to their instructional needs, and to provide cooperative learning experiences that enhance all students' social and academic skills.

Problem-Centered Learning

According to the NCTM Standards, problem solving should be the focus of the curriculum. Problem-solving activities in *Math Advantage* range from application problems to non-routine and open-ended problems to projects and activities that center the learning of mathematical content on a large problem. As the Professional Standards states, "Good tasks nest skill development in the context of problem solving" (NCTM, 1991, p. 32). The project with each chapter opener in Grades 3-8 presents, in a motivating context, a problem that promotes skill and concept development, problem-solving strategy development, cooperative learning, and communication skills development and practice. For some students, these motivating problems make mathematics relevant and answer the question "Why do I have to learn this?"

The wide variety of teaching strategies presented in *Math Advantage* will help you make your mathematics instructional program a rich experience for your students. You can choose instructional strategies based on your style of teaching and on the needs of your students with the assurance that research-based strategies for concept and problem-solving development and skill practice form the program's pedagogical approach.

References

Gardner, H. (1983). *Frames of mind: The theory of multiple intelligences.* New York: Basic Books.

Johnson, David W., Johnson, Roger T., and Holubec, Edythe Johnson (1994). *The new circles of learning: Cooperation in the classroom and school.* Alexandria, Virginia: Association for Supervision and Curriculum Development.

National Council of Teachers of Mathematics (1989). *Curriculum and evaluation standards for school mathematics.* Reston, Virginia.

National Council of Teachers of Mathematics (1991). *Professional standards for teaching mathematics.* Reston, Virginia.

Developing Algebraic Thinking

by Dr. Terence H. Perciante

"One critical transition is that between arithmetic and algebra. It is thus essential that in grades 5-8, students explore algebraic concepts in an informal way to build a foundation for the subsequent formal study of algebra. Such informal explorations should emphasize physical models, data, graphs, and other mathematical representations rather than facility with formal algebraic manipulation. Students should be taught to generalize number patterns, to model, represent, or describe observed physical patterns, regularities, and problems. These informal explorations of algebraic concepts should help students to gain confidence in their ability to abstract relationships from contextual information and use a variety of representations to describe those relationships."
NCTM, 1989.

For too many students, algebra becomes the stumbling block to a successful high school experience, to acceptance into college, and to a career choice that they can make based on their strengths rather than on their weaknesses. Measures to prevent this failure and the consequent limiting of life choices for students must begin in the early years of school.

Algebra--A Way of Thinking

Algebra is really a language and a way of thinking. To develop algebraic thinking ability in students, the mathematics curriculum must build concepts before skills, relationship thinking before formal manipulation, and concrete meaning before the related symbolic notation. The fundamentals of algebra are not number and symbol manipulation, even though students' success in algebra depends upon a strong proficiency in number and operation concepts. Rather, the fundamentals of algebra are relationship thinking, the ability to represent problems in various formats, the ability to reason logically, and an understanding of the concept of equivalence.

Relationship thinking is the ability to "see" relationships such as those expressed in patterns, problems, and formulas. For example, if you know the total amount of a loan, including interest, and the number of months in which the loan must be repaid, you can relate these two quantities to find the monthly payment. Perhaps the best examples of relationship thinking are seen when you use the formulas for distance and interest. You can establish a relationship between distance, rate, and time or between interest, principal, rate, and time based on which quantity is unknown. The identification of patterns is based on relationship thinking. Developing rules for operating with integers, determining the next number in a pattern, and examining numbers that produce geometric patterns are examples of expressing relationships in a pattern.

Representational thinking is focused on translating a word problem into an algebraic equation or a visual representation. Examples of representations in algebra include variables, expressions and equations, tables and graphs of linear equations, systems of equations, and inequalities. The ability to produce multiple representations of a relationship is a key to success in algebra. For example, if students can use concrete objects to model an equation on a simple balance, can use words to describe that relationship, and can then represent that model with numbers and symbols, they are developing their ability to represent their thinking in multiple ways.

Logical reasoning is evident when students can engage in "If. . . then" thinking. For example, when students think, "If I am going to travel 150 mi at 50 mi per hour, then I will be on the road for 3 hr," they are engaging in logical reasoning. If students had needed to find the rate of speed, they would have reasoned, "If I know the distance and the time, then I can divide the distance by the time to find the rate, or miles per hour."

Equivalence is thinking based on a comparison of quantities to determine whether or not they are the same. Students test for equivalence when they express a given quantity as a decimal, a fraction, and a percent. They maintain the equivalence relationship in an equation as they work through the steps to solve it. They use scientific notation as an equivalent representation of a number. They test their understanding of the concept of equivalence when they identify nonequivalent relationships.

In *Math Advantage* these fundamental ideas are embedded in the instructional approach. New concepts are introduced through the use of manipulative materials. Students are asked to record their thinking about new concepts by drawing representations of their models and by describing their thinking. Skills are developed following a careful sequence of concept development at the concrete level, followed by visual representations of the procedure in a step-by-step model. Symbolic notation is developed as a recording of the physical model and the visual representation. This developmental process is based on logical reasoning, on linking concepts and procedures to problem-solving experiences, and on representing problem-solving experiences in expressions, equations, tables, and graphs. The number and operation strand in the program is focused on developing the concept of equivalence. Students carefully develop their ability to test for equivalence or nonequivalence as they record names for numbers, compare quantities, represent a quantity as a fraction, decimal, and percent, and carry out other procedures.

Preparation for Algebra–It Begins in Kindergarten

Every essential concept and technique treated in algebra finds its first and most substantial development within the curriculum of the early grades from kindergarten onward. These concepts include the properties of numbers, the meaning of the operations, the notions associated with comparison symbols, the connection of numerical representation to graphical representation, and ideas associated with equivalent expressions and equations. Simply put, algebra symbolically and algorithmically summarizes that which students know concretely and numerically from the elementary grades.

Within the last decade, a heightened emphasis upon concept development and student articulation of emerging ideas has found concrete expression in journal activities. Commenting on her own experience in using journal assignments for these purposes with her own students, Kathleen Chapman wrote: "I found these journal assignments extremely valuable for diagnosing and troubleshooting misconceptions . . . . The journal was not only proving to be a valuable window on thinking but was also being used by the students as a forum for their frustrations" (Chapman, 1996, p. 589).

In the same article, Chapman referred to a natural concern that such journal assignments may absorb excessive instructional time, thereby detracting from the teaching process rather than delivering long term benefits. However, Chapman suggests that the opposite is true, citing the assertion that instructional time is saved both by having students explain what they mean and by examining their stated ideas.

Writing forces a kind of analysis and synthesis of ideas that go beyond memorization of facts. It is an intentional articulation of mathematical processes. In *Math Advantage* the Write About It activities found in many of the practice sets provide students with this opportunity for articulation of mathematical processes.

In *Math Advantage* the development of algebraic thinking truly begins in kindergarten as children work with the idea of testing for equivalence by establishing a one-to-one correspondence between groups of objects and as they build their understanding of a number as being one more than or one less than another number. This same careful development of algebraic thinking permeates the entire program. In the lesson plans for many lessons, a section titled Developing Algebraic Thinking describes how the content of the lesson is preparation for algebra. As you work through the program with your students, you can be confident that the fundamentals of algebra-relationship thinking, representational thinking, logical reasoning, and the concept of equivalence–are built into the very fiber of the program.

References

Chapman, K. P. (1996). Journals: Pathways to thinking in second-year algebra. *Mathematics Teacher*, 89, (October 1996), 588-590.

National Council of Teachers of Mathematics (1989). *Curriculum and evaluation standards for school mathematics*. Reston, Virginia.

Problem Solving–
The Focus of Mathematics Instruction

by Dr. Muriel Burger Thatcher

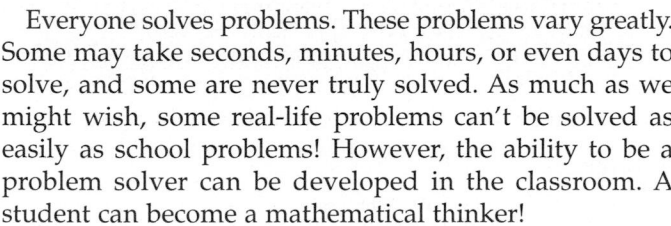

> **"When problem solving becomes an integral part of classroom instruction and children experience success in solving problems, they gain confidence in doing mathematics and develop persevering and inquiring minds. They also grow in their ability to communicate mathematically and use higherlevel thinking processes." NCTM, 1989.**

Everyone solves problems. These problems vary greatly. Some may take seconds, minutes, hours, or even days to solve, and some are never truly solved. As much as we might wish, some real-life problems can't be solved as easily as school problems! However, the ability to be a problem solver can be developed in the classroom. A student can become a mathematical thinker!

Memorization of facts, while a key part of building mathematical power, is not sufficient to ensure that students can solve meaningful problems. In fact, as early as 1980, NCTM, in its *An Agenda for Action* (1980, p. 2), stated, "Problem solving must be the focus of school mathematics." The 1989 Standards identified problem solving as the number-one goal for Grades K-12 and stated, "To develop such abilities, students need to work on problems that may take hours, days, and even weeks to solve. Although some may be relatively simple exercises to be accomplished independently, others should involve small groups or an entire class working cooperatively" (NCTM, 1989, p. 6).

This focus on problem solving implies a change in philosophy and practice for mathematics teaching. The Professional Teaching Standards describes this change by stating, "This is a major shift from learning mathematics as accumulating facts and procedures to learning mathematics as an integrated set of intellectual tools for making sense of mathematical situations" (NCTM, 1991, p. 2).

In order to help teachers focus on this shift in emphasis in the classroom, the 1989 Standards set the following goals for students:

- Learning to value mathematics–its role in and impact upon our technological society, historically and at the present time

- Becoming confident in one's own ability–having varied experiences which create trust in one's own mathematical thinking and power

- Becoming a mathematical problem solver–exploring and reflecting upon varied, unique, and often open-ended problems while working in individual and cooperative group experiences

- Learning to communicate mathematically–using the language of mathematics naturally in discourse, writing, and clarifying ideas

- Learning to reason mathematically–empowering logical thinking and conjecturing, which stresses the process approach to learning (NCTM, 1989, pp. 5-6)

In order to make these goals a reality for all students, problem-solving experiences and hands-on experiences designed to develop both conceptual and procedural understanding must be balanced with practice that develops skill proficiency. The successful student will be proficient with the skills and procedures of mathematics and will be able to apply those skills confidently in problem-solving situations. This competency frees students to focus on the process of problem solving,". . . the process by which students experience the power and usefulness of mathematics in the world around them" (NCTM, 1989, p. 75).

Problem Solving–A Process

Problem solving involves much more than exercises that use and practice algorithms. It consists of applying prior knowledge to find a possible solution for which there is no obvious path. "To solve a problem is to find a way where no way is known off-hand, to find a way out of a difficulty, to find a way around an obstacle, to attain a desired end that is not immediately attainable, by appropriate means" (NCTM, 1989, p. 75).

George Polya (1971) devised a problem-solving approach that guides students through the thinking process necessary for effective problem solving. His process includes the following steps:

- **Understand the Problem**–Students restate the problem in their own words, identify the information needed to solve the problem, and determine what question they are being asked.

- **Make a Plan**–Students choose a strategy by which they can reach a solution to the problem. In addition, they predict what answer they might get, or they estimate to determine a reasonable solution to problems requiring a numerical result.

- **Carry Out the Plan**–Students apply the strategy they have chosen. They evaluate their results in light of their plan and either formalize their solution or choose another strategy if their first choice did not work.

- **Look Back**–Students reflect on their choice of strategy and their solution to determine whether the solution is reasonable in light of the information given in the problem and whether the solution answers the question stated in the problem.

In *Math Advantage,* this process approach to problem solving forms the basis for the problem-solving strand. Using the key words–**Understand**, **Plan**, **Solve**, and **Look Back**–and the Problem-Solving Think Along®, students are guided through this thinking process, thereby ensuring that they develop the thinking and reasoning skills necessary for successful problem solving.

Problem-Solving Strategies–Developing, Practicing, Choosing and Justifying

Strategies are general approaches to solving a problem that require reasoning. Examples of strategies include *act it out, draw a diagram, find a pattern, guess and check, make a model, make a table, solve a simpler problem, use a formula, work backward,* and *write an equation.* As students develop their ability to use strategies, they will find them useful in clarifying the relationships in the problem and in presenting their solution.

In *Math Advantage*, problem solving is integrated into every lesson and forms the basis for developing conceptual and procedural understanding and proficiency. Students have opportunities to do the following:

- **develop strategies** as they work through lessons that present and model each of the problem-solving strategies

In Grades 1-2, children are using the four-step plan to solve problems using appropriate strategies. In Grades 3-5, one part of a lesson in the chapter is devoted to using the chapter math content in a problem-solving situation for which an appropriate strategy is modeled. In Grades 6-8, a section at the beginning of the book, titled "Focus on Problem Solving," reviews and models all of the strategies developed in the preceding grades. Students are referred to this section as they work through problem-solving situations throughout the year. In addition, a lesson focusing on each strategy is developed using appropriate math content.

- **practice strategies** as they solve problems presented throughout the lessons

In Grades 1 and 2, problem-solving activities are presented in which children practice the strategies that have been developed in the program. A workspace encourages children to use manipulatives, to draw pictures, and to record numbers and symbols as they work through the problem. In Grades 3-8, Mixed Applications and Problem-Solving Applications sets of problems in math content lessons provide practice in applying strategies to a wide variety of types of problems. The Problem of the Day in the lesson plan for each lesson provides opportunities for students to solve a wide variety of types of problems using the strategies they have learned.

- **choose strategies** as they solve a wide variety of types of problems

In the problem-solving strategy lessons, the Mixed Applications are designed to help students further develop their understanding of the problem-solving strategy presented in the lesson by presenting a wide variety of types of problems for which students must choose from a list of strategies. This practice in choosing an appropriate strategy is a critical step in the learning of a new strategy because it helps students compare strategies and understand what types of problems are best solved using a given strategy. They continue to practice choosing strategies as they work through the Mixed Applications in each lesson.

- **justify their choice of strategies** as they describe their thinking orally or in writing

As students reflect on the process of solving problems and look back at their solutions, they justify their choice of strategies and their solutions. They decide whether their answers are reasonable and whether they could have solved the problem by using a different strategy.

Problem solving is the focus of instruction in *Math Advantage*. The careful balance between the development of problem-solving abilities and skill proficiency ensures that students can confidently use their math skills as they think, reason, and apply the process of problem solving to relevant and interesting problems. Students learn that the skills they have learned are effective tools for problem solving.

References

National Council of Teachers of Mathematics (1980). *An agenda for action.* Reston, Virginia.

National Council of Teachers of Mathematics (1989). *Curriculum and evaluation standards for school mathematics.* Reston, Virginia.

National Council of Teachers of Mathematics (1991). *Professional standards for teaching mathematics.* Reston, Virginia.

Polya, G. (1971). *How to solve it.* Princeton, New Jersey: Princeton University Press.

Practicing for Proficiency

by Dr. Loye Y. (Mickey) Hollis

"Good tasks are ones that do not separate mathematical thinking from mathematical concepts or skills, that capture students' curiosity, and that invite them to speculate and to pursue their hunches." NCTM, 1991.

Drill is different from practicing for proficiency, both in theory and in application. Drill theory proposes that the student repeat a concept or skill until a correct response is automatic. The emphasis is on a correct response, and memorization is an important instructional strategy. Drill theory was criticized in 1935 by Brownell, who proposed an alternative: meaning theory (Kilpatrick, 1992, p. 19). More recently Wakefield states, "When children learn to get answers with memorized algorithms without understanding why these 'tricks' work, their confidence in their own ability for problem solving is undermined" (Wakefield, 1997, p. 105).

There is no "practice theory" of teaching mathematics. Rather, the concept of practice is embedded within other instructional strategies. Hiebert and Carpenter describe a theory based on learning mathematics with understanding. They define understanding in terms of the way information is represented and structured. They state, "A mathematical idea or procedure or fact is understood if it is part of an internal network. More specifically, the mathematics is understood if its mental representation is part of a network of representations" (Hiebert & Carpenter, 1992, p. 67).

This network of representations is used to make connections in several ways. A connection may be with a previously learned mathematical idea, procedure, or fact; an idea, procedure, or fact previously learned in another subject; a mathematical problem; or a nonmathematical problem. In other words, networking helps learners connect the mathematics being learned with existing mathematical knowledge. Practice facilitates making and reinforcing these connections. Proficiency is being able to apply the mathematics in a different context and to use it when solving problems.

Standard 1 of the NCTM Professional Standards recommends that teachers of mathematics should pose tasks that

- develop students' mathematical understandings and skills.
- stimulate students to make connections and develop a coherent framework for mathematical ideas.
- display sensitivity to and draw on students' diverse background experiences and dispositions (NCTM, 1991, p. 25).

Practicing for proficiency is consistent with the above recommendations.

The practice in *Math Advantage* is designed to build proficiency—a balance between the understanding of concepts and automaticity with skills. This balance is critical to ensure that students can use mathematics as a tool for problem solving and can achieve success in school mathematics, including the ability to score well on standardized tests and other assessment instruments.

Paper-and-pencil practice is a key component of every lesson in the Pupil's Editions. Practice sets are carefully designed to help students first reinforce the concepts underlying a skill and then practice the entire procedure or skill for the purpose of getting correct answers. In Kindergarten through Grade 4, there is a strong emphasis on developing the basic fact strategies and memorizing the basic facts for all four operations. Additional practice is available in the More Practice section of the Student Handbook in the Pupil's Editions for Grades 3-8. Practice designed to meet the individual needs of students is available in the Reteaching, Practice, and Enrichment Workbooks for Grades 1-8. This abundance of paper-and-pencil practice ensures that your students can be successful.

Practice–Methods Beyond Paper and Pencil

Educators have known for some time that students need more than paper-and-pencil practice to ensure success. They know that the use of manipulatives can facilitate learners' understanding of mathematical concepts and development of skills. These concrete materials can also be pictured in a textbook or drawn by the learner. Their use is an important and necessary step along the road to proficiency. Each step takes time and requires practice, more for some learners than for others.

Practice with manipulatives should precede paper-and-pencil practice and should be paired with opportunities to discuss, ask questions, and record in words and/or symbols the actions taken with the manipulatives.

In *Math Advantage*, practice with manipulatives is the basis for concept development. Throughout the Pupil's Edition lessons in Kindergarten through Grade 2, there are workspaces on the lesson pages and suggestions for the use of workmats in all concept development lessons. In the Hands On lessons in Grades 3-5 and in the Lab lessons in Grades 6-8, practice with manipulatives is used when new concepts are introduced. Math Fun activities, found at the end of some chapters, also include manipulative activities.

In addition to the use of manipulatives in Pupil's Edition lessons where new concepts are introduced, you will find the following opportunities for hands-on practice in the Teacher's Editions:

Learning Centers   Practice Games
Idea Files

Another key practice methodology in *Math Advantage* is the use of technology. The following technology products are tightly correlated to program content and provide practice designed to reinforce concept development and skills practice:

Zoo Zillions™   Carnival Countdown™
(Grades K-3)   (Grades K-3)

Number Heroes™   Calculating Crew™
(Grades 3-6)   (Grades 3-6)

Cosmic Geometry™   Astro Algebra™
(Grades 6-8)   (Grades 6-8)

E-Lab™
(Grades 3-8)

Anytime Is Math Time–Practice Throughout the Day

Integrating the content from other subject areas is an effective strategy for building networks and connections. In *Math Advantage* you will find cross-curricular connections in the Pupil's Editions for Grades 3-8 and in the Idea Files, Literature Connections, and Learning Centers of the Teacher's Editions. In Kindergarten through Grade 2, cross-curricular connections are also found at the beginning of each chapter, in Anytime Is Math Time. Cultural Connections in the Teacher's Edition Idea Files and in the Pupil's Editions for Grades 6-8 also provide practice opportunities in which students learn about other cultures.

Review–Keeping Skills Sharp

Teachers are often dismayed at how little their students remember after summer vacation or, with some students, after a weekend. They know that systematically reviewing concepts and skills improves the learner's retention and serves as a diagnostic tool.

In *Math Advantage*, Reviews and Mixed Reviews in the Pupil's Edition lessons provide review of all critical skills. In the Teacher's Editions for Grades 6-8, the Quick Check in each lesson plan provides review of previously taught skills critical for success in the lesson. In addition,

the Problem of the Day at the beginning of each lesson plan provides review of problem-solving strategies and skills.

Of particular importance in providing review is Tab B of the Teacher's Editions for Grades 1-8. The Practice Activities in Tab B provide opportunities for you to help students who have not mastered critical skills from the previous grade level. They are designed for learning centers, for individual practice, for cooperative groups, and for home practice and are activity-based–giving students a new and motivating way to practice previously taught skills.

The key to successful review is the variety of types of problems and methods which students practice. In *Math Advantage*, paper-and-pencil practice, games, learning centers, cross-curricular activities, and computer activities provide the variety that keeps students interested in and excited about mathematics.

It is by engaging in various types of practice that students build the critical balance between their understanding of concepts and their skill competence. Practice helps build students' confidence in their ability to do mathematics. Practice that builds both understanding and skill proficiency is the pathway to students' success.

References

Hiebert, J., & Carpenter, T. P. (1992). Learning and teaching with understanding. In D. A. Grouws (ed.), *Handbook of research on mathematics teaching and learning* (pp. 65-97) (A project of the National Council of Teachers of Mathematics). New York: Macmillan Publishing Company.

Kilpatrick, J. (1992). A history of research in mathematics education. In D. A. Grouws (Ed.) *Handbook of research on mathematics teaching and learning* (pp. 3-38) (A project of the National Council of Teachers of Mathematics). New York: Macmillan Publishing Company.

National Council of Teachers of Mathematics (1991). *Professional standards for teaching mathematics*. Reston, Virginia.

Wakefield, A. P. (1997). Thinking about math thinking. *Kappa Delta Pi record, 33*, 103-105.

Integrating Technology

by Dr. Howard C. Johnson

*"The relatively slow mechanical means of communication–
the voice and the printed page–have been supplemented by
electronic communication, enabling information to be
shared almost instantly with persons–or machines–anywhere.
Information is the new capital and the new material, and
communication is the new means of production. The impact
of this technological shift is no longer an intellectual abstraction.
It has become an economic reality. Today, the pace of economic
change is being accelerated by continued innovation in
communications and computer technology." NCTM, 1989.*

Mathematics teachers are at the forefront of integrating computers and calculators in the school curriculum. They have recognized the potential of these powerful educational tools and are fast becoming proficient in using this technology in their classrooms. In fact, computer-assisted instruction was pioneered by mathematics educators. Today the mathematics curriculum stands as the first of the traditional school curricula to undergo change due to the use of computers and calculators.

Computers and calculators are effective tools for illustrating or representing mathematical concepts and skills. They can be used for problem solving with real data, and they free students from calculations with numbers larger than those they can handle on their own. As a result, you can more easily engage students in difficult problems they may encounter in the world–problems that can be approached with a variety of strategies, and problems that may have more than one solution.

The National Council of Teachers of Mathematics (NCTM) has taken the following position with respect to the use and availability of computers and calculators:

- Appropriate calculators should be available to all students at all times.
- A computer should be available in every classroom for demonstration purposes.
- Every student should have access to a computer for individual and group work.
- Students should learn to use the computer as a tool for processing information and performing calculations to investigate and solve problems (NCTM, 1989, p. 9).

Research Results

One of the major impacts of technology on the mathematics curriculum is that it is now possible to introduce some topics simply because technology frees us from computation. Technology enables students to focus on the problem. A commonly held misconception is that if students are allowed to use computers and calculators, they will not be able to perform calculations or understand mathematical reasoning without the aid of technology. Research clearly dispels this myth. More than 200 studies over the last 12 years have focused on the effects of the use of calculators, and the evidence is strongly in favor of their use. Hembree and Dessart (1986) identified general trends from 79 calculator studies. Here are their conclusions:

- Students who use calculators in concert with traditional instruction maintain their paper-and-pencil skills without apparent harm.
- The use of calculators in testing produces much higher achievement scores than paper-and-pencil efforts, both in basic operations and in problem solving.
- Students using calculators possess both a better attitude toward mathematics and a better self-concept in mathematics than do students who do not use calculators. This statement applies across all grades and ability levels.

Jensen and Williams (1993) report that using calculator and computer technology has a positive effect on "both problem-solving achievement and attitudes toward the activity of problem solving."

Using calculators • The research evidence is clear concerning calculators. When provided with calculators–the least expensive and most prevalent technological tools available–students perform better on problem-solving tasks. Students are better able to manage the overall problem-solving process when they are expending less energy on its computational aspects. The *guess and check* strategy also becomes a more realistic approach when technology removes the drudgery of recalculating results with different inputs. Calculators let students solve challenging problems that would otherwise take too much time with paper and pencil.

Using computers • Research evidence shows the computer to be an invaluable tool that helps the problem solver focus on generating ideas, trying out various approaches, and checking hunches. Computers allow students to test their conjectures rather painlessly, since the machines can produce an endless stream of output that students can use as data for testing and revising their conjectures.

Using Technology in the Classroom

Careful planning is essential in using technology in classrooms. Campbell and Stewart (1993) offer the following guidelines:

- Use technology to enhance your curriculum goals, not for its own sake.
- Use technology to supplement manipulative activities, not to replace them.
- Demand interactive software that capitalizes on the potential of the microcomputer.
- Encourage students to question and help one another.
- Help students to determine when technology is appropriate and when other approaches are more appropriate or efficient.

Technology and *Math Advantage*

Technology is an integral part of the curriculum of *Math Advantage* and is correlated to specific lessons. The technology products in the program provide the following:

Concept reinforcement and practice The *Mighty Math* series of CD-ROM products offers drill and practice in motivating settings and allows students to work with multiple representations of mathematical concepts through the use of Virtual Manipulatives™. *Stanley's Sticker Stories* offers children in Grades K-2 the opportunity to make their own number books, thereby allowing them to produce a record of their development of number and operation concepts. This CD-ROM product encourages creativity and problem solving in the young learner.

Tools *Graph Links Plus* and *Data ToolKit* are software programs that allow students to manipulate and analyze data. *Graph Links Plus* allows students in Grades 2-5 to enter their data into a table and then graph that data in a pictograph, bar graph, line graph, or circle graph. *Data ToolKit* allows students in Grades 5-8 to use a spreadsheet, find measures of central tendency, and graph data in a variety of ways, including line plots and box-and-whisker graphs. These tools allow students to quickly produce several different graphs that accurately reflect their data, thereby allowing them to determine which representations are the best choices.

Explorations *E-Lab* is a series of applications that link directly to the Hands On lessons in Grades 3-5 and to the Lab lessons in Grades 6-8. These applications reinforce and extend the activities in the Pupil's Editions, allowing students multiple opportunities to explore a concept, to extend the hands-on activity done with the lesson in order to deepen understanding, and to work with multiple representations of a math idea.

The development team of *Math Advantage* took seriously the importance of developing technology that is tightly linked to the content and methodology of the program. You can be sure that the time your students spend on the computer with the software developed to support *Math Advantage* will enhance your classroom instruction program and ensure a higher level of success for all students.

The use of the calculator is also integrated throughout *Math Advantage*. In Kindergarten through Grade 2, calculator explorations are in the Activity Options in appropriate lessons, in the lesson Idea Files, and in Math Fun activities in the Pupil's Editions. In Grade 3 in the Pupil's Edition Student Handbook, there is a section of lessons and activities that teach students how to use the function keys on the TI-108 and Casio SL-450. In Grade 4-5 the TI Explorer and Casio fx-55 are explained. In appropriate Pupil's Edition lessons in Grades 3-8, examples show calculator key sequences. In Middle School Books I-III, the Pupil's Edition Student Handbook includes lessons and activities to help students use the TI Explorer Plus and the Casio fx-65, a scientific calculator. Use of the calculator as a tool is assumed throughout the middle school program.

There is no longer a question as to whether computers and calculators should be used in the learning and teaching of mathematics. It is necessary that this technology be available for all students. In the mathematics classroom, computers and calculators should be used as freely and frequently as paper and pencil have been used in the past. These powerful technology tools will enhance mathematics learning and bring students into the next century with skills and knowledge they will need in order to be competitive in the world that awaits.

References

Campbell, P., & Stewart, E. (1993). Calculators and computers. In R. J. Jensen (ed.), *Research ideas for the classroom: Early childhood mathematics* (pp. 251-268). New York: Macmillan.

Hembree, R., & Dessart, D. J. (1986). Effects of hand-held calculators in precollege mathematics: A meta-analysis. *Journal for research in mathematics education*, 17(2), 83-99.

Jensen, R., & Williams, B. (1993). Technology: Implications for middle grades mathematics. In D.T. Owens (ed.), *Research ideas for the classroom: Middle grades mathematics* (pp. 225-243). New York: Macmillan.

National Council of Teachers of Mathematics (1989). *Curriculum and evaluation standards for school mathematics*. Reston, Virginia.

Making Mathematics Relevant

by Dr. Sonia M. Helton

"When children enter school, they have not segregated their learning into separate school subjects or topics within an academic area. Thus, it is particularly important to build on the wholeness of their perspective of the world and expand it to include more of the world of mathematics. This can be done in many ways, both within and outside the realm of mathematics." NCTM, 1989.

Learning is an intensely personal activity. Therefore, making the mathematics being taught in today's classroom personally relevant is critical to the success of each learner. Since teachers know their students' interests and can relate the mathematics to problems unique to those interests and to the region in which students live, the teacher becomes the key element in making mathematics relevant. You can make mathematics meaningful to your students when you plan lessons that link to these areas:

- Literature
- Real-life experiences
- Other subject areas
- Cultural experiences
- Interdisciplinary units
- Fun

In *Math Advantage* you will find many ways to make mathematics relevant to your students. One of the program's major goals is to help students understand not only what they are learning but *why* they are learning it. You can use this information to discuss students' ideas about why they are learning about a particular math topic and how they think they might use this knowledge in their everyday lives. In addition, specific features that link mathematics with literature, other subject areas, cultural experiences, and just plain fun are found throughout the program. The following sections describe some of these special features you can use to make math more meaningful and relevant to your students.

Real-World Links to Mathematics

Interesting problem contexts provide examples of how mathematics is useful. The photographs and art in the Pupil's Editions show students how math is used in many real-life contexts. Many of the chapter openers provide critical thinking questions, activities, and projects that encourage students to work together on problem situations that can be linked to everyday activities.

Literature Links to Mathematics

Using literature as a motivator and as a springboard to a math activity is a useful tool to help students integrate reading and mathematics, to facilitate the development of reading skills, and to enhance the development of math concepts. For centuries stories have been told to teach children the principles of life and to help them make analogies for solving real-life problems. Good stories help inspire creative thinking and stimulate the imagination, and they can be used to set mathematical experiences in a rich and interesting context.

In *Math Advantage* the Literature Links contain activities for all levels of learners that link to an authentic piece of literature. These Literature Links are found on the Meeting Individual Needs pages at the beginning of each chapter. In Grades 1 and 2, literature is used to set the stage for many of the chapters. This tie to literature continues throughout the chapter and will help you build a strong link from mathematics to language arts.

Cross-Curricular Links to Mathematics

Connecting mathematics to other subject areas helps students learn to value mathematics as a tool for solving many different kinds of problems. One of the most powerful kinds of cross-curricular links is to the language arts–reading, listening, speaking, writing, and representing. If students can communicate clearly about their mathematical ideas, then it is evident that they understand the ideas. Writing is a particularly powerful link, since students who write as part of their study of mathematics must do considerable thinking and organizing of their thoughts in order to crystallize in their minds what they have studied (Helton, 1992; Shepherd, 1993; Waywood, 1994).

A math journal is an effective tool for learning mathematical concepts, as it enables students to explain their thinking and the strategies they used to solve problems. Cobb et al. (1991) concluded that when students used math journals, their conceptual and application understanding of mathematics was superior to that of students who experienced a year of traditional math instruction without the inclusion of journal-writing opportunities. They suggested that the problem-centered approach facilitates students' construction of operations whose concepts are increasingly sophisticated. Other researchers have also found that keeping a journal helps students formulate, clarify, and relate math concepts; appreciate how mathematics relates to the world; and apply the problem-solving strategies that underlie the doing of mathematics (Clarke et al., 1993).

Problems from science, social studies, health, physical education, music, and language arts can be documented and solved in the math journal. The journal-keeping link enables students to develop mathematical ideas by helping them understand the following ideas:

- Math can be explored in meaningful contexts.

- Journal keeping in mathematics helps develop problem-solving skills in all curriculum areas.

- A range of math skills is necessary to function in real life. Learning to record processes is necessary.

- They must begin to think and function as mathematicians (Helton, 1992).

In *Math Advantage*, cross-curricular links make up a thread that runs throughout the program. In Grades K-2, these links are in the Pupil's Editions on some of the Math Fun pages; they can be found in the Teacher's Editions in the Idea Files in the lesson plans and on the Anytime Is Math Time page at the beginning of each chapter. In Grades 3-8, cross-curricular links are found in many of the lessons in the Pupil's Editions and in the Idea Files in the lesson plans in the Teacher's Editions. In addition, some of the projects and activities in the chapter openers have cross-curricular links.

Cultural Links to Mathematics

Every culture has both a mathematical history and a heritage of arts and crafts that express mathematical ideas. These ideas are usually expressed in the form of patterns and artwork found in clothing, fabrics, jewelry, pottery, weavings, and woodcarvings. Having students investigate how customs and art forms from various cultures are represented in their lives in the United States today is a powerful way to link mathematical experiences to an appreciation of the diversity among students.

In *Math Advantage* these cultural links may be found in some lessons and at the ends of some chapters in the Pupil's Editions for Grades 3-8. In the Teacher's Editions for Grades 1-8, cultural links are included in the Idea Files in the lesson plans.

Fun with Mathematics

Having fun with mathematics accomplishes three goals. It helps students

- develop a favorable attitude toward learning mathematics.
- use creative thinking in mathematics.
- develop an appreciation for mathematics.

Having fun with mathematics can inspire students' creativity–the ability to tap past experiences and come up with something new. You can provide opportunities for creative thinking by encouraging students to

- invent constructions.
- visualize a representation for a problem.
- think of new ways to solve a problem.
- explore art that reflects the application of a mathematical concept.

Infusing the mathematics program with opportunities for fun and creativity helps students appreciate mathematics and feel good about it. A child who can freely express his or her creativity and share experiences about mathematics maximizes his or her learning, which in turn builds appreciation (Taylor, 1993).

Many of the experiences students will be having in the math classroom set the tone for appreciating mathematics. In *Math Advantage*, activities that are creative and fun to do will build this appreciation. Math Fun pages provide activities and games for all levels of learners that develop key skills and reinforce concepts. In Kindergarten

through Grade 2, Gameboards with Pieces provide games that reinforce skills and concepts. In addition, Learning Center suggestions found in every chapter for Grades K-5 often include games. Practice Activities in Tab B in the back of the Teacher's Editions provide interesting and motivating activities of all types to help students develop proficiency in their areas of weakness. Teachers know that more paper-and-pencil practice is not the answer to remediating students' weaknesses and that students respond positively to activities that are fun and engaging.

Interdisciplinary Units

Each teacher comes to the classroom with a wealth of life experiences that becomes a pivotal point upon which mathematical ideas are structured. Routman (1994) suggests that in planning for integrated or interdisciplinary instruction a teacher needs to follow these steps:

- Identify important concepts that students should learn.
- Decide if these concepts foster critical and creative thinking.
- Identify the learning skills and teaching strategies.
- Set up a climate that encourages inquiry and choice.
- Determine how students' work will be evaluated.
- Determine the student attitude being fostered.

In the middle school program of *Math Advantage*, suggestions for interdisciplinary units are included on the Meeting Individual Needs pages at the beginning of each chapter. These suggestions will help you build a diverse and exciting classroom setting in which students apply mathematical concepts and skills in a problem-based learning environment and in which mathematical concepts can be developed in rich and interesting contexts.

Making mathematics relevant to students' everyday lives is critical to maintaining interest and motivation. Helping students understand not only what they are learning but why they are learning it will help them see mathematics as a tool for solving everyday problems–not just as a classroom exercise.

References

Clarke, D. J., Waywood, A., & Stephens, M. (1993). Probing the structure of mathematical writing. *Educational studies in mathematics*, 25, 235-250.

Cobb, P., Wood, T., Yackel, E., Nicholls, J., Wheatley, G., Trigatti, B., & Perlwitz, M. (1991). Assessment of a problem-centered second-grade mathematics project. *Journal for research in mathematics education*, 22, 3-29.

Helton, S. M. (1992). You can count on it. *The whole idea*. Vol. II. No. 3, Spring, 6-7.

National Council of Teachers of Mathematics (1989). *Curriculum and evaluation standards for school mathematics*. Reston, Virginia.

Routman, R. (1994). *Invitations: Changing as teachers and learners. K-12*. Portsmouth, New Hampshire: Heinemann, 276-280.

Shepherd, R. (1993). Writing for conceptual development in mathematics. *Journal of mathematical behavior*, 12, 287-293.

Taylor, L. (1993). Vygotskian influences in mathematics education, with particular reference to attitude development. *Focus on learning problems in mathematics*, 15, 3-17.

Waywood, A. (1994). Informal writing-to-learn as a dimension of a student profile. *Educational studies in mathematics*, 27, 321-340.

Meeting Individual Needs in the Mathematics Classroom

by Dr. Judith Mayne Wallis

"For all children, wise coaches act as bridges between unformed ideas and mutual understanding. That's what this 'conversation game' is all about." **Jane Healy, 1992.**

Think about the work that goes into spring and fall gardens. With shovels and forks, gardeners scoop up the earth and turn it before bringing in additional soil and nutrients. Gradually the ground is prepared to receive the young seedlings. As they are positioned into the cultivated bed, careful attention is paid to each plant's particular needs: the depth of the planting, the spacing between the plants, and the amount of sunlight each needs. In the following weeks, gardeners check regularly for growth. They use hoes and other tools to remove weeds and keep the soil loose.

Tending a garden is much like your work with children. Today's classrooms are filled with children who have requirements just as individual as those of seedlings. And like gardeners, you are asked to meet the special needs you face. One way you can do this is by teaching children to use language as a tool for learning and growth.

Students Acquiring English–ESL Strategies

The *National Council of Teachers of Mathematics Curriculum and Evaluation Standards* (NCTM, 1989) emphasizes the importance of using language to explore mathematical concepts. You have wonderful opportunities to foster communication among your students, especially those acquiring English. For example, when you have students make drawings and build models, you can help them not only construct mathematical knowledge but also express that knowledge. Several strategies can be used with your students to develop the language they need for discussing their ideas.

• Build an Idea

Manipulatives can be used to assist students in developing mathematical thinking. In addition to manipulatives designed especially for mathematics, small tiles, cubes, wooden blocks, beads, and buttons can be used to build an idea. Once students have modeled their ideas, talk becomes a bridge to further understanding.

• Draw an Idea

After initial instruction, students can make drawings that demonstrate or extend a new concept. For example, they might make a sketch to show their understanding of congruence. After students are done, use their drawings to foster conversation. This taps into students' natural use of metaphorical language to convey and interpret mathematical observations and understanding (Whiten & Whiten, 1997). Drawing an idea enables students to develop their thinking in pictorial ways.

• Teach Me

Most teachers know that the best way to learn more about a concept is to teach it. In the thought and preparation that go into planning how to instruct someone else, the teacher learns the concept thoroughly and in a personal way. Asking students to teach you or a small group of their peers offers them that same opportunity. A key to making "Teach Me" work with students is to provide time for them to prepare. After you complete initial instruction, ask ESL students to join you for further discussion. Set up situations in which you can provide additional modeling. When students appear to understand, explain that they will have an opportunity to teach the concept they learned. If you allow a few days to go by, this will also provide distributed practice. Establish the teaching time, and suggest that models and sketches might be useful.

ESL Strategies in Math Advantage

ESL Tips, found in the Idea Files of selected lessons, provide alternative teaching strategies that focus on concrete and visual methods. These will help students use the "Build an Idea" and the "Draw an Idea" strategies to develop conceptual understanding and proficiency with the language of mathematics.

Vocabulary Cards, found in the Teaching Resources for Grades K-2, show both terms and illustrations. These cards may be used in the classroom in a "Teach Me" session for introducing or reviewing key math terms, or they may be sent home with suggestions about how family members can use them with the children.

Word Power, a feature found at the top of lesson pages in the Pupil's Editions for Grades 3-8, lists the new terms that will be introduced, defined, and developed in the lesson. This careful focus on vocabulary is particularly useful to students acquiring English.

Talk About It questions encourage students to use their emerging English skills to describe their mathematical thinking. The questions also provide students the opportunity to listen to other students' descriptions of mathematical thinking.

Hands On or Lab Lessons, which involve cooperative learning and the use of manipulatives, provide students who are acquiring English the opportunity to work with native speakers in "Build an Idea" or "Draw an Idea" activities. These will build concepts and develop oral and written communication skills.

Models show step-by-step visual representations of mathematical procedures and a word description of each step. They facilitate understanding by providing scaffolds that link visual representations and written descriptions.

Inclusion–Every Student's Opportunity to Participate

Inclusion students often require a different kind of support. There are several strategies that assist you in providing for students with special needs. By observing students as they write, draw, talk, and question, you will find occasions when you need to scaffold and support the learning of students with individual needs.

Math Advantage includes many opportunities for you to support students with special needs. Throughout the program you will find

- activities that reach all modalities—auditory, kinesthetic, and visual.

- activities that develop the multiple intelligences as defined by Howard Gardner.

- cooperative group activities that allow students with special needs to benefit from and contribute to the work of the whole group.

- hands-on activities that focus on concept development and encourage communication.

- journal-writing prompts that provide opportunities for students to show what they are learning and to describe their attitudes toward mathematics.

- a variety of technology components that provide support for concept development and practice.

- assessment options that allow you to assess students in a variety of ways.

Just as the small seedlings planted in a garden have individual needs, so do children. Supporting and nurturing the variety of individual needs rather than homogeneity

(Eisner, 1991) encourages all children to blossom into capable mathematical thinkers and learners.

References

Eisner, E. (1991). What really counts in schools. *Educational Leadership*, 48, 10-17.

Healy, J. M. (1992). *Is your bed still there when you close the door? . . . and other playful ponderings: How to have intelligent and creative conversations with your kids.* New York: Doubleday.

National Council of Teachers of Mathematics (1989). *Curriculum and evaluation standards for school mathematics.* Reston, Virginia.

Whiten, P. E. & Whiten, D. J. (1997). Ice numbers and beyond: Language lessons for the mathematics classroom. *Language Arts*, 74, 108-115.

Meeting Individual Needs

by Dr. Evelyn M. Neufeld

"Goals such as learning to make conjectures, to argue about mathematics using mathematical evidence, to formulate and solve problems–even perplexing ones–and to make sense of mathematical ideas are not just for some group thought to be 'bright' or 'mathematically able.' Every student can–and should–learn to reason and solve problems, to make connections across a rich web of topics and experiences, and to communicate mathematical ideas." NCTM, 1991.

Mathematics is an elegant and fascinating tool for learning to think reasonably and responsibly. It relates to the activities of everyday living and to the world of work. Therefore, children's work in mathematics is tied to real life–the life of childhood, adulthood, work, and home. Through the mathematics curriculum, every student has the opportunity to construct his or her mathematical knowledge by using methodologies and instructional materials that are undergirded by sound theory and research and that have withstood the test of time. However, students have different talents, achievement levels, learning styles, and interests. This diversity of learning capabilities challenges teachers to bring together the student, mathematical content, and instructional methodology in such a way that mathematical knowledge is constructed.

Curriculum specialist J. Cecil Parker remarked in his curriculum class, "We can create the best schools, hire the best teachers, adopt the best curricula, and then all the wrong children show up" (Neufeld, 1976). The children who show up are not miniature adults; rather, they have unique characteristics that set them apart from adults, from children who are at different levels of thinking, and even from children who generally seem to be at a similar level of development.

Each individual child, according to Jean Piaget (1967), brings to the subject matter and to the instructional methodologies a different genetic structure and different physical experiences, social experiences, and motivations. These four factors are said to govern the rate of progress through four developmental levels of cognition, which Piaget found to be age-related, not age-dependent. These four developmental levels and their approximate age ranges are as follows:

Sensory Motor Period of Intelligence (birth to 2 years)

Preoperational Period of Thought (2 to 7 or 8 years)

Concrete Operational Period of Thought (8 to young adulthood)

Formal Operational Period of Thought (12 to adulthood)

Lowery et al. (1986) found a wide range of differences in the ages at which children demonstrate conservation abilities–abilities to identify the invariance of a particular factor when other factors are changed. While most children conserved number at six or seven years of age, some children conserved as early as four and others not until nearly eight years of age. The following table shows the age ranges at which children's conservation abilities appear:

Lowery's Strand of Conservation Abilities:
Sequence of Appearances

Strand	Age of Appearance for Most Students
	3 4 5 6 7 8 9 10 11 12 13 14 15 16 17
Number	·········· ··· (4–7)
Continuous Length	·········· —— ···
Solid Amount	·········· —— ···
Liquid Amount	·········· —— ···
Discontinuous Length	·········· —— ···
Area	·········· —— ···
Weight	········ ——— ···
Solid Volume	········ ———— ···
Displaced Volume	·········· —— ··········

Table by Lawrence Lowery from *Learning about Learning: Conservation Abilities.* Published by University of California School of Education, Berkeley, 1981. Reprinted by permission of Lawrence Lowery.

Asher (1988) contends that mathematical concepts can be grasped by most children if the concepts are presented in a form that addresses the right side of the brain. He states, "If asked a question, the left brain can talk to explain what it has experienced. If information goes into the right brain, however, and a question is asked . . . the right brain cannot communicate in words what it has experienced. Although the right brain cannot communicate in language, it can express itself if asked to act out, to draw, to point, to spell, to sing or physically demonstrate with body movements such as pantomime" (Asher, p. 37). Asher's work confirms teachers' experiences in the classroom which show that instructional methodologies emphasizing auditory, visual, and kinesthetic modalities are more likely to meet learners' needs.

Howard Gardner (1982) found the differences in children's learning patterns to be extreme. Through his research, he identified recurrent patterns in children's approaches to learning. He labeled the types of learners as follows:

Verbalizers produced copious amounts of language.

Visualizers plunged directly into drawing or building.

Self-starters required but the barest stimulus to begin.

Completers displayed considerable reluctance to begin work. Yet when given a product to finish, assemble, or copy, completers took off, often constructing a work more appropriate to the demands of the task and even more inventive than the abundant, but often undisciplined, productions of the self-starters.

Person-centered learners used forms or symbols that emphasized communication or creation.

Object-centered learners constructed works that featured physical elements and machines. Their efforts were more private and seldom modulated by another individual's presence.

Based on his research, Gardner contends, "Far from representing a deterrent to decisive research, however, these individual differences may turn out to be thought-provoking. We have seen that the cognitivist is so preoccupied with grouping subjects according to developmental level that the enormous differences across children within each level have tended to elude him. For his part, the person concerned with the affective development has been so struck by personal characteristics and idiosyncratic qualities of specific youngsters that he has often overlooked profound continuities across children in skills of production and comprehension. We believe that, in the last analysis, the fullest understanding of the young child's work will come from setting aside this dichotomy, from thinking instead in terms of the intersection of multiple factors" (Gardner, 1982, pp. 121-122).

Gardner (1983), in his later work, named seven intelligences that he believed all learners have. Teachers recognize the importance of his work and are now implementing curriculum and instructional strategies that provide opportunities for children to develop multiple intelligences. See "Strategies for Teaching Mathematics" by Grace Burton for a description of the ways in which activities in *Math Advantage* are designed to develop multiple intelligences (Tab D, page D7).

In *Math Advantage* the authors and editors focused on designing the content and the instructional approach to reflect the variety of ways in which children learn. As you work through the program, you will quickly see how Piaget's ideas and Lowery's further development of those ideas about learning form the basis for the instructional approach. Lessons and activities have been developed to incorporate the results of research on brain-based learning and to enhance the development of multiple intelligences as defined by Gardner. Technology products have been developed to provide a wide range of instructional methodologies and to reflect the findings of research on the ways children learn.

In addition, the following elements of *Math Advantage* incorporate these ideas and focus on meeting the needs of all learners:

Intervention Strategies–In Grades 3-8, there are lessons that present prerequisite concepts and skills necessary for success with grade-level content. These lessons reflect the research on the various ways children learn and are designed not just to remediate weaknesses, but rather to prevent failure.

Extension Strategies–In Grades 3-8, there are lessons that extend the content of a chapter or provide students with a different approach to a key concept or skill developed in the chapter. These lessons are designed to stretch the thinking of the students who show understanding and proficiency with the chapter content.

In *Math Advantage* the focus of the instructional design is to build in strategies and to develop components that touch on the various ways research has shown children learn. This careful attention to the needs of all learners will assist you in planning a rich and varied mathematics program that ensures success for all students.

References

Asher, J. J. (1988). *Brainswitching: A skill for the 21st century*. Los Gatos, California: Sky Oaks Productions, Inc.

Gardner, H. (1982). *Art, mind, and brain: A cognitive approach to creativity*. New York: Basic Books.

Gardner, H. (1983). *Frames of mind: The theory of multiple intelligences*. New York: Basic Books.

Lowery, L. et al. (1986). *It's the thought that counts: A sourcebook of mathematics activities related to conservation abilities*. Palo Alto, California: Dale Seymour Publications.

National Council of Teachers of Mathematics (1989). *Curriculum and evaluation standards for school mathematics*. Reston, Virginia.

National Council of Teachers of Mathematics (1991). *Professional standards for teaching mathematics*. Reston, Virginia.

Neufeld, E. (1976). *Homework!* New Rochelle, New York: Cuisenaire.

Piaget, J. (1967). *Six psychological studies*. New York: Random House.

Assessing What Children Know and Can Do

by Dr. George W. Bright

"Assessment is the process of gathering evidence about a student's knowledge of, ability to use, and disposition toward mathematics and of making inferences from that evidence for a variety of purposes." NCTM, 1995.

Assessment is a tool that helps you understand what students know about mathematics and what they can do. Some effective ways to assess students' mathematical understanding, skill proficiency, and problem-solving abilities are paper-and-pencil tests, interviews, projects, hands-on tasks, and observations of daily work. Asking students to evaluate their own progress and understanding and their attitudes and dispositions toward mathematics is also a vitally important way to understand how to best meet your students' needs. Gathering data from multiple types of assessment will help you plan instruction that builds on what each of your students knows and can do.

Critical thinking questions are a key way to assess how students think. Asking students to explain, verify, and support their answers enhances their oral and written communication skills. Effective communication requires that students develop an understanding of math terminology and the ability to use that terminology as they communicate about their thinking. These critical thinking questions will help ensure that assessment is "a communication process in which assessors . . . learn something about what students know and can do and in which students learn something about what assessors value" (NCTM, 1995, p. 13).

Assessing Prior Knowledge—Building a Bridge to Success

The first step in planning effective lessons is deciding what new knowledge is important for students to know and what previously taught skills are critical for success with that topic. By quickly assessing whether students are proficient in these previously taught skills, you can plan instruction that meets the needs of students who lack the necessary skills. This important piece of your assessment program will ensure that more of your students are successful in learning new skills and concepts.

Assessment—A Complete Picture

Increasingly, research on students' knowledge of mathematics suggests that many students do not understand mathematics in the same ways that their teachers do. Teachers are now expected to choose assessment methodologies that elicit students' thinking and to interpret students' responses in light of instructional goals.

Carpenter and Fennema (Carpenter et al., 1989; Fennema et al., 1993, 1996) have shown that teachers can acquire very specific information about students' mathematical thinking and then use that information to plan instruction on increasingly sophisticated problem-solving strategies. Students in these settings score higher on both standardized tests and problem-solving tests.

Cobb and his colleagues (e.g., Nicholls, Cobb et al., 1990) have also focused on helping teachers understand students' thinking in order to adapt instruction. Students in programs designed to focus on students' thinking performed better on tests and developed better dispositions toward doing mathematics and engaging in mathematical tasks.

> For a teacher who seeks to foster students' higher-order mathematical thinking, no conventional achievement test is likely to be of immediate help. What teachers can use for this enterprise is more-or-less constant feedback on how students interpret the problems before them. (Nicholls et al., 1990, pp. 137–138)

Teachers who use many different techniques have rich feedback on students' thinking. Just as snapshots, portraits, and videos contribute to a complete picture of children's physical growth and change, multiple assessment methodologies jointly form a complete record of students' mathematical growth and development. The assessment program in *Math Advantage* is designed to provide this complete record. Assessment options in the program include the following:

Assessing Prior Knowledge

For each chapter, an activity (Grade K) or a worksheet (Grades 1–8) assesses each skill that is a prerequisite for the chapter's content. For students who have weaknesses in any of the skill areas tested, troubleshooting aids—references to specific activities or components of the program—are given.

Integrated Assessment

In each lesson plan, Problem of the Day (Grades K-8) assesses students' abilities to solve a wide variety of problem types and to choose and apply strategies. Quick Check (Grades 6–8) begins a lesson with a review of prerequisite skills. Assessment Tip (Grades K–8) checks students' understanding of the lesson's "big idea." Lesson Check (Grades 1–5) or Daily Quiz (Grades 6–8) assesses the lesson's skills. Critical thinking questions throughout the program provide ongoing opportunities to observe students' progress.

Multiple-Choice and Free-Response Tests

These tests assess students' concept development, skill proficiency, and problem-solving abilities. There are tests for each chapter, cumulative tests for all content up to the test, and multiple chapter tests for larger "chunks" of instruction so that you can see how well students are retaining material taught over time.

Performance Assessment

For each grade level, there are four performance tasks, with each task designed to cover one fourth of the grade-level content. To assist you in scoring these tasks, a rubric and scored student samples with explanations of the scores are provided. In addition, in the Pupil's Editions for Grades 3–8, performance assessment items are included on the What Did I Learn? pages. On each of these pages, a rubric and scored student samples are provided for one of the tasks.

Interview/Task Tests

These test items require students either to use manipulatives to model mathematical concepts or to explain the steps in a mathematical procedure. Checklists for scoring are provided.

Portfolio Suggestions are given throughout the Teacher's Edition for specific work samples that you may wish students to include in their portfolios. The *Assessment Guide* provides suggestions for evaluating students' portfolios, for talking with students about their work samples, and for sharing the portfolios with parents. Forms for student evaluation, parent evaluation, and teacher evaluation are included.

Student Self-Assessment Forms are included in the *Assessment Guide* for students to assess their work in a cooperative group, to assess their attitudes and dispositions toward mathematics, and to assess their portfolio work samples. Projects and activities that open each chapter in Grades 3–8 include a checklist for the members of cooperative groups to use for evaluating their own work before sharing their results.

Using Assessment Results

Your primary consideration is how to use the data collected from the various types of assessment. According to the Professional Teaching Standards, "When teachers have a good understanding of what their students know and can do, they are able to make appropriate instructional decisions" (NCTM, 1995, p. 45).

The following are among the instructional strategies you may choose on the basis of assessment results.

- Individualizing: You can adjust the numbers in or the structure of problems to make them easier or more difficult for individual students.

- Choosing problems: Problem choices can be based on your knowledge of the types of thinking the students are using. Choosing problems that elicit different kinds of thinking allows students to develop a repertoire of different strategies.

- Grouping: If there are wide discrepancies in students' thinking, you can form groups in which there is more similarity of processing.

- Using alternative strategies: Sometimes you may need to suggest alternative strategies so that students are exposed to a wide range of techniques. You can describe a different strategy as "a way I saw another student solve this problem." This presents the new strategy as just another way to solve a problem and allows students to accept or reject it on the basis of their personal understanding.

Learning how to get useful information about students' thinking and skill proficiency takes time. You have to decide what kinds of information are most useful and what techniques are effective for obtaining this information. However, you will have the tremendous payoff of helping your students understand mathematics more deeply and feeling personal satisfaction about the quality of your instructional program. A comprehensive assessment program is the cornerstone in helping you make sound instructional decisions.

References

Carpenter, T. P., Fennema, E., Peterson, P. L., Chiang, C-P., & Loef, M. (1989). Using knowledge of children's mathematics thinking in classroom teaching: An experimental study. *American Educational Research Journal*, 26, 499–531.

Fennema, E., Carpenter, T. P., Franke, M. L., Levi, L., Jacobs, V. R., & Empson, S. B. (1996). A longitudinal study of learning to use children's thinking in mathematics instruction. *Journal for Research in Mathematics Education*, 27, 403–434.

Fennema, E., Franke, M. L., Carpenter, T. P., & Carey, D. A. (1993). Using children's mathematical knowledge in instruction. *American Educational Research Journal*, 30, 555–584.

National Council of Teachers of Mathematics (1995). *Assessment standards for school mathematics*. Reston, Virginia.

National Council of Teachers of Mathematics (1991). *Professional standards for teaching mathematics*. Reston, Virginia.

Nicholls, J. G., Cobb, P., Yackel, E., Wood, T., & Wheatley, G. (1990). Students' theories about mathematics and their mathematical knowledge: Multiple dimensions of assessment. In G. Kulm (ed.), *Assessing higher order thinking in mathematics* (pp. 137–154). Washington, D.C.: American Association for the Advancement of Science.

Building the School-Home Connection

by Vicki Newman

"Programs that involve parents from the planning through the implementation stages also foster a strong sense of mutual trust between homes and schools." Teaching Children Mathematics (November 1996), pp. 118–121.

Building the school-home connection is a critical component in ensuring a successful mathematics program. Children benefit when they learn mathematics together with parents and teachers. If parents and teachers stress that math is used in their own daily lives and show how everyday math experiences can be enjoyed together, then children learn to value mathematics. They see a reason for studying math at school.

Since families play a vital role in developing the child's attitude toward mathematics, it is important to encourage family involvement. Effective school-home communication is the key. When communication between school and home is ongoing, parents are provided with a clear picture of what is happening in their child's classroom and how they can offer support at home. The following suggestions may help you build the school-home connection and provide this valuable information to parents.

- **Identify the goals of the mathematics** program to give parents an overview of their child's mathematics curriculum.

- **Explain your mathematics program** during parent conferences by sharing your instructional strategies. Allow children to become involved in the communication process by sharing classroom projects and daily work at classroom and school exhibits or at a schoolwide math event. Parents enjoy hearing their children talk about their work.

- **Suggest ways parents can support mathematics learning** at school and outside the classroom.

- **Share with parents samples of student work** that represent growth in their child's math ability.

- **Listen to questions, concerns, and suggestions** from parents and the community. Parent meetings and parent surveys provide opportunities for parents to ask questions about the school's mathematics program and help them feel they are taking an active role in their child's education.

The School-Home Connection in *Math Advantage* will help you implement these suggestions as you share the following materials:

School-Home Connections

- *Math Advantage Parent Preview*, a brochure for parents, provides an overview of the major topics that will be studied during the year, suggestions for ways parents can share their use of mathematics in everyday experiences in order to help children understand why they need to learn mathematics, and some activities that parents and their children can do together to reinforce important math skills. This brochure may be sent home at the beginning of the year.

- *Math at Home* letters describe the mathematics taught in each chapter, provide tips on how to help with homework, new vocabulary, and how to help the child be more successful on tests, and present activities that relate to the chapter content. These letters provide practical information that will help you involve parents as partners in supporting their child's growth in mathematical proficiency.

In the *Pupil's Edition*

- *Home Notes* on each lesson in Grades K–2 provide suggestions for ways parents can reinforce the concepts and skills taught.

- *Math Fun* activities in some of the chapters in Grades 3–8 include suggestions for sharing the activity with family members.

In the *Assessment Guide*

- Practical tips for building math portfolios include suggestions for possible work samples to include and forms that you, the child, and parents can use for comments on the work samples.

Mathematics Outside the Classroom

One of the key goals of *Math Advantage* is to help children see the relevance of mathematics to their daily lives. Parents can work with teachers to accomplish this goal by pointing out how math is used at the post office, the grocery store, the bank, and in other everyday settings. They can also help their child with projects and activities that are at the beginning of each chapter in the Pupil's Editions, Team-Up Time (Grades 3–5) and Team-Up Project (Grades 6–8). They can discuss the information in the Why Learn This? section at the top of each lesson in Grades 3–8. Many times both you and the child's parents can think of additional answers to the question "Why learn this?"

The focus on relevance in *Math Advantage* will help children see that mathematics is everywhere—not just in the classroom.

Math Homework—A Family Affair

Parents usually want to help their child with homework assignments. However, many times the way that mathematics is taught today is different from the parents' school experience, and their efforts to help tend to frustrate both them and the child. The information in the *Parent Preview* brochure and in the *Math at Home* letters can help parents better understand the mathematics their child is learning and how to best help with homework assignments. In addition, the explanations and models in the Pupil's Editions will provide help for parents in understanding their child's mathematical experiences.

Parents can be powerful partners with you in ensuring the success of their child. They can offer support at home and show how math is used every day in ways that can be linked directly to the mathematics you are teaching in school.

E Bibliography/Index

The following bibliography contains references to:

- fiction and nonfiction books for students
- technology resources
- professional books and magazines

These materials will assist you in creating an interesting learning environment. The references to literature are provided to help you work through your media center to acquire literature selections that you can use with Math Advantage. The math activities developed to correlate to these books will help you build a math curriculum to meet the needs of all students.

The index contains information for both Pupil's Editions and Teacher's Editions. The italicized items are found in the Teacher's Edition.

Bibliography

BOOKS FOR STUDENTS

All Things Bright and Beautiful.   Herriot, James. St. Martin's Press, 1974.

"Analysis of Baseball" by May Swenson in *American Sports Poems.*   Knudson, R.R., and May Swenson. Franklin Watts, 1988.

"Hockey" by Scott Blaine in *American Sports Poems.*   Knudson, R.R., and May Swenson, eds. Orchard Books, 1995.

"A Man Who Had No Eyes" in *Author's Choice: 40 Stories.*   Kantor, MacKinlay. Coward-McCann, Inc., 1944.

"Mother and Daughter" in *Baseball in April and Other Stories.*   Soto, Gary. Harcourt Brace & Company, 1990.

"The Treasure of Lemon Brown" by Walter Dean Myers in *Boys' Life Magazine,* March 1983.

Casey at the Bat.   Thayer, Ernest Lawrence. South China Printing Company, 1988.

"Up the Slide" in *The Complete Short Stories of Jack London.*   London, Jack. Stanford University Press, 1993.

The Crazy Horse Electric Game.   Crutcher, Chris. William Morrow & Company, 1987.

"The Force of Luck" by Rudolfo A. Anaya in *Cuentos: Tales from the Hispanic Southwest: Based on Stories Originally Collected by Juan B. Rael.*   Griego y Maestas, José, ed. Museum of New Mexico Press, 1980.

"I Watched an Eagle Soar" in *Dancing Teepees: Poems of American Indian Youth.* Sneve, Virginia Driving Hawk. Holiday House, 1989.

The Diary of Anne Frank.   Goodrich, Frances, and Albert Hackett. Heinemann, 1995.

"A Rendezvous with Ice" in *Earth's Changing Climate.*   Gallant, Roy. Four Winds Press, 1979.

Exploring the Titanic.   Ballard, Robert D. Scholastic, Inc., 1988.

Face on the Milk Carton.   Cooney, Caroline B. Laureleaf, 1994.

A Fall from the Sky: The Story of Daedalus. Serraillier, Ian. H. Z. Walck, 1966.

"My Mother Pieced Quilts" by Teresa Palma Acosta in *Fiesta in Aztlan: Anthology of Chicano Poetry.*   Empringham, Toni, ed. Capra Press, 1981.

From the Mixed-Up Files of Mrs. Basil E. Frankweiler.   Konigsburg, E. L. Atheneum, 1987.

"The Monsters Are Due on Maple Street" in *From the Twilight Zone.*   Serling, Rod. Nelson Doubleday, Inc., 1962.

"Gentleman of Río en Medio."   Sedillo, Juan A. A. *The New Mexico Quarterly*, August 1939.

"The Smallest Dragonboy" in *Get Off the Unicorn.*   McCaffrey, Anne. Random House, 1983.

The Gilded Cat.   Dexter, Catherine. William Morrow & Company, 1992.

The Girl Who Cried Flowers and Other Tales. Yolen, Jane. Thomas Y. Crowell, 1974.

Go Figure! The Numbers You Need for Everyday Life.   Hopkins, Nigel J.; John W. Mayne; and John R. Hudson. Gale Research Inc., 1992.

Good-bye, Mr. Chips.   Hilton, James. Little, Brown, 1986.

"Raymond's Run" in *Gorilla, My Love.* Bambara, Toni Cade. Random House, 1992.

"The Circuit" by Francisco Jiménez in *Growing Up Chicana/o: An Anthology.* López, Tiffany Ana, ed. William Morrow & Company, 1993.

Gulliver's Travels. Swift, Jonathan. Running Press Book Publishers, 1992.

Hello, My Name Is Scrambled Eggs. Gilson, Jamie. William Morrow & Company, 1985.

Hoops. Myers, Walter Dean. Dell, 1981.

I Have A Dream: The Life and Words of Martin Luther King, Jr. Haskins, James. Millbrook Press, 1992.

The Indian in the Cupboard. Banks, Lynne Reid. Avon, 1980.

Island of the Blue Dolphins. O'Dell, Scott. Houghton Mifflin Company, 1990.

Julie of the Wolves. George, Jean Craighead. HarperCollins, 1987.

King of the Wind. Henry, Marguerite. Macmillan, Inc., 1991.

"Thank, You M'am" in *The Langston Hughes Reader: The Selected Writings of Langston Hughes.* Hughes, Langston. George Braziller, Inc., 1958.

"In Search of Cinderella" in *A Light in the Attic: Poems and Drawings* by Shel Silverstein. Silverstein, Shel. Harper Collins, 1981.

Little Farm in the Ozarks. MacBride, Roger Lea. HarperCollins, 1994.

"The School Play" in *Local News.* Soto, Gary. Harcourt Brace & Company, 1993.

Lost at Sea (formerly titled *The Voyage of the Frog*). Paulsen, Gary. Orchard Books, 1989.

The Maltese Cat. Kipling, Rudyard. Creative Education, Inc., 1991.

"Tranquillity Base" in *Men from Earth.* Aldrin, Buzz. Bantam Books, 1991.

"The Dog on the Roof in Twelfth Street" by Joan Aiken in *The Methuen Book of Animal Tales.* Methuen, 1983.

"The Road Not Taken" in *Mountain Interval.* Frost, Robert. Henry Holt and Company, 1916.

"At Last I Kill a Buffalo" in *My Indian Boyhood by Chief Luther Standing Bear, Who Was the Boy Ota K'te (Plenty Kill).* Bear, Luther Standing. University of Nebraska Press, 1988.

A New Way of Life (formerly titled *The Vietnamese in America*). Rutledge, Paul. Lerner Publications Company, 1991.

Not for a Billion, Gazillion Dollars. Danziger, Paula. Delacorte Press, 1992.

Nothing But the Truth. Avi. Avon, 1991.

Number the Stars. Lowry, Lois. Houghton Mifflin Company, 1989.

One Two Three . . . Infinity: Facts and Speculations of Science. Gamow, George. Dover Publications, Inc., 1988.

Paul Revere's Ride. Longfellow, Henry Wadsworth. William Morrow & Company, 1985.

The Phantom Tollbooth. Juster, Norton. Random House, 1989.

Phoebe and the General. Griffin, Judith Berry. Coward, McCann & Geoghegan, Inc., 1977.

"The Parakeet Named Dreidel" in *The Power of Light: Eight Stories for Hanukkah.* Singer, Isaac Bashevis. Farrar, Straus & Giroux, 1980.

"The Forecast" in *Prairie Schooner.* Jaffe, Daniel F. University of Nebraska Press, 1964.

The Red Pony.   Steinbeck, John. Viking Penguin, 1992.

Robinson Crusoe.   Defoe, Daniel. Running Press Book Publishers, 1990.

Roll of Thunder, Hear My Cry.   Taylor, Mildred D. Bantam Books, 1989.

The Secret Garden.   Burnett, Frances Hodgson. HarperCollins, 1987.

Secrets of the Shopping Mall.   Peck, Richard. Dell, 1991.

She Flew No Flags.   Manley, Joan B. Houghton Mifflin Company, 1995.

Skinnybones.   Park, Barbara. Bullseye Books, 1995.

Snow Bound.   Mazer, Harry. Dell, 1990.

"What Are You Going to Be When You Grow Up?" in *Spaceships & Spells: A Collection of New Fantasy and Science-Fiction Stories.*   Yolen, Jane; Martin H. Greenberg; and Charles G. Waugh, eds. HarperCollins, 1987.

The Star Fisher.   Yep, Laurence. William Morrow & Company, 1991.

Sticks.   Bauer, Joan. Delacorte Press, 1996.

"The Emperor's New Clothes" in *Tales from Hans Christian Andersen.*   Andersen, Hans Christian. Simon & Schuster, 1987.

"How Winter Man's Power Was Broken" in *Tales of the Cheyennes.*   Penny, Grace Jackson. Houghton Mifflin Company, 1981.

"Dog of Pompeii" by Louis Untermeyer in *Teach Your Children Well: A Parent's Guide to the Stories, Poems, Fables, and Tales That Instill Traditional Values.*   Allison, Christine, ed. Delacorte Press, 1993.

"The Great Peanut Puzzle" in *Test-Tube Mysteries.*   Haines, Gail Kay. Dodd Mead, 1982.

That Was Then, This Is Now.   Hinton, S. E. Viking, 1980.

Too Soon a Woman (formerly titled *A Wonderful Woman*).   Johnson, Dorothy M. Dodd Mead, 1967.

Very Far Away From Anywhere Else.   LeGuin, Ursula K. Bantam Books, 1982.

"Emergency Landing" by Ralph Williams in *Visiting Mrs. Nabokov and Other Excursions.*   Amis, Martin, ed. Vintage Books, 1995.

The Walrus and the Carpenter.   Carroll, Lewis. Penguin Books USA Inc., 1992.

Water Sky.   George, Jean Craighead. HarperCollins, 1989.

What Hearts.   Brooks, Bruce. HarperCollins, 1992.

"A Mother in Mannville" in *When the Whippoorwill —.*   Rawlings, Marjorie Kinnan. Charles Scribner's Sons, 1940.

Where the Red Fern Grows: The Story of Two Dogs and a Boy.   Rawls, Wilson. Doubleday, Inc., 1961.

Where the Sidewalk Ends: The Poems and Drawings of Shel Silverstein.   Silverstein, Shel. HarperCollins, 1974.

The Whipping Boy.   Fleischman, Sid. William Morrow & Company, 1986.

The Widow and the Parrot.   Woolf, Virginia. Harcourt Brace & Company, 1992.

"Whoopie Ti Yi Yo, Git Along Little Dogies" in *Woody Guthrie Sings Folk Songs with Cisco Houston and Sony Terry, Vol. 2.*   Folkways Records, 1964.

Bibliography

COMPUTER SOFTWARE

Alge-blaster 3. Davidson & Associates, 1995.

Algebra CD. Expert Software, 1995.

Balancing Act. Ventura Math Power Series.

"Level B, Understanding Fractions and Percents: Lesson 6," in *CornerStone Mathematics.* SkillsBank Corporation, 1996.

"Level B, Understanding Numbers: Lesson 3," in *CornerStone Mathematics.* SkillsBank Corporation, 1996.

"Level B, Understanding Numbers: Lesson 4," in *CornerStone Mathematics.* SkillsBank Corporation, 1996.

"Level B, Using Decimals: Lesson 4," in *CornerStone Mathematics.* SkillsBank Corporation, 1996.

"Level B, Using Fractions and Percents: Lesson 1," in *CornerStone Mathematics.* SkillsBank Corporation, 1996.

"Level B, Using Fractions and Percents: Lessons 2 and 3," in *CornerStone Mathematics.* SkillsBank Corporation, 1996.

"Level B, Using Fractions and Percents: Lessons 4 and 5," in *CornerStone Mathematics.* SkillsBank Corporation, 1996.

"Level B, Using Whole Numbers: Lessons 5 and 7," in *CornerStone Mathematics.* SkillsBank Corporation, 1996.

"Level B, Using Whole Numbers: Lesson 8," in *CornerStone Mathematics.* SkillsBank Corporation, 1996.

"Level B, Using Whole Numbers: Lesson 9," in *CornerStone Mathematics.* SkillsBank Corporation, 1996.

"Level B, Working with Data: Lesson 2," in *CornerStone Mathematics.* SkillsBank Corporation, 1996.

"Level B, Working with Data: Lesson 4," in *CornerStone Mathematics.* SkillsBank Corporation, 1996.

"Level C, Using Fractions and Percents: Lesson 1," in *CornerStone Mathematics.* SkillsBank Corporation, 1996.

"Level C, Working with Data: Lessons 1–3," in *CornerStone Mathematics.* SkillsBank Corporation, 1996.

Counting on Frank: A Math Adventure Game. Electronic Arts, 1994.

Fraction Attraction. Sunburst Communications, 1996.

Graphers. Sunburst Communications, 1997.

Hot Dog Stand: The Works. Sunburst Communications, 1996.

How the West Was Negative One. Sunburst Communications, 1997.

Math Heads. Theatrix Interactive, Inc., 1996.

Mystery Math Island. Lawrence Productions, Inc., 1995.

The Quarter Mile Integers and Equations. Barnum Software, 1991.

Safari Search. Sunburst Communications, 1986.

Shape Up! Sunburst Communications, 1995.

TesselMania! MECC, 1995.

Bibliography

PROFESSIONAL BIBLIOGRAPHY

About Teaching Mathematics: Burns, Marilyn. Math Solutions Publications, 1993.

About Teaching Mathematics: A K–8 Resource. Burns, Marilyn. Math Solutions Publications, 1992.

"Adding to One," in *About Teaching Mathematics: A K–8 Resource.* Burns, Marilyn. Math Solutions Publications, 1992.

"Closest to 0, .5, or 1?" in *About Teaching Mathematics: A K–8 Resource.* Burns, Marilyn. Math Solutions Publications, 1992.

"Multiplication," in *About Teaching Mathematics: A K–8 Resource.* Burns, Marilyn. Math Solutions Publications, 1992.

"A Sample Activity — Introducing the Metric System," in *About Teaching Mathematics: A K–8 Resource.* Burns, Marilyn. Math Solutions Publications, 1992.

Activity Math: Using Manipulatives in the Classroom. Bloomer, Anne, and Phyllis Carlson. Addison-Wesley, 1993.

Addition and Subtraction. Losq, Christine. Great Source Education Group, 1998.

"Bats Incredible! Grades 2–4," in *AIMS Activities.* Project AIMS. AIMS Education Foundation, 1993.

"Fall Into Math and Science: K–1," in *AIMS Activities.* Project AIMS. AIMS Education Foundation, 1987.

"Glide Into Winter With Math and Science: K–1 Book 2," in *AIMS Activities.* Project AIMS. AIMS Education Foundation, 1987.

"Hardhatting in a Geo-World," in *AIMS Activities.* Project AIMS. AIMS Education Foundation, 1996.

"Mostly Magnets: Grades 2–8," in *AIMS Activities.* Project AIMS. AIMS Education Foundation, 1991.

"Pieces and Patterns: Grades 5–9 Book 8," in *AIMS Activities.*   Project AIMS. AIMS Education Foundation, 1986.

"Primarily Bears: Grades K–6, A Collection of Elementary Activities," in *AIMS Activities.*   Project AIMS. AIMS Education Foundation, 1987.

"Primarily Physics: Investigations in Sound, Light & Heat Energy, Grades K–3," in *AIMS Activities.*   Project AIMS. AIMS Education Foundation, 1990.

"Primarily Plants: K–3," in *AIMS Activities.* Project AIMS. AIMS Education Foundation, 1990.

"Seasoning Math and Science, Books A and B: Grades K–4," in *AIMS Activities.* Project AIMS. AIMS Education Foundation, 1987.

"Sense-able Science: Exploring and Discovering Our Five Senses, K–1," in *AIMS Activities.*   Project AIMS. AIMS Education Foundation, 1994.

"Spring Into Math and Science: K–1 Book 3," in *AIMS Activities.*   Project AIMS. AIMS Education Foundation, 1987.

"Water Precious Water: A Collection of Elementary Water Activities, Grades 2–6," in *AIMS Activities.*   Project AIMS. AIMS Education Foundation, 1988.

"Decimal Dash" by Francis (Skip) Fennell and David Williams in *The Arithmetic Teacher.* National Council of Teachers of Mathematics, 1986.

"Product Pairs" by Francis (Skip) Fennell and David Williams in *The Arithmetic Teacher.* National Council of Teachers of Mathematics, 1986.

"Symmetry the Trademark Way" by B. Renshaw in *The Arithmetic Teacher*. National Council of Teachers of Mathematics, 1986.

"Teaching Division of Fractions with Understanding" by Charles Thompson in *The Arithmetic Teacher, 26 (January 1979)*. National Council of Teachers of Mathematics.

Base-Ten Block Activities. Norton-Wolf, Sherry. Learning Resources, 1990.

Basic Math Games. Bright, George, and John Harvey. Dale Seymour Publications, 1987.

Between Never and Always. Singer, Margie, et al. Dale Seymour Publications, 1997.

Books You Can Count On: Linking Mathematics and Literature. Griffiths, Rachel, and Margaret Clyne. Heinemann, 1988.

Box It or Bag It Mathematics: Teachers Resource Guide, First–Second. Burk, Donna; Allyn Snider; and Paula Symonds. The Math Learning Center, 1988.

"Arithmetic," in *Box It or Bag It Mathematics: Teachers Resource Guide, First–Second.* Burk, Donna; Allyn Snider; and Paula Symonds. The Math Learning Center, 1988.

"Arithmetic, Facts to 10," in *Box It or Bag It Mathematics: Teachers Resource Guide, First–Second.* Burk, Donna; Allyn Snider; and Paula Symonds. The Math Learning Center, 1988.

"Extended Number Patterns—Beginning Multiplication and Division," in *Box It or Bag It Mathematics: Teachers Resource Guide, First–Second.* Burk, Donna; Allyn Snider; and Paula Symonds. The Math Learning Center, 1988.

"Money," in *Box It or Bag It Mathematics: Teachers Resource Guide, First–Second.* Burk, Donna; Allyn Snider; and Paula Symonds. The Math Learning Center, 1988.

"Numerals 0–10," in *Box It or Bag It Mathematics: Teachers Resource Guide, First–Second.* Burk, Donna; Allyn Snider; and Paula Symonds. The Math Learning Center, 1988.

"Part One: Seasonal Mathematics," in *Box It or Bag It Mathematics: Teachers Resource Guide, First–Second.* Burk, Donna; Allyn Snider; and Paula Symonds. The Math Learning Center, 1988.

"Part Three: The Calendar," in *Box It or Bag It Mathematics: Teachers Resource Guide, First–Second.* Burk, Donna; Allyn Snider; and Paula Symonds. The Math Learning Center, 1988.

"Pattern," in *Box It or Bag It Mathematics: Teachers Resource Guide, First–Second.* Burk, Donna; Allyn Snider; and Paula Symonds. The Math Learning Center, 1988.

"Place Value Addition and Subtraction," in *Box It or Bag It Mathematics: Teachers Resource Guide, First–Second.* Burk, Donna; Allyn Snider; and Paula Symonds. The Math Learning Center, 1988.

"Place Value Counting," in *Box It or Bag It Mathematics: Teachers Resource Guide, First–Second.* Burk, Donna; Allyn Snider; and Paula Symonds. The Math Learning Center, 1988.

"Sorting and Graphing," in *Box It or Bag It Mathematics: Teachers Resource Guide, First–Second.* Burk, Donna; Allyn Snider; and Paula Symonds. The Math Learning Center, 1988.

"Understanding Measuring," in *Box It or Bag It Mathematics: Teachers Resource Guide, First–Second*. Burk, Donna; Allyn Snider; and Paula Symonds. The Math Learning Center, 1988.

A Calculator Tutorial. Merrill, William L. Dale Seymour Publications, 1996.

Change Tiles. Cook, Marcy. Marcy Cook Math, 1985.

A Collection of Math Lessons from Grades 1–3. Burns, Marilyn, and Bonnie Tank. Math Solutions Publications, 1987.

A Collection of Math Lessons from Grades 6–8. Burns, Marilyn. Math Solutions Publications, 1990.

Connections: Linking Manipulatives to Mathematics, Grade 4. Charles, Linda Holden, and Micaelia Randolph Brummett. Creative Publications, 1989.

Constructing Ideas About Large Numbers. Brodie, Julie Pier. Creative Publications, 1995.

Create-a-Timeline. Dale Seymour Publications, 1997.

Creative Puzzles of the World. Van Delft, Peter, and Jack Botermans. Abrams, 1978.

Critters: K–6 Life Science Activities. AIMS Education Foundation, 1989.

"Developing Number Sense in the Middle Grades," by Frances R. Curcio, in *Curriculum and Evaluation Standards for School Mathematics Addenda Series*. National Council of Teachers of Mathematics, 1991.

Developing Graph Comprehension: Elementary and Middle School Activities. Curcio, F. National Council of Teachers of Mathematics, 1989.

"Activity 21: The Raisin Experiment," in *Developing Graph Comprehension: Elementary and Middle School Activities*. Curcio, F. National Council of Teachers of Mathematics, 1989.

Developing Skills with Tables and Graphs. Murphy, Elaine C. Dale Seymour Publications, 1981.

The Elementary Math Teacher's Book of Lists. Helton, S. M., and S. J. Micklo. Simon & Schuster, 1998.

Estimation Destinations. Threewit, Fran. Cuisenaire, 1994.

Every Day Counts • Grade 3. Kanter, Patsy, and Janet Gillespie. Great Source Education Group, 1994.

"Calendar Patterns," in *Family Math*. Stenmark, J.; V. Thompson; and R. Cossey. University of California, 1986.

"Five Containers," in *Family Math*. Stenmark, J., and R. Cossey. University of California, 1986.

"The Magnificent Inch," in *Family Math*. Stenmark, J., and R. Cossey. University of California, 1986.

"Perfect People," in *Family Math*. Stenmark, J., and R. Cossey. University of California, 1986.

"Perimeter Variations," in *Family Math*. Stenmark, J.; V. Thompson; and R. Cossey. University of California, 1986.

"Simple Symmetries," in *Family Math*. Stenmark, J.; V. Thompson; and R. Cossey. University of California, 1986.

"Some Survey Topics," in *Family Math*. Stenmark, J.; V. Thompson; and R. Cossey. University of California, 1986.

"Sorting and Classifying," in *Family Math*. Stenmark, J.; V. Thompson; and R. Cossey. University of California, 1986.

"Tic-Tac-Toe," in *Family Math*. Stenmark, J.; V. Thompson; and R. Cossey. University of California, 1986.

"What Time Is It? What Time Will It Be?" in *Family Math*. Stenmark, J., and R. Cossey. University of California, 1986.

"Find the Missing Link," in *Fifth–Grade Book: Addenda Series, Grades K–6*. Burton, G., and Levia, M., et al. National Council of Teachers of Mathematics, 1991.

"Comparing Fractions," in *50 Problem-Solving Lessons, Grades 1–6*. Burns, Marilyn. Math Solutions Publications, 1996.

Fraction Circle Activities. Berman, Barbara, and Fredda Friederwitzer. Dale Seymour Publications, 1983.

Fractions with Pattern Blocks. Zullie, Mathew E. Creative Publications, 1988.

Great Graphing. Lee, Martin, and Marcia Miller. Scholastic Professional Books, 1993.

Hands-On Base-Ten Blocks, K–3. Creative Publications.

"Heavy Work," in *Hardhatting in a Geo-World*. AIMS Education Foundation, 1996.

Helping Your Child with Math. Weiss, Sol. Prentice Hall, 1987.

"How Far Down Fraction Hill?" in *Ideas from the Arithmetic Teacher: Grades 4-6*. Fennell, S., and D. Williams. National Council of Teachers of Mathematics, 1986.

"Time for Tiling," in *Ideas from the Arithmetic Teacher: Grades 4-6*. Fennell, F., and D. Williams. National Council of Teachers of Mathematics, 1986.

"Undercover Shapes," in *Ideas from the Arithmetic Teacher: Grades 4–6*. Fennell, F., and D. Williams. National Council of Teachers of Mathematics, 1986.

"Circle Graph," in *IDEAS: Grades 4–6*. Fennell, F., and D. Williams. National Council of Teachers of Mathematics, 1986.

"Draw the Shapes," in *IDEAS: Grades 4–6*. Fennell, F., and D. Williams. National Council of Teachers of Mathematics, 1986.

"Merry Measuring," in *IDEAS: Grades 1–4*. Immerzeel, G., and M. Thomas. National Council of Teachers of Mathematics, 1982.

Integrating Beginning Math & Literature. Rommel, Carol A. Incentive Publications, Inc., 1991.

"Graphing," in *Janice Van Cleave's Math for Every Kid: Easy Activities That Make Learning Math Fun*. Van Cleave, Janice Pratt. Wiley, 1991.

Kaleidoscope Symmetry. Dale Seymour Publications, 1993.

Kaleidoscope Symmetry Poster. Dale Seymour Publications, 1993.

Kid Cash: Creative Money-Making Ideas. Lamancusa, Joe. TAB Books, 1993.

Macmillan Visual Almanac. Glassman, Bruce S., ed. Blackbirch Press, 1996.

Making Sense of Data. National Council of Teachers of Mathematics, 1993.

The Master Revealed–A Journey with Tangrams. Han, S.T., and Barbara E. Ford. Cuisenaire.

Mathematics: A Way of Thinking. Baratta-Lorton, Robert. Addison-Wesley, 1977.

Mathematics: A Way of Thinking: Division. Baratta-Lorton, Robert. Addison-Wesley, 1977.

Math and Literature (K–3). Burns, Marilyn. Math Solutions Publications, 1992.

Math By All Means: Division (Grades 3 and 4). Burns, Marilyn. Math Solutions Publications, 1994.

Math By All Means: Division, Grades 3 and 4. Ohanian, Susan. Math Solutions Publications, 1995.

Math By All Means: Multiplications Grade 3. Burns, Marilyn. Math Solutions Publications, 1994.

Math By All Means: Probability, Grades 3 and 4. Burns, Marilyn. Math Solutions Publications, 1994.

Math for Every Kid. Van Cleave, J. John Wiley, 1991.

Mathematics Their Way: An Activity-Centered Mathematics Program for Early Childhood Education. Baratta-Lorton, Mary. Addison-Wesley Publishing Company, Inc., 1976.

"Application and Extension of Place Value—Measuring," in *Mathematics Their Way: An Activity-Centered Mathematics Program for Early Childhood Education.* Baratta-Lorton, Mary. Addison-Wesley Publishing Company, Inc., 1976.

"Comparing," in *Mathematics Their Way: An Activity-Centered Mathematics Program for Early Childhood Education.* Baratta-Lorton, Mary. Addison-Wesley Publishing Company, Inc., 1976.

"Counting," in *Mathematics Their Way: An Activity-Centered Mathematics Program for Early Childhood Education.* Baratta-Lorton, Mary. Addison-Wesley Publishing Company, Inc., 1976.

"Number at the Concept Level, Word Problems," in *Mathematics Their Way: An Activity-Centered Mathematics Program for Early Childhood Education.* Baratta-Lorton, Mary. Addison-Wesley Publishing Company, Inc., 1976.

"Number at the Connecting Level," in *Mathematics Their Way: An Activity-Centered Mathematics Program for Early Childhood Education.* Baratta-Lorton, Mary. Addison-Wesley Publishing Company, Inc., 1976.

"Pattern One," in *Mathematics Their Way: An Activity-Centered Mathematics Program for Early Childhood Education.* Baratta-Lorton, Mary. Addison-Wesley Publishing Company, Inc., 1976.

Mental Math: Computation Activities for Anytime. Piccirilli, Richard S. Scholastic Professional Books, 1996.

Mental Math in Junior High. Hope, J.; B. Reys; and R. Reys. Dale Seymour Publications, 1987.

Mental Math in the Middle Grades. Hope, J., et al. Dale Seymour Publications, 1987.

The Money Book. Campbell, June H. Dale Seymour Publications, 1978.

Multicultural Math: Hands-On Math Activities from Around the World. Zaslavsky, Claudia. Scholastic Professional Books, 1994.

Multicultural Mathematics Materials. Krause, M. NCTM, 1993.

Name That Portion (Fractions, Percents, and Decimals). Akers, Joan; Cornelia Tierney; Claryce Evans; and Megan Murray. Dale Seymour Publications, 1997.

"Oh How We've Changed!" in *NCTM Addenda Series: Fourth Grade.* Leiva, M., et al. National Council of Teachers of Mathematics, 1992.

"Developing Measurement and NumberSense," in *New Directions for Elementary School Mathematics: NCTM 1989 Yearbook*. Shaw and Cliatt. National Council of Teachers of Mathematics, 1989.

One Hand at a Time. Smith, Patricia E. Dale Seymour Publications, 1987.

100 Activities for the Hundred Number Board. Clarkson, Sandra Pryor. Ideal School Supply Company, 1985.

"Fractions on a Geoboard," in *Opening Eyes to Mathematics, Volume 3*. The Math Learning Center, 1995.

Packages and Groups (Multiplication and Division). Economopoulos, Karen; Cornelia Tierney; and Susan Jo Russell. Dale Seymour Publications, 1995.

The Patchwork Quilt. Flournoy, Valerie, et al. Scholastic, 1996.

The Place Value Connection: Primary Activities and Games to Teach Place Value, Money, and More. D'Aboy, Diana A. Dale Seymour Publications, 1985.

Poster Tessellations. Dale Seymour Publications, 1997.

Powers of Ten. Morrison, Philip and Phylis. W. H. Freeman, 1982.

Probability, Grades 2–3. Burns, Marilyn. Math Solutions Publications, 1994.

Probability Model Masters. Seymour, Dale. Dale Seymour Publications, 1990.

Problem Solving Focus: Time and Money. Greenes, Carole, and G. Immerzeel. Dale Seymour Publications, 1993.

"From Head to Toe," in *Project AIMS*. (Grades 5–9 series). AIMS Education Foundation, 1986.

Seeing Fractions. Russell, Susan Jo, and Rebecca Corwin. California State Department of Education, 1990.

"Making and Using Fraction Strips," in *Seeing Fractions*. California Department of Education, 1991.

"It's the Last Straw," in *The Sky's the Limit*. AIMS Education Foundation, 1994.

Statistics: Middles, Means, and In-Betweens. Friel, Susan N.; Jancie R. Mokros; and Susan Jo Russell. Dale Seymour Publications, 1992.

Step-by-Step: Decimals. Bazik, Edna; Janet Barnard; and Carol Thornton. DLM–McGraw Hill, 1987.

Step-by-Step: Division. Bazik, Edna; Janet Barnard; and Carol Thornton. DLM–McGraw Hill, 1987.

Step-by-Step: Fractions. Bazik, Edna; Janet Barnard; and Carol Thornton. DLM–McGraw Hill, 1987.

Step-by-Step: Subtraction with Whole Numbers. Bazik, Edna; Janet Barnard; and Carol Thornton. DLM–McGraw Hill, 1987.

Symmetry Posters. Dale Seymour Publications, 1997.

Tangramath. Seymour, Dale. Creative Publications, 1971.

Teaching Primary Math with Music. Mendlesohn, Esther. Dale Seymour Publications, 1990.

Telling Time Bingo. Frank Schaeffer Publications, Inc., 1991.

Things That Come in Groups: Multiplication and Division. Tierney, Cornelia, et al. Dale Seymour Publications, 1996.

This Book Is About Time. Burns, Marilyn. Yolla Bolly Press, 1978.

Timelines of African-American History: 500 Years of Black Achievement. Cowan, Thomas, and Jack Maquire. Berkley Publishing Group, 1994.

20 Thinking Questions for Base-Ten Blocks, Grades 3–6. Stewart, Kelly, and Kathryn Walker. Creative Publications, 1995.

Uncovering Mathematics with Manipulatives and Calculators. Schielack, Jane F., and Dinah Chancellor. Texas Instruments, 1995.

Understanding Division. Losq, Christine. Great Source Education Group, 1998.

"Division as Sharing," in *Understanding Division.* Losq, Christine. Great Source Education Group, 1998.

"Division as Sharing Using Objects," in *Understanding Division.* Losq, Christine. Great Source Education Group, 1998.

Understanding Fractions. Losq, Christine. Great Source Education Group, 1998.

"Act Out Fractions," in *Understanding Fractions.* Losq, Christine. Great Source Education Group, 1998.

"Act Out Pie Graphs," in *Understanding Fractions.* Losq, Christine. Great Source Education Group, 1998.

"Make Rulers," in *Understanding Fractions.* Losq, Christine. Great Source Education Group, 1998.

"Problems with Brownies," in *Understanding Fractions.* Losq, Christine. Great Source Education Group, 1998.

"Show Me a Fraction," in *Understanding Fractions.* Losq, Christine. Great Source Education Group, 1998.

"Solve Problems About Brownies," in *Understanding Fractions.* Losq, Christine. Great Source Education Group, 1998.

"Who Has More?" in *Understanding Fractions.* Losq, Christine. Great Source Education Group, 1998.

Understanding Fractions, Decimals and Percent. Losq, Christine, and Robin Levy. Great Source Education Group, 1998.

Understanding Geometry. Franco, Betsy, et al. Great Source Education Group, 1998.

"Geometry Concentration," in *Understanding Geometry.* Franco, Betsy, et al. Great Source Education Group, 1998.

Understanding Multiplication. Losq, Christine. Great Source Education Group, 1998.

"Developing a Packaging Plan," in *Understanding Multiplication.* Losq, Christine. Great Source Education Group, 1998.

Used Numbers. Russell, Susan Jo, et al. Dale Seymour Publications, 1993.

Used Numbers: Measurement: From Paces to Feet. Corwin, Rebecca B., and Susan Jo Russell. Dale Seymour Publications, 1990.

Used Numbers: Statistics: Prediction and Sampling. Corwin, Rebecca B., and Susan N. Friel. Dale Seymour Publications, 1990.

Used Numbers: Statistics: The Shape of the Data. Russell, Susan Jo, and Rebecca B. Corwin. Dale Seymour Publications, 1989.

Using Big Numbers. Losq, Christine, and Robin Levy. Great Source Education Group, 1998.

Using Fractions, Decimals and Percent. Losq, Christine, and Robin Levy. Great Source Education Group, 1998.

"Read Decimals," in *Using Fractions, Decimals and Percent.*   Losq, Christine, and Robin Levy. Great Source Education Group, 1998.

Using Geometry.   Losq, Christine, and Robin Levy. Great Source Education Group, 1998.

Using Rates and Scales.   Losq, Christine, and Robin Levy. Great Source Education Group, 1998.

What Are My Chances? Book A.   Shulte, Albert P., and Stuart A. Choate. Creative Publications, 1977.

"Patterns with a Point" in *What's Next? Vol. 1,* AIMS Education Foundation, 1995.

"Patterns with a Point" in *What's Next? Vol. 2,* AIMS Education Foundation, 1995.

The Wonderful World of Mathematics: A Critically Annotated List of Children's Books in Mathematics.   Thiessen, Diane, and Margaret Mathias. National Council of Teachers of Mathematics, 1992.

Index

*Pupil's Edition–**Roman***   Teacher's Edition–***Italics***

D

E

F

G

J K

L

M

S

U

V

W

X

Y

Z